DYNAMIC ASPECTS OF
MICROTUBULE BIOLOGY

ANNALS OF THE NEW YORK ACADEMY OF SCIENCES
Volume 466

DYNAMIC ASPECTS OF MICROTUBULE BIOLOGY

Edited by David Soifer

The New York Academy of Sciences
New York, New York
1986

Cover photo: Transverse section through concentric ordered arrays of MTs alternating with layers of macromolecular material in a portion of neuritic extension near the soma. See p. 735.

Library of Congress Cataloging-in-Publication Data

Dynamic aspects of microtubule biology.

(Annals of the New York Academy of Sciences, ISSN 0077-8923; v. 466)
   Bibliography: p.
   Includes index.
   1. Microtubules.   I. Soifer, David, 1937–
II. Series.
Q11.N5 vol. 466   500 s   86-12592
[QH603.M44]   [574.87′34]
ISBN 0-89766-327-6
ISBN 0-89766-328-4 (pbk.)

SP
Printed in the United States of America
ISBN 0–89766–327–6 (Cloth)
ISBN 0–89766–328–4 (Paper)
ISSN 0077–8923

ANNALS OF THE NEW YORK ACADEMY OF SCIENCES

Volume 466
June 30, 1986

DYNAMIC ASPECTS OF MICROTUBULE BIOLOGY[a]

Editor and Conference Chairman
DAVID SOIFER

Organizing Committee
B.R. BRINKLEY and LESLIE WILSON

CONTENTS

[a]This volume is the result of a conference entitled Conference on Dynamic Aspects of Microtubule Biology, held in New York City on December 3–6, 1984, by the New York Academy of Sciences.

Part II. Microtubule-Associated Proteins

Part III. Properties of Tubulins and the Assembly and Disassembly of Microtubules

Part IV. Drugs Affecting Microtubule Assembly

Part V. Interactions of Microtubules with Other Cytoplasmic Components

Part VI. Microtubule Function

Financial assistance was received from:
- A. H. ROBINS COMPANY
- AMERSHAM CORPORATION
- DUPONT—NEN
- LILLY RESEARCH LABORATORIES
- MERRELL DOW RESEARCH INSTITUTE
- NATIONAL HEART, LUNG & BLOOD INSTITUTE—NIH
- NATIONAL INSTITUTE OF GENERAL MEDICAL SCIENCES—NIH
- NATIONAL INSTITUTE OF NEUROLOGICAL, COMMUNICATIVE
  DISEASES, AND STROKE—NIH
- NATIONAL INSTITUTE ON AGING—NIH
- NATIONAL SCIENCE FOUNDATION
- SANDOZ INC.
- SCHERING CORPORATION
- SCHLEICHER AND SCHUELL INC.
- STUART PHARMACEUTICALS, DIVISION OF ICI AMERICAS INC.
- UNITED STATES OFFICE OF NAVAL RESEARCH

Preface

DAVID SOIFER

Institute for Basic Research in Developmental Disabilities
Staten Island, New York 10314

In May 1974, representatives of nearly every laboratory in the world, presently engaged in research on any aspect of the biology of cytoplasmic microtubules, participated in the first New York Academy of Sciences Conference on this subject. There were few strangers at the conference. Nearly every participant was known to most of the others. The presentations from the platform during those three days spanned the work in the field, from the first attempts to understand microtubule assembly and the microtubule-associated proteins (MAPs), to studies of effects of antimitotic drugs on a variety of cell functions. Research on microtubules has come a long way in the ensuing ten years. Microtubules is a generic term for a class of structures. We now talk of many forms of tubulins, of families of genes coding for different tubulins, of kinetic models of tubulin assembly, and of a variety of microtubule proteins. Monoclonal antibodies and domain-specific polyclonal antibodies are used to dissect tubulin molecules whose amino acid sequence is known and whose properties continue to surprise us. The study of microtubules is carried out at most basic research centers in the United States and Western Europe. The sheer number of investigators of microtubules and of their constituent proteins makes it impossible to convene all of them at a conference on the general topic of microtubules. When it was decided to hold a second New York Academy of Sciences Conference on microtubules ten years after the first, it was clear that the subject would have to be more limited and that the Academy would have to plan for a larger conference.

The program for the conference was organized around the factors that regulate the presence of microtubules at any given time in the life of a cell. These factors ranged from the regulation of the expression of tubulin genes to the properties of MAPs and the kinetics of assembly. To help us maintain perspective, we added some discussion of what microtubules do in cells, but even so, we just touched on the vast amount of work on the function of microtubules.

Many colleagues participated in the development of the program. In particular, the members of the conference organizing committee, Bill Brinkley and Les Wilson, were always available for consultation and provided suggestions and assistance ranging from helping with the organization of the program to raising some of the funds that were necessary for the support of this meeting. Drs. Ellen Borenfreund, Muriel Feigelson, Olga Greengard, and Phil Siekovitz, of the Academy's conference committee, offered numerous constructive suggestions, both with respect to the program itself as well as with other aspects of the organization of the conference. Ellen Marks and the members of her staff simplified the onerous task of organizing a conference in many ways. Many others contributed to the success of the conference and this volume through their suggestions. Pat Casiano, of the New York State Institute for Basic Research in Developmental Disabilities, was the person who kept the entire organization together, handling much of the correspondence and endless revisions of the program. The organizations listed following the Table of Contents provided financial assistance for the conference. Drew Gomes, of Dupont/NEN, was particularly helpful in obtaining a significant grant that was used to defray some of the expenses incurred by speakers at the conference.

On behalf of the conference participants I should like to thank all of those who contributed to the success of the conference. I should also like to thank the New York Academy of Sciences for bringing us together for this productive meeting and for the publication of this monograph.

Factors Regulating the Presence
of Microtubules in Cells

DAVID SOIFER[a]

Institute for Basic Research in Developmental Disabilities
Staten Island, New York 10314

When microtubules are required by a cell for a particular function, microtubules assemble in the appropriate region of the cell, with the necessary orientation. As microtubules are no longer needed, they depolymerize. The word, microtubules, describes a class of similar structures, formed of specific proteins called tubulins. In different microtubules, the tubulins are copolymerized with any of a variety of proteins collectively known as microtubule-associated proteins, or MAPs. Both the tubulins and at least some of the MAPs may be coded by more than a single gene; both tubulins and MAPs are modified in various ways after they are synthesized. Thus the determination of what sort of microtubule will be present at a particular time in a particular cell depends, at least in part, upon the synthesis and processing of a number of proteins. But the presence of the right proteins, while necessary, is not sufficient for assembly to occur. The assembly of many, if not all microtubules depends on the presence and orientation of microtubule organizing centers, which may include a centriole or analogous structure. It is not clear what determines the length of a microtubule or what factors enable a cell to depolymerize its microtubules as required for normal cell function. In a living cell, one population of microtubules may be elongating at the same time as another population is disassembling.

The organizing principle for the conference on Dynamic Aspects of Microtubule Biology, and for this volume, is our understanding of the factors regulating the appearance and disappearance of microtubules in cells. All of the contributions to this conference deal with aspects of this fundamental feature of cell biology. As a point of departure for considering these problems, some basic features of what is known about factors regulating the presence of microtubules in at least some cells will be reviewed in this chapter, which will serve as an introduction to the proceedings of the conference.

What factors determine the presence of assembled microtubules in cells? Microtubule assembly may be seen to depend upon the size of the pool of subunits available for assembly; the microenvironment—the state of the cytoplasm; the presence of specific modulators of assembly such as MAPs, for example; and the mechanisms of the assembly and disassembly processes themselves. In simplest terms, microtubule assembly may be described by the following reaction:

$$\text{subunit pool} \rightleftharpoons \text{microtubules} \tag{1}$$

Modification of the concentration of functional subunits may affect this relationship. Changes in the stability of assembled microtubules or in the properties of the ends of microtubules may also modify this reaction. Alterations of the local environment, such as local variations of pH, of Ca^{++} concentration, of levels of the various MAPs, or of

[a] Send correspondence to David Soifer, PhD, Head, Laboratory of Cell Biology, Institute for Basic Research in Developmental Disabilities, 1050 Forest Hill Road, Staten Island, N.Y. 10314.

1

the amount or organization of other components of the cytoskeleton, can shift this reaction to favor assembly or disassembly. Other articles in this volume contain extensive discussions of assembly kinetics and the assembly reaction itself. I should like to focus on the left side of equation **1**.

First, the state of the subunit pool depends upon the availability of assembly-competent tubulin subunits. These are, presumably, heterodimeric tubulin-guanosine triphosphate (GTP) and oligomeric tubulin (FIGURE 1). The tubulins of the subunit pool are either recycled, by depolymerization from existing microtubules (the left directed arrow of equation **1**), or originate from nondimeric tubulin (FIGURES 2, 1). Regulation of any of the steps between the initiation of transcription of specific tubulin genes and the formation of tubulin dimer is potentially a site for regulation of assembly itself, by regulation of the size of the subunit pool. This group of reactions may be considered in two parts: first, the transcription of tubulin genes and the synthesis and posttranslational modification of tubulin monomer (FIGURE 2, reactions 1–4, 7); second, the dimerization and oligomerization of tubulins to form the pool of assembly-competent subunits (FIGURE 1).

The first group of reactions (FIGURE 2) may be regulated at several levels. Because most organisms express at least three α-tubulin and three β-tubulin genes, it is reasonable to consider that subunit concentration may be regulated at the level of transcription. Sequence analysis of two rat α-tubulin cDNA clones reveals close homology between the coding regions of these DNAs but considerable divergence between their 3'-noncoding regions.[1] Similar relationships have been reported among chicken,[2] and rat[3] β tubulins. Some tubulins are expressed only in specific tissues [see, for example, references 1–6], or at unique stages of development.[1,3,6] [See also reference 7 for evidence linking specific tubulins to specific functions in *Aspergillus*]. A possible function for multiple genes with great similarities in their coding regions but differing in noncoding sequences is to allow regulation of the same gene products by different regulators.[6] It is important to note the presence of some sequence heterogeneity within the amino acid sequences of different β tubulins,[1,2] especially within one of the regions that appear to be involved in the binding of MAPs.[1] There is also considerable diversity in noncoding regions of tubulin genes from such organisms as *Chlamydomonas,*[8] although the tubulins, in this organism at least, are coordinately expressed.

The size of the subunit pool itself has been shown to exert a feedback regulation on the synthesis of tubulin (FIGURE 2), presumably at a posttranscriptional level.[9,10] In this autoregulatory system, an increase in the pool of depolymerized tubulin results in a decrease in the amount of tubulin mRNA without turning off the transcription of tubulin genes. Self-adjustment of tubulin gene expression has now been demonstrated in mouse cells transfected with a chicken β_2-tubulin gene.[10] Colchicine treatment of cells that express these foreign genes, down-regulates both the host cell tubulin genes and the transfected genes. At this time, the precise site of the presumably posttranscriptional regulation for this feedback regulation system is not known (FIGURE 2, reaction 7).

The translation of the tubulin message represents another point at which the size of the subunit pool may be regulated. Although there is no clear evidence for translational regulation of the tubulins in chordates, what happens to the translation product is of great importance for this discussion. α Tubulins may, or may not be detyrosinated at their carboxyterminals. The detyrosinated α tubulin is absent from the astral rays of mitotic spindles[11] and from mature cerebellar granule cell axons (parallel fibers)[12] and generally has a limited intracellular distribution (but see reference 13). Nascent tubulins may be phosphorylated, methylated, and acetylated. Newly synthesized tubulin may be sequestered from the subunit pool by such events as insertion into

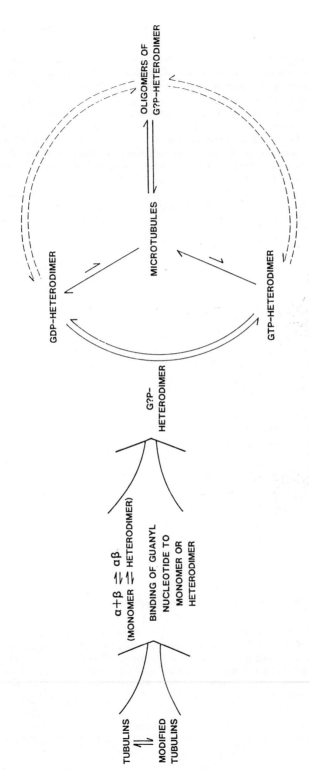

FIGURE 1. The subunit pool for microtubule assembly.

membranes. It is not clear to what extent the substrate tubulins for these modification reactions are monomeric, or to what extent the modifications are carried out on dimeric or oligomeric tubulins. It is also not clear whether the presence of a guanyl nucleotide ligand on the tubulins will affect any of the other posttranslational modifications (*i.e.* whether the modifications take place prior to guanylation of the tubulin).

All of the above represent possible levels of regulation of the subunit pool by regulation of the amount of newly synthesized tubulin that can be made available for assembly. The subunit pool might also be regulated by modifying the competence of tubulin already in the pool to assemble, by regulation of the guanylation of tubulin, or of $\alpha\beta$-heterodimer formation. The various models for the kinetics of microtubule assembly have taken the analysis of Oosawa as a point of departure and consider the

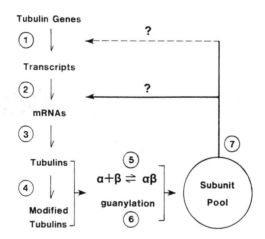

FIGURE 2. Points of regulation of assembly competent tubulin.
1. Transcription
2. Processing
3. Translation
4. Posttranslational modification
5. Monomer-heterodimer equilibrium
6. Binding of guanyl nucleotide to tubulins
7. Feed-back regulation of gene expression by size of subunit pool

rate of assembly (dn/dt) as a function of the concentration of dimeric tubulin that is competent to be incorporated into microtubules [see, for example, references 14–18]. Extending that analysis to consideration of GTP-tubulin capping of microtubule ends, the concentration of GTP-tubulin heterodimer is effectively the concentration of assembly competent tubulin. Because GTP-tubulin is necessary for assembly, several additional points for regulation of assembly are apparent. These include regulation of the effective concentration of guanyl nucleotide for binding to the tubulin, the effective concentration of NTP to serve as a source of high energy phosphate to maintain the bound nucleotide of the tubulin in the triphosphate form, and the phosphate exchange reactions allowing the conversion of bound dinucleotide to trinucleotide.

Numerous approaches to microtubule formation by tubulin in cell-free systems give a critical concentration of tubulin dimer on the order of 1 μM. Estimates of the dissociation constant for the monomer $\rightleftharpoons$ dimer equilibrium[19] are of the same order of magnitude as the critical concentration. These observations imply that, if the subunit for microtubule assembly is heterodimeric tubulin, regulation of the dimerization reaction will change the size of the subunit pool, thus changing the amount of subunit available for assembly.

Regulation of the synthesis of proteins or other substances that modify assembly may serve to vary the amount of assembled microtubules in the face of subunit pools of constant size. Assembly-promoting MAPs serve effectively to lower the critical concentration of subunit for assembly. Thus, by varying the real size of the subunit pool or by varying the critical concentration, thereby changing the minimal pool size required for assembly, the cell has numerous ways to control the assembly of microtubules.

As is made apparent by many of the papers presented in this volume, the diversity of the microtubule proteins, both tubulins and MAPs, means that not all microtubules are identical! In fact, we do not really know if each given microtubule is a homopolymer of identical heterodimeric subunits. Some tubulin genes are expressed in all cells; others are limited to certain types of cells and/or to certain stages of development. The modifications of tubulin discussed above in considering the regulation of the tubulin subunit pool, lead to a greatly varied intracellular tubulin population, representing in the brain, for example, more than twenty different tubulin isomorphs[20] [see also reference 21]. There is considerable heterogeneity in the MAPs, as well [see, for example, references 22–31]. Thus, an organism has the possibility of generating a large assortment of structures, all of which are microtubules.

A clear example of microtubule heterogeneity within a single cell is seen in examination of the spatial distribution of microtubule proteins in individual neurons. The axon, despite its small diameter, includes the bulk of the cytoplasm of the cell. There is minimal protein synthesis in the axon, however, so the protein of the axonal cytoplasm is constantly replenished from the perikaryon. Individual proteins and other structures that require rapid repletion are transported along the axon by a microtubule-dependent mechanism. The microtubules of dendrites do not extend for nearly as great a distance as those of axons and may differ in function from axonal microtubules. The MAPs of neurons are not homogeneously distributed through single cells. Some MAPs, such as MAP-2, are characteristic of dendrites.[22–24] Others appear to be components of axonal microtubules.[22–24,32] There is also evidence for inhomogeneous distribution of tubulins in given neurons. As already mentioned, tyrosinated tubulin is not present in the microtubules of the axons of cerebellar granule cells, the parallel fibers of the cerebellar molecular layer, but is a constituent of the microtubules of Purkinje cell dendrites and radial glia of the outer cerebellar cortex.[12] With appropriate antibodies, we should be able to determine the intracellular distributions of at least some of the individual tubulin isomorphs, as is being done for the mitotic spindle.[11] The many forms of tubulins and MAPs and the multitude of possible microtubules that may be formed, in theory, in a single cell, mean that the regulation of the presence of microtubules in cells includes presumably separate regulation of at least some of the different types of microtubules. Each potential regulatory point discussed for microtubules in general can serve for each possible type of microtubule.

What might be called the "ecology" of microtubule proteins, the interactions of microtubule proteins with elements of their environment, must also play a great role in the fine regulation of the state of microtubules in cells. Microtubule proteins, especially MAPs, interact with other components of the cytoskeleton in cell-free systems [see, for example, references 33, 34]. Intracellular levels of Ca^{++} and of cyclic adenosine monophosphate (AMP) also affect microtubule proteins, especially by activation of MAP phosphorylation, but perhaps by other mechanisms as well.[35,36] Local changes in pH have been reported to be effective regulators of microtubule assembly.[37]

The importance of microtubules to the life of eukaryotic cells is evidenced by the limited variability in the genes coding for these highly conserved proteins. Yet they do figure in human pathology in various ways. Immotile-cilia syndromes result from defective dynein or from other nontubulin mutations.[38] The altered microtubules described in Purkinje cell dendrites of developmentally disabled infants[39] appear to result from defective MAPs. The toxicity of microtubule-specific drugs attests to the range of activities that require functioning microtubules. As we come to understand more about microtubule function we find ourselves in a position to understand more about the normal functioning of cells as well as about the cell biology of disease.

Against this background, let us consider the following papers, with the realization that we are not discussing a single subcellular structure, but a class of multifunctional related structures whose roles are central to the operation of all eukaryotic organisms. The papers are grouped to consider the expression of tubulin genes, MAPs, microtubule assembly, effects of drugs on microtubules, interactions of microtubules with other components of the cytoskeleton, and some aspects of microtubule function.

REFERENCES

1. GINZBURG, I., A. TEICHMAN & U. Z. LITTAUER. 1986. Ann. N.Y. Acad. Sci. **466:** 31–40. This volume.
2. SULLIVAN, K. F., J. C. HAVERCROFT & D. W. CLEVELAND. 1984. *In* Molecular Biology of the Cytoskeleton. G. G. BORISY, D. W. CLEVELAND & D. B. MURPHY, Eds.: 321–332. Cold Spring Harbor Press.
3. FARMER, S. R., G. S. ROBINSON, D. MBANGKOLLO, J. F. BOND, G. B. KNIGHT, M. J. FENTON & E. M. BERKOWITZ. 1986. Ann. N.Y. Acad. Sci. **466:** 41–50. This volume.
4. MORRISON, M. R. & W. S. T. GRIFFIN. 1986. Ann. N.Y. Acad. Sci. **466:** 51–74. This volume.
5. ROTHWELL, S. W., W. A. GRASSER & D. B. MURPHY. 1986. Ann. N.Y. Acad. Sci. **466:** 103–110. This volume.
6. RAFF, E. C. 1984. J. Cell. Biol. **99:** 1–10.
7. MORRIS, N. R., G. S. MAY, J. A. WEATHERBEE, M. L.-S. TSANG & J. GAMBINO. 1986. Ann. N.Y. Acad. Sci. **466:** 13–17. This volume.
8. SILFLOW, C. D. & J. YOUNGBLUM. 1986. Ann. N.Y. Acad. Sci. **466:** 18–30. This volume.
9. BEN ZE'EV, A., S. R. FARMER & S. PENMAN. 1979. Cell **17:** 319–325.
10. LAU, J. T. Y., M. F. PITTENGER, J. C. HAVERCROFT & D. W. CLEVELAND. 1986. Ann. N.Y. Acad. Sci. **466:** 75–88. This volume.
11. GUNDERSON, G. G., M. H. KALNOSKI & J. C. BULINISKI. 1984. Cell **38:** 779–89.
12. CUMMING, R., R. D. BURGOYNE & N. A. LYTTON. 1984. J. Cell Biol. **98:** 347–351.
13. WEHLAND, J., H. C. SCHRÖDER & K. WEBER. 1986. Ann. N.Y. Acad. Sci. **466:** 609–621. This volume.
14. HILL, T. L. & M. W. KIRSCHNER. 1982. Int. Rev. Cytol. **78:** 1–125.
15. MITCHISON, T. & M. W. KIRSCHNER. 1984. Nature **312:** 237–242.
16. PANTALONI, D. & M-F. CARLIER. 1986. Ann. N.Y. Acad. Sci. **466:** 496–509. This volume.
17. WILSON, L. & K. W. FARRELL. 1986. Ann. N.Y. Acad. Sci. **466:** 690–708. This volume.
18. SOIFER, D. & K. MACK. 1984. Handb. Neurochemistry **7:** 245–280.
19. DETRICH III, H. W. & R. C. WILLIAMS JR. 1978. Biochemistry. **17:** 3900–3907.
20. LEE, J. C., D. J. FIELD, H. J. George & J. Head. 1986. Ann. N.Y. Acad. Sci. **466:** 111–128. This volume.
21. GOZES, I. & K. J. SWEADNER. 1981. Nature (London) **294:** 477–480.

22. VALLEE, R. B., G. S. BLOOM & F. C. LUCA. 1986. Ann. N.Y. Acad. Sci. **466:** 134–144. This volume.
23. BINDER, L. I., A. FRANKFURTER & L. I. REBHUN. 1986. Ann. N.Y. Acad. Sci. **466:** 145–166. This volume.
24. MATUS, A. & B. RIEDERER. 1986. Ann. N.Y. Acad. Sci. **466:** 167–179. This volume.
25. WICHE, G., H. HERMANN, J. M. DALTON, R. FOISNER, F. E. LEICHTFRIED, H. LASSMANN, C. KOSZKA & E. BRIONES. 1986. Ann. N.Y. Acad. Sci. **466:** 180–198. This volume.
26. FELLOUS, A., R. OHAYON, J. C. MAZIE, F. ROSA, R. F. LUDUENA & V. PRASAD. 1986. Ann. N.Y. Acad. Sci. **466:** 240–256. This volume.
27. DRUBIN, D., S. KOBAYASHI & M. W. KIRSCHNER. 1986. Ann. N.Y. Acad. Sci. **466:** 257–268. This volume.
28. OLMSTED, J. B., C. F. ASNES, L. M. PARYSEK, H. D. LYON & G. M. KIDDER. 1986. Ann. N.Y. Acad. Sci. **466:** 292–305. This volume.
29. MARGOLIS, R. L., D. JOB, M. PABION & C. T. RAUGH. 1986. Ann. N.Y. Acad. Sci. **466:** 306–321. This volume.
30. SOLOMON, F. 1986. Ann. N.Y. Acad. Sci. **466:** 322–327. This volume.
31. BLOOM, G. S., F. C. LUCA, C. A. COLLINS & R. B. VALLEE. 1986. Ann. N.Y. Acad. Sci. **466:** 328–339. This volume.
32. BRADY, S. T. & M. M. BLACK. 1986. Ann. N.Y. Acad. Sci. **466:** 199–217. This volume.
33. WILLIAMS JR., R. C. 1986. Ann. N.Y. Acad. Sci. **466:** 798–802. This volume.
34. SELDEN, S. C. & T. D. POLLARD. 1986. Ann. N.Y. Acad. Sci. **466:** 803–812. This volume.
35. KEITH, C. H., A. S. BAJER, R. RATAN, F. R. MAXFIELD & M. L. SHELANSKI. 1985. Ann. N.Y. Acad. Sci. **466:** 375–391. This volume.
36. VALLANO, M. L., J. R. GOLDENRING, R. S. LASHER & R. J. DELORENZO. 1986. Ann. N.Y. Acad. Sci. **466:** 357–374. This volume.
37. BERLIN, R. D., C. S. REGULA & J. R. PFEIFFER. 1980. *In* Microtubules and Microtubule Inhibitors 1980. M. DEBRABANDER & J. DEMEY, Eds.: 145–160. Elseiver/North-Holland. Amsterdam.
38. AFZELIUS, B. 1979. Int. Rev. Exp. Pathol. **19:** 1–43.
39. PURPURA, D., N. BODICK, K. SUZUKI, I. RAPIN & S. WURZELMANN. 1982. Dev. Brain Res. **5:** 287–297.

Evolutionary Aspects of Tubulin Structure

MELVYN LITTLE,[a,c] GÜNTER KRÄMMER,[a]
MONIKA SINGHOFER-WOWRA,[a] AND
RICHARD F. LUDUENA[b]

[a]*Institute of Cell and Tumor Biology*
German Cancer Research Center
Heidelberg, Federal Republic of Germany

[b]*Department of Biochemistry*
University of Texas Health Science Center
San Antonio, Texas 75284

INTRODUCTION

In 1981, the publication of tubulin sequences from the brains of chicks, pigs, and rats[1-4] suggested that this protein was extremely conservative. Particularly striking was the presence of only one amino acid difference between the brain α tubulins of chicks and pigs, organisms that diverged about 280 million years ago.[5] Since then, however, the elucidation of other tubulin sequences has shown that the mutation rates of both α and β tubulin can vary quite considerably. For example, the sequences of two yeast α tubulins[6] and one yeast β tubulin[7] contain more than a hundred differences compared to all the published sequences of other α and β tubulins. The α tubulin from another nonanimal organism, the slime mold *Physarum polycephalum,* also contains relatively many differences to other α tubulins (unpublished results, see below).

AMINO ACID SEQUENCE HOMOLOGIES

α Tubulins

TABLE 1 shows the percent homology between all the α tubulins for which extensive sequence data is available. All the animal tubulins are at least 97% homologous to one another. They show markedly less homology, however, to an α tubulin expressed in the plasmodia of *Physarum polycephalum* (83%) and even less to the α tubulins from the yeast *Schizosaccharomyces pombe* (76%). The least homology is between the yeast and *Physarum* α tubulins (70%). It appears, therefore, that both the yeast and *Physarum* α tubulins have a much higher rate of mutation than the animal α tubulins. Indeed, the large structural constraints imposed upon the animal α tubulins of TABLE 1 appear to allow only very few sequence changes.

β Tubulins

The largest sequence differences among the β tubulins are also between animal and nonanimal β tubulins, although only one nonanimal β tubulin, that of the yeast

---

[c]Send correspondence to Dr. M. Little, Institute of Cell and Tumor Biology, German Cancer Research Center, Im Neuenheimer Feld 280, D-6900 Heidelberg 1, Federal Republic of Germany.

8

TABLE 1. α-Tubulin Amino Acid Sequence Homologies (percent). The Number of Sequenced Amino Acids Is Shown in Parenthesis.

	Human brain (451)	Pig brain (451)	Rat brain (451)	Chick brain (411)	Sea urchin (161)	Physarum (423)	Yeast A 1 (455)	Yeast A 2 (451)
Human brain[8]	100	99.3	99.8	99.0	98.1	83.0	76.2	76.5
Pig brain[2]	99.3	100	99.6	99.3	97.5	82.5	76.0	76.1
Rat brain[4]	99.8	99.6	100	99.3	98.1	83.0	76.2	76.5
Chick brain[1]	99.0	99.3	99.3	100	97.5	83.0	75.9	75.4
Sea urchin[9]	98.1	97.5	98.1	97.5	100	86.3	78.3	77.6
Physarum	83.0	82.5	83.0	83.0	86.3	100	70.2	71.0
Yeast[6] A 1	76.2	76.0	76.2	75.9	78.3	70.2	100	87.6
Yeast[6] A 2	76.5	76.1	76.5	75.3	77.6	71.0	87.6	100

Schizosaccharomyces pombe,[7] was available at the time of writing (TABLE 2). The animal β tubulins, however, show many more differences among themselves than are found among the α tubulins. A large proportion of these differences are present at the C terminus, which appears to be particularly heterogeneous. By analogy to the α tubulin subunit, this part of the β tubulin molecule might be expected to protrude from the microtubule surface where it could be involved in binding regulatory molecules such as microtubule-associated proteins (MAPs).

CHORDATE BRAIN β TUBULINS

Particularly surprising is the high degree of homology (99.1%) between one of the chick brain tubulins and the corresponding variant in pig brain tubulin. By contrast, there is only 95% homology between the brain β tubulins of the more closely related humans and pigs. Because the human brain β tubulin sequences were derived from mRNAs present in fetal brain tissue, it is possible that different β tubulins are expressed at different stages of development. It seems highly likely, moreover, that these different β tubulins have specific functions, as indicated by the highly conserved amino acid sequences of corresponding β tubulins from quite different organisms. For example, all of the twelve amino acid differences found between the major and minor β tubulin components of bovine brain (β_1 and β_2, respectively) during an investigation on β_2-specific peptides,[13] correspond exactly to twelve of the thirty-six differences between

TABLE 2. β-Tubulin Amino Acid Sequence Homologies (percent). The Number of Sequenced Amino Acids Is Shown in Parenthesis.

	Human brain (444)	Human brain (444)	Pig brain (445)	Chick brain (445)	Chick (449)	Sea urchin (177)	Yeast (450)
Human brain[10]	100	94.6	95.3	94.8	91.7	93.2	73.4
Human brain[11]	94.6	100	96.2	95.7	91.2	91.5	74.3
Pig brain[3]	95.3	96.2	100	99.1	92.6	94.9	73.9
Chick brain[1]	94.8	95.7	99.1	100	91.6	94.9	73.7
Chick[5]	91.7	91.2	92.6	91.6	100	90.4	73.3
Sea urchin[9]	93.2	91.5	94.9	94.9	90.4	100	71.2
Yeast[7]	73.4	74.3	73.9	73.7	73.3	71.2	100

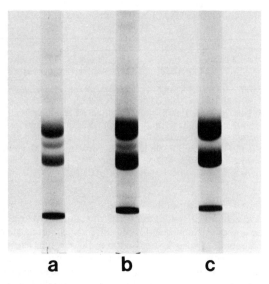

a   b   c

FIGURE 1. SDS-PAGE electrophoresis of carboxymethylated tubulin samples. The samples were run on 6% polyacrylamide gels using the system of Laemmli[16] and stained with Coomassie blue. a) calf brain tubulin; b) dogfish brain tubulin; c) sperm tail tubulin from sea urchin *Strongylocentrotus purpuratus*.

two chick β tubulins.[1,12] Some of these differences were first observed as heterogeneities during the sequencing of pig brain tubulin that consisted of a mixture of β_1 and β_2 tubulins.[3] Systematic sequencing of the separated mammalian β_1 and β_2 tubulins or their corresponding mRNAs will probably reveal a similar number of amino acid differences as found for the two chick β tubulins. In addition to a large number of amino acid differences at the C terminus,[12] a highly conserved cluster of amino acid differences is present around position 55.[3,12,13]

The two β tubulins of chordate brain are easily recognized by their different electrophoretic mobilities on sodium dodecyl sulfate polyacrylamide gel electrophoresis (SDS-PAGE) after carboxymethylation.[14] They also elute differently from hydroxyapatite. Although a wide variety of organisms were tested for the presence of β_2, it was only found in chordate brain; none was found in squid brain.

The relative amounts of β_1 and β_2, however, vary from organism to organism. In mammalian brain, β_2 accounts for approximately 25% of the β tubulin but for only about 8% of dogfish brain[14,15] (FIGURE 1). Perhaps the most striking difference between β_1 and β_2 tubulins is their different reactivity towards the potent microtubule inhibitor ethylene bis (iodoacetamide) (EBI), a bifunctional sulfhydryl reagent; two cysteines of β_1, but not β_2, can be specifically cross-linked. The presence of only one molecule of EBI cross-linked in a β_1 subunit prevents its assembly into microtubules.[15] The two cysteines were found to be in positions 239 and 354.[13] Although they are separated by 115 amino acid residues along the β_1 chain, they are maximally 9 Å apart in the tertiary structure.

In the presence of colchicine and podophyllotoxin, the two cysteines cannot be cross-linked, suggesting that colchicine binds on or near this region of β tubulin. Supporting evidence that this is indeed the case was provided by Thomas and

I   Ia   II

FIGURE 2. Peptide patterns of α tubulins after limited proteolytic digestion with *S. aureus* protease and electrophoresis on 15% acrylamide gels. Only the major peptides have been sketched that are consistent characteristics of these patterns. Type I: animal axonemal α tubulins; Type Ia: animal cytoplasmic α tubulins. The double-headed arrow denotes the position of a group of peptides showing some pattern variation; Type II: nonanimal α tubulins.

Botstein,[17] who showed that an arginine in position 241 was substituted by histidine in a benomyl-resistant mutant of the yeast *Saccharomyces cerivisae*. Benomyl-like compounds have been shown to compete for the colchicine binding site.[18,19] Moreover, a comparison of the sequence of yeast β tubulin around position 241 to mammalian brain β tubulins shows that it is fairly conservative, except for position 239, which is a serine in yeast and a cysteine in brain β_1 tubulin. It is possible, therefore, that the presence of a highly reactive cysteine in position 239 is important for colchicine binding. If this is indeed the case, then chordate brain β_2 tubulin should not bind colchicine as well as brain β_1 tubulin. Although the highly reactive cysteine 239 is obviously not required for polymerization, it may be involved in a sulfhydryl-mediated mechanism of microtubule assembly.

TUBULIN PEPTIDE MAPS

The *Staphylococcus aureus* protease one-dimensional peptide maps of 28 different α tubulins from a wide variety of organisms and organelles suggest that the large differences between the sequences of animal and nonanimal α tubulins is a general phenomenon.[20-22] The β-tubulin maps showed an even distribution of small to large peptides that all appeared to be fairly similar. The α-tubulin maps, in contrast, were very distinct and could be correlated with particular types of tubulin. It was thus possible to show that functionally similar tubulins from quite different organisms could be more related to one another than tubulins from the same organism.[20] This hypothesis has since been confirmed using two-dimensional tryptic peptide maps of both α and β tubulins (unpublished results). The strongest evidence, however, for this hypothesis is the sequence data discussed above for the brain β tubulins.

The major differences in the *S. aureus* protease one-dimensional maps were between the animal α-tubulin peptide maps, which all shared a basic similarity, and the nonanimal α-tubulin maps (FIGURE 2). Twenty animal tubulins gave a peptide pattern that was quite distinct from that of seven nonanimal tubulins (FIGURE 2). A summary of this data is given by Little *et al.*[22]

EVOLUTION OF METAZOAN TUBULINS

The data from tubulin sequence comparisons combined with that of the peptide maps suggests that the mutation rate of animal tubulins, particularly that of the α

tubulins, underwent a drastic reduction during the evolution of the metazoa. Indeed, the structural constraints imposed on some of the animal tubulins is so great, that their mutation rate appears to be comparable to those of the most conservative histones.

A possible explanation for the apparently very low mutation rate of many of the animal tubulins is that they are all involved to a greater or lesser extent in one or more functions specific to metazoan organisms that require a very precise tubulin structure. One obvious feature common to most metazoan organisms is the presence of nerve cells with a high tubulin content. It would thus be very interesting to examine the tubulins of sponges, organisms that diverged very early from the rest of the metazoa and that do not contain highly specialized nerve and muscle cells. Similar comparisons of primitive metazoan and nonanimal tubulins may help to identify domains in the tubulin molecule that are important for particular functions.

REFERENCES

1. VALENZUELA, P., M. QUIROGA, J. ZALDIVAR, W. J. RUTTER, R. W. KIRSCHNER & D. W. CLEVELAND. 1981. Nature (London) **289:** 650–655.
2. PONSTINGL, H., E. KRAUHS, M. LITTLE & T. KEMPF. 1981. Proc. Natl. Acad. Sci. USA **78:** 2757–2761.
3. KRAUHS, E., M. LITTLE, T. KEMPF, R. HOFER-WARBINEK, W. ADE & H. PONSTINGL. 1981. Proc. Natl. Acad. Sci. USA **78:** 4156–4160.
4. LEMISCHKA, J. R., S. FARMER, U. R. RACANIELLO & P. A. SHARP. 1981. J. Mol. Biol. **151:** 101–120.
5. DAYHOFF, M. O. 1978. Atlas of protein sequences and structure (National Biomedical Research Foundation, Washington D.C.) Vol. 5, Suppl. 3.
6. TODA, T., Y. ADACHI, Y. HIRAOKA & M. YANAGIDA. 1984. Cell **37:** 233–242.
7. NEFF, N. F., J. H. THOMAS, P. GRISAFI & D. BOTSTEIN. 1983. Cell **33:** 211–219.
8. COWAN, N. J., P. R. DOBNER, E. V. FUCHS & D. W. CLEVELAND. 1983. Mol. Cell. Biol. **3:** 1738–1745.
9. ALEXANDRAKI, D. & J. V. RUDERMAN. 1983. J. Mol. Evol. **19:** 397–410.
10. HALL, J. L., L. DUDLEY, P. R. DOBNER, S. A. LEWIS & N. J. COWAN. 1983. Mol. Cell. Biol. **3:** 854–862.
11. GWO-SHU LEE, M., C. LOOMIS & N. J. COWAN. 1984. Nucleic Acids Res. **12:** 5823–5836.
12. SULLIVAN, K. F. & D. W. CLEVELAND. 1984. J. Cell Biol. **99:** 1754–1760.
13. LITTLE, M. & R. F. LUDUEÑA. 1985. EMBO J. In press.
14. LITTLE, M. 1979. FEBS Lett. **108:** 283–286.
15. LUDUEÑA, R. F., M. C. ROACH, P. P. TRCKA, M. LITTLE, P. PALANIVELU & P. BINKLEY. 1982. Biochemistry **21:** 4787–4794.
16. LAEMMLI, U. K. 1970. Nature (London) **227:** 680–685.
17. THOMAS, J. H. & D. BOTSTEIN. 1984. *In* Molecular biology of the cytoskeleton. Cold Spring Harbor Laboratory Meeting, p. 9 (Abstract).
18. HOEBEKE, J., G. VAN NIJEN & M. DE BRABANDER. 1976. Biochem. Biophysics Res. Commun. **69:** 319–324.
19. CORTESE, F., B. BHATTACHARYYA & J. WOLFF. 1977. J. Biol. Chem. **252:** 1134–1140.
20. LITTLE, M., R. F. LUDUEÑA, G. M. LANGFORD, C. F. ASNES & K. FARRELL. 1981. J. Mol. Biol. **149:** 95–107.
21. LITTLE, M., R. F. LUDUEÑA, R. KEENAN & C. F. ASNES. 1982. J. Mol. Evol. **19:** 80–86.
22. LITTLE, M., R. F. LUDUEÑA, R. C. MOREJOHN, C. ASNES & E. HOFFMAN. 1984. Origins Life **13:** 169–176.

Genetic and Functional Analysis of Beta Tubulin in *Aspergillus nidulans*[a]

N. RONALD MORRIS, GREGORY S. MAY, JAMES A. WEATHERBEE, MONICA LIK-SHING TSANG, AND JOHN GAMBINO

Department of Pharmacology
UMDNJ-Rutgers Medical School
Piscataway, New Jersey 08876

In *Aspergillus,* as in most other eukaryotes, there are a number of tubulin polypeptides that are coded by a family of related genes. Two alpha-tubulin polypeptides, alpha 1 and alpha 3, are coded by the *tub* A gene and a third alpha tubulin, alpha 2, by another gene, *tub* B.[1] Similarly, two beta-tubulin polypeptides, beta 1 and beta 2, are coded by the *ben* A gene, and a third polypeptide, beta 3, is coded by *tub* C.[2] A number of mutations have been identified in the *ben* A and *tub* A genes that have been helpful in analyzing the functions of the *ben* A and *tub* A gene products; but, heretofore, no mutations have been described in the *tub* B or *tub* C genes.

Mutations in *ben* A have been shown to confer resistance to growth inhibition by antimicrotubule drugs, for example, benomyl, and to alter the electrophoretic behavior on 2-D gels of the beta-1 and beta-2 (but not the beta 3) polypeptides.[2,3] One such mutation, *ben* A33, causes a temperature dependent hyperstabilization of microtubules that has allowed us to demonstrate that *ben* A gene product functions in both chromosome and nuclear movement and is a component of both spindle and cytoplasmic microtubules.[4,5] Mutations in tub A were initially isolated as indirect suppressors of temperature sensitive *ben* A mutations, apparently by causing microtubule destabilization and thereby reversing *ben* A33-induced microtubule hyperstabilization.[1,4,5] These mutations, which affect the electrophoretic mobility of alpha 1 and alpha 3 tubulins (but not alpha-2 tubulin) have been used to demonstrate that the *tub* A gene product also functions in both chromosome and nuclear movement and is a component of both spindle and cytoplasmic microtubules.[4,5]

In this preliminary description of recent results, we describe a series of ultraviolet (UV)-induced mutations that affect both asexual sporulation (conidiation) and the production of beta-3 tubulin. Whether these mutations are in *tub* C, the structural gene for beta-3 tubulin, or in some other gene that affects the level of beta-3 tubulin is not yet known. We also demonstrate how recombinant DNA technology and DNA-mediated transformation can be used to generate authentic mutations in *tub* C, and we use one such artificially generated mutation to show that beta-3 tubulin plays a role in conidiation.

RESULTS

Identification of Conidiation-Resistant Mutants

Strains of *Aspergillus* carrying the *ben* A22 mutation[2] grow in the presence of concentrations of benomyl that are inhibitory to wild-type strains, but they do not form

[a]This work was supported by a Grant from the National Institutes of Health, GM29228. G.S. May was supported by a fellowship (600) from the Anna Fuller Fund.

asexual spores (conidia).[6] This observation suggested to us that conidiation might involve a benomyl-sensitive tubulin different from the benomyl-resistant beta tubulin that is involved in vegetative growth. To test this hypothesis, *ben* A22 was mutagenized with UV light, and mutants able to conidiate in the presence of benomyl were selected. These mutants were given the designation CR⁻ for conidiation resistance. Twenty independent CR⁻ mutants were selected.

The tubulin of these CR⁻ mutants was studied by 2D gel electrophoresis. Because the *ben* A22 mutation causes a 2+ electrophoretic charge shift in beta-1 and beta-2 tubulins,[2,3] beta-3 tubulin, which is normally occluded by beta 1 and beta 2, can be seen in this strain.[3] All twenty CR⁻ mutants were examined. In each of these mutants the beta-3 tubulin spot was absent. To determine whether the CR⁻ phenotype and the absence of beta 3 were the result of the same mutation, two CR⁻ mutants were crossed to two different wild-type strains, and the progeny was analyzed for CR⁻ and beta 3. The CR⁻ phenotype cosegregated with the absence of beta 3, demonstrating that the effects were caused by the same mutation. This result suggests strongly that absence of the beta-3 gene product is in all probability responsible for the CR⁻ phenotype in these mutants.

Cloning of the Beta Tubulin Genes

Two different beta tubulin clones, beta 5 and beta 14, were isolated from a Charon 4A *Aspergillus* library by hybridization with both chicken cDNA[7] and yeast genomic[8] beta-tubulin probes. Although the beta-5 and the beta-14 sequences both hybridized with chicken and yeast beta-tubulin genes at low stringency, they failed to hybridize with each other at high stringency, indicating that there is considerable sequence divergence between these genes. The beta-5 and beta-14 clones were subcloned into M13 and partially sequenced to verify their identity as beta tubulins. The tubulin sequences were then used to construct *Aspergillus*-transforming plasmids that also contained the *Neurospora* gene for orotidylate decarboxylase, *pyr* 4⁺, as a selective marker. Three such plasmids were constructed: AlpGM1 (*Aspergillus* Integrative plasmid GM1), which contains the beta 14 sequence, AlpGM4, which contains the beta-5 sequence, and AlpGM6, which contains an internal fragment of the beta-14 sequence.

Transformation with Tubulin Genes

A uridine auxotroph *pyr* G⁻ strain of *Aspergillus* (G191) was the initial recipient for transformation by AlpGM1 and AlpGM4. Protoplasts of G191 were transformed by the method of Balance, Buxton, and Turner,[9] and transformants to uridine prototrophy were selected. Transformation with AlpGM1 and AlpGM4 was integrative and site specific with each of the two plasmids integrating into a different genomic sequence. The fact that integration was site specific allowed us to determine which of the two beta-tubulin sequences was the *ben* A gene by using the plasmid-linked *pyr* 4⁺ (*Neurospora*) gene to map the site of integration of each of the plasmids with respect to the *ben* A locus. A site specific AlpGM1 (beta 14) *pyr* 4⁺ transformant and a site specific AlpGM4 (beta 5), *pyr* 4⁺ transformant were respectively crossed to a *pyr* G⁻, *ben* A15, benomyl-resistant mutant strain; the progeny were analyzed for both uridine prototrophy and benomyl resistance. In the case of the AlpGM4 (beta 5) transformant, there was essentially no recombination between the *pyr* 4+ marker and the *ben* A marker, indicating that this plasmid had integrated at the *ben* A locus. In the case of the AlpGM1 (beta 14) plasmid, the markers recombined freely, showing that the beta 14 gene was not closely linked to the *ben* A locus.

Additional evidence that AlpGM4 (beta 5) contained the *ben* A gene came from an experiment in which both AlpGM1 and AlpGM4 were used to transform a temperature-sensitive, *ben* A33, *pyr* G⁻ mutant. Because the *ben* A33 mutation is recessive for both temperature sensitivity and for benomyl resistance when heterozygous with wild-type (*ben* A33/+),[10] (TABLE 1), it seemed reasonable that integration of a wild-type copy of the *ben* A gene (which should result in a tandemly duplicated gene with one copy mutant and the other copy wild-type[11]) might have the same effect. In the case of the AlpGM4 (beta 5) transformants this was the case. The majority of the AlpGM4 transformants were not growth inhibited at 42° and were less resistant to benomyl than *ben* A33. All of the AlpGM1 (beta 14) transformants had the parental phenotype.

Disruption of the Tub C Gene

Site specific integration by DNA-mediated recombination of internal gene fragments lacking both the 5′ and 3′ ends of a gene is known to result in two incomplete tandem gene copies that, since neither contains the complete sequence of the gene,

TABLE 1. The Effect of Beta Tubulin Composition on Sensitivity to Benomyl of Vegetative Growth and Conidiation

Growth	Haploid Beta 1/2^R, Beta 3^S	Diploid Beta 1/2^R, Beta 3^S Beta 1/2^S, Beta 3^S	Haploid Beta 1/2^R, Beta 3°
Vegetative	R	S	R
Conidial	S	S	R

effectively constitute a gene disruption or "knock out."[11] We have used the AlpGM6 plasmid in this way to "knock out" the gene that corresponds to the beta-14 cloned sequence. Because we had previously identified the beta-5 sequence as corresponding to the *ben* A gene (see above), it seemed likely that the beta-14 sequence would correspond to the *tub* C gene, which codes for beta-3 tubulin. To be able to visualize (by 2-D gel electrophoresis) the effect of knocking out the *tub* C gene on its putative beta-3 tubulin gene product, we prepared a *ben* A22, *pyr* G⁻ recipient strain, which we then transformed with the AlpGM6 (beta 14) disrupter plasmid. Seven such transformants have been analyzed by Southern blotting and by 2-D gel electrophoresis of tubulin polypeptides. Integration was site specific in each of these, and in each, beta-3 tubulin was missing from the 2-D gel electropherogram. By analogy with the CR⁻ mutations, which cause the loss of beta-3 tubulin and also cause conidiation resistance, we expected that conidiation in these strains might be resistant to benomyl. This was tested and proved to be correct. All seven disrupted strains were conidiation resistant to benomyl.

DISCUSSION

Aspergillus nidulans is a relatively complex, simple eukaryote with a variety of different cell types and a well characterized pattern of development. Its tubulins have been characterized biochemically,[1-3] and during vegetative growth its microtubules resemble those of higher cells. There is a complex network of cytoplasmic microtubules present during interphase that disappears at mitosis and is replaced by a well-defined mitotic spindle with astral microtubules at the spindle poles.[5]

This laboratory has previously taken advantage of mutations affecting *Aspergillus* tubulins to analyze microtubule function in vegetative cells. As noted above, the *ben* A and *tub* A tubulins have been shown to be components of both cytoplasmic and spindle microtubules and to function in both chromosome and nuclear movement.[5] This type of analysis, however, has been limited by the fact that up to now we have been able to find mutations in only two of the four known *Aspergillus* tubulin genes. In the present report, we extend our capabilities by describing UV-induced and integrative-plasmid-induced mutations in the *tub* C gene.

The UV-induced CR⁻ mutations were selected in *ben* A22, in which vegetative growth is resistant to benomyl. Under the same conditions conidiation (asexual sporulation) is sensitive to benomyl. This differential sensitivity of vegetative growth and conidiation in *ben* A22 (CR⁺) strains suggests that the two beta-tubulin genes are used differentially, that only the *ben* A beta-tubulin gene product is needed for vegetative growth, but both beta-tubulin gene products are needed for conidiation. By this model, in the *ben* A22 mutant, the benomyl-resistant beta-1 and/or beta-2 tubulin gene products would confer resistance to inhibition of vegetative growth by benomyl, but the presence of a benomyl-sensitive beta-3 subunit would cause conidiation to be sensitive to the drug. The observation that heterozygous *ben* A⁻/+ diploids are sensitive to benomyl[10] is consistent with these data and indeed suggests this explanation. The data are summarized in TABLE 1. In the CR⁻ derivatives of *ben* A22, in which there is no beta-3 tubulin to cause sensitivity, conidiation becomes resistant to benomyl. An alternative explanation of the data is that only the *tub* C gene product functions in normal conidiation, but if beta-3 tubulin is absent, this causes the induction of beta-1 and/or beta-2 tubulin.

All twenty independently isolated CR⁻ mutants lacked beta-3 tubulin. Our previous experience with the *ben* A and *tub* A mutations would have led us to expect that some of these mutants would exhibit electrophoretic variants of beta 3 rather than a complete absence of the gene product.[1-3] Thus, either there exists no missense mutation of beta 3 that can cause benomyl resistance in the presence of the *ben* A22 mutation or such mutations are lethal. We also do not know whether the CR⁻ mutations are in *tub* C, the structural gene for beta 3, or whether they are in genes that affect the level of beta-3 transcription, translation, or stability. Nor do we yet know whether they represent a single class of mutation or several different mutations representing the gamut of above-mentioned, possible mechanisms.

We have cloned two genomic sequences that code for beta tubulin and by using DNA-mediated transformation have identified one as corresponding to the *ben* A gene by showing that it integrates at the *ben* A locus. The other beta tubulin clone has been identified as the *tub* C gene not only by virtue of the fact that it does not integrate at the *ben* A locus but most convincingly because the AlpGM6 disruptor plasmid containing an internal fragment of this gene ablates the beta-3 tubulin spot from 2-D gel electropherograms. Elimination of beta-3 tubulin by gene disruption in a *ben* A22 background also causes conidiation to become resistant to benomyl, thereby supporting the hypothesis that beta-3 tubulin plays a specific role in conidiation.

We consider the experiments described in this paper to be paradigmatic. By cloning the alpha-tubulin genes and doing gene disruption experiments similar to the one described here, it should be possible to determine which alpha-tubulin genes are essential for several cellular and developmental processes. Obviously, the same procedures, and variations on the procedure using *in vitro* mutagenesis of cloned sequences,[11] can in principle be used to analyze the function of any cloned gene in *Aspergillus*.

In summary, the many features of *Aspergillus* that we have described previously, its good cytology, ease of handling, and excellent genetics, are now complemented by DNA-mediated transformation to make it one of the most amenable systems for experimental analysis of the cell biology of microtubules and other proteins of the cytoskeleton.

ACKNOWLEDGMENT

We thank Happy Smith for excellent technical assistance, particularly with the transformation experiments.

REFERENCES

1. MORRIS, N. R., M. LAI & C. E. OAKLEY. 1979. Cell **16:** 437–42.
2. SHEIR-NEISS, G., M. LAI & N. R. MORRIS. 1978. Cell **15:** 639–47.
3. WEATHERBEE, J. A. & N. R. MORRIS. 1984. J. Biol. Chem. **259:** 15452–15459.
4. OAKLEY, B. R. & N. R. MORRIS. 1981. Cell **24:** 837–45.
5. GAMBINO, J., L. G. BERGEN & N. R. MORRIS. 1984. J. Cell Biol. **99:** 830–38.
6. OAKLEY, C. E. & N. R. MORRIS. Unpublished observation.
7. CLEVELAND, D. W., M. A. LOPATA, R. J. MACDONALD, N. J. COWAN, W. J. RUTTER & M. W. KIRSCHNER. 1980. Cell **20:** 95–105.
8. NEFF, N. F., J. H. THOMAS, P. GRISAFI & D. BOTSTEIN. 1983. Cell **33:** 211–19.
9. BALANCE, D. J., F. P. BUXTON & G. TURNER. 1983. Biochem. Biophys. Res. Comm. **112:** 284–9.
10. MORRIS, N. R. Unpublished observations.
11. BOTSTEIN, D. & R. MAURER. 1982. Annu. Rev. Genet. **16:** 61–83.

Chlamydomonas reinhardtii Tubulin Gene Structure

CAROLYN D. SILFLOW AND JAMES YOUNGBLOM

Department of Genetics and Cell Biology
University of Minnesota
St. Paul, Minnesota 55108

INTRODUCTION

The unicellular green alga *Chlamydomonas reinhardtii* is a useful model system for studies of microtubule diversity and function. The cells contain several functionally and spatially distinct types of microtubules, including those found in the mitotic apparatus, cytoplasmic rootlets, basal bodies, and flagella.[1,2] Heterogeneity of tubulin components of microtubules has been demonstrated by two-dimensional (D) gel analysis,[3-5] and may contribute to structural or functional differences among the various microtubule classes. Microheterogeneity of alpha tubulin in *C. reinhardtii* is due to both gene-encoded variation and posttranslational modification.[3-7]

C. reinhardtii has also been a useful system for studying the regulation of tubulin gene expression. Changes in the level of gene expression occur during the cell cycle, and in response to a number of experimental manipulations that alter the state of the microtubules within the cell. The relative rate of tubulin synthesis increases during the period of the cell cycle just prior to cytokinesis,[8] decreases during flagellar shortening (resorption), which can be induced by a number of chemical treatments,[9,10] and increases in response to the loss or shortening of flagella caused by deflagellation or chemically induced resorption.[9,11,12] Increases or decreases in the relative rate of tubulin synthesis correspond qualitatively to changes in the level of tubulin mRNA as determined by *in vitro* translation,[3,11] or hybridization of radioactive DNA probes to RNA blots.[13-17] Regulation of mRNA levels is controlled at least partly by increased transcription of tubulin genes in response to deflagellation and during the cell cycle.[18-20] In addition, alteration of the stability of tubulin transcripts appears to contribute to the regulation of mRNA levels after deflagellation.[19]

Tubulin gene regulation in *C. reinhardtii* is coordinate, in contrast to the regulation seen in many higher eukaryotes, where members of the tubulin gene family are expressed differentially during development (*e.g.* references 22–24). Two size classes of alpha-tubulin transcripts and two size classes of beta-tubulin transcripts are distinguished on RNA gel blots.[7,13,16,21] Using heterologous or homologous cDNA (complementary DNA) probes that hybridize to both of the alpha-, or to both of the beta-tubulin transcripts, the different transcripts appear to be regulated coordinately under a variety of conditions.[13,16] Hybridization of cloned cDNA probes prepared from each of the transcripts to blots of RNA and blots of genomic DNA has indicated that each transcript is produced from one of the four tubulin genes present in the haploid *C. reinhardtii* genome.[7,13,16,21] The *Chlamydomonas* system presents an opportunity to study both the molecular mechanisms for coordinating tubulin gene expression, and the function in the cell of the protein product produced by each member of the tubulin gene family.

18

RESULTS AND DISCUSSION

To determine the heterogeneity of tubulins encoded by the two alpha- and two beta-tubulin genes of *C. reinhardtii*, and to find sequences that may be important for the coordinate expression of the genes, we have sequenced the DNA of all four tubulin genes.[7,21] Our studies, in addition to those of Brunke *et al.*[25] on the 5′ promoter region of each gene, give a complete picture of the tubulin genes in this organism. Because our sequences from each of four different full-length cDNA clones were compared with those from genomic clones, our data support earlier evidence that each tubulin gene is transcribed, and produces a single major transcript.[25]

Sequence of Alpha-Tubulin Genes

The DNA sequence of the two alpha-tubulin cDNA clones (pcf 10-2, which hybridizes preferentially to a 1.9 kilobase (kb) transcript, and pcf 4-2, which hybridizes preferentially to a 1.8 kb alpha-tubulin transcript) was determined. Partial sequencing of two cloned genomic DNA fragments, and comparison with the cDNA sequence, identified the alpha-1 gene as the source of the 1.9 kb transcript, and the alpha-2 gene as the source of the 1.8 kb transcript.[7] These comparisons also determined the locations (after amino acid 15 and within amino acid 90) of the two intervening sequences in each of the genes.

The coding region of each cDNA clone was determined by comparison of sequence data with a rat alpha-tubulin sequence.[26] The coding region of each *C. reinhardtii* alpha-tubulin gene contains 1356 base pairs (bp) encoding a protein of 451 amino acids (FIGURE 1). The sequences are 98% conserved, with only 27 nucleotide differences between the two genes. The majority of the nucleotide differences are found at the third position of a codon, and do not change the predicted amino acid. Two nucleotide differences, however, at positions 923 and 1097, occur at the second position of a codon. These differences predict an arginine and a leucine at position 308 in the alpha-1 and alpha-2 genes, respectively. At amino acid 366, the alpha-1 tubulin contains a glycine, whereas the alpha-2 tubulin contains a valine. Thus, the sequence data predicts that two different alpha-tubulin primary gene products are produced in *C. reinhardtii*, and that the two proteins will differ by one net charge.

The two predicted alpha tubulins may correspond to the two alpha tubulins resolved on 2-D gels of *in vitro* translation products.[3,27] If this is the case, we predict that the alpha-1 protein, with a pI of approximately 5.7, is the product of the alpha-1 gene, whereas the alpha-3 protein, with a pI of 5.5, is the alpha-2 gene product. It is not known whether the amino acid differences specify functional differences between the proteins, or are simply the result of functionally unimportant sequence divergence of a duplicated gene. It is interesting to note, however, that the arginine at amino acid 308 is conserved in all other alpha tubulins and beta tubulins sequenced to date (*i.e.* references 26, 28–31), suggesting that the alpha-2 polypeptide is divergent at an amino acid important for the normal function of the tubulin polypeptides. The glycine at amino acid 366 is conserved in alpha tubulins from vertebrates,[26,28,29] although it is replaced by a serine in alpha tubulins from *Schizosaccharomyces pombe*.[31] These comparisons suggest that the alpha-1 gene is more closely related to the ancestral alpha-tubulin gene, and that alpha 2 is the more divergent gene.

Further studies will be needed to determine the function of the two alpha tubulins. An important question is whether both proteins are assembled into flagellar microtu-

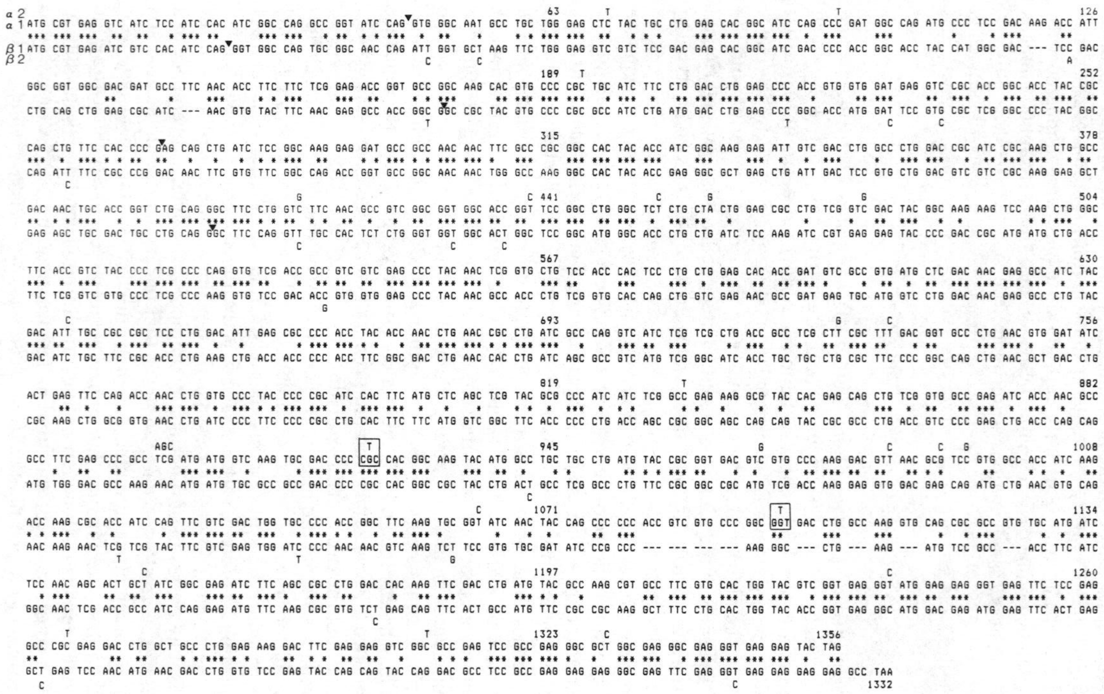

FIGURE 1. Comparison of nucleotide sequences of alpha- and beta-tubulin gene coding regions. Sequences were determined by dideoxynucleotide termination of DNA synthesis using cloned cDNA or genomic DNA templates.[7,21] Sequences read from 5′ to 3′ with respect to the mRNA. The full sequences of the alpha-1 and the beta-1 gene are shown. Only those nucleotides that differ from the alpha-1 or beta-1 gene sequence are shown for the alpha-2 and beta-2 genes. Asterisks indicate nucleotides conserved between the alpha-1 and beta-1 genes. Dashes are inserted into the beta-tubulin sequence as necessary to maximize the homology at the amino acid level (see FIGURE 2). Arrows mark the locations of intervening sequences within the genes. These locations are identical for both alpha-tubulin genes and for both beta-tubulin genes.

bules. The major flagellar alpha tubulin (alpha 3) has a more acidic pI (5.5) than the major cell body alpha tubulin (alpha 1, pI 5.7), and accumulates by a posttranslational acetylation of a lysine residue in the precursor alpha tubulin.[3,32] This modification would be expected to decrease the pI of the modified protein. In an experiment in which cells were labeled with [^{14}C] acetate during flagellar regeneration in the presence of cycloheximide, only one labeled flagellar alpha-tubulin species (alpha 3) was detected on 2-D gels.[5] This result suggests that only one of the alpha tubulins (presumably the more basic protein, alpha 1) is a substrate for the modification. If the alpha-3 gene product were acetylated, it should migrate with a pI more acidic than 5.5. An alternative explanation could be that both proteins are modified, but that only one acetylated tubulin is transported to the flagella. Additional experiments will be needed to determine whether both the alpha-1 and alpha-2 gene products are modified, and whether they are both assembled into flagellar microtubules.

Sequence of Beta-Tubulin Genes

Two beta-tubulin genomic clones were sequenced, and the coding regions were identified by comparison with a chicken beta-tubulin sequence.[28] The sequences were compared with partial sequences of two full-length beta-tubulin cDNA clones to determine that the beta-1 gene codes for a 2.1 kb transcript, and the beta-2 gene codes for a 2.0 kb transcript.[21] The coding regions of each gene contain 1332 bp, interrupted by three intervening sequences: the first is located after amino acid 8, the second interrupts amino acid 57, and the third interrupts amino acid 132 (FIGURE 1). The coding regions contain 19 nucleotide differences between the 2 genes. All of the differences occur at the third position of an amino acid codon, and do not change the predicted amino acid sequence. Thus, both beta-tubulin genes encode an identical protein of 443 amino acids. The sequencing results, together with the observed coordinate expression of the two beta-tubulin genes,[16,21] suggest that the gene products are assembled interchangeably into all of the different microtubules in the cell. Further studies will be needed to determine why the cells maintain two copies of the beta-tubulin genes. One possibility is that two genes provide for the optimal rate of beta-tubulin synthesis in response to deflagellation. Another possibility is that gene number is maintained by an unknown mechanism that selects for equal numbers of alpha-tubulin and beta-tubulin genes, a situation that has been noted in a number of different systems (reviewed in references 33, 34).

Comparison of Alpha-Tubulin and Beta-Tubulin Nucleotide and Amino Acid Sequences

The probable origin of alpha and beta tubulin from a common ancestral gene has been supported by amino acid sequence comparisons in several systems.[28,35] The predicted amino acid sequences of the C. reinhardtii proteins are aligned in FIGURE 2 to maximize their homology. Ten amino acids are inserted into the alpha-tubulin sequence relative to the beta-tubulin sequence. The locations of the additional amino acids in the alpha-tubulin sequence are the same as those used to maximally align the alpha and beta tubulins of pig and chicken.[28,35] Two insertions are located in the amino-terminal end of the sequence, at amino acid 40 and 49; eight insertions are located near the carboxyl-terminus, at positions 361–364, and at 367, 369, 371, and 375. The conservation of identical amino acids is 43%, a number that increases to 73% when conservative substitutions are considered[45]. This degree of conservation between the subunits is very similar to that observed in higher eukaryotes, and indicates the stability of these genes through eukaryotic evolution.[28,35]

The comparison of the *C. reinhardtii* alpha- and beta-tubulin genes at the nucleotide level is shown in FIGURE 1, with the same adjustments made as in FIGURE 2 to maximize the homology. The sequence conservation at this level is 60%, significantly higher than the 50% conservation noted between chicken alpha- and beta-tubulin genes.[28] The high degree of nucleotide conservation of the *C. reinhardtii* genes is probably related to the unusually high guanosine plus cytidine (G + C) content (64%) of the coding region. The G + C content is comparable to that reported for the genome as a whole,[36] but is significantly higher than the G + C content of 5' and 3' noncoding portions of the transcripts (54% G + C), or of the intervening sequences (59% G + C).

Codon Usage

The unusually high G + C content of the *C. reinhardtii* tubulin genes is related to the very biased codon usage found in these four genes (TABLE 1). For a total of 1788

```
                  20                  40                  60                  80
α MREVISIHIGQAGIQVGNACWELYCLEHGIQPDGQMPSDKTIGGGDDAFNTFFSETGAGKHVPRCIFLDLEPTVVDEVRT
  ***|| *| ** * *|*    **| | **** * *   |* |      |   * |*|*| |*| *** *||**** |* **|
β MREIVHIQGGQCGNQIGAKFWEVVSDEHGIDPTGTYHGD-SDLQLERI-NVYFNEATGGRYVPRAILMDLEPGTMDSVRS
                                                                                 78

                  100                 120                 140                 160
α GTYRQLFHPEQLISGKEDAANNFARGHYTIGKEIVDLALDRIRKLADNCTGLQGFLVFNAVGGGTGSGLGSLLLERLSVD
  *|* *|*|*| || *| |*|** *|**** * *||* |** |** *||* **** * |||*******|*|**| || ||
β GPYGQIFRPDNFVFGQTGAGNNWAKGHYTEGAELIDSVLDVVRKEAESCDCLQGFQVCHSLGGGTGSGMGTLLISKIREE
                                                                                 158

                  180                 200                 220                 240
α YGKKSKLGFTVYPSPQVSTAVVEPYNSVLSTHSLLEHTDVAVMLDNEAIYDICRRSLDIERPTYTNLNRLIAQVISSLTA
  *  |  * *|* ***|** |******| ** * *|*||*| ||*****|**** *|* |  **| |**|**| *|*||*
β YPDRMMLTFSVVPSPKVSDTVVEPYNATLSVHQLVENADECMVLDNEALYDICFRTLKLTTPTFGDLNHLISAVMSGITC
                                                                                 238

                  260                 280                 300    (L)          320
α SLRFDGALNVDITEFQTNLVPRIHFMLSSYAPIISAEKAYHEQLSVAEITNAAFEPASMMVKCDPRHGKYMACCLMYR
  |*** * **|*| ||   **|*|**|** | |||*| |     *|*|*|* || |** *****|*|| | ||*
β CLRFPGQLNADLRKLAVNLIPFPRLHFFMVGFTPLTSRGSQQYRALTVPELTQQMWDAKNMMCAADPRHGRYLTASALFR
                                                                                 318

                  340                 360   (V)          380                 400
α GDVVPKDVNASVATIKTKRTIQFVDWCPTGFKCGINYQPPTVVPGGDLAKVQRAVCMISNSTAIGEIFSRLDHKFDLMYA
  * | |*|*|| | |||||* |  **|* *| |*||| **    * * *    * *|***** *|* *|  |* *|
β GRMSTKEVDEQMLNVQNKNSSYFVEWIPNNVKSSVCDIPP----KG-L-K-MSA-TFIGNSTAIQEMFKRVSEQFTAMFR
                                                                                 390

                  420                 440        451
α KRAFVHWYVGEGMEEGEFSEAREDLAALEKDFEEVGAESAEGAGEGEGEEYΩ
  ||**|*** ****|* **|**  || |*| |||| ||***||** ****
β RKAFLHWYTGEGMDEMEFTEAESNMNDLVSEYQQYQDASAEEEGEFEGEEEEAΩ
                  443
```

FIGURE 2. Comparison of predicted amino acid sequences of alpha- and beta-tubulins from *C. reinhardtii*. The complete amino acid sequence predicted from the alpha-1 gene is compared with the sequence predicted by both beta-tubulin genes. The two amino acids that differ in the predicted alpha-2 protein are shown above the alpha-1 sequence at positions 308 and 366. Asterisks indicate amino acids conserved between alpha- and beta-tubulin; | bars indicate conservative substitutions based on the criteria described by Doolittle.[45] Dashes are inserted into the beta-tubulin sequence to maximize the homology.

TABLE 1. Codon Usage Comparison

	α	β		α	β		α	β		α	β		α	β
Phenylalanine			**Alanine**			**Serine**			**Tyrosine**			**Cysteine**		
UUU	1	0	GCU	5	10	UCU	1	5	UAU	0	0	UGU	0	0
UUC	39	46	GCC	64	42	UCC	24	24	UAC	34	28	UGC	22	20
Leucine			GCA	0	0	UCA	0	1	**Histidine**			**Tryptophan**		
UUA	0	0	GCG	5	2	UCG	24	20	CAU	0	2	UGG	6	10
UUG	0	0	**Valine**			AGU	0	0	CAC	24	18	**Arginine**		
CUU	2	1	GUU	2	1	AGC	7	8	**Glutamine**			CGU	6	4
CUC	6	28	GUC	32	28	**Proline**			CAA	0	0	CGC	35	40
CUA	1	0	GUA	0	0	CCU	1	1	CAG	28	42	CGA	0	0
CUG	64	31	GUG	35	31	CCC	37	33	**Asparagine**			CGG	0	0
Isoleucine			**Aspartic Acid**			CCA	0	0	AAU	2	0	AGA	0	0
AUU	7	5	GAU	12	5	CCG	0	4	AAC	30	42	AGG	0	0
AUC	47	31	GAC	40	43	**Threonine**			**Lysine**			**Glycine**		
AUA	0	0	**Glutamic Acid**			ACU	4	6	AAA	0	0	GGU	25	13
Methionine			GAA	0	0	ACC	46	47	AAG	38	30	GGC	54	61
AUG	22	40	GAG	70	74	ACA	0	0				GGA	0	0
						ACG	0	1				GGG	0	0
												Stop		
												UAA	0	2
												UAG	2	0
												UGA	0	0

amino acids, only 42 or 69% of the available codons are used. Approximately 90–92% of each tubulin protein is encoded by only 21 different codons. Most of the codons (93%) used in *C. reinhardtii* tubulins are synonyms that use G or C in the third position. In addition, an almost total exclusion of A at the third position has been noted; only two such codons are found in the four genes. That *C. reinhardtii* has an extremely low tolerance for A at the third position of codons is supported by examining third position substitutions between the two alpha tubulins and between the two beta-tubulin genes (FIGURE 3). A total of 42 such substitutions, which presumably have occurred since the genes were duplicated, are found: 12 are transversions and 30 are transitions. Of the transitions, 29 are C–T changes in the antisense or coding strand, whereas only one is an A–G change. Because these transitions presumably occur with equal frequency, the data suggest that G to A transitions are not tolerated in the third position. The one third position A to G difference occurs at amino acid 153 in the alpha-tubulin genes. We have previously suggested that the alpha-1 gene is the ancestral gene, based on its amino acid similarity with higher eukaryotes at positions 308 and 366. The third position A at amino acid 153 is found in the alpha-1 gene; the G is found in the alpha-2 gene. Thus, we can speculate that for this codon, the more ancestral third position A may have been replaced by a G in the divergent gene. Of the 12 third position transversion differences between the genes, the same bias against transversions involving A exists; only one such change, a C to A substitution in the beta-tubulin genes is found.

The lack of information on other genes in the *C. reinhardtii* genome precludes comparisons with the tubulin genes to determine whether the biased codon usage described here is typical of all *C. reinhardtii* genes or is more pronounced in the tubulin gene family. It has been suggested, for example, that abundant "housekeeping" proteins such as tubulins have a relatively high use of G or C in the third position of a codon, when compared with other less highly expressed genes.[37] Further examination of nuclear gene sequences in this organism will allow such comparisons.

Intervening Sequences

The ten intervening sequences (IVS) found in the four *C. reinhardtii* tubulin genes provide information on the structure of typical intervening sequences in this organism and on the divergence of noncoding portions of duplicated genes (FIGURE 4). The IVS occur in the 5′ one-third of the gene, as is true for most tubulin genes in other organisms.[38–40] The exact locations of the IVS, however, differ from those in other organisms and also differ between the alpha- and beta-tubulin genes. The IVS are relatively short, ranging in size from 126 to 359 bp. All 10 IVS contain consensus splice junctions (5′ G/GT (G/C) (A/C) G .... (C/A) AG/ (G/A) 3′) similar to those in other eukaryotic genes.[41] In the consensus regions, nucleotides that differ from the norm in one IVS tend to be conserved at the same location in the duplicate gene as well. In addition to the splice junction consensus sequences, all 10 IVS contain an adenosine (A) residue located in a consensus sequence similar to the one suggested recently to be involved in the splicing reaction in higher eukaryotes: Py-X-Py-T-Pu-A-Py.[42] The sequence begins 30 to 54 bp upstream of the 3′ end of each IVS in *C. reinhardtii*.

Because of the high degree of homology found in the coding regions of the duplicate genes, we examined conservation of sequence in the IVS. Computer matrix comparisons were used to detect regions of homology between the IVS. These searches showed no homology of sequences longer than 10 bp between corresponding first or second IVS in the beta-tubulin genes or between the first or second IVS in the alpha-tubulin genes. A striking homology between the third IVS in the beta-tubulin genes was detected, however. When the two IVS are aligned as shown in FIGURE 5, the sequences are 89% homologous. The high degree of homology between the third IVS in the beta-tubulin genes suggests that the sequence element may have a function in controlling gene transcription or transcript splicing. It has been suggested, for example, that the conserved IVS found in brain-specific genes may have a role in regulating the expression of those genes.[43] If the *C. reinhardtii* IVS 3 is important for the regulation of beta-tubulin gene expression, it might also be found in association with other genes regulated in a similar manner, such as the alpha-tubulin genes. We compared the IVS 3 from both the beta-1 and beta-2 genes to alpha-tubulin gene IVS and found limited homology with the second IVS in the alpha-tubulin genes. For example, 16 of 19 nucleotides in the alpha-1 IVS 2 are homologous with the beta-1 IVS 3 starting 79 bp downstream of the 5′ end of IVS 3 (sequences underlined in FIGURE 4, B and E). Also, 37 of 47 nucleotides of alpha-2 IVS 2 are homologous with beta-2 IVS 3, starting at nucleotide 77 of IVS 3. Because these homologous regions do not extend throughout the length of IVS 3 and because the homologies are limited, it is difficult to determine their biological significance. Further experiments will be necessary to determine

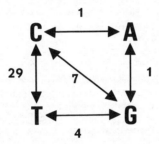

FIGURE 3. Categories of silent substitutions between *C. reinhardtii* tubulin genes. The two alpha-tubulin genes contain 23 silent substitutions at the third position of a codon; the two beta-tubulin genes contain 19 such substitutions. The categories of these substitutions in the coding strand are indicated: arrows for transitions are vertical; arrows for transversions are horizontal or diagonal.

whether the IVS 3 elements have a role in gene regulation. For example, it will be possible to determine whether IVS 3-like elements are associated with other flagellar genes regulated coordinately with the beta-tubulin genes.

The beta-tubulin IVS 3 elements were examined for open reading frames that might encode proteins. When the normal reading frame of the beta-1 or beta-2 gene is read into IVS 3, a stop codon occurs 57 bp into IVS 3. For beta 2 only, an initiation codon 56 bp downstream of the 5′ end begins an open reading frame that continues to the end of the tubulin coding sequence. Because no open reading frames extend throughout the length of the conserved IVS 3, it seems unlikely that the element functions as a coding sequence.

5′ and 3′ Untranslated Regions

We determined the DNA sequence of the 5′ ends of the four tubulin cDNA clones and compared the data with similar sequences from the genomic clones, and with the data reported by Brunke *et al.*[25] The 5′ untranslated portion of the alpha-1 cDNA

A α1, IVS 1

```
        10        20        30        40        50        60        70
GTCAGTTGACAGAGCGAGAGGGCTGGGACGTCGTTGCCGAGTTTGAATCGCCTTGCCTCCCCCCCGCGGCCTCCAGCTGT
        90       100       110       120       130       140       150
GTCCGATTAGCCTTGCTACTTTGCTGCTTGCTGTGGCCATGAGTTGCTTCTTTCAAGCGCTTGTGCCAAGTCGCGGCGAG
       170       180       190       200       210       220       230
CTACACCCCCCGCGTTCAAACTTCGAGGAACTTGGCGCTGACATTGCCCGTCCCGCTTCCTCCCTTGCCCGGACAG
```

α2, IVS 1

```
        10        20        30        40        50        60        70
GTCAGTCGCCTGCGAACGAGAGGGGAGCGGGACCGCGGGAAGATTCGTAGCGCCGATTCGAACTTGCCGCGAACCTGGC
        90       100       110       120       130       140       150
CCATTTGCTCGCTTTCTTCCTGGGGCACTCGCAAATTTGATCAGCAGCGTTCACAGCTTGCTGGCCTTGCCTCCTGCGCC
       170       180       190       200       210       220       230
CGTGCGCTTTGGGCGCGGATCTTCCCCCCTGACCTTGCTTGACTTTGCCGCTTCTGCCTCCTGCGTTCCCGCAAAG
```

B α1, IVS 2

```
        10        20        30        40        50        60        70
GTGAGTGACGCCAATTCCACATGTCACGCGACGTCGAGCTTCTGTTGAGCAGCTGGCCCATTGTTAAATGTGCATGCTTA
        90       100       110       120       130       140       150
GGCGCTCGACAGCCAAATGCCCATGGGCGTGCATATCTTACAGCCCAGTTTTCGTACCCTTGACTGCGTTTCCTCGTCTC
       170       180       190
GCTCTTAACATTGCCTTTACCCCCTTCCCGCCCACAG
```

α2, IVS 2

```
        10        20        30        40        50        60        70
GTGAGTTCGCATGTGCTTCGAACTTGTGTGCGTGCGTTCTAAAAGGGCTTCTCTTGGTGTTCGATCTGGGCTTCTCGCTT
        90       100       110       120       130       140       150
GCTATTGCAGTCATATGTTGGCCTTCGGCTGTGTCAGCATTCTTCAGCCAGTGCCTTGAAGTCCGCCCAGCTTCCGTCCT
       170       180       190       200       210
GCGCCCTTTCCTTGCCCTGTGCTGACTTCGCCTTCCCCCCGTTCCCTTCACGCTCGCAG
```

C β1, IVS 1

```
        10        20        30        40        50        60        70
GTGCGCAAGCTTCGGTCCTGGCTTAGGCTGGTCGGGGGAAAAGGCCGCGCTCGCAAGATCTTCGGAGTTCCGGTCGTCTC
        90       100       110       120
GCGCTCACCTCAATTTCGCCTTCTCGCTCCCCAATTGCATCCTCAG
```

β2, IVS 1

```
        10        20        30        40        50        60        70
GTGCGTTGAAGCGCTTAGCGCATTGGCTGAGGGCTAGCGCAGTCAAGGGGCGCGGGGTCGTGGCTACACCCCGCGGCTC
        90       100       110       120       130       140
AATTTCAAACCTGTTTCCGACTTCGAGGCTCATCGTCGCTCCGCCTGCTTGCGCCTTTACATCCACAG
```

D β1, IVS 2

```
        10        20        30        40        50        60        70
GTGAGTCAACTCAGTATGGTCGCTGGCGTGTCACGCGGGGTCTTTTGCTCGGCAACATTCCTTGAGTTGACCGCTGGTCA
        90       100       110       120       130       140       150
GCTGCGTGGGCTCGCCACGTTGTCCACAAGGCCGGTTGTGCGGCTAGCTTGGCTTCTTGTCCAAAACAGTCAGCTTCGTC
       170       180       190       200       210       220       230
GTGGCTGCTTAGCGCTAGCGCTGTGGGGCCGCGCACATGTGAACTTCACAGTCAGTAGTCAGAGTCAACCGCGTTGCGAC
       250       260       270       280       290       300       310
ACCCGAACTGTGCTAGTACGTGGTCCAGCGCTCGTGTGGATGCCCGTCGCGTGCTAGTCAGCACCTTCTCTCTGTTCCGA
       330       340       350
CACCACATGCGCTGACAACAATTCTGTCTTGCGCTTCAG
```

β2, IVS 2

```
        10        20        30        40        50        60        70
GTGAGTTGCCTGAAGAGTTCCTGTTGGGGCTGTAGCGAGCCGCGGGCAGGTTGCTAGCGCTGATGCTACTGTTGTGCTAG
        90       100       110       120       130       140       150
CGCACGACTTGTGTTCCTTGGAGCGCGCGTTTTTCGGCACCGCCGTTCCCAGCCCAATGCTTGATAATGTTTTTGCCGTCA
       170       180       190       200       210       220       230
TGGTTTGGTGCTTTTGCTCAGCCTGATGTCTGCTCCTCTTCCTTGCTCGGACATGCGGGCGAATGCCTGACCTTCTTCCC
       250
TTCCCTTTTGCTCCTCAG
```

E β1, IVS 3

```
        10        20        30        40        50        60        70
GTGAGCAGTTTTGCACGGGCCGCCGGGTGTCGGGGTGTGGACCGCAGGGCTTGGCATGACGGGCTGCT--GGGTCTGTGC
*********  *********** ******** * *********** *********** ***** *** ****** 
GTGAGCAGTTCTGCACGGGCCGTCGGGTGTCAGTGTATGGACCGCAGGGCTCGGCATGACGGGCTGCTTTGGG-CTGTGC
        10        20        30        40        50        60        70
```
β2, IVS 3

```
        90       100        110       120       130
TCCTCGTCGTCTTAACG-GTTGCTGACTTCTCTC-CTCTTGTTGTTCCCTTCCCTCAG
** * ******* **  **************** * * **** ****** *******
GCCCCTTCGTCTTCACACGTTGCTGACTTCTCTCTCCCCTGTTCTTCCCTACCCTCAG
        90       100       110       120       130
```

FIGURE 4. Nucleotide sequence of tubulin gene IVS. Sequences read from 5′ to 3′ with respect to the DNA coding strand. **A:** the first IVS in the alpha-tubulin genes, located after codon 15 in both genes. **B:** the second IVS, located in codon 90 in both alpha-tubulin genes. Underlined region in the alpha-1 gene has homology with beta-1, IVS 3. Underlined region in the alpha-2 gene has homology with beta 2, IVS 3. **C:** the first IVS in the beta-tubulin genes, located after codon 8. **D:** the second IVS in the beta-tubulin genes, located within codon 57. **E:** the third IVS in the beta-tubulin genes, located in codon 131. Homology between the two IVS is indicated by asterisks. Dashes are inserted into the sequence to maximize the homology. Within each IVS, the putative splicing reaction consensus sequence[42] is indicated by dashed lines.

A

α 1, 5′
```
        10        20        30        40        50        60        70
GCATTCACAGCTCTGCTTTTACTCCCCTTAACCTTACATTATCATCCCAGTTGGGACGCGGGTCATTACCCCTCGCGAAC
        90       100       110       120       130
CCAGCGTAGGAAAAGCTCCCTTCTCCTTCCCCTGATTTTCTCCGTCTCCTCCCTTCGCAACC
```

α 2, 5′
```
        10        20        30        40        50        60        70
AAGCCTAAAGCTTCTTACTCCCAACAACACTCCACAAACACAAGTACCGTTGGTACGCAGGTCATTAAGCCTCGCGTCAC
        90       100       110       120       130       140
CCCCGGTTAGCAGACTCCCCTTTCGCTCCCCTTCTCTTCTCCGTATCCAGTCCCTTCGCAACC
```

β 1, 5′
```
        10        20        30        40        50        60        70
ATTCTCAACCCTCAAGCACATACTTCATACTTTAAACAAAGGGCTGCCAGGACCTTCATACGTGAAGGCCTTTGACTAAA
        90       100       110       120       130       140       150
GCAGGCGAGACTTCGTCGCTCCCGCGTCCCAGTCCCCCCGCCTCTTCGGCTCCTCGCTTCGGCGCAGCCCCACAGCAAAC
```

β 2, 5′
```
        10        20        30        40        50        60        70
ATTATAGCGAGCTACCAAAGCCATATTCAAACACCTAGATCACTACCACTTCTACACAGGCCACTCGAGCTTGTGATCGC
        90       100       110       120       130
ACTCCGCTAAGGGGGCGCCTCTTCCTCTTCGTTTCAGTCACAACCCGCAAAC
```

B

α 1, 3′
```
        10        20        30        40        50        60        70
ACTCCTGGCATGGCATCAGGGTAGCAGCAAGTGTAGTGCTTTGCTGTTTGCCCGTCAAGATGAACAGCGGTTGCTAGGTT
.        90       100       110       120       130       140       150
GTGGATCGAAGGATCTAGCTAATGACGTTGCATCGTTAGCGTGGACGGGCACCCGTGCATGAGTCTGAACACATACTTGC
        170       180       190       200       210       220       230
TGCCTGAACAAACAGCAGTGGCTGCTAGCTTACAAGTGGACGAAGGGTCTAGCTAGTGACGTGACATGCGCAAGTGTTG
        250       260       270       280       290       300       310
CGGACACCGTGCATGAGTCTGAACATTGGCTGCTGCCCACCGGAACAATGGTGCTGTAATGGACTTGGAAGTCA
```

α 2, 3′
```
        10        20        30        40        50        60        70
AGTATTCCCGCACTCCAACCCACGAGCTAGCATGTGAGCGTGTCCTGTGCGGCGTGATCTGAGGCTTCGTTGGAAAAGCG
        90       100       110       120       130       140       150
TATTGTGGCTGTACTTATGCTAGCTTGGTCAAATGATGATGCACGACTGCAAGTTACATCATGGGTCGTGAACATCCACG
        170       180       190
ACCGTGCCTCTATTGGCTTGCTGTAAAGGATGAGAAG
```

β 1, 3′
```
        10        20        30        40        50        60        70
ATGCCTTGCCTACCTATCCACCAATGCGTGCATGTGTGCGCATGATTAGGTGATGTGCATACCTTACTCGTGATGTGCAT
        90       100       110       120       130       140       150
CACGGCTATTTAGGCGAGCGGTCTAGGGGCTTCGTGCGGGTGTGGTGGTTTGGTCCCACAGGCAAACGAAGAAGTGAAGC
        170       180       190       200       210       220       230
GTGGCGTGTGATAGCAATGGCGCAAGCAGGAGCTGTGTTCGTTTTTGTGGATTACTTCAGGCACATAGTCTTGACGTTAA
        250       260       270       280       290       300       310
TAAGCTCGTTTATGACTGCTGTACGAACACAAAACTATGACATGAGGGCACATCATTGTGTACCTATCATAGCATTTTTAC
        330       340       350       360       370       380       390
CGCTGGCGCTCTCCTTCGGCGCTCCATGGGCAGGCGACTAACGAGTGACTGCGGAGGAGGGAATACCTGAACTTAAGGCA
        410       420       430       440       450       460       470
CTCGTAGGTGCACGCGCTGCTCCGGGGTTTGACGGTAGTAGCAAGTGGCTGCTGAGGCTGACGAAGGACATGCTGGAAGC
        490       500       510       520       530       540       550
GTTGGCTGGGCGCGGCGATTGCCGGGCCAGGTGCCCCTTGCAGGATGCCTGAGGCAATCTCTGTAATACCATTCATG
```

β 2, 3′
```
        10        20        30        40        50        60        70
ATGCCGGCACCTCCATGCGCCACTGAACGTGTAGCGTGACTGTGGCGGCCTTGGCAGTTTTGACCGTGACTGACCCTGGA
        90       100       110       120       130       140       150
CAAAGGATCCCTGACTGAAGACAACTTGACATGTGATTGCCATTTGACGCTTTGGTGTGGAGGCGGATTGTGAGATGGGA
        170       180       190       200       210       220       230
GGGGGGGCCCATTGCCTTCGTGACCATAACGACATCGAATTTCATACATGTGAACAGTTCAGCATGGACATTCATCTCGTC
        250       260       270       280       290       300       310
GGATTAGCTCTTGTGTGATAGGCCATAGCAGCTGGACTGTTGTGAGCTCTCGATCTGCGTAGCTACTGGCTGTGATTGTG
        330       340       350       360       370       380       390
CTTCAGGCGGCAGGGGCAGGTAACTGCCCTGAACGTAAAGGTGCAGCAGCAGACAGCGGATGTGCAGAACGAATAGCGCA
        410       420       430       440       450
GTGGATAAGGTTGATGGGTGGCCACACACTCGTGCACGTGTAATAAGATACACG
```

FIGURE 5. Nucleotide sequences of tubulin transcript 5′ and 3′ untranslated regions. Sequences read from 5′ to 3′ with respect to the coding strand. A, sequences from the 5′ untranslated regions of the genes. Homologous sequences are underlined. B, sequences from the 3′ untranslated regions. A conserved putative polyadenylation signal is underlined for each transcript.

begins at a G residue 142 bp upstream of the translation start site, whereas the alpha-2 cDNA begins at an A residue 143 bp upstream (FIGURE 5A). The beta-1 and beta-2 transcripts begin at A residues and extend 160 bp and 132 bp upstream of the initiation codon, respectively. The 5' ends of the cDNAs are located at the same base (for the beta-tubulin clones) or two bases upstream (for the alpha-tubulin clones) of the transcription start site identified by Brunke et al. by primer extension.[25] These results suggest that the cDNA clones represent full-length tubulin transcripts. The sequence comparisons also indicate that no intervening sequences are located in the 5' untranslated region of the genes. The strong homology found in the coding regions of the alpha- and beta-tubulin genes extends upstream into the 5' noncoding sequence a short distance, as noted by Brunke et al.[25] A high degree of homology between the two alpha-tubulin transcripts is found throughout the final 25 bp of the 5' noncoding region.

The 3' noncoding regions of the tubulin cDNA clones range in length from 557 nucleotides for the beta-1 gene to 197 nucleotides for the alpha-2 gene. Because the cDNA clones were constructed by an mRNA-cDNA hybridization method with no S1 nuclease,[17] the clones probably represent the full 3' extent of the beta-tubulin transcripts. Thus, the differences in transcript length noted earlier are accounted for almost entirely by differences in the length of the 3' untranslated region. No homology of sequences longer than 10 bp was found when comparing the 3' untranslated region from each of the transcripts. A sequence TAATA occurs 13 bp upstream of the polyadenosine tracts in the two beta-tubulin cDNAs and is similar to the polyadenylation signal found in higher eukaryotes.[44] The only sequence conserved, however, between all four genes in this region is TGTAA (underlined in FIGURE 5B), starting 16–20 bp upstream of the polyadenosine tracts. Thus it is possible that this sequence may be the polyadenylation signal.

SUMMARY

DNA sequencing studies have provided a picture of the total information available at the gene level for tubulin production in C. reinhardtii. The data indicates that diversity at the gene level is very limited and that all the microtubules in the cell are composed of a very similar set of tubulins. These studies contrast with similar studies of S. pombe alpha-tubulin genes and chicken beta-tubulin genes that show much heterogeneity among members of the same gene family.[31,40] Further studies will be needed to investigate whether the high degree of conservation of tubulin genes is unique or common among lower eukaryotes, and what mechanisms are used to maintain homogeneity in C. reinhardtii tubulin gene families. Our DNA sequence analysis, in addition to the work of Brunke et al.,[25] has provided information on the noncoding, and possibly regulatory, portions of the tubulin genes. For example, the promoter regions of the 4 tubulin genes share a consensus sequence of 16 nucleotides upstream of the TATA box.[25] This sequence could be involved in regulating the coordinate expression of the genes. Although little homology exists generally in the noncoding region of the genes, striking homology between the third IVS in each beta-tubulin gene is observed. Small elements homologous to the beta-tubulin IVS 3 also exist in the second IVS of each alpha-tubulin gene. In addition, considerable homology in the 5' noncoding portion of the alpha-tubulin transcripts has been noted. These homologies may be the result of recent gene conversion events, and may not have functional significance. The possibility, however, must also be considered in future experiments that these elements may play a role in regulating the expression of the tubulin genes.

REFERENCES

1. RINGO, D. L. 1967. J. Cell Biol. **33:** 543–571.
2. GOODENOUGH, U. W. & R. L. WEISS. 1978. J. Cell Biol. **76:** 430–438.
3. LEFEBVRE, P. A., C. D. SILFLOW, E. D. WIEBEN & J. L. ROSENBAUM. 1980. Cell **20:** 469–477.
4. MCKEITHAN, T. W., P. A. LEFEBVRE, C. D. SILFLOW & J. L. ROSENBAUM. 1983. J. Cell Biol. **96:** 1056–1063.
5. L'HERNAULT, S. W. & J. L. ROSENBAUM. 1983. J. Cell Biol. **97:** 258–263.
6. BRUNKE, K. J., P. S. COLLIS & D. P. WEEKS. 1982. Nature (London) **297:** 516–518.
7. SILFLOW, C. D., R. L. CHISHOLM, T. W. CONNER, L. P. W. RANUM & J. STAUBUS. 1985. Mol. Cell. Biol. **5:** 2389–2398.
8. WEEKS, D. P. & P. S. COLLIS. 1979. Dev. Biol. **69:** 400–407.
9. LEFEBVRE, P. A., S. A. NORDSTROM, J. E. MOULDER & J. L. ROSENBAUM. 1978. J. Cell Biol. **78:** 8–27.
10. COLLIS, P. S. & D. P. WEEKS. 1978. Science **202:** 440–442.
11. WEEKS, D. P. & P. S. COLLIS. 1976. Cell **9:** 15–27.
12. WEEKS, D. P., P. S. COLLIS & M. A. GEALT. 1977. Nature (London) **268:** 667–668.
13. SILFLOW, C. D. & J. L. ROSENBAUM. 1981. Cell **24:** 81–88.
14. MINAMI, S. A., P. S. COLLIS, E. E. YOUNG & D. P. WEEKS. 1981. Cell **24:** 89–95.
15. ARES, M. & S. HOWELL. 1982. Proc. Natl. Acad. Sci. USA **79:** 5577–5581.
16. BRUNKE, K. J., E. E. YOUNG, B. U. BUCHBINDER & D. P. WEEKS. 1982. Nucleic Acids Res. **10:** 1295–1310.
17. SCHLOSS, J. A., C. D. SILFLOW & J. L. ROSENBAUM. 1984. Mol. Cell. Biol. **4:** 424–434.
18. KELLER, L. R., J. A. SCHLOSS, C. D. SILFLOW & J. L. ROSENBAUM. 1984. J. Cell Biol. **98:** 1138–1143.
19. BAKER, E. J., J. A. SCHLOSS & J. L. ROSENBAUM. 1984. J. Cell Biol. **99:** 2074–2081.
20. DALLMAN, T., M. ARES & S. H. HOWELL. 1983. Mol. Cell. Biol. **3:** 1537–1539.
21. YOUNGBLOM, J., J. A. SCHLOSS & C. D. SILFLOW. 1984. Mol. Cell. Biol. **4:** 2686–2696.
22. BOND, J. F. & S. R. FARMER. 1983. Mol. Cell. Biol. **3:** 1333–1342.
23. KALFAYAN, L. & P. C. WENSINK. 1982. Cell **29:** 91–98.
24. LOPATA, M. A., J. C. HAVERCROFT, L. T. CHOW & D. W. CLEVELAND. 1983. Cell **32:** 713–724.
25. BRUNKE, K. J., J. G. ANTHONY, E. J. STERNBERG & D. P. WEEKS. 1984. Mol. Cell. Biol. **4:** 1115–1124.
26. LEMISCHKA, I. R., S. FARMER, V. R. RACANIELLO & P. A. SHARP. 1981. J. Mol. Biol. **151:** 101–120.
27. SILFLOW, C. D., P. A. LEFEBVRE, T. W. MCKEITHAN, J. A. SCHLOSS, L. R. KELLER & J. L. ROSENBAUM. 1982. Cold Spring Harbor Symp. Quant. Biol. **46:** 157–169.
28. VALENZUELA, P., M. QUIROGA, J. ZALDIVAR, W. J. RUTTER, M. W. KIRSCHNER & D. W. CLEVELAND. 1981. Nature (London) **289:** 650–655.
29. PONSTINGL, H., E. KRAUHS, M. LITTLE & T. KEMPF. 1981. Proc. Natl. Acad. Sci. USA **78:** 2757–2761.
30. NEFF, N. F., J. H. THOMAS, P. GRISAFI & D. BOTSTEIN. 1983. Cell **33:** 211–219.
31. TODA, T., Y. ADACHI, Y. HIRAOKA & M. YANAGIDA. 1984. Cell **37:** 233–242.
32. L'HERNAULT, S. W. & J. L. ROSENBAUM. 1985. Biochemistry. **24:** 473–478.
33. CLEVELAND, D. W. 1983. Cell **34:** 330–332.
34. RAFF, E. C. 1984. J. Cell Biol. **99:** 1–10.
35. KRAUHS, E., M. LITTLE, T. KEMPF, R. HOFER-WARBINEK, W. ADE & H. PONSTINGL. 1981. Proc. Natl. Acad. Sci. USA **78:** 4156–4160.
36. CHIANG, K.-S. & N. SUEOKA. 1967. Proc. Natl. Acad. Sci. USA **81:** 713–717.
37. IKEMURA, T. 1985. Mol. Biol. Evol. **2:** 13–34.
38. LEMISCHKA, I. R. & P. A. SHARP. 1982. Nature (London) **300:** 330–335.
39. LEE, M. G., S. A. LEWIS, C. D. WILDE & N. J. COWAN. 1983. Cell **33:** 477–487.
40. SULLIVAN, K. F. & D. W. CLEVELAND. 1984. J. Cell Biol. **99:** 1754–1760.
41. MOUNT, S. M. 1982. Nucleic Acids Res. **10:** 459–472.
42. RUSKIN, B., A. KRAINER, T. MANIATIS & M. GREEN. 1984. Cell **38:** 317–338.

43. MILNER, R. J., F. E. BLOOM, C. LAI, R. A. LERNER & J. G. SUTCLIFFE. 1984. Proc. Natl. Acad. Sci. USA **81:** 713–717.
44. PROUDFOOT, N. J. & G. G. BROWNLEE. 1976. Nature (London) **263:** 211–214.
45. DOOLITTLE, R. F. 1979. *In* The proteins. H. Neurath & R. L. Hill, Eds.: **4:** 1–118. Academic Press, New York.

Isolation and Characterization of Two Rat Alpha-Tubulin Isotypes[a]

I. GINZBURG,[b] A. TEICHMAN, AND U.Z. LITTAUER

Department of Neurobiology
The Weizmann Institute of Science
Rehovot, Israel

INTRODUCTION

Microtubules are cytoskeletal filaments present in all eukaryotic cells and are involved in various cellular functions such as cell division, cell migrations, cell shaping, and secretion.[1-4] In the nervous system, microtubules are also involved in cell differentiation and synaptic transmission. The major protein component of microtubules is tubulin, which is a heterodimer composed of alpha and beta subunits having a molecular mass of approximately 50 kilodaltons. In addition to tubulin, various microtubule-associated proteins (MAPs) have been identified in preparations of cytoplasmic microtubules. These include high molecular weight proteins (MAP-1 and MAP-2) and tau factors of which five to six isoforms have been separated by gel electrophoresis.[5-7] Early during brain development, microtubules participate in spindle formation.[2] At later stages of differentiation, microtubules are involved in cell migration[8] and are essential in creating the basic asymmetry of neurons that distinguishes between axons and dendrites.[9,10] In mature neurons, axon outgrowth, synapse formation, and axoplasmic transport depend on microtubule integrity.[11,12] Each of these functions is performed in one or another of the nerve cell population. It appears that many microtubule types are involved in several neuronal functions that may be controlled by the presence of various tubulin isoforms and MAP forms. Our results[13-15] and those of others[16,17] show that the expression of mRNAs coding for tubulin and for tau factors[18] are modulated during brain development. These results could be a consequence of changes occurring similarly in each cell type or independent regulation of microtubule composition in different cell populations upon assumption of a specific task. The latter alternative would also suggest that the various isotubulins and MAP forms are not randomly distributed within a given microtubular system, but rather that their organization is directly coupled to their function.

Tubulin, like many other eukaryotic genes is encoded by a multigene family. Using a rat alpha-tubulin cDNA probe, we have shown that about 15 to 20 copies of tubulin genes exist in the rat genome;[14] the number of functional genes is unknown, however, and indeed several rat pseudogenes were identified and sequenced.[14,19] In this report we describe the isolation of several rat alpha-tubulin cDNA clones and their use for studying specific gene transcripts.

[a]This work was supported in part by grants from the United States-Israel Binational Science Foundation and from the Forcheimer Center for Molecular Genetics.
[b]Send correspondence to I. Ginzburg, Department of Neurobiology, The Weizmann Institute of Science, P.O. Box 26, Rehovot 76100, Israel.

RESULTS

Isolation and Sequence Analysis of Two Rat Alpha-Tubulin cDNA Clones

A more precise insight into the involvement of the multiple tubulin genes in microtubule function requires the construction of specific DNA probes for the individual gene transcripts. From the limited data available it appears that the various isotubulin mRNAs are distinguished by their highly divergent 3'-untranslated regions, whereas the coding regions show extensive sequence homology.[20–22] It was clear that only high efficiency cloning techniques may yield tubulin cDNA clones derived from the less abundant mRNA species. For this purpose, several libraries were constructed in lambda gt10 phage. These include cDNA libraries from rat brain at different developmental stages and cDNA libraries from various neuroblastoma and glioma cell lines. The latter cell lines have been previously shown to possess a somewhat more limited subset of tubulin isoforms as compared with the complex pattern observed in the brain.[23] FIGURE 1A shows a Southern blot analysis of EcoR1 fragments isolated from mouse neuroblastoma N18TG-2 and rat glioma C6BU-1 DNA that was hybridized to labeled alpha-tubulin cDNA clone pT25.[20] The hybridization pattern shows a complex tubulin gene family closely similar to that obtained with the parental DNA isolated from mouse or rat brain respectively. Thus, most of the tubulin genes are not lost during establishment of these cell lines in culture, and the somewhat reduced expression of tubulin isotypes appears to be controlled at the mRNA level.[24] The mRNAs were isolated from rat brain or cell cultures and converted into dscDNA according to the method of Gubler and Hoffmann.[25] This method has been modified from the procedure of Okayama and Berg[26] and renders large size cDNA molecules with an intact 3'-noncoding region. The double stranded cDNA molecules thus derived were cloned in the high efficiency phage lambda gt10 cloning system using EcoR1 linkers. In contrast to genomic libraries that also contain pseudogenes, the cDNA libraries score only sequences derived from mRNA transcripts. The tubulin cDNA clones were identified by *in situ* colony hybridization to ^{32}P-labeled alpha-tubulin cDNA. By these methods, over 100 tubulin cDNA clones were identified. Because we were interested in clones containing the 3'-ends, the tubulin cDNA clones were further hybridized to short synthetic oligonucleotide sequences (20 nucleotide long) that are homologous to the tubulin 3'-coding regions. By this selection, we limited our analysis to putative clones that also contain the adjacent 3'-noncoding regions.

FIGURE 2 shows the sequence of the 3'-noncoding regions of two rat alpha-tubulin clones. pTa26 is a cDNA clone identical to our previously described pT25 coding for alpha tubulin[20] except that it contains the complete 3'-noncoding regions up to and including the poly(A) tail. The second clone isolated and sequenced is pTa1, which is similar to the clone previously described by Lemischka *et al.*[21]

Whereas there is a close homology along the coding region between these clones, there is high divergence in the 3'-noncoding regions. Clone pTa26 contains 156 base pairs and its polyadenylation signal is AAGTAAA, which precedes by nine bases the poly(A) tract. On the other hand, the noncoding region of clone pTa1 is longer and contains 194 base pairs; its polyadenylation signal is AATAAA, and is located 14 bases before its poly(A) tract.

Interspecies Homology of Alpha-Tubulin Genes

It has been previously noted that while within a given species individual isotubulin genes are totally dissimilar, each in turn shares very high interspecies homolo-

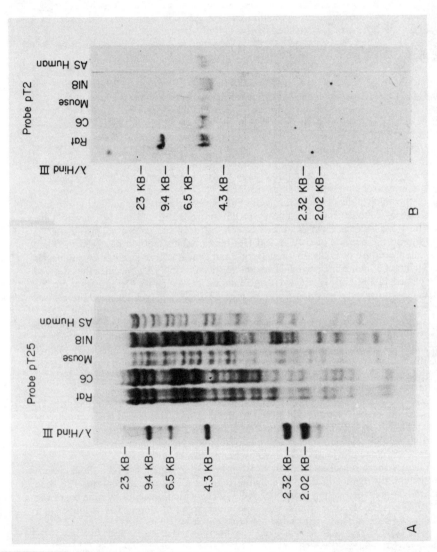

FIGURE 1. Hybridization of tubulin cDNA probes to EcoR1 digested genomic DNA. **A.** Hybridization with ^{32}P-labeled rat pT25 alpha-tubulin cDNA probe. **B.** Hybridization with ^{32}P-labeled rat pT2 3'-noncoding cDNA subclone derived from clone pT25. DNA was isolated from rat, C6 rat glioma, mouse, N18TG-2 mouse neuroblastoma, and SK-N-AS human neuroblastoma.

3'-Untranslated Regions of Rat Alpha-Tubulin cDNA clones

```
              10        20        30        40        50        60        70
pTa26: TTCACTCCTGCAGTCCCTGTATCATGTCAAACTCAACTCCAGCTCCAGCACTAGCTGCAGGCATCGATGC
pTa1 : ATTAAATGTCACAAGGTGCTGCTTCACAGGGATGTTATTCTGGTCCAACATAGAAAGTTGTGGGCTGA

              80        90       100       110       120       130       140
pTa26: TTCTATGCTGTTTCCCTTCTGTGATCATGTCTTCTCCATGTGTACCTCTTAAGTTTCCATGACGTCTCA
pTa1 : TCAGTTAATTTGTATGTGGCAATGTGTGCTTTCATACAGTTACTGACTTTAAGTGTGAATGATTTGTCAG

             150       160       170       180       190
pTa26: AAGTAAAAGCTTTAAGA(A)n
pTa1 : AGACCCGAGCCGTCCACTTCACTGATGGGTTTAAATAAAATACTCCCTGTCTT(A)n
```

FIGURE 2. Sequence of the 3'-noncoding regions of rat alpha-tubulin cDNA clones pTa26 and pTa1.

gy.[19,21,22,27] The homology in this region is evident when the polyadenylation signal is used as an alignment point for mammalian isotubulin sequences. Clearly, the rat-pTa26, the human-kα1,[22] and hamster-I genes[27] are closely related, having AAGTAAA as the polyadenylation signal and being derived from a common ancestral gene. On the other hand, rat-pTa1, human-bα1, and hamster-II genes with AATAAA as the polyadenylation signal are derived from a second different gene. It is interesting to note that the length of the 3'-noncoding regions of the mRNAs coding for isotypic species is similar within the groups. The length of human kα1, hamster-I, and rat-pTa26 3'-noncoding regions is 172, 172, and 156 base pairs, respectively, and the homology between the species ranges between 64% and 69 percent. The length of human-bα1, hamster-II, and rat-pTa1 3'-noncoding regions is 216, 202, and 194 base pairs, respectively, and the homology ranges between 75% and 88 percent. Moreover, it appears that the changes in the nucleotide sequence along the 3'-noncoding regions seem to be nonrandom. A higher conservation is observed towards the 3'-end of the noncoding region around the polyadenylation signal. Whether this extensive interspecies homology is related to any function remains to be determined.

Identification of the Gene Coding for a Unique Alpha-Tubulin

The conserved nature of the tubulin coding sequences allowed the estimation of the tubulin gene number in various species by using chick or rat cDNA clones.[28,29,14] In these studies, multiple copies of genomic tubulin sequences were detected in a variety of species. Thus, by employing labeled rat alpha-tubulin cDNA clones, we were able to observe between 10 and 20 hybridizable bands in DNA isolated from *Tetrahymena*, plant, mouse, rat, and human cells (reference 14 and FIGURE 1A). All these studies indicated that each of the genes coding for alpha- and beta-tubulin species constitutes a large multigene family. These DNA segments could represent either functional genes or nonfunctional pseudogenes. Sequence analysis of genomic recombinant fragments has shown that a significant number of human[30,31] and rat[19,14] alpha- and beta-tubulin genes are indeed pseudogenes. Thus, the number of functional tubulin genes is more limited than initially inferred from the earlier hybridization experiments.

Using a subclone of the 3'-noncoding region of pT25, namely pT2, we were able to detect its homologous gene (FIGURE 1B). The analysis was performed on the same blot shown in FIGURE 1A. One hybridization fragment was observed in DNA isolated from rat, C6 rat glioma, mouse, N18 mouse neuroblastoma, and SK-N-AS human neuroblastoma. Using a similar approach, subclones of the 3'-untranslated regions or synthetic nucleotide sequences specific for a given subtype may aid in the identification of the expressed genes of the tubulin gene family.

In situ *Hybridization Studies*

We have previously shown that the microheterogeneity found in the brain is developmentally determined, increasing from five to six isotubulins prenatally to seven isotubulins postnatally, and reaching a value of nine to eleven distinct components during early brain maturation.[32,33] The question therefore arises as to whether changes occur in the relative proportion of the isotubulines upon assumption of different roles within the same nerve cell. Alternatively, the increase in tubulin microheterogeneity might arise from changes in the brain cell population during brain development, which unlike other organs is composed of many cell types. The only way to differentiate between the above possibilities is to use the *in situ* hybridization histochemistry

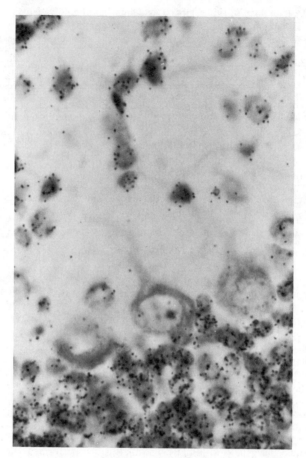

FIGURE 3. *In situ* hybridization of ^{3}H-labeled pT25 cDNA with a section from ten-day-old rat cerebellum.

technique that measures both aspects at the cellular level, that is, changes in gene expression, cell type, and number. FIGURE 3 shows the *in situ* hybridization of the ^{3}H-labeled pT25 rat alpha-tubulin cDNA clone to cerebellum slices of 10-day-old rats. It was observed that there is relatively more of the total alpha-tubulin mRNA in mitotically active external granule layer cells than in the internal granule layer cells. These results show that migration and differentiation of the granule cells is accompanied by a decrease in their alpha-tubulin mRNA level. Furthermore, the relative levels of alpha-tubulin mRNA both in the prenatally formed Purkinje cells and the postnatally formed stellate cells is less than that observed in the differentiated granule cells. We are now developing methods that will allow the use of synthetic oligonucleotides for *in situ* hybridization experiments. These specific probes will enable us to study the changes in the expression of the various isotubulin genes during normal brain development and to compare them to cases of neurological disorders.

DISCUSSION

The dynamic nature of microtubules and the broad spectrum of cellular processes in which they are involved suggests that their assembly and growth is controlled not only temporally but also spatially in all cells. The detailed mechanisms by which microtubules regulate this process are not yet well understood.[34-36] The great variety of microtubule functions can only be explained by the large number of interactions between tubulin and other proteins such as MAPs and is possibly governed by the microheterogeneity of the microtubule proteins. Microtubule heterogeneity can arise from its specific assembly from its various components, that is, the diverse tubulin isoforms and the families of MAPs. It appears that neither genetic diversity nor any other single mechanism may account for cellular control over the enormous versatility of microtubule functions. Among the possible different mechanisms, one may involve the presence of multiple genes for both tubulin and MAPs. Second, the variety of isotypes may be generated by posttranslational modification of one or the other protein subunits. And finally, the different interactions that may take place between all of these components may have a functional significance.

We have previously shown the presence of 15 to 20 alpha-tubulin genes in the rat.[14] Similarly, multiplicity of tubulin genes was observed in other mammals.[22,30,31,33] The number of expressed genes, however, remains to be determined. In this work we describe the sequence of two alpha-tubulin cDNA clones that differ markedly in their 3'-untranslated regions. In contrast to the high nucleotide divergence observed between the two rat species, a high interspecies conservation is observed when compared with sequences isolated from human[22] and Chinese hamster ovary cells.[27] The interspecies homology is also based on the polyadenylation signal that is different in the two species, that is, AATAAA and AAGTAAA, as well as the size conservation of the 3'-untranslated regions, being between 156–172 base pairs and 194–216, respectively. The conservation of the unique AAGTAAA polyadenylation signal is noteworthy and may be of functional significance. It will also be interesting to determine whether a complementary sequence exists downstream from this sequence on the genomic level, which may form hairpin loops with the different polyadenylation signals. It has been suggested that the formation of such hairpin loops is required for correct polyadenylation.[37,38]

From the sequence homology data it is also clear that the 3'-end of the 3'-untranslated region is more conserved than other parts of the noncoding regions. The strong conservation of 3'-noncoding regions of isotypic mRNAs is not limited to the tubulin gene family. A homology of 60% was determined for rat and chick beta actin, whereas 85% homology was calculated for the human and rat isotypes.[39] Similarly, the rat and human cardiac actins show 84% homology in their 3'-untranslated regions,[40,41] and the chick and rat skeletal muscle mRNA show over 75% homology.[42] The rate of accumulation of silent substitution in the nucleotide sequence of DNA during evolution has been estimated to be in the order of 1% per one million years.[43] The accumulation of these changes has been used as a biological clock to study the evolutionary history of genes.[43-45] The degree of homology expected between the rat and human, having diverged about 75 million years ago, is less than 50 percent. On the other hand, the high degree of sequence homology observed in the 3'-untranslated regions of mRNA coding for tubulin isotypes in distantly related mammals indicates a very strong evolutionary constraint to preserve these sequences.

Whereas the alpha-tubulin isotypes show a stringent conservation of the C-terminal region, the beta-tubulin isotypes show variability within the 20 carboxy-terminal amino acids. These differences may contribute to changes in tubulin binding sites towards its various ligands, thus generating functionally different microtubules. This hypothesis is supported by our recent studies in which we have developed a specific binding assay for the interaction of [125I]MAPs with tubulin or its cleavage peptides. To identify the tubulin-binding domains to MAPs, we have examined, in collaboration with Drs. H. Ponstingl and M. Thierauf (German Cancer Research Center, Heidelberg), the binding of rat brain [125I]MAP-2 or [125I]tau factors to 60 cleavage peptides derived from pig alpha- and beta-tubulin. Our results show that MAP-2 will specifically interact with only two peptides located at the C-terminus of beta tubulin, between positions 392–445 and 416–445. Strong binding sites for tau factors were also located at the C-terminus of beta-tubulin, between positions 392–445 and 416–445. In addition, tau factors, but not MAP-2, interacted with two peptides derived from alpha tubulin. The first peptide is located near the N-terminus of alpha tubulin between positions 1–75 and a second weaker binding site located between positions 314–377.[34]

For tubulin isotypes, there has been much speculation about the possibility of specific gene functions. The above results indicate that several isotubulin genes differ in a limited number of base pairs in the coding regions, and differ markedly in the 3'-untranslated region. One should consider, however, additional possibilities for the expression of a multigene family. It is possible that several genes will have an identical coding region, but will differ in their nontranslated regions. Another alternative would imply nonidentical coding regions with similar noncoding regions. Although such isotubulin genes have not yet been described, they would imply alternative control mechanisms leading to tissue-specific expression of these proteins as well as their differential expression during cell differentiation.

ACKNOWLEDGMENTS

We thank Dr. E. M. Elliot for providing her data[27] prior to publication and Dr. L. Helson for the SK-N-AS human neuroblastoma cell line.

REFERENCES

1. BRAY, D. 1973. Nature (London) **224:** 93–96.
2. DUSTIN, D. 1978. Microtubules. Springer-Verlag. New York.
3. BRAY, D. & D. GILBERT. 1981. Annu. Rev. Neurosci. **4:** 505–523.
4. ROBERTS, K. & J. HYAMS, Eds. 1979. Microtubules. Academic Press. New York.
5. MURPHY, D. B. & G. G. BORISY. 1975. Proc. Natl. Acad. Sci. USA **72:** 2696–2700.
6. WEINGARTEN, M. D., A. H. LOCKWOOD, S.-Y. HWO & M. W. KIRSCHNER. 1975. Proc. Natl. Acad. Sci. USA **72:** 1858–1862.
7. KIRSCHNER, M. W., R. C. WILLIAMS, N. D. WEINGARTEN & J. C. GERHART. 1974. Proc. Natl. Acad. Sci. USA **72:** 1159–1163.
8. OLMSTEAD, J. B. & G. G. BORISY. 1973. Annu. Rev. Biochem. **42:** 567.
9. MATUS, A., G. HUBER & R. BERNHARDT. 1983. C.S.H. Symposia on Quantitative Biology **48:** 775–782.
10. DE CAMILLI, P., P. E. MILLER, F. NAVONI, W. E. THEURKAVE & R. B. VALLEE. 1984. Neurosci. **11:** 819–846.
11. DANIELS, M. P. 1972. J. Cell Biol. **53:** 164–176.
12. MARCHISIO, P. C., K. WEBER & M. OSBORN. 1980. In Tissue culture in neurobiology. E. Giacobini, A. Vernadakis & A. Shahar, Eds.: 99–109. Raven Press. New York.
13. GINZBURG, I., S. RYBACK, Y. KIMHI & U. Z. LITTAUER. 1983. Proc. Natl. Acad. Sci. USA **80:** 4243–4247.

14. GINZBURG, I., T. SCHERSON, S. RYBAK, Y. KIMHI, D. NEUMAN, M. SCHWARTZ & U. Z. LITTAUER. 1983a. Cold Spring Harbor Symposia on Quantitative Biology, vol. XLVIII: Molecular Neurobiology, C.S.H. Laboratory, 783–790.
15. GOZES, I., A. DE BAETSELIER & U. Z. LITTAUER. 1980. Eur. J. Biochem. **103:** 13–20.
16. MORRISON, M. R., S. PARDUE & W. S. T. GRIFFIN. 1981. J. Biol. Chem. **256:** 3550–3556.
17. MORRISON, M. R., S. PARDUE & W. S. T. GRIFFIN. 1983. J. Neurogenet. **1:** 105.
18. I. GINZBURG, T. SCHERSON, D. GIVEON, L. BEHAR & U. Z. LITTAUER. 1982. Proc. Natl. Acad. Sci. USA **79:** 4892–4896.
19. LEMISCHKA, I. & P. A. SHARP. 1982. Nature (London). **300:** 330–335.
20. GINZBURG, I., L. BEHAR, D. GIVOL & U. Z. LITTAUER. 1981. Nucleic Acids Res. **9:** 2691–2697.
21. LEMISCHKA, I. R., S. FARMER, V. R. RACANIELLO & P. A. SHARP. 1981. J. Mol. Biol. **151:** 101–120.
22. COWAN, N. J., P. R. DOBNER, E. V. FUCHS & D. W. CLEVELAND. 1983. Mol. Cell Biol. **3:** 1738.
23. GOZES, I., D. SAYA & U. Z. LITTAUER. 1979. Brain Res. **171:** 171.
24. LITTAUER, U. Z., A. ZUTRA & I. GINZBURG. 1985. The expression of tubulin and various enzyme activities during neuroblastoma differentiation. Third Conference on Advances in Neuroblastoma Research, Philadelphia, Pa, Allan Liss, Inc. New York. 193–208.
25. GUBLER, U. & B. J. HOFFMAN. 1983. Gene **2:** 262.
26. OKAYAMA, H. & P. BERG. 1982. cDNA. Mol. Cell Biol. **2:** 101.
27. ELLIOT, E. M., H. OKAYAMA, F. SARANGI, G. HENDERSON & V. LING. 1985. Mol. Cell. Biol. **5:** 236–241.
28. CLEVELAND, D. W., M. A. LOPATA, R. J. MACDONALD, N. J. COWAN, W. J. RUTTER & M. W. KIRSCHNER. 1980. Cell **20:** 95–105.
29. SANCHEZ, F., J. E. NATZLE, D. W. CLEVELAND, M. W. KIRSCHNER & B. J. MCCARTHY. 1980. Cell **22:** 845–854.
30. COWAN, N. J., C. D. WILDE, L. T. CHOW & F. C. WEFALD. 1981. Proc. Natl. Acad. Sci. USA **78:** 4877–4881.
31. HALL, J. L., L. DUDLEY, P. R. DOBNER, S. A. LEWIS & N. J. COWAN. 1983. Mol. Cell Biol. **3:** 854–862.
32. GOZES, I. & U. Z. LITTAUER. 1978. Nature (London). **276:** 411–413.
33. LITTAUER, U. Z., A. DE BAETSELIER, I. GINZBURG & I. GOZES. 1980. Neurotransmitters and their receptors. U. Z. Littauer *et al.*, Eds.: 547–557. J. Wiley and Sons, Ltd. New York.
34. GINZBURG, I. & U. Z. LITTAUER. 1984. The expression and cellular organization of microtubule proteins: Brain specific probes in the study of differential expression during brain development. *In* Molecular Biology of the Cytoskeleton. Cold Spring Harbor Laboratory. 357–366.
35. SCHERSON, T., T. E. KREIS, J. SCHLESSINGER, U. Z. LITTAUER, G. G. BORISY & B. GEIGER. 1984. Dynamic interactions of fluorescent MAPs in living cells. J. Cell Biol. **99:** 425–434.
36. LITTAUER, U. Z. & I. GINZBURG. 1985. Expression of microtubule proteins in brain. *In* Gene Expression in Brain. C. Zomsely-Neurath and W. A. Walker, Eds. J. Wiley & Sons. New York. 125–156.
37. MCDEVITT, M., M. IMPERIALE, H. ALI & J. R. NEVINS. 1984. Cell. **37:** 993–999.
38. BENECH, P., G. MERLIN, M. REVEL & J. CHEBATH. 1984. Nucleic Acids Res. **13:** 1267–1281.
39. HANUKOGLU, I., N. TANESE & E. FUCHS. 1983. J. Mol. Biol. **163:** 673–678.
40. MAYER, Y., H. CZOSNEK, P. E. ZEELON, D. YAFFE & U. NUDEL. 1984. Nucleic Acids Res. **12:** 1987–1100.
41. NUDEL, U., Y. MAYER, R. ZAKUT, M. SHANI, H. CZOSNEK, B. ALONI & D. YAFFE. 1984. *In* Experimental Biology and Medicine. H. M. Eppenberger & J. C. Perriard, Eds.: **9:** 219–227. Kanger Press. Basel.
42. ORDAL, C. P. & T. A. COOPER. 1983. Nature (London) **303:** 348–349.
43. MIYATA, T., H. HAYASHIDA, M. HASEGAWA, M. KOBAYASHI & K. KOIKE. 1982. J. Mol. Evol. **19:** 28–35.

44. SHEN, S. I., J. L. SLIGHTOM & O. SMITHIES. 1981. Cell **26.** 191–203.
45. EFSTRATIADIS, A., J. W. POSAKONY, T. MANIATIS, R. M. LAWN, C. O'CONNELL, R. A. SPITZ, J. K. DE RIEL, B. G. FORGET, S. M. WEISSMANN, J. L. SLIGHTOM, A. E. BLECHL, D. SMITHIES, F. E. BARALLE, C. C. SHOULDERS & N. J. PROUDFOOT. 1980. Cell **21:** 653–668.

Differential Expression of the β-Tubulin Multigene Family during Rat Brain Development[a]

STEPHEN R. FARMER, GREGORY S. ROBINSON, DAVID
MBANGKOLLO, JULIAN F. BOND, GLENN B. KNIGHT,
MATTHEW J. FENTON, AND ELLEN M. BERKOWITZ

Department of Biochemistry
Boston University Medical School
Boston, Massachusetts 02118

INTRODUCTION

The differentiation of nerve cells both *in vivo* and *in vitro* requires the intimate involvement of the various cytoskeletal elements. The microtubules, in particular, have been shown to play crucial roles in neurite extension and axoplasmic transport.[1] Qualitative and quantitative changes in the expression of tubulin, both at the mRNA level and in terms of protein microheterogeneity, have been demonstrated during brain maturation.[2-5] Similarly, other studies have revealed that the regulation of tubulin mRNA production responds to the levels of unpolymerized tubulin in the cell.[6,7] At the outset of our present studies we were interested in assessing whether the changes in tubulin gene expression, accompanying nerve cell development, were responsive to the polymerization state of the microtubules. To address this question we embarked on a study of tubulin mRNA production in the rat brain.[8] At this time, other studies were beginning to reveal that tubulin genes within most higher eukaryotic cells were comprised of a complex multigene family.[9-12] This led to speculation that many of the genes may code for different tubulin isotypes, some of which may construct function and/or cell-type-specific microtubules, that is, neurite microtubules. Therefore, we have expanded our initial studies to include an analysis of the differential expression of the rat β-tubulin multigene family during nerve cell development. We have isolated and characterized three different cDNA clones; these have been sequenced and shown to correspond to three β-tubulin isotypes. Two of these are present only in cells of neurological origin, and their expression is differentially regulated during rat brain development.

RESULTS

Our initial studies were directed towards analyzing the expression of α tubulin, β tubulin, and actin mRNAs using both *in vitro* translation and cDNA hybridization techniques.[8] During brain maturation (0–80 days postnatal), these mRNA species undergo a dramatic decrease in abundance, in particular, a 1.8 kilobase (kb) and a 2.9

[a]This work was supported by U.S. Public Health Service Grants AG-00001 and GM-29630 from the National Institutes of Health.

kb β-tubulin mRNA species decrease by 90–95% in the cerebrum after day 11 and a decrease of 80% in the cerebellum after day 16. Coincident with this response is an increased production of a 2.5 kb β-tubulin mRNA. This change in mRNA expression occurs at a time of significant morphological differentiation (*i.e.*, cessation of cell division and neurite extension) in the brain.

To address the question of whether these β-tubulin mRNAs are transcribed from different genes, we isolated and characterized three different cDNA clones corresponding to β-tubulin mRNAs. The three clones are referred to as RBT1, RBT2, and RBT3, and each was obtained from a different cDNA library: RBT1 was derived from neonatal brain mRNA, RBT2 from adult cerebellum mRNA, and RBT3 from neonatal spleen mRNA.

Tissue Specific Expression of the β-Tubulin mRNAs

DNA sequence analysis reveals that the coding region of each mRNA is highly homologous, whereas the 3'-untranslated segments are significantly diverged. This suggests that the specificity of the different mRNA species resides in the 3' ends. Therefore, in order to study the tissue distribution of each sequence, we have subcloned these 3' regions and hybridized them to equal amounts of RNA isolated from different rat tissues and cells. TABLE 1 reveals that RBT1 and RBT2 are expressed only in cells of neurological origin, whereas RBT3 appears to be present in all cell types analyzed.

Beta-Tubulin mRNA Expression during Brain Development

In order to analyze the expression of the various β-tubulin mRNAs during brain development in more detail, we hybridized these three cDNA clones to RNA isolated

TABLE 1. Expression of the Different β-Tubulin mRNAs in Various Rat Tissues and Cells[a]

	RBT1[b]	RBT2	RBT3
Tissues			
Brain	+ + + + +	+ + +	+ +
Spleen	—	—	+ +
Lung	—	—	+
Liver	—	—	+
Kidney	—	—	+
Heart	—	—	+
Cells			
Neuroblastoma (B50, B103)	+ +	ND[c]	ND
Pheochromocytoma (PC12)	+	ND	ND
Fibroblast	—	—	+
Epithelial	—	—	ND

[a]Data was derived from reference 13, and Mbangkollo, Robinson, and Farmer, (unpublished data).

[b]Tissue analyses of RBT1 and RBT3 were performed separately on 5-day-old rats, and of RBT3 on 35-day-old rats.

[c]ND: no determination performed.

TABLE 2. Differential Expression of the Three β-Tubulin mRNAs during Postnatal Rat Cerebellum Development[a]

	Age (Days)					
	0	7	14	21	35	48
RBT1	+++++	+++++	++	—	—	—
RBT2	+	+	+++	+++	+++	+++
RBT3	ND[b]	++	ND	ND	ND	+

[a]Data obtained from reference 13 and Mbangkollo, Robinson, and Farmer, (unpublished data).
[b]ND: no determination performed.

from cerebellum at various ages of postnatal development. TABLE 2 shows that RBT1 is expressed in very high abundance during the first week of development. Its production then declines rapidly, such that by day 21 there are no detectable sequences. This clone corresponds only to a 1.8 kb mRNA species. RBT2 sequences can also be detected at early stages; the levels, however, are very low. The increased expression of this mRNA occurs between 7 and 14 days, the period when RBT1 expression is beginning to decline. RBT2 mRNA corresponds to the 2.5 kb species, previously shown to increase in abundance during later stages of postnatal development. The RBT3 specific subclone hybridizes to both a 1.8 and 2.9 kb mRNA, and these two species decrease only slightly in abundance during the entire postnatal period (see reference 13 for further details).

Analysis of a Genomic Clone Corresponding to the RBT2 β-Tubulin mRNA

We have screened a rat genomic library (kindly given to us by John Tamkun of M.I.T.) for β-tubulin gene sequences, and using the 3'-untranslated region subclone of RBT2 as the hybridization probe, we have isolated a specific genomic clone, 207. FIGURE 1 shows a restriction map of 207 and various subclones generated from it. Parallel restriction digests of 207 and total rat genomic DNA were compared by Southern analysis to test for rearrangements within the clone (data not shown). In all cases, the DNA fragments are identical, suggesting that no gross rearrangements, additions, or deletions have occurred during cloning. The extreme 5' end of this clone 207 (a Sal I—Bgl II fragment) was subcloned into the vector pSP64 to give rise to the plasmid pSP64-207-32. This clone, which contains 5' untranscribed sequences, probably the promoter, and two small exons, were used to generate specific 5' RNA transcripts to probe the Southern blots in DNase I sensitivity studies.

Tissue–Specific DNase I Sensitivity of the RBT2 Gene

We have employed limited digestion with the enzyme, pancreatic DNase I, to assess the potential for transcriptional activity of the RBT2 gene. Whole brain nuclei from 0 day old rats were treated with increasing concentrations of DNase I. The DNA was purified, digested with BAM HI, and analyzed by hybridization to a runoff transcript of pSP64-207-32 digested with EcoRI. This transcript does not contain exon sequences. After the lower stringency wash (FIGURE 2A), 4 bands of hybridization were observed. Two of these bands (4.5 and 3.9 kb) are DNase I sensitive indicating

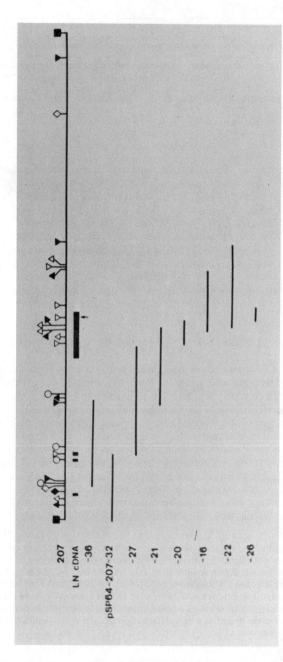

FIGURE 1. Restriction map of the RBT2 genomic clone (207) and its respective subclones. The top line is a map of 207 with restriction sites indicated by the following symbols: ■ = Sal I; ▲ = Bam HI; △ = Ava I; ◆ = Sma I; ◇ = EcoRI; O = PvuII; ▼ = Hind III; ▽ = Bg1 II. The locations of the regions corresponding to the cDNA (LNcDNA) are shown on the second line. The subclone pSP64-207-32 was used in the DNase I sensitivity experiments described in FIGURES 2 and 3.

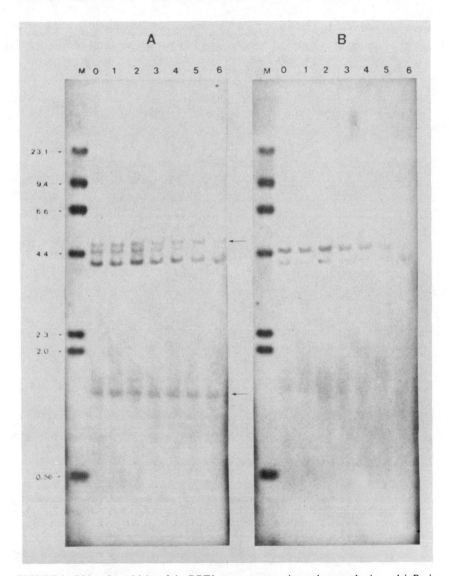

FIGURE 2. DNase I sensitivity of the RBT2 gene sequences in newborn rat brain nuclei. Brain nuclei from 0 day old rats were digested with increasing concentrations of DNase I. DNA was then purified from these nuclei, digested with Bam HI, fractionated on a 1% agarose gel, Southern blotted, and hybridized to a RNA transcript of the pSP64-207-32 subclone (see FIGURE 1). Lanes 0-6 correspond to 0, 0, 4, 8, 12, 16, and 20 U/ml of DNase I, respectively. M = Hind III digested λ DNA size markers. Panel A shows a low stringency wash resulting in four bands of hybridization. Panel B shows the blot after a high stringency wash: the arrows indicate the bands that are removed by this treatment.

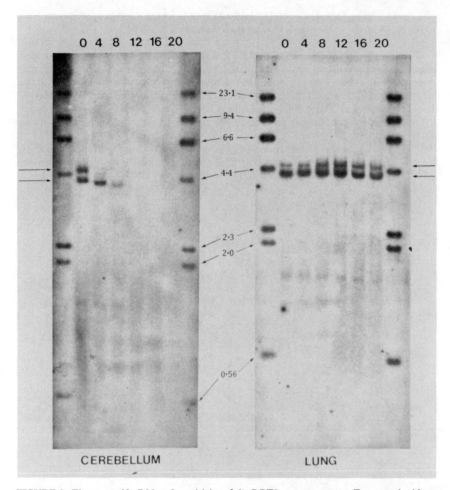

FIGURE 3. Tissue-specific DNase I sensitivity of the RBT2 gene sequences. Two-month-old rat cerebellar and lung nuclei and the resultant DNA were treated as described in FIGURE 2. Lanes 1–6 correspond to 0, 4, 8, 12, 16, and 20 U/ml DNase I. The Hind III λ DNA size markers are in lane M.

that these fragments are portions of the transcriptionally active RBT2 gene. The other two bands are DNase I insensitive and thus correspond to a pseudogene. The probe can easily be washed from these two pseudogene fragments at higher stringency (FIGURE 2B), suggesting some degree of divergence of nucleotide sequences between the active gene and inactive pseudogene. FIGURE 3 shows a comparison of the DNase I sensitivity of the RBT2 gene in adult rat cerebellum and lung employing the same procedures as described in FIGURE 2. We observe that the two BamHI fragments (4.5 and 3.9 kb) are only sensitive to DNase I digestion in cerebellar nuclei. This observation correlates with the neural specific expression of RBT2 mRNA sequences (see TABLE 1 and reference 13).

Comparison of the Amino Acids in the Most Divergent Regions of Different Vertebrate β Tubulins

The three rat β-tubulin cDNAs (RBT1, RBT2, and RBT3) have been sequenced using the Sanger dideoxy chain terminator method. A comparison of the nucleotide sequences with those of other β tubulins reveals a high degree of homology within the coding regions. The predicted amino acids have also been compared, and, as expected, they are conserved throughout evolution. There are, however, some regions of divergence among the various vertebrate proteins, and these are clustered within at least four domains. These are as follows: amino acid numbers a) 35–57 b) 216–235 c) 288–365, and d) the last fifteen amino acids of the carboxyl terminus. When analyzing the first three domains we noticed that RBT1 was identical to a chicken β tubulin CB2,[16] and that a pig sequence[15] and RBT2 were also very similar. All four of these β tubulins were derived from brain tissue. RBT3 is more closely related, however, within

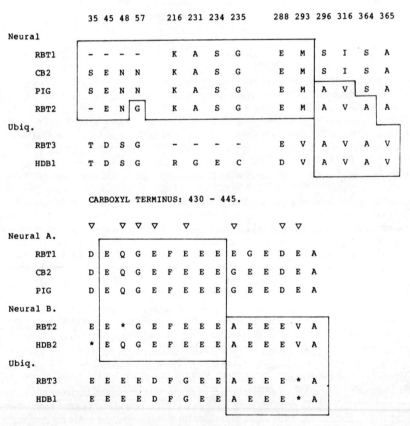

FIGURE 4. Comparison of the amino acids in the most divergent regions of different vertebrate β tubulins. Data were derived from DNA sequencing of RBT1, RBT2, and RBT3 (data not shown), and 14 (for human, HDB1 and HDB2), 15 (for pig), and 16 (for chicken, CB2). Numbers correspond to amino acid positions along each β-tubulin polypeptide.

these regions, to a human β tubulin HDB1[14] than to its rat counterparts. Both RBT3 and HDB1 are expressed in several different tissues. A comparison of the carboxy terminal amino acids appears to be the most useful parameter for grouping together the various β tubulins into similar isotypes. This region of RBT1 is almost identical to CB2 and the corresponding pig sequences. On the basis of this and the previous comparison, we propose that these three isotypes are evolutionarily related and are the same neural specific isotype (FIGURE 4, Neural A). The carboxyl terminus of RBT3 is significantly diverged from that of RBT1 in that 9 of 15 amino acids are different. This sequence is identical with that of HDB1: we propose that these two are also related and correspond to ubiquitous isotypes. Finally, the carboxyl terminus of RBT2 differs somewhat from both the neural A and ubiquitous groups: it is closely related, however, to another human β tubulin, HDB2.[14] This suggests that HDB2 is also a neural isotype and that these two sequences represent another evolutionarily conserved gene (FIGURE 4, Neural B). It is interesting to note that the RBT2 isotype appears to be a combination of both neural and ubiquitous-specific amino acids. This can most prominantly be seen within the carboxyl terminus, but it is also apparent at amino acids 296, 316, 364, and 365.

DISCUSSION

The studies described here reveal that there is differential expression of the rat β-tubulin multigene family during rat brain development. Our conclusions are derived from the identification of three different cDNAs (RBT1, RBT2, and RBT3) corresponding to different β-tubulin mRNAs. The RBT1 clone hybridizes to a 1.8 kb mRNA, RBT2 hybridizes to a 2.5 kb mRNA, and both mRNAs are present only in tissues of neurological origin. On the other hand, RBT3 hybridizes to both a 1.8 kb and a 2.9 kb mRNA found in most tissues. The expression of the two neural-specific mRNAs (RBT1 and RBT2) is regulated during brain development in a very interesting manner. RBT1 mRNA is present in high abundance during the first week of postnatal brain development. Then, during the second week, the level of this mRNA decreases dramatically, such that, by day 21 there are no detectable RBT1 sequences present. On the other hand, RBT2 is expressed only at very low levels during the first two weeks of development, but interestingly there is a rapid increase in abundance between 7 and 15 days postnatal. It appears, therefore, that there is a switch in expression of these two neural-specific β-tubulin genes during this developmental period. It should be noted, however, that the abundance of RBT2 in the mature brain is only a fraction of the abundance of RBT1 in the neonatal brain.

Within the time points studied, two major developmental events involving microtubules are occurring in the cerebellum: rapid cell proliferation, followed by terminal differentiation, involving extension of neurite processes and synapse formation.[17] The proliferative stage of differentiation occurs during the first two weeks after birth. It is during this time that we observe the concomitant high levels of RBT1 mRNA. Neuronal cell differentiation commences with the cessation of cell division followed by the extension of processes to give rise to synapses with other cell types.

The result of these processes, occurring between 7 and 21 days after birth, is a reduction in the outer proliferative zone and formation of well differentiated inner cell layers. It is interesting that the switch in expression of RBT1 and RBT2 genes correlates closely to the change from the proliferative stage to the neuritic outgrowth stage of development. It is conceivable that RBT1 synthesizes a β tubulin that is produced in large amounts, stockpiled, and then mobilized during the process formation to build axonal and dendritic classes of microtubules. The expression of RBT2 later in development may be required to complete the formation of these specific microtubules. This β tubulin, however, may be synthesized to participate in a completely different set of microtubules, for instance, those involved in synaptogenesis. On the other hand, it is also possible that RBT1 and RBT2 are each expressed in separate neural cell types that are changing in their relative abundance during development. We should be able to address this latter possibility by performing *in situ* hybridization studies on brain slices taken from rats at various ages, using the radiolabeled pSP6 RNA transcripts as probes.

RBT3 appears to be a ubiquitous isotype of β tubulin whose expression is not regulated to any major extent during brain development. Furthermore, from hybridization analysis, it appears that this sequence is closely related to a 2.9 kb mRNA species. This observation has previously been noted for the human β-tubulin gene family.[12] These investigators demonstrated the transcription of both a 1.8 kb and a 2.6 kb mRNA from the same gene (HDB1). Our sequence data (FIGURE 4) suggests that RBT3 and the human HDB1 are the same gene: thus it is not surprising that the RBT3 specific subclone hybridizes to a larger mRNA.

By comparing the predicted amino acid sequences of the entire β-tubulin polypeptides we have been able to group these vertebrate β tubulins into two major isotype classes (neural and ubiquitous). A comparison of the last fifteen amino acids of the carboxyl terminus has further allowed us to propose that the neural class comprises two subclasses, that is, neural A containing RBT1, which is expressed early in brain development, and neural B containing RBT2, which is expressed later.

It is possible that these regions of divergence correlate with important functional domains within the β-tubulin polypeptide? The conserved parts of the protein may be responsible for maintaining subunit interactions within the microtubule, whereas the divergent regions may be important for specialized functions such as binding of ligands and other proteins, that is, microtubule-associated proteins (MAPs). Although there is very little information on the structural/functional features of β tubulin, investigators have determined the major secondary structures of the protein.[15] The carboxyl terminus, which is a region of significant divergence, is highly acidic and has potential to adopt an α-helical conformation. It has been suggested that this region may have a specialized function in the binding of cationic proteins.[15] Isotypic amino acid differences may subtly alter the α-helical structure of this region to allow binding to different cellular components. A comparison of many different β-tubulin sequences resulting in the various isotype groups as presented here should also allow the design of interesting future experiments. For instance, production of domain specific antibodies and expression of recombinant tubulin sequences lacking specific domains will help us to relate structure and function in a precise manner.

REFERENCES

1. SHELANSKI, M. L. & H. FEIT. 1972. Filaments and tubules in the nervous system. G. H. Bourne, Ed.: **6:** 47–80. *In* Structure and function of the nervous tissue. Academic Press, Inc. New York.
2. SCHMITT, H., I. GOZES & U. Z. Littauer. 1977. Decrease in levels and rates of synthesis of tubulin and actin in developing rat brain. Brain Res. **121:** 327–342.
3. GOZES, I. & U. Z. LITTAUER. 1978. Tubulin microheterogeneity increases with rat brain maturation. Nature (London) **276:** 411–413.
4. GOZES, I., A. BAETSELIE & U. Z. LITTAUER. 1980. Translation *in vitro* of rat brain mRNA coding for a variety of tubulin forms. Eur. J. Biochem. **103:** 13–20.

5. MORRISON, M. R., S. PARDUE & S. T. GRIFFIN. 1981. Developmental alterations in the levels of translationally active messenger RNAs in the postnatal rat cerebellum. J. Biol. Chem. 256: 3550–3556.
6. BEN-ZE'EV, A., S. R. FARMER & S. PENMAN. 1979. Mechanisms of regulating tubulin synthesis in cultured mammalian cells. Cell 17: 319–325.
7. CLEVELAND, D. W., M. A. LOPATA, P. SHERLINE & M. W. KIRSCHNER. 1981. Unpolymerized tubulin modulates the level of tubulin mRNAs. Cell 25: 537–546.
8. BOND, J. F. & S. R. FARMER. 1983. Regulation of tubulin and actin mRNA production in rat brain: expression of a new β-tubulin mRNA with development. Mol. Cell. Biol. 3: 1333–1342.
9. COWAN, N. J., C. D. WILDE, L. T. CHOW & F. C. WEFALD. 1981. Structural variation among human β-tubulin genes. Proc. Natl. Acad. Sci. USA 78: 4877–4881.
10. WILDE, C. D., C. E. CROWTHER & N. J. COWAN. 1982. Diverse mechanisms in the generation of human β-tubulin pseudogenes. Science 217: 549–552.
11. WILDE, C. D., C. E. CROWTHER, T. P. CRIPE, M. GWO-SHU LEE & N. J. COWAN. 1982. Evidence that a human β-tubulin pseudogene is derived from its corresponding mRNA. Nature (London) 297: 83–84.
12. GWO-SHU LEE, M., S. A. LEWIS, D. C. WILDE & N. J. COWAN. 1983. Evolutionary history of a multigene family: an expressed human β-tubulin gene and three processed pseudogenes. Cell 33: 477–487.
13. BOND, J. F., G. S. ROBINSON & S. R. FARMER. 1984. Differential expression of two neural cell-specific β-tubulin mRNAs during rat brain development. Mol. Cell. Biol. 4: 1313–1319.
14. HALL, J. L., L. DUDLEY, P. L. DOBNER, S. A. LEWIS & N. J. COWAN. 1983. Identification of two human β-tubulin isotypes. Mol. Cell. Biol. 3: 854–862.
15. KRAUHS, E., M. LITTLE, T. KEMPF, R. HOFER-WARBINEK, W. ADE & H. PONSTINGL. 1981. Complete amino acid sequence of β-tubulin from porcine brain. Proc. Natl. Acad. Sci. USA 78: 4156–4160.
16. VALENZUELA, P., M. QUIROGA, J. ZALDIVAR, W. J. RUTTER, M. W. KIRSCHNER & D. W. CLEVELAND. 1981. Nucleotide and corresponding amino acid sequences encoded by α and β tubulin mRNAs. Nature (London) 289: 650–655.
17. JACOBSON, M. 1978. Developmental neurobiology, 2nd ed., p. 76–88. Plenum Publishing Corp. New York.

Quantitation and *in Situ* Localization of Tubulin mRNA in the Mammalian Nervous System[a]

MARCELLE R. MORRISON AND W. SUE T. GRIFFIN

Departments of Neurology and Cell Biology
University of Texas Health Science Center at Dallas
Dallas, Texas 75235

Microtubules have a variety of cytoskeletal functions and are present in all cell types.[1] They are especially abundant in neuronal cells where they are found not only in the cell body but also as a major component of axons and dendrites.[2]

The major proteins found in microtubules are the α and β tubulins. Each of these proteins is composed of several isotypes. Neuronal cells contain many more discrete α- and β-tubulin isotypes[3,4] than do other cells.[5] Each isotype is present in individual neurons,[6] but their relative abundance varies in different brain areas and during development.[5] Some of the tubulin isotypes result from posttranslational modification of proteins synthesized from one messenger RNA (mRNA),[7] whereas others may be translation products of mRNAs that have a different primary structure. Several expressed genes for the α and β tubulins have been identified in rat, chick, and human.[8–13] During rat brain development, at least three of the β-tubulin genes are differentially expressed.[12,13]

We have shown that four major tubulin mRNAs, two α and two β, are expressed in mouse neuroblastoma cells, rat brain, and human brain. The two neuroblastoma β-tubulin mRNAs can be differentiated by cytoplasmic distribution, efficiency of translation, and degree of polyadenylation. These two mRNAs are also differentially regulated during the development of the rat cerebellum and cortex. *In vitro* translation and slot-blot hybridization to a cDNA probe demonstrated that the levels of the less abundant brain β-tubulin mRNA decrease concomitant with the cessation of cell division and with the extension of neuronal processes. Quantitation of this β-tubulin mRNA by *in situ* hybridization showed that it is differentially expressed in the individual cell types of the 14-day-old rat cerebellum; it is relatively more abundant in individual granule cells than it is in postmigratory cells or other neuronal cell types.

DIFFERENTIAL REGULATION OF TUBULIN mRNAs IN MOUSE NEUROBLASTOMA CELLS

The mouse neuroblastoma cell is a tumor cell line derived from neurons of neural crest origin.[14] The cloned cell lines display many properties characteristic of differentiated neurons when treated with a variety of differentiating agents. The cellular homogeneity of these neuronal cell lines enables us to characterize the mRNA populations present in undifferentiated and differentiated cells.[15–19]

[a]This work was supported by NIH Grants HD 14886 (M.R. Morrison) and AI 14663 (W.S.T. Griffin) and by the Leland Fikes Organization (MRM).

51

The tubulins and actin are among the most abundant neuroblastoma proteins. FIGURE 1 shows the actin and tubulin area of a two-dimensional gel of total neuroblastoma proteins. The tubulins were identified by coelectrophoresis with brain tubulins that were isolated by several cycles of polymerization and depolymerization and by precipitation with vinblastine. Two α-tubulin isotypes, α_1 and α_2, and two β-tubulin isotypes, β_1 and β_2, were resolved by two-dimensional gel electrophoresis. *In vitro* translation products of polyadenylated (poly(A)$^+$) neuroblastoma polysomal RNAs were then analyzed to determine if the tubulin isotypes were the products of different mRNAs or if they were formed by posttranslational modification. Among the *in vitro* translation products were four proteins that comigrated and copolymerized with the unlabeled tubulin isotypes (FIGURE 2A), indicating that the proteins are synthesized by mRNAs in a wheat germ cell-free system.

In order to differentiate between primary translation products and proteins that have been posttranslationally modified *in vitro*,[20] it is necessary to demonstrate that the levels of the isotypes synthesized *in vitro* can be independently regulated. In the course of our studies on translational regulation, we have obtained proof for the independent regulation of β_1 and β_2 mRNAs in undifferentiated neuroblastoma cells.[17] Approximately 30% of neuroblastoma cytoplasmic mRNAs are not present on polysomes, but are found in nontranslated messenger ribonucleoprotein (mRNP) particles. Our quantitation of *in vitro* translation products of mRNAs isolated from polysomes and from mRNP particles demonstrates that several of the abundant mRNAs are differentially distributed between the two cytoplasmic fractions.[17] Several mRNAs, including those that encode the β_1 tubulin, are present in the polysome (FIGURE 2A) but not the mRNP fraction (FIGURE 2B). Other mRNAs, including those that code for actin and the β_2 tubulin, are enriched in the polysome fraction. By contrast, the mRNAs for the α tubulins and vimentin are relatively more abundant in the mRNP fraction. These results demonstrate that mRNAs isolated from the polysome and mRNP cell fractions synthesize different relative amounts of the β_1 and β_2 tubulins when translated in a cell-free system. Therefore, each isotype must be translated from a unique mRNA.

This conclusion is strengthened by comparing translational efficiencies of the different mRNAs under conditions where ribosomes and initiation factors are rate

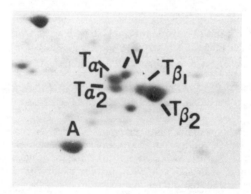

FIGURE 1. Two-dimensional gel electrophoresis of unlabeled proteins from neuroblastoma S-20 cells. Only the tubulin area of the gel is shown in this and subsequent electrophoretograms. The isoelectric focusing dimension is from left to right. The molecular weight dimension is from top to bottom. A = actin, V = vimentin; T_{α_1}, T_{α_2} = α_1 and α_2 tubulins; T_{β_1}, T_{β_2} = β_1 and β_2 tubulins.

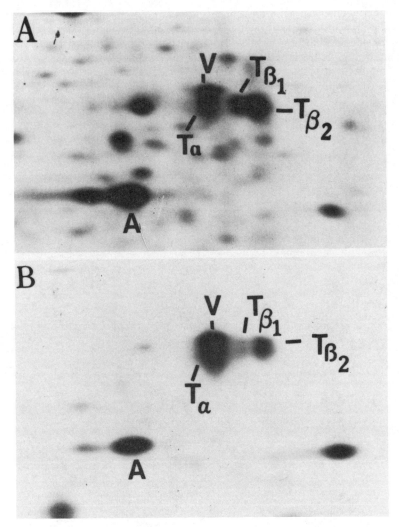

FIGURE 2. *In vitro* translation products of poly(A)$^+$ mRNAs isolated from neuroblastoma cells. Two-dimensional gel electrophoresis of the *in vitro* translation products of poly(A)$^+$ RNAs isolated from neuroblastoma polysomes (**A**) and from the postribosomal mRNP fraction (**B**). Translation was performed under subsaturating conditions in a wheat germ *in vitro* protein synthesizing system using [^{35}S] methionine as the radiolabel.[16]

limiting.[21] Under these conditions, the β_2 isotype is preferentially synthesized, whereas synthesis of the β_1 isotype is significantly reduced (FIGURE 3B) when compared to translation under normal, nonsaturating conditions (FIGURE 3A). Because the translational efficiency of the β_1 and β_2 mRNAs *in vitro* does not correlate with their cytoplasmic distribution, factors other than primary structure must regulate the distribution of β_1 and β_2 mRNAs between polysomes and mRNP particles *in vivo*.

In addition to cytoplasmic distribution and translational efficiency, the neuroblastoma β_1- and β_2-tubulin mRNAs can be distinguished by their degree of adenylation. Unlike the actin mRNAs, 90% of the cytoplasmic α- and β-tubulin mRNAs are adenylated in neuroblastoma cells.[16] Two-dimensional gel electrophoresis of the proteins encoded by the nonadenylated mRNAs, that is, those that do not bind to

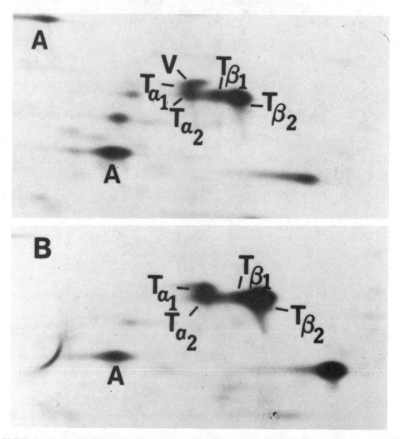

FIGURE 3. *In vitro* translation products of neuroblastoma poly(A)$^+$ RNAs translated under subsaturating and supersaturating conditions. Two-dimensional gel electrophoresis of wheat germ, ^{35}S-labeled *in vitro* translation products of mRNAs translated under subsaturating (**A**) and supersaturating (**B**) conditions.

oligo(dT) cellulose, shows that these mRNAs contain higher levels of the β_2-tubulin mRNA than does the adenylated mRNA fraction (FIGURE 4). This result suggests that the poly(A) regions of the tubulin mRNAs are processed at different rates and/or that the nonadenylated β_2-tubulin mRNA is more stable than the others.[16]

It is clear from these results that the neuroblastoma β_1 and β_2 tubulins are

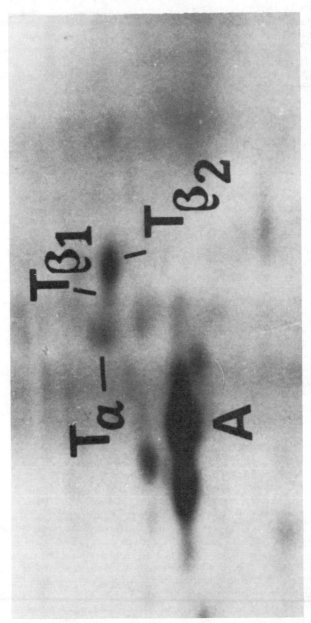

FIGURE 4. *In vitro* translation products of neuroblastoma nonadenylated mRNAs. Two-dimensional gel electrophoresis of wheat germ, ^{35}S-labeled *in vitro* translation products of mRNAs that do not hybridize to oligo (dT) cellulose at 4°C and that therefore contain less than 6 As.[16]

synthesized by different mRNAs and that these mRNAs are differentially translated and processed. It is important to note that tubulin isotypes with similar isoelectric points and molecular weights might not be resolved in our two-dimensional gel system. Thus, the β tubulins synthesized *in vitro* are encoded by a minimum of two mRNAs.

Some aspects of neuronal differentiation can be mimicked by increasing intracellular cyclic AMP (cAMP) levels in neuroblastoma. In the cholinergic line S-20, cholineacetyltransferase and cAMP-binding protein levels increase several fold as cells differentiate.[22] After treating cells with dibutyryl cAMP, there is an early transient increase in tubulin mRNA.[23] After three days, there is a 1.4–1.7-fold increase in an mRNA that encodes a cAMP-binding protein, and a 10% decrease in tubulin mRNA levels.[19] These results show that increasing intracellular cAMP levels in neuroblastoma has little long-term effect on tubulin mRNA levels.

TUBULIN mRNA LEVELS IN CEREBELLUM AND CORTEX

The tubulin proteins are involved in a number of developmental events including cell division, axon and dendrite formation, and axoplasmic transport.[1] Their levels in whole brain decrease during development.[24–27] This decrease is accompanied by alterations in the relative amounts of specific isotypes[5,28] and their encoding mRNAs. Neither the functions, however, nor the intra- and intercellular distributions, nor the factors that regulate the synthesis and posttranslational modifications of the different tubulin isotypes in the brain is known.[26,29] To correlate different tubulin mRNA levels with specific developmental events, we have quantitated these levels at defined stages of cortex and cerebellum development.

The postnatal development of the rat cerebellum has been well characterized anatomically, electrophysiologically, and neurochemically.[30–32] The major cell types in the cerebellum of newborn rats are the prenatally formed Purkinje and Golgi cells (15%), the glial cells (15%), and the germinal cells of the external granular layer (EGL) (70%).[31,33–36] Granule cell proliferation is maximal at days 6–8, whereas postmitotic granule cells are migrating into their final positions in the internal granular layer (IGL).[37] By day 14, granule cell proliferation has declined by 60%,[37] and the rate of synaptogenesis and myelination is high.[31,38] Cell division has virtually ceased by day 21,[37] but even after day 30 there is continued arborization of Purkinje cell dendrites.[33] In the adult, greater than 90% of cells are the granule cell interneurons of the IGL.[36]

Two-dimensional gel analysis of unlabeled cerebellar proteins demonstrated that the steady state levels of the tubulin isotypes decreased during cerebellar development.[39] Short-term labeling studies *in vivo* showed that this decrease was due to a developmental decrease in the *de novo* synthesis of these isotypes.[39] Two-dimensional electrophoresis of *in vivo* translation products of cerebellar mRNAs isolated early in postnatal development resolves clearly two α- and two β-tubulin isotypes. Quantitation of the relative levels of mRNAs encoding the tubulin isotypes demonstrated correlations with specific developmental stages. The mRNAs for both α tubulins decreased postnatally 25% between days 2 and 6. A further decrease to adult levels occurred between days 14 and 30. The β_1-tubulin mRNA is never present at more than 25% of the level of the β_2 mRNA during any developmental stage. The postnatal levels of the β_2- and the β_1-tubulin mRNAs both decreased after 14 days. The levels of the mRNA encoding the β_2 isotype remained relatively high, even in the adult (50% of day 2), whereas the β_1-mRNA levels had decreased more than 75 percent. These results indicate that the levels of α- and β-tubulin mRNAs in cerebellum are independently regulated as are the mRNAs encoding the two β-tubulin isotypes.

Correlations between tubulin mRNA levels and human brain development are also possible. First, we established that postmortem degradation of mRNAs in rat and human brain is essentially random.[40] The abundant mRNAs isolated from human and rat cerebellum kept 16 hours postmortem are undegraded and translationally active. These mRNAs are present in the same relative amounts as in rat cerebellar RNA isolated immediately postmortem.[40,41] Isolation and *in vitro* translation of poly(A)$^+$ RNAs from one month postnatal and adult human cerebella shows that the β_1-tubulin mRNA is still present at one month, but is significantly reduced in the adult, whereas the levels of the β_2 mRNA are relatively high in both (FIGURE 5).

The developmental timetable of the rat cortex is different from that of the cerebellum. Cortical cell division takes place between gestational days 16 and 21. The

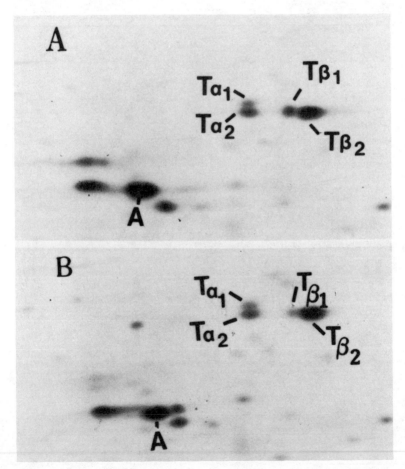

FIGURE 5. *In vitro* translation products of poly(A)$^+$ RNAs isolated from one month postnatal and adult human cerebellum. Two-dimensional gel electrophoresis of wheat germ ^{35}S-labeled *in vitro* translation products of poly(A)$^+$ mRNAs isolated from one month human (**A**) and adult human (**B**) cerebella.

earliest formed cells migrate within two days to the outer layers of the cortex.[42] Cells formed between gestational days 19–21 take 3–10 days to migrate to the inner cortical layers.[43] Axons elongate concomitant with cell migration, whereas the bulk of dendritic arborization and synaptic connections occur between days 12 and 20 postnatal.[44] As in other brain areas such as the cerebellum, large cells develop first. Several studies have shown that the chromatin repeat length[45,46] and the content of chromatin-associated proteins in cortex and cerebellum[47,48] change concomitant with the arrest of cell division and the beginning of terminal differentiation. This occurs between postnatal days 1–7 in cortex and days 7–30 in the cerebellum.

Developmental changes in translationally active cortical tubulin mRNAs were compared with those in the cerebellum. *In vitro* translation of total RNAs isolated from fetal rat cortex showed that the synthesis of the minor tubulin subunit (β_1) was

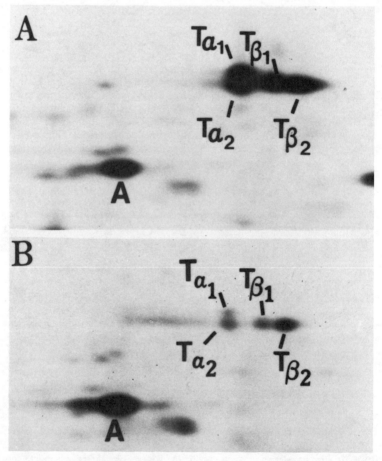

FIGURE 6. *In vitro* translation products of rat fetal cortex RNAs. Two-dimensional gel electrophoresis of wheat germ ^{35}S-labeled *in vitro* translation products of poly(A)$^+$ RNAs (**A**) and poly(A)$^-$ RNAs (**B**) isolated from rat fetal cortex.

nearly equivalent to that of the major (β_2) cerebellar form.[49,50] Separation and translation of the poly(A)$^+$ and poly(A)$^-$ RNAs showed that the β_1-tubulin mRNA was present in the poly(A)$^+$ mRNA fraction at higher levels than was the β_2-tubulin mRNA (FIGURE 6A), whereas the β_2-tubulin mRNA was present at higher levels in the poly(A)$^-$ RNAs (FIGURE 6B). These results parallel the distribution of the β_1- and β_2-tubulin mRNAs in the nonadenylated mRNA fraction in neuroblastoma, as noted above. Although the β_1-tubulin mRNA was present at much higher levels in fetal rat cortex than at any stage in cerebellar development, its levels in the 10 day and adult rat cortex were similar to those in 10 day and adult rat cerebellum, respectively. Again, this developmental pattern in rat cortex was mirrored in human fetal and adult cortex.[41]

Using our recently devised RNA microisolation procedure, we have isolated undegraded, translationally active whole-cell RNAs from as little as 5–10 mg of wet weight brain tissue (Ilaria *et al.*, manuscript in preparation). Using this procedure, multiple comparisons (up to 36 samples per day) of β-tubulin mRNA levels in cortex and cerebellum were done at one time. The chick β-tubulin clone of Cleveland *et al.*[51] was used as a probe for Northern analysis of mRNAs at various stages of cerebellar development. At early stages of cerebellar development, the probe hybridizes to a 1.8 kilobase (kb) mRNA (FIGURE 7). After postnatal day 21, hybridization to the 1.8 kb mRNA was significantly decreased, and additional hybridization to a 2.5 kb mRNA was seen. These results are similar to the developmental pattern of cerebellar RNAs hybridizing to the brain-specific $R\beta T_1$ clone of Bond *et al.*[13] They contrast with the hybridization patterns observed with their $R\beta T_3$ clone. This cDNA hybridizes to a 2.9 kb and a 1.8 kb mRNA at early developmental stages.[13] In the adult, the 2.9 kb mRNA is not expressed, but the 1.8 kb mRNA is still present at relatively high levels.[13] The chick β-tubulin cDNA, therefore, corresponds to the developmentally regulated rat β-tubulin mRNA, $R\beta T_1$. Furthermore, the chick β-tubulin clone does not cross hybridize with the major adult rat β-tubulin mRNA, $R\beta T_3$. We conclude that the chick β-tubulin cDNA is a selective probe for the rat $R\beta T_1$ mRNA at early stages of cerebellar development. In the adult, the chick probe will also recognize the developmentally induced 2.5 kb β-tubulin mRNA corresponding to the $R\beta T_2$ clone of Bond *et al.*[13] Another group has also shown that the chick β-tubulin clone does not cross hybridize under stringent conditions with the major rat brain tubulin mRNA (Ginzburg, personal communication).

Total RNAs were isolated from the cortex and cerebellum of rats at different stages of development. Slot blot hybridizations, using the chick β-tubulin probe, demonstrated that the complementary mRNAs were several fold more abundant in fetal cortex and early postnatal cortex than at any stage of cerebellar development. The highest cerebellar levels of this mRNA were at 4–6 days postnatal. At 10 days postnatal, cortex levels were similar to those in cerebellum. After day 10, the mRNA levels in both brain areas declined with a similar time course and had reached a plateau by day 21 (Morrison *et al.*, manuscript in preparation).

The hybridization pattern of this β-tubulin mRNA exactly parallels the levels of the β_1 tubulin identified and quantitated by two-dimensional electrophoresis of *in vitro* translation products (see above). Like the β_1 tubulin, the levels of the β-tubulin mRNA are highest in early postnatal cortex and become equivalent to those in cerebellum only after 10 days postnatal. We tentatively conclude that the major translation product of the mRNA hybridizing to the chick β-tubulin probe is the β_1-tubulin isotype. This conclusion would be substantiated by positive hybrid selection of this mRNA[52] and two-dimensional analysis of its translation products.

The tubulin mRNA levels peak at the time of maximal cell division in the cerebellum (days 6–8) and decline concomitant with the decline in cell division of the

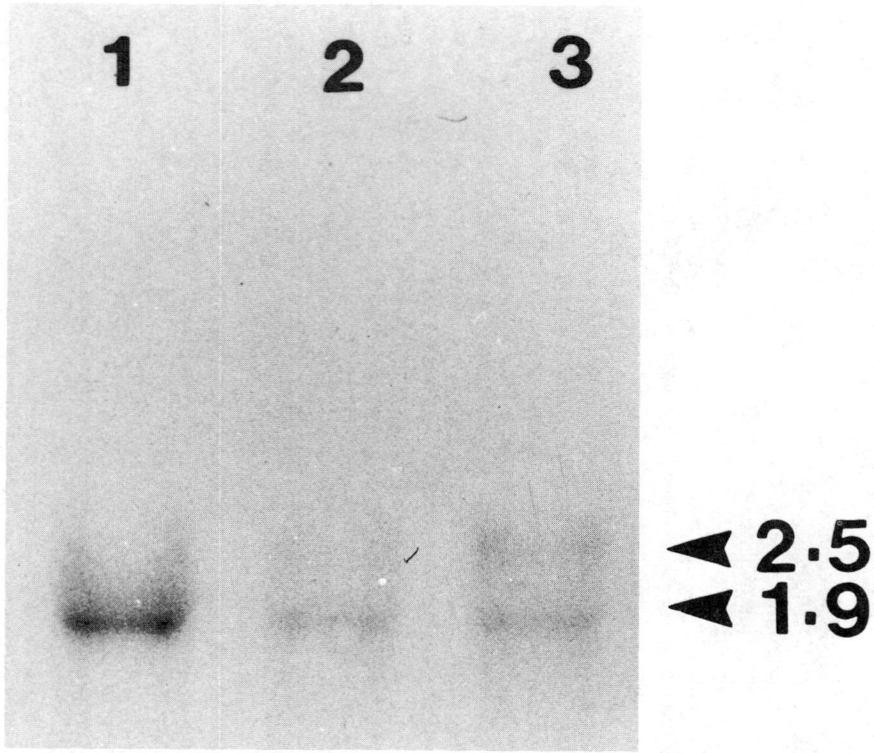

FIGURE 7. Northern analysis of 14 and 21 day total cerebellar RNAs hybridized to ^{32}P-labeled chick β-tubulin DNA. Four μg total cerebellar RNAs were electrophoresed on denaturing agarose gels,[57] transferred to Zeta-probe (Bio-Rad) and hybridized to ^{32}P-chick β-tubulin cDNA labeled by nick translation.[58] Before exposure to film, filters were washed at high stringency (final washes were done in 0.5 × SSC at 55°C). Lane 1 = day 14; lane 2 = day 21; lane 3 = day 28.

granule cells in the external granule layer. In the cortex, the postnatal β-tubulin mRNA levels are higher several days after the cessation of cell division (4 days postnatal) than they are perinatally (although we have not yet quantitated mRNA levels prior to fetal day 18). Only a small percentage of cortical cells are dividing at the perinatal days that were measured. This suggests that the high β-tubulin mRNA levels are related to neuronal migration and extensive process formation in early postnatal cortex rather than to cell division. A requirement for the β-tubulin mRNA in process formation would also explain the relatively low levels of this mRNA in cerebellum, as there is minimal arborization of postnatally formed cells in this brain region. High levels in cortex might also be a function of cell type; differentiating glia or large neurons (both relatively more abundant in cortex than in cerebellum) might require more β tubulin than do differentiating granule cells.

Quantitation of tubulin mRNA levels in individual cells during development would determine whether the β-tubulin mRNA is more abundant in specific cell types. It would also determine the point in development where the levels of this mRNA decrease in specific cell types. The developing rat cerebellum, with its discrete layers and well-delineated cell types, is an ideal area of the central nervous system for determining the relationship between β-tubulin mRNA levels and cell types and/or cell stage. The technique of *in situ* hybridization was used to quantitate the steady state levels of the β-tubulin mRNAs in specific cell types during rat cerebellar development.

Quantitation of the exact copy number of a specific mRNA in cells by *in situ* hybridization cannot be performed unless a known amount of the mRNA can be added to the cells and a standard curve generated from grain counts after *in situ* hybridization. Because this is not feasible in brain, we have chosen to calculate specific mRNA levels relative to the levels of total poly(A)$^+$ mRNA in each cell population of interest. Poly(A)$^+$ mRNA levels are measured by hybridization to [^{3}H]polyuridylic acid (poly(U)). The relative amount of specific mRNA present in each cell type can then be determined by quantitating grain counts in adjacent sections hybridized either to a ^{3}H-labeled recombinant DNA probe or to [^{3}H]poly(U).[53] This circumvents possible variations in total mRNA content in different cell types as well as possible variations in probe access to or RNA leakage from specific cell types.[53] Conditions of hybridization were such that the poly(U) was present in 100-fold excess. Each 5 μm section of 14 day rat cerebellum contains 5 ng of total poly(A)$^+$ mRNA, and, assuming an average steady state poly(A) tail size of 80 nucleotides,[16] each section contains 0.2 ng of poly(A). Twenty ng of [^{3}H]poly(U) was applied to each section. Similarly, the β-tubulin probe was also present in 100-fold excess, assuming β-tubulin mRNAs constitute 10% of the total mRNAs in all cerebellar cell types.

The probes were hybridized *in situ* to adjacent 5 μm sections of 14 day rat cerebellum that had been fixed in Bouin's fluid, embedded in paraffin, deparaffinized, and permeabilized with 0.2N HCl.[53,54] After stringent washing, autoradiography, and staining, autoradiographic grains were counted over individual neuronal cells of the cerebellum.

We found that the majority of grains for both the poly(U) and the β-tubulin probes were localized over cells (FIGURES 8 and 9). Few, if any, grains were seen over glomeruli in the IGL. This was expected because glomeruli contain axon terminals[55] and hence should contain no mRNA.

Two controls were performed to show that hybridization was specific. First, poly(U) did not hybridize to sections pretreated with RNase A or RNase T2 (results not shown), demonstrating that the poly(U) was hybridizing to poly(A). Second, few grains were seen over sections hybridized to the [^{3}H]pBR322 vector, demonstrating that the tubulin DNA insert was hybridizing specifically to complementary sequences in the cerebellar sections. The grain counts over each cell type were highly reproducible

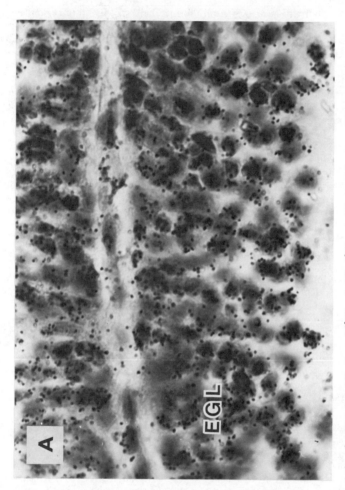

FIGURE 8. Autoradiographs of *in situ* hybridization of [³H]-poly(U). [³H]-poly(U) (3 Ci/mole; 39,500 cpm/section) was hybridized to cerebellar sections from 14-day-old rats. **A:** Grain density over cells in the EGL. **B:** Grain density in Purkinje cells in the EGL. **C:** Grain density in granule cells of the IGL. Note the paucity of grains over the mossy fiber axon terminals in the glomerulus (G). **D:** Grain density in Purkinje cells and granule cells of the IGL. The low density of grains over cells in **D** are due to pretreatment with RNAse A before hybridizations. Autoradiographic exposure was for five days.[53]

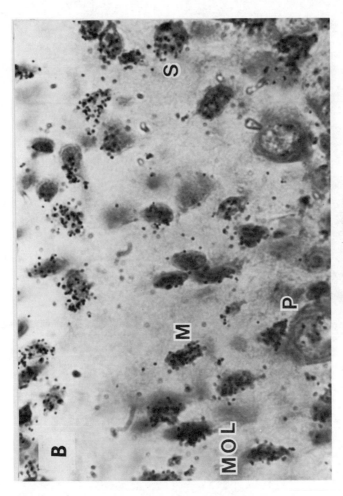

FIGURE 8B. Legend on p. 62.

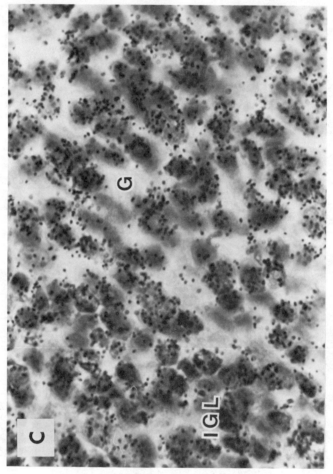

FIGURE 8C. Legend on p. 62.

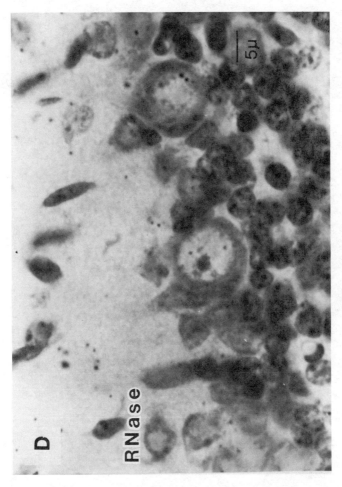

FIGURE 8D. Legend on p. 62.

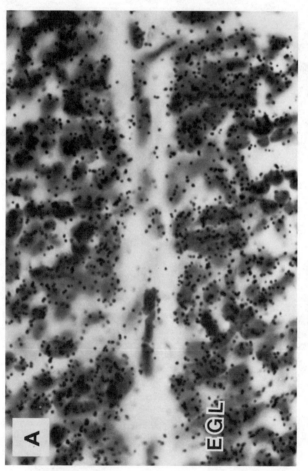

FIGURE 9. Autoradiographs of *in situ* hybridization of [³H]beta tubulin probe and [³H]native pBR322. [³H]beta tubulin probe (60,000 cpm/section), (A–C), and native pBR322 (60,000 cpm/section), (D) were hybridized to cerebellar sections from 14-day-old rats. **A:** Grain density over cells in the EGL. **B:** Grain density in Purkinje cells (P) as well as in the stellate (S) and basket (B) cells in the molecular layer. **C:** Grain density in granule cells of the IGL. The low density of grains over cells in **D** shows the lack of hybridization of radiolabeled native pBR322 to cerebellar cells.[53]

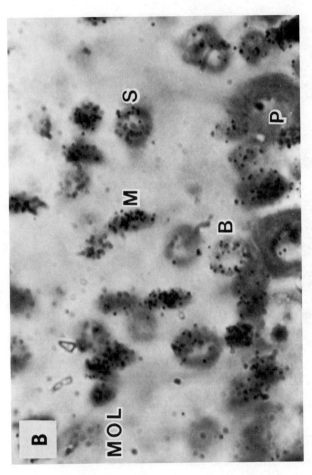

FIGURE 9B. Legend on p. 66.

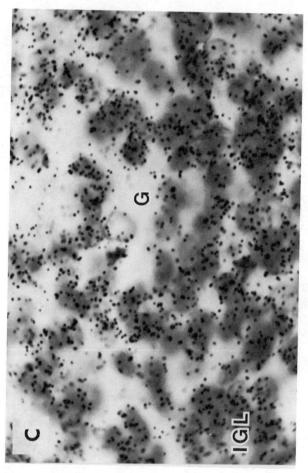

FIGURE 9C. Legend on p. 66.

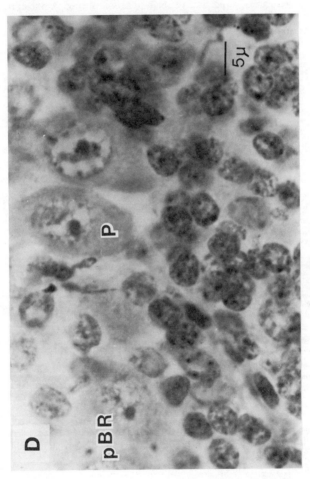

FIGURE 9D. Legend on p. 66.

with a standard deviation of less than 10%; each cell type had a distinct complement of both poly(A)$^+$ mRNAs and β-tubulin mRNAs (reference 53 and TABLE 1).

The ratios of grains hybridizing to poly(A)$^+$ mRNAs and tubulin mRNAs showed that the relative amount of β-tubulin mRNA in cerebellum of 14-day-old rats was approximately twofold greater in granule cells of the EGL than it was in the IGL. This correlates with the developmental decrease in this β-tubulin mRNA that we see in both rat[39,49,50] and human cerebellum. The functional significance of this decrease may be related to the different metabolic requirements of granule cells during mitosis and axon formation in the EGL and during their final maturation in the IGL. In other systems, axon regrowth is accompanied by an increase in tubulin mRNA levels,[56] indicating that normal axon growth, particularly in parallel fibers that contain

TABLE 1. *In Situ* Hybridization of β-Tubulin cDNA and [^{3}H]Poly(U) to 14-Day-Cerebellar Neurons[a]

	Grains per Cell	Relative Levels of β-Tubulin mRNA
Purkinje Cells		
β tubulin	14.7 ± 0.3	0.48
Poly(U)	30.1 ± 3.4	
Stellate Cells		
β tubulin	24.7 ± 1.8	0.35
Poly(U)	70.6 ± 2.6	
EGL Granule Cells		
β tubulin	53.0 ± 2.0	2.30
Poly(U)	23.0 ± 2.4	
IGL Granule Cells		
β tubulin	33.0 ± 2.4	1.36
Poly(U)	24.0 ± 3.0	

[a]The distribution of relative levels of tubulin mRNAs and total poly(A) mRNAs over Purkinje and stellate cells and the granule cells in the EGL and IGL.

Grains were counted over cells such as those shown in FIGURES 8 and 9. The 5 μm cerebellar sections, examined at 1000 diameters magnification, yielded "grains/cell" (mean ± SD of at least 100 cells of each cell type from three different experiments). Grains were counted by focusing at every level of the emulsion-coated slide. The relative level of tubulin mRNA in each cell type is the ratio of grains/cell type in sections hybridized to the [^{3}H]beta tubulin probe to the grains/cell type in sections hybridized to [^{3}H]poly(U).[53]

microtubules and no neurofilaments,[55] may similarly require increased synthesis of tubulin. The β-tubulin isotype may be required specifically for one or more developmental tasks, including elaboration of axons and/or dendrites.

Comparison of the relative levels of this β-tubulin mRNA in the postmigratory, differentiated IGL granule cells with that in the other neuronal types (TABLE 1), suggests that the elaboration of dendrites in the 14 days stellate and Purkinje cells requires less tubulin synthesis than does maintenance of the long parallel fibers by the small cell bodies of the granule cells.

Our results indicate that the β-tubulin mRNA is present in all cerebellar neurons. To determine whether this mRNA is also present in glia, we have hybridized the [^{3}H]β-tubulin cDNA probe to white matter. The results show that the β-tubulin mRNA levels are 2.5-fold higher in glial cells (previously identified as such by staining

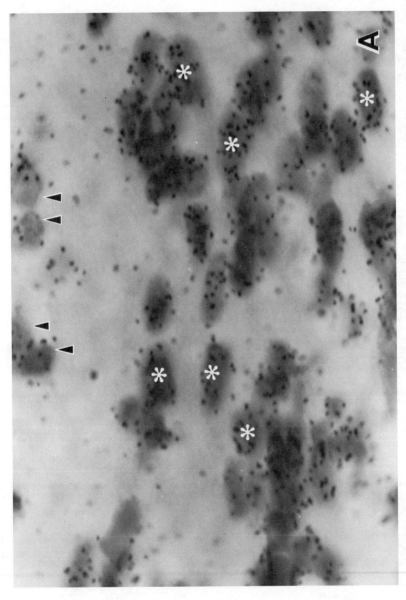

FIGURE 10. Autoradiographs of *in situ* hybridization of [³H]beta tubulin probe to the arbor vitae of the cerebellum and to the deep cerebellar nuclei. [³H]beta tubulin probe (60,000 cpm/section) was hybridized to cerebellar sections from four-day-old rats. **A:** Grain density over cells (white asterisks) in the arbor vitae (white matter) and over cells in the adjacent IGL (black arrow tips). **B:** Grain density of cells of similar size (black arrow tips) in the deep cerebellar nuclei.

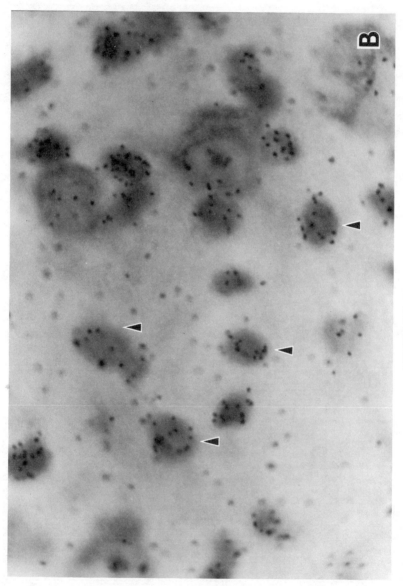

FIGURE 10B. Legend on p. 71.

with antibody to glial fibrillary acidic protein) (FIGURE 10A) relative to adjacent small and large neurons in the deep cerebellar nuclei (FIGURE 10B). Thus, this β-tubulin mRNA is an example of an mRNA that is brain-specific but is found in both neurons and glia.

ACKNOWLEDGMENTS

The technical assistance of Sibile Pardue, Riki Ison, and Mike Alejos, the editorial assistance of Katherine Miller, and the secretarial assistance of Cheryl Beisert are gratefully acknowledged.

REFERENCES

1. DUSTIN, P. 1978. Microtubules. Springer-Verlag. Berlin.
2. SHELANSKI, M. L., H. FEIT, R. W. BERRY & M. P. DANIELS. 1972. Filaments and tubules in the nervous system. *In* The Structure and Function of Nervous Tissue. G. H. Bourne, Ed.: VI: 47–80. Academic Press. New York.
3. MAROTTA, C. A., J. L. HARRIS & J. M. GILBERT. 1978. J. Neurochem. **30:** 1431–1440.
4. GEORGE, H. J., L. MISRA, D. J. FIELD & J. C. LEE. 1981. Biochem. **20:** 2402–2409.
5. GOZES, I. & U. Z. LITTAUER. 1978. Nature (London) **276:** 411–413.
6. GOZES, I. & K. J. SWEADNER. 1981. Nature (London) **294:** 477–480.
7. EDDE, B., C. JEANTET & F. GROS. 1981. Biochem. Biophysics Res. Commun. **103:** 1035–1043.
8. GINZBURG, I., L. BEHAR, D. GIVOL & U. Z. LITTAUER. 1981. Nucleic Acids Res. **9:** 2691–2697.
9. LEMISCHKA, I. R., S. FARMER, V. R. RACANIELLO & P. A. SHARP. 1981. J. Mol. Biol. **151:** 101–120.
10. LOPATA, M. A., J. C. HAVERCROFT, L. T. CHOW & D. W. CLEVELAND. 1983. Cell **32:** 713–724.
11. COWAN, N. J., P. R. DOBNER, E. V. FUCHS & D. W. CLEVELAND. 1983. Mol. Cell. Biol. **3:** 1738–1745.
12. BOND, J. F. & S. R. FARMER. 1983. Mol. Cell. Biol. **3:** 1333–1342.
13. BOND, J. F., G. S. ROBINSON & S. R. FARMER. 1984. Mol. Cell. Biol. **4:** 1313–1319.
14. AMANO, T., E. RICHELSON & M. NIRENBERG. 1972. Proc. Natl. Acad. Sci. USA **69:** 258–263.
15. MORRISON, M. R., F. BASKIN & R. N. ROSENBERG. 1977. Biochim. Biophys. Acta **476:** 228–237.
16. MORRISON, M. R., R. BRODEUR, S. PARDUE, F. BASKIN, C. L. HALL & R. N. ROSENBERG. 1979. J. Biol. Chem. **254:** 7675–7683.
17. CROALL, D. E. & M. R. MORRISON. 1980. J. Mol. Biol. **140:** 549–564.
18. MORRISON, M. R., C. L. HALL, S. PARDUE, R. BRODEUR, F. BASKIN & R. N. ROSENBERG. 1980. J. Neurochem. **34:** 50–58.
19. MORRISON, M. R., S. PARDUE, N. PRASHAD, D. E. CROALL & R. BRODEUR. 1980. Eur. J. Biochem. **106:** 463–472.
20. GARRELS, J. I. & T. HUNTER. 1979. Biochim. Biophys. Acta **564:** 517–525.
21. LODISH, H. F. 1974. Nature (London) **251:** 385–388.
22. PRASHAD, N. & R. N. ROSENBERG. 1978. Biochim. Biophys. Acta **539:** 459–469.
23. GINZBURG, I., S. RYBAK, Y. KIMHI & U. Z. LITTAUER. 1983. Proc. Natl. Acad. Sci. USA **80:** 4243–4247.
24. FELLOUS, A., J. FRANCON, A. VIRION & J. NUNEZ. 1975. FEBS. Lett. **57:** 5–8.
25. SCHMITT, H., I. GOZES & U. Z. LITTAUER. 1977. Brain Res. **121:** 327–342.
26. SABORIO, J. L., E. PALMER & I. MEZA. 1978. Exp. Cell Res. **114:** 365–373.
27. DENOULET, P., B. EDDE, C. JEANTET & F. GROS. 1982. Biochimie **64:** 165–172.

28. DAHL, J. L. & V. J. WEIBEL. 1979. Biochem. Biophysics Res. Commun. **86:** 822–828.
29. GOZES, I., A. DE BAETSELIER & U. Z. LITTAUER. 1980. Eur. J. Biochem. **103:** 13–20.
30. WOODWARD, D. J., B. J. HOFFER & L. W. LAPHAM. 1969. *In* Neurobiology of Cerebellar Evolution and Development. R. Llinas, Ed.: 725–744. American Medical Association Press. Chicago.
31. ALTMAN, J. 1972. J. Comp. Neurol. **145:** 353–398.
32. GILAD, G. M. & I. J. KOPIN. 1979. J. Neurochem. **33:** 1195–1204.
33. ADDISON, W. H. F. 1911. J. Comp. Neurol. **21:** 459–485.
34. GRIFFIN, W. S. T., D. J. WOODWARD & R. CHANDA. 1977. J. Neurochem. **28:** 1269–1279.
35. GRIFFIN, W. S. T., J. R. HEAD & M. F. PACHECO. 1979. Brain Res. Bull. **4:** 313–317.
36. ZAGON, I. S. & P. J. MCLAUGHLIN. 1979. Brain Res. **170:** 443–457.
37. CLARK, B. R., M. E. WEICHSEL, JR. & R. E. POLAND. 1978. Biol. Neonate **34:** 209–216.
38. CAMPAGNONI, C. W., G. D. CAREY & A. T. CAMPAGNONI. 1978. Arch. Biochem. Biophys. **190:** 118–125.
39. MORRISON, M. R., S. PARDUE & W. S. T. GRIFFIN. 1981. J. Biol. Chem. **256:** 3350–3356.
40. MORRISON, M. R. & W. S. T. GRIFFIN. 1981. Anal. Biochem. **113:** 318–324.
41. MORRISON, M. R., S. PARDUE & W. S. T. GRIFFIN. 1983. J. Neurogenet. **1:** 105–111.
42. BERRY, M., A. W. ROGERS & J. T. EAYRS. 1964. Nature (London) **203:** 591–593.
43. HICKS, S. P. & C. J. D'AMATO. 1968. Anat. Rec. **160:** 619–634.
44. EAYRS, J. T. & B. GOODHEAD. 1959. J. Anat. **93:** 385–402.
45. BROWN, I. R. 1978. Biochem. Biophysics Res. Commun. **84:** 285–292.
46. JAEGER, A. W. & C. C. KUENZLE. 1982. EMBO J. **1:** 811–816.
47. HEIZMANN, C. W., E. M. ARNOLD & C. C. KUENZLE. 1980. J. Biol. Chem. **255:** 11504–11511.
48. HEIZMANN, C. W., E. M. ARNOLD & C. C. KUENZLE. 1982. Eur. J. Biochem. **127:** 57–61.
49. MORRISON, M. R. & W. S. T. GRIFFIN. 1984. Developmental expression of specific messenger RNAs in rat and human brain. *In* Symposium on Neuronal Communications. B. J. Meyer & S. Kramer, Eds.: 107–113. Balkema Press. Cape Town.
50. MORRISON, M. R. & W. S. T. GRIFFIN. 1985. Molecular biology of the mammalian brain. *In* Gene Expression in Brain. C. Zomzely-Neurath & W. A. Walker, Eds.: 57–98. Wiley Press.
51. CLEVELAND, D. W., M. A. LOPATA, R. J. MACDONALD, N. J. COWAN, W. J. RUTTER & M. W. KIRSCHNER. 1980. Cell **20:** 95–105.
52. RICCIARDI, R. P., J. S. MILLER & B. E. ROBERTS. 1979. Proc. Natl. Acad. Sci. USA **76:** 4927–4931.
53. GRIFFIN, W. S. T., M. ALEJOS & M. R. MORRISON. 1985. J. Cell Biochem. **27:** 205–211.
54. GRIFFIN, W. S. T., M. ALEJOS & M. R. MORRISON. 1983. Brain Res. Bull. **10:** 597–601.
55. PALAY, S. & V. CHAN-PALAY. 1974. *In* Cerebellar Cortex. 153–171. Springer-Verlag. New York.
56. GINZBURG, I., T. SCHERSON, S. RYBAK, Y. KIMHI, D. NEUMAN, M. SCHWARTZ & U. Z. LITTAUER. 1983. Cold Spring Harbor Symp. Quant. Biol. **XLVIII:** 783–790.
57. THOMAS, P. S. 1980. Proc. Natl. Acad. Sci. USA **77:** 5201–5205.
58. RIGBY, P. W. J., M. DIECKMANN, C. RHODES & P. BERG. 1977. J. Mol. Biol. **113:** 237–251.

Reconstruction of Tubulin Gene Regulation in Cultured Mammalian Cells[a]

JOSEPH T. Y. LAU, MARK F. PITTENGER, JANE C. HAVERCROFT, AND DON W. CLEVELAND

Department of Biological Chemistry
The Johns Hopkins University School of Medicine
Baltimore, Maryland 21205

INTRODUCTION

Microtubules, which are comprised principally of dimeric subunits of one α- and one β-tubulin polypeptide, participate in a diverse spectrum of cellular functions including the establishment of programmed modifications of cell shape during morphogenesis, formation of mitotic and meiotic spindles, and establishment of cilia- and flagella-dependent cell motility. Given these important functions, it is not unexpected that the synthesis of tubulin should be a closely regulated process. Thus, as Ben Ze'ev et al.[1] initially reported, it was not too surprising to find that the marked alterations in the morphology of cultured animal cells following colchicine-induced microtubule depolymerization were accompanied by depression of new tubulin synthesis. What has emerged less expectedly from this work and from our own subsequent efforts[2-6] is the realization that the synthesis of tubulin is apparently established in these cells by an autoregulatory pathway closely linked to the pool size of unpolymerized subunits. Thus, as we document below, elevation of the level of free subunits, either by treatment of cells with microtubule disrupting agents, such as colchicine, or by direct microinjection of purified tubulin subunits, results in dramatically lowered levels of new tubulin polypeptide synthesis.

Although the molecular mechanism through which this autoregulation of tubulin is achieved has not yet been unambiguously identified, we document that the down-regulation of tubulin synthesis in response to an increased pool size of free subunits is accompanied by a rapid loss of tubulin mRNAs. The apparent rates of tubulin gene transcription, however, in nuclei isolated from cells with normal or elevated pools of tubulin subunits are indistinguishable.

To further dissect the precise molecular events that underlie this autoregulation, we have also sought to define the important control sequences that are carried on a tubulin gene and/or its corresponding mRNA. To this end, we have used DNA transfection to transiently introduce a heterologous chicken β-tubulin gene into cultured mouse L cells. We show that the mouse cells correctly express and process the RNA transcript copied from this gene. Moreover, upon elevation of the intracellular pool of unpolymerized tubulin subunits following drug-induced microtubule depolymerization, we demonstrate that the level of the transfected chicken tubulin gene RNA transcripts is down-regulated coordinately with the endogenous mouse α- and β-tubulin RNAs.

[a]J.T.Y. Lau and M.F. Pittenger were recipients of NIH postdoctoral and predoctoral support, respectively, during the early stages of this work. This work has been supported by Grants from the NIH, American Heart Association, and the March of Dimes to D.W. Cleveland, who is also the recipient of an NIH Research Career Development Award.

METHODS

Two-Dimensional Gel Electrophoresis

Analysis of ^{35}S-methionine-labeled protein samples was achieved by two-dimensional polyacrylamide gel electrophoresis according to the method of O'Farrell.[7] The resultant pattern of polypeptides was visualized by fluorography[8] using Kodak XAR X-Omat film.

Microinjection of Cultured Cells

Each cell seeded onto a small glass cover chip (about 20–50 cells per chip) was serially microinjected using a fine-tipped glass needle using a modification[5] of the technique of Graessmann.[9] Following injection, the cells were returned to a CO_2 incubator. At an appropriate time, the cover chip was removed and placed in a small drop of serum-free, methionine-free medium supplemented with 50 μCi of [^{35}S]methionine. After a one-hour labeling period, total proteins were solubilized in sodium dodecyl sulfate (SDS) gel sample buffer[10] and analyzed by two-dimensional gel electrophoresis and fluorography.

RNA Electrophoresis and Blotting

RNA samples were separated according to apparent molecular weight using electrophoresis on 1% agarose gels containing 2.2 M formaldehyde.[11] RNA was transferred to and immobilized on nitrocellulose as described by Thomas.[12] Specific RNAs were detected by autoradiography following hybridization of these immobilized RNAs with ^{32}P-labeled DNA probes copied from cloned sequences corresponding to α tubulin, β tubulin, or actin. Labeled probes were prepared with the random priming method of Shank et al.[13] For details of the hybridization reactions, see reference 4.

Nuclear Transcription and RNA Isolation

RNA transcription in nuclei and subsequent RNA isolation was performed in 100 μl reactions by the method of McKnight and Palmiter[14] as modified by Groudine and Weintraub.[15] Typically, using [α-^{32}P]uridine triphosphate (UTP) of specific activity 400 Ci/mmol, $3–5 \times 10^7$ cpm were incorporated per reaction into trichloroacetic acid-precipitable material.

DNA Transfection and S1 RNA Analysis

Reintroduction of cloned DNAs into cultured mouse L cells was performed using the diethylaminoethyl (DEAE)-dextran mediated DNA transfection protocol as detailed by Lopata et al.[16] RNA was recovered from transfected cells by the guanidine thiocyanate/CsCl centrifugation method of Chirgwin et al.[17] RNA was analyzed for the presence of transcripts derived from the transfected gene using an S1 nuclease protection protocol under R-looping conditions.[18]

RESULTS

Tubulin Synthesis in Cultured CHO Cells Is Regulated by the Apparent Pool Size of Depolymerized Subunits

To demonstrate that the level of new tubulin polypeptide synthesis is specifically depressed by antimicrotubule drugs that induce microtubule depolymerization and a consequent increase in the pool size of depolymerized subunits, we treated cultured chinese hamster ovary (CHO) cells with a level of colchicine (10μM) sufficient to induce rapid microtubule depolymerization. After six hours of incubation, newly synthesized proteins were labeled with [^{35}S]methionine, whole cell protein was solubilized with Laemmli gel sample buffer,[10] and the polypeptides were separated by two-dimensional gel electrophoresis. FIGURE 1A displays the pattern of proteins derived from untreated, control cells, whereas the corresponding pattern from colchicine-treated cells is shown in FIGURE 1B. Among the large number of protein species that are visible in either part, the four marked with arrows represent the cytoskeletal proteins α tubulin, β tubulin, actin, and vimentin. Careful inspection of the figure reveals that although the overall pattern of protein synthesis is not substantially affected by colchicine treatment, a marked and specific repression of new α- and β-tubulin synthesis is apparent. Quantitation of the level of depression using a two-dimensional gel scanner (Loats Associates, Inc.) further reveals that this repression is approximately 10 fold for both α tubulin and β tubulin.

Qualitatively similar data are obtained in parallel experiments using nocodazole,[4] a drug that like colchicine induces microtubule depolymerization and an obligatory increase in the pool of depolymerized subunits. However, taxol and vinblastine, additional antimicrotubule drugs whose mechanisms of action lower the pool size of subunits,[19–22] have been found to induce modest increases in the level of new synthesis (reference 4; Pachter and Cleveland, unpublished).

The sum of these data is thus consistent with the hypothesis that cultured cells contain an autoregulatory control mechanism that monitors the pool of depolymerized tubulin subunits and that adjusts the rate of new tubulin synthesis in response to suboptimal levels of unassembled subunits.

Elevation of Tubulin Levels by Microinjection Suppresses New Tubulin Synthesis

It should be noted, however, that interpretation of these initial experiments in the context of such an autoregulatory control mechanism rests entirely on the presumptive effects of the various antimicrotubule drugs. This caveat is nontrivial, as the drugs obligatorily induce gross morphological alterations, and the detailed mechanisms of action and specificities are not known with certainty.

To investigate the effect of elevated levels of tubulin subunits on the rate of new tubulin synthesis in the absence of drug treatments and their concomitant morphological changes, we have microinjected purified tubulin subunits into mammalian cells in culture.[5] Each of approximately 30 cells attached to a glass cover chip were serially microinjected either with purified tubulin subunits at an initial concentration of 6.6 mg/ml or with buffer alone. The cells were then returned to an incubator for three hours, and newly synthesized proteins were labeled by incubation of the chips for one additional hour in media containing [^{35}S]methionine. Total cellular protein was then analyzed by two-dimensional gel electrophoresis. The results of such an experiment are shown in FIGURE 2. Panel A of the figure displays the newly synthesized proteins

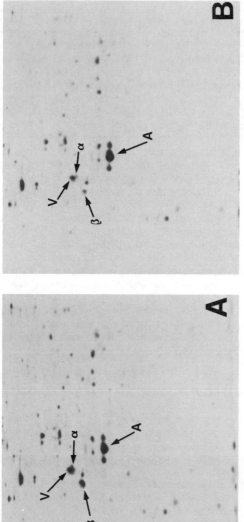

FIGURE 1. Specific depression in synthesis of α- and β-tubulin polypeptides following colchicine-induced microtubule depolymerization. Duplicate dishes of CHO cells were seeded in parallel. Six hours prior to harvesting, colchicine was added to one dish to a final concentration of 10μM. Thirty minutes prior to harvesting, newly synthesized proteins in each dish were labeled by addition of [^{35}S]methionine to the media. Total cellular protein was solubilized in Laemmli sample buffer and subjected to two-dimensional gel electrophoresis. A fluorograph of the resultant patterns is shown. **A:** Newly synthesized proteins from control cells; **B:** newly synthesized proteins from cells incubated for six hours in colchicine. Spots whose identities are known are labeled. $\alpha = \alpha$ tubulin; $\beta = \beta$ tubulin; A = actin; and V = vimentin. The acidic end of the isoelectric dimension is on the left.

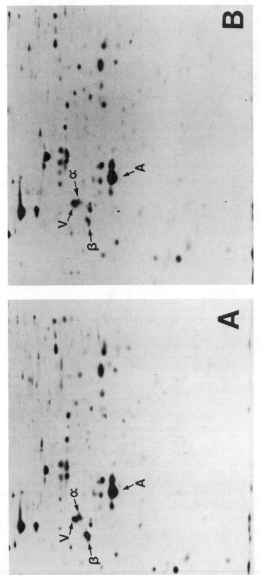

FIGURE 2. Elevation of tubulin levels by microinjection supresses new tubulin synthesis. Approximately 30 cells attached to each of two glass cover chips was serially microinjected either with buffer (50 mM PIPES, pH 6.7, 0.1 mM EDTA, 0.5 mM MgCl$_2$, 1 mM β-mercaptoethanol) or with purified tubulin at 6.6 mg/ml in the same buffer. Following injection, both cover chips were returned to a CO$_2$ incubator. Three hours later, the chips were removed from their original dishes, placed in methionine-free media supplemented with 50 μCi of [^{35}S]methionine, and returned to the CO$_2$ incubator. After a one hour labeling period, the cover chips were placed in Laemmli gel sample buffer,[10] and newly synthesized proteins were analyzed by two-dimensional gel electrophoresis and fluorography. **A:** Newly synthesized proteins in buffer-injected (mock-injected) cells; **B:** newly synthesized proteins in cells injected with tubulin. $\alpha = \alpha$ tubulin; $\beta = \beta$ tubulin; A = actin; and V = vimentin.

produced in mock (buffer) injected cells. As expected, this pattern is qualitatively identical to that obtained from uninjected cells (*e.g.,* see FIGURE 1A). Hamster cells injected with exogenous hog brain tubulin subunits (FIGURE 2B) display a pattern of protein synthesis that is also remarkably similar to that of mock-injected, control cells with the noteworthy exception that synthesis of α- and β-tubulin polypeptides is markedly repressed. More specifically, quantitation of the depression in new β-tubulin synthesis by excision of the appropriate spots from the dried gels, followed by scintillation counting, reveals a level of β-tubulin synthesis in tubulin-injected cells of only 1/10th the rate of control cells.

Tubulin constitutes roughly 2–3% of total cell protein in most cultured mammalian cells,[23] and its intracellular concentration has been estimated to be about 2 mg/ml[23] of which approximately 50% is in the unpolymerized form at any specific time.[23–25] The volume routinely injectable into cultured cells without loss of viability has been empirically determined by previous workers and by ourselves to be about 1/10th of the initial cell volume.[9,26] Thus, in the present example, injection of exogenous, depolymerized tubulin at an initial concentration of 6.6 mg/ml will increase the cellular tubulin content by roughly 50% and initially increase the tubulin subunit pool by 100 percent. As shown in FIGURE 2B, such microinjection of tubulin results in a rapid and specific depression of new tubulin synthesis, thus clearly demonstrating that cultured cells can and do monitor tubulin content and adjust the rate of new synthesis accordingly. Moreover, companion experiments in which the injected subunits were prebound to colchicine strongly support the hypothesis that it is the subunit form of tubulin that is monitored.[5]

Autoregulation of Tubulin Synthesis Is Not Achieved through a Reversible, mRNA Sequestration Mechanism

To begin to determine the molecular mechanism through which tubulin synthesis is specified, we initially tested whether the specific loss of synthesis of new tubulin subunits resulted from the loss of tubulin mRNA sequences from the cell cytoplasm or whether the tubulin mRNAs were somehow translationally inactivated or sequestered. Cytoplasmic RNA was prepared from parallel dishes of CHO cells that had, or had not, been incubated for six hours in colchicine. Equivalent amounts of recovered RNA were then analyzed by RNA blot analysis for the presence of α tubulin, β tubulin, or actin mRNAs using ^{32}P-labeled probes constructed from appropriate cloned DNAs. Autoradiograms of the resultant hybridization patterns obtained on triplicate blots are shown in FIGURE 3A,B,C for α tubulin, β tubulin, and actin, respectively. Clearly, both α- and β-tubulin mRNAs decline in amount as a consequence of colchicine treatment (compare the − and + lanes of each part). Moreover, that this loss of tubulin RNAs is not due to unintentional RNA preparation/loading artifacts is demonstrated in FIGURE 3C that documents that, as expected, actin RNA levels are not similarly depressed. (On the contrary, actin RNAs show a mild and reproducible elevation in amount.) We conclude that tubulin RNAs are specifically lost from the cell following colchicine treatment. Hence, translational sequestration of intact mRNAs cannot be the mechanism responsible for controlling tubulin expression.

Regulation of Tubulin Synthesis Is Probably Not Achieved through a Transcriptionally Derived Control Mechanism

For most cellular genes thus far studied in detail in higher eukaryotes, control of expression has been demonstrated to be achieved primarily at the level of RNA transcription (*e.g.*, globin,[14,27] ovalbumin,[13,28] and a variety of randomly selected RNAs from liver[29]). To test whether tubulin expression as a function of the apparent pool size of tubulin subunits is also regulated transcriptionally, we isolated nuclei from control cells and from cells depleted in tubulin mRNAs as the result of treatment with colchicine. [^{32}P]UTP was added to aliquots containing equivalent numbers of either sample of nuclei. Under conditions that do not permit new transcriptional initiation, transcription of the *in vivo*-initiated nascent transcripts was allowed to proceed for 5–30 minutes. At this point the heterogeneous, radiolabeled RNAs resulting from these *in vitro* transcription/elongation reactions were then isolated. To determine the fraction of newly labeled RNA that is in α- or β-tubulin sequences relative to that

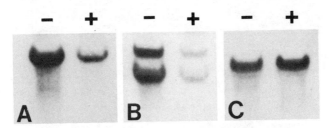

FIGURE 3. Measurement of cytoplasmic α tubulin, β tubulin and actin RNAs in control and colchicine-treated cells. Cytoplasmic RNA was prepared from control and colchicine-treated cells, and equal amounts of RNA (10µg) were electrophoresed on a denaturing gel. After transfer to nitrocellulose, α tubulin, β tubulin, or actin RNAs were detected by hybridization to appropriate ^{32}P-labeled probes prepared from clones pT1, pT2, and pA1 of reference 32. Autoradiograms of those hybridizations are shown. **A:** Hybridization to the α-tubulin probe; **B:** hybridization to the β-tubulin probe; **C:** hybridization to the actin probe. Slots labeled ($-$) are from control cells; slots labeled ($+$) are from cells incubated for six hours in colchicine.

which represents actin transcripts, we prepared nitrocellulose replicas containing cloned actin, α tubulin, or β tubulin cDNAs. Duplicate filters were then hybridized with RNA transcribed from colchicine-treated (FIGURE 4B and D) or control nuclei (FIGURE 4A and C). Hybridization conditions were determined to be in DNA excess and to be carried to saturation.[2] Under these conditions, the relative or absolute numbers of radiolabeled transcripts derived from α- or β-tubulin genes and or from actin genes can be determined by quantitation of the resultant autoradiographic signals.

The results of two independent experiments are shown in Figure 4. Panel **A** of the figure displays an autoradiogram of the transcription products of control nuclei. Slots 1, 2, and 3 represent the apparent, relative transcription rates of actin, α-tubulin, and β-tubulin RNAs, respectively. Panel **B** displays the corresponding experiment for transcripts isolated from nuclei derived from drug-treated cells. No significant differences in the relative or absolute rates of α- or β-tubulin gene transcription are apparent in nuclei from control and colchicine-treated cells.

Moreover, when parallel *in vitro* transcription reactions were performed in the presence of heparin [which remove histones and most other chromosomal proteins from the DNA, but leaves initiated RNA polymerases still bound to DNA and capable of elongation, but not new initiation (see discussion of reference 14)], a qualitatively and quantitatively similar rate of tubulin transcription was again seen in nuclei derived from control- and colchicine-treated cells (compare parts **C** and **D** of FIGURE 4). The observed transcription signals do appear to represent authentic transcription products of RNA polymerase II, because transcription reactions performed in the presence of 2 μg/ml of α amanitin yield no detectable signals for the tubulins or actin (FIGURE 4E). This level of α amanitin is known to be sufficient to fully inhibit RNA polymerase II.[30]

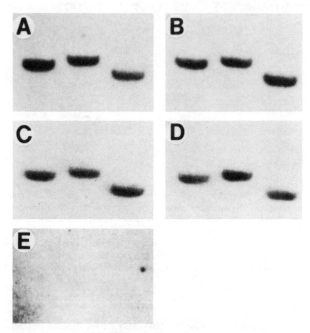

FIGURE 4. Relative transcription in CHO nuclei isolated from control cells or from colchicine-treated cells. ^{32}P-labeled runoff RNA from nuclei of control or colchicine-treated cells was hybridized for 4 days to filters containing cloned human cDNAs specific for (lane 1) actin, (lane 2) α tubulin, and (lane 3) β tubulin. Parts **A** and **B** represent hybridization of 5×10^6 cpm of RNA derived from 30 min of *in vitro* transcription of control and colchicine-treated nuclei, respectively. Parts **C** and **D** represent duplicates of **A** and **B** except that the *in vitro* runoff reactions were carried out in the presence of one mg/ml heparin. Part **E** displays the hybridization pattern of 10^7 cpm of *in vitro* transcribed RNA isolated from control CHO nuclei to which α amanitin had been added to a final concentration of 2μg/ml. The human cDNA clones used were as previously described.[2]

If we employ the reasonable assumption that the cytoplasmic appearance of new tubulin RNAs can be described by a zero order rate constant encompassing transcription/RNA processing/transport and that the cytoplasmic rate of RNA degradation can be described by a first-order decay process, then in order for a transcriptional control mechanism to achieve a specified lowering of the cytoplasmic RNA level, there must be a proportional reduction in the transcription/processing/

transport rate. As we have seen, although cytoplasmic tubulin RNA levels fall up to 10-fold (FIGURE 3), no difference in the transcription rates of tubulin genes in control and colchicine-treated cells can be detected (FIGURE 4). With the explicit assumption that the *in vitro* transcription reaction in isolated nuclei faithfully mirrors the *in vivo* situation [as has been found for a variety of previously studied genes (*e.g.*, references 13,14,30)], the sum of the present data argue clearly that the principal mechanism modulating tubulin synthetic rates in cultured cells does not derive from a transcriptional level.

Expression of the Chicken β_2 Gene in Mouse L Cells

To further dissect the precise molecular events that underlie the autoregulation of tubulin synthesis, we have now sought to define the important control sequences that are carried on a tubulin gene and/or its corresponding mRNA. To this end, we have used DEAE-dextran mediated DNA transfection to transiently introduce a heterologous chicken β-tubulin gene into cultured mouse L cells. With this technique, exogenous DNA can be introduced into cultured cells, and if appropriate transcription signals are present on the DNA, it will be expressed. The DNA is not, however, integrated into the host cell chromosomes, but rather remains in plasmid form.

The gene we have chosen for transfection is the chicken β_2 gene. The complete nucleotide sequence starting 275 bases 5' to the presumptive cap site for RNA transcription through 270 bases 3' to the site of polyadenylation has been determined and has been presented elsewhere (Sullivan, Lau, and Cleveland, submitted). Thus, the major structural features of this gene, including the four protein coding exon sequences, the presumptive TATA promoter sequence, the 5' and 3' untranslated regions, and the ACATAAA signal for polyadenylation are known precisely. As described previously,[31] the entire β_2 tubulin gene has been isolated on a 4.5 kilobase (kb) Eco RI fragment and subcloned into the unique Eco RI site of pBR322 in an orientation such that transcription would proceed clockwise in the normal presentation of pBR322.

The ability of mouse L cells to correctly express the β_2 tubulin gene and to process its nascent RNA transcript following introduction by transfection was demonstrated by using S1 nuclease protection experiments and by RNA blotting. In the former method, two S1 probes were prepared. The first of these (probe 1) was isolated from a subclone of pT2[32,33] (named pT2-H3) that contains a nearly full length cloned copy of the mature β_2 mRNA transcript. Plasmid pT2-H3 was opened at the unique Bgl II site that lies 313 nucleotides from the start of the cDNA sequence contained in the subclone. A second probe (probe 2) was constructed directly from our original β-tubulin cDNA plasmid pT2.[32,33] Plasmid pT2 contains 318 nucleotides of cDNA 5' to the Bgl II site. After 5'-end labeling with polynucleotide kinase and γ-^{32}P-ATP, both probes were hybridized to various RNAs, the hybridization mixtures digested with the single-strand specific nuclease S1 and the probe fragments protected from digestion by hybridization to RNA detected by autoradiography following electrophoresis on DNA sequencing gels. As clearly demonstrated in FIGURE 5, lane 7, an S1 experiment using probe 1 and authentic chicken brain mRNA yielded the expected 313 nucleotide probe fragment. Similarly, after hybridization to probe 2, chicken RNA protected a 318 base fragment from S1 digestion (FIGURE 5, lane 8). (The weaker 265 nucleotide signal in lanes 7 and 8 is derived from the presence of an mRNA encoded by the chicken β_1 gene, a highly homologous sister gene to β_2 that diverges from the β_2 sequence 265 bases 5' to the Bgl II site [Sullivan, Lau, and Cleveland, submitted].) Both S1 probes are, however, specific for the RNA transcripts copied from chicken tubulin genes

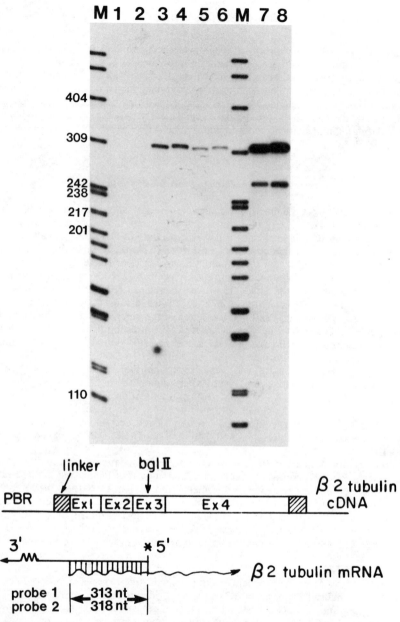

FIGURE 5. Effects of colchicine-induced microtubule depolymerization on the level of expression of the chick β_2 tubulin gene introduced by transfection. Duplicate dishes of mouse L cells were transfected in parellel with the β_2 gene. Twenty-six hours posttransfection, colchicine was added to one dish to a final concentration of 10μM. Three hours later, total RNA was prepared from both dishes, and equal amounts of RNA from each were assayed with two different S1 probes as described in the text. Lanes 1 and 2: fragments protected using probes 1 and 2 hybridized to RNA from mock transfected cells; lane 3: fragments protected using probe 1 hybridized to RNA from control cells; lane 4: fragments protected using probe 2 hybridized to RNA from control cells; lane 5: fragments protected using probe 1 hybridized to RNA from colchicine-treated cells; lane 6: fragments protected using probe 2 hybridized to RNA from colchicine-treated cells; lanes 7 and 8: fragments protected using probes 1 and 2 hybridized to authentic chick brain RNA. Size markers in nucleotides are shown in lanes labeled M.

because endogenous mouse L cell RNAs do not protect any portion of either probe from S1 digestion (FIGURE 5, lanes 1 and 2).

When RNA prepared from mouse L cells 26 hours after transfection with the β_2 gene was analyzed, the presence of authentic β_2 mRNA was demonstrated by the presence of the 313 nucleotide fragment protected in probe 1 (FIGURE 5, lane 3) and by the 318 nucleotide fragment protected in probe 2 (FIGURE 5, lane 4). Moreover,

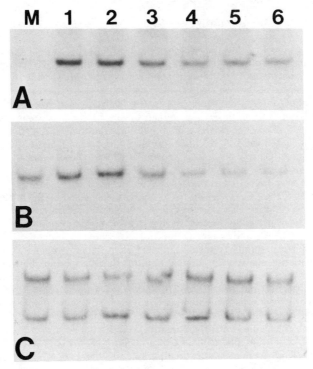

FIGURE 6. Effects of colchicine-induced microtubule depolymerization on the level of expression of a transfected chick β_2 tubulin gene and the endogenous mouse tubulin genes as assayed by RNA blotting. Identical dishes of cells that had been transfected with the β_2 gene were exposed to colchicine for 0, 0.5, 1.5, 3.0, 4.5, and 6.0 hours (lanes 1–6, respectively). Equal amounts of RNA from each dish were then analyzed by blot analysis. Part **A**: Blot analysis using a probe specific for the 3'-untranslated region of the chick β_2 gene; part **B**: analysis using the coding region from a β-tubulin cDNA clone (pT2 of reference 32) that hybridizes to both the endogenous mouse and transfected chicken β-tubulin RNAs; part **C**: analysis using a probe (from plasmid pXLr101A) that is specific for ribosomal RNAs. Lane M contains RNA from mock transfected mouse cells.

because both probes were end-labeled at a site within the third exon (amino acid position 84), the presence of only a 313 (or a 318) nucleotide-protected fragment (and the absence of shorter fragments) indicated that correct excision of the first and second introns occurred at high efficiency for the transfected gene transcript.

To further demonstrate the correct expression and processing of the heterologous chick tubulin mRNA in the L cells, RNA from transfected cells was analyzed by RNA blotting using a probe specific for the 3'-untranslated region of chick β_2 RNA. As shown in FIGURE 6A, lane M, this probe did not hybridize to endogenous L-cell transcripts. A single 1800 base β_2 specific RNA species, however, was detected in transfected cells (FIGURE 6A, lane 1), and moreover, this RNA species was indistinguishable in size from the authentic β_2 transcripts present in chicken RNA (not shown).

Regulation of Chicken β_2 Expression in Mouse L Cells

We next sought to determine whether the level of expression of the transfected chicken β_2 gene was sensitive to colchicine-induced microtubule depolymerization. To test this, parallel dishes of cells were transfected with the β_2 gene, and 26 hours posttransfection, colchicine was added to some dishes to a final concentration of 10 μM. Three hours later, total RNA was prepared from each dish, and equivalent amounts of RNA from each were examined by the S1 protocol using either of the probes described above. The results using probe 1 are shown in FIGURE 5, lanes 3 and 5 for β_2 RNAs in control and colchicine-treated transfected cells, respectively. The corresponding results using probe 2 are shown in FIGURE 5, lanes 4 and 6. Remarkably, using either probe, the level of expression of the transfected gene is seen to be appropriately down-regulated in response to colchicine.

The results of the S1 experiment have also been confirmed by RNA blot analysis. Again, multiple dishes of cells were transfected in parallel. Colchicine was then added to each dish for a period of zero to six hours prior to harvesting. Total RNA was isolated from each dish, and equal amounts of RNA were electrophoresed and blotted. FIGURE 6 displays autoradiograms of triplicate blots. The first of these (part A) was hybridized with a probe specific for the 3'-untranslated region of the transfected β_2 gene, whereas the second blot (FIGURE 6B) was hybridized with a probe that detects both the endogenous and transfected β-tubulin mRNAs. Lanes 1–6 of each part represent RNAs from cells incubated for 0, 0.5, 1.5, 3, 4.5, or 6 hours in colchicine. Clearly, the RNAs encoded by the transfected β_2 gene (FIGURE 6A) decline upon treatment with colchicine in a fashion qualitatively similar to that of the endogenous β-tubulin RNAs (FIGURE 6B). A final blot probed for rRNA sequences (FIGURE 6C) reveals that rRNA contents were comparable in the two RNA samples, thus demonstrating that the loss of tubulin RNA sequences in the colchicine-treated sample could not be due to inadvertent differences in RNA quantitation/preparation in control and colchicine-treated cells.

DISCUSSION

Most animal cells rapidly depress synthesis of new α- and β-tubulin polypeptides in response to microtubule inhibitors that cause microtubule depolymerization and presumably increase the intracellular concentration of free subunits. Additional drugs that interfere with microtubule function but that lead to a decrease in the subunit pool size result in a mild increase in the rate of new tubulin synthesis.[1,4] These findings are thus consistent with the hypothesis that the level of tubulin gene expression in cultured animal cells is established through an autoregulatory mechanism that monitors the pool size of depolymerized tubulin subunits. Although interpretation of the drug experiments must be tempered with caution because the drugs obligatorily induce gross morphological changes and the ultimate specificities and mechanisms of action cannot be known with certainty, microinjection into cells of a level of exogenous tubulin subunits comparable to that which would be liberated by endogenous microtubule depolymerization has revealed that elevation of the cellular tubulin content in the absence of such drug treatments results in a rapid and specific repression of new tubulin synthesis (FIGURE 2). Collectively, these experiments show quite clearly that cultured cells can and do monitor their tubulin subunit content and adjust the rate of new synthesis accordingly.

With regard to the molecular mechanism responsible for such apparent autoregulation, we have demonstrated (FIGURE 3) that tubulin RNAs are specifically lost from the cell cytoplasm following colchicine-induced microtubule depolymerization; hence, translational sequestration of intact mRNAs cannot be the mechanism responsible for controlling tubulin expression. *A priori,* the loss of tubulin RNAs may be the result of suppression of new tubulin RNA transcription, of failure of newly synthesized tubulin RNAs to be properly processed or transported from the nucleus, or of an increased rate of cytoplasmic tubulin RNA degradation. Although transcriptional regulation has been demonstrated for most cellular eukaryotic genes thus far investigated in detail, we have found that the apparent rates of tubulin RNA transcription are essentially identical in isolated nuclei derived from colchicine-treated or control cells. This finding strongly suggests that transcription is not the principal level at which control of tubulin synthesis is exercised. Overall, the sum of the present data point to a relatively novel autoregulatory control mechanism that establishes tubulin synthetic rates in higher cells and that is probably operative at the level of tubulin mRNA processing/transport efficiency or mRNA stability.

To continue our efforts to determine in detail the molecular mechanism responsible for tubulin autoregulation, we have begun to test whether exogenous tubulin genes introduced transiently into cells by DNA transfection are expressed correctly and if so, whether the level of expression of the transfected genes is subject to the same regulatory controls as are the endogenous tubulin gene sequences. Our present data (FIGURE 5) have demonstrated that mouse cells transfected with a chicken β-tubulin gene do indeed express and correctly process the RNA transcripts copied from the chicken gene. While this finding is not particularly surprising given the many eukaryotic genes that have been successfully expressed by this kind of methodology, it remains considerably more remarkable that following colchicine-induced microtubule depolymerization, the levels of transfected and endogenous tubulin RNAs are coordinately down-regulated. Hence, the level of expression of the transfected chicken gene is subject to the same pattern of control as that of the chromosomal mouse α- and β-tubulin genes. The sum of these transfection data mandate the somewhat surprising conclusion that the requisite recognition signal(s) for appropriate tubulin gene regulation by the apparent pool size of depolymerized tubulin subunits must reside directly in the primary tubulin gene DNA sequences.

The demonstration that proper tubulin gene regulation can be achieved on a cloned tubulin gene that has been reintroduced into cultured cells should now afford a powerful and potentially tractable tool for further investigation of the DNA/RNA sequences required to specify regulated tubulin expression. As a consequence, it seems very likely that by construction and analysis of hybrid genes containing tubulin promoter, coding, intron, or flanking regions sequences that the requisite regulatory sequences and pathway of regulation will be identified in the near future.

REFERENCES

1. BEN ZE'EV, A., S. R. FARMER & S. PENMAN. 1979. Cell 17: 319–325.
2. CLEVELAND, D. W. & J. C. HAVERCROFT. 1983. J. Cell Biol. 99: 919–924.

3. CLEVELAND, D. W. & M. W. KIRSCHNER. 1982. Cold Spring Harbor Symp. Quant. Biol. **46:** 171–183.
4. CLEVELAND, D. W., M. A. LOPATA, P. SHERLINE & M. W. KIRSCHNER. 1981. Cell **25:** 537–546.
5. CLEVELAND, D. W., M. F. PITTENGER & J. R. FERAMISCO. 1983. Nature (London) **305:** 738–740.
6. CLEVELAND, D. W., M. A. LOPATA & M. F. PITTENGER. 1983. J. Submicrosc. Cytol. **15:** 353–358.
7. O'FARRELL, P. O. 1975. J. Biol. Chem. **250:** 4007–4021.
8. BONNER, W. M. & R. A. LASKEY. 1976. Eur. J. Biochem. **46:** 83–88.
9. GRAESSMANN, A., M. GRAESSMANN & C. MUELLER. 1980. Methods Enzymol. **65:** 816–825.
10. LAEMMLI, U. K. 1970. Nature (Lodnon) **227:** 680–685.
11. BOEDTKER, H. 1971. Biochim. Biophys. Acta **240:** 448–453.
12. THOMAS, P. S. 1980. Proc. Natl. Acad. Sci. USA **77:** 5201–5205
13. SHANK, P. R., S. H. HUGHES, H. J. KUNG, J. E. MAJORS, N. QUINTRELL, R. V. GUNTAKA & H. E. VARMUS. 1978. Cell **15:** 1383–1395
14. MCKNIGHT, G. S. & R. PALMITER. 1979. J. Biol. Chem. **254:** 9050–9058.
15. GROUDINE, M., M. PERETZ & H. WEINTRAUB. 1981. Mol. Cell. Biol. **1:** 281–288.
16. LOPATA, M. A., D. W. CLEVELAND & B. SOLLNER-WEBB. 1984. Nucleic Acids Res. **12:** 5705–5717.
17. CHIRGWIN, J. M., A. E. PRZYBYLA, R. J. MACDONALD & W. J. RUTTER. 1979. Biochemistry **18:** 5294–5299.
18. CASEY, J. & N. DAVIDSON. 1977. Nucleic Acids Res. **4:** 1539–1552.
19. SCHIFF, P. B. & S. B. HORWITZ. 1980. Proc. Natl. Acad. Sci. USA **77:** 1561–1565.
20. SCHIFF, P. B., J. FANT & S. B. HORWITZ. 1979. Nature (London) **277:** 665–667.
21. BRYAN, J. 1971. Exp. Cell Res. **66:** 129–136.
22. FUJIWARA, K. & L. G. TILNEY. 1975. Ann. N.Y. Acad. Sci. **253:** 27–50.
23. HILLER, G. & K. WEBER. 1978. Cell **14:** 796–804.
24. SPIEGELMAN, B. M., S. M. PENNINGROTH & M. W. KIRSCHNER. 1977. Cell **12:** 587–600.
25. OLMSTED, J. B. 1981. J. Cell Biol. **89:** 418–423.
26. STACEY, D. W. & V. G. ALLFREY. 1976. Cell **9:** 729–737.
27. LANDES, G. M. & H. G. MARTINSON. 1982. J. Biol. Chem. **257:** 1102–1107.
28. NGUYEN-HUU, M. C., K. J. BARRETT, K. GIESECKE, T. WURTZ, A. E. SIPPEL & G. SCHUTZ. 1978. Hoppe-Seyler's Z. Physiol. Chem. **359:** 1307–1313.
29. DERMAN, E., K. KRAUTER, L. WALLING, C. WEINGERGER, M. RAY & J. E. DARNELL. 1981. Cell **23:** 731–739.
30. KEDINGER, C., M. GNIAZDOWSKI, J. L. MANDEL, R. GISSINGER & P. CHAMBON. 1970. Biochem. Biophysics Res. Commun. **38:** 165–171.
31. LOPATA, M. A., J. C. HAVERCROFT, L. T. CHOW & D. W. CLEVELAND. 1983. Cell **32:** 713–724.
32. CLEVELAND, D. W., M. A. LOPATA, R. J. MACDONALD, W. J. RUTTER & M. W. KIRSCHNER. 1980. Cell **20:** 95–105.
33. VALENZUELA, P., M. QUIROGA, J. ZALDIVAR, W. J. RUTTER, M. W. KIRSCHNER & D. W. CLEVELAND. 1981. Nature (London) **289:** 650–655.

In Vitro Studies of the Biosynthesis of Brain Tubulin on Membranes[a]

JEFFREY M. GILBERT AND PAOLA STROCCHI

Neurochemistry Laboratory
Laboratories for Psychiatric Research
Mailman Research Center
McLean Hospital
Belmont, Massachusetts 02178
and
The Department of Psychiatry
Harvard Medical School
Boston, Massachusetts 02114

INTRODUCTION

Research from numerous laboratories has indicated that membranous elements of brain tissue contain significant amounts of tubulin.[1-14] Previous studies from our laboratory[15,16] have provided evidence that relatively large amounts of alpha- and beta-tubulin subunits are present in smooth microsomes and plasma membranes isolated from central nervous system (CNS) tissue. There is little known about how membrane-associated tubulin is synthesized and incorporated into membranous structures. In this communication we present evidence that there are qualitative differences in tubulin subunits synthesized by free polysomes and rough endoplasmic reticulum (membrane-bound polysomes). Some of the tubulin subunits synthesized by the free polysome population are found in membranes. *In vitro* synthesized tubulin subunits from free polysomes have hydrophobic characteristics that may be an important factor in the association of the proteins with membranous elements of the cell.

THE CHARACTERIZATION OF CNS-SOLUBLE AND MEMBRANE TUBULIN

Soluble proteins from rat forebrain were prepared and analyzed by two-dimensional gel electrophoresis (2DGE) by a modification[15] of the O'Farrell procedure;[17] the resulting electrophoretogram was stained by Coomassie blue and photographed as shown in FIGURE 1. The identity of the tubulin subunits was confirmed by biochemical purification, microtubule aggregation-disaggregation, and colchicine affinity chromatography.[15] Other major acidic, soluble brain proteins (FIGURE 1) include actin and the 68K 5.6 protein (having a molecular mass of 68,000 daltons and an isoelectric point of 5.6) that comigrates with a 68K microtubule-associated protein.[15,29] This protein can be purified by calmodulin affinity chromatography (unpublished results), consistent with its identification as a Tau factor.[19] The 68K 5.6 protein is also found in intermediate filament preparations[16] and among proteins found in the slow phase of axonal transport in the rat optic nerve.[20]

[a]This work was supported by NSF Grant BNS8416618 and the Marion Benton Trust.

89

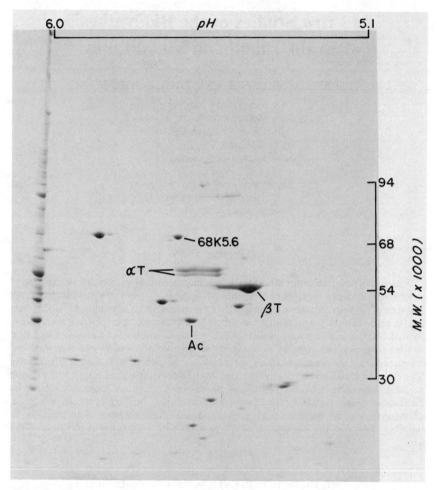

FIGURE 1. 2-DGE of soluble proteins (cytosol from rat forebrain). The method of 2DGE has been described.[15] The isoelectric focusing gel was loaded with 150 micrograms of protein as assayed by Coomassie blue.[36] The 2D electrophoretogram was stained with Coomassie brilliant blue and photographed.

Relatively large amounts of tubulin were also detected in the 2-DGE analysis of proteins in membrane fractions enriched in smooth microsomes (FIGURE 2) and synaptic membranes (FIGURE 3). The identity of these species as tubulin subunits was confirmed by peptide mapping of the proteins taken from the 2D electrophoretograms.[15] High salt washing of the membrane fractions (0.5M KCl) did not remove the alpha- or beta-tubulin subunits. Also, limited proteolytic digestion of these membrane fractions prior to analysis did not degrade the major beta-tubulin subunits (unpublished observation); thus, it is unlikely that these proteins were adhering to the outside of the membrane vesicles.

CELL-FREE BIOSYNTHESIS OF TUBULIN SUBUNITS

Highly purified free and membrane-bound polysomes were prepared from rat forebrain and translated in a cell-free system containing high specific activity [35S]methionine and protein synthesis factors from either rat forebrain or rabbit reticulocytes.[21-23] Radiolabeled translation products were analyzed by 2-DGE followed

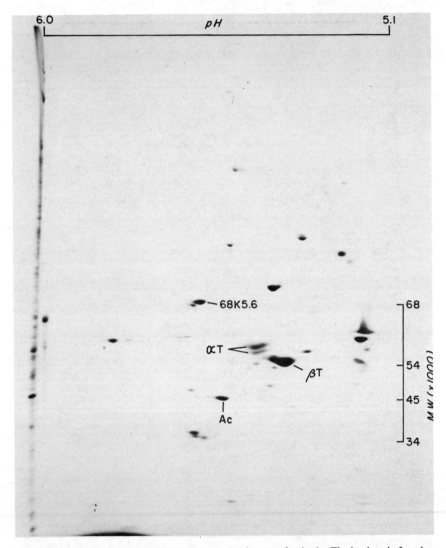

FIGURE 2. 2-DGE of smooth microsomal proteins from rat forebrain. The isoelectric focusing gel was loaded with 150 micrograms of protein. The 2D electrophoretogram was stained with Coomassie brilliant blue and photographed.

by fluorography of the 2D electrophoretograms. The free polysome translation products are shown in FIGURE 4 and include all the major alpha- and beta-tubulin subunits. The *in vitro* synthesized tubulin subunits were greatly enriched if the translation products were purified by three cycles of microtubular aggregation-disaggregation (FIGURE 5A) or immunopurified with a polyclonal antibody to tubulin (FIGURE 5B) generously provided by Dr. K. Fugiwara.[24] Membrane-bound polysomes

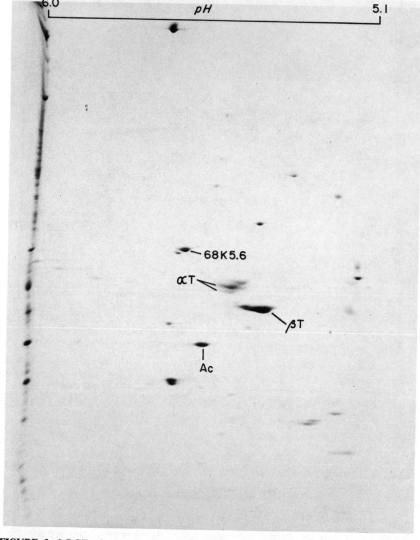

FIGURE 3. 2-DGE of plasma membrane fraction enriched in synaptic membranes. The isoelectric focusing gel was loaded with 150 micrograms of protein.

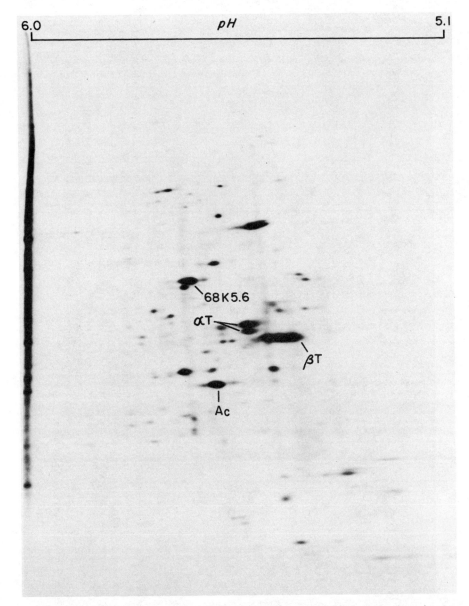

FIGURE 4. 2-DGE analysis of free polysome translation products. 0.6 ml reaction mixture containing 9.0 A_{260nm} units of rat forebrain free polysomes, [^{35}S] methionine (specific activity 1200 Ci/mmole), and factors required for protein synthesis from rabbit reticulocytes[21] was incubated for 30 min at 30°C. There were 226.8×10^6 cpm of radioactive amino acid incorporated into hot trichloroacetic acid insoluble material. An aliquot of 2.1×10^6 cpm was analyzed by 2DGE and fluorography for 24 hours.

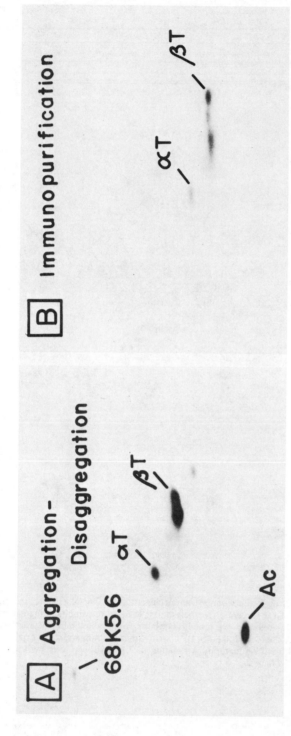

FIGURE 5. 2-DGE analysis of free polysome translation products purified for tubulin by three cycles of aggregation-disaggregation or immunoprecipitation. **A:** Aggregation-disaggregation—three cycles. 242 × 10⁶ cpm of free polysome translation products labeled with [³⁵S] methionine (1100 Ci/mmole) were added to a fresh cytosolic extract as previously described.²² After three cycles of aggregation-disaggregation there were 1.29 × 10⁶ cpm in the tubulin fraction. An aliquot of 110,000 cpm was analyzed by 2DGE. The electrophoretogram was fluorographed for six days. The positions of tubulin subunits (αT and βT) and actin (Ac) and the 68K 5.6 protein¹⁵ are shown. **B:** Immunoprecipitation. Ten microliters of tubulin antisera²⁴ were added to 37.2 × 10⁶ cpm of free polysome translation products labeled with [³⁵S] methionine (1000 Ci/mmole). The immune complexes were isolated by protein A sepharose as previously described.³⁵ The protein A sepharose was eluted with lysis buffer resulting in a fraction containing 672,000 cpm. One half of this sample was analyzed by 2DGE followed by fluorography for one day.

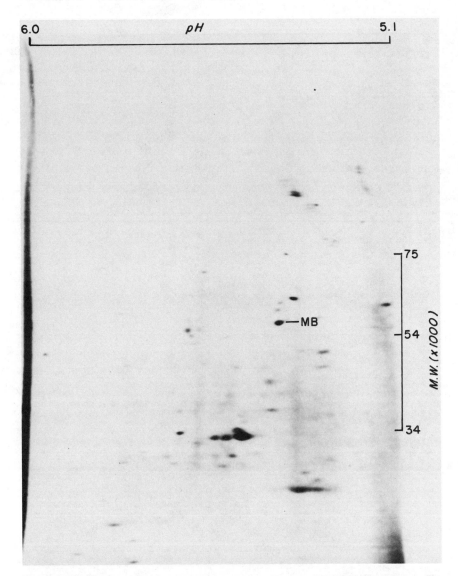

FIGURE 6. 2-DGE analysis of polypeptides synthesized by membrane-bound polysomes. The reaction mixture (0.50 ml) contained 3.2 A_{260nm} units of polysomes, 0.80 mg of reticulocyte factors, 0.50 mCi of [^{35}S]methionine (950 Ci/mmol), and other components required for protein synthesis are listed in experimental procedures. There was 51×10^6 cpm of hot trichloroacetic acid-precipitable material after the 30 min incubation. An aliquot of 2.0×10^6 cpm was analyzed by 2DGE followed by fluorography.

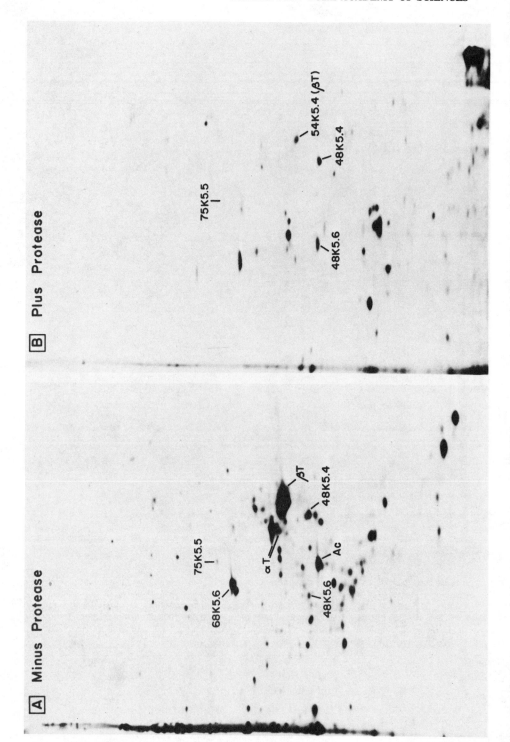

FIGURE 7. 2-DGE analysis of free polysome translation products added to smooth microsomes. **A:** Free polysome translation products added to smooth microsomes. 900 micrograms of the smooth microsome fraction was added to the reaction mixture described above (see Legend to FIGURE 4) containing free polysome translation products plus puromycin to inhibit further protein synthesis. This reaction mixture was incubated for 30 min at 30°C and then layered over a discontinuous sucrose gradient (SW36 Beckman rotor) containing 3.0 ml 2.0 M sucrose and 4.0 ml 1.3 M sucrose. Centrifugation was for 22 hrs at 35,000 rpm. The 1.3 M sucrose-supernatant interface was removed, and the smooth microsomes in one-third of this fraction were collected by centrifugation at $100,000 \times g \times 1$ hour. The pellet was washed with 0.50 M KCl, 1 mM MgCl$_2$. There were 8.69×10^6 cpm in the membrane pellet. An aliquot of 1.4×10^6 cpm was analyzed by 2-DGE and fluorography for three days. **B:** 2DGE analysis of translation products added to smooth microsomes with subsequent trypsin and chymotrypsin treatment. Smooth microsomes were reisolated from a reaction mixture containing free translation products as described above. The smooth microsome fraction from two-thirds of this reaction mixture was treated with trypsin and chymotrypsin.[25] The microsomes were collected by centrifugation, and there were 3.37×10^6 cpm in the membrane pellet. An aliquot of 1.12×10^6 cpm was analyzed by 2DGE and fluorography for six days. In the description of proteins, the first two or three digits represent the apparent molecular weight, with K representing thousands, and the last two digits representing the isoelectric point.

(rough endoplasmic reticulum) were also prepared from rat forebrain, and products of translation were analyzed by 2DGE as shown in FIGURE 6. Only trace amounts of tubulin subunits were detectable (FIGURE 6) as compared with the product analysis of the free polysomes (FIGURE 4). The membrane-bound polysomes synthesized a product denoted as "MB" in FIGURE 6 that had similar (but not identical) molecular weight and isoelectric point to alpha tubulin. Also the peptide map of the MB translation product was similar, but not identical, to the peptide map of alpha tubulin,[22] and MB copurified with tubulin subunits after microtubular aggregation-disaggregation.[22] MB synthesized by intact rough microsomes was resistant to proteolytic digestion,[23] suggesting that after synthesis it is inside the membrane or vesicle. In conclusion, the membrane-bound polysome synthesize a protein (MB) very similar to alpha tubulin: many of the major tubulin subunits, however, including some of those found in membrane fractions, are synthesized by the free polysome population.

STUDIES ON THE ASSOCIATION OF *IN VITRO*-SYNTHESIZED TUBULIN WITH CNS MEMBRANES AND HYDROPHOBIC LIGANDS

Free polysomes were incubated with [35S]methionine and other components required for cell-free protein synthesis. After 30 min incubation, protein synthesis was inhibited with puromycin or pancreatic ribonuclease. A membrane fraction enriched in smooth microsomes was added to the reaction mixture. After an additional period of incubation, the membranes were repurified by discontinuous sucrose gradient centrifugation in high salt, and the membrane fraction proteins were analyzed by 2DGE and fluorography as shown in FIGURE 7A. Significant amounts of newly synthesized alpha- and beta-tubulin subunits were associated with the membrane fractions (FIGURE 7A). Results identical to the analysis in FIGURE 7A were obtained if the smooth microsomes were present in the reaction mixture during protein synthesis and without subsequent incubation after inhibition of protein synthesis (data not shown). After incubation with radiolabeled translation products, the smooth microsomes were subjected to trypsin-chymotrypsin digestion,[25] followed by 2DGE analysis and fluorography as shown in FIGURE 7B. Of all the tubulin subunit translation products added to the smooth microsomes, only one species in the beta-tubulin region (54K 5.4) survived proteolytic digestion. Proteolysis of native proteins in smooth microsome fractions also caused degradation of proteins in the tubulin subunit complex, with the exception of one of the beta-tubulin subunits (data not shown); this protease resistant beta tubulin had identical molecular weight and isoelectric point with the protease-resistant beta-tubulin translation product shown in FIGURE 7. Our data is consistent with the conclusion that at least one of the tubulin subunits found in CNS membranes is synthesized by the free polysome population and then becomes posttranslationally associated with membrane elements. Our results are consistent with the findings of Soifer and Czosnek[11] who showed that newly synthesized tubulin subunits from free polysomes could posttranslationally associate with microsomal membranes;[11] the association of tubulin subunits from disassembled microtubule protein and acid-precipitated tubulin did not occur, however,[10] suggesting that a posttranslational change may take place preventing a tubulin-membrane interaction.

One factor responsible for the association of *in vitro*-synthesized tubulin with membranes may be an interaction between a hydrophobic domain(s) in tubulin subunits with hydrophobic membrane components, such as the lipid bilayer. The translation products of free polysomes were purified by hydrophobic chromatography using ethane-sepharose.[26] Radiolabeled products tightly binding to the affinity matrix

FIGURE 8. 2-DGE analysis of free polysome translation products purified by hydrophobic affinity chromatography with ethane-agarose. 80.9 × 10⁶ cpm of free polysome translation products labeled with [³⁵S]methionine (specific activity 850 Ci/mmole) were applied to a ethane-agarose column (2.0 ml bed volume).[26] The final 8 M urea elution contained 1.72 × 10⁶ cpm, and an aliquot of 200,000 cpm was analyzed by 2-DGE. Fluorography was for two days. The positions of tubulin subunits (αT and βT) and actin (Ac) are shown.

were eluted with urea and analyzed by 2DGE as shown in FIGURE 8. There was significant enrichment of alpha- and beta-tubulin subunits and actin isomers. Only trace amounts of these proteins were detectable if sepharose (without ethane) was used in the affinity purification (data not shown). A similar purification starting with native soluble brain proteins did not result in an enrichment of tubulin subunits (data not shown) suggesting that posttranslational processing may result in a loss of tubulin hydrophobicity.

SUMMARY AND CONCLUSIONS

Membrane elements in brain tissue contain relatively large amounts of alpha- and beta-tubulin (FIGURES 2 and 3). We have investigated the subcellular sites of tubulin biosynthesis in order to determine the origin of this membrane-associated tubulin. Free and membrane-bound polysomes from rat forebrain were separated by differential centrifugation, and the products of translation from these polysome populations were analyzed by 2DGE (FIGURES 4 and 6). Alpha- and beta-tubulin subunits were synthesized by the free polysome population (FIGURES 4 and 5A and B). The membrane-bound polysome fraction synthesized a protein with similar (but not identical) characteristics to alpha-tubulin (denoted as "MB" in FIGURE 6), including isoelectric point, molecular weight, peptide map, and copurification with microtubules after aggregation-disaggregation. Tubulin subunits synthesized *in vitro* by free polysomes could associate posttranslationally with a microsome fraction (FIGURE 7A). The association of the tubulin translation products with membranes was not disrupted by high salt; the associated tubulin, however, was susceptible to proteolytic digestion, with the exception of one of the beta-tubulin subunits (FIGURE 7B). There was an identical protease-resistant beta-tubulin subunit among the native proteins of the smooth microsome fractions. Our data is consistent with the conclusion that at least one beta subunit of membrane-associated tubulin is synthesized by free polysomes and becomes posttranslationally added to membrane structures. It is unlikely that a cotranslational mechanism is responsible, in which there is a signal-mediated insertion of a growing polypeptide chain to membrane.[27,28] Our results, however, are consistent with a "membrane trigger" mechanism proposed by Wickner[18] in which the membrane lipid bilayer triggers the folding of a polypeptide into a configuration that allows integral membrane insertion. The association of tubulin with membranes may also be secondary to the interaction of hydrophobic elements. The amino acid sequence of beta tubulin is known to contain several hydrophobic domains.[30] Tubulin can be incorporated into phospholipid vesicles[31-33] and various subcellular membrane elements.[34,10,11] In our studies, *in vitro* synthesized tubulin from free polysome was found to be purified by hydrophobic affinity chromatography with ethane-sepharose (FIGURE 8). Thus, the hydrophobic characteristics of newly synthesized tubulin could be partially responsible for the posttranslational association of tubulin subunit with membranes. Native tubulin in a soluble fraction of CNS tissue was not purified by hydrophobic affinity chromatography. Therefore, posttranslational modification of tubulin may occur, resulting in a covering over of hydrophobic domains.

REFERENCES

1. BABITCH, J. A. & L. A. BENAVIDES. 1979. Comparison of synaptic plasma membrane and synaptic vesicle polypeptides by two-dimensional polyacrylamide gel electrophoresis. Neurosci. **4**: 603–613.

2. BLITZ, A. L. & R. E. FINE. 1974. Muscle-like contractile proteins and tubulin in synaptosomes. Proc. Natl. Acad. Sci. USA **71:** 4472–4476.
3. ESTRIDGE, M. 1977. Polypeptides similar to the α and β subunits of tubulin are exposed on the neuronal surface. Nature (London) **268:** 59–63.
4. FEIT, H. & S. BARONDES. 1970. Colchicine-binding activity in particulate fractions of mouse brain. J. Neurochem. **17:** 1355–1364.
5. GOZES, I. & U. Z. LITTAUER. 1979. Tubulin microheterogeneity increases with rat brain maturation. Nature (London) **276:** 411–413.
6. KELLY, P. T. & C. W. COTMAN. 1978. Synaptic proteins: Characterization of tubulin and actin and identification of a distinct postsynaptic density polypeptide. J. Cell Biol. **79:** 173–183.
7. KORNGUTH, S. E. & E. SUNDERLAND. 1975. Isolation and partial characterization of a tubulin-like protein from human and swine synaptosomal membranes. Biochim. Biophys. Acta **393:** 100–114.
8. LAGNADO, J. R., C. LYONS & G. WICKREMASINGHE. 1971. The subcellular distribution of colchicine-binding protein (microtubule protien) in rat brain. FEBS Lett. **15:** 254–258.
9. ROSTAS, J. A. P., P. T. KELLY, R. H. PESIN & C. W. COTMAN. 1979. Protein and glycoprotein composition of synaptic junctions prepared from discrete synaptic regions and different species. Brain Res. **168:** 151–167.
10. SOIFER, D. & H. CZOSNEK. 1980. Association of newly synthesized tubulin with brain microsomal membranes. J. Neurochem. **35:** 1128–1136.
11. SOIFER, D. & H. CZOSNEK. 1980. The possible origin of neuronal plasma membrane tubulin. *In* Microtubules and Microtubule Inhibitors, M. DeBrabender & J. DeMey, Eds.: 429–447. Elsevier/North Holland Biomedical Press. Amsterdam.
12. WALTERS, B. B. & A. I. MATUS. 1975. Tubulin in postsynaptic junctional lattice. Nature (London) **257:** 496–498.
13. WANG, Y. J. & H. R. MAHLER. 1976. Topography of the synaptosomal membrane. J. Cell Biol. **71:** 639–658.
14. ZISAPEL, N., M. LEVI & I. GOZES. 1980. Tubulin: An integral protein of mammalian synaptic vesicle membrane. J. Neurochem. **34:** 26–32.
15. STROCCHI, P., B. A. BROWN, J. A. YOUNG, J. A. BONVENTRE & J. M. GILBERT. 1981. The characterization of tubulin in CNS membrane fractions. J. Neurochem. **37:** 1295–1307.
16. STROCCHI, P., D. DAHL & J. M. GILBERT. 1982. Studies on the biosynthesis of intermediate filament proteins in the rat CNS. J. Neurochem. **39:** 1132–1141.
17. O'FARRELL, P. H. 1972. High resolution two-dimensional electrophoresis of proteins. J. Biol. Chem. **250:** 4007–4021.
18. WICKNER, W. 1979. The assembly of proteins into biological membranes: The membrane trigger hypothesis. Annu. Rev. Biochem. **48:** 23–45.
19. SOBUE, K., M. FUJITA, Y. MUMMOTO & S. KAKIUCHI. 1981. The calmodulin-binding protein in microtubules is tau factor. FEBS Letts. **132:** 137–140.
20. STROCCHI, P., J. M. GILBERT, L. I. BENOWITZ, D. DAHL & E. R. LEWIS. 1984. Cellular origin and biosynthesis of rat optic nerve proteins: A two-dimensional gel analysis. J. Neurochem. **43:** 349–357.
21. GILBERT, J. M. 1974. Differences in the translation of rat forebrain messenger RNA dependent on the source of protein synthesis factors. Biochim. Biophys. Acta **340:** 140–146.
22. GILBERT, J. M., P. STROCCHI, B. A. BROWN & C. A. MAROTTA. 1981. Tubulin synthesis in rat forebrain: Studies with free and membrane-bound polysomes. J. Neurochem. **36:** 839–846.
23. GILBERT, J. M. & P. STROCCHI. 1983. Studies on the cell-free biosynthesis of CNS membrane proteins. J. Neurochem. **40:** 153–159.
24. FUGIWARA, K. & T. D. POLLARD. 1978. Simultaneous localization of myosin and tubulin in human tissue culture cells by double antibody staining. J. Cell Biol. **77:** 182–195.
25. SCHEELE, G., R. JACOBY & T. CARNE. 1980. Mechanism of compartmentation of secretory proteins: Transport of exocrine pancreatic proteins across the microsomal membranes. J. Cell Biol. **87:** 611–628.
26. SHALTIEL, S. 1974. Hydrophobic chromatography. *In* Methods in Enzymology. W. B. Jakoby & M. Wilchek, Eds.: Vol. XXXIV: 126–140. Academic Press. New York.

27. BLOBEL, G. & B. DOBBERSTEIN. 1975. Transfer of proteins across membranes. J. Cell Biol. **67:** 835–851.
28. MILSTEIN, C., G. G. BROWNLEE, T. M. HARRISON & M. B. MATHEWS. 1972. A possible precursor of immunoglobin light chains. Nature (London) New Biol. **239:** 117–120.
29. WHATLEY, S. A., C. HALL, A. N. DAVISON & L. LIM. 1984. Alterations in the relative amounts of specific mRNA species in the developing human brain in Down's syndrome. Biochem. J. **220:** 179–187.
30. KRAUHS, E., M. LITTLE, T. KEMPT, R. HOFER-WARBINCK, W. ADE & H. PONSTINGL. 1981. Complete amino acid sequence of beta-tubulin from porcine brain. Proc. Natl. Acad. Sci. USA **78:** 4156–4160.
31. CARON, J. M. & R. D. BERLIN. 1979. Interaction of microtubule proteins with phospholipid vesicles. J. Cell Biol. **81:** 665–671.
32. KLAUSNER, R. D., N. KUMAR, J. N. WEINSTEIN, R. BLUMENTHAL & M. FLAVIN. 1981. Interaction of tubulin with phospholipid vesicles. Association with vesicles at the phase transition. J. Biol. Chem. **256:** 5879–5885.
33. REAVEN, E. & S. AZHAR. 1981. Effect of various hepatic membrane fractions on microtubule assembly with special emphasis on the role of membrane phospholipids. J. Cell Biol. **89:** 300–308.
34. BERNIER-VALENTIN, F., D. AUNIS & B. ROUSSET. 1983. Evidence for tubulin-binding sites on cellular membranes: plasma membranes, mitochondrial membranes and secretory granule membranes. J. Cell Biol. **97:** 209–216.
35. SAPIRSTEIN, V. S., M. WISLOWSKI, P. STROCCHI & J. M. GILBERT. 1983. Characterization and biosynthesis of soluble and membrane-bound carbonic anhydrase in brain. J. Neurochem. **40:** 1251–1261.
36. SEDMAK, J. J. & S. E. GROSSBERG. 1977. A rapid, sensitive and versatile assay for protein using Coomassie brilliant blue G250. Anal. Biochem. **79:** 544–552.

Tubulin Variants Exhibit Different Assembly Properties

STEPHEN W. ROTHWELL, WILLIAM A. GRASSER, AND
DOUGLAS B. MURPHY

The Johns Hopkins University
School of Medicine
Baltimore, Maryland 21205

INTRODUCTION

The discovery that tubulin is encoded by multiple, unique genes in vertebrates and other eukaryotes has stimulated new interest in the idea that biochemical variants of tubulin may perform different functions in the cell.[1,2] Biochemical variants of tubulin that are derived from different genes[3-6] or are generated by posttranslational modifications such as tyrosinolation[7] or acetylation[8] have been isolated and characterized, and in many instances these variants have been shown to be associated with specific tissues, cell types, or with specific subcellular organelles. In most cases, however, it has not been possible to demonstrate whether the tubulin subunits themselves also exhibit differences in their assembly properties and thus may account for the specificity of their expression.

Recently we demonstrated that tubulins isolated from chicken brain and erythrocytes represent two distinct variants that differ in primary structure as determined by peptide mapping and amino acid composition.[9] The differences in their primary structures are reflected in differences in their electrophoretic mobilities, isoelectric points, solubilities, and drug-binding abilities.[10,11] In previous studies we also demonstrated differences in the *in vitro* polymerization of the tubulin variants with respect to nucleation and elongation, critical concentration, and oligomer formation.[9,12,13] In this report we show that the basis for the differences in polymerization is due to the fact that the two tubulin variants exhibit different association and dissociation rate constants at the microtubule ends when measured under identical *in vitro* conditions. Comparison of the rate constants allows us to predict that the treadmilling potential and rates of subunit exchange at the microtubule ends are significantly different for brain and erythrocyte microtubules. The findings suggest that tubulin variants with different polymerization rate constants exhibit different assembly properties and may perform different functions in the cell.

MATERIAL AND METHODS

Preparation of Chicken Brain and Erythrocyte Tubulin

Chicken brain tubulin was isolated by the method of Dentler *et al.*[14] in 0.1 M piperazine-N,N'-bis-(2-ethanesulfonic acid) (PIPES) buffer containing 1 mM MgCl$_2$, 2 mM ethylene glycol-bis-(β-amino ethyl ether) N,N'-tetra-acetic acid (EGTA), 1 mM guanosine triphosphate (GTP), and 4 M glycerol.

Chicken erythrocyte tubulin was prepared from chicken blood by the method of

103

Murphy and Wallis.[12] Tubulin was purified free of microtubule-associated proteins (MAPs) by ion exchange chromatography using Whatman P-11 phosphocellulose (PC)[15] and cycled once prior to use. Unless stated otherwise, microtubule assembly buffer was 0.1 M Na-PIPES pH 6.94 containing 1 mM MgCl$_2$, 1 mM GTP (PMG), and 5% glycerol.

Erythrocyte Tubulin Antibody

A rabbit polyclonal antibody was produced against the beta subunit of the erythrocyte tubulin dimer. The preparation of the antigen and the affinity purification of the antibody has been previously described.[16]

Preparation of Microtubule Seeds and Subunits

Microtubule seeds were prepared by polymerizing PC erythrocyte tubulin at 37°C in 0.1 M PIPES pH 6.94 containing 10 mM MgCl$_2$, 1 mM GTP, and 20% glycerol, and sedimenting and resuspending the polymers to 5–13 mg/ml in assembly buffer. Polymers were sheared by 10 passes through a 27-gauge needle. Examination of the microtubule lengths in the seed preparation revealed no redistribution in seed length for up to five hours. PC tubulin subunits (either brain or erythrocyte) were diluted to the desired concentration in the same buffer at 5°C. The subunit preparations were lightly sonicated prior to use to disperse any oligomers.[12]

Preparation of Heteropolymers for Measurement of Initial Rates of Elongation

Microtubule elongation (subunit addition onto the ends of exogenous, preformed microtubules) was performed using the guidelines established previously by Johnson and Borisy.[17] Tubulin polymerization was monitored by recording the change in optical density at 350 nm in a Gilford spectrophotometer equipped with a water-jacketed cuvette holder adjusted to 32°C. Typically, a 0.8 ml aliquot of subunits (either brain or erythrocyte tubulin) was placed in a cuvette and warmed to polymerization temperature (30°C). The temperature increased exponentially, with the time required for half-maximal equilibration being less than 5 seconds. After preincubation of the subunits at 30°C for 60 sec, polymerization was initiated by the addition of 15 μl of seeds of the other microtubule type. The optical density at 350 nm was recorded continuously within 3 sec of mixing, and 20 μl samples of the elongating polymers were taken at 15, 30, 45, and 60 sec for electron microscopic examination.

We determined that the criteria established by Johnson and Borisy[17] were met by our protocol for measuring initial rates: (1) no microtubules were formed in the absence of seeds; (2) the initial rate of assembly was directly proportional to the number concentration of seeds added (determined by the measurements of the rates of turbidity development); (3) the change in optical density following the addition of seeds exhibited a single pseudo-first-order exponential.

Labeling of Microtubule Heteropolymers with Tubulin Antibody—Protein A Gold

Electron microscope (EM) grids containing glutaraldehyde-fixed microtubules were rinsed successively with phosphate-buffered saline (PBS), 10 mM NaBH$_4$ to reduce the free aldehyde,[18] PBS, and incubated with the erythrocyte tubulin antibody at 50 μg/ml for 10 minutes. The grids were rinsed with PBS again and incubated with gold colloid coated with protein-A.[19-21] This complex was fixed with 5% glutaraldehyde in PMG + 5% glycerol, reduced with NaBH$_4$ and washed with PBS before negative staining with 1% uranyl acetate. The fixation after incubation with protein A-gold was required to prevent the redistribution of label. Measurements of the lengths of elongated brain or erythrocyte tubulin polymers were made using a Zeiss EM 10A electron microscope. At each concentration, the mean lengths of the elongated portions were plotted versus time, yielding an initial rate of elongation.

Bergen and Borisy[22] showed that the rate of elongation of a microtubule could be related to its association (k$^+$) and dissociation rate constant (k$^-$) by the following equation:

$$dL/dt = k^+[S] - k^-$$

In a plot of initial rate of elongation versus tubulin subunit concentration, the slope, the y intercept, and the x intercept will yield the association rate constant, dissociation rate constant, and the critical concentration, respectively. Because the EM assay measures the lengths of the elongated portions of individual microtubules, the quantity representing the microtubule number concentration is dropped from the equation.[17]

Sodium Dodecyl Sulfate (SDS)-Polyacrylamide Gel Electrophoresis and Immunoblotting

SDS slab gels containing 12% acrylamide (pH 9.1) were prepared as previously described.[10] For immunoblotting, the protein was electrophoretically transferred to nitrocellulose strips.[23] The strips were then incubated with antiserum to beta erythrocyte tubulin diluted 1/500, followed by an incubation with [^{125}I]protein-A.

Determination of Protein Concentration

Total protein concentrations were determined by the Bradford protein assay.[24] Bovine serum albumin was used as a standard.

Biochemical Materials

PIPES and sodium salt, was obtained from Calbiochem. Other chemicals and nucleotides were obtained from Sigma Chemical Co.

RESULTS

Demonstration of Antibody Specificity for Erythrocyte Beta Tubulin

The specificity of the erythrocyte beta tubulin antibody is demonstrated in Figure 1A in an immunoblot of chicken brain and erythrocyte tubulin. Only the beta subunit of the erythrocyte tubulin was labeled. This antibody was also specific when used in an immuno-gold labeling procedure for electron microscopy. Erythrocyte microtubules were heavily stained compared to brain microtubules, which allowed the two types of microtubules to be distinguished (FIGURE 1B, 1C).

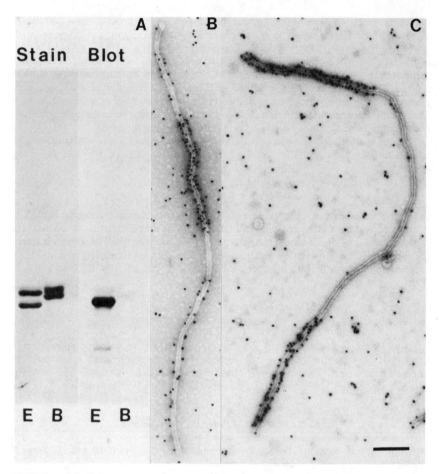

FIGURE 1. Specificity of rabbit antibody to erythrocyte beta tubulin. **A:** Immunoblot. Microtubule proteins from chicken brain (lower **B**) and erythrocyte (**E**) were fractionated by SDS gel electrophoresis at pH 9.1 to resolve the alpha and beta subunits in both preparations. The figure shows Coomassie stain and corresponding immunoblot after incubation in the presence of erythrocyte tubulin antiserum (1/500 dilution) and iodinated protein-A. **B** and **C:** Immunoelectron microscopy. **B:** Erythrocyte seeds initiate the elongation of brain tubulin subunits. Growing heteropolymers were fixed and labeled with antibody and protein A-gold colloid to identify microtubule domains containing erythrocyte tubulin. Magnification × 50,000. Bar = 0.2 μm. **C:** Brain microtubule seeds were used to initiate the elongation of erythrocyte tubulin subunits at microtubule ends.

Determination of Rate Constants

Plots of the initial rates of elongation at various subunit concentrations for each tubulin isoform were prepared in order to determine the association and dissociation rate constants for microtubule assembly. The electron microscope assay allowed us to calculate the rate constants for each end of the microtubules (FIGURE 2). We found

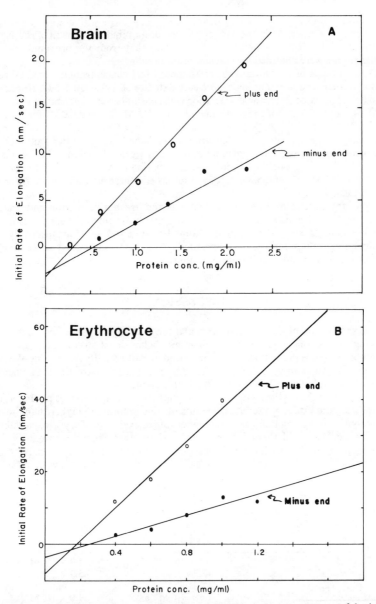

FIGURE 2. Determination of the association and dissociation rate constants of brain and erythrocyte tubulin. The initial rate of elongation of microtubule seeds were measured at different tubulin subunit concentrations. Microtubule seeds (10 μl, 5 mg/ml) were added to PC tubulin subunits (0.8 mg/ml) and monitored spectrophotometrically at 350 nm to confirm that the elongation kinetics were linear. Small aliquots (10 μl) were periodically fixed, labeled with antibody, and examined by EM in order to construct a plot of length of elongated polymers versus time. Initial rates of elongation were determined for both the plus and minus ends of the microtubules from these measurements. **A:** Brain tubulin subunit addition off erythrocyte tubulin seeds. **B:** Erythrocyte tubulin subunit addition off brain tubulin seeds.

that for each tubulin type there was an asymmetry in growth, with one end consistently shorter than the other. As in the case of porcine brain tubulin,[21] growth of erythrocyte tubulin from the distal (+) end was faster than that from the proximal (−) end. Similar results were obtained for chicken brain tubulin.

From analysis of the initial rate plots the critical concentrations for erythrocyte tubulin at the plus and minus ends were calculated to be 1.6 and 1.8 μM, respectively (TABLE 1). The association and dissociation rate constants were determined to be 8.4 × 10^6 $M^{-1}s^{-1}$ and 13.9 s^{-1} for the plus end and 2.5 × 10^6 $M^{-1}s^{-1}$ and 6.0 s^{-1} for the minus end.

These parameters were also measured for chicken brain tubulin (TABLE 1). The critical concentrations for the plus and minus ends were calculated to be 2.7 μM and 4.7 μM, respectively. The association and dissociation rate constants for the plus end were 1.9 × 10^6 $M^{-1}s^{-1}$ and 5.3 s^{-1}; for the minus end these constants were 1.0 × 10^{-6} $M^{-1}s^{-1}$ and 4.7 s^{-1}.

From the rate constants, we calculated the treadmilling rate and the flux parameter "s".[25] The flux rates for erythrocyte and brain tubulin were determined to be 0.38 subunits sec^{-1} (0.8 $\mu m/hr$) and 1.8 subunits sec^{-1}, respectively. The corresponding values for the flux parameter "s" were determined to be 0.07 for erythrocyte tubulin and 0.13 for brain tubulin.

DISCUSSION

From the values of the rate constants and the critical concentrations we made the following predictions regarding the dynamic behavior of erythrocyte and brain microtubules. These predictions are summarized in FIGURE 3. By calculating Wegner's "s" parameter we estimate a difference of a factor of two in the treadmilling efficiency, with the brain tubulin being the more efficient variant.

Even though the erythrocyte tubulin treadmilling rate, however, is predicted to be low, the absolute values of the rate constants for this isoform are higher than those for brain tubulin, indicating that there is a higher rate of exchange of erythrocyte subunits per unit time. We would, therefore, predict that the erythrocyte tubulin would elongate faster than the brain tubulin. Using the increase in optical density of polymerizing tubulin as an independent assay for elongation, we observed that this was, in fact, the case.[10]

TABLE 1. Association and Dissociation Rate Constants of Brain and Erythrocyte Tubulin[a]

		Erythrocyte	Brain
Plus end	k^+	8.4 × 10^6 $M^{-1}s^{-1}$	1.9 × 10^6 $M^{-1}s^{-1}$
	k^-	13.9 s^{-1}	5.3 s^{-1}
	c	1.6 μm	2.7 μm
Minus end	k^+	2.5 × 10^6 $M^{-1}s^{-1}$	1.0 × 10^6 $M^{-1}s^{-1}$
	k^-	6.0 s^{-1}	4.7 s^{-1}
	c	1.8 μm	4.7 μm
Flux		.38 s^{-1}	1.8 s^{-1}
		(.8 μm hr^{-1})	(3.9 μm hr^{-1})
"s"		.07	.13

[a]The association rate constants (k^+) and the dissociation rate constants (k^-) and the critical concentrations (c) for the plus and minus ends were obtained from initial rate plots shown in FIGURE 2. The rate of flux ($k^+c - k^-$) and the flux parameter "s" were determined from the rate constants as described by Bergen and Borisy.[22]

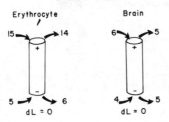

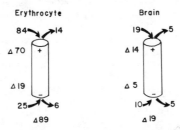

FIGURE 3. Dynamics of subunit equilibrium at steady state and during elongation. The number of association-dissociation events occurring per second are shown for erythrocyte and brain tubulin at steady state (dL = 0) in the top panel and during elongation in 10 μm tubulin in the bottom panel. The number of events and total events are indicated (Δ).

The high values of the erythrocyte rate constants may also be related to our independent observations of microtubule fusion *in vitro,* which show that erythrocyte microtubules fuse at a faster rate than brain microtubules. We observed that the microtubule number concentration decreases rapidly with time at steady state. By EM observation of microtubule chimeras containing erythrocyte and brain tubulin, we were able to determine that the rate of decrease in microtubule number was precisely correlated with the rate of end-on-end fusion of microtubules.[13] Although we do not have an explanation for the preferential fusion of erythrocyte microtubules, the differences in the values of the association rate constants may, in part, account for the observation. It is possible that erythrocyte microtubule fusion may constitute a repair mechanism for maintaining long microtubule polymers in the marginal band of the chicken erythrocyte.

In summary these studies establish that tubulin variants can exhibit different assembly properties that can be related to differences in their critical concentrations and assembly rate constants. It is possible that the differences in assembly properties are related to differences in their functions *in vivo.*

REFERENCES

1. STEPHENS, R. E. 1975. *In* Molecular and Cell Movement. S. INOUE & R. E. STEPHENS, Eds.: 181–206. Raven Press. New York.
2. FULTON, C. & P. A. SIMPSON. 1976. Cold Spring Harbor Conference on Cell Proliferation. Vol. 3: 987–1005.

3. RAFF, E. C. & M. T. FULLER. 1984. *In* Molecular Biology of the Cytoskeleton. G. G.
    BORISY, D. W. CLEVELAND & D. B. MURPHY, Eds.: 293–304. Cold Spring Harbor
    Laboratory, N.Y.
4. RAFF, E. B. 1984. J. Cell Biol. **99:** 1–10.
5. SULLIVAN, K. F., J. C. HAVERCROFT & D. W. CLEVELAND. 1984. *In* Molecular Biology of
    the Cytoskeleton. G. G. BORISY, D. W. CLEVELAND & D. B. MURPHY, Eds.: 321–332.
    Cold Spring Harbor Laboratory.
6. FARMER, S. R., J. F. BOND, G. S. ROBINSON, D. MBANGKOLLO, M. J. FENTON & E. M.
    BERKOWITZ. 1984. *In* Molecular Biology of the Cytoskeleton. G. G. Borisy, D. W.
    Cleveland & D. B. Murphy, Eds.: 333–342. Cold Spring Harbor Laboratory.
7. KUMAR, N. & M. FLAVIN. 1982. Eur. J. Biochem. **128:** 215–223.
8. L'HERNAULT, S. W. & J. L. ROSENBAUM. 1985. Biochemistry **24:** 473–478.
9. MURPHY, D. B., K. T. WALLIS & W. A. GRASSER. 1984. *In* Molecular Biology of the
    Cytoskeleton. G. G. BORISY, D. W. CLEVELAND & D. B. MURPHY, Eds.: 59–70. Cold
    Spring Harbor Laboratory.
10. MURPHY, D. B. & K. T. WALLIS. 1983. J. Biol. Chem. **258:** 7870–7875.
11. LUDUENA, R. F., M. C. ROACH, M. A. JORDAN & D. B. MURPHY. 1985. J. Biol. Chem.
    260: 1257–1264.
12. MURPHY, D. B. & K. T. WALLIS. 1983. J. Biol. Chem. **258:** 8357–8364.
13. ROTHWELL, S. W., W. A. GRASSER & D. B. MURPHY. 1984. J. Cell Biol. **99:** 350a.
14. DENTLER, W. L., S. GRANETT & J. L. ROSENBAUM. 1975. J. Cell Biol. **65:** 237–241.
15. ROOBOL, A., C. I. POGSON & K. GULL. 1980. Biochem. J. **189:** 305–312.
16. ROTHWELL, S. W., W. A. GRASSER & D. B. MURPHY. 1985. J. Cell Biol. In press.
17. JOHNSON, K. A. & G. G. BORISY. 1977. J. Mol. Biol. **117:** 1–31.
18. OSBORNE, M. & K. WEBER. 1982. Methods Cell Biol. **241:** 97–132.
19. GEOGHEHAN, W. D. & G. A. ACKERMAN. 1977. J. Histochem. Cytochem. **25:** 1187–1200.
20. HORRISBERGER, M. 1979. Biol. Cell. **36:** 253–258.
21. SLOT, J. W. & H. J. GEUZE. 1981. J. Cell Biol. **90:** 533–535.
22. BERGEN, L. G. & G. G. BORISY. 1980. J. Cell Biol. **84:** 141–150.
23. TOWBIN, H., T. STAEHELIN & J. GORDON. 1979. Proc. Natl. Acad. Sci. USA **76:** 4350–
    4354.
24. BRADFORD, M. 1976. Anal. Biochem. **72:** 248–254.
25. WEGNER, A. 1976. J. Mol. Biol. **108:** 139–150.

Biochemical and Chemical Properties
of Tubulin Subspecies[a]

JAMES C. LEE, DEBORAH J. FIELD,[b]

HENRY J. GEORGE,[c] AND JOSEPH HEAD

Edward A. Doisy Department of Biochemistry
St. Louis University School of Medicine
St. Louis, Missouri 63104

Microtubules are found in all eukaryotes and are apparently involved in a variety of cellular functions including mitosis, maintenance of cell shape, cell motility, intracellular transport, and secretion.[1] This suggests an immediate question: Is one protein responsible for all of these diverse functions, or are these functions maintained by a family of proteins that, although exhibiting gross structural similarity, actually possess subtle chemical differences that designate them for particular functional roles? To address this issue, a study was initiated to define the extent of microheterogeneity in tubulin from vertebrate brains.

AUTHENTICITY OF TUBULIN SUBSPECIES

The concept of multiple subspecies in the tubulin family is well established by protein and DNA sequencing.[2-6] The exact number of subspecies is still in doubt. There have been an increasing number of reports implying that brain tubulin exhibits extensive heterogeneity when subjected to isoelectric focusing.[7-13] The reported number of subspecies varies from three to nine. The presence of multiple peaks during isoelectric focusing, however, does not necessarily indicate heterogeneity of the protein sample. Cann and coworkers[14-18] have elegantly demonstrated by computer simulation that even for a homogeneous sample of macromolecule, a pattern showing multiple well-resolved peaks can be observed if the system undergoes a reversible carrier ampholyte-macromolecule interaction, for example, $P + A \rightleftharpoons PA$. Such an interaction may include ligand-induced isomerization, association-dissociation of the macromolecule, or pH-dependent conformational transitions. The number of peaks then reflects the presence of various ampholyte-macromolecule complexes, but not necessarily the intrinsic heterogeneity of the macromolecule. Several examples are found in the literature, and these include an acidic protein from wool,[19] bovine serum albumin,[20,21] myoglobin,[22] and transfer RNA (tRNA).[23,24] Hence, it is imperative to establish that the subspecies identified by electrophoretic techniques indeed reflect authentic tubulin heterogeneity. Thus, control experiments were conducted.

[a]This work was supported by Grants from the National Institutes of Health, NS-14269 and AM-21489.

[b]Present address: Department of Medical Genetics, University of Toronto, Toronto, Ontario M5S 1A1, Canada.

[c]Present address: Molecular Genetics, Inc., Edina, Minnesota 55343.

111

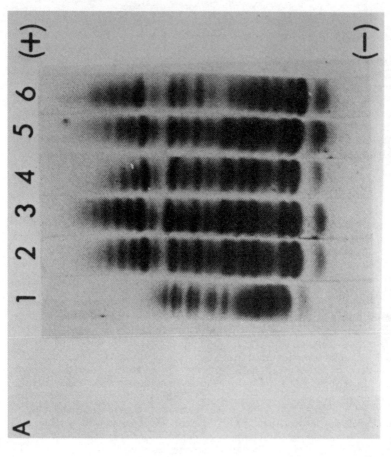

FIGURE 1. Isoelectric focusing patterns of tubulin. Experimental conditions were the following: (1) Total duration of experiment was 18 h with an Ampholine mix of 1.2% (v/v), pH range 4–6, and 0.8% (v/v) Ampholine mix of pH range 5–7; (2–6) total duration of experiments was 72 h with a 2% (v/v) Ampholine mix of pH range 4–6 at 8°C. A 100 μg sample of protein at 3.0 mg/ml was applied to each gel. The protein samples are as follows: (1 and 2) carboxyamidomethylated W-tubulin; (3) W-tubulin; (4) carboxyamidomethylated PC-tubulin; (5) PC-tubulin; (6) fluorescent-labeled tubulin.

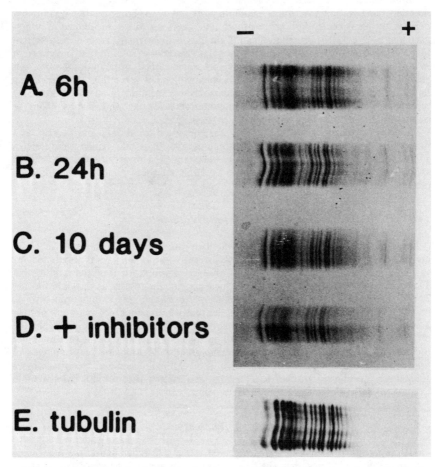

FIGURE 2. Evaluation by isoelectric focusing of possible modification of tubulin during purification. Total soluble protein was prepared from calf brain by homogenization in 10 m*M* sodium phosphate/0.5 m*M* MgCl$_2$/0.24 *M* sucrose, pH 7.0, followed by centrifugation at 12,000 × g for 30 min (the first step of the standard Weisenberg purification). Aliquots of the supernatant were incubated at 4°C for 6 hr (lane A), 24 hr (lane B), or 10 days (lane C), measured from the time of animal death. A separate supernatant (lane D) was prepared with protease inhibitors included in the homogenization buffer as follows: 30 m*M* KF, 2 m*M* benzamidine, 2 m*M* phenylmethylsulfonyl fluoride, and α_2-macroglobulin at 3 mg/ml. After incubation, supernatants were carboxyamidomethylated, dialyzed against distilled deionized water at 4°C., and lyophilized. Proteins were resolved by isoelectric focusing and stained with Coomassie blue. Purified tubulin was also included on the gel (lane E). Approximately 100–350 μg of protein was loaded in each lane.

Effect of Purification Procedure

Brain tubulin was isolated by either the Weisenberg (W-tubulin) or polymerization-depolymerization procedure.[25] Tubulin prepared by the latter method was further purified by phosphocellulose chromatography according to the procedure of Weingarten et al.,[26] and it is designated as PC-tubulin. The polymerization-depolymerization method conceivably could selectively isolate the pool of tubulin that is competent in forming microtubules, and because glycerol was included in the buffer system, it is possible that contaminants in glycerol may chemically modify tubulin,[27,28] thus leading to the presence of multiple resolvable components. On the other hand, the Weisenberg procedure uses conventional protein fractionation principles and is most likely to isolate proteins with similar physiochemical properties.

Similar isoelectric focusing patterns were observed for tubulin purified by both procedures, as shown in FIGURE 1. The same number of protein bands was resolved, and the relative intensity for each band was unchanged. This result indicates that the same population of tubulin was isolated by either of the well-adopted purification procedures of tubulin.

Effect of Protease Inhibitors

Tubulin heterogeneity may be generated during preparation due to proteolytic digestion. To resolve this issue, total soluble calf brain protein was incubated for various periods of time in the presence or absence of a mixture of protease inhibitors. The samples were analyzed by isoelectric focusing, and the results, as shown in FIGURE 2, demonstrate that the isoelectric focusing patterns are very similar and give no indication of time-dependent changes in the relative amounts of any of the protein bands focusing in the region of tubulin subspecies. This strongly suggests that the observed heterogeneity of tubulin is not produced by modification of the tubulin under the conditions used for purification. This result assumes greater significance in view of the recent report by Sackett et al.,[29] that proteolytically modified tubulin is more efficient in forming microtubules. Thus, the assumption that tubulin competent in forming microtubules is native protein is no longer valid.

Effect of Ampholytes

If the multiple number of peaks observed was a true reflection on the number of heterogeneous subspecies, then the total number of peaks and the relative area under each peak should remain unchanged upon perturbation of either protein concentration, total protein amount, or ampholyte concentration. If, however, there are indeed ligand-induced changes in the macromolecule, then on the basis of the LaChatelier's principle, the number of peaks resolvable by isoelectric focusing and the relative area under each peak should vary as a function of protein or ampholyte concentration. Isoelectric focusing experiments were, therefore, carried out either as a function of initial protein concentration or ampholyte concentration.

Typical results of these experiments show that the number of resolvable bands remains the same, and the relative intensity of each band remains approximately the same. Thus, the results seem to indicate that the presence of these peaks is a true reflection on the number of heterogeneous species in tubulin. The quantitative validity of these results, however, is compromised by the technical complication that not all of the protein that was layered onto the gel migrated into the gel matrix; hence, the exact protein amount that migrated into the gel was not defined. Due to the failure to obtain unequivocal results by using the experimental tests described, two-dimensional isoelectric focusing experiments were conducted. Briefly, isoelectric focusing in tube gels constitutes the first dimension, and it is followed by focusing in slab gels under identical conditions for the second dimension. If ampholyte-induced changes in the

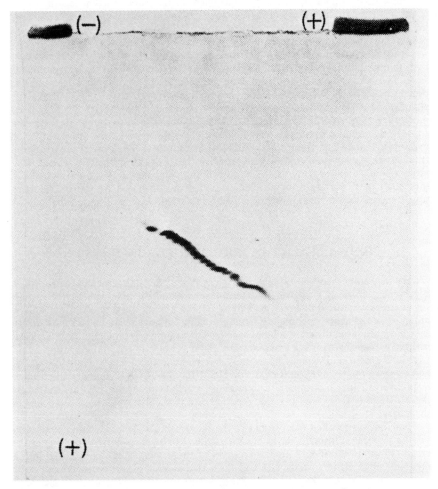

FIGURE 3. Two-dimensional isoelectric focusing of tubulin. 100 μg of carboxyamidomethylated-W-tubulin was subjected to isoelectric focusing under the same experimental conditions as those in FIGURE 1. Isoelectric focusing in the second dimension was conducted under the same conditions on a slab gel (15.5 × 14.5 × 0.1 cm) for 24 hours.

macromolecules are the cause of the presence of multiple bands in the first dimension, then it is expected that each band would further resolve into multicomponents in the second dimension, leading to a pattern of multiple bands on either side of the diagonal on the slab gel. Conversely, if the multiple peaks do indeed represent genuine polymorphic components of brain tubulin, then the pattern should show the bands distributing along the diagonal on the slab gel. Carboxyamidomethylated-tubulin was subjected to two-dimensional isoelectric focusing, and the results are shown in FIGURE 3. It is evident that no additional bands were generated in the second dimension, and the protein bands were distributed along the diagonal on the slab gel. The pH gradient generated on a slab gel was usually not as linear as in a tube gel, thus leading to the slight curvatures in the gel pattern. When the experiment was conducted with a steeper pH gradient or for a shorter duration, the pattern became straighter with less resolution between protein bands.

A conclusive demonstration of heterogeneity is to isolate individual bands, concentrate to initial protein concentrations, and repeat the isoelectric focusing experiments for each isolated band under identical experimental conditions. Six representative protein bands were sliced from eight identical gels. These were electrophoretically eluted, exhaustively dialyzed against deionized water, and lyophilized. These isolated protein fractions were resuspended in sample buffer and adjusted to a concentration equal to the initial unfractionated protein sample. Subsequently, they were subjected to isoelectric focusing; the results are shown in FIGURE 4. Each of these six tubulin fractions shows no indication of heterogeneity. The isoelectric points of the isolated protein bands are the same as the corresponding components in the unfractionated sample. It may, therefore, be concluded that, at least for the six protein fractions tested, these components represent genuine polymorphic species of brain tubulin.

HETEROGENEITY IN VERTEBRATE BRAIN TUBULIN

Having established that tubulin heterogeneity as detected by isoelectric focusing is a genuine feature, it is logical to determine the extent of heterogeneity in various vertebrate brain tubulin, because individual laboratories have selected not only a different experimental technique to study tubulin heterogeneity, but, in most cases, they have also used different animal species. For example, Denoulet et al.[30] has used isoelectric focusing to study mouse brain tubulin; Krauhs et al.[2] and Ponstingl et al.[3] have used amino acid sequencing to study porcine brain tubulin, while Wilde et al.[31,32] and Lopata et al.[33] have studied the number of tubulin genes in human and chicken, respectively. Thus, it is important to establish which common features are shared by these systems so that an observation reported on a particular system can be applied to the others.

Tubulin was purified from brains of the six vertebrate species commonly used in tubulin research. FIGURE 5 shows one-dimensional slab gel isoelectric focusing of the vertebrate brain tubulins. Tubulins from all of these sources can be resolved into a minimum of 15 distinct bands, although the actual number of subspecies is likely to be greater: the most basic bands are unusually broad and do not sharpen with increased time of focusing. Close inspection of the gels suggests that these bands are composed of two, and in some cases three, nearly comigrating proteins. Comparison of the isoelectric focusing patterns among the six species indicates that corresponding subspecies have very similar, if not identical, isoelectric points. Differences in the relative intensities of some of the corresponding subspecies bands are evident, but the overall distribution is very similar. No species-specific forms of tubulin were identified by this one-dimensional isoelectric focusing system. In spite of this good resolution of tubulin species, the one-dimensional isoelectric focusing procedure cannot identify the subspecies as being an α or β subunit. A two-dimensional system with high resolution was developed and applied to the six brain systems, as shown in FIGURE 6. In general,

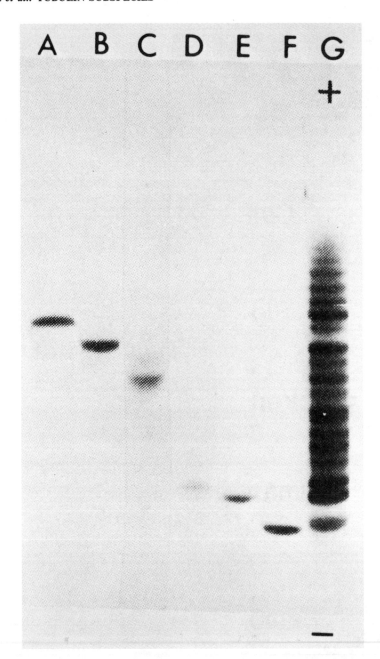

FIGURE 4. Isoelectric focusing of isolated subspecies of tubulin. The experimental conditions were the same as those in FIGURE 1. A–F: Isolated subspecies 14, 12, 10, 3, 2, and 1, respectively. A 10 μg sample of each subspecies at 3.0 mg/mL was applied. 45 μg of carboxyamidomethylated-W-tubulin at 3.0 mg/mL was applied to gel G.

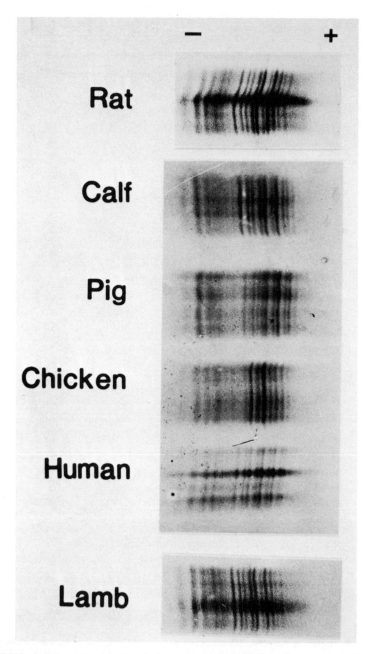

FIGURE 5. One-dimensional isoelectric focusing gels of purified vertebrate brain tubulins. Gels were stained with Coomassie blue. Approximately 100 μg of protein was loaded per lane.

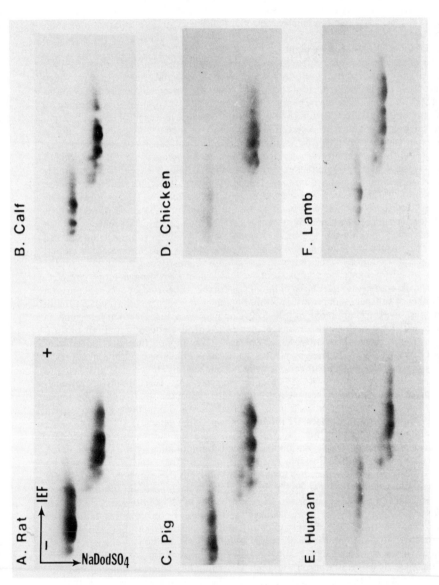

FIGURE 6. Two-dimensional isoelectric focusing/NaDodSO₄ electrophoretic analysis of purified vertebrate brain tubulins. Gels were stained with Coomassie blue. Approximately 12–18 µg of protein was applied to each gel. IEF = isoelectric focusing.

two-dimensional gel analyses of higher vertebrate species reveal that the α subunits are highly conserved, whereas the most basic subspecies of the β subunit exist in two to three subpopulations of different apparent molecular weight. The presence of these subpopulations is apparently species dependent. Rat, human, and calf β subunits are very similar, whereas pig tubulin contains an additional subgroup of β tubulin. Chicken tubulin subspecies appear as a cluster of proteins with very similar apparent molecular weight rather than as two or three distinguishable subpopulations of proteins with different apparent weights.

The subspecies of higher apparent molecular weight are most likely the same ones identified in porcine brain tubulin by Little.[34] The present study has further identified them as basic β subspecies. In addition to this subgroup, porcine brain tubulin contains two other subgroups. One corresponds to what seems to be a common group among all vertebrate brain tubulin tested in this study; a third group, however, of lower apparent molecular weight can also be observed. The existence of such a component can only be observed by the developed two-dimensional procedure. The identity of this subgroup as members of the β subunit was verified by peptide mapping.[35] This β subgroup is apparently also present in chicken brain tubulin, although the ones with higher apparent molecular weight are missing. Hence, there are detectable differences in the composition of the basic β subspecies among the vertebrate species tested. The observed difference in molecular weight among β-tubulin subspecies most likely is indeed a reflection on the size difference among these polypeptides. Recently, Sullivan and Cleveland[6] reported the nucleotide sequence of a β-tubulin subunit from chicken brain. This β-tubulin gene encodes a polypeptide that is four amino acids longer. Hence, results obtained from either studying protein chemistry or genomic structure of tubulin will most likely be complementary and aid in providing a more complete picture of the complexity in tubulin. This view is further supported by recent reports on the number of subspecies in brain β tubulin. Lopata et al.[33] reported that four β tubulin (or at most five) are sufficient to allow development of a higher eukaryote such as chicken; as shown in FIGURES 5 and 6, however, chicken and mammalian brain β tubulins resolve into apparently 12 subspecies. This implies that the majority of β-tubulin subspecies are formed by posttranslational modifications, although efforts to identify posttranslational modifications were not successful. Hence, it is gratifying to note that Sullivan and Cleveland[6] recently reported that three to five additional divergent chicken β-tubulin sequences are identified. Thus, the relative contribution of posttranslational modification to tubulin heterogeneity is diminished, and at least a majority of β-tubulin subspecies are products of functional genes.

Results from this study may also help to resolve some apparent contradictory immunological reports. Using antibodies to rat tubulin, Joniau et al.[36] reported full cross-reactivity among tubulin from rat, rabbit, or pig. Morgan et al.[37] performed a similar experiment using antibodies to chicken tubulin, but these authors reported only very weak cross-reactivity between chicken and mouse or lamb brain tubulin. If they used antibodies to mouse tubulin instead of antibodies to chicken tubulin, they found complete cross-reactivity between chicken and mouse brain tubulin. Hiller and Weber[37a] confirm the results of Morgan et al.[37] and demonstrate that there is only weak cross-reactivity between mouse or pig brain tubulin where antibody to chicken tubulin is used as the probe. These results suggest that tubulin from different species contain different mass densities of shared antigenic determinants, whereas chicken contains highly antigenic determinants unique to chicken. The present study corroborates these immunologic findings and shows that the tubulin distribution of basic β subspecies is unique for chicken. If one considers the large degree of conservation among tubulin from all other vertebrate brains studied, then one can expect that the polyclonal antibodies generated against rat tubulin will be directed towards these common determinants. Consequently, a high degree of cross-reactivity among different tubulins will be observed using antibodies to rat tubulin.[36] On the other hand, the polyclonal antibodies to chicken tubulin may be made to unique and highly antigenic determinants in addition to the less antigenic, common ones. In conclusion, if one uses antibodies to chicken tubulin, then it is reasonable to expect only weak cross-reactivity between chicken and mouse tubulin, as reported by Morgan *et al.*[37] and good cross-reactivity between tubulin from rat, pig, and chicken in using antimouse antibody. The explicit assumption is that the unique isoelectric focusing pattern in chicken reflects the presence of unique highly antigenic determinants.

HETEROGENEITY IN NONNEURONAL TISSUES

An understanding of the significance of tubulin heterogeneity ultimately requires the determination of functional differences among each of the subspecies. Having successfully applied isoelectric focusing procedures using immobilized pH gradients (IPG-IEF) in resolving brain tubulin subspecies,[38,39] this technique was employed to provide high resolution of individual subspecies in tubulin populations from nonneuronal sources. The goal is to compare the identity of tubulin subspecies in these tissues with that of the neuronal ones.

FIGURE 7 shows a one-dimensional IPG-IEF gel of tubulin purified from nonneuronal tissues and whole brain. It is evident that the nonneuronal tissues have a simpler distribution of major protein bands than does brain. In order to identify these bands as tubulin, a strip from this gel was overlayed with V8 protease and then electrophoresed in a second dimension to produce peptide maps of each band. In all of the peptide maps, all the subspecies are identifiable. There were no "lanes" of unique peptide patterns that would have indicated contamination by nontubulin proteins, thus supporting the conclusion that these proteins are tubulins.

In order to identify individual bands as α or β tubulin, strips from the one-dimensional IPG-IEF gel were cut from each lane and electrophoresed in a second dimension in the presence of sodium dodecyl sulfate (SDS) and urea, as shown in FIGURE 8.

In agreement with previous reports, FIGURES 7 and 8 show that the nonneuronal tissues have a simpler distribution of major tubulin subspecies than does brain.[7,30,40] The tubulin populations of all of the nonneuronal tissues are very similar, consisting predominantly of a few of the most basic α subspecies and a group of β subspecies with isoelectric points near the middle of the range of β found in brain. This similarity was even apparent among tissues from different species: calf kidney tubulin has a subspecies distribution that is essentially indistinguishable from that of lamb liver (FIGURE 8, C, E). Based on isoelectric points, the same major tubulins appear to be present in each of the nonneuronal tissues (FIGURE 7). Several other subspecies are also present in much smaller amounts (detectable by silver staining, FIGURE 8), but many of the more acidic subspecies seen in brain are not detectable in the nonneuronal tissues. No identified subspecies were unique to any of the nonneuronal tissues. A previous report has suggested that rat liver contains a basic α tubulin not found in brain.[30] The present data show that lamb brain contains subspecies with isoelectric points corresponding to even the most basic α tubulins found in lamb liver FIGURE 7. It has recently been shown that these very basic subspecies are only minor components of the tubulin populations of several vertebrate brains,[35] and they may have been overlooked previously.

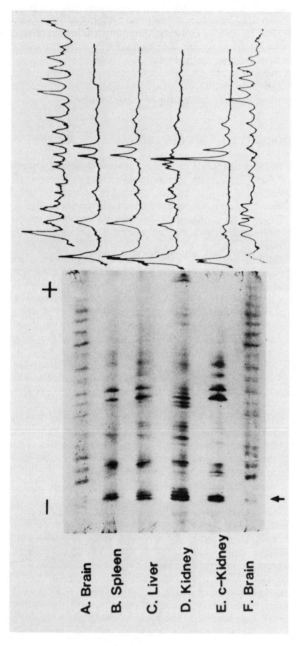

FIGURE 7. One-dimensional IPG-IEF gels (pH 5.8–6.4) and densitometer scans of tubulin isolated from five tissues. Approximately 400 μg was loaded per lane. Gels were stained with Coomassie brilliant blue G-250. The arrow points to the most basic α tubulin. Scanning was performed on an LKB 2202 Ultroscan laser densitometer equipped with an LKB 2220 recorder integrator. Scan length was limited to 105 mm; therefore, scans are aligned to show comigrating tubulins (B–E). Scans for brain tubulin are skewed to show the large amount of acidic β tubulin; the first α tubulin in brain (see arrow) is not shown on these scans. Lanes A and F are identical samples of brain tubulin; they are both included to demonstrate the reproducibility of the technique. Lanes B–E are as labeled on the figure. Tubulin was purified by published procedures.[39] Calf kidney medulla tubulin was kindly donated by L. Barnes of The University of Texas, San Antonio, Texas.

The large differences in subspecies distribution between brain and nonneuronal tissues, as well as previous Ampholine-IEF of tubulin from neonatal brain suggests that the extreme heterogeneity found in adult brain might only be present in fully developed and differentiated neuronal cells.[7,12,41-44] To investigate this possibility, tubulin purified from neonatal brain and undifferentiated neuroblastoma cells was examined. FIGURES 9A and 9B show two-dimensional gels of purified brain tubulin

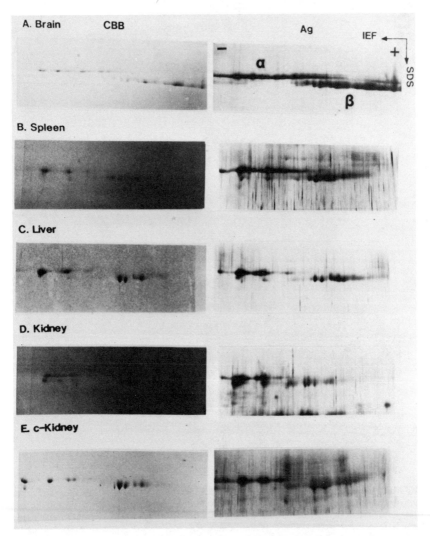

FIGURE 8. Two-dimensional IPG-IEF/NaDodSO₄ electrophoresis analysis of tissue tubulins. Tubulin sources are as labeled in the figures. Gels are aligned as in FIGURE 7; only the tubulin region is shown in each gel. Approximately 50–100 μg was loaded per gel.[38] Gels were sequentially stained with Coomassie blue G-250 (CBB) and silver (Ag).

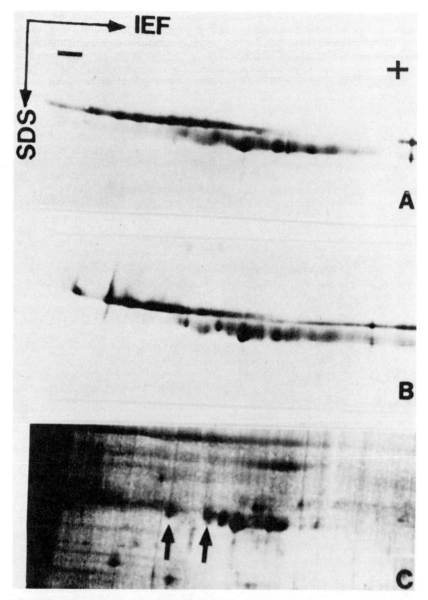

FIGURE 9. Two-dimensional IPG-IEF/NaDodSO$_4$ electrophoresis analysis of tubulin from neonate brains (**A** and **B**) and mouse N-18 neuroblastoma cells. **A:** purified brain tubulin from 10-day-old human female; **B:** purified brain tubulin from a 3-day-old rat; **C:** total soluble protein from N-18 neuroblastoma cells; only the tubulin region is shown. Approximately 100 μg was loaded per gel. **A** and **B** stained with Coomassie brilliant blue G-250; **C** stained with silver. Neonate human brain and N-18 neuroblastoma cells were kindly donated by W. Frey and T. Rustan of St. Paul-Ramsey Medical Center, St. Paul, Minnesota, and A. Howlett, St. Louis University, St. Louis, Missouri, respectively.

from neonate human (10 day) and rat (3 day). The IPG-IEF data show improved resolution of individual subspecies and are generally consistent with previous reports[7,12,42-44] using lower resolution one- and two-dimensional gels, which showed that neonate brain contains fewer β tubulins than does adult brain. As in the nonneuronal tissues (see above), the β populations of neonate brain are deficient in the more acidic subspecies.

FIGURE 9C shows the tubulin region of a two-dimensional gel of soluble proteins from undifferentiated mouse N-18 neuroblastoma cells. The tubulin distribution in this neoplastic neuronal cell line is much simpler than in adult or even neonate brain and only slightly more complex than in any of the nonneuronal tissues (*cf.* FIGURE 8). As in the nonneuronal tissues, the more acidic β tubulins are absent. A striking difference between neuroblastoma cell tubulin and nonneuronal cell tubulins is the presence of two prominent basic β tubulins (indicated by arrows in FIGURE 9). These are absent or only present in very small amounts in the nonneuronal tissues, but present in neonate and adult brain. This suggests that they may be specific for neuronal cells.

In summary, the present study shows that tubulins isolated from spleen, liver, and kidney all have a very similar subspecies composition. In all cases, the tubulin population is much less complex than in brain and consists of almost the same small subset of proteins, which is markedly deficient in the more acidic subspecies found in brain. Although the cells in these tissues are functionally quite different, they do perform a common set of functions, including mitosis, secretion, and formation of a cytoskeleton. Microtubules are involved in these functions. Thus, this simple subspecies composition may not be strictly tissue specific, but rather may represent the minimal tubulin population required for these basic functions. This is in contrast to cells that contain microtubules involved in specialized functions that may require morphologically distinct arrays of microtubules, such as formation of the marginal band in erythrocytes,[45,46] synaptic junctions, and axoplasmic transport in neuronal cells,[47] and sperm flagella in testes.[48,49] The present finding is in agreement with preliminary evidence using lower resolution isoelectric focusing systems demonstrating that neuronal tissues and cells,[7,41] erythrocyte,[45] and sperm cells[50] contain tubulin composed of discrete subpopulations of tubulin subspecies. In a few instances, these changes in subspecies population have been shown to correlate with differences in relative amounts of expression of different tubulin genes.[42,48,49,51,52]

BIOCHEMICAL PROPERTIES OF TUBULIN SUBSPECIES

As a consequence of structural differences, tubulin subspecies may exhibit different solution properties. Hence, antitubulin drugs were employed as perturbants to probe the system. One of the drugs is nocodazole (methyl[5-(2-thienylcarbonyl)-1H-benzimidazol-2-yl]) carbamate. The interaction between nocodazole and calf brain tubulin in 10^{-2} M sodium phosphate, 10^{-4} M guanosine triphosphate (GTP), and 12% (v/v) dimethylsulfoxide at pH 7.0 was studied. The number of binding sites for nocodazole was shown to be one per tubulin monomer of 50,000 as a result of equilibrium binding studies by gel filtration and spectroscopic techniques. At 25°C, the apparent equilibrium constant is $(4 \pm 1) \times 10^{-5}$ M^{-1}, and the equilibrium constant is not significantly affected by temperature.[53]

The mechanism of nocodazole-tubulin interaction was studied by stopped-flow spectroscopy. The time dependence absorbance change of mixing 2×10^{-6} M tubulin and 1.8×10^5 M nocodazole at 25°C is shown in FIGURE 10A. The data were analyzed by pseudo-first order reaction scheme, and clearly the data show curvature and can

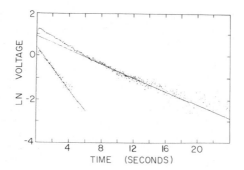

FIGURE 10A. Change in voltage upon formation of nocodazole-tubulin complex in 10^{-2} M sodium phosphate, 10^{-4} M GTP, and 12% (v/v) dimethylsulfoxide at pH 7.0 and 25°C. The concentrations of tubulin and nocodazole were 2×10^{-6} M and 1.8×10^{-5} M, respectively. The dots represent experimental data points, and the solid curve represents the fitting, using the parameters obtained in **B.**

only be analyzed with a consideration of two phases, as shown in FIGURE 10B. Inclusion of two phases yields a good fit of the data, as shown in FIGURE 10A. The effect of nocodazole concentration on the apparent rate constants of the fast and slow phases at various temperatures was monitored. There is an increase in the apparent rate constants with increasing nocodazole concentration in both the slow and fast phases. The simplest interpretation of the kinetic data that is consistent with the thermodynamic and structural data is that the fast and slow phases are treated as independent, parellel reactions, each of which involves a simple two-step mechanism:

$$T + N \underset{k_2}{\overset{k_1}{\rightleftharpoons}} TN \underset{k_4}{\overset{k_3}{\rightleftharpoons}} \overset{*}{T}N$$

k_1, k_2 are the forward and reverse rate constants of the binding reaction; k_3 and k_4 are the forward and reverse rate constants of the ligand-induced conformational change in

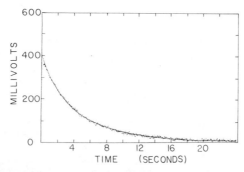

FIGURE 10B. Logarithmic plot of change in voltage upon nocodazole binding to tubulin. The linear portion of the upper curve was analyzed with LN $(T - T_\infty)$ = LN B $- \beta t$, and the curved part by LN $(T - T\infty - Be^{-\beta t})$ = LN A $- \alpha t$, where $\alpha = 0.45$ sec^{-1} and $\beta = 0.15$ sec^{-1}.

tubulin. This conclusion implies that there are at least two populations of tubulin, each of which reacts with nocodazole in a simple two-step mechanism of biomolecular binding followed by a unimolecular isomerization step.[53] Because the ratio of amplitudes for these two phases does not change with varying ligand concentration or temperature, it implies that the reactants in the fast and slow phases are not linked by rapid equilibrium that is highly sensitive to nocodazole concentration, temperature, or Mg^{++} ion. It most likely is a reflection of heterogeneity in the solution properties of tubulin subspecies. Such an interpretation is most tentative, based on the limited nature of results; it is, however, consistent with the kinetic studies of colchicine-derivative tubulin interaction.[54]

In conclusion, the present knowledge on tubulin heterogenetiy, as revealed by protein chemical studies, indicates that most, if not all, of β-tubulin subspecies from brain tissues are primary gene products. The heterogeneity in brain β tubulin not only involves charge differences, but also involves size; there are detectable species differences in the distribution of basic β subunit. These subspecies do not only have unique chemical composition, but preliminary data show that they quantitatively exhibit different solution properties.

REFERENCES

1. ROBERTS, K. & J. S. HYAMS, Eds. 1979. Microtubules. Academic Press. London.
2. KRAUHS, E., M. LITTLE, T. KEMPF, R. HOFER-WARBINEK, W. ADE & H. PONSTINGL. 1981. Proc. Natl. Acad. Sci. USA 78: 4156–4160.
3. PONSTINGL, H., E. KRAUHS, M. LITTLE & T. KEMPF. 1981. Proc. Natl. Acad. Sci. USA 78: 2757–2761.
4. HALL, J. L., L. DUDLEY, P. R. DOBNER, S. A. LEWIS & N. J. COWAN. 1983. Mol. Cell. Biol. 3: 854–862.
5. VALENZUELA, P., M. QUIROGA, J. ZALDIVAR, W. J. RUTTER, M. W. KIRSCHNER & D. W. CLEVELAND. 1981. Nature (London) 289: 650–655.
6. SULLIVAN, K. F. & D. W. CLEVELAND. 1984. J. Cell. Biol. 99: 1754–1760.
7. GOZES, I. & U. Z. LITTAUER. 1978. Nature (London) 276: 411–413.
8. GOZES, I. & U. Z. LITTAUER. 1979. Brain Res. 171: 171–175.
9. MAROTTA, C. A., J. L. HARRIS & J. M. GILBERT. 1978. J. Neurochem. 30: 1431–1440.
10. FEIT, H., V. NEUDECK & F. BASKIN. 1977. J. Neurochem. 28: 697–706.
11. FORGUE, S. T. & J. L. DAHL. 1979. J. Neurochem. 32: 1015–1025.
12. DAHL, J. L. & V. J. WEIBEL. 1979. Biochem. Biophysics Res. Commun. 86: 822–828.
13. NELLES, L. P. & J. R. BAMBURG. 1979. J. Neurochem. 32: 477–481.
14. HARE, D. J., D. I. STIMPSON & J. R. CANN. 1978. Arch. Biochem. Biophys. 187: 274–275.
15. CANN, J. R. & D. I. STIMPSON. 1977. Biophys. Chem. 7: 103–114.
16. STIMPSON, D. I. & J. R. CANN. 1977. Biophys. Chem. 7: 115–119.
17. CANN, J. R., D. I. STIMPSON & D. J. COX. 1978. Anal. Biochem. 86: 34–49.
18. CANN, J. R. & K. J. GARDINER. 1979. Biophys. Chem. 10: 203–210.
19. FRATER, R. 1970. J. Chromatogr. 50: 469–474.
20. KAPLAN, L. J. & J. F. FOSTER. 1971. Biochemistry 10: 630–636.
21. WALLEVIK, K. 1973. J. Biol. Chem. 248: 2650–2655.
22. FELGENHAUER, K., D. GRAESSLIN & B. D. HUISMANS. 1971. Protides Biol. Fluids Proc. Colloq. 19: 575–578.
23. DRYSDALE, J. W. & P. G. RIGHETTI. 1972. Biochemistry 11: 4044–4052.
24. GALANTE, E., T. CARVAGGIO & P. G. RIGHETTI. 1976. Biochem. Biophys. Acta 442: 309–315.
25. WILLIAMS, R. C., JR. & J. C. LEE. 1982. Methods Enzymol. 85: 376–385.
26. WEINGARTEN, M. D., A. H. LOCKWOOD, S. Y. HWO & M. W. KIRSCHNER. 1975. Proc. Natl. Acad. Sci. USA 72: 1858–1862.

27. DETRICH III, H. W., S. A. BERKOWITZ, H. KIM & R. C. WILLIAMS JR. 1976. Biochim.
     Biophysics Res. Commun. **68:** 961–968.
28. BELLO, J. & H. R. BELLO. 1976. Arch. Biochem. Biophys. **172:** 608–610.
29. SACKETT, D. L., B. BHATTACHARYYA & J. WOLFF. 1985. J. Biol. Chem. **260:** 43–45.
30. DENOULET, P., C. JEANET & F. GROS. 1982. Biochem. Biophysics Res. Commun.
     105: 806–813.
31. WILDE, C. D., C. E. CROWTHER, T. P. CRIPE, M. G.-S. LEE & N. J. COWAN. 1982. Nature
     (London) **297:** 83–84.
32. LEE, M.G.-S., S. A. LEWIS, C. D. WILDE & N. J. COWAN. 1983. Cell **33:** 477–487.
33. LOPATA, M. A., J. C. HAVERCROFT, L. T. CHOW & D. W. CLEVELAND. 1983. Cell
     32: 713–724.
34. LITTLE, M. 1979. FEBS Lett. **108:** 283–286.
35. FIELD, D. J., R. A. COLLINS & J. C. LEE. 1984. Proc. Natl. Acad. Sci. USA **81:** 4041–
     4045.
36. JONIAU, M., M. DEBRABANDER, J. DEMEY & J. HOEBEKE. 1977. FEBS Lett. **78:** 307–
     312.
37. MORGAN, J. L., C. R. HOLLADAY & B. S. SPONNER. 1978. Proc. Natl. Acad. Sci. USA
     75: 1414–1417.
37a. HILLER, G. & K. WEBER. 1978. Cell **14:** 795–804.
38. FIELD, D. J. & J. C. LEE. 1985. Anal. Biochem. **144:** 584–592.
39. FIELD, D. J. 1984. Doctoral Dissertation, St. Louis University.
40. DIEZ, J. C., M. LITTLE & J. AVILA. 1984. Biochem. J. **219:** 277–285.
41. GOZES, I., D. SAYA & U. Z. LITTAUER. 1979. Brain Res. **171:** 171–175.
42. GOZES, I., A. DE BAETSELIER & U. Z. LITTAUER. 1980. Eur. J. Biochem. **103:** 13–20.
43. MOURA NETO, V., M. MALLAT, C. JEANET & A. PROCHIANTA. 1983. EMBO J. **2:** 1243–
     1248.
44. SULLIVAN, K. F. & J. WILSON. 1984. J. Neurochem. **42:** 1363–1371.
45. MURPHY, D. B. & K. T. WALLIS. 1983. J. Biol. Chem. **258:** 7870–7875.
46. MURPHY, D. B. & K. T. WALLIAS. 1983. J. Biol. Chem. **258:** 8357–8364.
47. GOZES, I. & U. Z. LITTAUER. 1982. Scand. J. Immunol. **15:** 299–316.
48. DISTEL, R. J., K. C. KLEENE & N. B. HECHT. 1984. Science **224:** 68–70.
49. KEMPHUES, K. J., E. C. RAFF, R. A. RAFF & T. C. KAUFMAN. 1980. Cell **21:** 445–451.
50. LITTLE, M., C. ROHRICHT & D. SCHROETER. 1983. Exp. Cell. Res. **147:** 15–22.
51. BOND, J. F., & S. R. FARMER. 1983. Mol. Cell. Biol. **3:** 1333–1342.
52. BOND, J. C., G. S. ROBINSON & S. R. FARMER. 1984. Mol. Cell. Biol. **4:** 1313–1319.
53. HEAD, J., L. L.-Y. LEE, D. J. FIELD & J. C. LEE. 1985. J. Biol. Chem., **260:** 11060–11061.
54. ENGELBORGHS, Y. & T. FITZGERALD. 1986. Ann. N.Y. Acad. Sci. **466:** 709–717. This
     volume.

Changes in Tubulin mRNAs during Differentiation of a Parasitic Protozoan *Leishmania mexicana*[a]

DUNNE FONG[b] AND KWANG-POO CHANG[c]

[b]*Departments of Biology and Microbiology*
University of Alabama at Birmingham
Birmingham, Alabama 35294

[c]*Department of Microbiology and Immunology*
University of Health Sciences
Chicago Medical School
North Chicago, Illinois 60064

Leishmania mexicana is the causative agent for human cutaneous leishmaniasis. This trypanosomatid protozoan has two developmental stages: an extracellular, motile promastigote form in the alimentary canal of the sandfly vector and an intracellular, nonmotile amastigote form in the phagolysosomes of macrophages of the mammalian host. Because of the presence of a variety of microtubules, namely the flagellar, the subpellicular and the mitotic types, tubulin is the most abundant protein in this protozoan, and we are studying tubulin genes and proteins both as a model for eukaryotic gene regulation and as a potential target for chemotherapeutic attack against leishmaniasis.

A change in tubulin biosynthesis during cell differentiation of *L. mexicana* was initially reported by Fong and Chang.[1] Promastigotes were grown in culture medium, and amastigotes were maintained intracellularly in a mouse macrophage cell line. [^{35}S]methionine incorporation into protein was analyzed during leishmanial differentiation. Tubulin biosynthesis was found to increase in amastigote-to-promastigote transformation and to decrease in the reverse differentiation. These biosynthetic changes corresponded well with the morphological changes of the parasites. Thus tubulin biosynthesis is developmentally regulated in leishmanias.

Tubulin biosynthesis was also characterized by *in vitro* translation using a rabbit reticulocyte lysate cell-free system (Wallach, Fong, and Chang[2]). Unexpectedly, poly A$^+$ RNA from both stages were found to translate tubulin protein equally well *in vitro*. The result implied the presence of similar amounts of tubulin mRNA between the two leishmanial stages. Because we have previously shown that promastigotes synthesized more tubulin protein that amastigotes *in vivo*, a posttranscriptional control for tubulin biosynthesis during leishmanial differentiation is indicated.

Leishmanial tubulin mRNA was further characterized using tubulin-specific cDNA probes derived from *Chlamydomonas* and chick brain (Fong *et al.*[3]). The results of dot blot hybridization showed similar amounts of α- and β-tubulin mRNA between amastigotes and promastigotes. Leishmanial tubulin mRNA was also analyzed by northern blot hybridization. With the chick brain α-tubulin probe, a mRNA species of 2100 nucleotides in size was detected in amastigote and promastigote total

[a]This work was supported in part by a Graduate School Faculty Research Grant to D. Fong and National Institutes of Health Grant AI20486 to K. P. Chang.

129

FIGURE 1. Northern blot hybridization of leishmanial RNA. Equal amounts (6.5 μg) of total RNA from amastigotes (A) and promastigotes (P) were run on formaldehyde gel, blotted to nitrocellulose filter, and hybridized with a chicken brain β-tubulin cDNA probe.

RNA in roughly equal amounts. With the β-tubulin probe, however, promastigotes were shown to have three mRNA species, at 2800, 3600, and 4400 nucleotides in size, with the smallest being the most predominant; amastigotes had only one species at the location of 3600 nucleotides (FIGURE 1). Because the major β-tubulin mRNA species are of different sizes in amastigotes and promastigotes, the regulation of β-tubulin gene expression appears to be different between the two developmental stages of this leishmanial species.

For *Leishmania enriettii*, which only infects guinea pigs, Landfear and Wirth[4] reported that promastigotes contain four- to fivefold more tubulin mRNA than amastigotes, and that a single species of β-tubulin mRNA is present in both stages, but differs in the amount between them. Thus their results point to transcriptional control, whereas ours indicate posttranscriptional control. Additional study is needed to determine whether the discrepancy is due to species variation.

REFERENCES

1. FONG, D. & K. P. CHANG. 1981. Proc. Natl. Acad. Sci. USA **78:** 7624–7628.
2. WALLACH, M., D. FONG & K. P. CHANG. 1982. Nature (London) **299:** 650–652.
3. FONG, D., M. WALLACH, J. KEITHLY, P. W. MELERA & K. P. CHANG. 1984. Proc. Natl. Acad. Sci. USA **81:** 5782–5786.
4. LANDFEAR, S. M. & D. F. WIRTH. 1984. Nature (London) **309:** 716–717.

Tubulin Heterogeneity in the Ciliate, *Tetrahymena thermophila*

K. A. SUPRENANT, E. HAYS, E. LeCLUYSE,
AND W. L. DENTLER.

Department of Physiology and Cell Biology
University of Kansas
Lawrence, Kansas 66045

Ciliary and cytoplasmic tubulins were isolated from *Tetrahymena thermophila* [strain SB 711[1]] and analyzed by 2-D sodium dodecyl sulfate (SDS)-polyacrylamide gel electrophoresis (PAGE) and by 1-D peptide mapping in order to begin to determine whether the assembly of a specific microtubule-containing organelle is regulated by the availability of a specific tubulin. Separation of *Tetrahymena* tubulins by 1-D SDS-PAGE[2] was a function of the pH of the separating gel with maximum resolution of α and β tubulins occurring at pH 8.25–8.50. The α subunit was identified as the fastest migrating tubulin under these conditions. In 1-D separating gels with a pH greater than 8.7, the α and β tubulins merged into a single band that comigrated with chick and bovine brain β tubulin.

High resolution isoelectric focusing (IEF) and 2-D SDS-PAGE[3] indicated that the ciliary axoneme was composed of a major α (α_3) and two minor α tubulins (α_1, α_2), and a major β (β_2) and a minor β (β_1) tubulin. The minor axonemal β_1 was found to be the major β tubulin in solubilized central pair microtubules, obtained by dialysis against Tris-EDTA.[4] By contrast, β_2 was the major β tubulin in outer doublet microtubules. *Tetrahymena* cytoplasmic tubulin purified and assembled *in vitro*[5] and was composed of two unique tubulin subunits, α_4 and α_5 that did not comigrate with any of the axonemal tubulins and a β tubulin that comigrated with axonemal β_2. The major cytoplasmic α, α_5, was the fastest migrating on SDS-PAGE and had the most basic pI on IEF gels of all the tubulin subunits.

Axonemal and cytoplasmic α and β tubulins, resolved by 1-D SDS-PAGE, were cut out of these gels, and compared with one another after 1-D limited proteolytic digestion with *Staphylococcus aureus* V8 protease.[6,7] The resulting peptides were analyzed by electrophoresis in pH 8.25 and 8.80 separating gels. No differences were observed in the peptide maps of axonemal and cytoplasmic β tubulin at either pH 8.25 or 8.80. By contrast, the pattern of peptides resolved from cytoplasmic α tubulin at pH 8.25 was different than the axonemal α cleavage products. The peptide maps of axonemal and cytoplasmic α tubulins were identical in pH 8.80 gels.

In summary, we have identified seven organelle-specific tubulins by IEF, 1-D and 2-D SDS-PAGE, and by 1-D peptide mapping (see TABLES 1 and 2). It is likely that

TABLE 1. Isoelectric Points of Tubulin Subunits

Subunit	pI
$\beta_{1,2}$	5.1–5.2
$\alpha_{1,2,3}$	5.3–5.4
α_4	5.5
α_5	5.55

TABLE 2. Distribution of Tubulin Isotypes

	β_1	β_2	α_1	α_2	α_3	α_4	α_5
Cytoplasm		x				x	
Axoneme	x	x	x	x	x		
Central pair	x		x				
Outer doublet		x		x	x		

some of the tubulin isotypes have arisen by posttranslational modification, because only one α-tubulin gene[8] has been identified at this time.

1. ORIAS, E., M. FLACKS & B. H. SATIR. 1983. J. Cell Sci. **64:** 49.
2. LAEMMLI, U. K. 1970. Nature (London) **227:** 680.
3. O'FARRELL, P. 1975. J. Biol. Chem. **250:** 4007.
4. GIBBONS, I. R. 1965. Arch. Biol. Liege **76:** 317.
5. MAEKAWA, S. & H. SAKAI. 1978. J. Biochem. (Tokyo) **83:** 1065.
6. BORDIER, C. & A. CRETTOL-JARVINEN. 1979. J. Biol. Chem. **254:** 649.
7. CLEVELAND, D. W., S. G. FISCHER, M. W. KIRSCHNER & U. K. LAEMMLI. 1977. J. Biol. Chem. **252:** 1102.
8. CALLAHAN, P. C., G. SHALKE & M. A. GOROVSKY. 1984. Cell **36:** 441.

Differential Structure and Distribution of the High Molecular Weight Brain Microtubule-Associated Proteins, MAP-1 and MAP-2

RICHARD B. VALLEE, GEORGE S. BLOOM,[a] AND
FRANCIS C. LUCA

Cell Biology Group
Worcester Foundation for Experimental Biology
Shrewsbury, Massachusetts 01545

When the first meeting of this kind was held ten years ago, some of the earliest evidence was presented indicating that cytoplasmic microtubules are biochemically complex. It is now well established that these structures consist not only of tubulin, but, in addition, a variety of microtubule-associated proteins, or MAPs. The MAPs were originally identified as a pair of extremely high molecular weight proteins,[1] now referred to as MAP-1 and MAP-2[2] ($M_r \sim 300,000$). A considerable number of additional MAPs have since been identified, as discussed elsewhere in this volume.

Of all the MAPs, MAP-1 and MAP-2 have received the most attention. They are the most abundant and most readily isolated. In addition, they have proven to be structurally interesting. They appear as fine, fibrous projections on the surface of microtubules,[3-5] suggesting a role in mediating the interaction of microtubules with other cellular organelles. Electron microscopic images of similar fibers cross-linking microtubules with other organelles in cells have been obtained repeatedly (for example, see references 6 and 7), and a considerable body of physiological and cytological evidence has accumulated indicating that these interactions occur in the living cell. It seems likely that MAP-1 and MAP-2 play a significant role in these interactions.

These and the other MAPs are known to have an additional property, that of stimulating microtubule assembly. Some evidence exists that MAPs increase in abundance during neurite outgrowth,[8-10] consistent with a role in the formation or stabilization of new microtubules in the cell.

COMPARISON OF PROPERTIES OF MAP-1 AND MAP-2

It has been possible to separate the high molecular weight MAPs into MAP-1 and MAP-2 fractions,[5,11-13] corresponding to the electrophoretic species discussed above. This has allowed us to compare the properties of these proteins, though, as we discuss below, we now know these fractions to be biochemically complex. Superficially, the MAP-1 and MAP-2 preparations seem to be similar in their properties. They are composed of proteins of similar large size. When combined with tubulin, both

[a]Present address: Department of Cell Biology, University of Texas Health Science Center at Dallas, 5323 Harry Hines Blvd., Dallas, Texas 75235.

preparations promote microtubule assembly and show fine, regularly spaced projections on the microtubule surface. Both preparations contain proteins that are, for the most part, extremely sensitive to protease digestion.[14,15]

Here, however, the similarities end. MAP-1 and MAP-2 both contain lower molecular weight polypeptide components, but the nature of these polypeptides and the stoichiometry of the complex is different in the two cases. MAP-2 preparations contain a substoichiometric level of an associated type II cyclic AMP dependent protein kinase,[13,16] whereas preparations of MAP-1 contain a pair of low molecular weight polypeptides—the "MAP-1 light chains"—each of approximately 1:1 stoichiometry with the high molecular weight protein.[5] MAP-2 can be readily phosphorylated *in vitro* by its associated kinase,[2,13,16] and this represents another major difference from MAP-1. Two reports have indicated that some component of MAP-1 is phosphorylated *in vivo*,[2,9] but phosphorylation *in vitro* has not been achieved.

MAP-1 and MAP-2 have quite distinct cellular distributions (see below). MAP-2 is highly enriched in neuronal dendrites.[17-21] While the protein has also been detected in a number of other tissues and cells,[22,23] the levels are considerably lower than in the dendrite. In contrast to this situation, the component polypeptides of MAP-1 have a more general distribution both within neuronal cells and in other cell types[24-27] (see below).

MAP-1 and MAP-2 have distinct solution properties. MAP-1 has a tendency to aggregate. MAP-2 also shows this property, but to a much lower extent. This has allowed us to purify MAP-2 chromatographically from mixtures of the MAPs.[13] MAP-1 could not be isolated in pure form from standard MAP preparations. Instead, we found it necessary to use as starting material MAPs obtained from white matter, which contains relatively low amounts of MAP-2.[5,17]

One of the most striking differences between MAP-1 and MAP-2 is their stability to elevated temperatures.[11,12] When purified microtubules are exposed to elevated temperature, MAP-2 remains soluble, whereas tubulin and MAP-1 precipitate. We believe that this behavior is an extreme version of the tendency of MAP-1 to aggregate, observed under less severe conditions. Elevated temperature apparently serves to promote the aggregation process.

To determine whether MAP-1 was inherently less stable to denaturation than MAP-2, we exposed unfractionated MAP to 8 M urea and 6 M guanidine HCl.[5] Neither treatment affected the ability of MAP-1 or MAP-2 to bind to microtubules. These experiments suggested that the two proteins were not nearly as distinct as was indicated by their relative stability to elevated temperature.

MAP-1 IS MORE THAN ONE PROTEIN

Several reports have noted that the high molecular weight MAPs consist of multiple polypeptides. Purified MAP-2 was found to contain two polypeptides with similar biochemical properties.[12] The relationship between the various additional MAP species has not been clear. Do some represent proteolytic fragments of the naturally occurring proteins? Are the various polypeptides related at all in primary structure?

A number of lines of investigation in our laboratory have led us to the conclusion that the high molecular weight MAPs are more complex than originally suspected. We have noted that, in preparations of microtubules obtained from bovine white matter (FIGURE 1, and see reference 24), the multiplicity of the high molecular weight MAPs is much more apparent than in traditional microtubule preparations, for which whole brain or cerebral cortex has been used as starting material. In the white matter

FIGURE 1. Four percent sodium dodecyl sulfate (SDS)-urea electrophoretic gel of calf brain white matter microtubules.[5,17,24] Five high molecular weight MAP polypeptides may be seen. Arrows denote top and dye front of gel. Tubulin ran in the dye front. Some gray matter was included in the white matter preparation to increase MAP-2 to a readily detectable level (Bloom et al.[26]).

TABLE 1. Antibodies to High Molecular Weight Brain MAPs

Antibody Name[a]	Reacts with	Properties	References
Anti-MAP-2	MAP-2A and MAP-2B	Rabbit polyclonal	16,18,19,29
MAP-2-1	MAP-2A and MAP-2B	Mouse monoclonal IgG_{2b}	36
MAP-1A-1	MAP-1A	Mouse monoclonal IgG_1	24,25
MAP-1B-1	MAP-1B[b]	Mouse monoclonal IgG_1	27,28
MAP-1B-2	MAP-1B	Mouse monoclonal IgG_1	27,28
MAP-1B-3	MAP-1B[c]	Mouse monoclonal IgM	28
MAP-1B-4	MAP-1B	Mouse monoclonal IgG_1	27,28
MAP-1B-5	MAP-1B	Mouse monoclonal IgG_1	27,28

[a] Hybridoma clones and antibodies were named for the immunogenic protein species, followed by a number designating the order of isolation. For example, MAP-1B-1 was the first anti-MAP-1B antibody isolated.

[b] Also reacts with centosomes (see text and FIGURE 4).

[c] Reacts with phosphorylated form of MAP-1B and shows cross-reactivity with MAP-1A (see text and FIGURE 5).

preparations we can reproducibly identify five high molecular weight MAP species. Two species, which we refer to as MAP-1B and MAP-1C, are specifically enriched in these preparations, suggesting that these proteins may be more concentrated in axons or in glial cells than the other MAPs.

Peptide maps of the MAP-1 polypeptides were very different, indicating that even these proteins may not be closely related.[27,28] Peptide maps of the two MAP-2 polypeptides, MAP-2A and MAP-2B, revealed that these polypeptides were closely related to each other, but not to the MAP-1 species.[27]

We have raised monoclonal antibodies to the two most abundant MAP-1 polypeptides, MAP-1A and MAP-1B (TABLE 1). With one possible and interesting exception (see below) these antibodies do not show evidence of cross-reactivity between MAP-1A and MAP-1B, adding further support to our contention that the two polypeptides are structurally distinct.

PROPERTIES OF THE MAP-1 POLYPEPTIDES

Cellular and Subcellular Distribution

In calf brain tissue MAP-1A[24] and MAP-1B[27] both differ from MAP-2 in that they are clearly detectable in axons and cell bodies of neurons (TABLE 2). Glial cells were also strongly stained by anti-MAP-1A and anti-MAP-1B antibodies. In cultured cell

TABLE 2. Distribution of MAPs[26]

MAP Name	Distribution within Neurons	Distribution in Other Cells
MAP-1A[24,25]	Dendrites, Cell bodies, Axons	Glia, General
MAP-1B[27,28]	Dendrites, Cell bodies, Axons	Glia, General
MAP-1C	?	?
MAP-2A and -2B[17-22]	Dendrites, Cell bodies	Occasional, at low level

preparations, both antigens were found to be widespread.[24,28] This was in contrast to MAP-2, which had a highly restricted distribution in brain and in cultured cells.

In primary cultures of rat brain cells, MAP-1B was found to be prominent in very long, varicose processes (FIGURE 2). These processes did not react with anti-MAP-2.[27] These data are consistent with the identification of the processes as axons.

In earlier work we found MAP-2 immunoreactivity to be codistributed with intermediate filaments in primary cultured brain cells.[29] None of the anti-MAP-1 antibodies show this pattern of immunoreactivity.[26]

Microtubule Binding Characteristics

FIGURE 3 shows the distribution of MAP-1B in cultured TC-7 kidney epithelial cells. The protein shows a fibrous distribution, which we found by double-labeling with antitubulin, to correspond to microtubules. The use of colchicine, vinblastine, and taxol confirmed that MAP-1B was associated with microtubules in the cell.

With the availability of specific antibodies to the individual MAP, it is possible to examine the partitioning of these proteins into various subcellular fractions during purification. We have used immunoblot analysis to determine how efficiently the individual MAPs associate with microtubules during the initial isolation of these structures from cytosolic extracts. With the exception of MAP-1B, all of the MAPs we have examined in brain, as well as in sea urchin eggs,[30,31] cosediment completely with microtubules during the initial stages of microtubule purification. MAP-1B has the unique property that only a fraction of the protein is recovered with the microtubules, whereas a significant fraction remains in the supernate after microtubule sedimentation.[27] That fraction of MAP-1B that does sediment remains strongly associated with the microtubules, however, as evidenced by its cosedimentation in later purification steps. The basis for this behavior is not certain, though some preliminary evidence indicates that it may result from inefficient competition of MAP-1B for the microtubule surface with the other MAPs in the preparation.

Whatever the explanation for these results, they indicate that determination of the abundance of MAP-1B by examination of purified microtubules is subject to considerable error. We believe, in particular, that MAP-1B is likely to be one of the major MAPs—if not the major MAP—in neuronal as well as nonneuronal cells. This feature is, so far, unique to MAP-1B among the known MAPs. It is, of course, worth considering that other proteins may exist that are similarly underrepresented in purified microtubules as currently prepared.

REACTION OF ANTI-MAP-1B WITH THE CENTROSOME

One of our anti-MAP-1B antibodies (MAP-1B-1, TABLE 1) shows a reaction with the centrosomal region of the cell (FIGURE 4). Similar results have been obtained by Sherline and Mascardo using a polyclonal antibody to MAP-1,[32] and by De Mey et al. using a monoclonal antibody to MAP-1.[33] It is not certain whether the different antibodies reacted with the same MAP-1 polypeptide species. Our results indicate, however, that MAP-1B, in particular, may have a common epitope with some component of the centrosome. Whether the centrosome actually contains MAP-1B is difficult to determine and must await further work.

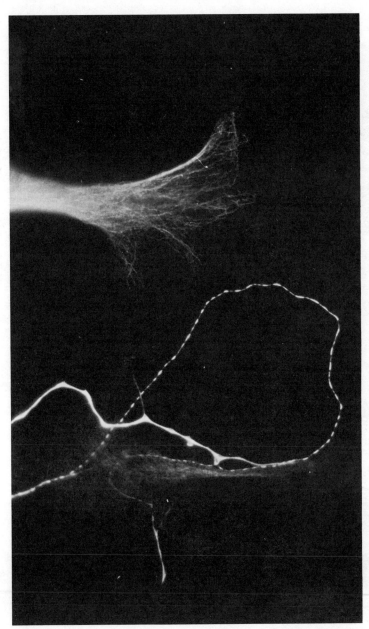

FIGURE 2. Immunofluorescence microscopy of a primary culture of newborn rat brain cells[24] stained with antibody MAP-1B-4 (TABLE 1). Long fine varicose processes such as the looping process at left are among the most striking features stained by the anti-MAP-1B antibodies in these preparations. In addition, thicker processes (left) are stained that are, in general, also MAP-2 and MAP-1A positive, and that may correspond to dendrites. At right, microtubules are seen to be stained by the antibody in a cell of nonneuronal morphology.

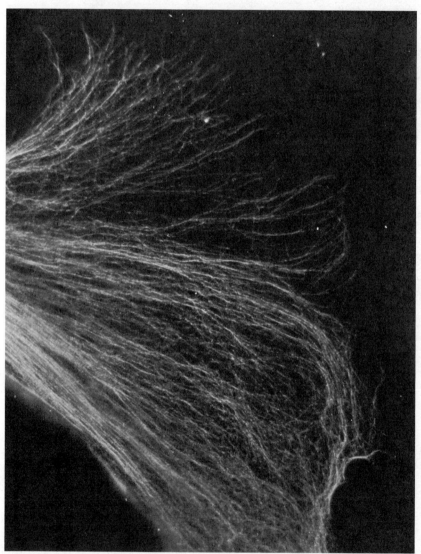

FIGURE 3. Immunofluorescence microscopy of TC-7 kidney epithelial cells stained with antibody MAP-1B-2 (TABLE 1). Microtubules are clearly stained in this, and a wide variety of other nonneuronal cell types, indicating that, as for MAP-1A,[24,25] MAP-1B has a widespread cellular distribution.

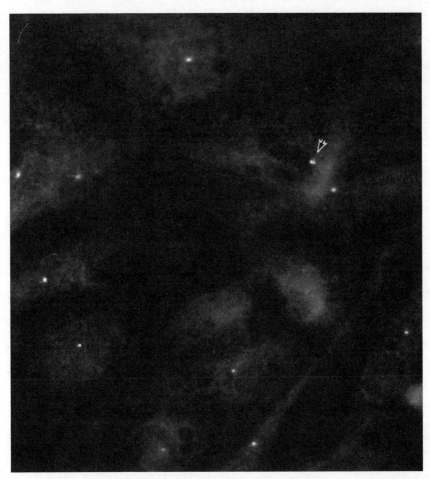

FIGURE 4. Immunofluorescence microscopy of PTK-2 cells stained with antibody MAP-1B-1 (TABLE 1). Staining of bright single or double spots (arrowhead) on the surface of the nucleus may be seen in all cells. In other cell types, such as Chinese hamster ovary (CHO), strong microtubular staining may also be seen. The microtubules in those cells emanate from the brightly stained juxtanuclear spots, identifying them as centrosomes.

AN ANTIBODY THAT REACTS WITH THE PHOSPHORYLATED FORM
OF MAP-1B

During screening of our anti-MAP-1B antibodies, we found that one antibody (MAP-1B-3, TABLE 1) showed a number of properties quite distinct from the other antibodies. This particular antibody showed some reaction with higher molecular weight material in both calf and rat brain microtubules. In calf, discrete reaction with a polypeptide at the position of MAP-1A was observed (FIGURE 5A), indicating that, despite the structural and biochemical differences in the MAP, they may, in fact, have some limited sequence homology. The antibody also showed unique staining characteristics in brain tissues sections.

The unique features of this antibody led us to suspect that it might recognize a modified form of the MAPs. To determine whether this might be the case, we transferred microtubule proteins to nitrocellulose paper and incubated strips of the paper with a protein phosphatase. The strips were subsequently incubated in antibody and processed to visualize the antibody-antigen reaction as usual.

As may be seen in FIGURE 5B, immunoreactivity was abolished by the phosphatase treatment. To be certain that the effect of the phosphatase preparation was specific for the phosphate group, we included 50 mM sodium phosphate in the phosphatase reaction mixture as a competitive inhibitor of the enzyme. Immunoreactivity persisted in this case (FIGURE 5C), confirming that the epitope for MAP-1B-3 included a phosphate group.

These results have several implications. First, they are the only evidence that MAP-1B is a phosphoprotein as isolated, and strongly imply that the protein is phosphorylated *in vivo*. They also indicate that MAP-1A is a phosphoprotein. Because the shared epitope is a phosphorylation site, this suggests that MAP-1A and MAP-1B may be phosphorylated by way of a common mechanism, and, perhaps, a common protein kinase.

WHY MULTIPLE MAPS?

Unlike the tubulins, which are described earlier in this volume, the high molecular weight MAPs appear to represent a structurally and biochemically diverse class of proteins.

Conceivably, by analogy with the variety of intermediate filaments that have been described,[34,35] the observed diversity of the high molecular weight MAPs masks a fundamental structural similarity. Thus, the MAPs may have regions of structural similarity, or even identity, that are relatively nonimmunogenic, and have, therefore, so far eluded detection. Furthermore, their overall morphology, like that of the various intermediate filaments, so far appears to be similar.

A particularly apt analogy could be the relationship between neurofilaments and vimentin filaments. The neurofilaments contain multiple subunits, each of which is distinct from the subunits of vimentin filaments. Yet the basic structure of the two types of filament is very similar. The functional basis for the difference in filament composition is not yet fully understood.

How similar this situation is to that of the MAP-1 and MAP-2 polypeptides remains to be seen. Unlike the intermediate filament case, it is already known that microtubules are involved in a wide variety of cellular functions, and it is certainly tempting to speculate that the MAPs are the basis for this diversity. We have already found that in certain primary brain cells, MAP-2 immunoreactivity is associated with intermediate filaments, whereas MAP-1A and MAP-1B immunoreactivity is not.[26,29] This suggested that MAP-2 has a binding site for intermediate filaments, whereas MAP-1A and MAP-1B do not.

It should be noted that, in our earlier study,[29] MAP-2 was found to colocalize with vimentin-type intermediate filaments. It is conceivable that MAP-1A and MAP-1B

FIGURE 5. Reaction of antibody MAP-1B-3 with phosphorylated form of MAP-1B and MAP-1A. Immunoblot analysis was performed as follows.[24] Calf brain white matter microtubule proteins were separated by sodium dodecyl sulfate (SDS)-urea electrophoresis on a 4% gel as in FIGURE 1. The proteins were transferred to nitrocellulose paper, which was cut into strips. The strips were pretreated to assay for protein-bound phosphate. The treatments were A: none; B: 2.5 hr at 37°C in the presence of 43 μg/ml calf intestinal alkaline phosphatase; and C: same as B, with addition of 50 mM sodium phosphate as competitive inhibitor of phosphatase. Two features of MAP-1B-3 immunoreactivity are noteworthy. First, the antibody reacts with both MAP-1A and MAP-1B. In addition, immunoreactivity is abolished specifically by phosphatase, indicating that the antibody reacts with a phosphorylated epitope on MAP-1A and MAP-1B.

will prove to have a similar interaction with intermediate filaments, but of a different type. Thus, the presence of the MAP-1 polypeptides in axons makes them more suitable candidates for an interaction with neurofilaments, for example.

In this case, the roles of the MAPs would overlap. Clearly, much further work will be needed to determine if this is the case, of if, alternatively, each MAP plays a totally distinct function in the cell.

REFERENCES

1. BORISY, G. G., J. M. MARCUM, J. B. OLMSTED, D. B. MURPHY & K. A. JOHNSON. 1975. Ann. N.Y. Acad. Sci. **253:** 107–132.
2. SLOBODA, R. D., S. A. RUDOLPH, J. L. ROSENBAUM & P. GREENGARD. 1975. Proc. Natl. Acad. Sci. USA **72:** 177–181.
3. MURPHY, D. B. & G. G. BORISY. 1975. Proc. Natl. Acad. Sci. USA **72:** 2696–2700.
4. DENTLER, W. L., S. GRANETT & J. L. ROSENBAUM. 1975. J. Cell Biol. **65:** 237–241.
5. VALLEE, R. B. & S. E. DAVIS. 1983. Proc. Natl. Acad. Sci. USA **80:** 1342–1346.
6. SMITH, D. S., U. JARLFORS & B. F. CAMERON. 1975. Ann. N.Y. Acad. Sci. **253:** 472–506.
7. HIROKAWA, N. 1982. J. Cell Biol. **94:** 129–142.
8. OLMSTED, J. B. & H. D. LYON. 1981. J. Biol. Chem. **256:** 3507–3511.
9. GREENE, L. A., R. K. H. LIEM & M. L. SHELANSKI. 1983. J. Cell Biol. **96:** 76–83.
10. DRUBIN, D., S. KOBAYASHI & M. KIRSCHNER. 1986. Ann. N.Y. Acad. Sci. **466:** 257–268. This volume.
11. HERZOG, W. & K. WEBER. 1978. Eur. J. Biochem. **92:** 1–8.
12. KIM, H., L. BINDER & J. L. ROSENBAUM. 1979. J. Cell Biol. **80:** 266–276
13. VALLEE, R. B., M. J. DiBARTOLOMEIS & W. E. THEURKAUF. 1981. J. Cell Biol. **90:** 568–576.
14. VALLEE, R. B. & G. G. BORISY. 1977. J. Biol. Chem. **252:** 377–382.
15. VALLEE, R. B. 1980. Proc. Natl. Acad. Sci. USA **77:** 3206–3210.
16. THEURKAUF, W. E. & R. B. VALLEE. 1982. J. Biol. Chem. **257:** 3284–3290.
17. VALLEE, R. B. 1982. J. Cell Biol. **92:** 435–442.
18. MILLER, P., U. WALTER, W. E. THEURKAUF, R. B. VALLEE & P. DE CAMILLI. 1982. Proc. Natl. Acad. Sci. USA **79:** 5562–5566.
19. DE CAMILLI, P., P. MILLER, F. NAVONE, W. E. THEURKAUF & R. B. VALLEE. 1984. Neurosci. **11:** 819–846.
20. HUBER, G. & A. MATUS. 1984. J. Neurosci. **4:** 151–160.
21. CACERES, A., L. I. BINDER, M. R. PAYNE, P. BENDER, L. REBHUN & O. STEWARD. 1984. J. Neurosci. **4:** 394–410.
22. VALDIVIA, M. M., J. AVILA, J. COLL, C. COLACO & I. V. SANDOVAL. 1982. Biochem. Biophys. Res. Commun. **105:** 1241–1249.
23. WEATHERBEE, J. A., P. SHERLINE, R. N. MASCARDO, J. G. IZANT, R. B. LUFTIG & R. R. WEIHING. 1982. J. Cell Biol. **92:** 155–163.
24. BLOOM, G. S., T. A. SCHOENFELD & R. B. VALLEE. 1984. J. Cell Biol. **98:** 320–330.
25. BLOOM, G. S., F. C. LUCA & R. B. VALLEE. 1984. J. Cell Biol. **98:** 331–340.
26. BLOOM, G. S., F. C. LUCA & R. B. VALLEE. 1985. Ann. N.Y. Acad. Sci. **455:** 18–31.
27. BLOOM, G. S., F. C. LUCA & R. B. VALLEE. 1984. Proc. Natl. Acad. Sci. USA **82:** 5404–5408.
28. LUCA, F. C., G. S. BLOOM, S. SCIAVI & R. B. VALLEE. 1984. J. Cell Biol. **99:** 189a.
29. BLOOM, G. S. & R. B. VALLEE. 1983. J. Cell Biol. **96:** 1523–1531.
30. VALLEE, R. B. & G. S. BLOOM. 1983. Proc. Natl. Acad. Sci. USA **80:** 6259–6263.
31. BLOOM, G. S., F. C. LUCA, C. A. COLLINS & R. B. VALLEE. 1986. Ann. N.Y. Acad. Sci. **466:** 328–339. This volume.
32. SHERLINE, P. & MASCARDO, R. 1982. J. Cell Biol. **95:** 316–322.
33. DEMEY, J., F. AERTS, M. MOEREMANS, G. GEUENS, G. DANEELS & M. DE BRABANDER. 1984. J. Cell Biol. **99:** 447a.
34. WANG, E., D. FISCHMAN, R. K. H. LIEM & T. T. SUN, Eds. 1985. Ann. N.Y. Acad. Sci. **455.**
35. STEINERT, P. M., W. W. IDLER & R. D. GOLDMAN. 1980. Proc. Natl. Acad. Sci. USA **77:** 4534–4538.
36. VALLEE, R. B. & BLOOM, G. S. 1984. *In* Modern Cell Biology. B. Satir, Ed.: **3:** 21–75. Alan R. Liss. New York.

Differential Localization of MAP-2 and Tau in Mammalian Neurons *in Situ*[a]

LESTER I. BINDER, ANTHONY FRANKFURTER,
AND LIONEL I. REBHUN

Department of Biology
University of Virginia
Charlottesville, Virginia 22901

INTRODUCTION

Microtubules are one of the major constituents of the neuronal cytoskeleton. Although it is generally accepted that these linear polymers play an important role in regulating cell morphology and intracellular transport processes, the mechanisms that dictate their assembly, their interactions with each other, as well as their interactions with other cytoplasmic structures are poorly understood. In addition to their dimeric tubulin subunits, microtubules contain numerous accessory proteins known collectively as microtubule-associated proteins (MAPs).[1] It is thought that the MAPs control the dimer-polymer equilibrium[2] and mediate the interaction of microtubules with other cytoskeletal elements and cytoplasmic organelles.[3–7] Therefore, it is likely that the MAP composition of microtubules will affect not only microtubule stability but also the structure and stability of the surrounding cytoplasmic elements with which these microtubules interact.

Most of what is currently understood regarding microtubule stability stems from studies performed on microtubules isolated by repetitive cycles of *in vitro* assembly[8] from soluble adult brain extracts. Initial homogenization of adult whole brain results in the simultaneous mixing of many regions, cell types (neurons and glia), and cell compartments (dendrites and axons) that in turn, results, upon purification by assembly-disassembly procedures, in a composite microtubule polymer containing as many as 21 tubulin isoforms[9] and numerous MAPs.[1,10]

In vitro analyses of brain microtubule assembly have definitively demonstrated that two MAPs, MAP-2 ($M_r > 300,000$)[11] and tau (M_r 55–62,000)[12] stimulate assembly stoichiometrically and can displace each other from binding sites on microtubules (Kim *et al.*, this volume). Results such as these lead one to speculate whether or not these two proteins are present on the same microtubule *in situ* and raise the possibility that they may be differentially compartmentalized in brain tissue. In this regard, MAP-2 is known to be a predominantly dendritic protein,[13–18] whereas the distribution of tau in brain tissue has not been previously reported. Because different neuronal cell compartments perform different tasks, an understanding of microtubule function requires the precise localization of MAP-2 and tau to specific cell compartments and subclasses of microtubules. To this end, the present monograph outlines our initial immunohistochemical and biochemical localization experiments using monoclonal antibodies to MAP-2 and tau in both adult and developing brain.

[a]This work was supported by Grant NS 17588 to A. Frankfurter and by Grant J-30 from the Jeffress Memorial Trust to L. I. Rebhun.

MATERIAL AND METHODS

Monoclonal Antibodies

The MAP-2 and tubulin monoclonal antibodies used in this study were those described by Caceres et al.[14] The production, selection, and characterization of the tau monoclonal antibody (tau-1) is described elsewhere (reference 39 and Binder, Frankfurter, and Rebhun, submitted for publication).

Protein Determination, Sodium Dodecyl Sulfate-Polyacrylamide Gel Electrophoresis, and Immunoblotting Procedures

Protein determinations were performed by the method of Lowry et al.[19] as modified.[20] Sodium dodecyl sulfate (SDS)-polyacrylamide slab gels ($17 \times 11.5 \times 0.15$ cm) were of the formulation of Laemmli.[21] The separating gel consisted of a 32 ml, linear acrylamide or urea-acrylamide gradient. The specific gradient used is described in each gel figure legend. Electrophoresis was performed at 20 mA constant current until the tracking dye reached the bottom of the gel (ca 7 hours). After electrophoresis, gels were either stained with Coomassie brilliant blue R or transferred to nitrocellulose for subsequent immunoblot analysis.[22] Blotted nitrocellulose sheets were blocked by gentle agitation in 5% nonfat dry milk (w/v) in 50 mM Tris-HCl, pH 7.6, 0.2 M NaCl (Tris-Saline) for 45 min at room temperature[23] prior to overnight incubation at 4°C in Tris-Saline containing the monoclonal antibody. After primary antibody incubation, blots were washed in Tris-Saline and blocked again for 15 min in 5% nonfat dry milk, Tris-Saline, and then placed in the same solution containing a 1/5000 dilution of peroxidase conjugated goat antimouse IgG (Hyclone, Logan, Utah) for 2–3 hrs at 4°C. At this time the blots were washed again 3–4 times in Tris-Saline and placed in a substrate solution containing 0.0075% (v/v) H_2O_2, 0.04% (w/v) diaminobenzidine in 50 mM Tris, and 10 mM Imidazole-HCl, pH 7.6.[28] The reaction was stopped after sufficient color had developed (ca 2–3 min) by immersing the blot in deionized water. After air drying on filter paper, the blots were photographed using a dark blue filter.

Microtubule Protein Purification

Microtubules were purified by the assembly-disassembly procedure of Shelanski et al.[8] Brain tissue was dissected from either cerebral cortex or caudate nucleus (gray matter) or internal capsule (white matter) and suspended at a ratio of 1 g tissue/ml of polymerization buffer containing 100 mM PIPES, pH 6.8, 1 mM MgSO$_4$, 2 mM EGTA, and 0.1 mM GTP. Homogenization, centrifugation, and cycling of microtubule protein were as described by Kim et al.[24,11] Tubulin was purified from twice-cycled microtubules using phosphocellulose chromatography.[25] MAP-2 and tau were purified by a modification of the method of Kim et al.[11] using heat precipitation and chromatography of the heat-stable fraction in high salt on a BioGel A-1.5m column.

Papain Cleavage of MAP-2

MAP-2 (ca 0.2 mg/ml) was incubated at 37°C in the presence of 1 μg/ml papain in 100 mM PIPES buffer, pH 6.8 with NaOH. After 2 min incubation, the reaction was stopped by addition of an equal volume of $2\times$ electrophoresis sample buffer.[21] The entire mixture was then boiled for 4 min and loaded onto a slotless SDS-urea polyacrylamide gel. Electrophoretic transfer to nitrocellulose and immunostaining with anti-MAP-2 monoclonal antibodies were performed as outlined above.

Quantitative Enzyme-Linked Immunosorbent Assay (ELISA)

The molar ratio of tau to tubulin was determined in microtubules assembled from both gray matter and white matter. This was accomplished using a competitive ELISA.[26] Purified tau or tubulin was attached to microtiter plates at a concentration of 0.01 mg/ml. The plates were washed and blocked in a solution containing 20 mM sodium phosphate buffer, pH 7.4, 0.13 M NaCl, 0.05% Tween-20, and 0.4% bovine serum albumin (PBS-TA). Initial serial antibody dilutions were performed to determine the antibody concentration that yielded an absorbance value of 60% saturation. This antibody concentration was held constant in the presence of a serial dilution of unbound protein (either purified tubulin, tau, or an unknown sample). The antibody-protein suspension was preincubated at 37°C for 1 hr and then plated on microtiter plates that contained preattached tubulin or tau and were then allowed to incubate an additional 1 hr at 37°C. After incubation, the plates were washed in PBS-TA and incubated in the presence of a second antibody peroxidase conjugate (Hyclone: goat antimouse IgG-peroxidase conjugate; 1/3000 dilution) for 1 hr at room temperature. The plates were again washed with PBS-TA, and an equal volume of substrate solution containing a citrate-phosphate buffer, 0.25 mg/ml o-phenylenediamine and 0.003% H_2O_2, was added to each microtiter well. After 20 min, the reaction was stopped with an equal volume of 0.65 M H_2SO_4, and the plates were read immediately on a Titertek Multiskan ELISA reader at 492 nanometers. Standard curves were constructed by competing known amounts of tau or tubulin in the soluble phase with bound tau or tubulin. The amount of tau or tubulin in unknown samples was determined by extrapolation of the linear regions of the competition curves to the linear portions of the standard curves.

Immunohistochemical Procedures

Anesthetized rats were intracardially perfused with 500 ml of phosphate buffer (96 mM K_2PO_4, 24 mM NaH_2PO_4, pH 7.4) containing 20 μM $CaCl_2$, 2% paraformaldehyde, 0.25% glutaraldehyde, and 15% (v/v) of a saturated picric acid solution. The temperature of the perfusate was adjusted to 41°C immediately prior to perfusion. Following perfusion, the animal was decapitated, and the head was immersed in fixative for 3–4 hours. At this time, the brain was removed and stored in 4% paraformaldehyde in phosphate buffer at 4°C for 24 hours. The following day, blocks of tissue were transferred to 50 mM Tris-HCl, 130 mM NaCl, 5 mM KCl, 5 mM $MgSO_4$, 1 mM EGTA, and 0.05% NaN_3, pH 7.6 at 25°C (TBS), containing 100 mM DL-lysine and stored at 4°C with continuous agitation for 12–18 hours. All procedures described below were performed in this TBS solution unless otherwise specified. The tissue blocks were embedded in 5% molten agar, sectioned to a thickness ranging from 40–60 μm with a vibratome, and the resultant sections were placed in ice cold TBS and stored at 4°C for no longer than 24 hours.

The unlabeled antibody method[27] was used to identify the distribution of anti-beta-tubulin, anti-MAP-2 and anti-tau binding sites. Free-floating sections were blocked for 3 hrs at 30°C in TBS containing 5% bovine serum albumin (BSA) and 1% normal rabbit serum prior to being placed in various dilutions of primary antibody in

TBS containing 0.5% BSA and 0.1% normal rabbit serum. Upon removal from primary antibody, sections were washed extensively in TBS prior to unlabeled antibody treatment as described by Sternberger.[27] After removal from peroxidase-antiperoxidase (PAP) complex, the sections were again exhaustively washed in TBS and then immersed in chromogen solution containing 50 mM Tris-HCl, pH 7.6 at 25°C, 10 mM imidazole, 0.04% 3,3' diaminobenzidine (DAB) tetrahydrochloride, and 0.0075% H_2O_2.[28] The reaction was stopped by immersing the sections in ice cold TBS. After a number of further washes, the sections were mounted on gelatinized slides, air-dried overnight and dehydrated in ethanol, cleared in xylene, and mounted in permount.

Transmission Electron Microscopy

Anesthetized rat pups, ages 5 and 10 days postnatal, were intracardially perfused for 30 min with 1% paraformaldehyde, 1% glutaraldehyde in phosphate buffer (77 mM K_2HPO_4, 19 mM NaH_2PO_4) containing 20 uM $CaCl_2$ at 41°C. This is a slightly hypotonic solution when compared to a standard perfusion mixture used for adult animals. Following perfusion, the head was removed and immersed in fixative at room temperature for 3 to 4 hours. The skull overlying the cerebellum was then removed and 2–3 sagital slices approximately 1 mm in thickness were made in the cerebellum with a razor blade. The head was reimmersed in fresh fixative and stored at 4°C. After 24–36 hours, the slices of cerebellum were postfixed for 3 hrs in 2% OsO_4 in phosphate buffer, containing 7% dextrose and 20 uM $CaCl_2$, *en bloc* stained with 0.5% aqueous uranyl acetate, dehydrated in methanol, and embedded in Maraglas. Thin sections were stained with uranyl acetate and lead citrate and examined in a Hitachi (HU12-A) electron microscope. The reader is referred to Palay and Chan-Palay[29] for a complete description of the preparative procedures for electron microscopy outlined in the preceding paragraph.

RESULTS

MAP-2 Distribution in Adult Brain Tissue

Examination of adult and developing brain tissue has been performed using three monoclonal antibodies to MAP-2 (AP-9, -13, -14). An immunoblot demonstrating their specificity in whole rat brain is shown in FIGURE 1, lanes B–D. In all cases, the MAP-2 monoclonal antibodies bind to a doublet of polypeptides at $M_r > 300,000$, which we[30] and others[31,32] have designated MAP-2a and 2b. Also shown are the specificities of the two beta-tubulin monoclonal antibodies employed in our experiments (FIGURE 1, lanes E, F). All of the MAP-2 and tubulin monoclonal antibodies used are of the IgG_1 subclass.

Preliminary experiments have been performed that indicate that all three of our MAP-2 monoclonal antibodies bind to different epitopes on the MAP-2 molecule. When proteolyzed MAP-2 (see METHODS) is loaded across the top of a slotless gel, transferred to nitrocellulose and adjacent strips challenged with the different monoclonal antibodies, unique patterns are revealed (FIGURE 2). Even though the MAP-2 fragment pattern must be the same on each strip of nitrocellulose, different fragments are recognized by the different antibodies, proving that each antibody recognizes a unique epitope.

In the molecular layer of the cerebellum, the MAP-2 antibodies stain both neuronal cell bodies and dendrites (FIGURE 3A) but do not stain the numerous parallel

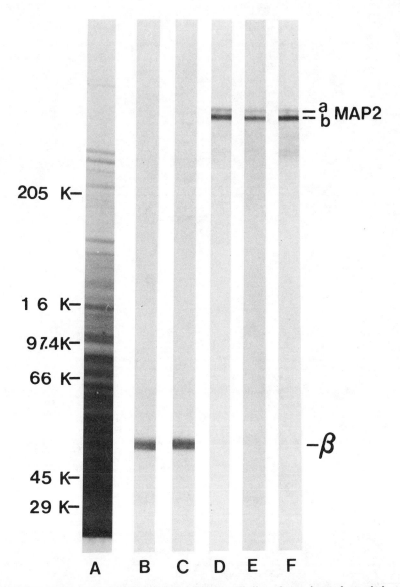

FIGURE 1. Immunoblot of an SDS extract of whole cerebellum electrophoresed on a slotless gel prior to transfer to nitrocellulose. Adjacent strips were cut and treated as described below: (A) amido black stain; (B) immunostain with the anti-beta tubulin monoclonal, Tu9B; (C) immunostain using the anti-beta tubulin monoclonal, Tu27B; (D) immunostain using, respectively, the anti-MAP-2 monoclonal antibodies AP-9, AP-13, and AP-14. The numbers to the left of Panel A denote molecular weight ($\times 10^{-3}$). The gel employed was composed of a 3–10% linear acrylamide gradient superimposed on a 1–8 M linear urea gradient.

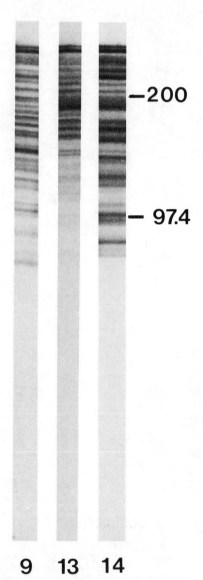

FIGURE 2. MAP-2 cleaved with papain prior to electrophoresis on a slotless gel and transfer to nitrocellulose. Adjacent strips were challenged with the MAP-2 monoclonal antibodies indicated at the bottom of each strip. Each monoclonal illuminates a unique fragment pattern proving that they are binding to unique epitopes. The numbers to the right of the third panel denote molecular weight ($\times 10^{-3}$). The gel was composed of linear gradients of 4–16% acrylamide and 1–8 M urea.

fiber axons that give a peppery appearance when observed in cross sections stained with either of our anti-beta tubulin monoclonal antibodies (FIGURE 3B). Also, apparent in FIGURE 3B but not in FIGURE 3A are the processes of Bergmann glia that are well stained by the anti-beta tubulin monoclonals but remain unstained by any of our anti-MAP-2 antibodies. Indeed, in the molecular layer of the cerebellum, no axon or glial processes are stained with any of our three MAP-2 monoclonal antibodies. These results alone suggest that MAP-2 is exclusively localized in cell bodies and dendrites of neurons. In other areas of the nervous system, however, axonal staining can be detected with two of our three MAP-2 monoclonal antibodies. The axons of the internal capsule (FIGURE 4A) contain detectable MAP-2 immunoreactivity, whereas staining is also apparent in the hypoglossal axons as they course through the medulla (FIGURE 4B).

These observations support previous reports of MAP-2 immunoreactivity in axons of peripheral neurons[5] as well as in the white matter of the cerebellum.[14] In order to validate our immunocytochemical findings, however, we sought confirmation by independent biochemical methods. To accomplish this, small punches of tissue were taken from the internal capsule of bovine brain, homogenized in a microtubule polymerization buffer, and carried through two cycles of temperature-dependent assembly and disassembly (see METHODS). After electrophoresis and transfer to nitrocellulose, the blots were probed with all three of our MAP-2 monoclonal antibodies. In all cases, a doublet of immunoreactivity was observed corresponding to MAP-2a and 2b (data not shown). When equal loadings of gray matter and white matter microtubule preparations were blotted, however, considerably more MAP-2 was present in the gray matter preparations than in those prepared from the internal capsule (FIGURE 5, lanes A, B). From these experiments, we conclude that, in brain, MAP-2 is primarily a neuronal protein that is highly concentrated in cell bodies and dendrites; detectable quantities of MAP-2, however, can be found in certain axonal populations.

MAP-2 and Tubulin Localization in Developing Cerebellum

We have examined the distribution of MAP-2 and tubulin early in postnatal development of the cerebellum. Immunocytochemical localization of MAP-2 using our monoclonal antibodies demonstrated intense dendritic staining during all stages of Purkinje cell dendritic differentiation examined. As is shown in FIGURE 6B, at postnatal day 10, many Purkinje cells display extensive dendritic arbors that are strongly immunoreactive against all MAP-2 monoclonal antibodies tested. Further-more, by contrast to the results obtained by Bernhardt and Matus,[33] dendrites at this stage of development and earlier also reacted strongly against anti-beta tubulin monoclonal antibodies (FIGURE 6A). Routine electron microscopic analysis confirmed the presence of large numbers of microtubules throughout the length of the Purkinje cell dendrites as early as postnatal day 5 (FIGURE 7A, B). From these results, we conclude that tubulin and MAP-2 enter the growing dendrite in a nearly simultaneous fashion. The notable exception to this statement is the presence of MAP-2 in immature spines and dendritic growth cones in the apparent absence of microtubules or tubulin. Growth cones and immature spines appear as large protuberances projecting from the main dendritic shafts at early postnatal times. These structures stain intensely with anti-MAP-2 antibodies (FIGURE 6B), but not with antitubulin antibodies (FIGURE 6A). Electron microscopic examination confirms the presence of immature spines in these structures and further indicates that no microtubules are present in the dendritic growth cone proper (data not shown). It is important to note that this absence of

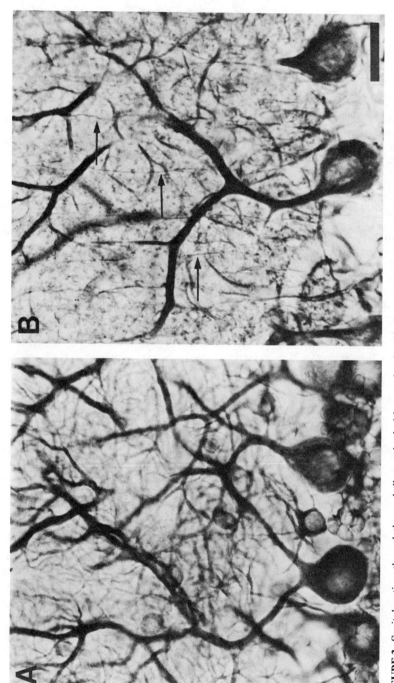

FIGURE 3. Sagital sections through the cerebellum stained with monoclonal antibodies to (A) MAP-2 and (B) beta tubulin. In panel A, only neuronal cell bodies and dendrites are stained. In panel B, the anti-beta tubulin antibody stains, in addition to neuronal cell bodies and dendrites, processes of Bergmann glia (arrows) and parallel fiber axons (the dot-like profiles in the molecular layer). Bar = 20 μm.

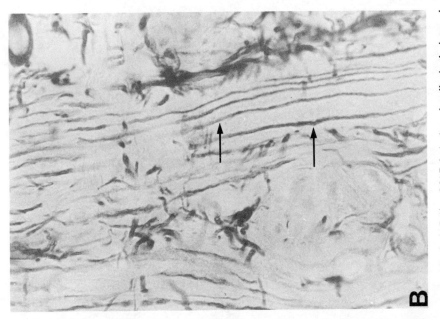

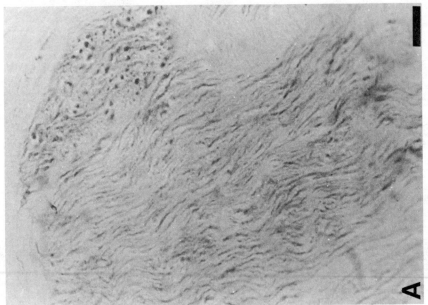

FIGURE 4. Axons of the central nervous system stained with a monoclonal antibody (AP-9) to MAP-2. A: Axon bundles in the internal capsule, and B: hypoglossal axons (arrows) in the medulla. Bar = 20 μm.

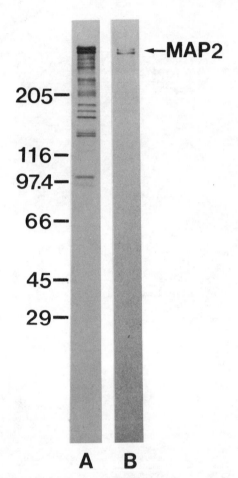

FIGURE 5. Immunoblot of equal amounts of bovine (A) gray matter microtubules and (B) white matter microtubules probed with the same concentration of AP-14, a MAP-2 monoclonal antibody. Note that there is more MAP-2 in the gray matter microtubule preparation than in the white matter microtubule preparation. The numbers to the left of panel A represent molecular weight ($\times 10^{-3}$). The gel was composed of a linear 5–12.5% acrylamide gradient. Numerous immunoreactive bands are present beneath the MAP-2 doublet in A. These are proteolytic fragments of MAP-2.

tubulin and microtubules in spines persists into adulthood as does the presence of MAP-2 in these structures (FIGURE 8A–C). These results demonstrate that MAP-2 can exist *in vivo* in the apparent absence of tubulin or microtubules.

Tau Localization in Adult Brain

We have also determined the localization of the heterogeneous tau proteins in mammalian brain. This was accomplished using an IgG_{2a} monoclonal antibody (tau-1)

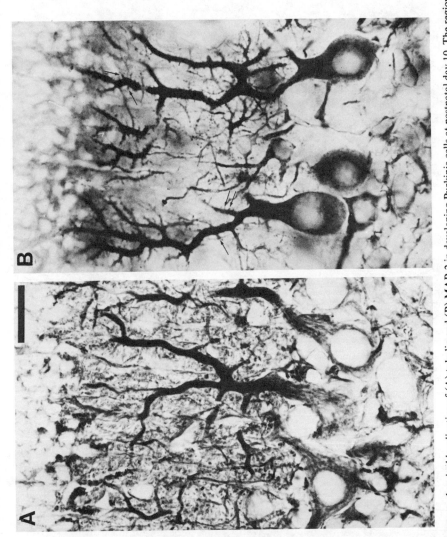

FIGURE 6. Immunocytochemical localization of (A) tubulin and (B) MAP-2 in developing Purkinje cells at postnatal day 10. The region of the neuropil surrounding the dendrites that is stained in A but remains unstained in B represents the parallel fiber axons that contain tubulin but not MAP-2. A few of the numerous immature dendritic spines and presumptive growth cones stained in B are indicated (arrows). Bar = 20 μm.

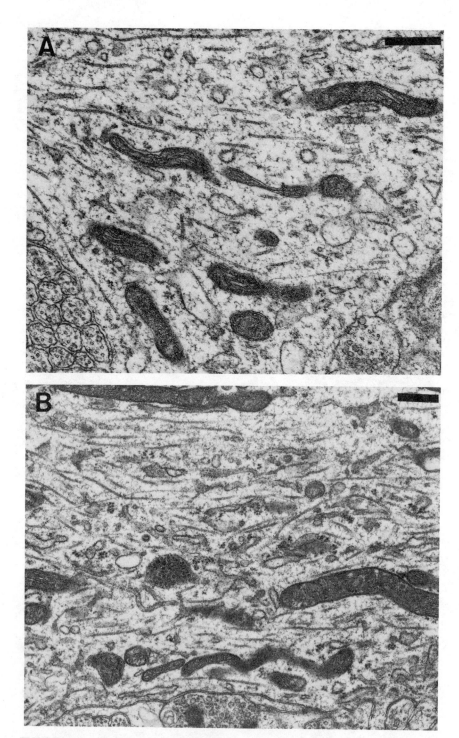

FIGURE 7. Transmission electron micrographs of sagital sections through developing Purkinje cell dendrites at (A) postnatal day 5, and (B) postnatal day 10. Note the presence of numerous microtubules in both sections. Bars = 0.5 μm.

that recognizes all known electrophoretic species of tau in both rat and bovine brain. When reacted against blots of whole bovine brain extracts, a minimum of five immunoreactive species are apparent (FIGURE 9, lanes A, D), whereas one of these is lost during purification of microtubules from such an extract by cycles of assembly and disassembly (FIGURE 9, lanes B, E).

Immunohistochemical studies were performed on vibratome sections taken from adult rat brain tissue fixed as described in METHODS. The unlabeled antibody

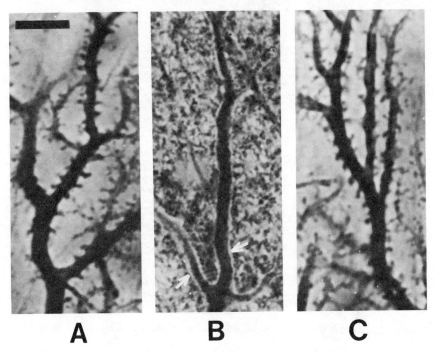

FIGURE 8. Immunocytochemistry of the spiny branchlets of adult Purkinje cell dendrites using an anti-MAP-2 monoclonal antibody (A and C) and an anti-beta tubulin antibody (B). Note the periodic appearance of the anti-MAP-2 immunoreactive spines along the lengths of the dendritic shafts in A and C. No spine staining is present in B, but instead, the dendrites are surrounded by a nonstained glial sheath that appears as a "halo" (arrows). Bar = 5 μm.

procedure was used to examine the distribution of tau immunostaining at the level of the light microscope. Sections through the substantia nigra (FIGURE 10A–C) have been stained with tubulin, MAP-2, and tau monoclonal antibodies. As expected, an anti-beta tubulin antibody stains neuronal cell bodies and dendrites in the substantia nigra as well as axons of the nigra and the underlying cerebral peduncle (FIGURE 10A). By contrast, our MAP-2 monoclonal antibodies give localization patterns typified by FIGURE 10B, in which the majority of the staining is found in the cell bodies and dendrites of the substantia nigra. The small nigral axon bundles and the axons of the cerebral peduncle contain little MAP-2 immunoreactivity. When reacted with tau-1, however, only axonal staining is observed (FIGURE 10C), whereas both neuronal cell bodies and dendrites appear unstained. This general staining pattern is confirmed in a

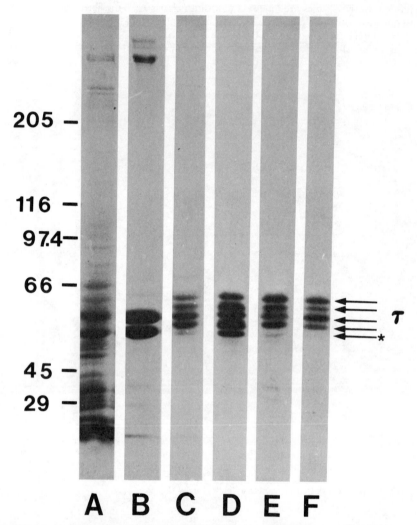

FIGURE 9. Purification of tau from bovine cerebral cortex: stained with Coomassie blue (A–C) and immunostained with tau-1 following transfer to nitrocellulose (D–F). (A) Crude extract of bovine brain cerebral cortex, (B) twice cycled microtubule preparation from bovine cerebral cortex, (C) purified, heat-stable tau-factor, (D) immunoblot of the crude extract shown in A, (E) immunoblot of the twice cycled microtubule preparation shown in B, and (F) immunoblot of 25 ng of purified tau shown in C. Tau* represents the immunoreactive polypeptide present in extracts of whole cortex but lost during purification of microtubules. The numbers to the left represent molecular weight ($\times 10^{-3}$).

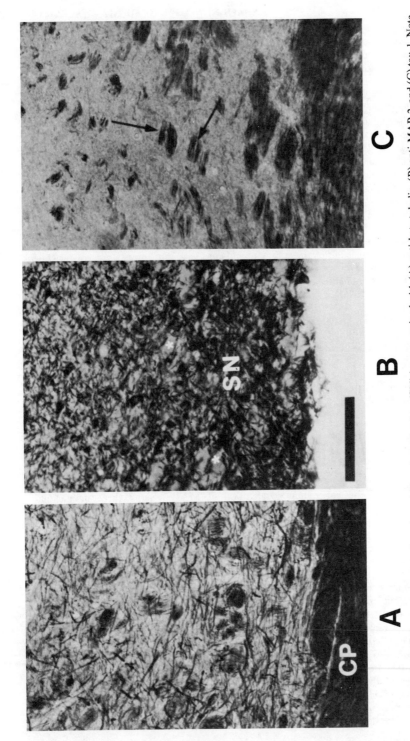

FIGURE 10. Low power of coronal sections through the substantia nigra (SN) immunostained with (A) anti-beta tubulin, (B) anti-MAP-2, and (C) tau-1. Note the cerebral peduncles (CP) are stained intensely by the anti-beta tubulin antibody (A) and tau-1 (C), whereas only light staining is observable upon staining with an anti-MAP-2 antibody (B). Anti-tubulin and anti-MAP-2 antibodies stain the dendrites and cell bodies (A and B); these are unstained by tau-1 (C). The arrows point to fiber bundles in the substantia nigra, which are stained with tau-1 (C); a few of the corresponding areas that are not stained by anti-MAP-2 (B) are starred (*). Bar = 8 μm.

higher magnification of the cerebellar molecular layer (FIGURE 11A, B). Here, the neuronal cell bodies and dendrites appear in negative relief to the densely stained parallel fiber axons that give rise to dark stippled areas in the surrounding neuropil. Also not stained by tau-1 are the processes of Bergmann glia (FIGURE 11B), indicating that, like MAP-2, tau is primarily confined to neurons.

Although the results described above suggest that tau is confined to axons, they were obtained using only one monoclonal antibody. It is conceivable that this antibody recognizes a site on the tau molecule that is only available in axons, being masked in cell bodies and dendrites. For this reason, additional experiments were performed on microtubules purified from white matter and gray matter-enriched brain regions. Immunoblots of sodium dodecyl sulfate (SDS) gels and competitive ELISAs were performed to qualitatively and quantitatively determine the relative amounts of tau associated with microtubules from different brain regions. When crude tissue extracts from the caudate nucleus (gray matter) were compared to an equal amount of protein from a similar extract from internal capsule (white matter), the white matter preparation contained more tau (FIGURE 12A, B). Similarly, immunoblots of microtubules purified from each of these regions indicate that more tau is present in white matter microtubules than in gray matter microtubules (FIGURE 12C, D). In addition, microtubules assembled from white matter contain more of the lower molecular weight tau species (tau*) than do those purified from gray matter (FIGURE 12C, D) even though tau* appears to be in a similar ratio to the other tau species in crude extracts of both gray and white matter tissue (FIGURE 12A, B). That white matter microtubules contained more tau than those from gray matter was confirmed by competitive ELISA (see METHODS). The results of the tau:tubulin ratios obtained from these assays are shown in TABLE 1, which indicates that white matter microtubules contain a threefold greater ratio of tau:tubulin than gray matter microtubules. Taken together, the immunocytochemistry and biochemistry reported here suggest that tau is primarily confined to axons.

DISCUSSION

The Distribution of MAP-2

The cellular and subcellular localization of MAP-2 using three monoclonal antibodies, for the most part, agrees with other published accounts of MAP-2's distribution.[13,15,17] We find MAP-2 to be mostly concentrated in dendrites and neuronal cell bodies and undetectable in most glia. Furthermore, we have not found MAP-2 in the parallel fiber axons in the molecular layer of the cerebellum. In contrast to numerous published reports, however, we have demonstrated the presence of MAP-2 in certain axonal subpopulations in agreement with Caceres et al.[14] and Papazomenos et al.[5] The amount of MAP-2 in axons is small in comparison to dendrites and neuronal cell bodies, but its presence has been confirmed biochemically by immunoblots of twice cycled microtubules from white matter. Here, our biochemical determination agrees with our immunocytochemical localizations, in that much more MAP-2 is present in gray matter microtubules, but an easily detectable amount is also present in microtubules from white matter. Attempts are currently underway to quantify the MAP-2 to tubulin ratios in white versus gray matter microtubules and soluble extracts by competitive ELISA.

Using a picric acid-aldehyde fixation procedure (see METHODS) the presence of MAP-2 immunoreactivity has been unequivocally demonstrated in Purkinje cell dendritic spines at the light microscopic level. This confirms an early report by Caceres

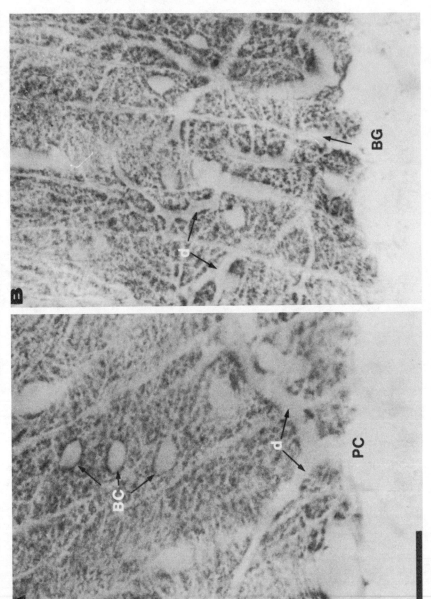

FIGURE 11. High power of sagital sections through the Purkinje cell and molecular layers of the cerebellum immunostained with tau-1. (A) Section through an unstained Purkinje cell body (PC) showing continuity with its unstained dendrites (d) extending into the molecular layer towards the pial surface. Also unstained are the basket cell bodies (BC). (B) Section showing unstained processes of Bergmann glia (BG plus arrow), as well as numerous unstained Purkinje cell dendrites (d). The stippled appearance provided by the positive staining in the molecular layer is due to the presence of parallel fiber axons that obviously contain abundant tau.

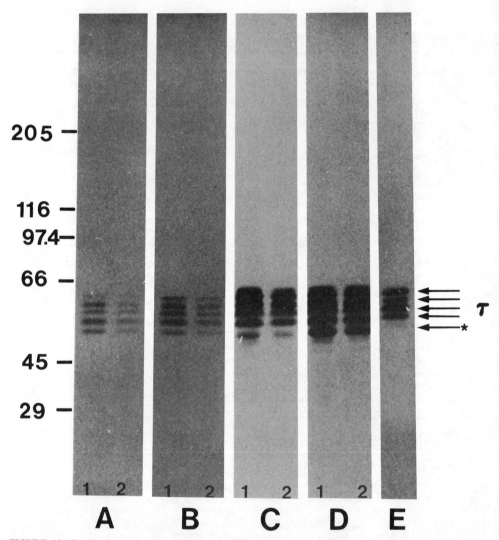

FIGURE 12. Qualitative determination of the relative amounts of tau in gray matter versus white matter extracts and twice cycled microtubules. Immunoblots of a soluble extract from gray matter, (A) and a soluble extract from white matter, (B). The amount of protein loaded in Panel A, lane 1 equals the amount loaded in Panel B, lane 1, and the amount of protein loaded in Panel A, lane 2 equals the amount loaded in Panel B, lane 2. Immunoblots of twice-cycled microtubule preparations from (C) the gray matter extract shown in A, and (D) the white matter extract shown in B. The same amounts of protein were loaded in both Panel C, lane 1 and in Panel D, lane 1. The amount of protein loaded in Panel C, lane 2 equals the amount of protein loaded in Panel D, lane 2. E: 25 ng of a purified bovine tau standard. Note the enrichment of tau* in D as compared to C, and also note that there is qualitatively more tau in both white matter extracts and microtubules than in corresponding preparations from gray matter. The numbers to the left represent molecular weight ($\times 10^{-3}$).

et al.[34] who presented similar evidence using electron microscopic localization. Such findings are in opposition to those reported by DeCamilli *et al.*[15] and Bernhardt and Matus,[13] both of which failed to detect MAP-2 localization in dendritic spines. It should be noted, however, that no tubulin immunoreactivity is present in dendritic spines (FIGURE 8), thus arguing against the nonspecific migration of reaction product from the dendritic shaft into the spines. Furthermore, the absence of tubulin in spines agrees with reports that demonstrate ultrastructurally that microtubules do not enter the spine proper (*e.g.*, Caceres *et al.*, reference 34).

The most important aspect of the presence of MAP-2 in adult dendritic spines and in developing spines and presumptive growth cones is that it occurs in the apparent absence of tubulin and microtubules. This suggests that MAP-2 may have functions that are distinct from its association with microtubules. The colocalization of MAP-2 with actin in dendritic spines[34] was the first *in vivo* indication that the work demonstrating MAP-2–actin interactions *in vitro*[4,6,35] may, indeed, have biological significance. Also, others have demonstrated that MAP-2 can interact with intermediate filaments[3] and neurofilaments[5] *in situ*. An interesting future direction in MAP-2 research will be the determination of the mechanisms controlling the relative affinities of MAP-2 for microtubules and other cytoskeletal elements.

TABLE 1.[a]

	White Matter MT	Gray Matter MT
τ (mg)	0.19 ± 0.01	0.08 ± 0.01
β tubulin (mg)	2.19 ± 0.19	2.77 ± 0.10
Molar Ratio (τ/tubulin)	1/12.8	1/38.5

[a]The tau/tubulin ratios obtained from competitive ELISAs performed on twice-cycled microtubules from both gray and white matter extracts. The indicated amounts of protein (in mg) have been standardized to one milliliter. The data is expressed as an average (+) or (−) one standard deviation. The molar ratios were calculated using molecular weights of 55,000 for beta tubulin and 61,000 for tau.

We have also studied the distribution of MAP-2 and tubulin in the developing cerebellum. This work was prompted by a report of Bernhardt and Matus,[33] which demonstrated that MAP-2 preceded microtubules and tubulin in developing Purkinje cell dendrites by as much as 21 days, leading them to suggest that MAP-2 was a "determinant" of dendritic differentiation. We have been unable, however, to confirm these findings. When proper fixation methods are employed (see METHODS), MAP-2 and tubulin appear to enter the dendrites simultaneously with dendritic elongation. Furthermore, in contrast to another report,[32] much of the tubulin present in developing dendrites is in the form of microtubules as early as postnatal day five. The only evidence of noncolocalization of MAP-2 and tubulin was obtained in presumptive dendritic growth cones and both mature and immature dendritic spines. These structures stained intensely with our anti-MAP-2 antibodies, but were devoid of tubulin immunoreactivity.

The Distribution of Tau in the Adult Brain

A monoclonal antibody that binds to the tau polypeptides was described. This antibody binds to five polypeptides in extracts of bovine cerebral cortex, but after two cycles of assembly-disassembly from either extracts of cerebral cortex (FIGURE 9) or caudate nucleus (FIGURE 12), the resultant microtubules contain little or no tau*, the lowest molecular weight tau species. This result is identical to that reported by Drubin et al.[36] using a polyclonal antibody to tau. By contrast, microtubules purified from white matter do not lose tau* (FIGURE 12), but instead, all of the tau species seem to maintain the same ratios to each other as are present in the initial crude extract. The reasons for these differences are not known, but it is safe to conjecture that the differential binding of tau* is either due to the presence of different tubulins or different MAPs on gray versus white matter microtubules. For a discussion of these possibilities, see Kim et al. (this volume).

The localization of tau in the mammalian central nervous system was also reported. Immunocytochemical experiments indicated that tau was exclusively axonal in its distribution. No localization was seen in glial cells or neuronal cell bodies and dendrites. The reason that tau localization was absent in neuronal cell bodies is not known. Certainly, tau must be synthesized in the cell body, because there is no evidence for protein synthesizing activity in axons. Either the tau in cell bodies is in a form not recognized by our monoclonal antibody, or it is synthesized very near the initial segment of the axon and transported immediately down the axon.

Competitive ELISAs performed on twice-cycled microtubules from bovine gray and white matter also indicated that most tau was axonal in that the tau:tubulin molar ratios were threefold higher in microtubules assembled from white matter than in those from gray matter. That there was tau in gray matter microtubules was not surprising because gray matter contains as much as 40% axons by volume.[37] White matter, on the other hand, contains virtually no neuronal cell bodies or dendrites and, as would be expected, contains very little MAP-2 (FIGURE 5 and reference 38). Although we have reported that certain axonal subpopulations contain detectable amounts of MAP-2, the distribution of tau and MAP-2 in the central nervous system can be thought of, to a limited degree, as complementary. The molar ratio of 1 tau per 12.8 tubulin dimers in twice-cycled microtubules from white matter is very close to the MAP-2:tubulin ratio of 1:12 reported for taxol-stabilized microtubules from gray matter.[38] We report here that the tau to tubulin ratio in gray matter microtubules is low (1:38), whereas our immunoblot evidence indicates that the amount of MAP-2 in white matter microtubules is correspondingly low. This work combined with our immunocytochemical localization studies suggest that there is a minimum of two subclasses of microtubules in neurons, those in the cell bodies and dendrites that are defined by MAP-2 and those in the axon that are defined by tau.

ACKNOWLEDGMENT

Most of the MAP-2 and tubulin monoclonal antibodies reported in this monograph were isolated in collaboration with Dr. Michael R. Payne, Department of Anatomy, New York Medical College, Valhalla, New York. Dr. Payne's continued advice and support are greatly appreciated.

REFERENCES

1. SLOBODA, R. D., S. A. RUDOLPH, J. L. ROSENBAUM & P. GREENGARD. 1975. Cyclic AMP-dependent endogenous phosphorylation of a microtubule-associated protein. Proc. Natl. Acad. Sci. USA 72: 177–181.

2. MURPHY, D. B., K. A. JOHNSON & G. G. BORISY. 1977. Role of tubulin-associated proteins in microtubule nucleation and elongation. J. Mol. Biol. **117:** 33–52.

3. BLOOM, G. S. & R. B. VALLEE. 1983. Association of microtubule-associated protein 2 (MAP2) with microtubules and intermediate filaments in cultured brain cells. J. Cell Biol. **96:** 1523–1531.

4. GRIFFITH, L. M. & T. D. POLLARD. 1982. The interaction of actin filaments with microtubules and microtubule-associated proteins. J. Biol. Chem. **257:** 9143–9151.

5. PAPASOZAMENOS, S. CH., L. I. BINDER, P. K. BENDER & M. R. PAYNE. 1985. Microtubule-associated protein 2 (MAP2) within axons of spinal motor neurons: associations with microtubules and neurofilaments in normal and beta, beta'-iminodipropionitrile-treated axons. J. Cell Biol. **100:** 74–85.

6. SELDON, S. C. & T. D. POLLARD. 1983. Phosphorylation of microtubule-associated proteins regulates their interaction with actin filaments. J. Biol. Chem. **258:** 7064–7071.

7. SUPRENANT, K. A. & W. L. DENTLER. 1982. Association between endocrine pancreatic secretory granules and *in vitro*-assembled microtubules is dependent upon microtubule-associated proteins. J. Cell Biol. **93:** 164–174.

8. SHELANSKI, M. L., F. GASKIN & C. R. CANTOR. 1973. Microtubule assembly in the absence of added nucleotides. Proc. Natl. Acad. Sci. USA **70:** 765–768.

9. DENOULET, P., B. EDDE, C. JEANTET & F. GRAS. 1982. Evolution of tubulin heterogeneity during mouse brain development. Biochimie **64:** 165–172.

10. SLOBODA, R., W. DENTLER & J. L. ROSENBAUM. 1976. Microtubule-associated proteins and the stimulation of tubulin assembly *in vitro*. Biochemistry **15:** 4497–4505.

11. KIM, H., L. I. BINDER & J. L. ROSENBAUM. 1979. The periodic association of MAP2 with brain microtubules *in vitro*. J. Cell Biol. **80:** 266–276.

12. CLEVELAND, D. W., S. Y. HWO & M. W. KIRSCHNER. 1977. Physical and chemical properties of purified tau factor and the role of tau in microtubule assembly. J. Mol. Biol. **116:** 227–247.

13. BERNHARDT, R. & A. MATUS. 1984. Light and electron microscopic studies of the distribution of microtubule-associated protein 2 in rat brain: a difference between dendritic and axonal cytoskeletons. J. Comp. Neurol. **226:** 203–221.

14. CACERES, A., L. I. BINDER, M. R. PAYNE, P. BENDER, L. REBHUN & O. STEWARD. 1983. Differential subcellular localization of tubulin and the microtubule-associated protein MAP2 in brain tissue as revealed by immunocytochemistry with monoclonal hybridoma antibodies. J. Neurosci. **4:** 394–410.

15. DECAMILLI, P., P. E. MILLER, F. NAVONE, N. E. THEURKAUF & R. B. VALLEE. 1984. Distribution of microtubule-associated protein 2 in the nervous system of the rat studied by immunofluorescence. Neurosci. **11:** 819–846.

16. FRANKFURTER, A., L. I. BINDER, M. R. PAYNE & L. I. REBHUN. 1983. Immunohistochemical localization of tubulin and the high molecular weight microtubule-associated proteins (MAP1 and MAP2) in the developing cerebellum. Neurosci. (Abstract) **9:** 851.

17. HUBER, G. & A. MATUS. 1984. Differences in the cellular distributions of two microtubule-associated proteins, MAP1 and MAP2, in rat brain. J. Neurosci. **4:** 151–160.

18. MATUS, A., R. BERNHARDT & T. HUGH-JONES. 1981. High molecular weight microtubule-associated proteins are preferentially associated with dendritic microtubules in brain. Proc. Natl. Acad. Sci. USA **78:** 3010–3014.

19. LOWRY, O. H., N. J. ROSEBROUGH, A. L. FARR & R. J. RANDALL. 1951. Protein measurement with the Folin phenol reagent. J. Biol. Chem. **193:** 265–275.

20. BENSADOUN, A. & D. WEINSTEIN. 1976. Assay of protein in the presence of interfering materials. Anal. Biochem. **70:** 241–250.

21. LAEMMLI, U. K. 1970. Cleavage of structural proteins during the assembly of the head of bacteriophage T_4. Nature (London) **227:** 680–685.

22. TOWBIN, H., T. STAEHELIN & J. GORDON. 1979. Electrophoretic transfer of protein from polyacrylamide gels to nitrocellulose sheets: procedure and some application. Proc. Natl. Acad. Sci. USA **76:** 4354–4356.

23. JOHNSON, D. A., J. W. GAUTSCH, J. R. SPORTSMAN & J. H. ELDER. 1984. Improved technique utilizing nonfat dry milk for analysis of proteins and nucleic acids transferred to nitrocellulose. Gene Anal. Technol. **1:** 3–8.

24. KIM, H. 1983. Brain Microtubule Structure *In Vitro*. Ph.D. Thesis. University of Virginia, Charlottesville, Va.
25. WEINGARTEN, M. D., A. H. LOCKWOOD, S. Y. HWO & M. W. KIRSCHNER. 1975. A protein factor essential for microtubule assembly. Proc. Natl. Acad. Sci. USA **72:** 1858–1862.
26. VOLLER, A., A. BARTLETT & D. E. BIDWELL. 1978. Enzyme immunoassays with special reference to ELISA techniques. J. Clin. Pathol. **31:** 507–520.
27. STERNBERGER, L. A. 1979. Immunocytochemistry. A Wiley Medical Publication. New York.
28. STRAUS, W. 1982. Imidazole increases the sensitivity of the cytochemical reaction for peroxidase with diaminobenzidine at a neutral pH. J. Histochem. Cytochem. **30:** 491–493.
29. PALAY, S. L. & V. CHAN-PALAY. 1974. Cerebellar Cortex: Cytology and Organization. Springer-Verlag. New York.
30. BINDER, L. I., A. FRANKFURTER, H. KIM, A. CACERES, M. R. PAYNE & L. I. REBHUN. 1984. Heterogeneity of microtubule-associated protein 2 during rat brain development. Proc. Natl. Acad. Sci. USA **81:** 5613–5617.
31. BLOOM, G. S., T. A. SCHOENFELD & R. B. VALLEE. 1984. Widespread distribution of the major polypeptide component of MAP1 (Microtubule-Associated Protein 1) in the nervous system. J. Cell Biol. **98:** 320–330.
32. BURGOYNE, R. D. & R. CUMMING. 1984. Ontogeny of microtubule-associated protein 2 in rat cerebellum: differential expression of the doublet polypeptides. Neurosci. **11:** 157–167.
33. BERNHARDT, R. & A. MATUS. 1982. Initial phase of dendritic growth: Evidence for the involvement of high molecular weight microtubule-associated proteins (HMWP) before the appearance of tubulin. J. Cell Biol. **92:** 589–593.
34. CACERES, A., M. R. PAYNE, L. I. BINDER & O. STEWARD. 1983. Immunocytochemical localization of actin and microtubule-associated protein (MAP2) in dendritic spines. Proc. Natl. Acad. Sci. USA **80:** 1738–1742.
35. SATTILARO, R. F., W. L. DENTLER & E. L. LECLUYSE. 1981. Microtubule-associated proteins (MAPs) and the organization of actin filaments *in vitro*. J Cell Biol. **90:** 467–473.
36. DRUBIN, D. G., D. CAPUT & M. W. KIRSCHNER. 1984. Studies on the expression of the microtubule-associated protein, tau, during mouse brain development, with newly isolated complementary DNA probes. J. Cell Biol. **98:** 1090–1097.
37. CACERES, A. & O. STEWARD. 1983. Dendritic reorganization in the denervated dentate gyrus of the rat following entorhinal cortical lesions: A Golgi and electron microscopic analysis. J. Comp. Neurol. **214:** 387–403.
38. VALLEE, R. B. 1982. A taxol-dependent procedure for the isolation of microtubules and microtubule-associated proteins. J. Cell Biol. **92:** 435–442.
39. BINDER, L. I., A. FRANKFURTER & L. I. REBHUN. 1984. A monoclonal antibody to tau-factor localizes predominately in axons. J. Cell Biol. **99:** 191a.

Microtubule-Associated Proteins in the Developing Brain

ANDREW MATUS AND BEAT RIEDERER

The Friedrich Miescher Institute
4002 Basel, Switzerland

Brain microtubules repolymerized *in vitro* contain a rich variety of nontubulin peptides known as the microtubule-associated proteins (MAPs). Until recently the properties of these molecules were poorly documented because their low abundance made it impractical to isolate them. There were two exceptions—the proteins known as MAP-2 (molecular weight circa 280,000) and tau (molecular weight circa 70,000) whose resistance to heat denaturation made them relatively easy to isolate.[1,2] Both of these proteins are effective promoters of tubulin polymerization *in vitro*,[1-4] and this has, to some extent, fostered an expectation that a true MAP should contribute to microtubule assembly *in vitro*.

The investigation of these proteins has been radically changed by the introduction of monoclonal antibodies. By way of their agency several novel kinds of data are being rapidly accumulated. First, the cytological localization of different MAP species is being established. Second, subforms of previously identified MAPs are being discriminated and made accessible to study. Third, novel MAPs are being identified and characterized.

With respect to the function of MAPs in brain, the growing body of cytological data has acquired a particular significance, for it has transpired that each MAP species has a characteristic distribution between different categories of brain cells as well as within the cytoplasm of individual types of neurons.[5] Thus we have observed that MAP-1 and MAP-2 in adult brain are expressed in neurons but not in glia, occur at very different concentrations in different neuronal cell types, and in at least some neurons, differ in cytoplasmic distribution within the same cell.[5-7]

CHEMICAL CHARACTERIZATION OF MAP SPECIES

Our recent experiments have used libraries of monclonal antibodies made against the MAP fraction from repolymerized brain microtubules. These contain antibodies against the previously identified proteins MAP-1 and MAP-2 and against several novel species including MAP-3 and MAP-5, which are discussed here.

Some of the Apparent Heterogeneity among MAP Peptides Can Be Traced to Proteolytic Fragments

A striking feature of all these antibodies was that on sodium dodecyl sulfate (SDS)-gel blots of microtubule proteins they reacted with more than one peptide band. These fall into patterns that are reproducible and characteristic for each class of antibody. Thus, for example, antibodies that react with MAP-1 invariably also react

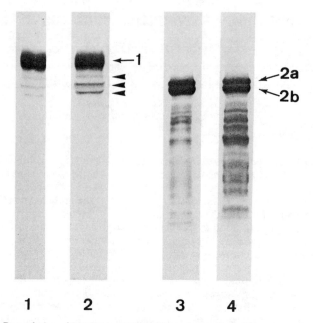

FIGURE 1. Degradation of MAP-1 and MAP-2 by endogenous brain proteases. 0.4 mg of brain microtubule proteins and 1 mg of brain supernatant fraction (100,000 × g for 60 min), final volume 120 μl, were incubated for 2 h at 37°C. These samples (strips 2 and 4) were run side by side with unincubated controls (strips 1 and 3) in SDS-gels (40 μg per channel), which were blotted onto nitrocellulose and stained with either anti-MAP-1 (strips 1 and 2) or anti-MAP-2 (strips 3 and 4). The native components of MAP-1 (1) and MAP-2 (2a and 2b) are indicated together with those breakdown products of MAP-1 that comigrate with MAP-2 (arrowheads, strip 2).

with three minor bands that in SDS-gels comigrate with the conventionally designated MAP-2 band (FIGURE 1, slots 1 and 2). Similarly, all our antibodies against MAP-2 react with both components of the 280,000 molecular weight MAP-2 doublet (MAP-2 a and b) and with a complex pattern of lower molecular weight species (FIGURE 1, slots 3 and 4).

It is well known that MAPs 1 and 2 are susceptible to degradation by endogenous brain proteases.[8,9] So the question arose, Which of these bands are independent molecules and which are proteolytic fragments of larger peptides? This was investigated by incubating microtubule proteins with brain supernatant and using antibodies to selectively reveal the pattern of breakdown products for MAP-1 (FIGURE 1, slot 2) or MAP-2 (FIGURE 1, slot 4). This suggests that all the lower anti-MAP-1 reactive bands are produced by degradation of the broad 350,000 molecular weight band. In the case of MAP-2, the two components of the 280,000 doublet appear to be independent species with the lower bands being breakdown products of this high molecular weight doublet. Both calcium-stimulated and calcium-independent proteases appear to be involved in mediating this degradation (B. Riederer and A. Matus, unpublished observations).

Novel MAP Species Contribute to Overall MAP Heterogeneity

Although proteolytic fragments can account for a good many of the bands seen with any one antibody, it is also clear that novel species, such as MAP-3,[5,10] possess their own characteristic pattern of bands, none of which cross-react with antibodies against peptide bands of other MAP species. Thus MAP-3 consists of two apparently native peptides, molecular weight *circa* 180,000, and a series of smaller antibody-reactive doublet bands apparently derived by the clipping of equally sized fragments from the primary pair of components (reference 10, FIGURE 2, panel labeled 3). A second novel MAP species we have characterized, MAP-5, consists of a broad band migrating between MAP-1 and MAP-2 and has a major breakdown product that migrates just below MAP-2 (FIGURE 2, panel labeled 5).

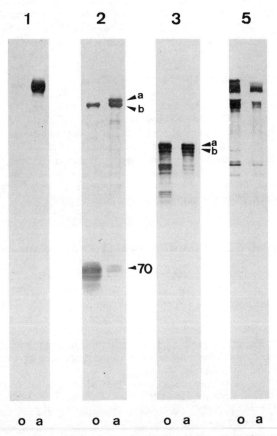

FIGURE 2. Western blots of SDS-gels in which microtubule proteins from newborn (0) and adult (a) were separated in adjacent channels. Each panel of two slots was stained with a different monoclonal antibody against MAPs 1, 2, 3, or 5 as indicated above. Otherwise, they are identical in content and treatment. All channels were loaded with 10 µg protein.

CELLULAR DISTRIBUTION OF DIFFERENT MAPS
IN THE ADULT BRAIN

The initial observation that stimulated our interest in MAPs was that an antiserum against high molecular weight MAPs stained only neuronal dendrites in brain sections.[11] It transpired that this antiserum reacts selectively with MAP-2.[5] There is now a considerable body of evidence, using both monoclonal and polyclonal antibodies, showing that MAP-2 is much more highly concentrated in neuronal dendrites than in any other cellular compartment.[6,7,12-14] In fact according to our results, MAP-2 is exclusively associated with dendrites both in the adult brain[6,7] and throughout development.[15,16]

This conclusion is supported by the observation that MAPs 1, 3, and 5, all of which are less abundant than MAP-2, can nevertheless be readily detected in axons. Furthermore, the pattern of axonal staining is itself characteristic for each of these three MAP species. Thus the only axons in adult brain in which MAP-3 appears are those rich in neurofilaments,[5,10] whereas MAP-1 occurs in myelinated fibers (at reproducibly lower levels than MAP-3) and in some but not all granule cell axons of the cerebellum[7,14,17,18] Similarly, the absence of MAPs 1, 2, and 5 from glia contrasts with the strong association of MAP-3 with glia throughout the brain in adult animals.[5,10] Each of the four MAPs that we have examined by immunohistochemistry with monoclonal antibodies thus presents a cellular staining pattern reproducibly different from the others (TABLE 1). That these are real differences and not artifacts is indicated by the fact that the patterns characteristic of each MAP are seen in serial sections cut from a single tissue block stained side by side with the same reagents, differing only in the particular anti-MAP anitbody with which they were treated. Furthermore, the antibodies themselves are all monclonal IgG_1's, so artifacts due to differences in physicochemical properties of the antibodies are unlikely. The characteristic staining pattern of each MAP is also maintained under two very different conditions of tissue preparation, the one involving cardiac perfusion with aldehydes, the other, acetone postfixation of rapidly frozen fresh tissue.[7]

DEVELOPMENTAL CHANGES IN MAP EXPRESSION IN BRAIN

It has been well demonstrated that the growth of neuronal processes is absolutely dependent on the assembly of microtubules.[19] This raises the possibility that proteins such as MAP-2 and tau, which can promote microtubule assembly *in vitro,* may play a role in regulating process formation during neuronal differentiation. The selective association of MAP-2 with dendrites further suggests that individual microtubule proteins may be specifically involved in the structural differentiation of a particular morphological compartment or microdomain of the developing neuron, a process we call neuronal microdifferentiation.[5]

Immunocytological staining indeed shows that MAP-2 in developing dendrites is highly concentrated in a core region behind the growth cone.[5,15] On this basis we have hypothesized that MAP-2 may be involved in mediating the transition from the flexible phase of growth exemplified by the growth cone into the fixed form of the established dendritic tree. More recently, we have compared the developmental appearance of MAPs 1, 2, and 3 in the cerebellum[15] and cerebral cortex.[16] In both tissues these three MAPs show the same basic characteristics. MAP-1 is associated with both axons and dendrites throughout development and is more highly concentrated in developing axons than in axons in the adult brain. MAP-2 is, according to our observations, exclusively associated with dendrites throughout development and is highly concentrated in the distal growing branches of the dendritic tree in both cerebellar Purkinje cells[15] and cortical pyramidal cells.[16] MAP-3 is strongly expressed in neurons during early development in both tissues, but after the second postnatal week in the rat, rapidly disappears from all neuronal compartments[15,16] other than neurofilament-rich axons and is consequently mainly found in glial cells in mature brain tissue.[5,10]

From the point of view of quantitative expression of MAPs, equally striking changes occur during early brain development.[16] These are also reflected in the different chemical compositions of microtubules prepared from newborn and adult brain (FIGURE 2). MAP-1 levels are very low in the newborn brain and increase progressively to reach adult levels in the fourth postnatal week. MAP-2 levels show a small increase over the same period, whereas MAP-3 shows the opposite phenomenon, being present in higher amounts in the newborn brain than in adult tissue.[16] MAP-5 is also more abundant in newborn brain than in the adult (B. Riederer and A. Matus, in preparation).

TABLE 1. Properties of Microtubule-Associated Proteins in Adult Brain

Name	Chemical Composition[a]	Cellular Distribution	References
MAP-1	broad band M_r 350,000	neurons, stronger in dendrites than axons	6,17
MAP-2	two peptides M_r 280,000	dendrites	6,7
MAP-3	two peptides M_r 180,000	neurofilament-rich axons and glia	5,10 [b]
MAP-5	broad band M_r 320,000	neurons, axons, and dendrites	

[a]See FIGURE 2.
[b]B. Riederer and A. Matus, in preparation.

These changes are not simply the result of an increase or decrease in the gross levels of the MAP proteins; changes in the abundance of individual peptide subspecies also take place.[16] The most striking of these occur in the expression of MAP-2 and MAP-3. In the adult brain, both of these proteins are present as pairs of peptides that appear as doublet bands on SDS-gel blots,[16] (FIGURE 2). In each case only one of the two bands is present in the neonatal brain. In the case of MAP-2, it is the faster migrating component, MAP-2b, which is present at birth (references 16, 20, and 21 and FIGURE 2). In addition, three smaller peptides (molecular weights around 70,000) that bear the MAP-2 antigen are markedly more abundant in the newborn brain. These are apparently distinct species from the 280,000 MAP-2 bands, and they are not generated when MAPs are exposed to proteolytic enzymes present in brain supernatant. In the adult, the 70,000 material appears as a doublet. The level of 70,000 MAP-2 in the developing brain drops sharply at the same time as the high molecular weight MAP-2a band makes its appearance, between postnatal days 10 and 15 (FIGURE 3). There is, in addition, an interesting change in the endogenously generated proteolytic fragments from MAP-2 280,000 during this period. Thus, fragments evidently generated from the MAP-2b component (b 1–3, FIGURE 3) are joined by fragments apparently derived from the MAP-2a component (a 1 and 2, FIGURE 3). These differences between their endogenous peptide maps suggest that the two 280,000 components are independent protein species rather than one being a posttranslational modification of the other.

In the case of MAP-3, it is the upper band, MAP-3a, which is alone present in embryonic brain. The faster migrating component, MAP-3b, appears at about the time of birth and then increases in level until it is more abundant than MAP-3a at 10 days

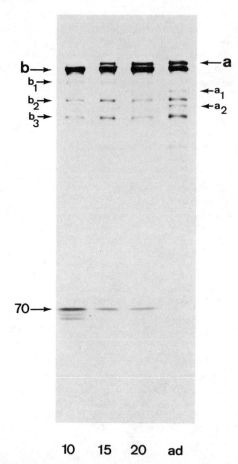

                          10   15   20   ad

FIGURE 3. Development of MAP-2 peptides in neonatal rat brain. An anti-MAP-2 stained nitrocellulose blot of a single slab gel in which 10 μg samples of rat brain supernatant proteins were run in adjacent channels. The ages of the animals from which the samples were taken is shown below in days after birth (ad=adult). For other labeling see text.

postnatal.[16] After this, there is a rapid and dramatic fall in the levels of both components, which occurs simultaneously with the disappearance of MAP-3 antigen from most neurons throughout the brain (references 15, 16 and FIGURE 7).

For each of the MAPs we have investigated, all of the peptides detected by the antibody reassemble with microtubules through several cycles of polymerization (see FIGURE 3 for data on MAPs 1, 2, and 3, data for MAP-5, B. Riederer and A. Matus, in preparation). This confirms that all of these molecules, despite their differing cytological localizations and patterns of expression during development, possess microtubule-binding sites.

Developmental Changes in Cytological Distribution of MAPs

During the period when these changes in quantity and peptide composition of the various MAPs are occurring, their cellular distribution within the developing brain also undergoes striking alterations. It is important to note that during this neonatal period, when these changes occur, all of the MAPs coassemble selectively with microtubules (FIGURE 4) as they do in the adult. In the case of MAPs 1, 2, and 3, the changes that occur in cerebellum[15] and cerebral cortex[16] share characteristics found throughout the brain. FIGURES 5–8 illustrate the cytological pattern for each of the four MAPs in the hippocampus at birth (A), at postnatal day 10 (B) and in the adult (C). Thus MAP-1 shows a steady increase in neurons from the barely detectable levels present at birth. At no stage is it detectable in glia (FIGURE 5). MAP-2 is abundant throughout the developmental sequence, like MAP-1 being expressed only in neuron and, in addition, being exclusively associated with dendrites (FIGURE 6). MAP-3 shows the most striking change, at first accumulating in neurons from which it then vanishes leaving a purely glial pattern of staining (FIGURE 7). This change also occurs in cerebellum and cortex;[15,16] in the hippocampus, the neurofilament-rich mossy fiber tract that innervates the CA3 region also continues to express MAP-3 in the adult.[10] MAP-5, like MAPs 1 and 2, shows a neuron-specific distribution pattern in hippocampus and is present in both axons and dendrites throughout development (FIGURE 8).

WHAT DO WE KNOW ABOUT MAP FUNCTION?

These studies with monoclonal antibodies have improved our knowledge of the nontubulin brain microtubule proteins and have at the same time opened up new questions. For example, it is clear that known proteins such as MAP-2 occur in multiple forms whose expression may be independently regulated during development. How this relates to their cellular function and particularly their highly selective association with growing dendrites is not known. Even in the adult brain the existence of two or more related MAP-2 peptides, all of which selectively bind to microtubules *in vitro*, suggests functional heterogeneity, which is also of unknown significance at present. The same obviously applies to the doublet peptides of MAP-3 in adult brain and particularly with respect to their different time courses of appearance during development.

Another important question concerns the functional nature of the relationship between the MAP proteins and microtubules *in vivo*. These molecules are known as MAPs (microtubule-associated proteins) because they bind to microtubules *in vitro*. Providing nothing further is assumed about them, this is a reasonable and useful categorization. The diversity of their cellular distribution and the striking differences in their developmental patterns of expression, however, suggest that the binding to microtubules is only one facet of the function of these molecules in living cells.

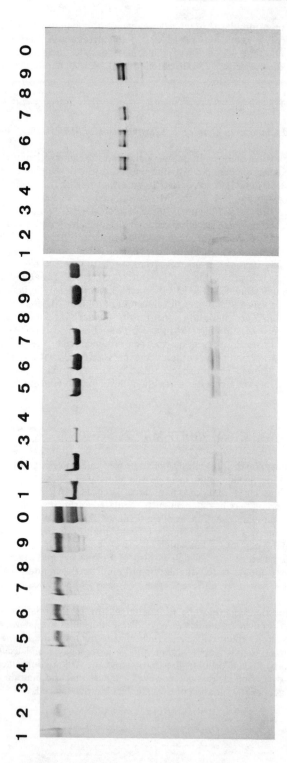

FIGURE 4. Selective assembly of MAPs 1, 2, and 3 with microtubules. Samples containing 10 μg of protein taken at each stage during the isolation of microtubules from newborn rat brain were run on SDS-gels, blotted onto nitrocellulose and stained with one or other anti-MAP antibody as indicated. Samples are as follows: 1, homogenate; 2, first supernatant; 3, first cold pellet; 4, supernatant to first warm cycle pellet; 5, first cycle warm pellet (microtubules); 6, cold supernatant from no. 5 (depolymerized microtubules); 7, cold insoluble pellet from no. 5; 8, supernatant to second cycle warm pellet; 9, second cycle warm pellet (microtubules); 0, adult second cycle microtubules.

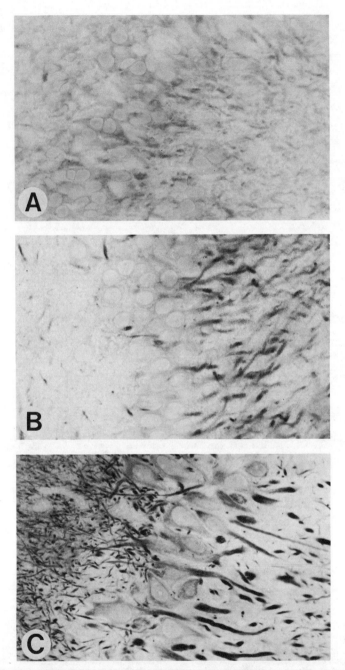

FIGURE 5. Sections of rat hippocampus taken from all three ages. All the sections in the **A** panels are from the same block of newborn rat brain, all of the **B** panels are from the same block of 10-day old tissue, and all of the **C** panels are from a single block of adult hippocampus. Staining: FIGURE 5, anti-MAP-1; FIGURE 6, anti-MAP-2; FIGURE 7, anti-MAP-3; FIGURE 8, anti-MAP-5. Magnification in all micrographs is ×220.

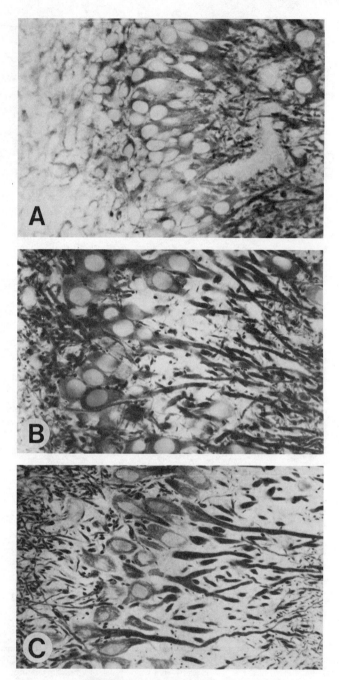

FIGURE 6. For legend, see FIGURE 5.

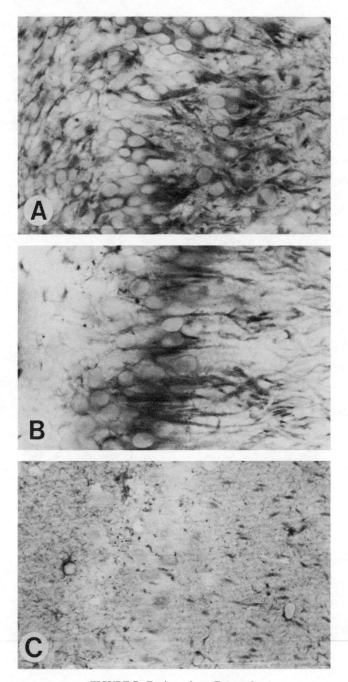

FIGURE 7. For legend, see FIGURE 5.

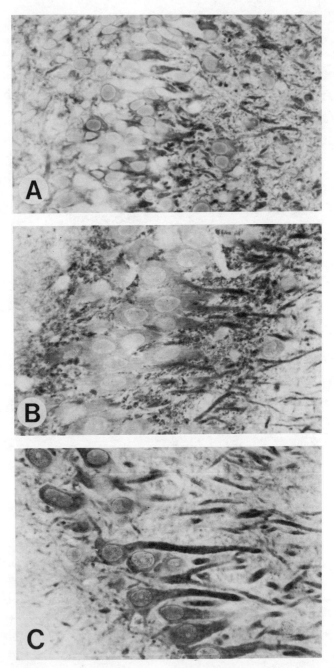

FIGURE 8. For legend, see FIGURE 5.

REFERENCES

1. FELLOUS, A., J. FRANCON, A.-M. LENNON & J. NUNEZ. 1977. Microtubule assembly *in vitro*. Eur. J. Biochem. **78:** 167–174.
2. HERZOG, W. & K. WEBER. 1978. Fractionation of brain microtubule-associated proteins. Isolation of two different proteins which stimulate tubulin polymerisation *in vitro*. Eur. J. Biochem. **92:** 1–8.
3. MURPHY, D. B. & G. G. BORISY. 1975. Association of high molecular weight proteins with microtubules and their role in microtubule assembly *in vivo*. Proc. Natl. Acad. Sci. USA **72:** 2696–2700.
4. WEINGARTEN, M. D., A. H. LOOKWOOD, S.-Y. HWO & M. W. KIRSCHNER. 1975. A protein factor essential for microtubule assembly. Proc. Natl. Acad. Sci. USA **72:** 1858–1862.
5. MATUS, A., G. HUBER & R. BERNHARDT. 1983. Neuronal microdifferentiation. Cold Spring Harb. Symp. Quant. Biol. **48:** 775–782.
6. BERNHARDT, R. & A. MATUS. 1984. Light and electron microscopic studies on the distribution of microtubule-associated protein 2 in rat brain: A difference between dendritic and axonal cytoskeletons. J. Comp. Neurol. **226:** 203–219.
7. HUBER, G. & A. MATUS. 1984. Differences in the cellular distributions of two microtubule-associated proteins, MAP 1 and MAP 2, in rat brain. J. Neurosci. **4:** 151–160.
8. SANDOVAL, I. G. & K. WEBER. 1978. Calcium-induced inactivation of microtubule formation in brain extracts. Eur. J. Biochem. **92:** 463–470.
9. SCHERSON, T., B. GEIGER, Z. ESHHAR & U. LITTAUER. 1982. Mapping of distinct structural domains of microtubule-associated protein 2 by monoclonal antibodies. Eur. J. Cell Biol. **129:** 295–302.
10. HUBER, G., D. ALAIMO-BEURET & A. MATUS. 1985. MAP 3: Characterization of a novel microtubule-associated protein. J. Cell Biol. **100:** 496–507.
11. MATUS, A., R. BERNHARDT & T. HUGH-JONES. 1981. HMWP proteins are preferentially associated with dendritic microtubules in brain. Proc. Natl. Acad. Sci. USA **78:** 3010–3014.
12. CACERES, A., L. I. BINDER, M. R. PAYNE, P. BENDER, L. REBHUN & O. STEWARD. 1984. Differential subcellular localization of tubulin and the microtubule-associated protein MAP 2 in brain tissue as revealed by immunocytochemistry with monoclonal hybridoma antibodies. J. Neurosci. **4:** 394–410.
13. DE CAMILLI, P., P. E. MILLER, F. NAVONE, W. E. THEURKAUF & R. B. VALLEE. 1984. Distribution of microtubule-associated protein 2 in the nervous system of the rat studied by immunofluorescence. Neuroscience **11:** 819–846.
14. WICHE, G., E. BRIONES, H. HIRT, R. KREPLER, U. ARTLIEB & H. DERK. 1983. Differential distribution of microtubule-associated proteins MAP-1 and MAP-2 in neurons of rat brain and association of MAP-1 with microtubules of neuroblastoma cells (clone N2A) EMBO J. **2:** 1915–1920.
15. BERHARDT, R., G. HUBER & A. MATUS. 1985. Differences in the developmental patterns of three microtubule-associated proteins in the rat cerebellum. J. Neurosci. **5:** 977–991.
16. RIEDERER, B. & A. MATUS. 1985. Differential expression of distinct microtubule-associated proteins during brain development. Proc. Natl. Acad. Sci. USA **82:** 6006–6009.
17. HUBER, G. & A. MATUS. 1984. Immunocytochemical localization of microtubule-associated protein 1 in rat cerebellum using monoclonal antibodies. J. Cell Biol. **98:** 777–781.
18. BLOOM, G. S., T. A. SCHONFELD & R. B. VALLEE. 1984. Widespread distribution of the major polypeptide component of MAP 1 (Microtubule-associated protein 1) in the nervous system. J. Cell Biol. **98:** 320–330.
19. DANIELS, M. P. 1975. The role of microtubules in the growth and stabilization of nerve fibers. Ann. N. Y. Acad. Sci. **253:** 535–544.
20. BURGOYNE, R. D. & R. CUMMING. 1984. Ontogeny of microtubule-associated protein 2 in rat cerebellum: differential expression of the doublet polypeptides. Neuroscience **11:** 157–167.
21. BINDER, L. I., A. FRANKFURTER, H. KIM, A. CACERES, M. R. PAYNE & L. I. REBHUN. 1984. Heterogeneity of microtubule-associated protein 2 during rat brain development. Proc. Natl. Acad. Sci. USA **81:** 5613–5617.

Molecular Aspects of MAP-1 and MAP-2: Microheterogeneity, *in Vitro* Localization and Distribution in Neuronal and Nonneuronal Cells[a]

GERHARD WICHE, HARALD HERRMANN, J. MITCHELL
DALTON, ROLAND FOISNER, FRANZ E. LEICHTFRIED,
HANS LASSMANN[b], CHRISTIANE KOSZKA, AND
ESTHER BRIONES

Institute of Biochemistry and [b]Neurological Institute
University of Vienna Medical School
Vienna, Austria

INTRODUCTION

Microtubule-associated proteins (MAP)-1 and MAP-2, aside from tubulin, are the major components of microtubules polymerized *in vitro* from mammalian brain. As recently shown, using various, mainly immunological methods, MAP-1[1,2] as well as MAP-2[2] are apparently of widespread occurrence not only in neuronal, but also nonneuronal cells and tissues. It is generally assumed that MAP-1 and MAP-2 play an important role as cytoskeletal connecting links between microtubules and a variety of different cell organelles, including other cytoskeletal filaments (for a recent review see reference 3). Over the last few years we have studied the structure and function of high M_r MAPs with a special focus on possible differences between MAPs from neuronal versus nonneuronal sources. Here we report on the microheterogeneity of MAP-1 and MAP-2 in mammalian brain, their *in vitro* localization, and their distribution in regenerating peripheral nerve and nonneuronal cultured cell lines.

MICROHETEROGENEITY

To prepare MAP-1 and MAP-2 from hog brain by repeated rounds of microtubule polymerization/depolymerization, we followed the protocol of Karr *et al.*[4] by which MAP-1 and MAP-2, contrary to other procedures, are isolated in roughly equal amounts. When analyzed on high-percentage polyacrylamide gels, MAP-1 and MAP-2 each migrated as single bands with an apparent M_r of around 350,000 and 300,000. Upon analysis on low percentage gels (5%), both MAP-1 and MAP-2 split into a number of bands (FIGURE 1). The major components visualized by Coomassie blue staining (FIGURE 1, lane 1) were designated MAP-1A, -1B, and -1C, and

[a]This work was supported by Grant 5263 from the Austrian Science Research Fund (Österreichischer Fonds zur Förderung der Wissenschaftlichen Forschung) and Grant 18.853/ 2-10/83 from the Austrian Ministry of Science and Research (Bundesministerium für Wissenschaft and Forschung).

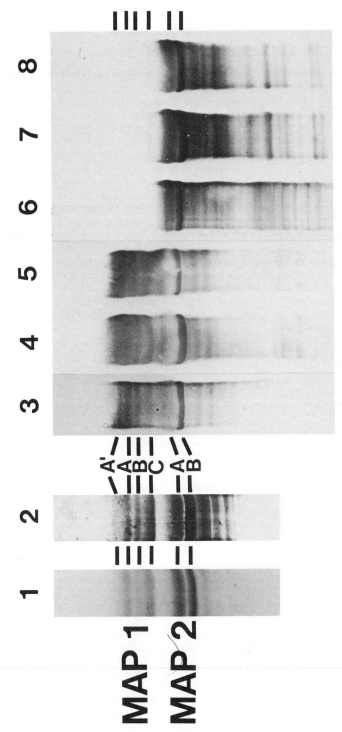

FIGURE 1. Electrophoresis and immunoblots of MAP-1 and MAP-2 on low percentage gels. Microtubule proteins prepared by two (lanes 1, 2, 5, and 8) or one round of polymerization/depolymerization (lanes 4 and 7) and soluble cell extracts containing cold-instable microtubule proteins (lanes 3 and 6) were run on 5% sodium dodecyl sulfate (SDS) polyacrylamide gels containing 25% glycerol.[9] Lane 1, Coomassie brilliant blue staining; lane 2, silver staining;[10] lanes 3–5, immunoblot using antiserum to MAP-1; lanes 6–8, immunoblot using antiserum to MAP-2. Only the upper parts of gels were processed.

MAP-2A and -2B, according to Bloom et al.,[5] who observed similar band splitting with high-M_r MAPs from calf brain. In addition, a minor band migrating slightly above MAP-1A, termed MAP-1A', and a number of bands migrating ahead of MAP-2B were observed after silver staining (FIGURE 1, lane 2).

All four MAP-1 bands were immunoreactive with anti-MAP-1 antibodies[6,7] raised against the entire MAP-1 group (FIGURE 1, lanes 3–5). In addition, MAP-2B was prominently stained, whereas MAP-2A was hardly stained at all. An antiserum raised to a mixture of MAP-2A and -2B[6,7] reacted, as expected, with both of these polypeptide bands (FIGURE 1, lanes 6–8). The group of bands below MAP-2 revealed by silver staining (FIGURE 1, lane 2) was also intensively stained on immunoblots with antiserum to MAP-2 (FIGURE 1, lanes 6–8), whereas it was only slightly reactive with antiserum to MAP-1 (FIGURE 1, lanes 3–5). Because these bands were hardly detectable by conventional protein staining (FIGURE 1, lane 1), their relative amounts were clearly overestimated by immunoblotting.

When immunoblots of MAPs present in fractions obtained early during repeated rounds of polymerization/depolymerization (FIGURE 1, lanes 3 and 6) were compared to those of later fractions (FIGURE 1, lanes 4, 5, 7, and 8), no significant changes in number and relative proportion of bands immunostained by antisera to MAP-1 and to MAP-2 were observed. Similar patterns were observed on immunoblots obtained from brain extracts boiled directly in SDS solutions.[8] Thus, the multiplicity of the major bands in the MAP-1 and MAP-2 group definitely was not a result of the microtubule polymerization procedure.

To compare the major subcomponents of MAP-1 and MAP-2 by proteolytic fingerprinting, individual MAP bands were radioiodinated using chloramine T and digested with trypsin. The fragments were analyzed in two dimensions.[11] Peptide patterns generated from MAP-1A (FIGURE 2A) and MAP-1C (FIGURE 2B) were very similar. Those from MAP-2A and MAP-2B were virtually identical (FIGURES 2C and D). A certain degree of homology became apparent also by comparing the patterns of MAP-1s with those of MAP-2s, supporting a previous report on partial structural homology between MAP-1 and MAP-2.[7]

After in vitro radiolabeling of microtubule proteins with $[\gamma - {}^{32}P]$ adenosine triphosphate (ATP), limited digestion (10 ng V8 protease) of MAP-2A and MAP-2B and analysis on 10% gels revealed five major fragments of M_r 106,000, 76,000, 71,000, 64,000, and 54,000 (FIGURE 3A, lanes 1 and 2). When the amount of protease was increased to 50 ng, undigested MAP-2 as well as the 106,000 fragment disappeared, and a new band at 42,000 appeared (FIGURE 3A, lanes 3 and 4). At 250 ng of protease the 54,000 and 42,000 fragments became most prominent in addition to material at the front possessing a M_r less than 13,700 (FIGURE 3A, lanes 5 and 6). When a similar experiment was performed on 7.5% gels, an additional fragment of M_r around 160,000 appeared. MAP-2s phospholabeled in the presence of cyclic adenosine monophosphate (cAMP) (FIGURE 3B, lanes 3 and 4) showed a very similar pattern when compared with MAP-2s from unstimulated incubations (FIGURE 3B, lanes 1 and 2), except for an enhanced labeling of two major fragments around 35,000 and above 100,000 (FIGURE 3B, stars). Under all conditions tested, the fragments generated from MAP-2A and MAP-2B were of undistinguishable size and of similar proportions, again indicating that these two polypeptides must be virtually identical in structure. The MAP-2 fragmentation patterns were also similar to those of MAP-1s run in parallel. When digestions were performed under the conditions of FIGURE 3B, lanes 1 and 2, the seven fragments indicated by arrowheads were generated also from MAP-1s, although with differences in relative proportions. A direct comparison of MAP-2A with MAP-1C at a low degree of digestion is shown in FIGURE 3C. With the exception of the two

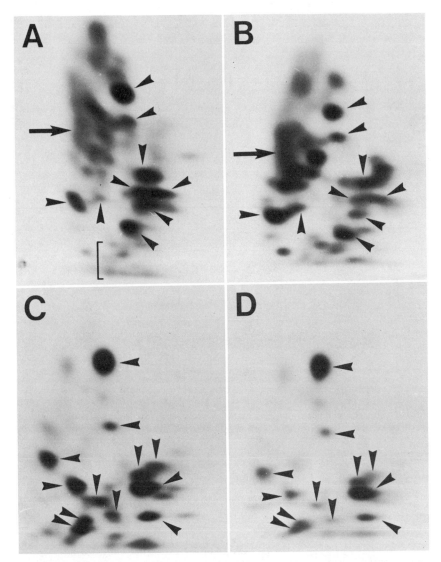

FIGURE 2. Two-dimensional peptide mapping of MAP-1 and MAP-2 subcomponents. After electrophoresis on 5% gels, proteins were stained with 0.2% Coomassie brilliant blue in 50 mM Tris (pH 7.2). Bands to be analyzed were excised and radioiodinated with chloramine T[11] followed by extensive digestion with trypsin. **A**: MAP-1A; **B**: MAP-1C; **C**: MAP-2A; **D**: MAP-2B. Electrophoresis was from left to right and chromatography from bottom to top. Arrowheads in **A, B, C,** and **D** point to common peptides of MAP-1A and 1C, and MAP-2A and 2B, respectively. Arrows in **A** and **B** indicate common ring-like cluster of peptides. The bracket in **A** indicates peptides of low mobility in the second dimension that are only weakly labeled by radioiodination but that are highly labeled after *in vitro* phosphorylation of MAPs (see FIGURE 5).

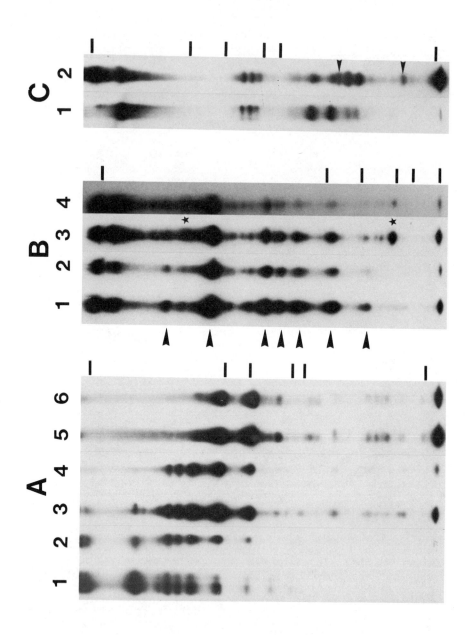

fragments indicated by arrowheads, the sizes of all major fragments generated from both polypeptide species were identical.

Partial digestion with V8 protease of MAP-1A, -1B, and -1C, phospholabeled in the absence of cAMP, gave rise to breakdown products of identical M_r, although individual fragments varied in relative proportions (FIGURE 4, lanes 1–3). Two new prominently labeled fragments were generated from all three MAP-1s, and three additional fragments from MAP-1B and MAP-1C after phosphorylation in the presence of cAMP (FIGURE 4, lanes 4–6). When these patterns were compared with those of MAP-2 analyzed in parallel (FIGURE 4, lanes 7 and 8), the four major fragments of MAP-2A and -2B were found in all MAP-1 digests, also (FIGURE 4, arrows); four more fragments common to MAP-1s and MAP-2s labeled in the presence of cAMP became clearly apparent only after long exposure times (FIGURE 4, arrowheads).

The similarity of γ^{32}P-labeled phosphopeptides generated from individual MAP-subcomponents became evident also after fingerprint analysis of tryptic digests. MAP-1A and MAP-1C yielded virtually identical peptide maps consisting of eight major radioactive spots (FIGURE 5, A and B), and MAP-2A and MAP-2B were hardly distinguishable (FIGURE 5, C and D). Moreover, MAP-1 and MAP-2 patterns strongly resembled one another. In the second dimension (chromatography) in general, phosphopeptides were less mobile than the major peptides observed after radioiodination (FIGURE 2). It is likely that the phosphopeptides shown in FIGURE 4 correspond to a group of spots, which were only weakly labeled by iodination (see bracket in FIGURE 2A). These fragments most likely contain no tyrosine and hence were radiolabeled by the chloramine T method to a low extent on other amino acids.

To examine whether phosphorylation affected the electrophoretic mobility and hence, the apparent microheterogeneity of high M_r MAPs, phospholabeling was performed under various conditions. When microtubule proteins were labeled with $[\gamma^{32}\text{P}]$ATP for 10 min under cAMP stimulation, 90% of the total radioactivity resided in MAP-2. MAP-2A and MAP-2B were labeled in a ratio of 2.5:1 (FIGURE 6, lanes 1, 5, and 6). Postincubation ("chasing") with unlabeled guanosine triphosphate (GTP) for 20 min induced a slight shift of both the MAP-2A and the MAP-2B gel bands to an apparent higher molecular weight. This shift was not accompanied by a reduction in label indicating that the radioactive P_i on MAPs was not exchanged by GTP (FIGURE 6, lane 2). Simultaneous incubation of microtubules with $[\gamma\text{-}^{32}\text{P}]$ATP and 1 m$M$

FIGURE 3. Autoradiography of one-dimensional peptide maps of MAP-2 subcomponents phosphorylated *in vitro:* comparison with MAP-1. **A:** Microtubule proteins were labeled with 2.5 μM $[\gamma\text{-}^{32}\text{P}]$ATP in the absence of cAMP as described,[7] and excised MAP bands were partially digested[12] with 10 ng (lanes 1 and 2), 50 ng (lanes 3 and 4), and 250 ng (lanes 5 and 6) of *Staphylococcus aureus* V8 protease. Lanes 1, 3, and 5: MAP-2A; lanes 2, 4, and 6: MAP-2B. **B:** Microtubule proteins were labeled with 2.5 μM $[\gamma\text{-}^{32}\text{P}]$ATP in the absence (lanes 1 and 2) or presence (lanes 3 and 4) of 10 μM cAMP, and excised bands were digested with 10 ng SV8 protease. Lanes 1 and 3: MAP-2A; lanes 2 and 4: MAP-2B. Arrowheads point to common fragments obtained from MAP-1A, 1B, and 1C run in parallel lanes (not shown). Stars indicate fragments prominently phosphorylated only after cAMP stimulation. **C:** Microtubule proteins were labeled with 2.5 μM $[\gamma\text{-}^{32}\text{P}]$ATP in the presence of 10 μM cAMP, and excised bands were digested with 100 ng SV8 protease. Lane 1: MAP-2B; lane 2: MAP-1C. Arrowheads point to fragments unique to MAP-1C. In **A**, separation gels consisted of 10%, in **B** of 7.5% polyacrylamide, and in **C** of 10% polyacrylamide plus 7 M urea. Bars on the right indicate positions of M_r marker proteins: MAP-2 (300,000), tubulin (53,000), 3-phosphoglycerate kinase (44,000), glyceraldehyde-3-phosphate dehydrogenase (36,000), DNase I (31,000), RNase A (13,700). Full length gels are shown.

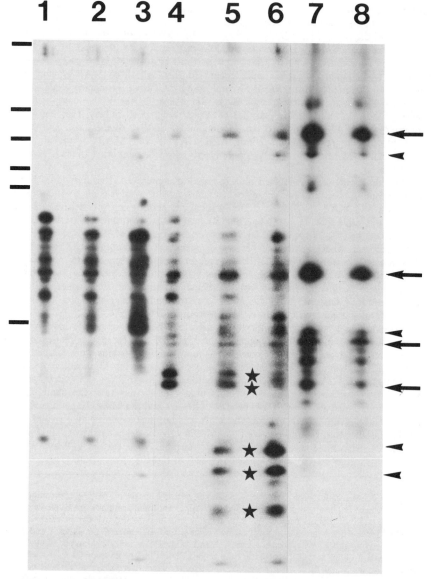

FIGURE 4. Autoradiography of one-dimensional peptide maps of MAP-1 subcomponents phosphorylated *in vitro:* comparison with MAP-2. Microtubule proteins were incubated with 2.5 μM [γ-^{32}P]ATP, as described.[7] Samples analyzed in lanes 4–8 were labeled in the presence of 10 μM cAMP. Digestion was performed with 250 ng V8 protease, and fragments separated in parallel on 10% gels. Lanes 1 and 4, MAP-1A; lanes 2 and 5, MAP-1B; lanes 3 and 6, MAP-1C; lane 7, MAP-2A; lane 8, MAP-2B. M_r markers (bars) are as in FIGURE 3. Major fragments common to MAP-2s and MAP-1s are indicated by arrows; minor common fragments are indicated by arrowheads. Stars indicate fragments of MAP-1s labeled only under cAMP stimulation.

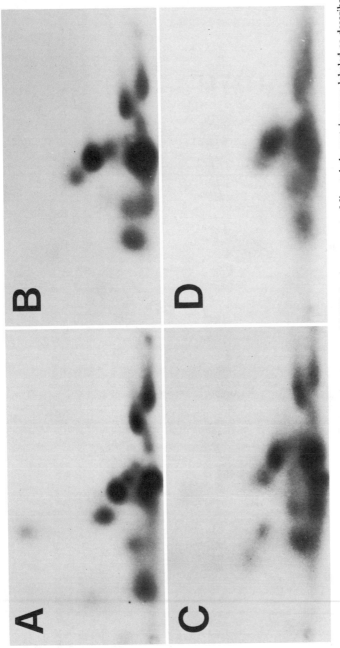

FIGURE 5. Two-dimensional peptide mapping of ^{32}P-labeled MAP-1 and MAP-2 subcomponents. Microtubule proteins were labeled as described in FIGURE 3, and digestion of MAPs in gel bands was performed using trypsin. Electrophoresis was from left to right, chromatography from bottom to top. **A:** MAP-1A; **B:** MAP-1C; **C:** MAP-2A; **D:** MAP-2B.

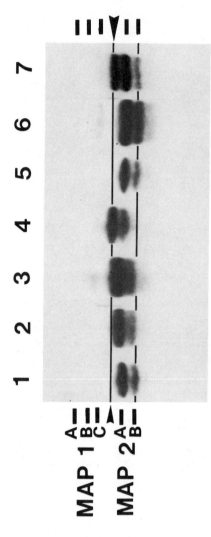

FIGURE 6. Effect of phosphorylation on the electrophoretic mobility of MAP-2 subcomponents. Microtubule proteins cycled for two times were preincubated with 10 μM cAMP for 3 min and then labeled with 2.5 μM [γ-^{32}P]ATP for 5 minutes. Reactions were either stopped by addition of electrophoresis sample buffer (lanes 1, 5, and 6) or continued for 20 min in the presence of 1 mM GTP (lane 2) or 1 mM ATP (lane 4) prior to addition of SDS. In lane 3, 1 mM GTP was present from the beginning and phosphorylation stopped after 25 min by addition of SDS. Lane 7 is a mixture of equal aliquots from lanes 1 and 4. Electrophoresis was on 5% polyacrylamide gels; in lane 6, the protein loading was three times higher than in all other lanes. Arrowheads and upper line indicate new position of MAP-2A; lower line indicates the position of original MAP-2B.

unlabeled GTP for 25 min did not inhibit incorporation of the label into MAPs, and shifting of MAP-2A and MAP-2B was slightly enhanced possibly due to the longer incubation with GTP. Both MAP-2A and MAP-2B, however, were not resolved into discrete bands, probably because their phosphorylation was not complete (FIGURE 6, lane 3). By contrast, "chasing" with 1 mM ATP instead of GTP resulted in a more pronounced upward shift of well-resolved MAP-2A and MAP-2B in unison (FIGURE 6, lane 4). The radiolabel residing in both MAP-2A and MAP-2B was not reduced compared to the controls (FIGURE 6, lanes 1, 5, and 6), indicating that also in this case, radiolabeled P_i-residues were not exchanged. Under all these conditions, neither a change in phosphorylation of the MAP-1 group subcomponents, nor a shift of these subcomponents to higher M_r was observed. These results allow the following conclusions to be drawn: Increased phosphorylation of MAP-2s reduces their mobility on SDS gels. As previously shown,[13] MAP-2 is phosphorylated more efficiently by ATP than by GTP. With GTP, other sites on MAP-2 must become phosphorylated as compared with ATP, because unlabeled GTP at a four-hundred-fold higher concentration over labeled ATP did not inhibit the incorporation of label from ATP. Phosphatase activity seems to be low in our preparations, because phosphate once incorporated is not "chased" out. MAP-1s, as isolated, might be saturated with phosphate in contrast to MAP-2s.

IN VITRO LOCALIZATION

To visualize the spatial arrangement of MAPs, hog brain microtubules reconstituted from phosphocellulose-purified tubulin and MAP-fractions were decorated with rabbit-antibodies to MAP-1 or MAP-2 followed by secondary antibody labeled with colloidal gold particles. With both antibody preparations, all microtubules visualized on the grids were uniformly decorated (FIGURE 7A). At higher magnification, a helical arrangement of antibody/antigen complexes and, thus, of MAP-1 and MAP-2 became clearly apparent (FIGURE 7, **B** and **C**). The periodicity of helical turns was 100 nm in both cases. Using mouse anti-MAP-1 and rabbit anti-MAP-2 antibodies, and secondary antibodies labeled with colloidal gold particles of two different sizes, double labeling experiments were carried out. Again, helically arranged surface extensions, in this case apparently consisting of both antigens, were visualized (FIGURE 7D). On microtubules polymerized in the presence of taxol, helical arrangements of MAP-1 and MAP-2 were less frequently observed (FIGURE 7E), and microtubules were decorated less uniformly. Strikingly, however, bulky extensions protruding from the polymer's surface were heavily decorated with antibodies to MAP-1 (FIGURE 7E) and antibodies to MAP-2 (not shown). Often these extensions seemed to interconnect adjacent microtubules. Thus, taxol treatment seemed to alter the molecular arrangement of both MAPs on the surface of microtubules assembled *in vitro*.

DISTRIBUTION IN REGENERATING PERIPHERAL NERVE

The differential distribution of MAP-1 and MAP-2 to microtubules of neuronal tissues has been demonstrated by immunocytochemistry using polyclonal[2] as well as monoclonal [5,16] antibodies. There is general agreement that MAP-2 is exclusively found in dendrites and cell bodies of neurons, and that MAP-1 is also found in axons. Moreover, in developing dendrites, MAP-2 was found at distal ends of dendrites before microtubules could be detected there.[17] Studying the cytoskeleton of de- and regenerat-

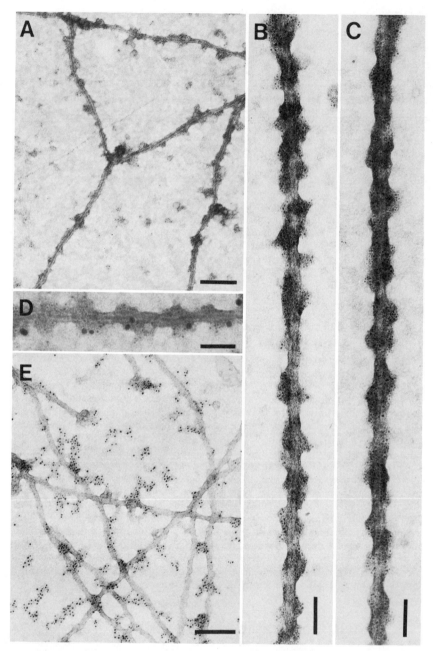

FIGURE 7. Immunolocalization of MAPs on microtubules assembled *in vitro*. For assembly in the absence (**A–D**) or presence of 20 μM taxol (**E**), microtubule protein preparations[4] fractionated by phosphocellulose chromatography[14] were used. Immunodecoration of microtubules with antibodies to MAPs was performed following protocols given in reference 15. For double immunodecoration (**D**), specimens were incubated, in sequence, with mouse antibodies to MAP-1 (undiluted serum), rabbit antibodies to MAP-2 (diluted 1/40), goat anti-rabbit IgG labeled with 5 nm gold particles, and goat anti-mouse IgG labeled with 20 nm gold particles. **A** and **C**: Antibodies to MAP-2; **B**: antibodies to MAP-1; **D**: double immunodecoration (see above); **E**: antibodies to MAP-1. Bars: 200 nm (**A** and **E**); 100 nm (**B**, **C**, and **D**).

ing peripheral nerve, we observed another case of drastic differences in the distribution of high M_r MAPs and tubulin. In normal spinal ganglion cells, the nucleus was in the middle of the nerve cell body (FIGURE 8, A–D). Structures immunoreactive with antibodies to tubulin (FIGURE 8C), to MAP-1 (FIGURE 8A), to MAP-2 (FIGURE 8B), and to neurofilament M_r = 68,000 protein (FIGURE 8D) were evenly distributed over the entire cytoplasm. When the peripheral nerve was cut near the spinal ganglion, "central chromatolysis" was observed, a phenomenon where the cell body of affected spinal ganglion cells is enlarged and the cell nucleus is dislocated to the cell periphery (FIGURE 8, E–H). In these neurons, MAP-1 (FIGURE 8E) and MAP-2 (FIGURE 8F) were still evenly distributed throughout the cytoplasm, whereas tubulin was located

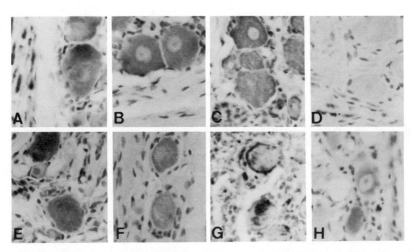

FIGURE 8. Immunolocalization of MAP-1, MAP-2, tubulin and 68.000 neurofilament protein in spinal ganglion. A–D: nerve cells in normal ganglia; E–H: nerve cells in spinal ganglia 14 days after nerve transection. A, E: antiserum to MAP-1; B, F: antiserum to MAP-2; C, G: monoclonal antibodies to tubulin; D, H: antibodies to 68,000 neurofilament protein. Peroxidase-labeled secondary antibodies were used. Note presence of MAP-1, MAP-2, and tubulin throughout the cytoplasm of nerve cell perikarya in A, B, E, and F, but preferential location of tubulin at the cell periphery and in some spotted areas in G. Neurofilament is mainly present in axons with little staining of nerve cell perikarya in D; neurofilament, however, is found predominantly in the central portion of reactive nerve cells in H. Cell nuclei were counterstained with hematoxylin. Magnifications: ×260.

mainly at the cell periphery or in discrete regions of the cells (FIGURE 8G). Neurofilaments were located mainly in the cell center (FIGURE 8H).

CULTURED CELL MAP

The isolation of MAP-1 and MAP-2 from cultured cells by rounds of repeated *in vitro* polymerization/depolymerization is ineffective, largely because of proteolytic breakdown and partial retention of both MAPs in insoluble cell fractions. Microtubules polymerized from isotonic cell extracts of cultured cells by taxol, however, were found to be enriched in both MAPs.[2,15] Upon electrophoresis of taxol-polymerized microtubule preparations from glioma C6 cells on low percentage acrylamide gels, multiple protein bands in the M_r-region of MAPs became apparent (FIGURE 9A, lane 2). The bands migrating in the MAP-1 region were immunoreactive with antibodies to neuronal MAP-1, as revealed by immunoblotting (FIGURE 9B, lane 1). Protein bands migrating at the position of brain MAP-2A were immunoreactive with antibodies to neuronal MAP-2 (FIGURE 9B, lanes 2 and 3) and with antibodies to plectin[21] (FIGURE 9B, lane 4). Depending on the preparation, often additional bands of lower M_r, presumably breakdown products of MAP-2, were immunoreactive with antibodies to MAP-2 (FIGURE 9B, lane 2); these bands were not reactive with antibodies to plectin. Thus, we conclude that a number of distinct polypeptides all of which have M_rs similar

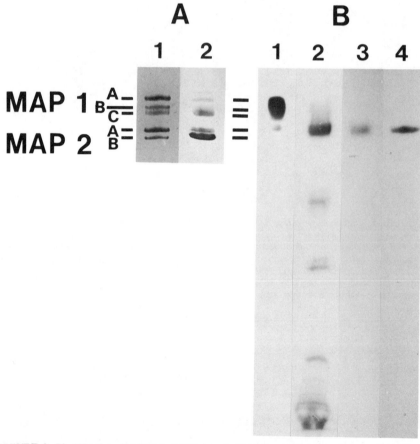

FIGURE 9. Identification of MAP-1, MAP-2 and plectin in taxol-polymerized microtubule preparations from glioma C6 cells. Taxol-polymerized microtubules were prepared as described.[2,15] A: Silver staining.[10] Hog brain standard (lane 1); C6 cell taxol-pellet (lane 2). Only high M_r region of gel is shown. B: Autoradiography of immunoblots using antibodies to MAP-1 (lane 1), MAP-2 (lanes 2 and 3) and plectin (lane 4). Full length gels are shown. Five percent polyacrylamide gels containing 25 percent glycerol were used.

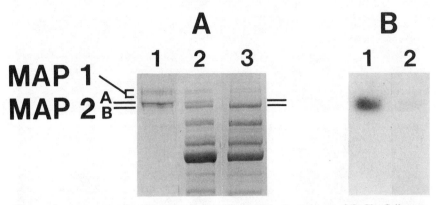

FIGURE 10. Absence of MAP-2 in taxol-pellets prepared in the presence of $CaCl_2$. Cells were homogenized under isotonic conditions, 5 mM $CaCl_2$ were added to the homogenate, and taxol pellets were prepared as described.[2,15] **A**: Coomassie brilliant blue staining. Lanes: 1, hog brain microtubule proteins; 2, control (taxol-pellet prepared without $CaCl_2$); 3, taxol-pellet prepared in the presence of $CaCl_2$. **B**: Autoradiography of immunoblots using antibodies to MAP-2. Lanes: 1, control; 2, taxol-pellet prepared in the presence of $CaCl_2$. Only high M_r regions of gels are shown.

to MAP-1s and MAP-2s from mammalian brain cosediment with microtubules assembled from a nonneuronal cultured cell line.

The isolation of undegraded high M_r polypeptides, particularly MAP-2, from cultured cells was to a large extent dependent on the removal of free Ca-ions. When taxol polymerized microtubules were prepared from glioma C6 cells under strict exclusion of free Ca-ions, a band at the position of MAP-2A was clearly revealed on gels by regular Coomassie blue staining (FIGURE 10A, lane 2). This band, clearly identified as a MAP-2 by immunoblotting (FIGURE 10B), was absent in taxol-pellets prepared in the presence of 5 mM $CaCl_2$.

DISTRIBUTION IN NONNEURONAL CULTURED CELLS

There are two recent reports that establish the widespread occurrence of MAP-1 in nonneuronal cells and tissues. In one of them,[1] the association of MAP-1A with microtubule structures of various cultured cell lines was shown using a monoclonal antibody. In the other,[2] polyclonal antibodies were used to demonstrate the widespread occurrence of MAP-1 in a variety of tissues and cultured cells. The widespread distribution of MAP-2 in nonneuronal cultured cells and tissues is comprehensively shown only in one study.[2] Several other studies, including a recent one,[18] claim MAP-2 to be neuron-specific. Here, we present further evidence for the *in vivo* association of MAP-2 with various microtubule structures of nonneuronal cell types. FIGURE 11A shows the MAP-2 specific staining of cytoplasmic filaments in BALB/c 3T3 cells. These filaments were characterized as microtubules by their sensitivity to colcemid (FIGURE 11D and E). Tubulin paracrystals formed in BALB/c 3T3 cells were unreactive with antibodies to MAP-2 (FIGURE 11F) as formerly observed in primary cell cultures from brain.[19] MAP-2 antibodies stained also spindle microtubules of fibroblast cells in meta- (FIGURE 11B, C) and all other phases of mitosis (data not

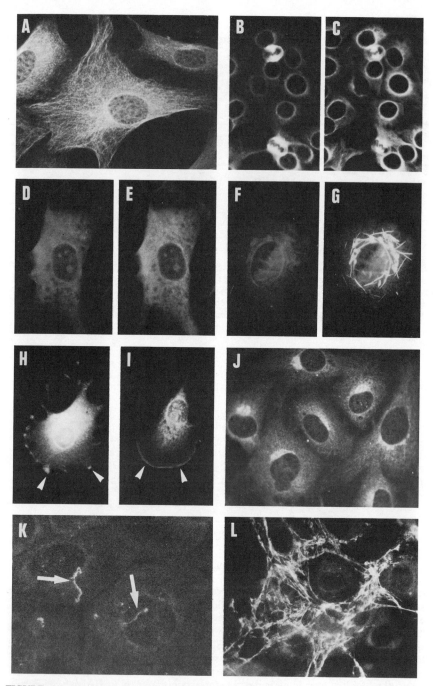

FIGURE 11. Immunolocalization of MAP-2 and MAP-1-related antigens in nonneuronal cultured cells. Cells in A–J were treated with methanol for 5 min at −20°C, and those in **K** and **L** with 3.5% paraformaldehyde for 30 min at room temperature. Primary and secondary antibodies, and their respective dilutions for immunostaining have been described previously.[2] A–G, J, and L: BALB/c 3T3 cells; **H**: HeLa cells; **I** and **K**: MDBK cells. **B, C; D, E;** and **F, G**: double immunofluorescence microscopy. **A, J,** antiserum to MAP-2, texas red optics; **B, D, F, H, I,** affinity purified antibodies to MAP-2, fluorescein optics; **C, E, G,** monoclonal antibodies to α and β tubulin, texas red optics; **K,** antiserum to MAP-1, rhodamine optics; and **L,** affinity purified antibodies to MAP-1, rhodamine optics. Magnifications: **A, J, L:** ×460; **B, C, H, I:** ×350; **F, G:** ×550; **D, E:** × 750; **K:** ×960.

shown). Besides at typical microtubule structures, MAP-2 was specifically located at the leading edge of motile cells (FIGURE 11I), and the peripheral regions of spreading cells (FIGURE 11H). Furthermore, a conspicuous accumulation of MAP-2 was observed in the pericentriolar region of confluent fibroblast cells (FIGURE 11 J), whereas in this case, cytoplasmic microtubules were barely stained. The staining of microtubule structures with antibodies to MAP-2 was generally less pronounced than that with antibodies to MAP-1. For instance, the staining of primary cilia (FIGURE 11K) was only observed with antibodies to MAP-1.

MAPS ARE IMMUNOLOGICALLY RELATED TO AN EXTRACELLULAR MATRIX PROTEIN

One quite unexpected location of a MAP-related antigen, namely in the extracellular matrix of cultured fibroblast cells (FIGURE 11L), was revealed by affinity-purified antibodies to MAP-1. After radiolabeling BALB/c 3T3 fibroblast cells with various radioactive precursors and immunoprecipitation of secreted polypeptides, the antigen related to MAP-1 was identified as a sulfated glycoprotein of $M_r = 205,000$ (FIGURE 12A, and reference 20). This extracellular antigen was found in the spent growth medium as well as in preparations of extracellular matrix proteins.[20] To establish the immunological relationship between the intracellular MAP and the extracellular $M_r = 205,000$ protein on the molecular level, antibodies present in the antiserum to MAP-1 were affinity purified by adsorption to the $M_r = 205,000$ protein band after its transfer to nitrocellulose sheets.[2] These affinity-purified antibodies were immunoreactive with MAP-1 and MAP-2 from brain (FIGURE 10B, lane 3). Thus, they showed a similar specificity as anti-MAP-1 antibodies affinity purified on hog brain MAP-1 (FIGURE 12B, lane 2; and reference 2). (For the cross-reactivity of our anti-MAP-1 antibodies with MAP-2, see also references 2, 6, and 7). We conclude, that the $M_r = 205,000$ protein found in the exterior of mouse fibroblast cells shares immunological determinants with the intracellular proteins MAP-1 and MAP-2.

SUMMARY AND CONCLUSIONS

We have studied various aspects of MAP-1 and MAP-2 from neuronal as well as nonneuronal sources. MAP-1 and MAP-2 polymerized from brain were resolved into a number of subcomponents upon electrophoresis on low percentage gels. Based on peptide mappings performed under a variety of different conditions, we conclude that the three major subcomponents of MAP-1 have very similar, though not identical structures. The two major MAP-2 subcomponents might have identical structure, because their peptide maps were hardly distinguishable. The apparent microheterogeneity of high M_r MAPs is not yet understood on a molecular basis. Proteolysis during isolation or a different degree of phosphorylation, however, seems to be an unlikely cause for microheterogeneity.

When localized on microtubules polymerized *in vitro* by electron microscopy, both MAP-1 and MAP-2 polypeptides apparently form helical arrays on the polymer's surface with periodicities of 100 nm. In the presence of taxol, MAPs form irregular and bulky extensions.

Both MAPs are found to be widespread in neuronal as well as nonneuronal cells. MAP-1- and MAP-2-related polypeptides, together with other high M_r proteins, such as plectin, were associated with microtubules polymerized by taxol from extracts of a nonneuronal cultured cell line. MAP-2 from cultured cells was found to be extremely sensitive to proteolysis, in particular in the presence of free Ca-ions.

MAP-1 and MAP-2 generally were found associated with typical microtubule structures such as interphase and spindle microtubules and primary cilia. A differential distribution of MAP-1 and MAP-2 was clearly evident in neural tissues, where MAP-2 was restricted to cell bodies and dendrites, whereas MAP-1 was present also in

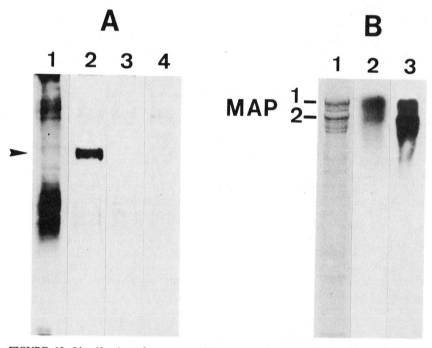

FIGURE 12. Identification of an extracellular antigen immunoreactive with antibodies to MAP-1. **A:** BALB/c 3T3 cells were labeled with [35S]sulfate and proteins secreted into the growth medium analyzed by gel electrophoresis.[20] Fluorographies of gels are shown. Lanes: 1, total proteins released into the medium; 2 and 3, proteins immunoprecipitated with antibodies to MAP-1 and MAP-2, respectively; 4, control (nonimmunoserum). Samples were run on a 7.5% polyacrylamide gel. Arrowhead indicates the position of the $M_r = 205,000$ antigen. **B:** A hog brain microtubule preparation was run on a 6.25% polyacrylamide gel and transferred to nitrocellulose. Transferred proteins were immunostained[2] using antibodies to MAP-1, affinity purified[2] either on nitrocellulose-immobilized MAP-1 bands (lane 2) or on the extracellular $M_r = 205,000$ protein (lane 3). Lane 1, Coomassie blue staining (hog brain microtubule proteins); lanes 2 and 3, autoradiographies.

axons. Moreover, a differential distribution of MAPs and tubulin was observed in de-and regenerating peripheral nerve, and in a few occasions, also with nonneuronal cells. A quite unexpected result was the identification of a protein in the extracellular matrix of cultured fibroblast cells, which has antigenic determinants in common with MAP-1 and MAP-2 from brain.

As a whole, the data presented support a concept in which a family of structurally homologous, though not identical, high M_r polypeptides constitute the crosslinking elements between microtubules and various other cellular components. The structural diversity of these polypeptides might play a role in the development and dynamic changes in the cytoskeletal architecture.

REFERENCES

1. BLOOM, G. S., F. C. LUCA & R. B. VALLEE. 1984. Widespread cellular distribution of MAP-1A (microtubule-associated protein 1A) in the mitotic spindle and on interphase microtubules. J. Cell Biol. **98:** 331–340.
2. WICHE, G., E. BRIONES, C. KOSZKA, U. ARTLIEB & R. KREPLER. 1984. Widespread occurrence of polypeptides related to neurotubule-associated proteins (MAP-1 and MAP-2) in non-neuronal cells and tissues. EMBO J. **3:** 991–998.
3. WICHE, G. 1985. High molecular weight microtubule associated proteins (MAPs): A ubiquitous family of cytoskeletal connecting links. Trends Biochem. Sci. **10:** 67–70.
4. KARR, T. L., H. D. WHITE & D. L. PURICH. 1979. Characterization of brain microtubule proteins prepared by selective removal of mitochondrial and synaptsomal components. J. Biol. Chem. **254:** 6107-6111.
5. BLOOM, G. S., T. A. SCHOENFELD & R. B. VALLEE. 1984. Widespread distribution of the major polypeptide component of MAP 1 (microtubule-associated protein 1) in the nervous system. J. Biol. Chem. **98:** 320–330.
6. WICHE, G., E. BRIONES, H. HIRT, R. KREPLER, U. ARTLIEB & H. DENK. 1983. Differential distribution of microtubule-associated proteins MAP-1 and MAP-2 in neurons of rat brain and association of MAP-1 with microtubules of neuroblastoma cells (clone N_2A). EMBO J. **2:** 1915–1920.
7. HERRMANN, H., R. PYTELA, J. M. DALTON & G. WICHE. 1984. Structural homology of microtubule-associated proteins 1 and 2 demonstrated by peptide mapping and immunoreactivity. J. Biol. Chem. **259:** 612–617.
8. HERMANN, H., J. M. DALTON & G. WICHE. 1985. Microheterogeneity of microtubule-associated proteins, MAP-1 and MAP-2, and differential phosphorylation of individual subcomponents. J. Biol. Chem. **260:** 5797–5803.
9. CARRARO, U. & C. CATANI. 1983. A sensitive SDS-PAGE method separating myosin heavy chain isoforms of rat skeletal muscles reveals the heterogeneous nature of the embryonic myosin. Biochem. Biophys. Res. Commun. **116:** 793–802.
10. MARSHALL, T. 1984. Detection of protein in polyacrylamide gels using an improved silver stain. Anal. Biochem. **136:** 340–346.
11. ELDER, J. H., R. A. PICKETT II, J. HAMPTON & R. A. LERNER. 1977. Radioiodination of proteins in single polyacrylamide gel slices. J. Biol. Chem. **252:** 6510–6515.
12. CLEVELAND, D. W., S. G. FISCHER, M. W. KIRSCHNER & U. K. LAEMMLI. 1977. Peptide mapping by limited proteolysis in sodium dodecyl sulfate and analysis by gel electrophoresis. J. Biol. Chem. **252:** 1102–1106.
13. JAMESON, L., T. FREY, B. ZEEBERG, F. DALLDORF & M. CAPLOW. 1980. Inhibition of microtubule assembly by phosphorylation of microtubule-associated proteins. Biochemistry **19:** 2472–2479.
14. WEINGARTEN, M. D., A. H. LOCKWOOD, S.-Y. HWO & M. W. KIRSCHNER. 1975. A protein factor essential for microtubule assembly. Proc. Natl. Acad. Sci. USA **72:** 1858–1862.
15. KOSZKA, C., F. E. LEICHTFRIED & G. WICHE. 1985. Identification and spatial arrangement of high molecular weight proteins (300,000–330,000) co-assembling with microtubules from a cultured cell line (glioma C6). Eur. J. Cell Biol. **38:** 149–156.
16. HUBER, G. & A. MATUS. 1984. Immunocytochemical localization of microtubule-associated protein 1 in rat cerebellum using monoclonal antibodies. J. Cell Biol. **98:** 777–781.
17. BERNHARDT, R. & A. MATUS. 1982. Initial phase of dendrite growth: Evidence for the involvement of high molecular weight microtubule-associated proteins (HMWP) before the appearance of tubulin. J. Cell Biol. **92:** 589–593.
18. BERNHARDT, R. & A. MATUS. 1984. Light and electron microscopic studies of the

distribution of microtubule-associated protein 2 in rat brain: A difference between dendritic and axonal cytoskeletons. J. Comp. Neurol. **226:** 203–221.

19. BLOOM, G. S. & R. B. VALLEE. 1983. Association of microtubule associated protein 2 (MAP-2) with microtubules and intermediate filaments in cultured brain cells. J. Cell Biol. **95:** 1523–1531.

20. BRIONES, E. & G. WICHE. 1985. M_r 205,000 sulfoglycoprotein in extracellular matrix of mouse fibroblast cells is immunologically related to high molecular weight microtubule-associated proteins. Proc. Natl. Acad. Sci. USA. **82:** 5776–5780.

21. WICHE, G., H. HERRMANN, F. E. LEICHTFRIED & R. PYTELA. 1982. Plectin: a high-molecular weight cytoskeletal polypeptide component that copurifies with intermediate filaments of the vimentin type. Cold Spring Harbor Symp. Quant. Biol. **46:** 475–482.

Axonal Transport of Microtubule Proteins: Cytotypic Variation of Tubulin and MAPs in Neurons[a]

SCOTT T. BRADY[b] AND MARK M. BLACK[c]

[b]*Department of Cell Biology*
University of Texas Health Science Center at Dallas
Dallas, Texas 75235

[c]*Department of Anatomy*
Temple University Medical School
Philadelphia, Pennsylvania 19140

The fundamental importance of microtubules in cell organization and function is reflected in the high degree of conservation for structure and sequence of the tubulins, so that tubulins from sources as evolutionarily distant as sea urchin eggs and primate brain will copolymerize.[27,28,34,61,86] Evidence continues to accumulate, however, that not all microtubules are equivalent. They may differ with respect to ultrastructure, polypeptide composition, resistance to destabilizing treatments, and cellular distribution. Understanding specializations and adaptations of the cytoskeletal elements in different cell types and subcellular structures is essential for understanding the roles of cytoskeletal elements in cell function.

The majority of studies on the properties of tubulin and microtubules have used vertebrate brain as a source of microtubule protein.[27,34,42,59] Tubulin is the major protein of vertebrate brain, representing 10–26% of total brain protein.[1,2,48] The amount of tubulin in brain is even more striking when compared with the tubulin content of other tissues, which average only 1–3% of tissue protein.[1,48] Less than 5% of the protein in cells derived from glia and other nonneuronal cells of the brain is tubulin,[48] so most brain tubulin is derived from neurons. This is consistent with the number of microtubules present in the axons and dendrites of neurons.[84,108,109] The large amount of tubulin in neurons suggests that microtubules are particularly important in neurons. The list of likely microtubule-associated functions includes organization of the cytoplasm,[18,21,46,50,90,100] generation and maintenance of neuronal morphologies,[38,49,95] and axonal transport,[14,44,64,93] as well as other less well-defined functions. How microtubules are involved in each of these functions remains to be understood.

As the resolution of sensitivity of analytic techniques have improved, however, a considerable amount of biochemical complexity has been noted in microtubule protein from whole brain.[35,36,39,41,42,94] It is clear from both biochemical and immunohistochemical studies that neuronal microtubules vary considerably in composition. This heterogeneity could reflect either different functional roles for populations of microtubules or a need for microtubules with different biochemical properties. The number of polypeptides associated with microtubules in brain continues to increase, and multiple forms of both tubulin and microtubule-associated proteins (MAPs) can be found

[a]The work described in this manuscript was supported in part by Grants from the National Institutes of Health to S.T. Brady (NS 18361) and M.M. Black (NS 17682) and a Research Center Development Award to M.M. Black.

199

within a single neuron.[43,82,83] The tubulins are a multigene family,[22,28] but the number of variants noted for tubulin exceeds the number of gene copies. At least 20 isoforms of brain tubulin[35,36] and 9 distinct classes of brain MAPs[7,25,79,83] have been described. Moreover, characterizations of brain tubulin usually do not include the substantial fraction of total brain tubulin that fails to be solubilized by conventional methods for preparation of microtubule protein from brain.[2,17,55,73,106] Thus, there is considerable heterogeneity of both tubulin and MAPs in vertebrate brain.

Although biochemical heterogeneity in neuronal microtubules was recognized in preparations of microtubule protein from whole brain homogenates, little could be said about the cellular and subcellular origin or about the physiological significance of a particular isotype of tubulin or MAP. The ability to label and analyze microtubule proteins in given cell types and subcellular regions is necessary for studying microtubule specialization. Recently, immunohistochemical and biochemical studies have shown that certain MAPs[5,19,20,30,53,54,103] or isoforms of tubulin[17,29,97] may be enriched in particular regions of the nervous system. Two levels of heterogeneity for microtubule proteins must be considered: 1) differences among the various cell types of the nervous system and 2) differences among populations of microtubules within a single cell. Both types of heterogeneity have been demonstrated in neurons by a variety of approaches.

VARIATION OF TUBULINS AMONG AND WITHIN NEURONS

The existence of cell-type specific tubulins was first clearly demonstrated in *Drosophila*[55-57] with a testis-specific beta subunit. Studies on vertebrate tubulin genes[12,22,28,69] have also shown that different tubulin isotypes are expressed preferentially, though not exclusively, in different tissues. Two lines of evidence indicate that neurons express tubulin subunits differentially. First, immunocytochemical studies with monoclonal antibodies specific to different isotypes of tubulin reveal different staining patterns for antibodies with different specificities.[29] Second, axonal transport studies reveal several types of heterogeneity among tubulins (references 17, 97–99 and next section). The use of axonal transport in these latter studies has provided a method for direct biochemical analysis of microtubule protein heterogeneity in neurons.

Axonal Transport and Tubulin Heterogeneity in Neurons

Axonal proteins can be specifically labeled *in vivo* through axonal transport processes,[8,9,16,44,51,64,70,101,102,107] allowing examination of the biochemistry and properties of these proteins *in situ*. When properly used, the axonal transport method is a valuable complement to traditional biochemical and physiological approaches for studying the cytoskeleton.[15] Studies based on axonal transport provided the first identification of the subunits of neurofilaments,[51] as well as the identities or properties of a number of axonal proteins associated with the cytoskeleton.[8,15,16,40,102] TABLE 1 briefly lists properties for the major rate components of axonal transport. Proteins are conveyed in the axon at discrete rates, and each rate component is composed of a specific set of proteins that, with few exceptions, are not present in other rate components (see FIGURE 1 and references 9, 15, 16, 64, 66, 67).

Axonal proteins are moved in the form of cytological structures rather than individual proteins,[9,66,101] and assembly of these cytological structures appears to occur primarily, perhaps exclusively, in the cell body.[6] As a result, studies on the axonal transport of microtubule protein permit labeling of axonal microtubules in a specific class of neurons.[17,97] The axonal microtubule proteins can then be compared with microtubule protein from whole brain, from nonneuronal sources, or from the axons of other classes of neurons. Both physiological and biochemical properties of axonal microtubules can be identified in such studies because the use of axonal transport provides information about composition and dynamic properties of axonal microtubules *in situ*.

In optic nerve of guinea pig and rat, all of the tubulin in axonal transport moves with the slowest rate component of axonal transport, SCa.[9,16,17,75] Tubulins are among the most heavily labeled proteins in SCa, accounting for some 35–45% of the radioactivity moving with SCa in rat optic nerve with most of the remaining radioactivity being associated with the neurofilament triplet proteins.[9,16,17] When axonally transported tubulin is extracted in the cold under conditions used to cycle microtubules from whole brain homogenates, some 60% of the axonal tubulin fails to be solubilized (reference 17 and TABLE 2). This cold insoluble tubulin is also resistant

TABLE 1. Major Rate Components of Axonal Transport[a]

Rate Component	Rate (mm/day)	Composition	Cytological Structure
Fast			
Anterograde	250–400	Glycoproteins, lipids, transmitter-related materials	Vesicles, Tubulovesicular structures
Mitochondria	50	F1-ATPase[70]	Mitochondria
Retrograde	100–200	Nerve, growth factor, lysosomal enzymes	Prelysosomal structures, multivesicular bodies
Slow			
SCb	2–4	Actin, clathrin, glycolytic enzymes	Cytoplasmic matrix
SCa	0.2–1	Tubulin, MAP, Neurofilament triplet	Microtubule-neurofilament network

[a]Axonal transport rate components fall into two general categories: 1) fast transport, which includes the movement of membranous organelles in both anterograde and retrograde directions and 2) slow transport, which represents the movement of cytoskeletal elements and associated proteins.[66] Some distinguishing features of the major axonal transport rate component are summarized below, and further detail can be found in a number of recent reviews (see, for example, references 9, 15, 16, 44, 64, 66, 67). The rates given are typical of the rates seen in rodent visual system axons. Different rates may be observed in other types of nerves[67,75,80] in the same animals or in other species.[44]

to solubilization by extraction with millimolar calcium and 2 *M* guanidine HCl (see FIGURE 1).

Two-dimensional polyacrylamide gel electrophoretic (2-D PAGE) (see FIGURE 2) and one-dimensional peptide mapping analyses make several features of axonally transported tubulin apparent.[17] Both the soluble and insoluble beta subunits comigrate with the beta subunit of cycled microtubules from whole brain homogenates, but a very different result is obtained with the alpha subunits. The soluble fraction has a typical alpha subunit that comigrates with the alpha subunit of cycled tubulin from whole brain homogenates in 2-D PAGE. When corrected for differential labeling, the ratio of alpha to beta in the soluble fraction gives the expected one to one stoichiometry in 2-D PAGE. By contrast, relatively little of the typical alpha subunit is seen in 2-D PAGE of the insoluble fraction, although the one-dimensional electrophoretic analysis (reference 17 and see FIGURE 1) indicated that the expected amount of a protein with the

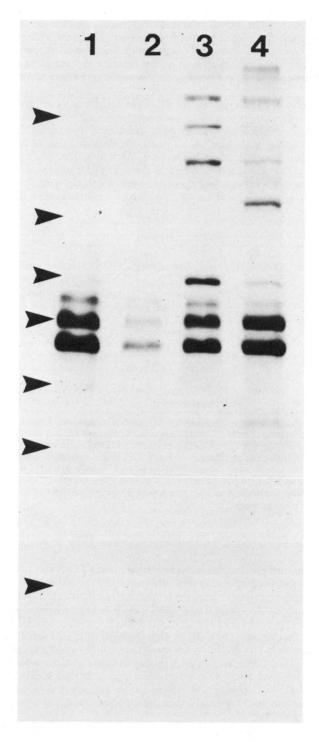

TABLE 2. Fractionation of Axonally Transported Tubulin[a]

Rodent Optic Axons—CSa Tubulin			
	S1	S2	P2
Alpha	41.88 ± 4.45	9.83 ± 1.37	48.29 ± 4.83
Beta	41.32 ± 4.45	9.40 ± 1.43	49.27 ± 3.94
n = 10, ±SEM			

Rat Dorsal Root Ganglion Cell Axons			
	S1	S2	P2
SCa Tubulin[b]	33.96	9.74	56.30
SCb Tubulin+	56.41 ± 4.97	10.48 ± 2.11	31.86 ± 5.5

[a]In all experiments, axonal proteins were labeled by axonal transport following an injection of [^{35}S]methionine into the region of the cell bodies (vitreous of the eye or dorsal root ganglion). Intervals were chosen such that only the appropriate rate component was labeled in the region of the nerve analyzed. The fractions were generated as described in reference 17. Briefly, the nerve was homogenized in MT buffer (100 mM MES, pH 6.4, 1 mM EGTA, 1 mM guanosine triphosphate (GTP), and 0.5 mM MgCl$_2$) at 0–4 °C, then centrifuged at 130,000 × g. The proteins in this supernatent correspond to S1 (see FIGURE 1 and 2) and represent cold soluble proteins. The pellet was resuspended in CMT buffer (same as MT except 5 mM CaCl$_2$ replaces 1 mM EGTA) at room temperature and centrifuged again at 130,000 × g. The supernatant, S2, represents those proteins in the cold insoluble fraction that are solubilized by millimolar levels of Ca^{2+}. The pellet, P2, is cold and Ca^{2+}-insoluble material, including both neurofilament proteins and a substantial amount of tubulin. As seen in FIGURE 1, a substantial fraction of this insoluble tubulin remains insoluble in the presence of 2 M guanidine HCl, although neurofilament proteins are largely solubilized by this treatment. After analysis in SDS gel electrophoresis and fluorography to locate labeled proteins, the tubulin bands are excised from the gel and radioactivity associated with proteins of interest determined. Data are expressed as percent of total radioactivity of the appropriate molecular weight in the three fractions. For molecular weights of 57,000 and 53,000 in these studies, 2-D PAGE indicate that only alpha and beta tubulin are significantly labeled. The data given are for guinea pig optic nerve,[17] but similar values are found for rat optic nerve.

Preliminary studies comparing tubulin moving with SCa to tubulin moving with SCb in dorsal root ganglion cells indicate that a smaller proportion of tubulin moving at SCb rates is insoluble. Injection-sacrifice intervals were chosen so that SCa- and SCb-labeled materials were in similar segments of the nerve, so that differences represent differences in the labeled material only.

[b]n = 2 (replicates ± 2–4%); +n = 4, ± SEM

molecular weight of alpha tubulin is present in the insoluble fraction. Analysis of the behavior of axonally transported tubulin in isoelectric focusing (IEF) revealed that a significant fraction of the alpha tubulin in axonal transport in optic nerve migrated anomalously in IEF, failing to focus in the gel.[17] Peptide maps of this insoluble material following limited proteolysis by the Cleveland method[23] show that the insoluble proteins are alpha and beta tubulins,[17] but reveal subtle differences in the alpha

FIGURE 1. Fractionation of axonal tubulin labeled by axonal transport. Optic nerve of rat containing labeled SCa proteins was homogenized and centrifuged at 0–4°C as described in reference 17 to give a cold extractable fraction (lane 1). The cold insoluble pellet was resuspended in a buffer containing millimolar Ca^{2+} and recentrifuged to give a Ca^{2+} extractable fraction (lane 2). In this experiment, the cold- and Ca^{2+}-insoluble pellet was suspended once again in a buffer containing 2 M guanidine HCl. The guanidine soluble fraction (lane 3) includes some tubulin and most of the neurofilament triplet, but nearly a third of the labeled axonal tubulin remains in the insoluble pellet (lane 4) (unpublished data, S. Brady). The samples have been analyzed on a SDS polyacrylamide gradient gel (6–17.5%) using a modification of the Laemmli method.[62] Arrowheads indicate molecular weight markers: 200, 94, 68, 57, 43, 30, and 14.

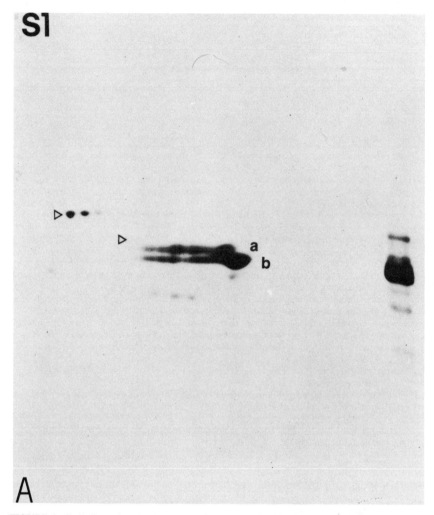

FIGURE 2. Two-dimensional gel electrophoresis of axonal tubulin fractions. Guinea pig optic nerve containing labeled SCa proteins were fractionated as described in reference 17 (also see TABLE 2 and FIGURE 1 for a brief description), then analyzed in 2-D PAGE according to a modification of the method of O'Farrell.[81] **A:** S1 represents the cold extractable labeled proteins of the axon. In S1, the tubulin pattern (alpha [a] and beta [b]) is very similar to the Coomassie blue stained pattern, when cycled microtubule protein from whole brain is run in the same gel. **B:** The cold- and Ca^{2+}-insoluble material (P2), however, has a very different tubulin pattern. In this exposure, no label is detected in the region where alpha (a) tubulin should be seen. Comigration with carrier microtubule protein prepared from whole brain indicates that little label comigrating with alpha tubulin can be detected in the P2 fraction. The cold insoluble alpha tubulin was found to migrate anomalously in 2-D PAGE. (S. Brady et al.[17] With permission from the *Journal of Cell Biology.*)

tubulins of the two fractions. The biochemical basis of this anomalous migration for the cold insoluble alpha tubulin is uncertain, but appears to be the result of a posttranslational modification, expression of a different alpha tubulin gene, or a combination of both.

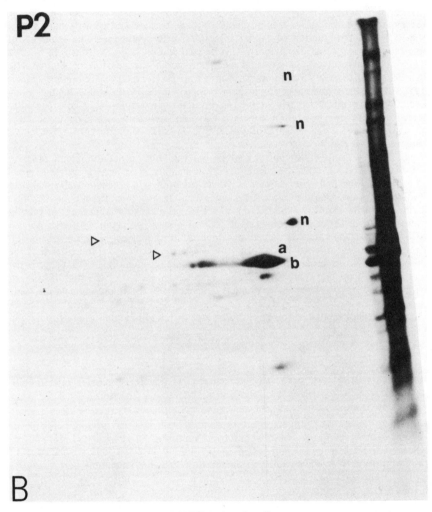

FIGURE 2. (continued)

Independently, Tashiro and Komiya[97,98] have described a unique subunit composition for axonally transported tubulin in the vagal nerve of guinea pig. They also find a difference in the alpha subunit of transported tubulin[97] and altered solubility properties of transported tubulin under a different set of conditions.[99] Based on the effects of the drug β,β'-iminodiproprionitrile, Tashiro et al.[99] suggested that the insolubility results from an interaction of the insoluble tubulin with neurofilaments. Extraction, however, of the insoluble fraction of axonally transported tubulin from rat optic nerve with 2 M guanidine HCl solubilizes the 80–90% of the neurofilament protein, but approximately 30% of the axonal tubulin (half of the insoluble fraction) continues to pellet (FIGURE 1 and unpublished observation, S. Brady). This suggests that some property of the tubulin itself is responsible for its lack of solubility.

The unusual solubility properties of axonal tubulin raised questions about the morphological correlate for this insoluble fraction. Sahenk and Brady[88] have shown that a substantial fraction of axonal tubulin is in the form of stable segments of microtubules that resist depolymerization by cold and antimitotic drugs (see FIGURE 3). Stable microtubules have also been identified in the neurites of sympathetic neurons grown *in vitro*[10] and in other types of neurons (for example see references 47, 77). These short segments of microtubules appear to represent stable regions along axonal microtubules.[88] Clearly, stable regions in a microtubule would be expected to alter the dynamics of microtubule assembly and disassembly, because the presence of such a region at the end of a microtubule would be expected to prevent further loss of subunits at that end.[17]

In addition to the optic[17] and vagus[97] nerves, the axons of the dorsal root ganglion (DRG) cell (reference 99 and see TABLE 2) and axons of the ventral motor neurons (VMN) (unpublished observations, S. Brady) also contain a substantial amount of insoluble tubulin. One difference between axonal transport of tubulin in the optic axons and in the axons of the DRG or VMN is that a substantial amount of tubulin moves ahead of the neurofilaments and SCa at a rate consistent with transport in SCb,[51,60,75,76,80] (the rate component in which actin and glycolytic enzymes move). Fractionation of transported tubulin in the VMN and DRG axons indicates that the proportion of SCa-associated tubulin that is insoluble is comparable to that seen in rat optic nerve (TABLE 2), but a significantly smaller proportion of the tubulin moving with the SCb fraction is insoluble (TABLE 2). Changes in the amount of cold insoluble tubulin have also been noted during development and aging.[2,13] The significance of these differences is not certain, but is consistent with a regulation of heterogeneity in axonal tubulin.

VARIATIONS IN MICROTUBULE-ASSOCIATED PROTEINS OF NEURONS

MAPs are thought to stabilize microtubules and organize them with respect to each other and to other components of the cytoskeleton. As with tubulin, much of our current understanding of MAPs stems from studies of MAPs isolated from brain, where they were originally identified as nontubulin proteins that copurified with tubulin through successive cycles of temperature-dependent assembly and disassembly.[3,25,79] The *in vivo* association of MAPs with microtubules was subsequently demonstrated in studies with specific antibodies prepared against several MAPs species (for examples see references 11, 54, 74; reviewed in reference 59; also see other chapters in this volume) and in biochemical studies using sequential extraction assays.[7,33,83,103]

As a result of many studies, a catalog of neuronal MAPs now exists. TABLE 3 lists MAPs identified in pure sympathetic neurons grown in tissue culture. The relative homogeneity of the neuron population in these cultures suggests that the MAP profile observed in total cultures is indicative of that in individual neurons. This possibility is supported by immunostaining analyses which show that MAP-1a, MAP-1b, MAP-2, and at least some of the tau species are present in all neurons of individual cultures. A

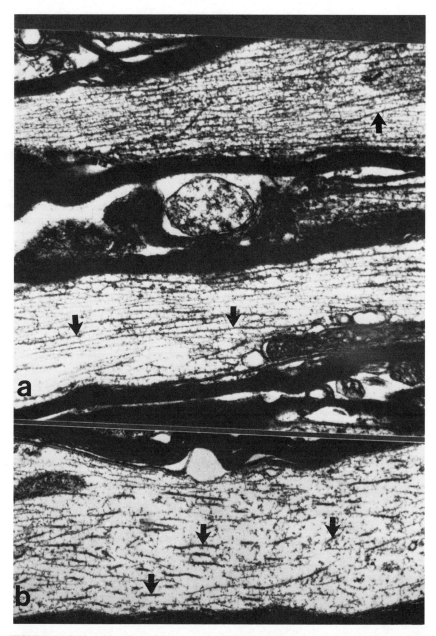

FIGURE 3. Cold stable microtubules in retinal ganglion cell axons. a) Axons incubated and processed for electron microscopy under control conditions. b) Axon incubated for one hour at 0–2°C for one hour and then processed for electron microscopy in the cold. These are longitudinal sections chosen for measurement of the mean length of microtubules in the section. The mean length of microtubule profiles in cold treated axons is significantly shorter than the mean length in control axons.[88] Similar short microtubule segments were observed in nerves treated with podophyllotoxin or vincristine, and addition of taxol to cold treated axons increases the mean length the microtubule segments. Morphometric analyses of longitudinal and transverse sections of axons demonstrate that the cold-insoluble tubulin of axons is largely in the form of short stable segments of microtubules.[88]

similar number of neuronal MAPs has also been identified in brain (for example, see references 3, 25, 26, 54, 79 and other chapters in this volume) consistent with the idea that neurons contain a considerable variety of MAPs.

Variation in MAP Composition between Neurons

Several lines of evidence indicate that the MAP composition of different neurons differs both quantitatively and qualitatively. For example, the tau proteins of brain (a structurally related family of polypeptides 55,000 to 70,000 in molecular weight),[25,31] appear to be present at different levels in immunoblots of different regions of the brain (reference 31 and L. Binder, personal communication). Because antibodies that recognize the full complement of brain tau localize primarily to neuronal elements of brain by immunostaining, this would suggest that the profile of tau varies among different populations of neurons. More dramatic differences in tau composition are observed when tau of sympathetic neurons is compared to that of brain. The most abundant tau species in sympathetic neurons both *in vivo* and *in vitro* has a molecular

TABLE 3. Microtubule-Associated Proteins of Cultured Sympathetic Neurons[a]

MAP-1a
MAP-1b
MAP-2
210K MAP
MAP-3
130K MAP
Tau (2–3 species of molecular weights 100–110 and 5 species with molecular weights 55–68)
60–76K MAPs (4 species of MAPs in the molecular weight range of 60–76 that are biochemically and immunologically distinct from tau proteins)
32K Map

[a]This table lists MAPs that have been identified using biochemical and immunological procedures in rat sympathetic nerves in primary culture. Compiled from references 7, 82, 83.

weight of 100–110K, and on appropriate gels is composed of two or three species.[83] Similar high molecular weight (HMW) tau proteins are the principal tau species in PC12 pheochromocytoma cells,[32] which are considered models for sympathetic neurons, cultured dorsal root ganglion neurons (unpublished observation, M. Black), and also in the optic nerve and tract (L. Binder, personal communication). By contrast, this high molecular weight tau is present at only trace levels in brain. Similarly, the composition of HMW MAPs varies for different regions of the brain (see other chapters in this volume and references 11, 20, 30, and 53). Both tau and HMW MAPs have also been found to change during development.[4,31,37,72]

The significance of variation in MAP content among different neuronal types is presently unknown, although consideration of the properties of MAPs as determined by *in vitro* experiments provides possible clues. MAPs bind to microtubules *in vitro* and thereby stabilize them by decreasing the rate at which tubulin subunits dissociate from the microtubule.[24,25,59,78,91] Also, the various MAPs are apparently specialized with respect to their effects on microtubules (see references 71, 110 for examples). If MAPs also stabilize microtubules *in vivo,* then regional variation in the composition of MAPs among different neuronal types may be indicative of a corresponding variation in microtubule stability among these neurons.

Axonal MAPs and Heterogeneity of MAPs within Neurons

Results from several laboratories indicate that MAPs are not uniformly distributed among the various domains of a neuron, such as the cell body, dendrites, and axons. The first clue for this stemmed from studies in which the principal high molecular weight MAPs, MAP-1 and MAP-2, were not observed in axonal transport, although tubulin and other MAPs were transported along the axon.[102] In the optic nerve of rat and guinea pig, two MAPs with molecular weights of 55,000 and 66,000 were detected in axonal transport studies, whereas at least four polypeptides in the molecular weight range of 55,000–70,000 are detectable in the retina.[102] It is likely that these proteins correspond to some of the tau proteins because immunoblots of rat optic nerve with antibodies to the tau fraction similarly show only two polypeptides in the 55,000–70,000 molecular weight range, corresponding to the labeled polypeptides (unpublished observation, S. Brady; personal communication, L. Binder). Other MAPs, however, unrelated to the taus in the molecular weight range of 60,000–76,000 have also been detected in axons (see TABLE 3). Both quantitative and qualitative differences have been found between the MAPs present in axonal transport in optic axons and those being transported in the axons of the DRG and VMN,[80] indicating that the exact complement of axonal MAPs varies among axons from different types of neurons.

Immunohistochemical studies have also been valuable for demonstrating the partitioning of MAPs within neurons. The apparent absence of MAP-2 from most axons was confirmed by immunostaining[19,20,30,52] and biochemical[82,83,103] studies, where it was localized to neuron cell bodies and their dendrites, but not to axons. MAP-1, on the other hand, has been detected in axons as well as in cell bodies and dendrites by immunostaining and by biochemical procedures (references 11, 53, and see below). There are several possible explanations for the failure to detect MAP-1 in axonal transport studies. For example, MAP-1 levels are relatively low in axons, and the procedures used to detect axonally transported proteins may not have been sensitive enough to detect these levels of MAP-1. Immunostaining analyses using a monoclonal antibody to tau indicate that this class of MAPs is highly enriched in axons (reference 5 and see below). Also, a recently described brain MAP of 180,000 kD, designated MAP-3, localizes by immunostaining to relatively large myelinated axons, but not to cell bodies and dendrites.[54] These considerations suggest that neuronal MAPs can be classed as dendrite-specific, axon-specific, and present in both axons and dendrites.

One difficulty with the immunostaining data is to correctly interpret the absence of staining. This issue is especially relevant in the case of staining with anti-tau antibodies, which localize to axons, but not to dendrites or cell bodies. Axonal tau proteins are synthesized in the neuron cell body and then transported into the axon. Thus, tau is clearly present in the cell body, even if not detected by immunostaining: similar arguments must also apply to MAP-3). There are many possible explanations for the absence of tau immunoreactivity in cell bodies, some of which have potential biological interest. For example, tau may be posttranslationally modified to an immunochemically distinct form as it is transported into the axon. A precedent for this mechanism is the phosphorylation of neurofilament protein.[96]

A somewhat different approach, based on the axonal transport paradigm, has been taken to identify axonal MAPs and to look at the distribution of MAPs in cultured neurons. This approach uses explant cultures of pure sympathetic neurons. Based on morphological and physiological criteria, sympathetic neurons in culture produce two types of neurites: one resembling axons and the other resembling dendrites.[19,63,104,105] The axon-like neurites attain lengths of many millimeters, whereas the dendrite-like neurites are relatively short, having lengths of only 50–100 microns. In explant cultures, pieces of sympathetic ganglia, 0.5 mm in diameter and consisting of thousands of neurons, are plated onto culture dishes and allowed to extend neurites. Cultures are freed of nonneuronal cells by treatment with various poisons, and after approximately three weeks, neurites extend out from the ganglionic explant or cell body mass (CBM) for 4–6 mm in all directions. The dendrite-like neurites remain confined to the CBM because their length is short relative to the diameter of the CBM. By contrast, the axon-like neurites extend out from the CBM for several millimeters. Thus, explant cultures can be used to obtain fractions that are pure axon-like neurites and fractions containing cell bodies and dendrites by cutting the CBM away from the culture with a scalpel blade, leaving behind the pure axon-like neurites. The CBM, which is enriched in cell bodies and dendrites, and the pure axon-like neurites are then analyzed for microtubule proteins. That cell bodies and dendrites are effectively separated from axons is demonstrated by the observation that MAP-2, which is a highly specific marker for cell bodies and dendrites, is only detectable in CBM preparation of explant cultures (see FIGURE 4).

Both immunoblotting and immunohistochemical analyses demonstrate differential distributions of MAPs in the various domains of these cultured sympathetic neurons. Monospecific monoclonal antibodies against MAP-1a and MAP-1b (these antibodies were provided by Dr. Richard Vallee), MAP-2 and tau (these antibodies were provided by Dr. Lester Binder), and tubulin were used to establish the location of these polypeptides. Alpha and beta tubulin were present in the CBM and axon-like neurites in roughly equal amounts (see FIGURE 4). Similarly, MAP-1a and MAP-1b were present in both the axon-like neurites and the CBM (I. Peng and M. Black, unpublished observations). By contrast, MAP-2 and tau exhibited a very different pattern of distribution between the CBM and axon-like neurites. MAP-2 was only detected in the CBM, indicating the MAP-2 is restricted in its distribution to the cell bodies and dendrites. Immunohistochemistry confirmed this observation as antibodies to MAP-2 stained all neuron cell bodies, but only some of the neurites. All of the MAP-2 positive neurites appeared to extend only a short distance from the neuron cell bodies (I. Peng and M. Black, unpublished observations). In favorable preparations, some neurites elaborated by a given neuron were MAP-2 negative, whereas others were MAP-2 positive. In constrast, tau was greatly enriched in the axon-like neurites, with only trace amounts in the CBM (see FIGURE 4), suggesting that tau is preferentially associated with axonal microtubules compared to microtubules in cell bodies and dendrites. Comparable results were also obtained by identifying MAPs of CBM and pure axon-like neurites from metabolically labeled explant cultures by coassembly with unlabeled carrier brain microtubule proteins in the presence of taxol. These latter experiments showed that the 210K MAP was also enriched in the axon-like neurites relative to the CBM.

On the basis of these and other data, it is clear that MAPs are differentially distributed among the microtubules of the neuron. Three distinct patterns of MAP distribution in neurons can be discerned: dendrite-enriched, axon-enriched, and present in all domains of neurons with a distribution pattern similar to that of tubulin. The consequences of selective partitioning for MAPs in neurons are still being explored, but MAPs influence microtubule stability and also cross-link microtubules with each other as well as with other cytoskeletal structures.[45,58,68,87,89] Moreover, the various MAPs appear to differ with respect to these parameters. For example, MAP-2, but not tau, is able to cross-link microtubules with actin filaments.[85] If the *in vitro* observations reflect the function of MAPs *in vivo,* then selective partitioning of MAPs may represent a mechanism for differentially regulating the stability and organization of microtubules in different regions of the neuron. In this regard, it is well documented that the spatial organization of microtubules in axons differs from that in dendrites.[108]

In dendrites, microtubules are uniformly distributed across the dendrite, whereas in axons, microtubules occur in small clusters that are irregularly arrayed within axoplasm. In addition, the center-to-center spacing of neighboring microtubules in axons differs from that in dendrites. It seems reasonable to suggest that the differences in the organization of microtubules observed between axons and dendrites are in part

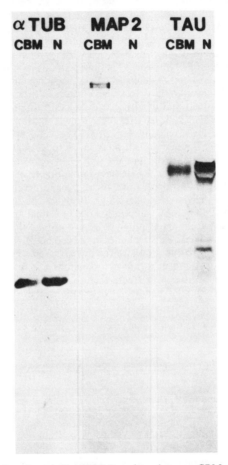

FIGURE 4. Distribution of α tubulin, MAP-2, and tau between CBM and axon-like neurites. CBM and axon-like neurite (N) preparations from individual explant cultures were probed by immunoblotting with monoclonal antibodies against α tubulin (right), MAP-2 (middle), or tau (left). Tubulin is relatively evenly distributed between CBM and axon-like neurites. By contrast, MAP-2 is enriched in the CBM fraction, whereas tau is enriched in the axon-like neurites.

due to the differences in the MAPs associated with the microtubules in these neurites.

The mechanisms by which this partitioning of cytoskeletal element is accomplished require further study. There are, however, a number of reasons to think that the segregation begins in the cell body.[65] By electron microscopy, two ultrastructurally distinct domains are observed in the neuron soma.[65,92] One domain is characterized by an abundance of polysomes and other structures associated with protein synthesis. The other domain is characterized by an abundance of microtubules and neurofilaments, and a paucity of polysomes. These microtubule-neurofilament-rich domains are apparently continuous into the axon and dendrites,[92] raising the possibility that the cytoskeletal networks of these neurites originate in the cell body. Further support for this possibility stems from our studies on the time interval between the synthesis and assembly of tubulin and neurofilament proteins.[6] These proteins are synthesized as soluble precursors that are incorporated into their respective polymers relatively rapidly, such that newly synthesized tubulin and neurofilament proteins reach the *in vivo* steady state monomer-polymer distribution within 1–2 h (tubulin) or 15–30 min (neurofilament protein) after synthesis. This time period is too short for these proteins to be transported into the axons or dendrites. Thus, the initial incorporation of tubulin and neurofilament proteins into the cytoskeleton occurs in the cell body.

If segregation of axonal and dendritic cytoskeletal proteins is begun in the cell body, then it is possible that multiple sites for cytoskeletal assembly exist in the soma: one for axons and at least one for dendrites. The concept of multiple sites for cytoskeletal assembly in the neuron cell body has implications for the regulation of neuronal morphology as well as for partitioning of cytoskeletal proteins. Individual neurons have several morphologically distinct domains, most notably the cell body, axon, and dendrites. We have proposed that the regional differences in the morphology of individual neurons reflect an underlying difference in the internal cytoskeleton.[17,65,102] As discussed above, the composition of the axonal and dendritic cytoskeletons differ substantially from one another. We have argued that these compositional differences can be traced back to distinct sites in the cell body where newly synthesized cytoskeletal proteins are incorporated into the cytoskeleton. In this view, the specific components of the axonal and dendritic cytoskeletons are specified at the sites for cytoskeletal assembly. If this is correct, then by regulating which proteins are incorporated into the axonal and dendritic cytoskeletons, these sites will influence the structure of these cytoskeletons and thereby the external shape of the axon and dendrites.

CONCLUSIONS AND SUMMARY

Axonal microtubules represent a highly specialized population of cytoplasmic microtubules with an unusual stability and biochemical composition. A large fraction of axonal tubulin cannot be solubilized by conventional methods for preparation of microtubules from whole brain. The MAPs of axonal microtubules also represent a subset of the MAPs described in the whole brain homogenates. Microtubules in axons from different populations of neurons may differ in rates of movement, in metabolism, and in polypeptide composition. Recognition of this heterogeneity in the cytoskeletal elements of the neuron and consideration of physiological roles of these specializations may provide important insights into the development and function of the nervous system.

ACKNOWLEDGMENTS

The authors would like to thank Isaac Peng for the use of FIGURE 4 and Zarife Sahenk for the micrographs in FIGURE 3.

REFERENCES

1. ANDERSON, P. 1979. The structure and amount of tubulin in cells and tissues. J. Cell Biol. **254:** 2168–2171.
2. BAMBURG, J., E. SHOOTER & L. WILSON. 1973. Developmental changes in microtubule protein of chick brain. Biochemistry **12:** 1476–1482.
3. BERKOWITZ, S., J. KATAGIRK, H. BINDER & R. WILLIAMS. 1977. Separation and characterization of microtubule proteins from calf brain. Biochemistry **16:** 5610–5617.
4. BINDER, L. I., A. FRANKFURTER, H. KIM, A. CACERES, M. PAYNE & L. REBUHN. 1984. MAP2 heterogeneity during rat brain development. Proc. Nat. Acad. Sci. USA **81:** 5613–5617.
5. BINDER, L. I., A. FRANKFURTER & L. REBUHN. 1984. A monoclonal antibody to tau factor localizes predominantly in axons. J. Cell Biol. **99:** 191a.
6. BLACK, M., P. KEYSER & E. SOBEL. 1985. Interval between the synthesis and assembly of cytoskeletal proteins in cultured neurons. J. Neurosci. In press.
7. BLACK, M. & J. KURDYLA. 1983. Microtubule associated proteins of neurons. J. Cell Biol. **97:** 1020–1028.
8. BLACK, M. & R. LASEK. 1979. Axonal transport of actin: Slow component b is the principal source of actin for the axon. Brain Res. **171:** 410–413.
9. BLACK, M. & R. LASEK. 1980. Slow components of axonal transport: Two cytoskeletal networks. J. Cell Biol. **86:** 616–623.
10. BLACK, M., M. COCHRAN & J. KURDYLA. 1984. Solubility properties of neuronal tubulin: Evidence for extractable and stable microtubules. Brain Res. **295:** 255–263.
11. BLOOM, G., T. SCHOENFELD & R. VALLEE. 1984. Widespread distribution of MAP1 (microtubule associated protein 1) in the nervous system. J. Cell Biol. **98:** 320–330.
12. BOND, J., G. ROBINSON & S. FARMER. 1984. Differential expression of two neural cell-specific B-tubulin mRNAs during rat brain development. Mol. Cell. Biol. **4:** 1313–1319.
13. BRADY, S. 1984. Increases in cold-insoluble tubulin during aging. Soc. Neurosci. Abstr. **10:** 273.
14. BRADY, S. 1984. Basic properties of fast axonal transport and the role of fast transport in axonal growth. *In* Axonal Transport in Neuronal Growth and Regeneration. J. Elam & P. Cancalon, Eds.: 13–29. Plenum. New York.
15. BRADY, S. & R. LASEK. 1982. Axonal transport: A cell biological method for studying proteins that associate with the cytoskeleton. Methods Cell Biol. **25:** 366–398.
16. BRADY, S. & R. LASEK. 1982. The slow components of axonal transport: Movements, compositions, and organization. *In* Axoplasmic Transport D. G. Weiss, Ed. Springer Verlag. Berlin.
17. BRADY, S., M. TYTELL & R. LASEK. 1984. Axonal transport and axonal tubulin: Biochemical evidence for cold-stability. J. Cell Biol. **99:** 1716–1724.
18. BURTON, P. & J. PAIGE. 1981. Polarity of axoplasmic microtubules in the olfactory nerve of the frog. Proc. Natl. Acad. Sci. USA **78:** 3269–3273.
19. CACERES, A., G. BANKER, O. STEWARD, L. BINDER & M. PAYNE. 1984. MAP2 is localized to the dendrites of hippocampal neurons which develop in culture. Dev. Brain Res. **13:** 314–318.
20. CACERES, A., L. BINDER, M. PAYNE, P. BENDER, L. REBUHN & O. STEWARD. 1984. Differential subcellular localization of tubulin and the microtubule associated protein MAP-2 in brain tissue as revealed by immunocytochemistry with monoclonal hybridoma antibodies. J. Neurosci. **4:** 394–410.
21. CHALFIE, M. & J. THOMPSON. 1979. Organization of neuronal microtubules in the nematode *Caenorhabditis elegans*. J. Cell Biol. **82:** 278–289.
22. CLEVELAND, D., 1983. The tubulins: From DNA to RNA and back again. Cell **34:** 330–332.
23. CLEVELAND, D., S. FISHER, M. KIRSCHNER & U. LAEMMLI. 1977. Peptide mapping by limited proteolysis in sodium dodecyl sulfate and analysis by gel electrophoresis. J. Biol. Chem. **252:** 1102–1106.
24. CLEVELAND, D., S. HWO & M. KIRSCHNER. 1977. Physical and chemical properties of purified tau factor and role of tau in microtubule assembly. J. Mol. Biol. **116:** 227–247.

25. CLEVELAND, D., S. HWO & M. KIRSCHNER. 1977. Purification of tau, a microtubule associated protein which induces assembly of microtubles from purified tubulin. J. Mol. Biol. **116:** 207–225.
26. CLEVELAND, D., B. SPIEGELMAN & M. KIRSCHNER. 1979. Conservation of microtubule associated proteins: Isolation and characterization of tau and the high molecular weight microtubule associated proteins from chick brain and rat fibroblast and comparison to the corresponding mammlian brain proteins. J. Biol. Chem. **254:** 12670–12678.
27. CORREIA, J. & R. WILLIAMS. 1983. Mechanisms of assembly and disassembly of microtubules. Ann. Rev. Biophys. Bioeng. **12:** 211–235.
28. COWAN, N. & L. DUDLEY. 1983. Tubulin isotypes and the multigene tubulin families. Intl. Rev. Cytol. **85:** 147–173.
29. CUMMING, R., R. BURGOYNE & N. LYTTON. 1983. Axonal subpopulations in the central nervous system demonstrated using monoclonal antibodies against a-tubulin. Eur. J. Cell Biol. **31:** 241–248.
30. DECAMILLI, P., P. MILLER, F. NAVONE, W. THEURKAUF & R. VALLEE. 1984. Distribution of microtubule associated protein 2 (MAP-2) in the nervous system of the rat studied by immunofluorescence. Neurosci. **11:** 819–846.
31. DRUBIN, D., D. CAPUT & M. KIRSCHNER. 1984. Studies on the expression of the microtubule associated protein, tau, during mouse brain development, with newly isolated complementary probes. J. Cell Biol. **98:** 1090–1097.
32. DRUBIN, D., S. FEINSTEIN & M. KIRSCHNER. 1985. Induction of microtubule-associated tau protein by nerve growth factor during neurite outgrowth in PC12 cells. *In* Molecular Biology of the Cytoskeleton. G. Borisy, D. Cleveland & D. Murphy, Eds.: 343–355. Cold Spring Harbor Press.
33. DUERR, A., D. PALLAS & F. SOLOMON. 1981. Molecular analysis of cytoplasmic microtubules *in situ:* Identification of both widespread and specific proteins. Cell **24:** 203–211.
34. DUSTIN, P. 1978. Microtubules. Springer Verlag. Berlin.
35. FIELD, D., R. COLLINS & J. LEE. 1984. Heterogeneity of vertebrate brain tubulins. Proc. Natl. Acad. Sci. USA **81:** 4041–4045.
36. FIELD, D. & J. LEE. 1985. Isoelectric focusing and two dimensional electrophoresis of tubulin using immobilized pH gradients under denaturing conditions. Anal. Biochem. **144:** 584–592.
37. FRANCON, J., A. LENNON, A. FELLOUS, A. MAVECK, M. PIERRE & J. NUNEZ. 1982. Heterogeneity of microtubule-associated proteins and brain development. Eur. J. Biochem. **129:** 465–471.
38. FRIEDE, R. & T. SAMORAJSKI. 1970. Axon caliber related to neurofilaments and microtubules in sciatic nerve fiber of rats and mice. Anat. Rec. **167:** 397–387.
39. FULTON, C. & P. SIMON. 1976. Selective synthesis and utilization of flagellar tubulin: The multitubulin hypothesis. *In* Cell Motility. R. Goldman, T. Pollard & J. Rosenbaum, Eds.: 988–1005. Cold Spring Harbor Laboratory. Cold Spring Harbor, New York.
40. GARNER, J. & R. LASEK. 1981. Clathrin is axonally transported as part of slow component b: the axoplasmic matrix. J. Cell Biol. **88:** 172–178.
41. GEORGE, H., L. MISRA, D. FIELD & J. LEE. 1981. Polymorphism of brain tubulin. Biochemistry **20:** 2402–2409.
42. GOZES, I. 1982. Tubulin in the nervous system. Neurochem. Int. **4:** 101–120.
43. GOZES, I. & K. SWEADNER. 1981. Multiple tubulin forms are expressed by a single neuron. Nature (London) **294:** 477–480.
44. GRAFSTEIN, B. & D. FORMAN. 1980. Intracellular transport in neurons. Physiol. Rev. **60:** 1167–1283.
45. GRIFFITH, L. & T. POLLARD. 1982. The interaction of actin filaments with microtubules and microtubule-associated proteins. J. Biol. Chem. **27:** 9143–9151.
46. HEIDEMANN, S., J. LANDERS & M. HAMBORG. 1981. Polarity orientation of axonal microtubules. J. Cell Biol. **91:** 661–665.
47. HEIDEMANN, S., M. HAMBORG, S. THOMAS, B. SONG, S. LINDLEY & D. CHIU. 1985. Spatial organization of axonal microtubules. J. Cell Biol. **99:** 1289–1295.
48. HILLER, G. & K. WEBER. 1978. Radioimmunoassay for tubulin: A quantitative comparison of the tubulin content of different established tissue culture cells and tissues. Cell **14:** 795–804.

49. HILLMAN, D. 1980. Neuronal shape parameters and substructure as a basis for neuronal form. *In* The Neurosciences, 4th Study Program. (F. Schmitt & F. Worden, Eds.: 477–498. MIT Press. Cambridge, Mass.

50. HIROKAWA, N. 1982. Cross-linker system between neurofilaments, microtubules, and membranous organelles in frog axons revealed by the quick-freeze, deep-etching method. J. Cell Biol. **94:** 129–142.

51. HOFFMAN, P. & R. LASEK. 1975. The slow component of axonal transport. Identification of major structural polypeptides of the axon and their generality among mammalian neurons. J. Cell Biol. **66:** 351–366.

52. HOFFMAN, P., R. LASEK, J. GRIFFIN & D. PRICE. 1983. Slowing of the axonal transport of neurofilament proteins during development. J. Neurosci. **3:** 1694–1700.

53. HUBER, G. & A. MATUS. 1984. Differences in the cellular distribution of two microtubule associated proteins, MAP-1 and MAP-2, in rat brain. J. Neurosci. **4:** 151–160.

54. HUBER, G., D. ALAIMO-BEURET & A. MATUS. 1985. MAP3: characterization of a novel microtubule associated protein. J. Cell Biol. **100:** 496–507.

55. JOB, D. & R. MARGOLIS. 1982. Recycling of cold-stable microtubules: Evidence that cold stability is due to substoichiometric polymer blocks. Biochemistry **21:** 509–515.

56. KEMPHUES, K., R. RAFF, T. KAUFMAN & E. RAFF. 1979. Mutation in a structural gene for a β-tubulin specific to testic in *Drosophila melanogaster.* Proc. Natl. Acad. Sci. USA **76:** 3991–3995.

57. KEMPHUES, K., T. KAUFMAN, R. RAFF & E. RAFF. 1982. The testis specific β-tubulin in *Drosophila melanogaster* has multiple functions in spermatogenesis. Cell **31:** 655–670.

58. KIM, H., L. BINDER & J. ROSENBAUM. 1979. The periodic association of MAP2 with brain microtubules *in vitro.* J. Cell Biol. **80:** 266–276.

59. KIRSCHNER, M. 1978. Microtubule assembly and nucleation. Int. Rev. Cytol. **54:** 1–72.

60. KOMIYA, Y. 1980. Slowing with age of the rate of slow axonal flow in bifurcating axons of rat dorsal root ganglion cells. Brain Res. **183:** 477–480.

61. KRAUHS, E., M. LITTLE, T. KEMPT, R. HOFER-WARBINEK, W. ADE & H. PONSTINGL. 1981. Complete amino acid sequence of beta tubulin from porcine brain. Proc. Natl. Acad. Sci. USA **78:** 4156–4160.

62. LAEMMLI, U. 1970. Cleavage of structural proteins during the assembly of the head of bacteriophage T4. Nature (London) **227:** 680–685.

63. LANDIS, S. 1977. Morphological properties of the dendrites and axons of dissociated rat sympathetic neurons. Neurosci. Abstr. **3:** 525.

64. Lasek, R. 1981. The dynamic ordering of neuronal cytoskeletons. Neurosci. Res. Program Bull. **19:** 7–32.

65. LASEK, R. & S. BRADY. 1982. The axon: A prototype for studying expressional cytoplasm. Cold Spring Harbor Symp. Quant. Biol. **46:** 113–124.

66. LASEK, R. & S. BRADY. 1982. The structural hypothesis of axonal transport: Two classes of moving elements. *In* Axoplasmic Transport. D. Weiss, Ed.: 397–405. Springer Verlag. Berlin.

67. LASEK, R., I. MCQUARRIE & S. BRADY. 1983. Transport of cytoskeletal and soluble proteins in neurons. *In* Biological Structures and Coupled Flows. A. Silberberg, Ed.: 329–347. Academic/Balaban Press. Rehovet, Israel. In press.

68. LETERRIER, J., R. LIEM & M. SHELANSKI. 1982. Interactions between neurofilament and microtubule associated proteins: a possible mechanism for intra-organelle bridging. J. Cell Biol. **95:** 982–986.

69. LOPATA, M., J. HAVERCROFT, L. CHOW & D. CLEVELAND. 1983. Four unique genes required for beta tubulin expression in vertebrates. Cell **32:** 713–724.

70. LORENZ, T. & M. WILLARD. 1978. Subcellular fractionation of intra-axonally transported polypeptides in the rabbit visual system. Proc. Natl. Acad. Sci. USA **75:** 505–509.

71. LUDUENA, R., A. FELLOUS, J. FRANCON, J. NUNEZ & L. MCMANUS. 1981. Effect of tau on the vinblastine-induced aggregation of tubulin. J. Cell Biol. **89:** 680–684.

72. MARECK, A., A. FELLOUS, J. FRANCON & J. NUNEZ. 1978. Changes in composition and activity of microtubule associated proteins during brain development. Nature (London) **284:** 353–355.

73. MARGOLIS, R. & C. RAUCH. 1981. Characterization of rat brain crude extract microtubule assembly: Correlation of cold stability with the phosphorylation state of a microtubule associated 64K protein. Biochemistry **20:** 4451–4458.

74. MATUS, A., R. BERNHARDT & T. HUGH-JONES. 1981. High molecular weight microtubule associated proteins are preferentially associated with dendritic microtubules in brain. Proc. Natl. Acad. Sci. USA **78:** 3010–3014.

75. MCQUARRIE, I., S. BRADY & R. LASEK. 1980. Polypeptide composition and kinetics of SCa and SCb in sciatic nerve motor axons and optic axons of the rat. Soc. Neurosci. Abstr. **6:** 501.

76. MORI, H., Y. KOMIYA & M. KUROKAWA. 1979. Slowly migrating axonal polypeptides: Inequalities in their rate and amount of transport between two branches of bifurcating axons. J. Cell Biol. **82:** 174–184.

77. MORRIS, J. & R. LASEK. 1982. Stable polymers of the axonal cytoskeleton: The Axoplasmic Ghost. J. Cell Biol. **92:** 192–198.

78. MURPHY, D., K. JOHNSON & G. BORISY. 1977. Role of tubulin associated proteins in microtubule nucleation and elongation. J. Mol. Biol. **117:** 33–52.

79. MURPHY, D., R. VALLEE & G. BORISY. 1977. Identity and polymerization stimulatory activity of the non-tubulin proteins associated with microtubules. Biochemistry **26:** 2598–2605.

80. OBLINGER, M., S. BRADY, I. MCQUARRIE & R. LASEK. 1985. Cytotypic differences in the protein composition of the axonally transported cytoskeleton in mammalian neurons. J. Neurochem. Submitted for publication.

81. O'FARRELL, P. 1975. High resolution two-dimensional electrophoresis of proteins. J. Biol. Chem. **250:** 4007–4021.

82. PENG, I., L. BINDER & M. BLACK. 1986. Biochemical and immunological analyses of cytoskeletal domains of neurons. J. Cell Biol. In press.

83. PENG, I., L. BINDER & M. BLACK. 1985. Cultured neurons contain a variety of microtubule associated proteins. Brain Res. **361:** 200–211.

84. PETERS, A., S. PALAY & H. WEBSTER. 1976. The fine structure of the nervous system: The neurons and supporting cells. Saunders. Philadelphia.

85. POLLARD, T., S. SELDEN & P. MAUPIN. 1984. Interaction of actin filaments with microtubules. J. Cell Biol. **99:** 33s–37s.

86. PONSTINGL, H., E. KRAUHS, M. LITTLE & T. KEMPF. 1981. Complete amino acid sequence of alpha tubulin from procine brain. Proc. Natl. Acad. Sci. USA **78:** 2757–2761.

87. RUNGE, M. & R. C. WILLIAMS. 1982. Formation of an ATP dependent microtubule-neurofilament complex *in vitro.* Cold Spring Harbor Symp. Quant. Biol. **46:** 483–493.

88. SAHENK, Z. & S. BRADY. 1983. Morphologic evidence for stable regions on axonal microtubules. J. Cell Biol. **97:** 210a.

89. SATTILARO, R., W. DENTLER & E. LECLUYSE. 1981. Microtubule-associated proteins (MAPs) and the organization of actin filaments *in vitro.* J. Cell Biol. **90:** 467–473.

90. SCHNAPP, B. & T. REESE. 1982. Cytoplasmic structure in rapid frozen axons. J. Cell Biol. **94:** 667–679.

91. SLOBODA, R. & J. ROSENBAUM. 1979. Decoration and stabilization of intact, smooth-walled microtubules with microtubule-associated proteins. Biochemistry **18:** 48–55.

92. SMITH, D. 1973. The location of neurofilaments and microtubules during the postnatal development of Clark's nucleus in the kitten. Brain Res. **55:** 41–53.

93. SMITH, R. 1073. Microtubule and neurofilament densities in amphibian spinal root fibers: Relationship to axoplasmic transport. Can J. Physiol. Pharmacol. **51:** 798–806.

94. SOIFER, D. & K. MACK. Microtubules in the nervous system. *In* Handbook of Neurochemistry. A. Lajtha, Ed.: **7:** 245–280.

95. SOLOMON, F. 1984. Determinants of neuronal form. Trends Neurosci. **7:** 17–20.

96. STERNBERGER, L. & N. STERNBERGER. 1983. Monoclonal antibodies distinguish phosphorylated and nonphosphorylated forms of neurofilaments *in situ.* Proc. Natl. Acad. Sci. USA **80:** 6126–6130.

97. TASHIRO, T. & Y. KOMIYA. 1983. Subunit composition specific to axonally transported tubulin. Neurosci. **4:** 943–950.

98. TASHIRO, T. & Y. KOMIYA. 1983. Two distinct components of tubulin transport in sensory axons of the rat recognized by dimethyl sulfoxide treatment. Biomed. Res. **4:** 443–450.

99. TASHIRO, T., M. KUROKAWA & Y. KOMIYA. 1983. Two populations of axonally transported tubulin diferentiated by their interactions with neurofilaments. J. Neurochem. **43:** 1220–1225.

100. Tsukita, S. & H. Ishikawa. 1981. The cytoskeleton in myelinated axons: a serial section study. Biomed. Res. **2:** 424–437.
101. Tytell, M., M. Black, J. Garner & R. Lasek. 1981. Axonal transport: Each of the major rate components consist of distinct macromolecular complexes. Science **214:** 179–181.
102. Tytell, M., S. Brady & R. Lasek. 1984. Axonal transport of subclass of tau proteins: Evidence for the regional differentiation of microtubules in neurons. Proc. Natl. Acad. Sci. USA **81:** 1570–1574.
103. Vallee, R. 1982. A taxol dependent procedure for the isolation of a microtubule and microtubule associated proteins (MAPs). J. Cell Biol. **92:** 435–442.
104. Wakshull, E., M. Johnson & H. Burton. 1979. Postnatal rat sympathetic neurons in culture. I. A comparison with embryonic neurons. J. Neurophysiol. **42:** 1410–1425.
105. Wakshull, E., M. Johnson & H. Burton. 1979. Postnatal rat sympathetic neurons in culture. II. Synaptic transmission by postnatal neurons. J. Neurophysiol. **42:** 1426.
106. Webb, B. & J. Wilson. 1980. Cold stable microtubules from brain. Biochemistry **19:** 1993–2001.
107. Willard, M., W. Cowan & P. Vagelos. 1974. The polypeptide composition of intraaxonally transported proteins: Evidence for four transport velocities. Proc. Natl. Acad. Sci. USA **71:** 2183–2187.
108. Wuerker, R. & J. Kirkpatrick. 1972. Neuronal microtubules, neurofilaments, and microfilaments. Int. Rev. Cytol. **33:** 43–75.
109. Zenker, W., R. Mayr & H. Gruber. 1975. Neurotubules: different densities in peripheral motor and sensory nerve fibers. Experientia **31:** 318–320.
110. Zingsheim, H., W. Herzog & K. Weber. 1979. Differences in surface morphology of microtubules reconstituted from pure brain tubulin using two different microtubule-associated proteins: The high molecular weight MAP-2 proteins and tau proteins. Eur. J. Cell Biol. **19:** 175–183.

The Binding of MAP-2 and Tau on Brain Microtubules *in Vitro:* Implications for Microtubule Structure[a]

HELEN KIM,[b,e] CYNTHIA G. JENSEN,[c]
AND LIONEL I. REBHUN[d]

[b]*Department of Microbiology*
University of Virginia
Medical School
Charlottesville, Virginia 22908

[c]*Department of Anatomy*
University of Auckland
Medical School
Auckland, New Zealand

[d]*Department of Biology*
University of Virginia
Charlottesville, Virginia 22901

INTRODUCTION

A MAP, or microtubule-associated protein, was originally defined as a nontubulin polypeptide that stoichiometrically copurified with the microtubule subunit tubulin through cycles of *in vitro* assembly and disassembly.[1] MAP-2, a 300,000 dalton polypeptide, and tau, a heterogeneous group of polypeptides of average molecular weight, 60,000, have both been shown to fulfill that definition of a MAP (reference 1 and Binder *et al.,* this volume). Both MAP-2 and tau have been shown to be heat-stable polypeptides that can independently stimulate microtubule assembly *in vitro.*[2-4] To better understand how these two MAPs interact with tubulin and with the microtubule polymer, our experiments have addressed the following specific questions: What is the specific distribution of these MAPs on a microtubule? Do MAP-2 and tau compete for the same binding sites on microtubules? and How is the identity and the distribution of a MAP-binding site on the microtubule polymer specified?

We showed previously that brain microtubules reassembled *in vitro* were saturable with MAP-2 in the sense that further addition of MAP-2 to a polymerized sample of microtubules did not result in a further increase in turbidity at 350 nm.[4] By fixing such MAP-2-saturated microtubule pellets in the presence of tannic acid prior to thin

[a]Much of this work was accomplished while H. Kim was a graduate student in the Biophysics Program at the University of Virginia. During that time she was supported by Pratt Funds from the University of Virginia, by an NIH training grant to the Biochemistry Department, University of Virginia, and by funds from the NIH and NSF to L. I. Rebhun. C. G. Jensen was supported by funds from the Medical Research Council of New Zealand.

[e]Present address: Department of Cell Biology and Anatomy, University of Alabama at Birmingham, Birmingham, Alabama 35294.

section electron microscopy, we were able to show that microtubules with MAP-2 were heavily decorated with filamentous projections along their lengths. Whereas we were able to enhance for an average periodicity of 32 nm for the MAP-2 projections along the microtubule length, direct measurement of spacings between projections along any particular length of microtubule often yielded a variety of other distances.[5] To obtain a more complete understanding of the distribution of MAP-2 projections on MAP-2–saturated microtubules, thin section electron micrographs of the latter were subjected to microdensitometer-computer autocorrelation analysis as described by Jensen and Smaill elsewhere in this volume.

Comparative *in vitro* analysis of MAP-2–tau–tubulin interactions indicates that the two MAPs interact differently with tubulin. It has been shown that vinblastine induces different effects on tubulin, depending on whether MAP-2 or tau is present.[6] Similarly, tubulin preincubated with colchicine to suppress assembly has different numbers of free sulfhydryls accessible for alkylation, depending on whether MAP-2 or tau is present; tau does not affect the number of sulfhydryls on tubulin, whereas MAP-2 suppresses the extent of alkylation by about 30 percent.[7] Although both MAP-2 and tau stoichiometrically promote the polymerization of microtubules when either is recombined with tubulin, quantitative differences in *in vitro* behavior of the resulting different microtubule populations have been documented (Bender and Rebhun, this volume). MAP-2–microtubules are about twofold more stable to calcium depolymerization than tau microtubules.[8] Moreover, MAP-2 has been shown to induce a tubulin-critical concentration of 0.05–0.08 mg/ml,[3,4] whereas tau induces a higher critical concentration of 0.25 mg/ml.[3,4] In addition to the biochemical differences conferred on tubulin and the microtubule polymer by MAP-2 and tau, there is increasing evidence that the two MAPs may be differentially compartmentalized within neurons; MAP-2 appears to be predominantly dendritic, whereas tau appears to be localized in axons (Binder *et al.*, this volume). For all these reasons, it was of interest to determine whether MAP-2 and tau could compete for the same sites on brain microtubules *in vitro*.

High resolution isoelectric focusing of the tubulin in microtubule protein obtained by standard procedures from whole brain extract has resolved 17–20 different isotypes.[9] In light of the differences in affinity between MAP-2 and tau for the microtubules polymerized from such heterogeneous tubulin, as well as the apparent segregation of MAP-2 from tau within a single cell, it was of interest to determine whether there was any corellation between tubulin isotype composition and MAP composition among microtubule populations. To that end, gray matter microtubules and white matter microtubules were analyzed for tubulin isotype composition, and the results are described here.

Finally, in an effort to relate tubulin isotype composition and heterogeneity with the specific distribution of MAP-binding sites found on microtubules *in vitro*, a model for a nonrandom distribution of tubulin isotypes within a microtubule is presented.

MATERIAL AND METHODS

Preparation of Microtubule Protein

Twice-cycled microtubule protein from bovine brain was obtained according to the Berkowitz *et al.*[10] modifications of the method of Shelanski *et al.*,[11] except that the second polymerization did not include glycerol.[4]

Purification of MAP-2, Tau, and Tubulin

MAP-2 and tau were purified from twice-cycled microtubule protein by molecular sieve chromatography of the heat-stable fraction on Bio-Gel A1.5M as described previously,[4] with the following modifications: all protein samples were concentrated by precipitation with 50% ammonium sulfate, and the column load volume was always 1% or less of the total column bed volume. This last modification was essential for optimal resolution between MAP-2 and tau.

Tubulin dimers were purified from twice-cycled microtubule protein by phosphocellulose chromatography of twice-cycled microtubule protein as described previously.[5] After obtaining the tubulin from the phosphocellulose column, the protein was concentrated by precipitation with 50% ammonium sulfate, resuspended in phosphocellulose column buffer (PCCB): 50 mM PIPES, 0.5 mM $MgSO_4$, 1 mM EGTA, 0.1 mM GTP, pH 6.9, desalted by centrifugation through Sephadex G-25 equilibrated in PCCB,[12] and stored in small (100 μl or less) aliquots at $-80°C$.

Determination of Protein Concentration

The concentration of total protein in appropriate samples was determined according to a modification[13] of the Lowry procedure.[14]

MAP-2–Tau Displacement Experiments

The polymerization of microtubules saturated with either MAP-2 or tau was monitored spectrophotometrically; phosphocellulose-purified tubulin (PC6S) was mixed with either MAP-2 at a mass ratio of 1:1, or with tau, at a tau:tubulin mass ratio of 0.25:1.0, and allowed to polymerize by raising the temperature to 37°C. To ensure that the sample was saturated for capacity to assemble, the final (equilibrium) turbidity value was compared with that of a sample containing the same concentration of PC6S, but a MAP concentration 10–15% higher than in the first sample. This was carried out because of slight variation from preparation to preparation in the capacity of both MAPs to stimulate microtubule assembly.

After equilibrium in the polymerization reaction was reached, as assayed spectrophotometrically, 50 μl aliquots of the polymer mixture were overlaid onto 200 μl of a 20% sucrose cushion containing an appropriate concentration of the second MAP in PCCB. The microtubule samples were centrifuged through the sucrose cushions in 6 × 25 mm glass tubes (W.C. Thomas Co.) at 30,000 × g for 1 hour at 30°C, in an SS-34 rotor. Glass tubes were used because the glass allowed easy localization of the microtubule pellets after the centrifugation with the aid of a polarimeter. After centrifugation, the microtubule pellets were washed three times by filling the tube with warm PCCB and aspirating. After the washes, the microtubule pellets were overlaid with cold (4°C) PCCB and allowed to sit in an ice bath for 10–15 min, and then monitored with the polarimeter to ensure that depolymerization was complete, as indicated by the lack of birefringence. A small aliquot of each depolymerized microtubule sample was saved for Lowry determination. Another aliquot was diluted 1/25 with antigen dilution buffer (AgB) (50 mM PIPES, 1.0 mM $MgCl_2$, 1 mM EGTA, 0.3 M NaCl, 0.02% NaN_3, and 1 mg/ml BSA, pH 7.0), and analyzed for MAP-2 and tau by radioimmunoassay (RIA), as described by Bender,[8,15] using an IgG-enriched fraction of a rabbit polyclonal antiserum, which contained antibodies to both MAP-2 and to tau. This antiserum has been characterized elsewhere.[15]

Isoelectric Focusing

Isoelectric focusing of samples was carried out in slab gels (0.75 mm thick × 15 cm long × 10 cm wide) essentially as described by Field *et al.*,[9] except that after prefocusing at 8 watts for 30 min, electrofocusing was carried out at 1000 volts for 12 hours. After electrofocusing, the gel was fixed and washed in 10% acetic acid overnight, and stained with Coomassie brilliant blue R-250.

Sodium Dodecyl Sulfate (SDS) Polyacrylamide Gel Electrophoresis

SDS-urea polyacrylamide slab gel electrophoresis was carried out as described previously,[4] incorporating gradients in acrylamide and in urea. The gels (20 × 15 × .15 cm) were stained with Coomassie brilliant blue R-250.

Microtubule Polymerization Assay

Microtubule polymerization was monitored by continuously recording the increase in turbidity at 350 nm[16] in a Gilford spectrophotometer equipped with a temperature-controlled cuvette chamber. All polymerization was carried out in 50 mM PIPES, 1.0 mM EGTA, 0.5 mM MgSO$_4$, pH 6.9, with 1 mM GTP. Components for the reactions were mixed in the appropriate cuvettes at 4°C, and polymerization was initiated by raising the temperature to 37°C.

Thin Section Electron Microscopy

Thin section electron microscopy was carried out as described previously,[4] using a modification of procedures developed by Tilney *et al.*[17] and by Begg *et al.*[18] Electron microscopy was carried out with a Phillips 201 electron microscope, calibrated with a carbon replica of an optical diffraction grating.

Preparation of Gray Matter and White Matter Microtubule Protein

Bovine cerebral cortex or caudate nucleus and internal capsule were used as the sources of gray and white matter, respectively. Tissue was homogenized at a ratio of 1 g tissue per ml of buffer in homogenization buffer consisting of 100 mM PIPES, 1 mM MgSO$_4$, 2 mM EGTA, and 0.1 GTP, pH 6.9. Further processing of the extracts was as described previously for obtaining twice-cycled microtubule protein (see *Preparation of Microtubule Protein*, this section).

Microdensitometer-Computer Correlation Analysis of Spacings between MAP-2 Projections on MAP-2–Saturated Microtubules

The details of this technique are described elsewhere in this volume (see Jensen and Smaill).

RESULTS

Purification of MAP-2 and Tau from Cycled Microtubule Protein

During initial purifications of MAP-2 from the heat-stable fraction of cycled microtubule protein, the degree of purity of the non-MAP-2 fractions was not of much concern, because our efforts were focused on MAP-2.[4] We found, however, that simply by keeping the volume of the column load under 1% of the column volume, reproducibly clean MAP-2 and tau fractions could be obtained.[5] FIGURE 1 shows a typical elution profile from a Bio-gel A1.5M molecular sieve column, whereas FIGURE 2 is an SDS-polyacrylamide gel of the corresponding fractions, confirming the resolution of the MAP-2–containing fractions from the tau-containing fractions.

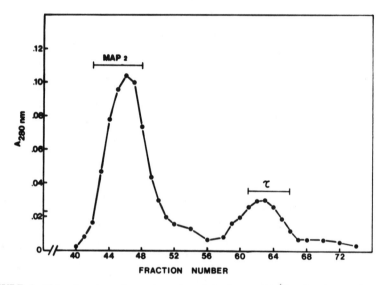

FIGURE 1. Molecular sieve chromatography of the heat-stable fraction from twice-cycled bovine brain microtubule protein on Bio-Gel A1.5M.

FIGURE 3 shows the typical tubulin (lane A), tau (lane B), and MAP-2 (lane C) fractions used in the experiments described here.

Ultrastructural Analysis of MAP-2–Saturated Microtubules

Thin section electron microscopy of MAP-2–saturated microtubules yielded micrographs such as shown by FIGURE 4. It was apparent that (a) all the microtubules were decorated with filamentous projections along their length, presumably due to the presence of MAP-2, and (b) that these MAP-2 projections were attached at nonrandom intervals along the length and all around the microtubules. Quantitative analysis of the spacings between MAP-2 projections along many microtubules revealed an average spacing of 32 nm.[4] Whereas we were able to enhance for an average periodicity

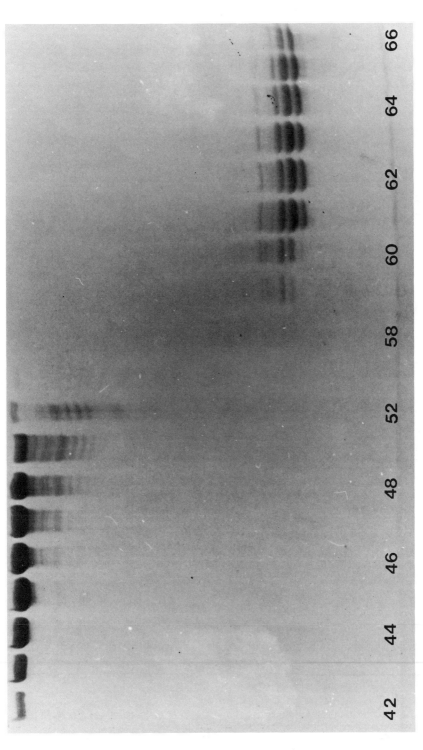

FIGURE 2. SDS-polyacrylamide gel electrophoresis of the fractions from the Bio-Gel A1.5M column loaded with the heat-stable fraction from twice-cycled brain microtubule protein. The numbers correspond to the column fraction; see FIGURE 1. This is a 7.5% acrylamide gel, according to the Laemmli[32] formulation, stained with Coomassie blue.

FIGURE 3. SDS-urea polyacrylamide gel electrophoresis of the protein components used in the experiments described here. A: 25 μg of phosphocellulose-purified tubulin; B: 40 μg of tau; C: 10 μg of MAP-2.

of 32 nm using Markham techniques, direct measurement of spacings between MAP-2 projections along any particular length of microtubule in an electron micrograph often yielded a variety of distances,[5] indicating that MAP-2 was not distributed in a strictly periodic manner along the microtubule long axis, but probably in a more complicated helical distribution around the microtubule as well. To determine all the different spacings between MAP-2 projections, as well as to determine whether the arrangement of MAP-2 projections could be described by a symmetric distribution, microdensitometer-computer correlation analysis was carried out (Jensen and Smaill, this volume)

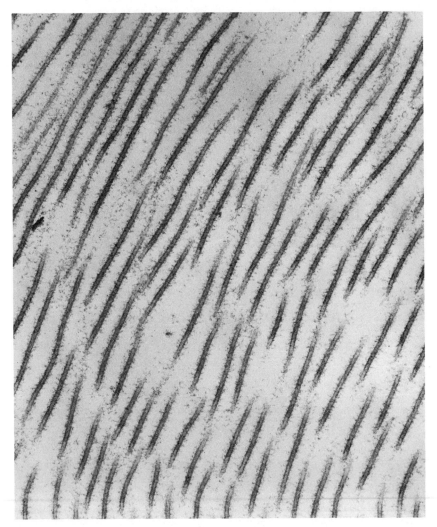

FIGURE 4. Thin section electron micrograph of microtubules saturated with MAP-2. Microtubules were prepared and processed for thin section electron microscopy as described in the MATERIAL AND METHODS.

using electron micrographs similar to FIGURE 4. The observed spacings between MAP-2 projections along the microtubule were compared with various models for MAP arrangement as initially described by Amos.[19] The arrangement of MAP-2 on MAP-2–saturated microtubules correlated best with the 12-dimer superlattice model[19] at a confidence level greater than 98% (Jensen and Smaill, this volume). FIGURE 5 shows schematically the 12-dimer superlattice of MAP binding sites superimposed on the unrolled 13-protofilament microtubule cylinder. In this model, there is one MAP every 12 tubulin dimers, along a single protofilament; biochemically, then, this model requires one MAP molecule, or perhaps more accurately, one MAP projection, every 12 tubulin dimers. The helical nature of the MAP superlattice also predicts that in any random longitudinal plane, such as might be seen by thin section electron microscopy, a number of nonidentical distances will be seen between apparently adjacent MAP projections. These spacings are indicated in FIGURE 6, which is a three-dimensional representation of the 12-dimer MAP superlattice. All of the distances between MAP sites predicted by the 12-dimer superlattice were found on MAP-2–saturated microtubules (Jensen and Smaill, this volume). The probability that the observed spacings between MAP-2 projections would fit a 12-dimer superlattice distribution due to random binding of MAP molecules onto the microtubule surface was 1.8% (Jensen and Smaill, this volume). Additionally, however, other distances between MAP-2 projections were detected by the microdensitometric scans, in particular, an 11 nm minimal

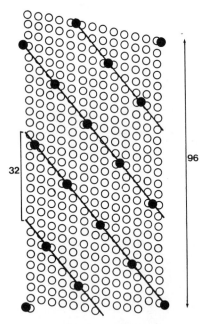

FIGURE 5. Schematic diagram of the 12-dimer MAP superlattice proposed by Amos (1977). For simplicity, all monomers are represented equivalently here, by the open circles. The black dots indicate the MAP-binding sites located helically, along every double 8-start dimer helix, indicated by the lines connecting the dots. The 32 indicates the 32 nm or 4-dimer distance between "ribbons" of MAP-binding sites. The 96 indicates the 96 nm or 12-dimer distance between identical MAP sites along a single protofilament. This is redrawn from Amos.[19]

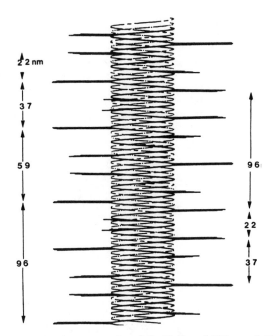

FIGURE 6. Three-dimensional representation of the 12-dimer MAP superlattice. The numbers 22, 37, 59, and 96 indicate the spacings between MAP projections predicted by the 12-dimer superlattice and that would be observed in random longitudinal sections of microtubules such as might be seen by thin section electron microscopy.

distance between projections, as well as a 48 nm, and an 85 nm spacing (Jensen and Smaill, this volume). These spacings are incompatible with a 12-dimer superlattice; they are, however, predicted by a 6-dimer superlattice arrangement—that is, where there is 1 MAP projection every 6 tubulin dimers along a single protofilament. This is shown schematically in FIGURE 7, where the MAP sites predicted by the 12-dimer superlattice are indicated by the round black dots, and the additional MAP sites predicted by the 6-dimer superlattice are indicated by the black squares. For comparison, three-dimensional representations of both the 6-dimer MAP superlattice and the 12-dimer superlattice are shown in FIGURE 8.

MAP-2 and Tau Binding Sites on in Vitro-Assembled Microtubules

It has been shown that microtubules depolymerized to different extents in the same concentration of calcium, depending on whether the microtubules had MAP-2 or tau associated with them (reference 8 and Bender and Rebhun, this volume). Moreover, it has been shown that the sensitivity of tubulin to certain drugs can vary, depending on which MAP is associated with the tubulin.[7] Thus, a logical question that arose from these observations was, Did both MAP-2 and tau bind to the same site on the microtubule polymer surface? The approach we took was to polymerize microtubules *in vitro* saturated with one MAP, expose these microtubules to saturating amounts of the second MAP, and determine whether the latter would displace any of the former, and/or bind to the microtubules.

At a PC6S (phosphocellulose-purified tubulin; see MATERIAL AND METHODS) concentration of 1 mg/ml, where there was no self-assembly, increasing concentrations of MAP-2 stoichiometrically promoted microtubule formation.[4] Between 0.75–1.0 mg/ml MAP-2, there was no further increase in polymer formation, indicating that the microtubules were saturated with respect to MAP-2. As with MAP-2, increasing concentrations of tau stimulated microtubule formation (FIGURE

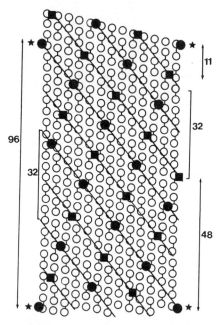

FIGURE 7. Schematic diagram of the 6-dimer MAP superlattice. (All numbers are in nanometers.) The 11 and the 48 indicate additional spacings between MAP-2 projections detected by the microdensitometer correlation analysis in longitudinal thin sections of MAP-2–saturated microtubules. The round black dots indicate the MAP-binding sites predicted by the 12-dimer MAP superlattice originally proposed by Amos,[19] for a 1:12 MAP:tubulin molar ratio. The squares indicate the proposed positions of the additional MAP-binding sites on a microtubule with a MAP-2–tubulin molar ratio of 1:6. The 32 on the right indicates how the distance between the "ribbons" of square MAP sites is the same as the distance between the original ribbons of MAP sites, represented by the 32 on the left. The starred dots indicate the 96 nm distance between identical sites along a single protofilament.

9), until the system was saturated above a tau concentration of 0.25 mg/ml. Similar results have been reported by Herzog and Weber[3] and by Bender.[8] Our initial determination of the MAP-2: tubulin molar ratio in MAP-2–saturated microtubules was 1:9, taking into account only the high molecular weight doublet on SDS gels, and using an average molecular weight of 300,000. It has since been shown that most of the minor polypeptides that migrate beneath the MAP-2 doublet on SDS gels and that copurify with MAP-2 are proteolytic fragments of MAP-2.[20] Taking this into account, quantitative densitometry of Fast Green-stained SDS gels revealed a MAP-2:tubulin molar ratio of 1:6.5[5] for MAP-2–saturated microtubules. When microtubules satu-

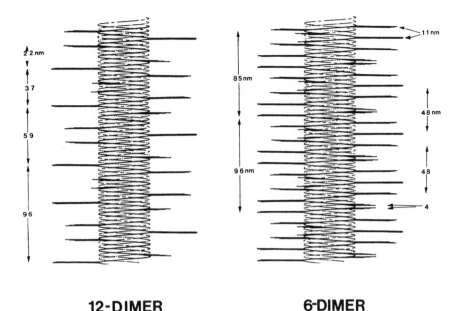

12-DIMER   **6-DIMER**

FIGURE 8. Three-dimensional comparison of the 12-dimer MAP superlattice and the 6-dimer MAP superlattice. The 11 nm, 48 nm, and 4 nm in the 6-dimer superlattice indicate the additional spacings between MAP-2 projections that the 6-dimer model predicts would be detected by the microdensitometer analysis.

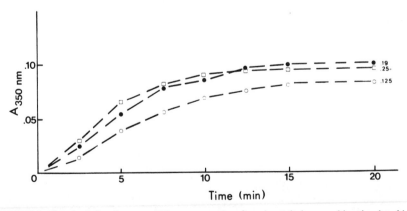

FIGURE 9. Graph of absorbance at 350 nm versus time for microtubule assembly stimulated by purified tau. The tubulin concentration was held constant in all samples at 1 mg/ml. The tau concentration in the different samples is indicated by the numbers next to the different turbidity curves. All samples were mixed together at 4°C in cuvettes in a thermostatically controlled cuvette chamber. Assembly was initiated by raising the temperature to 37°C.

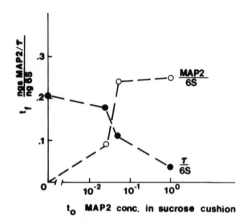

FIGURE 10. Graph of t_f (final) ratio of either MAP-2/6S or tau/6S versus t_0 (initial) concentration of MAP-2 in a 20% sucrose cushion. This graph shows the changes in amounts of MAP-2 and of tau, relative to tubulin, in microtubules initially saturated with tau, and centrifuged through different amounts of MAP-2 in a 20% sucrose cushion. Note that the abscissa is a log scale; thus, to go from 10^{-2} mg/ml to 10^0 mg/ml covers a 100-fold range in MAP-2 concentration.

rated with tau (as indicated by turbidimetric assay) were similarly analyzed, the molar ratio of tau:tubulin in tau-saturated microtubules ranged from 1:3–1:5, using an average molecular weight of 60,000 for tau.[5]

To determine whether MAP-2 or tau could displace each other from microtubules saturated with either tau or MAP-2, samples of microtubules polymerized with either saturating amounts of tau or with MAP-2 were centrifuged through 20% sucrose cushions containing varying concentrations of the other MAP. When tau-saturated microtubules were exposed to varying amounts of MAP-2, approximately 75% of the tau was displaced by MAP-2, as shown in FIGURE 10. When MAP-2–saturated microtubules were exposed to varying amounts of tau, however, only 25% of the MAP-2 was displaced, with an equivalent amount of tau becoming associated with the microtubules.[5] These results are summarized in TABLE 1.

Isoelectric Focusing of Tubulin from Gray Matter and from White Matter

It has been shown that white matter microtubules differ in their MAP content from gray matter microtubules (reference 21 and Binder *et al.*, this volume). To determine

TABLE 1. Summary of MAP-2–Tau Displacement Data

$t_0\dfrac{\text{moles MAP}}{\text{mole 6S}}$	$t_f\dfrac{\text{moles MAP}}{\text{mole 6S}}$	$t_f\dfrac{\text{moles total MAPs}}{\text{mole 6S}}$
(1) MAP-2/6S = 1/6, τ/6S = 0	MAP-2/6S = 1/8, τ/6S = 1/30	1/6.3
(2) τ/6S = 1/3, MAP-2/6S = 0	MAP-2/6S = 1/10, τ/6S = 1/15	1/6.0

whether there were any differences in tubulin isotype composition in microtubule populations that were known to differ in their MAP composition, bovine gray matter and white matter microtubule protein was subjected to isoelectric focusing, as described in MATERIAL AND METHODS. The results, shown in FIGURE 11, indicate that there are qualitative and quantitative differences in tubulin isotype composition between the two microtubule populations. There were 17 tubulin isotypes in gray matter microtubule protein. White matter microtubules lacked two of the isotypes contained in gray matter (indicated by arrows and *'s), and contained detectably lesser amounts of two additional isotypes (indicated by arrows only).

DISCUSSION

Ultrastructural Analysis of MAP-2–Saturated Microtubules

We have shown by microdensitometer-computer correlation analysis that the projections on MAP-2–saturated microtubules are not arranged randomly; instead, the arrangement is highly compatible with the 12-dimer MAP superlattice originally proposed by Amos.[19] Although the additional spacings observed between MAP-2 projections are predicted by the 6-dimer MAP superlattice, statistical analysis indicates that the probability that the match to the 6-dimer model is due to random binding is greater than 90% (Jensen and Smaill, this volume). This would be predicted, however, if not all the MAP sites on the 6-dimer superlattice were filled; this observation is consistent with the molar ratio of MAP-2:tubulin in these microtubules ranging between 1:6.5 to 1:9 (see RESULTS). We suggest that the 6–dimer MAP superlattice more completely describes the total set of MAP-binding sites on the microtubule lattice. The statistical fit of the data to the model may be poor for two reasons. The additional sites specified by the 6-dimer superlattice may be sites where shorter or less visible proteolytic fragments of MAP-2 are bound. Alternatively, the MAP-2 bound at the additional sites could be chemically and conformationally distinct from the MAP-2 bound at sites specified by the 12-dimer superlattice, resulting in a less easily detectable molecule. For example, the MAP-2's bound at the 6-dimer superlattice sites could be in a different state of phosphorylation. Clearly, further experimentation is required to resolve the question of whether the 6-dimer model is correct. In support of the 6-dimer MAP superlattice, interprojection spacings that are predicted by the 6-dimer superlattice have been detected in microtubules *in situ*.[22]

MAP-2 and Tau Binding Sites on Brain Microtubules in Vitro

The results presented here show that *in vitro* assembled brain microtubules are saturable with either MAP-2 or tau alone, but that approximately 25% of the MAP-2 on MAP-2–saturated microtubules can be displaced by tau, and that about 75–85% of the tau on tau-saturated microtubules can be displaced by MAP-2. In both cases, the resulting microtubule populations contain approximately the same ratio of MAP-2 to tau. More importantly, the ratio of *total* MAPs to tubulin in both cases is the same, one MAP per approximately six tubulin dimers, which is the same as the MAP-2:tubulin molar ratio in microtubules saturated with MAP-2 alone. These data indicate that *in vitro* assembled microtubules polymerized from whole brain (predominantly gray matter) extract contain a pool of binding sites for MAP-2 and tau, which consists of two classes, defined by different affinities for each MAP. Thus, the MAP-2 sites prefer

GM   WM

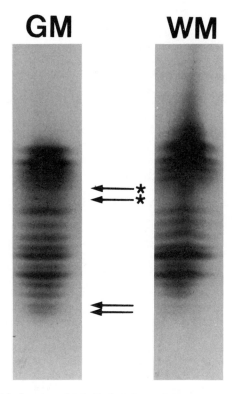

FIGURE 11. Isoelectric focusing of tubulin in twice-cycled gray matter (GM) and in white matter (WM) microtubule protein. The two different microtubule protein samples were prepared as described in MATERIAL AND METHODS from freshly dissected bovine brain tissue. Because MAP-2 electrofocuses with a pI similar to some of the tubulin isotypes (5.3–5.5), the gray matter tubulin was purified away from the MAPs by phosphocellulose chromatography as described in MATERIAL AND METHODS. Isoelectric focusing was carried out essentially as described by Field et al.[9]

to bind MAP-2, but will bind tau in the exclusive presence of the latter; the tau sites prefer to bind tau, but will bind MAP-2 in the exclusive presence of the latter.

In considering what general ways MAP-microtubule interaction could be described, three simple models can be hypothesized, shown schematically in FIGURE 12. In model A, each MAP has a unique binding site, and its binding to the microtubule is independent of the binding of other MAP at their sites. In model B, all the MAP-binding sites are identical, and have equal affinity for all the MAPs. Here, the binding of a particular MAP in the presence of other MAPs would be a function of the relative concentrations of the other MAPs. In model C, there are different classes of MAP-binding sites, defined by different affinities for the different MAPs. TABLE 2 shows for each model in FIGURE 12 the conditions it imposes on MAP-microtubule interaction, and whether the data described here satisfy those conditions. Model A, where each MAP binds to a unique site, independent of other MAPs, predicts that binding of a second MAP should not change the amount of the first MAP, relative to a constant number of tubulin dimers in the microtubule. It follows, however, that with binding of the second MAP, the total number of MAPs bound to the microtubule per constant number of tubulin dimers should increase. Model B, where all the MAP sites have equal affinity for each MAP, predicts that a second MAP in solution should displace a MAP already bound on the microtubule, with the degree of displacement being proportional to the concentration of the second MAP. This model, unlike model A, predicts that the total number of MAPs per tubulin dimer should not change with binding of a second MAP. Model C, where the sites have different affinities for each MAP, requires that binding of a second MAP displace the first MAP, to a saturable point, and that the total number of MAPs bound per tubulin dimer should not increase, with binding of a second MAP. The results presented here have demonstrated (1) a maximum number of MAP-binding sites on *in vitro* assembled microtubules, (2) that either of the two MAPs, MAP-2 or tau, can by itself saturate those sites, (3) that both MAPs can be displaced by each other, but to different extents, and (4) that the total number of MAPs bound per tubulin dimer does not change, with one MAP being displaced by the other. As summarized in TABLE 2, the only model for which all the conditions are satisfied by the data presented here is model C. Thus, we suggest that microtubules reassembled *in vitro* from extracts of whole brain contain a pool of MAP-binding sites that consists of at least two classes—one that has higher affinity for

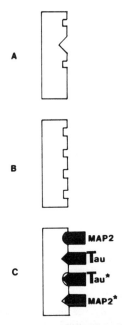

FIGURE 12. Models of MAP-microtubule interaction. In model A, the square and the triangular indentations in the microtubule wall represent different MAP binding sites, unique for each MAP. In model B, the identical square indentations represent MAP binding sites that can bind different MAPs with equal affinity. In model C, the rounded indentations represent sites that have higher affinity for MAP-2, but that will bind tau; the triangular indentations represent sites that have higher affinity for tau than for MAP-2, but that will bind the latter. MAP-2* and tau* represent molecules bound at their lower affinity sites.

MAP-2, but that will bind tau, and another that has higher affinity for tau, but that will bind MAP-2.

It should be noted that in the experiments described here, the state of phosphoryl-ation of either MAP-2 or tau was not ascertained; both species were isolated from heat-stable fractions of brain microtubule protein, and used as is. Both MAP-2 and tau are phosphorylated *in vitro*,[17,23] and MAP-2 has been shown to be phosphorylated *in vivo* in chick brain.[17] If microtubules do bind both MAP-2 and tau *in vivo* as they appear to *in vitro*, perhaps the difference between MAPs that are displaced and those that are not displaced by a second MAP lies in the state of phosphorylation. It is not clear whether phosphorylated MAP-2 and dephosphorylated MAP-2 are incorporated identically into microtubules.[24,25] Selden and Pollard[26] have shown, however, that

TABLE 2. Real Versus Predicted Data for MAP-Microtubule Interaction

Model	Conditions Imposed by Model	Conditions Met by Data
A: Unique, independent MAP-binding sites	(1) No change in amount of first MAP, with binding of second MAP	No
	(2) Total number of MAPs should increase with binding of second MAP	No
B: Equivalent MAP-binding sites, with equal affinity for each MAP	(1) Increasing concentrations of second MAP should displace increasing amounts of first MAP until complete displacement	No
	(2) Total number of MAPs bound should not increase with binding of second MAP	Yes
C: MAP-binding sites can bind either MAP, but with different affinities	(1) Displacement of first MAP should not be complete, even in the presence of excess second MAP	Yes
	(2) Total number of MAPs bound should not increase with binding of second MAP	Yes

phosphorylation of MAP-2 inhibits binding of MAP-2 to actin filaments. It is not yet known whether phosphorylation affects incorporation of tau into brain microtubules, although it has been noted that tau can be highly phosphorylated *in vitro*.[23] There is precedent in other cytoskeletal systems for the function of a protein being regulated by phosphorylation; when the myosin light chains are phosphorylated, the actin-activated ATPase activity of smooth muscle myosin increases.[27] Thus, it may be important to reexamine the competition of MAP-2 and tau for binding sites on microtubules when different states of phosphorylation for the two proteins can be defined.

The results and interpretations here extend the observations of Sandoval and Vanderkerckhove,[28] who quantitated MAP-2–tau competition for binding sites, and obtained data compatible with the results reported here.

Is there now sufficient information from the data presented here as well as by others to describe the arrangement of different MAPs on the microtubule lattice *in vitro?* The microdensitometer-computer correlation analysis of MAP-2–saturated microtubules indicates that the 12-dimer superlattice originally proposed by Amos[19] may specify one set of MAP-2 sites. Preliminary analysis of the spacings between tau molecules on tau-saturated microtubules indicates that the same 12-dimer superlattice may describe the distribution of tau sites (Binder, unpublished data). What is not known is how tau and MAP-2 are distributed on the microtubule when both are bound at the same time. We can make some statements, however; (1) the results presented here show that more than half of the MAP-2 originally on MAP-2–saturated microtubules remains associated with the microtubule in the presence of excess tau; (2) the distribution of MAP-1, the highest molecular weight brain MAP, on microtubules saturated *in vitro* with MAP-1, appears to resemble that of MAP-2 on MAP-2–saturated microtubules (Jensen, personal communication); it is also known that the MAP-1 to tubulin molar ratio in these microtubules is 1 MAP-1 per every 6.7 tubulin dimers[29]; thus, the 6-dimer superlattice may specify these MAP-1 sites, when microtubules are saturated with MAP-1 alone; (3) the helical repeat of 96 nm or 1 MAP every 12 dimers is observed on microtubules that have been cycled with all the MAPs, (MAP-1, MAP-2, and tau),[19] and that are probably not saturated with respect to any of the three. In light of all the above, a possible arrangement for the three MAPs on the same microtubule would be to have the MAP-2's on one 96 nm repeat helix, and have the other MAPs (MAP-1 and tau) occupy sites on the other helix.

Tubulin Isotype Composition in Gray Matter and White Matter Microtubules

It has been shown that microtubules polymerized *in vitro* from bovine gray matter differ in their MAP composition from those microtubules polymerized from bovine white matter (reference 21 and Binder *et al.,* this volume). It has also been shown that MAP-2 immunoreactivity is much more intense in dendrites than axons, whereas tau immunoreactivity appears to be restricted to axons (Binder *et al.,* this volume). To determine whether the differences in MAP content were reflected in the tubulin isotype composition, twice-cycled microtubules from gray matter and from white matter were prepared under identical conditions, and the tubulin in both preparations was subjected to isoelectric focusing, as described in MATERIAL AND METHODS. Seventeen tubulin isotypes were resolved from gray matter microtubules, and 15 from white matter microtubules (see FIGURE 11). In addition to lacking two of the isotypes contained in gray matter, the white matter tubulin contained less of two other isotypes (see FIGURE 11). These results indicate that white matter tubulin differs qualitatively and quantitatively from gray matter tubulin. The results presented here probably represent a low estimate of the difference(s) between white matter and gray matter tubulin isotypes, because it has been suggested that a significant proportion of axonal (white matter) tubulin is cold stable, and is lost during cycles of assembly and disassembly (Brady *et al.,* this volume). If one were to hypothesize that tubulin isotype composition were a factor in determining MAP composition, then the results presented here are consistent with the hypothesis, because white matter microtubules are a subset of gray matter microtubules, and one would expect white matter microtubules to contain a subset of the tubulin isotypes contained in gray matter microtubules. Previous reports have described changes in complexity of tubulin isotypes[30] and in MAP content[31] during mammalian brain development. We suggest that these sets of changes are not unrelated.

If one considers the MAP-2–tau displacement data in light of the data indicating differences in tubulin isotype content between gray and white matter microtubules, we may speculate that the brain microtubules reconstituted *in vitro* in our experiments may be "hybrid" microtubules with respect to their tubulin isotype content, and that the heterogeneity in MAP-binding sites may result from tubulin isotypes that are segregated from each other *in vivo,* polymerizing together *in vitro.* An obvious experiment will be to determine how different MAPs compete for binding sites on a more pure population of microtubules, such as microtubules from white matter. It is already known that these microtubules contain significantly higher amounts of tau protein than MAP-2, after two cycles of assembly *in vitro* (Binder *et al.,* this volume). In light of the data discussed here, we predict that MAP-2 would not displace tau from white matter microtubules to the same extent as from gray matter microtubules.

Hypothesis: A Nonrandom Distribution of Tubulin Isotypes Determines the Arrangement of MAP-Binding Sites in a Microtubule

We have presented data here that indicate that (1) MAPs bind to the microtubule polymer at sites that are nonrandomly distributed on the microtubule lattice; (2) microtubules polymerized *in vitro* contain a heterogeneous population of MAP-binding sites; (3) microtubule samples that could be said to represent different neuronal cell compartments (gray matter or dendritic and cell body versus white matter or axonal microtubules) contain subtle but detectable differences in their tubulin isotype compositions, in addition to the clear differences in their MAP contents. The question yet to be answered is, How is a MAP-binding site specified within a microtubule? We suggest that the identity and distribution of MAP-binding sites might be specified within a microtubule by the composition and nonrandom distribution of tubulin isotypes. FIGURE 13 shows two general ways in which tubulin isotypes could be nonrandomly distributed within a microtubule. In both models 1 and

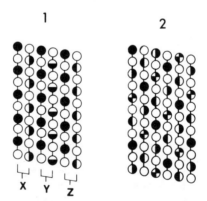

FIGURE 13. Models for nonrandom distribution of tubulin isotypes in the microtubule. For the sake of argument, the alpha tubulin monomers will be identical. In both 1 and 2, the open circles represent identical alpha tubulin monomers; the variously filled-in circles represent different beta tubulin isotypes. Also in both models, the monomers are connected vertically to indicate the alignment of monomers to form protofilaments. In model 1, the different betas are distributed such that each protofilament consists of only one type of beta. Thus, X sites are specified along the X pair of protofilaments, the Y sites are specified along the Y pair of protofilaments, and so forth. In model 2, the different betas are distributed in repeating groups down each protofilament, such that all the protofilaments in the microtubule are identical, and helical asymmetry is introduced by the staggering of adjacent protofilaments.

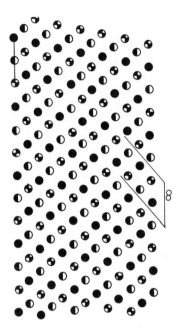

FIGURE 14. Schematic diagram of a microtubule lattice consisting of identical protofilaments, each consisting of linear repeats of the three different beta tubulin isotypes connected by the lines in the diagram. For simplicity, again, the alphas have been kept identical, and, in fact, have been omitted from this diagram. By comparing this diagram with FIGURES 5 and 7, it can be seen that this particular staggering of protofilaments can specify either the 12-dimer or the 6-dimer MAP superlattice. The 8 indicates the 8-start double dimer helix.

2, the open circles arbitrarily represent alpha tubulin monomers that will remain identical. The variously filled-in circles then represent different beta tubulin monomers. In model 1, the protofilaments differ from each other by consisting of identical alphas but different betas. Thus, protofilament pair X differs from protofilament pair Y, which differs from protofilament pair Z. If a MAP-binding site is now defined by a unique beta tubulin isotype, it can be seen that in model 1, identical MAPs would only bind axially along the protofilaments. For example, MAP X would bind along protofilament pair X, MAP Y would bind along protofilament pair Y, and MAP Z would bind along protofilament pair Z. For a given MAP, the spacings between binding sites should occur strictly periodically along the microtubule long axis.

The ultrastructural and biochemical analysis presented here, however, strongly indicates that at least one class of MAP-binding site, the MAP-2 site, is distributed helically around the microtubule. Is there another way the tubulin isotypes could be distributed nonrandomly in the microtubule to generate a helical distribution of MAP-binding sites? Model 2 in FIGURE 13 shows a nonrandom distribution of beta tubulin isotypes that is qualitatively different from that depicted in model one. Here, all the protofilaments are identical, and consist of groups of different beta tubulin isotypes (and identical alphas) repeating linearly down the protofilament. To introduce helical asymmetry, the protofilaments need only be staggered relative to each other. In this model, then, a unique MAP-binding site would repeat helically around the microtubule cylinder, resulting in each unique MAP having a "ribbon" of binding sites helically winding around the microtubule, possibly intertwining with other "ribbons" of binding sites for other MAPs. It should be noted that in both models 1 and 2, the basic monomer helical pattern is identical.

Given the constraints for distribution of MAP-2 binding sites defined by the ultrastructural analysis presented here and elsewhere (Jensen and Smaill, this volume), FIGURE 14 depicts schematically how a microtubule consisting of three different beta tubulin isotypes (indicated by the three linked together in the first protofilament) could specify a 6- or 12-dimer superlattice of MAP-binding sites. Again, for simplicity, the alpha tubulins have been kept identical, and have been omitted from this schematic representation.

SUMMARY

We have presented data that indicate that MAP-2 associates with brain microtubules at nonrandomly distributed sites, whose distribution on the microtubule polymer can best be described by the 12-dimer MAP superlattice originally described by Amos[19]; because of the additional spacings, however, between MAP-2 projections observed on MAP-2–saturated microtubules, we suggest that the 6-dimer MAP superlattice, or what we will call the double Amos superlattice, more completely specifies the total set of MAP-binding sites on cytoplasmic microtubules. Second, we have shown that brain microtubules reassembled *in vitro* contain a heterogeneous population of MAP-binding sites, which differ in their affinities for the two MAPs, MAP-2 and tau. Third, we have shown that microtubule populations that differ in their MAP content have subtle, but detectable differences in their tubulin isotype composition. Based on all the data presented here, we have presented the idea of a nonrandom distribution of tubulin isotypes within a microtubule as a means by which a cell could specify both the identity and the distribution of MAP-binding sites.

REFERENCES

1. SLOBODA, R. D., S. A. RUDOLF, J. L. ROSENBAUM & P. GREENGARD. 1975. Cyclic-AMP dependent endogenous phosphorylation of a microtubule-associated protein. Proc. Natl. Acad. Sci. USA **72:** 177–181.
2. FELLOUS, A., J. FRANCON, A. LENNON & J. NUNEZ. 1977. Microtubule assembly *in vitro*. Purification of assembly-promoting factors. Eur. J. Biochem. **78:** 164–174.
3. HERZOG, W. & K. WEBER. 1978. Fractionation of brain microtubule-associated proteins. Isolation of 2 different proteins which stimulate tubulin polymerization *in vitro*. Eur. J. Biochem. **92:** 1–8.
4. KIM, H., L. I. BINDER & J. L. ROSENBAUM. 1979. The periodic association of MAP 2 with brain microtubules *in vitro*. J. Cell Biol. **80:** 266–276.
5. KIM, H. 1983. Brain microtubule structure *in vitro*. Ph.D. Dissertation. University of Virginia, Charlottesville, Va.
6. LUDUENA, R. F., A. FELLOUS, J. FRANCON, J. NUNEZ & L. MCMANUS. 1981. Effect of tau on the vinblastine-induced aggregation of tubulin. J. Cell Biol. **89:** 680–683.
7. LUDUENA, R. F., M. C. ROACH, P. BINKLEY & M. A. JORDAN. 1982. Alkylating agents as probes for the binding of microtubule-associated proteins (MAPs) to tubulin. J. Cell Biol. **95:** 347a.
8. BENDER, P. K. 1982. Microtubule-associated proteins. Ph.D. Dissertation. University of Virginia.
9. FIELD, D., & J. LEE. 1984. Heterogeneity of vertebrate brain tubulins. Proc. Natl. Acad. Sci. **81:** 4041–4045.

10. BERKOWITZ, S. A., J. KATAGIRI, H. K. BINDER & R. C. WILLIAMS, JR. 1977. Separation and characterization of microtubule protein from calf brain. Biochemistry **16:** 5610–5617.
11. SHELANSKI, M. L., F. GASKIN & C. R. CANTOR. 1973. Microtubule assembly in the absence of added nucleotides. Proc. Natl. Acad. Sci. USA **70:** 765–768.
12. NEAL, M. W. & J. R. FLORINI. 1973. A rapid method for desalting small volumes of solutions. Anal. Biochem. **55:** 328–330.
13. BENSADOUN, A. & D. WEINSTEIN. 1976. Assay of protein in the presence of interfering materials. Anal. Biochem. **70:** 241–250.
14. LOWRY, O. H., N. J. ROSEBROUGH, A. L. FARR & R. J. RANDALL. 1951. Protein measurement with the folin phenol reagent. J. Biol. Chem. **193:** 265–275.
15. BENDER, P., L. I. REBHUN & D. C. BENJAMIN. 1982. Analysis of the relationship of bovine high molecular weight protein-2 and tau protein using radioimmunoassay. Biochim. Biophys. Acta **708:** 149–159.
16. GASKIN, F., C. R. CANTOR & M. L. SHELANSKI. 1974. Turbidimetric studies of the *in vitro* assembly and disassembly of porcine neurotubules. J. Mol. Biol. **89:** 737–755.
17. TILNEY, L. G., J. BRYAN, D. J. BUSH, K. FUJIWARA, M. MOOSEKER, D. B. MURPHY & D. H. SNYDER. 1973. Microtubules: evidence for thirteen protofilaments. J. Cell Biol. **59:** 267–275.
18. BEGG, D., R. RODEWALD & L. I. REBHUN. 1978. The visualization of actin filament polarity in thin sections. Evidence for the uniform polarity of membrane-associated filaments. J. Cell Biol. **79:** 846–852.
19. AMOS, L. A. 1977. Arrangement of high molecular weight associated proteins on purified mammalian brain microtubules. J. Cell Biol. **72:** 642–654.
20. SLOBODA, R. D., W. L. DENTLER & J. L. ROSENBAUM. 1976. Microtubule-associated proteins and the stimulation of tubulin assembly *in vitro*. Biochemistry **15:** 4497–4505.
21. VALLEE, R. B. 1982. A taxol-dependent procedure for the isolation of microtubules and microtubule-associated proteins. J. Cell Biol. **92:** 435–442.
22. JENSEN, C. G. 1982. The arrangement of cross-bridges and side arms in cells of *Haemanthus* endosperm exposed to taxol. J. Cell Biol. **95:** 335a.
23. CLEVELAND, D. W., S. Y. HWO & M. W. KIRSCHNER. 1977. Purification of tau, a microtubule-association protein which induces assembly of microtubules from purified tubulin. J. Mol. Biol. **116:** 227–248.
24. JAMESON, L., T. FREY, B. ZEEBERG, F. DALLDORF & M. CAPLOW. 1980. Inhibition of microtubule assembly by phosphorylation of microtubule-associated proteins. Biochemistry **19:** 2472–2478.
25. MANSO-MARTINEZ, R., A. VILLASANTE & J. AVILA. 1980. Incorporation of the high molecular weight microtubule-associated protein (MAPs) into microtubules at steady-state *in vitro*. Eur. J. Biochem. **105:** 307–313.
26. SELDEN, S. C. & T. D. POLLARD. 1983. Phosphorylation of microtubule-associated proteins (MAPs) regulates their interaction with actin filaments. J. Biol. Chem. **258:** 7064–7071.
27. CHACKO, S., M. A. CONTI & R. S. ADELSTEIN. 1977. Effect of phosphorylation of smooth muscle myosin on actin activation and Ca^{+2} regulation. Proc. Natl. Acad. Sci. USA **74:** 129–133.
28. SANDOVAL, I. V. & J. S. VANDERKERCKHOVE. 1981. A comparative study of the *in vitro* polymerization of tubulin in the presence of the microtubule-associated proteins MAP 2 and tau. J. Biol. Chem. **256:** 8795–8800.
29. VALLEE, R. B. & S. E. DAVIS. 1983. Low molecular weight microtubule-associated proteins are light chains of microtubule-associated protein 1 (MAP 1). Proc. Natl. Acad. Sci. USA **80:** 1342–1346.
30. DENOULET, P., B. EDDE, C. JEANTET & F. GRAS. 1982. Evolution of tubulin heterogeneity during mouse brain development. Biochimie **64:** 165–172.
31. FRANCON, J., A. M. LENNON, A. FELLOUS, A. MARECK, M. PIERRE & J. NUNEZ. 1982. Heterogeneity of microtubule-associated proteins and brain development. Eur. J. Biochem. **129:** 564–571.
32. LAEMMLI, U. K. 1970. Cleavage of structural proteins during the assembly of the head of bacteriophage T4. Nature (London) **227:** 680–685.

Tau Microheterogeneity:
An Immunological Approach with
Monoclonal Antibodies[a]

ARLETTE FELLOUS,[b] RENÉE OHAYON,[b]
JEAN-CLAUDE MAZIE,[c] FREDERIC ROSA,[c]
RICHARD F. LUDUENA,[d] AND VEENA PRASAD[d]

[b]Unité INSERM
Bicêtre, France

[c]Department of Immunology
Pasteur Institute
Paris, France

[d]Department of Biochemistry
University of Texas Health Science Center
San Antonio, Texas 78284

INTRODUCTION

Microtubules, ubiquitous in eukaryotic cells, are involved in several specialized functions, including cell shape regulation, mitosis, intracellular translocation, cell motility, and secretion.[1,2] Within cells of the nervous system they are conspicuous components of the cytoskeleton and play a crucial role in several cellular events. During development, microtubules appear to be directly involved in the growth and (or) maintenance of nerve cell processes and may also have some indirect role in cell migration. In the stabilized axons and dendrites they probably participate in intracellular transport[3] as well as in other neuronal functions. The mechanisms that regulate the involvement of microtubules in a wide variety of cellular functions are still largely unknown. It is likely, however, that the heterogeneity of microtubular components controls the geometric organization of microtubules within the cytoplasm, their connections with other cytoskeletal structures and other cell organelles, and consequently their functional properties. Brain tubulins have been shown to exhibit considerable microheterogeneity;[4-8] most of these tubulin isoforms are present at the same time in a single neuron.[9] It is difficult, however, to envisage how the large number of functions ascribed to microtubules might be regulated by tubulin microheterogeneity. Microtubule associated proteins (MAPs), a variety of accessory proteins that copurify in constant stoichiometry to tubulin, are probably closely related to the multiplicity of microtubule functions. MAPs are divided into two major classes of polypeptides, one with molecular weights greater than 270,000, called the high molecular weight proteins,[10,11] the other with molecular weights from about 53,000 to 67,000, called tau proteins.[12,13] Both the 270,000 molecular weight polypeptide complex referred to as MAP-2[14] and the tau proteins have been the subject of extensive

[a]Support was provided by Grant CA 26276 from the National Cancer Institute to R.F. Luduena.

structural and functional studies. The two classes of MAPs differ by their amino acid composition, peptide maps, and antigenic properties.[13] Both classes, however, promote *in vitro* the assembly of tubulin into microtubules.[14,15,19] Furthermore, both proteins have been shown, by using specific antibodies, to be present along the length of microtubules *in situ*.[16,17] These observations suggest that MAP-2 and tau factor do have some role to play in the assembly and (or) the function of microtubules *in vivo*. But this role is not yet clearly established. The examination of MAP-2 and tau function is considerably complicated by the heterogeneity that exists in both classes of proteins.[13,18,19] The apparent molecular weights of tau proteins were found to be particularly heterogeneous. The amino acid composition and peptide maps of these proteins, however, presented sufficient similarity for them to be considered isoforms. In this report, we have confirmed the high level of tau microheterogeneity and have improved the resolution of tau subspecies. Developmental changes of tau composition in rat cerebrum and cerebellum were shown to be very similar, suggesting that the appearance of new tau isoforms during brain maturation, which has been shown to be partially controlled by thyroid hormones, might reflect the different steps that occur during neuronal differentiation. In an attempt to investigate the biological significance of tau heterogeneity, *in vitro* functional properties of tau polypeptides were analyzed, and both differences and similarities in these properties were demonstrated. Several lines of evidence raise the possibility that the multiple subspecies of tau might induce at variable rates the formation of different subclasses of microtubules. The functional similarities of tau isoforms might be ascribed to the conserved nature of these very widely distributed microtubule-associated proteins.[20] Using an immunological approach, we have confirmed the extensive evolutionary conservation of tau polypeptides. Monoclonal antibodies raised against tau factor recognize epitopes conserved not only in different tissues but also in different species.

CONSIDERABLE TAU HETEROGENEITY

The microheterogeneity that characterizes tau proteins purified from mammalian brain tissue has been extensively described in a previous report[19] and confirmed by several authors.[18,20] Previous data have demonstrated the plurality of isoforms at every stage of brain development and the similarity of the physical properties of these isoforms. Furthermore, tau composition changes considerably when analyzed at different stages of brain development. Developmental changes in tau activity have also been described: tau factor isolated from rat brain at early stages of development was found to be much less active in promoting pure tubulin assembly than the adult factor. These observations have led to the hypotheses that tau heterogeneity may be related to specific cellular events that occur during brain development, that is, cell division, cell migration, or cell differentiation. Another possibility is that tau complexity might also be related to different cell types present in different areas of the brain. In the present report, we present the evidence that tau factor appears much more heterogeneous than previously demonstrated when analyzed by using high resolution one-dimensional sodium dodecyl sulfate (SDS)-gel electrophoresis. Brain tau factor was resolved into more than seven entities when analyzed in rats 21 days after birth (FIGURE 1c). The highest molecular weight isoform was the most abundant component. At an earlier stage of brain development, that is, 6 days after birth, tau factor comprised two polypeptide families: one, slowly migrated and was mainly composed of three entities; the other, migrated quickly and was probably composed of several isoforms that could not be resolved into distinct entities even under high resolution gel electrophoresis,

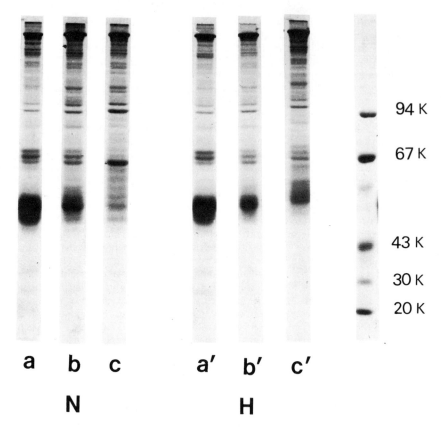

94 K

67 K

43 K

30 K

20 K

a   b   c   a′   b′   c′

N   H

FIGURE 1. Changes in MAP composition at different stages of rat brain development. MAPs were prepared by the method of Fellous *et al.*[21] from brain microtubules of normal rats (N) or hypothyroid rats (H) sacrificed at day 6 (a and a′), day 15 (b and b′), and day 21 (c and c′) after birth. Hypothyroidism was induced from the beginning of development as described by Fellous *et al.*[22] Aliquots of MAPs were analyzed by SDS/8.5% polyacrylamide gel electrophoresis performed according to Laemmli's procedure[23] slightly modified. The Coomassie blue staining is represented.

because their electrophoretic mobilities were so similar (FIGURE 1a). Tau composition was also analyzed in hypothyroid rats. Several authors have reported that brain maturation is impaired when thyroid activity is blocked from the beginning of rat brain development.[24–26] We have demonstrated previously that a lack of thyroid hormones impairs tubulin polymerizing activity during a long period after birth and that this alteration might be related to MAPs composition in hypothyroid animals.[22] FIGURE 1c′ shows that the immature pattern of tau factor was present for a longer period after birth in hypothyroid than in normal rats. This observation lends support to the hypothesis that tau function is closely related to specific cellular events during the different steps of brain maturation. This hypothesis is also strengthened by the finding that the composition of tau factor is equally complex in both the cerebrum and cerebellum (FIGURE 2). Moreover, the electrophoretic pattern of tau family at different stages of development in cerebellum and cerebrum resembled each other closely, suggesting that tau heterogeneity is more related to the stage of development than to differences in cell type distribution.

The extent of tau heterogeneity was evaluated by a two-dimensional SDS-gel electrophoresis. FIGURE 3A clearly shows that tau in adult rat brain was resolved into eight sets of proteins with different molecular weights. Each set consisted of several proteins that differed in isoelectric point and, to a lesser degree, in apparent molecular weight. The highest molecular weight set comprised a doublet of very basic, more abundant proteins. These basic polypeptides appear to be specific to the rat. FIGURE 3B shows a completely different electrophoretic pattern for cow brain tau, confirming the previously described diversity[19] of tau factor among different species.

In spite of very large differences in their apparent molecular weight and isoelectric points, the different tau isoforms present certain identical chemical and physical properties that were described in a previous report[19] and by other authors:[20] heat stability, similar peptide distribution, amino-acid composition, sedimentation coefficient, and capacity to undergo phosphorylation.[28-34] Both the structural heterogeneity and the similarities of tau factor raise several questions. The first question concerns the level at which tau heterogeneity is generated. It was demonstrated that the major differences in the composition of tau factor are controlled at the mRNA level: different

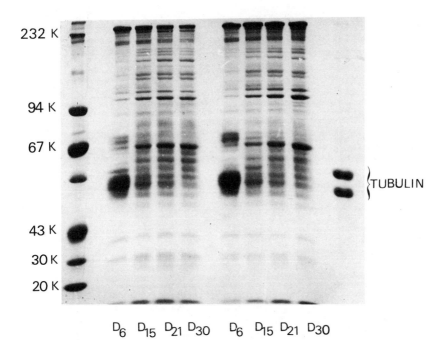

FIGURE 2. Changes in cerebrum and cerebellum MAP composition during rat brain development. MAPs were prepared from cerebrum and cerebellum microtubules of rats sacrificed at day 6 (D_6), day 15 (D_{15}), day 21 (D_{21}), and day 30 (D_{30}) and analyzed by SDS/8.5% polyacrylamide gel electrophoresis. The Coomassie blue staining is represented.

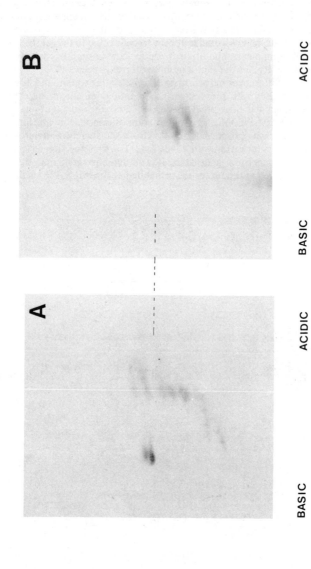

FIGURE 3. Two-dimensional gel electrophoresis of rat and cow tau factor. The tau factor isolated from adult rat brain (**A**) or cow brain (**B**) was analyzed by the procedure referred to as nonequilibrium pH gradient electrophoresis (NEPHGE) as described by Patricia O'Farrell *et al.*[27] Tau factor was prepared as described by Fellous *et al.*[21]

mRNA seem to exist for different tau species.[32,33] It has also been reported that mRNA species specific for mature tau isoforms appear in the brain before their corresponding proteins accumulate,[32] suggesting that a translational control mechanism may regulate the expression of stage-specific tau proteins. In addition to this genetic regulation of tau heterogeneity, it is likely that different posttranslational modifications act to generate several tau isoforms. *In vitro* phosphorylation of tau proteins was demonstrated in a previous report[19] and was largely documented by several authors.[28–31] Phosphorylation of tau proteins has also been found to occur *in vivo*.[34] Whereas the phosphorylation state of tau was reported to have a significant effect upon the rate and extent of microtubule polymerization *in vitro,* the effect *in vivo* has not yet been determined. In any case, a certain state of tau phosphorylation appears to diminish the electrophoretic mobility[31] and probably modifies the isoelectric points of the entities affected by phosphorylation. Tau proteolysis, which, as shown here, produces some lower molecular weight tau entities with modified activity, may also generate tau heterogeneity. Whatever the origin of tau heterogeneity, a second question arises concerning the biological significance of this heterogeneity. That tau polypeptides are also highly conserved raises a third question: whether or not the conserved sequences present not only in different mammals and chicken,[20] but probably also in frog and *Drosophila,*[33] are responsible for a common function, different from that of other MAPs.

A COMMON FUNCTIONAL PROPERTY SHARED BY DIFFERENT SUBCLASSES OF TAU PROTEINS: TWO POSSIBLE MECHANISMS

Previous data[35,36] have demonstrated that the capacity to induce extensive aggregation of tubulin into paracrystalline arrays in the presence of vinblastine is restricted to tau proteins. It was demonstrated that MAP-2 proteins were unable to catalyze the same process under the same conditions. The interaction of both tau factor and vinblastine with purified tubulin promotes the formation of a large number of long and tight spirals that have the tendency to bind together and form large aggregates as revealed by negative staining (FIGURE 4B). MAP-2 induced polymers, analyzed by the same procedure, were detectable as small aggregates, rings and short spirals (FIGURE 4C). The whole set of tau polypeptides, which copolymerizes with tubulin in normal microtubules (FIGURE 4A) was shown to be copurified with the vinblastine-induced spirals, suggesting that the different tau isoforms isolated from cow brain can catalyze vinblastine-induced paracrystalline arrays.

In order to determine whether this specific property of cow brain tau might be ascribed to one or several subspecies of polypeptides identified as tau factor, several experiments using different populations of tau were performed. Tau proteins isolated either at an early stage of development or at the adult stage, differ in their ability to promote tubulin assembly.[19] Both classes of tau proteins, however, were found to have the same property to promote vinblastine-induced paracrystals as revealed by the large increase in turbidity that always follows the formation of dense aggregates (FIGURE 5A). The different "adult" tau isoforms as well as the "young" fast migrating tau participate to the formation of paracrystals (FIGURE 5C). Since the "young" slow migrating tau is present in small amounts among the tau protein extracted from the spirals (FIGURE 5C$_2$), either this "young" tau is very easily degraded into smaller polypeptides that migrate at the rate of the fast migrating component or its function is markedly different from that of other tau components. "Young" MAP-2, like "adult" MAP-2, could not induce paracrystalline arrays as shown by the decrease of turbidity

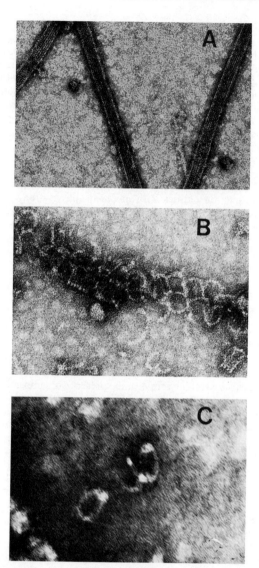

FIGURE 4. Electron micrographs of negatively stained microtubules and vinblastine-induced aggregates. Tubulin, tau and MAP-2 were purified from cow brain. Samples of 10 μM tubulin were incubated in the presence of 1.5 μM tau (A), 1.5 μM tau, and 60 μM vinblastine (B), and 1.8 μM MAP-2 and 60 μM vinblastine (C). Negative staining was performed according to Luduena et al.[36] On the right side of figures **A** and **B** are represented the electrophoretic patterns of the tau isoforms copurified with microtubules and vinblastine-induced spirals obtained during the incubation of samples **A** and **B**. The copurified tau was extracted according to the method of Fellous et al.[21]

induced by vinblastine (FIGURE 5B). New tau isoforms may be produced enzymatically *in vitro,* and similar modifications may occur *in vivo* as the result of posttranslational mechanisms. When tau was submitted to exogenous phosphorylation, all the tau isoforms including the phosphorylated components, were copurified with the spirals as shown in FIGURE 5(C_5). Tau phosphorylation, which is reported to be mediated by several kinases, including a cAMP kinase, a calcium and calmoduline-dependent protein kinase, led to an inhibition of microtubule assembly.[30,31] The phosphorylated tau, however, have the capacity either equal to or greater than tau, which is not phosphorylated, to produce vinblastine-mediated paracrystalline arrays (results not shown). The degree of phosphorylation of tau isoforms may induce variability in the rate and extent of microtubule assembly, but does not alter spiral formation. Mild proteolysis represents another kind of enzymatic reaction that may modify MAPs properties *in vivo.* A-chymotrypsin treatment of tau appears to markedly reduce tau polymerizing activity, but appears to enhance significantly the paracrystalline formations with vinblastine (FIGURE 6) as shown by the large increase in turbidity that follows the addition of vinblastine. This kind of observation strongly suggests that the cleavage of tau polypeptides into very short fragments is sufficient to inhibit certain tubulin interactions required for normal polymerization and to favor other tubulin interactions required for spiral formation.

Such experiments lead to the conclusion that although tau isoforms display different degrees of efficiency in microtubule promotion, they all induce spiral formation in the presence of vinblastine. Furthermore, different tau isoforms, whether they are produced at a transcriptional or posttranscriptional level, may share a common biological function that, at present, is still unknown. The following two hypotheses may be proposed to describe tau interaction with tubulin (FIGURE 7). In the first hypothesis, A, tau is a linear element that binds to several tubulin dimers and produces a protofilament. This protofilament may interact laterally and normally with other filaments to form normal microtubules. In the presence of vinblastine, it becomes helical and forms a spiral. Abnormal lateral interactions may explain the formation of paracrystalline arrays. In the second hypothesis, B, tau binding to tubulin changes the conformation of the tubuline molecule, which in turn modifies the two sites of longitudinal and lateral interactions. Strong longitudinal interactions and weak lateral interactions between tubulin subunits appear to result from the interaction of tau with tubulin. Longitudinal interactions may be consolidated and lateral interactions weakened even more by enzymatic reactions that maintain or enhance the specific spiral-forming property of tau proteins. Thus, the whole family of tau subspecies, generated either at the transcriptional or at the posttranscriptional level, share not only several physical properties, but also the capacity to strengthen longitudinal interactions between tubulin dimers, a prerequisite for vinblastine-induced paracrystalline arrays. These functional and structural similarities might be related to the highly conserved nature of tau factor. It is tempting to speculate that the conserved sequences of tau polypeptides are responsible for a specific change of tubulin conformation induced by tau binding. The nonconserved domains that cause tau microheterogeneity might explain the ability of tau factor to perform different biological functions.

AN IMMUNOLOGICAL APPROACH WITH MONOCLONAL ANTIBODIES RAISED AGAINST TAU FACTOR

Somatic cell hybridization performed as described by Kholer and Milstein[37] has provided eight clones secreting immunoglobulins that bind specifically to tau protein.

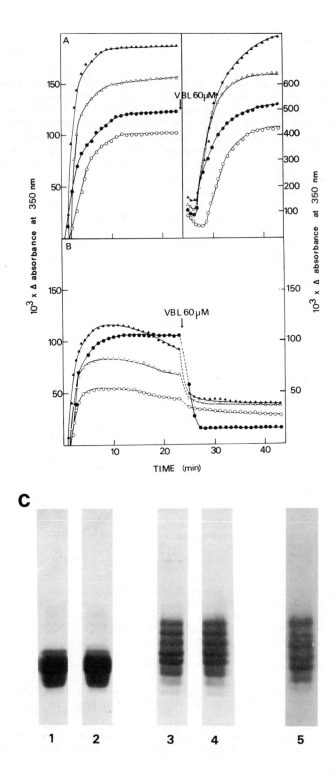

Two clones, H7-9 and H15-14, were shown to recognize epitopes among the multiple tau isoforms. H7-9, for example, reacts with the different tau polypeptides isolated from rat cerebrum at day 6,15,21, and 30 (FIGURE 8) and from cerebellum of rats sacrificed at the same ages (results not shown). No reactivity, however, was observed with the slowly migrating component of tau factor in six-day-old preparations. This tau component may not belong to the tau family or does not contain the conserved sequences recognized by H7-9 antibody. Further chemical and structural analysis is required to determine which of these two possibilities is correct. H7-9 appears to react less intensively with the highest molecular weight component of adult tau factor, suggesting that the slow migrating tau entities contain either tau isoforms not recognized by H7-9 monoclonal antibody or other MAPs not yet identified. By using a biochemical and immunological approach, we are investigating these two hypotheses. Despite some discrepancies in the reactivity of the monoclonal antibody with the different tau isoforms, the data summarized in FIGURE 8 provide additional arguments that tau polypeptides contain conserved sequences, and also demonstrate that the monoclonal antibody H7-9 can recognize them. We have demonstrated that these conserved sequences are distributed in species other than rat, as shown in FIGURE 9. H7-9 reacts with the different tau isoforms isolated from cow brain. By using indirect immunofluorescence techniques, we have confirmed the wide distribution of tau proteins in different tissues and in different species. We have also noticed their absence in certain cell types. The visualization by indirect immunofluorescence procedure of tau proteins with monoclonal antibody H15-14 in a skin dermal fibroblastic cell line from human origin (FIGURE 10A) has demonstrated that the conserved sequences identified in rat or cow brain tau isoforms are maintained in tau polypeptides of human and nonneuronal tissues. By immunoblotting procedure, H15–14 was shown to also react with the multiple isoforms of tau factor (result not shown). When embryonic neurones isolated from the striatum of 13-day-old rat embryo were induced to grow under the experimental conditions established by Prochiantz *et al.*[40] and were examined with the antibodies H7-9 and H15-14, a significant staining of neuronal cell bodies were observed, whereas very few processes appeared stained. FIGURE 10B shows the staining pattern observed with H15-14. The processes that are not stained by the anti-tau monoclonal antibody appear very well stained when the monoclonal antibody H12-5, specific for MAP-2, was used (FIGURE 10C). These observations suggest that the embryonic neurones develop under the experimental conditions that have been

---

FIGURE 5. Effect of vinblastine on microtubules assembled in the presence of a series of tau and MAP-2 isolated from adult and young rat brain. **A:** Samples of 10 μM rat brain tubulin were incubated at 37°C in the presence of 3.45 μM tau isolated from adult rat brain (●—●), 3.45 μM (O—O), 5.17 μM (△—△), and 6.90 μM (▲—▲) tau isolated from six-day-old rat brain. Sixty μM vinblastine was added to the four samples after 24 min of incubation. **B:** Samples of 10 μM rat brain tubulin were incubated at 37°C in the presence of 2.6 μM MAP-2 isolated from adult rat brain (●—●), 1.48 μM (O—O), 2.22 μM (△—△), and 2.96 μM (▲—▲) MAP-2 isolated from six-day-old rat brain. Sixty μM vinblastine was added to the four samples after 24 min of incubation. **C:** Electrophoretic pattern of a series of tau copurified with microtubules and vinblastine-induced spirals. Tau isoforms copurified with microtubules, and vinblastine-induced spirals were extracted as in FIGURE 4 and subjected to SDS/8.5% polyacrylamide gel electrophoresis. The gels were then stained by Coomassie blue. Electrophoretic pattern of tau extracted from microtubules catalyzed by six-day-old rat brain tau (slot 1), from vinblastine-induced spirals in the presence of six-day-old rat brain tau (slot 2), from microtubule-catalyzed by adult rat brain tau (slot 3), from vinblastine-induced spirals in the presence of adult rat brain tau (slot 4), or ^{32}P-labeled phosphorylated adult rat brain tau. (Slot 5 represents the autoradiography of the gel).

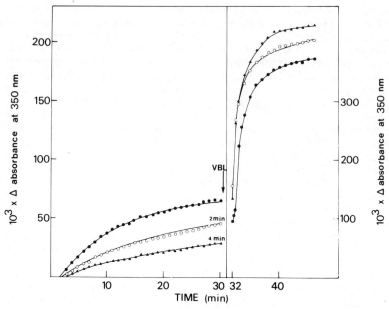

FIGURE 6. Effect of tau proteolysis on tau-dependent microtubule assembly and on vinblastine-induced tau dependent tubulin aggregation. Three samples of 10 μM bovine brain tubulin were incubated at 37°C in the presence of 5.7 μM tau, preincubated with α chymotrypsin (0.7 μg/ml) for 2 min (O——O), 4 min (▲——▲), or without α chymotrypsin (●——●). Sixty μM of vinblastine were added to the three samples after 30 min of incubation.

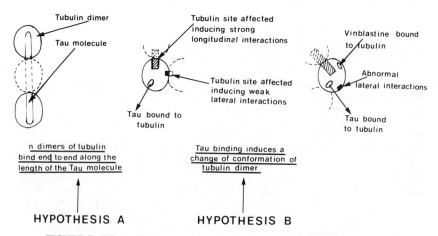

FIGURE 7. Effect of tau interaction with tubulin: a model with two hypotheses.

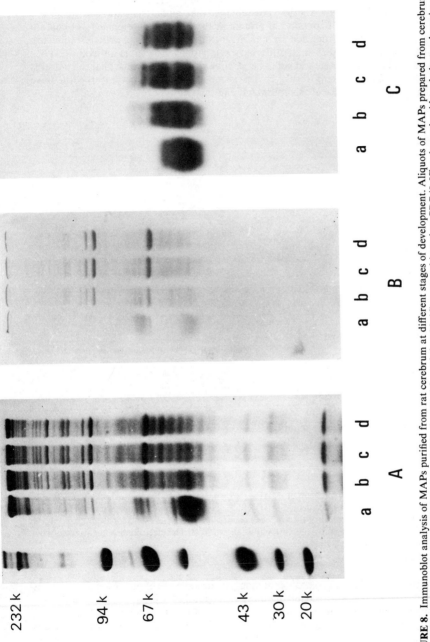

FIGURE 8. Immunoblot analysis of MAPs purified from rat cerebrum at different stages of development. Aliquots of MAPs prepared from cerebrum of rats sacrificed at day 6 after birth (a), day 15 (b), day 21 (c), and day 30 (d) were subjected to SDS/8.5% polyacrylamide gel electrophoresis; after electrophoresis the gel was stained by Coomassie blue (A). The same PAGE was repeated in duplicate. After electrophoresis, the separated proteins were transferred from the two gels onto nitrocellulose sheets using the method of Towbin *et al.*,[38] slightly modified. Half of the blot was stained by Coomassie blue (B), and the other half was incubated with the monoclonal antibody H7-9, washed, and then incubated with ^{125}I-labeled goat anti-mouse IgG. After washing and drying the blot was exposed to X-ray film (C).

chosen, numerous dentritic-like processes, which contain a large amount of MAP-2.[41-45] The same embryonic neurones appear to extend very few axonal processes known to contain tau factor.[46,47] Another possiblity is that numerous axon-like fibers are formed but either do not contain tau protein at this stage of development, or contain tau polypeptides not recognized by our monoclonal antibodies. Further studies are required to choose between these different explanations. The same antibodies, H7-9, H15-14, and H12-5, were found not to be reactive with rat embryonic glial cell cultures (results not shown). This result is not surprising because several reports have demonstrated the absence of tau and MAP-2 in this cell type.[43,45,48]

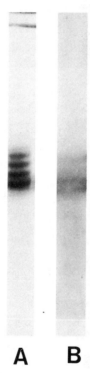

A   B

FIGURE 9. Immunoblot analysis of MAPs purified from cow brain. An aliquot of MAPs prepared from cow brain was subjected to SDS/8.5% polyacrylamide gel electrophoresis, and the same procedure as described in FIGURE 8 was carried out using the monoclonal antibody H7-9. The polypeptides recognized by the antibody is shown in B. The electrophoretic pattern of cow brain tau is shown in A.

In conclusion, immunological techniques, such as the use of monoclonal antibodies that have the property of recognizing most of the tau isoforms, appear very useful for investigating not only the presence of tau factor in different cell types but also for investigating its precise cellular location. In cells that undergo morphological changes in response to external stimuli, for instance, to hormones, or to growth factors, these techniques will make it possible to measure precisely the increase or decrease in the expression of tau factor and to explore the changes that may occur in the cellular

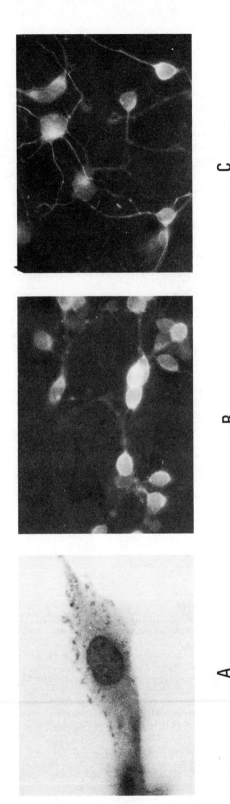

A   B   C

FIGURE 10. Indirect immunofluorescence staining of human skin fibroblasts and rat embryonic neurons. Human skin fibroblasts grown on coverslips in normal culture medium were treated with the monoclonal antibody H15-14, washed, then incubated with the fluorescein-conjugated goat antimouse IgG. After washing, the coverslips were mounted on a drop of glycerol solution containing *p*-phenylenediamine according to the procedure of Oriol *et al.*[39] and viewed in a Zeiss photomicroscope equipped with filters for fluorescein (**A**). Embryonic neurons isolated from the stratum of rat embryos sacrificed at day 13 were grown on coverslips in culture medium as described by Prochiantz *et al.*[40] One coverslip was treated with the monoclonal antibody H15-14 (**B**) and another one with monoclonal antibody H12-5 (**C**). These two coverslips were mounted on a drop of PBS-glycerol.

location of this MAP. *In vitro* studies have strongly suggested that tau factor is involved in promoting tubulin assembly and in stabilizing the microtubules by strengthening longitudinal interactions. The immunological approach will provide further information about tau function. It has been proposed that tau may be involved in connecting microtubules to neurofilaments,[49] microfilaments,[29] and coated vesicles.[50] It has also been suggested that tau may play a role in axonal transport,[47] in nerve regeneration,[51] and in the spindle formation during mitosis.[52] The differential subcellular localization of tau as revealed by immunocytochemistry with monoclonal antibodies may suggest certain biological functions of tau factor in different cellular events.

SUMMARY

The family of tau polypeptides purified from mammalian brain exhibit both extensive heterogeneity and large similarities in their chemical, physical, and functional properties.

All the tau isoforms generated at a transcriptional or posttranscriptional level share the property of interacting with tubulin dimers in a specific manner. They strengthen longitudinal interactions between tubulin dimers and thus may stabilize microtubules once they are formed. Mild proteolysis or phosphorylation does not remove but only modulates the tau specific function that is probably related to the conserved sequences of the molecules.

Monoclonal antibodies raised against tau were found to recognize epitopes conserved not only between species but also in different tissues. Using indirect immunofluorescence, a specific staining pattern was observed on rat neuronal cells and also on human skin fibroblasts. The same antibodies did not recognize glial cells, suggesting that these cells either do not contain detectable levels of tau or contain tau molecules different from the neuronal ones.

These data suggest that tau protein is widely distributed, highly conserved, and may be preferentially associated with special subclasses of microtubules.

ACKNOWLEDGMENTS

We thank M.A. Jordan for taking electron micrographs, A. Prochiantz and B. Chamak for preparing embryonic neuronal and glial cell cultures, and A. Guedec, C. Sais, and M. Bahloul for assistance in manuscript preparation.

REFERENCES

1. ROBERTS, K. & J. S. HYAMS, Eds. 1979. Microtubules. Academic Press, London & New York.
2. SOIFER, D., Ed. 1975. The biology of cytoplasmic microtubules. Ann. N.Y. Acad. Sci. Vol. 253.
3. TYTELL, M., S. T. BRADY & R. LASER. 1984. Proc. Natl. Acad. Sci. USA 81: 1570–1574.
4. FEIT, H., U. HEUDECK & F. BASKIN. 1977. J. Neurochem. 28: 697–706.
5. GEORGE, H. J., L. HISRA, O. J. FIELD & J. C. LEE. 1981. Biochemistry 20: 2402–2409.
6. GOZES, I. & U. Z. LITTAUER. 1978. Nature (London) 276: 411–413.
7. MAROTTA, C. A., P. STROCCHI & J. M. GILBERT. 1979. Brain Res. 167: 93–106.
8. NELLS, L. P. & J. R. BAMBURG. 1979. J. Neurochem. 32: 477–489.
9. GOZES, I. & K. J. SWEDNER. 1981. Nature (London) 294: 477–480.

10. MURPHY, O. B. & G. G. BORISY. 1975. Proc. Natl, Acad. Sci. USA **72:** 2696–2700.
11. MURPHY, D. B., K. A. JOHNSON & G. G. BORISY. 1977. J. Mol. Biol. **117:** 33–52.
12. WEINGARTEN, M. D., A. H. LOCKWOOD, S. Y. HWO & M. W. KIRSCHNER. 1975. Proc. Natl. Acad. Sci. USA **72:** 1858–1862.
13. CEVELAND, D. W., S. Y. HWO & M. W. KIRSCHNER. 1977. J. Mol. Biol. **116:** 227–247.
14. BORISY, G. G., J. M. MARCUM, J. B. OLMSTED, D. B. MURPHY & K. A. JOHNSON. 1975. Ann. N.Y. Acad. Sci. **253:** 107–132.
15. CLEVELAND, D. W., S. Y. HWO & M. W. KIRSCHNER. 1977. J. Mol. Biol. **116:** 207–225.
16. CONNOLLY, J. A., V. I. KALNINS, D. W. CLEVELAND & M. W. KIRSCHNER. 1977. Proc. Natl. Acad. Sci. USA **74:** 2437–2240.
17. CONNOLLY, J. A., V. I. KALNINS, D. W. CLEVELAND & M. W. KIRSCHNER. 1978. J. Cell Biol. **76:** 781–786.
18. BINDER, L. I., A. FRANKFURTER, H. KIM, A. CACERES, M. R. PAYNE & L. I. REBHUN. 1984. Proc. Natl. Acad. Sci. USA **81:** 5613–5617.
19. FRANCON, J., A. M. LENNON, A. FELLOUS, A. MARECK, M. PIERRE & J. NUNEZ. 1982. Eur. J. Biochem. **129:** 465–471.
20. CLEVELAND, D. W., B. M. SPIEGELMAN & M. W. KIRSCHNER. 1979. J. Biol. Chem. **254:** 12670–12678.
21. FELLOUS, A., J. FRANCON, A. M. LENNON & J. NUNEZ. 1977. Eur. J. Biochem. **78:** 167–174.
22. FELLOUS, A., A. M. LENNON, J. FRANCON & J. NUNEZ. 1979. Eur. J. Biochem. **101:** 365–376.
23. LAEMMLI, U. K. 1970. Nature (London) **227:** 680–685.
24. LEGRAND, J. 1972. Arch. Anat. Microsc. Morphol. Exp. **56:** 291–308.
25. NICHOLSON, J. L. & J. ALTMAN. 1972. Brain Res. **44:** 13–23.
26. REBIERE, A. & J. LEGRAND. 1972. Arch. Anat. Micros. Morphol. Exp. **61:** 105–126.
27. O'FARRELL, P. Z., H. J. GOODMAN & P. H. O'FARRELL. 1977. Cell **12:** 1133–1142.
28. PIERRE, M. & J. NUNEZ. 1983. Biochem. Biophysics. Res. Commun. **115:** 212–219.
29. SELDEN, S. C. & T. D. POLLARD. 1983. J. Biol. Chem. **258:** 7064–7071.
30. YAMAMOTO, H., K. FUKUNAGA, E. TAHAKA & E. MIYAMOTO. 1983. J. Neurochem. **41:** 1119–1125.
31. LINDWALL, G. & R. D. COLE. 1984. J. Biol. Chem. **259:** 5301–5305.
32. GINZBURG, I., T. SCHERSON, D. GIVEON, L. BEHAR & U. Z. LITTAUER. 1982. Proc. Natl. Acad. Sci. USA **79:** 4892–4896.
33. DRUBIN, D. G., D. CAPUT & M. W. KIRSCHNER. 1984. J. Cell Biol. **98:** 1090–1097.
34. PALLAS, D. & F. SALOMON. 1982. Cell **30:** 407–414.
35. LUDUENA, R. F., A. FELLOUS, J. FRANCON, J. NUNEZ & L. MCMANUS. 1981. **89:** 680–683.
36. LUDUENA, R. F., A. FELLOUS, L. MCMANUS, M. A. JORDAN & J. NUNEZ. 1984. J. Biol. Chem. **259:** 12890–12898.
37. KOHLER, G. & C. MILSTEIN. 1975. Nature (London) **256:** 495–497.
38. TOWBIN, H., T. STAEHELIN & J. GORDON. 1979. Proc. Natl. Acad. Sci. USA **76:** 4350–4354.
39. ORIOL, R. & R. MANCILLIA-JIMENEZ. 1983. J. Immunol. Methods **62:** 185–192.
40. PROCHIANTZ, A., U. DIPORZIO, A. KATO, B. BERGER & J. GLOWINSKI. 1979. Proc. Natl. Acad. Sci. USA **76:** 5387–5391.
41. BERNHARDT, R. & A. MATUS. 1982. J. Cell Biol. **92:** 589–593.
42. MATUS, A., R. BERNHARDT & T. HUGH-JONES. 1981. Proc. Natl. Acad. Sci. USA **78:** 3010–3014.
43. HUBER, G. & A. MATUS. 1984. J. Neurosci. **4:** 151–160.
44. CACERES, A., G. BANKER, O. STEWARD, L. BINDER & M. PAYNE. 1984. Dev. Brain Res. **13:** 314–318.
45. CACERES, A., L. I. BINDER, M. R. PAYNE, P. BENDER, L. REBHUN & O. STEWARD. 1984. J. Neurosci. **4:** 394–410.
46. BLACK, M. M. & J. T. KURDYLA. 1983. J. Cell Biol. **97:** 1020–1028.
47. TYTELL, M., S. T. BRADY & R. J. LASEK. 1984. Proc. Natl. Acad. Sci. USA **81:** 1570–1574.

48. IZANT, J. G. & J. R. MCINTOSH. 1980. Proc. Natl. Acad. Sci. USA **77:** 4741–4745.
49. SHELANSKI, M. L., J. F. LETERRIER & R. K. H. LIEM. 1981. Neurosci. Res. Program Bull. **19:** 32–43.
50. PFEFFER, S. R., D. G. DRUBIN & R. B. KELLY. 1983. J. Cell Biol. **97:** 40–47.
51. NEUMANN, D., T. SHERSON, I. GINZBURG, U. Z. LITTAUER & M. SCHWARTZ. 1983. FEBS Lett. **162:** 270–276.
52. CONNOLLY, J. A., V. I. KALNINS, D. W. CLEVELAND & M. W. KIRSCHNER. 1977. Proc. Natl. Acad. Sci. USA **74:** 2437–2440.

Association of Tau Protein with Microtubules in Living Cells

DAVID DRUBIN, SUMIRE KOBAYASHI,
AND MARC KIRSCHNER

Department of Biochemistry
University of California, San Francisco
San Francisco, California 94143

INTRODUCTION

Several well-characterized microtubule-associated proteins (MAPs) coassemble with microtubules *in vitro* and promote the assembly of purified tubulin. These include the tau proteins of 55,000–68,000 kD,[1,2] MAPs of 120,000 and 210,000 kD,[3] and MAPs of approximately 300,000 kD, called MAP-1 and MAP-2.[4-6] The relevance of *in vitro* studies on MAPs rests first upon direct demonstration that the MAPs associate with microtubules in living cells, and second upon direct demonstration of their *in vivo* functions. Immunocytochemical studies have supported the conclusion that tau,[7,8] the 210,000 kD MAP,[9] and the 300,000 kD MAPs[8,10-14] bind microtubules in living cells. Immunocytochemical experiments alone, however, can be ambiguous, as evidenced by inconsistencies in reported staining patterns. For example, in our own studies with a highly specific affinity-purified antiserum raised against bovine brain tau protein, we find vanishingly low levels of tau protein in cells previously reported to be stained by tau antibodies (see discussion below). In addition, other investigators have noted discrepancies in high molecular weight MAP staining.[12]

In this report, we reexamine the distribution of tau protein in established cell lines and in various mouse organs. We find that tau protein is found predominantly in cells of neuronal origin. We also use both biochemical and immunocytochemical techniques to demonstrate conclusively that tau protein is bound to microtubules in living cells. Finally, we discuss ways in which MAP function can be studied *in vivo*.

RESULTS

Tau Distribution in Mouse Organs and Established Cells Lines

Tau protein was originally characterized as 4–5 brain polypeptides of 55–68 kD that coassemble with purified tubulin and promote microtubule assembly *in vitro*.[1] These polypeptides are closely related by physical properties and by primary structure.[2] In addition, a MAP of 125 kD in rat pheochromocytoma and in mouse neuroblastoma cells was shown to be closely related to the 55–68 kD taus, by several criteria.[15] In order to determine how widely tau protein is distributed outside of brain tissue where it was originally identified, we performed immunoblots[16] with a highly specific affinity-purified tau antiserum[17,18] on equal amounts of protein extracted from various mouse and rat tissues. The results of this study are summarized in TABLE 1. In some tissues the tau antibodies reacted with proteins of 42 kD and of 200 kD. We have not characterized these proteins further, but we note the results because the reaction with the tau antibodies appears to be highly specific.

TABLE 1. Tau Distribution in Mouse and Rat Tissues

Mouse tissue	Molecular Mass of Tau Species			
	42 kD	46–68 kD	125 kD	200 kD
cerebrum	− [a]	+ + + [b]	± [c]	−
cerebellum	−	+ + [d]	±	−
brain stem	−	+ +	±	−
spinal cord	−	+ +	+ +	−
7-day superior cervical ganglia (rat)	−	+ +	+	−
adult superior cervical ganglia (rat)	−	+ [e]	+ +	−
dorsal root ganglia (rat)	−	−	+	−
sciatic nerve (rat)	−	±	−	±
liver	−	−	−	−
spleen	+ +	±	−	−
thymus	+ +	±	−	−
heart	−	±	−	−
lung	−	±	−	−
pancreas	−	±	−	−
adrenal gland	−	−	−	+ + +
kidney	−	−	−	−
skeletal muscle	−	±	−	−
small intestine	−	±	−	−
lymph node	−	−	−	+ + +
esophagus	−	−	−	−
trachea	−	±	−	+

[a] − = not detectable
[b] + + + = high
[c] ± = trace
[d] + + = intermediate
[e] + = low

The well-characterized taus of 46–68 and 125 kD are most abundant in nervous tissue in general and in central nervous tissue in particular. The 125 kD tau protein first observed in established cell lines[15] is not an artificially produced protein unique to transformed cell lines, but is actually the major tau species in spinal cord, superior cervical ganglia, and dorsal root ganglia. In nonnervous tissue, we detect only trace amounts of taus in the molecular weight ranges typical of well-characterized taus. We can not distinguish between the possibility that tau proteins are low in abundance in these tissues and the possibility that our antibodies recognize neuron-specific antigenic determinants.

We have also examined tau distribution in established cell lines. The results of these analyses are summarized in TABLE 2. With the exception of HeLa cells, tau proteins are found only in cells of neuronal origin. (PC12 cells display many features of nerve cells and derive from the neural crest.[19]) The presence of tau in HeLa cells may be normal for epithelial cells, or tau could be expressed inappropriately, as is vimentin,[20] in some established cell lines. These results are inconsistent with earlier reports of tau distribution,[7,8] based on immunofluorescence studies. The earlier studies found tau protein localized on microtubules in C6 glial and 3T3 fibroblast cells. When we use our affinity-purified tau antiserum that recognizes only tau proteins on immunoblots[18] containing total brain protein extracts, we do not detect any tau in C6 or 3T3 cells. Moreover, we have performed immunofluorescence studies on tau protein in 3T3 cells and several other fibroblast cell lines and detect no tau staining (see tau microinjection section below). These data do not exclude the possibility that tau protein is expressed in fibroblasts, but at levels that are undetectable with our affinity-purified tau antiserum. Indeed, tau protein was detected biochemically in 3T3 cells after coassembly of radiolabeled 3T3 proteins with brain tubulin.[21,22] We now believe, however, that tau protein is expressed predominantly in cells of neuronal origin. The trace amounts of tau detected in various nonneuronal tissues (TABLE 1) could be due to low levels of expression of tau in these tissues, or simply due to innervation of these tissues. We note that tau protein has been unambiguously detected in one untransformed nonneuronal cell, the chicken erythrocyte.[23]

Tau Immunofluorescence

In order to study the subcellular localization of tau protein, we performed immuno-fluorescence studies on glutaraldehyde-fixed cells with our affinity-purified tau antiserum. As mentioned above, in initial studies with mouse 3T3 cells we observed no tau staining. We therefore decided to study tau distribution in PC12 cells differentiated for five days by nerve growth factor because we had determined by immunoblotting that these cells contain high levels of tau protein.[15] FIGURE 1, panel **a** shows a PC12 cell neurite stained with a mouse monoclonal α-tubulin antibody and a fluorescein-conjugated secondary antibody and visualized in the fluorescein channel. The neurite that is densely filled with microtubules[24] stains brightly, and individual microtubule fibers can be seen in the growth cone where microtubules splay outward. The same neurite was stained with the rabbit affinity-purified tau antiserum and a rhodamine-conjugated secondary antibody (FIGURE 1, panel **b**). Notice that the neurite body stains brightly as with the tubulin antibody, and that the same microtubules are stained in the growth cone as are stained with the tubulin antibody (FIGURE 1, panel **a**). In control experiments, staining with tau or tubulin antibodies alone did not affect the staining pattern. Furthermore, the drug colchicine, which depolymerizes microtubules, abolished the tau staining pattern, and the drug vinblastine, which causes tubulin and tau to form paracrystalline aggregates *in vitro*,[25] caused both the tubulin and tau staining to localize to ellipsoid aggregates in the cytoplasm. In

TABLE 2. Tau Distribution in Established Cell Lines

| | Molecular Mass of Tau Species | |
Cell line	55–68 kD	125 kD
Rat 6 glial	— [a]	—
Mouse neuroblastoma N₂A	±	±
Mouse neuroblastoma N115 undifferentiated	±	±
Mouse neuroblastoma N115 differentiated	+	+ + +
Mouse 3T3 fibroblasts	—	—
Mouse BC₃HI brain tumor	±	—
Mouse NCTC L929 connective tissue	—	—
Mouse NCTC 1469 liver	—	—
Rat PC12 pheochromocytoma	+	+
Rat PC12 pheochromocytoma differentiated	+ + +	+ + +
Hamster CHO ovary	—	—
Human HeLa epithelium	+	—

[a]See TABLE 1 for definition of symbols.

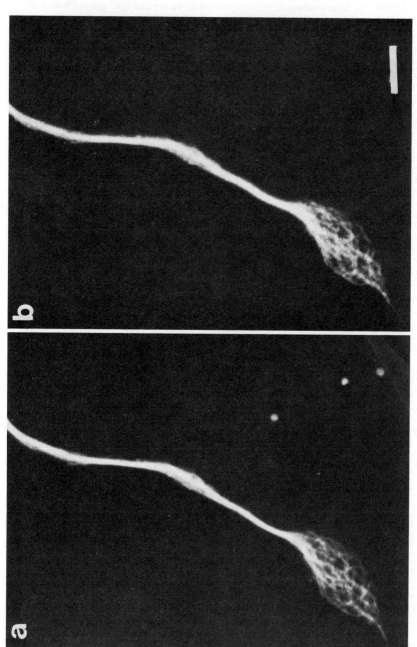

FIGURE 1. Tubulin and tau immunofluorescent staining of a PC12 cell neurite. Double-labeling was performed with monoclonal anti-α tubulin and fluorescein-conjugated goat anti-mouse IgG (**a**) and polyclonal rabbit anti-tau and rhodamine goat anti-rabbit IgG (**b**). Cells were washed in phosphate-buffered saline (PBS), fixed (10 min in 0.3% glutaraldehyde, 1% Empigen BB (alkyl betaine, Allbright and Wilson, Whitehaven, Cumbria, England), 80 m*M* PIPES, 5 m*M* EGTA, 1 m*M* MgCl$_2$, pH 6.8), washed in PBS, incubated 7 min in 1 mg/ml NaBH$_4$ in PBS, and then washed 5 times in PBS, 1% bovine serum albumin, 0.1% Tween 20. Primary antisera were incubated 30 min and washed as above; then secondary antisera were applied and washed as for primaries. Bar, 5 *μ*m.

addition, cells that had low levels of tau protein, as determined by immunoblotting, were devoid of tau immunofluorescence staining, showing that the immunofluorescence pattern is dependent on the presence of tau protein in the cell, and is not due to adventitious binding of tau antibodies to glutaraldehyde-fixed microtubules. We conclude that tau protein is bound to microtubules in differentiated PC12 cells. We have also determined by immunofluorescence that MAP-1 is bound to microtubules in differentiated PC12 cells, consistent with findings for other cell types.[14]

Microinjection of Tau Protein into Rat Fibroblasts

Whereas immunofluorescence studies on tau distribution in PC12 cells demonstrated that tau protein is present on PC12 cell microtubules, the small size of PC12 cells and their round morphology make it difficult to determine if tau protein is localized exclusively on microtubules, particularly in the cell body. To address this question we microinjected purified tau protein into RAT1 fibroblast cells, whose flat morphology is ideal for visualizing microtubules and other cellular organelles, and looked at the tau localization by immunofluorescence. FIGURE 2, panel **a**, shows a field of RAT1 cells stained with the DNA-binding dye Hoechst. Normally these cells, like other fibroblasts we have examined, contain no detectable tau protein. The three cells marked with arrows were microinjected with 2 mg/ml tau protein, then the entire field was fixed with glutaraldehyde and stained with tau antiserum and rhodamine-conjugated secondary antibody. As shown in FIGURE 2, panel **b**, it is clear that tau staining is seen only in the microinjected cells and that the staining is localized exclusively on the microtubules that can be seen to emanate from centrosomes. Thus, in living cells, it appears that tau protein associates predominantly with microtubules and not with other organelles.

Biochemical Demonstration of Tau Association with Microtubules

Whereas the immunofluorescence experiments described above strongly suggest that tau protein is associated exclusively with microtubules in living cells, the remote possibility exists that the glutaraldehyde-fixation employed causes tau protein to become artificially cross-linked to microtubules. To dispense with this possibility, we sought a third criterion to demonstrate that tau is an *in vivo* MAP. One method used to identify MAPs, independent of their *in vitro* properties, is to prepare detergent-extracted cytoskeletons under microtubule-stabilizing conditions from normal cells, or cells treated with colchicine, to selectively remove microtubules from their cytoskeletons, and to compare the nontubulin proteins that selectively fractionate with tubulin.[26] FIGURE 3 shows the results of such an analysis. An immunoblot was probed with our affinity-purified tau antiserum and an iodinated secondary antiserum. The first lane shows the tau protein found in a differentiated PC12 cytoskeleton. The third lane shows the extracted tau protein. Tau protein remains almost exclusively in the cytoskeleton fraction. If the microtubules, however, are selectively depolymerized for 35 min with colchicine before lysis, most of the tau protein is lost from the cytoskeleton fraction (second lane). The tau protein is now in the extracted fraction (fourth lane). These results show biochemically that tau protein is associated almost exclusively with assembled microtubules in PC12 cells, and overcome the concern that immunofluorescent localization of tau to microtubules is due to glutaraldehyde fixation, artificially causing tau to associate with microtubules. We have obtained similar results for

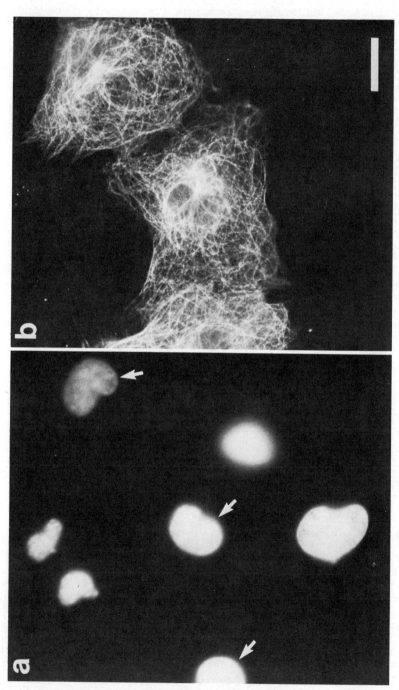

FIGURE 2. Microinjection of tau protein into rat fibroblast cells. **a:** RAT1 cells stained with DNA-binding Hoechst dye. Cells marked with arrows were microinjected with 2 mg/ml tau protein in 90 mM KCl, and 10 mM NaPO$_4$, pH 7.0. Forty minutes later the cells were fixed and treated for tau immunofluorescence as described in FIGURE 1. **b:** Tau immunofluorescence of same cells shown in **a**. Bar, 15 μm.

MAP-1, supporting earlier conclusions that MAP-1 binds to microtubules in living cells.

Tau Form Dependent on Interaction with Microtubules

We have shown both biochemically and immunocytochemically that tau protein associates with microtubules in living cells. We would like to report one additional observation that perhaps demonstrates this association most strongly and also suggests a role for posttranslational modifications in regulating the association of tau protein with microtubules. Using the extraction procedure described in the preceding section, Pallas and Solomon showed that the extent of phosphorylation of some MAPs *in vivo* is dependent on whether they interact with microtubules.[27] If the phosphorylation state of tau protein is dependent on its interaction with microtubules, then treating cells with drugs affecting microtubule assembly should alter the phosphorylation state of tau protein. Because Lindwall and Cole showed that dephosphorylation of tau protein *in vitro* causes it to run with an increased mobility on SDS gels,[28] it might then be possible to detect changes in tau phosphorylation, caused by microtubule drugs on one-dimensional gels. FIGURE 4 shows an immunoblot probed with our affinity-purified tau antiserum. The first lane shows the tau pattern observed in differentiated PC12 cell extracts without drug treatment. Treatment with colchicine to depolymerize microtubules causes an increase in the mobility of all three tau species (second lane), similar to that observed upon treatment of purified tau with phosphatase.[28] Taxol, which binds microtubules and shifts the tubulin monomer/polymer equilibrium to favor assembly, also causes this mobility shift (third lane). In addition, nocodazole, which has the same affect on microtubule assembly as colchicine, and vinblastine, which causes tubulin and tau to coaggregate, also cause the same mobility shift (not shown). Because these drugs, which specifically affect microtubule assembly in different ways, all cause tau mobility shifts when applied to living cells, tau protein must interact directly with microtubules in cells. This result circumvents shortcomings of the approaches described in previous sections, where fixation or detergent-extraction might cause an adventitious association of tau with microtubules, because cells were lysed directly in sodium dodecyl sulfate (SDS) and boiled immediately. Although we have not formally demonstrated that the tau mobility shift is due to dephosphorylation, we believe that this is the case because this mobility shift is exactly as described *in vitro*.[28] Because tau protein form can be modified by changing the state of microtubules in a cell, it stands to reason that the cell may normally alter tau form to change the state of its microtubules.

DISCUSSION

We have examined the cellular distribution and subcellular localization of tau protein. As the result of an extensive analysis of tau levels in different tissues and different established cell lines, we have come to view tau as a predominantly nervous-tissue-specific protein. In the course of these studies it became clear that the complexity of tau-related protein species is far greater than was originally described. Our highly specific affinity-purified tau antiserum recognizes polypeptides ranging from 42 kD to 200 kD in different tissues (TABLE 1). We have shown that the 125 kD taus observed in nervous tissues are closely related to the previously well-characterized taus.[15] Further analysis will be required to prove that other proteins that react with our

polymer monomer

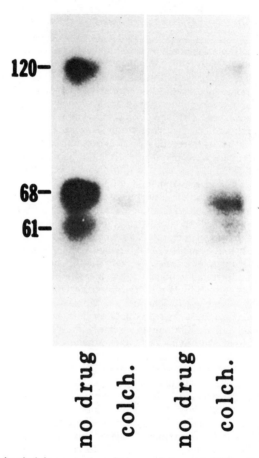

FIGURE 3. Biochemical demonstration of tau association with PC12 cell microtubules. Detergent-extracted cytoskeletons were prepared from differentiated PC12 cells as described in reference 26, and nonextracted or extracted tau proteins were analyzed by immunoblotting with tau antiserum. The left two lanes show tau proteins in PC12 cytoskeletons from untreated cells, or cells treated for 35 min with 10 μm colchicine to depolymerize the microtubules, respectively. The right two lanes show the extracted tau proteins from nontreated and colchicine-treated PC12 cells respectively.

tau antibodies are related to tau protein. The exact way in which the taus are related can only be determined by thorough analysis of tau cDNA clones. It is intriguing that we detect only one or a few tau genes in the mouse genome.[18] The ability to translate multiple tau species from mouse brain mRNA in cell-free translation systems[18,29] suggests that a single tau gene may give rise to multiple tau mRNAs as has been seen for several genes (see reference 30, for example).

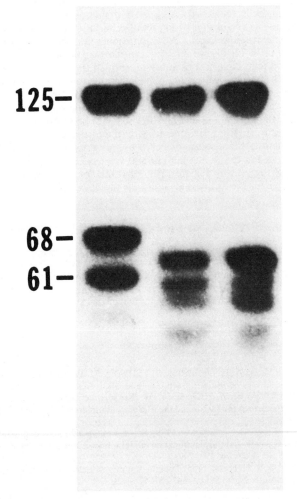

FIGURE 4. Effect of colchicine and taxol on tau form. tau protein in extracts from PC12 cells treated with no drug, 10 μM colchicine for 6 hr, or 10 μM taxol for 6 hr was analyzed on an immunoblot probed with tau antiserum.

Because tau protein characterization was based on its *in vitro* association with microtubules, it was important to demonstrate that the tau behaves like a MAP in living cells. We have demonstrated the association of tau with microtubules in cells in four ways. (1) Immunofluorescence using tau antibodies showed that tau protein binds microtubules in PC12 cells. Furthermore, all microtubules appear to be coated with tau protein. It remains possible that different tau subspecies react with different microtubules to specify functionally distinct microtubules. We will only be able to determine if this is the case if we can generate antibodies that distinguish tau subspecies. (2) By microinjecting tau protein into fibroblast cells that normally lack tau protein, we were able to verify that tau associates with microtubules in living cells. In addition, the flat morphology of the fibroblasts allowed us to determine that tau protein associates exclusively with microtubules in cells. (3) By preparing detergent-extracted cytoskeletons with or without microtubules, we were able to show biochemically that tau is associated with the microtubule network. (4) Treatment of PC12 cells with any of four microtubule drugs changes tau form, showing that, independent of fixing or extracting cells, tau is still associated with microtubules.

These experiments establish a strong set of criteria to demonstrate that a protein associates with microtubules in cells. New approaches must be found to determine the role of MAPs in cells. Ideally we might use a pharmacological agent or a mutation specific for tau protein. No drugs have been identified that specifically interfere with tau action. RAT1 fibroblast cells contain no detectable tau protein, however, and are therefore analogous to genetic tau null mutants. We can introduce tau protein into these cells, and we know that it will bind exclusively to microtubules (FIGURE 2). We plan to compare tubulin on and off rates in cells with or without tau by monitoring treatment, or recovery from treatment, with microtubule depolymerizing drugs. Microtubule dynamics can also be compared by introducing derivatized tubulin into cells[31] containing or not containing tau protein. We have extended *in vitro* studies on tau protein into living cells by demonstrating that tau associates with microtubules *in vivo*. The challenge for the future is to determine the *in vivo* function of tau protein.

ACKNOWLEDGMENTS

We thank David Gard for his assistance in mouse dissections. We are grateful to Stuart Feinstein for supplying PC12 cells and nerve growth factor, Steve Blose for supplying tubulin antibodies, David Asai for supplying MAP-1 antibodies, and Cynthia Cunningham-Hernandez for preparation of the manuscript.

REFERENCES

1. WEINGARTEN, M. D., A. H. LOCKWOOD, S. Y. HWO & M. W. KIRSCHNER. 1975. A protein factor essential for microtubule assembly. Proc. Natl. Acad. Sci. USA **72:** 1858–1862.
2. CLEVELAND, D. W., S. Y. HWO & M. W. KIRSCHNER. 1977. Physical and chemical properties of purified tau factor and the role of tau in microtubule assembly. J. Mol. Biol. **116:** 227–247.
3. BULINSKI, J. C. & G. G. BORISY. 1979. Self-assembly of microtubules in extracts of cultured HeLa cells and the identification of HeLa microtubule-associated proteins. Proc. Natl. Acad. Sci. USA **76:** 293–297.
4. KEATES, R. A. & R. H. HALL. 1975. Tubulin requires an accessory protein for self assembly into microtubules. Nature (London) **257:** 418–420.

5. MURPHY, D. B. & G. G. BORISY. 1975. Association of high molecular weight proteins with microtubules and their role in microtubule assembly *in vivo*. Proc. Natl. Acad. Sci. USA **72:** 2696–2700.

6. SLOBODA, R. D., S. A. RUDOLPH, J. C. ROSENBAUM & P. GREENGARD. 1975. Cyclic-AMP-dependent endogenous phosphorylation of a microtubule-associated protein. Proc. Natl. Acad. Sci. USA **72:** 177–181.

7. CONNOLLY, J. A., V. I. KALNINS, D. W. CLEVELAND & M. KIRSCHNER. 1977. Immuno-fluorescent staining of cytoplasmic and spindle microtubules in mouse fibroblasts with antibody to τ protein. Proc. Natl. Acad. Sci. USA **74:** 2437–2440.

8. CONNOLLY, J. A. & V. I. KALNINS. 1980. The distribution of tau and HMW microtubule-associated proteins in different cell types. Exp. Cell. Res. **127:** 341–350.

9. DEBRABANDER, M., J. C. BULINSKI, G. GEUENS, J. DE MAY & G. G. BORISY. 1981. Immunoelectron microscopic localization of the 210,000-mw microtubule-associated protein in cultured cells of primates. J. Cell Biol. **91:** 438–445.

10. SHERLINE, P. & K. SCHIAVONE. 1977. Immunofluorescence localization of proteins of high molecular weight along intracellular microtubules. Science **198:** 1038–1040.

11. CONNOLLY, J. A., V. I. KALNINS, D. W. CLEVELAND & M. W. KIRSCHNER. 1978. Intracellular localization of the high molecular weight microtubule accessory protein by indirect immunofluorescence. J. Cell Biol. **76:** 781–786.

12. IZANT, J. G. & J. R. MCINTOSH. 1980. Microtubule-associated proteins: a monoclonal antibody to MAP2 binds to differentiated neurons. Proc. Natl. Acad. Sci. USA **77:** 4741–4745.

13. WEATHERBEE, J. A., P. SHERLINE, R. N. MASCASDO, J. G. IZANT, R. B. LUFTIG & R. R. WEIHING. 1982. Microtubule-associated proteins of HeLa cells: heat stability of the 200,000 mol. wt. HeLa MAPs and detection of the presence of MAP2 in HeLa cell extracts and cycled microtubules. J. Cell Biol. **92:** 155–163.

14. BLOOM, G. S., F. C. LUCA & R. B. VALLEE. 1984. Widespread cellular distribution of MAP-1A (microtubule-associated protein 1A) in the mitotic spindle and on interphase microtubules. J Cell Biol. **98:** 331–340.

15. DRUBIN, D., S. FEINSTEIN & M. KIRSCHNER. 1984. Microtubule-associated tau protein induction by nerve growth factor during neurite outgrowth in PC12 cells. *In* Molecular Biology of the Cytoskeleton. Cold Spring Harbor. pp. 343–355.

16. BURNETTE, W. N. 1981. "Western Blotting": electrophoretic transfer of proteins from sodium dodecyl sulfate-polyacrylamide gels to unmodified nitrocellulose and radio-graphic detection with antibody and radioiodinated protein A. Anal. Biochem. **112:** 195–203.

17. PFEFFER, S. R., D. G. DRUBIN & R. J. KELLY. 1983. Identification of three coated vesicle components as alpha and beta tubulin linked to a phosphorylated 50,000 dalton polypeptide. J. Cell Biol. **97:** 40–47.

18. DRUBIN, D. G., D. CAPUT & M. W. KIRSCHNER. 1984. Studies on the expression of the microtubule-associated protein, tau, during mouse brain development with newly isolated complementary DNA probes. J. Cell Biol. **98:** 1090–1097.

19. GREENE, L. A. & A. S. TISCHLER. 1976. Establishment of a nonadrenergic clonal line of rat adrenal pheochromocytoma cells which respond to nerve growth factor. Proc. Natl. Acad. Sci. USA **73:** 2424–2428.

20. LAZARIDES, E. 1980. Intermediate filaments as mechanical integraters of cellular space. Nature (London) **283:** 249–256.

21. SPIEGELMAN, B. M. 1978. Doctoral dissertation. Princeton University.

22. CLEVELAND, D. W., B. M. SPIEGELMAN & M. W. KIRSCHNER. 1979. Conversation of microtubule associated proteins. J. Biol. Chem. **254:** 12670–12678.

23. MURPHY, D. B. & K. T. WALLIS. 1983. Isolation of microtubule protein from chicken erythrocytes and determination of the critical concentration for tubulin polymerization *in vitro* and *in vivo*. J. Biol. Chem. **258:** 8357–8364.

24. LUCKENBILL-EDDS, L., C. VAN HORN & L. A. GREENE. 1979. Fine structure of initial outgrowth of processes induced in a pheochromocytoma cell line (PC12) by nerve growth factor. J. Neurocytol. **8:** 493–511.

25. LUDUENA, R. G., A. FELLOUS, L. MCMANUS, M. A. JORDAN & J. NUNEZ. 1981. Contrasting roles for tau and microtubule-associated protein 2 in the vinblastine-induced aggregation of brain tubulin. J. Biol. Chem. **259:** 12890–12898.
26. SOLOMON, F., M. MAGENDANTZ & A. SALZMAN. 1979. Identification with cellular microtubules of one of the co-assembling microtubule-associated proteins. Cell **18:** 431–438.
27. PALLAS, D. & F. SOLOMON. 1982. Cytoplasmic microtubule-associated proteins: phosphorylation at novel sites is correlated with their incorporation into assembled microtubules. Cell **30:** 407–414.
28. LINDWALL, G. & R. D. COLE. 1984. Phosphorylation affects the ability of tau protein to promote microtubule assembly. J. Biol. Chem. **259:** 5301–5305.
29. GINZBURG, I., T. SCHERSON, D. GIVEON, L. BEHAV & U. Z. LITTAUER. 1982. Modulation of mRNA for microtubule-associated proteins during brain development. Proc. Natl. Acad. Sci. USA **79:** 4892–4896.
30. SCHWARTZBAUER, J. E., J. W. TAMKUN, I. R. LEMISCHKA & R. O. HYNES. 1983. Three different fibronectin mRNAs arise by alternative splicing within the coding region. Cell **35:** 421–431.
31. SAXTON, W. M., D. L. STEMPLE, R. J. LESLIE, E. D. SALMON, M. ZAVORTINK & J. R. MCINTOSH. 1984. Tubulin dynamics in cultured mammalian cells. J. Cell Biol. **99:** 2175–2186.

Dynein as a Microtubule-Associated Protein[a]

PETER SATIR AND JOCK AVOLIO

Department of Anatomy and Structural Biology
Albert Einstein College of Medicine
Bronx, New York 10461

INTRODUCTION

The potential for association between dynein, the axonemal ATPase, and axonemal microtubules has been recognized since the original work of Gibbons.[1] The localization of dynein to the arms extending from subfiber A of each axonemal doublet microtubule supported the early speculations of Afzelius[2] regarding arm-microtubule interactions. With the demise of the contractile models, and the growing evidence in favor of the sliding microtubule model of axonemal activity powered by dynein arms (Satir;[3,4] Summers and Gibbons;[5] Sale and Satir[6]), these interactions have assumed increased significance.

An *in vitro* system for examining such interactions has been pioneered by Takahashi and Tonomura.[7] In this system, dynein is extracted from axonemes to give armless microtubules and then readded in an appropriate incubation buffer. As seen by negative stain procedures, in the absence of adenosine triphosphate (ATP), dynein decorates both subfiber A and subfiber B of each doublet microtubule, much as heavy meromyosin may be used to decorate actin microfilaments. Satir *et al.*[8] have confirmed these results and in addition have demonstrated that the dynein that decorates also cosediments with the microtubules, and that the binding site for the *Tetrahymena* dynein arm is common to other axonemal doublets and to brain microtubules. Similar results have been obtained by Haimo and her colleagues, mainly using *Chlamydomonas reinhardtii* dynein and thin section techniques (cf. Haimo and Telzer[9]).

Addition of ATP releases arms, but if sufficient time elapses so that ATP is hydrolyzed, the arms will rebind to the microtubules. In the presence of ATP, the released arms do not cosediment with the microtubules and are found in the supernate.

Based on the Takahashi and Tonomura[7] and Satir *et al.*[8] assays, Nasr and Satir[10] have developed an association-dissociation procedure, alloaffinity filtration, for purifying certain dyneins. The *Tetrahymena* dynein that is routinely purified by this procedure has the characteristics of 30S *Tetrahymena* dynein prepared by conventional methods. Alloaffinity purified *Tetrahymena* dynein is a complex molecule composed of heavy (H), intermediate (I), and light (L) chains. The major H peak is at M_r about 3×10^5; the I region usually consists of 1–3 major peaks of M_r about 6–9×10^4, whereas the L group profile has peaks at M_r 23, 17, 15, and 14×10^3. According to Johnson and Wall,[11] the particle weight of isolated 30S *Tetrahymena* dynein molecules is 1.95×10^6. This would be consistent with an alloaffinity dynein composition of 6H: 3I: 6L chains per molecule. Although some estimates of the true molecular weight of the H chain are about 50% higher than the value used here and some estimates of the

[a]This work was supported by a Grant from the USPHS (HL22560).

particle weight are about one-third lower, there is clearly more than one H chain per dynein molecule, as is also the case for sea urchin and *Chlamydomonas reinhardtii* dynein arms. The H chains are the likely sites of ATPase activity, and they presumably form the major part of the compact globular domains that are the signature of isolated dynein (FIGURE 1a). Avolio *et al.*[12] have concluded that *in situ* the major domains of dynein forming an arm extend transversely across the interdoublet gap from doublet N to doublet N + 1 (FIGURE 1b). We discuss this conclusion further in this article.

RESULTS AND DISCUSSION

The *in situ Tetrahymena* arm is composed of a head, body, and cape. The head supposedly swivels around its articulation with the body, so that in one position the whole arm does not extend completely across the gap, whereas in another it does. In the latter position, the head can attach to subfiber B of doublet N + 1. Avolio *et al.*[12] have

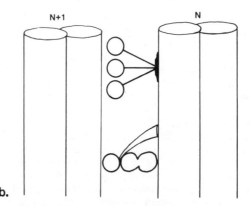

FIGURE 1. (a) Representation of isolated 30S *Tetrahymena* dynein as seen by scanning transmission microscopy. The isolated molecule appears as a bouquet containing three major globular domains (Johnson and Wall[11]). (b) Orientation of dynein molecule *in situ*. Above: Vertical disposition of subunits as proposed by Johnson and Wall.[11] Below: Horizontal disposition of subunits as proposed by Avolio *et al.*[12] In this model, the globular domains composed in part of dynein H chains attach to doublet N as the body of the arm and to doublet N + 1 as the head. A cape, possibly formed by the stalk-like extensions from the globular domains, makes a second attachment to doublet N.

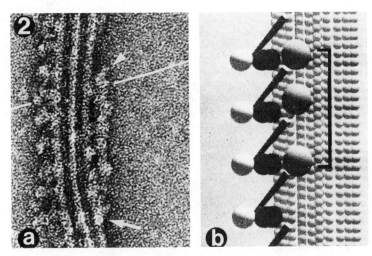

FIGURE 2. (a) Negative stain image of splayed *Tetrahymena* axoneme in AMP-PCP. During splaying, arms have been torn from doublet N but still attach to subfiber B of doublet N + 1 by their heads (arrow); barbed ends are free (arrowhead). Lines indicate arms in the trans-orientation defined by Avolio et al.[12] Magnification × 260,000. (b) Computer reconstruction of AMP-PCP image using arms composed of three globular domains with horizontal *in situ* orientation. A radial spoke group is indicated by bracket.[15]

demonstrated this attachment in the presence of the nonhydrolyzable ATP analog, AMP-PCP (β,γ methylene adenosine 5'-triphosphate). In the presence of this analog, doublets will not slide, but they will sometimes splay apart, probably under the influence of forces generated during the preparative procedures. In many instances, the dynein arms remain firmly attached to subfiber B by their head ends. Avolio et al.[12] refer to this type of image as the *trans*-configuration. A doublet whose subfiber B is decorated with arms in the *trans*-configuration is shown in FIGURE 2a. The head attachment, demonstrated in FIGURE 2a, which does not rely on decoration experiments, is a critical step in a proposed mechanochemical cycle of the arm (cf. Avolio et al.[12] for further discussion).

Viewed from the side, the tail or barbed end of the arm is V-shaped. One side of the V is formed by the body, the other by the cape. An *in situ* arm attached by its head end to subfiber B of doublet N + 1 is also attached by its barbed end to subfiber A of doublet N. In this configuration the body is oriented toward the base of the microtubule, the cape toward the tip. Attachments between both ends of the V and subfiber A are formed. FIGURE 3 is a stereo image showing a series of arms in AMP-PCP where the subunit construction and attachment across the interdoublet gap are evident. As can be seen, there are two usual attachment sites on doublet N at the barbed end of the dynein arm. In profile, the body-microtubule attachment site spans about three tubulin subunits, whereas the cape-microtubule attachment site spans a single subunit, about three subunits more proximally along the subfiber. Throughout the mechanochemical cycle that produces sliding of N + 1 relative to N, at least one of these sites remains attached at all times; therefore, at least one, but not necessarily both, of these sites is ATP-insensitive. It is only during splaying in AMP-PCP that, unphysiologically, both attachments to doublet N are sometimes lost, whereas head attachment to doublet N + 1 persists. In FIGURE 2a, the barbed ends of arms attached to subfiber B are free.

The stringency of assembly of the *in situ* arm with respect to the microtubule doublet is well known. The *in situ* arm never normally decorates subfiber B by its barbed end, nor is it ever associated with central pair microtubules. Moreover, the rows of arms on subfiber A generally end before subfiber B terminates (Satir[4]). The arm periodicity is 24 nanometers. Two rows of arms assemble only along specific protofilaments of subfiber A. In the nomenclature of Warner and Satir,[13] the inner arm

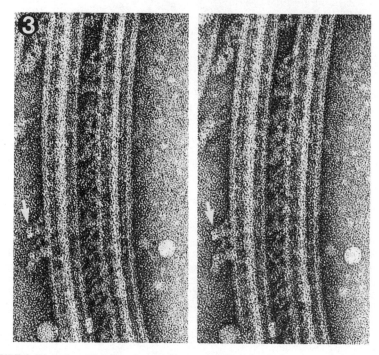

FIGURE 3. Negative stain stereo images of arms in splayed *Tetrahymena* axoneme in AMP-PCP. Arrowhead indicates individual arm showing subunit construction. Note that both the cape and the body are attached to subfiber A on this doublet. To the right, a row of arms spans the interdoublet gap. Magnification × 240,000.

assembles along protofilaments 2–4, the outer arm along 6–9. Recently, we (Avolio and Satir[14] and Avolio *et al.*[15]) have demonstrated that the two rows are in exact alignment along the *Tetrahymena* doublets, not staggered as was previously thought, and are separated by about one protofilament (no. 5). A computer model of rows of inner and outer arms with the appropriate periodicity, placement, and subunit construction is shown in FIGURE 2b. In the reconstruction, the body is composed of two of the three globular domains of isolated *Tetrahymena* dynein; the head is the third domain. The computer image can be rotated to produce detailed comparisons to stereo negative stain images such as FIGURE 3.

The dynein arm orientation found after decoration of axonemal microtubules with isolated crude or alloaffinity purified *Tetrahymena* dynein is much more promiscuous and less stringent than that seen *in situ*. The structure of the arm is somewhat altered by extraction and readdition. Although the barbed end of the arm can readily be identified by the cape, body, and characteristic V shape, head subunits are not normally seen.

Takahashi and Tonomura[7] showed that along a single doublet where subfiber B as well as subfiber A decorates with added dynein, both rows of arms tilt toward the base, forming a series of arrowheads. Avolio et al.[12] refer to this as the *cis* orientation. The *cis* orientation can also be produced by rebinding of released arms in the absence of ATP to doublets whose *in situ* arms remain intact. Although there is no single unique *cis* orientation, the binding to subfiber B usually occurs by the V-shaped barbed end of the arm, with the cape oriented tipward, implying that many regions of the microtubule lattice have the potential to bind dynein by its barbed end. What specifies *in situ* binding so precisely in the face of this potential is unknown; *in situ* binding probably depends partially, but dramatically, on the requirement that the *in situ* medium contains sufficient ATP to cause dissociation of ATP-sensitive binding sites on the arm. Because neither the morphology of the arm nor the stringency of the attachment to microtubules has yet been retained in the decoration experiments, some caution seems appropriate when using these experiments to uncover the principles of arm-microtubule interactions.

Attachment in decoration experiments, unlike that *in situ,* is completely sensitive to ATP under the ionic conditions specified in Avolio et al.[12] This might indicate that the barbed end of the dynein arm contains an ATP-sensitive site, which is the predominant binding site in the absence of ATP. The relative strengths of the two attachment sites at the barbed end of the arm might be determined by an interplay between ionic strength and ATP concentration. Such interplay seems critical in purifying dynein by an association-disassociation method such as the alloaffinity procedure. In this regard, quite different conditions pertain during *in situ* assembly from those that are used *in vitro.*

The finding that the barbed end of the arm contains an ATP-sensitive site suggests that one or more of the ATPase H chains of the arm is associated with either the cape or, more likely, the body. Because of the apparent role of the head subunit of the *in situ* arm during mechanochemical force transduction, a second ATPase is probably located in the head. In *Chlamydomonas reinhardtii,* Witman et al.[16] have found one dynein (12S) from the outer arm appears to have dimensions similar to those of the head, whereas another (18S) resembles the body and the cape. The transverse orientation of the *in situ* arm across the interdoublet gap that we demonstrate may imply that complex interactions involving ATP are possible at both ends of the arm, which, under given ionic conditions, might be important in the activation of a particular arm at a particular time in the beat cycle.

ACKNOWLEDGMENTS

We thank Dr. M. E. J. Holwill and Dr. A. N. Glazzard (Queen Elizabeth College, London) for help with the computer reconstructions.

REFERENCES

1. GIBBONS, I. R. 1963. Studies on the protein components of cilia from *Tetrahymena pyriformis*. Proc. Natl. Acad. Sci. USA **50:** 1002–1010.

2. AFZELIUS, B. A. 1959. Electron microscopy of the sperm tail: results obtained with a new fixative. J. Biophys. Biochem. Cytol. **5:** 269–278.

3. SATIR, P. 1965. Studies on cilia. II. Examination of the distal region of the ciliary shaft and the role of filaments in motility. J. Cell Biol. **26:** 805–834.

4. SATIR, P. 1968. Studies on cilia. III. Further studies on the cilium tip and a "sliding filament" model of ciliary motility. J. Cell Biol. **39:** 77–94.

5. SUMMERS, K. E. & I. R. GIBBONS. 1971. Adenosine-triphosphate-induced sliding of tubules in trypsin-treated flagella of sea urchin sperm. Proc. Natl. Acad. Sci. USA **63:** 3092–3096.

6. SALE, W. S. & P. SATIR. 1977. Direction of active sliding of microtubules in *Tetrahymena* cilia. Proc. Natl. Acad. Sci. USA **74:** 2045–2050.

7. TAKAHASHI, M. & Y. TONOMURA. 1978. Binding of 30S dynein with a B-tubule of the outer doublet of axonemes from *Tetrahymena pyriformis* and adenosine triphosphate-induced dissociation of the complex. J. Biochem. (Tokyo) **84:** 1339–1355.

8. SATIR, P., J. WAIS-STEIDER, S. LEBDUSKA. A. NASR & J. AVOLIO. 1981. The mechano-chemical cycle of the dynein arm. Cell Motility **1:** 303–327.

9. HAIMO, L. T. & B. R. TELZER. 1981. Dynein microtubule interactions: ATP-sensitive dynein binding and the structural polarity of mitotic microtubules. Cold Spring Harbor Symp. Quant. Biol. **46:** 207–217.

10. NASR, A. & P. SATIR. 1985. Alloaffinity filtration: A general approach to the purification of dynein and dynein-like molecules. Anal. Biochem. **151:** In press.

11. JOHNSON, K. A. & J. S. WALL. 1983. Structure and molecular weight of the dynein ATPase. J. Cell Biol. **96:** 669–678.

12. AVOLIO, J., S. LEBDUSKA & P. SATIR. 1984. Dynein arm substructure and the orientation of arm-microtubule attachments. J. Mol. Biol. **173:** 389–401.

13. WARNER, F. D. & P. SATIR. 1973. The substructure of ciliary microtubules. J. Cell Sci. **12:** 313–326.

14. AVOLIO, J. & P. SATIR. 1984. Negative stain stereo images of ciliary doublet microtubules: three dimensional arrangement of arms, spokes and links. J. Cell Biol. **99:** 47a.

15. AVOLIO, J., A. N. GLAZZARD, M. E. J. HOLWILL & P. SATIR. 1986. Structures attached to doublet microtubules of cilia: computer modeling of thin section and negative stained stereo images. Proc. Natl. Acad. Sci. USA In press.

16. WITMAN, G. B., K. A. JOHNSON, K. K. PFISTER & J. S. WALL. 1983. Fine structure and molecular weight of the outer arms of dyneins of *Chlamydomonas*. J. Submicrosc. Cytol. **15:** 193–197.

Dynein-like Cytoplasmic Microtubule Translocators[a]

DAVID J. ASAI,[b] ROGER J. LESLIE, AND
LESLIE WILSON

Department of Biological Sciences
University of California
Santa Barbara, California 93106

INTRODUCTION

Microtubule-mediated motility is a common and important phenomenon in eukaryotic cells. It includes eukaryotic flagellar and ciliary beating, chromosome movements during mitosis, and particle migration in axoplasmic transport. Microtubules have also been implicated in other motile processes such as rearrangements of cell surface receptors, pigment granule movement, and pronuclear migration. Despite the many examples of microtubule-mediated motility that have been described, there is little understanding at present of the mechanisms driving the motility.

The only documented mechanism of microtubule-based movement is that of axonemal outer doublet microtubule sliding in cilia and flagella.[1,2] The geometry of the axoneme and control of active sliding produce the observed bending that is perpendicular to the direction of microtubule sliding.[3] Active sliding is powered by a complex of force-producing proteins called dynein.[4] Dynein has only been described in axonemes, probably in part because cilia and flagella are such favorable organelles for the study of motility. For example, cilia and flagella have a well-ordered axonemal ultrastructure, they are often easy to visualize and isolate, and it has been possible to demembranate and reactivate their normal movement (*e.g.,* references 5, 6). These same advantages are not shared by other microtubule-mediated motility systems, although several kinds of cytoplasmic microtubule-based movements do appear to involve an active sliding of elements along microtubules. Because the axoneme offers a well-characterized paradigm, and because cytological observations often suggest microtubule sliding, it has been attractive to propose the existence of nonaxonemal microtubule translocators, or "cytoplasmic dyneins," which, by analogy to axonemal dynein, might mediate microtubule-based movements by an adenosine triphosphate (ATP)-dependent attachment-detachment mechanism.

This paper addresses four main points. First, we discuss the properties of axonemal dynein and relate those properties to our expectations for a nonaxonemal dynein-like translocator. Second, we briefly review some of the evidence for a cytoplasmic dynein-like activity. Third, we present some thoughts about cytoplasmic microtubule motors, especially in the context of mitosis. And fourth, we summarize the results from our laboratory that characterize a dynein-like activity from the cytoplasm of unfertilized sea urchin eggs.

[a]This work was supported by NIH Grants HD16956 to D. J. Asai, GM09657 to R. J. Leslie, and NS13560 to L. Wilson.
[b]Send correspondence to Dr. David J. Asai, Department of Biological Sciences, Lilly Hall of Science, Purdue University, West Lafayette, Indiana 47907.

275

PROPERTIES OF DYNEIN

Axonemal Dynein

Axonemal dyneins are the translocators that constitute the cross-bridge arms on axonemal outer doublet microtubules. Dynein arms remain structurally anchored to the A subfiber of each outer doublet microtubule and cyclically attach and detach to the B subfiber of the adjacent outer doublet microtubule. The cross-bridge cycle, analogous to the myosin cross-bridge cycle in striated muscle, is mediated by $MgATP^{2-}$.[7,8] Detachment of dynein from the B subfiber occurs upon the addition of $MgATP^{2-}$ and prior to ATP hydrolysis.[9,10] Either during the hydrolysis of the ATP or the subsequent release of products ((adenosine diphosphate) ADP and (inorganic phosphate) P_i), the dynein arm undergoes a conformational change and rebinds to a

TABLE 1. Properties of Dynein

Generally	Sea Urchin Sperm Flagellar Dynein (after Gibbons *et al.*)
1. ATPase activity.	1. Mg^{2+}/Ca^{2+} ATPase; two ATPases per particle.
2. Asymmetric: one end binds to a microtubule and the other end binds to a matrix that is fixed in space relative to the microtubule.	2. "A-end" is structural binding to A subfiber microtubule; "B-end" is ATP-sensitive binding to B subfiber microtubule. (A subfiber represents the "matrix").
3. Work: interacts with microtubules to produce mechanochemical work.	3. Outer dynein arms slide outer doublet microtubules in a controlled fashion to produce axonemal bending.
4. Inactive in uncoupled form. Activated by coupling with microtubules.	4. Latent activity dynein is activated by rebinding to microtubules, by nonionic detergent, by thiourea. Latent activity dynein rebinds to outer doublet microtubules and restores sliding function.
5. Particle 10-30S.	5. 21S latent activity dynein (approximately 1.3×10^6 daltons); and a 10-14S species.
6. Multimeric protein, including high molecular weight proteins.	6. 21S latent activity dynein: two high molecular weight proteins, A_α and A_β (approx. 420,000 daltons); three intermediate chains (85,000–115,000 daltons), and four light chains (15,000–25,000 daltons).

new site on the B subfiber so that the two outer doublet microtubules are translocated with respect to each other. This ATP-dependent active sliding occurs in only one direction with respect to any pair of outer doublet microtubules,[11] and it is this unidirectionality that helps to generate the observed bending pattern.

The examination of axonemal function, then, reveals several important features of axonemal dynein: (1) it is a MgATPase; (2) it is asymmetric in that it binds to microtubules in two different ways: a structural attachment to the A subfiber lattice that is not sensitive to ATP and a "B end" binding that is reversed by ATP; and (3) it performs work with microtubules *in vivo*. The properties of dynein have been delineated by Ian Gibbons and his coworkers,[12,13] and we summarize the salient points in TABLE 1.

Another important aspect of axonemal dynein is its latency. From the sea urchin flagellum, a functional form of dynein has been isolated as a 21S particle and termed latent activity dynein, or LAD.[14] Latent activity dynein can be irreversibly activated *in vitro* by various treatments including incubation with nonionic detergent, gentle heating, or mild chaotropic agents such as thiourea.[15] Upon *in vitro* activation, the dynein often exhibits an apparent MgATPase-specific activity tenfold higher than the LAD. Latent activity dynein can also be activated upon rebinding to axonemal microtubules.[14,16] In this regard, LAD is analogous to myosin in the absence of filamentous actin.[17] Latent activity dynein can be considered the physiologically important form of the enzyme, because it is capable of reattaching to axonemes that have been stripped of their endogenous dynein, with restoration of active sliding function as measured by the return of reactivated beat frequency.[18,19] Gibbons and his coworkers have demonstrated that sea urchin spermatozoan LAD is a 21S particle containing 9 polypeptides (2 heavy chains, 3 intermediate chains, and 4 light chains), and they have provided convincing evidence that the ATPase site(s) resides on the high molecular weight polypeptides species.[20–22]

Dynein Binding to Axonemal Microtubules Is Asymmetric

An important functional aspect of axonemal dynein is its binding to microtubules in order to produce work *in vivo*. The dynein binding to microtubules must be asymmetric in that each end of the dynein must interact with a microtubule (or other structure) in a kinetically nonidentical way. The axoneme reveals this asymmetry in that the dynein arm binds the A subfiber permanently and binds the B subfiber transiently and in an ATP-sensitive fashion (see reference 23). Given the structural complexity of the dynein arm, both in terms of its polypeptide composition[22] and its morphology (*e.g.*, references 24, 25), it is reasonable to conclude that a significant portion of the asymmetry in the interaction between the dynein and the microtubules is encoded in the dynein arm structure. Another aspect of the asymmetry is found in the surface lattices of the A and B subfibers, that is, the surface lattices of the two microtubules are different.[26] It is these surface lattices that are important when considering dynein-microtubule interactions. The A lattice appears to be unique to the axoneme, and the dissimilarity between the two subfibers presumably allows the dynein arm to bind with the appropriate orientation. It is important to note, then, that although the axonemal outer doublet microtubules all display the same structural polarity (parallel arrangement of the nine outer doublet microtubules), the outer doublet active sliding that is observed in axonemes is not strictly parallel microtubule sliding but sliding between two different surface lattice elements.

Non-Axonemal Dynein-Like Translocators

Based upon our understanding of axonemal dynein, a cytoplasmic microtubule translocator should fulfill three basic criteria: it must bind to microtubules; it must be asymmetric in its binding to microtubules and to other structures; and it must interact with microtubules *in vivo* to perform work. In addition, by analogy to axonemal dynein, we might expect such cytoplasmic microtubule translocators to derive their energy by the hydrolysis of ATP, to bind to microtubules in an ATP-sensitive fashion, and to be composed of proteins, at least some of which are high molecular weight ATPases.

As we discussed above, in order to perform work, the translocator must display an asymmetry. One end of the translocator must be able to bind to a microtubule and work against the microtubule, perhaps in an ATP-dependent manner. The other end of the translocator must be structurally dissimilar from the first end so that its interaction with either another microtubule or with a nonmicrotubular structure is kinetically different than the interaction with the microtubule at the first end. For example, it is conceivable that a cytoplasmic microtubule translocator might not cross-bridge two adjacent microtubules, but might instead be imbedded in a membranous structure or attached to a non-microtubule filament or bound to the cytomatrix.

A CYTOPLASMIC DYNEIN?

A number of different microtubule-mediated movements appear to involve translocation of particles along microtubules. Sometimes, as in anaphase mitotic movement, overlapping microtubules appear to slide against each other. In other cases, as in axoplasmic transport, vesicles and membranous organelles appear to actively track along microtubules. It has been of considerable interest to hypothesize the existence of a dynein-like translocator that might cause elements to slide along cytoplasmic microtubules. Indeed, previous work over the past quarter century from several laboratories has established the existence of a dynein-like MgATPase in unfertilized echinoderm eggs.[27-35]

Work from our laboratory,[72] as we summarize here, has identified a 20S enzyme that resembles axonemal dynein in its *in vitro* properties. It is a MgATPase, is isolated as a 20S latent activity enzyme, and is composed, at least in part, of high molecular weight polypeptides that appear to contain the ATPase activity. It binds microtubules *in vitro*, and the binding is accompanied by a striking enrichment of the MgATPase activity. The binding to microtubules is ATP-sensitive, and it shows immunological homology to axonemal dynein.

Nevertheless, the enzyme we have characterized as well as all of the other cytoplasmic dynein-like enzymes thus far described fall short of one very important criterion for a true dynein. A dynein must interact with microtubules to perform work *in vivo*. Until a cytoplasmic enzyme is shown unequivocally to be used by the living cell to move microtubules (or to move elements along microtubules), the various dynein-like MgATPases heretofore described can only be considered promising candidates. It is important to remember the possibility that cytoplasmic dynein does not exist, in spite of how attractive the axonemal paradigm. If there is a dynein-like microtubule translocator, it may be sufficiently different from axonemal dynein to be unrecognizable as a dynein.

Strategies to Identify Cytoplasmic Dynein

Two strategies have been used to attempt to assign an axonemal dynein-like mechanism to nonaxonemal movements: drug inhibitor studies and antibody localization experiments. Although these approaches are somewhat useful, they are based on the assumption that a cytoplasmic microtubule translocator will turn out to be very homologous to axonemal dynein. Each are also plagued by difficulties.

The axonemal dynein inhibitors vanadate and erythro-9-[3-(2-hydroxynonyl)]adenine (EHNA) have been extensively used to infer the existence of dynein-like enzymes in several cytoplasmic processes (*e.g.,* references 36–41). Although vanadate has been very useful in elucidating the mechanism of action of axonemal dynein when the purified enzyme or flagellum has been available,[42-44] the utility of vanadate as a specific probe for all dyneins remains tenuous. Because vanadate inhibits other ATPases[45,46] and because the mechanism of inhibition of a putative cytoplasmic dynein is not known, vanadate inhibition of a particular movement is not mechanistically diagnostic. Similarly, although EHNA at high concentrations inhibits flagellar dynein ATPase activity,[47] its mechanism of inhibition of a putative cytoplasmic dynein is unknown and EHNA by itself is not diagnostic for dynein. Recent evidence underscores this point by suggesting that EHNA, when introduced into cells under conditions previously thought to be specific for dynein inhibition, will inhibit movement by inhibiting actin function.[48]

The second strategy to identify putative cytoplasmic dyneins has involved the use of antibodies made against axonemal dynein. Many antiaxonemal dynein antibodies appear to cross-react with cytoplasmic components (*e.g.*, references 49, 50). For example, the antifragment 1A antibody prepared by Ogawa and Mohri[51] has been found to immunofluorescently stain mitotic spindles[52] and to inhibit the anaphase-like movements of glycerol-stabilized mitotic apparatuses.[53] In the absence of enzyme isolation, however, and the demonstration that the antibody reacts specifically with the purified dynein-like enzyme, it is possible that antibody localization and inhibition studies such as those listed above rely on nondynein peptide cross-reactivity by the antibody. Indeed, the immunofluorescent staining of mitotic spindles by antifragment 1A was largely eliminated if the antibody was first absorbed with pig brain microtubule protein,[52] suggesting that either the pig brain microtubule protein preparation contained significant quantities of dynein or a dynein-like material, or that the antibody staining of the spindle was due, at least in part, to activity against nondynein components, such as tubulin.

MITOSIS AND DYNEIN

Chromosome Movements

Dynein has been implicated in mitosis. Vanadate and EHNA, known inhibitors of axonemal dynein ATPase, inhibited anaphase motion in lysed PtK1 cells.[36,37] There is also some biochemical evidence for a molecule like axonemal dynein heavy chain in sea urchin embryos during mitosis.[31,49] These exciting results were somewhat surprising because the structural details of mitotic apparatuses differ from those of ciliary and flagellar axonemes. For example, dynein arms in axonemes cross-bridge microtubules of parallel polarity, whereas the microtubules of the mitotic interzone are antiparallel.[54,55] The motor(s) of mitosis and axonemes differ from each other in functional details in addition to the differences in their microenvironments. Whereas spindle elongation (anaphase B) is a unidirectional process, thus possibly produced by a dynein-like motor, prometaphase chromosome movement is clearly bidirectional.[56]

There is some controversy about how such bidirectional movement is produced. Ostergren[57] and others[58,59] have proposed that chromosomes move in two directions during prometaphase because of opposing forces exerted on the sister kinetochores. Ostergren believed that the amount of force exerted on the kinetochore is proportional to the distance between the kinetochore and the pole to which it is directed. A chromosome that is not at the metaphase plate would have unequal opposing forces acting on it and thus bidirectional movement would occur as a result of these opposing forces. Others[60-62] propose that chromosomes move in two directions during prometaphase because there are two motors in operation. Both motors would be unidirectional, one requiring energy and moving kinetochores along microtubules away from the poles (antipolar movement) and the other not requiring energy and moving kinetochores to the poles (polar movement). A dynein-like enzyme would be a good candidate for the motor causing antipolar movement, whereas Porter's microtrabecular lattice might be involved in poleward movement. Evidence using inhibitors suggests that a dynein-like enzyme acts during bidirectional translocation of pigment granules in erythrophores.[40] Experiments are needed to distinguish between the balance of forces model and the two-motor model of chromosome motion. Note that once chromosomes have attached to the spindle, both models predict that a balance of forces would result in metakinesis.

Because of their morphological prominence, microtubules have remained central in thought concerning mitosis for the past 20 years. Recent observations concerning microtubule polarity in cytoplasmic structures[54,55,63] suggest an important role for microtubule polarity in determining the polarity of kinetochore movements.[64] A functional role for microtubule polarity in kinetochore attachment to the spindle has not been established. Another possibility is that movement polarity is built into the chromosome so that sister kinetochores point in opposite directions. If the motors are associated with the kinetochores or with the fibrous coronae surrounding the kinetochores,[65] the inherent polarity of the chromosomes would favor an opposite pole

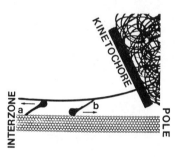

FIGURE 1. Dynein-like molecules in prometaphase. A diagram of proposed interactions between microtubules, dynein-like translocators, and fibers associated with the kinetochores.[65] For simplicity, a single fiber is shown associated with one microtubule. Translocators might bind permanently to the microtubule (a) or to the filament (b), with the globular ATP-sensitive binding site at the opposite end. The geometry of the binding is such that the power stroke of the dynein-like cross-bridge would move the kinetochore toward the interzone.

attachment for sister kinetochores. Syntelic orientation of sister kinetochores (sister kinetochores attached to the same pole) would be disfavored because of the stress that this orientation would place on the chromosome. Reorientation of syntelic chromosomes has been described.[66–68] Pratt[30] has proposed that microtubule-related movement of cytoplasmic vesicles might be the result of dynein-like molecules associated with the vesicles. An analogous model might be proposed for kinetochores.

Switchable Forces

As chromosomes attach to a spindle and as the spindle grows during mitosis, the direction and nature of the forces exerted on the spindle change. For example, as chromosomes attach to the prophase spindle they exert a compressional force on the interpolar microtubules of the spindle. This compressional force is present continuously from prometaphase until chromatid separation at the beginning of anaphase

A.[58,59,69,70] The forces that cause the spindle to elongate are present during prometaphase and anaphase B but do not appear during metaphase.[69] Forces that result in spindle elongation are discontinuous and thus must be switched on and off. The forces that cause antipolar movement of chromosomes during prometaphase do not act during anaphase A and so must also be switched on and off.

The most likely role of a cytoplasmic dynein in mitosis would be to produce these switchable forces. Switching mechanisms have also been proposed for the dynein cross-bridge activity in axonemes.[71] An axonemal bend in one direction occurs because dyneins on one cross-sectional half of the axoneme in that region are switched on, while those on the opposite side are switched off. A better understanding of the factors that control dynein activity in the axoneme might provide clues to the control of similar motors that may function in the mitotic apparatus.

We have constructed a conceptual framework around which the possibility of dynein-like cytoplasmic microtubule translocators might be considered; should we impose, however, such constraints on our search for motors in mitosis? The structural and functional differences between the axoneme, in which dynein is known to operate, and the mitotic apparatus, in which dynein is proposed to work, are significant. In the presence of such dissimilarities, strong homologies in molecular details would be surprising. It is possible that a molecule analogous to the dynein heavy chain could be modular and its association with the axoneme or with the mitotic apparatus mediated by other molecules specific to each microtubule framework. An example of a modular protein is calmodulin, the calcium-binding polypeptide that apparently associates with a variety of enzymes. A modular model of translocator function predicts differences in the proteins that associate with the heavy chain ATPase polypeptide. A comparison of purified axonemal dynein and dynein-like complexes from mitotic apparatuses would be useful in this regard.

CHARACTERIZATION OF A CYTOPLASMIC DYNEIN-LIKE MgATPase

A 20S Latent Activity MgATPase

We have recently isolated a 20S MgATPase from unfertilized sea urchin (*Strongylocentrotus purpuratus*) eggs.[72] Cytoplasmic tubulin was removed from the egg homogenate by absorption with diethylaminoethyl (DEAE) Sephacel, leaving a crude MgATPase preparation whose specific activity was typically 10 nmol phosphate released/mg protein/minute. The egg MgATPase was stimulated by the nonionic detergent Triton X-100, and activity was inhibited by ethylenediamine tetraacetic acid (EDTA). Magnesium or calcium ions were required for activity, with an apparent preference for Mg^{2+}.

Sedimentation of the DEAE-absorbed sea urchin egg supernatant through 5–20% sucrose density gradients revealed a broad peak of MgATPase activity, and the faster-sedimenting portion of the peak was stimulated by incubation with Triton X-100 (see FIGURE 2). Triton-stimulation was usually 6- to 9-fold, and the detergent-activated egg species was found to migrate slightly slower than axonemal latent activity dynein, at approximately 20S. Sucrose gradient-purified flagellar 21S dynein and egg 20S MgATPase were prepared to approximately the same specific activity, and both were shown to be sensitive to vanadate. The egg MgATPase, however, was considerably less sensitive to vanadate ion than was flagellar 21S dynein. TABLE 2 compares some of the properties of 21S flagellar dynein and 20S egg MgATPase.

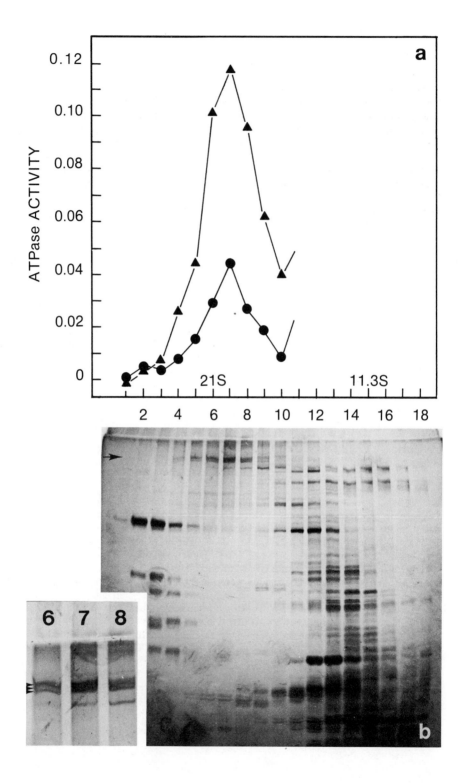

Polypeptide Composition of Egg 20S MgATPase

Polyacrylamide gel electrophoresis of sucrose gradient fractions from several different egg preparations consistently identified a group of high molecular weight polypeptides that sedimented with the 20S MgATPase activity (FIGURE 2). Close inspection of high resolution polyacrylamide gels revealed that there were three high molecular weight proteins—called polypeptides 1, 2, and 3—and that polypeptides 1 and 2 appeared to be very similar to spermatozoan axonemal $A\alpha$ and $A\beta$ heavy chains (FIGURE 3). In addition, the egg 20S activity peak often contained two intermediate-sized polypeptides: polypeptide 4 of M_r circa 130 K and polypeptide 5 of M_r circa 100 K. The intermediate polypeptides did not appear to coelectrophorese with the intermediate chains of 21S flagellar dynein. Resedimentation experiments were performed in which the major MgATPase activity again migrated as a 20S species and the five polypeptides cosedimented with the 20S activity.

The 20S Egg MgATPase Binds to Microtubules in Vitro

An important property of a microtubule translocator is that it must be able to bind microtubules *in vitro*. In addition, by analogy to axonemal dynein, we would expect that a dynein-like translocator binding to microtubules be reversed by $MgATP^{2-}$. Two different kinds of microtubule binding experiments were performed that demonstrate that the egg 20S MgATPase polypeptides are able to bind selectively to reassembled microtubules *in vitro*, and that the binding is sensitive to $MgATP^{2-}$.

A non-equilibrium microtubule affinity assay was designed to identify the components in a crude egg homogenate that were able to bind rapidly and with high affinity to the surfaces of reassembled microtubules. The nonequilibrium affinity assay was carried out by sedimenting reassembled sea urchin egg microtubules through discontinuous sucrose gradients that contained crude extracts of the egg MgATPase.[72] The egg microtubules were prepared by reassembly of cytoplasmic sea urchin egg tubulin (without taxol) and were completely free of any high molecular weight (HMW) associated proteins.[73] As we have shown elsewhere,[72] only two HMW polypeptides were isolated by the nonequilibrium affinity method, and the binding was completely reversed if the bottom sucrose layer contained 1 mM $MgATP^{2-}$. The two polypeptides corresponded to polypeptides 1 and 2 of the egg MgATPase. Binding to the reassembled egg microtubules was very striking because the two components identified represented a very minor fraction of the crude egg homogenate that was available for binding to the microtubules.

The second method, a preparative microtubule affinity assay, was designed to maximize binding and to provide enough material so that accurate MgATPase activity and protein determinations could be performed. Recently, a similar method has been

---

FIGURE 2. Electrophoretic analysis through a sucrose gradient of egg MgATPase. A DEAE-absorbed sea urchin egg homogenate was sedimented through a 5–20% sucrose gradient, and the MgATPase activity in each fraction was determined, in the absence of nonionic detergent (circles), and after a five-minute incubation with 0.2% (v/v) Triton X-100 (triangles). An ATPase activity measurement of 0.01 at 660 nm represents approximately 7.6 nmol phosphate released by 100 μl of the fraction/min. Equal volumes of each fraction were electrophoresed, and the gel was stained with silver. A cluster of HMW polypeptides was identified that sedimented with the 20S MgATPase (arrow and inset). The positions of 21S flagellar latent activity dynein and 11.3S catalase in matched gradients are indicated. The bottom of the gradient is to the left (fraction 1).

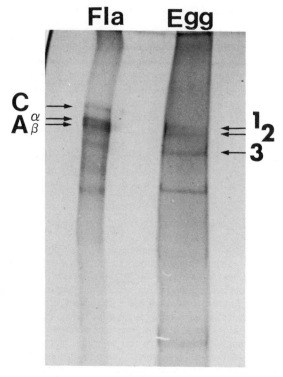

FIGURE 3. The HMW polypeptides of flagellar dynein and of egg 20S MgATPase. The heavy chains of sea urchin spermatozoan flagellar dynein (Fla) and the egg 20S MgATPase HMW polypeptides (Egg) are compared. Egg polypeptides 1 and 2 closely resemble the A_α and A_β bands of flagellar dynein. Both samples were run on the same gel (stained with silver).

termed alloaffinity chromatography.[74] The preparative affinity assay used high concentrations of taxol-stabilized bovine brain microtubules as the substrate for binding. Bovine brain microtubules, containing HMW microtubule-associated proteins (MAPs), were assembled with GTP at 37° and then stabilized with 20 μM taxol (from Dr. Matthew Suffness, National Cancer Institute). The microtubules were pretreated with 1 mM MgATP^{2-} and washed through 50% sucrose by centrifugation in order to deplete any microtubule components that might be removed by ATP. The taxol-stabilized microtubules were then incubated with various sea urchin egg cytoplasmic fractions for 30 min at 0°. Samples were collected by centrifugation through 50% sucrose, and the pellets were treated in either of two ways. Some pellets were resuspended in ATPase buffer and analyzed for MgATPase activity and protein concentration. Other pellets were resuspended in taxol-containing buffer supplemented with 1 mM MgATP^{2-} and recentrifuged. Supernatants were then analyzed by discontinuous polyacrylamide gel electrophoresis to identify the components that initially bound to the microtubules and that were subsequently eluted with ATP.

The preparative affinity method provided striking results. This one-step purification scheme increased the apparent specific MgATPase activity approximately 15-fold: for example, in one experiment, activity increased from 10.1 nmol phosphate released/mg protein/min in the crude egg homogenate to 153 nmol phosphate released/mg protein/min after correcting for the phosphate and protein contributions by the microtubules. In addition, polyacrylamide gel electrophoresis (FIGURE 4) indicated that proteins corresponding to polypeptides 1 and 2 of 20S egg MgATPase were able to bind to the microtubules and then were eluted with the $MgATP^{2-}$. Polypeptides 1 and 2 were identified if the starting material was the crude egg homogenate (FIGURE 4, lane b), or if the starting material was the sucrose gradient-

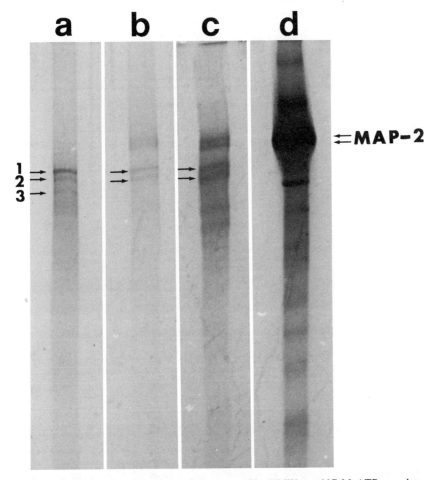

FIGURE 4. Preparative microtubule affinity assay. The HMW egg 20S MgATPase polypeptides 1, 2, and 3 (lane a) are compared to the ATP supernatants of a preparation in which a crude egg homogenate was the starting material (lane b) and a preparation in which sucrose gradient-purified 20S MgATPase was the starting material (lane c). In both cases, egg polypeptides 1 and 2 were identified as binding to the taxol-stabilized microtubules and subsequently being eluted with 1 m*M* $MgATP^{2-}$. Significant amounts of bovine MAP-2 were also removed from the microtubules by the ATP. Lane d is of the microtubule pellet remaining after ATP elution of sample b. All four samples were run on the same gel (stained with silver).

purified 20S MgATPase (FIGURE 4, lane c). Significant amounts of bovine MAP-2 were also eluted from the microtubules by the ATP treatment. The preparative affinity assay was also performed with taxol-stabilized microtubules prepared from MAP-depleted (by phosphocellulose chromatography) bovine brain tubulin with similar results.

The two different microtubule affinity experiments strongly suggest that the egg 20S MgATPase polypeptides 1 and 2 contain the MgATPase active site(s) and are directly involved in the ATP-sensitive binding to microtubules. The egg polypeptides 3, 4, and 5 were not identified in either of the microtubule affinity assays.

Egg 20S MgATPase and Flagellar Dynein Are Immunologically Related but Not Identical

Two different kinds of antibody probes have been used to assess the relatedness of the egg 20S enzyme to axonemal dynein. The first kind of probe, polyclonal antibodies made against flagellar dynein—including antifragment 1A[51] and antidynein 1[75]—demonstrated that the egg and sperm enzymes were immunologically related. Immunological cross-reaction has been shown by solid-phase binding assays, by competitive binding assays, and by immunoblot analyses. On the other hand, a group of several

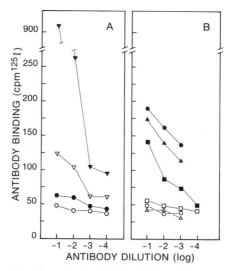

FIGURE 5. Solid-phase binding assay. The binding of a polyclonal antibody to flagellar dynein (antifragment 1A, in panel A) and three different monoclonal antibodies to egg cytoplasmic MgATPase (panel B) is expressed as the deposition of [^{125}I]protein A as a function of antibody dilution. Panel A: antiflagellar dynein binding to egg MgATPase (filled triangles) and to flagellar dynein (open triangles); antiflagellar dynein preimmune serum binding to egg MgATPase (filled hexagons) and to flagellar dynein (open hexagons). Panel B: three different monoclonal anticytoplasmic MgATPase antibodies binding to egg MgATPase (filled symbols) and to flagellar dynein (open symbols). The binding of antifragment 1A to the egg MgATPase indicates that the egg enzyme is antigenically related to flagellar dynein. The lack of binding by the monoclonal antiegg MgATPase antibodies, however, also indicates that the two enzymes are not identical.

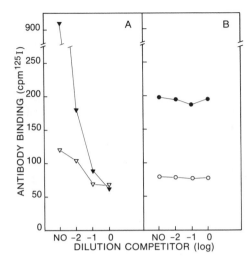

FIGURE 6. Competitive solid-phase binding assay. The binding of polyclonal antifragment 1A antibody (panel A) and monoclonal antiegg MgATPase antibody (panel B) is expressed as the deposition of [^{125}I]protein A as a function of increasing quantities of soluble, heterologous competitor. Panel A: antifragment 1A binding to egg MgATPase is competed by increasing amounts of flagellar dynein (filled triangles); antifragment 1A binding to flagellar dynein is competed by increasing amounts of cytoplasmic MgATpase (open triangles). Panel B: monoclonal antibody binding to egg MgATPase is unaffected by increasing amounts of flagellar dynein (filled circles); monoclonal antibody binding to flagellar dynein is negligible and is not affected by increasing amounts of cytoplasmic MgATPase.

monoclonal antibodies prepared against the egg MgATPase and reactive with the HMW polypeptides 1 and 2 of the 20S MgATPase,[50] although able to bind the egg MgATPase, failed to recognize any component in spermatozoan dynein preparations. FIGURES 5 and 6 summarize the immunological partial cross-reactivity between the two enzymes.

A Noncytoplasmic "Cytoplasmic Dynein?"

The term "cytoplasmic dynein" conveys two implications: first, the enzyme is implied to be a true dynein, fulfilling all of the criteria set forth above; and second, the enzyme is supposedly cytoplasmic, interacting with cytoplasmic microtubules to perform work. It is very important, then, to carefully study the fate of a cytoplasmic dynein-like molecule, and, in the case of the sea urchin egg, the added complication of the expression of blastula-stage embryonic cilia must be considered. Transcriptional and translational inhibitor studies have demonstrated that the sea urchin embryo possesses a precursor pool of ciliary proteins,[76] and this suggests the possibility that a cytoplasmic dynein-like enzyme such as we have described is a ciliary dynein precursor and is not involved in cytoplasmic movements.

The question of whether the sea urchin egg dynein-like enzyme is a ciliary precursor remains, and must await further studies. Nevertheless, this very real possibility underscores the fact that an enzyme is not fully understood until its *in vivo* function is elucidated. In short, an enzyme is not a "cytoplasmic dynein" until we know what it does in the cell. Restraint in our current nomenclature should avoid confusion in the future.

The Sea Urchin as a Model System

The developing sea urchin embryo displays a number of microtubule-mediated movements that might involve dynein-like mechanisms, including pronuclear migration, mitosis, and blastula-stage expression of motile cilia. A superficial indication might be that the various movements obscure the characterization of any particular translocator; for example, the presence of ciliary dynein precursors in the egg might interfere with identifying a mitotic dynein-like molecule. Closer inspection, however, suggests just the opposite: the sea urchin system is an excellent one for identifying and comparing several cytoplasmic microtubule translocators. The advantages are as follows: (1) large quantities of gametes and embryos can be obtained for biochemical

TABLE 2. Comparison of 21S Synein and 20S MgATPase

Flagellar 21S Dynein	Egg 20S MgATPase
1. Mg^{2+}/Ca^{2+} ATPase.	1. Mg^{2+}/Ca^{2+} ATPase.
2. Asymmetric: "A-end" structural and "B-end" ATP-sensitive.	2. "B-end" like binding; ATP-sensitive binding to reassembled microtubules.
3. Work: slides outer doublet microtubules.	3. Function *in vivo* unknown.
4. Latent activity dynein.	4. Latent; stimulated 6- to 9-fold by Triton X-100.
5. 21S latent activity dynein.	5. Sediments as a 20S latent activity MgATPase.
6. 9 proteins: 2 heavy molecular weight (HMW) A_α and A_β chains; 3 intermediate chains; 4 light chains.	6. 2 HMW proteins very similar to flagellar A_α and A_β; 1 other HMW protein; 2 intermediate-sized proteins; no light chains (?)
7. Vanadate inhibition: 50% inhibition at approximately 7 μM; 96% inhibition at 10 μM.	7. Not as sensitive to Vanadate as 21S dynein: 50% inhibition at approximately 20 μM; 89% inhibition at 50 μM.
8. Antibodies: anti-flagellar dynein antibodies react with both 21S flagellar dynein and 20S egg MgATPase.	8. Anti-egg MgATPase antibodies do not cross-react with flagellar dynein.

studies, and therefore large amounts of particular molecules are available for antigen; (2) large-scale experiments can be performed with synchronous populations of embryos, allowing for stage-specific isolations and characterizations of translocator molecules; (3) embryos can be manipulated; their large size allows eggs and early embryos to be easily microinjected with antibody and other inhibitors, and surface cilia can be repeatedly removed and regenerated; (4) axonemal dynein from sea urchin spermatozoan flagella is extremely well characterized, and this true dynein provides the biochemical criteria by which putative cytoplasmic dynein-like molecules can be examined (see *e.g.* references 12, 13). The fact that antibodies can distinguish among different dyneins and dynein-like molecules allows us to use the complex sea urchin embryo; an immunological approach will allow us to differentiate between a ciliary precursor and a cytoplasmic dynein-like molecule. Because that distinction is feasible, the sea urchin system becomes extremely informative. By exploiting the variety of microtubule movements in the sea urchin, we might be able to identify a whole family of dynein-like translocators whose members are all related in that they share certain common structures, but who are also each slightly modified for their specialized function.

SUMMARY

The eukaryotic flagellum presents an excellent predictive model of microtubule-mediated motility: movement is caused by microtubule translocators, called dyneins, which actively slide outer doublet microtubules against each other. Cytoplasmic movements, such as certain aspects of mitotic motion, may also be powered by dynein-like molecules. It is important, then, to carefully assess potential cytoplasmic dynein-like translocators by applying criteria defined by the properties of axonemal dynein. A cytoplasmic microtubule translocator may be only partially homologous to axonemal dynein; a modular construction may provide the translocator with domains that are shared with dynein, and with other domains that give it functional specificity. Finally, it is important to consider the possibility that a dynein-like enzyme that is found in the cytoplasm may not function in the cytoplasm but rather is awaiting incorporation into an axoneme.

ACKNOWLEDGMENTS

This paper is dedicated to the memory of Dr. Chris Bell, who was instrumental in biochemically defining the axonemal dynein particle, and whose thoughts about dynein will remain as the criteria for evaluating putative, nonaxonemal, dynein-like molecules.

REFERENCES

1. SATIR, P. 1968. J. Cell Biol. **39:** 77–94.
2. SUMMERS, K. E. & I. R. GIBBONS. 1971. Proc. Natl. Acad. Sci. USA **68:** 3092–3096.
3. BROKAW, C. J. 1972. Science **178:** 455–462.
4. GIBBONS, I. R. & A. J. ROWE. 1965. Science **149:** 424–426.
5. BROKAW, C. J. 1961. Exp. Cell Res. **22:** 151–162.
6. GIBBONS, B. H. 1982. *In* Methods in Cell Biology. L. Wilson, Ed. Vol. 25: 253–271.
7. GIBBONS, B. H. & I. R. GIBBONS. 1974. J. Cell Biol. **63:** 970–985.
8. WARNER, F. D. 1978. J. Cell Biol. **77:** R19–R26.
9. PORTER, M. E. & K. A. JOHNSON. 1983. J. Biol. Chem. **258:** 6582–6587.
10. JOHNSON, K. A. 1983. J. Biol. Chem. **258:** 13825–13832.
11. SALE, W. S. & P. SATIR. 1977. Proc. Natl. Acad. Sci. USA **74:** 2045–2049.
12. GIBBONS, I. R., E. FRONK, B. H. GIBBONS & K. OGAWA. 1976. *In* Cell Motility. R. Goldman, T. Pollard & J. Rosenbaum, Eds.: 915–932. Cold Spring Harbor Laboratory.
13. BELL, C. W. 1982. *In* Biological Functions of Microtubules and Related Structures. H. Sakai, H. Mohri & G. G. Borisy, Eds.: 137–150. Academic Press. New York.
14. GIBBONS, I. R. & E. FRONK. 1979. J. Biol. Chem. **254:** 187–196.
15. BROKAW, C. J. & B. BENEDICT. 1971. Arch. Biochem. Biophys. **142:** 91–100.
16. OMOTO, C. K. & K. A. JOHNSON. 1984. J. Cell Biol. **99:** 350a.
17. MOOS, M. 1973. Cold Spring Harbor Symp. Quant. Biol. **XXXVII:** 137–143.
18. GIBBONS, B. H. & I. R. GIBBONS. 1976. Biochem. Biophysics Res. Commun. **73:** 1–6.

19. GIBBONS, B. H. & I. R. GIBBONS. 1979. J. Biol. Chem. **254:** 197–201.
20. TANG, W.-J. Y., C. W. BELL, W. S. SALE & I. R. GIBBONS. 1982. J. Biol. Chem. **257:** 508–515.
21. BELL, C. W. & I. R. GIBBONS. 1982. J. Biol. Chem. **257:** 516–522.
22. BELL, C. W., E. FRONK & I. R. GIBBONS. 1979. J. Supramol. Struct. **11:** 311–317.
23. HAIMO, L. T. & R. D. FENTON. 1984. Cell Motility **4:** 371–385.
24. JOHNSON, K. A. & J. S. WALL. 1983. J. Cell Biol. **96:** 669–678.
25. WITMAN, G. B., K. A. JOHNSON, K. K. PFISTER & J. S. WALL. 1983. J. Submicrosc. Cytol. **15:** 193–197.
26. AMOS, L. A., R. W. LINCK & A. KLUG. 1976. *In* Cell Motility. R. Goldman, T. Pollard & J. Rosenbaum, Eds.: 847–867. Cold Spring Harbor Laboratory.
27. MAZIA, D., R. R. CHAFFEE & I. M. IVERSON. 1961. Proc. Natl. Acad Sci. USA **47:** 788–790.
28. MIKI, T. 1963. Exp. Cell Res. **29:** 92–101.
29. WEISENBERG, R. & E. W. TAYLOR. 1968. Exp. Cell Res. **53:** 372–384.
30. PRATT, M. M. 1980. Dev. Biol. **74:** 364–378.
31. PRATT, M. M., T. OTTER & E. D. SALMON. 1980. J. Cell Biol. **86:** 738–745.
32. HISANAGA, S. & H. SAKAI. 1980. Dev. Growth Differ. **22:** 373–384.
33. HISANAGA, S. & H. SAKAI. 1983. J. Biochem. **93:** 87–98.
34. HOLLENBECK, P. J., F. SUPRYNOWICZ & W. Z. CANDE. 1984. J. Cell Biol. **99:** 1251–1258.
35. SCHOLEY, J. M., B. NEIGHBORS, J. R. MCINTOSH & E. D. SALMON. 1984. J. Biol. Chem. **259:** 6516–6525.
36. CANDE, W. Z. 1982. Nature (London) **295:** 700–701.
37. CANDE, W. Z. & S. M. WOLNIAK. 1978. J. Cell Biol. **79:** 573–580.
38. FORMAN, D. S., K. J. BROWN & D. R. LIVENGOOD. 1983. J. Neurosci. **3:** 1279–1288.
39. FORMAN, D. S., K. J. BROWN & M. E. PROMERSBERGER. 1983. Brain Res. **136:** 194–197.
40. BECKERLE, M. C. & K. R. PORTER. 1982. Nature (London) **295:** 701–703.
41. SCHATTEN, G., R. BALCZON, C. CLINE & H. SCHATTEN. 1982. J. Cell Biol. **95:** 166a.
42. OKUNO, M. 1980. J. Cell Biol. **85:** 712–725.
43. SHIMIZU, T. & K. A. JOHNSON. 1983. J. Biol. Chem. **258:** 13833–13840.
44. SHIMIZU, T. & K. A. JOHNSON. 1983. J. Biol. Chem. **258:** 13841–13846.
45. GOODNO, C. C. 1979. Proc. Natl. Acad. Sci. USA **76:** 2620–2624.
46. HANSEN, O. 1979. Biochim. Biophys. Acta **568:** 265–269.
47. BOUCHARD, P., S. M. PENNINGROTH, A. CHEUNG, C. GAGNON & C. W. BARDIN. 1981. Proc. Natl. Acad. Sci. USA **78:** 1033–1036.
48. SCHLIWA, M., R. M. EZZELL & U. EUTENEUER. 1984. Proc. Natl. Acad Sci. USA **81:** 6044–6048.
49. PIPERNO, G. 1984. J. Cell Biol. **98:** 1842–1850.
50. ASAI, D. J. & L. WILSON. 1982. J. Cell Biol. **95:** 329a.
51. OGAWA, K. & H. MOHRI. 1975. J. Biol. Chem. **250:** 6476–6483.
52. IZUTSU, K., K. OWARIBE, S. HATANO, K. OGAWA, H. KOMADA & H. MOHRI. 1979. *In* Cell Motility: Molecules and Organization. S. Hatano, H. Ishikawa & H. Sato, Eds.: 621–638. University of Tokyo Press. Tokyo.
53. SAKAI, H., I. MABUCHI, S. SHIMODA, R. KURIYAMA, K. OGAWA & H. MOHRI. 1976. Dev. Growth Differ. **18:** 211–219.
54. TELZER, B. R. & L. T. HAIMO. 1981. J. Cell Biol. **89:** 373–378.
55. EUTENEUER, U. & J. R. MCINTOSH. 1981. J. Cell Biol. **89:** 338–345.
56. TIPPIT, D. H., J. D. PICKETT-HEAPS & R. J. LESLIE. 1980. J. Cell Biol. **86:** 402–416.
57. OSTERGREN, G. 1950. Hereditas **36:** 1–18.
58. HAYS, T. S., D. WISE & E. D. SALMON. 1982. J. Cell Biol. **93:** 374–382.
59. MCNEILL, P. A. & M. W. BERNS. 1981. J. Cell Biol. **88:** 543–553.
60. LUBY, K. J. & K. R. PORTER. 1980. Cell **21:** 13–23.
61. MCINTOSH, J. R. 1981. *In* International Cell Biology, 1980–1981. H. G. Schweiger, Ed.: 359–368. Springer-Verlag. New York.
62. PICKETT-HEAPS, J. D., D. H. TIPPIT & K. R. PORTER. 1982. Cell **29:** 729–744.
63. HEIDEMANN, S. R. & J. R. MCINTOSH. 1980. Nature (London) **286:** 517–519.
64. PICKETT-HEAPS, J. D. & T. D. SPURCK. 1982. Eur. J. Cell Biol. **28:** 83–91.

65. SCHIBLER, M. J. & J. D. PICKETT-HEAPS. 1980. Eur. J. Cell Biol. **22:** 687–698.
66. NICKLAS, B. R. & C. A. STAEHLY. 1967. Chromosoma **21:** 1–16.
67. NICKLAS, B. R. & C. A. KOCH. 1969. J. Cell Biol. **43:** 40–50.
68. JANICKE, M. A. & J. R. LAFOUNTAIN. 1984. J. Cell Biol. **98:** 859–869.
69. LESLIE, R. J. & J. D. PICKETT-HEAPS. 1983. J. Cell Biol. **96:** 548–561.
70. IZUTSU, K. 1961. Mie Med. J. **11:** 213–232.
71. BROKAW, C. J. 1982. Symp. Soc. Exp. Biol. **35:** 313–338.
72. ASAI, D. J. & L. WILSON. 1985. J. Biol. Chem. **260:** 699–702.
73. DETRICH, H. W., III & L. WILSON. 1983. Biochemistry **22:** 2453–2462.
74. NASR, A. & P. SATIR. 1985. Anal. Biochem. **151:** In press.
75. OGAWA, K., D. J. ASAI & C. J. BROKAW. 1977. J. Cell Biol. **73:** 182–192.
76. AUCLAIR, W. & B. W. SIEGEL. 1966. Science **154:** 913–915.

Distribution of MAP-4 in Cells and in Adult and Developing Mouse Tissues[a]

J.B. OLMSTED,[b] C.F. ASNES,[b] L.M. PARYSEK,[b]
H.D. LYON,[b] AND G.M. KIDDER[c]

[b]Department of Biology
University of Rochester
Rochester, New York 14627

[c]Department of Zoology
University of Western Ontario
London, Ontario N6A 5B7

INTRODUCTION

Over the last decade, a number of proteins have been identified that coassemble with microtubules *in vitro*. As outlined in this volume and elsewhere,[1–4] these include low molecular weight species from brain (tau), and high molecular weight proteins from brain ((microtubule-associated protein) MAP-1, MAP-2) and cultured cells (210 kD HeLa MAP, MAP-4). Antisera against all of these MAPs stain interphase arrays in cultured cells, and most also react with mitotic spindles. Although studies on the distribution of these proteins in tissues are being initiated, the function of any of these MAPs *in vivo* is still largely unknown.

We have been investigating the occurrence of MAP-4[3] in a variety of systems. This protein was originally identified as a 215 kD MAP in neuroblastoma cells, the synthesis of which appeared to be induced upon neurite differentiation.[2,5] The distribution of MAP-4 in extracts[3] and semi-thin sections[6] of mouse tissues has recently been described. This paper outlines the species distribution of MAP-4, further analyses on the complexity of this MAP, and the occurrence of MAP-4 during brain development and in early mouse embryos.

MATERIAL AND METHODS

The identification of MAP-4 in cells and tissues, the preparation of antisera to this MAP, and procedures used for immunoblotting and immunofluorescence have all been described previously.[3] Immunoelectron microscopy of microtubules formed *in vivo* was performed by fixing the samples with 1.0% glutaraldehyde in 0.1 M PIPES, pH 6.9, and placing the solutions on formvar and carbon-coated nickel grids. The grids were then inverted and treated at room temperature with the following solutions: phosphate buffered saline (PBS), 10 min; 1 mg/ml $NaBH_4$ in water, 15 min; 1% BSA in TBS, pH 7.6, 30 min; 1/20 MAP-4 or tubulin antibodies in TBS, pH 7.6 with 1% BSA, 1 hr; TBS, 1 mg/ml BSA, pH 8.3, 5 × 5 min rinses; goat anti-rabbit serum conjugated with 20 nm colloidal gold (1/20 in TBS, pH 8.3 with 1% BSA), 1 hr; TBS, pH 8.3, 5 × 5

[a]This work was supported by NIH Grant GM 22214 to J.B. Olmsted.

min; and 1% glutaraldehyde in TBS, pH 8.3, 5 minutes. Grids were then rinsed successively with 3–4 drops each of water, 1 mg/ml cytochrome c, water, and 1% aqueous uranyl acetate, and air dried.

Peptide mapping was carried out using a modification[7] of the method of Cleveland *et al.*[8] Samples containing MAP-4 were resolved in a 15 cm 4–6% linear gradient acrylamide gel. A small portion of the lane was reserved for staining and an adjacent strip placed horizontally across a second SDS gel made with a 1.5 cm stacking gel (3.1% acrylamide) and a 14 cm running gel (12–18% linear gradient of acrylamide). The gel strip was overlaid with 0.5 ml sample buffer lacking mercaptoethanol and containing 0.5 μg/ml *Staphylococcus aureus* V8 enzyme. Electrophoresis was carried out until the tracking dye reached the front of the stacking gel, and the current turned off for 30 min; electrophoresis was then continued until the tracking dye reached the bottom of the running gel. Gels were silver-stained for proteins and peptides.[9]

Mouse embryos on days 1–4 of gestation were collected from the reproductive tract of superovulated females (Ha/ICR strain mated with CB6/F$_1$J males) using previously published procedures.[10] Day 5 embryos were obtained by culturing day 4 embryos overnight in modified Eagle's medium supplemented with 10% serum.[11] Embryos were washed through five drops of PBS containing 0.3% polyvinylpyrrolidone, and then were quick-frozen in dry ice-methanol prior to storage at −70°C. Frozen embryos were boiled in 2× SDS sample buffer prior to electrophoresis.

RESULTS AND DISCUSSION

Distribution of MAP-4 on Microtubules in Vitro *and* in Vivo

MAP-4 was initially identified as a MAP in neuroblastoma cells by the partitioning of this protein with tubulin through multiple cycles of *in vitro* assembly.[2,5] As shown in FIGURE 1A, antibodies reacting with MAP-4 label the surface of microtubules formed *in vitro;* the distribution of label appears periodic along the microtubule length. Upon treatment of taxol-stabilized microtubules with 0.35 M NaCl,[12] the microtubules are no longer labeled with MAP-4 antibody (FIGURE 1B); this is consistent with blot analyses that indicate MAP-4 is removed from the microtubules by treatment with salt, as is also true for other MAPs.[12] The salt-treated microtubules, however, are still labeled with tubulin antisera (FIGURE 1C). These data demonstrate MAP-4 is associated with the surface of microtubules, and the interrupted pattern of labeling suggests that MAP-4 is regularly spaced on the microtubule lattice. Both thin sections and negatively stained preparations have demonstrated that MAP-1[13] and MAP-2[14,15] have regular spacings on brain microtubules formed *in vitro*. Immunoelectron microscopy of whole cells incubated with 210 kD antiserum and ferritin-conjugated second antibody also suggest this MAP is periodically distributed *in situ*.[16] The significance of these spacings, however, in the function of the MAP has not yet been elucidated.

In cultured cells, immunolocalization studies have shown that MAP-4 antibody labels interphase arrays in patterns that mimic those seen with tubulin antibody.[2,3] Additional MAP-4 positive structures, however, are observed when neuroblastoma cells are treated with concentrations of taxol (10 μM for 1 hr) sufficient to cause microtubule bundling. As shown in FIGURE 2, taxol-treated cells stained with tubulin antisera contain prominent arrays of microtubules. As opposed to the pattern in untreated cells, the neurite microtubules in taxol-treated cells often appear to extend to the extreme end of the neurite and double back (FIGURE 2B). These observations are consistent with the data indicating taxol promotes the polymerization and bundling of microtubules *in vivo.*[17,18] Neurite microtubule arrays are also strongly labeled in cells immunostained for MAP-4 (FIGURE 2A). These cells also show small perinuclear spots from which the MAP-4–positive fibers emanate (FIGURE 2C); these spots are not visible in untreated neuroblastoma cells. These data suggest that taxol may induce the aggregation of at least some of MAP-4 into discrete areas around the nucleus. Whereas it will be necessary to use higher resolution immunomicroscopy to identify the structures being stained, observations from the work of DeBrabander *et al.*[18] suggest a possible origin for this pattern. In electron microscopic analyses of taxol-treated cells, they found that microtubules converged on points at which no centrioles (centrosomes) existed. We postulate that in the presence of taxol, MAP-4 in neuroblastoma cells may

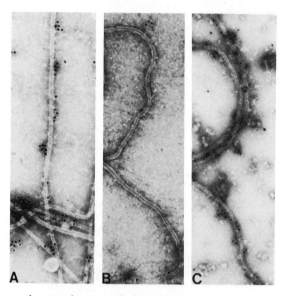

FIGURE 1. Immunoelectron microscopy of microtubules prepared by taxol treatment. Microtubules were polymerized from differentiated neuroblastoma cell extracts with taxol, pelleted through sucrose, washed twice with taxol-containing buffer, and stained with antibodies either before or after release of MAP by treatment with 0.35 M NaCl.[12] **A:** Staining with MAP-4 antiserum before 0.35 M NaCl treatment. **B:** Staining with MAP-4 antiserum after 0.35 M NaCl treatment. **C:** Staining with tubulin antiserum after 0.35 M NaCl treatment. Magnification 40,000×.

undergo rearrangement into discrete areas, perhaps in association with amorphous tubulin, and that this may be material around which the microtubules focus. The question remains as to whether this arrangement represents an aberrant assemblage, or corresponds to one that occurs naturally, but to a lesser degree, at other points in the cell. The existence of aggregations of MAP not associated with microtubules has been inferred from several other immunocytochemical studies. For example, it was found that brain high molecular weight MAPs were organized into discrete submembranous patches in developing neurons, and that this organization antedated the assembly and organization of microtubules, presumably at these points, during dendrite formation.[19,20] Recently, monoclonal antibodies raised to brain MAP-1 have been found to

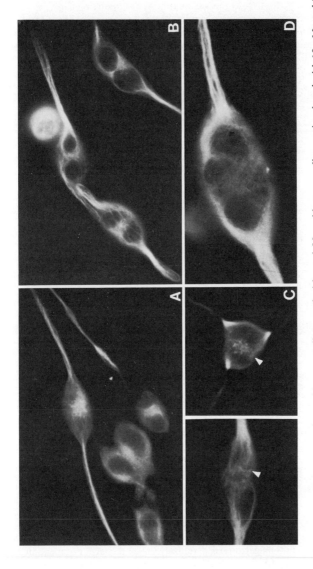

FIGURE 2. Indirect immunofluorescence of neuroblastoma cells treated with taxol. Neuroblastoma cells were incubated with 10 μM taxol for 1 hr before being fixed and stained with either MAP-4 (**A, C**) or tubulin (**B, D**) antisera. Arrows in FIGURE 2C indicate punctate foci stained with MAP-4 antibody. Magnification 400× (**A, B**); 500× (**C**); 800× (**D**).

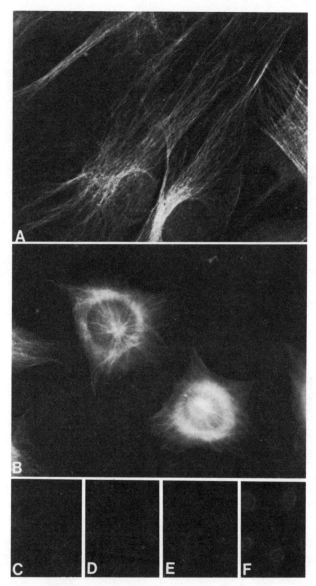

FIGURE 3. Indirect immunofluorescence of various cell lines labeled with MAP-4 antibody. Cell lines of murine (**A**: 3T3 cells; **B**: L cell) or nonmurine [**C**: HeLa (human); **D**: BHK (hamster kidney); **E**: C_6 (rat glioma); **F**: PtK$_1$ (marsupial kidney)] origin were incubated with MAP-4 antibody and fluorescein-conjugated second antibody.[3] Note the lack of staining in nonmurine cells. Magnification $1500\times$ (**A, B**); $380\times$ (**C–F**).

label kinetochores and centrosomes as well as spindle fibers.[21] Given these results, it is tempting to speculate that the location of some MAPs, including MAP-4, in regions other than the surface of formed microtubules, may be important for the organization of microtubules during various cellular events.

To examine whether MAP-4 might exist in cell lines other than those of mouse origin, immunological analyses of cultured cells of different species were undertaken. As shown in FIGURE 3, microtubule patterns visualized with MAP-4 antibody are evident in L and 3T3 cells, but are absent in PtK$_1$ (marsupial), BHK (hamster), C$_6$ (rat), or HeLa (human) cells. Similar results were obtained when microtubules formed

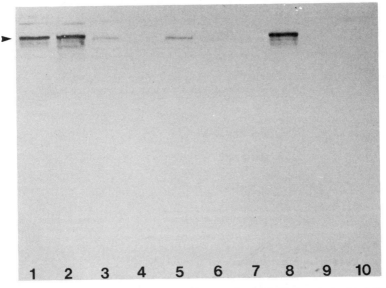

FIGURE 4. Immunoblot analysis of extracts from various cell lines probed with MAP-4 antibody. 1: mouse neuroblastoma (Nb2a-HL-1); 2: mouse fibroblast (3T3); 3: baby hamster kidney (BHK); 4: marsupial kidney (PtK$_1$); 5: hamster fibroblast (NIL); 6: embryonic *Drosophila* (SL2); 7: human carcinoma (ME180); 8: mouse epithelium (L); 9: human epitheloid carcinoma (HeLa); 10: rat glioma (C$_6$).

Cell extracts were prepared from a variety of cell lines and analyzed by immunoblotting with MAP-4 antibody. Note the reaction of MAP-4 with a triplet of proteins in lanes 1, 2, and 8, weak reaction with hamster lines (lanes 3, 5), and lack of reaction with other cell types. (Arrow: MAP-4).

in extracts from each of these lines were analyzed. As assessed by immunoelectron microscopy, microtubules derived from murine cell lines were decorated with MAP-4 antibody, whereas those of other species origin were not (unpublished results). The species-specific reactivity of the antibody was confirmed using immunoblot analysis of cell extracts of a large number of cell lines. As shown in FIGURE 4, mouse lines of a variety of tissue origins react with MAP-4; in each of these lines, the reaction of MAP-4 antibody with a complex of polypeptides between 220–240 kD is evident (see discussion below, and reference 3). There are some mouse lines, however, myeloma lines SP2 and P3, that appear to contain no MAP-4. The antiserum shows little or no reaction with cell lines derived from *Drosophila,* rat, hamster, marsupial, or human tissues. In addition, no reaction of MAP-4 antibody is seen with microtubules from rat liver, rat lung, or rat or hog brain (unpublished results).

Our data on MAP-4 antibody are complementary to the findings on the limited species reactivity of antibodies to a 210 kD MAP derived from HeLa cells. Polyclonal[22] antisera against this MAP have been demonstrated to react only with human or primate cell lines, although some monoclonal antisera to the 210 kD MAP also react with marsupial cells.[23] As discussed previously,[3] the 210 kD protein from HeLa shares properties of thermostability and complexity of polypeptides with MAP-4, and it seems likely that these proteins are species-specific homologs. A number of proteins of

TABLE 1. Distribution of MAP-4 in Tissues

Tissue	Cell Type Containing Microtubules	MAP-4
All tissues	blood vessel endothelia and adventitia, lymphatic endothelium	+
CNS	neuronal elements	−
	Astrocytes, oligodendrocytes, Bergmann glia, tanycytes, Müller cells	+
Testis	Sertoli cells	+
	spermatogonia, spermatocytes	−
	mature sperm	−
	spermatid manchettes	+
Liver	hepatocytes	−
	Kuppfer cells, endothelial cells	+
Small intestine	absorptive cells (columnar epithelia)	−
	connective tissue of villus core	+
Muscle	cardiac	+
	smooth	+
	skeletal	
	eye muscle fibers	+
	anterior thigh muscle fibers	−
	connective tissue around myofibrils	+
Kidney	absorptive cells of nephronic tubule	−
	connective tissue around tubules	+
	podocytes and glomerular epithelium	+
Bone marrow	developing blood cells	−
	endothelia, adventitial reticular cells	+

approximately 220 kD have been identified as MAPs by obtaining fractions from selectively solubilized cells that coassemble with exogenous tubulin.[24] MAPs of approximately 205 kD have also recently been identified in *Drosophila.*[25] Because there may be a number of MAPs with similar properties in this molecular mass range, it will be interesting to establish whether regions of the molecule fixed to the tubulin lattice are conserved, whereas other regions are variable from species to species.

Whereas our analysis of cultured cells demonstrated MAP-4 associated with microtubules *in vivo,* an investigation of the occurrence of this protein in mouse tissues was undertaken as an approach for assessing the possible functions of this MAP. Extracts and/or microtubule fractions prepared from brain, heart, liver, lung, and thymus all contained polypeptides that reacted with MAP-4 antiserum, whereas skeletal muscle from thigh, sperm, and peripheral blood contained no reactive species.[3] A subset of tissues, kidney, spleen, and stomach, contained a protein, band 4, that was immunologically related to MAP-4, but that did not bind to microtubules *in vitro.* To examine where the immunoreactive species were *in situ,* immunocytological analyses of semi-thin sections of mouse tissue were undertaken. As summarized in TABLE 1, and detailed elsewhere,[6,25a] MAP-4 is present in association with microtubules in a number of cell types. This MAP is highly restricted in distribution, however, being found primarily in supportive cell types within each tissue. For example, in the central nervous system, MAP-4 is found in glial cells. This distribution contrasts with that of other major brain MAPs. MAP-2,[26–29] tau,[30] and some types of MAP-1[29,31,32] seem localized to neurons alone, whereas MAP-1A[33] and MAP-3[32] are found in both glia and neurons. Interestingly, we have found that the distribution of MAP-4 in tissues does not always mimic that which would have been deduced from analysis of cell lines of various tissue origins. For example, MAP-4 was originally isolated from neuroblastoma cells. This cell line was derived from a spinal ganglion tumor[34] and possesses neuronal-like properties, including enzymes for neurotransmitter production.[35] MAP-4 is absent, however, in neurons *in vivo.* These data suggest that the synthesis of MAP-4 may be induced in cells that are transformed, or during adaptation of cells to continuous culture. Therefore, whereas the presence of MAP-1A,[36] MAP-1B,[37] and 210 kD HeLa MAP[22] has been demonstrated in a number of cell lines, whether these MAPs occur as widely in normal tissues remains to be examined.

MAP-4 Is a Complex of Related Polypeptides

Although the initial gel analyses on MAPs from neuroblastoma cells demonstrated MAP-4 (designated at 215 kD in reference 5) was a single polypeptide, immunoblotting[3] and silver staining (FIGURE 5) have shown that this MAP actually comprises three polypeptides. This complex migrates between 220 and 240 kD, and the components have been designated as MAP-4A, 4B, and 4C in order of increasing mobility on gels.[3] All three of the polypeptides have been shown to coassemble with microtubules *in vitro,*[2,3] and all are thermostable and antigenically related.[3] As shown in FIGURE 5, one-dimensional peptide maps also demonstrate similarities of the components of MAP-4. Whereas MAP-4A and 4C show extensive homology, MAP-4B shows some differences in smaller molecular weight peptides. A number of other MAPs have also been found to consist of more than one protein. Tau was initially identified as having four polypeptides,[38] and a monoclonal antibody to tau has recently been shown to react with all of these species.[30] MAP-1 has been resolved into three polypeptides; MAP-1C is more resistant to proteolysis,[33] and MAP-1B shows different cellular distribution[39] than MAP-1A[33] in adult brain. MAP-2 also appears to be composed of at least two distinct species,[20,39,40] although these are antigenically similar.[39,40] A 205 kD MAP from *Drosophila* cells also consists of four related polypeptides (reference 25 and L. Goldstein and J.R. McIntosh, personal communication). As discussed below, changes in the complexity of the MAP-4 polypeptides during development suggest that the different components may reflect alterations important in the function of these MAPs during tissue maturation.

MAP-4 Polypeptides Change during Development

To assess whether the complexity of MAP-4 might reflect heterogeneity that is important in tissue function, an analysis of this MAP at various stages of brain development was undertaken. As shown in FIGURE 6, the complexity of MAP-4 increases with age. In extracts from 13-day-old fetal brain, only MAP-4A is dominant, whereas in 17-day-old brain, MAP-4B is also present. At birth, all three bands of the triplet can be identified. Beyond the age of three weeks, MAP-4C diminishes. Because incubation of brain extract from 13-day-old or 17-day-old fetal mice for prolonged periods at 37°C did not cause the production of any bands of lesser molecular weight, it

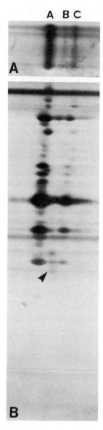

FIGURE 5. One-dimensional peptide mapping of MAP-4 polypeptides. Microtubules were prepared from neuroblastoma extracts by taxol-driven assembly, and boiled to prepare thermostable MAPs (see ref. 3 for procedure). The thermostable fraction was electrophoresed on a 4–6% gradient gel, and parallel strips either stained for protein (**A**), or digested and run in a second dimension before staining (**B**). Note the similarity of the patterns of MAP-4A and C peptides, and the different peptides seen in MAP-4B (arrow). The continuous streak at the top of the gel in FIGURE 5B is the *S. aureus* V8 protease used for digestion. The portion of the gel above the *S. aureus* V8 line, which was blank, is not shown.

appears that the changes in MAP-4 composition observed on blots are not the result of *in vitro* proteolysis. By contrast to the changes in MAP-4, the protein reacting with antibodies to MAP-1 (FIGURE 6, two dots) appeared constant over the same time period. Analyses of other MAPs have also shown changes in the proteins during brain

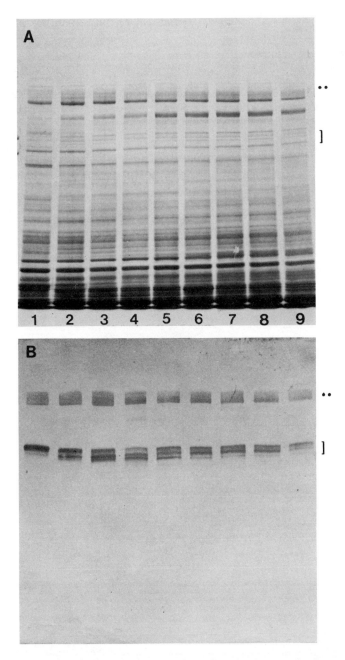

FIGURE 6. Complexity of MAP-4 during mouse brain development. 1: 13 day (prenatal); 2: 17 day (prenatal); 3: newborn; 4: one week; 5: 2 weeks; 6: 3 weeks; 7: 4 weeks; 8: 5 weeks; 9: 6 weeks (adult). Four to six percent gradient gels (**A**) of brain extracts taken from mice of various ages were blotted and probed with antisera containing both MAP-1 and MAP-4 antibodies (**B**). Note that MAP-4 increases in complexity with age, although after the age of three weeks, MAP-4C begins to diminish. MAP-1 appears constant throughout all stages. Two dots: MAP-1. Bracket: MAP-4.

development. For example, Francon et al.[41] found that tau proteins in young rat, mouse, and guinea pig brain were gradually replaced by other tau proteins characteristic of the adult state, but that little change was seen in the amount or peptide map of young and adult MAP-2. Other workers[20,40] have found, however, that MAP-2 appears first as a single band and then as a doublet. Binder et al.[40] have suggested that this change occurs because of a cAMP-dependent phosphorylation event that begins at day 15 of postnatal rat brain development. Whereas it does not appear that MAP-4 is phosphorylated in cultured cells (unpublished results), we do not yet know whether changes in the polypeptides of MAP-4 during development arise from specific posttranslational processing events, or from the expression of different genes for each of the components. Whether the appearance of MAP-4B and C reflect the maturation of specific glial functions in developing brain will be the subject of future studies.

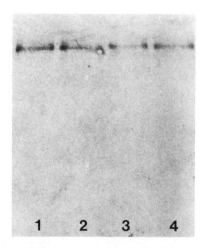

FIGURE 7. MAP-4 in early mouse embryos. 1: day 2 (90% 2 cells, 10% 3–4 cell; 411 embryos, approximately 11 μg protein); 2: day 3 (8 cells; 575 embryos, approximately 13 μg protein); 3: day 4 (early blastocyst; 254 embryos, approximately 6 μg protein); 4: day 5 (late blastocyst; 273 embryos, approximately 6 μg protein).

Embryo lysates were analyzed by immunoblotting with MAP-4 antibody. All stages examined contain a protein corresponding to MAP-4A; this appears to remain constant in amount per embryo.

To determine whether MAP-4 might be important in very early developmental events, mouse embryos were obtained between the two cell and late blastocyst stage (day 2 to day 5 of gestation) and analyzed by immunoblotting. As shown in FIGURE 7, MAP-4 antibodies react with a single band, corresponding in position to MAP-4A, in all of the embryonic stages examined. To a first approximation, this MAP appears to be present in constant amount per embryo, indicating that no net synthesis of this protein occurs over the first five days of development. Whereas the origin of this protein is still unknown, our previous analyses using either immunostaining[6] or immunoblotting[3] have shown no reaction of MAP-4 antibodies with mature sperm. Further, studies have failed to detect the synthesis of a MAP protein in the molecular weight range of MAP-4 from either blastocyst or early postimplantation embryos.[11] Because of the presence of MAP-4 in the two-cell stage embryos, it seems likely that

MAP-4 is either an oogenetic product, or is synthesized shortly after fertilization. The presence of MAP-4 in one-cell zygotes, also revealed by immunoblotting (results not shown) argues for the former possibility. Using immunolocalization, it should be possible to establish whether MAP-4 is present in developing and mature oocytes, and how this protein is distributed during early formation of germ layers. By combining these analyses with studies on the expression of MAP-4 gene(s), it may be possible to understand how the regulation of this protein is involved in determining the function of MAPs in tissues.

ACKNOWLEDGMENT

The taxol used in these studies was provided by the National Products branch of the National Cancer Institute.

REFERENCES

1. KIRSCHNER, M. W. 1978. Microtubule assembly and nucleation. Int. Rev. Cytol. **54:** 1–71.
2. OLMSTED, J. B., J. V. COX, C. F. ASNES, L. M. PARYSEK & H. D. LYON. 1984. Cellular regulation of microtubule organization. J. Cell Biol. **99:** 28s–32s.
3. PARYSEK, L. M., C. F. ASNES & J. B. OLMSTED. 1984. MAP 4: occurrence in mouse tissues. J. Cell Biol. **99:** 1309–1315.
4. VALLEE, R. B., G. S. BLOOM & W. E. THEURKAUF. 1984. Microtubule-associated proteins: molecular components of the cytomatrix. J. Cell Biol. **99:** 38s–44s.
5. OLMSTED, J. B. & H. D. LYON. 1981. A microtubule-associated protein (MAP) specific to differentiated neuroblastoma cells. J. Biol. Chem. **256:** 3507–3511.
6. PARYSEK, L. M., J. WOLOSEWICK & J. B. OLMSTED. 1984. MAP 4: a microtubule-associated protein specific for a subset of tissue microtubules. J. Cell Biol. **99:** 2287–2296.
7. BORDIER, C. & A. CRETTOL-JÄRVINEN. 1979. Peptide mapping of heterogeneous protein samples. J. Biol. Chem. **254:** 2565–2567.
8. CLEVELAND, D. W., S. G. FISCHER, M. W. KIRSCHNER & U. K. LAEMMLI. 1977. Peptide mapping by limited proteolysis in sodium dodecyl sulfate and analysis by gel electrophoresis. J. Biol. Chem. **252:** 1102–1106.
9. MORRISSEY, J. H. 1981. Silver stain for proteins in polyacrylamide gels: a modified procedure with enhanced uniform sensitivity. Anal. Biochem. **117:** 307–310.
10. MCLACHLIN, J. R., S. CAVENEY & G. M. KIDDER. 1983. Control of gap junction formation in early mouse embryos. Dev. Biol. **98:** 155–164.
11. BATES, W. R. & G. M. KIDDER. 1984. Synthesis of putative microtubule-associated proteins by mouse blastocysts during early outgrowth *in vitro*. Can. J. Biochem. Cell Biol. **62:** 885–893.
12. VALLEE, R. B. 1982. A taxol-dependent procedure for the isolation of microtubules and microtubule-associated proteins. J. Cell Biol. **92:** 435–442.
13. VALLEE, R. B. & S. B. DAVIS. 1983. Low molecular weight microtubule-associated proteins are light chains of microtubule-associated protein 1 (MAP 1). Proc. Natl. Acad. Sci. USA **80:** 1342–1346.
14. KIM, H., L. I. BINDER & J. L. ROSENBAUM. 1979. The periodic association of MAP 2 with brain microtubules *in vitro*. J. Cell Biol. **80:** 266–276.
15. VOTER, W. A. & H. P. ERICKSON. 1982. Electron microscopy of MAP 2 (microtubule-associated protein 2). J. Ultrastruc. Res. **80:** 374–382.
16. DEBRABANDER, M., J. C. BULINSKI, G. GEUENS, J. DEMEY & G. G. BORISY. 1981. Immunoelectronmicroscopic localization of 210,000 mol. wt. microtubule associated proteins in cultured cells of primates. J. Cell Biol. **91:** 438–445.
17. HORWITZ, S. B., J. PARNESS, P. B. SCHIFF & J. J. MANFREDI. 1981. Taxol: a new probe for

studying the structure and function of microtubules. Cold Spring Harbor Symp. Quant. Biol. **46**(1): 219–227.

18. DeBrabander, M., G. Geuens, R. Nuydens, R. Willebords & J. DeMey. 1981. Taxol induces the assembly of free microtubules in living cells and blocks the organizing capacity of the centrosomes and kinetochores. Proc. Natl. Acad. Sci. USA. **78**: 5608–5612.

19. Bernhardt, R. & A. Matus. 1982. Initial phase of dendrite growth: evidence for the involvement of high molecular weight microtubule-associated proteins (HMWP) prior to the appearance of tubulin. J. Cell Biol. **92**: 589–593.

20. Burgoyne, R. D. & R. Cumming. 1984. Ontogeny of microtubule-associated protein 2 in rat cerebellum: Differential expression of the doublet polypeptides. Neurosci. **11**: 157–167.

21. DeMey, J., F. Aerts, M. Moermans, G. Geuens, G. Daneels & M. DeBrabander. 1984. Anti-MAP 1 reacts with centrosomes, kinetochores, midbody and spindle of mitotic PtK$_2$ cells. J. Cell Biol. **99**: 447a.

22. Bulinski, J. C. & G. G. Borisy. 1980. Widespread distribution of 210,000 mol. wt. microtubule-associated protein in cells and tissues of primates. J. Cell Biol. **87**: 802–808.

23. Izant, J. G., J. A. Weatherbee & J. R. McIntosh. 1983. A microtubule associated protein antigen unique to mitotic spindle microtubules in PtK$_1$ cells. J. Cell Biol. **96**: 424–434.

24. Duerr, A., D. Pallas & F. Solomon. 1981. Molecular analysis of cytoplasmic microtubules *in situ:* Identification of both widespread and specific proteins. Cell **24**: 203–212.

25. Goldstein, L. S. B. & J. R. McIntosh. 1984. A microtubule-associated protein from *Drosophila:* Identification, production of specific antibodies, and isolation of gene(s). J. Cell Biol. **99**: 238a.

25a. Parysek, L. M., M. del Cerro & J. B. Olmsted. 1985. Microtubule-associated protein 4 antibody: a new marker for astroglia and oligodendroglia. Neurosci. **15**: 869–875.

26. Caceres, A., M. R. Payne, L. I. Binder & O. Steward. 1983. Immunocytochemical localization of actin and microtubule-associated protein MAP 2 in dendritic spines. Proc. Natl. Acad. Sci. USA. **80**: 1738–1742.

27. Caceres, A., L. I. Binder, M. R. Payne, P. Bender, L. Rebhun & O. Steward. 1984. Differential subcellular localization of tubulin and the microtubule associated protein MAP 2 in brain tissue as revealed by immunocytochemistry with monoclonal hybridoma antibodies. J. Neurosci. **4**: 394–410.

28. DeCamilli, P., P. E. Miller, F. Navone, W. E. Theurkauf & R. B. Vallee. 1984. Distribution of microtubule associated protein 2 in the nervous system of the rat studied by immunofluorescence. Neurosci. **11**: 819–846.

29. Wiche, G., E. Briones, H. Hirt, U. Artleib & H. Denk. 1983. Differential distribution of microtubule-associated proteins MAP-1 and MAP-2 in neurons of rat brain, and association of MAP-1 with microtubules of neuroblastoma cells (clone N$_2$A). EMBO J. **2**: 1915–1920.

30. Binder, L. I., A. Frankfurter & L. I. Rebhun. 1985. The distribution of tau in the mammalian central nervous system. J. Cell Biol. **101**: 1371–1378.

31. Huber, G. & A. Matus. 1984. Immunocytochemical localization of microtubule-associated protein 1 in rat cerebellum using monoclonal antibodies. J. Cell Biol. **98**: 777–781.

32. Matus, A., G. Huber & R. Bernhard. 1981. Neuronal microdifferentiation. Cold Spring Harbor Symp. Quant. Biol. **48**: 775–782.

33. Bloom, G. S., T. A. Schoenfeld & R. B. Vallee. 1984. Widespread distribution of the major polypeptide component of MAP 1 (microtubule-associated protein 1) in the nervous system. J. Cell Biol. **98**: 320–330.

34. Klebe, R. J. & F. H. Ruddle. 1969. Neuroblastoma: cell culture analysis of a differentiating stem cell system. J. Cell Biol. **43**: 69a.

35. Augusti-Tocco, G. & G. Sato. 1969. Establishment of functional clonal lines of neurons from mouse neuroblastoma. Proc. Natl. Acad. Sci. USA **64**: 311–315.

36. Bloom, G. S., F. C. Luca & R. B. Vallee. 1984. Widespread cellular distribution of MAP

1A (microtubule-associated protein 1A) in the spindle and on interphase microtubules. J. Cell Biol. **98:** 331–340.

37. LUCA, F. C., G. S. BLOOM, S. SCHIAVI & R. B. VALLEE. 1984. Variation in MAP 1B immunoreactivity among cultured cells and as a function of PC12 cell differentiation. J. Cell Biol. **99:** 189a.

38. CLEVELAND, D. W., S. Y. HWO & M. W. KIRSCHNER. 1977. Purification of tau, a microtubule associated protein that induces assembly of microtubules from purified tubulin. J. Mol. Biol. **116:** 207–226.

39. BLOOM, G. S., F. C. LUCA & R. B. VALLEE. 1984. Microtubule-associated protein 1B: Identification of a major component of the neuronal cytoskeleton. Proc. Natl. Acad. Sci. USA **82:** 5404–5408.

40. BINDER, L. I., A. FRANKFURTER, H. KIM, A. CAERCES, M. PAYNE & L. I. REBHUN. 1984. Heterogeneity of MAP2 during rat brain development. Proc. Natl. Acad. Sci. USA **81:** 5613–5617.

41. FRANCON, J., A. LENNON, A. FELLOUS, A. MYRECK, M. PIERRE & J. NUNEZ. 1982. Heterogeneity of microtubule-associated proteins and brain development. Eur. J. Biochem. **129:** 465–471.

Sliding of STOP Proteins on Microtubules: A Model System for Diffusion-Dependent Microtubule Motility[a]

ROBERT L. MARGOLIS,[b] DIDIER JOB,[c] MICHEL PABION,[c]
AND CHARLES T. RAUCH[b]

[b]The Fred Hutchinson Cancer Research Center
Seattle, Washington 98104
and
[c]Laboratoire des Regulations Endocrines
INSERM U244
Department de Recherche Fondamentale
Centre d'Etudes Nucléaires
38041 Grenoble, France

INTRODUCTION

Microtubules exhibit an impressive repertoire of activities within the living cell. They may function to create skeletal structure or an apparatus for motility, or perhaps both at once, as in flagella.[1] To create subcellular structure, microtubules often stabilize against dissassembly and cross-link with other microtubules or other skeletal elements. In their motility role, microtubules are often highly labile and transient structures. The molecular mechanisms that underlie the movement of materials on cytoplasmic microtubules have remained obscure, as are the mechanisms that would cause microtubules to align, stabilize, and cross-link into particular arrays.

We have been studying a microtubule associated protein, STOP (stable tubule only polypeptide), which has both microtubule stabilization and motility functions embodied within it. This unique protein stabilizes microtubules against disassembly when present in extremely low concentrations,[2] and also, interestingly, slides along the polymer to which it is bound and from which it cannot dissociate.[3] The STOP protein thus has a distinctive and highly intriguing duality to its nature. It stabilizes microtubules, and yet it moves on them. It is possible, as we will discuss, that both aspects of this duality may be exercised at once, yielding a subtle and efficient transport machine.

MATERIAL AND METHODS

Material

All reagents not otherwise identified were from Sigma Chemicals, Inc., and were the purest grades available. [γ^{32}P]adenosine triphosphate (ATP) and [^{3}H]guanosine triposphate (GTP) were obtained from New England Nuclear, podophyllotoxin, (PLN) was a gift from Dr. L. Wilson, diethylaminoethyl (DEAE)-cellulose (DE-52)

[a]This work was supported by Grants (GM 28189) from the NIH, and from the Ministere de la Recherche et de l'Industrie, as well as supporting funds from INSERM. M. Pabion is a fellow of the Association pour la Recherche sur le Cancer.

was from Whatman, and the calmodulin affinity column matrix was purchased from BioRad. The buffer used throughout (designated MME) was composed of 100 mM 2-(N-morpholino) ethane sulfonic acid (MES), 1.0 mM MgCl$_2$, 1.0 mM ethylene glycol-bis(β-amino ethyl ether)N,N,N',N'-tetraacetic acid (EGTA), 0.02% NaN$_3$, pH 6.75.

Purification of Microtubule Protein

Beef brain microtubules were purified by three assembly/disassembly cycles according to Margolis and Wilson[4] and Asnes and Wilson[5] except that all procedures were carried out in MME buffer. After the third cycle of polymerization, the preparation was cleaned by centrifugation of microtubules through 50% (wt/vol) sucrose in MME buffer (200,000 × g, 2.0 hr Beckman (Palo Alto, Calif.) 70.1 ti rotor, 25° C). Cold-stable microtubules were purified from the brains of adult rats (inbred strains W/FU and Sprague-Dawley) according to published procedures.[2,6] After a sucrose cushion sedimentation step, cold-stable microtubules were depolymerized by rapid and repeated passage at 0° C through a syringe fitted with a 25-gauge needle in MME buffer with an added 1.0 mM free calcium. After 10 minutes on ice, the depolymerized microtubule protein was centrifuged at 12,000 × g for 30 minutes at 4° C, and the supernatant was used for further separatory procedures.

Column Separatory Procedures

DEAE Cellulose Column

Routinely, the one-cycle purified microtubule protein from eight rat brains was depolymerized in 0.75 ml of MME buffer with 1 mM free Ca^{++} and centrifuged as indicative above. The supernatant protein was run into a 4.5 ml DE-52 column. The flow-through fraction was then eluted in a step-wise fashion with successive 0.5 ml Ca^{++}/MME buffer additions and collected in plastic tubes. In some experiments 0.1 M NaCl was included during protein preparation and column elution. Peak fractions were combined and used for stability assay after addition of 2 mM EGTA, or were used directly for further separatory procedures.

Calmodulin Affinity Column

The flow through eluate peak fractions from a DE-52 column were run into a 1.5 ml calmodulin column (BioRad) in MME buffer plus 1.0 mM free calcium by stepwise addition of 0.5 ml aliquots at 10 minute intervals (4° C). After binding, the column was washed with Ca^{++}/MME buffer, then eluted with 6.0 ml of MME buffer containing 0.3 M KCl and 1.0 mM Ca^{++}. After another Ca^{++}/MME buffer wash, the calmodulin-specific polypeptides were eluted in MME buffer containing 0.1 M KCl and collected in 0.5 ml aliquots. The peak of activity usually eluted at approximately 2.5 ml.

Filter Assay

The filter assay, to determine radioactive regions of microtubules, labeled with [^{3}H]GTP, was performed essentially as previously described.[3,7] Assay conditions, unless otherwise noted, were protein at 2.0 mg/ml; GTP, 50 μM; [^{3}H]GTP, 10 μCi/ml; and acetate kinase, 0.05 U/ml. Acetyl phosphate, when used to maintain constant GTP concentration, was 10 mM. Time points were taken for filter assay by removing 50 μl aliquots into 500 μl of a stop buffer. This buffer, composed of MME, 10% DMSO, 25% glycerol, and 50 μM PLN, was always maintained at 30° C. The buffer was designed to prevent further assembly or disassembly reactions of the polymer after taking a time point. Samples in STOP buffer were applied to GF/C filters under negative pressure and processed as previously reported.[3,7] All measurements were the average of three replicates, and are reported as net values after subtraction of blanks. The blanks, usually about 1000 cpm each, were generated by incubating microtubule protein with PLN from the time assembly was initiated, to prevent assembly, or were residual counts after disassembling microtubules (10 min at 7° C in MME buffer plus 2 mM CaCl$_2$). The blanks obtained by both methods are equivalent.

Turbidimetric Measurement

Assays of assembly, disassembly, and cold stable levels of polymer were performed with 1.0 ml samples in semimicro quartz cuvettes. Changes in optical density (which linearly covary with the microtubule assembly state[8] were followed at 350 nm and 30° C using a Varian Cary 219 recording spectrophotometer (Los Altos, Calif.) equipped with a constant temperature chamber and a cell programmer for assaying five samples successively. Assembly was initiated by the addition of 1.0 mM GTP and warming to 30° C. Spectrophotometric disassembly assays were performed as previously described.[9,10]

Other Procedures

Protein concentrations were determined by the method of Bradford.[11] SDS-polyacrylamide gel electrophoresis was performed using the procedures of Schier-Neiss et al.[12] All polyacrylamide gels were 8% acrylamide unless noted otherwise.

RESULTS

Purification of STOP

The first step in purifying the protein factor responsible for stabilizing microtubules was to effect the disassembly of the stable polymers and solubilize their constituent proteins for purification procedures. We determined that the disassembly of cold stable microtubules could be effected by use of Ca^{2+}-calmodulin at micromolar concentrations,[9] by the combination of millimolar calcium and 0° C[3,10,13] (this represents a synergistic behavior because neither calcium nor cold temperature is effective alone), or by mild shearing of the stable polymers.[2] The fact that shearing was so effective in disassembling the stable microtubules suggested that the stabilizing factor was acting in a highly substoichiometric manner and protecting the inner regions of polymers from end-wise disassembly.[2]

When the soluble protein, derived from cold-stable microtubules, was passed through a DEAE ion exchange column, the major microtubule proteins (tubulin, MAPs, and tau) were all retained, whereas a minor protein fraction with stabilizing activity could be isolated in the flow-through eluate.[2,14]

Further purification of the stabilizing protein took advantage of the apparent calmodulin binding activity of the stabilizing moiety. To demonstrate that the active protein interacted with calmodulin directly, we applied the DEAE column flow-through eluate to calmodulin affinity columns.

When the protein was run through such a column in the presence of calcium, the activity was retained in the column, whereas passage of protein in the absence of calcium caused no loss of activity in the flow-through eluate (FIGURE 1). When the calmodulin affinity column was sequentially run with calcium and then with excess EGTA, we could elute a specific protein fraction that had all the stabilizing activity. There was one predominant polypeptide, of 145 kD, which we designated STOP$_{145}$ (FIGURE 2).

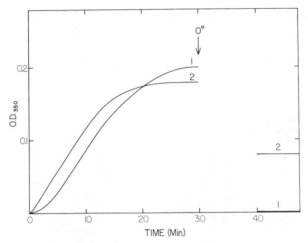

FIGURE 1. Conversion of cold-stable and cold-labile microtubule protein by passage through a Ca^{2+}-calmodulin affinity column. Purified cold-stable microtubules were depolymerized, centrifuged (40,000 g, 30 min, 4° C), and the supernatant divided in two. One fraction was passed through a 2 ml calmodulin affinity column in MME buffer plus 0.5 mM excess CaCl$_2$ (line 1), and the other in MME buffer alone (line 2). Both eluates were brought to identical buffer conditions by addition of Ca^{2+} or EGTA, equilibrated to 30° C in a spectrophotometer, and then assembled with 0.1 mM GTP. At the arrow, both samples were chilled for 10 min at 0° C in the presence of podophyllotoxin (25 μM). Line 2, representing the sample run through the column in the absence of Ca^{2+}, is 40% cold stable; line 1, the sample run in the presence of Ca^{2+}, has no cold stability.

There were also several minor polypeptides present that could possibly account for activity (FIGURE 2). To discount this possibility, we subsequently established that the STOP$_{145}$ accounted for all the activity by raising an antibody against STOP$_{145}$ and forming an antibody-linked affinity column. Passage of the active fraction from a DEAE column through this antibody column caused the retention of only the STOP$_{145}$ and the simultaneous retention of all the STOP protein activity (Margolis and Rauch, submitted for publication).

Response of the STOP Protein to Physiological Signals

From the calmodulin-binding properties of STOP, and the ability of Ca^{2+}-calmodulin to cause the rapid disassembly of cold stable microtubules, we have

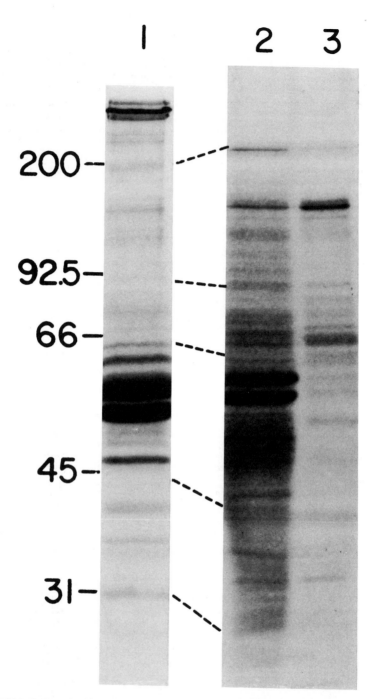

FIGURE 2. Polyacrylamide gel analysis of STOP protein at various stages of purification. Total cold-stable microtubule protein (lane 1) is compared with the DEAE column flow-through eluate fraction, containing STOP protein activity (lane 2) and with the further purified Ca^{2+}-calmodulin column specific fraction, containing the highly purified 145 kD STOP protein (lane 3). The SDS-polyacrylamide gels were 8% in acrylamide, and were stained with Coomassie blue R.

inferred that calmodulin may have a role in the physiological regulation of STOP protein activity. Calmodulin can act alone,[9] but it appears to have an additional role, when ATP is present, in the modulation of STOP protein activity.[15] It activates a protein kinase that causes phosphorylation of several specific proteins while microtubules are released from cold stability through an apparent phosphorylation reaction. This response requires the presence of ATP, Ca^{2+}, and calmodulin and is inhibited by trifluoperazine (FIGURE 3A). Note that stability is not released by ATP or Ca^{2+} in the absence of calmodulin (FIGURE 3A). The observed calmodulin-dependent activation is peculiar to this particular preparative stage. For this experiment, whole two-cycle purified cold stable microtubules from rat brain were assayed.

Curiously, the purest STOP protein fraction we have yet obtained, the specific eluate from a calmodulin affinity column, has a substantially different activity when used to reconstitute cold stable microtubules. Cold stable microtubules reconstituted with this fraction, exhibit a relatively slow lability to either millimolar ATP or to micromolar calcium (FIGURE 3B) in the absence of calmodulin. Both agents together give a substantial synergistic effect; they cause a very rapid stable microtubule disassembly (FIGURE 3B). It would seem that the synergistic response might be attributable to a calcium-dependent protein kinase, yet we find the presence of calcium causes no change in the phosphorylation pattern of the active protein fraction (FIGURE 3B). The reason for this calcium-dependent activation with ATP is unclear; a calcium dependent ATPase activity has not been ruled out. The reason for the change in activity with stage of purification also remains to be elucidated. Perhaps a STOP–associated protein that modifies the ATP dependent response has been removed by purification of the STOP activity.

Sliding of STOP Protein on Microtubules

From the above evidence, as elaborated elsewhere (Margolis and Rauch, submitted for publication), we conclude that $STOP_{145}$ protein is responsible for creating the large population of stable microtubules in the brain. It is remarkable for its ability to stabilize microtubules when present in extremely low concentration. Not only do the data require a substoichiometric blocker to endwise disassembly, but also this blocker must add to the polymer irreversibly. An equilibrium of STOP with the microtubule, accompanied by random STOP rebinding would effect no protection of microtubules to disassembly at low STOP protein concentrations.

We have found that STOP protein, added back to microtubules, binds irreversibly under our experimental conditions to the polymers.[3] Part of the evidence demonstrating a lack of equilibrium is shown in FIGURE 4. Cold stable microtubules were mixed at steady state with cold labile microtubules whose subunits had been labeled with [^{3}H]GTP. Cold-stable microtubules represented three-quarters of the total microtubule population. If equilibrium exchange of STOP occurred, we would expect to see a corresponding stabilization of the labile microtubules. In fact, we observed no stabilization (FIGURE 4). These experiments were analyzed by the specific retention of ^{3}H-GTP labeled microtubules on glass fiber filters. This procedure has been shown to yield a quantitative representation of the assembly state of a labeled subfraction of the microtubule population.[3,7] The reverse experiment was also done. Cold labile microtubules were mixed in large excess with labeled cold-stable microtubules, and we observed no loss of cold stability with time.

Having established that there was no equilibrium exchange of STOPs between microtubules, we asked whether there was any migration or sliding of the STOP protein from one position to another on its host polymer. The assay we used took advantage of the fact that STOP proteins bind randomly on the polymer and create the cold stability of any random portion of the polymer that may lie between two STOPs. It is thus protected from end-wise disassembly. The experimental design is shown in FIGURE 5A. Again we used the [³H]GTP and glass fiber filter assay.

If a cold-stable microtubule population is assembled in the presence of [³H]GTP, each microtubule will be labeled randomly through its entire length. Upon chilling to 7° C, only cold-stable regions remain (FIGURE 5A). Rewarming this microtubule

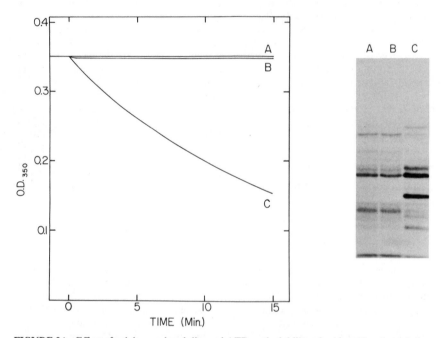

FIGURE 3A. Effect of calcium, calmodulin, and ATP on the lability of cold-stable microtubules. Response of purified (two cycles) cold-stable microtubules, assayed by turbidity change. To the sample protein (2.5 mg/ml) was added at time 0: 200 μM ATP alone (line A); 200 μM ATP, 9.0 μM free calcium, 0.5 μM calmodulin, 25 μM trifluoperazine (line B); or same as B but without trifluoperazine (line C). Podophyllotoxin (25 μM) was present, and the assay was performed at 30° C in MME buffer containing 10 μM free calcium. Final protein concentrations were 1.25 mg/ml microtubule protein. (Right) Gel autoradiograph showing the phosphorylation pattern of the stabilizing fraction, isolated as a DEAE column flow-through eluate after phosphorylation. Lanes: A: with ATP only; B: with ATP, calmodulin, and trifluoperazine; C: with ATP and calmodulin. Twenty micrograms of protein were loaded in each lane.

population in the presence of a 30-fold GTP chase will leave the cold-stable regions of the polymers uniquely labeled. One can then ask if the STOP protein displaces on the polymer relative to the labeled subunits by challenging these microtubules with cold temperature at time intervals. If the STOPs have displaced from the labeled subunits, they will become increasingly cold labile with time.

The result we have obtained clearly demonstrates sliding of STOPs on microtubules. Controls show that the microtubule assembled state does not change in the course of this experiment (FIGURE 5B), nor does the overall cold-stable level of the

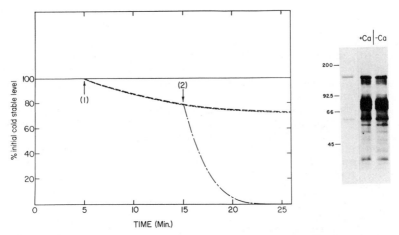

FIGURE 3B. Effect of calcium, calmodulin, and ATP on the lability of cold-stable microtubules. Response of cold-stable microtubules produced by admixture of cold-labile microtubule protein with the Ca^{2+}-calmodulin column specific $STOP_{145}$ protein. Cold-labile microtubule protein was assembled, mixed with the CaM column-specific protein, and assayed turbidometrically for disassembly response as detailed above. The control line (————) shows no disassembly with time. At the point indicated (1), separate microtubule samples were exposed to either 200 μM ATP (———) or to 10 μM free calcium (---------). Final protein concentrations were 1.25 mg/ml microtubule protein; 10 μg/ml of CaM column-specific protein was used. (Right) Gel autoradiograph showing the phosphorylation pattern of calmodulin column-specific proteins, phosphorylated as indicated in the presence or absence of Ca^{2+}. A Coomassie stained gel of the same protein is shown for comparison.

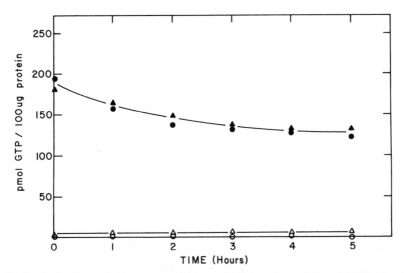

FIGURE 4. Assay of the exchange of STOPs between microtubules. Control: cold-labile (3X cycled) microtubules (2 mg/ml) were assembled for 50 min at 30° C in the standard conditions for filter assay, including 50 μM [^{3}H] GTP. At steady state (time 0), 10 μM podophyllotoxin and 1.0 mM GTP were added. Total (●——●) and cold stable (○——○) levels were filter assayed at the indicated time points (50 μl/ filter). Competition: at steady state, preassembled, cold-stable microtubules at 2 mg/ml were added, 3 volumes to 1, along with 10 μM podophyllotoxin. Final concentrations of all materials were the same as for controls. Aliquots (200 μl/ filter) were assayed at time points for total label (▲——▲) and for cold-stable levels (△——△).

microtubule population. For those microtubules labeled uniquely in their cold-stable regions, we find there is no loss of label with time when assayed at warm temperature. The only parameter that changes with time is the cold stability of the original cold stable region. The cold stability of this region drops (FIGURE 5B) presumably because the STOP protein is sliding on the polymer relative to the labeled subunits.

The reverse experiment was also conducted,[3] in which the cold labile regions of the polymers were uniquely labeled by a [³H]GTP pulse upon rewarming chilled cold-stable microtubules. We found that the originally labile regions became progressively stable with time, in a manner apparently reciprocal to the labilization observed in the reverse experiment. Because we have shown that STOPs do not jump from one polymer to another, the displacement of STOPs relative to subunits must be occurring within the context of a single polymer.

Sliding and Treadmilling in a Coordinate System

We have described here a protein that both stabilizes microtubules and exhibits motility behavior on them. Previously we described a motility system intrinsic to the equilibrium behavior of the microtubule: treadmilling.[4,16] Treadmilling involves the net flux of subunits from one end of the polymer to the other, through the favored addition of tubulin subunits to one polymer end. This molecular machine could, in theory, produce motility phenomena in the cell.[16,17]

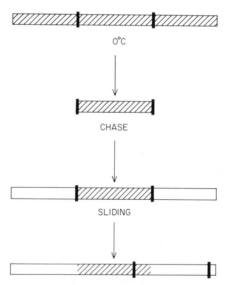

FIGURE 5A. Assay of sliding of STOP proteins on microtubules. Diagram of the experimental protocol and expected result, if sliding occurs. Cold-stable microtubules were assembled with [³H]GTP, disassembled at cold temperature to leave only residual cold stable regions, and reassembled in 20-fold excess GTP so that only cold-stable regions were labeled. If STOP proteins move relative to labeled subunits, the subunits that come to lie in regions of the polymer external to STOPs will become cold labile. The heavy vertical lines delimit the cold-stable region of the polymer; the diagonal lines indicate the position of the labeled subunits.

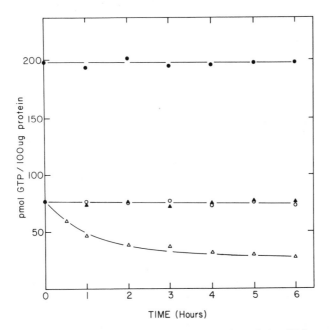

FIGURE 5B. Assay of the sliding of STOP proteins on microtubules. Sliding experiment. Cold-stable microtubules (2 mg/ml) were assembled in [³H]GTP under standard conditions for filter assay. After 50 min at 30° C, the protein was exposed to 7° C for 40 min and separated into two parts. The first sample, the control, was reassembled under the same conditions (the polymer remained fully labeled), and filter assayed at the indicated time points (time zero being the time of rewarming) for total label incorporation (●——●), and for cold stability (○——○) after cooling aliquots to 7° C for 20 minutes. The second part, the experimental, was reassambled at the indicated time 0 in the presence of 2 m*M* unlabeled GTP. At the indicated time points, the total label at 30° C (corresponding to the initial cold-stable part) was assayed (▲——▲), and the residual cold stability of the labeled region was assayed (△——△) after cooling aliquots to 7° C for 20 minutes.

We wished to ask how the binding of STOP proteins to microtubules influences microtubule treadmilling behavior. STOP protein, after all, prevents subunit loss due to rapid shocks such as application of calcium or cold temperature, and could be suspected of blocking subunit loss to treadmilling as well. On the other hand, STOPs slide on the polymer and apparently at approximately the rate of treadmilling (unpublished observations). So it is possible that sliding keeps pace with treadmilling, and treadmilling is unhindered.

Microtubule protein was assembled to steady state, then mixed either with a STOP protein active fraction or with a similar amount of buffer as a control. After 10 minutes, [³H]GTP was introduced as a pulse label to measure treadmilling. The result shows that the presence of STOP protein does not alter at all the rate of treadmilling (FIGURE 6). The STOP containing microtubule population was 67% cold stable, and label traversed 15% of the average polymer over the course of this experiment. Therefore, by the end of the pulse, approximately 50% of the polymers should have had STOP proteins at their net disassembly ends in the absence of sliding.

The observations that the STOP proteins are not lost from the polymer, that they

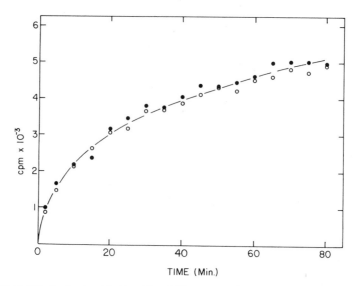

FIGURE 6. Effect of microtubule cold stability on treadmilling. Cold-labile (3X cycled) microtubule protein was assembled to steady state (30 min, 30° C) in MME buffer containing 0.1 mM GTP, 5 mM acetylphosphate, and 0.1 IU/ml acetate kinase. At steady state, the microtubule protein was split into two aliquots: one received an equal volume of prewarmed calmodulin column specific STOPs (85 μg); the other received an equal volume of prewarmed MME buffer. After ten minutes, each aliquot received 20 μCi of [^{3}H]GTP. The indicated time zero is the time of label addition. Time points were then taken for filter assay on glass fiber filters. Label incorporation represents treadmilling addition of tubulin subunits at steady state. At the time the pulse began, the sample containing the CaM-specific fraction was 67% cold stable. By the last time point, label incorporation represented approximately 15% of fully labeled controls (assembled in the presence of [^{3}H]GTP). Microtubule protein concentration was 1.05 mg/ml during the pulse. Solid circles, sample containing STOPs; open circles, sample without STOPs.

slide on the polymer, that sliding keeps pace with treadmilling, and that STOPs do not hinder treadmilling, all suggest that STOP sliding and treadmilling occur with complementarity. STOPs may therefore be floating on the polymer (perhaps along with other associated proteins) as if they were a moving sleeve that is never lost from the microtubule flowing beneath.

DISCUSSION

There are two obvious locales where cold-stable microtubules may play some role in cytoplasmic motility: the neuronal system from which our STOP protein has been isolated, and the mitotic apparatus of the typical dividing cell.

Cold stable microtubules are abundant in the mammalian neuronal system.[18,19] Further, they appear ubiquitous to the different cell regions and regions of the brain.[18] Cold-stable microtubules have also been localized uniquely to the kinetochore-to-pole fibers of the mitotic spindle,[20] though of course it remains to be seen if this stabilization is due to the same STOP protein.

The STOP protein has two distinct activities, stabilization and sliding. We will propose here that these two activities are possibly coordinate, and are both important to STOPs' intracellular function. Through these two activities and in conjunction with treadmilling, STOP protein may produce a subtle and economical machine for the generation of intracellular motility. Furthermore, the STOP protein is undoubtedly not unique in its sliding properties on microtubules. It should therefore serve as an interesting model system to describe other examples of microtubule–dependent motility.

A Basic Mechanism of Motility

Although the STOP protein is the first cytoskeletal protein to exhibit sliding behavior, it joins company with a number of such proteins associated with chromatin or with RNA (see references in Pabion *et al.*[3]). A model explaining the behavior of these sliding chromatin proteins has been proposed.[21] Their movement on the polymer is produced by their association with an isopotential "track" on the polymer. They may slide from position to position neither gaining nor losing a portion of their binding site, because it is dispersed indefinitely along the track. On the other hand, loss from the polymer is blocked by a considerable thermodynamic barrier.[21]

Assuming a free diffusion of STOP protein on the microtubule surface, but a substantial barrier to equilibrium loss, both aspects of which are in accord with our findings, then the sliding can be made unidirectional by the mechanics of the polymer treadmilling beneath the STOP. Suppose that a STOP protein is bound near the microtubule's net disassembly end. It can only slide in one direction, toward the net assembly end, and it cannot be lost from the net disassembly end due to a thermodynamic barrier. As it slides toward the net assembly end, it exposes tubulin subunits for disassembly (FIGURE 7A). If the sliding of STOP occurs more readily than the loss of subunits to treadmilling equilibria, then sliding and treadmilling will occur coordinately, as we have found (FIGURE 6).

A similar behavior of STOP protein at the assembly end is possible (for example, at

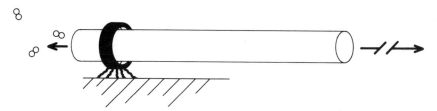

FIGURE 7A. Model of sliding-/treadmilling-dependent motility. Depiction of the general mechanism of sliding-/treadmilling-induced unidirectional movement of microtubules. The STOP protein (or similar sliding protein) is depicted as attached simultaneously near the net disassembly end of a microtubule and to a cellular structure. It is free to slide on the polymer in either direction through diffusional mechanisms. Its motion, however, is constrained to be largely toward the net assembly end due to the thermodynamic barrier to its loss from the net disassembly end, and due to the net loss of microtubule subunits exposed beyond the STOP protein at the net disassembly end of the polymer. The STOP protein is represented by a dark collar around the microtubule, and net subunit loss is shown by an arrow and double circles (representing dimeric subunits). The large broken arrow represents the inability of the microtubule to move in a net manner toward its net assembly end, due to the STOP protein that cannot be lost from the net disassembly end.

the centromere), always diffusing toward the subunits most recently added. If, as depicted in FIGURE 7A, the STOP protein is bound to a subcellular structure, then the combination of treadmilling and STOP sliding can coordinate to form a very effective unidirectional machine.

The Neuronal System

The long neuronal cell processes, axons and dendrites, contain abundant microtubules, which are responsible for maintenance of the cell morphology and synapse function through microtubule-dependent axonal transport.[22]

Axonal microtubules are interesting in many ways: they are abundantly cold stable,[18,19] they are uniformly oriented with their net assembly ends distad (pointed in the direction of their movement toward the synapse),[19,23] and they are almost all discontinuous, that is, both their net assembly and disassembly ends are free and exposed to the cytosol.[24-26]

Cold stability, as we have seen from our *in vitro* reconstitution system, does not infer a removal of microtubules from equilibria. Treadmilling continues unhindered,

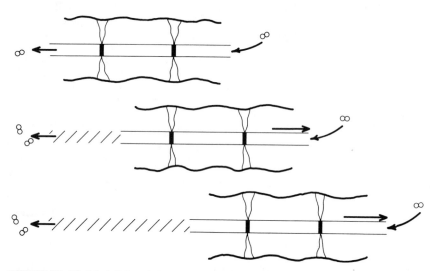

FIGURE 7B. Model of sliding-/treadmilling-dependent motility. The neural system. A sliding/ treadmilling mechanism is applied to model the net distal movement of microtubules and their attached neurofilaments in axons. A microtubule is shown advancing toward a synapse by net addition of subunits at one end, and net loss at the other, due to an intrinsic treadmilling process. At the same time, STOP and other sliding proteins are constrained to remain on the polymer and, therefore, advance forward as it treadmills. As a result, the microtubule advances in slow axonal transport by treadmilling and neurofilaments, attached to it by sliding proteins, advanced by a coordinate sliding mechanism. STOP or a similar linker sliding protein is represented by a dark collar on the microtubule. The thin wavy lines represent neurofilaments and their microtubule linkers. Arrows and double circles represent net entry and loss of dimeric tubulin subunits through treadmilling. Diagonal lines depict that portion of the polymer in the original frame that has been lost to treadmilling.

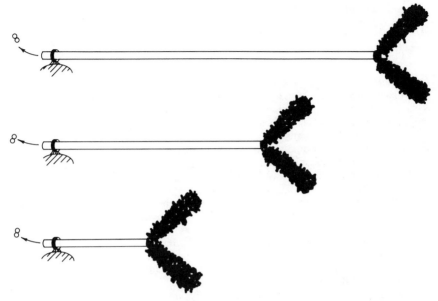

FIGURE 7C. Model of sliding-/treadmilling-dependent motility. The mitotic apparatus. The sliding/treadmilling principle is here applied to chromosome movement in anaphase. The kinetochore-to-pole microtubule is here depicted as linking a chromatid to an anchored sliding protein in the pericentriolar region. In anaphase, these cold-stable microtubules must remain attached at both the kinetochore and at the pole and must simultaneously shorten. The microtubule is shown here shortening due to the net loss of subunits (double circles) at the pole, whereas an anchored sliding protein continually migrates inward toward the centromere.

and STOP proteins are free to slide. Further, interaction with calmodulin or with ATP may quickly modulate the observed stabilization.

If the *in vitro* system is reflecting correctly the *in vivo* situation, then axonal microtubules, whether cold stable or not, would be free to migrate toward synapses by treadmilling reactions, because both their ends are free and treadmilling could therefore cause the net addition of subunits at the end pointed toward the synapse. Note that treadmilling would not be observed in this case by fluorescent photobleaching experiments, because each subunit of the polymer remains stationary: only the ends are moving.

Because STOP proteins slide on the treadmilling polymer in a manner apparently coordinate with treadmilling, they will be passively transported toward the synapse. Anything bound to the STOPs will be transported in like manner. It is probable that proteins other than STOP are capable of migrating on microtubules in a sliding manner.

Intermediate filaments of the neuronal type (neurofilaments) migrate toward synapses almost coordinately with microtubules, and their movement is dependent on the presence of microtubules and on their linkage to microtubules.[22]

We propose that the linker protein is sliding on microtubules. Thus, neurofilaments may migrate by sliding on microtubules that are independently migrating toward the synapse by treadmilling. This interaction and its consequences are depicted schematically in FIGURE 7B.

The Mitotic Apparatus

A coordinate system of treadmilling and sliding, as outlined above for the neuronal system, can also be envisaged as producing the work required for the poleward migration of chromatids in anaphase.

Chromatids are connected to the pericentriolar region through kinetochore-to-pole microtubules, which are uniquely cold stable.[20] In anaphase there are two requirements for the behavior of these microtubules. They must remain firmly bound to the pole and to the kinetochore, and they must shorten the distance between the two.

Sliding proteins, like STOP, could produce microtubules with the required characteristics. If a sliding protein were anchored to the polar region, its unidirectional sliding behavior (as described above) would cause subunits to be lost from microtubules while the polymers remained firmly anchored. The net result would be shortening of the kinetochore-to-pole distance and constant attachment of microtubules to both elements (FIGURE 7C).

A similar sliding protein at the kinetochore could allow for net assembly of microtubules at the end attached to the kinetochore. The difference in behavior would be attributable solely to the orientation of the polymers. Their net assembly ends are attached at the kinetochores, and their net disassembly ends are at the poles.[27]

The mechanism is in general accord with the data so far available on mitosis. We have proposed a model of mitotic behavior based in part on treadmilling of the microtubules,[17] and refined recently to include anchored lateral elements capable of binding to microtubules and sliding on them as they treadmill.[16] We predicted a unique orientation of microtubules in the spindle, which turned out to be correct.[27] We predicted that kinetochores would seed assembly of microtubules with their net assembly ends attached. This appears to be confirmed.[28]

Now we have isolated and characterized a protein that would have the requisite unidirectional sliding behavior on cold-stable microtubules. Preliminary evidence, obtained with antibody to $STOP_{145}$, suggests that a STOP protein is present uniquely on the cold-stable kinetochore-to-pole microtubules of cells in culture (Margolis and Rauch, in preparation). We note that calmodulin, which regulates STOP protein behavior *in vitro,* is also localized uniquely to the kinetochore-to-pole microtubules in the spindle.[29] This fact suggests a calcium-regulated sliding behavior in the spindle.

Our findings, and the mechanisms modeled on them, suggest an intracellular machinery that is energy efficient and subtle. We will be working to determine the correctness of our hypotheses based on a sliding/treadmilling system.

SUMMARY

STOP proteins, of 145 kD, act substoichiometrically to block end-wise disassembly of microtubules. STOPs bind to microtubules either during microtubule assembly or when added at steady state, and when binding to the polymers is apparently irreversible. They are not measurably lost from polymers under competition conditions, and there is no measurable exchange between polymers. Nonetheless, STOP proteins exhibit an extraordinary behavior: they "slide" laterally on the surface of the microtubule. Displacement is assayed by forming hybrid microtubules in which cold stable or cold labile region subunits are labeled. Displacement of STOPs on the polymer with time will cause labeled subunits of cold-stable regions to become increasingly cold labile in a manner reciprocal to cold stabilization of previously cold-labile subunits. Because equilibrium exchange of STOP proteins onto and off the polymers can be ruled out, the displacement of STOPs relative to subunits can only be explained by lateral diffusion or "sliding." Axonal transport and mitotic mechanisms were discussed as implications of such a lateral translocation mechanism for microtubule-dependent motility.

ACKNOWLEDGMENT

We thank Dr. Douglas Palmer for critically reading the manuscript.

REFERENCES

1. DUSTIN, P., 1973. Microtubules. Springer-Verlag. New York.
2. JOB, D., C. T. RAUCH, E. H. FISCHER & R. L. MARGOLIS. 1982. Biochemistry **21:** 509–15.
3. PABION, M., D. JOB & R. L. MARGOLIS. 1984. Biochemistry **23:** 6642–6648.
4. MARGOLIS, R. L. & L. WILSON. 1978. Cell **13:** 1–8.
5. ASNES, C. F. & L. WILSON. 1979. Anal. Biochem. **98:** 64–73.
6. MARGOLIS, R. L. & C. T. RAUCH. 1981. Biochemistry **20:** 4451–8.
7. WILSON, L., K. B. SNYDER, W. C. THOMPSON & R. L. MARGOLIS. 1982. Methods Cell Biol. **24:** 145–58.
8. GASKIN, F., C. R. CANTOR & M. L. SHELANSKI. 1975. Ann. N.Y. Acad. Sci. **253:** 133–46.
9. JOB, D., E. H. FISCHER & R. L. MARGOLIS. 1981. Proc. Natl. Acad. Sci. USA **78:** 4679–82.
10. PIROLLET, F., D. JOB, E. H. FISCHER & R. L. MARGOLIS. 1983. Proc. Natl. Acad. Sci. USA **80:** 1560–4.
11. BRADFORD, M. M., 1976. Anal. Biochem. **72:** 248–54.
12. SCHIER-NEISS, G., M. H. LAI & N. R. MORRIS. 1978. Cell **15:** 639–47.
13. JOB, D. & R. L. MARGOLIS. 1984. Biochemistry **23:** 3025–31.
14. WEBB, B. L. & L. WILSON. 1980. Biochemistry **19:** 1993–2001.
15. JOB, D., C. T. RAUCH, E. H. FISCHER & R. L. MARGOLIS. 1983. Proc. Natl. Acad. Sci. USA **80:** 3894–8.
16. MARGOLIS, R. L. & L. WILSON. 1981. Nature (London) **292:** 705–11.
17. MARGOLIS, R. L., L. WILSON & B. I. KIEFER. 1978. Nature (London) **272:** 450–2.
18. MORRIS, J. R. & R. J. LASEK. 1982. J. Cell Biol. **92:** 192–8.
19. HEIDEMANN, S. R., M. A. HAMBORG, S. J. THOMAS, B. SONG, S. LINDLEY & D. CHU. 1984. J. Cell Biol. **99:** 1289–95.
20. BRINKLEY, B. R. & J. CARTWRIGHT. 1975. Ann. N.Y. Acad. Sci. **253:** 428–39.
21. BERG, O. G., R. B. WINTER & P. H. VON HIPPEL. 1982. Trends Biochem. Sci. **7:** 52–5.
22. LASEK, R. J., 1982. Proc. R. Soc. London **B299:** 313–27.
23. HEIDEMANN, S. R., J. M. LANDERS & M. A. HAMBORG. 1981. J. Cell Biol. **91:** 661–5.
24. CHALFIE, M. & J. N. THOMSEN. 1979. J. Cell Biol. **82:** 278–89.
25. BRAY, D. & M. B. BUNGE. 1981. J. Neurocytol. **10:** 589–605.
26. TSUKITA, S. & H. ISHIKAWA. 1981. Biomed. Res. **2:** 424–37.
27. EUTENEUER, U. & J. R. MCINTOSH. 1981. J. Cell Biol. **89:** 338–45.
28. EUTENEUER, U., H. RIS & G. G. BORISY. 1983. J. Cell Biol. **97:** 202–8.
29. WELSH, M. J., J. R. DEDMAN, B. R. BRINKLEY & A. R. MEANS. 1978. Proc. Natl. Acad. Sci. USA **75:** 1867–71.

What Might MAPs Do? Results of an *in Situ* Analysis

FRANK SOLOMON

Department of Biology and Center for Cancer Research
Massachusetts Institute of Technology
Cambridge, Massachusetts 02129

The apparent multiple roles of microtubules in cellular physiology have been listed many times. We still do not know how these structures, with their highly conserved morphology and their highly conserved major component, can participate in so many different functions and be organized into so many different organelles. One hypothesis holds that the information that specifies the forms of this diversity resides not in the tubulin protein and its rather subtle variations, but instead in minor components, or microtubule associated proteins (MAPs). In this view, noncovalent interaction between tubulin in the microtubule lattice and other proteins can modulate the state of assembly and the interactions with other structures.

Our laboratory has been testing this hypothesis for the last six years. The premise that underlies our approach is that microtubule structure can best be analyzed by examining isolated cytoplasmic organelles, rather than *in vitro* reconstituted structures, and that that analysis can reveal small but significant differences between microtubules in different structural and functional contexts. In this article, I will describe this approach, including its strengths and weaknesses; the results obtained with it, to date, results that support the hypothesis that associated proteins can specify microtubule diversity; recent results that suggest a role for associated proteins in specifying microtubule form in great detail; and finally, prospects for further study of microtubule structure and function *in vivo*. This work has been done with Margaret Magendantz, Alan Salzman, Ann Duerr, David Pallas, Gary Zieve, Lorraine Pillus, Ellen Sutherland, Molly Miller, and Judith Swan.

AN *in SITU* ANALYSIS OF MICROTUBULES

To analyze microtubule structures, we adapted extant detergent extraction procedures[1-3] to conditions that preserve assembled microtubules.[4] At about the same time, Osborn and Weber[5] showed that such techniques could be used to enhance visualization of cytoplasmic microtubules by immunofluoresence, because they removed so much of the cytoplasmic protein including unassembled tubulin. We showed that these preparations contained tubulin, detectable on autoradiograms of one- or two-dimensional gels. That tubulin represented the assembled microtubules seen in the microscope, and only that material; all of the unassembled pool of tubulin was released. The pool of assembled microtubules could be manipulated either *in vivo* before extraction, with microtubule depolymerizing drugs, or *in vitro* after extraction, by changing the extraction buffer by addition of excess calcium ions. Those properties of the extraction procedure defined our assay for microtubule components. A microtubule component remains behind after detergent extraction if and only if there are intact microtubules, but not if the cells have been preincubated with depolymerizing drugs; they were quantitatively released from such preparations when the microtubules were depolymerized by subsequent extraction in the calcium-containing buffers. In practice, the microtubule components are detected by comparing the microtubule depolymerizing extract from the two populations of cells—drug treated and untreated. Both extracts do contain a number of background proteins, but the extract from untreated cells alone contains tubulin and the associated proteins. These species can now be detected by one- or two-dimensional gel analysis of the extracts, and they also coassemble to constant specific activity with carrier brain microtubule protein. So do other proteins, reenforcing again what others have claimed: that *in vitro* assembly is not a stringent criterion for association *in vivo*.[6]

Implicit in this assay is an operative definition for a MAP. We will detect only those proteins that are integrated into the microtubule structure in such a way that their association with the detergent extracted cytoskeleton is dependent upon microtubule integrity. We will miss two possible classes of microtubule components. First are those that can be removed from the microtubules without causing them to fall apart in the extraction buffer. These will be released in the initial extraction and not heard from again. A second class are those that can remain behind even after the microtubules are gone because they are firmly anchored elsewhere. These last are of considerable interest, because they may be the connectors between microtubules and other cytoskeletal elements. Later in this article, we will describe evidence that such proteins have been detected, and an assay that has been devised for them.

The advantages of this approach are, first and foremost, its fidelity to the *in vivo* situation. Within the strictly defined limits stated above, the assay detects MAPs without relying upon *in vitro* associations under irrelevant conditions. Second, the assay allows us to look at many sorts of cells, even those whose extracts do not support (or support only poorly) *in vitro* assembly. Third, different microtubule structures within the same cell type can be analyzed (for example, cytoplasmic arrays and mitotic arrays) and their components compared, because the assay begins with the assembled structures in the absence of unassembled components. Fourth, the assay separates the assembled pool and unassembled pools of microtubule components, so that we can compare them quantitatively and qualitatively. Fifth, these preparations have created a method for looking at MAPs attached to other structures. Each of these advantages has been exploited in our work.

The disadvantage of this sort of analysis is that the amounts of material in hand are small. For example, it is unlikely that these experiments could produce enough protein to permit studies of the physical properties of the associated proteins. Using recently developed techniques, however, we now know that they do produce enough to permit preparation, screening, and purification of antibodies,[7] and that those reagents can in turn, be used to clone the genes that code for the associated proteins. These advances will enable us to approach many of the problems that, by standard methodologies, would have required much more material.

SPECIFICITY OF MAPs

The ability to look at different sorts of microtubules under essentially identical conditions has enabled us to identify minor components that are common to all situations, as well as some that are unique. These proteins have all been described in published papers.[4,8–11]

All rodent cells examined have a 69 kD and a 220 kD protein. The latter is unlikely to be related to the protein of similar molecular weight identified from mouse neuroblastoma by Olmsted and colleagues,[12] because unlike their protein, it is present in both differentiated and undifferentiated cells. Both these proteins are present in both interphase and mitotic cell microtubules.

Two different proteins, of 120 kD and 200 kD, are found in primate cells. They are likely to be identical to the proteins of similar molecular weight identified by *in vitro* assembly by Bulinski and Borisy,[13] judged by cross-reaction with their antisera.

A protein of 150 kD is present in the microtubules of mitotic rodent cells, but not in the interphase arrays of those cells.

Cells of neural origin contain species of 72 kD and 80 kD not found in fibroblastic cells of the same species. A very similar set of proteins has been identified by Black and his colleagues from primary neurons, using this approach.[14] In fact, these molecular masses are only nominal, because each of these proteins is in fact a group of polypeptides of varying molecular weight and isoelectric point. Analysis of complete tryptic peptides demonstrates that these proteins and the 69 kD protein are homologous. In particular, all the methionine-containing tryptic peptides of the 69 kD protein are present in the 80 kD protein, as well as at least three others. The isoelectric variants are generated partly, but not solely, by phosphorylation at serine residues. We have called this family of proteins the chartins. Two aspects of the chartins are particularly intriguing. First, the correlation between additional sequence in the 80 kD chartin and their limited distribution to neural cells raises the possibility that the extra sequence may have functional significance. Second, as discussed in the next section, it has been possible to suggest a biological correlate for the covalent modification as well.

ANALYZING POOLS OF MICROTUBULE COMPONENTS

The original extraction in microtubule stabilization buffer releases unassembled tubulin and the unassembled pool of associated proteins as well. These last can be identified by position on two-dimensional gels, and that identification confirmed by peptide mapping. In the case of the chartins, analysis of the two pools, assembled and unassembled, reveals that the more basic, or even once phosphorylated variants, are distributed between assembled and unassembled states stoichiometrically with tubulin. The highly phosphorylated variants, however, are compartmentalized in the assembled pool and are not detectable in the unassembled pool.[11] This result could mean that the location of this protein is specified by the covalent modification. It could also mean that the heavily phosphorylated forms are extent limiting for microtubule assembly in the cytoplasm. In subsequent experiments,[15] we have been able to show that the phosphorylated variants increase in some, but not all, circumstances where microtubule assembly is increased. The correlation between phosphorylation and state of assembly, however, has remained intact. The possible roles of phosphorylation and the mechanism of the localization are now being studied.

ASSOCIATED PROTEINS AND THE FORMATION
OF MICROTUBULE ORGANELLES

We have recently turned to analysis of a novel microtubule organelle, the marginal band of avian erythrocytes. Others have shown long ago that these microtubules are confined to the periphery, and to one plane, of these lentiform cells, and that their number is characteristic of the species. The cells are easily obtained in large quantities, for example from chicken blood, each one displaying the identical structure. These features make the marginal band a particularly accessible one for the study of elements that specify microtubule form.

The microtubules of the marginal band can be detected by immunofluorescence or by standard electron microscopy. Detergent extracted preparations, made under the stabilization conditions described above, also contain microtubules that can be seen by whole mount electron microscopy. In addition, an assembled pool of tubulin can be detected in these cells, remaining after the extraction. When cells are incubated in the cold for 60 minutes, all the microtubules are disassembled: no profiles are seen by any of the microscopy techniques, and the pool of assembled tubulin is also no longer present. In short, these cells respond to cold the same way that other animal cells respond to microtubule-depolymerizing drugs (for some reason, they do not respond to the drugs). But this temperature-mediated depolymerization is completely reversible. Cells returned to the warm will reform their microtubules into a marginal band that is quantitatively and qualitatively indistinguishable from that in untreated controls.

Because this phenomenon has such clearly defined beginning and end points, it can be treated as a sort of *in vivo* assembly reaction.[16] In particular, because the reforming microtubules can be seen with such clarity in the whole mount preparations of the detergent extracted cells, and because their contour length can be measured accurately, it is possible to follow the kinetics of microtubule reformation and to look for intermediates. We simply take time points by extracting the cells within a few minutes after shift-up. When analogous experiments are done in other animal cells that contain radial microtubular arrays, the intermediates appear to contain several microtubules starting from (usually) a single point near the cell center.[17-19] But in chicken red blood cells, a quite different pattern emerges. At early times after cold-treated cells are shifted to 39°, single microtubules can be seen extending part or all of the way around the circumference of the cell. Only very rarely is more than one microtubule profile seen before the entire circumference of the cell is ringed. After one circumferential profile is formed, more follow in smooth fashion. These results are consistent with one or two growing ends within the cell, rather than the multiple growing ends visualized in other cells. The kinetic data are also consistent with this conclusion. The total microtubule length increases linearly with time after shift to warm temperature; there is no evidence for an increase in the number of available growing ends with time. Calculations of tubulin concentration in these cells, and values of the pseudo-first-order rate constant for microtubule assembly taken from the work of others, lead to the conclusion that between one and two growing ends are available per cell. Finally, no evidence, phenomenological, morphological, or immunological can be obtained for the presence of a microtubule organizing center in these cells. Therefore, some other cell component must be responsible for specifying the marginal bands' characteristics.

We have pursued this matter further,[20] using an experimental paradigm first developed by Brinkley and his colleagues.[21] They showed that 6S tubulin, added to cells that had been preincubated with drugs so that all the endogenous tubulin was gone, would repolymerize in an orderly fashion. That polymerization, like the *in vivo* regrowth experiments described above, appeared to proceed from single sites near the cell center, and in that sense shared an element in common with the normal cytoplasmic organelle. Other characteristics, however, are not displayed *in vitro:* the regrown microtubules are straight, rather than curvilinear, and they do not stop at the cell margin. We had repeated these experiments with neuroblastoma cells, and seen only radially symmetrical arrays of microtubules: there was no evidence of the asymmetry of these cells represented by their neurites.

We purified 6S tubulin from calf brain microtubules, and showed it to be contaminated with less than 0.5% nontubulin proteins, as assayed by silver staining of sodium dodecyl sulfate polyacrylamide gel electrophoresis (SDS-PAGE) gels. This material is not competent to assemble even at high concentrations in the microtubule stabilization buffer. When the protein is added back to detergent-extracted chicken red blood cells that have been preincubated in the cold, new microtubules are formed in about 90% of the cells. In almost all cases, the microtubules are confined to the periphery. They are not straight; they curve where the cell curves. In addition, the number of microtubules reformed is the same as were present in the original cell. Even doubling the concentration of tubulin, and increasing the length of incubation threefold, does not produce an increase in the number of microtubules.

We conclude that the elements that specify marginal band morphology are in the detergent-extracted preparations. They are not in the tubulin itself, even though there is evidence that chicken red blood cell beta tubulin is distinct,[22] because all these properties of the marginal band can be recapitulated using calf brain tubulin. And they are not in any organizing center, because no such center can be detected. Whatever the regulatory species are, they have eluded our standard definition for an associated protein. They have remained behind even though all the microtubules are gone. Above, we pointed out that this was one class of associated protein we might never see. Now we have seen it, at least in the sense that we have a functional assay for it. Our effort now is to identify it, or them, and to determine to what else they are connected.

PROSPECTS

Each of the results above has raised as many questions as have been solved. The mechanism of localization with phosphorylation, or the fate of spindle proteins once mitosis is complete, and several other issues need to be resolved. In broader terms, however, two major questions remain. First, to what extent can tubulin variations participate in specifying microtubule form and function? The existence of specific associated proteins in no way precludes the existence of specific tubulins for them to contact. Second, what are the precise roles of associated proteins *in vivo?* Data consistent with various functions have been gathered. More conclusive findings will probably depend upon the development of techniques to interfere with them in the cytoplasm. These are the questions we are now attacking, using techniques described above as well as other approaches.

REFERENCES

1. BROWN, S., W. LEVINSON & J. A. SPUDICH. 1976. Cytoskeletal elements of chick embryo fibroblasts revealed by detergent extraction. J. Supra. Mol. Struct. **5:** 119–130.
2. HYNES, R. O. & A. T. SESTREE. 1978. 10nm filaments in normal and transformed cells. Cell **13:** 151–163.
3. LENK, R., L. RANSOM, Y. KAUFMANN & S. PENMAN. 1977. A cytoskeletal structure with associated polyribosomes obtained from HeLa cells. Cell **10:** 67–78.
4. SOLOMON, F., M. MAGENDANTZ & A. SALZMAN. 1979. Identification with cellular microtubules of one of the co-assembling microtubule-associated proteins. Cell **18:** 431–438.
5. OSBORN, M. & K. WEBER. 1977. The display of microtubules in transformed cells. Cell **12:** 561–571.
6. LEE, J. C., N. TWEEDY & S. TIMASHEFF. 1978. *In vitro* reconstitution of calf brain microtubules: effects of macro-molecules. Biochemistry **17:** 2783–2790.

7. MAGANDANTZ, M. & F. SOLOMON. 1985. Analyzing the components of microtubules: Antibodies against chartins, associated proteins from cultured cells. Proc Natl. Acad. Sci. USA **82:** 6581–6585.

8. DUERR, A., D. PALLAS & F. SOLOMON. 1981. Molecular analysis of cytoplasmic microtubules *in situ:* identification of both widespread and specific species. Cell **24:** 203–212.

9. ZIEVE, G. & F. SOLOMON. 1982. Proteins specifically associated with the microtubules in the mammalian mitotic spindle. Cell **27:** 233–242.

10. ZIEVE, G. & F. SOLOMON. 1984. Direct isolation of neuronal microtubule skeletons. Mol. Cell. Biol. **4:** 371–374.

11. PALLAS, D. & F. SOLOMON. 1983. Cytoplasmic microtubule associated proteins: phosphorylation at novel sites is correlated with their incorporation into assembled microtubules. Cell **30:** 407–414.

12. OLMSTED, J. & H. D. LYON. 1981. A microtubule associated protein specific to differentiated neuroblastoma cells. J. Biol. Chem. **256:** 3507–3511.

13. BULINSKI, J. C. & G. G. BORISY. 1979. Self-assembly of microtubules in extracts of cultured HeLa cells and the identification of HeLa microtubule-associated proteins. Proc. Natl. Acad. Sci. USA **76:** 293–297.

14. BLACK, M. M. & J. T. KURDYLA. 1983. Microtubule-associated proteins of neurons. J. Cell. Biol. **97:** 1020–1028.

15. SUTHERLAND, E., J. A. SWAN, D. PALLAS & F. SOLOMON. 1985. Cellular regulation and phosphorylation of a microtubule component: an *in vivo* analysis. Submitted for publication.

16. MILLER, M. & F. SOLOMON. 1984. Kinetics and intermediates of marginal band reformation: evidence for peripheral determinants of microtubule organization. J. Cell. Biol. **99:** 70S–75S.

17. OSBORN, M. & K. WEBER. 1976. Cytoplasmic microtubules in tissue culture cells appear to grow from an organizing structure towards the plasma membrane. Proc. Natl. Acad. Sci. USA **73:** 857–861.

18. SPEIGELMAN, B. M., M. A. LOPATA & M. W. KIRSCHNER. 1979. Multiple sites for the initiation of microtubule assembly in mammalian cells. Cell **16:** 239–252.

19. SPEIGELMAN, B. M., M. A. LOPATA & M. W. KIRSCHNER. 1979. Aggregation of microtubule initiation sites preceding neurite outgrowth in mouse neuroblastoma cells. Cell **16:** 253–263.

20. SWAN, J. A. & F. SOLOMON. 1984. Reformation of the marginal band of avian erythrocytes *in vitro* using calf-brain tubulin: peripheral determinants of microtubule form. J. Cell Biol. **99:** 2108–2113.

21. BRINKLEY, B. R., S. M. DOX, D. A. PEPPER, L. WIBLE, S. L. BRENNER & R. L. PARDUE. 1981. Tubulin assembly sites and the organization of cytoplasmic microtubules in cultured mammalian cells. J. Cell Biol. **90:** 554–562.

22. MURPHY, D. B. & K. T. WALLIS. 1983. Brain and erythrocyte microtubules from chicken contain different beta tubulin polypeptides. J. Cell Biol. **258:** 7870–7875.

Isolation of Mitotic Microtubule-Associated Proteins from Sea Urchin Eggs[a]

GEORGE S. BLOOM,[b] FRANCIS C. LUCA,
CHRISTINE A. COLLINS, AND RICHARD B. VALLEE

Cell Biology Group
Worcester Foundation for Experimental Biology
Shrewsbury, Massachusetts 01545

INTRODUCTION

Sea urchin eggs have been one of the most important systems for the investigation of mitosis during the past thirty years. They have had particular value for studies involving the manipulation and isolation of the mitotic spindle. They also offer significant potential for the biochemical investigation of mitosis, because of the large amount of material that can be obtained from fertilized eggs highly synchronized in mitosis. Tubulin has been purified from the sea urchin egg, and its *in vitro* assembly properties have been characterized.[1-3] A "cytoplasmic dynein" has also been identified in the egg,[4-6] but its role in mitosis is uncertain.

In the interest of identifying additional components of the mitotic machinery, we sought to isolate microtubules containing both tubulin and microtubule-associated proteins (MAPs) directly from crude extracts of sea urchin eggs. To this end, we have adopted an approach that combines the use of taxol to purify microtubules and MAPs, and monoclonal antibodies for further analysis of the proteins.[7,8] Here, we summarize our use of these antibodies to examine purified microtubules, as well as those found in the mitotic spindle of the fertilized egg, and in cilia, flagella, and the interphase cytoplasm of differentiated sea urchin cells.

MATERIAL AND METHODS

Microtubules were prepared with the aid of taxol from unfertilized eggs of *Lytechinus variegatus, Strongylocentrotus purpuratus,* and *Arbacia puctulata* as described previously.[7] Heat stable MAPs were obtained from resuspended pellets of purified microtubules by adding DTT to 10 mM and NaCl to 1M, heating the solutions for 5 minutes at 90–100 degrees, and centrifuging at 30,000 g for 25 minutes at 4 degrees.[9] Those proteins remaining soluble were classified as heat stable. Sodium dodecyl sulfate (SDS) polyacrylamide gel electrophoresis was performed according to the method of Laemmli.[10]

Monoclonal antibodies to sea urchin MAPs have been described.[7,8] Immunofluo-

[a]This work was supported by NIH Grant GM 32977 and March of Dimes Grant 5-388 to Richard B. Vallee, and by the Mimi Aaron Greenburg Fund.
[b]Present address: Department of Cell Biology and Anatomy, University of Texas Health Science Center at Dallas, Dallas, Texas 75235.

rescence microscopy was performed on cells obtained from *L. variegatus* and *S. purpuratus*. Mitotic zygotes were stained as described earlier.[7] To stain coelomocytes, undiluted coelomic fluid was plated onto poly(L-lysine)-coated coverslips, and the cells were allowed to attach for one hour. The coverslips were then rinsed in artificial sea water, immersed in methanol at −20 degrees for 5 minutes, and processed for indirect immunofluorescence by the same protocol used for cultured mammalian cells.[11] Immunoblotting was performed as detailed previously, using peroxidase-labeled sheep anti-mouse IgG and 4-chloro-1-naphthol.[7]

Cilia were isolated from blastula stage embryos of *S. purpuratus* as described by Auclair and Siegel.[12] Flagella were isolated from *S. purpuratus* sperm. The cells were washed with artificial sea water, and the sperm heads were dislodged from the tails by gentle homogenization in a Dounce tissue grinder. A 5-second centrifugation step at 15,000 g in a microcentrifuge was used to pellet the heads and the few intact sperm that persisted. The free flagella remaining in the supernate were then collected by centrifugation at 5,000 g for 30 minutes.

RESULTS AND DISCUSSION

Identification of Sea Urchin Egg MAPs

The first step in the identification of mitotic spindle MAPs was to purify microtubules from unfertilized eggs, which was accomplished by a taxol-based procedure. We had shown earlier that the addition of taxol to cytosolic extracts of mammalian tissue or cultured cells induced the formation of microtubules, which could then be collected by centrifugation.[13] The microtubules were composed of tubulin plus several other proteins, the latter of which were identical in composition to the MAPs present in microtubule preparations obtained by temperature cycling procedures. We also found that the MAPs could be dissociated from taxol-stabilized microtubules by exposure to elevated salt concentrations, implying that ionic bonds were important in the interaction of MAPs with microtubules.

Application of the taxol technique to cytosolic extracts of sea urchin eggs also yielded microtubules containing a diversity of nontubulin proteins that represented presumptive MAPs. Microtubules did not form when taxol was omitted, probably because of microtubule assembly inhibitors present in the extracts.[14,15] We have successfully purified microtubules from five species of sea urchin representing three genera: *Lytechinus variegatus*, *L. pictus*, *Strongylocentrotus purpuratus*, *S. droebachiensis*, and *Arbacia punctulata*.

An example of the complete purification procedure using extracts obtained from *A. punctulata* eggs is shown in FIGURE 1. The first lane of this Coomassie blue-stained gel shows the unfractionated cytosolic extract. Taxol was added to a final concentration of 20 micromolar, inducing the spontaneous assembly of microtubules. Centrifugation at this point yielded a postmicrotubule supernate (lane 2) and the first microtubule pellet (lane 3). The major protein in the pellet was tubulin, but several putative MAPs were found as well. The most conspicuous of these was a protein of M_r 77,000. A number of high molecular weight (HMW) species ranging from about 200,000 to 350,000 were also observed, and several less abundant proteins were found throughout the gel. To remove cytosolic proteins that may have been entrapped nonspecifically in the first pellet, the microtubules were resuspended in taxol-containing buffer and centrifuged again. The few proteins released from the microtubules by this washing step remained in the supernate (lane 4), whereas the protein composition of the new pellet (lane 5) was nearly indistinguishable from that of the previous one (see lane 3). To determine whether the nontubulin proteins in the second pellet could be dissociated from the microtubules by high salt, as was found for mammalian MAPs,[13] the microtubule pellet was resuspended in taxol-containing buffer, NaCl was added to 0.4 M, and the preparation was centrifuged again. As can be seen in lane 6, nearly all of the

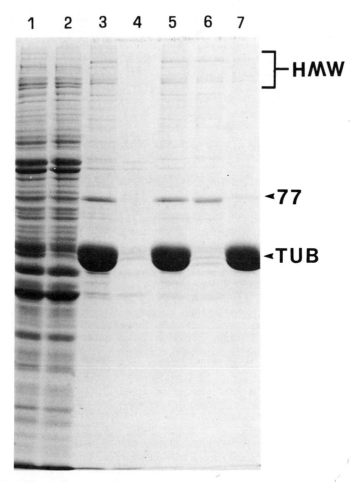

FIGURE 1. Electrophoretic analysis of microtubules purified from unfertilized *A. punctulata* eggs. Taxol was added to a cytosolic extract (lane 1) to stimulate microtubule assembly. Centrifugation yielded a postmicrotubule supernate (lane 2) and the first microtubule pellet (lane 3). Microtubules were resuspended in taxol-containing buffer and centrifuged again, producing a supernate (lane 4) containing proteins nonspecifically entrapped in the first pellet and a second washed pellet of microtubules (lane 5). After the pellet was resuspended in taxol-containing buffer, NaCl was added to 0.4 M, and the microtubules were centrifuged once more. MAPs remained in the supernate (lane 6), and the final microtubule pellet was composed of nearly pure tubulin (lane 7). The positions of tubulin (TUB), the 77,000 dalton (77), and high molecular weight (HMW) MAPs are indicated.

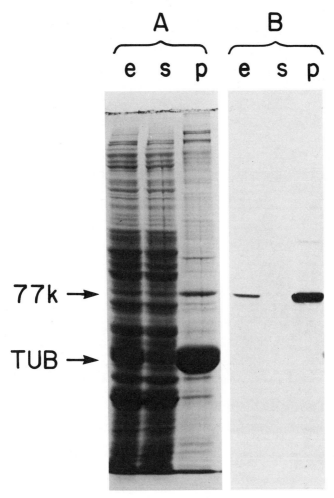

FIGURE 2. Immunoblot analysis of the first three steps in the purification of microtubules from unfertilized eggs of *S. purpuratus*. (A) Coomassie blue-stained gel. (B) Corresponding immunoblot stained with the monoclonal antibody to the 77,000 dalton MAP. The lanes are the cytosolic extract (e), postmicrotubule supernate (s), and first microtubule pellet (p). The positions of tubulin (TUB) and the 77,000 dalton MAP (77k) are shown.

presumptive MAPs, and very little of the tubulin were solubilized by salt. The final microtubule pellet (lane 7), by contrast, was composed of nearly pure tubulin.

Similar results have been obtained with additional species of sea urchins we[7] and others[16] have examined, suggesting that sea urchin eggs contain a variety of MAPs representing a broad spectrum of molecular weights. The M_r 77,000 protein, which is probably identical to an 80,000 dalton mitotic spindle protein described by Keller and Rebhun,[17] is the major nontubulin protein in microtubules purified from all species

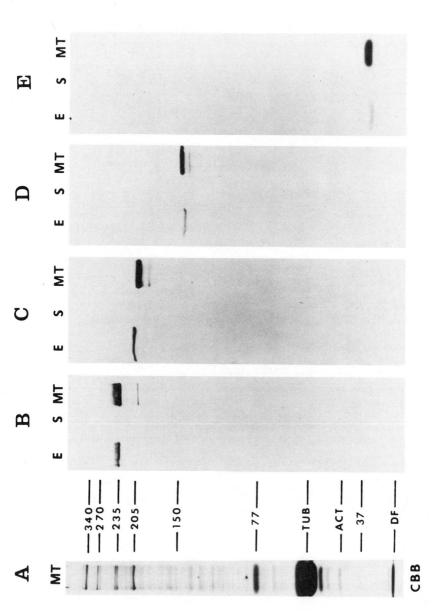

FIGURE 3. Immunoblot analysis of *L. variegatus* microtubules with monoclonal antibodies. A: Coomassie blue-stained gel of the first microtubule pellet isolated with taxol from unfertilized eggs. B–E: Nitrocellulose immunoblots of the cytosolic extract (lanes E), postmicrotubule supernate (lanes S), and first microtubule pellet (lanes MT). The immunoblots were stained with monoclonal antibodies to MAPs of 235,000 (B), 205,000 (C), 150,000 (D), and 37,000 (E) daltons. The size in kilodaltons of several MAPs is shown to the right of panel A, as are the locations of tubulin (TUB) and actin (ACT). The arrows to the left of panel A indicate the positions of the four immunoreactive MAPs. From Vallee and Bloom.[7]

examined. Whereas HMW proteins have also been observed consistently, the precise composition of these appears to be species-specific.

Electron microscopy (not shown) revealed that purified microtubules containing nontubulin proteins are covered along their lengths with small projections. Longer, less numerous projections that seemed to bridge adjacent microtubules were also observed.[7] Both classes of surface decoration are presumably formed by the nontubulin proteins. These projections are similar in appearance to microtubule cross-bridges observed in the mitotic spindle, where they may be involved in regulating the architecture and motile machinery of the mitotic apparatus.[18-23]

To further our investigation of the nontubulin proteins we prepared monoclonal antibodies using two types of immunogens: the proteins solubilized from *L. variegatus* microtubules by high salt (as in FIGURE 1, lane 6) and the 77,000 dalton band purified by preparative gel electrophoresis from isolated *S. purpuratus* egg microtubules.[7,8] Hybridomas were screened for the production of specific antibodies by immunofluorescence microscopy of fertilized mitotic eggs, and by immunoblotting of purified sea urchin egg microtubules. Several independent clones secreting antibodies that stain the spindle and react with nontubulin proteins have been obtained in this manner.

Five distinct proteins (37, 77, 150, 205, and 235 kilodaltons) were recognized by these clones. Consistent with their identification as MAPs, each of these proteins was shown by immunoblotting to copurify with microtubules out of crude cytosolic extracts (see FIGURES 2 and 3). These proteins were undetectable in postmicrotubule supernates and were highly enriched in the first microtubule pellets. Hence, virtually the total pool of each of these proteins in the egg cytosol was able to coassemble with egg tubulin into microtubules. On Coomassie blue-stained gels of purified microtubules, three of these proteins, the M_r 77,000, 205,000, and 235,000 species, appeared as major bands, whereas the 37,000 and 150,000 dalton proteins were barely detectable. These results suggest that many, and perhaps most of the nontubulin bands in the taxol-purified microtubules, regardless of their level of abundance, are likely to be true sea urchin egg MAPs.

All of the monoclonal antibodies labeled the spindle throughout mitosis, and none of the immunoreactive proteins appeared to be restricted to specific regions within the spindle. FIGURE 4 shows examples of fertilized eggs stained with two of the antibodies. The distributions of the 77,000 dalton MAP at the onset of anaphase and the 235,000 dalton MAP at telophase are illustrated in panels **a** and **b**, respectively.

The five immunoreactive proteins therefore behave biochemically as would be expected for MAPs and are specifically associated with microtubules *in vivo*. On the basis of these results it seems clear that the five proteins are, indeed, bona fide mitotic spindle MAPs in the sea urchin. That they represent both major and minor proteins present in purified egg microtubules indicates that the sea urchin mitotic spindle contains an extremely complex set of both abundant and scarce MAPs that presumably regulate the activities of microtubules during mitosis.

Heat-Stable MAPs in Sea Urchin Eggs

One unusual property of some mammalian MAPs is their solubility following brief (~5 minutes) exposure to temperatures of 90–100 degrees. The heat-stable mammalian MAPs described to date include MAP-2,[9,24] the tau proteins,[25] and several apparently related proteins ranging from 200,000–240,000 daltons.[26,27] We have found that at least two sea urchin egg MAPs are also heat stable. When microtubules purified from *L. variegatus* eggs were boiled, immunoblotting indicated that all of the

M_r 150,000 MAP and much of the M_r 235,000 species remained soluble (FIGURE 5). The other MAPs identified with monoclonal antibodies and by Coomassie blue staining were precipitated by this treatment (not shown). Whether these findings are indicative of functional homologies between thermally stable MAPs from sea urchins and mammals remains to be determined.

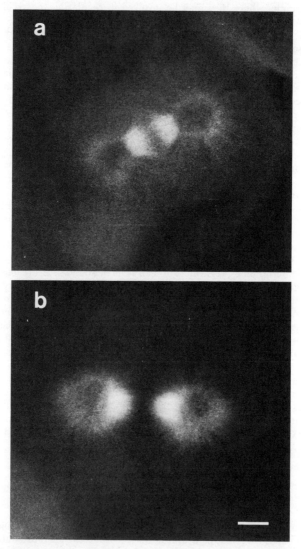

FIGURE 4. Localization of MAPs in mitotic fertilized eggs. The distributions of the 77,000 dalton MAP in a *S. purpuratus* embryo at the first anaphase (a) and the 235,000 MAP in a *L. variegatus* embryo at the first telophase (b) are shown by immunofluorescence microscopy. The bar in panel b equals 10 micrometers.

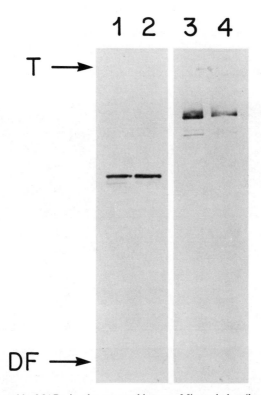

FIGURE 5. Heat-stable MAPs in the sea urchin egg. Microtubules (lanes 1 and 3) and heat-stable MAPs (lanes 2 and 4) were subjected to SDS polyacrylamide gel electrophoresis and transferred electrophoretically to nitrocellulose. The blots were then stained with monoclonal antibody to the 150,000 (lanes 1 and 2) or 235,000 (lanes 3 and 4) dalton MAP.

Presence of Immunoreactive MAPs in Other Sea Urchin Microtubules

There are several possible explanations for the presence of the five immunoreactive MAPs in the mitotic spindle of fertilized eggs. These proteins may mediate mitosis-specific functions of microtubules. They may also regulate interphase properties of microtubules. Finally, they may represent precursor proteins of the ciliary axonemes that form at the blastula stage of development approximately one day after fertilization. To distinguish among these possibilities, we used the monoclonal antibodies specific for the five MAPs to examine different classes of microtubules found in sea urchin cells.

None of the five MAPs were detectable by either immunoblotting or immunofluorescence microscopy in cilia isolated from blastula, or in sperm flagellar axonemes. Thus, these MAPs do not appear to be ciliary precursors that are stockpiled in the egg, as has been suggested for "cytoplasmic dynein."[4,28] In fact, the five MAPs seem to be generally absent from axonemal microtubules.

Circulating cells from the coelomic cavity were used to assess whether the five immunoreactive MAPs were present on cytoplasmic microtubules of differentiated, postmitotic cells. Among the numerous cell types present in the coelomic cavity, several, including the well-known petaloid/filopodial cells, will attach readily to glass coverslips and assume a well-spread morphology. Using antitubulin immunofluorescence, Edds[29] has shown that these cells typically contain arrays of cytoplasmic microtubules that emanate from a juxtanuclear organizing center. When we processed

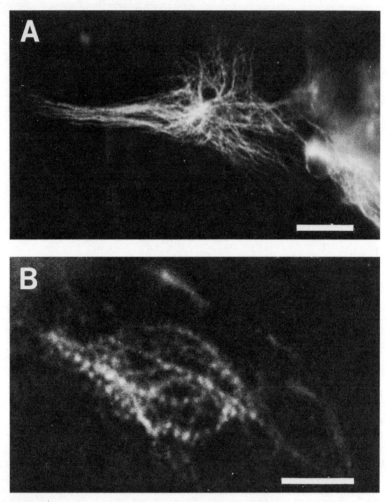

FIGURE 6. Localization of MAPs in coelomocytes. The distributions of the 235,000 dalton MAP in a *L. variegatus* cell (**A**) and the 150,000 dalton MAP in a *S. purpuratus* cell (**B**) are shown. The bars in panels **A** and **B** equal 10 and 5 micrometers, respectively.

coelomocyte cultures for immunofluorescence microscopy with the five monoclonal anti-MAP antibodies, colchicine-sensitive fibers that presumably represented microtubules were observed in most attached cells. An example of a cell stained with the antibody to the 235,000 dalton MAP is shown in FIGURE 6A. Numerous fibers can be seen to radiate from the vicinity of the nucleus and extend throughout the cytoplasm.

This general pattern of immunofluorescence was also observed when antibody to the 37, 77, or 205 kilodalton MAP was used (not shown). The antibody to the M_r 150,000 MAP, however, reacted in a unique and novel manner with coelomocyte microtubules, as illustrated in FIGURE 6B. This antibody stained apparent bundles of microtubules in a striking pattern of periodic striations consistently spaced about 0.8 micrometers apart. Thus, at least five mitotic spindle MAPs in the sea urchin are also found on interphase microtubules of differentiated, postmitotic cells.

Possible Functions of Sea Urchin MAPs

A few general conclusions can be made regarding the functions of the five immunoreactive sea urchin MAPs. Earlier studies from other laboratories had indicated that unfertilized sea urchin eggs contain abundant tubulin,[30] but no microtubules.[31] We have demonstrated that unfertilized eggs contain high levels of multiple MAP species, as well. This indicates that the mere presence of these proteins is not a sufficient condition to drive microtubule assembly in the egg. Other factors must also be involved, perhaps including posttranslational modification of MAPs.

Because both the mitotic spindle and the axoneme constitute microtubule-based motile systems, elucidating the physical organization and molecular composition of these structures is of fundamental importance for understanding how they function. Extensive progress has been made in this regard for the axoneme. There, dynein ATPases, in concert with numerous additional proteins, generate the forces and regulate the architecture that enable outer doublet microtubules to slide past one another in the precisely controlled manner that causes the cilium or flagellum to beat.[32] Although dynein-like molecules have been reported in mitotic spindles isolated from sea urchin zygotes,[4-6] the spindle is very poorly understood by comparison. It is significant, therefore, that none of the immunoreactive spindle MAPs described here seem to be associated with axonemal microtubules in the sea urchin. Whereas these results certainly do not eliminate the possibility that fundamental similarities underlie the mechanochemical systems within the axoneme and the mitotic spindle, they strongly suggest that spindle and axonemal MAPs are essentially distinct sets of proteins. Whether any essential mitotic spindle MAPs are structurally or functionally homologous to proteins in the axoneme remains to be demonstrated.

The finding that all five immunoreactive MAPs are common to microtubules in both the mitotic spindles of fertilized eggs and the interphase cytoplasm of coelomocytes suggests that these proteins perform functions of general importance for cytoplasmic microtubules in the sea urchin. Few clues for what those functions may be currently exist. One possible function for the 150,000 dalton MAP, however, has been suggested by its distribution in coelomocytes. The preferential association of that MAP with apparent microtubule bundles raises the possibility that the protein may be involved in cross-linking microtubules to one another, as has been observed for mitotic spindle microtubules not only in sea urchins,[23] but in other species as well.[18-22] We hope that further studies will clarify the functions of this protein and the other sea urchin MAPs during mitosis and interphase.

SUMMARY

We have used a taxol-based microtubule purification procedure and monoclonal antibodies to isolate and characterize the MAPs of mitotic spindle microtubules in the fertilized sea urchin egg. In so doing, we hope to have identified some of the essential working parts of the mitotic apparatus, namely those proteins that regulate the assembly, disassembly, organization and mechanochemical properties of spindle microtubules.

The results of this effort strongly suggest that a rich diversity of polypeptides associate with mitotic spindle microtubules. Whether each of these represents an individual protein species is not currently known. It is possible, for example, that particular spindle MAPs comprise multiple, distinct subunits. This would not be surprising in light of the facts that both MAP-1[33] and MAP-2[34,35] contain lower molecular weight subunits, and that axonemal dyneins are complex assemblies of several polypeptide species.[6,36-39] Our future efforts with the sea urchin system will be to determine how the various mitotic spindle MAPs we have identified function individually and in concert, and how those functions contribute to the mechanochemical properties of the spindle.

REFERENCES

1. KURIYAMA, R. 1977. Biochemistry **81:** 1115–1125.
2. SUPRENANT, K. A. & L. I. REBHUN. 1983. J. Biol. Chem. **258:** 4518–4525.
3. DETRICH III, H. W., & L. WILSON. 1983. Biochemistry **22:** 2453–2462.
4. WEISENBERG, R. & E. W. TAYLOR. 1968. Exp. Cell. Res. **53:** 372–384.
5. PRATT, M. M., T. OTTER & E. D. SALMON. 1980. J. Cell Biol. **86:** 738–745.
6. PIPERNO, G. 1984. J. Cell Biol. **98:** 1842–1850.
7. VALLEE, R. B. & G. S. BLOOM. 1983. Proc. Nat. Acad. Sci. USA **80:** 6259–6263.
8. BLOOM, G. S., F. C. LUCA, C. A. COLLINS & R. B. VALLEE. 1985. Cell Motility **5:** 431–446.
9. KIM, H., L. I. BINDER & J. L. ROSENBAUM. 1979. J. Cell Biol. **80:** 266–276.
10. LAEMMLI, U. K. 1970. Nature (London) **227:** 680–685.
11. BLOOM, G. S. & R. B. VALLEE. 1983. J. Cell Biol. **96:** 1523–1531.
12. AUCLAIR, W. & B. W. SIEGEL. 1966. Science **154:** 913–915.
13. VALLEE, R. B. 1982. J. Cell Biol. **92:** 435–442.
14. BRYAN, J., B. W. NAGLE & K. H. DOENGES. 1975. Proc. Nat. Acad. Sci. USA **72:** 3570–3574.
15. NARUSE, H. & H. SAKAI. 1981. J. Biochem. (Tokyo) **90:** 581–587.
16. SCHOLEY, J. M., B. NEIGHBORS, J. R. McINTOSH & E. D. SALMON. 1984. J. Biol. Chem. **259:** 6516–6525.
17. KELLER III, T. C. S., & L. I. REBHUN. 1982. J. Cell Biol. **93:** 788–796.
18. WILSON, H. J. 1969. J. Cell Biol. **40:** 854–859.
19. BRINKLEY, B. R. & J. CARTWRIGHT, JR. 1971. J. Cell Biol. **50:** 416–431.
20. McINTOSH, J. R. 1974. J. Cell Biol. **61:** 166–187.
21. INOUE, S. & H. RITTER. 1975. *In* Molecules and Cell Movement. S. Inoue & R. E. Stephens, Eds.: 3–30. Raven Press. New York.
22. WITT, P. L., H. RIS & G. G. BORISY. 1981. Chromosoma **83:** 523–540.
23. SALMON, E. D. & R. R. SEGALL. 1980. J. Cell Biol. **86:** 355–365.
24. FELLOUS, A., J. FRANCON, A. LENNON & J. NUNEZ. 1977. Eur. J. Biochem. **78:** 167–174.
25. WEINGARTEN, M. D., A. H. LOCKWOOD, S. HWO & M. W. KIRSCHNER. 1975. Proc. Nat. Acad. Sci. USA **72:** 1858–1862.
26. WEATHERBEE, J. A., P. SHERLINE, R. M. MASCARDO, J. G. IZANT, R. B. LUFTIG & R. R. WEIHING. 1982. J. Cell Biol. **92:** 155–163.
27. PARYSEK, L. M., C. F. ASNES & J. B. OLMSTED. 1984. J. Cell Biol. **99:** 1309–1315.
28. STEPHENS, R. E. 1972. Biol. Bull. (Woods Hole, Mass.) **142:** 489–504.
29. EDDS, K. T. 1984. Cell Motility **4:** 269–281.
30. BORISY, G. G. & E. W. TAYLOR. 1967. J. Cell Biol. **34:** 535–548.
31. BESTOR, T. H. & G. SCHATTEN. 1981. Dev. Biol. **88:** 80–91.
32. SATIR, P. 1968. J. Cell Biol. **39:** 77–94.
33. VALLEE, R. B. & S. E. DAVIS. 1983. Proc. Nat. Acad. Sci. USA **80:** 1342–1346.

34. VALLEE, R. B., M. J. DiBARTOLOMEIS & W. E. THEURKAUF. 1981. J. Cell Biol. **90:** 568–576.
35. THEURKAUF, W. E. & R. B. VALLEE. 1982. J. Biol. Chem. **257:** 3284–3290.
36. GIBBONS, I. R. & E. FRONK. 1979. J. Biol. Chem. **254:** 187–196.
37. PFISTER, K. K., R. B. FAY & G. B. WITMAN. 1982. Cell Motility **2:** 525–547.
38. TANG, W.-J.Y., C. W. BELL, W. S. SALE & I. R. GIBBONS. 1982. J. Biol. Chem. **257:** 508–515.
39. HISANAGA, S. & H. SAKAI. 1983. J. Biochem. **93:** 87–98.

Modulation of the Kinetic Parameters of Microtubule Assembly by MAP-2 Phosphorylation, the GTP/GDP Occupancy of Oligomers, and the Tubulin Tyrosylation Status

ROY G. BURNS AND KHALID ISLAM

Biophysics Section
Department of Physics
Imperial College of Science and Technology
London, England SW7 2BZ

INTRODUCTION

Microtubule assembly has been proposed[1] to conform to the linear condensation model first postulated for actin,[2] with the change in the unpolymerized subunit[a] concentration being described by:

$$-d[S]/dt = k_{+1}[M][S] - k_{-1}[M]$$

in which [S] is the concentration of free "subunit," [M] is the number concentration of microtubule ends, and k_{+1} and k_{-1} are the respective association and dissociation rate constants. One clear consequence of this analysis, and also of the subsequent consideration of the assembly and disassembly at the two polymer ends,[3,4] is that the cell could regulate the rate of assembly by controlling either the effective subunit concentration of k_{+1}, but could only regulate the rate of disassembly by controlling k_{-1}.

Whereas pure tubulin dimers can be assembled,[5,6] the observed critical concentration (C_o, equal to k_{-1}/k_{+1}) is high compared with the tubulin concentration found in most cells. Microtubules purified through successive cycles of assembly and disassembly, however, contain stoichiometric amounts of microtubule associated proteins (MAPs). The addition of such MAPs reduces C_o by lowering k_{-1},[7] so that k_{-1} can only be treated in the above model as a constant, provided that polymerization does not preferentially reduce either the MAP or the tubulin concentrations.

Immunofluorescence studies have demonstrated that MAPs are associated with microtubules *in vivo*,[8-12] and a large number of presumptive MAPs have been identified[13-17] in addition to the more highly characterized brain MAP-2 and tau. If the MAP:tubulin affinity depends upon the particular MAP species or tubulin isoform, the cell could alter the assembly kinetics by synthesizing new MAPs or tubulin subunits. This would be temporally insensitive, but may be important in cells with "stable" cytoskeletons.

---

[a]The term subunit describes the functional unit added to, or lost from, a microtubule end. Its identity, a tubulin dimer or a MAP-2:tubulin oligomer, will later be shown to depend upon the precise assembly conditions.

The system could, however, respond more rapidly if the MAP:tubulin affinity could be altered by modifying preexisting MAPs or tubulins. We shall show that subtle, but physiologically significant, alterations to the assembly conditions cause immediate, specific, and graduated changes in k_{+1} and k_{-1}, and can also affect the effective tubulin concentration. Early studies identified a number of factors affecting the rate or the maximum extent of assembly, both indicative of changes to C_o or to the effective subunit concentration. Several, such as temperature and ionic strength, are probably not significant physiological controls. The sensitivity, however, to the calcium and nucleotide triphosphate concentrations are of particular interest in view of the importance of calcium:calmodulin and the adenylate charge upon a wide variety of cellular activities.

We have concentrated on the effects of nucleotides on microtubule assembly as follows. (a) Brain MAP-2, tau, and many of the presumptive MAPs are substrates for a protein kinase(s),[14,18-20] and as phosphorylation affects a wide variety of protein:protein interactions, the phosphorylation of MAPs may alter the kinetics of microtubule assembly. Furthermore, as the gene products of various oncogenes have been shown to resemble certain protein kinases,[21] the marked changes in the microtubule cytoskeleton that occur in certain cell lines on transformation might reflect a change in the pattern of phosphorylation. (b) Tubulin dimers bind one mole guanosine triphosphate (GTP) · mole^{-1} at a nonexchangeable site,[22] a further mole · mole^{-1} at an exchangeable site on β tubulin,[23] which is hydrolyzed to GDP following assembly,[24] and adenosine triphosphate (ATP) at a third site on α tubulin.[25,26] Consequently, changing either the ATP or GTP concentrations, or their relative concentrations by altering the nucleoside diphosphate kinase (NDP kinase) activity could affect the competency of tubulin to assemble.

Before considering the specific effects of the nucleotide conditions of k_{+1} and k_{-1}, and the concentration of effective tubulin, it is necessary to describe the microtubule protein used in our studies.

THE MICROTUBULE PROTEIN

The microtubule protein used was purified from the brains of a day-old chick by successive cycles of assembly and disassembly.[18,27] The buffer contained, unless otherwise stated, 0.1 M morpholinoethane sulfonic acid, 2.5 mM ethyleneglycol-bis-(β aminoethyl ether) N,N' tetraacetic acid (EGTA), 1.0 mM dithiothreitol, 0.5 mM MgCl$_2$, and 0.1 mM ethylenediamine tetraacetic acid (EDTA), pH 6.4 with KOH, including leupeptin (10 μg · ml^{-1}) during the initial homogenization to minimize MAP-2 degradation. The pelleted microtubules were dissociated at 4°C in assembly buffer lacking added nucleotide for 2 h after the second and subsequent cycles to facilitate the complete dephosphorylation of the MAP-2 by the copurifying phosphoprotein phosphatase(s).[28-30]

MAP-2 and tubulin are the principle components present after two cycles of assembly, although some preparations contain trace amounts of MAP-1 and tau.[30,31] As MAP-2 is found primarily or exclusively in dendrites,[32,34] the *in vitro* phosphorylation and polymerization studies relate to the properties of dendritic microtubules. A number of enzymatic activities are also present including the phosphoprotein phosphatase(s), protein kinase(s), and NDP kinase.[27-29,35-37]

Quantitative sodium dodecyl sulfate (SDS)-polyacrylamide gel electrophoresis of highly purified microtubules yields a molar ratio of one MAP-2 molecule: 12 tubulin dimers,[30] applying M_r of 280,000[31] and 100,000[38] respectively. This is in excellent agreement with the model based on image reconstruction in which the MAP-2 sidearms are helically ordered on the tubulin lattice with a 96 nm periodicity along each protofilament.[39] A similar periodicity was observed with negatively stained microtubules decorated with DEAE-dextran.[40] There is therefore a sidearm every twelve tubulin dimers, each consisting of a single MAP-2 molecule.

The cold dissociated microtubule protein contains, as previously reported (e.g. references 41–43), numerous MAP-2:tubulin rings, and approximately 55% of the total tubulin dimer exists in this oligomeric form.[44]

MAP-2 PHOSPHORYLATION

MAP-2 can be phosphorylated by a protein kinase (M_r 35,000[35]) that copurifies with microtubules through cycles of assembly and disassembly.[18,35] We have shown that this activity labels MAP-2 in vitro using $[\gamma\text{-}^{32}P]ATP^{18}$ or $[\gamma\text{-}^{32}P]GTP^{45}$ to twelve moles · mole^{-1}, and that whereas ATP is the preferred donor, both nucleotide triphosphates label the same sites.[45] This stoichiometry has been confirmed by direct measurement of the difference between the phosphate content of acid phosphatase-treated MAP-2 and after rephosphorylation with ATP.[46]

Both the initial rate (FIGURE 1) and the final extent[18,45] of MAP-2 phosphorylation are dependent upon the phosphoryl donor concentration. The observation that the final extent of phosphorylation is directly proportional to the ATP concentration (3–300 μM) is unusual. This donor concentration dependency is not due to substrate depletion by the copurifying GTPase and ATPase activities, and indicates that the $k_{m,ATP}$ may differ for each of the twelve phosphorylation sites.

Early reports suggested that MAP-2 could only be phosphorylated to two moles · mole^{-1} and that one site was cyclic adenosine monophosphate (cAMP)-dependent.[47,48] The chick brain microtubule protein prepared at pH 6.4, however, lacks detectable cAMP-dependency, is insensitive to the heat-stable inhibitor,[30] yet permits phosphorylation to twelve moles · mole^{-1}.

Some preparations of chick brain microtubule protein purified at pH 6.9 do exhibit cAMP sensitivity, indicating the copurification of the regulatory subunit under certain assembly conditions. The same initial rate of phosphorylation, using such protein, is observed at a low ATP concentration and 10 μM cAMP as at a higher ATP concentration but without cAMP (FIGURE 1). Binding of cAMP to the regulatory subunit[49] therefore appears to enhance the effect of the donor concentration on the number of sites available for phosphorylation.

MAP-2 contains two functional domains, one being the ultrastructural sidearm, whereas the other is associated with the microtubule wall.[50] The sidearm portion is 80–90 nm long by rotary shadowing, whereas the purified MAP-2 is about 165–185 nm long.[51] The difference is consistent with a 96 nm domain interacting with twelve tubulin dimers along each protofilament between successive MAP-2 projections. As MAP-2 increases the lateral tubulin:tubulin spacing of zinc-induced sheets by about 0.5 nm without affecting the longitudinal periodicity,[52] the tubulin-binding domain has apparent dimensions of 96 × 0.5 nm.

Partial α-chymotryptic digestion and cleavage with nitrothiocyanobenzoic acid (NTCB) shows that at least ten of the twelve phosphorylation sites lie towards the end of the MAP-2 primary sequence, forming the tubulin-binding domain.[30] There is recent evidence for two MAP-2 isoforms, MAP-2a and MAP-2b,[32,34] and further analysis of the NTCB digest indicates that two MAP-2 isoforms are present with similar molecular weights and occurring in approximately equal amounts. In each case the phosphorylation sites are clustered towards one end of the primary sequence.[53]

The tubulin-binding domain includes a 25–35,000 M_r proteolytic fragment at one end of the primary sequence,[48,50] but it may extend into the 255,000 M_r "sidearm" domain as some of the most readily phosphorylated residues lie outside the terminal fragment.[31,37,48] This domain interacts with twelve tubulin dimers and contains most or all of the twelve phosphorylation sites, so that there is a 1:1 equivalence between the number of phosphorylation sites and tubulin dimers. The extent of phosphorylation, itself determined by the cAMP and phosphoryl donor concentrations, may therefore regulate the MAP-2:tubulin interaction.

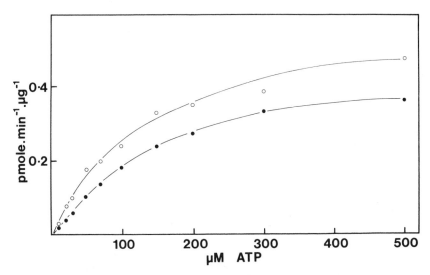

FIGURE 1. Stimulation by cAMP of the initial rate of phosphorylation as a function of the ATP concentration. Microtubule protein was purified through two cycles of assembly and disassembly using reassembly buffer[18,27] at pH 6.9. The protein (1.1 mg.ml^{-1}) was challenged in the absence (●) or presence of (○) 10 μM cAMP for 10 min at 37°C with increasing concentrations of [γ-^{32}P]ATP. The assay buffer, in the pH 6.9 reassembly buffer, also contained 20 mM NaF and 200 μM hypoxanthine. The reaction was terminated by addition of 10% trichloroacetic acid, and the samples were processed as previously described in detail.[18]

EFFECT OF MAP-2 PHOSPHORYLATION ON MICROTUBULE ASSEMBLY

The affinity of MAP-2 for assembled tubulin can be assessed by salt titration of taxol-stabilized microtubules.[54] Increasing the level of phosphorylation significantly reduces this affinity,[55,56] and the profile indicates that the level of phosphorylation exerts a cooperative effect on the MAP-2:tubulin interaction.[55]

This indirect approach, however, cannot establish whether phosphorylation has any specific effect on the kinetics of microtubule assembly. We have examined the kinetics of seeded assembly as a function of the initial protein concentration in order to determine k_{+1} and k_{-1}, using protein prephosphorylated to differing levels and then separated from free nucleotides by gel filtration. The microtubules were assembled with 100 μM GTP, a concentration that results in minimal phosphorylation[45] and, as we shall show later, minimizes the direct addition of MAP-2:tubulin oligomers to the microtubule end.

Raising the level of prephosphorylation increased C_o, determined from both the initial rate and the final extent of assembly, from about $0 \ \mu M$ for the unphosphorylated protein to about 1.5 μM tubulin dimer for protein prelabeled to 10 moles $\cdot$ mole^{-1} MAP-2.[55] There was, however, no effect on k_{+1}: the increase in C_o being due solely to

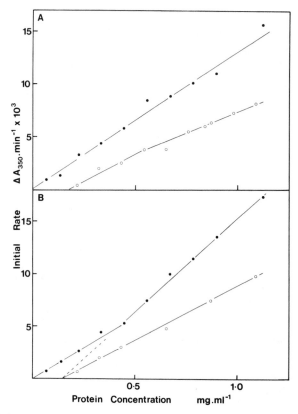

FIGURE 2. Initial rate of seeded microtubule assembly as a function of the concentration of unphosphorylated and prephosphorylated protein at 50 μM and 500 μM GTP. Microtubule protein was purified through two cycles of assembly and disassembly[18,27] using the pH 6.4 reassembly buffer. The protein (8.56 mg.ml^{-1}) was incubated with either 500 μM ATP or without added nucleotide and 20 mM NaF for 2 h at 37°C, and was then eluted through a Sephadex G-50 column (0.9 × 9 cms) preequilibrated with reassembly buffer 20 mM NaF. The void fractions were used to determine the extent of MAP-2 phosphorylation by SDS-polyacrylamide gel electrophoresis (see reference 18 for details) and for the assembly studies. The unphosphorylated (●) and prephosphorylated protein (O : 5.8 mole phosphate $\cdot$ mole^{-1} MAP-2) was diluted with reassembly buffer (pH 6.4), degassed, transferred to the temperature-controlled cuvette of a Beckman DU-8 spectrophotometer, and warmed to 37°C for 5 minutes. There was no assembly during this period. Microtubule assembly was initiated by the simultaneous addition of either (A) 50 μM GTP or (B) 500 μM GTP and preassembled microtubule seeds.[55] The assembly was monitored by the increase in A_{350nm}, and the initial rate is plotted against the initial protein concentrations.

an alteration in k_{-1}, such that the assembly of unphosphorylated protein was essentially irreversible ($k_{-1} = 0$), but k_{-1} was increased by prephosphorylation to 10 $\sec^{-1}$. Intermediate values were observed for protein prelabeled to lower stoichiometric levels, and, again, an apparently cooperative effect was observed.[55]

Significantly, microtubule assembly does not affect the initial rate of phosphorylation, and MAP-2 is labeled with the same kinetics before and after assembly.[55] Therefore, although the tubulin-binding domain of MAP-2 is located in the groove between adjacent protofilaments, the substrate residues remain available to the protein kinase.

The availability of the phosphorylation sites and the specific effect on k_{-1} imply that phosphorylating a preassembled microtubule should lead to its dissociation. This has been tested by assembling microtubules to equilibrium with 100 μM GTP and then challenging with increasing ATP concentrations. The addition of ATP immediately initiates microtubule disassembly, at rates that are directly proportional to the rates of MAP-2 phosphorylation.[55]

In summary, our results show that the extent of MAP-2 phosphorylation is determined, in part, by the phosphoryl donor concentration, and that this phosphorylation specifically modulates k_{-1} without affecting k_{+1}. This differential effect demonstrates that assembly and disassembly are not kinetically equivalent, and implies that the binding of MAP-2 to the incoming dimer occurs subsequent to dimer addition to the elongating microtubule. As MAP-2 can be phosphorylated to twelve moles · mole^{-1}, this novel graduated control of k_{-1} may have thirteen integer steps.

MAP-2 phosphorylation, however, has been shown by two other laboratories to affect both k_{+1} and k_{-1} with little effect on C_o.[56–58] As these studies used significantly higher GTP concentrations to promote assembly, we have compared the initial rate of seeded assembly at 50 and 500 μM GTP of unphosphorylated microtubule protein and protein containing MAP-2 prephosphorylated to 5.8 moles · mole^{-1}. The assembly at 50 μM GTP confirmed our earlier finding that k_{-1} is approximately zero for the unphosphorylated protein, whereas the phosphorylated protein exhibited a higher C_o (0.175 mg.ml^{-1}), due to a specific increase in k_{-1} (FIGURE 2A).

The kinetics at 500 μM GTP were significantly different. A plot of the initial rate as a function of the unphosphorylated protein concentration was distinctly biphasic, with critical concentrations equal to 0 and 0.15 mg.ml^{-1} (FIGURE 2B), but was linear with an apparent C_o of 0.15 mg.ml^{-1} for the phosphorylated protein. The GTP concentration, therefore, has a profound effect on the apparent values of k_{+1} and k_{-1}, and the discrepancies between the reported results from different laboratories can be attributed to the particular assembly conditions. The interpretation of the apparent effect of phosphorylation on both k_{+1} and k_{-1} at 500 μM GTP will be considered after examining the effect of the GTP concentration on the assembly of unphosphorylated microtubule protein.

MICROTUBULE ASSEMBLY AS A FUNCTION OF THE GTP CONCENTRATION

The classic study demonstrating that microtubule assembly conforms to the linear condensation model, considered only a single subunit species[1] as the protein had been centrifuged to remove oligomeric material. Most other studies, however, have used unfractionated protein, containing an equilibrium mixture of MAP-2:tubulin oligomers, tubulin dimers, and free MAP-2. Significantly, both temperature-jump ultracentrifugation[59] and time-resolved x-ray diffraction studies[43] on the polymerization of such protein have shown that oligomers can be added directly to the microtubule end.

Whereas polymerization has generally been induced by raising the temperature and adding 1 mM GTP, we have recently shown[44] that tubulin dimers and free MAP-2 are preferentially added at 20 μM GTP, whereas the MAP-2:tubulin oligomers are also added directly at 1 mM GTP. The maximal extent of assembly, however, is independent of the GTP concentration, so that the assembly at 20 μM GTP must involve the initial dissociation of the oligomers to dimers and free MAP-2 prior to their addition to the microtubule end. This GTP concentration dependency is also evident in the assembly of unphosphorylated microtubule protein at 50 and 500 μM GTP, in that the initial rate is directly proportional to the protein concentration at 50 μM GTP (FIGURE 2a), but is markedly biphasic at 500 μM GTP (FIGURE 2B). Significantly, the initial rate at 50 μM GTP tends to deviate from linearity at higher initial protein concentrations.

The dissociation constant for GTP to the tubulin dimer ($k_{d,GTP}^{dimer}$) is about 0.1 μM,[60] whereas much higher GTP concentrations are required to detect any binding to the MAP-2:tubulin oligomers,[44,61] although the absolute oligomer GTP and GDP binding constants have not been determined. Calculations using values of $k_{d,GTP}$ and $k_{d,GDP}$ of 0.1 and 0.2 μM[60] show that, under our experimental conditions,[44] the dimer approaches saturation by 20 μM GTP,[62] so that there can be no further increase in k_{+1}^{dimer} (defining k_{+1}^{dimer} and k_{-1}^{dimer} as the association and dissociation rate constants for dimer binding to the microtubule end).

The observed value of k_{+1} is, however, proportional to ln[GTP][62] and only approaches a maximum at 500 μM GTP. This rise must, therefore, reflect an increase in the direct addition of oligomers to the microtubule end and result from the increased GTP occupancy of the relatively nonexchangeable E site on the oligomers. The $k_{d,GTP}^{oligo}$ therefore appears to control the rate of addition of oligomers. The GTP occupancy of the oligomers appears to modulate k_{+1}^{oligo} (*i.e.*, the association rate constant for oligomer addition to the microtubule end) in a way comparable to the control exerted by MAP-2 phosphorylation on k_{-1}. The number of integer steps in this control of k_{+1} is currently unknown but will relate to the number of tubulin subunits per oligomer and the maximum GTP occupancy.

The initial rate of assembly of unphosphorylated microtubule protein at 500 μM GTP yields two values for C_o (0 and 0.15 mg.ml^{-1}, FIGURE 2B), and the higher C_o must relate to the addition and loss of oligomers at the microtubule end. As the maximum extent of assembly is independent of the GTP concentration, it can only apply to the elongation phase and not to steady state conditions. Furthermore, the higher value of C_o^{oligo} means that the subunit species added under steady state conditions must be the tubulin dimer and free MAP-2. Consequently, initial rate measurements at high protein or high GTP concentrations cannot be used to assess the stability or rate of disassembly of preassembled microtubules.

The C_o^{oligo} therefore only describes the probability that a MAP-2:tubulin oligomer newly added to an elongating microtubule end will not be lost before its incorporation is stabilized by the binding of additional subunits. Any analysis under conditions that permit direct oligomer addition or loss (*e.g.* measurements of the initial rates of assembly at high protein and high GTP concentrations) must take account of the complex subunit composition. Furthermore, assembly conditions that permit the addition of oligomers introduce a new variable, in that the oligomeric rate constants and the structural integrity of the oligomers could both be affected. Changes in the latter would have the effect of altering the apparent subunit concentration.

We have examined whether phosphorylation alters the structural integrity of the oligomers by comparing the proportion of total tubulin of unphosphorylated and prephosphorylated protein eluting as oligomers (FIGURE 3). The fraction is significantly decreased from 58% for the unphosphorylated protein to 30% after labeling to 10.8 moles · mole^{-1} MAP-2. This decrease could either be due to a reduced number of oligomers of constant structure or to a smaller number of dimers in each oligomer. The former possibility would imply, as the MAP-2 is multiply-phosphorylated, that only a fraction of the protein is labeled. Phosphorylation, however, affects the mobility of certain of the NTCB peptides, yielding a "ladder" of isoforms differing by a single phosphorylated residue, with no evidence for a specific unlabeled MAP-2 subset.[53]

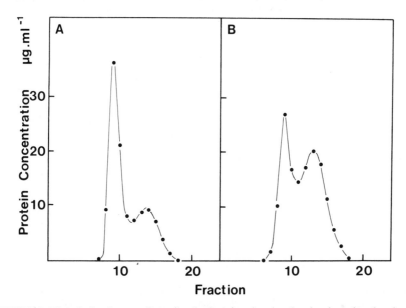

FIGURE 3. The tubulin oligomer:dimer distribution of unphosphorylated and prephosphorylated microtubule protein. Twice-recycled microtubule protein was incubated in reassembly buffer (pH 6.4) containing 20 mM NaF with or without 500 μM [γ-^{32}P] ATP for 2 h at 37°C and eluted through a Sephadex G-50 column (see FIGURE 2 for details). An aliquot was removed for determining the MAP-2:tubulin ratio and the stoichiometry of MAP-2 phosphorylation by SDS-polyacrylamide gel electrophoresis.[18,30] The remainder was warmed to 37°C for 5 min, lightly fixed with glutaraldehyde, cooled to 4°C, and fractionated on a Sephacryl S-300 column (0.9 × 9 cm) as previously described.[44] The volume (mean = 450 μl) and protein concentration of each fraction was determined and, together with the MAP-2:tubulin ratio (19:81), was used to calculate the tubulin distribution assuming that all the MAP-2 (M_r of 280,000) eluted in the void fractions. These calculations show that for the unphosphorylated protein (A), 58% of the total tubulin coeluted with the MAP-2, but that after prephosphorylation to 10.8 moles phosphate · mole^{-1} MAP-2 (B), the value fell to only 30 percent.

Phosphorylation therefore probably affects the number of tubulin dimers per oligomer, although this awaits direct confirmation.

If the number of tubulin dimers per oligomer is defined as c, then phosphorylation decreases the value of c and so reduces the extent of assembly resulting from the addition of a single oligomer. This smaller oligomer remains an effective subunit species, and whereas the molar concentration of oligomeric subunits is unchanged, the concentration of free tubulin dimers is increased.

Correcting the initial rate of assembly as a function of the protein concentration (FIGURE 2) for c of unphosphorylated and prephosphorylated protein (FIGURE 3) suggests (although the levels of phosphorylation differed), that phosphorylation has no significant effect on either k_{+1}^{oligo} or k_{-1}^{oligo}. Phosphorylation, therefore, appears to specifically affect k_{-1} dimer and to alter the identity of the oligomeric subunit species, but has little or no affect on k_{+1}^{oligo} or k_{-1}^{oligo}. Furthermore, the slight deviation from linearity of the initial rate of assembly of the unphosphorylated protein at 50 μM GTP (FIGURE 3A) suggests that there is some oligomer incorporation at low concentrations, but at a rate that is slower than that for dimers.

KINETICS OF ASSEMBLY WITH ATP

This description of the utilization of dimers and oligomers at increasing GTP concentrations and the effect of MAP-2 phosphorylation fails to account for one significant observation. The addition of ATP to preassembled microtubules induces an immediate depolymerization at a rate that is proportional to the rate of phosphorylation.[55] Depolymerization, however, continues after the C_0 characteristic of the extent of phosphorylation has been attained. Therefore, although the rate of depolymerization is governed by the rate of phosphorylation, it does not determine the new equilibrium.

ATP directly induces microtubule assembly and does not require the generation of GTP by the copurifying NDP kinase charging up-bound GDP.[63] The maximum extent of assembly, however, is 25–35% lower than that observed with GTP, and is independent of the ATP concentration (20–500 μM). The addition of 50 μM GTP to microtubules assembled to equilibrium with 50 μM ATP fails to reverse this inhibition, whereas adding ATP to microtubules preassembled with GTP causes depolymerization. The decreased extent of assembly cannot be accounted for by an increased C_0 or MAP-2 phosphorylation. Electron microscopy shows that samples assembled with GTP contain microtubules and unidentified globular particules (FIGURE 4A). The ATP samples also contain microtubules, but in addition, there are also short filaments that are not seen following assembly with GTP. In favorable preparations these filaments associate with the globular particles to form "aster-like" complexes (FIGURE 4B). The filaments are not observed following extended assembly with 1 mM ATP, possibly as a result of MAP-2 phosphorylation weakening the MAP-2:tubulin interaction. Therefore, only a fraction of the protein that is assembly-competent with GTP can be polymerized into microtubules in the presence of ATP, whereas the remainder is channeled into assembly-incompetent filaments.

This fraction (25–35% depending upon the preparation) might represent a specific tubulin isoform or posttranslational modification. Purified microtubule protein contains a mixture of α-tubulin subunits that are either tyrosylated or detyrosylated at the C terminus.[64,65] The primary α-tubulin gene product is tyrosylated,[38] but this terminal residue can be removed by a specific tubulin-tyrosine carboxyperptidase (TTC).[66] The tubulin, though, can be retyrrosylated by tubulin-tyrosine ligase (TTL),[67,68] a reaction requiring ATP hydrolysis. The tyrosine status of tubulin has no apparent effect on the rate or extent of microtubule assembly with 1 mM GTP,[69] and whereas the MAP-2:tubulin stoichiometry is affected, this is probably due to MAP-2 phosphorylation. The partial assembly with ATP may reflect, as α tubulin binds ATP (k_d = 200 μM^{25}), a differential effect on detyr- and tyr-tubulin.

This has been examined by pretreating chick brain microtubule protein with α carboxypeptidase to remove the terminal tyrosine and then to assemble the protein with ATP or GTP. Microtubule assembly with 50 μM GTP was unaffected by the

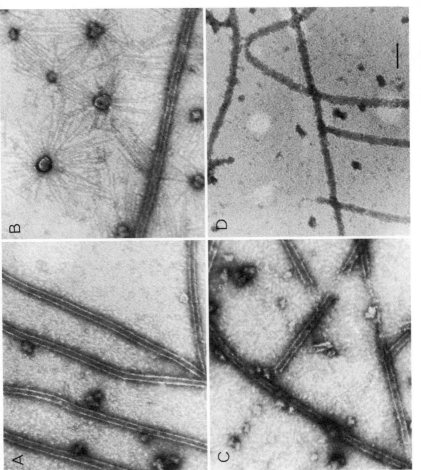

FIGURE 4. Electron microscopy of untreated and carboxypeptidase-digested microtubule protein assembled with GTP and with ATP. Twice-recycled microtubule protein was assembled with 500 μM GTP for 10 min at 37°C using reassembly buffer lacking dithiothreitol. The assembled microtubules were incubated for a further 10 min with or without 0.55 $\mu g.ml^{-1}$ α carboxypeptidase (53 units.mg^{-1}), at which time 15 mM dithiothreitol was added to terminate the digestion. The assembled microtubules were then pelleted through a 30% sucrose cushion, cold dissociated, clarified, and eluted through a Sephadex G-50 column[44] (0.9 × 9 cm). The untreated (**A** and **B**) or carboxypeptidase-treated protein (**C** and **D**) was diluted to 1 mg · ml^{-1} and assembled to equilibrium at 37°C with either 50 μM GTP (**A** and **C**) or 50 μM ATP (**B** and **D**). The samples were fixed with 1% glutaraldehyde, diluted tenfold, negatively stained with 1% uranyl acetate, and viewed with a Philips 301 electron microscope. Bar = 100 nm.

pretreatment (FIGURE 4C), whereas no microtubules were observed after assembly with 50 μM ATP. There were, however, a small number of long ribbons that lacked the characteristic structure of microtubules (FIGURE 4D), indicating that pretreatment with carboxypeptidase enhances the ATP discrimination. Attempts to reverse the effects of carboxypeptidase pretreatment by incubating detyr-tubulin with tyrosine and a crude preparation of TTL have only been partially successful, possibly because only a rather low level of retyrosylation has been achieved. These results suggest that assembly with ATP may, unlike GTP-assembly,[67,69] discriminate between tyr- and detyr-tubulin.

LINEAR CONDENSATION: EFFECTS OF MAP-2 PHOSPHORYLATION AND GTP CONCENTRATION

It is clear from these results that the usual description[1] of the linear condensation model can only apply to the assembly of microtubule protein under strictly limited conditions. In particular, the model only applies to the assembly of pure tubulin or tubulin dimers and free MAP-2 present in stoichiometric amounts, studied either at low protein and low GTP concentrations or under conditions, such as inclusion of 75 mM NaCl,[44] which dissociate the oligomers. In addition, the level of phosphorylation must not change during assembly; yet this condition is not normally satisfied, as GTP is a donor for the protein kinase.[45] Finally, the GTP concentration must be adequate to fully saturate the tubulin dimer E-site such that there is not a mixed population of GTP- and GDP-dimers.

Under such conditions, phosphorylation only affects k_{-1}^{dimer}, so that the rate of elongation and the conditions at steady state can be defined as:

$$-d[Tu^{dimer}]/dt = k_{+1}^{dimer}[Tu^{dimer}][M] - \sum_{1-o}^{1-12} k_{-1}^{dimer,i}[M]$$

in which $k_{-1}^{dimer,i}$ defines the dissociation rate constants relating to i levels of MAP-2 phosphorylation, and where [S] of the original equation is replaced by the tubulin dimer concentration $[Tu^{dimer}]$.

This relationship must be extended if the assembly conditions permit the direct addition and loss of oligomers, to include terms defining the oligomer species and concentration, and the association and dissociation rate constants for each level of GTP occupancy:

$$-d[Tu^{dimer}]/dt = k_{+1}^{dimer}[M][Tu^{dimer}] - \sum_{1-o}^{1-12} k_{-1}^{dimer,i}[M]$$
$$+ [M] \sum_{j-o}^{j-x} (c\, k_{+1}^{oligo,j}[Tu^{oligo,j}]) - [M] \sum_{j-o}^{j-x} (c\, k_{-1}^{oligo,j}).$$

$[Tu^{oligo,j}]$ defines the concentration of each of the oligomer species with a GTP occupancy of $0 - x$ moles GTP $\cdot$ mole^{-1} oligomer: $k_{+1}^{oligo,j}$ and $k_{-1}^{oligo,j}$ are the respective association and dissociation rate constants for these oligomeric species, and c is the number of tubulin dimers per oligomer (a function modulated by the MAP-2 phosphorylation status).

This analysis is restricted to the instantaneous elongation rate. It does not describe either the full kinetics or the rate of dissociation, as the dimer:oligomer composition changes with time due to the preferential usage of oligomers during the earlier stages of assembly. Furthermore, the analysis does not consider the consequences of GTP hydrolysis[24] following subunit addition. Hydrolysis has no direct effect on k_{+1}^{dimer}, as the tubulin dimer is saturated above $\simeq 20$ μM GTP, and higher concentrations merely establish the GTP occupancy of the oligomers. Hydrolysis of GTP will markedly affect the dissociation kinetics by determining whether GTP dimers or GDP dimers are lost,[70,71] yielding $k_{-1}^{GTP,dimer,i}$ and $k_{-1}^{GDP,dimer,i}$.

The rate of attachment or loss of a subunit from a microtubule end depends upon the number of protein:protein interactions, and this number reflects the precise geometry of the microtubule end.[72] Consequently, each of the dimeric and oligomeric association and dissociation rate constants must be a number-weighted average describing the characteristics of the microtubule end. Finally, this analysis of the utilization of dimers and oligomers has not yet been extended to describe the kinetic events at the two ends of the microtubule, although they have been shown to differ (*e.g.* references 3, 4, 66) with the resulting treadmilling or head-to-tail polymerization.

IN VIVO CONSEQUENCES

The nucleotide regime affects microtubule assembly *in vitro* in three distinct ways. (1) The concentration of ATP, and GTP but less efficiently, modulates the extent of MAP-2 phosphorylation, and this in turn affects k_{-1}^{dimer} and the number of dimers per oligomer (c). This modulation can be potentiated by cAMP. In addition, recent studies suggest that MAP-2 can also be multiply phosphorylated by a Ca^{++}/calmodulin-dependent protein kinase.[73–76] It is unclear, however, whether the same residues are labeled by this and the cAMP-dependent/independent kinase activities, as direct measurements of the MAP-2 phosphate content indicate that certain residues cannot be labeled by the latter activity[46] (2) ATP directly affects the extent of microtubule assembly, possibly by specifically inactivating detyr-tubulin, whereas both detyr- and tyr-tubulin can be assembled *in vitro* by GTP. (3) The GTP concentration affects the assembly competency of MAP-2:tubulin oligomers, with a graduated increase in the net association rate constant with increasing GTP concentrations. The precise structure of these oligomers, which is also sensitive to the extent of MAP-2 phosphorylation, is currently unknown, but high resolution electron microscopy coupled with synchratron x-ray diffraction suggests that they are short fragments of microtubule protofilaments derived from the partial dissociation of the cold-stable rings.[43]

The key question is whether these effects are important in controlling microtubule assembly *in vivo*. The ATP concentration affects MAP-2 phosphorylation over the range 3–500 μM, whereas the change in oligomer utilization occurs over the range of 100–500 μM GTP. Neuronal tissue contains 4–7 times more ATP than GTP,[77,78] and assuming that *in vivo* ATP concentration is 1–2 mM, this yields an *in vivo* GTP concentration of about 100–500 μM, that is, exactly the range observed *in vitro* for the modulation of oligomer addition. The ATP concentrations, however, that alter the extent of MAP-2 phosphorylation *in vitro* are somewhat lower than the assumed cellular concentration. This may be due to the unphysiological buffer conditions, such as the use of 0.1 M 4-morpholino sulphonic acid or the omission of ADP. By contrast, both the optimal cAMP and Ca^{++} sensitivities lie with the predicted physiological ranges.

The *in vitro* effects of the nucleotide concentrations would, of course, only apply if MAP-2 is phosphorylated *in vivo*, and if the MAP-2:tubulin oligomers exist in cells. MAP-2 has been labeled by intracerebral injection of [^{32}P]phosphate,[47] although not necessarily at the same residues as *in vitro*. Currently, there is no direct evidence for oligomers *in vivo*, but it is probable that they occur as the specific, and similar, protein:protein interactions required for microtubule assembly are satisfied both *in vivo* and *in vitro*.

If the *in vitro* kinetics accurately reflect microtubule assembly *in vivo*, then it is clear that subtle changes in the metabolism of the cell should profoundly affect the microtubule cytoskeleton. For instance, raising the level of MAP-2 phosphorylation, either by elevating the ATP concentration or by cAMP-activation, would lead to an increase in k_{-1}^{dimer} (FIGURE 5). This would result in microtubules shortening if the size of the effective subunit pool remains unchanged and if the number of microtubules is governed by the microtubule organizing center (MTOC). Shortening would continue until the size of the subunit pool had increased to the C_0 appropriate for the level of phosphorylation. Similarly, decreasing the extent of phosphorylation, by lowering the ATP and/or cAMP concentrations or by increasing the phosphoprotein phosphatase(s) activity, would result in the lengthening of preexisting microtubules.

Whereas the rate of microtubule dissociation can only be modulated by varying k_{-1}, the rate of elongation could be regulated by controlling either k_{+1} or the subunit concentration. Altering the GTP concentration changes the GTP occupancy of the oligomers, and hence the identity of the effective subunit species. The GTP concentration alters k_{+1} (FIGURE 5), but although C_0^{dimer} and C_0^{oligo} (FIGURE 2B) differ, the maximum extent of assembly is unaffected, as the oligomers exist in equilibrium with the dimers. Consequently, increasing the cellular GTP concentration would result in a higher GTP occupancy of the oligomers and hence a higher rate of elongation rate without affecting the maximal extent of assembly. Raising the oligomer GTP occupancy also increases k_{-1}^{oligo}, that is, it increases the probability that an oligomer added to a microtubule end will be lost before becoming stabilized by the addition of further subunits.

A general increase in cellular metabolism would be expected to increase both the ATP and GTP concentrations. The net association rate constant ($k_{+1}^{dimer} + k_{+1}^{oligo}$) would therefore be predicted to be greater, but there would also be increases in both k_{-1}^{dimer} and k_{-1}^{oligo}, and c would be smaller. Consequently, the rate of assembly would be faster, the extent at equilibrium reduced, and the microtubules would be far more dynamic than at a lower metabolic state.

It should be noted, however, that as the *in vitro* studies have been confined to MAP-2:tubulin microtubules, this model only strictly applies to dendritic microtubules. It may, however, have a broader application, as a number of presumptive MAPs can be phosphorylated,[14] and tau can be labeled at multiple sites.[19,20] Significantly, conditions that lead to the depletion of the ATP pool result in the stabilization of the microtubule cytoskeleton,[79] whereas the addition of ATP, but not AMPPCP, to detergent-extracted cells enhances the rate of microtubule depolymerization.[80] Lowering the ATP concentration would reduce the extent of MAP-2 phosphorylation *in vitro* and hence k_{-1}, whereas the addition of ATP should increase phosphorylation and k_{-1}.

These studies on nonneuronal cell lines suggest that MAP phosphorylation may indeed be a general control mechanism. For example, microtubule assembly in lightly permeabilized 3T3 cells is markedly affected by the nucleotide conditions.[81] In particularly, 1 m*M* GTP or uridine triphosphate (UTP) resulted in the random assembly of short microtubules, whereas longer microtubules were assembled in the presence of 1 m*M* ATP. Nucleation was only restricted to the MTOCs on incubation with 1 m*M* ATP-0.3 m*M* GDP-10 μM bromo-cAMP at pH 7.6, conditions that should, from the *in vitro* observations, yield a high rate of MAP phosphorylation and a low GTP concentration, that is, conditions favoring a high C_0 but slow elongation, and (as we shall argue below) a lowered concentration of assembly-competent tubulin.

Numerous ultrastructural and immunofluorescent studies have indicated that the

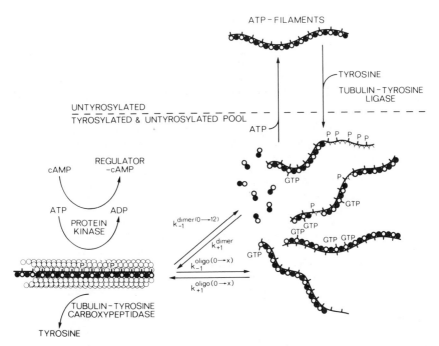

FIGURE 5. Proposed model of the effects of MAP-2 phosphorylation, the oligomer GTP occupancy, and the tyrosylation/detyrosylation of α tubulin upon the kinetics of microtubule assembly. The FIGURE shows an assembled microtubule composed of tubulin dimers and MAP-2, in which the tubulin-binding domain of the MAP-2 molecule interacts with twelve dimers along the protofilament axis. One such MAP-2:12 tubulin dimer complex is drawn in bold to emphasize this specific interaction. The sidearm portion of the MAP-2 molecule is not shown. MAP-2 can be phosphorylated *in vitro* to twelve moles · mole^{-1} MAP-2, and at least ten of these sites fall within the tubulin-binding domain. The microtubule is drawn such that a phosphorylation site is associated with each of the twelve tubulin dimers. Microtubule assembly does not affect the kinetics of MAP-2 phosphorylation by the associated protein kinase. The extent of MAP-2 phosphorylation is determined by the phosphoryl donor concentration, and this effect is potentiated by cAMP. The extent of MAP-2 phosphorylation specifically modulates the dissociation rate constant ($k_{-1}^{dimer(0 \to 12)}$), but has no effect upon the dimeric association rate constant k_{+1}^{dimer}, giving rise to the term $\Sigma_{i-0}^{i-12} k_{-1}^{dimer,i}$. Both tubulin dimers and MAP-2:tubulin oligomers can participate as direct assembly intermediates, at rates defined by k_{+1}^{dimer} and k_{+1}^{oligo}. The oligomeric association rate constant is modulated by the GTP occupancy of the oligomers, to yield rate constants $k_{+1}^{oligo,(0 \to x)}$ that give rise to the term $\Sigma_{j-0}^{j-x} k_{+1}^{oligo,j}$ in which x defines this occupancy. Oligomers with varying GTP occupancies are shown. The oligomeric dissociation rate constants ($k_{-1}^{oligo(0 \to x)}$ and giving rise to the term $\Sigma_{j-0}^{j-x} k_{-1}^{oligo,j}$) describe the loss of oligomers from the elongating microtubule and before they are stabilized by the addition of other subunits. For simplicity, the illustration shows a microtubule with a 'smooth' end, although the end would be predicted to be 'ragged'. MAP-2 phosphorylation also affects the stability of the MAP-2:tubulin oligomers. This is shown diagrammatically by the loss of bound tubulin dimers from those MAP-2 sites that are phosphorylated. The number of tubulin dimers per oligomer (defined as c) added to a microtubule end, therefore, depends upon the extent of MAP-2 phosphorylation. ATP inhibits the maximum extent of microtubule assembly by promoting filament formation. This effect is enhanced by pretreatment with α carboxypeptidase. It is postulated that, at physiological ATP concentrations, filament formation reduces the size of the microtubule assembly-competent tubulin pool, that these filaments are specifically composed of detyr-tubulin, and that this protein can only be restored to the assembly-competent pool by retyrosylation. The addition of tyrosine to α tubulin by the tubulin-tyrosine ligase is shown as regulating, together with any tyr-tubulin released by net microtubule disassembly, increases in the size of the assembly-competent pool. Similarly, the tubulin-tyrosine carboxypeptidase, which is specific for polymerized microtubules, is shown as regulating (together with net microtubule assembly) decreases in the assembly-competent pool size by raising the concentration of detyr-tubulin, which on binding ATP assembly, forms the ATP filaments.

size of the assembly-competent tubulin subunit pool is controlled. Temperature-jump studies on the birefringence of sea urchin spindles[82] have shown that this control is exercised at preprophase and is temperature-sensitive. The molecular basis of the activation of the tubulin pool has not emerged from the kinetic studies of GTP-induced microtubule assembly.[72]

The predominant nucleotide triphosphate *in vivo* is, however, ATP. The extent of microtubule assembly *in vitro* with ATP is lower than with GTP, apparently resulting from the specific inactivation of detyr-tubulin. This occurs *in vitro* at all ($\geq 20 \ \mu M$) ATP concentrations, so that detyr-tubulin would be expected not to assemble into microtubules *in vivo*. TTC has a specific requirement for assembled tubulin,[83] so that the assembled cytoskeleton should contain both tyr- and detyr-tubulin. Tyrosylation may constitute a cellular control of the effective subunit pool size, complementing the effects of phosphorylation on k_{-1} and the GTP concentration on k_{+1}.

The monoclonal antibody YL 1/2 against α tubulin is specific for tyr-tubulin.[84] Immunofluorescence shows that this antibody cross-reacts with the cytoskeletons of a wide variety of cells,[85-87] indicating that this epitope is highly conserved and that cellular microtubules can contain tyr-Tu. There is no difference, however, in the ratio of tyr:detyr tubulin between exponentially growing and mitotically arrested HeLa cells,[88] which might imply that tyrosylation does not activate the assembly-competent tubulin subunit pool. Alternatively, as interphase HeLa cells contain an extensive microtubule cytoskeleton, the size of the assembly-incompetent pool may not change significantly. By contrast, there is a marked increase in tyrosylation associated with the differentiation of cultured neuroblastoma-glioma hybrid cells.[89]

The *in vivo* tyr-tubulin pool size may be increased by *de novo* tubulin synthesis or by changing the TTL activity. Significantly, the primary gene transcript is tyrosylated[38] so that newly synthesized tubulin would be predicted to be assembly-competent. In addition, TTL activity has been reported for various organisms[90] and varies with development.[67] Similarly, the pool size may be decreased by net microtubule assembly or by increasing the activity of TTC, and a dynamic detyrosylation-tyrosylation cycle is present in muscle.[91]

The size of the effective subunit pool may, therefore, reflect the balance between increases in the tyr-tubulin concentration resulting from TTL activity and net microtubule disassembly, and reductions due to ATP-inactivation of detyr-tubulin generated by the activity of TTC and to microtubule assembly (FIGURE 5).

The linear condensation model for microtubule assembly permits the control of k_{-1}, k_{+1}, and the subunit concentration. We have shown that the nucleotide conditions exert specific effects on each of these three possible controls. In summary, the ATP concentration affects the extent of MAP-2 phosphorylation, and this modulates k_{-1}^{dimer} and the identity of the oligomeric species. The GTP concentration modulates the utilization of the MAP-2:tubulin oligomers by changing the GTP occupancy. The observed inactivation of microtubule protein by ATP suggests that tyrosylation may control the size of the tubulin pool. These studies describe, for the first time, aspects of the molecular linkage between the metabolic state of the cell and the extent and stability of the microtubule cytoskeleton.

ACKNOWLEDGMENTS

We wish to thank the Medical Research Council, the University of London, and the Cancer Research Campaign for their support during various stages of this work.

REFERENCES

1. JOHNSON, K. A. & G. G. BORISY. 1977. J. Mol. Biol. **117:** 1–31.
2. OOSAWA, F. & M. KASAI. 1962. J. Mol. Biol. **4:** 10–21.
3. MARGOLIS, R. L. & L. WILSON. 1978. Cell **13:** 1–8.
4. COTE, R. H. & G. G. BORISY. 1981. J. Mol. Biol. **150:** 577–602.
5. HERZOG, W. & K. WEBER. 1977. Proc. Natl. Acad. Sci. USA **74:** 1860–1864.
6. FARRELL, K. W., J. A. KASSIS & L. WILSON. 1979. Biochemistry **18:** 2642–2647.
7. MURPHY, D. B., K. A. JOHNSON & G. G. BORISY. 1977. J. Mol. Biol. **117:** 33–52.
8. OCHS, R. L. & M. E. STEARNS. 1981. Biol. Cell **42:**19–28.
9. CONNOLLY, J. A., V. I. KALNINS, D. W. CLEVELAND & M. W. KIRSCHNER. 1978. J. Cell Biol. **76:** 781–786.
10. IZANT, J. G. & J. R. MCINTOSH. 1980. Proc. Natl. Acad. Sci. USA **77:** 4741–4745.
11. SHERLINE, P. & K. SCHIAVONE. 1977. Science **198:** 1038–1040.
12. DE BRABANDER, M., J. C. BULINSKI, G. GEUENS, J. DE MEY & G. G. BORISY. 1981. J. Cell Biol. **91:** 438–445.
13. BULINSKI, J. C. & G. G. BORISY. 1979. Proc. Natl. Acad. Sci. USA **76:** 293–297.
14. ZIEVE, G. & F. SOLOMON. 1982. Cell **28:** 233–242.
15. DUERR, A., D. PALLAS & F. SOLOMON. 1981. Cell **24:** 203–211.
16. VALLEE, R. B. & G. S. BLOOM. 1983. Proc. Natl. Acad. Sci. USA **80:** 6259–6263.
17. BERKOWITZ, S. A., J. KATAGIRI, H. BINDER & R. C. WILLIAMS. 1977. Biochemistry **16:** 5610–5617.
18. ISLAM, K. & R. G. BURNS. 1981. FEBS Lett. **123:** 181–185.
19. PIERRE, M. & J. NUNEZ. 1983. Biochem. Biophysics Res. Commun. **115:** 212–219.
20. SELDEN, S. C. & T. D. POLLARD. 1983. J. Biol. Chem. **258:** 7064–7071.
21. LAND, H., L. F. PARADA & R. A. WEINBERG. 1983. Science **222:** 771–778.
22. PENNINGROTH, S. M. & M. W. KIRSCHNER. 1977. J. Mol. Biol. **115:** 643–673.
23. GEAHLEN, R. L. & B. E. HALEY. 1979. J. Biol. Chem. **254:** 11982–11987.
24. DAVID-PFEUTY, T., H. P. ERICKSON & D. PANTALONI. 1977. Proc. Natl. Acad. Sci. USA **74:** 5372–5376.
25. ZABRECKY, J. R. & R. D. COLE. 1982. Nature (London) **296:** 775–776.
26. ZABRECKY, J. R. & R. D. COLE. 1983. Arch. Biochem. Biophys. **225:** 475–481.
27. BURNS, R. G. & K. ISLAM. 1981. Eur. J. Biochem. **117:** 515–519.
28. PRUS, K. & M. WALLIN. 1983. FEBS Lett. **151:** 54–58.
29. SHETERLINE, P. & J. G. SCHOFIELD. 1975. FEBS Lett. **56:** 297–302.
30. BURNS, R. G. & K. ISLAM. 1984. Eur. J. Biochem. **141:** 599–608.
31. BURNS, R. G. & K. ISLAM. 1982. Eur. J. Biochem. **122:** 25–29.
32. BURGOYNE, R. D. & R. CUMMING. 1983. Eur. J. Cell Biol. **30:** 154–158.
33. MATUS, A., R. BERNHARDT & T. HUGH-JONES. 1981. Proc. Natl. Acad. Sci. USA **78:** 3010–3014.
34. BINDER, L. I., A. FRANKFURTER, H. KIM, A. CACERES, M. R. PAYNE & L. REBHUN. 1984. Proc. Natl. Acad. Sci. USA **81:** 5613–5617.
35. VALLEE, R. B., M. DIBARTOLOMEIS & W. THEURKAUF. 1981. J. Cell Biol. **90:** 568–576.
36. SHETERLINE, P. 1977. Biochem. J. **168:** 533–539.
37. NICKERSON, J. A. & W. W. WELLS. 1978. Biochem. Biophysics Res. Commun. **85:** 820–826.
38. VALENZUELA, P., M. QUIROGA, J. ZALDIVAR, W. J. RUTTER, M. W. KIRSCHNER & D. W. CLEVELAND. 1981. Nature (London) **289:** 650–655.
39. AMOS, L. A. 1977. J. Cell Biol. **72:** 642–654.
40. BURNS, R. G. 1978. J. Ultrastruc. Res. **65:** 73–82.
41. SCHEELE, R. B. & G. G. BORISY. 1978. J. Biol. Chem. **253:** 2846–2851.
42. BAYLEY, P. M., P. A. CHARLWOOD, D. C. CLARK & S. R. MARTIN. 1982. Eur. J. Biochem. **121:** 579–585.
43. BORDAS, J., E.-M. MANDELKOW & E. MANDELKOW. 1983. J. Mol. Biol. **164:** 89–135.
44. BURNS, R. G. & K. ISLAM. 1984. FEBS Lett. **173:** 67–74.
45. BURNS, R. G. & K. ISLAM. 1984. Biochem. J. **224:** 623–627.
46. THEURKAUF, W. E. & R. B. VALLEE. 1983. J. Biol. Chem. **258:** 7883–7886.

47. SLOBODA, R. D., S. A. RUDOLF, J. L. ROSENBAUM & P. GREENGARD. 1975. Proc. Natl. Acad. Sci. USA **72:** 177–181.
48. VALLEE, R. B. 1980. Proc. Natl. Acad. Sci. USA **77:** 3206–3210.
49. THEURKAUF, W. E. & R. B. VALLEE. 1982. J. Biol. Chem. **257:** 3284–3290.
50. VALLEE, R. B. & G. G. BORISY. 1977. J. Biol. Chem. **252:** 377–382.
51. VOTER, W. A. & H. P. ERICKSON. 1982. J. Ultrastruct. Res. **80:** 374–382.
52. MCEWEN, B. F., T. A. CASKA, R. H. CREPEAU & S. J. EDELSTEIN. 1983. J. Mol. Biol. **166:** 119–140.
53. BURNS, R. G. 1984. J. Cell Biol. **99:** 191a.
54. VALLEE, R. B. 1982. J. Cell Biol. **92:** 435–442.
55. BURNS, R. G., K. ISLAM & R. CHAPMAN. 1984. Eur. J. Biochem. **141:** 609–615.
56. MURTHY, A. S. N. & M. FLAVIN. 1983. Eur. J. Biochem. **137:** 37–46.
57. JAMESON, L., T. FREY, B. ZEEBERG, F. DALLDORF & M. CAPLOW. 1980. Biochemistry **19:** 2472–2479.
58. JAMESON, L. & M. CAPLOW. 1981. Proc. Natl. Acad. Sci. USA **78:** 3413–3417.
59. WEISENBERG, R. C. 1974. J. Supramol. Struct. **2:** 451–465.
60. JACOBS, M. & M. CAPLOW. 1976. Biochem. Biophysics Res. Commun. **68:** 127–135.
61. CAPLOW, M. & B. ZEEBERG. 1980. Arch. Biochem. Biophys. **203:** 404–411.
62. ISLAM, K. & R. G. BURNS. 1985. **466:** 639–641. This volume.
63. ISLAM, K. & R. G. BURNS. 1984. FEBS Lett. **178:** 264–270.
64. RODRIGUEZ, J. A. & G. G. BORISY. 1978. Biochem. Biophysics Res. Commun. **83:** 579–586.
65. PONSTINGL, H., M. LITTLE, E. KRAUHS & T. KEMPF. 1979. Nature (London) **282:** 423–424.
66. ARGARANA, C. E., H. S. BARRA & R. CAPUTTO. 1978. Mol. Cell. Biochem. **19:** 17–21.
67. RABIN, D. & M. FLAVIN. 1977. Biochemistry **16:** 2189–2194.
68. MUROFUSHI, H. 1980. J. Biochem. (Tokyo) **87:** 979–984.
69. KUMAR, N. & M. FLAVIN. 1982. Eur. J. Biochem. **128:** 215–222.
70. CARLIER, M.-F. & D. PANTALONI. 1978. Biochemistry **17:** 1908–1915.
71. FARRELL, K. W., R. H. HIMES, M. A. JORDAN & L. WILSON. 1983. J. Biol. Chem. **258:** 14148–14156.
72. WEISENBERG, R. C. 1980. J. Mol. Biol. **139:** 660–677.
73. YAMAUCHI, T. & H. FUJISAWA. 1982. Biochem. Biophysics Res. Commun. **109:** 975–981.
74. YAMAUCHI, T. & H. FUJISAWA. 1983. Biochem. Biophysics Res. Commun. **110:** 287–291.
75. SCHULMAN, H. 1984. J. Cell Biol. **99:** 11–19.
76. GOLDENRING, J. R., B. GONZALES, J. S. MCGUIRE & R. J. DELORENZO. 1983. J. Biol. Chem. **258:** 12632–12640.
77. FRANKLIN, T. J. & P. A. TWOSE. 1977. Eur. J. Biochem. **77:** 113–117.
78. BROSTROM, C. O., S. B. BOCCKINO & M. A. BROSTROM. 1983. J. Biol. Chem. **258:** 14390–14399.
79. DE BRABANDER, M., G. GEUENS, R. NUYDENS & J. DE MAY. 1982. Biol. Cell **45:** 429.
80. BERSHADSKY, A. D. & V. I. GELFAND. 1981. Proc. Natl. Acad. Sci. USA **78:** 3610–3613.
81. DEERY, W. J. & B. R. BRINKLEY. 1983. J. Cell Biol. **96:** 1631–1641.
82. STEVENS, R. E. 1972. Biol. Bull. Mar. Biol. Lab. (Woods Hole) **142:** 145–149.
83. ARCE, C. A. & H. S. BARRA. 1983. FEBS Lett. **157:** 75–78.
84. WEHLAND, J., M. C. WILLINGHAM & I. V. SANDOVAL. 1983. J. Cell Biol. **97:** 1467–1475.
85. KILMARTIN, J. V., B. WRIGHT & C. MILSTEIN. 1982. J. Cell Biol. **93:** 576–582.
86. WEHLAND, J. & M. C. WILLINGHAM. 1983. J. Cell Biol. **97:** 1476–1490.
87. CUMMING, R., R. D. BURGOYNE & N. A. LYTTON. 1984. Neurosci. **12:** 775–782.
88. BULINSKI, J. C., J. RODRIGUEZ & G. G. BORISY. 1980. J. Biol. Chem. **255:** 1684–1688.
89. NATH, J. & M. FLAVIN. 1979. J. Biol. Chem. **254:** 11505–11510.
90. KOBAYASHI, T. & M. FLAVIN. 1981. Comp. Biochem. Physiol. B. **69B:** 387–392.
91. THOMPSON, W. C., G. G. DEANIN & M. W. GORDON. 1979. Proc. Natl Acad. Sci. USA **76:** 1318–1322.

Association of Calcium/Calmodulin-Dependent Kinase with Cytoskeletal Preparations: Phosphorylation of Tubulin, Neurofilament, and Microtubule-Associated Proteins[a]

MARY LOU VALLANO,[b,d] JAMES R. GOLDENRING,[b]
ROBERT S. LASHER,[c] AND ROBERT J. DELORENZO[b,e]

[b]Department of Neurology
Yale University School of Medicine
New Haven, Connecticut 06510

[c]Department of Anatomy
University of Colorado Medical School
Denver, Colorado 80220

The neuronal cytoskeleton is responsible for the development and maintenance of cell shape and position. Microtubules and associated cytoskeletal elements have been implicated in the regulation and coordination of such dynamic cellular processes as neuroplasmic transport, mitosis, and secretion. Therefore, the endogenous factors that rapidly modulate cytoskeletal function are of considerable interest. Successful repolymerization of tubulin into microtubules *in vitro* was initially achieved in buffer containing sufficient ethyleneglycol-bis-(β-aminoethyl ether)N,N'-tetraacetic acid (EGTA) to chelate free calcium ions,[1] suggesting that calcium reduces microtubule stability. More recently, a number of investigations have demonstrated the importance of the calcium binding protein, calmodulin, in mediating calcium effects on the dynamics of microtubule function. *In situ* and *in vitro* studies indicate that low concentrations of calcium produce rapid disassembly of microtubules in a calmodulin-dependent manner.[2-4] A subpopulation of microtubules, designated as cold-stable, are especially sensitive to the depolymerizing effects of low concentrations of adenosine triphosphate (ATP), calcium, and calmodulin.[5-7] Microtubule preparations contain calmodulin-binding proteins[8-10] and calmodulin localizes on cold-stable microtubules in cultured cells.[11-12] These data suggest that calmodulin-regulated enzyme systems are associated with microtubules and may mediate calcium effects on the cytoskeleton.

[a]This research was supported by United States Public Health Service Grant NS 13532, Air Force Office of Scientific Research Grant 82-0284 (R. J. DeLorenzo) and Biomedical Research Grant BRSG-05357 (R. S. Lasher). J. R. Goldenring is a recipient of a Medical Scientist Training Program Fellowship.

[d]Present address: Department of Pharmacology, SUNY-Upstate Medical Center, 766 Irving Avenue, Syracuse, New York 13210.

[e]Present address: Department of Neurology, Medical College of Virginia, Richmond, Virginia 23298.

Evidence indicates that calcium/calmodulin-dependent phosphorylation of the microtubule-associated protein 2 (MAP-2) may play a role in microtubule assembly/disassembly dynamics. Phosphorylation of MAP-2, with a calcium/calmodulin-dependent kinase purified from brain cytosol inhibits MAP-2-stimulated microtubule assembly.[13] Moreover, addition of purified calcium/calmodulin-dependent kinase to microtubules produces rapid disassembly of microtubules only under conditions in which the MAP-2 kinase is activated.[14] Calcium/calmodulin-dependent kinase phosphorylates MAP-2 with a high specific activity in reconstituted systems[13,15] and in microtubule preparations that contain endogenous kinase and MAP-2.[10,16] Thus, this calcium/calmodulin-dependent kinase may play an important role in regulating cytoskeletal dynamics through phosphorylation of MAP-2.

Calcium/calmodulin-dependent phosphorylation of MAP-2 and other cytoskeletal proteins may also affect the interaction between microtubules and neurofilaments. *In vivo* studies have demonstrated extensive cross-bridging between neurofilaments and microtubules,[17,18] and microtubule preparations are a rich source of neurofilaments.[19,20] Neurofilaments isolated from microtubule preparations contain the neurofilament triplet polypeptides (210000, 160000, 70000 Da), tubulin and MAP-2. The presence of MAP-2 and tubulin on neurofilaments has been correlated with globular decorations projecting from the neurofilament surface.[19] MAP-2 binds to neurofilaments with high affinity,[21] and its presence may impart the neurofilament protein with the ability to promote tubulin polymerization into microtubules.[22] In cultured brain cells, MAP-2 preferentially associates with intermediate filaments, an analogue of neurofilaments, when a microtubule-disrupting agent is introduced into the cell.[23] Neurofilament and microtubule proteins are transported together in the slowest component of axonal transport.[24] More recently, a specific population of tubulin that is resistant to standard solubilization procedures has been identified in association with neurofilaments during axoplasmic transport and may be related to the tubulin in cold-stable microtubules.[25] These experiments suggest that microtubules and neurofilaments are associated *in vivo* and that MAP-2 serves as a cross-bridging protein. A possible mechanism for regulation of neurofilament/microtubule interaction is reversible phosphorylation of the cross-bridge proteins. Consistent with this hypothesis, both neurofilament proteins[26] and MAP-2[27] are phosphorylated *in vivo,* and ATP regulates the interactions between these two cytoskeletal elements *in vitro.*[28]

Our laboratory has isolated and characterized a calcium/calmodulin-dependent protein kinase from brain cytosol that phosphorylates MAP-2 and tubulin.[29] This kinase is representative of a group of identical or similar multimeric kinases, designated as Type II calcium/calmodulin-dependent kinase (CaM kinase II).[29-34] We have recently identified this kinase in microtubule preparations.[10,35] Both neurofilaments[19,20] and CaM kinase II are enriched in cold-stable microtubules. These data suggest that a significant amount of the kinase may be associated with neurofilaments and may regulate interactions between microtubules and neurofilaments.

In this report, we present evidence that CaM kinase II is present in microtubules prepared under conditions to preserve kinase activity. The enzyme phosphorylates endogenous MAP-2 with a high specific activity, and its subunits are the major calmodulin-binding proteins in microtubule preparations. In addition to endogenous MAP-2 and tubulin, synapsin I and neurofilament proteins are substrates for the kinase. A microtubule fraction that is enriched in the kinase demonstrates depolymerization in an ATP-dependent, calmodulin-dependent manner. When microtubule protein is resolved by gel filtration chromatography, the kinase is enriched in a high molecular weight fraction that also contains neurofilaments. Finally, an antibody to the kinase labels microtubules and filamentous structures in neurons. On the basis of these data, we suggest that CaM kinase II, like MAP-2, is shared between neurofilaments and microtubules and may play a role in regulating the dynamics of microtubule and cytoskeletal function in the neuron.

EXPERIMENTAL PROCEDURES

Material

[γ-^{32}P]ATP (5–10 Ci/mmol) was purchased from New England Nuclear. [^{125}I]calmodulin, produced by Bolton-Hunter labeling of fluphenazine-purified calmodulin was a gift from Dr. F. Gorelick (Department of Cell Biology, Yale University School of Medicine). A polyclonal antibody to tau proteins was a gift from Dr. M. Kirschner (Department of Biochemistry, University of California, San Francisco). A filament preparation from rat spinal cord that is enriched for the neurofilament polypeptides was a gift from Dr. F.-C. Chiu (Department of Neurology, Albert Einstein College of Medicine, Bronx, New York). Molecular weight standards were obtained from Pharmacia. Calmodulin was purified from calf brain by chromatography on diethylaminoethyl (DEAE) cellulose (DE-52) and Affigel fluphenazine (Bio-Rad).[36] All other chemicals used in the experiments were reagent grade and were obtained from commercial sources.

METHODS

Preparation of Microtubule Protein

Microtubules were prepared by cycles of temperature-dependent assembly/disassembly[37] with modifications to preserve kinase activity. Rat brains were rapidly excised and homogenized in less than 15 sec in ice-cold buffer A (100 mM 2-(N-morpholino)ethane sulfonic acid (MES), 1 mM EGTA, 0.5 mM MgCl$_2$, 0.3 mM phenylmethylsulfonyl fluoride, pH 6.75, 2 ml/brain). The homogenate was centrifuged at 100000 $\times$ g for 60 min at 4°C. The supernatant was supplemented with glycerol and guanosine triphosphate (GTP) to obtain final concentrations of 3.4 M and 1 mM, respectively, and transferred to a shaking water bath for 15 min at 37°C. The incubated mixture was centrifuged at 75000 $\times$ g for 25 min at 37°C. The microtubule pellet was then gently suspended in buffer A (1–4 mg/ml), placed on ice for 20 min to solubilize cold-labile microtubules, and centrifuged at 75000 $\times$ g for 25 min at 4°C to obtain a cold-stable microtubule pellet. The supernatant was supplemented with glycerol and GTP, incubated at 37°C for 15 min, and centrifuged at 75000 $\times$ g for 25 min at 37°C to obtain twice-cycled microtubules. In some experiments, the cycling procedure was repeated to obtain thrice-cycled microtubules. Cold-stable microtubules were further resolved into two primary peaks of protein by gel filtration chromatography on Bio-Gel A 15m (Bio-Rad). Throughout the experimental protocol, the formation or depolymerization of microtubules was monitored by negative stain electron microscopy.[38]

Determination of Calcium/Calmodulin-Dependent Kinase Activity

Calmodulin-dependent kinase activity associated with microtubule protein was assayed as previously described[29] except that 20 mM MES replaced piperazine-N,N-bis-(2-ethane sulfonic acid) (PIPES) as the buffer. Unless otherwise indicated, samples were incubated for 1 min at 37°C with 7 μM ATP. Microtubule protein was resolved by one-dimensional sodium dodecyl sulfate-polyacrylamide gel electrophoresis (SDS-PAGE) using 8.5% or 7% acrylamide as described,[39] stained with Coomassie blue or silver,[40] and exposed for autoradiography on XAR-5 (Kodak) film. Phosphorylation of specific substrates was assessed both by excision of specific bands from SDS-PAGE gels and scintillation counting[41] or by densitometric scanning of gels and autoradiographs.[42] [^{125}I]calmodulin binding to polypeptides in denaturing gels was performed as previously described.[29,43] Protein was determined by the method of Bradford using bovine serum albumin as a standard.[44]

Preparation of Antibodies against CaM Kinase II

Monoclonal antibodies against the 52000 Da subunit of CaM kinase II were prepared. The 52000 Da subunit of purified kinase was separated from the 63000 Da subunit by SDS-PAGE and blotted on nitrocellulose. Pieces of nitrocellulose containing the 52000 Da subunit were used to immunize an F1 female (C57B1/6J female X A/J male) mouse, and spleen cells from this mouse were then fused with PAI-O myeloma cells (a variant of P3 × 63-Ag 8.653 cells) using polyethylene glycol 4000. The initial colonies and subsequent clones obtained by limiting dilution were screened against both the major postsynaptic density protein and CaM kinase II. An IgM monoclonal antibody (39-29:X) with high affinity for the 52000 Da subunit of the kinase was selected for use in the experiments. Specific labeling of neuronal structures in the cerebral cortex and hippocampus were examined in Vibratome sections of rat brains fixed by perfusion with either 5% paraformaldehyde (PF) alone or combined with 2% glutaraldehyde (PF+G). Incubation with primary and secondary antibodies was carried out in the absence or presence of 0.3% Triton X-100, respectively. The secondary antibody used was a horseradish peroxidase-conjugated, affinity-purified goat anti-mouse IgG (TAGO). After osmication, the Vibratome sections were embedded in PolyBed 812, thin-sectioned, and examined without counterstaining in a Philips EM 300.

RESULTS

Calcium/Calmodulin-Dependent Kinase Activity in Microtubule Preparations

Using appropriate preparation conditions, calmodulin-dependent kinase activity was observed in microtubule preparations. FIGURE 1 shows the protein pattern and autoradiograph of a thrice-cycled microtubule protein that was incubated under standard phosphorylation conditions and subsequently resolved on one-dimensional SDS-PAGE. In the presence of calcium and calmodulin, specific incorporation of radioactive phosphate was observed in several proteins with major bands corresponding to 52000 Da, 80000 Da, and 280000 Da. Inclusion of 10 μM trifluoperazine, a calmodulin inhibitor, in the reaction tubes blocked calmodulin stimulation. The identity of the 52000–63000 Da phosphoproteins as alpha tubulin, beta tubulin, and the autophosphorylated rho and sigma subunits of the calmodulin-dependent kinase, and of the 280000 Da phosphoprotein as MPA-2 has been shown previously by two-dimensional isoelectric focusing/SDS-PAGE and phosphopeptide mapping.[10,35] The 80000 Da doublet has recently been identified as synapsin I on the basis of mobility in nonequilibrium pH gradient electrophoresis, phosphopeptide mapping, and comigration with purified synapsin I (Goldenring et al., manuscript in preparation).

It was imperative to use rapid brain excision and homogenization in the presence of protease inhibitors in conjunction with rapid microtubule preparation procedures in order to preserve calmodulin-dependent kinase activity in microtubule preparations. The *postmortem* lability of calcium/calmodulin-dependent protein kinases is well documented.[45,46] We have also observed a gradual loss of calcium/calmodulin-

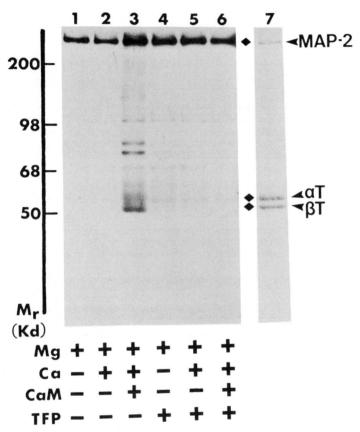

FIGURE 1. Calmodulin-dependent kinase activity in microtubules. Autoradiograph (lanes 1–6) and protein pattern (lane 7) of thrice-cycled, phosphorylated microtubule protein resolved by one-dimensional SDS-PAGE (8.5% acrylamide). Protein was incubated for 1 min at 37°C in MES buffer containing the additions indicated on the figure. Mg = magnesium; Ca = calcium; CaM = calmodulin; TFP = trifluoperazine. The positions of MAP-2, alpha tubulin (αT) and beta tubulin (βT) are depicted.

dependent kinase activity if the brains are not rapidly homogenized after excision. In addition, incubation of brain cytosolic protein at 37°C to allow polymerization of microtubules inactivates a significant fraction of the kinase present in cytosol (FIGURE 2). Thus, standard methods of microtubule preparation result in a loss of approximately 80% of the calcium/calmodulin-dependent kinase activity after 15 min

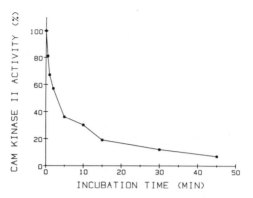

FIGURE 2. Inactivation of calmodulin-dependent kinase activity. Brain cytosol was incubated at 37°C for 0–45 min and subsequently examined for calmodulin-dependent kinase activity by the standard phosphorylation assay described above. Phosphoproteins were excised from one-dimensional SDS-PAGE, and incorporated phosphate was quantitated by liquid scintillation counting. The relative amount of autophosphorylation of the rho subunit of the calmodulin-dependent kinase is shown.

incubation of cytosol at 37°C. This observation may account for the lack of information on the effects of calcium/calmodulin-dependent kinases in microtubule preparations. The precise nature of the enzyme inactivation is not known, but in the case of MAP-2, proteolytic digestion appears to be partially responsible for the observed loss of MAP-2 phosphorylation.

As described previously, the kinase was present in multiply-cycled microtubules and was enriched in cold-stable microtubules.[10,35] Approximately 56% of the calmodulin-dependent MAP-2 kinase activity in brain cytosol was recovered in the cold-stable fraction, and the specific activity of the enzyme was enriched by 16-fold. We have recently shown that the calmodulin-dependent kinase in cold-stable microtubule preparations is identical by all criteria tested to CaM kinase II purified from brain cytosol.[10]

Previous studies have demonstrated that cold-stable microtubules are especially sensitive to the depolymerizing effects of calcium and calmodulin, and that this effect is ATP-dependent.[5–7] In order to confirm that the CaM kinase II in our preparation was effective in mediating the depolymerization of microtubules, we examined the effects of calcium, calmodulin, and ATP on cold-stable microtubules using spectrophotometry and electron microscopy to follow microtubule depolymerization. We observed depolymerization of microtubules in a calcium, calmodulin, and ATP-dependent manner. These data suggest that microtubule depolymerization may be mediated by activation of endogenous CaM kinase II in this preparation.

Calmodulin-Binding Proteins in Microtubules

Earlier studies have demonstrated that calmodulin localizes on microtubules *in situ,* and that it specifically labels cold-stable microtubules.[11,12] *In vitro* studies have demonstrated that microtubule preparations contain several calmodulin-binding proteins, including MAP-2 and tau proteins.[8–10] Our laboratory has observed that the subunits of CaM kinase II in microtubule preparations are also major calmodulin-binding proteins.[10,35] FIGURE 3 shows that microtubule preparations contain two major calmodulin-binding proteins that comigrate with the rho and sigma subunits of the kinase. Because the molecular masses of the enzyme subunits (52000–63000 Da) are in the same range as tau proteins (55000–62000 Da), one might argue that the calmodulin-binding proteins that we have indentified as CaM kinase II are tau proteins. Two approaches were used to resolve this issue. First, two-dimensional isoelectric focusing/SDS-PAGE gel electrophoresis demonstrated that the major calmodulin-binding proteins in microtubule preparations comigrated with the subunits of the purified kinase. Next, we examined whether a polyspecific antibody prepared against tau proteins reacted with purified CaM kinase II. Based on immunoblot analysis, there was no immunoreactivity between CaM kinase II and tau antibody, suggesting that these proteins are distinct.

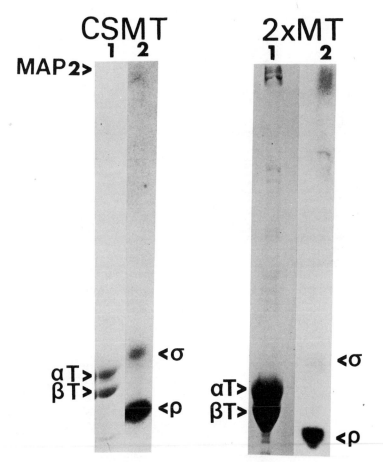

FIGURE 3. Calmodulin-binding proteins in microtubules. High-resolution SDS-PAGE showing protein pattern (lane 1) and autoradiograph (lane 2) of [^{125}I]calmodulin-binding proteins in cold-stable (CSMT) and twice-cycled (2xMT) microtubules. The positions of alpha tubulin (αT), beta tubulin (βT), rho (ρ), sigma (σ), and MAP-2 are depicted.

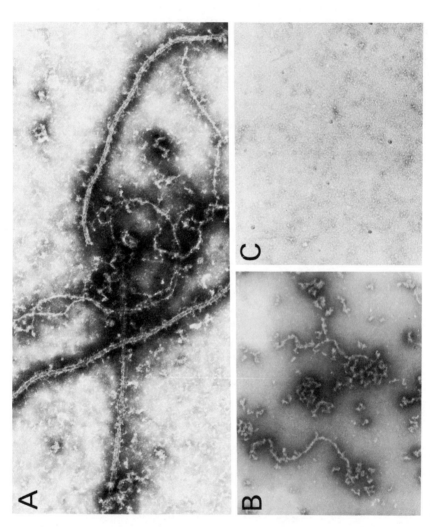

FIGURE 4. Microtubules and 10 nm filaments in a cold-stable microtubule preparation. **A:** Negative stain electron micrograph of a cold-stable microtubule preparation showing 10 nm filaments (neurofilaments) and microtubules. **B:** Void fraction. **C:** Back peak fraction. (Original magnification: 50,000×; reduced by 16 percent.)

Enrichment of CaM Kinase II in a Neurofilament Fraction
Derived from Microtubules

When microtubules are prepared by multiple cycles of temperature-dependent assembly/disassembly, CaM kinase II activity,[10,35] neurofilaments,[19] calmodulin-binding proteins, and calcium/calmodulin sensitivity[3,5–7] are gradually lost from the microtubule fractions. CaM kinase II[10,35] and neurofilaments[19,20] are enriched in cold-stable microtubules. FIGURE 4A shows that neurofilaments are present in a preparation of cold-stable microtubules. The parallel loss of neurofilaments and CaM kinase II from microtubule preparations suggests that a significant amount of this enzyme is associated with neurofilaments in these preparations. In view of the evidence that microtubules and neurofilaments are linked in the cell, we examined whether CaM kinase II coelutes with neurofilaments during chromatography of microtubules. A cold-stable microtubule preparation containing neurofilaments was separated into two primary peaks of protein by gel filtration chromatography on Bio-Gel A 15. The

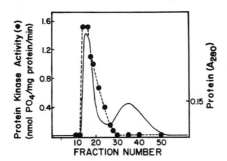

FIGURE 5. Gel filtration chromatography of microtubules. Cold-stable microtubules were prepared, fragmented by repeated passage through a syringe fitted with a 26-gauge needle, and chromatographed on Bio-Gel A 15m. Fractions from the column were phosphorylated under standard conditions and resolved on one-dimensional SDS-PAGE (7% acrylamide). The elution of protein was monitored at 280 nm (———), and fractions were assayed for calmodulin-dependent kinase activity (●----●). Column buffer contained 10 mM MES, 0.1 mM MgCl$_2$, 200 mM NaCl, pH 6.7.

front peak of protein, designated the void peak, contained numerous 10 nm filaments (FIGURE 4B), whereas the trailing peak of protein contained only small globular particles (FIGURE 4C). All of the recovered calmodulin-dependent kinase activity eluted in the void protein peak, along with 10 nm filaments (FIGURE 5). The kinase activity in the void fraction was enriched 2-fold by chromatography of cold-stable microtubules, representing an enrichment of 32-fold over cytosol.

The protein pattern and autoradiograph for cold-stable microtubules, void protein peak, and trailing protein peak are shown in FIGURE 6. For comparison, the neurofilament triplet polypeptides present in a cytoskeletal preparation are shown. These proteins were present in cold-stable microtubules, enriched in the void fraction, but the trailing protein peak did not contain neurofilament proteins. These data demonstrate that CaM kinase II activity is enriched in a neurofilament fraction from cold-stable microtubules, and further suggest that standard methods of microtubule preparation removes this potentially important regulator of cytoskeletal function from multiply-cycled microtubules. The possibility that CaM kinase II functions solely in

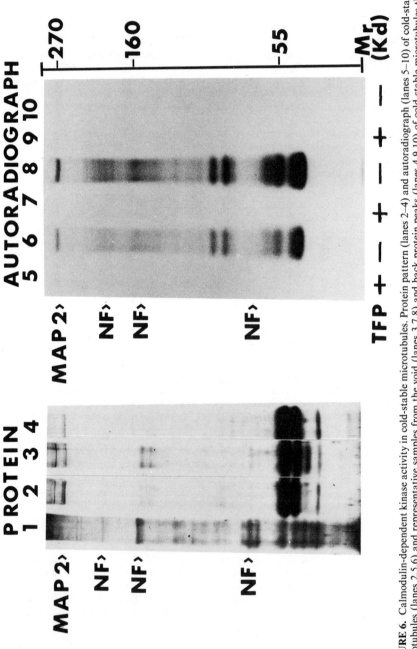

FIGURE 6. Calmodulin-dependent kinase activity in cold-stable microtubules. Protein pattern (lanes 2–4) and autoradiograph (lanes 5–10) of cold-stable microtubules (lanes 2,5,6) and representative samples from the void (lanes 3,7,8) and back protein peaks (lanes 4,9,10) of cold-stable microtubules that were chromatographed on Bio-Gel A 15m as described in FIGURE 5. For comparison, a cytoskeletal fraction enriched in neurofilament proteins is shown (lane 1). The positions of MAP-2 and the three neurofilament proteins are depicted. The phosphorylation assay was performed under standard conditions in the presence of magnesium, calcium, and calmodulin ± 100 µM trifluoperazine (TFP).

the dynamic regulation of the cold-stable subpopulation of microtubules and their interaction with neurofilaments, however, is a viable alternative.

Neurofilament Protein Phosphorylation by CaM Kinase II

Neurofilaments are phosphorylated *in vivo*,[26] and evidence suggests that phosphorylation stabilizes neurofilaments.[47] Therefore, it was of interest to determine whether the neurofilament polypeptides are substrates for CaM kinase II. A cytoskeletal preparation that is enriched for the neurofilament triplet polypeptides was

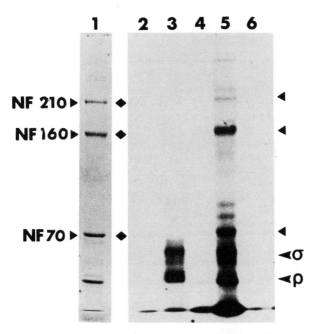

FIGURE 7. Neurofilament protein phosphorylation by CaM kinase II. Cytoskeletal protein (lane 1) was incubated under standard phosphorylation conditions with CaM kinase II purified from brain cytosol and subsequently resolved on one-dimensional SDS-PAGE (7% acrylamide). The autoradiograph shows CaM kinase II alone in the absence (lane 2) and presence (lane 3) of calmodulin; CaM kinase II plus cytoskeletal protein in the absence (lane 4) and presence (lane 5) of calmodulin; and cytoskeletal protein alone in the presence of calmodulin (lane 6).

examined for calmodulin-dependent phosphorylation using purified cytosolic CaM kinase II. FIGURE 7 shows the protein pattern and autoradiograph of samples that were phosphorylated and resolved on one-dimensional SDS-PAGE. These data demonstrate that all three neurofilament proteins are phosphorylated by the enzyme in a calmodulin-dependent manner. Under standard conditions, the stoichiometry of phosphate incorporation into neurofilament protein by the kinase ranged from 0.5–1.0 mol PO_4/mol protein. A careful study, however, using phosphatases to remove endogenous phosphate from neurofilament proteins is required before an accurate determination can be made.

CaM Kinase II Antibody Labeling of Cytoskeletal Structures

In order to determine whether CaM kinase II is localized on cytoskeletal structures in neurons, monoclonal antibody 39-29:X to the enzyme was prepared, and its intracellular distribution was examined. As seen in FIGURE 8, antibody 39-29:X specifically labels microtubules (FIGURE 8A–C) and neurofilaments (FIGURE 8C) in dendrites in the rat cerebral cortex. Antigen also appears to be associated with filamentous material surrounding microtubules (FIGURE 8A). In some areas, antigen apparently on microtubules is in the form of 25 nm diameter spheres having a center-to-center spacing of about 35 nm (FIGURE 8A). These data are consistent with another report describing microtubule labeling in neurons by a monoclonal antibody against CaM kinase II.[48]

DISCUSSION

The data presented here and elsewhere are consistent with an emerging view that calmodulin-dependent kinases modify the cytoskeleton. A significant amount of neuronal CaM kinase II appears to be fixed in the fibrous matrix of the postsynaptic density or associated with the neurofilament/microtubule axis.[48–51] The presence of the kinase on the cytoskeleton may provide a mechanism for rapid modulation of the interaction between these structures by autophosphorylation or phosphorylation of associated cytoskeletal proteins.

Several proteins are potential candidates for mediating the effects of calcium on cytoskeletal function. Phosphorylation of MAP-2, tubulin, neurofilament protein, synapsin I, and the enzyme subunits occurs rapidly and is dependent upon calcium and calmodulin.[10,35] Most of this information has been obtained recently, and further studies are needed to determine the physiological role of these substrates in the regulation of cytoskeletal dynamics. Phosphorylation of MAP-2 by both cAMP-dependent[23] and calmodulin-dependent kinases,[13–15] however, has received considerable attention as a possible mechanism mediating microtubule assembly/disassembly and microtubule/neurofilament interactions.

The MAP-2 molecule consists of an assembly-promoting domain, which interacts with the microtubule, and a projection portion that appears to cross-bridge microtubules with other microtubules, neurofilaments, and membrane.[23] MAP-2 is phosphorylated *in vivo*[27] and a type II cAMP-dependent kinase is tightly bound to a fraction of the MAP-2 that copurifies with microtubules.[23] The cAMP-dependent kinase phosphorylates both the assembly-promoting and projection portion of the MAP-2 molecule, suggesting that it may regulate both assembly/disassembly dynamics and interaction between microtubules and other cytoskeletal elements. *In vitro* studies indicate that approximately 13 of the 22 available sites on MAP-2 are phosphorylated in a cAMP-dependent manner, whereas the remaining sites are phosphorylated in a cAMP-independent fashion.

CaM kinase II is a good candidate for regulating the phosphorylation state of MAP-2[10,29,34,35] and the dynamics of microtubule assembly/disassembly.[13–15] CaM kinase II is present in microtubule preparations and it phosphorylates endogenous MAP-2 rapidly and with a high specific activity.[10,35] Phosphopeptide-mapping studies indicate that CaM kinase II may account for all of the cAMP-independent phosphorylation sites on MAP-2.[52] Moreover, because cAMP-dependent kinase and CaM kinase II phosphorylate distinct sites on MAP-2,[52,53] they may differentially regulate MAP-2

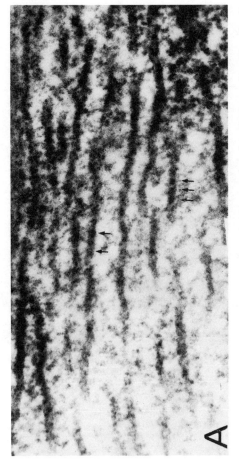

FIGURE 8 A, B, and **C.** CaM kinase II antibody labeling of microtubules. Electron micrographs of thin sections through a dendrite from the superficial layer of the rat cerebral cortex. **A:** Longitudinal section. Arrows indicate location of antigen on microtubules in the form of 25 nM diameter spheres. (Fix: PF + G; magnification :64,000×) **B:** Cross to oblique section. (Fix: PF + G; Magnification: 64,000×). **C:** Cross section of a small dendrite in the hippocampus. Both microtubules (large arrows) and filaments (small arrows) are covered with reaction product. (Fix: PF; Magnification: 118,500×). No label was observed in control sections.

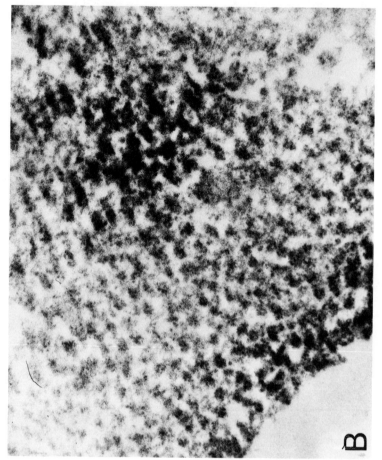

FIGURE 8B. See legend on page 369.

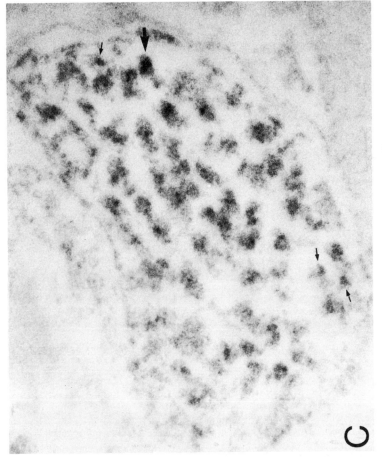

FIGURE 8C. See legend on page 369.

function. MAP-2 consists of two separate proteins, designated MAP-2A and MAP-2B. We are currently examining the possibility that one of these MAPs and CaM kinase II are specifically associated with the neurofilaments that are present in microtubule preparations. Possibly, MAP-2 and CaM kinase II serve as detachable cross-bridges between microtubules and neurofilaments.

FIGURE 9 is a simplified schematic model that illustrates the possible locations of CaM kinase II on microtubules and neurofilaments. *In situ* antibody labeling studies suggest that CaM kinase II is associated with microtubules and filamentous structures in apposition to the microtubule axis (FIGURE 8).[48] *In vitro* isolation experiments indicate that the kinase preferentially associates with neurofilament fractions during the separation of neurofilaments from microtubule preparations (FIGURES 5,6).[35] These data support the hypothesis that CaM kinase II is associated either directly with microtubules and neurofilaments or with the side arms that project from these

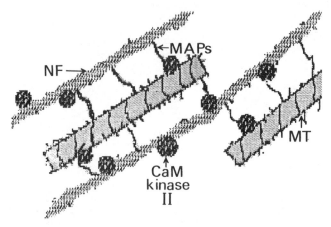

FIGURE 9. Possible locations of CaM kinase II on the cytoskeleton. Schematic representation showing the possible locations of CaM kinase II on neurofilaments (NF), microtubules (MT), and microtubule-associated proteins (MAPs).

cytoskeletal structures (*e.g.* MAP-2). Further studies, however, are required to demonstrate the precise location of CaM kinase II on the cytoskeleton.

Most of the evidence available to date indicates that phosphorylation of MAP-2 diminishes its ability to promote tubulin assembly into microtubules by reducing the interaction between MAP-2 and tubulin. The effect of MAP-2 phosphorylation on its interactions with other cytoskeletal and membrane elements remains to be determined. It is tempting to hypothesize that phosphorylation might transiently induce detachment of the MAP-2 cross-bridge and thereby reduce interactions between cytoskeletal elements. Such alterations in the synaptic cytoarchitecture might provide a biochemical mechanism for the rapid transfer of intraneuronal and interneuronal information. In the axon, this process may subserve transport of proteins and organelles. In the synaptic terminal, modification of the cytoskeletal postsynaptic density may be necessary for insertion of neurotransmitter receptors or other important transmembrane proteins into the postsynaptic membrane.

SUMMARY

Calcium and calmodulin have been implicated in the regulation of cytoskeletal function. In this report, we demonstrate that microtubule preparations from rat brain contain a calcium/calmodulin-dependent protein kinase that phosphorylates endogenous MAP-2, tubulin, synapsin I, and neurofilament proteins. This cytoskeletal-associated kinase has been biochemically characterized and shown to be identical to Type II calcium/calmodulin-dependent protein kinase (CaM kinase II). The subunits of CaM kinase II represented major calmodulin-binding proteins in cytoskeletal preparations. A monoclonal antibody against the 52000 Da subunit of CaM kinase II specifically labeled cytoskeletal elements in cortical neurons. These results indicate that CaM kinase II is associated with the neuronal cytoskeleton and may play a role in mediating some of the effects of calcium on cytoskeletal function.

ACKNOWLEDGMENT

The authors are grateful to Mr. John Albert for assistance with the preparation of this manuscript.

REFERENCES

1. WEISENBERG, R. C. 1972. Science. **177:** 1104–1105.
2. MARCUM, J. M., J. R. DEDMAN, B. R. BRINKLEY & A. R. MEANS. 1978. Proc. Natl. Acad. Sci. USA **75:** 3771–3775.
3. NISHIDA, E., H. KUMAGAI, I. OHTSUKI & H. SAKAI. 1979. J. Biochem. (Tokyo) **85:** 1257–1266.
4. KEITH, C., M. DIPAOLA, F. R. MAXFIELD & M. L. SHELANSKI. 1983. J. Cell Biol. **97:** 1918–1924.
5. JOB, D., E. H. FISCHER & R. L. MARGOLIS. 1981. Proc. Natl. Acad. Sci. USA **78:** 4679–4682.
6. PIROLLET, F., C. T. RAUCH, E. H. FISCHER & R. L. MARGOLIS. 1983. Proc. Natl. Acad. Sci. USA **80:** 1560–1565.
7. JOB, D., C. T. RAUCH, E. H. FISCHER & R. L. MARGOLIS. 1983. Proc. Natl. Acad. Sci. USA **80:** 3894–3898.
8. SOBUE, K., M. FUJITA, Y. MURAMOTO & S. KAKIUCHI. 1981. FEBS. Lett. **132:** 137–140.
9. LEE, Y. C. & J. WOLFF. 1984. J. Biol. Chem. **259:** 1226–1230.
10. LARSON, R. E., J. R. GOLDENRING, M. L. VALLANO & R. J. DELORENZO. 1985. J. Neurochem. **44:** 1566–1574.
11. WELSH, M. J., J. R. DEDMAN, B. R. BRINKLEY & A. R. MEANS. 1979. J. Cell Biol. **81:** 624–634.
12. DEERY, W. J., A. R. MEANS & B. R. BRINKLEY. 1984. J. Cell Biol. **98:** 904–910.
13. YAMAMOTO, H., K. FUKUNAGA, E. TANAKA & E. MIYAMOTO. 1983. J. Neurochem. **41:** 1119–1125.
14. YAMAUCHI, T. & H. FUJISAWA. 1983. Biochem. Biophysics Res. Commun. **110:** 287–291.
15. YAMAUCHI, T. & H. FUJISAWA. 1982. Biochem. Biophysics Res. Commun. **109:** 975–981.
16. VALLANO, M. L. & R. J. DELORENZO. 1986. Ann. N.Y. Acad. Sci. **466:** 453–456. This volume.
17. WUERKER, R. 1970. Tissue Cell **2:** 1–19.
18. HIROKAWA, N. 1982. J. Cell Biol. **94:** 129–142.
19. BERKOWITZ, S. A., J. KATAGIRI, H. K. BINDER & R. C. WILLIAMS, Jr. 1977. Biochemistry **16:** 5610–5617.

20. THORPE, R. A., A. DELACOURTE, M. AYERS, C. BULLOCK & B. H. ANDERTON. 1979. Biochem. J. **181:** 275–284.
21. LETERRIER, J.-F., R. K. H. LIEM & M. L. SHELANSKI. 1982. J. Cell Biol. **95:** 982–986.
22. LETERRIER, J.-F., J. WONG, R. K. H. LIEM & M. L. SHELANSKI. 1984. J. Neurochem. **43:** 1385–1391.
23. VALLEE, R. B., G. S. BLOOM & W. E. THEURKAUF. 1984. J. Cell Biol. **99:** 38s–44s.
24. BLACK, M. M. & R. J. LASEK. 1980. J. Cell Biol. **86:** 616–623.
25. ERICKSON, P. F., K. B. SEAMON, B. W. MOORE, R. S. LASHER & L. N. MINIER. 1980. J. Neurochem. **35:** 242–248.
26. JULIEN, J.-P. & W. E. MYSHYNSKI. 1981. J. Neurochem. **37:** 1579–1585.
27. SLOBODA, R. D., S. A. RUDOLPH, J. L. ROSENBAUM & P. GREENGARD. 1975. Proc. Natl. Acad. Sci. USA **72:** 177–181.
28. MINAMI, Y., H. MUROFUSHI & H. SAKAI. 1982. J. Biochem. **92:** 889–898.
29. GOLDENRING, J. R., B. GONZALEZ, J. S. MCGUIRE, JR. & R. J. DELORENZO. 1983. J. Biol. Chem. **258:** 12632–12640.
30. FUKUNAGA, K., H. YAMAMOTO, K. MATSUI, K. HIGASHI & E. MIYAMOTO. 1982. J. Neurochem. **39:** 1607–1617.
31. BENNETT, M. K., N. E. ERONDU & M. B. KENNEDY. 1983. J. Biol. Chem. **258:** 12735–12744.
32. MCGUINNESS, T. L., Y. LAI, P. GREENGARD, J. R. WOODGETT & P. COHEN. 1983. FEBS Lett. **163:** 329–334.
33. YAMAUCHI, T. & H. FUJISAWA. 1983. Eur. J. Biochem. **132:** 15–21.
34. SCHULMAN, M. 1984. J. Cell Biol. **99:** 11–19.
35. VALLANO, M. L., J. R. GOLDENRING, T. M. BUCKHOLZ, R. E. LARSON & R. J. DELORENZO. 1985. Proc. Natl. Acad. Sci. USA **82:** 3202–3206.
36. DELORENZO, R. J. & J. R. GOLDENRING. 1984. Brain Receptor Methodologies. pp. 191–207. Academic Press. New York
37. SHELANSKI, M. L., F. GASKIN & C. R. CANTOR. 1973. Proc. Natl. Acad. Sci. USA **70:** 765–768.
38. ATKINSON, M. A. L., J. S. MORROW & V. T. MARCHESI. 1982. Cell Biochem. **18:** 143–155.
39. LAEMMLI, U. K. 1970. Nature (London) **227:** 680–685.
40. MERRIL, C. R., D. GOLDMAN, S. A. SEDMAN & M. H. EBERT. 1981. Science **211:** 1437–1438.
41. BURKE, B. E. & R. J. DELORENZO. 1981. Proc. Natl. Acad. Sci. USA **78:** 991–995.
42. DELORENZO, R. J., G. P. EMPLE & G. H. GLASER. 1977. J. Neurochem. **28:** 21–30.
43. CARLIN, R. K., D. J. GRAB & P. SIEKEVITZ. 1981. J. Cell Biol. **89:** 449–455.
44. BRADFORD, M. M. 1976. Anal. Biochem. **72:** 248–254.
45. DUNKLEY, P. R. & P. J. ROBINSON. 1981. Biochem. J. **199:** 269–272.
46. JUSKEVICH, J. C., D. M. KUHN & W. LOVENBERG. 1982. Biochem. Biophysics Res. Commun. **108:** 24–30.
47. STERNBERGER, L. A. & N. H. STERNBERGER. 1983. Proc. Natl. Acad. Sci. USA **80:** 6126–6130.
48. OUIMET, C. C., T. L. MCGUINNESS & P. GREENGARD. 1984. Proc. Natl. Acad. Sci. USA **81:** 5604–5608.
49. KENNEDY, M. B., M. K. BENNETT & N. E. ERONDU. 1983. Proc. Natl. Acad. Sci. USA **80:** 7357–7361.
50. KELLEY, P. T., T. L. MCGUINNESS & P. GREENGARD. 1984. Proc. Natl. Acad. Sci. USA **81:** 945–949.
51. GOLDENRING, J. R., J. S. MCGUIRE, JR. & R. J. DELORENZO. 1984. J. Neurochem. **42:** 1077–1084.
52. GOLDENRING, J. R., M. L. VALLANO & R. J. DELORENZO. 1985. J. Neurochem. **45:** 900–905.
53. SCHULMAN, H. 1984. Mol. Cell. Biol. **4:** 1175–1178.

Calcium and Calmodulin in the Regulation of the Microtubular Cytoskeleton[a]

C. H. KEITH,[b] A. S. BAJER,[c] R. RATAN,[d] F. R.
MAXFIELD,[d] AND M. L. SHELANSKI[d]

[b]Department of Zoology
University of Georgia
Athens, Georgia 30602

[c]Department of Biology
University of Oregon
Eugene, Oregon 97403

[d]Department of Pharmacology
New York University School of Medicine
New York, New York 10016

Calcium has been found to act as an intracellular second messenger in a wide variety of cellular systems, as recently reviewed by Campbell.[1] In most cases, calmodulin has been found to act as an intracellular receptor in the translation of calcium ion fluxes into enzyme activation or inhibition. For example, calcium-saturated calmodulin activates the brain cyclic nucleotide phosphodiesterase (Kakiuchi and Yazamaki,[2] Cheung[3]), and the red blood cell calcium-activated ATPase (Vincenzi et al.[4]), among a great many others.

Because Weisenberg's demonstration that microtubules in crude extract are sensitive to low levels of free calcium (Weisenberg[5]), this ion has been suspected to act as a physiological regulator of the dynamic equilibrium of microtubules. In the test tube, microtubule polymerization is found to be sensitive to calcium in the range of 10–500 micromolar, depending on the purity of the microtubule preparation (Rosenfeld et al.[6]; Olmsted and Borisy[7]). The mitotic spindle, which is clearly the site of extensive microtubule polymerization and depolymerization, is sensitive to free calcium in the range of a few micromolar, both in the cell (Kiehart[8]; Izant[9]) and after partial (Cande[10]) or complete isolation (Silver et al.[11]; Salmon and Segall[12]) of the mitotic apparatus. The progress of mitosis is also sensitive to extracellular calcium (Chai and Sandberg[13]; Wagenaar[14]), and to caffeine (Harris[15]), a drug that is known to cause release of calcium from the sarcoplasmic reticulum of muscle cells. In addition, the mitotic spindle has been found to contain membrane-bound compartments that have high internal calcium levels (Wick and Hepler[16]; Schatten et al.[17]), and, after isolation, can actively concentrate calcium from the medium (Silver et al.[11]). Finally, in a number of other different microtubule-associated motile systems, such as retinal cone elongation (Gilson and Burnside[18]) and pigment granule migration (Schliwa[19]; Luby-Phelps and Porter[20]), micromolar calcium has been shown to be inhibitory, suggesting that the motile machinery in these systems is sensitive to low levels of calcium.

[a]This work was supported by Grants from the American Cancer Society (CD-129) and the NIH (NS 15076) to M. L. Shelanski, from the NIH (AM 27083) and the Irma T. Hirschl Charitable Trust to F. R. Maxfield, and from the NSF (PCM83-10016) to A. S. Bajer.

375

A number of observations have implicated calmodulin in the transduction of calcium's effect to microtubules. In the first place, the mitotic spindle has been shown to have high concentrations of calmodulin in the chromosome-to-pole region, both in immunofluorescence (Welsh et al.[21]) and fluorescent analogue cytochemical localizations (Zavortnik et al.[22]). Calmodulin can also augment the in vitro calcium sensitivity of microtubules to the micromolar range when present in considerable excess over tubulin. Finally, calmodulin is found to be localized along the cold-stable microtubules of lysed cell models (Deery et al.[23]) and can interact with the stable tubule only polypeptide (STOP) proteins (which stabilize microtubules to cold) to sensitize microtubules to physiological calcium levels (Margolis[24]).

The above observations have lent rather strong circumstantial support to the notion that calcium and calmodulin can act as regulators of microtubule dynamics in the cell. In order to make a stronger case for the regulation of microtubules by calcium and calmodulin, we need to know how microtubules interact with calcium-saturated and calcium-free calmodulin in the intact cell. We would also like to know the pattern of calcium variation in a microtubule-associated motile system in which there is evidence for regulation by calmodulin. We have undertaken experiments designed to address both of these issues. First, in order to elucidate the effect, both extent and stoichiometry, of calmodulin on cellular microtubules in situ, we have injected calcium-saturated and calcium-free calmodulin into living fibroblasts, and visualized the consequences of this perturbation on cellular microtubules by antitubulin immunohistochemical staining. We have found that calcium-saturated calmodulin is capable of disrupting cellular microtubules in a localized fashion at a 2:1 excess over tubulin. This disruption is specific to the calcium-saturated form of calmodulin and persists for at least ninety minutes after injection. Second, we have measured both cell-wide and local calcium ion concentrations in the cytoplasm of mitotic cells. We have found that cellular calcium levels are lowered in the cell as a whole during mitosis, but during anaphase they are specifically and locally elevated at the spindle poles. These experiments are consistent with models in which calcium and calmodulin act in concert to locally regulate the behavior of the mitotic spindle.

MICROINJECTION OF CALMODULIN INTO INTERPHASE CELLS

Calcium-saturated calmodulin ($0.1-0.25$ mM), mixed with a known volume of FITC-labeled bovine serum albumin, was injected into interphase gerbil fibroma cells in an injection buffer consisting of 10 mM sodium phosphate and 137 mM KCl. The visualization of microtubules by immunofluorescence techniques revealed that this treatment caused the rapid loss of polymerized microtubules around the site of injection (FIGURE 1). This disruption of microtubules is sharply delimited, with free ends of microtubules visualized at the border of the region of disruption. Microtubules outside the depolymerized area appear normal. Immunofluorescence staining of calmodulin in injected cells reveals a similar pattern of localization around the site of injection (FIGURE 2), providing that the cell's internal calcium is maintained high by use of a calcium ionophore. The effect of calcium-calmodulin injection on microtubules persists for at least ninety minutes; at later time points, immunofluorescence staining of microtubules reveals the initiation of new microtubule growth from the presumptive microtubule organizing center, even while the peripheral fringe of old microtubules persists (FIGURE 3). Stress fibers are also disrupted around the site of calcium-calmodulin injection, albeit in a more limited region than the microtubules (FIGURE 4). Intermediate filaments of microinjected cells, revealed by antivimentin immunofluo-

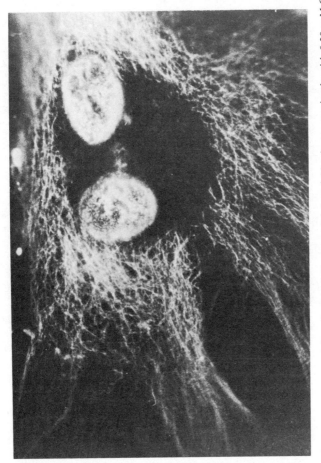

FIGURE 1. Antitubulin immunofluorescence staining of a gerbil fibroma cell injected 20 minutes previously with 0.20 mM Ca$_4$-calmodulin. Note the sharply delimited disruption of microtubules around the site of injection, between the nuclei of this binucleate cell. ×440. (C. H. Keith *et al.*[31] With permission from the *Journal of Cell Biology*.)

rescent staining, are intact throughout the cell, although their arrangement may be somewhat disturbed (FIGURE 5).

The disruption of microtubules in injected cells is specific to the calcium-saturated form of calmodulin: when cells were injected with a solution containing 0.25 mM calmodulin and 1.2 mM [ethylene-bis-(oxyethylene nitrilo)]-tetracetic acid (EGTA), their microtubules were undisturbed (FIGURE 6). In addition, in our injection buffer, 1.0 mM calcium chloride had no discernible effect on the microtubules of injected cells (FIGURE 7).

In addition to driving the disruption of microtubules, the calcium saturation of microinjected calmodulin controls its localization within the cell. As mentioned above,

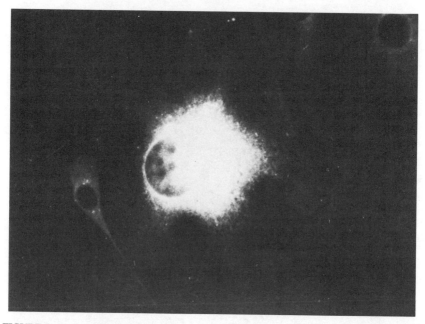

FIGURE 2. Anticalmodulin immunofluorescence staining of a gerbil fibroma cell, injected 20 minutes previously with 0.25 mM Ca$_4$-calmodulin. Calcium saturation of the cytoplasm was maintained throughout incubation and fixation by the inclusion of the calcium ionophore A23187 and 100 μM CaCl$_2$ in the medium. ×375. (C. H. Keith *et al.*[31] With permission from the *Journal of Cell Biology.*)

immunofluorescence staining of microinjected calmodulin only reveals a localized distribution when the calmodulin is maintained in the calcium-saturated state. In addition, when cells are injected with EGTA-calmodulin (1 mM EGTA + 0.25 mM calmodulin) and later treated with a calcium ionophore in the presence of 25 micromolar calcium, their microtubules are disrupted throughout the cell, rather than locally (FIGURE 8A). Uninjected sister cells in the same dish showed normal displays of microtubules, demonstrating that the calcium ionophore itself did not disturb the microtubules in the presence of normal levels of cellular calmodulin (FIGURE 8B). Finally, cells that had been microinjected with calcium-saturated calmodulin still showed local microtubule disruption (FIGURE 8C). This last experiment illustrates that

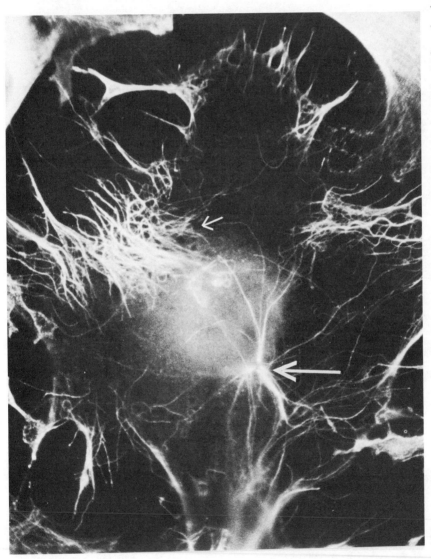

FIGURE 3. Antitubulin immunofluorescence staining of a gerbil fibroma cell ninety minutes after its injection with 0.2 mM Ca$_4$-calmodulin. Note that new microtubule initiation is occurring from the microtubule organizing center (heavy arrow), even while the fringe of old microtubules (light arrow) persists at the periphery of the cell. ×440.

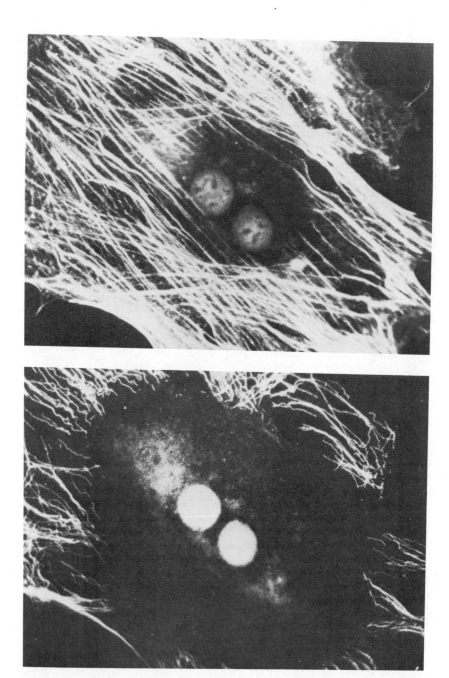

FIGURE 4. (Top) NBD-*N*-(7-nitrobenz-2-oxa-1,3-diazole-4-yl)-phallacidin and (bottom) anti-tubulin immunofluorescence staining of a gerbil fibroma cell 20 minutes after its injection with 0.25 m*M* Ca_4-calmodulin. Note that, like the microtubules, stress fibers are disrupted by the microinjection of calcium-saturated calmodulin, but the radius of the effect is smaller. (C. H. Keith *et al.*[31] With permission from the *Journal of Cell Biology*.)

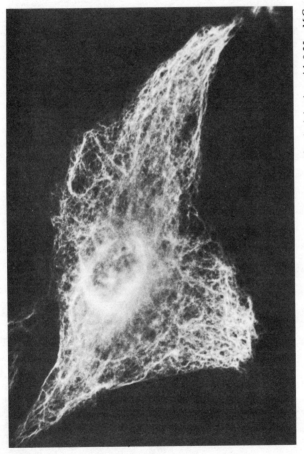

FIGURE 5. Antivimentin immunofluorescence staining of a gerbil fibroma cell, 20 minutes after its injection with 0.25 mM Ca$_4$-calmodulin. Note that the intermediate filaments are intact, eliminating the possibility that cytoskeletal disruption by calmodulin is due to a contaminating or cellular calcium-activated protease. ×375.

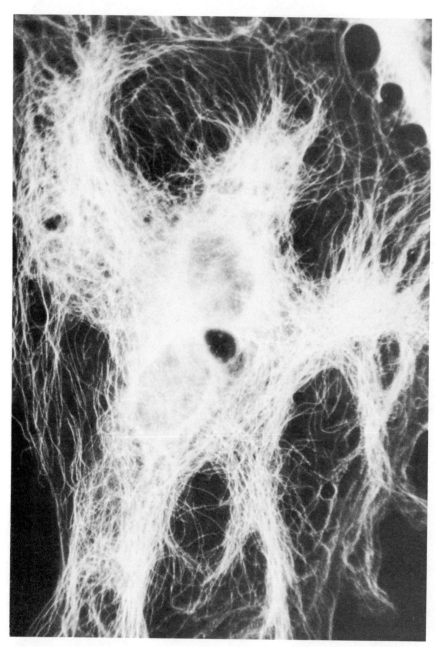

FIGURE 6. Gerbil fibroma cell injected 20 minutes previously with 0.25 mM calmodulin + 1.2 mM Na$_2$-EGTA. Note that the microtubules are largely intact; the small (2 micron diameter) region immediately around the injection site is a region where upper and lower cell membranes were fused during microinjection. ×440; reduced by 6 percent. (C. H. Keith *et al.*[31] With permission from the *Journal of Cell Biology*.)

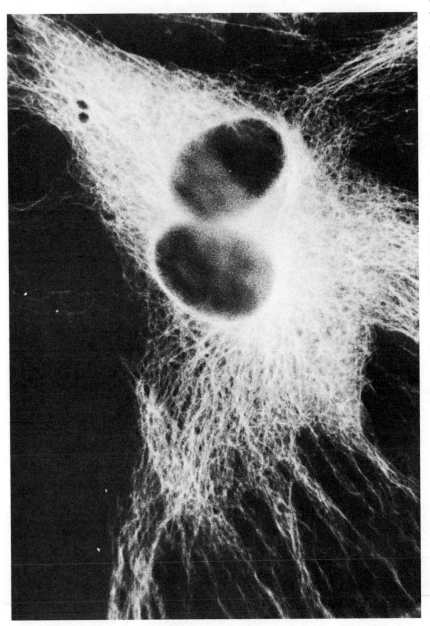

FIGURE 7. Gerbil fibroma cell injected 20 minutes previously with 1.0 mM CaCl$_2$. Note that the microtubules are intact at the resolution of immunofluorescence microscopy. ×440; reduced by 6 percent. (C. H. Keith *et al.*[31] With permission from the *Journal of Cell Biology*.)

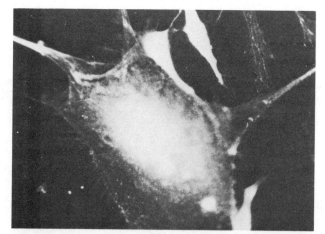

a.

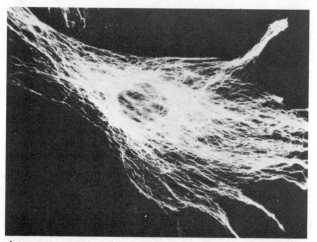

b.

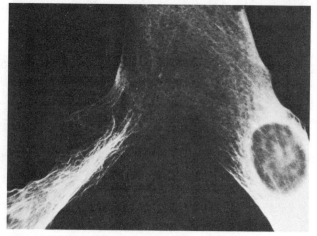

c.

ionophore does not somehow drive the dispersal of microinjected calmodulin, but that its mobility is a consequence of its state of calcium saturation. Thus, we can conclude that calcium ion levels control not only the activity, but also the localization of microinjected calmodulin.

An experimental protocol derived from the one just described also allowed the determination of the amount of calmodulin required to disrupt a cell's microtubules. Briefly, cells were injected with EGTA-calmodulin, and later permeabilized to extracellular calcium with calcium ionophore. After immunocytochemical processing to visualize microtubules, the tracer fluorescence within injected cells was measured, and they were scored for the extent to which their microtubules were disrupted. In this manner, the amount of tracer fluorescence corresponding to the amount of calmodulin necessary to fully disrupt a cell's microtubules was determined; by measuring the fluorescence of the injection mixture, this quantity could be related to a volume of solution, and thereby to an amount of calmodulin. In this manner, we could determine that the introduction of approximately 0.4 femtomoles of calcium-calmodulin into a gerbil fibroma cell is sufficient to disrupt all its microtubules that can be visualized by immunofluorescence techniques. The average tubulin content of these cells is measured by the tritiated colchicine filter binding assay (Borisy[25]) to be 0.2 femtomoles, so that we can calculate that a twofold excess of calmodulin is sufficient to disrupt polymerized microtubules in the living cell.

FREE INTRACELLULAR CALCIUM CONCENTRATIONS DURING MITOSIS

Mitosis is a process that requires considerable regulation of microtubule polymerization equilibria. By any account, microtubules must depolymerize in a controlled manner in at least three stages of the mitotic process: during early prophase, when the interphase microtubules must break down to be replaced by the spindle; during late anaphase, when the chromosome to pole distance decreases significantly; and during telophase, when the interzonal microtubules must break down. If calcium is a physiological regulator of microtubule polymerization equilibria within the cell, we might expect to see calcium concentrations vary either cell-wide or locally during mitosis, in a manner that is consistent with the pattern of microtubule assembly and disassembly.

The calcium indicator dye Quin2 (Tsien[26]; Tsien *et al.*[27]) is well suited to the measurement of intracellular calcium ion concentrations during the course of mitosis. The fluorescence emission at 492 nm of the free acid is highly dependent on calcium levels when excited at 340 nm, but is independent of calcium when excited at 360 nm (Tsien *et al.*[27]). These spectral characteristics mean that the ratio of Quin2 fluorescence when excited at these two wavelengths is a quantity that increases with the free calcium ion concentration, but is independent of the amount of dye in the optical path

---

FIGURE 8. Antitubulin immunofluorescence staining of gerbil fibroma cells treated for 5 minutes with the calcium ionophore A23187 in the presence of 25 micromolar $CaCl_2$. The cell in **a** was injected with 200 micromolar calmodulin + 1.0 mM EGTA 15 minutes before the start of ionophore treatment, and the cell in **c** was injected with 240 micromolar Ca_4-calmodulin 20 minutes before ionophore treatment. The cell in **b** is a control, uninjected cell. Note that microtubules are disrupted throughout the cytoplasm in the case of injection with calcium-free calmodulin **a**, but are still disrupted locally in the case of injection with calcium-saturated calmodulin (**c**) ×390. (C. H. Keith *et al.*[31] With permission from the *Journal of Cell Biology*.)

(Kruskal *et al.*[28]; Keith *et al.*[29]). The fluorescence ratio method is therefore ideally suited for the study of local and global free calcium levels in individual cells on the stage of the fluorescence microscope (Kruskal *et al.*[28]), particularly during a process such as mitosis, where cells change in size and shape as they go through the process (Keith *et al.*[29]). Quin2 may be added to cells as the acetoxymethyl ester, which, being uncharged, readily permeates the plasma membrane; once inside the cell it is cleaved to the free acid, which is the indicator form, and is trapped inside the cytoplasm. We have found that a variety of mitotic cells take up sufficient quantities of Quin2 to give readily visible fluorescence under excitation at either 340 or 360 nm.

We have used the Quin2 fluorescence ratio method to follow the free intracellular calcium ion levels of mitotic PtK2 cells on the stage of a fluorescence microscope, equipped with UV-transparent optics and a photometer. After measuring the average autofluorescence at both wavelengths of unloaded cells, unsynchronized cells were loaded with Quin2 on the microscope stage; loading conditions used were such that cells appeared normal and proceeded through mitosis with their usual timing. On the basis of visual examination, cells were chosen as being in a given stage of mitosis, and their fluorescence was measured on excitation at 340 and 360 nm. After subtracting the average autofluorescence at each wavelength, a 340/360 fluorescence ratio was calculated. By reference to a calibration curve of fluorescence ratios versus the free calcium levels of a series of Quin2-containing calcium-EGTA buffer solutions (Kruskal *et al.*[28]; Keith *et al.*[29]), these ratios were converted to free intracellular calcium ion concentrations. In this manner, a large number of observations of free calcium levels at each stage of mitosis were accumulated; these observations were averaged at each mitotic stage to give a picture of free calcium variation during mitosis. In a second set of experiments, the free calcium levels of seven individual cells were followed through mitosis, to confirm that they followed the overall pattern accurately.

When we followed the change of intracellular calcium concentration in mitotic PtK2 cells, we found a pattern that, at first glance, is rather surprising, in that intracellular calcium dropped during the course of mitosis, only to rise again at the end of the process. Specifically, interphase calcium levels of approximately 50 nM dropped to 25 nM in metaphase, and remained at this low level until late telophase, when they floated back up toward interphase levels (FIGURE 9). An identical pattern was observed in individual cells followed through the course of mitosis, although in the latter case, the scatter of measurements was considerably higher than in the averaged cells.

In order to measure local calcium levels during mitosis, we used a computer-based image processor to effectively duplicate the action of a photometer at each of 262,144 picture elements (pixels) within a video image. Using the image processor, the intensities within a 340 nm (excitation) fluorescence image of a Quin2-loaded cell were measured (the image was digitized) and stored. Then a 360 nm fluorescence image of the same cell was digitized, and the 340 nm image intensities were divided point-by-point by the corresponding 360 nm intensities. The intensities at any point within the resulting 340/360 ratio image should then indicate the free calcium concentration at that point in the cell, averaged through the thickness of the cell.

The cells chosen for the investigation of local calcium levels during mitosis were the cell-wall free endosperm cells of the African globe lily, *Haemanthus katherinae Baker*. These cells were chosen for three reasons: a) Their spindles are of the order of three to five times the size of most mammalian spindles in each linear dimension, so that the chance of resolving local variations in calcium levels are correspondingly better than they would be in mammalian cells. b) They are known to have significant stores of calcium in membrane-bounded vesicles (Wick and Hepler[16]); the calcium within such a compartment would not be sampled by Quin2, but might act as a local

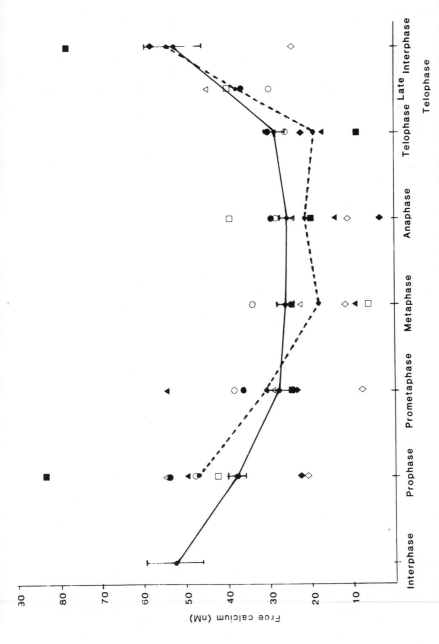

FIGURE 9. Intracellular free calcium levels in PtK2 epithelial cells during the course of mitosis, as determined by the Quin2 ratio method. The solid line represents the averaged calcium levels of a number of cells at each stage of mitosis. The error bars on these measurements represent the standard error of the mean calcium value at that stage. The individual symbols (circles, squares, triangles, diamonds) represent calcium levels measured from individual cells as they pass through mitosis. The dashed line represents the average of these seven individual time course measurements.

source for calcium release into the cytoplasm. c) They proceed normally through mitosis in the absence of extracellular calcium; if the baseline level of calcium in the cytoplasm is proportional to extracellular calcium, lowered calcium levels outside the cell should improve the "contrast" of local calcium release above background.

When local calcium levels in Quin2-loaded *Haemanthus* endosperm cells are studied in this manner, two features of the apparent calcium distribution are immediately apparent. First, through prometaphase and metaphase, the calcium distribution throughout the cell is relatively uniform, except that the chromosomes appear dark, probably because of some effect of local environment on Quin2. In anaphase, however, the intracellular calcium distribution becomes remarkably nonuniform, with the spindle poles having a 340/360 ratio, almost twofold higher than the interzone (FIGURE 10). The region of high fluorescence ratio, and therefore of high calcium, becomes quite tightly localized around the spindle poles during late anaphase. A second region of high calcium is also seen in some cells at the side of the midzone, but this is not a consistent feature among the images analyzed and will require further analysis.

Second, during interphase, prophase, and late telophase, the cell nucleus shows a significantly lower fluorescence ratio than the surrounding cytoplasm. This lowered ratio is not an artifact of Quin2 exclusion from the nucleus, as it is seen even in telophase cells that were loaded with Quin2 during metaphase. It is not clear, however, that this lowered ratio is necessarily due to lowered calcium ion concentrations in the nucleus, because the ionic environment in the nucleus is substantially different from that in the bulk cytoplasm, due to the presence of high concentrations of DNA and histones.

CONCLUSION

The experiments we have described strengthen the case for the involvement of calcium and calmodulin in the regulation of microtubule assembly equilibria in the intact cell. In the first place, we have demonstrated that calmodulin can act to cause the depolymerization of microtubules locally and at approximately unit stoichiometry in response to calcium in the living cell. We have also shown that calcium ion levels control not only the activity but also the localization of calmodulin within the cell, thus demonstrating a mechanism that could be responsible for the observed immobilization of both endogenous (Welsh *et al.*[21]; Anderson *et al.*[30]) and microinjected (Zavortnik *et al.*[22]) calmodulin within the mitotic spindle. We have also demonstrated that, against a generally lowered background of free calcium, calcium ion concentration is elevated by a factor of two to three at the spindle poles of anaphase cells. These factors combine to suggest that it is the local release of calcium ion at the anaphase spindle pole that is responsible for the observed localization of calmodulin within the anaphase half-spindle of mitotic cells.

Our current state of knowledge of the chemistry of the regulation of microtubule equilibria by calcium and calmodulin is thus at a transitional stage. From our experiments it is clear that purified tubulin and calmodulin make an inadequate model of the regulatory system acting on microtubules in the living cell. Our observations of microtubule disruption by microinjected calcium-calmodulin, and of calcium fluxes during mitosis, make a suggestive case for the regulation of microtubules by these factors in the living cell. We do not yet know, however, what the local concentration of calmodulin in the spindle is, so we do not know the level of calcium saturation of spindle calmodulin, and we do not know the nature of the calmodulin-microtubule interaction

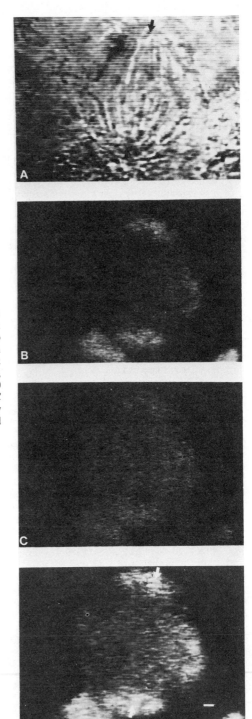

FIGURE 10. Local calcium levels in a late anaphase *Haemanthus* endosperm cell, as determined by the Quin2 ratio method. **A:** Brightfield image. **B:** 490 nm fluorescence image, with excitation at 340 nm. **C:** 490 nm fluorescence image, with excitation at 360 nm. **D:** 340/360 ratio image, formed by dividing **B** by **C** element by element. Note high intensity at spindle poles (arrows). ×400; reduced by 23%. (Bar = 10 μM)

that pertains in the cell. The current state of the technology is such that these questions can be addressed in the living cell by microinjection and quantitative fluorescence microscopy, so that we should be able to further dissect the regulatory process in the near future.

REFERENCES

1. CAMPBELL, A. K. 1983. Intracellular calcium: its universal role as regulator. John Wiley and Sons. New York.
2. KAKIUCHI, S. & R. YAZAMAKI. 1970. Calcium-dependent phosphodiesterase activity and its activating factor (PAF) from brain. Studies on cyclic 3'–5' nucleotide phosphodiesterase (III) Biochem. Biophysics Res. Commun. 41: 1104–1110.
3. CHEUNG, W. Y. 1970. Cyclic 3',5'-nucleotide phosphodiesterase: demonstration of an activator. Biochem. Biophysics Res. Commun. 38: 533–538.
4. VINCENZI, F. F., T. R. HINDS & B. U. RAESS. 1980. Calmodulin and the plasma membrane calcium pump. Ann. N.Y. Acad. Sci. 356: 292–301.
5. WEISENBERG, R. C. 1972. Microtubule formation in vitro in solutions containing low calcium concentration. Science 177: 1104–1105.
6. ROSENFELD, A. C., R. V. ZACKROFF & R. C. WEISENBERG. 1976. Magnesium stimulation of calcium binding to tubulin and calcium-induced depolymerization of microtubules. FEBS Lett. 65: 144–147.
7. OLMSTED, J. & G. G. BORISY. 1975. Ionic and nucleotide requirements for microtubule formation in vitro. Biochemistry 14: 2996–3005.
8. KIEHART, D. P. 1981. Studies on the in vivo sensitivity of spindle microtubules to calcium ions and evidence for a vesicular calcium-sequestering system. J. Cell Biol. 88: 604–617.
9. IZANT, J. G. 1983. The role of calcium during mitosis. Calcium participates in the anaphase trigger. Chromosoma 88: 1–10.
10. CANDE, W. Z. 1981. Physiology of chromosome movement in lysed cell models. In International Cell Biology 1980–1981. Springer Verlag, Berlin. 593–608.
11. SILVER, R. B., R. D. COLE & W. Z. CANDE. 1980. Isolation of mitotic apparatus containing vesicles with calcium sequestration activity. Cell 19: 505–516.
12. SALMON, E. D. & R. R. SEGALL. 1980. Calcium-labile mitotic spindles isolated from sea urchin eggs (Lytechinus variagatus) J. Cell Biol. 86: 355–367.
13. CHAI, L. S. & A. A. SANDBERG. 1983. Effect of divalent cations on metaphase to telophase progression and nuclear envelope formation in chinese hamster cells. Cell Calcium 4: 237–252.
14. WAGENAAR, E. B. 1983. Increased free Ca^{++} levels delay the onset of mitosis in fertilized and artificially activated sea urchin eggs. Exp. Cell. Res. 148: 72–83.
15. HARRIS, P. 1978. Triggers, trigger waves, and mitosis: a new model. In Monographs in Cell Biology. 25–104. Academic Press. New York.
16. WICK, S. B. & P. K. HEPLER. 1980. Localization of Ca^{++}-containing antimonate precipitates during mitosis. J. Cell. Biol. 86: 500–513.
17. SCHATTEN, G., H. SCHATTEN & C. SIMERLY. 1982. Detection of sequestered calcium during mitosis in mammalian cell cultures and in mitotic apparatus isolated from sea urchin zygotes. Cell Bio. Int. Rep. 6: 717–724.
18. GILSON, C. & B. BURNSIDE. 1984. Calcium inhibition of reactivated elongation in cone mobile models. J. Cell Biol. 99: 182a.
19. SCHLIWA, M. 1976. The role of divalent cations in the regulation of microtubule assembly. In vivo studies on the microtubules of the helizoan axapodium using the ionophore A23187. J. Cell Biol. 70: 527–540.
20. LUBY-PHELPS, K. & K. R. PORTER. 1982. The control of pigment migration in isolated erythrophores of Holocentrus ascensionis (Osbeck). II. The role of calcium. Cell 29: 441–450.
21. WELSH, M. J., J. R. DEDMAN, B. R. BRINKLEY & A. R. MEANS. 1978. Calcium-dependent regulator protein: localization in the mitotic apparatus of eukaryotic cells. Proc. Natl. Acad. Sci. USA 75: 1867–1871.

22. ZAVORTNIK, M., M. J. WELSH & J. R. MCINTOSH. 1983. The distribution of calmodulin in living mitotic cells. Cell Res. **149**: 375–385.
23. DEERY, W. J., A. R. MEANS & B. R. BRINKLEY. 1984. Calmodulin-microtubule association in cultured mammalian cells. J. Cell Biol. **98**: 904–910.
24. MARGOLIS, R. L. 1983. Calcium and microtubules. *In* Calcium and Cell Function, **4**: 313–335.
25. BORISY, G. G. 1972. A rapid method for quantitative determination of microtubule protein using DEAE-cellulose filters. Anal. Biochem. **50**: 373–385.
26. TSIEN, R. Y. 1980. New calcium indicators and buffers with high selectivity against magnesium and protons: design, synthesis, and properties of prototype structures. Biochemistry **19**: 2396–2404.
27. TSIEN, R. Y., T. POZZAN & T. J. RINK. 1982. T-cell mitogens cause early changes in cytoplasmic Ca^{++} and membrane potential in lymphocytes. Nature (London) **295**: 68–71.
28. KRUSKAL, B., C. H. KEITH & F. R. MAXFIELD. 1984. TRH-induced changes in intracellular $[Ca^{2+}]$ measured by microspectrofluorometry on individual quin2-loaded cells. J. Cell Biol. **99**: 1167–1172.
29. KEITH, C. H., F. R. MAXFIELD & M. L. SHELANSKI. 1985. Intracellular free calcium levels are reduced in mitotic PtK2 epithelial cells. Proc. Natl. Acad. Sci. USA **82**: 800–804.
30. ANDERSON, B., M. OSBORN & K. WEBER. 1978. Specific visualization of the distribution of the calcium dependent regulatory protein of cyclic nucleotide phosphodiesterase (modulator protein) in tissue culture cells by immunofluorescence microscopy: mitosis and intracellular bridge. Cytobiologie **17**: 354.
31. KEITH, C. H., M. DIPAOLA, F. R. MAXFIELD & M. L. SHELANSKI. 1983. Microinjection of Ca^{++}-calmodulin causes a localized depolymerization of microtubules. J. Cell Biol. **97**: 1918–1924.

The Calcium Sensitivity of MAP-2 and Tau Microtubules in the Presence of Calmodulin[a]

PATRICK K. BENDER AND LIONEL I. REBHUN

Department of Biology
University of Virginia
Charlottesville, Virginia 22901

INTRODUCTION

Microtubules isolated from bovine brain tissue by temperature-dependent polymerization-depolymerization contain, in addition to the tubulin subunits, several microtubule-associated proteins (MAPs). The MAPs are composed of a 340,000 molecular weight protein (MAP-1), a 300,000 molecular weight protein (MAP-2), and several proteins of molecular weights between 53,000 and 68,000, collectively designated tau.[1,2] All three groups of proteins have been shown to lower the critical concentration necessary for tubulin polymerization and, thus, are possibly involved in regulating the intracellular equilibrium between tubulin polymer and monomer. Calcium has also been implicated in regulating tubulin polymerization.[3,4] The calcium concentrations necessary to cause significant *in vitro* depolymerization of microtubules, however, are several orders of magnitude greater than *in vivo* concentrations, unless the calcium-binding proteins troponin C, S-100, or calmodulin are included in the assay.[5] Of these proteins, calmodulin is of particular interest because it and tubulin are present in all nucleated eukaryotic cells.

The mechanism by which calmodulin augments the calcium sensitivity of microtubules is complicated by the number of different proteins present in a microtubule preparation. Kumagi and Nishida[6] report that calmodulin interacts with tubulin, whereas Sobue *et al.*[7] report that calmodulin interacts with tau. Jemiolo[8] reported that MAP-2 was specifically bound to a calmodulin-sepharase affinity matrix in a calcium-dependent manner and could be competed away from the calmodulin by tubulin, whereas Lee and Wolfe[9] showed that tau and MAP-2 could both bind to calmodulin. These results are partially contradictory, and as a result, the calmodulin interaction or combination of interactions responsible for the increased calcium sensitivity of microtubules is not resolved. We investigated the question by using purified samples of tubulin, MAP-2, and tau to prepare reconstituted microtubules and measure the calcium and calcium-calmodulin sensitivity of these microtubules directly. Using this experimental approach, the calcium-calmodulin sensitivity of MAP-2-tubulin and tau-tubulin microtubules can be compared. We show that (1) MAP-2 microtubules and tau microtubules are relatively insensitive to calcium in the absence of calmodulin. MAP-2 microtubules, however, are considerably less sensitive to calcium than tau microtubules. (2) The change in calcium sensitivity in the presence

[a]This work was supported by Grant 2RO1GM26784 from the NIH and PCM8011676 from the NSF.

of calmodulin is greater for the MAP-2-tubulin system than for the tau-tubulin system. (3) Nucleation of MAP-2-tubulin and tau-tubulin microtubules is affected at lower Ca^{2+}-calmodulin concentrations than elongation. (4) Comparison of the MAP-2 and tau-tubulin systems with the cycle-purified system indicates that there are additional MAPs that affect the calcium and calmodulin sensitivity of microtubules. Some of these results are similar to those reported by Berkowitz and Wolfe[10] and Lee and Wolfe.[9,11]

MATERIAL AND METHODS

Reagents

1,4-Piperazinediethanesulfonic acid (PIPES), ethylenediaminetetraacetic acid (EDTA), guanosine 5'-triphosphate (GTP), and dithiothreitol were purchased from Sigma Biochemicals (St. Louis, Mo.). All other reagents, unless reported otherwise, were from J.T. Baker Chemical Company.

Preparation of Protein Samples

The preparation of cycle-purified microtubule protein, MAP-2, and tau has been previously reported.[12]

Purified 6S tubulin was prepared from the cycle-purified microtubule protein by chromatography on phosphocellulose (P-11, Whatman), according to procedures previously reported.[1,13] This preparation is referred to as PC tubulin in this text.

Lyophilized bovine brain calmodulin was a generous gift from Dr. Wilson Burgess and Dr. Robert Kretsinger (Department of Biology, University of Virginia). The purification procedure for this calmodulin has been previously reported.[14] The lyophilized calmodulin was resuspended in buffer A (0.1 M PIPES-NaOH pH = 6,8, 1.0 mM ethylene-bis(oxy-ethylenenitrile)tetraacetic acid (EGTA), 1.0 mM $MgCl_2$) containing 1.0 mM $CaCl_2$ and dialyzed against the same. Microtubule protein, MAP-2, tau, and 6S tubulin were all in buffer A.

Polymerization Studies

Purified tubulin was mixed with either high molecular weight (HMW)-2 or tau, and GTP (50 mM stock solution in H_2O, pH = 7.0) was added to a concentration of 1.0 mM while the sample was at 4°C. Three hundred μl aliquots of the mixtures were then transferred to masked micro-cuvettes, and the absorbance at 340 nm was measured in a Perkin Elmer model 526 spectrophotometer equipped with a temperature controlled cuvette chamber, 5-cell cuvette changer, and strip chart recorder. The samples were maintained at 15°C until polymerization was initiated by shifting the temperature to 37°C. The cuvette chamber was purged with dry air to prevent condensation. $CaCl_2$ was added from a 25 mM or 50 mM stock solution in H_2O using a 10 μl Hamilton syringe. In all experiments the dilution due to $CaCl_2$ addition was less than 2.5 percent. For experiments using calmodulin, the samples of purified tubulin with either HMW-2 or tau, or the cycle-purified samples were prepared at concentrations such that after transfer to the cuvettes and addition of calmodulin the concentration would be as reported. For experiments using various concentrations of calmodulin, the volume of calmodulin added to each cuvette was different. The total volume in each cuvette and the volume added of buffer A containing 1.0 mM CaCl$_2$, however, was maintained the same in each cuvette by addition of the calmodulin dialysate buffer.

Miscellaneous Techniques

Polyacrylamide gel electrophoresis in the presence of SDS was according to the method of Laemmli[15] except that SDS was omitted from the resolving and stacking gels. The stacking gel was 3% acrylamide, and the resolving gel was a 5% to 12% acrylamide gradient.

The concentration of protein in each sample was determined by a modification of the Lowry method.[16] The illustrated polymerization curves are tracings from the spectrophotometer chart recorder. Where indicated, the 15°C baseline absorbance of some samples was manually shifted on the chart recorder to aid in illustration. In these cases the absorbance scale is relative. Formation of microtubules was confirmed by transmission electron microscopy of negatively stained aliquots of the various samples.

RESULTS

Stimulation of Microtubule Polymerization by MAPs

Purified samples of PC tubulin, MAP-2, and tau were prepared from cycle-purified microtubule protein as described in MATERIAL AND METHODS. Aliquots of these proteins were electrophoresed in NaDodSO$_4$ on polyacrylamide gels. The results are illustrated in FIGURE 1, where samples were deliberately overloaded to reveal minor species. The cycle-purified preparation (3X) contains, in addition to the α- and β-tubulin subunits, the microtubule-associated proteins MAP-1, MAP-2, and tau. Most of the tau proteins, however, are not visible due to the similar molecular weights of tubulin and tau and the abundance of the tubulin subunits. The PC tubulin sample contains only the α- and β-tubulin subunits. No tau proteins were detectable in this sample by 2-dimensional gel electrophoresis (data not shown). The MAP-2 sample contains primarily a 300,000 molecular weight protein plus several polypeptides of lower molecular weights that are thought to be proteolytic fragments of MAP-2.[17] The tau sample contains several proteins of molecular weights between 53,000 and 68,000. No tubulin was detectable in this sample by two-dimensional gel electrophoresis (data not shown).

These samples were used to prepare reconstituted microtubules by mixing different concentrations of either MAP-2 or tau with PC tubulin at 1 mg/ml. Under these conditions, tubulin would not polymerize in the absence of either MAP-2 or tau. The absorbance at 340 nm of each reconstituted sample was recorded at 15°C, and tubulin polymerization was initiated by increasing the temperature to 37°C. The formation of microtubules was quantitated by the change in absorbance at 340 nm (ΔA_{15-37}). The results are plotted in FIGURE 2 as the ΔA_{15-37} versus concentration of either MAP-2 or tau. The stimulation of polymerization by these factors reaches a maximum at 0.9 mg/ml for MAP-2 and 0.25 mg/ml for tau. These results are in agreement with previously published results characterizing the affects of MAP-2 and tau on the stimulation of tubulin polymerization.[18]

Calcium Sensitivity of Microtubules

To determine the calcium sensitivity of the microtubules formed with either MAP-2 or tau we used concentrations of MAP-2 and tau that were slightly less than

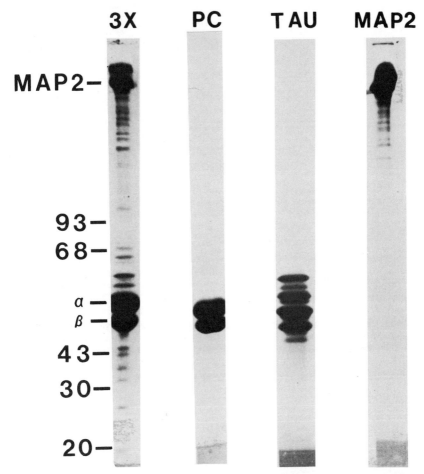

FIGURE 1. Samples of cycled-purified microtubule protein (3X), phosphocellulose-purified 6S tubulin (PC), microtubule-associated protein of 300,000 molecular weight (MAP-2), and microtubule-associated proteins-tau were electrophoresed on polyacrylamide gels in the presence of NaDodSO₄. The mobilities of five protein standards with known molecular weights are also illustrated.

the levels needed for maximum tubulin polymerization because the microtubules thus formed are more sensitive to perturbations that might involve MAP-2 or tau interactions. Therefore, in the following experiments samples of PC tubulin at 1.0 mg/ml in 1.0 mM EGTA buffer were prepared with either 0.7 mg/ml MAP-2 or 0.2

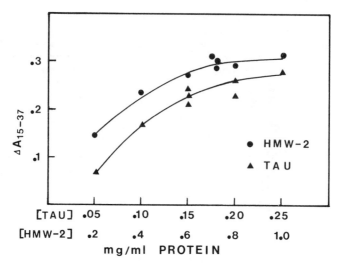

FIGURE 2. Tubulin polymerization was measured by the change in absorbance at 340 nm after shifting the temperature from 15 to 37°C (ΔA_{15-37}). The samples contained purified tubulin at 1.0 mg/ml with either HMW-2 (●) or tau (▲) at the indicated concentrations.

mg/ml tau. For comparison, cycle-purified tubulin was used at two different concentrations. Polymerization was initiated by shifting the temperature to 37°C, and after polymerization reached steady state, concentrated $CaCl_2$ was added to the microtubule samples in various amounts to obtain several molar ratios of $Ca^{2+}/EGTA$. The amount of free Ca^{2+} in each sample was determined using the equations of Portzehl et al.[19] and assuming that EGTA was the primary Ca^{2+} binding species. This procedure is justified because the association constants of Mg^{2+} for EGTA is five orders of magnitude lower than that of Ca^{2+} for EGTA and that of GTP for Ca^{2+} is seven orders of magnitude lower.[20,21] The amount of polymer lost after calcium addition was measured by the decrease in absorbance at 340 nm of the sample. The A_{340} nm was then used to calculate the percent depolymerization at various free Ca^{2+} levels by the following formula:

$$\frac{A_{37} - A_{37}^{Ca}}{A_{37} - A_{15}} \times 100$$

A_{15} = the absorbance of the sample at 15°C before
    polymerization was initiated.
A_{37} = the absorbance of the sample at 37°C after
    polymerization and steady state was obtained.
A_{37}^{Ca} = the absorbance of the sample at 37°C after
    addition of calcium and steady state was obtained.
$A_{37} - A_{15}$ = ΔA_{15-37}, a measure of the initial microtubule
    concentration before Ca^{2+} addition.[22]

Therefore, the percent depolymerization measures the ratio of the amount of microtubules lost after calcium addition to the total amount of microtubules initially present. The results are shown in FIGURE 3, which includes a table of the percent depolymerization for the four samples at three calcium levels. Comparison of the percent depolymerization between the MAP-2-tubulin, tau-tubulin, and cycle-purified microtubules, demonstrates that the MAP-2-tubulin microtubules have the lowest calcium sensitivity. It requires approximately a sevenfold higher concentration of Ca^{2+} to depolymerize MAP-2 microtubules to the same extent as tau microtubules. The observed calcium sensitivity for the three systems varies in the order: MAP-2 microtubules $\ll$ tau microtubules $\leq$ cycle-purified microtubules. Furthermore, comparison of the percent depolymerization of the two cycle-purified samples at different concentrations demonstrates, within the concentration range tested, that their calcium sensitivities are approximately the same, even though the concentrations of microtubule polymer differ by more than a factor of two.

Ca^{2+}-Calmodulin Effects on Tubulin Polymerization

In the following experiments the effects of calmodulin on the three microtubule systems are determined. To aid in comparing the results, the calmodulin concentration used in various experiments is normalized to the concentration of tubulin and expressed as the molar ratio of calmodulin to tubulin. For the MAP-2-tubulin and tau-tubulin systems this is readily calculated because the concentration of tubulin is accurately known. Cycle-purified microtubule protein is reported to be approximately 75–80% tubulin by weight as was confirmed for our preparation by densitometric

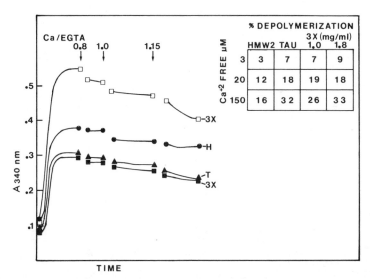

FIGURE 3. Tubulin polymerization was measured by the change in absorbance at 340 nm after shifting the temperature of the samples from 15 to 37°C. The polymerization curves are tracings from the spectrophotometer chart recorder. $CaCl_2$ was added from a 50 mM stock solution to each sample at the times indicated by the arrows. The molar ratios of Ca^{2+} to EGTA after $CaCl_2$ addition are indicated. The samples polymerized are cycle-purified microtubule protein (3X) at 1.8 mg/ml ($\square$) and 1.0 mg/ml ($\blacksquare$). Samples of purified tubulin at 1.0 mg/ml were prepared with either 0.7 mg/ml HMW-2 ($\bullet$) or 0.2 mg/ml tau ($\blacktriangle$). The percent depolymerization and concentration of free Ca^{2+} were calculated as explained in the text.

measurements of a cycle-purified sample after electrophoresis on a polyacrylamide gel and staining with fast green (data not shown). Therefore, in experiments using calmodulin and the cycle-purified system, the molar ratio of calmodulin to tubulin was calculated assuming 80% of the protein was tubulin.

Cycle-Purified Microtubules and Calmodulin

The effect of calmodulin on the calcium sensitivity of cycle-purified microtubules is shown in FIGURE 4. For these experiments, the calmodulin sample was previously

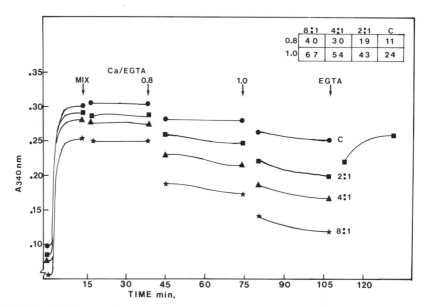

FIGURE 4. Polymerization curves illustrating the effect of calmodulin on the calcium-induced depolymerization of cycle-purified microtubule protein. Cycle-purified microtubule protein at a concentration of 1.25 mg/ml was polymerized in the absence of calmodulin (●) and with calmodulin at a calmodulin to tubulin molar ratio of 8:1 (★), 4:1 (▲), and 2:1 (■). After polymerization, the samples were mixed, and $CaCl_2$ was added to give a Ca^{2+} to EGTA molar ratio of 0.8 in each sample. The resulting change in A_{340nm} was recorded, and a second addition of $CaCl_2$ was made to give a Ca^{2+} to EGTA ratio of 1.0. At the end of the experiment EGTA was added in excess of Ca^{2+} to the 2:1 sample. The percent depolymerization as a result of calcium addition is tabulated in the figure. The absorbance scale is absolute for the control sample (absence of calmodulin) and relative for the other three samples because their 15°C baseline was manually lowered on the chart recorder.

dialyzed against a buffer containing equal milli-molar EGTA and $CaCl_2$ (Ca^{2+}/ EGTA = 1.0). An aliquot of the calmodulin sample was then added to the cycle-purified microtubule protein in an EGTA buffer to obtain the calmodulin to tubulin molar ratios indicated. Polymerization was initiated by increasing the temperature to 37°C. After polymerization reached steady state, $CaCl_2$ was added from a stock solution to obtain the molar ratios of Ca^{2+} to EGTA indicated. In the calculations of the Ca/EGTA ratios, the contribution from Ca^{2+} dissociation from the calmodulin has not been considered. At the Ca^{2+} to EGTA ratio of 1.0, this contribution, however, should be negligible, because the calmodulin was previously dialyzed against a buffer with the same ratio. The 0.8 ratio may be in error due to some Ca^{2+} dissociation; the data, however, obtained at this ratio is still valid for comparison between experiments with the same concentration of calmodulin. A table of the Ca^{2+}-induced percent depolymerization is shown on the graph for the three calmodulin concentrations used and a control sample without calmodulin. The data demonstrates that calmodulin increases the calcium sensitivity of cycle-purified microtubules as previously reported (Marcum, et al.).[5] Furthermore, by addition of EGTA in excess of Ca^{2+} the A_{37} returns to within 5% of its initial value as indicated for one of the samples, but that is true for all the samples. In addition, after calcium is added, the samples can be further depolymerized by incubation at 4°C for 15 minutes, and after a second incubation at 37°C will repolymerize to within 10% of their calcium-induced plateau value (A_{37}^{Ca}) before depolymerization. This indicates that the plateau values recorded after calcium addition represent steady state and are independent of whether calcium is added to polymer, and depolymerization is recorded or added to monomer, and polymerization is recorded. The ability to repolymerize after cold depolymerization is observed in all the samples except that in which the Ca^{2+}-calmodulin-induced depolymerization is greater than 50 percent. For example, when the sample containing 8:1 calmodulin to tubulin at a Ca^{2+}/EGTA ratio of 1.0 is cold depolymerized, there is a 10–20% increase in its cold baseline (A_{15}), indicating formation of some light scattering aggregates. Rewarming this sample to 37°C results in poor polymerization, returning to only 20% of its A_{37}^{Ca} value before depolymerization. Whereas polymerization can be induced by addition of EGTA, it is less than that observed in the other samples after EGTA addition. These results indicate that there may be some irreversible denaturation due to the presence of Ca^{2+} calmodulin at high ratios to tubulin during the incubation period at 37°.

Tau-Tubulin Microtubules

The effects of calmodulin on the calcium sensitivity of tau-tubulin microtubules are shown in FIGURE 5. Three concentrations of calmodulin have been used, and a control sample without calmodulin is included for comparison. The calculated percent depolymerization values are also tabulated in FIGURE 5. The effect of calmodulin on the tau-tubulin microtubules is similar to that observed for the cycle-purified microtubules in that the presence of calmodulin increases the calcium sensitivity as demonstrated by the increase in the percent depolymerization at a given calcium concentration. Similar to the cycle-purified samples, the Ca^{2+}-calmodulin effect on the tau-tubulin microtubules is reversible by addition of EGTA. Furthermore, after calcium addition, the samples can be cold depolymerized and repolymerized (37°C) to their former plateau values (A_{37}^{Ca}). This is observed in all the samples except for that containing a calmodulin to tubulin molar ratio of 8:1 and a Ca^{2+}/EGTA ratio of 1.0. This sample will not repolymerize after cold depolymerization unless EGTA is added. This is different than observed in the cycle-purified sample that will polymerize at the same Ca^{2+} and calmodulin concentrations although partially inhibited due to denaturation. This will be further investigated in a later section.

MAP-2-Tubulin Microtubules

The effect of calmodulin on the calcium sensitivity of MAP-2-tubulin microtubules is shown in FIGURE 6, which also contains a table showing the percent depolymeriza-

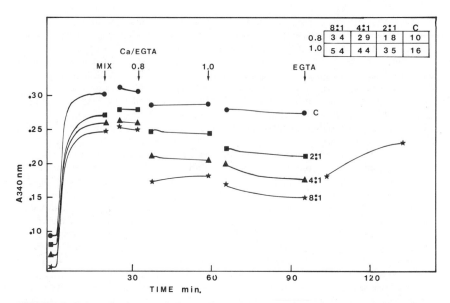

FIGURE 5. Polymerization curves illustrating the effect of calmodulin on the calcium-induced depolymerization of samples containing 1.0 mg/ml purified tubulin and 0.2 mg/ml tau. The samples were polymerized either in the absence (●) or presence of calmodulin at a calmodulin to tubulin molar ratio of 8:1 (★), 4:1 (▲), and 2:1 (■). After polymerization, the samples were mixed, and CaCl₂ was added at different times to give the Ca^{2+}/EGTA molar ratios indicated. At the end of the experiment, EGTA was added in excess of Ca^{2+} to the 8:1 sample. The percent depolymerization as a result of calcium addition is tabulated in the figure. The absorbance scale is absolute for the control sample and relative for the other three.

tion for the various calcium and calmodulin concentrations. Increasing the concentration of calmodulin results in increasing calcium sensitivity of MAP-2-tubulin microtubules. The reversal of the Ca^{2+}-calmodulin effect on the MAP-2-tubulin microtubules by addition of EGTA occurs in a manner similar to that of tau-tubulin and cycle-purified tubulin. The ability of the MAP-2-tubulin microtubules to be cold depolymerized after calcium addition and subsequently repolymerized after rewarming is similar to that observed for the tau-tubulin samples, in that all the samples will repolymerize to their former plateau values (A_{37}^{Ca}), except that containing an 8:1 calmodulin to tubulin molar ratio and a 1.0 Ca^{2+}/EGTA ratio. This sample will not repolymerize unless EGTA is added.

Calmodulin increased the calcium sensitivity of all three samples. The MAP-2-tubulin system, however, exhibits the largest increase in its calcium sensitivity. At the calmodulin to tubulin molar ratio of 2:1, the MAP-2-tubulin microtubules depolymerize by 16% at the 0.8 Ca^{2+}/EGTA ratio. This percent depolymerization is greater than that observed at the Ca^{2+}/EGTA ratio of 1.15 for a duplicate sample without calmodulin. Calculating the free Ca^{2+} concentration from the Ca^{2+}/EGTA ratio indicates that the MAP-2-tubulin samples show a greater than 50-fold increase in their calcium sensitivity at a 2:1 calmodulin to tubulin molar ratio. Similar calculations for the tau-tubulin and cycle-purified tubulin samples show an approximate 7-fold increase in their calcium sensitivity in the presence of a 2:1 calmodulin to tubulin ratio. Even in the presence of calmodulin, however, MAP-2-tubulin is still less sensitive to Ca^{2+} than is either tau-tubulin or cycle-purified tubulin. The differences in calcium sensitivity between the three systems is less than that observed in the absence of calmodulin.

Effect of Calcium and Calmodulin on Nucleation of Microtubules

The inability of the MAP-2-tubulin and tau-tubulin samples at 8:1 calmodulin to tubulin ratios to repolymerize after cold treatment in the absence of EGTA suggests a possible effect of Ca^{2+}-calmodulin on nucleation. To investigate this possibility, three samples of each tubulin system were prepared with calmodulin. Two of the samples contained a calmodulin to tubulin molar ratio of 8:1, whereas the third sample contained a 4:1 calmodulin to tubulin molar ratio. Calcium was added to the 4:1 and to one of the 8:1 samples to a Ca^{2+}/EGTA ratio of 1.0 prior to polymerization. The duplicate 8:1 sample to which no exogenous calcium was added contained approximately $10^{-7}M$ free calcium (resulting from the calmodulin addition). All three samples were then warmed to 37°C, and polymerization was monitored by turbidity. After the $10^{-7}M$ calcium, 8:1 molar ratio sample polymerized to steady state, calcium was added to a Ca^{2+}/EGTA ratio of 1.0, and depolymerization was recorded. The difference between the two samples containing the 8:1 ratio of calmodulin to tubulin was that calcium was added before polymerization in one sample and after polymeriza-

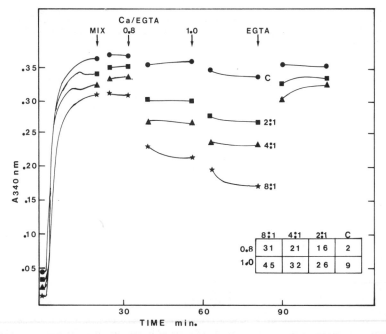

FIGURE 6. Polymerization curves illustrating the effect of calmodulin on the calcium-induced depolymerization of a sample containing 1.0 mg/ml purified tubulin and 0.7 mg/ml HMW-2. The experimental design and $CaCl_2$ additions were the same as illustrated in FIGURES 4 and 5. At the end of the experiment, EGTA was added to all of the samples except the 8:1.

tion in the other. If Ca^{2+}-calmodulin does not affect nucleation, then these two samples should reach the same concentration of microtubules. This result is obtained for the cycle-purified tubulin illustrated in FIGURE 7. Thus, addition of calcium (Ca/EGTA = 1.0) before polymerization to the 4:1 and 8:1 calmodulin to cycle-purified tubulin samples allows polymerization to approximately the same A_{37}^{Ca} as duplicate samples that had calcium added after polymerization. These results indicate that nucleation of the cycle-purified sample is not significantly inhibited by Ca^{2+}-calmodulin at these levels.

Different results, however, are obtained for tau and MAP-2 reconstituted microtubules as illustrated for the tau-tubulin samples shown in FIGURE 8. The 4:1 calmodulin to tubulin sample with calcium added before the temperature shift polymerizes poorly (FIGURE 8, open squares), although it eventually approaches the A_{37}^{Ca} value observed in a previous experiment (FIGURE 5). Polymerization of the 8:1 calmodulin to tubulin sample with calcium added before the temperature shift, however, is almost completely inhibited, whereas the duplicate 8:1 sample with calcium added after polymerization contains approximately 45% of its microtubules (FIGURE 8, closed circles versus closed triangles). The difference in microtubule concentration between the two 8:1 samples may occur because (1) an inhibition of nucleation occurs in the sample with calcium added prior to polymerization, or (2) calcium-calmodulin-induced depolymerization is a two-step process, and microtubules will slowly depolymerize after the fast initial decrease. To test these two possibilities, the two samples were mixed in equal proportions at the time indicated on the graph, and the A_{37}^{Ca} of the mixture was monitored. If 1 above is correct the mixture should contain the necessary nucleation sites added from the polymerized sample and, therefore, the A_{37}^{Ca} of the mixture should increase and reach plateau at the same value as the 8:1 sample that had calcium added after polymerization, because both samples have the same concentrations of tubulin,

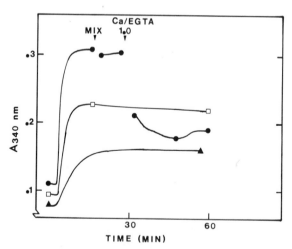

FIGURE 7. The effects of calcium and calmodulin on the polymerization of cycle-purified microtubule protein. Before polymerization was initiated by shifting the temperature to 37°C, the three samples of 1.25 mg/ml cycle-purified protein contained either calmodulin at a calmodulin to tubulin molar ratio of 8:1 and no calcium (●), or a calmodulin to tubulin ratio of 8:1 with a Ca^{2+}/EGTA ratio of 1.0 (▲), or a calmodulin to tubulin ratio of 4:1 and a Ca^{2+}/EGTA ratio of 1.0 (□). After polymerization, calcium was added only to the 8:1 sample polymerized without calcium. The absorbance scale is absolute only for this sample and relative for the other two.

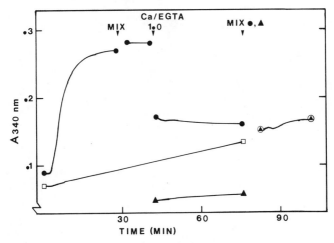

FIGURE 8. The effects of calcium and calmodulin on the polymerization of 1.0 mg/ml purified tubulin with 0.2 mg/ml tau. Before polymerization was initiated, the three samples contained either calmodulin at an 8:1 molar ratio to tubulin without calcium (●), or a calmodulin to tubulin molar ratio of 8:1 with a Ca^{2+}/EGTA ratio of 1.0 (▲), or a calmodulin to tubulin molar ratio 4:1 and a Ca^{2+}/EGTA ratio of 1.0 (□). At the time indicated, $CaCl_2$ was added to the sample, which was polymerized without calcium. At this time the other 8:1 sample with calcium was warmed to 37°C. After 30 minutes, the two 8:1 samples were mixed in equal proportions, and the absorbance was recorded (▲).

tau, and calmodulin. If 2 above is correct, then the mixture would initially have 50% of the A_{37}^{Ca} value of the polymerized 8:1 sample and would then decrease due to Ca^{2+} depolymerization. The results in FIGURE 8 demonstrate that the mixture polymerizes to the A_{37}^{Ca} value observed for the 8:1 sample that had calcium added after polymerization. Thus it appears that inhibition of nucleation prevents polymerization in the sample with calcium added before the polymerization is initiated by the temperature shift.

Similar results were obtained for the HMW-2-tubulin system shown in FIGURE 9. The 4:1 sample polymerizes slowly, but does reach the expected plateau (A_{37}^{Ca}). The 8:1 sample with calcium added before the temperature shift does not polymerize, whereas the 8:1 sample with calcium added after polymerization retains approximately 50% of its microtubules. When these two samples are mixed in 1:1 proportions, the resulting mixture polymerizes to the same A_{37}^{Ca} previously observed for the 8:1 sample with calcium added after polymerization. These results indicate that nucleation in the HMW-2-tubulin sample is inhibited at these levels of calcium and calmodulin.

Comparison of these results indicate that cycle-purified tubulin contains a factor(s) that is either responsible for nucleation and unaffected by Ca^{2+}-calmodulin or relieves the inhibition of Ca^{2+}-calmodulin on nucleation.

Alternatively, the combination of MAP-2 and tau in the cycle-purified preparation may affect nucleation differently than either by itself. To investigate the latter possibility, a 1.0 mg/ml sample of tubulin was prepared with 0.17 mg/ml MAP-2 and 0.05 mg/ml tau, the approximate concentrations of these proteins in a 1.25 mg/ml cycle-purified sample. This reconstituted sample was polymerized, and the calcium-induced percent depolymerization was determined in the absence and presence of a 4:1 molar ratio of calmodulin to tubulin. The percent depolymerization was 10.6% in the

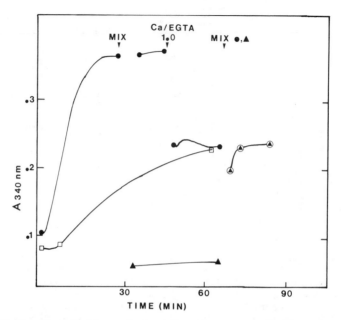

FIGURE 9. The effects of calcium and calmodulin on the polymerization of 1.0 mg/ml purified tubulin with 0.7 mg/ml HMW-2. Before polymerization was initiated, the three samples contained either calmodulin at an 8:1 molar ratio to tubulin without calcium (●), or a calmodulin to tubulin ratio of 8:1 with a Ca^{2+}/EGTA of 1.0 (▲), or a calmodulin to tubulin molar ratio of 4:1 and a Ca^{2+}/EGTA ratio of 1.0 (□). At the time indicated, $CaCl_2$ was added to the sample polymerized without calcium. At this time the other 8:1 sample with calcium was warmed to 37°C. After 30 minutes the two 8:1 samples were mixed in equal proportions, and the absorbance was recorded (▲ in circle).

absence and 33.9% in the presence of calmodulin at a Ca^{2+}/EGTA ratio of 1.0. These values are only slightly higher than those previously determined for the MAP-2-tubulin system containing 0.7 mg/ml MAP-2. Furthermore, the reconstituted sample containing calmodulin would not repolymerize after calcium addition and cold depolymerization unless EGTA was added. This is consistent with the results previously illustrated (FIGURES 8 and 9) except that at these lower concentrations of MAP-2 and tau, nucleation is inhibited at a 4:1 calmodulin to tubulin ratio instead of an 8:1 ratio. The cycle-purified tubulin, however, does polymerize at this calcium concentration even at an 8:1 calmodulin to tubulin ratio (refer to FIGURE 7). Consequently, the combination of tubulin with both MAP-2 and tau at concentrations similar to those found in the cycle-purified system does not mimic the calcium or calmodulin sensitivity of the cycle-purified system.

Ca^{+2}-Calmodulin Effects at Different Critical Concentrations

It is possible to modify the critical concentration of tubulin by using different ratios of MAPs to tubulin. In these experiments we have used three samples of MAP-2-tubulin prepared with 0.35, 0.70, and 1.4 mg/ml MAP-2 containing 1.0 mg/ml PC tubulin and a 4:1 calmodulin to tubulin molar ratio. The ratios of calmodulin to

MAP-2 (8:1, 4:1, and 2:1) are tabulated in FIGURE 10. The samples were polymerized, calcium was added, and the percent depolymerization was determined. The results are shown in FIGURE 10. The percent depolymerization of all three samples is approximately the same. A similar experiment was done using the tau-tubulin system with tau at 0.1, 0.2, and 0.4 mg/ml and a calmodulin to tubulin ratio of 4:1. These samples have the same percent depolymerization at a Ca^{2+}/EGTA ratio 1.0 (data not shown). Thus, increasing tau concentration also does not significantly affect the percent calcium-calmodulin-induced depolymerization of microtubules.

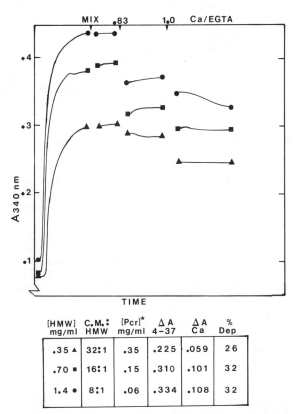

[HMW] mg/ml	C.M.: HMW	[Pcr]* mg/ml	ΔA 4-37	ΔA Ca	% Dep
.35 ▲	32:1	.35	.225	.059	26
.70 ■	16:1	.15	.310	.101	32
1.4 ●	8:1	.06	.334	.108	32

FIGURE 10. Samples of purified tubulin at 1.0 mg/ml, containing calmodulin at a 4:1 calmodulin to tubulin ratio, were polymerized with either 0.35 (▲), 0.7 (■), or 1.4 (●) mg/ml HMW-2. $CaCl_2$ was added at the times indicated to give 0.83 and 1.0 ratios of Ca^{2+} to EGTA. Below the curves are tabulated the tubulin critical concentration for the samples (*Pcr), the change in absorbance due to the temperature shift (ΔA_{15-37}), the total change in absorbance after both calcium additions ($A_{37} - A_{37}^{Ca} = \Delta A_{Ca}$), and the percent depolymerization after both calcium additions (% Dep.).

*The critical concentration of 0.06 mg/ml is the average of the reported values 0.05 and 0.08. The critical concentration of 0.15 mg/ml was determined experimentally, and the value of 0.35 mg/ml was calculated assuming that absorbance was proportional to concentration of polymer in the three samples.

DISCUSSION

Understanding the interaction of calcium with microtubules is likely to be key in the understanding of a variety of physiological processes. The system is, however, highly complex because the specific conditions of interaction determine very different outcomes. For example, purified 6S brain tubulin is highly sensitive to depolymerization by calcium[8,10,23,24] as is tubulin isolated from spindles[25] or eggs[26] of the sea urchin. Cycle-purified brain tubulin containing MAPs, however, is at least 100-fold less sensitive to calcium than purified 6S tubulin. Similarly, the addition of MAPs to purified sea urchin tubulin also greatly decreases its calcium sensitivity.[26] In the brain system, the presence of calmodulin partially restores the calcium sensitivity to a level approximating that for purified tubulin. The mechanism by which Ca^{2+} and Ca^{2+} calmodulin depolymerize microtubules, however, is likely to be different. For example, previous work in this laboratory has shown that the ultracentrifuge patterns of Ca^{2+}-calmodulin depolymerized tubulin is different from that of calcium-depolymerized tubulin in that the latter contains both 6S and 30–36S peaks (ring structures), whereas the former contains only the 6S peak.[27] Because the 30–36S peak is an oligomer of tubulin and MAPs, it is likely that calcium-calmodulin affects MAP-tubulin interaction more severely than does calcium alone.

Other differences between calcium-induced depolymerization versus that resulting from calcium-calmodulin interaction with microtubules concerns the effects of these two systems on the critical concentration for polymerization. It is clear from the work of Karr,[28] that calcium acting on cycled microtubules induces an endwise depolymerization. It is stated that the critical concentration of calcium-treated tubulin under their conditions was double that of control tubulin. Weisenberg and Deery[29] point out, however, that critical concentration cannot be simply derived from the intercept of the line relating polymer concentration to total protein concentration as is usually done (Oosawa and Kasai),[30] but must take into account the amount of inactive protein present. When this is done there is no difference in critical concentration between untreated and calcium-treated tubulin.[29] Thus, the mechanism by which calcium alone appears to cause depolymerization of intact cycled microtubules involves an endwise depolymerization or destabilization along the length of the microtubule, which inactivates tubulin for further incorporation into microtubules without affecting its critical concentration.

For the case of calcium-calmodulin-induced depolymerization of microtubules, consider the data illustrated in FIGURE 10. Three samples of MAP-2-tubulin were polymerized with different critical concentrations. If calcium-calmodulin simply inactivated the tubulin protomer pool without changing the critical concentration, then under the conditions of the experiment, the sample with the highest critical concentration should suffer the greatest loss of polymer due to calcium-calmodulin. In the three cases investigated, however, all percentages of depolymerization were independent of the critical concentrations. Similar results were obtained for tau microtubules. Thus, the extent of depolymerization of microtubules in the presence of calcium-calmodulin is proportional to the amount of microtubules present. Considering the observation that calcium-calmodulin disrupts ring structures, its affect on microtubule depolymerization may be to destabilize MAP-tubulin interactions, leading to an increase in critical concentration.

Let us consider the data in FIGURE 10 in the light of a model in which the critical concentration increases as a function of the presence of calcium-calmodulin. Let c_i be the critical concentration of sample i and Δ_i be the increase in critical concentration due to calcium-calmodulin. Then the fractional decrease in polymer concentration due to calcium-calmodulin is

$$\frac{T - (c_i + \Delta i)}{T - c_i} = 1 - \frac{\Delta i}{T - c_i}$$

where T is the concentration of total polymerizable tubulin and $T - c_i$ is the concentration of tubulin in the form of polymer before calcium addition. If calcium-calmodulin acts on polymer rather than on monomers in the steady state, then it is likely that Δ_i is proportional to the amount of polymer present, that is, $\Delta_i = k(T - c_i)$ where k is a proportionality constant. In this case $1 - \Delta i/T - c_i = 1 - k$, that is, the fractional (and therefore, percent) depolymerization is independent of critical concentration as is the case experimentally. No other simple model yields this result, and the model does account for the data.

The discussion so far involves experiments in which calcium or calcium-calmodulin is added to polymer at steady state. We now examine experiments in which calcium is added prior to polymerization, that is, in which nucleation must occur to initiate polymerization. This has been reported in some detail for sea urchin tubulin obtained from isolated spindles[25] or directly from unfertilized eggs.[26] In both cases, it was clear that microtubule elongation was considerably less sensitive to calcium inhibition than was the initiation of polymerization. Addition of exogenous seeds speeded polymerization of calcium-inhibited tubulin solutions, and steady-state levels ultimately reached those of solutions to which calcium had not been added. Thus, it is clear that in these samples of highly purified, MAP-free tubulins, nucleation is considerably more calcium sensitive than is depolymerization from the microtubule state. Similar results were inferred from experiments done with calf brain microtubules.[10] In the latter paper, a critical concentration was obtained for calcium-treated cycled microtubules that was higher than that obtained for untreated cycled tubulin, thus differing from the report of Weisenberg and Deery.[29] From examination of the figure legends in that paper, however, it appears that the experiments were done by application of calcium prior to polymerization, compared to those obtained by Weisenberg and Deery, in which calcium was applied to steady-state polymer. It is not known whether this accounts for the differing conclusions obtained by these separate authors, but in light of our discussion so far, this is a strong possibility.

In the presence of calcium-calmodulin, nucleation of purified MAP-2 tubules or purified tau tubules is more sensitive to calcium-calmodulin than is depolymerization from the steady state, supporting the results discussed above for sea urchin and brain tubulin. It was pointed out, however, that in cycled brain tubulin containing whole MAPs, little difference in the effects of calcium-calmodulin occurred when it was added prior to polymerization compared to those obtained after polymerization. In purified MAP-2 tubulin or tau tubulin, or in a combination of the two, there is a clearly greater effect of calcium-calmodulin on nucleation compared to its effect on depolymerization from the steady state. This suggests the presence in whole MAPs of some factor that regulates the effects of Ca^{+2} calmodulin on nucleation of MAP-tubulin.

It is clear from the totality of these results that the effects of calcium on tubulin is highly complex and that experiments should be set up with specific components and orders of addition of reagents clearly outlined. Thus, effects on initiation of polymerization, on steady state depolymerization (presumably an effect on treadmilling), or on elongation may be different depending upon which MAP factors are present, their phosphorylation state, and how the experiment is done. The cell may also use some of these properties in its physiology. For example, large transients of calcium can occur in response to hormone stimulation[31] or nerve impulses. In such cases, it may be of considerable advantage to the cell to maintain its microtubule complement intact, for example to maintain cell shape. In addition, suppression of spontaneous nucleation by calcium may allow the cell to better specify where it will assemble microtubules by the distribution of nucleating centers.[32] The difference in effect of calcium on nucleation versus elongation allows this to be accomplished.

Finally, we have found a clear difference on nucleation versus depolymerization of reconstructed microtubules (MAP-2 tubules and tau tubules) compared with cycled tubules. Because we now know that MAP-2 and tau often occur in different compartments within the same neuron (Binder, *et al.,* this volume, p. 145–166) it is clear that experiments with microtubules should be done with purified reconstructed tubules in addition to cycled tubules if proper models concerning physiology are to be made.

REFERENCES

1. SLOBODA, R. D., W. L. DENTLER & J. L. ROSENBAUM. 1976. Biochemistry **15:** 4497–4505.
2. CLEVELAND, D. W., S. HWO & M. W. KIRSCHNER. 1977. J. Mol. Biol. **116:** 207–225.
3. WEISENBERG, R. C. 1972. Science **177:** 1104–1105.
4. OLMSTEAD, J. B. & G. G. BORISY. 1975. Biochemistry **14:** 2996–3004.
5. MARCUM, J. M., J. R. DEDMAN, B. R. BRINKLEY & A. R. MEANS. 1978. Proc. Natl. Acad. Sci. USA **75:** 3771–3775.
6. KUMAGI, H. & E. NISHIDA. 1980. Biomed. Res. **1:** 223–229.
7. SOBUE, K., M. FUJITA, Y. MURAMOTO & S. KAKIUCHI. 1981. FEBS Lett. **132:** 137–140.
8. JEMIOLO, D. K., W. H. BURGESS, L. I. REBHUN & R. H. KRETSINGER. 1980. J. Cell Biol. **87:** 248a.
9. LEE, Y. C. & J. WOLFF. 1981. J. Biol. Chem. **259:** 1226–1230.
10. BERKOWITZ, S. & J. WOLFF. 1981. J. Biol. Chem. **256:** 11216–11223.
11. LEE, Y. C. & J. WOLFF. 1982. J. Biol. Chem. **257:** 6306–6310.
12. BENDER, P. K., L. I. REBHUN & D. C. BENJAMIN. 1982. Biochim. Biophys. Acta **708:** 149–159.
13. WILLIAMS, R. C. & W. DETRICK. 1979. Biochemistry **18:** 2499–2503.
14. BURGESS, W. H., D. K. JEMIOLO & R. H. KRETSINGER. 1980. Biochim. Biophys. Acta **623:** 257–270.
15. LAEMMLI, U. K. 1970. Nature (London) **227:** 680–685.
16. BENSADOUN, A. & D. WEINSTEIN. 1976. Anal. Biochem. **70:** 241–250.
17. BINDER, L., M. PAYNE, H. KIM, V. SHERIDAN, D. SCHROEDER, C. WALKER & L. REBHUN. 1982. J. Cell Biol. **95:** 349a.
18. SANDOVAL, I. V. & J. S. VANDEKERCKHOVE. 1981. J. Biol. Chem. **256:** 8795–8800.
19. PORTZEHL, H., P. C. CALDWELL & J. C. RUEGG. 1964. Biochim. Biophys. Acta **79:** 581–591.
20. BLINKS, J., G. WIER, P. HESS & F. PREDERGAST. 1982. Prog. Biophys. Mol. Biol. **40:** 1–114.
21. GOLDSTEIN, D. A. 1976. Biophys. J. **26:** 235–242.
22. GASKIN, F., C. R. CANTOR & M. L. SHELANSKI. 1975. J. Mol. Biol. **89:** 737–758.
23. BENDER, P. K. 1982. Microtubule Associated Proteins. Ph.D. Dissertation, University of Virginia.
24. JEMIOLO, D. K. 1981. Calcium Control of Tubulin Assembly. Ph.D. Dissertation, University of Virginia.
25. KELLER, T., D. JEMIOLO, W. BURGESS & L. I. REBHUN. 1982. J. Cell Biol. **93:** 797–803.
26. SUPRENANT, K. A. & L. I. REBHUN. 1984. Cell Motility **4:** 333–350.
27. REBHUN, L. I., D. JEMIOLO, T. KELLER, W. BURGESS & R. KRETSINGER. 1980. Calcium, calmodulin and control of assembly of brain and spindle microtubules. *In* Microtubules and Microtubule Inhibitors. M. DeBrabander & J. DeMay, Eds.: 243–252. Elsevier/North-Holland Biomedical Press. Amsterdam.

28. KARR, T. L., D. KRISTOFFERSON & D. L. PURICH. 1980. J. Biol. Chem. **255:** 11853–11856.
29. WEISENBERG, R. C. & W. J. DEERY. 1981. Biochem. Biophys. Res. Commun. **102:** 924–931.
30. OOSAWA, R. & M. KASAI. 1962. J. Mol. Biol. **4:** 10–21.
31. RASMUSSEN, H., I. KOJIMA, K. KOJIMA, W. JAWALICH & W. APFELDORF. 1984. *In* Advances in Cyclic Nucleotides and Protein Phosphorylation Research. P. Greengard & G. Robison, Eds. **18:** 159–194. Raven Press. New York.
32. CACERES, A., P. BENDER, L. SNAVELY, L. REBHUN & O. STEWARD. 1983. Neurosci. **10:** 449–461.

Microtubule-Associated Proteins

In Vitro Isolation Versus *in Vivo* Function[a]

DAVID J. ASAI,[b] WILLIAM C. THOMPSON,[c]
DANIEL L. PURICH,[d] AND LESLIE WILSON[e]

[b]*Department of Biological Sciences*
Purdue University
West Lafayette, Indiana 47907

[c]*Department of Human Biological Chemistry and Genetics*
University of Texas Medical Branch
Galveston, Texas 77550

[d]*Department of Biochemistry and Molecular Biology*
University of Florida
Gainesville, Florida 32610

[e]*Department of Biological Sciences*
University of California
Santa Barbara, California 93106

Proteins that copurify with tubulin through repetitive cycles of microtubule polymerization and depolymerization *in vitro* are commonly called microtubule-associated proteins or MAPs. Three classes of MAPs have been identified and partially characterized from mammalian brain: a very high molecular weight group of polypeptides called MAP-1 (>350 kD), a cluster of polypeptides designated MAP-2 (280 kD), and a group of polypeptides known as tau (55–65 kD). All three classes of MAPs apparently stimulate the polymerization of tubulin into microtubules *in vitro* and, taken together with the coisolation data, the *in vitro* reassembly results suggest that the MAPs may be important in the regulation of microtubule polymerization in cells.[1-4]

Immunological localization studies using antibodies to MAP-2[5] and to tau[6] have revealed that the two *in vitro* MAPs appear to be found in association with microtubules in a variety of cells. The immunofluorescence data has been interpreted to mean that both MAP-2 and tau participate in some important way in the cellular functions of microtubules. Antibodies to MAP-2 have also been reported to associate with neurofilaments in axons,[7] suggesting that MAP-2 may be involved in interactions between microtubules and neurofilaments.

In contrast to the immunolocalization studies with MAP-2 and tau, antibodies to MAP-1 have yielded significantly different results among the several laboratories studying MAP-1 distribution. Monoclonal antibodies to a MAP-1 polypeptide (MAP-1A) stain a variety of microtubule structures in a wide distribution of mammalian cells.[8,9] Two other antibodies to MAP-1, however, specifically stained dendritic processes in mammalian brain sections, and did not appear to react with structures in other nonneural cells.[10,11] A third laboratory reported that a monoclonal antibody to

[a]The authors were supported by NIH Grants HD16956 (D. J. Asai), GM24958 (D. L. Purich), and NS13560 (L. Wilson), and by a Guion Pool Keating Endowment (W. C. Thompson).

MAP-1 produced a punctate pattern of intranuclear spots in interphase mammalian cells, as well as centrosome staining; that antibody did not stain microtubules in the cells.[12]

Work from our laboratory[13] has also shown that a monoclonal antibody to MAP-1 does not stain microtubules but rather stains stress fibers and speckles the interphase nuclei in fixed and permeabilized mammalian cells. We have further demonstrated that the anti-MAP-1 associates with MAP-1-containing microtubules that have been reassembled *in vitro*. Our results clearly indicate that the MAP-1 polypeptides recognized by the antibody are able to tightly adhere to microtubules *in vitro*, but appear to not be associated with microtubules *in vivo*.

The differential staining behavior observed in our studies suggests that a more specific nomenclature is necessary to differentiate among proteins displaying microtubule association under *in vitro* versus *in vivo* conditions. It is reasonable, for example, that a protein might have a substantial affinity for the microtubule surface lattice, and be able to bind to the microtubule surface and therefore copolymerize repeatedly with tubulin through cycles of assembly and disassembly *in vitro* without an *in vivo* functional relationship between the *in vitro* MAP and microtubules. Such a nonspecific adventitious interaction *in vitro* could profoundly affect the assembly and disassembly kinetics of microtubules by conferring a large stability to the polymer.[14,15] With the immunocytochemical methods available, it is now possible to carefully assess the cellular distribution of molecules that behave as *in vitro* MAPs. That assessment should yield two classes of *in vitro* MAPs—those molecules that are truly *in vivo* MAPs in that they are capable of associating with microtubules in the cell, and those molecules that are not.

A second and related consideration emerges from the various immunological studies: the chemical compositions of the high molecular weight *in vitro* MAP-1 and MAP-2 polypeptides are probably much more complex than previously suspected. Given the variety of staining patterns with the four different antibodies to MAP-1 mentioned above, it seems likely that each of the "MAP-1" proteins used was different, and that the antibodies to the various molecules are revealing the heterogeneity. Until better chemical characterization of the high molecular weight proteins is achieved, and until more specific antibody probes are established, the entire high molecular weight population of *in vitro* MAPs, and the attendant terminology, must await clarification.

REFERENCES

1. SLOBODA, R. D., W. L. DENTLER, R. A. BLOODGOOD, B. R. TELZER, S. GRANETT & J. L. ROSENBAUM. 1976. *In* Cell Motility. R. Goldman, T. Pollard & J. Rosenbaum, Eds.: 1171–1212. Cold Spring Harbor Laboratory.
2. HERZOG, W. & K. WEBER. 1978. Eur. J. Biochem. **92:** 1–8.
3. KUZNETSOV, S. A., V. I. RODIONOV, V. I. GELFAND & V. A. ROSENBLAT. 1981. FEBS Lett. **135:** 241–244.
4. VALLEE, R. B. & S. E. DAVIS. 1983. Proc. Natl. Acad. Sci. USA **80:** 1342–1346.
5. SHERLINE, P. & K. SCHIAVONE. 1977. Science **198:** 1038–1040.
6. CONNOLLY, J. A., V. I. KALNINS, D. W. CLEVELAND & M. W. KIRSCHNER. 1977. Proc. Natl. Acad. Sci. USA **71:** 2268–2272.
7. PAPASOZOMENOS, S. CH., L. I. BINDER, P. BENDER & M. R. PAYNE. 1982. J. Cell Biol. **95:** 341a.
8. BLOOM, G. S., F. C. LUCA & R. B. VALLEE. 1984. J. Cell Biol. **98:** 331–340.
9. BLOOM, G. S., T. A. SCHOENFELD & R. B. VALLEE. 1984. J. Cell. Biol. **98:** 320–330.
10. HUBER, G. & A. MATUS. 1984. J. Neurosci. **4:** 151–160.
11. HUBER, G. & A. MATUS. 1984. J. Cell Biol. **98:** 777–781.

12. SATO, C., K. NISHIZAWA, H. NAKAMURA, Y. KOMAGOE, K. SHIMADA, R. UEDA & S. SUZUKI. 1983. Cell Struct. Function **8:** 245–254.
13. ASAI, D. J., W. C. THOMPSON, L. WILSON, C. F. DRESDEN, H. SCHULMAN & D. L. PURICH. 1985. Proc. Natl. Acad. Sci. **82:** 1434–1438.
14. TIMASHEFF, S. N. 1979. Trends Biochem. Sci. **4:** 61–65.
15. LEE, J. C., N. TWEEDY & S. N. TIMASHEFF. 1978. Biochemistry **17:** 2783–2790.

Protein-Protein Interactions in Microtubules as Determined by Reversible Protein Cross-Linking[a]

ROGER D. SLOBODA AND DONNA GOTTWALD

Department of Biological Sciences
Dartmouth College
Hanover, New Hampshire 03755

Tubulin and microtubule associated proteins (MAPs) interact with each other in a highly specific manner to produce microtubules with side-arm projections having a characteristic periodicity. To determine the component of the microtubule subunit lattice to which MAPs bind, microtubules composed predominantly of MAP-2 and tubulin were reacted with the reversible protein cross-linker 2-iminothiolane.[1,2] In these experiments, free sulfhydryls on the proteins in question are first blocked with a suitable reducing agent (*e.g.* mercaptoethanol) prior to reaction with the cross-linker. Then, when 2-iminothiolane binds to an available amino group, the cross-linker decyclizes to generate an oxidizable sulfhydryl. Thus, proteins having accessible E-amino groups within the cross-linking distance (1.6 nm) of the 2-iminothiolane can potentially be cross-linked after oxidation of the newly inserted sulfhydryls using H_2O_2. The progress of the cross-linking reaction as well as analysis of the subunit composition of the cross-linked aggregates is performed using sodium dodecyl sulfate (SDS)-polyacrylamide gel electrophoresis. First, the cross-linked aggregates are resolved in the presence of urea and SDS under nonreducing conditions, and then the bands corresponding to the cross-linked aggregates of interest are excised, incubated in reducing solution, and run on a second gel in the presence of SDS, urea, and mercaptoethanol.

When three-times-cycled microtubules containing tubulin and MAPs are subjected to the preceding cross-linking protocol, high molecular weight aggregates form, as determined by SDS polyacrylamide gel electrophoresis under nonreducing conditions. One aggregate migrates with an apparent molecular weight of approximately 120,000 whereas the second is much larger and barely enters the gel. When the first aggregate is excised, reduced, and run on a second dimension gel, it can be determined that this aggregate is composed solely of equimolar amounts of alpha and beta tubulin. This aggregate is thus formed by 2-iminothiolane-induced cross-linking of the alpha-beta tubulin heterodimers that constitute the microtubule wall. A similar analysis of the aggregate at the top of the gel reveals that it is composed of MAPs as well as alpha and, to a lesser extent, beta tubulin. Such a result, however, does not demonstrate that MAPs were cross-linked to tubulin in this aggregate because the aggregate is too large to be resolved sufficiently by the separating gel (composed of linear gradients of 3–15% acrylamide and 12–48% urea).

Thus, in order to show that MAPs can indeed be cross-linked to tubulin using these procedures, another approach was used, based on the observation that MAPs can be

[a]This work was supported by American Cancer Society Grant No. IN00157 and National Science Foundation Grant PCM 80-21976 to R.D. Sloboda.

removed from taxol-stabilized microtubules by treatment with high salt.[3] FIGURE 1 shows the results of such an experiment. When control microtubules not treated with 2-iminothiolane are stabilized in taxol and then treated with high salt the MAPs can be readily separated from tubulin. MAPs, however, cannot be removed from microtubules cross-linked with 2-iminothiolane, presumably because the salt-sensitive, noncovalent interactions that normally keep the MAPs and tubulin associated are further stabilized

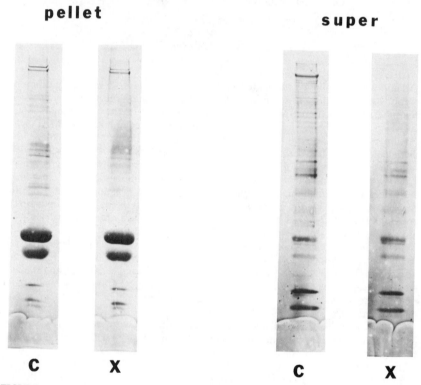

pellet   **super**

C   X   C   X

FIGURE 1. Purified microtubules containing tubulin and MAPs were cross-linked with 2-iminothiolane, stabilized with taxol, and then treated with high salt.[3] The solubilized MAPs (super) were then separated from the microtubules (pellet) by centrifugation. The samples were then analyzed by SDS-polyacrylamide gel electrophoresis under reducing conditions. C: control, non-cross-linked microtubules; X: an identical sample cross-linked with 2-iminothiolane. Note the lack of solubilized MAPs in the supernatant of the cross-linked sample.

by the presence of introduced disulfide bonds due to the cross-linking activity of the 2-iminothiolane.

An analysis of the time course of the cross-linking reaction reveals that alpha tubulin reacts preferentially with MAPs. In FIGURE 2, panels A and B show the results of such a time course. It is evident that as the time of the cross-linking reaction increases, there is a preferential and selective loss of MAPs and alpha, and to a lesser extent, beta tubulin. When the aggregate at the center of the gel (at a position of 40–50

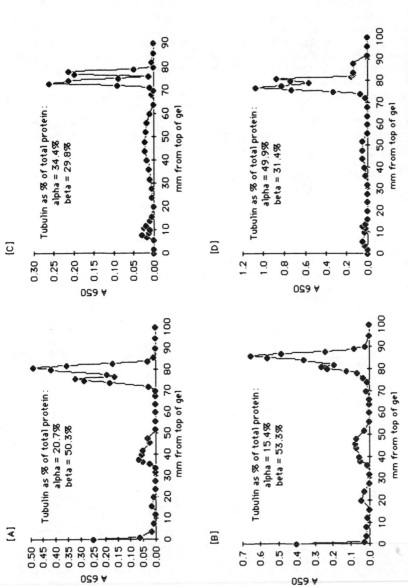

FIGURE 2. Gel scans of microtubules cross-linked for (A) 7 minutes and (B) 10 minutes. The cross-linked moieties at the top of the gel were then reduced and run on a second gel under reducing conditions (C, D). Note the greater loss of alpha tubulin relative to beta in A and B, and the corresponding increase in alpha relative to beta in C and D.

mm) is excised, reduced, and run on a second dimension gel it can be shown to be composed of equimolar amounts of alpha and beta tubulin (data not shown). If the aggregates at the top of the gel (at a position of 0–4 mm) are excised, reduced, and run on a second dimension gel, the gel profiles appear as shown in panels C and D. As was suggested by the results presented in panels A and B, more alpha than beta tubulin cross-links to the MAPs.

Such experiments suggest, although they do not conclusively prove, that MAPs (particularly MAP-2) associate with the tubulin subunit lattice by way of a direct interaction with the alpha subunit of the lattice. The results, however, presented would also occur if alpha tubulin had more reactive groups than beta tubulin. In support of the conclusion presented here, however, is the recent result of Serrano et al.[4] who reported that controlled proteolysis of tubulin liberated a C-terminal domain from alpha tubulin that was capable of binding MAP-2. Furthermore, tubulin digested in this manner was capable of assembly into sheets of protofilaments that could not incorporate MAP-2. Thus the evidence presented here as well as that summarized above suggest that MAP-2 binds to the microtubule wall through a direct interaction with alpha tubulin.

REFERENCES

1. JUE, R., J. M. LAMBERT, L. R. PIERCE & R. R. TRAUT. 1978. Biochemistry **17**: 5399–5406.
2. LAMBERT, J. M., R. JUE & R. R. TRAUT. 1978. Biochemistry **17**: 5406–5416.
3. VALLEE, R. B. 1982. J. Cell Biol. **92**: 435–442.
4. SERRANO, L., J. AVILA & R. B. MACCIONI. 1984. Biochemistry **23**: 4675–4681.

A Technique for Analyzing the Spatial Organization of Microtubular Arms and Bridges (MAPs)[a]

C. G. JENSEN[b] AND B. H. SMAILL[c]

[b]Department of Anatomy
[c]Department of Physiology
University of Auckland School of Medicine
Auckland, New Zealand

Projections from the walls of microtubules (MTs), present as side arms in isolated MTs and as cross bridges extending between MTs or between MTs and other organelles, are seen in electron micrographs from a variety of organisms. These projections appear to be the morphological correlates of a group of high molecular weight polypeptides collectively referred to as microtubule-associated proteins (MAPs), which includes MAP-1, MAP-2, and tau. Difficulties in identifying arms and bridges in micrographs, angulation of the projections, uncertainties in magnification, and the absence of long straight stretches of MTs for analysis have hindered past attempts to determine the exact arrangement of arms and bridges. Thus, there is presently only limited information on their arrangement. In this respect, McIntosh[1] has suggested that MT arms and bridges are attached to strictly periodic sites on the MT wall that are related to its substructure. Conversely, Amos[2] has suggested that the MAP-2 projections associated with brain microtubules are arranged according to a symmetrical helical superlattice that repeats after 12 tubulin dimers.

Neither of the above models has been adequately tested on data from electron micrographs. We have therefore developed a technique to quantitatively and objectively analyze the arrangement of MT arms and bridges. We use a microdensitometer to scan the area next to a longitudinally sectioned MT (FIGURE 1a and b). Standard digital signal enhancement procedures are then applied to the densitometric data, resulting in a filtered binary data sequence (FIGURE 1c). The filtered data are then analyzed in each of two ways. (1) Autocorrelograms are formed by correlating the array of a whole record with a template array consisting of the first half of the record. Correlogram peaks from many scans are then pooled to produce a peak frequency histogram. (2) The filtered densitometric data are crosscorrelated with various computer-generated model templates, and a best-fit match is obtained (FIGURE 1d). The significance of the match, as compared with random matches to the same model, is then determined for a group of scans.

We have analyzed the MAP-2 projections associated with brain MTs (micrographs kindly supplied by H. Kim[3]) using the two techniques described above. We have found that the match of MAP-2 to a 12-dimer superlattice is highly significant and more than one order of magnitude better than any other model tested (see TABLE 1). We have also found that some longitudinal spacings between MAP-2 projections are compatible

[a]This research was supported by grants from the Medical Research Council of New Zealand, the Auckland Medical Research Foundation, and the University of Auckland Research Committee.

417

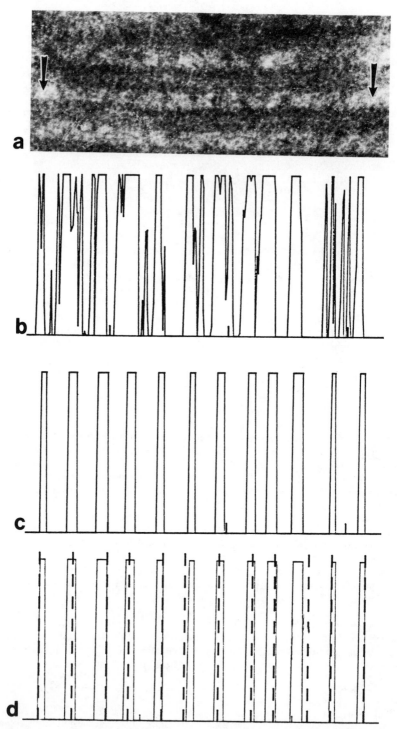

FIGURE 1 **a:** *Trypanosoma* periplast MTs. Arrows indicate the area scanned to produce the densitometer data in **b;** ×360,000; reduced by 7%; **b:** Unfiltered densitometer data. **c:** Filtered densitometer data. **d:** Best-fit match of filtered densitometer data to the 12-dimer model (indicated by dashed lines).

with a 6- rather than a 12-dimer model. We thus propose that additional, less-preferred sites of a 6-dimer superlattice may also be occupied by MAP-2.

We similarly analyzed the cross bridges of *Trypanosoma* periplast MTs (micrographs kindly supplied by H. Fuge) and of cold-treated endosperm spindle MTs. The match for both types of cross bridges with the 12-dimer model is also highly significant, showing a probability of random match of only 0.7% for *Trypanosoma* MT bridges and of 1.4% for cold-treated MT bridges. Our analysis, however, also reveals preferred longitudinal spacings between cross bridges that approximate either a 16 nm or a 21 nm periodicity. Our interpretation of this latter result is that steric constraints on the possible associations between projections from neighboring MTs result in a distortion of the 12-dimer superlattice arrangement.

TABLE 1. Best-Fit Matches of 26 MAP-2 Densitometric Records to Various Models

Possible Superlattice Model	Probability that the Match to the Model is Due to Random Arrangement
6-Dimer	93.2%
7-Dimer	28.1%
9-Dimer	53.2%
12-Dimer	1.8%
14-Dimer	20.6%
15-Dimer	34.1%
17-Dimer	18.1%
Periodic (24 nm)	14.5%

In summary, we have developed a powerful tool for a quantitative, comparative analysis of the arrangement of microtubular arms and bridges from a variety of organisms. Using this technique we have verified the 12-dimer superlattice arrangement for MAP-2 and for MT bridges.

REFERENCES

1. McIntosh, J. R. 1974. J. Cell Biol. **61:** 166.
2. Amos, L. A. 1977. J. Cell Biol. **72:** 642.
3. Kim, H. *et al.* 1979. J. Cell Biol. **80:** 266.

Age-Related Modifications of MAP-2

K. S. KOSIK, S. BAKALIS, L. GALIBERT, D. J. SELKOE,
AND L. K. DUFFY

Harvard Medical School
McLean Hospital
Belmont, Massachusetts 02178

The microtubule-associated protein 2 (MAP-2) copurifies with tubulin and several other proteins when brain microtubules are prepared by temperature-dependent phases of assembly and disassembly.[1] Ultrastructurally, MAP-2 molecules appear as projections extending outward from the microtubule wall.[2-5] By contrast to most other MAPs, MAP-2 is highly enriched in the dendritic and somatic portions of neurons, whereas axons and nonneuronal elements appear free of the antigen as seen in brain tissue sections by immunohistochemical techniques.[6,7] Because of the extensive literature concerning the complexities of dendritic morphology, development, plasticity, and attrition, a biochemical marker for the region of the neuron opens a number of investigative avenues. Previously, we studied MAP-2 during senescence in the human condition Alzheimer's disease.[8] We now present data regarding the expression of MAP-2 during fetal development and maturity in the rat.

MAP-2-enriched samples were prepared from rat forebrain using the taxol method[9] to isolate the microtubules. Rats at the following developmental time points were used: E17, E19, P0, (0 = zero), P14, P21, and mature animals. For each of these time points a suspension of the taxol-prepared microtubules was made 0.75 M NaCl and placed in a boiling H_2O bath to precipitate tubulin and maintain heat-stable MAPs in the supernatant.[2,5,10] By P14, MAP-2 appears as a doublet with a more heavily stained lower molecular weight (MW) component (MAP-2b) and a more lightly stained higher MW component (MAP-2a). The doublet is resolved in a 5–15% sodium dodecyl sulfate polyacrylamide gel electrophoresis (SDS-PAGE) system by loading ~50 μg of protein (FIGURE 1). A ladder of faintly staining putative degradation products can often be visualized beneath the doublet. At all fetal time points, studied and at birth, MAP-2 is resolved as a single band that migrates between MAP-2a and MAP-2b. These fetal proteins were shown to be MAP-2 not only by their M_r, their heat stability in a microtubule preparation, but also by their immunoreactivity with our monospecific monoclonal antibodies (mAb) against MAP-2 (FIGURE 1). These antibodies, designated 5F9 and 4F7, have been characterized previously.[8] MAP-2 is thus present in high yield at a time when neurons begin migrating into areas that are destined to become their ultimate architectonic fields. Within the first two weeks of postnatal life, MAP-2 takes its mature form as a doublet. The data suggest that MAP-2 expression begins with the onset of neuronal differentiation and forms a doublet at its cessation, for example, at the time of terminal differentiation. A similar ontogeny has been described in the cerebellum for the plasma membrane-associated cytoskeletal protein, spectrin.[11]

REFERENCES

1. SLOBODA, R. D., S. A. RUDOLPH, J. L. ROSENBAUM & P. GREENGARD. 1975. Proc. Natl. Acad. Sci. USA **72:** 177–181.

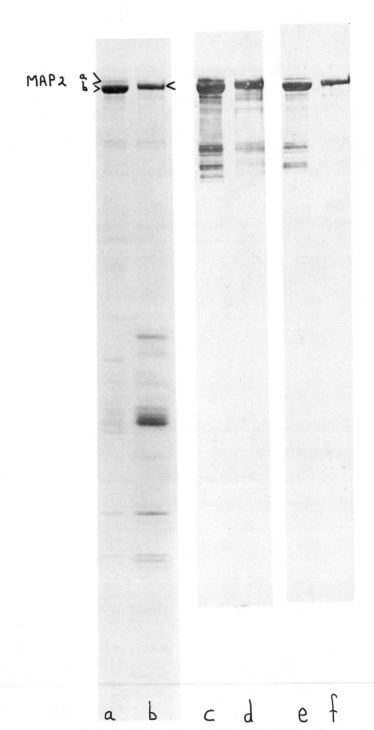

FIGURE 1. Lane a: Coomassie brilliant blue-stained SDS-PAGE of taxol-prepared MAP-2 from mature rat forebrain. Lane b: Coomassie brilliant blue-stained SDS-PAGE of taxol-prepared MAP-2 from fetal rat forebrain (E17). Lane c: Immunoblot of sample from lane a with 5F9. Lane d: Immunoblot of sample from lane b with 5F9. Lane e: Immunoblot of sample from lane a with 4F7. Lane f: Immunoblot of sample from lane b with 4F7.

2. HERZOG, W. & K. WEBER. 1978. Eur. J. Biochem. **92:** 1–8.
3. ZINGSHEIM, H. P., W. HERZOG & K. WEBER. 1979. Eur. J. Cell Biol. **19:** 175–183.
4. VOTER, W. A. & H. P. ERICKSON. 1982. J. Ulstruct. Res. **80:** 374–382.
5. KIM, H., L. I. BINDER & J. L. ROSENBAUM. 1979. J. Cell Biol. **80:** 266–276.
6. MATUS, A., R. BERNHARDT & T. HUGH-JONES. 1981. Proc. Natl. Acad. Sci. USA **78:** 3010–3014.
7. MILLER, P., V. WALTER, W. E. THEURKAUF, R. B. VALLEE & P. DECAMILLI. 1982. Proc. Natl. Acad. Sci. USA **79:** 5562–5566.
8. KOSIK, K. S., L. K. DUFFY, M. M. DOWLING, C. ABRAHAM, A. MCCLUSKEY & D. J. SELKOE. 1984. Proc. Natl. Acad. Sci. USA **81:** 7941–7945.
9. VALLEE, R. B. 1982. J. Cell Biol. **92:** 435–442.
10. FELLOUS, A., J. FRANCON, A. LENNON & J. NUNEZ. 1977. Eur. J. Biochem. **78:** 167–174.
11. LAZARIDES, E. & W. J. NELSON. 1983. Science **222:** 931–933.

Estramustine Phosphate Inhibits Microtubule Assembly by Binding to the Microtubule-Associated Proteins

MARGARETA WALLIN,[a] JOHANNA DEINUM,[b]
AND BERYL HARTLEY-ASP[c]

[a]Department of Zoophysiology
[b]Department of Medical Physics
University of Göteborg, S-400 33 Göteborg, Sweden
and
[c]AB Leo Research Laboratories
S-251 09 Helsingborg, Sweden

Estramustine phosphate, estradiol-3,[N-bis-(2-chloroethyl) carbamate] 17β-phosphate, (Estracyt) is active against advanced prostatic carcinoma.[1] It exhibits several classical mitotic inhibitory characteristics: it is cytotoxic and induces mitotic arrest at metaphase.[2] Its mechanism of action, however, has been unknown. We have therefore investigated the effect of the drug on microtubule proteins.[3,4]

EXPERIMENT

Estramustine phosphate was synthesized by AB Leo, Helsingborg, Sweden. Microtubule proteins were prepared from bovine brain cortex in the absence of glycerol by three cycles of assembly-disassembly, but in the presence of 0.5 mM MgSO$_4$.[3,4] The microtubule-associated proteins (MAPs) were separated from tubulin by ion exchange chromatography on Mg(II)-treated phosphocellulose by one-step elution with 0.6 M NaCl in buffer.[4] Microtubule proteins were assembled in 0.1 M piperazinediethane sulfonic acid (PIPES), 0.5 mM MgSO$_4$, 1 mM ethylenediaminetetracetic acid (EDTA), and 1 mM GTP as described in the legends to the figures. EDTA was present to avoid precipitation of the insoluble Mg(II)-estramustine phosphate complex and was found not to affect assembly of the microtubule proteins.

RESULTS AND CONCLUSION

Estramustine phosphate (0.01–0.5 mM) inhibited the assembly of brain microtubule proteins *in vitro* and disassembled preformed microtubules, see FIGURE 1. In the presence of estramustine phosphate the minimum microtubule-protein concentration sufficient for the assembly of microtubules was increased. Furthermore, low concentrations of taxol, 20 μM, completely reversed the inhibition of assembly by estramustine phosphate (see FIGURE 1). The content, however, of the MAPs was reduced in these taxol reversed estramustine phosphate-inhibited microtubules.

The effects were specific to estramustine phosphate because neither estradiol-17β-phosphate, the hormonal moiety of the drug, nor nornitrogen mustard, the alkylating moiety, had any effect on assembly.

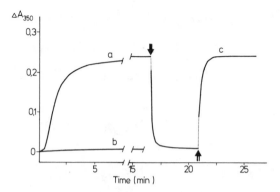

FIGURE 1. Time course of assembly of brain microtubules. Estramustine phosphate (0.42 mM) was added initially, (trace A) or at steady state level of assembly (trace B at the arrow). After complete disassembly, 20 μM taxol was added (trace B at the arrow). Microtubule assembly, (1.8 mg/ml protein) was started by raising the temperature from 10 to 37°C and monitored by the increase in absorbance at 350 nm against time. The assembly buffer contained 0.1 M PIPES at pH 6.8, 0.5 mM MgSO$_4$, 1 mM GTP, and 1 mM EDTA. (Kanje et al.[3] With permission from *Cancer Research*.)

As shown in FIGURE 2, additional MAPs relieved the inhibition of assembly by estramustine phosphate. In agreement, we have found that tritiated estramustine phosphate bound predominantly to the MAPs. Estramustine phosphate seems thus to inhibit microtubule assembly by binding to the MAPs.

In conclusion, our results suggest that the cytotoxic action of estramustine

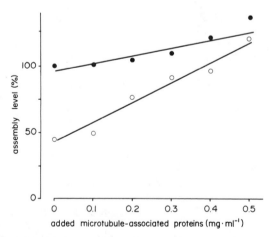

FIGURE 2. Effect of estramustine phosphate on microtubule assembly in the presence of additional MAPs. The level of assembly is given as percent of the absorbance difference at 350 nm of the control versus the additionally added MAPs (mg/ml) present during assembly. ● = no estramustine phosphate added; ○ = after incubation at 4°C with 0.2 mM estramustine phosphate. The conditions were as described in FIGURE 1. The microtubule protein concentration was originally 2.1 mg/ml.

phosphate could partially be dependent on an interaction with microtubules by binding to the MAPs. Furthermore, as most mitotic inhibitors bind to tubulin,[5] estramustine phosphate belongs to a new class of microtubule inhibitors.

REFERENCES

1. JÖNSSON, G., B. HÖGBERG & T. NILSSON. 1977. Treatment of advanced prostate carcinoma with estrasmustine phosphate (Estracyt®). J. Scand. Urol. Nephrol. **11:** 231–238.
2. HARTLEY-ASP, B. 1984. Estramustine induced mitotic arrest in two human prostatic carcinoma cell-lines Du145 and PC-3 Prostate **5:** 93–100.
3. KANJE, M., J. DEINUM, M. WALLIN, P. EKSTRÖM, A. EDSTRÖM & B. HARTLEY-ASP. 1985. The effect of estramustine phosphate on the assembly of isolated brain microtubules and fast axonal transport. Cancer. Res. **45:** 2234–2239.
4. WALLIN, M., J. DEINUM & B. FRIDÉN. 1985. Interaction of estramustine phosphate with microtubule-associated proteins. FEBS Lett. **179:** 289–293.
5. WALLIN, M. & J. DEINUM. 1983. Tubulin. *In* Handbook of Neurochemistry. A. Laitha, Ed.: Vol. 5: 101–126. Plenum Press. New York.

In Vivo Taxol Treatment Alters the Solubility Properties of Microtubule-Associated Proteins (MAPs) of Cultured Neurons[a]

MARK M. BLACK AND ISAAC PENG

Department of Anatomy
Temple University School of Medicine
Philadelphia, Pennsylvania 19140

Taxol is a drug that induces microtubule (MT) assembly *in vivo* and *in vitro*,[1,2] and also stabilizes MTs to a variety of depolymerization-promoting conditions.[1] Taxol induces abnormal MT arrangements *in vivo* that are characterized by bundles of very closely spaced MTs.[3] Several lines of evidence indicate that microtubule-associated proteins (MAPs) influence the spacing between MTs.[4] Thus, MTs in taxol-induced bundles may have an altered composition of MAPs. We have tested this possibility using cultured sympathetic neurons as a model system. These neurons contain a variety of MAPs.[5,6] The present report concerns a group of four MAPs that range in molecular weight from 60K to 76K (these MAPs are distinct from tau). Our results indicate that under *in vivo* conditions, taxol interferes with the incorporation of the 60–76K MAPs into MTs.

Cultured sympathetic neurons were prepared as described previously.[5] To determine whether *in vivo* taxol treatment alters the MAPs assembled into MTs, gel electrophoresis was used to compare fractions containing assembled MAPs obtained from taxol-treated cultures with similarly prepared fractions from control cultures and cultures depleted of MT by exposure to podophyllotoxin. In these experiments, cultures were labeled with [^{35}S]methionine, and the drug treatments were coincident with the labeling period. Briefly, cultures were extracted with a Triton X-100 containing MT-stabilizing buffer (without taxol) under conditions that solubilized unpolymerized MT proteins, whereas most or all MT remained insoluble. Triton X-100 insoluble residues from control and taxol-treated cultures contain MTs, whereas those from podophyllotoxin-treated cultures do not. The Triton X-100-insoluble residues were then extracted with a Ca^{++}-containing buffer that solubilized the majority of their MTs. Ca^{++}-soluble fractions, which contain the proteins assembled into MT at the time of extraction, were prepared by centrifugation and analyzed on 1-D sodium dodecyl sulfate (SDS) gels.

FIGURE 1 shows a fluorograph depicting the labeled peptides in Ca^{++}-soluble fractions from the various cultures. The tubulin subunits ($\rightarrow$) are major components of the material from control, taxol, and DMSO-treated cultures, but are greatly diminished in the material from podophyllotoxin-treated cultures. Note also that, as expected, the 60–76K MAPs (>) were absent from Ca^{++}-soluble fractions from MT-depleted cultures compared to controls. Unexpectedly, these MAPs were also greatly diminished in the material from taxol-treated cultures. That this effect is

[a]This work was supported by NIH Grants NS17681 and NS00698.

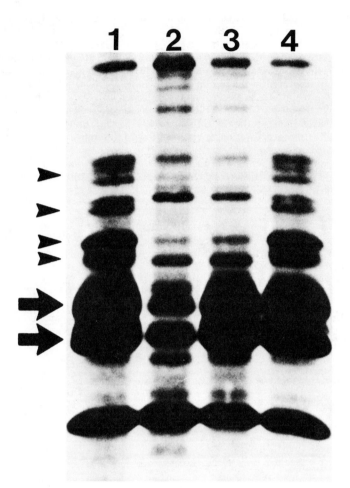

FIGURE 1. A portion of a fluorograph of a 1-D gel showing the labeled peptides in Ca^{++}-soluble fractions prepared from control cultures (lane 1) and cultures treated for 22 h with 4.5 μM podophyllotoxin (lane 2), 10 μM taxol (lane 3), and 0.1% DMSO (lane 4). The fractions were obtained by sequentially extracting cultures with a Triton X-100-containing MT-stabilizing buffer followed by a Ca^{++}-containing buffer as described in the text and in reference 5. The resulting Ca^{++}-soluble fractions are enriched in the proteins (tubulin and MAP) assembled into MTs at the time of extraction. The arrows identify the α and β tubulins, whereas the arrowheads identify the four MAPs in the molecular weight range of 60–76K.

Cultures treated with podophyllotoxin contain few or no MTs. Consequently, Ca^{++}-soluble fractions from such cultures contain little or no tubulin and MAPs (compare lanes 1 and 2). Control cultures and cultures treated with taxol or DMSO contain MTs as evidenced by the abundance of tubulin in Ca^{++}-soluble fractions prepared from such cultures. Ca^{++}-soluble fractions from control and DMSO-treated cultures contain comparable levels of 60–76K MAPs, whereas these MAPs are greatly diminished in similarly prepared fractions from taxol-treated cultures. These results suggest that the 60–76K MAPs are incorporated into MTs in control and DMSO-treated neurons, but not in neurons treated with taxol.

specific for taxol, and not DMSO, the solvent for taxol, is demonstrated by analyses of Ca^{++}-soluble fractions from DMSO-treated cultures (FIGURE 1, lane 4), in which the 60–76K MAPs are present at control levels. Additional experiments have shown that the levels of labeled 60–76K MAPs in taxol-treated cultures are indistinguishable from those in control cultures, and that the 60–76K MAPs of taxol-treated cultures are quantitatively recovered in the Triton X-100-soluble fraction, which contains unassembled MT proteins. These observations show that *in vivo* taxol treatment alters the partitioning of the 60–76K MAP in the extraction assay for assembled MAPs.

One possible interpretation for these results is that *in vivo* taxol treatment interferes with the incorporation of the 60–76K MAPs into microtubules. In this regard, *in vitro* experiments indicate that the 60–76K MAPs coassemble very poorly with brain microtubule proteins in the presence of taxol. Perhaps the unusual MT formations in taxol-treated neurons result, in part, from taxol interfering with the incorporation of the 60–76K MAPs into MTs.

REFERENCES

1. DeBrabender, M., G. Geuens, R. Nuydens, R. Willebrords & J. DeMey. 1981. Proc. Natl. Acad. Sci. USA **78:** 5608–5612.
2. Parness, J. & S. B. Horowitz. 1981. J. Cell Biol. **91:** 479–487.
3. Masurovsky, E. B., E. R. Peterson, S. M. Crain & S. B. Horowitz. 1983. Neurosci. **10:** 491–509.
4. Kim, H., L. I. Binder & J. L. Rosenbaum. 1979. J. Cell Biol. **80:** 266–276.
5. Black, M. M. & J. T. Kurdyla. 1983. J. Cell Biol. **97:** 1020–1028.
6. Peng, I., L. I. Binder & M. M. Black. 1986. Brain Res. In press.

Retinoic Acid Induces MAP-2-Containing Neurites in Mouse Neuroblastoma Cells

ITZHAK FISCHER,[a,b] THOMAS B. SHEA,[a,b] KENNETH S.
KOSIK,[c,d] AND VICTOR S. SAPIRSTEIN[a,b]

[a]Department of Biochemistry
Eunice Kennedy Shriver Center
Waltham, Massachusetts 02254

Department of [b]Biological Chemistry and [c]Neurology
Harvard Medical School
Boston, Massachusetts 02115

[d]The Mailman Research Center
Mclean Hospital
Belmont, Massachusetts 02178

We have studied the process of neuronal differentiation using mouse neuroblastoma cells (NB2a) as a model system. We found that whereas control cells possessed only short neurites (FIGURE 1A), retinoic acid (RA) induced an elaborate network of highly branched neurites (FIGURE 1C) that dramatically differed from neurites induced by db cAMP (FIGURE 1E) or serum deprivation (FIGURE 1G), which were mostly bipolar and unbranched. Examination by electron microscopy showed that the RA-induced neurites contained polysomes and microtubules, but not intermediate filaments, and their surface membrane had an irregular contour with spines, all of which are typical of dendrites. By contrast, the dibutyryl cyclic adenosine monophosphate (db cAMP)-induced neurites were without spines or polysomes, contained intermediate filament bundles and long microtubules, and even formed what appeared to be synaptic densities, all of which are typical of axons.

To study the properties of these neurites, we used a monoclonal antibody highly specific to microtubule-associated protein 2 (MAP-2), which in brain appears to have a distribution limited to neuronal cell bodies and dendrites. By immunofluorescence microscopy we found that the cell bodies and neurites induced by RA treatment stained with MAP-2 antibody (FIGURE 1D). Cultures with neurites induced by db cAMP treatment (FIGURE 1F) or serum deprivation (FIGURE 1H) showed similar staining of the cytoplasm; however, no neurite staining was observed. The short neurites of control cells were also unreactive (FIGURE 1B), but the cytoplasm did stain. Immunoprecipitation of pulse-labeled proteins indicated that there was a 2–3-fold increase in the rate of MAP-2 synthesis 24 hr after treatment of the cells with RA, which returned to control levels by 72 hours. The transient increase in MAP-2 was not reflected in an increase in the steady state amounts of MAP-2 as analyzed by immunoblots. Changes in the rate of synthesis of other cytoskeleton proteins following the RA treatment were also observed; the affected proteins were distinct from those whose synthesis was affected by db cAMP.

Both the morphological and cytochemical data suggest that the neurites induced by RA have unique dendritic properties different from the well-studied axon-like neurites induced by db cAMP or serum deprivation. The transient increase in synthesis of MAP-2 early during the RA treatment may be required for the initial formation of such dendrite-like neurites. The small changes in the steady state amounts of MAP-2 during that process, however, suggests that it is the specific distribution of MAP-2 that is probably the important event that defines the structure and nature of neurites.

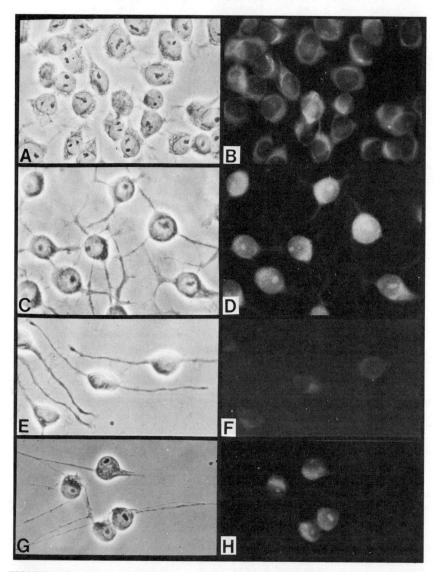

FIGURE 1. Distribution of MAP-2 in neuroblastoma cells under various conditions. NB2a cells induced to form neurites were examined by phase contrast microscopy and stained with a monoclonal antibody against MAP-2 using indirect immunofluorescence. (**A, B**) phase contrast and corresponding immunofluorescence of control cells; (**C, D**) RA-treated cells; (**E, F**) db cAMP-treated cells; (**G, H**) serum-deprived cells.

Studies on Sea Urchin Egg Cytoplasmic ATPases of Possible Significance for Microtubule Functions[a]

A. S. DINENBERG, J. R. McINTOSH, AND J. M. SCHOLEY

Department of Molecular, Cellular, and Developmental Biology
University of Colorado
Boulder, Colorado 80309

Cytoplasmic extracts of unfertilized sea urchin eggs contain ATPase activity with dynein-like properties.[1] We have examined the association of the ATPase present in such egg extracts with taxol-assembled microtubules (MTs).[2] We observed that as little as 14%[2] and as much as 45% (TABLE 1) of the ATPase copelleted with the MTs (whereas more than 90% tubulin pelleted), giving rise to approximately a 10–30-fold increase in specific activity in MTs over extracts. A dynein-like ATPase (column ATPase) can be prepared by gel filtration and hydroxyapatite chromatography from the MT-depleted extract.[2] We have investigated whether this latter ATPase would copellet with MTs upon addition of bovine brain tubulin to the egg extract prior to MT assembly (TABLE 1). In the absence of MT assembly, negligible quantities of ATPase pelleted. Approximately 3 mg MT protein and 44% ATPase pelleted from egg extracts to which no tubulin was added. When 3 mg bovine brain tubulin was added to an aliquot of the extract 6 mg MT protein pelleted, but there was no significant increase in the amount of ATPase recovered in the pellet. Gel-chromatographed cytoplasmic dynein binds efficiently to bovine brain taxol MTs,[3] so factors other than lack of availability of tubulin-binding sites in our extracts must explain why the column ATPase does not copellet with MTs.

Both the MTs isolated from egg extracts and the column purified ATPase contain polypeptide(s) of about 350 kilodaltons (kD), which may correspond to cytoplasmic dynein (FIGURE 1). On high resolution gels[5] of these protein mixtures, one sees polypeptides that coelectrophorese with A-alpha and A-Beta subunits of flagellar dynein, but the column ATPase contains more A-Beta than A-alpha on stained gels (FIGURE 1). These preparations have been analyzed by immunoblotting with a rat antiserum raised against sea urchin sperm flagellar dynein, revealing an immunological cross-reaction between 350 kD flagellar and cytoplasmic polypeptides.

We have compared the enzymatic properties of the column ATPase with those of flagellar dynein. Under conditions where flagellar dynein has specific activity of between 1.0 and 2.5 micromoles per min per mg, the column ATPase has specific activity of between 150 and 300 nanomoles per min per mg. The effects of various ATPase inhibitors, and changing the pH monovalent cation concentration, divalent cation, or nucleotide substrate on the activity of the two enzymes was similar. For example, the column ATPase possessed (a) 0% maximal activity at pH 5.0, 55% at pH 6.0, 62% at pH 7.0, 100% at pH 8.0, and 65% at pH 9.0 (b) 100% activity in Mg^{2+}, 27%

[a]J. M. Scholey was supported by a Medical Research Council traveling fellowship and a British-American exchange fellowship of the British Heart Foundation and American Heart Association. This work was supported by National Institutes of Health Grant GM 30213 and American Cancer Society Grant CD8 to J. R. McIntosh.

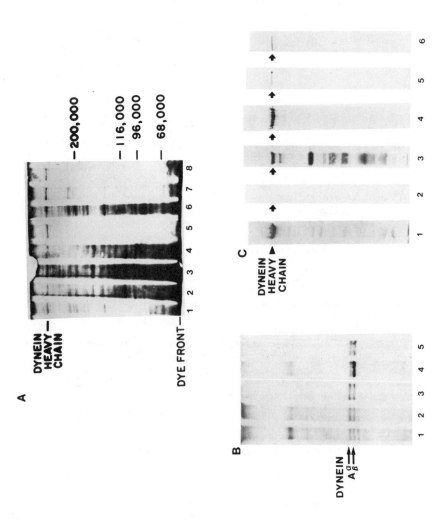

FIGURE 1. SDS gel analysis of high molecular weight polypeptides present in egg cytoplasmic ATPase and flagellar dynein preparations. **A:** Five percent SDS polyacrylamide gels (silver stained) showing various *S. purpuratus* ATPase fractions including (a) sperm dynein prepared by high salt extraction of flagellar axonemes followed by biogel A-5M gel filtration chromatography. The major heavy chain corresponds to "dynein A." The A-α and A-β polypeptides are not resolved. (b) ATP MAPs, prepared by differential centrifugation of egg taxol MTP in $5mM$ ATP. (c) Urchin egg "column ATPase" prepared by $(NH_4)_2SO_4$ fractionation, Biogel A-5M chromatography and hydroxyapatite chromatography. From left to right, the lanes show (1) column ATPase; (2) column ATPase + ATP MAPs; (3) ATP MAPs; (4) ATP MAPs + sperm dynein; (5) sperm dynein; (6) column ATPase; (7) column ATPase + sperm dynein; (8) sperm dynein. The migration positions of molecular weight standards are indicated on the right. Note that all the *S. purpuratus* ATPase fractions contain a prominent polypeptide (M_r approx. 350,000) that comigrates with the major heavy chain of sperm flagellar dynein. **B:** Silver-stained 3.2% polyacrylamide gel run in 0–8M urea[5] showing (1) MAPs; (2) flagellar dynein + MAPs; (3) flagellar dynein; (4) column ATPase; (5) column ATPase + flagellar dynein. Arrows indicate the position of the dynein A-alpha chain (the upper band) and A-beta (the lower). **C:** Immunoblotting analysis of the ATPases with a rat antiserum raised against flagellar dynein (lane 1). Preimmune (lane 2) and immune sera (lane 3) were analyzed on nitrocellulose blots of 6% SDS gels of flagella dynein. Antibody was affinity-purified on blots of electrophoretically separated dynein heavy chain[7] and analyzed on blots of flagella dynein (lane 4), MAPs (lane 5) and column ATPase (lane 6).

in Ca^{2+}, and 3% in EDTA (c) 73% activity in $0.1M$ NaCl, 84% in $0.2M$ NaCl, and 100% in $0.5M$ NaCl (d) 100% activity in ATP and 5% in GTP, and (e) 26% activity in $50\mu M$ Na_3VO_4. The column ATPase activity thus resembles cytoplasmic dynein.[4]

We have compared the enzymatic properties of the column ATPase with those of the MT-associated ATPase. The MTs contain polypeptides related to dynein heavy chains (FIGURE 1), and most of the MT-associated ATPase cofractionates with these polypeptides.[2] The ATPase, however, that copellets with MTs differs from dynein in

TABLE 1. The Effect of Microtubule Assembly on the Pelleting of MgATPase Activity from Cytoplasmic Extracts of Unfertilized *Strongylocentrotus purpuratus* Eggs

Fraction	Volume (ml)	Concen- tration (mg/ml)	Protein (mgs)	ATPase; Specific Activity (nanomol/min/ mg)	ATPase; Total Activity (nanomol/min)	ATPase; (percent)
1. Extract + No Taxol No GTP						
S	8	9	72.0	1.5	108	98%
P	1	0.6	0.6	3.6	2	2%
2. Extract + 20 μM Taxol + 1 mM GTP						
S	8	14	112.0	0.8	90	56%
P	1	2.8	2.8	25.4	71	44%
3. Extract + 20 μM Taxol + 1 mM GTP + 3 mg bo- vine brain tubulin						
S	8	13	104.0	1.0	104	53%
P	1	6.4	6.4	14.0	90	46%

Bovine brain tubulin was purified from thrice-cycled microtubule protein using diethylamino-ethyl (DEAE)-sephadex chromatography, and found to be MAP-free on Coomassie-stained sodium dodecyl sulfate (SDS) gels. Sea urchin eggs were extracted in $0.1M$ piperazinediethane-sulfonic acid (PIPES) pH 6.95, 2.5 mM Mg $(CH_3COO)_2$, 5 mM ethylene glycol bis (β-amino ethyl ether) N,N' tetraacetic acid (EGTA), 0.1 mM ethylenediamine tetraacetic acid (EDTA), 0.9 M glycerol, 0.5 mM dithiothreitol (DTT), 10^{-4} M phenylmethyl sulfonyl fluoride (PMSF), 10 μg/ml para-tosyl-arginine-methyl ester (TAME)·HCl, 10 μg/ml aprotinin, 1 μg/ml leupep-tin, and 1 μg/ml pepstatin as described previously.[2] Aliquots of the extract were incubated (1) with no taxol or GTP (2) with 20 μM taxol plus 1 mM GTP (3) with 20 μM taxol, 1 mM GTP plus 3 mg 6S bovine brain tubulin, for 5 mins at 37° C, then cooled on ice for 15 minutes and centrifuged for 45 minutes at 23,000 g through a 15% sucrose cushion in a swinging bucket rotor. The pellets were resuspended then assayed for protein concentration and ATPase activity (by following liberation of P^{32} from γ $[P^{32}]$ATP). S = supernatant, P = pellet.

possessing lower specific activity,[2] in showing lower substrate specificity (100% activity in ATP, 71% in GTP), and in its response to changes in monovalent cation concentration (84% activity in $0.1M$ NaCl, 100% in $0.2M$ NaCl, and 38% in $0.5M$ NaCl). Our working hypothesis is that an ATPase that differs from cytoplasmic dynein also copellets with our MT preparations. This activity does not, however, seem to be the same as the bovine brain MT-associated ATPase that is derived from membrane vesicles.[6] In immunoblotting experiments with an antibody that recognizes this 50 kD brain ATPase,[6] we detected a 50 kD immunoreactive polypeptide in egg homogenates, but not in our cytoplasmic extracts or MT pellets.

In summary, we are studying two cytoplasmic ATPase preparations of possible significance for microtubule function; namely the MT-associated ATPase and the column ATPase. Our aim is to clarify the relationship between the dynein-like polypeptides and ATPase activities in these preparations, and to eludicate their significance for microtubule-mediated processes within cells.

ACKNOWLEDGMENTS

We wisk to thank our colleagues in the McIntosh laboratory for advice and encouragement, and Dr. D. B. Murphy for a generous gift of brain 50 kD ATPase antibody.

REFERENCES

1. PRATT, M. M. 1984. Int. Rev. Cytol. **87:** 83–105.
2. SCHOLEY, J. M., B. NEIGHBORS, J. R. MCINTOSH & E. D. SALMON. 1984. J. Biol. Chem. **259:** 6516–6525.
3. HOLLENBECK, P. J., F. SUPRYNOWICZ & W. Z. CANDE. 1984. J. Cell Biol. **99:** 1251–1258.
4. HISANAGA, S. & H. SAKAI. 1983. J. Biochem. **93:** 87–98.
5. PIPERNO, G. & D. J. L. LUCK. 1979. J. Biol. Chem. **254:** 3084–3090.
6. MURPHY, D. B., K. T. WALLIS & R. R. HIEBSCH. 1983. J. Cell Biol. **96:** 1306–1315.
7. OLMSTED, J. B. 1981. J. Biol. Chem. **256:** 11955–11957.

Presumptive MAPs and "Cold-Stable" Microtubules from Antarctic Marine Poikilotherms[a]

ROBLEY C. WILLIAMS JR.[b,d] AND
H. WILLIAM DETRICH III[c]

[b]Department of Molecular Biology
Vanderbilt University
Nashville, Tennessee 37235
and
[c]Department of Biochemistry
University of Mississippi Medical Center
Jackson, Mississippi 39216

Most studies of microtubule (MT) assembly *in vitro* have been performed with MT protein from the brains of warm-blooded vertebrates. Relatively little is known about the composition and mechanism of assembly of MTs from poikilothermic organisms. Fish living in Antarctic waters are of particular interest because their MTs are stable at habitat (and body) temperatures in the range $-1.8°C$ to $\sim +2°C$, conditions under which mammalian MTs become dissociated.

Tubulin (Tb) was isolated from brains of *Pagothenia borchgrevinki* by two cycles of temperature-dependent (0°C and 37°C) assembly/disassembly. Although disassembly was incomplete at 0°C (see below), the critical concentration was high enough to allow a good harvest of MTs. These MTs contained nearly indetectable amounts of MAPs.[1] Their Tb binds colchicine, polymerizes to make normal MTs, and comigrates with bovine brain Tb in sodium dodecyl sulfate (SDS)-polyacrylamide electrophoresis. MTs prepared in a similar way from brains of *Notothenia coriiceps* and *Chaenocephalus aceratus* also contained only Tb and negligible quantities of MAPs.

FIGURE 1 shows the temperature dependence of the critical concentration (C_{CRIT}) of Tb from *P. borchgrevinki* over the range $-1.2°C$ to 35°C. At 35°C, corresponding roughly to mammalian body temperatures, $C_{CRIT} = 0.04$ mg/ml, a value almost 100-fold smaller than that expected for pure mammalian Tb under identical conditions. The extrapolated value of C_{CRIT} at $-1.8°C$, the habitat temperature of the fish, is 0.74 mg/ml, a number close to the physiologically relevant concentration. Thus, the cold stability of these MTs is largely a property of their Tb, rather than one conferred by MAPs or other stabilizing factors.[2] At a given temperature, the critical concentration of Antarctic fish Tb is simply much smaller than that of mammalian Tbs under the same conditions.

Although MAPs do not appear to be required for the assembly of cold-stable MTs *in vitro*, they may nonetheless be associated with MTs *in vivo*. The taxol-dependent protocol of Vallee[3] was used to isolate microtubule proteins from brains of two Antarctic cods (*N. coriiceps* and *N. gibberifrons*) and an ice fish (*Chionodraco*

[a]This work was supported in part by NSF Grant DPP-8317724 to H.W. Detrich III.
[d]Present address: Department of MCD Biology, Campus Box 347, University of Colorado, Boulder, Colorado 80309.

hamatus). As shown in FIGURE 2, MTs prepared from these fish (lanes C, D, F, and G) contained Tb and a complement of presumptive MAPs. Particularly prominent among them were previously unreported proteins of unusually high apparent molecular weight, 415K to 430K (K = 1000) as well as proteins with molecular weights of 220–280K, 140–155K, 85–95K, 40–50K, and 32–35K. As far as we know, MAPs of molecular weight above 400,000 have not been observed in temperate fish.[4,5] Some of these proteins may represent functional homologs of mammalian MAPs, and the general similarity of their electrophoretic patterns suggests that they may be functionally or structurally related to each other. Finally, we note that many of the MAPs present in the taxol preparations from *N. coriiceps* and from *N. gibberifrons* comigrated with proteins observed in MTs from the first cycle of assembly/disassembly preparations (compare lane B with C & D and lane E with F), and with those present in second cycle MTs prepared from *N. coriiceps* by a Ca^{2+}-EGTA procedure.[6] The failure to recover MAPs from the second cycle MTs from these fish

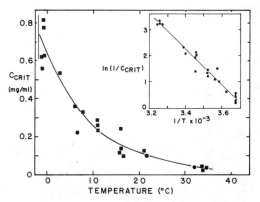

FIGURE 1. Critical concentration of Antarctic fish Tb as a function of temperature. (■) Results obtained by measurement of turbidity at 350 nm; (●) results obtained by centrifugation of MTs and measurement of concentration of supernatant. Inset: The same data represented as a van't Hoff plot, from which apparent thermodynamic quantities are obtained for addition of a tubulin dimer to a MT: $\Delta G^0 = -6.4$ kcal/mol; $\Delta H^0 = +13.7$ kcal/mol; $\Delta S^0 = +74$ cal/(deg mol). Assembly is thus entropy-driven.

indicates that they do not copurify efficiently with Tb during temperature-dependent assembly/disassembly.

We conclude that a major part of the cold stability of MTs of these organisms is due to adaptive properties of their Tbs. Novel MAPs are present, and they may contribute further to cold stability and to functional properties of the MTs.

REFERENCES

1. WILLIAMS JR., R. C., J. J. CORREIA & A. L. DEVRIES. 1985. Formation of microtubules at low temperature by tubulin from antarctic fishes. Biochemistry **24:** 2790–2798.
2. JOB, D., C. T. RAUCH, E. H. FISCHER & R. L. MARGOLIS. 1982. Recycling of cold-stable microtubules: Evidence that cold stability is due to substoichiometric polymer blocks. Biochemistry **21:** 509–515.

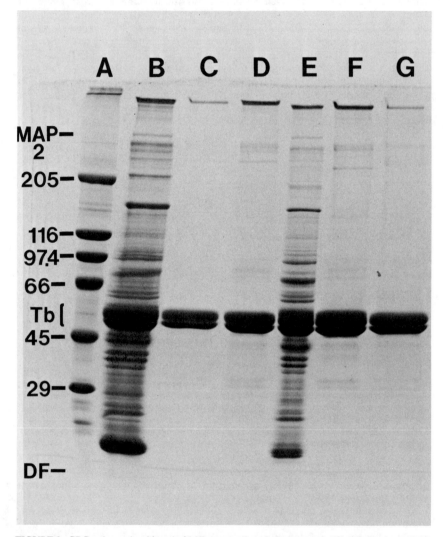

FIGURE 2. SDS-polyacrylamide gel of MT preparations from Antarctic fish. Molecular weights are indicated on the left in thousands, and the approximate positions of Tb and bovine MAP-2 are indicated. Lane A = standards; lane B = MT proteins from *N. coriiceps* prepared by one cycle of assembly/disassembly; C and D = two different taxol-dependent preparations of MT proteins from *N. coriiceps;* E = MT proteins from *N. gibberifrons* prepared by one cycle of assembly/ disassembly; F = taxol-dependent preparation of MT proteins from *N. gibberifrons;* G = taxol-dependent preparation of MT proteins from *C. hamatus.*

3. VALLEE, R. B. 1982. A taxol-dependent procedure for the isolation of microtubules and microtubule-associated proteins. J. Cell Biol. **92:** 435–442.
4. LANGFORD, G. M. 1978. *In vitro* assembly of dogfish brain tubulin and the induction of coiled ribbon polymers by calcium. Exp. Cell. Res. **111:** 139–151.
5. MACCIONI, R. B. & W. MELLADO. 1981. Characteristics of the *in vitro* assembly of brain tubulin of *Cyprinus carpio.* Comp. Biochem. Physiol. **70B:** 375–380.
6. WEBB, B. C. & L. WILSON. 1980. Cold-stable microtubules from brain. Biochemistry **19:** 1993–2001.

Microtubule-Associated Proteins (MAPs) of Dogfish Brain and Squid Optic Ganglia[a]

GEORGE M. LANGFORD, ERROL WILLIAMS, AND
DARRYL PETERKIN

Department of Physiology
School of Medicine
University of North Carolina
Chapel Hill, North Carolina 27514
and
Marine Biological Laboratory
Woods Hole, Massachusetts 02543

Experiments were designed to determine the types of high molecular weight (HMW) MAPs present in brain tissue of the smooth dogfish shark (*Mustelus canis*) and the optic ganglia of the squid (*Loligo pealei*). We previously reported[1] that microtubule proteins (MTPs) from dogfish brain, purified by temperature cycling, showed a lack of HMW MAPs. In subsequent studies, we have found that the temperature at which the microtubules are assembled significantly influences the retention of MAPs during purification. We found that when microtubules are assembled at 30°C rather than at 18–22°C (the physiological temperatures for these organisms), HMW MAPs were consistently observed after two or three cycles of purification. Therefore, we have used 30°C as the assembly temperature in these studies.

Samples of dogfish brain MTPs, after one cycle of assembly-disassembly, were run on sodium dodecyl sulfate (SDS) polyacrylamide gels. Three prominent HMW bands were seen (FIGURE 1, lanes 3 and 4). The slowest migrating HMW band ran slightly slower than cow brain MAP-1 (FIGURE 1, lanes 5 and 6) and is presumed to be homologous to MAP-1. The fastest migrating HMW band ran faster than cow brain MAP-2 and is presumed to be its homologue. The middle band was a minor HMW protein that migrated between MAPs 1 and 2.

MAP-1 of dogfish brain (FIGURE 1, lane 10) was found to be heat labile after boiling at 100°C for 5 min (FIGURE 1, lanes 11 and 12). Tau-like proteins that comigrated with the tau proteins of cow brain (FIGURE 1, lanes 8 and 9) were seen on gels of dogfish brain MTP samples (FIGURE 1, lanes 11 and 12) after heat treatment.

Experiments were also designed to determine if differences in the relative amounts of the HMW MAPs were apparent in different regions of the dogfish central nervous system (CNS). Three regions of the CNS (designated brain part I, brain part II, and the spinal cord) were investigated. Brain part I (the telen-, mesen-, and diencephalon) was highest in gray matter. Brain part II (rhombencephalon and corpus cerebelli) had similar amounts of gray and white matter, and the spinal cord (a 10–12 cm segment from the anteriormost portion of the cord) was highest in white matter.

Densitometic scans of the gel lanes (FIGURE 2) showed that after one cycle of assembly, brain part I had a higher amount of MAP-2 than MAP-1 (FIGURE 2, lane 1). The high amount of MAP-2 in brain part I may reflect the high percentage of gray matter in this part of the brain. The relative amounts of MAPs 1 and 2 in brain part II

[a]This work was supported by NIH Grant GM-28107.

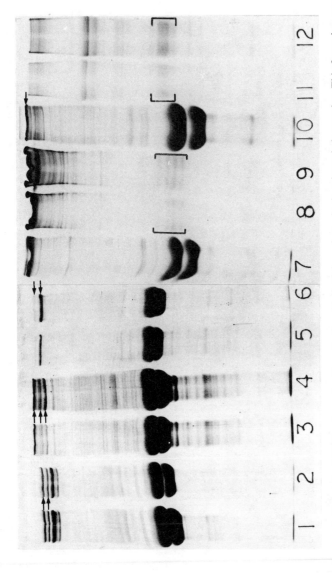

FIGURE 1. Coomassie blue-stained SDS-polyacrylamide gels of squid, dogfish, and cow brain microtubule proteins. This figure is a composite of two different gels. The acrylamide concentration is 10% in lanes 1–6 and 7.5% in lanes 7–12. Cow brain MAPs 1 and 2 are indicated by arrows at the top of lanes 5 and 6 and cow brain tau MAPs are shown in brackets in lanes 8 and 9. The samples in each of the lanes are as follows: 1 = squid second cycle cold supernatant (C2S); 2 = squid third cycle warm pellet (H3P); 3 = dogfish C1S; 4 = dogfish H1P; 5 = cow C2S; 6 = cow H2P; 7 = cow H3P; 8 = cow heat stable MAPs; 9 = cow heat stable MAPs (higher loading); 10 = dogfish H2P; 11 = dogfish heat stable MAPs; 12 = dogfish heat stable MAPs (higher loading).

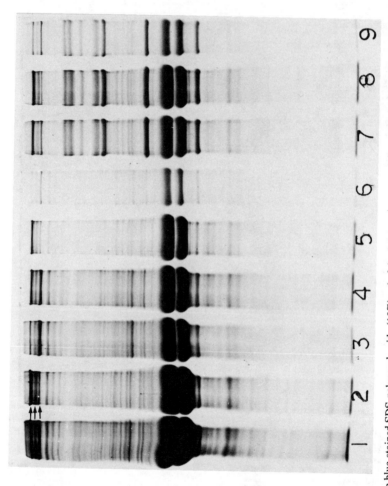

FIGURE 2. Coomassie blue-stained SDS-polyacrylamide (10%) gel of dogfish brain microtubule proteins. The three major bands at the tops of the lanes are the high molecular weight MAPs. The samples on each of the lanes are as follows: 1 = brain part I H1P; 2 = brain part I C1S; 3 = brain part I C1P; 4 = brain part II H1P; 5 = brain part II C1S; 6 = brain part II C1P; 7 = spinal cord H1P; 8 = spinal cord C1S; 9 = spinal cord C1P.

were found to be roughly equivalent (FIGURE 2, lane 4), thereby paralleling the concentrations of white and gray matter in this region. The spinal cord had a higher amount of MAP-1 than MAP-2 (FIGURE 2, lane 7).

The amount of MAP-2 in samples from each of the three areas of the CNS was reduced relative to MAP-1 (FIGURE 2, lanes 2, 5, and 8) after the first cold centrifugation step. MAP-2 depletion occurred because it remained bound to material that sedimented during this centrifugation step (FIGURE 2, lanes 3, 6, and 9). Because MAP-2 is depleted during cycle purification, the first microtubule pellet is presumed to be more representative of the relative concentrations of MAPs 1 and 2 in the tissues.

Squid MTPs, purified by the same procedure and assembled at the same temperature, contained 4 HMW proteins (FIGURE 1, lanes 1 and 2). The two upper bands comigrated with MAPs 1 and 2 of cow brain (FIGURE 1, lanes 5 and 6). These two proteins are most likely to be homologous to MAPs 1 and 2. The third band from the top of the gel is presumed to be a contaminant because it is found in very high concentration in the initial extract where the tubulin concentration is relatively low. The fourth protein is not thought to be a MAP based on its lower molecular weight.

REFERENCE

1. LANGFORD, G. M. 1978. Exp. Cell. Res. **111:** 139–151.

Proteins Associated with an Erythrocyte Marginal Band[a]

J. A. JOSEPH-SILVERSTEIN AND W. D. COHEN

Department of Biological Sciences
Hunter College
New York, New York 10021

The cytoskeletal system of "blood clam" (*Noetia ponderosa*) erythrocytes (FIGURE 1a) consists of a marginal band (MB) of microtubules (MTs) (FIGURE 1b), and a cell surface-associated cytoskeleton (SAC). As in vertebrate erythrocytes,[1] the MB in this invertebrate erythrocyte functions to maintain shape in mechanically perturbed cells.[2] The MB MTs are linked to one another by means of cross-bridges. In living blood clam erythrocytes, the MB is cold labile, disassembling at 0°C and reassembling in association with a centriole pair upon rewarming to room temperature.[3] Thus, this is a good system for studying a relatively simple microtubule bundle that is generated by a centriole-containing organizing center. The ability to manipulate the MB by temperature cycling has provided us with a useful tool for identifying proteins associated with the MB, one or more of which may be cross-bridge proteins. When cytoskeletons prepared from these cells by Triton extraction are analyzed by sodium dodecyl sulfate polyacrylamide gel electrophoresis (SDS-PAGE), the major proteins observed comigrate with human erythrocyte α spectrin (clam 240K), actin, and tubulin. The band comigrating with actin has been identified as such by Western blotting using polyclonal anti-actin. SDS-PAGE comparison of cytoskeletal proteins from cells at room temperature (with MBs), cells incubated at 0°C (no MBs), and rewarmed cells (with reassembled MBs) revealed that two minor bands with M_r 105,000 and M_r 80,000, together with tubulin, cycled with the MB. Clam 240K and actin remained constant, supporting their identification as components of the cold stable SAC rather than the cold labile MB. To determine whether the presence of the M_r 105,000 and M_r 80,000 proteins in the erythrocyte cytoskeletons was dependent upon the presence of an intact MB, we used the MT inhibitors nocodazole and colchicine and the MT-stabilizing drug taxol to prepare cells with and without MBs at the same temperature. Ten μg/ml nocodazole and 0.1mM colchicine blocked MB reassembly at room temperature following 0°C incubation of erythrocytes, whereas 10μg/ml taxol prevented MB disassembly at 0°C. The cytoskeletal composition of erythrocytes with MBs was compared to that of erythrocytes without MBs by SDS-PAGE. The relative amounts of the M_r 80,000 and M_r 105,000 proteins, as well as tubulin, were decreased in cytoskeletal preparations from cells lacking MBs regardless of the drug or temperature used to prepare the cells (FIGURE 1c). Therefore, the behavior of the M_r 80,000 and M_r 105,000 proteins is consistent with that expected for MB-associated proteins.

We have developed a Brij-lysed system with a cold-labile MB in order to identify proteins released into the soluble phase upon MB disassembly. Brij, hemoglobin, and other soluble components can be washed out of the cytoskeletal preparations prior to MB disassembly, simplifying analysis of the disassembly products. The use of Brij-58

[a]This work was supported by NSF Grant PCM-8409159.

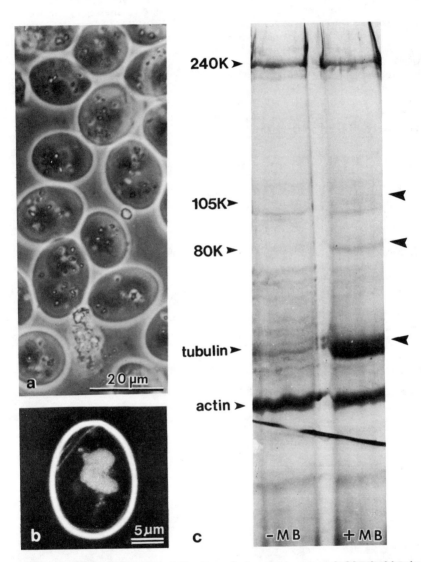

FIGURE 1. a: Living erythrocytes of *Noetia ponderosa*, phase contrast. **b:** Marginal band of erythrocyte cytoskeleton visualized by anti-tubulin immunofluorescence. (Note: nuclear fluorescence is nonspecific effect of secondary antibody). **c:** SDS-PAGE of cytoskeletons prepared from cells incubated at 0°C in the presence of taxol (+ MB) or absence of taxol (− MB). Note absence of the M_r 105,000 and M_r 80,000 proteins (as well as tubulin) in the cytoskeletons lacking MBs.

facilitated repeated resuspension of the cytoskeletons following washes. MB disassembly in Brij cytoskeletons at 0°C was verified by indirect immunofluorescence using monoclonal anti-tubulin. The supernate resulting from MB disassembly was concentrated in Centricon microfilters and analyzed by SDS-PAGE. Large amounts of tubulin were present, verifying release of soluble proteins from Brij-extracted cells. The M_r 80,000 and M_r 105,000 proteins constituted 12% of the total supernate protein, whereas tubulin constituted 78 percent. Actin was also present in significant amounts, but we conclude that it was not MB-associated because (a) actin does not cycle together with MB tubulin in living cells, and (b) in MB-containing cytoskeletons, polyclonal anti-actin binds to the SAC but not to the MB, as determined by indirect immunofluorescence. The Brij-lysed system developed in this work provides sufficient quantities of the MB proteins for biochemical and immunological studies.

REFERENCES

1. JOSEPH-SILVERSTEIN, J. A. & W. D. COHEN. 1984. The cytoskeletal system of nucleated erythrocytes. III. Marginal band function in mature cells. J. Cell Biol. **98:** 2118–2125.
2. JOSEPH-SILVERSTEIN, J. A. & W. D. COHEN. 1985. Role of the marginal band in an invertebrate erythrocyte: evidence for a universal mechanical function. Can. J. Biochem. Cell Biol. **63:** 621–630.
3. NEMHAUSER, I., J. A. JOSEPH-SILVERSTEIN & W. D. COHEN. 1983. Centrioles as microtubule organizing centers for marginal bands of molluscan erythrocytes. J. Cell Biol. **96:** 979–989.

Comparison of Cytoplasmic and Flagellar Dynein by One-Dimensional Peptide Mapping[a]

M. M. PRATT

University of Miami School of Medicine
Miami, Florida 33101

Unfertilized sea urchin eggs contain a Mg^{++}ATPase that is very similar to axonemal dynein. The egg enzyme shows the unique ionic and nucleotide specificities and sensitivity to inhibitors (sodium vanadate and erythro-9-2,3-hydroxynonyladenine) characteristic of sperm flagellar dynein. In addition, the native enzyme is large, exhibiting a sedimentation coefficient of 12-14S, similar to some forms of axonemal dynein. As determined by sodium dodecyl sulfate-polyacrylamide gel electrophoresis (SDS-PAGE), the egg ATPase is composed at least in part of high molecular weight (HMW) polypeptides (A, D and B; see reference 1) that comigrate with flagellar subunits. Recently, these polypeptides have been shown to bind ATP specifically.[2] In order to examine the primary structural homology between the egg ATPase and a known dynein, the HMW polypeptides of the two enzymes prepared from sea urchin eggs and sperm flagella were compared by one-dimensional peptide mapping.

The egg ATPase and sperm flagellar dynein were isolated from the gametes of *Strongylocentrotus purpuratus* according to standard procedures. Peptide maps were generated by the method of Cleveland *et al.*[3] with minor modifications. The HMW dynein subunits were fractionated by SDS-PAGE on 5–10% gradient gels containing a 30:0.3 ratio of acrylamide:bis-acrylamide.[4] After brief staining with Coomassie blue, the A, D, and B bands were excised and soaked in sample buffer. Each band was loaded into the well of a second gel containing a 7.5–15% gradient of polyacrylamide alternating egg and flagellar bands across the gel. The slices were overlaid with sample buffer and then with either 1–500 ng of *Staphylococcus* V8 protease or 0.5–5 μg of α chymotrypsin. The mapping gels were run at 15 mA, and in the case of chymotrypsin digestion, the current was turned off for 30 min when the protein was 0.5 cm from the bottom of the stacking gel. Mapping gels were visualized with a silver nitrate stain.

Adjacent lanes where egg and flagellar HMW subunits were digested under identical conditions were compared visually, and 35 to 40 peptides were examined on each map. No more than four differences (absent or substantially decreased peptides) were found between the egg and flagellar dynein subunits. This indicates that each polypeptide component of the egg ATPase is 85–90% homologous to its flagellar counterpart. When the flagellar subunits A, B, and D were compared one to another, 80–85% homology was observed by the same criteria.

These data support the identification of the egg-derived polypeptides as components of a "cytoplasmic dynein". Because these polypeptides are ATP binding,[2] it is likely that the active sites of egg and flagellar dynein are highly conserved. The high degree of similarity between cytoplasmic and axonemal dynein may support an early

[a]This work was supported by National Science Foundation Grants PCM-82-45321 and PCM-84-03006.

hypothesis that the egg ATPase serves as a precursor for embryonic ciliary dynein. This important question is currently under investigation.

REFERENCES

1. GIBBONS, I. R., E. FRONK, B. H. GIBBONS & K. OGAWA. 1976. *In* Cell Motility. R. Goldman, T. Pollard & J. Rosenbaum, Eds. Vol. C: 915–932. Cold Spring Harbor Press. New York.
2. PRATT, M. M. 1984. J. Cell Biol. **99**(4):46a.
3. CLEVELAND, D. W., S. G. FISCHER, M. W. KIRSCHNER & U. K. LAEMMLI. 1977. J. Biol. Chem. **252**(3): 1102–1106.
4. PORTER, M. & K. JOHSON. 1983. J. Biol. Chem. **258**(10): 6575–6581.

Interaction of the Regulatory Subunit (R$_{II}$) of cAMP-Dependent Protein Kinase with Tissue-Specific Binding Proteins Including Microtubule-Associated Proteins[a]

SUZANNE M. LOHMANN,[b] ULRICH WALTER,[b] AND
PIETRO DeCAMILLI[c]

[b]*Departments of Physiological Chemistry and Medicine*
University of Würzburg
Wüzburg, West Germany
and
[c]*CNR Center of Cytopharmacology and*
Department of Medical Pharmacology
University of Milan
Milan, Italy

One mechanism of signal transfer from hormones and neurotransmitters to intracellular regulators of physiological functions is by way of cyclic adenosine monophosphate (cAMP) and cAMP-dependent protein kinase.[1-5] In eukaryotes the cAMP-dependent protein kinase consists of a regulatory subunit (R$_I$ or R$_{II}$, for the type I and type II protein kinases, respectively) and a catalytic subunit (C).[1,6,7] The catalytic subunit is inhibited by the regulatory subunit and becomes active and able to phosphorylate substrate proteins only after the holoenzyme is dissociated by cAMP. Certain evidence, however, has led us to believe that the regulatory subunit may have other roles in addition to inhibition of C. (1) Forms of life (both prokaryotic and eukaryotic) early in evolution have cAMP binding proteins analogous to R, which are not associated with C or kinase activity, for example, *E.coli* catabolite activator protein (CAP), a DNA regulatory protein,[8] and *Dictyostelium discoideum* cell surface R, which is apparently a component of the amoeba's chemotactic response to cAMP.[9] (2) Under certain conditions, the concentration of regulatory subunits can be selectively increased several fold in many cell types, without any change in the concentration of C (reviewed in reference 10). (3) Protein kinase association with certain cellular structures, such as membranes,[11-14] microtubules,[15,16] and a calmodulin-binding complex,[17] appear to be mediated by the R subunit. Additionally, the use of cDNA probes has demonstrated cAMP-mediated increases in specific mRNAs coding for several proteins (reviewed in references 10, 18), raising the possibility that the R subunit of cAMP-dependent protein kinase could directly function as a DNA-regulatory protein in enkaryotes, analogous to CAP in prokaryotes.

In our studies, two independent methods were used to identify the tissue-specific cellular binding proteins with which the regulatory subunit interacts. High affinity binding of R$_{II}$ to a number of cellular proteins distinct from C were demonstrated by experiments in which these proteins copurified with R$_{II}$ under nondenaturing condi-

[a]This work was supported by the Deutsche Forschungsgemeinschaft (Heisenberg Program and Wa 366/3-2) and by the Italian Consiglio Nazionale delle Ricerche.

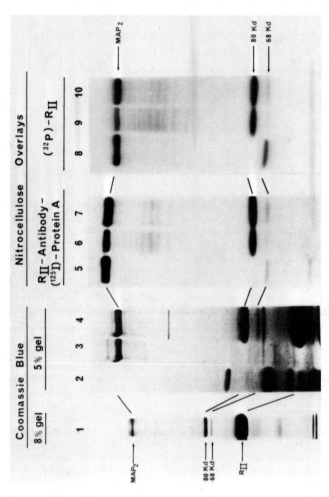

FIGURE 1. Comparison of the proteins that copurified with bovine brain R_{II} on AHA-cAMP Sepharose (lanes 1 and 4: Coomassie blue stain of 8% and 5% gels) and those proteins of bovine cerebrum cytosol (lanes 6, 7, 9, 10) transferred from 5% gels to nitrocellulose, which bound R_{II} present in an overlay medium. The same "R_{II} binding proteins" on nitrocellulose were visualized by autoradiography if $[^{32}P]R_{II}$ was used or if nonlabeled R_{II} was subsequently labeled by monospecific antibody against R_{II} and $[^{125}I]$Protein A. Lane 2: M_r protein standards; lanes 3, 5, 8: MAP-2 (16 μg); lanes 6, 7, 9, 10: bovine cerebrum cytosol (300 μg). In other experiments,[18] the proteins in bovine cerebrum cytosol that bind R_{II} on nitrocellulose were shown to be identical to those proteins that coeluted with R_{II} from the cAMP affinity column.

tions and were labeled on Western blots treated with R_{II} overlays. The comparison of the proteins identified by both methods is shown in FIGURE 1. In bovine cerebrum cytosol, a specific interaction between R_{II} and proteins of M_r 300, 80, and 68 kD were found. The 300 and 68 kD proteins were identified respectively as microtubule-associated protein 2, and a lower molecular mass protein that copurified with it.[18] In contrast to these proteins that are primarily found in the brain cytosol, the 80 kD protein was almost equally distributed between soluble and membrane fractions of brain (FIGURE 2). Future work is designed to identify the 80 kD protein and determine whether it is also a microtubule-associated or cytoskeletal protein. The interaction of R_{II} with binding proteins was not affected by R_{II}phosphorylation (FIGURE 1), nor by

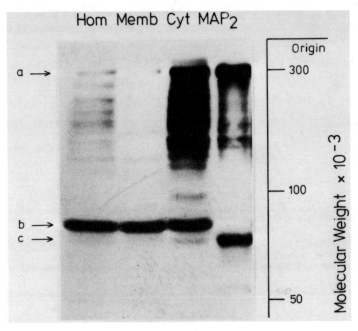

FIGURE 2. Subcellular distribution of R_{II} binding proteins in the soluble and membrane fractions of bovine cerebrum obtained by centrifugation of homogenates at $100,000 \times g$ for 1 hour. MAP-2 (300 kD) and the 68 kD protein are cytosolic proteins, whereas the 80 kD protein is associated with both soluble and membrane fractions.

the presence of cAMP (not shown). Several binding proteins distinct from those shown in FIGURE 1 were found in bovine heart.[18] Many of the R_{II} binding proteins from brain and heart served as substrates for the purified catalytic subunit of cAMP-dependent protein kinase (not shown). One hypothesis of the significance of the protein kinase regulatory subunit interaction with cellular binding proteins is that this may control the protein kinase holoenzyme localization and thereby define the substrate targets most accessible for phosphorylation by the active C subunit. MAP-2 is a known substrate protein for both cAMP-dependent protein kinase[19] and Ca^{2+}/calmodulin protein kinase,[5] and the net effect of phosphorylation of MAP-2 appears to favor microtubule disassambly.[20,21] Alternatively, R_{II} binding to a variety of cellular proteins may regulate their function, that is, R_{II} could be a regulator for multiple proteins in addition to C.

REFERENCES

1. KREBS, E. G. & J. A. BEAVO. 1979. Annu. Rev. Biochem. **48:** 923–939.
2. COHEN, P. 1982. Nature (London) **296:** 613–620.
3. INGEBRITSEN, T. S. & P. COHEN. 1983. Science **221:** 331–338.
4. GREENGARD, P. 1981. Harvey Lect. **75:** 277–331.
5. NESTLER, E. J. & P. GREENGARD. 1983. Nature (London) **305:** 583–588.
6. FLOCKHART, D. A. & J. D. CORBIN. 1982. CRC Crit. Rev. Biochem. **12:** 133–186.
7. BEAVO, J. A. & M. C. MUMBY. 1982. *In* Handbook of Experimental Pharmacology. J. A. Nathanson & J. W. Kebabian, Eds. Vol. 58I: 363–392. Springer. New York.
8. WEBER, I. T., K. TAKIO, K. TITANI & T. A. STEITZ. 1982. Proc. Natl. Acad. Sci. USA **79:** 7679–7683.
9. GERISCH, G. 1982. Annu. Rev. Physiol. **44:** 535–552.
10. LOHMANN, S. M. & U. WALTER. 1984. Adv. Cyclic Nucleotide Res. **18:** 63–117.
11. WALTER, U., P. KANOF, H. SCHULMAN & P. GREENGARD. 1978. J. Biol. Chem. **253:** 6275–6280.
12. CORBIN, J. D., P. H. SUGDEN, T. M. LINCOLN & S. L. KEELY. 1977. J. Biol. Chem. **252:** 3854–3861.
13. HOFMANN, F., P. J. BECHTEL & E. G. KREBS. 1977. J. Biol. Chem. **252:** 1441–1447.
14. RUBIN, C. S., J. ERLICHMAN & O. M. ROSEN. 1972. J. Biol. Chem. **247:** 6135–6139.
15. VALLEE, R. B., M. J. DIBARTOLOMEIS & W. E. THEURKAUF. 1981. J. Cell Biol. **90:** 568–576.
16. MILLER, P., U. WALTER, W. E. THEURKAUF, R. B. VALLEE & P. DECAMILLI. 1982. Proc. Natl. Acad. Sci. USA **79:** 5562–5566.
17. HATHAWAY, D. R., R. S. ADELSTEIN & C. B. KLEE. 1981. J. Biol. Chem. **256:** 8183–8189.
18. LOHMANN, S. M., P. DECAMILLI, I. EINIG & U. WALTER. 1984. Proc. Natl. Acad. Sci. USA **81:** 6723–6727.
19. SLOBODA, R. D., S. A. RUDOLPH, J. L. ROSENBAUM & P. GREENGARD. 1975. Proc. Natl. Acad. Sci. USA **72:** 177–181.
20. MURTHY, A. S. N. & M. FLAVIN. 1983. Eur. J. Biochem. **137:** 37–46.
21. BURNS, R. G., K. ISLAM & R. CHAPMAN. 1984. Eur. J. Biochem. **141:** 609–615.

Separation of Microtubule-Associated cAMP and Calmodulin-Dependent Kinases that Phosphorylate MAP-2

MARY LOU VALLANO[a] AND ROBERT J. DeLORENZO[b]

Yale University Medical School
New Haven, Connecticut 06510

Microtubule-associated protein 2 (MAP-2) is a proposed regulator of microtubule assembly dynamics and the interaction between microtubules and other cytoskeletal elements and membrane.[1] A possible mechanism for rapid modulation of cytoskeletal function is phosphorylation of MAP-2. MAP-2 is phosphorylated *in vivo*,[2] and reconstitution studies indicate that phosphorylation of MAP-2 with exogenous cyclic adenosine monophosphate (cAMP)-dependent kinase[3,4] or calmodulin-dependent kinase[5] diminishes the ability of the molecule to promote tubulin assembly into microtubules. A cAMP-dependent kinase/MAP-2 complex is endogenous to microtubule preparations and phosphorylates 13 of the 22 available sites on MAP-2.[1] The remaining sites are phosphorylated in a cAMP-independent manner by an unidentified kinase. In the present report, we demonstrate that microtubule preparations also contain an endogenous calmodulin-dependent kinase that phosphorylates endogenous MAP-2 with a high specific activity.

Depolymerized microtubule protein was resolved into two primary peaks of protein by gel filtration chromatography, designated 30–36 S and 6 S tubulin.[6] The cAMP-dependent kinase activity coelutes with the 30–36 S ring form of tubulin, because it is tightly bound to MAP-2, which is part of the ring structure.[1,6] All of the calmodulin-dependent kinase activity also coelutes with the rings near the void volume of the column (FIGURE 1A). The autoradiographs of samples from the void and trough fractions are shown in FIGURE 2A.

NaCl was added to the depolymerized microtubule protein prior to gel filtration chromatography in order to determine whether the endogenous calmodulin-dependent kinase activity was associated with the rings, or another high molecular weight complex in the void fraction. Treatment of microtubules with NaCl disassembles rings into a cAMP-dependent kinase/MAP-2 complex, MAP-2, and tubulin.[1,6] FIGURE 1B shows the protein elution profile, calmodulin and cAMP-dependent kinase activities in NaCl treated chromatographed microtubules. Autoradiographs of samples from the void and trough fractions are shown in FIGURE 2B. All of the recovered calmodulin-dependent kinase activity eluted in the void fraction, representing an enzyme enrichment of 10-fold for MAP-2 phosphorylation over twice-cycled microtubules. The cAMP-dependent kinase activity was shifted to the lower molecular weight trough fraction. The endogenous calmodulin-dependent kinase in the void fraction incorporated at least 3 mol PO_4/mol endogenous MAP-2 and was identical by all criteria

---

[a]Present address: Department of Pharmacology, SUNY-Upstate Medical Center, Syracuse, New York 13210.
[b]Present address: Department of Neurology, Medical College of Virginia, Richmond, Virginia 23298.

examined to a previously purified calmodulin-dependent kinase from brain cytosol.[7] Calmodulin-dependent kinase and MAP-2 from the void fraction also cosedimented during sucrose-density gradient centrifugation and coeluted during chromatography on calmodulin-affinity resin, suggesting that they may be associated. Thus, two distinct MAP-2 fractions were prepared, one fraction associated with cAMP-dependent kinase and the other with calmodulin-dependent kinase. These studies

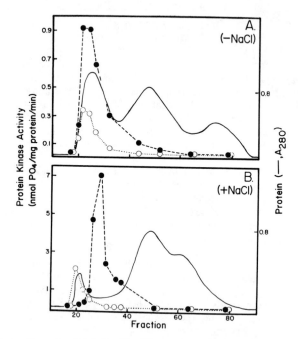

FIGURE 1. Separation of calmodulin-dependent and cAMP-dependent kinase activities in microtubules. **A:** Twice-cycled microtubule protein was chromatographed on Bio-Gel A 15m and phosphorylated under standard conditions.[7] Fractions were assayed for calmodulin-dependent protein phosphorylation (O · · · · O) and cAMP-dependent protein phosphorylation (●----●). The elution of protein was monitored at 280 nm (——). Column buffer contained 10 mM 2-(N-morpholino) ethanesulfonic acid (MES), 0.1 mM MgCl$_2$, pH 6.75. **B:** The experimental conditions were as described in **A**, except that the microtubules were treated with NaCl prior to column chromatography. Column buffer contained 10 mM MES, 0.1 mM MgCl$_2$, 500 mM NaCl, pH 6.75.

indicate that microtubule-associated calmodulin-dependent MAP-2 kinase may play a role in mediating calcium effects on microtubule function.

REFERENCES

1. VALLEE, R. B., G. S. BLOOM & W. E. THEURKAUF. 1984. J. Cell Biol. **99:** 38s–44s.
2. SLOBODA, R. D., S. A. RUDOLPH, J. L. ROSENBAUM & P. GREENGARD. 1975. Proc. Natl. Acad. Sci. USA **72:** 177–181.

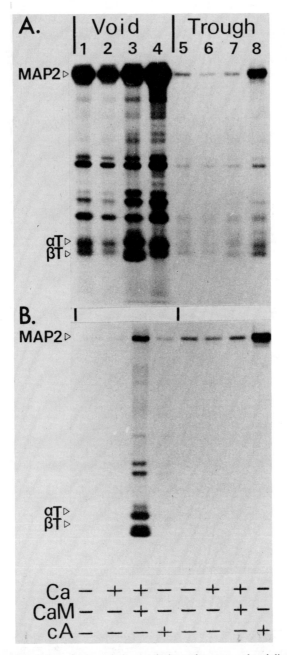

FIGURE 2. Phosphorylation of microtubule protein by endogenous calmodulin-dependent and cAMP-dependent kinases in microtubules. **A:** Autoradiograph of microtubule protein corresponding to samples from the profile shown in Figure 1A. Void = fraction #24; Trough = fraction #32. **B:** Autoradiography of microtubule protein corresponding to samples from the profile shown in FIGURE 1B. Void = Fraction #22; Trough = fraction #34. All samples (**A** and **B**) were phosphorylated under standard conditions[7] in buffer containing the additions indicated on the figure (Ca-calcium; CaM-calmodulin; cA-cAMP), and subsequently resolved on one-dimensional sodium dedecyl sulfate polyacrylamide gel electrophoresis (SDS-PAGE) (7% acrylamide). Magnesium was present in all samples. The positions of MAP-2, alpha tubulin (αT) and beta tubulin (βT) are shown.

3. JAMESON, L., T. FREY, B. ZEEBURG, F. DALLDORF & M. CAPLOW. 1980. Biochemistry
    19: 2472–2479.
4. JAMESON, L. & M. CAPLOW. 1981. Proc. Natl. Acad. Sci. USA **78:** 3413–3417.
5. YAMAMOTO, H., K. FUKANAGA, E. TANAKA & E. MIYAMOTO. 1983. J. Neurochem.
    41: 1119–1125.
6. VALLEE, R. B. & G. G. BORISY. 1978. J. Biol. Chem. **253:** 2834–2845.
7. VALLANO, M. L., J. R. GOLDENRING, R. S. LASHER & R. J. DELORENZO. 1986. Ann. N.Y.
    Acad. Sci. **466:** 357–374. This volume.

Phosphorylation of MAP-2 at Distinct Sites by Calmodulin- and Cyclic AMP-Dependent Kinases

JAMES R. GOLDENRING AND ROBERT J. DeLORENZO

Department of Neurology
Yale University School of Medicine
New Haven, Connecticut 06510

Microtubule-associated protein 2 (MAP-2) is highly phosphorylated endogenously.[1,2] Theurkauf and Vallee[2] have shown that there are at least 22 phosphorylated sites on MAP-2 of which 13 are cyclic adenosine monophosphate (cAMP)-dependent and at least 8 are cAMP-independent. Theurkauf and Vallee[3] have also documented that cAMP-dependent protein kinase and MAP-2 exist as a complex. The nature of the 8 cAMP-independent sites has not been well understood until recently. We and others have recently purified a calmodulin-dependent kinase from rat brain cytosol that phosphorylates MAP-2 as one of its major substrates.[4-6] This kinase is representative of a family of calmodulin-dependent kinases now designated as CaM kinase II. This kinase can be isolated in association with tubulin[7] and is endogenous to microtubule preparations.[8] We have therefore sought to compare the phosphorylation of MAP-2 by microtubule-associated cAMP-dependent kinase with that by both purified and microtubule-associated CaM kinase II.

MAP-2 was phosphorylated *in vitro* by endogenous microtubule-associated cAMP-dependent protein kinase, purified cytosolic CaM kinase II, and microtubule-associated calmodulin-dependent kinase. The phosphorylated MAP-2 was excised from sodium dodecyl sulfate polyacrylamide gel electrophoresis (SDS-PAGE) gels and digested to completion with trypsin. Phosphopeptides were resolved with two-dimensional cellulose thin layer electrophoresis/chromatography.[9] Major phosphopeptide species were excised from the thin layer plate and hydrolyzed in 6N HCl. Phosphoamino acids were resolved in one-dimension with cellulose thin layer electrophoresis.[4,9,10]

FIGURE 1 displays the resulting phosphopeptide maps. Eleven major phosphopeptides were phosphorylated by cAMP-dependent kinase (FIGURE 1A). Five major phosphopeptides were phosphorylated by both endogenous microtubule-associated calmodulin-dependent kinase (FIGURE 1B) and purified brain CaM kinase II (FIGURE 1C). Comigration studies carried out between phosphopeptides phosphorylated by both kinases are illustrated in FIGURE 2. They revealed that two peptides, B6 and D9, were indistinguishable by migration. Thus, amino acid analysis was performed on the major phosphopeptides. All eleven of the cAMP-dependently phosphorylated peptides contained only phosphoserine. Among the calmodulin-dependent phosphopeptides, however, A, B, C, and E contained 20, 50, 50, and 100 percent phosphothreonine residues, respectively. Thus, only the D9 phosphopeptide was indistinguishable by both phosphopeptide migration and phosphoamino acid analysis. By this combination of techniques, we observed 11 cAMP-dependent and 8 calmodulin-dependent phosphorylation sites in close agreement with the results of Theurkauf and Vallee.[2]

The results demonstrate that cAMP-dependent and calmodulin-dependent kinases have markedly different patterns of phosphorylation on MAP-2. Several investigators

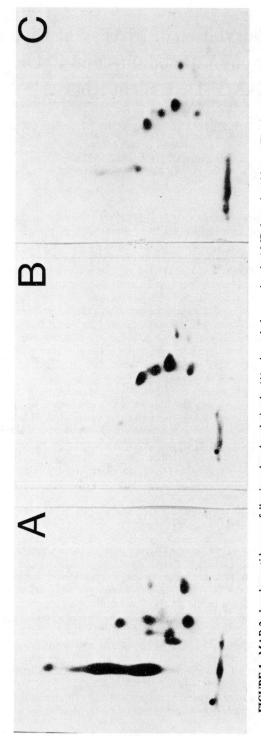

FIGURE 1. MAP-2 phosphopeptide maps following phosphorylation by (**A**) microtubule associated cAMP-dependent kinase, (**B**) microtubule-associated calmodulin-dependent kinase, and (**C**) purified CaM kinase II.

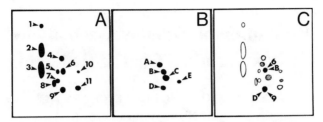

FIGURE 2. Comigration analysis of MAP-2 phosphopeptides phosphorylated by **(A)** cAMP-dependent kinase (phosphopeptides 1–11) and **(B)** calmodulin-dependent kinase (phosphopeptides A–E). **C:** Comigration showed two peptides, B6 and D9, with indistinguishable migration.

have indicated that phosphorylation of MAP-2 by both cAMP-dependent[11] and calmodulin-dependent[12] kinases can alter the ability of MAP-2 to promote microtubule polymerization. Further investigations will attempt to define this common pathway and its relationship to physiological stimuli in nerve cells.

REFERENCES

1. VALLEE, R. B. 1980. Proc. Natl. Acad. Sci. USA **77**: 3206–3210.
2. THEURKAUF, W. C. & R. B. VALLEE. 1983. J. Biol. Chem. **258**: 7883–7886.
3. THEURKAUF, W. C. & R. B. VALLEE. 1982. J. Biol. Chem. **257**: 3284–3290.
4. GOLDENRING, J. R., B. GONZALEZ, J. S. MCGUIRE JR. & R. J. DELORENZO. 1983. J. Biol. Chem. **258**: 12632–12640.
5. BENNETT, M. K., N. E. ERONDU & M. B. KENNEDY. 1983. J. Biol. Chem. **258**: 12735–12744.
6. SCHULMAN, H. 1984. J. Cell Biol. **99**: 11–19.
7. GOLDENRING, J. R., J. E. CASANOVA & R. J. DELORENZO. 1984. J. Neurochem. **43**: 1669–1679.
8. LARSON, R. E., J. R. GOLDENRING, M. L. VALLANO & R. J. DELORENZO. 1985. J. Neurochem. **44**:1566–1576.
9. AXELROD, N. 1978. Virology **87**: 366–383.
10. HUNTER, T. & B. H. SEFTON. 1980. Proc. Natl. Acad. Sci. USA **77**: 1311–1315.
11. JAMESON, L., T. FREY, B. ZEEBERG, F. DALLDORF & M. CAPLOW 1980. Biochemistry **19**: 2472–2479.
12. YAMAMOTO, H., K. FUKUNAGA, E. TANAKA & E. MIYAMOTO. 1983. J. Neurochem. **41**: 1119–1125.

Promotion of Tubulin Assembly by Carboxyterminal Charge Reduction

DAN L. SACKETT, B. BHATTACHARYYA,[a]

AND J. WOLFF

*National Institute of Arthritis, Diabetes, and
Digestive and Kidney Diseases
National Institutes of Health
Bethesda, Maryland 20205*

The ability of tubulin to assemble into microtubules is facilitated by numerous agents including glycerol, dimethyl sulfoxide, taxol, high Mg^{++} concentrations, certain polycations, and microtubule-associated proteins (MAPs). MAPs have been considered basic proteins on the basis of their behavior on ion exchange resins,[1] electrophoretic mobility,[2] the resemblance of their effects on tubulin polymerization to those produced by polycations,[1,3–8] and the apparent competition for MAPs by polyanions such as RNA or DNA.[9,10] In addition, the considerable sensitivity of microtubule protein to salt suggests the possibility of charge-charge interactions in the polymerization process.[11–13] When the amino acid sequence of α and β tubulin was solved,[14,15] it immediately became apparent that the carboxy terminal regions of both subunits were highly acidic and contained 17/24 excess anionic groups of the protein in their last 20 amino acid residues. These might thus be the acidic domains interacting with MAPs or polylysine. At the pH customarily used for polymerization, the glu and asp carboxyls would be dissociated, forming random coils presumably extending into the solvent.[17] They would also tend to repel neighboring carboxy termini in the polymer and might be expected to impede the polymerization process. Selective removal of the acidic domains of α and β tubulin might thus facilitate polymerization and diminish the need for associated proteins. In an effort to test this hypothesis, we have explored the effects of limited proteolysis and the properties of tubulin and have found that subtilisin can, under the appropriate incubation conditions, cleave small carboxy terminal fragments from both the α and β subunits of rat or cow brain tubulin that remove sufficient negative charge from the remaining protein to have profound effects on the facility with which polymerization occurs.[18,19] Similar attempts have been made by Serrano *et al.*,[20,21] and the results are in general agreement with ours.

In the present studies we have used pure 6S rat or cow brain tubulin, purified by phosphocellulose chromatography, that is devoid of MAPs on sodium dodecyl sulfate (SDS) polyacrylamide gel electrophoresis (PAGE). The time course of subtilisin-mediated cleavage of calf brain tubulin is depicted in FIGURE 1 where the left panel depicts SDS-PAGE and the right panel agarose electrophoresis in which the tubulin is maintained in the undenatured state. At early times the major cleavage occurs in the β subunit with the production of a large derivative that is smaller than the starting material by ~2 kD. This reaction is virtually complete by 15 minutes. The α subunit is hydrolyzed more slowly, again producing a derivative ~2 kD smaller than the starting material on SDS gels. Similar results have been obtained with rat brain (6S) tubulin as

[a]On leave from the Bose Institute, Calcutta, India.

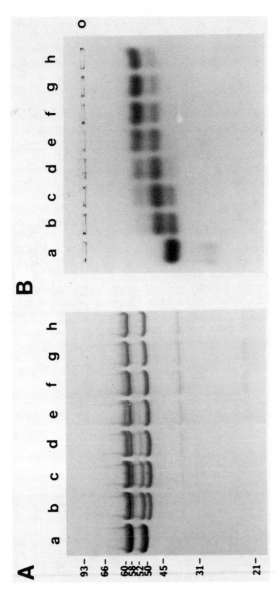

FIGURE 1. Time course of digestion of calf brain tubulin by subtilisin. Microtubule protein was prepared from fresh calf brains by cycles of temperature-dependent polymerization and depolymerization using MES assembly buffer (0.1 *M* 2-(*N*-morpholino)ethanesulfonic acid (MES), 1 m*M* MgCl₂, 1 m*M* EGTA, 1 m*M* GTP, pH 6.7). Tubulin was freed of MAPs by phosphocellulose chromatography, concentrated to >5 mg/ml by ultrafiltration, drop frozen in liquid nitrogen, and stored at −70° C or in liquid nitrogen. For digestion, the protein was diluted to 2 mg/ml in assembly buffer, and subtilisin BPN (solution at 1 mg/ml stored at −70° in aliquots that were thawed once only) was added to 1% (w/w). Incubation was at 30° C. Aliquots were removed at specified times, and 1% by volume of phenylmethylsulfonyl chloride (PMSF, 1% in DMSO) was added. All aliquots were kept on ice until the end of the experiments, when samples were processed for electrophoresis.

A: SDS electrophoresis was performed in 10% acrylamide gels using the tris glycine system of Laemlli, modified as described in reference 18. Molecular mass standards were phosphorylase B (93 kD), bovine serum albumin (66 kD), ovalbumin (45 kD), carbonic anhydrase (31 kD), and soybean trypsin inhibitor (21 kD). The apparent molecular masses are indicated for α (60 kD) $α_s$ (58 kD), β (52 kD), and $β_s$ (50 kD). Time points shown are (in minutes) a) 0, b) 2, c) 5, d) 10, e) 20, f) 40, g) 80, h) 105.

B: Nondenaturing agarose electrophoresis was performed in 1% agarose gels prepared in assembly buffer. Samples were prepared by addition of glycerol and bromphenol blue. The gel was run in the cold, submerged in assembly buffer. Further details are presented in reference 19. Time points are identical to **A.** 0 = origin, and direction of migration is toward the anode.

substrate.[18] The differences in the rates of cleavage of α and β tubulin are depicted in FIGURE 2 and are not dependent on the species studied. Thus, the half times for α- and β-tubulin hydrolysis were 10 min and 3 min, respectively. The charge changes, depicted in the right hand panel of FIGURE 1, reveal that hydrolysis proceeds by two stages with the formation of an intermediate that exhibits only partial reduction in negative charge; this eventually proceeds to a metastable state exhibiting further reduction of negative charge. These stages correspond in time to the changes seen in the SDS gels—first in the β subunit leading to the charge intermediate and then to cleavage of the α subunit as well, leading to further charge reduction in the doubly cleaved dimer. We have called the doubly cleaved product tubulin S (by analogy to ribonuclease S) and the cleaved subunits α_s and β_s. Thus, the intermediate is viewed as the dimer $\alpha\beta_s$ and tubulin S as $\alpha_s\beta_s$.[19]

It is also apparent from the native gels of FIGURE 1 (right lanes) that the small peptide cleaved off the α and β subunits can no longer be bound to the parent protein because no charge reduction would have occurred in the native electrophoretic pattern. This behavior differs substantially from hydrolysis products obtained with trypsin or chymotrypsin, which migrate like the original tubulin under nondenaturing conditions (FIGURE 3, right panel), whereas cleavage deep within the protein is apparent under denaturing conditions (FIGURE 3, left panel). This behavior was first surmised by Brown and Erickson,[22] who reported that cleavage of tubulin with trypsin and chymotrypsin yielded products that could polymerize with constant stoichiometry for three warm/cold cycles. Cleavage in the center of the tubulin subunits occurs late following subtilisin treatment (FIGURE 1) and yields major fragments with apparent molecular masses of approximately 39 kD and 22 kD (FIGURE 3). Moreover, these products have a markedly diminished ability to polymerize (FIGURE 4).

Tubulin S has retained its ability to form polymers and can be cycled repeatedly. The structure of the polymer formed is largely in the form of sheets of protofilaments

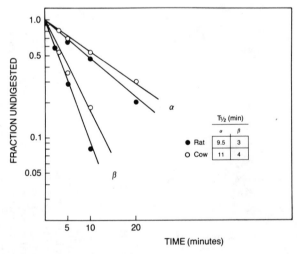

FIGURE 2. Quantitation of digestion time course. The gel in FIGURE 1A and a gel from a similar experiment using rat brain tubulin were scanned with an integrating densitometer. The results were analyzed according to Ueno and Harrington,[16] and presented as a semi-log plot. The half times for digestion are indicated in the inset.

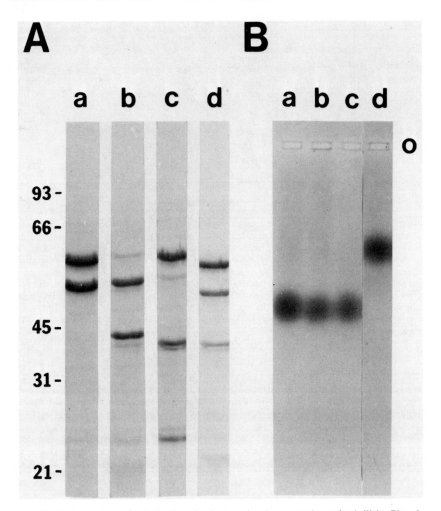

FIGURE 3. Comparison of tubulin digestion by trypsin, chymotrypsin, and subtilisin. Phosphocellulose-purified rat brain tubulin at 1 mg/ml in assembly buffer was digested for 1 hour at 30° C by addition of 1% w/w trypsin, chymotrypsin, or subtilisin. Following incubation, digestion was terminated by addition of PMSF to 0.05% w/v in the case of chymotrypsin and subtilisin, and aprotinin to 50 μg/ml in the case of trypsin. Samples were processed, and electrophoresis was performed as described in FIGURE 1. **A** = SDS electrophoresis; **B** = agarose electrophoresis. a) no enzyme, b) trypsin, c) chymotrypsin, d) subtilisin.

with only rare microtubules present.[18,19] Similar polymers have also been reported by Serrano *et al.*[20,21] During the early stages of digestion with subtilisin, there is formation of a polymer (increased OD) that is transient and diminishes as digestion proceeds. This light scattering peak corresponds to the formation of the $\alpha\beta_s$ intermediate (FIGURE 1) and has been detailed elsewhere.[19] Continued digestion leads to formation of tubulin S, which polymerizes in a manner analogous to untreated 6S rat brain

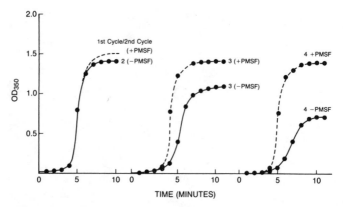

FIGURE 4. Temperature-dependent cycles of polymerization of tubulin-S. Rat brain tubulin (1.2 mg/ml in assembly buffer) was digested for 45 minutes at 30° C with 1% w/w subtilisin. Following digestion, the sample was placed on ice for 15 minutes and divided into two cuvettes. One cuvette received an aliquot (1% by volume) of PMSF (1% w/v in DMSO) to stop digestion. Both cuvettes were then transferred to a Cary Model 219 spectrophotometer, thermostatically controlled to 37° C. Turbidity at 350 nm was monitored with time. Following development of a plateau of turbidity, the cuvettes were returned to ice to depolymerize the polymer, and returned to the spectrophotometer to monitor the next cycle of assembly. The first cycle of assembly was the same for both samples. The numbers on the curves refer to the number of times the sample has been sequentially cycled.

tubulin (FIGURE 4). The polymer is cold sensitive and can be repolymerized at least three times in a temperature-dependent manner, provided further proteolysis is arrested by the addition of PMSF (FIGURE 4). When this precaution is not taken, further hydrolysis leads to products (FIGURE 1) that polymerize with progressively decreasing efficiency (FIGURE 4).

The regulatory function of the carboxy termini of tubulin is demonstrated by the very low critical concentration (C_c) for polymerization exhibited by tubulin S.[18] As shown in TABLE 1, the C_c for the native 6S tubulin was 2 mg/ml in the absence of MAPs. By contrast, the extrapolated C_c of tubulin S was ~0.04 mg/ml under the same conditions. This is as low or lower than the C_c attained by 6S tubulin in the presence of MAPs (0.2 mg/ml) and suggests that these proteins may interact with tubulin at the

TABLE 1. Critical Concentration for Assembly of Tubulin, Tubulin S,[a] and Mixtures of Tubulin and Tubulin S

Protein	Apparent critical concentration
	mg/ml
Tubulin	2.00
Tubulin S	0.04
Tubulin:Tubulin S (97:3)	0.4
Tubulin:Tubulin S (93:7)	0.1

[a]Tubulin S, prepared by the digestion of tubulin with subtilisin (1% w/w), was mixed with tubulin at different proportions. The critical concentration (C_c) was then determined at each ratio of tubulin and tubulin S by plotting plateau absorbance versus total protein concentration and extrapolating to zero absorbance.

carboxyterminal end. This has also been proposed by Serrano *et al.*,[20] whose smallest peptide appeared, however, to be larger than ours. The C_c of their subtilisin-treated pig brain tubulin was lowered approximately to that seen in the presence of MAPs.

The low C_c for tubulin S suggested to us that this derivative might promote polymerization of native 6S tubulin below the normal C_c for that particular preparation. This proved to be the case as is shown in TABLE 1. Mole fractions as low as 0.03 led to a 5-fold reduction in the apparent critical concentration of the mixture, and it could be shown[19] that this was the result of copolymerization of both types of tubulin. Increases in the mole fraction of tubulin S in the mixture led to modest further reductions in the apparent critical concentration, but the important point is that the presence of low concentrations of tubulin S in mixtures had a remarkable enhancing effect on polymerization when the native tubulin concentration was too low to polymerize. The potential significance of this effect *in situ* is under investigation.

Many of the well-recognized properties of tubulin persist in subtilisin-treated preparations (TABLE 2). These include a requirement for GTP in polymerization,

TABLE 2. Properties of Tubulin and Tubulin S[a]

Properties	Tubulin	Tubulin S
GTP requirement for assembly[b]	Required	Required
Temperature sensitivity (0°C)	Sensitive	Sensitive
Calcium effect on assembly (1 mM)	*Inhibits*	*No effect*
Salt effect (240 mM NaCl)	*Inhibits*	*No effect*
Podophyllotoxin effect on assembly (50 μM)	Inhibits	Inhibits
GDP effect on assembly (5 mM)	Inhibits	Inhibits
Vinblastine-induced aggregation (100 μM)	Aggregates	Aggregates
Colchicine binding[c]	Binds	Binds

[a]Tubulin S was prepared by digesting purified 6S tubulin (1 mg/ml) with subtilisin at a concentration of 1% (w/w) to tubulin in assembly buffer containing 1 mM guanosine triphosphate (GTP) at 30°C for 45 minutes. The reaction was terminated by the addition 1% by volume of 1% (w/v) PMSF in dimethylsulfoxide.

[b]Assembly and vinblastine-induced aggregation were monitored by turbidity at 350 nm using a Cary Model 219 spectrophotometer. The sample chamber was thermostatically controlled to 37°C ± 0.1°C using a Lauda K2-R circulator.

[c]Colchicine binding was determined using the diethylaminoethyl (DEAE) filter disc assay as described by Williams and Wolff.[23] Incubation was for 1 hour at 37°C.

depolymerization in the cold, inhibition of assembly by podophyllotoxin and other antimitotic drugs, inhibition of polymerization by guanosine diphosphate (GDP), formation of aggregates by vinblastine, and the binding of colchicine. On the other hand, two important properties of the native protein are no longer present in tubulin S. These are the inhibition of polymerization by Ca^{2+} (1 mM) and by high ionic strength (240 mM NaCl).[19] Whereas these effects may be explicable on the basis of the charge properties of the carboxy termini, it is equally possible that the greater propensity for polymerization in tubulin S may effectively compete with these disaggregating actions of Ca^{2+} or NaCl through long range effects.

Ten of the twenty carboxy terminal amino acids of α tubulin and eleven of β tubulin are either aspartyl or glutamyl residues. At the pH used in polymerization, these would carry a negative charge, would tend toward random coil conformation, and would be expected to exert repulsive forces on neighboring carboxy termini in the microtubule. We have postulated that such forces would exert a major influence on microtubule assembly by rendering the polymerization process more difficult. In this view, any form of charge neutralization would facilitate polymerization and lower the critical concentration. One role of MAPs would thus be such a neutralizing one, and Serrano *et al.*[20] have proposed that this interaction occurs at the carboxy terminal ends of the subunits. Because of the large size of the associated proteins, it seems unlikely that their effects are due primarily to relieving steric hindrance, and a charge determined effect seems more probable. The results presented in the present paper show that abolition or reduction of the carboxy terminal charges by the use of subtilisin proteolysis accomplishes a similar "neutralization" that results in a polymerizable product (tubulin S) that has a very low critical concentration, markedly reduced Ca^{2+} and salt sensitivity, but otherwise resembles native tubulin in most respects (TABLE 2, FIGURE 4). Moreover, tubulin S has such a marked increase in the tendency to polymerize that very low mole fractions promote polymerization of native tubulin present at levels well below its critical concentration. We also suggest that drugs that modulate polymerization must act elsewhere than at the carboxy termini because their effects persist in the cleaved product (TABLE 2). We are currently attempting to learn whether nature has availed herself of this property of carboxy terminal cleaved tubulin in systems in which the tubulin dimer concentration or other factors are unfavorable for microtubule assembly.

REFERENCES

1. ERICKSON, H. P. & W. A. VOTER. 1976. Proc. Natl. Acad. Sci. USA **73:** 2813–2817.
2. BÄRMANN, M., K. MANN & H. FASOLD. 1982. Biochem. Biophysics Res. Commun. **105:** 653–658.
3. JACOBS, M., P. M. BENNETT & M. J. DICKENS. 1975. Nature (London) **257:** 707–709.
4. BEHNKE, O. 1975. Nature (London) **257:** 709–710.
5. LEVI, A., M. CIMINO, D. MERCANTI, J. S. CHEN & P. CALISSANO. 1975. Biochim. Biophys. Acta **399:** 50–60.
6. MURPHY, D. B., R. B. VALLEE & G. G. BORISY. 1977. Biochemistry **16:** 2598–2605.
7. LEE, J. C., N. TWEEDY & S. N. TIMASHEFF. 1978. Biochemistry **17:** 2783–2790.
8. BANERJEE, A., A. C. BANERJEE & B. BHATTACHARYYA. 1981. FEBS Lett. **124:** 285–288.
9. BRYAN, J., B. W. NAGLE & K. H. DOENGES. 1975. Proc. Natl. Acad. Sci. USA **72:** 3570–3574.
10. VILLASANTE, A., J. DE LA TORRE, R. MANSO-MARTINEZ & J. AVILA. 1980. Eur. J. Biochem. **112:** 611–616.
11. NISHIDA, E. 1978. J. Biochem. (Tokyo) **84:** 507–512.
12. OLMSTED, J. B. & G. G. BORISY. 1975. Biochemistry **14:** 2996–3005.
13. VALLEE, R. B. 1982. J. Cell Biol. **92:** 435–442.
14. PONSTINGL, H., E. KRAUHS, M. LITTLE & T. KEMPF. 1981. Proc. Natl. Acad. Sci. USA **78:** 2757–2761.
15. KRAUHS, E., M. LITTLE, T. KEMPF, R. HOFER-WARBINEKI & H. PONSTINGL. 1981. Proc. Natl. Acad. Sci. USA **98:** 4156–4160.
16. UENO, H. & W. F. HARRINGTON. 1984. J. Mol. Biol. **173:** 35–61.
17. JACOBSON, A. L. 1964. Biopolymers **2:** 237–244.
18. SACKETT, D. L., B. BHATTACHARYYA & J. WOLFF. 1985. J. Biol. Chem. **260:** 43–45.
19. BHATTACHARYYA, B., D. L. SACKETT & J. WOLFF. J. Biol. Chem. **260:** 10208–10216.
20. SERRANO, L., J. AVILA & R. B. MACCIONI. 1984. Biochemistry **23:** 4675–4681.
21. SERRANO, L., J. DE LA TORRE, R. MACCIONI & J. AVILA. 1984. Proc. Natl. Acad. Sci. USA **81:** 5989–5993.
22. BROWN, M. R. & H. P. ERICKSON. 1983. Arch. Biochem. Biophys. **220:** 46–51.
23. WILLIAMS, J. A. & J. WOLFF. 1972. J. Cell Biol. **54:** 157–165.

The Role of Lys 394
in Microtubule Assembly[a]

G. S. BLANK, M. B. YAFFE, J. SZASZ, E. GEORGE,
T. L. ROSENBERRY, AND H. STERNLICHT[b]

Department of Pharmacology
Case Western Reserve University
Cleveland, Ohio 44106

Carboxytermini of α and β tubulin are essential for microtubule assembly despite their potential to destabilize microtubules by way of electrostatic repulsion. These regions (residues 410–451 in α polypeptide and residues 400–445 in β polypeptide) typically contain ~40% of all the glutamates and ~20% of all the aspartates in tubulin.[1-3] Microtubule assembly is affected by divalent cations[4-6] and basic proteins, that is, microtubule-associated proteins (MAPs).[7,8] A number of these agents are thought to bind at the carboxytermini and affect assembly in part by modulating electrostatic interactions. Proteolytically cleaved tubulins that lack carboxytermini regions show a marked decrease in their ability to associate with MAP-2, and they undergo aberrant assembly characterized by open and twisted sheets.[9,10] Although it is generally known that cations and MAPs are important regulators of microtubule assembly, there is much less appreciation of the fact that basic residues in tubulin also play essential roles in assembly. We have used reductive methylation to probe the role of lysine residues in assembly[11,12] and have identified a highly reactive, essential lysine in the alpha polypeptide of tubulin that is likely to function by electrostatic interaction with an acidic residue(s) in the carboxyterminus. Our studies complement related studies by Maccioni *et al.*[13] and Mellado *et al.*[14] on the role of arginyl and lysyl residues in microtubule assembly.

MATERIAL AND METHODS

Protein Preparation

Microtubule protein (MTP) was isolated by repetitive cycles of assembly-disassembly from bovine brains following a procedure[15] modified from Gaskin *et al.*[16] Twice cycled preparations were used immediately or stored at $-20°C$ in pH 6.7 assembly buffer [2-(N-morpholino)ethane sulfonic acid (MES), 2mM ethylene bis (oxyethylene nitrilo)tetraacetic acid(EGTA) 0.1 mM ethylenediaminetetraacetic acid (EDTA), 2mM mercaptoethanol, and 0.5 mM MgCl$_2$] supplemented with 5M glycerol for subsequent use within two weeks. MTP was ~85% tubulin and ~15% MAPs as determined by gel electrophoresis.[17] Protein concentrations were established by the Lowry method[18] using bovine serum albumin as standard.

[a]This work was supported in part by American Cancer Society Grant CH-99D and CH-99E to Himan Sternlicht.
[b]To whom reprint requests should be addressed.

467

Reductive Methylation

MTP samples were dual-isotope labeled with [³H]- and [¹⁴C]HCHO in order to facilitate isolation of the cyanogen bromide (CNBr) peptide containing the highly reactive lysine (CNBr-HRL peptide). Reductive methylation procedures followed previously established protocols.[11,12]

Isolation of CNBr-HRL Peptide

Lyophilized, dual-isotope labeled MTP was dissolved to a final concentration of ~5 mg/ml in 70% formic acid/10 mg/ml CNBr and incubated for 18–36 hr at room temperature. The digestion reaction was quenched by the addition of 10 volumes of water, and the material dried under vacuum. Lyophillized digests were chromatographed on SP-Sephadex (18 × 100 mm column, 0–8% pyridine gradient in 25% acetic acid, 5 mM mercaptoethanol). Fractions enriched in the HRL were dissolved in 0.1% trifluoroacetic acid (TFA) and 2% isopropanol and subjected to high performance liquid chromatography (HPLC) on a C18 reverse phase 250 × 4 mm column (Synchropak RP-P, 5 μM beads, 300 °A pore diameter) using a 5–60% isopropanol gradient in 0.1% TFA (60 ml total volume, 0.5 ml/min flow rate).[19]

Secondary Structure Analysis

Predicted alpha polypeptide secondary structures based on Garnier et al.,[20] were obtained using a program written in Forth by M. Glynias and M. Yaffe for the Commodore 64 PC computer. Hydrophobicity assignments are based on Kyte and Doolittle algorithm.[21] The Garnier et al. method treats the four possible conformational states of a residue (i.e., helix, extended sheet, turn and coil) equivalently, and uses the concept of residue and interresidue "information content" estimated from statistical analysis of known structures to evaluate the contribution each amino acid residue makes to its conformation and that of its neighbors. An eight-residue window on each side of the residue of interest was used, and the conformational state assigned was the one with the highest summed probability. Helix net projections were obtained either on the Commodore 64 using Forth or on the North Star Advantage using G Basic.

Cross-Linking Studies

Freshly prepared 1-ethyl-3(3 dimethylaminopropyl) carbodiimide (EDC) from a 100 mM stock in MES buffer (25 mM MES, pH 6.6) was added to MTP or phosphocellulose-purified tubulin in MES buffer supplemented with 1.5 mM MgCl$_2$ (EDC final concentration: 2 mM) and reacted at room temperature for the times indicated. The cross-linking reaction was stopped by the addition of 20 μl of neat β-mercaptoethanol per ml of reaction mixture, or by addition of one-half volume pH 7.8 stop solution (9% sodium dodecyl sulfate (SDS), 30 mM Tris, 15% Glycerol, 0.05% bromphenol blue, supplemented with 18 mg DDT per ml of stop solution) followed by boiling for 1–2 minutes.

RESULTS

Background

Reductive methylation of proteins by HCHO and $NaCNBH_3$ is specific for α- and ϵ-amino group (FIGURE 1) and can be carried out at physiological pH, yielding mono- and dimethylated amino derivatives.[22] This methylation reaction is capable of replacing 1 or 2 hydrogens per amine group with methyls, nonbulky groups whose Van der Waal radii are only 0.8°A larger than that of an amine hydrogen,[23] and has little effect on animo residue pK_as.[11] Despite the conservative nature of the modification, >50% assembly competence is lost at 1 mM HCHO where ~5 methyls (out of a possible total of ca 34) are introduced into tubulin dimer.[11] By contrast, colchicine and GTP binding is largely retained at low HCHO concentrations (FIGURE 2).[11] Furthermore assembly-incompetent tubulin obtained by methylation is incapable of coassembling with

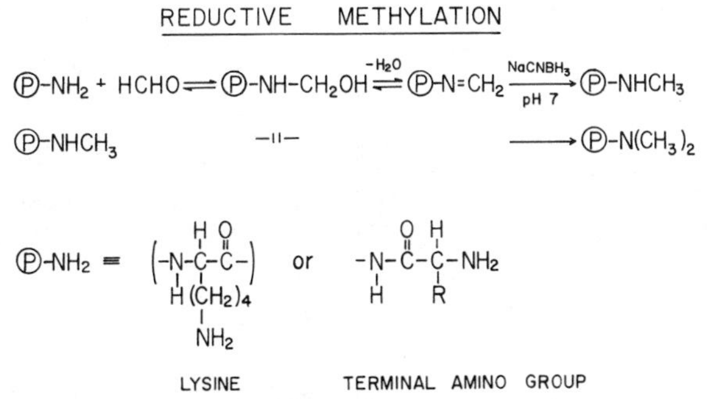

REDUCTIVE   METHYLATION

$$(P)\text{-}NH_2 + HCHO \rightleftharpoons (P)\text{-}NH\text{-}CH_2OH \underset{-H_2O}{\rightleftharpoons} (P)\text{-}N\text{=}CH_2 \xrightarrow[pH\ 7]{NaCNBH_3} (P)\text{-}NHCH_3$$

$$(P)\text{-}NHCH_3 \qquad -\text{"}- \qquad\qquad \longrightarrow (P)\text{-}N(CH_3)_2$$

LYSINE   TERMINAL AMINO GROUP

FIGURE 1. Reductive methylation: reaction scheme. Reductive methylation is highly specific for amino residues and is dependent on the pK_a of these residues.[22] α-amino groups of N-terminii are generally more reactive than ϵ-amino groups of lysines. Local environments, however, can significantly alter relative reactivities.[12] Reductive methylation involves a hydroxymethyl lysine intermediate that is converted to a Schiff base. Once the Schiff base is formed, reduction by $NaCNBH_3$ occurs rapidly.[12]

unmodified tubulin, suggesting that a critical domain for assembly had been modified.[11] Loss of assembly competence upon methylation results from a specific modification of tubulin, rather than from an alteration in MAP properties,[11] and correlates with the methylation of a highly reactive lysine (HRL) in the alpha chain.[12] By contrast, microtubules lack the HRL and microtubules methylated at low levels of HCHO (<2 mM HCHO) depolymerize and reassemble normally with little or no loss of assembly competence.[11,12]

Identification of the HRL

A small CNBr fragment containing the HRL (referred to below as the CNBr-HRL peptide) was isolated using HPLC chromatography (FIGURE 3). In this rapid isolation procedure, which took advantage of the enhanced reactivity of the HRL in native dimer, no attempt was made to separate α polypeptide from β polypeptide.

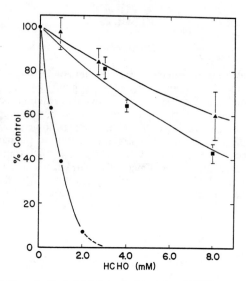

FIGURE 2. Methylation selectively inhibits assembly at low HCHO concentration (reproduced from reference 11). GTP-depleted microtubule protein preparations were reacted for 15 min at 37°C with varying concentrations of formaldehyde and NaCNBH₃, and assayed for extent of assembly (●) and for colchicine (▲) and GTP binding (■). Details are given in reference 11.

TABLE 1. Comparison of Amino Acid Composition of CNBr-HRL Peptide with that for Alpha Peptide 378-398

	CNBr-HRL	Peptide 378-398[a]
Lysine	6.6	5.0
Histidine	4.4	5.0
Arginine	4.5	5.0
CM-Cystine	ND[b]	0
Asparagine	12.0	15.0
Threonine	7.5	10.0
Serine	5.7	5.0
Glutamine	8.5	5.0
Proline	2.2	0.0
Glycine	3.7	0
Alanine	16.1	20.0
Valine	1.4	0
Methionine	~6[c]	5.0
Isoleucine	4.6	5.0
Leucine	9.7	15.0
Tyrosine	0.4	0.0
Phenylalanine	5.7	5.0
Tryptophan	N.D.	(0)[d]

[a]Ponstingl *et al.*[1]
[b]Not determined.
[c]Estimated from homoserine and homserine-lactone content.
[d]Peptide 378-398 has 1 Trp that was not included in mole percent.

Rather, CNBr digests of dual-isotope-labeled MTP (radiolabeled with [^{14}C]HCHO in the unpolymerized state and [^{3}H]HCHO in the microtubule state) were chromatographed on SP-Sephadex (FIGURE 3A), and fractions identified as enriched in HRL by their enhanced [^{14}C/^{3}H]$_{dpm}$ ratios were combined and chromatographed on HPLC

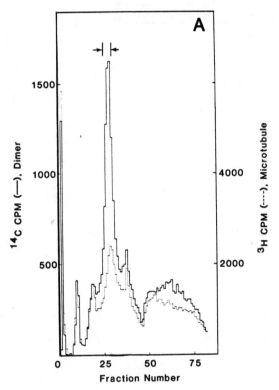

FIGURE 3 A and B. Isolation of CNBr-HRL peptide from dual-isotope-labeled tubulin. MTP (~15 mg/ml) (——) was reductively methylated with limiting concentrations of [^{14}C]HCHO (1mM HCHO, 6mM NaCNBH$_3$, 15 min, 37 °C) and mixed with an equal amount of microtubule polymer (----) that had been reductively methylated with [^{3}H]HCHO using identical reaction conditions (MATERIAL AND METHODS). The mixture was extensively dialyzed, CNBr digested, and resuspended in 1.5 ml of 25% acetic acid + 40 mM DTT. A: SP-Sephadex chromatography. Four ml fractions were collected and counted. Fractions 25–30 enriched in the CNBr-HRL (R$_f$ ~ 0.92 in tube gel electrophoresis[12]) emerged at ca 2.5% pyridine. B: Fractions 25–30 from A were combined, lyophilized, and resuspended in 0.1% TFA and 2% isopropanol for chromatography by reverse phase HPLC (MATERIAL AND METHODS). 0.5 ml fractions were collected and counted. Peak I (fractions 62–63) was identified as the CNBr-HRL fraction on the basis of its enhanced ^{14}C/^{3}H ratio.

(FIGURE 3B). Two peaks (I and II) displaying enhanced [^{14}C/^{3}H]$_{dpm}$ ratios emerged at gradient values indicative of moderately hydrophobic peptides (FIGURE 3B). Amino acid composition analysis of the major peak (peak I), which had the highest [^{14}C/^{3}H]$_{dpm}$ ratio and ran at the CNBr-HRL peptide position (R$_f$ ~ 0.92) on rod gels[12]

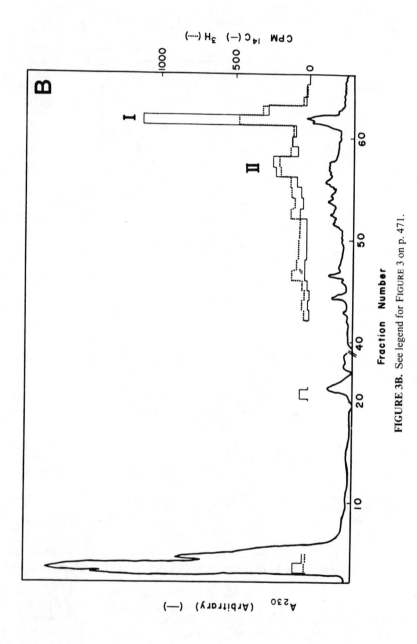

FIGURE 3B. See legend for FIGURE 3 on p. 471.

(not shown), indicated a high alanine content (~16 mole %), a high leucine/isoleucine ratio, and little proline, tyrosine or valine (TABLE 1). The amount of methionine conversion products, homoserine and homoserine lactone, corrected for loss during hydrolysis and expressed as percent methionine, indicated a polypeptide with 17–20 residues (TABLE 1). Analysis of the minor peptide, peak II, indicated a composition similar to that of peak I (data not shown) suggesting that peak II peptide was an artifact of the separation or, possibly, a modified form of peak I peptide. Comparisons of the composition of the major peptide with the known compositions of CNBr peptides of alpha tubulin[1] suggested that peak I peptide could be CNBr peptide 378–398 (TABLE 1), which had a leucine residue at its N-terminus and a single lysine residue at position 394. This assignment was confirmed by sequencing of CNBr-HRL peptide derived from dual-isotope labeled α tubulin, which indicated a leucine at the N-terminus and a highly reactive lysine at position 17, that is, at position 394 in the α chain (manuscript submitted).[24]

Secondary Structure Analysis

Secondary structure prediction studies (FIGURES 4 and 5) as well as preliminary cross-linking studies (below) were carried out to elucidate the local environment of Lys 394. Structure predictions based on Chou and Fassman rules[3,25,26] as well as our predictions based on the Garnier algorithm[20] (FIGURE 4; MATERIAL AND METH-ODS) suggest that residues 370–450 in alpha polypeptide are α helical. Although both predictive methods agree that carboxyterminus region 370–450 has strong helical propensity, these predictive methods are statistical in nature and ignore electrostatic interactions that can be significant in the carboxyterminus region. The degree of α helix in the glutamate-rich regions 410–450 may actually depend on external conditions such as pH, salt, and the presence or absence tyrosine[27,28] with the ultimate portion, residues 440–450, being random coil.[29]

If our assumption that Lys 394 is in an α helix is correct (FIGURE 4), then Lys 394 is part of a positively charged cluster consisting of Arg 390, His 393, and Lys 394 with acidic residues in the 385–402 region distant to Lys 394. Furthermore, Lys 394 is predicted to have a ring of hydrophobic near-neighbor residues in addition to its two positively charged near-neighbor residues, Arg 390 and His 393. A Corey-Pauling-Kolton space-filling model was constructed to illustrate the positively charged pocket consisting of Lys 394, His 393 and Arg 390 (FIGURE 5) and to investigate the steric constraints on these residues. These positively charged residues appear capable of close approach as required by our proposed mechanism for the enhanced reactivity of Lys 394 as a nucleophile (see DISCUSSION and FIGURE 9).

Cross-linking Studies

We were intrigued with the secondary structure prediction that Lys 394 is part of a positively charged cluster, possibly in close proximity to the highly negatively charged carboxyterminus region, residues 410–450, and sought to further probe the local environment of Lys 394 using EDC, a zero-length cross-linker, specific for cross-linking carboxylates to amine residues (FIGURE 6). Our results are still in the preliminary stages, but indicate the following: (1) When MTP or MTP-depleted tubulin is reacted with EDC, an intramolecular cross-linked α, that is, α^* is generated, which is electrophoretically altered and resolvable on SDS polyacrylamide gel electrophoresis (PAGE) (FIGURE 7). (2) This electrophoretically altered α is dimin-

```
              10        20        30        40        50        60        70        80        90       100
I   MRECISIH .         .         .         .         .PRAVFVDL.E .         .         .         .       
A            .    .MELY        HGI.QP QMP DK.T         V   PTVIDE   R  H E. .QLI KEDA.
B   G.QA     .    .                                   .VR    F P .
T        .GV .    QIGNAC.  CLE     DG  S . IGGGDDSFN.T FFSETGAGK.H      T.GTY QL      TG
C      .V .                                                                            .A.
H   ++++ ++++.++++++++ +. +++ + +       .         ++++++++++++ +++ .         +.

              10        20        30        40        50        60        70        80        90       200
I            .         .         .         .         .         .         .         .         .       
A   RGH  . K  L .  EIID VLD.PIRKLADQ      .FLVF    .LLMERLSV   .SKLEFS.            TTLEHSDC.
B   NNYA Y.      .        .              .T.       T.S         A QVSTA.   SI T.
T      TI.G      C .RLQG      HS.FGGGT GF     GS       .IYP P     .VEPYN L       .TH
C            .         .         .         D.YGKK    .         .V       .
H   +. ++++++++++. +         .         +++++++++++. +  +++++++++.  ++.+++++++++.+ ++++  +++  ++

              10        20        30        40        50        60        70        80        90       300
I   AFMVDNE  .         .         .         .         .         .         .         .         .       
A   AIY.     .E.     .NRL.ISQIV  A.S  .LRFDGALN .LTE       .ISAEK.AYHEQLSVAE.ITNACFER
B        .DICRRNLDI.  .SSIT.     V.D  F   .YPR HFPLA. YAPV              .AN.
T      .RPTY    TNL .              .QTNLV.F    I  .T
C            .         .         .         I    .         .         .
H   +++ ++ .       ++.++++++++++.+++++++++.++++++++ +.+    +  +++.++++++++.+ ++++

              10        20        30        40        50        60        70        80        90       400
I            .         .         .         .         .         .         .         .         .       
A   QM   . KYNAC R.G VV  NA.AIAT   R. FV      KVG  .QPP.T   .VQRAVCMLS . TAIAEAWAR.LDHKFDLHYA.
B   D    .CLL         .       T.IQ    .KV6      .DLAK.
T   VKC PSHG.  Y. D KDV    KT  DW PTG.F  INY  .PG6      .VV  .N.T
C            .         .         .         .         .         .         .         .         .
H   . +++++++++++.+ +++++ +.    ++++.++++ + .++++ +.+ ++++ .         .++ +++++ +  ++.

I   KRAFVHNYVG.EGMEEGEFSE.AREDMAALEK.DYEEVGVDSV.EGEGEEE
A            .         .         .GEE.Y
B            .         .
T            .         .
C            .         .
H            .         .
```

Lys 394 ←

ished at low HCHO concentrations (FIGURE 7**B** and 8) suggesting that an amino residue highly reactive to HCHO (*i.e.*, the *N*-terminus methionine or the HRL) may be directly involved in the cross-link, or that methylation at low HCHO concentrations generates a conformationally perturbed α that cross-links differently than unmethylated α. α^* has been isolated in this laboratory and is being analyzed to distinguish between these possibilities. Preliminary analyses indicate that *N*-terminus methionine is not cross-linked.

DISCUSSION

In this study we identified Lys 394 in the α polypeptide (FIGURE 3, TABLE 1) of tubulin dimer as the essential HRL previously implicated in assembly.[11,12] We proposed that Lys 394 is part of a positively charged pocket consisting of Lys 394, His 393, and Arg 390. (FIGURES 4, 5). We believe there are compelling reasons arguing for the pocket, although further experimentation is clearly needed to verify its existence. The model both rationalizes the enhanced nucleophilicity of Lys 394 and provides insights as to why the Lys 394 region is important for assembly that can be subjected to further testing.

Enhanced nucleophilicity observed for certain lysines in proteins is not fully understood, but is thought to involve a lowered pK_a induced by the local environment.[11,30,31] We suggest that positively charged near-neighbor residues, His 393 and Arg 390 (FIGURES 4 and 5), electrostatically interact with Lys 394 to lower its pK_a. Furthermore, because the positively charged cluster appears to be in a region of low dielectric constant (the cluster is surrounded by a hydrophobic ring), electrostatic repulsions between His 393, Arg 390, and Lys 394 should be enhanced. pH studies of the reactivity of the HRL relative to bulk lysines[24] indicate a bell-shaped profile with a reactivity maximum at pH $\sim$ 7.5 (HRL $\sim$ 20–30-fold more reactive than bulk lysine) and a minimum at pHs 5.5 and 9.0, the extremums of the study where HRL and bulk lysine reactivity are approximately equal. This profile is consistent with the titration of a histidine ($pK_a \sim 6.3$)[32] and a conformation change in tubulin at high pH.[33] We speculate that His 393 functions as a general acid-base catalyst to facilitate formation of the Schiff base intermediate (FIGURES 5 and 9).

Lys 394 is proximal in the α-tubulin sequence to the highly negatively charged carboxyterminus region (residues 410–451) previously implicated in assembly.[8–10] Our analysis further suggests that the positive charged cluster containing Lys 394 is at least partially in a "hydrophobic" environment, and thus not fully accessible to solvent water. Based on a survey of $\sim$500 lysines in globular proteins, Rashin and Honig[34]

FIGURE 4. Prediction of α helix (A), β-pleated sheet (B), turns (T), and random coil (C) regions in alpha tubulin. Conventional single letter codes are used to designate amino acid residues. Residue numbering proceeds from left to right and from top to bottom; each row consists of 100 residues that are further subdivided into sets of 10 (vertical dashes). H designates the hydrophobicity index and represents a running average over a 9 residue window as described by Kyte and Doolitle[21] with + indicating a positive hydrophobicity. Negative designations indicating hydrophilicity are not shown to facilitate the display. A number of residues at the *N*- and *C*-terminii could not be assigned, and these appear in the I, that is, indeterminate classification. Further refinements in the secondary structure predictions are clearly required in this region. One source of difficulty is the fact that secondary structure predictions neglect electrostatic interactions that are certainly important at the carboxyterminus (*cf*, helix-coil transition study of polyglutamic acid[39]).

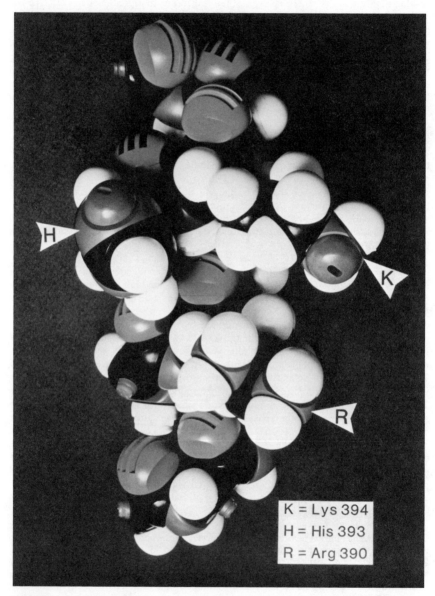

FIGURE 5. Space-filling model of the 383–400 region illustrating the positively charged pocket consisting of Lys 394, His 393, and Arg 390. Side chains from other residues have been omitted for clarity. In this model His 393 is sterically capable of interacting with the ϵ-amine group of Lys 394.

reported that ~3–4% of lysines can be expected to not be fully water accessible. Furthermore, >90% of these lysines pair with an apparent hydrogen-bonding partner and/or salt-bridged partner.[34] Charged residues interacting electrostatically have been identified as factors contributing to protein stability,[35,36] and as driving forces for protein association.[37,38] We suspect that intramolecular electrostatic interactions are

CARBODIIMIDE COUPLING

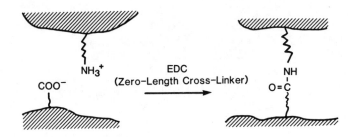

FIGURE 6. Carbodiimide coupling reaction scheme. EDC (R_1 = ethyl; R_2 = 3-dimethylamino-propyl), a zero-length cross-linker reacts with proteins in a two-step process; (a) the activation of carboxy groups, and (b) the displacement by the nucleophile, for example, NH_2-protein, which releases EDC as a soluble urea derivative.

also important for microtubule assembly. Because ~40% of the acidic residues are concentrated in the carboxyterminus, it seems reasonable to hypothesize that the positive pocket (FIGURES 4 and 5) interacts with the carboxyterminus. Although we do not know at present the specific nature of the interaction, we think it unlikely that Lys 394 forms a salt bridge with an acidic residue in the dimer because salt-bridge formation raises lysyl pKa and reduces rates of methylation relative to bulk lysines.[31]

Lys 394, however, has a more normal rate of reactivity in the microtubule state,[11,12] and we do not exclude the possibility that Lys 394 forms a salt bridge in the microtubule that is important for assembly. Thus, previous observations that methylation of the HRL abolishes microtubule assembly[11,12] may simply indicate interference with the

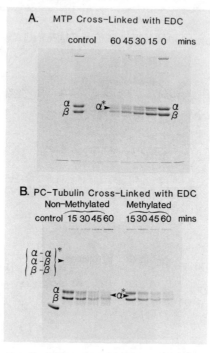

FIGURE 7. Cross-linking studies. **A:** Microtubule protein (2 mg/ml) was reacted with 2 mM EDC in 25 mM MES, pH 6.7, for various lengths of time and analyzed by gel electrophoresis. A band, whose intensity increased with reaction time, reaching an apparent maximum at 30 min, was detected immediately below the α band. Identification of this band, α^*, as an internal cross-linked α was done in an independent experiment that took advantage of the high specificity of tryosylating brain enzyme for α tubulin.[40] Phosphocellulose-purified tubulin (0.75 mg/ml) was enzymatically tyrosylated in the presence of [14C]tyrosine[40] and then cross-linked under identical conditions as in **A.** Radiolabel was incorporated only in α and α^*, and the extent of incorporation (cpm/peak height) of the two bands were similar (data not shown).

B: Methylation studies implicate Lys 394 as a possible site of cross-linking in α^*. Phosphocellulose-purified tubulin (~1.5 mg/ml) was divided into two aliquots and either immediately cross-linked or first methylated with 0.5 mM [14C]HCHO (15 min, 37°C, [NaCNBH$_3$]/[HCHO] = 6) or with 2 mM [14C]HCHO (FIGURE 8**B**) and then cross-linked with 2mM EDC for the times indicated. Visual inspection of Coomasie-blue-stained gels suggested that methylation affected α^* yields but had relatively little affect on α and β. Actin (~41 kD) (left lane *B*) was used as a molecular mass marker.

formation of an essential salt bridge involving Lys 394 neighbors, for example, Arg 390 or His 393, in the dimeric state or Lys 394 itself in the microtubule state.

The above discussion concerning the mechanism of Lys 394 methylation and its significance in assembly is largely speculative. We expect that additional studies

α* BAND IS SENSITIVE TO METHYLATION

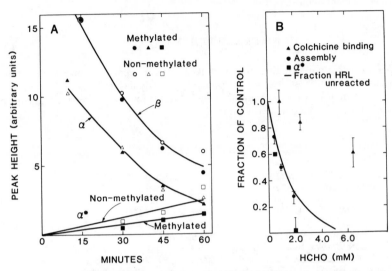

FIGURE 8. A: Coomasie-blue-stained gels of cross-linked nonmethylated and methylated phosphocellulose-purified tubulin (FIGURE 7B) were scanned spectrophotometrically, and peak heights of α, β, and α* at various times were plotted. Plots confirmed that methylation (0.5 m*M* HCHO) had little effect on α- and β-band intensities. **B:** α* yields as a function of HCHO concentration closely paralleled the amount of unmethylated HRL present (——). Data for fraction of unmethylated HRL, assembly (●) and colchicine binding (▲) as a function of HCHO, were taken from reference 12.

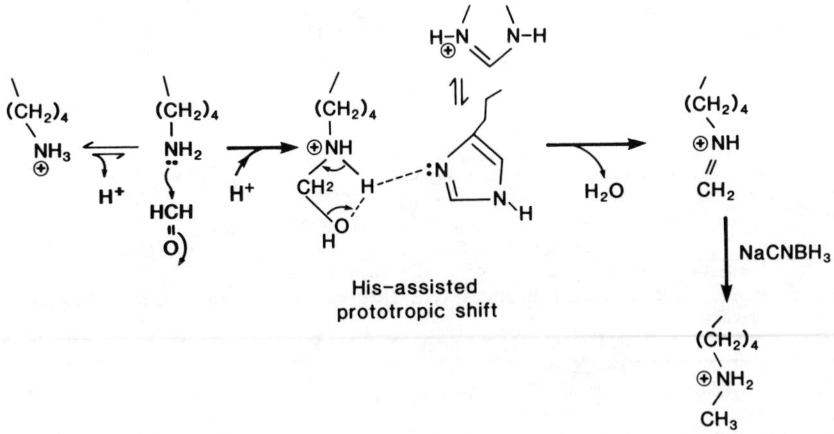

FIGURE 9. Speculative methylation scheme illustrating histidine-assisted prototropic shift.

currently in progress, for example, histidine modification, circular dichroism analysis of synthetic peptides spanning the Lys 394 region as well as cross-linking studies in the dimeric (FIGURES 7 and 8) and polymeric state will provide insights concerning the local environment about Lys 394.

ACKNOWLEDGMENTS

We wish to thank G. Sherman and B. Haas for technical support on several occasions during this study. We also thank M. Glynias for his assistance with secondary structure analysis of tubulin and J. Mieyal for enlightening discussions on the mechanism of reductive methylation.

REFERENCES

1. PONSTINGL, H., E. KRAUHS, M. LITTLE & T. KEMPF. 1981. Proc. Natl. Acad. Sci. **78:** 2757–2761.
2. KRAUHS, E., M. LITTLE, T. KEMPT, R. HOFFER-WARBINEK, A. WOLFGANG & H. PONSTINGL. 1981. Proc. Natl. Acad. Sci. **78:** 4156–4160.
3. VALENZUELA, P., M. QUIROGA, J. ZALDIVA, W. J. RUTTER, M. W. KIRSCHNER & D. W. CLEVELAND. 1981. Nature (London) **298:** 650–655.
4. SOLOMON, F. 1977. Biochemistry **16:** 358–363.
5. ROSENFELD, A., R. ZACKROFF & R. WEISENBERG. 1976. FEBS Lett. **65:** 144–147.
6. LARSSON, H., M. WALLIN & A. EDSTROM. 1976. Exp. Cell. Res. **100:** 104–110.
7. LEE, J. C., N. TWEEDY & S. N. TIMASHEFF. 1978. Biochemistry **17:** 2783–2790.
8. SCHEELE, R. B. & G. G. BORISY. 1979. *In* Microtubule Inhibitors. K. Roberts & J. Hyams, Eds: 176–253. Academic Press. New York.
9. MACCIONI, R. B., L. SERRANO & J. AVILA. 1984. Fed. Proc. Fed. Am. Soc. Exp. Biol. **43:** 2015a.
10. SERRANO, L., J. AVILA & R. B. MACCIONI. 1984. Biochemistry **23:** 4675–4681.
11. SZASZ, J., R. BURNS & H. STERNLICHT. 1982. J. Biol. Chem. **257:** 3697–3704.
12. SHERMAN, G., T. L. ROSENBERRY & H. STERNLICHT. 1983. J. Biol. Chem. **258:** 2148–2156.
13. MACCIONI, R. B., J. C. VERA & J. C. SLEBE. 1981. Arch. Biochem. Biophys. **207:** 248–255.
14. MELLADO, W., J. C. SLEBE & R. B. MACCIONI. 1982. Biochem. J. **203:** 675–681.
15. STERNLICHT, H. & I. RINGEL. 1979. J. Biol. Chem. **254:** 10540–10550.
16. GASKIN, F., C. R. CANTOR & M. L. SHELANSKI. 1974. J. Mol. Biol. **89:** 737–755.
17. LAEMLLI, U. K. 1970. Nature (London) **227:** 680–685.
18. LOWRY, O. H., N. J. ROSEBROUGH, A. L. FARR & R. RANDALL. 1951. J. Biol. Chem. **193:** 265–275.
19. MAHONEY, W. & M. HERMODSON. 1980. J. Biol. Chem. **255:** 11199–11203.
20. GARNIER, J., D. J. OSGUTHORPE & B. ROBSON. 1978. J. Mol. Biol. **120:** 97–120.
21. KYTE, J. & R. F. DOOLITLE. 1982. J. Mol. Biol. **157:** 105–132.
22. JENTOFT, N. & D. G. DEARBORN. 1979. J. Biol. Chem. **254:** 4359–4365.
23. PAULING, L. 1960. The Nature of the Chemical Bond. 3rd edit: 260–261. Cornell University Press. Ithaca, New York.
24. SZASZ, J., M. B. YAFFE, M. ELZINGA, G. S. BLANK, & H. STERNLICHT. 1985. Submitted for publication.
25. CHOU, P. Y. & G. D. FASMAN. 1978. Annu. Rev. Biochem. **47:** 251–276.
26. CLEVELAND, D., Personal communication.
27. PONSTINGL, H., M. LITTLE, E. KRAUHS & T. KEMPF. 1979. Nature (London) **282:** 423–424.
28. PONSTINGL, H., E. KRAUHS & M. LITTLE. 1983. Submicrosc. Cytol. **15:** 359–362.
29. RINGEL, I. & H. STERNLICHT. 1984. Biochemistry **23:** 5644–5653.

30. JENTOFT, J. E., T. A. GERKEN, N. JENTOFT & D. DEARBORN. 1981. J. Biol. Chem. **256:** 231–236.
31. GERKEN, T. A., J. E. JENTOFT, N. JENTOFT & D. DEARBORN. 1982. J. Biol. Chem. **257:** 2894–2900.
32. JARDETZKY, O. & G. C. K. ROBERTS. 1981. Nmr In Molecular Biology: 281–284. Academic Press. New York.
33. LEE, J. C., D. CORFMAN, R. P. FRIGON & S. N. TIMASHEFF. 1978. Arch. Biochem. Biophys. **185:** 4–14.
34. RASHIN, A. A. & B. HONIG. 1984. J. Mol. Biol. **173:** 515–521.
35. WADA, A. & H. NAKAMURA. 1981. Nature (London) **293:** 757–758.
36. FRIEND, S. H. & F. R. N. GURD. 1979. Biochemistry **18:** 4612–4619.
37. SIMONDSEN, R. P., P. C. WEBER, F. R. SALEMME & G. TOLLIN. 1982. Biochemistry **21:** 6366–6375.
38. POULOS, T. L. & J. KRAUT. 1980. J. Biol. Chem. **255:** 10322–10330.
39. APPLE, P. & J. T. YANG. 1965. Biochemistry **4:** 1244–1249.
40. FLAVIN, M., T. KOBAYASHI & T. M. MARTENSEN. 1982. *In* Methods in Cell Biol. L. Wilson, Ed. Vol. 24A: 257–263. Academic Press. New York.

Studies on the Exchangeable Nucleotide Binding Site of Tubulin[a]

JYOTI P. NATH, GEOFFREY R. EAGLE,
AND RICHARD H. HIMES[b]

Department of Biochemistry
University of Kansas
Lawrence, Kansas 66045

Tubulin, a dimer of nonidentical subunits, contains two guanine nucleotide binding sites, one of which exchanges with nucleotide in the medium.[1] To identify the subunit that contains the individual binding site and to compare the nucleotide binding domain of tubulin with other guanine nucleotide binding proteins, it is necessary to determine the primary sequence around the sites. In such studies use is often made of affinity labels that interact covalently with the protein.

To localize and characterize the exchangeable guanosine triphosphate (GTP) site, two photoaffinity analogues of GTP, 8-azido GTP and (3'-p-azido benzoyl)-GTP, as well as nonphotoaffinity analogues, the periodate oxidation product of GTP, 2-(guanylformylmethoxy)-3-(triphospho) propanal and 5'-p-fluorosulfonyl benzoyl guanosine, have been used (TABLE 1). Results from the use of such analogues have led to inconsistent conclusions. Gaehlen and Haley concluded that the 8-azido GTP binds to the β subunit, although significant nonspecific binding to the α subunit also occurred.[2] More recently, Haley *et al.*[3] reported that the α subunit was labeled exclusively. Maccioni and Seeds[4] found that the 3'-p-azido benzoyl GTP and 2-(guanylformylmethoxy)-3-(triphospho)propanal bound equally well to the α and β subunits, but the binding was saturable and competitive with GTP, indicating that it was due to a specific interaction at the exchangeable GTP site. On the other hand Kirsch and Yarbrough[5] concluded that labeling with the periodate oxidized GTP was nonspecific on the basis that they found no competition with GTP. Conflicting data has also been obtained with the use of 5'-p-fluorosulfonyl benzoyl guanosine, an analogue that is capable of reacting with sulfhydryl groups. In one case,[6] tubulin treated with this compound was capable of self-assembly, and the compound was incorporated into the β subunit. In another study[7] the analogue did not support assembly, and the subunit location of the site of interaction could not be identified. Although affinity analogues are useful in defining binding sites, results obtained with GTP analogues in attempts to identify the amino acid sequence at the exchangeable nucleotide binding site in tubulin have been ambiguous.

Direct photoaffinity labeling has the advantage of covalently cross-linking the natural nucleotide substrate or ligand to its binding site in proteins.[8–12] With this method, proteins are cross-linked to their natural ligands under the direct action of ultraviolet light, without the introduction of affinity labels on either of the reactants. The direct photoaffinity labeling technique has been used to covalently link nucleotides to a number of proteins, and nucleic acids to proteins;[13] very little, however is known about the photochemistry of these reactions and the nature of the adducts formed.[14] It

[a]This work was supported by National Institutes of Health Grant NS 11360.
[b]To whom correspondence should be addressed.

has been suggested that glycine, alanine, and tyrosine residues are involved in cross-linking DNA to alcohol dehydrogenase and salmine,[15] and a phenylalanine residue participates in the cross-linking of single stranded DNA to DNA binding protein.[16] In enzyme-ligand interactions, the direct photoaffinity approach has demonstrated that a glycine is involved in the labeling of isoleucine-tRNA synthetase with adenosine triphosphate (ATP),[8] and an isoleucine residue of ribonuclease A is cross-linked to pUp.[17] In this paper, we report the successful covalent cross-linking of GTP and guanosine diphosphate (GDP) to tubulin by the action of ultraviolet light and describe studies on some of the characteristics of this interaction. The ultimate aim of this work is to identify the exchangeable nucleotide binding site. Our studies reported here and earlier[18] indicate that the direct photoaffinity labeling technique should be useful for the determination of the amino acid sequence around the exchangeable nucleotide site.

TABLE 1. GTP Analogues Used to Label the Exchangeable Nucleotide Site

Analogue	Subunit Labeled	Reference
8-azido GTP	α—nonspecifically	
	β—specifically and nonspecifically	2
	α—exclusively	3
3'-*p*-azido benzoyl GTP	α—specifically	
	β—specifically	4
2-(guanylformylmethoxy)-	α—specifically	
3-(triphospho) propanal	β—specifically	4
	α—nonspecifically	
	β—nonspecifically	5
5'-*p*-fluorosulfonyl	primarily β, some α	6
benzoyl guanosine		
	two sites, not identified	7

MATERIAL AND METHODS

Tubulin Preparation

Bovine brain microtubule protein was prepared by the assembly-disassembly procedure of Shelanski *et al.*[19] After two cycles of assembly-disassembly the protein was suspended in a buffer containing 100 mM 2-(*N*-morpholino)ethane sulfonic acid (MES), 0.5 mM MgSO$_4$, 1 mM ethyleneglycol-bis-(β-aminoethyl ether)N,N'-tetraacetic acid (EGTA), and 2 M glycerol, pH 6.5, at a protein concentration of 10–15 mg/ml, and dialyzed against the same buffer for 4 hr at 0° C. The solution was then added dropwise to liquid nitrogen, and the resulting pellets were stored at −70° C. Further purification was done to obtain the tubulin dimer. Microtubule protein solution pellets were thawed, and an equal volume of 0.8 M piperazine diethane sulfonic acid (PIPES)—20% DMSO, pH 6.9, was added. The protein was then incubated at 37° C for 20 min in the presence of 0.5 mM GTP, and the polymerized protein was then collected by centrifugation. This procedure results in tubulin largely depleted of associated proteins.[20] The remaining associated proteins and buffer components were removed by passing the cold-solubilized protein through a 10 × 1.2 cm column of phosphocellulose (Whatman P 11) that was placed on top of a Biogel P-10 column (25 × 1.2 cm). The column had been washed and equilibrated with PED buffer; 100 mM PIPES, pH 6.9, containing 1 mM EGTA and 1 mM dithiothreitol (DTT), and the same buffer was used to elute tubulin. In some cases 1 mM MgSO$_4$ was included in the buffer (PEMD buffer). The fractions containing tubulin were pooled, concentrated, frozen dropwise in liquid nitrogen, and stored at $-70°$ C.

Tubulin depleted of nucleotide at the exchangeable site was prepared by charcoal treatment. Tubulin (5 mg/ml) in PEMD buffer was treated for 20 min at 5° C with a tenfold (w/w) excess of bovine serum albumin-saturated Norit charcoal.[21] In some cases a second treatment was performed. Charcoal was removed by two consecutive centrifugations at 40,000 g, and protein and bound nucleotide contents were determined in the supernatant. The amount of nucleotide remaining at the exchangeable site after charcoal treatment ranged from 0 to 0.3, that is, total bound nucleotide was 1.0 to 1.3.

Photoaffinity Labeling

Tubulin solutions (0.15–0.5 ml), after incubation for 45 min at 0° C with radiolabeled nucleotide, were placed in 4.3 × 4.3 × 1 cm plastic weighing boats on top of ice. Irradiation was performed by exposing the solution, from a distance of 6 cm, to a RPR-253.7 nm lamp equipped with a 2 mm thick vycor filter. The dosage was 600–700 μW/cm^2 as measured with a Blak-Ray J-225 ultraviolet intensity meter. The degree of covalent labeling was determined from the amount of radioactive nucleotide found in the protein after precipitation with 10% HClO$_4$. The precipitated protein was washed several times with 5% HClO$_4$ and H$_2$O until radioactivity was no longer found in the supernatant. The final precipitate was dissolved in 0.1 M NaOH, and the protein and radioactivity contents were determined.

Protein Nucleotide and Sulfhydryl Determinations

Protein concentrations were determined by the method of Bradford[22] or from the A_{275} value using a ϵ value of 1.13 mg^{-1} ml cm^{-1}. Noncovalently bound nucleotide was determined by precipitating the protein with 10% HClO$_4$ and measuring the absorbance of the supernatant at 254 nm using a molar extinction coefficient of 12,100 M^{-1} cm^{-1}. The free sulfhydryl content of tubulin was estimated with 5,5'-dithiobis (2-nitrobenzoic acid) using a molar extinction coefficient of 13,600 M^{-1} cm^{-1} at 412 nm.[23]

RESULTS

Cross-Linking of [^{3}H]GTP to Tubulin

Tubulin, depleted of microtubule-associated proteins by phosphocellulose chromatography, was irradiated at 0° C for various periods of time in the presence of [^{3}H]GTP. The amount of incorporated label was determined after precipitation of the protein with 10% HClO$_4$. A time-dependent incorporation occurred with a plateau value being reached within about 30–45 min (FIGURE 1). A maximum incorporation of 0.2–0.22 mole/mole of tubulin occurred under these conditions. In separate experiments gel filtration was used to determine the amount of [^{3}H]GTP that actually exchanged into tubulin used for the cross-linking experiments. The average value was 0.6 for a number of experiments. On the basis of this value, 33% of the [^{3}H]GTP that exchanged into the dimer became covalently incorporated upon irradiation. It has been noted that the efficiency of cross-linking by direct photoaffinity labeling is usually less than 25 percent.[16] The conclusion that the labeling was covalent was based on the stability of the label to $HClO_4$ precipitation, urea denaturation, and boiling in sodium dodecyl sulfate (SDS).

Effect of UV-Irradiation on Properties of Tubulin

Treatment with UV radiation can cause structural alterations in the protein as well as nucleotide cross-linking. Several experiments were done to assess the changes produced by the radiation treatment. Because cross-linking experiments were done in the absence of added GTP and in the presence of a 100-fold molar excess of GTP, these two conditions were used in examining for changes in the dimer. One way by which

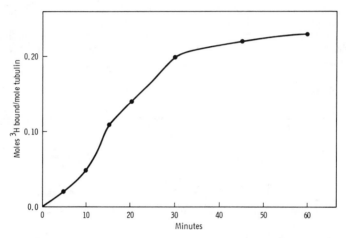

FIGURE 1. Effect of time of irradiation on the extent of cross-linking. Tubulin (15–45 μM) in PEMD buffer was irradiated in the presence of a 100-fold excess of [^{3}H]GTP and treated as described under MATERIAL AND METHODS.

radiation can bring about structural changes in the tubulin dimer is by causing oxidation of sulfhydryl groups. Indeed, it has been shown that tubulin sulfhydryl groups are oxidized as a result of ionizing radiation[24] and ultraviolet radiation.[25] We examined the effect of irradiation under different solution conditions: in the absence of added DTT or GTP, in the presence of DTT, and in the presence of DTT and GTP (TABLE 2). Tubulin sulfhydryls are clearly sensitive to the irradiation treatment; a loss of approximately 40% of the sulfhydryl groups occurs within 30 minutes. DTT significantly protected the sulfhydryl residues, reducing the loss to 10% in the same time period. In the presence of both DTT and a 100-fold excess of GTP, no detectable sulfhydryl loss occurred in the 30-min period. The latter conditions were chosen for most of the labeling studies conducted in this work.

Alterations in tubulin structure and sulfhydryl loss would be expected to result in decreased assembly competence of the protein. We measured assembly in 10% DMSO after irradiation, and the results are presented in FIGURE 2. Electron microscopy

TABLE 2. Effect of Irradiation on the Free Sulfhydryl Content of Tubulin[a]

Buffer Additions	Mole Sulfhydryl/Mole Tubulin		
	0 min	10 min	30 min
None	19.6	17.5	12.0
1 m*M* DTT	20.4	19.6	18.1
1 m*M* DTT and 4.5 m*M* GTP	19.5	19.2	19.2

[a]Tubulin at 5 mg/ml in 100 m*M* PIPES, pH 6.9, containing 1 m*M* EGTA, was irradiated for the times shown; then an aliquot was withdrawn and passed through a Biogel P-6DG desalting column (1.3 × 12 cm). The protein fractions were pooled and the free sulfhydryl content determined by the Ellman procedure.[23] Data from reference 18.

showed that microtubules (MTs) were the only polymerized product present, and the absorbance decreased to zero upon cooling on ice, indicating the absence of nonspecific aggregated products. In the absence of a large excess of GTP, almost complete inhibition of assembly occurred within 5 minutes. The rate of inactivation, however, was greatly reduced when 4.5 m*M* GTP was present. Loss of assembly competence is probably due to a combination of protein denaturation and cross-linking. That cross-linked tubulin GTP could not assemble was shown in the following way. After irradiation for 30 min in the presence of a 100-fold excess of [³H]GTP, the protein was assembled in the presence of 10% DMSO, centrifuged, and the amount of covalent labeling determined. Before assembly it was determined that 0.21 mole of nucleotide was incorporated per dimer. After assembly, the microtubules, collected by centrifugation, were found to contain only 0.03 moles of label per dimer, whereas in the

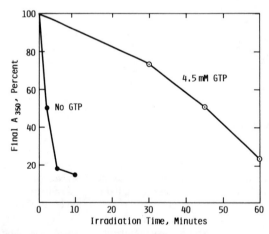

FIGURE 2. Effect of UV irradiation on the self assembly of tubulin. Tubulin (45 μ*M*) in PEMD buffer containing either no added GTP or 4.5 m*M* GTP was irradiated as described under MATERIAL AND METHODS for different periods of time in the absence and presence of excess GTP. Aliquots were taken, and assembly was performed at 1.5 mg/ml tubulin concentration in 10% DMSO at 37° C. The final concentrations of GTP during assembly were 0.5 m*M* for the sample irradiated in the absence of GTP and 2.0 m*M* for the sample irradiated in the presence of GTP. The maximum A₃₅₀ values for each irradiated aliquot are expressed as a percentage of the value of the control (no irradiation). (J. P. Nath *et al.*[18] With permission from *Biochemistry*.)

unassembled protein this figure was 0.16. It is concluded, therefore, that tubulin that contains the cross-linked nucleotide is not capable of self-assembly.

To examine for conformational changes that could result in the release of bound nucleotide from the exchangeable or nonexchangeable sites, we measured the nucleotide content after irradiation. The experiment was done under two conditions: in the absence and presence of excess GTP during the irradiation (TABLE 3). After irradiation, the protein was separated from unbound nucleotide by gel filtration and examined for bound-nucleotide content. When excess GTP was present, the decrease in nucleotide released upon addition of $HClO_4$ (0.18 mole/mole of tubulin) could be accounted for solely by the amount of cross-linking that occurred. When excess GTP was not present, the decrease in nucleotide released by $HClO_4$ was much larger. This was due to the fact that 0.45 mole of nucleotide per mole of tubulin dissociated from the dimer during irradiation, probably as a result of structural changes in the protein.

When fresh preparations of tubulin are allowed to age at 0° C, an aggregate forms, as observed by high performance liquid chromatography (HPLC)-size exclusion

TABLE 3. Loss of Bound Nucleotide After Irradiation[a]

Irradiation Time	GTP Added, mM	Nucleotide Content, Mole/Mole Tubulin	
Min		Protein fraction[b]	Nucleotide fraction[c]
0	0	2.01	0
30	0	1.27	0.45
0	4.5	1.97	ND[d]
30	4.5	1.79	ND

[a]Tubulin (5 mg/ml) in PED buffer received either no or 30 min irradiation, and 0.45 ml was passed through a Biogel P-6 DG column (1.2 × 12 cm). The fractions containing protein and free nucleotide, determined from absorbance measurements, were pooled.

[b]The protein was precipitated by adding an equal volume of 20% $HClO_4$, and the nucleotide content of the supernatant was determined from the A_{254} value.

[c]The fractions containing nucleotide were lyophilized, dissolved in 1 ml of H_2O, and the A_{252} value determined. A correction was made for buffer that had been treated in the same manner because DTT contributes to the absorbance. The values listed are the total nucleotide in this fraction divided by the amount of tubulin placed on the column.

[d]ND = Not determined. The presence of a large excess of nucleotide during the irradiation precluded this determination.

chromatography and by gel filtration on Sephadex G100. An extensive study of the aging of tubulin has been presented.[26] We noticed that after irradiation of a sample of tubulin in the presence of a 100-molar excess of GTP, the amount of tubulin in the aggregate fraction (the void volume) increased from 18% to 40% of the total. It is possible that in our irradiation experiments cross-linking is occurring only in an aggregated fraction, and thus the exchangeable GTP could be in a conformational state of the protein not found in the dimer. To examine for this possibility we isolated both the aggregates and dimer fractions after gel filtration, and after concentration, placed them on a desalting column in the presence of 8 M urea. Labeled exchangeable nucleotide, but not cross-linked nucleotide would be released by this procedure. Both the dimer and the aggregate fractions contained cross-linked labeled GTP. It is likely, therefore, that cross-linking occurs prior to increased aggregation of the protein.

Finally, it was shown that irradiation in the absence of excess GTP caused extensive protein covalent cross-linking. SDS-polyacrylamide gel electrophoresis

(PAGE) demonstrated that over a 90-min period, the monomer bands decreased considerably and higher and lower molecular weight bands became evident. A significant amount of material that would not enter the gel also appeared (FIGURE 3). In the presence of a large excess of GTP, the extent of protein cross-linking was greatly

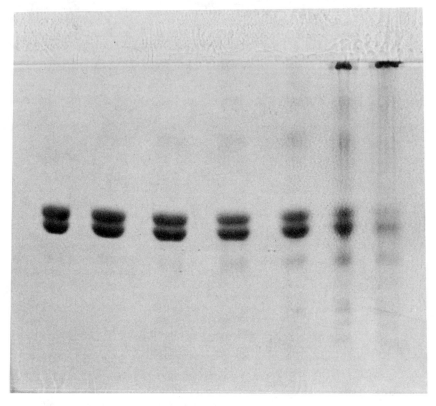

A B C D E F G

FIGURE 3. Effect of irradiation on the protein. Tubulin (5 mg/ml) was irradiated as described under MATERIAL AND METHODS in the absence and presence of 4.5 mM GTP. Aliquots were taken at different times for SDS-PAGE. A: no irradiation. B–D: with GTP present. E–G: GTP absent. Irradiation time for B and E: 30 min; for C and F: 60 min; for D and G: 90 minutes.

reduced, and in the 30-min time period used in most of this work, little covalent aggregation or protein cleavage occurred.

Specificity of Labeling

Several experiments were done to determine whether the covalent labeling was due to a specific interaction of GTP with tubulin. Nonspecific labeling results in a

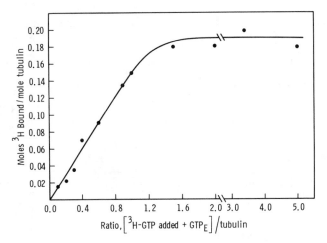

FIGURE 4. Photolabeling of charcoal-treated tubulin as a fraction of increasing [³H]GTP concentration. A 2 mg/ml solution of charcoal-treated tubulin in PEMD buffer was incubated for 30 min at 5° C with increasing amounts of [³H]GTP to yield varying ratios of GTP (remaining E site nucleotide plus added GTP)/tubulin. Irradiation and the determination of the amount of covalently bound ³H per mole tubulin were performed as described under MATERIAL AND METHODS. The specific activity of [³H]GTP used was corrected for the residual E site GTP present in tubulin. The results are the average values of four separate experiments. (J. P. Nath *et al.*[18] With permission from *Biochemistry*.)

nonsaturable incorporation of label. To examine the dependence of labeling on GTP concentration, it is necessary to remove nucleotide from the exchangeable site because this site is fully saturated as tubulin is isolated. Nucleotides were removed from the exchangeable site by charcoal treatment and UV irradiation was done in the presence of increasing concentrations of [³H]GTP. The results in FIGURE 4 show that covalent incorporation increased with increasing nucleotide concentration, and a plateau value was reached at a GTP/tubulin ratio of about 1.2–1.5. This value is consistent with the reported kD of tubulin · GTP of 22 nM.[27] Saturation would not be expected if nonspecific labeling were responsible for the incorporation of isotope.

In another series of experiments, the tubulin concentration was varied from 2.3 μM to 45 μM and the GTP/tubulin ratio from 1 to 100 (TABLE 4). Under these conditions the amount of label incorporated per tubulin dimer did not change. Specificity of

TABLE 4. Effect of Tubulin and GTP Concentrations on the Extent of Labeling[a]

Tubulin Concentration, μM	GTP Concentration, μM	Incorporation, Mole/Mole Tubulin
2.3	230	0.19
4.5	450	0.20
45	45[b]	0.20
45	4500	0.20

[a]Photo-induced cross-linking was performed for 30 min as described under MATERIAL AND METHODS.

[b]No excess GTP was added. The amount shown refers to the amount present in the exchangeable site.

TABLE 5. Specificity of Labeling[a]

Protein	Labeled Nucleotide	Incorporation, Mole/Mole Protein
Tubulin, 45 μM	ATP, 4500 μM	0.01
Tubulin, 45 μM	GDP, 4500 μM	0.19
Tubulin, 22.5 μM	GTP, 22.5 μM,[b] + GMP (unlabeled), 225 μM	0.20
Tubulin, 22.5 μM	GTP, 22.5 μM,[b] + GMP (unlabeled), 2250 μM	0.16
Tubulin, 45 μM	GTP, 450 μM, + GMP (unlabeled), 4500 μM	0.22
Tubulin, 45 μM	GMP, 4500 μM, + GTP (unlabeled), 45 μM[b]	0.01
Tubulin, urea-denatured, 40 μM	GTP, 4500 μM	0.01
Serum Albumin, 45 μM	GTP, 4500 μM	0.02

[a]Photoaffinity labeling was performed for 30 min as described under MATERIAL AND METHODS.

[b]No excess GTP was added. The amount shown refers to the amount present in the exchangeable site.

labeling was also examined by determining the incorporation of [³H]GDP, [³H] guanosine monophosphate (GMP), and [³H]ATP into tubulin, as well as the incorporation of [³H]GTP into denatured tubulin and into serum albumin (TABLE 5). Labeling by [³H]GDP is expected because this nucleotide binds at the exchangeable site with an affinity close to that of GTP.[27] The facts that [³H]GMP, a nucleotide that does not bind strongly to tubulin, was not cross-linked to the protein, even when present at a 100-fold excess, and did not inhibit labeling by [³H]GTP, represent strong evidence that incorporation of label was of a specific nature. Specificity was further indicated by the lack of incorporation of [³H]ATP into tubulin and [³H]GTP into serum albumin and urea-denatured tubulin.

Effect of UV Irradiation on GTP Exchange

To further support the conclusion that the cross-linking was indeed occurring at the exchangeable nucleotide binding site, we determined the effect of prior irradiation of tubulin, containing unlabeled GTP, on the subsequent exchange with [³H]GTP.

TABLE 6. Effect of UV Irradiation on [³H]GTP Exchange by Tubulin[a]

Irradiation Time Minutes	Mole[³H]GTP/Mole Tubulin	
	a	b
0	0.60	0.57
30	0.19	0.37

[a]Tubulin (5 mg/ml) in PED buffer was irradiated for 30 min as described under MATERIAL AND METHODS. After irradiation, [³H]GTP was added, and incubation was continued for 60 min at 0°C. The samples were passed through a Biogel P-6DG column (12 × 1.2 cm). The fractions containing protein were pooled and both protein and [³H]GTP determined. Column a: The irradiation was done in the absence of excess GTP. GTP (4.5 mM) was added with the [³H]GTP to measure exchange. Column b: The irradiation was done in the presence of 4.5 mM GTP. The values presented are the average of two experiments.

Tubulin was irradiated in the absence of excess GTP and in the presence of a 100-molar excess of GTP for 30 min followed by incubation with [³H]GTP and gel filtration. In the absence of GTP, the irradiation treatment caused a 68% reduction in the amount of GTP that could be exchanged (TABLE 6). When excess GTP was present during the irradiation, however, the value decreased by 35 percent.

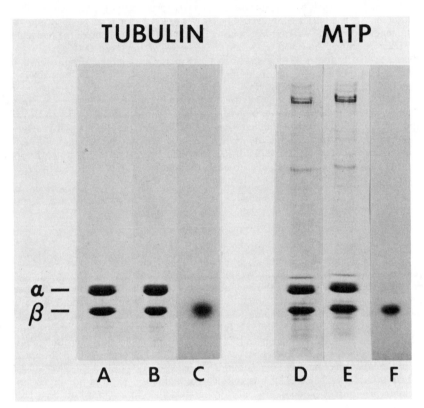

FIGURE 5. SDS-PAGE and autoradiography of carboxymethylated tubulin-α-[³²P]GTP after irradiation. Tubulin and twice-cycled microtubule protein were UV irradiated in the presence of a 100-fold excess of α-[³²P]GTP and prepared for SDS-PAGE and autoradiography. Protein precipitated and washed with HClO₄ was carboxymethylated under conditions described by Allen.[30] SDS-PAGE was done by the Laemmli procedure.[31] A and D: Coomassie stained sample without UV irradiation; B and E: Coomassie stained sample after 30-min UV irradiation; C and F: autoradiograph of 30-min UV irradiated samples. (J. P. Nath *et al.*[18] With permission from *Biochemistry*.)

Localization of the Site of Labeling

To localize the site of cross-linking, irradiation of tubulin was performed in the presence of α-[³²P]GTP followed by SDS-PAGE of the carboxymethylated protein and by autoradiography (FIGURE 5). All but a trace of the label was found in the β subunit. Incubation of tubulin with α-[³²P]GTP under similar conditions without irradiation followed by SDS-PAGE and autoradiography revealed no incorporation of the label. In other experiments, gel slices were cut out, digested, and counted. No more than 5% of the label was found in the α subunit. The results of experiments involving irradiation of microtubule protein are also shown in FIGURE 5. These results show that the presence of microtubule-associated proteins in such preparations does not change the labeling pattern.

Limited proteolysis by chymotrypsin is known to preferentially cleave the β subunit of tubulin.[28,29] The initial products of this cleavage were reported to be 35 K and 18 K fragments, with the 18 K fragment coming from the N-terminal end of the molecule.[29] At these meetings, however, E.-M. Mandelkow et al. reported that chymotrypsin cleaves primarily at Tyr 281 giving rise to a 30 K peptide from the N-terminal end and 20 K and 1 K peptides coming from the C-terminal end. We used this approach to try to narrow down the site of cross-linking in the β subunit. Results of these experiments are shown in FIGURE 6. In our hands the cleavage products are a fragment in the vicinity of 36 K and several lower molecular weight fragments, 19 K, 17 K, 15 K, 13 K, and smaller peptides. When the digestion of GTP-labeled tubulin with chymotrypsin was done as a function of time, the 36 K, 15 K, and 13 K fragments, as well as lower molecular weight fragments, were labeled. With increased time of digestion, the 36 K peptide was no longer labeled even though the amount of this peptide appeared to increase as judged by Coomassie blue staining.

DISCUSSION

The advantage of the direct photoaffinity labeling method over the use of photoaffinity or other chemically reactive analogues, is that the natural ligand is cross-linked to the protein, thus eliminating the possibility of the binding site undergoing a conformational change to accommodate a modified ligand. Our results indicate that the covalent incorporation of nucleotide was highly specific: (1) titration of charcoal-treated tubulin with GTP followed by irradiation produced a saturation-type curve; (2) the presence of a 100-fold excess of nucleotide over the tubulin-nucleotide complex did not increase the amount of label incorporated; (3) essentially all of the label was found in the β subunit; (4) incorporation of labeled GMP or ATP did not occur; (5) labeled GTP was not incorporated into denatured tubulin or serum albumin. The direct photoaffinity labeling method therefore appears to be a better technique than the use of photoaffinity and chemically reactive analogues of GTP.

The conditions we chose for the irradiation treatment minimized damage to the protein. When tubulin was irradiated in the absence of added nucleotide, the assembly competence was completely destroyed, reduction in free sulfhydryl content occurred, there was a 68% decrease in the quantity of GTP that could exchange into the exchangeable site, and there was a loss in nucleotide from the exchangeable site. The presence of a large excess of GTP prevented the loss in sulfhydryl groups and nucleotide and decreased the loss in assembly competence and GTP exchange activity. This effect possibly could be due to the absorbance of harmful radiation by the nucleotide solutions. Under these conditions approximately 0.2 mole of nucleotide became incorporated into the protein, determined either by measuring incorporation of label or decrease in the amount of nucleotide released by $HClO_4$ precipitation. Moreover, the amount of nucleotide that could subsequently exchange into tubulin (TABLE 6) was decreased by the same amount. The fact that approximately the same amount of cross-linking is achieved when measured by incorporation of labeled nucleotide (exchangeable site) or by reduction in nucleotide released on treatment with

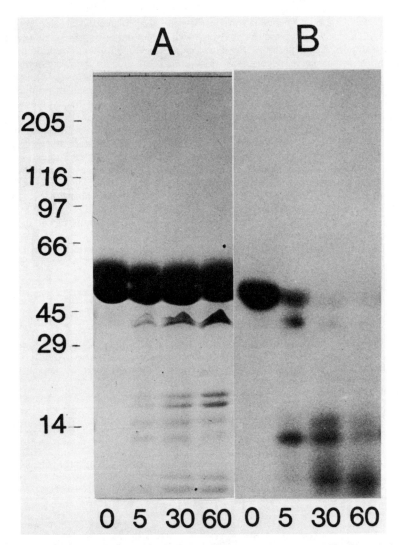

FIGURE 6. SDS-PAGE and autoradiography of tubulin cross-linked with GTP and subjected to limited proteolysis. Tubulin (5 mg/ml) was irradiated in the presence of a 100-fold molar excess of α-[^{32}P]GTP for 30 min at 0° C. After removal of the excess nucleotide by gel filtration, the protein was concentrated to 2.8 mg/ml with the use of an Amicon microconcentrator and incubated with α-chymotrypsin (tubulin: α chymotrypsin = 100:1 w/w) at 25° C for the time periods shown. PMSF (final concentration = 5 mM) was added to stop the reaction, and samples were treated for SDS-PAGE. A: Coomassie stained; B: autoradiograph.

$HClO_4$ (exchangeable and nonexchangeable sites) indicates that the nucleotide at the nonexchangeable site is not being covalently incorporated into the protein. This may mean that the conformations around the two sites are not identical.

On the basis of sequence comparisons with other nucleotide binding proteins, Krauhs et al.[32] have suggested that a glycine-rich cluster in the 140–146 residue region in the β subunit may represent a flexible loop in the exchangeable nucleotide binding domain. A similar cluster occurs in the α subunit. There are two contradictory reports on the initial sites of cleavage by chymotrypsin (see reference 29 and E.-M. Mandelkow et al., this volume, p. 645). E.-M. Mandelkow et al. propose that the site is at Tyr 281, which could give rise to a 30 K peptide containing the glycine cluster. Whether or not this is similar to the 36 K fragment is not clear. It is interesting to note that the label associated with the 36 K unit rapidly disappeared (as judged by autoradiography) even though this peptide and the β subunit were still present in substantial amounts as judged by Coomassie blue staining. An explanation of this result could be that the conformation of the labeled protein had been altered sufficient to make the protein much more susceptible to digestion than the unmodified protein. In an earlier report[33] we stated that the 36 K peptide did not become labeled. In the experiment in the earlier report, digestion was carried out for 30 min, and on the basis of the results presented in FIGURE 6, the labeled peptide probably had been digested further in this time period.

These studies indicate that the direct photoaffinity labeling technique may be useful in determining the amino acid sequence of the exchangeable nucleotide binding site in tubulin.

REFERENCES

1. WEISENBERG, R. C., G. G. BORISY & E. W. TAYLOR. 1968. Biochemistry **7:** 4466–4479.
2. GEAHLEN, R. L. & B. E. HALEY. 1979. J. Biol. Chem. **254:** 11982–11987.
3. HALEY, B. E., H. POSTINGL & K.-H. DOENGES. 1983. Hoppe-Seyler's Z. Physiol. Chem. **364:** 1137.
4. MACCIONI, R. B. & N. W. SEEDS. 1983. Biochemistry **22:** 1572–1579.
5. KIRSCH, M. & L. R. YARBROUGH. 1981. J. Biol. Chem. **256:** 106–111.
6. STEINER, M. 1984. Fed. Proc. Fed. Am. Soc. Exp. Biol. **43:** 2015.
7. PRASAD, A. R. S. & R. F. LUDUENA. 1984. Fed. Proc. Fed. Am. Soc. Exp. Biol. **43:** 1705.
8. YUE, V. T. & P. R. SCHIMMEL. 1977. Biochemistry **16:** 4678–4684.
9. CARROLL, S. F., S. LORY & R. J. COLLIER. 1980. J. Biol. Chem. **255:** 12020–12024.
10. CARAS, I. W., T. JONES, S. ERIKSSON & D. W. MARTIN. 1983. J. Biol. Chem. **258:** 3064–3068.
11. GOLDMAN, R. A., T. HASAN, C. C. HALL, W. A. STRYCHARZ & B. S. COOPERMAN. 1983. Biochemistry **22:** 359–368.
12. MUHN, P. & F. HUCHO. 1983. Biochemistry **22:** 421–425.
13. SMITH, K. C. 1976. In Photochemistry and Photobiology of Nucleic Acids. S. Y. Wang, Ed.: 187–218. Academic Press. New York.
14. SHETLAR, M. D. 1980. In Photochemical and Photobiological Reviews. K. C. Smith, Ed.: 105–197. Plenum Press. New York.
15. TOTH, B. & K. DOSE. 1976. Rad. Environ. Biophys. **13:** 105–113.
16. MERRILL, B. M., K. R. WILLIAMS, J. W. CHASE & W. H. KONIGSBERG. 1984. J. Biol. Chem. **259:** 10850–10856.
17. HAVRON, A. & J. SPERLING. 1977. Biochemistry **16:** 5631–5635.
18. NATH, J. P., G. R. EAGLE & R. H. HIMES. 1985. Biochemistry. **24:** 1555–1560.
19. SHELANSKI, M. L., F. GASKIN & C. R. CANTOR. 1973. Proc. Natl. Acad. Sci. USA **70:** 765–768.
20. HIMES, R. H., P. R. BURTON & J. M. GAITO. 1977. J. Biol. Chem. **252:** 6222–6228.

21. SANDOVAL, I. V., E. MACDONALD, J. L. JAMESON & P. CUATRECASAS. 1977. Proc. Natl. Acad. Sci. USA **74:** 4881–4885.
22. BRADFORD, M. M. 1976. Anal. Biochem. **72:** 248–254.
23. ELLMAN, G. L. 1959. Arch. Biochem. Biophys. **82:** 70–77.
24. ZAREMBA, T. G. & R. D. IRWIN. 1981. Biochemistry **20:** 1323–1332.
25. ZAREMBA, T. G., T. R. LEBON, D. B. MILLAR, R. M. SMEJKAL & R. J. HAWLEY. 1984. Biochemistry **23:** 1073–1080.
26. PRAKASH, V. & S. N. TIMASHEFF. 1982. J. Mol. Biol. **160:** 499–515.
27. ZEEBERG, B. & M. CAPLOW. 1979. Biochemistry **18:** 3880–3886.
28. BROWN, H. R. & H. P. ERICKSON. 1983. Arch. Biochem. Biophys. **220:** 46–51.
29. MACCIONI, R. B. & N. W. SEEDS. 1983. Biochemistry **22:** 1567–1572.
30. ALLEN, G. 1981. *In* Laboratory Techniques in Biochemistry and Molecular Biology. T. S. Work & R. H. Burdon, Ed.: 30–31. Elsevier/North-Holland Biomedical Press. New York.
31. LAEMMLI, U. K. 1970. Nature (London) **227:** 680–685.
32. KRAUHS, E., M. LITTLE, T. KEMPF, R. HOFER-WARINEK, W. ADE & H. PONSTINGL. 1981. Proc. Natl. Acad. Sci. USA **78:** 4156–4160.
33. NATH, J. P., G. R. EAGLE & R. H. HIMES. 1984. Fed. Proc. Fed. Am. Soc. Exp. Biol. **43:** 2015.

Involvement of Guanosine Triphosphate (GTP) Hydrolysis in the Mechanism of Tubulin Polymerization: Regulation of Microtubule Dynamics at Steady State by a GTP Cap

D. PANTALONI and M-F. CARLIER

Laboratoire d' Enzymologie
CNRS
91190 Gif-sur-Yvette, France

INTRODUCTION

During the past years our efforts have focused on the role of nucleotides in protein polymerization (glutamate dehydrogenase, microtubule and actin assembly), and more specifically the involvement of guanosine triphosphate (GTP) hydrolysis in tubulin polymerization. This issue is important because the fact that GTP hydrolysis accompanies polymerization,[1–5] and takes place in a steady-state fashion after polymerization, indicates that microtubules are steady-state polymers. The thermodynamics of assembly, therefore, can be modified as compared to a polymer undergoing reversible assembly and allow the possibility for special properties. For instance, the possibility of head-to-tail polymerization or "treadmilling" has been initially suggested[6–8] as resulting from a thermodynamic difference between the two ends of the polymer. More recent data[9–11] indicate that the highly dynamic behavior of microtubules in the cell can be accounted for by the involvement of GTP hydrolysis in tubulin polymerization.[12–17] In other words, GTP hydrolysis might be involved in the regulation of microtubule assembly in the cell. Our purpose is to understand the pathways through which this regulation may be accomplished. In this approach, we will address the following questions: How is GTP hydrolysis kinetically related to the polymerization reaction? How does the energetic state of the cell, that is, the GTP/GDP ratio (GDP = guanosine diphosphate), affect the rate of microtubule assembly? By what kinetic and thermodynamic features is a steady-state polymer, such as microtubules (or actin filaments) different from an equilibrium polymer such as hemoglobin S or GDH (or microtubules polymerized with nonhydrolyzable analogs of GTP)? How are the length fluctuations and the monomer-polymer exchanges affected? What is the role of GTP hydrolysis at steady-state? All these questions can be summarized in one: Why is it interesting for the cell to spend so much energy to form and maintain microtubules?

RESULTS

Uncoupling between Tubulin Polymerization and GTP Hydrolysis

The kinetic analysis of the correlation between the time courses of microtubule assembly *in vitro* and associated GTP hydrolysis[18] showed that these two reactions are not simultaneous, but that GTP hydrolysis follows polymerization and that the steady state of GTP hydrolysis is established later than the steady state of polymerization. Appreciable uncoupling was most conveniently observed when polymerization was fast, that is, when tubulin was polymerized at a high concentration, whereas GTP hydrolysis was more closely associated to polymerization when tubulin polymerized more slowly, at a low concentration, or close to the steady state of polymerization when the monomer concentration reached the critical concentration.

These observations eliminated the possibility that GTP hydrolysis and polymerization occur in a cycled compulsory order, but were rather in favor of a mechanism in which GTP tubulin first polymerizes and then the GTP bound to tubulin is hydrolyzed on the polymer, in a single turnover monomolecular process, as described by the following scheme.

$$MT_n (GDP_m, GTP_p) + T\text{-}GTP \underset{k_-}{\overset{k_+}{\rightleftharpoons}} MT_{n+1} (GDP_m, GTP_{p+1})$$

$$MT_n (GDP_m, GTP_p) \xrightarrow{k_h} MT_n (GDP_{m+1}, GTP_{p-1}) + P_i \qquad (1)$$

In this scheme, $n = m + p$ represents the number of subunits in a microtubule MT containing m T-GDP subunits and p T-GTP subunits; k_+ and k_- are the association and dissociation rate constants for a T-GTP subunit; and k_h is the first order rate constant for GTP hydrolysis on the microtubule.

FIGURE 1 represents the time courses of polymerization of one mole of tubulin and hydrolysis of one mole of GTP. A growing microtubule is drawn at different times of the reaction, showing that the proportion of GTP and GDP bound to the polymer changes with time, the polymerizing ends being always richer in GTP than the body that had polymerized previously.

Quantitative analysis of the data agreed with the proposed model. The following equation was fitted to the GTP hydrolysis curves:

$$P_i(t) = [T - GDP]_{MT}(t) = \frac{1}{k_1 - k_h} [k_1(1 - e^{-k_h t}) - k_h(1 - e^{-k_1 t})], \qquad (2)$$

where $k_1 = k_+ [MT]$

The value determined for k_h was found to be $0.004 s^{-1}$, independent of the tubulin concentration, in agreement with the model. The amount of GTP bound to microtubules is represented by the difference, at each time, between the polymerization curve and the GTP hydrolysis curve.

In order to get a clearer representation of the respective rates at which a microtubule grows and GTP is hydrolyzed on the polymer, let us consider a single microtubule growing in a solution of dimeric tubulin at a concentration c. In this experiment, the concentration c can be considered as constant, because the amount of tubulin incorporated in the polymer is negligible as compared to the amount of free subunits. The polymer is therefore growing indefinitely, in contrast with the polymerization experiment shown in FIGURE 1 where the concentration of dimeric tubulin decreases with time and finally reaches the critical concentration. The rate of growth is

$$\frac{dn_0}{dt} = k_+ (C - C_c) \qquad (3)$$

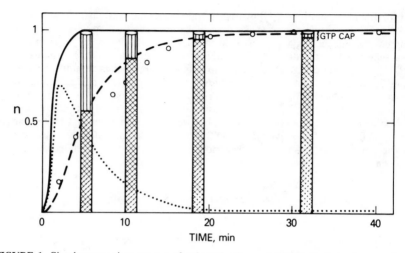

FIGURE 1. Simultaneous time courses of polymerization and GTP hydrolysis of one mole of GTP tubulin. Pure tubulin (34.6 μM) is polymerized at 37°C in a buffer consisting of 50 mM 2-(N-morpholino)ethane sulfonic acid (MES) pH 6.6, 0.5 mM EGTA, 6 mM MgCl$_2$, 3.4 M glycerol, and 0.1 mM [γ^{32}p] GTP. Solid line: polymer formed (from turbidity data). Symbols GTP hydrolyzed. Dashed line: theoretical curve for GTP hydrolysis according to equation **2**, with the following values of the parameters: $k_1 = 0.016$ s^{-1}, $k_h = 0.004$ s^{-1}. Dotted line: difference between the solid curve and the dashed curve, representing the evolution with time of polymerized GTP-tubulin complex. At different times of the polymerization process, a representative microtubule is shown. The GDP core is represented by hatched areas, whereas the top portion of the microtubule consists of GTP subunits. Note the persistence of a small GTP cap at steady state when the critical concentration is reached.

where n_0 is the number of subunits added to the "seed" microtubule that is added to the solution of dimeric tubulin, and C_c is the critical concentration. The rate of GTP hydrolysis on the polymer is

$$\frac{dn_0}{dt} = \frac{d(n_0 - n_T)}{dt} = k_h n_T \qquad (4)$$

and $n_0 = n_T + n_D$, n_T and n_D being the number of polymerized subunits having GTP or GDP bound respectively, at time t. Integration of these differential equations leads to

$$n_0 = k_+ (C - C_c) t$$

$$n_T = \frac{k_+}{k_h} (C - C_c) (1 - e^{-k_h t})$$

$$n_D = k_+ (C - C_c) t - \frac{k_+}{k_h} (C - C_c)(1 - e^{-k_h t}). \qquad (5)$$

FIGURE 2 shows the evolution of n_0 and n_D. The fact that GTP hydrolysis is uncoupled from polymerization results in the existence of the GTP cap at the end of the growing microtubule. Following a pre-steady-state period, a constant value of the GTP cap is maintained. At infinite time, the relative proportion of GTP polymerized subunits in the cap n_T/n_0 becomes a relatively negligible proportion of the total number of subunits in the polymer. It can be noted that the proportion of GTP subunits is independent of the tubulin concentration. Once the steady-state GTP is established, the rate at which GTP is hydrolyzed is equal to the rate of elongation. The size of the GTP cap at steady state is

$$n_T^{ss} = \frac{k_+}{k_h} (C - C_c).$$ (6)

The slower the rate of hydrolysis k_h, as compared to the rate of polymerization k_+, the bigger the size of the cap. Also, the size of the cap increases linearly with the concentration of dimeric tubulin.

At the critical concentration, the rate of growth is zero, and equation 6 indicates that the size of the GTP cap would be zero. This neglects, however, the statistical nature of the association-dissociation reactions that take place in equal proportions at steady state. Due to the delay between GTP hydrolysis and tubulin addition to microtubule ends, there actually is an average small cap of GTP subunits at the ends of microtubules at the critical concentration. For a given microtubule, the size of this cap fluctuates with time, and when it reaches zero, GDP will be present at the end of the

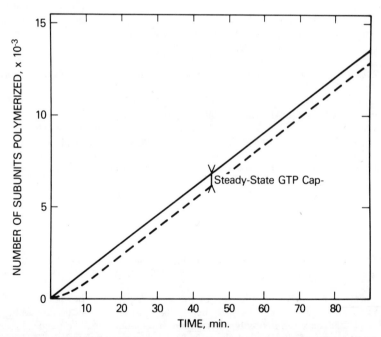

FIGURE 2. Growth of one microtubule in a solution of dimeric tubulin at a constant concentration, c. The total number of subunits, n_0, added to a virtual "seed" is represented by the solid straight line, of slope k_+ $(c - c_c)$ (equation 3). The dashed line represents the number of GDP subunits (equation 4). Note that following a pre-steady-state period, a steady GTP cap is maintained at the tip of the microtubule. The following parameters were used: $C - C_c = 5 \mu M$; $k_+ = 0.5 \mu M^{-1} s^{-1}$; $k_h = 0.004 s^{-1}$. The GTP cap here represents 625 subunits (equation 6).

microtubule. This event is rare when the microtubule grows, that is, at concentrations significantly above the critical concentration, because in this case the GTP cap is big enough to ensure a very low probability of having GDP at the end; it cannot be neglected, however, at concentrations close to the critical concentration. Therefore, it is important to understand how much difference it makes for the microtubule to have GTP or GDP bound to the end.

Interplay between GTP and GDP in Tubulin Polymerization

It was found early that GDP was an inhibitor of polymerization[2] and of nucleation[19] and caused microtubule depolymerization.[2,20-23] How GTP and GDP interact in microtubule assembly can be understood in detail by studying the rate of polymerization in the presence of controlled concentrations of both nucleotides. These experiments[22] showed that GDP acted as a competitive inhibitor with respect to GTP. As soon as GTP was present, only GTP tubulin was the polymerizing species, and no measurable amount of exogenous GDP was found incorporated in microtubules. This is not to suggest that GDP tubulin does not polymerize at all. In fact, elongation of GDP tubulin onto seeds can occur, although very slowly, almost two orders of magnitude more slowly than in the presence of GTP.[24] Therefore, even in the presence of high ratios GDP/GTP, essentially only polymerization of GTP tubulin is observed. In the same fashion, total depolymerization of microtubules was observed upon addition of GDP, but this may simply indicate that the critical concentration of GDP tubulin is higher than the tubulin concentration used in these experiments (10–20 μM range).

Dependence of the Rate of Microtubule Elongation on GTP-Tubulin Concentration

The peculiar properties of microtubules emphasized above have been taken into account to propose a polymerization model.[25] The analytical theory for microtubule elongation was developed by T. Hill.[12] The fact that the microtubule is a heterogeneous polymer with a GDP core and GTP caps at the ends, and that the presence of either GTP or GDP at the ends affects the rate parameters results in a nonlinear concentration dependence of the rate of elongation. Let us first recall that for an equilibrium polymer, as defined by Oosawa,[26] the elongation reaction can be described by the following equation:

$$J(c) = \frac{dc}{dt} = k_+ Mc - k_- M \tag{7}$$

where M is the concentration of elongating polymer ends, c is the monomer concentration, and k_+ and k_- the association and dissociation rate constants of the same monomer species T.

FIGURE 3 shows the J(c) plot for an equilibrium polymer. Incidentally, if GTP hydrolysis were tightly coupled to polymerization, that is, if GTP tubulin bound to a GDP end, and only GDP tubulin dissociated from microtubule (Wegner's model, reference 6), the same linear dependence J(c) would be observed; in this case k_+ is the association rate constant of GTP tubulin and k_- the dissociation rate constant of GDP tubulin. In the model we present, by contrast, the rate of elongation will be dependent on the nature of the nucleotide present at the ends, that is, it will reflect the involvement of GTP hydrolysis in the polymerization process. The steady-state analytical theory[12] shows that the J(c) plot is linear above the critical concentration, but exhibits a strong transition in slope at the critical concentration, and drops sharply below the critical concentration. At c = 0, the ordinate intercept gives the value of the dissociation rate constant for GDP tubulin. FIGURE 4 shows a typical theoretical plot and the influence on the shape of the plot of crucial parameters, namely the rate of nucleotide exchange at microtubule ends, and the rate of GTP-tubulin association to a GDP end, as compared to the rate of its association to a GTP end. Qualitatively, the shape of this plot can be explained as follows. Above the critical concentration, the polymer is growing with a GTP cap at the end, as seen above (FIGURE 2). The reactions taking place at the end, therefore, are essentially association and dissociation reactions of GTP tubulin. The system behaves approximately as a polymer undergoing reversible polymerization. Below the critical concentration, the polymer is shortening, and the depolymerization brings GDP subunits from the deep core to the polymer end. The possibility of direct GTP exchange for GDP at the end is also included in the model. Although we do not have direct evidence for such a reaction, it is logical to assume that

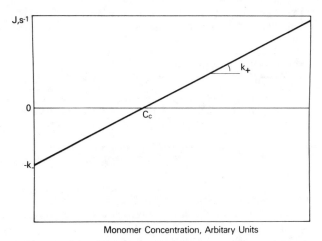

FIGURE 3. Monomer concentration dependence of the rate of growth of a polymer undergoing reversible polymerization. (Oosawa's model). According to equation 7, the rate of elongation, J, is a straight line of slope k_+ intercepting the abscissa at the critical concentration C_c. The ordinate at C = 0 indicates the value of the dissociation rate constant k_-.

it can occur, because nucleotide exchange takes place on dimeric tubulin at a rate that has been measured,[27,28] and nucleotide is blocked and nonexchangeable in the E-site of microtubules after hydrolysis, but not before.[2,18] It is reasonable to think, as proposed by Weisenberg,[29] that the rate of nucleotide exchange is inversely correlated with the number of interactions between the tubulin molecules in the microtubule; on the terminal subunit, the nucleotide might dissociate more slowly than on the free subunits, but faster than in the deep inside of the polymer. It is important to note that the theoretical plot in FIGURE 4 is calculated at steady state, that is, the amount of GTP subunits in the cap, at each tubulin concentration, is the one that exists at steady state (FIGURE 2).

The experimental J(c) plot has been obtained[14] and obviously is not linear but exhibits a sharp bend at the critical concentration (FIGURE 5). The ordinate intercept corresponds to a value for the dissociation rate constant about 25–30-fold larger than

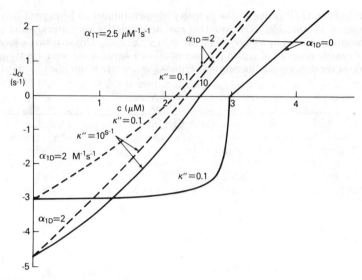

FIGURE 4. Theoretical steady state J(c) plot for a microtubule growing with a GTP cap and a GDP core. The microtubule is considered as a single helix growing at one end. Four curves are drawn to illustrate the influence of two important parameters on the shape of the J(c) plot; the sharpness of the transition at the critical concentration α_{1D} is the rate of GTP-tubulin association to a GDP end; k'' is the rate of nucleotide exchange on the terminal subunit of the helix. Solid lines: $\alpha_{1D} = 0$; $k'' = 0.1$ s^{-1} (right curve) and 10 s^{-1} (left curve). Dashes lines: $\alpha_{1D} = 2$ μM^{-1} s^{-1}; $k'' = 0.1$ s^{-1} and 10 s^{-1}. In all curves, α_{1T}, rate of GTP-tubulin association to a GTP end, is equal to 2.5 μM^{-1} s^{-1}. Note that the nonlinearity of the J(c) plot tends to vanish when α_{1D} is close to α_{1T} and k'' is large.

the value that would be determined by extrapolation of the linear upper branch of the plot. Therefore, the critical concentration is not the ratio between the dissociation and association rate constants measured far from steady state. Unpublished data by Ken Johnson (Ph.D. thesis, 1975) show evidence for a similar nonlinear J(c) plot, although no interpretation could be provided at that time. The same plot was obtained whether pure tubulin or whole microtubule protein preparations were studied. In good agreement with the model we propose, the J(c) plot experimentally obtained in the presence of GMPPNP, a nonhydrolyzable analog of GTP, is linear because, in this case, microtubules are equilibrium polymers (FIGURE 5, inset). The ordinate intercept of the plot in GTP corresponds to the dissociation rate constant of GDP tubulin. In agreement with this conclusion, the same rate of depolymerization was observed whether microtubules were diluted in buffer containing GTP or GDP (FIGURE 6) or in a solution containing fructose-6-phosphate and phosphofructokinase in order to transform all GTP into GDP. Direct comparison between the theoretical and the experimental plot is not feasible because the theory is calculated at steady state, whereas the experiments are transient in which the GTP cap of microtubules first undergoes a relaxation from the initial state (microtubules in the stock solution at the critical concentration) to the final state (microtubules in a solution of tubulin at the given concentration). The existence of these transients has been demonstrated,[13] and a good agreement has been obtained between the experimental plot and one particular transient.[14] These two experiments confirm that the peculiar shape of the J(c) plot for microtubules in GTP is due to the property of microtubules to be capped polymers, consisting of an unstable GDP core that can depolymerize very quickly and that is dynamically stabilized by a cap of GTP subunits at the ends. This structure guarantees a very dynamic behavior of such a polymer in the cell, because any signal that causes a decrease of the dimeric tubulin below its critical concentration (for instance, binding to

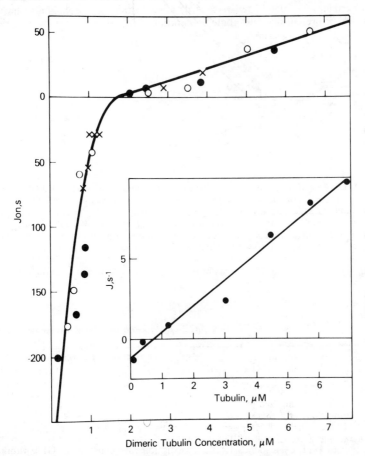

FIGURE 5. Experimental J(c) plot for microtubules growing in the presence of GTP. A solution of 13.6 μM polymerized tubulin at steady state was diluted 10-fold (●), 4-fold (○), and 2-fold (×) in solutions of dimeric tubulin at different concentrations. The initial rate of polymerization or depolymerization was measured from the turbidity change. The plot was normalized to the same number concentration of microtubules in each series of dilutions. The number concentration of microtubules in the assay was deduced from measurement of the average length of microtubules in the stock solution. Knowing the number concentration of microtubules and the turbidity change due to polymerization of a given amount of tubulin, the J(c) plot can be expressed in number of subunits polymerized per microtubule per second. Inset: Similar plot obtained with GMPPNP microtubules diluted in solutions of GMPPNP tubulin. Note that the plot is linear; the critical concentration is about one-half that in the presence of GTP; k_+ is about one-sixth that in the presence of GTP.

a sequestering protein) will result in a very rapid depolymerization of microtubules; any change in the ratio GTP/GDP will result in a smaller GTP cap at the end of microtubules and will cause a subsequent change in their stability. In this way, the energy state of the cell affects the stability of microtubules. We can also expect that protein binding to the ends of microtubules might modify the rate of exchange of nucleotides and therefore change the rate of polymerization by switching the elongating sites either in an "on" (actively elongating, GTP-bound) or "off" (GDP-bound, depolymerizing) position. In this regard, a nucleotide diphosphokinase has been reported to be present in microtubule preparations.[30]

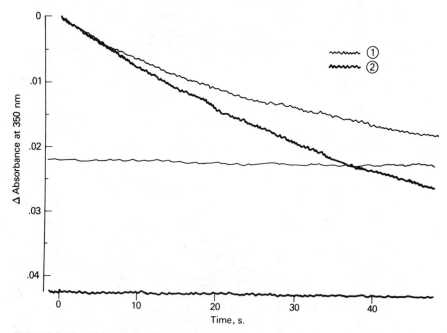

FIGURE 6. Depolymerization of microtubules upon dilution in the presence of GTP or GDP. A solution of 12 μM polymerized pure tubulin, at steady state in the presence of 200 μM GTP, 50° M GDP, was diluted 5-fold in assembly buffer containing either 1 mM GTP (1) or 1 mM GDP (2). The recordings of the decrease in turbidity with time exhibit the same initial rate, which is close to the rate of dissociation of GDP tubulin in both cases.

The critical concentration appears as a point of transition between two phases of the microtubule, as defined by T.L. Hill:[15-17] the GTP-capped, growing phase, and the uncapped, essentially depolymerizing phase. In the region of the critical concentration, microtubules oscillate between these two phases. If the state of a given microtubule (*i.e.* the size of the GTP cap) is followed with time in Monte-Carlo calculations, it can be observed that, although the number of events in each phase is the same, the microtubule spends much less time in the uncapped (depolymerizing phase) than in the capped (growing) phase, due to the high rate of GDP-subunit dissociation. This behavior is quite different from that of an equilibrium polymer at the critical concentration, which spends equal periods of time in the association and dissociation events.

Monomer-Polymer Exchange and Length Fluctuations of Microtubules at Steady State in GTP

The above considerations lead directly to the issue of the mechanism of the change in length distribution of microtubules and of the exchange between free and polymerized subunits in the presence of GTP. Let us consider first the case of an equilibrium polymer surrounded by monomers at the critical concentration. Because the rates of monomer association to and dissociation from the polymer end are equal ($k_+ c_c = k_-$), only stochastic incorporation of monomer into the polymer can take place. This process can be conveniently observed if the monomer is labeled, and if monomer and polymer can be easily separated, as is the case for microtubules. A quantitative solution to this problem, more accurate than the approximate treatment of Kasai and Oosawa,[31] has been brought by Kristofferson and Purich,[32] and tested on data obtained with microtubules. Whereas Zeeberg et al.[33] found that a similar model of diffusional exchange accounted fairly well for their data, Kristofferson and Purich found that after some time the label incorporated at a faster rate than expected within the simple equilibrium exchange, which was suggestive of some treadmilling mechanism.

All the above treatments of stochastic (diffusion-like) exchange have been made with the assumption that microtubules were polymerizing reversibly in the presence of GTP. Obviously the law of stochastic exchange at steady state should be different for a GTP-capped microtubule oscillating with time between the uncapped (rapidly depolymerizing) and capped (growing) configurations. Qualitatively we can expect that, at steady state, for a short period of time, a small number of microtubules are in the uncapped conformation and, to a large extent, depolymerize, while the major part of the population, in the GTP-capped conformation, slowly rebind the subunits liberated by the depolymerizing microtubules. Because of the large negative fluctuations involved in this mechanism, we can expect that this kind of "asymmetric" stochastic exchange will lead to a faster label uptake than the classical "symmetric" stochastic exchange for an equilibrium polymer, and may account for the available data. Interestingly, this mechanism predicts that this rapid diffusional incorporation will take place, in the presence of GTP, even if microtubules have only one free end; in contrast to what is expected in the treadmilling model, that is, directional label uptake, the two ends are not necessary. This should provide an easy means to check the validity of the GTP-cap model for microtubules.

Another related point is the question of length fluctuations at steady state. The same law of monomer-polymer exchange at steady state leads to the process of length redistribution from a Poisson to an exponential distribution, reported by Oosawa.[26] For very long polymers such as actin filaments or microtubules, this process is expected to be extremely slow, that is, beyond the realistic experimental periods of time. Careful studies by Kristofferson and Purich,[34] however, showed that a weak but measurable shift in length from 11.3 μM to 12.1 μM took place within three hours. Starting with sheared microtubules, that is, with a shorter length, the initial increase in length was faster. The same phenomenon of shift from short to long microtubules at steady state was observed in polymerizing microtubules with taxol[35] and was also attributed to a random walk process.

If we consider a population of polymers under reversible equilibrium, there is some probability that, at equilibrium, short polymers disappear, whereas others get longer. This increase in average length is accompanied by a decrease in number. (Note that this process is somewhat different from the size redistribution described by Oosawa, which takes place at a constant number concentration of polymers). This phenomenon has been observed and analyzed for F-actin at equilibrium in the presence of ADP.[36] Following fragmentation upon a pulse of sonication, the solution spontaneously returns to the initial distribution it had before fragmentation. Kinetic analysis of the process within the assumption of the random walk mechanism, and knowing the rate constant for ADP-actin dissociation from filaments fitted the data quite well. Therefore, length fluctuations are expected for an equilibrium polymer. For a GTP-capped microtubule at steady state, however, the kinetics of such a process is going to be affected by the phase transition that takes place around the critical concentration. Due to the high rate of depolymerization of GDP tubulin, microtubules in the uncapped conformation have a much higher probability of depolymerizing completely within a short interval of time, than microtubules with a GTP cap, which are more stable. Recently Mitchison and Kirshner[9,10] observed the behavior of individual microtubules and did find that two classes could be distinguished at steady state, a class of growing microtubules and a class of rapidly depolymerizing microtubules. Such a puzzling behavior is well accounted for by the model of the GTP cap. A complete mathematic description of this phenomenon has been developed by T.L. Hill.[17]

This model also accounts for the very high rate of microtubule turnover observed *in vivo* by Salmon[11] by the technique of fluorescence recovery after photobleaching. The very high rate of depolymerization measured in these experiments is compatible with the dissociation rate constant of 200–300 s^{-1} found for GDP tubulin in our experiments, which allow a microtubule 10 μM long to disappear in one minute.

GTP Hydrolysis at Steady State

It is well established that microtubules hydrolyze GTP at steady state.[2,3] Such a steady state GTPase activity is expected within the model of the GTP cap. Internal subunits within the cap hydrolyze GTP at the measured rate of 0.004 s^{-1}. The constant renewal of the cap is necessary to the maintenance of microtubules. When GTP is exhausted, microtubules depolymerize spontaneously.[19] The steady-state level of polymer is strictly a function of the GTP/GDP ratio in solution.[22]

The rate of GTP hydrolysis, however, that is measured at steady state is much faster than what is expected if only the GTP cap hydrolyzes GTP. Because the GTPase activity is proportional to the number of microtubule ends[37] and is competitively inhibited by GDP,[22] it was, therefore, proposed that the terminal tubulin subunit, at the very end of the microtubule, had a particular conformation which conferred to it a catalytic GTPase activity with a turnover rate constant of ~ 1 s^{-1}.[25] Another attractive possibility, which has not been demonstrated, however, is to assume that the rate at which GTP hydrolysis takes place on a polymerized subunit in the microtubule is dependent on whether GTP or GDP is bound to the neighboring subunits, as proposed by Hill (references 14 (appendix) and 17). For instance, a GTP-bound subunit surrounded by other GTP subunits would hydrolyze GTP at a much slower rate than when it is in contact with a GDP-bound subunit. Such a mechanism would make it possible to observe the slow rate of 0.004 s^{-1} for initial GTP hydrolysis on a GTP polymer; but once the interface between the GDP core and the GTP cap was established, the GTP subunit in contact with the most distal GDP subunit of the core would hydrolyze GTP at a faster rate. This latter reaction gives a GTPase rate proportional to the number of microtubules in solution, whereas the former is proportional to the concentration of polymerized GTP subunits. More experiments are needed to test this possibility, and are now underway.

Extended Model for Microtubule Polymerization in the Presence of GTP

In the model of the GTP-capped microtubule presented thus far,[12,14] microtubules have been considered as a polymer formed by five independent helices, that is, calculations were made for a single helix. In the real case, it may not be appropriate to consider the five helices as independent. Recent Monte-Carlo studies of the GTP cap in a five-start helix model of microtubule[38] show that in this more refined and realistic case, too, the presence of the GTP cap creates the same phase changes in the region of the critical concentration, and the model can be fitted to the data obtained by Mitchison and Kirschner.[9,10] If a microtubule is considered as made of five nonindepen-

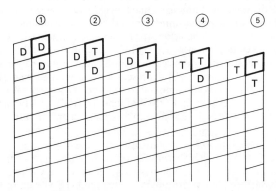

FIGURE 7. Simplified model for microtubule elongation in the presence of GTP. A flattened microtubule wall is represented. Elongation takes place onto the 5-start helices pictured. The association and dissociation parameters of the terminal subunit on each helix (pictured in bold lines) are different for all five cases represented. In analogy with the model presented for actin,[37] the nature of the nucleotide bound to the two adjacent subunits in each of the five cozy corners has to be considered.

dent helices, and if a subunit in the lattice is assumed to interact with four neighbors, with two longitudinal and two lateral interactions, it is clearly apparent that the tip of a growing microtubule is a very complex structure. FIGURE 7 tries to give a simplified representation of the possible configurations of the end of a microtubule growing in the presence of GTP. For sake of simplicity, it is assumed that growth takes place through additions of GTP subunits in "cozy corners" only.[39] The strength of the interaction of a GTP subunit with a "GTP-GDP" corner might be different depending on whether longitudinal or lateral interactions take place with the GTP or GDP subunit of the corner, respectively. Actually previous results suggest that longitudinal interactions are much stronger than lateral interactions.[19] Therefore, the rate constants of elongation must be different according to the nature of nucleotide bound to each of the subunits of the corner. The rate of GTP hydrolysis on polymerized tubulin also might be different according to the nature of the neighboring subunits, as emphasized by Chen and Hill.[38]

Due to the subdivision of rate constants, refined experiments will have to be designed to go further into the kinetic analysis of microtubule assembly. Fruitful comparisons are expected to be made with actin, which exhibits dynamic properties very similar to microtubules, but has a simpler single-helix structure.

Comparison between Actin and Microtubule Dynamics

Since actin filaments and microtubules are the major components of the cytoskeleton and share many structural and functional properties, in particular hydrolysis of nucleotide associated with polymerization, it was interesting to investigate whether in the case of actin the same relationship exists between adenosine triphosphate (ATP) hydrolysis and polymerization as for microtubules. It now appears[40] that actin filaments, similarly, are ATP-capped polymers and correlatively that the concentration dependence of the rate of actin filament growth is not linear, but shows a strong bend close to the critical concentration. Under most of the polymerization conditions, the rate of ADP-actin dissociation from filaments is about 10-fold larger than the rate of ATP-actin dissociation, as determined from the linear portion of the plot of the rate of filament growth versus monomer concentration, above the critical concentration. A model has been developed[41,42] to describe the involvement of nucleotide hydrolysis in the polymerization kinetics.

CONCLUSION

The results we have summarized above convey the following view of the role of nucleotides in the dynamics of microtubules. (The same is true for actin filaments.) GTP hydrolysis is not necessary for tubulin polymerization. The energy linked to the conformation of the GTP-tubulin complex is sufficient to allow it to polymerize. GTP hydrolysis uncoupled to polymerization promotes the existence of an unstable GDP-core polymer that is maintained dynamically stable by a GTP cap at the ends. The GTP cap provides a remarkable dynamic response of microtubules to changes in the environment and is fundamental to their role in cell motility. The same is true for actin filaments that have an ATP cap. The cell spends a lot of energy in the formation of microtubules and maintenance of the GTP cap, only to assure a much more dynamic response than what would be achieved with polymers in reversible equilibrium.

ACKNOWLEDGMENTS

I thank Dr. T. L. Hill, Dr. Y. Chen, Dr. M. W. Kirschner, and Dr. Tim Mitchison for making preprints of their work available to me and Dr. E. D. Korn for stimulating discussions.

REFERENCES

1. ARAI, T. & Y. KAZIRO. 1976. Biochem. Biophysics Res. Commun. **69:** 369–376.
2. WEISENBERG, R. C., W. J. DEERY & P. J. DICKINSON. 1976. Biochemistry **15:** 4248–4254.
3. DAVID-PFEUTY, T., H. P. ERICKSON & D. PANTALONI. 1977. Proc. Natl. Acad. Sci. USA **74:** 5372–5376.
4. SPIEGELMAN, B. M., S. M. PENNINGROTH & M. W. KIRSCHNER. 1977. Cell **12:** 587–600.

5. MACNEAL, R. K. & D. L. PURICH. 1978. J. Biol. Chem. **253:** 4683–4687.
6. WEGNER, A. 1976. J. Mol. Biol. **108:** 139–150.
7. MARGOLIS, R.L . & L. WILSON. 1978. Cell **13:** 1–8.
8. HILL, T. L. & M. W. KIRSCHNER. 1982. Int. Rev. Cytol. **78:** 1–125.
9. MITCHISON, T. & M. W. KIRSCHNER. 1984. Nature (London) **312:** 232–237.
10. MITCHISON, T. & M. W. KIRSCHNER. 1984. Nature (London) **312:** 237–242.
11. SALMON, E. & R. J. MACINTOSH. 1984. J. Cell Biol. **99:** 1067.
12. HILL, T. L. & M-F. CARLIER. 1983. Proc. Natl. Acad. Sci. USA **80:** 7234–7238.
13. CHEN, Y. & T. L. HILL. 1983. Proc. Natl. Acad. Sci. USA **80:** 7520–7523.
14. CARLIER, M-F., T. L. HILL & Y. CHEN. 1984. Proc. Natl. Acad. Sci. USA **81:** 771–775.
15. HILL, T. L. & Y. CHEN. 1984. Proc. Natl. Acad. Sci. USA **81:** 5772–5776.
16. HILL, T. L. 1984. Proc. Natl. Acad. Sci. USA **81:** 6728–6732.
17. HILL, T. L. 1985. Proc. Natl. Acad. Sci. USA. **82:** 431–435.
18. CARLIER, M-F. & D. PANTALONI. 1981. Biochemistry **20:** 1918–1924.
19. CARLIER, M-F. & D. PANTALONI. 1978. Biochemistry **17:** 1908–1915.
20. MARGOLIS, R. L. 1981. Proc. Natl. Acad. Sci. USA **78:** 1586–1590.
21. LEE, S-H., D. KRISTOFFERSON & D. L. PURICH. 1982. Biochem. Biophysics Res. Commun. **105:** 1605–1610.
22. CARLIER, M-F. & D. PANTALONI. 1982. Biochemistry **21:** 1215–1224.
23. ENGELBORGHS, Y. & A. VAN HOUTTE. 1981. Biophys. Chem. **14:** 195–202.
24. ZACKROFF, B., W. J. DEERY & R. C. WEISENBERG. 1980. J. Mol. Biol. **139:** 641–659.
25. CARLIER, M-F. 1982. Mol. Cell. Biochem. **47:** 97–113.
26. OOSAWA, F. & S. ASAKURA. 1975. *In* Thermodynamics of the Polymerization of Protein. Academic Press. New York.
27. ENGELBORGHS, Y. & J. ECCLESTON. 1982. FEBS Lett. **141:** 78–81.
28. BRYLAWSKI, B. P. & M. CAPLOW. 1983. J. Biol. Chem. **258:** 760–763.
29. WEISENBERG, R. C. 1980. J. Mol. Biol. **139:** 660–677.
30. HUITOREL, P., C. SIMON & D. PANTALONI. 1984. Eur. J. Biochem. **144:** 233–241.
31. KASAI, M. & F. OOSAWA. 1969. Biochim. Biophys. Acta **172:** 300.
32. KRISTOFFERSON, D. & D. L. PURICH. 1981. J. Theor. Biol. **92:** 85–96.
33. ZEEBERG, M., R. REID & M. CAPLOW. 1980. J. Biol. Chem. **255:** 9891–9899.
34. KRISTOFFERSON, D. & D. L. PURICH. 1981. Arch. Biochem. Biophys. **211:** 222–226.
35. CARLIER, M-F. & D. PANTALONI. 1983. Biochemistry **22:** 4816–4821.
36. CARLIER, M-F., D. PANTALONI & E. D. KORN. 1984. J. Biol. Chem. **259:** 9987–9991.
37. DAVID-PFENTY, T., J. LAPORTE & D. PANTALONI. 1978. Nature (London) **272:** 282–284.
38. CHEN, Y. & T. L. HILL. 1985. Proc. Natl. Acad. Sci. USA. **82:** 4127–4131.
39. ERICKSON, H. P. & D. PANTALONI. 1981. Biophys. J. **34:** 293–309.
40. CARLIER, M-F., D. PANTALONI & E. D. KORN. 1984. J. Biol. Chem. **259:** 9983–9986.
41. CARLIER, M-F., D. PANTALONI & E. D. KORN. 1984. J. Cell Biol. **99:** 30a.
42. PANTALONI, D., T. L. HILL, M-F. CARLIER & E. D. KORN. 1985. Proc. Natl. Acad. Sci. USA **82:** 7207–7211.

Location of the Guanosine Triphosphate (GTP) Hydrolysis Site in Microtubules[a]

MICHAEL CAPLOW

Department of Biochemistry
University of North Carolina
Chapel Hill, North Carolina 27514

In the earliest formulations of microtubule assembly it was assumed that the GTP in the tubulin E-site was hydrolyzed concomitant with subunit addition to microtubule ends. In recent studies, however, using purified tubulin, which had been separated from microtubule-associated proteins (MAPs) by diethylaminoethyl (DEAE) chromatography, it has been possible to obtain convincing evidence that GTP hydrolysis is not coincident with subunit addition.[1] The study of this problem is more difficult when MAPs are present, because the nonspecific nucleotide triphosphatase activity of the MAPs[2] can obscure the assembly-associated GTP hydrolysis. This problem has, however, been cleverly solved by using a high concentration of the nonhydrolyzable adenosine triphosphate (ATP) analogue adenyl β-imidodiphosphate to inhibit nonspecific GTP hydrolysis.[3] Results obtained under these conditions have been taken to indicate that GTP hydrolysis and microtubule assembly are not perfectly coincident. The difference in assembly and GTPase rates is, however, extremely small, and further study of this system appears to be warranted. It is of special interest to know if GTP hydrolysis is required for subunit incorporation with MAP-containing tubulin, for when assembly is promoted by MAPs (rather than by glycerol[1]) tubulin-guanosine diphosphate (GDP) (TuD) subunits do not readily elongate microtubules,[3–6] suggesting that the GTP hydrolysis is required for the reaction. As a result, it might be expected that GTP hydrolysis will be tightly coupled to subunit addition.

If GTP hydrolysis is not tightly coupled to microtubule assembly, microtubules will contain subunits with E-site GTP. Most of this GTP is eventually hydrolyzed, because the nucleotide in microtubules at steady state is predominately GDP.[7–10] We are interested in the mechanism for this hydrolysis reaction, specifically, whether the GTP hydrolysis rate in nonterminally located tubulin-GTP (TuT) subunits is influenced by the identity of the nucleotide in adjacent subunits. We next describe how this question can be answered, by analysis of the pre-steady state and steady state GTPase rates.

We first consider the case when the hydrolysis rate in TuT subunits is independent of adjacent subunits. Under these conditions GTP hydrolysis in TuT subunits is first-order and the observed rate for Pi release is proportional to the number of TuT subunits within the microtubule. This rate is greatest at some point during the pre-steady state phase of the assembly process, because when the TuT concentration is high, the second-order assembly rate can exceed the first-order hydrolysis rate, so that appreciable TuT subunits can transiently accumulate in the assembling microtubule. At later times, when the assembly reaches a steady state, the TuT content of the microtubules reaches a minimal value and the GTPase rate is proportionally decreased.

Very different kinetic behavior is predicted for a reaction scheme in which the

[a]This work was supported by a Grant from the National Institutes of Health (DE03246).

GTPase reaction in TuT subunits within the microtubule is enhanced by adjacent TuD subunits, especially if the GTPase activity is extremely low in the absence of this enhancement. In this case, the GTPase rate within a cap of TuT subunits at the microtubule's ends is low until the hydrolysis is activated by either: (a) a random GTPase event in a TuT subunit, or more likely, (b) a TuD subunit that was formed by a hydrolysis event during the nucleation phase for the assembly, or during TuT subunit addition to the microtubule end. If (a) is negligible, then the GTPase reaction will occur predominately at one or both ends of a core of TuD subunits within the body of the microtubule. Because the number of such sites remains approximately constant during microtubule assembly and at steady state (see below), the GTPase rate is expected to be about the same during the pre-steady-state and steady-state phase for assembly.

We describe here a kinetic study of GTP hydrolysis and microtubule assembly. The results are consistent with a mechanism in which the number of active GTPase sites does not change during the assembly process. Based on this result we propose a mechanism in which GTP hydrolysis occurs predominately at one or both interfaces of a core of TuD subunits and a cap of TuT subunits at the microtubule ends.

KINETICS FOR MICROTUBULE ASSEMBLY AND GTP HYDROLYSIS

The rates for microtubule assembly and the associated GTP hydrolysis are described in FIGURES 1 and 2. Following a brief lag period, the assembly rapidly proceeds to a steady state. To compare the rates for Pi release and assembly it is necessary to know the stoichiometry for guanine nucleotide incorporation into microtubules. This value is not equal to the number of moles of tubulin that is incorporated into microtubules, because following assembly with radioactive GTP, the radioactive GDP + GTP incorporated into microtubules/mole of tubulin dimer in microtubules is less than 1.0; values equal to 0.91,[10] 0.87,[13] 0.23,[14] 0.50–.64,[15] 0.8–.9,[8] 0.41–.47,[16] and 0.1[17] have been reported. To determine the number of moles of [³H]guanine nucleotide incorporated into microtubules under the conditions used here, we have carried out the assembly with [³H]GTP with varying specific activity, so that the specific activity can be corrected for the amount of E-site guanine nucleotide.[18] When microtubules were isolated by centrifugation (Airfuge, 3 min.) and the nucleotide was extracted, it was found that for the reaction shown in FIGURE 1, an amount of [³H]guanine nucleotide corresponding to 15.8 μM was in the pellet. The protein composition of the pellet corresponded to 18 μM.

As discussed above, if GTP hydrolysis is not tightly coupled to subunit incorporation, and the hydrolysis is a random first-order process, the GTPase rate will change during the course of the assembly. This is also true if the GTPase reaction is perfectly coupled to subunit addition, because the addition rate is faster during the pre-steady state phase of the reaction, as compared to when the reaction attains a steady state. There is one additional characteristic for a mechanism in which assembly and hydrolysis are perfectly coupled, and a mechanism in which assembly and hydrolysis are not coupled and microtubule-associated TuT subunits undergo first-order hydrolysis: the zero-time extrapolate of a plot of total Pi released as a function of time will correspond to the difference between the number of moles of E-site nucleotide incorporated into the microtubules ([³H]GTP + GDP, following assembly with [³H]GTP) and the number of moles of E-site GTP within subunits in the steady state microtubules; the latter value is believed to be relatively small.[7–10] By contrast, if the number of GTP hydrolysis sites does not change during the assembly and at steady

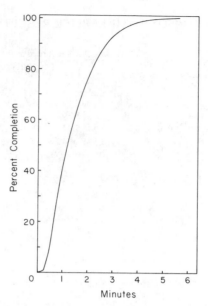

FIGURE 1. Kinetics for microtubule assembly. Assembly was measured spectrophotometrically (total absorbance change of 0.57), following the addition of 300 μM GTP to a mixture of 30 μM protein and 1 mM adenyl β-imidodiphosphate. At steady state the microtubules contained 15.8 μM [^{3}H]guanine nucleotide.

state, then the zero-time extrapolate of the steady state GTPase rate is zero. Thus, the three GTPase mechanisms (*i.e.,* assembly and hydrolysis are perfectly coupled; assembly and hydrolysis are not coupled and GTP hydrolysis is a random first-order process; assembly and hydrolysis are not coupled and the number of GTPase sites remains constant) can be distinguished by comparing the GTPase rate during the pre-steady state and steady state, and/or by determining the zero-time intercept of the steady-state GTPase rate. The latter method is preferable, because it is not ordinarily possibly to collect a sufficient number of data points of adequate accuracy during the brief pre-steady state phase of the reaction to precisely determine the pre-steady state GTPase rate. By contrast, the steady state phase of the reaction persists for a sufficient time so that the GTPase rate during this period can be unambiguously estimated. Equally important, it is possible to estimate the theoretical zero-time extrapolate of the steady-state GTPase rate from simultaneous determinations of the amount of [^{3}H]GTP + GDP incorporated into the microtubules following assembly with [^{3}H]GTP. As noted above, in the assembly reaction described in FIGURE 1, 15.8 μM [^{3}H]GTP + GDP is incorporated into microtubules. This is very much higher than the zero-time extrapolate of the steady state GTPase rate; the latter value is approximately zero, when the data points at later times are corrected for competitive inhibition by product GDP. As described above, the only mechanism that is consistent with an extrapolation of the rate to zero is one in which GTP hydrolysis is not tightly coupled to TuT subunit incorporation and the number of active GTPase sites does not change during the assembly reaction.

COMPARISON WITH EARLIER RESULTS

Although it was previously concluded that with microtubular protein the steady-state GTPase rate does not extrapolate to zero,[3] we have reanalyzed these results[18] and suggest that the previously derived nonzero intercept may result from GDP product-inhibition of the GTPase rate. Obviously, if the steady state GTPase rate is decreased by competitive inhibition, the progressively decreasing slope will generate a zero-time extrapolate different from zero. Thus, the results described here and previously[3] are consistent with a mechanism in which GTP hydrolysis is not tightly coupled to assembly and the number of GTPase sites remains constant during assembly.

With MAP-free tubulin, which was purified by chromatography on diethylamino-ethyl (DEAE) and assembled in a 4M glycerol-induced reaction, it was found that the assembly rate was faster than that for Pi release and the zero-time extrapolate of the steady state GTPase rate was equal to the moles of tubulin incorporated into microtubules.[1] If under these conditions the moles of tubulin incorporated into

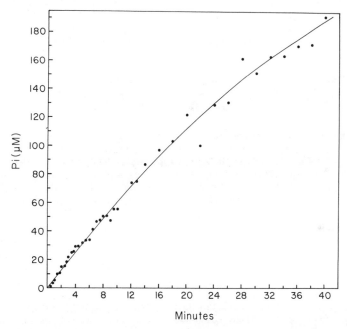

FIGURE 2. Time course for the hydrolysis of GTP by microtubular protein in the reaction shown in FIGURE 1. The solid line was calculated from the integrated rate expression describing product inhibition of a thermodynamically favorable reaction (eq. II.48 in reference 11), using K_{ms} equal to 2.2×10^{-8} M.[12] K_{mp} for GDP was taken to be 3.52×10^{-8} M, because the apparent inhibition disassociation constant for the GDP effect on the steady-state GTPase reaction is 1.6 times as large as K_{ms} for GTP.[3] V_{max} was estimated from the initial rate (0–5 min), with an appropriate correction (page 59 in reference 11) for the contamination of the initial reaction mixture by the 30 μM GDP, which is contained in the tubulin. It should be noted that the reasonably good fit of the data points to the theoretical curve means that if the results were corrected for competition inhibition, these would fit a linear plot with an extrapolate to the origin.

microtubules perfectly corresponds to moles of E-site nucleotide incorporated into microtubules (however, see above), then it may be concluded that under these conditions the mechanism is different from that in a MAP-induced assembly reaction, in that the number of GTPase sites varies during the reaction. It is not possible to implicate a similar mechanism for MAP-free tubulin purified by phosphocellulose chromatography, because with this material, the zero-time extrapolate of the rate for Pi release corresponds to 1.5–1.8 moles/mole of tubulin incorporated into microtubules.[1] We have obtained similar results in a study of the 4M glycerol-induced assembly of MAP-free tubulin dimer, purified by chromatography on Sepharose 6B (unpublished results from this laboratory). In this case the zero-time extrapolate of the steady state GTPase rate corresponds to 2.0 mole of Pi/mole of [^{3}H]GTP + GDP incorporated into microtubules. This extrapolate is not the result of a fall off in rate from competitive inhibition by product GDP, because a very high (600 μM) GTP concentration was used, so that the steady state rate was almost perfectly linear. Thus, some fraction, perhaps all, of the two moles of Pi released during the pre-steady state phase of the assembly process results from an as yet unidentified GTPase reaction. One possibility is that there is extensive GTPase activity during the nucleation process that precedes microtubule elongation. A similar process may account for the stoichiometry of 1.5–1.8 reported for phosphocellulose purified tubulin[1] and the burst-release of a stoichiometric amount of Pi seen with DEAE-purified tubulin.[1] If this is the case, then the above-described conclusions describing the mechanism for DEAE-purified MAP-free tubulin may be incorrect.

LOCATION OF THE GTPase SITE

If GTP hydrolysis is not coincident with subunit addition, it is possible to have two interfaces at each microtubule end, one between the end and solvent (*i.e.,* the tip), and one between a long core of TuD subunits constituting the body of the microtubule and a cap of TuT subunits at the end. The number of such interfaces can be expected to remain approximately constant during net microtubule assembly and at a steady state, and our observation that the GTPase rate remains approximately constant is taken to indicate that this reaction occurs predominately at one or more interface sites. We consider it unlikely that this site is at the microtubule's solvent interface, because if this were to be the principal hydrolysis site, it would be necessary to assume that this hydrolysis rate is enhanced, relative to that in a TuT subunit that is nonterminally located in the microtubule or in a free subunit in solution, by an interaction with the solvent that is not possible for free subunits. Instead, we suggest that the principal GTPase site is at the interface between a core of TuD subunits and cap of TuT subunits at one or both microtubule ends. The enhanced hydrolysis at the cap-core interface represents a "nucleation-elongation process," in which the core of TuD subunits in the microtubule alters the protein conformation of adjacent TuT subunits, so that the hydrolysis rate is enhanced. The resultant hydrolysis serves to elongate the TuD core. The enhancement in the GTPase rate provided by the TuD core would appear to be very large, as evidenced by the fact that GTP hydrolysis in free TuT subunits occurs at a negligible rate compared to that in assembling microtubules, and the fact that the enhancement is sufficient so that GTP hydrolysis is apparently coincident with TuT subunit addition, except when the TuT subunit concentration is extremely high.[1]

Alternate locations for the GTPase reaction have been proposed. For example, as described above, the observations that in the glycerol-induced assembly of DEAE-purified MAP-free tubulin, GTP hydrolysis is not coincident with microtubule assembly, and the pre-steady state GTPase rate is faster than the steady state GTPase rate, were taken to indicate that GTP hydrolysis in TuT subunits in the microtubule is a random first-order process within a TuT cap at the microtuble ends.[1] We believe, however, that this mechanism is suspect, because the steady state rate is about 67% as fast as the pre-steady state rate (FIGURE 1 in reference 1).[b] This fact is of concern because from consideration of the relative rates for the first-order incorporation of TuT subunits into microtubules and the subsequent first-order hydrolysis reaction, it was calculated that the initially formed microtubule is predominately composed of TuT subunits. (The results in FIGURE 4 of reference 19 suggest that as much as 60% of the microtubule consists of TuT subunits in the early stages of assembly.) Thus, because the steady state GTPase rate is about 67% as rapid as the pre-steady state rate, it could be concluded that about 67% of 60%, that is, 40% of the TuT subunits persist in steady state microtubules. This is, however, not believed to be the case: steady state microtubules contain only small amounts of GTP.[8,9]

A different site for the principal steady state GTPase site was proposed from subsequent studies in which it was found that the GTPase rate is dependent upon the GTP/GDP ratio in solution.[3] This dependence suggested that GDP is in rapid equilibrium with the active GTPase site. The location of the GTPase site was further delineated by the fact that added GDP is not able to exchange with GTP in TuT subunits within microtubules (FIGURE 4 in reference 3 and FIGURE 6 in reference 19). GDP exchange for GTP in the single ring of subunits at the microtubule's tips, however, could not be ruled out and it was, therefore, assumed that exchange occurs at this site. To account for the GDP inhibition of the GTPase rate it was suggested that the principal GTPase reaction occurs in TuT subunits at this location. Although this model appears to be the only one consistent with a mechanism in which GTP in TuT subunits within the microtubule is exchangeable with added GDP (*i.e.,* not ruled out by the results in references 3 and 19), these authors have analyzed an alternate model in which the principal GTPase site is located in the most internal TuT subunit in a string of TuT subunits at the microtubule end (see appendix to reference 20), and such a model was used in subsequent kinetic analyses. The basis for this new assignment is unclear.

The inhibition of the GTPase rate at low GTP/GDP ratios[3] is an important fact to be accounted for in defining the GTPase site. To this point, we have discussed this inhibition in terms of the model previously suggested; we believe, however, that this inhibition may, in fact, not represent simple competitive inhibition.[c] That is, we suggest that much of the previously observed inhibition at very high GDP concentrations results from loss of microtubule ends accompanying GDP-induced disassembly of microtubules. For example, when microtubules were polymerized with 100 μM GTP and 1.6 mM GDP was added at steady state, about 40% of the microtubule mass was disassembled (FIGURE 3 in reference 3). This disassembly is expected to result in a dramatic decrease in the microtubule number concentration, because short microtubules will fully disassemble, and as a result, the GTPase rate will be reduced. Thus, the observation that addition of 1.5 mM GDP to microtubules at steady state in the presence of 165 μM GTP results in a 67% decrease in the GTPase rate (FIGURE 2 in

[b]In an apparently equivalent reaction (FIGURE 6 in reference 19), however, the steady-state rate is only about 40% as fast as in the pre-steady-state phase of the reaction. This variability apparently results from the brevity of the pre-steady-state process, so that only limited data can be obtained.

[c]We refer here to the inhibition induced by addition of a large amount of GDP[3], not the relatively small accumulation of GDP in the assembly reaction (see FIGURE 2).

reference 3) may largely reflect a loss of microtubule ends, rather than competitive inhibition by GDP.

If, however, a significant fraction of the reported[3] GDP inhibition does not result from a decrease in microtubule ends, then we must be able to account for GDP inhibition in terms of the mechanism proposed here, in which the principal GTPase site is at the interface of a TuT cap and TuD core. As described above, this site would appear to be ruled out by the observation that GTP does not exchange into the E-site of nonterminal TuT subunits.[3,19] Our mechanism, however, differs from the one previously described[3] in that it does not include GDP exchange (or GTP exchange; see pp. 6751–6756 in reference 21) with microtubule-associated GTP. Rather, we have accounted for the effects of GDP on the assembly rate[3,6] and the steady state GTPase[3] rate by a mechanism involving TuD addition to microtubule ends, instead of GTP exchange into the microtubule. That is, to account for GDP inhibition of the assembly rate,[3,6] we propose that TuD subunits add nonproductively to microtubule ends, so as to generate an end that cannot undergo subsequent TuT subunit addition.[22] As noted previously (equation 8 in reference 3), this mechanism is kinetically indistinguishable with one in which GDP exchanges into the microtubule's terminal subunits, to generate an end that cannot undergo further elongation. To account for the effect of GDP on the steady state GTPase rate we propose that TuD subunits bind nonproductively to microtubule ends and thereby increase the time interval between productive TuT subunit additions. As a result, the hydrolysis at the cap-core interface has a greater chance of catching up with the ends, so that the fraction of ends without TuT caps is increased. An increase in the fraction of microtubule ends without TuT caps corresponds to a decrease in the number of active GTPase sites, so that the observed hydrolysis rate will be inhibited. Thus, the mechanism proposed here predicts GDP inhibition of the steady state GTPase rate, without GDP exchange into the E-site of tubulin subunit in the microtubule.[d]

The mechanism proposed here to account for the constancy of the GTPase rate (FIGURE 2) is predicated upon there being no change in the fraction of microtubule ends that are capped with TuT subunits during the reaction. One might, however, expect that even in the absence of the GDP inhibition, when the tubulin-GTP subunit concentration is reduced as the reaction reaches a steady state, the GTPase reaction will catch up with the subunit addition, so that the number concentration of microtubule ends having tubulin-GTP subunits at the end will be decreased. If this were the case, then the steady state GTPase rate would be lower than what is observed during net microtubule assembly. To evaluate this question, we have analyzed the nucleotide composition of pelleted microtubules, which had been polymerized to a steady-state condition in the presence of [³H]GTP. It was found, as reported previously,[8,9] that 5–10% of the radioactive nucleotide in the microtubules is [³H]GTP. Further evidence indicating that under steady-state conditions only a small fraction of microtubules contain fewer than one tubulin-GTP subunit at an end (and are according to the proposed mechanism, therefore, not actively catalyzing GTP hydrolysis) was derived from a theoretical analysis of this question.[22] We have derived a formula that describes the fraction of microtubule ends that do not have one or more

---

[d]We have not yet developed a quantitative analysis of GDP inhibition of the GTPase reaction for the mechanism proposed here, and the analysis of the result in FIGURE 2 in terms of simple competitive inhibition is only an approximation. In this reaction, however, the rate is linear, within experimented error, until the Pi (and GDP) concentration is 120 μM. If we consider these results only, our conclusion remains valid, that is, that the extrapolate of the rate is to zero molar Pi.

tubulin-GTP subunits at the end:

$$(f_0) = 1 - (C_1 k_{on})/(k_{off} + k_{hydrolysis}) \qquad (1)$$

C_1 represents the steady state tubulin-GTP subunit concentration; k_{on} is the second-order rate constant for tublin-GTP subunit addition; k_{off} is the first-order rate constant for subunit dissociation from the microtubule. When f_0 is close to zero, only a small fraction of the microtubule ends will have fewer than one tubulin-GTP subunit, so that GTP hydrolysis will be taking place at virtually all of the microtubule ends. In evaluating the formula for f_0, we suggest that $C_1 k_{on}$ is about equal to k_{off}, because the Wegner s-value for microtubule treadmilling ($s = C_1 k_{on} - k_{off}$ [at a given microtubule end]/$C_1 k_{on}$ [at both microtubule ends]) is very small.[23] The other requirement for f_0 to be near zero is k_{off} ($\simeq C_1 k_{on}$) $> k_{hydrolysis}$. We have determined k_{off} to be equal to about 120 sec^{-1} in studies of the dilution-induced disassembly of microtubules.[24] This rate constant represents k_{off} for tubulin-GDP subunits, because in a dilution-induced disassembly reaction the entire microtubule, which is composed primarily of tubulin-GDP subunits, undergoes disassembly. Although k_{off} for tubulin-GDP subunits is about equal to $k_{hydrolysis}$ (100 sec^{-1})[18,23] the k_{off} for tubulin-GTP subunits is significantly larger than that for tubulin-GDP subunits.[19] Thus, it would appear that k_{off} predominates in the denominator term of equation one. From this we conclude that $f_0 < 0.5$, so that the majority of steady state microtubules contains one or more tubulin-GTP subunits. As a result, the steady state and pre-steady state GTPase rates are expected to be equivalent for a mechanism in which GTP hydrolysis occurs predominately at an interface of a core of TuD subunits and a cap of TuT subunits at microtubule ends.

SUMMARY

The rate for GTP hydrolysis remains approximately constant during microtubule assembly from microtubular protein. This indicates that GTP hydrolysis does not accompany tubulin-GTP subunit addition to microtobule ends. We suggest that GTP, within tubulin-GTP subunits that are incorporated into microtubules, is hydrolyzed predominately at one or both microtubule ends at an interface of a cap of tubulin-GTP subunits and a core of tubulin-GDP subunits.

ACKNOWLEDGMENTS

The work described here was done by Bruna Pegoraro Brylawski and John Shanks. The author is grateful for helpful discussions with Ralph Reid, Barry Zeeberg, and Tom Hays.

REFERENCES

1. CARLIER, M. F. & D. PANTALONI. 1981. Biochemistry **20:** 1918–1924.
2. WHITE, H. D., B. A. COUGHLIN & D. L. PURICH. 1980. J. Biol. Chem. **255:** 486–491.
3. CARLIER, M. F. & D. PANTALONI. 1982. Biochemistry **21:** 1215–1224.
4. JAMESON, L. & M. CAPLOW. 1980. J. Biol. Chem. **255:** 2284–2292.
5. ZACKROFF, R. V., W. J. DEERY & R. C. WEISENBERG. 1980. J. Mol. Biol. **139:** 641–677.
6. ENGELBORGHS, Y. & A. VAN HOUTE. 1981. Biophys. Chem. **14:** 195–202.
7. KOBAYASHI, T. 1975. J. Biochem. (Japan) **77:** 1193–1197.

8. CAPLOW, M. & B. ZEEBERG. 1980. Arch. Biochem. Biophys. **203:** 404–411.
9. KIRSCH, M. & L. R. YARBROUGH. 1981. J. Biol. Chem. **256:** 106–111.
10. HAMEL, E., A. A. DEL CAMPO & C. M. LIN. 1983. Biochemistry **22:** 3664–3671.
11. SEGEL, I. H. 1975. Enzyme Kinetics. John Wiley and Sons. New York.
12. ZEEBERG, B. & M. CAPLOW. 1979. Biochemistry **18:** 3880–3886.
13. WEISENBERG, R. C., W. J. DEERY & P. J. DICKINSON. 1976. Biochemistry **15:** 4248–4254.
14. PENNINGROTH, S. M. & M. W. KIRSCHNER. 1977. J. Mol. Biol. **115:** 643–673.
15. MARGOLIS, R. L. & L. WILSON. 1978. Cell **13:** 1–8.
16. WILSON, L. K. B. SNYDER., W. C. THOMPSON & R. L. MARGOLIS. 1982. Methods Cell Biol. **24:** 159–169.
17. PABION, M., D. JOB & R. L. MARGOLIS. 1984. Biochemistry **23:** 6642–6648.
18. CAPLOW, M., J. SHANKS & B. P. BRYLAWSKI. 1985. Can. J. Biochem. Cell Biol. **63:** 422–429.
19. BONNE, D. & D. PANTALONI. 1982. Biochemistry **21:** 1075–1081.
20. CARLIER, M. F., T. L. HILL & Y. CHEN. 1984. Proc. Natl. Acad. Sci. USA **81:** 771–775.
21. CAPLOW, M., B. P. BRYLAWSKI & R. REID. 1984. Biochemistry. **23:** 6745–6752.
22. CAPLOW, M. & R. REID. 1985. Proc. Natl. Acad. Sci. USA. **82:** 3267–3271.
23. CAPLOW, M., G. M. LANGFORD & B. ZEEBERG. 1982. J. Biol. Chem. **257:** 15012–15021.
24. ZEEBERG, B., R. REID & M. CAPLOW. 1980. J. Biol. Chem. **255:** 9891–9899.

Several Metabolic Factors Governing the Dynamics of Microtubule Assembly and Disassembly

SUN-HEE LEE,[a] GREGORY FLYNN,[b]
PAUL S. YAMAUCHI,[b] AND DANIEL L. PURICH[b,c]

Department of Chemistry
University of California
Santa Barbara, California 93106

Cytoskeletal metabolism is a new field that presently lacks the coherence of typical metabolic pathways in which reactions may be sequentially organized by rules of organic chemistry. Thus, whereas nonmuscle cells have cytoskeletal networks comprised of such components as microtubules, intermediate filaments, and actin fibers, the interconnections and interactions among these components are far less well understood. This issue is made even more difficult to assess because cytoskeletal proteins are targets for enzymatic covalent modification. Tubulin, for example, can serve as a substrate for phosphorylation, acetylation, adenosine diphosphoribosylation, tyrosination, and proteolysis.[1,2] Likewise, other microtubule-associated proteins (MAPs), the metabolites guanosine triphosphate (GTP), guanosine diphosphate (GDP), and adenosine triphosphate (ATP), as well as a number of enzymes are currently linked, in some cases only circumstantially, to the control of the microtubule system. The cardinal difference between elucidating metabolic pathways and cytoskeletal metabolism is that the former has a chemically well defined product that submits to retro-biosynthetic analysis. By contrast, the latter involves structural and dynamic changes in the cytoplasmic architecture, and the rules for the design of such transient structures does not readily reveal the role of particular enzymes, proteins, and metabolites. Instead, the only available approaches rely on the ability of an agent to alter assembly/disassembly properties, tubule structure, or tubule interactions with other components. In this report, we consider two different modifying agents, the neurofilament protein triplet and phosphatidyl-inositol, and we discuss the involvement of a high molecular weight microtubule associated protein.

EXPERIMENTAL PROCEDURES

Microtubule protein was prepared from bovine brain by the assembly/disassembly protocols of Shelanski *et al.*[3] and Karr *et al.*[4] under hypotonic and isotonic extraction conditions, respectively. Cold-depolymerized, second-cycle protein was stored at −80°C in assembly buffer[4] containing 35% (v/v) glycerol, and a third assembly/

---

[a]Current address: Department of Biochemistry, University of California, Berkeley, California.
[b]Current address: Department of Biochemistry and Molecular Biology, Box J-245, University of Florida College of Medicine, Gainesville, Florida 32610.
[c]To whom correspondence should be sent.

disassembly cycle was performed prior to each experiment. Spinal cord neurofilament proteins were prepared as outlined elsewhere,[5] reassembled at 4°C by overnight dialysis against 25 mM imidazole, 0.15 M KCl, 5 mM magnesium sulfate, 2 mM dithiothreitol, 0.125 mM EGTA, 0.2 mM phenylmethane sulfonylfluoride (pH 7.1), maintained at 4°C in the same buffer with 2 M glycerol, and transferred into microtubule assembly buffer[4] by Sephadex G-25 gel filtration. MAPs were fractionated from tubulin on phosphocellulose chromatography;[6] heat-stable MAPs (primarily MAP-2) were obtained by the protocol of Kim et al.;[8] and salt-treated microtubule protein was prepared by the method of Delacourte et al.[7]

Polymerization was monitored turbidometrically at 350 nm (37°C) as described elsewhere.[4] Viscosity measurements were accomplished with a falling-ball viscometer constructed and calibrated as specified by Pollard.[8] Sodium dodecyl sulfate (SDS)

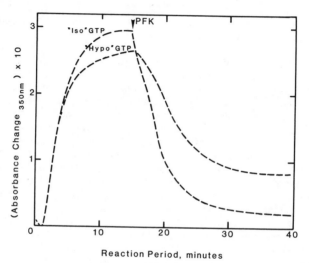

FIGURE 1. Phosphofructokinase-induced disassembly of assembled hypotonic (1.8 mg/ml) and isotonic (1.7 mg/ml) microtubules. Arrows indicate addition of PFK, and absorbance values correspond to turbidity changes.

acrylamide gel electrophoresis was carried out with 7.5% cross-linking as outlined by Laemmli,[9] and Coomassie brilliant blue stain revealed the protein banding pattern.

RESULTS

Evidence for Microtubule-Neurofilament Interactions

We have recently been interested in identifying more of the components that interact with microtubules *in vitro*. In one effort, we have attempted to compare the differences in the assembly-disassembly properties of microtubule proteins as prepared by standard hypotonic extraction protocols[3,10] with the microtubule proteins as isolated by our isotonic extraction procedure.[4] The need for such a comparative analysis

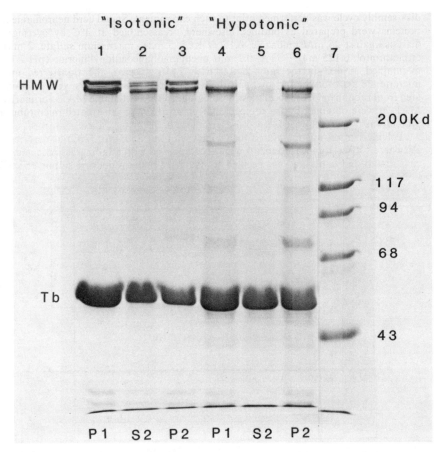

FIGURE 2. SDS gel electrophoretic analysis of hypotonic and isotonic microtubule protein before and after PFK-induced disassembly. Two identically treated samples were assembled, and aliquots were removed for centrifugation at 90,000 × g for 75 minutes. Protein samples were then analyzed by gel electrophoresis. Lanes 1, 2, and 3 correspond to the sedimented isotonic microtubules before PFK addition, the supernatant proteins after PFK addition, and sedimented proteins after PFK addition. Lanes 4, 5, 6 contain the corresponding samples with hypotonic microtubules. Lane 7 contains protein molecular weight standards.

TABLE 1. Cross-Reconstitution Experiments with Tubulin (Tb) and MAPs Fractions from Isotonic and Hypotonic Microtubule Preparations[a]

Conditions	Extent of Depolymerization
Iso-MAPs + "Hypo-Tb"	98% (95%)
Hypo-MAPs + "Iso-Tb"	68% (70%)

[a]After assembly of microtubules (1.4 mg/ml Tb and 0.8 mg/ml MAPs) in 0.2 mM GTP, 1 mM 5′-adenylyl-imidodiphosphonate (to inhibit endogenous brain nucleoside-5′-triphosphatase), and 10 mM fructose-6-P at 37°C, disassembly was induced by addition of PFK (0.3 IU/ml). The values in parentheses correspond to typical extents of disassembly in unfractionated isotonic and hypotonic protein fractions.

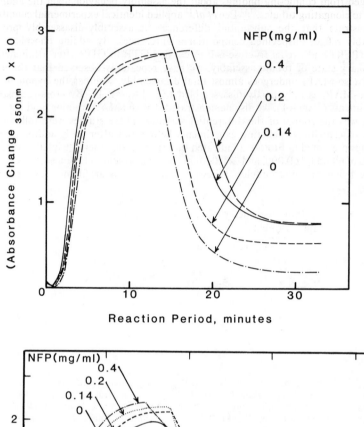

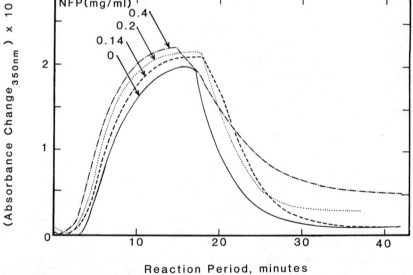

FIGURE 3. Effect of the presence of added neurofilaments on the assembly-disassembly properties of salt-treated hypotonic microtubules (*top*) and isotonic microtubules (*bottom*). Microtubule protein was 1.6 and 1.4 mg/ml, respectively, and neurofilament levels were as indicated in each panel.

emerged from conflicting findings about the assembly behavior upon the addition of GDP to elongating tubules.[11,12] Lee *et al.*[13] applied identical experimental conditions to demonstrate that the substantial differences in assembly-disassembly properties probably reflect unspecified compositional differences. By adding phosphofructokinase (PFK) and excess fructose-6-P to convert GTP to GDP after attainment of a stationary state of tubule assembly, we have repeatedly observed that the isotonic MAP (Iso-MAP) undergoes almost total disassembly, whereas the hypotonic MAP (Hypo-MAP) is only partly disassembled (See FIGURE 1). By cross-reconstitution experiments (TABLE 1), we also demonstrated that the MAP fraction, and not tubulin itself, was the source of the differential stability. These studies were followed by a series of centrifugal pelleting experiments before and after PFK addition, and the findings presented in FIGURE 2 indicated that the neurofilament "triplet" proteins (M_r values of 68,000, 160,000, and 210,000) are always present in the most stable fractions of the hypotonic microtubules. Likewise, when these neurofilaments were removed

TABLE 2. Falling-Ball Viscometry of Recombined Microtubule and Neurofilament Proteins[a]

Components Added					Apparent Viscosity (Centipoise)
Hypo-MTP	NFP	Tb	MAPs	ATP	
+	−	−	−	−	22
+	+	−	−	−	50
+	+	−	−	+	26
+	+	—	—	+[b]	10
−	−	+	+	−	87
−	+	+	+	−	Gel
−	+	+	+	+	146
−	+	+	+	+[b]	<10

[a]Hypotonic microtubule protein (Hypo-MTP), neurofilament protein (NFP), tubulin (Tb), and ATP (added at time of assembly) were 4.0 mg/ml, 0.6 mg/ml, 2.0 mg/ml, 0.8 mg/ml, and 0.25 mM, respectively, when present. Other components present were GTP (0.2 mM), 5'-adenylyl-imidodiphosphonate (AMPP (NH)P) (1 mM), additional magnesium sulfate (1 mM), and tubule assembly buffer.

[b]ATP and MAPs fraction were preincubated in the presence of 0.025 mM cyclic AMP for 15 min prior to assembly.

from three-cycle Hypo-MAP by the salt-treatment method of Delacourte *et al.*,[7] the salt-treated MAP had properties resembling Iso-MAP.

We have observed that salt-treated Hypo-MAP and untreated Iso-MAP provide an excellent means for demonstrating neurofilament-induced stabilization of microtubules. The turbidometric data in FIGURE 3 (top and bottom) are supported by SDS-gel electrophoretic and electron microscopical analysis. These results are in harmony with the observations of Runge *et al.*[14] and Leterrier *et al.*[15] Moreover, the MAP-2 component reinforces the neurofilament interaction with microtubules, but we also find that MAP-2 phosphorylation leads to substantially greater susceptibility of the microtubule-neurofilament complex to PFK-induced disassembly. To follow up on this observation, the data presented in TABLE 2 were obtained by use of the low-shear falling-ball viscometry method described by Pollard.[8] Whereas the combination of MAPs, tubulin, and isolated spinal cord neurofilaments lead to gel formation, even brief exposure to ATP leads to a much weaker interaction. The preincubation of MAPs and ATP in the presence of cyclic AMP is known to lead to MAP-2 phosphorylation,[16-18] and we observe that the viscosity of the tubulin, phosphorylated MAPs, and neurofilaments does not result in any significant cross-linking interaction between the tubule and neurofilament networks.

Evidence for Phosphatidyl-Inositol (PI) Destabilization of Microtubules in Vitro

In a search for agents that might serve to promote or inhibit microtubule formation *in vitro*, we have also turned our attention to low molecular weight metabolites. Many agents (*e.g.*, glycerol, dimethylsulfoxide, heparin, and polylysine) do alter the assembly properties of microtubules; likewise tubulin-directed drugs have both stabilizing and destabilizing actions. Yet, there has been little evidence that low molecular weight metabolites other than nucleotides can play such roles. In a series of screening tests, we have observed that PI and, to a lesser extent, other phosphatidylated substances can act as assembly inhibitors and microtubule-destabilizing agents. Because these actions

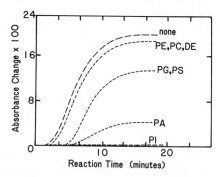

FIGURE 4. Comparison of the inhibitory action of phospholipids on the assembly of microtubule protein. Each sample contained isotonic microtubule protein (1.6 mg/ml), 0.5 mM GTP in reassembly buffer (pH 6.8). Additions were as follows: none, uppermost line; 0.14 mM phosphatidyl-ethanolamine (PE), 0.14 mM phosphatidyl-choline (PC), and 0.14 mM 1,2-diolein (DE), upper line; 0.14 mM phosphatidyl-glycerol (PG) and 0.14 mM phosphatidyl-serine (PS), middle line; 0.14 mM phosphatidic acid (PA), lower line; and 0.14 mM phosphatidyl-inositol (PI), lowest line.

were observed in the submillimolar concentration range near the levels thought to prevail within the cell, we have characterized this behavior more extensively.

As shown in FIGURE 4, addition of various phospholipids (at 140 μM) immediately prior to microtubule polymerization lead to four basic types of response. One group (including phosphatidyl-ethanolamine, phosphatidyl-choline, and 1,2-diacylglycerol (diolein)) has little effect as compared to the control sample assembled in the absence of any phospholipid. Another group (including phosphatidyl-glycerol and phosphatidyl-serine) behaves as indicated by the third curve from the top. Phosphatidic acid (fourth curve from top) exhibits stronger assembly inhibitor action, but phosphatidyl-inositol completely blocks assembly under these conditions. These observations indicate that either nucleation or elongation could be the target for such action, but the general destabilizing action is supported by the findings in FIGURE 5. Here, PI causes disassembly both during the course of elongation (upper panel) and after attainment of a stationary state of assembly. Furthermore, the critical concentration for microtubule

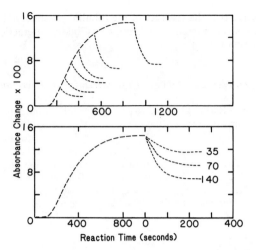

FIGURE 5. Effect of the addition of phosphatidyl-inositol (0.12 m*M*) on the elongation (upper panel) and assembled microtubule protein (lower panel). Protein concentration was 1.4 mg/ml.

polymerization was observed to depend upon the level of PI present (see FIGURE 6). With PI present, values as high as 1.5 mg/ml for the critical concentration were observed in several experiments. In companion experiments, we observed that phospholipase C could only partly relieve the PI effect, but taxol completely blocked inhibition by all of the phosphatidylated compounds.

Because PI is so potent in inhibiting tubule assembly, we investigated the effect of the tubule-stabilizing drug taxol. We observed that 20–50 micromolar levels of such

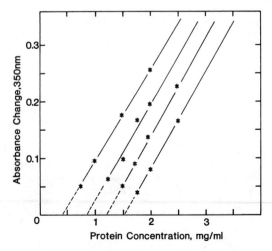

FIGURE 6. Critical concentration plots for microtubule protein at 0, 0.04, 0.08, and 0.12 m*M* phosphatidyl-inositol.

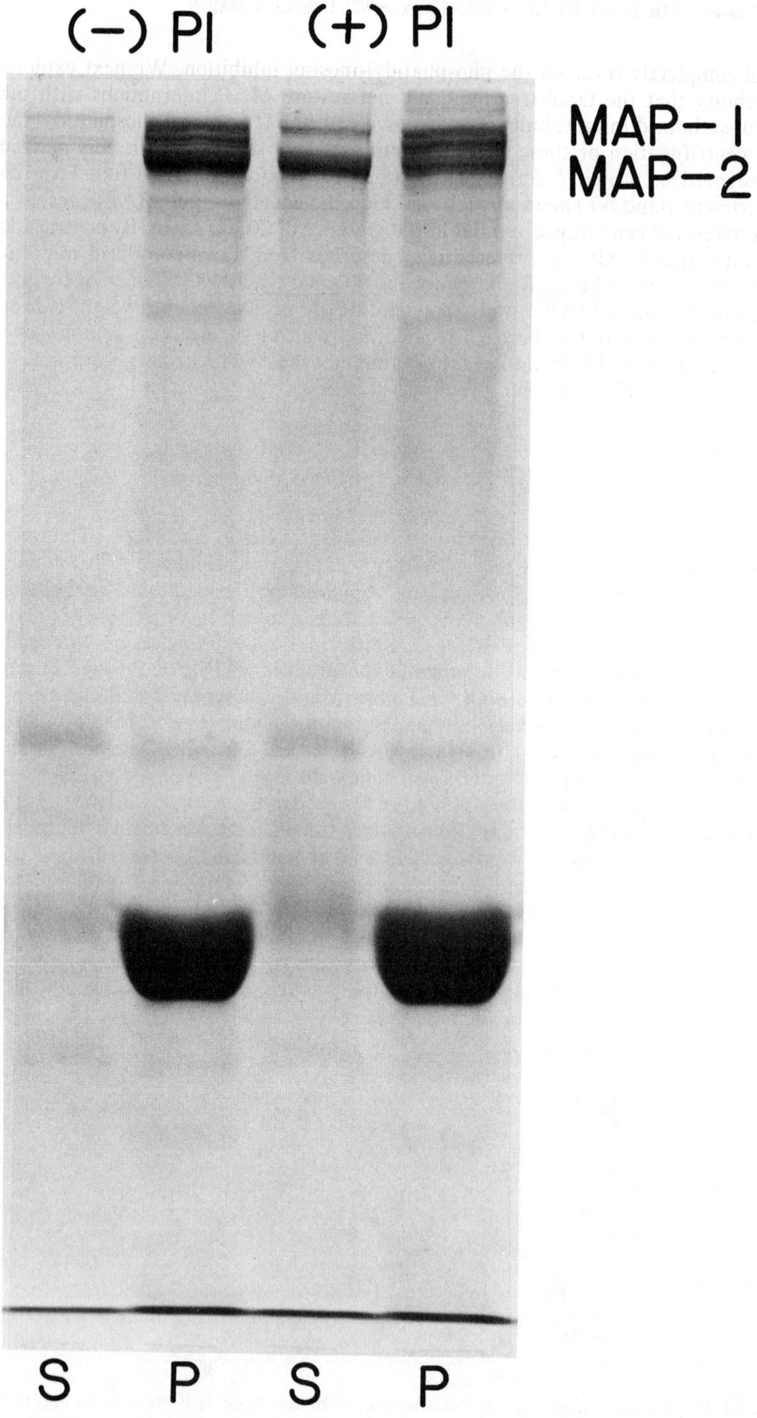

FIGURE 7. Electrophoretic analysis of supernatant (S) and pelleted (P) fractions of taxol-stabilized microtubules (1.5 mg/ml), the absence and presence of 0.12 mM phosphatidyl-inositol. Approximately 0.1 ml of supernatant and resuspended pellet fractions were applied to each lane.

taxol completely reversed the phosphatidyl-inositol inhibition. We next explored the possibility that the taxol-treatment did not restore MAP interactions with tubules. Taxol-stabilized microtubules were exposed to 120 μM levels of phosphatidyl inositol, and centrifugation of these samples permitted us to analyze both polymerized and unpolymerized protein fractions by SDS gel electrophoresis. In FIGURE 7, the control experiment (lane A) shows virtually no high molecular weight MAPs present in the supernatant of centrifuged tubules in the presence of 20 μM taxol. By contrast, lane C indicates that MAP-2 is preferentially desorbed from taxol-stabilized microtubules upon PI addition. (Lanes B and D are the polymer fractions sedimented by ultracentrifugation.) Some MAP-1 and a polypeptide of lower molecular weight (M_r around 70,000) are observed in both PI-treated and control supernatant samples of taxol-stabilized tubules. This reinforces the notion that the PI effect is largely directed at the MAP-2 interaction with microtubules.

DISCUSSION

Discovering the interconnections among cytoskeletal and other cellular components remains a challenging task. Whether cells are responding to a global realignment of cytoskeletal components (as in mitosis or meiosis) or responding to local changes (as in individual granule or organellar contact with tubules), there are likely to be many different control signals. Likewise, the status of the cytoskeleton may depend upon cellular metabolic conditions; we already know that the GTP/GDP ratio, for example, can profoundly alter microtubule assembly/disassembly behavior. Identifying other signals might be facilitated by an analysis of cytoskeletal systems in mutant cell lines with altered metabolic and/or structural properties. Until then, we are much more likely to rely on serendipity, as was apparently the case in the discovery of the original conditions for promoting tubule assembly.[10]

The neurofilament and phosphatidyl-inositol effects represent two opposing types of interaction. The former imposes a state of greater stability toward tubule depolymerization agents, whereas PI decreases tubule stability. Whereas there may be controversy on the significance of these interactions in cell processes, we are in the process of analyzing the PI effect on microtubule-neurofilament network formation. We are also interested in manipulating cellular PI metabolism to investigate the impact of PI in MAP-2-mediated connections. Likewise, we wish to learn whether PI acts by binding to MAP-2 or the microtubule binding sites for MAP-2. The answer to this question may offer insights about the importance and generality of such interactions.

REFERENCES

1. TERRY, B. J. & D. L. PURICH. 1982. Adv. Enzymol. Relat. Areas Mol. Biol. A. Meister, Ed.: **53:** 113–161. J. Wiley. New York.
2. PURICH, D. L. & R. M. SCAIFE. 1985. Curr. Top. Cell. Regul. R. L. Horecker & E. R. Stadtman, Eds.: **27:** 107–116.
3. SHELANSKI, M., F. GASKIN & C. R. CANTOR. 1973. Proc. Natl. Acad. Sci. USA **70:** 765–768.
4. KARR, T. L., H. D. WHITE & D. L. PURICH. 1979. J. Biol. Chem. **254:** 6107–6111.
5. CHIU, F.-C. & W. T. NORTON. 1982. J. Neurochem. **39:** 1252–1260.
6. WEINGARTEN, M. D., A. H. LOCKWOOD, S. Y. HWO & M. W. KIRSCHNER. 1975. Proc. Natl. Acad. Sci. USA **72:** 1858–1862.

7. DELACOURTE, A., G. FILLIATREAU, F. BOUTTEAU, G. BISERTE & J. SCHREVEL. 1980. Biochem. J. **191:** 543–546.
8. POLLARD, T. D. 1982. Methods Cell Biol. L. Wilson, Ed.: **24:** 31–312. Academic Press. New York.
9. LAEMMLI, U. K. 1970. Nature (London) **227:** 680–685.
10. WEISENBERG, R. C. 1972. Science **177:** 1104–1105.
11. WEISENBERG, R. C., W. J. DEERY & P. J. DICKINSON. 1976. Biochemistry **15:** 4248–4254.
12. KARR, T. L., A. E. PODRASKY & D. L. PURICH. 1979. Proc. Natl. Acad. Sci. USA **76:** 5475–5479.
13. LEE, S.-H., D. KRISTOFFERSON & D. L. PURICH. 1982. Biochem. Biophysics Res. Commun. **105:** 1605–1610.
14. RUNGE, M. S., T. M. LAUE, D. A. YPHANTIS, M. R. LIFSICS, A. SAITO, M. ALTIN, K. REINKE & R. C. WILLIAMS JR. 1981. Proc. Natl. Acad. Sci. USA **78:** 1431–1435.
15. LETTERIER, J.-F., R. K. H. LIEM & M. L. SHELANSKI. 1982. J. Cell. Biol. **95:** 982–986.
16. SLOBODA, R. D., S. A. RUDOLPH, J. L. ROSENBAUM & P. GREENGARD. 1975. Proc. Natl. Acad. Sci. USA **71:** 4472–4476.
17. COUGHLIN B. A., H. D. WHITE & D. L. PURICH. 1980. Biochem. Biophysics Res. Commun. **92:** 89–94.
18. THEURKAUF, W. E. & R. B. VALLEE. 1983. J. Biol. Chem. **258:** 7883–7886.

Mechanism of Assembly of Sea Urchin Egg Tubulin[a]

H. WILLIAM DETRICH III[b] and MARY ANN JORDAN[c]

[b]*Department of Biochemistry*
The University of Mississippi Medical Center
Jackson, Mississippi 39216-4505
and
[c]*Department of Biological Sciences*
University of California
Santa Barbara, California 93106

INTRODUCTION

The assembly and disassembly of cytoplasmic microtubules *in vivo* are spatially and temporally coupled to microtubule-dependent processes such as mitosis, nerve growth and regeneration, cytoplasmic transport, and the maintenance of cell shape.[1,2] Thus, the mechanisms of microtubule assembly, both *in vitro* and *in vivo,* have received considerable attention. Microtubule assembly *in vitro* displays the properties of a nucleated condensation polymerization reaction.[3] The detailed molecular mechanism of microtubule assembly, however, is not completely understood (reviewed by Correia and Williams[4]). For example, estimates of the number of tubulin dimers necessary to form the critical nucleus for polymer initiation vary from two to twelve.[5-8] Ring-shaped oligomers composed of tubulin and microtubule-associated proteins (MAPs) have been postulated to function as nucleation centers,[9,10] yet solutions of pure tubulin lacking rings will assemble microtubules *in vitro.*[11-14] Furthermore, Mandelkow *et al.*[15] demonstrated by X-ray diffractometry that rings break down into smaller structures prior to the start of microtubule assembly. Both Erickson and Pantaloni[16] and Voter and Erickson[8] suggest that sheet structures formed directly from tubulin dimers may function as the critical nuclei for microtubule assembly.

Most studies of microtubule assembly *in vitro* have been performed with microtubule proteins (tubulin plus MAPs) from vertebrate brain tissue[3,9,10,15] or with pure brain tubulins in the presence of high concentrations of nonphysiological solvents such as glycerol or dimethyl sulfoxide (DMSO).[6,7,11,17] Thus, differences in the protein preparations and solvent conditions employed may be responsible for the apparent discrepancies in assembly mechanisms described previously. By contrast, we have reported that highly purified tubulin from eggs of the sea urchin *Strongylocentrotus purpuratus* assembles readily to form microtubules in the absence of ring oligomers, MAPs, glycerol, DMSO, or high concentrations of magnesium ions.[14] Furthermore, we have observed an overshoot in the turbidity generated during the assembly of egg tubulin that is correlated with changes in the form and in the organization of the

[a]This work was supported by research Grants from the U.S. Public Health Service (NS13560) to L. Wilson and (GM23568) to R. C. Williams Jr., from the National Science Foundation (DPP-8317724) to H. W. Detrich, and by the Vanderbilt University Natural Sciences Fund (to R.C.W.). H. W. Detrich was also supported by U.S. Public Health Service National Research Service Award GM07182 (1980–1982).

products of assembly. Our results suggest that the assembly of microtubules from egg tubulin proceeds through, or in competition with, formation of tubulin sheets.

MATERIAL AND METHODS

Purification of Tubulin

Sea urchin egg tubulin was prepared from unfertilized eggs of *Strongylocentrotus purpuratus* as previously described.[14] Briefly, eggs were homogenized in PMI buffer [0.1 M piperazinediethanesulfonic acid (PIPES)-NaOH, 2 mM ethylene glycol bis (β-aminoethyl ether)-N, N, N', N'-tetraacetic acid (EGTA), 2 mM dithioerythritol, 1 mM MgSO$_4$, 1 mM p-tosyl-L-arginine methyl ester (TAME) (Sigma, St. Louis, Mo.), 0.1 mM guanosine 5'-triphosphate (GTP), and 0.02% (w/v) NaN$_3$, pH 6.82], and the homogenate was centrifuged at 210,000 g for 100 min at 4°C to prepare a high-speed supernatant (HSS). Tubulin was isolated by ion-exchange chromatography of the HSS on diethylaminoethyl (DEAE)-Sephacel.[14]

The partially purified tubulin obtained by ion-exchange chromatography (DEAE-tubulin) was further purified by n cycles of temperature-dependent polymer assembly and disassembly, where n = 1, 2, or 3. Solutions of egg tubulin in PB buffer [0.1 M PIPES-NaOH, 1 mM EGTA, 1 mM MgSO$_4$, and 0.02% (w/v) NaN$_3$, pH 6.82] containing 1 mM GTP were warmed to 37°C until assembly was complete as judged by turbidimetry, and assembled polymer was collected by centrifugation (40,000 g, 30 min, 35°C), producing a supernatant (H$_n$S) (discarded) and a microtubule-containing pellet (H$_n$P). The pellets from each cycle of assembly were resuspended in PB + 1 mM GTP, the resuspended polymer was depolymerized by incubation at 0°C for 30 min, and the depolymerized protein was centrifuged (40,000 g, 30 min, 4°C) to produce a cold pellet (C$_n$P) (discarded) and a cold supernatant (C$_n$S). The C$_n$S was then carried through further assembly-disassembly cycles or was frozen dropwise in liquid nitrogen. Unless otherwise noted, all experiments were performed with twice-cycled tubulin (C$_2$S).

Assembly Conditions

Experiments were performed in PB buffer containing GTP (0.1 or 1 mM) and a GTP-regenerating system (20 mM acetyl phosphate, 0.15 IU/ml acetate kinase),[18] except where noted.

Turbidimetric Measurement of Assembly

The assembly of sea urchin egg tubulin was monitored turbidimetrically at 350 nm with Cary Model 118c or Gilford Model 2400 recording spectrophotometers equipped with thermostatable cuvette chambers. Assembly reactions were started by rapid delivery of cold (0°C) solutions of tubulin into cuvettes prewarmed to the appropriate temperature.

Small-Angle Light Scattering Studies

Scattering of light at a solid angle of 6–7° by assembling solutions of egg tubulin was measured by the use of a Chromatix KMX-6 photometer (LDX/Milton-Roy,

Riviera Beach, Fla.) equipped with a thermostated sample cell. For a detailed description of these methods, see Williams.[19] Briefly, the scattering cell was carefully rinsed with Millex-GV-filtered (Millipore Corporation, Bedford, Mass.) PB buffer until fewer than three dust particles per ml could be detected. The tubulin sample was degassed at 0°C, introduced into a syringe, and placed in a refrigerated (4°C) syringe pump. The cold tubulin solution was then pumped into the prewarmed (18 or 37°C) 200-microliter scattering cell at 0.75 ml/min through the same Millex-GV filter employed for the buffer wash. Internal cell temperature was monitored by a thermistor. Scattering intensity at 6–7° was recorded on a strip-chart recorder, and calibrated neutral-density filters were introduced, as necessary, into the beam to maintain a measurable output. Measured intensities were converted to the Rayleigh ratio, R_θ, by $R_\theta = I_\theta/(I_0\sigma l)$, where I_0 is the intensity of the incident beam and I_θ is the intensity of the scattered light, both corrected for the filters. The factor σl is a correction for the scattering volume, accounting for the presence of a given field stop in the scattered beam.[20,21]

Electron Microscopy

Aliquots (100 μl) of egg tubulin (1.4–1.8 mg/ml in PB + 0.1 mM GTP + the GTP–regenerating system) were incubated separately at 37°C to initiate assembly. At intervals, individual aliquots were fixed for negative-stain electron microscopy by dilution (1:10 to 1:50) with aqueous glutaraldehyde (37°C) to produce a final glutaraldehyde concentration of 0.25 percent. Following 30 sec of fixation, single drops of a sample were deposited on collodion- or formvar-coated grids for 30 sec, excess solution was withdrawn with filter paper, and each grid was sequentially rinsed with cytochrome C (one drop for 15 sec) and with water (three drops). Excess water was drawn off with filter paper, and the samples were stained for 20–30 sec with 1% (w/v) aqueous uranyl acetate. Finally, the uranyl acetate solution was removed with filter paper, and the grids were allowed to air dry. For thin-section electron microscopy, aliquots of assembling egg tubulin were fixed in a solution of 0.3 to 1.1% glutaraldehyde and 6.4% tannic acid in 0.15 M sodium phosphate buffer (pH 6.7) for 30 min in suspension; occasionally, polymer pellets obtained by centrifugation (50,000 × g, 30 min, 25°C) were fixed overnight with the same solution. After washing once in phosphate buffer, pellets were postfixed in 1% OsO_4 in 0.15 M phosphate buffer or 0.1 M cacodylate buffer for one hour, dehydrated in acetone, embedded in Araldite, sectioned, and stained with methanolic uranyl acetate followed by lead citrate. Samples were observed with a Philips EM 300 electron microscope operated at 80 kV.

Electrophoresis

Sodium dodecyl sulfate-polyacrylamide gel electrophoresis (SDS-PAGE) was performed by the method of Laemmli[22] on slab gels containing linear gradients of acrylamide (4–16%) and of urea (1–8 M).[23] The gels were subsequently fixed and stained with Coomassie brilliant blue R-250.[14] For quantitative studies, gradient slab gels were fixed and stained with 1% Fast Green[24] in 50% methanol-7% acetic acid (v/v) for 24 h and destained in 5% methanol-7% acetic acid (v/v).[25] Appropriate lanes were cut out of the slabs and scanned at 640 nm with a Gilford Model 2400 spectrophotometer equipped with a linear transport. The amount of protein in a given electrophoretic band was determined by integration of the corresponding peak in the resulting scan.

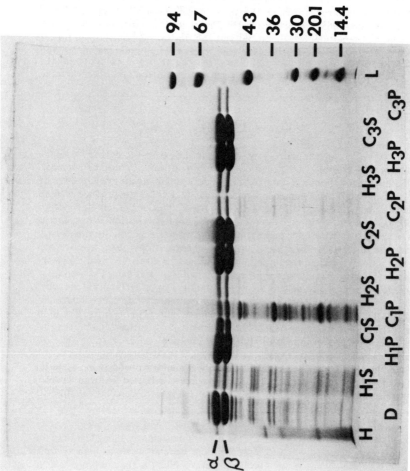

FIGURE 1. Electrophoretic analysis of fractions from three cycles of assembly and disassembly of DEAE ion-exchange-purified egg tubulin. Samples were electrophoresed in the presence of sodium dodecyl sulfate on urea-polyacrylamide gradient gels (top to bottom = 4–16% acrylamide, 1–8 M urea), and the gels were stained with Coomassie brilliant blue R-250 (see MATERIAL AND METHODS). Electrophoretic migration was from top to bottom. The molecular weights of standards (lanes H, L) are given in thousands on the vertical axis, and the approximate positions of the α and β tubulins are noted. Abbreviations: D=desalted DEAE-tubulin; H_nS, H_nP, C_nS, C_nP=purification fractions obtained after n cycles of assembly/disassembly, where H denotes fractions from 35°C centrifugations, C denotes fractions from 4°C centrifugations. S denotes supernatant, P denotes pellet.

RESULTS

Characteristics of Purified Sea Urchin Egg Tubulin

Tubulin was purified from unfertilized eggs by DEAE ion-exchange chromatography followed by cycles of polymer assembly and disassembly *in vitro* (*cf* references 14, 26). The protein compositions of the partially purified DEAE-tubulin and of the supernatants and pellets from three consecutive cycles of assembly and disassembly were analyzed by electrophoresis in the presence of sodium dodecyl sulfate on urea-polyacrylamide gradient gels (FIGURE 1). Proteins corresponding to the α and β tubulins (apparent molecular weights of 60,700 and 56,100, respectively, on this gel system) were progressively enriched in the warm pellet (H_nP) and cold supernatant

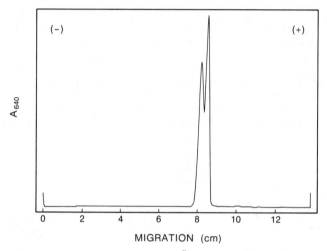

FIGURE 2. Densitometric scan of a Fast Green-stained, sodium dodecyl sulfate-urea-polyacrylamide gradient gel of an egg tubulin C_2S preparation (43 μg). Absorbance at 640 nm (arbitrary units) is plotted against distance of migration. The top and bottom of the gel are indicated by the discontinuities at the left (0 cm) and right (*ca* 14 cm), respectively. The dye front was located at about 13 cm.

(C_nS) fractions. Trace amounts of seven proteins with molecular weights less than the tubulin chains were detected when C_2S preparations were analyzed at high loadings (>50 μg) on these gels. No associated proteins corresponding to the tau proteins or to the high molecular weight MAPs of vertebrate brain,[27-29] however, have been observed in tubulin from sea urchin eggs.

The purity of six different C_2S preparations was examined by quantitative densitometry of gradient slab gels stained with Fast Green. The scan from one of these samples is presented in FIGURE 2. Proteins of molecular weight greater or less than the tubulin chains were not detected with this dye at loadings as high as 43 μg total protein. Based on the sensitivity of the Fast Green stain (*ca* 0.5 μg), we estimate that the α and β tubulins constitute greater than 98% of the C_2S.

When an egg tubulin C_3S was examined by two-dimensional electrophoresis, two α

tubulins (pI $\simeq$ 5.5) and a single β species (pI $\simeq$ 5.35) were observed.[14] Thus, the chemical complexity of tubulin from sea urchin eggs is quite low.

Assembly Characteristics

Microtubule assembly systems *in vitro* are characterized by a "critical" concentration of subunit proteins necessary to support polymerization. Below the critical concentration assembly does not occur, whereas above the critical concentration, the extent of polymerization is a linear function of the total protein concentration. We have shown that the critical concentration for assembly of purified egg tubulin (C_2S) is 0.12–0.15 mg/ml at 37°C[14] and 0.71 mg/ml at 18°C.[30] These values are similar to those reported by Suprenant and Rebhun.[26] Furthermore, the ultimate products of the assembly reaction are complete microtubules (FIGURE 6c, and d). Therefore, sea urchin egg tubulin assembles to form microtubules *in vitro* both at physiological and nonphysiological temperatures and at physiological protein concentrations.

When examined by turbidimetry, egg tubulin displayed an unusual, concentration-dependent overshoot in the scattering signal at concentrations above approximately 0.5–0.6 mg/ml. FIGURE 3 presents the temporal evolution of turbidity for three concentrations of egg tubulin (C_1S) at 37°C. No overshoot was observed at 0.46 mg/ml; turbidity developed slowly and reached a plateau at approximately 90 minutes. At a tubulin concentration of 0.69 mg/ml, turbidity increased to a peak at approximately 15 min, then declined slightly during the next 140 minutes. By contrast, a prominent overshoot in turbidity was observed at the highest tubulin concentration

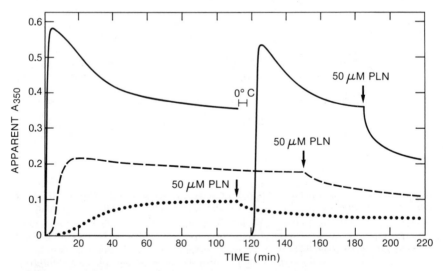

FIGURE 3. Assembly of sea urchin egg C_1S observed by turbidimetry. Three samples of egg C_1S in PB plus 1 mM GTP (no regenerating system) were warmed from 0 to 37°C at zero time, and assembly was monitored as the apparent absorbance at 350 nm. Protein concentrations: solid line, 1.4 mg/ml; dashed line, 0.69 mg/ml; dotted line, 0.46 mg/ml. The sample at 1.4 mg/ml was incubated on ice for 5 min where indicated. Podophyllotoxin (PLN) was added (arrows) to a final concentration of 50 μM. (H. W. Detrich & L. Wilson.[14] With permission from *Biochemistry*.)

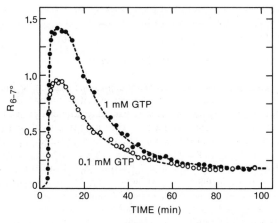

FIGURE 4. Assembly of urchin egg C_2S observed by small-angle light scattering. Samples of tubulin (1.0 mg/ml) in PB plus GTP (1 mM or 0.1 mM, as designated) plus the GTP-regenerating system were warmed from 0 to 37°C at zero time. The Rayleigh ratio observed at a solid angle of 6–7° ($R_{6-7°}$) is plotted on the ordinate.

(1.4 mg/ml). The turbidity generated by this sample increased rapidly to a peak upon warming, then decreased asymptotically toward a steady state value. Following brief exposure of the sample at highest concentration (1.4 mg/ml) to low temperature (0°C), the pattern of development and loss of turbidity during a second assembly reaction was nearly identical with that observed during the first assembly (FIGURE 3). Finally, each sample apparently depolymerized following addition of the antimitotic drug podophyllotoxin (arrows).

FIGURE 4 shows the assembly of egg tubulin (1.0 mg/ml, 37°C, GTP concentrations as indicated) observed by light scattering at a solid angle of 6–7°. Both samples displayed a pronounced overshoot in light scattering (5–8-fold), and both attained identical plateau values of the Rayleigh ratio by 100 minutes. Because a GTP-regenerating system was present in each sample, the overshoot cannot be attributed to polymer instability due to rapid accumulation of GDP during assembly. It is interesting, however, to note that the extent of the overshoot was dependent upon the concentration of GTP. This observation is not understood at the present time.

To determine whether a change in polymer mass during assembly was responsible for the overshoots in turbidity or in light scattering we employed a quantitative sedimentation assay.[30] Briefly, aliquots of twice-cycled egg tubulin (1.6 mg/ml in PB + 1 mM GTP) were incubated at 37°C for various intervals, and polymer of greater than 350 S was rapidly separated from unassembled tubulin by sedimentation in a Beckman Airfuge. Pelletable mass was found to increase rapidly to a stable plateau value; the assembled mass did not parallel the overshoot in turbidity observed for an identical companion sample. Thus, the overshoot phenomenon must result from changes in particle shape or in the organization of polymer in solution.

The wavelength-dependence of turbidity can yield important information concerning the shapes of scattering particles.[31,32] For example, arbitrarily long rods (i.e., microtubules) scatter light in proportion to λ^{-3},[33,34] whereas the wavelength exponent for scattering by large sheets is expected to be -2.[35] When we examined the wavelength-dependence of the turbidity produced by assembling solutions of egg tubulin, the wavelength exponents were observed to change from values near -2.5 during the initial stages of assembly to values somewhat more negative than -3 at later times.[30] Thus, these changes are consistent with, but do not prove, the initial formation of tubulin sheets followed by relaxation of the sheets toward microtubules.

Characterization of the Products of Assembly by Electron Microscopy

FIGURES 5 and 6 present representative electron micrographs of the polymers formed during assembly of egg tubulin at high concentration (1.4–1.8 mg/ml) and high temperature (37°C), conditions characterized by a pronounced overshoot in assembly-dependent turbidity (cf FIGURE 3). Nine seconds following the initiation of assembly, numerous small, sheet-like structures were observed (FIGURE 5a). Frequently, the sheets were branched in appearance (not shown). At intermediate times (45 sec–10 min) the sheets increased considerably in length (cf FIGURE 6a). When viewed in cross section, C-shaped profiles and S-shaped structures (wavy sheets) up to 180 nm in width predominated (FIGURES 5b, c); few closed microtubules were observed. Occasionally, wider sheets (up to 1 μm in profile length) lacking the wavy appearance (FIGURE 5c) were present. By 30 min, during the decay of the overshoot in turbidity, the major structures observed were microtubules, although hybrid structures composed of regions of open sheet and closed microtubule were also found (FIGURE 6b). Finally, following attainment of the steady-state level of turbidity at 120 min, almost 97% of the structures present were singlet microtubules (counted in thin sections); the remainder were doublet microtubules or C-shaped sheets (FIGURES 6c, d). Clearly, a structural transformation in the products of assembly from sheets to microtubules accompanies the overshoot in turbidity.

We have also measured the average length of representative samples of the polymers formed during assembly of egg tubulin at 18°C and at 37°C. Concentrations of egg tubulin both below and above the threshold for overshoot at each temperature were employed. In each case, average polymer length increased throughout the course of assembly.[30] For example, when egg tubulin (2.0 mg/ml) was assembled at 37°C, the average lengths of the formed elements present 9 sec, 4 min, 30 min, and 90 min after initiation of polymerization were approximately 1 μm, 3 μm, 7 μm, and 17 μm, respectively. These data, together with the results of the quantitative sedimentation assay, demonstrate that the population of tubulin polymers increases dramatically in mean length at constant polymer mass.

Polymer Alignment during Assembly

When solutions of assembling egg tubulin at high protein concentration were placed between crossed polarizers, birefringent striations were observed to develop at late times in the assembly reaction.[30] This observation is consistent with the formation of regions of partially aligned microtubules during the assembly process. As discussed below, progressive alignment of microtubule polymer in solution may also contribute to the overshoots in turbidity and in light scattering observed during assembly of egg tubulin.

DISCUSSION

We have isolated tubulin from unfertilized sea urchin eggs by DEAE ion-exchange chromatography followed by cycles of temperature-dependent assembly and disassem-

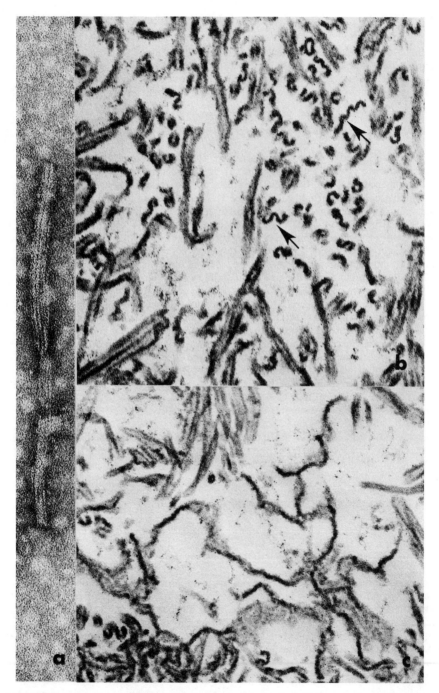

FIGURE 5. Electron micrographs of polymers formed during the initial stages of assembly of egg tubulin. Aliquots of egg tubulin (1.4–1.8 mg/ml in PB + 0.1 m*M* GTP + the GTP-regenerating system) were warmed from 0 to 37°C at zero time. At the times given below, individual aliquots were processed for negative-stain (**a**) or thin-section (**b** and **c**) electron microscopy as described in MATERIAL AND METHODS, (**a**) Sheet-like element formed during the first nine seconds of assembly, × 190,000. (**b**) Transverse sections (arrows) through wavy sheets (multiples of "C-" or "S-" shaped structures) present in preparation fixed 45 sec after the initiation of assembly, × 100,000. (**c**) Slightly oblique transverse section through large sheet polymers. Same preparation as (**b**), × 85,000.

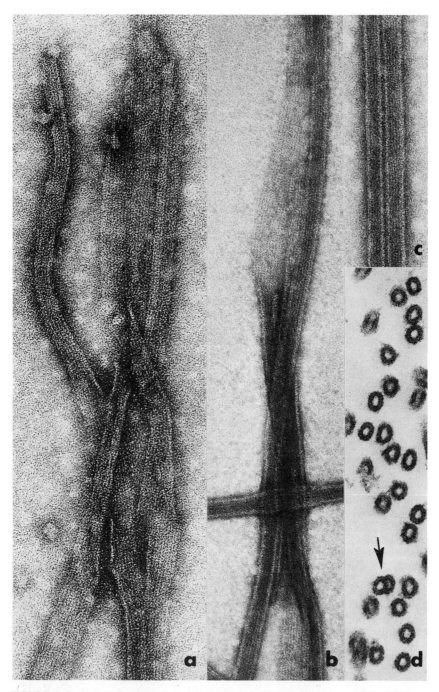

FIGURE 6. Representative formed elements present during decay of the turbidity overshoot. Aliquots of egg tubulin (see legend to FIGURE 5) were incubated at 37°C for the intervals indicated below, and then processed for negative-stain (**a–c**) or thin-section (**d**) electron microscopy. (**a**) Branched sheet polymer, preparation fixed 4 min after start of assembly, × 200,000. (**b**) Hybrid sheet-microtubule polymer, sample prepared 30 min after initiation of assembly, × 200,000. (**c**) Microtubules, sample fixed 120 min after initiation, × 190,000. (**d**) Cross sections of singlet microtubules and of one apparent doublet, sample prepared 120 min after start of assembly, × 200,000.

bly *in vitro*. Twice-cycled egg tubulin contains the α and β tubulins (>98% of the total protein) and trace amounts of several low molecular weight (<55,000) proteins.[14] No associated proteins comparable to the high molecular weight MAPs[28,29] or to the tau proteins[27] of vertebrate brain have been observed in these preparations.[14] Most, if not all, of the low molecular weight proteins present in the C_2S represent breakdown products of tubulin produced when tubulin samples are heated to 100°C in sodium dodecyl sulfate electrophoresis sample buffer.[14] Thus, the purity of our C_2S preparations is probably greater than that stated here. We have chosen to analyze microtubule assembly in this simple, MAP-free system in an effort to determine the principal apparent intermediates on the pathway to microtubule assembly.

The Assembly Overshoot

Detrich and Wilson[14] and Suprenant and Rebhun[26] have shown previously that purified egg tubulin assembles efficiently *in vitro* to form microtubules in the apparent absence of MAPs, ring oligomers, and high concentrations of nonphysiological solvents. Furthermore, we have observed a striking overshoot in assembly-dependent light scattering; at concentrations above 0.5 mg/ml at 37°C and above 1.2 mg/ml at 18°C the assembly of egg tubulin is accompanied by overshoots in turbidity and in light scattering at small angles.[30] Similar overshoots have been observed during assembly of other pure tubulins,[13,26,36,37] but the phenomenon has received little study.

Four potential causes for the overshoot can be eliminated based on the results of our studies. First, because repeated assembly reactions produce nearly identical overshoots in turbidity (FIGURE 3), the phenomenon cannot be due to proteolysis or denaturation of the tubulin. Second, the overshoot does not result from polymer instability produced by rapid depletion of GTP during assembly; the presence of a GTP-regenerating system has little, if any, effect on the magnitude of the overshoot.[14,30] Third, quantitative sedimentation analysis demonstrated that the overshoot is not due to changes in the mass of tubulin present in polymers.[30] Finally, changes in the distribution of microtubule lengths cannot contribute to the overshoot phenomenon. The last point deserves additional comment.

Overshoot in light scattering has also been observed in the temperature-induced polymerization of tobacco mosaic virus protein (TMVP).[38,39] During the initial stages of the TMVP-overshoot, a comparatively small number of long rods (up to 200 nm) form; at later times the polymer population relaxes toward a comparatively large number of small polymers (10–30 nm). Because the long rods scatter more light per unit mass than the short rods do, the turbidity overshoot reflects the transient formation and disappearance of the longer polymers. Two considerations render unlikely the contribution of a similar length-redistribution mechanism to the overshoot observed with egg tubulin. First, average polymer length increased monotonically throughout the course of assembly reactions performed at 18°C and at 37°C.[30] Second, the sizes of the assembled particles observed during the overshoot of egg tubulin (>5,000 nm) fall beyond the range where turbidity and light scattering are sensitive to rod length.[30] Therefore, other explanations must be sought for the observed overshoot.

Changes in Polymer Shape

Changes in the scattering form factors due to changes in polymer shape may be responsible for the turbidity and light scattering overshoots. The micrographs presented in FIGURES 5 and 6 demonstrate that sheet-like structures predominate early in the assembly reaction (*i.e.,* during the increase in scattering to its peak value), whereas the percentage of microtubules approaches 100% as the scattering signal declines toward its steady-state value. Sheet-like forms have frequently been observed as intermediates during polymerization of microtubule proteins,[9,10,15] and they have been invoked as intermediates in theoretical treatments of the mechanism of microtubule assembly.[16,40,41] Furthermore, changes in the wavelength-dependence of turbidity are consistent with the initial formation of tubulin sheets and the subsequent relaxation of the sheets toward microtubules.

Whereas the preceding arguments are compelling, formation and disappearance of sheet-like intermediates *per se* may not be the only process contributing to the overshoot. The reason for this uncertainty is that most of the sheets observed by electron microscopy (FIGURES 5 and 6) are less than 200 nm wide. Because variations in width in the range below the wavelength of the scattered light (633 nm for the small-angle light scattering studies) have little effect on the light scattering form factor,[42] the scattering properties of the ribbons may not be sufficiently different from those of microtubules to produce all of the large overshoot in light scattering at small angles. However, many of the intermediates observed during the assembly overshoot were complex structures composed of several ribbons in lateral association (FIGURE 6a). Aggregates of laterally associated or branched sheets could scatter more light per unit assembled mass than do populations of microtubules. Moreover, we note that some of the sheet structures formed by egg tubulin were quite wide (FIGURE 5c). The large sheets would be expected to exhibit light-scattering properties different from those of microtubules. Therefore, we suggest that the transient presence of ribbon aggregates and/or large sheets may also contribute to the overshoots in turbidity and in light scattering.

Polymer Alignment

We have found that solutions of egg tubulin develop birefringent striations as assembly proceeds; this observation is consistent with the progressive alignment of microtubule polymer to produce "bundles" of nearly parallel rods. Such ordering would reduce the independence of motion of individual microtubules, thereby decreasing the intensity of light scattering at all angles. Thus, polymer alignment may contribute to the decline of the turbidity and light scattering signals toward their steady-state values.

SUMMARY

Tubulin purified from eggs of the sea urchin *Strongylocentrotus purpuratus* assembles efficiently *in vitro* to form microtubules at physiological (18°C) and nonphysiological (37°C) temperatures. MAPs, ring oligomers, and high concentrations of nonphysiological solvents are not required for the assembly reaction. At concentrations above 1.2 mg/ml at 18°C and 0.5 mg/ml at 37°C a concentration-dependent overshoot in turbidity and in light scattering at small angles was observed: turbidity and scattering increased rapidly to a peak, then decreased asymptotically toward a steady-state value. Electron microscopic analysis demonstrated that tubulin sheets were prevalent during the initial stages of overshoot assembly, whereas complete microtubules were present at steady state. Qualitative observations of solution birefringence suggested that the polymer became progressively more aligned during assembly. The overshoot cannot be explained by proteolysis or denaturation of tubulin, by depletion of GTP, by a decrease in assembled mass, or by redistribution of polymer lengths. Taken together, the results suggest that changes in the form and/or in the organization of the assembling polymer are responsible for the overshoots in turbidity and in light scattering at small angles. Our results are consistent with models of microtubule assembly that postulate nucleation by tubulin sheets[16] and subsequent folding of the sheets to form mature microtubules.[16]

ACKNOWLEDGMENTS

We thank Dr. Leslie Wilson and Dr. Robley C. Williams Jr. for their generous support and encouragement and for their extensive intellectual contributions to this work. We also thank Mr. Herb Miller for excellent technical assistance.

REFERENCES

1. DUSTIN, P. 1978. Microtubules. Springer-Verlag. New York.
2. ROBERTS, K. & J. S. HYAMS. 1979. Microtubules. Academic Press. London.
3. JOHNSON, K. A. & G. G. BORISY. 1977. J. Mol. Biol. **117:** 1–31.
4. CORREIA, J. J. & R. C. WILLIAMS JR. 1983. Annu. Rev. Biophys. Bioeng. **12:** 211–235.
5. ENGELBORGHS, Y., L. C. M. DEMAEYER & N. OVERBERGH. 1977. FEBS Lett. **80:** 81–85.
6. CARLIER, M. -F. & D. PANTALONI. 1978. Biochemistry **17:** 1908–1915.
7. ROBINSON, J. & Y. ENGELBORGHS. 1982. J. Biol. Chem. **257:** 5367–5371.
8. VOTER, W. A. & H. P. ERICKSON. 1984. J. Biol. Chem. **259:** 10430–10438.
9. ERICKSON, H. P. 1974. J. Supramol. Struct. **2:** 393–411.
10. KIRSCHNER, M. W., L. S. HONIG & R. C. WILLIAMS. 1975. J. Mol. Biol. **99:** 263–276.
11. LEE, J. C. & S. N. TIMASHEFF. 1975. Biochemistry **14:** 5183–5187.
12. BINDER, L. I. & J. L. ROSENBAUM. 1978. J. Cell Biol. **79:** 500–515.
13. FARRELL, K. W. & L. WILSON. 1978. J. Mol. Biol. **121:** 393–410.
14. DETRICH III, H. W. & L. WILSON. 1983. Biochemistry **22:** 2453–2462.
15. MANDELKOW, E. -M., A. HARMSEN, E. MANDELKOW & J. BORDAS. 1980. Nature (London) **287:** 595–599.
16. ERICKSON, H. P. & D. PANTALONI. 1981. Biophys. J. **34:** 293–309.
17. HIMES, R. H., P. R. BURTON & J. M. GAITO. 1977. J. Biol. Chem. **252:** 6222–6228.
18. MACNEAL, R. K., B. C. WEBB & D. L. PURICH. 1977. Biochem. Biophys. Res. Commun. **74:** 440–447.
19. WILLIAMS JR., R. C. 1985. Methods Enzymol. In press.
20. KAYE, W. & A. J. HAVLICK. 1973. Appl. Opt. **12:** 541–550.
21. KAYE, W. 1973. Anal. Chem. **45:** 221A–225A.
22. LAEMMLI, U. K. 1970. Nature (London) **227:** 680–685.
23. KIM, H., L. I. BINDER & J. L. ROSENBAUM. 1979. J. Cell Biol. **80:** 266–276.
24. GOROVSKY, M. A., K. CARLSON & J. L. ROSENBAUM. 1970. Anal. Biochem. **35:** 359–370.
25. BERKOWITZ, S. A., J. KATAGIRI, H. K. BINDER & R. C. WILLIAMS JR. 1977. Biochemistry **16:** 5610–5617.
26. SUPRENANT, K. A. & L. I. REBHUN. 1983. J. Biol. Chem. **258:** 4518–4525.
27. WEINGARTEN, M. D., A. H. LOCKWOOD, S. -Y. HWO & M. W. KIRSCHNER. 1975. Proc. Natl. Acad. Sci. USA **72:** 1858–1862.
28. MURPHY, D. B. & G. G. BORISY. 1975. Proc. Natl. Acad. Sci. USA **72:** 2696–2700.
29. SLOBODA, R. D., S. A. RUDOLPH, J. L. ROSENBAUM & P. GREENGARD. 1975. Proc. Natl. Acad. Sci. USA **72:** 177–181.
30. DETRICH III, H. W., M. A. JORDAN, L. WILSON & R. C. WILLIAMS JR. 1985. J. Biol. Chem. **260:** 9479–9490.
31. CAMERINI-OTERO, R. D. & L. A. DAY. 1978. Biopolymers **17:** 2241–2249.

32. BERKOWITZ, S. A. & L. A. DAY. 1980. Biochemistry **19:** 2696–2702.
33. DOTY, P. & R. F. STEINER. 1950. J. Chem. Phys. **18:** 1211–1220.
34. BERNE, B. J. 1974. J. Mol. Biol. **89:** 755–758.
35. KRATKY, O. & G. POROD. 1940. J. Colloid. Sci. **4:** 35–70.
36. HERZOG, W. & K. WEBER. 1977. Proc. Natl. Acad. Sei. USA **74:** 1860–1864.
37. ANDREU, J. M. & S. N. TIMASHEFF. 1981. Arch. Biochem. Biophys. **211:** 151–157.
38. SCHUSTER, T. M., R. B. SCHEELE & L. H. KHAIRALLAH. 1979. J. Mol. Biol. **127:** 461–485.
39. SCHEELE, R. B. & T. M. SCHUSTER. 1974. Biopolymers **13:** 275–288.
40. KIRSCHNER, M. W. 1978. Int. Rev. Cytol. **54:** 1–71.
41. WEISENBERG, R. C. 1980. J. Mol. Biol. **139:** 660–677.
42. GUINIER, A. & G. FOURNET. 1955. Small-angle scattering of x-rays. John Wiley & Sons. New York. p. 40.

Kinetic and Steady State Analysis of Microtubule Assembly

RICHARD C. WEISENBERG

Department of Biology
Temple University
Philadelphia, Pennsylvania 19122

The study of microtubule assembly has been marked by many controversies. Such disputes are not unusual in science, but they should be based upon agreed definitions and analytical procedures. Unfortunately some of the most basic concepts relating to microtubule assembly have frequently been misused, resulting in unnecessary and unproductive disagreements. Certainly the most misused and abused concept is that of the critical subunit concentration. Other areas of research have suffered because of the failure to adequately define or specify kinetic parameters. It is the goal of this paper to provide a framework that may help future microtubule research to proceed more efficiently.

According to the nucleation-condensation model[1,2] helical polymers are stable only if the subunit concentration exceeds the concentration required for nucleation. The subunit concentration required for stable nuclei to form has been termed the critical subunit concentration. Once nucleation has occurred, assembly and disassembly most likely takes place between free subunit and the polymer ends, and the equilibrium, or steady state, reaction is described by

$$M + S_c = M \tag{1}$$

where M is the molar concentration of assembly and disassembly sites and S_c is the free subunit concentration at steady state. This relationship reflects the fact that assembly alters the length of the polymer, but not the concentration of assembly or disassembly sites. If a true equilibrium is reached the equilibrium constant for polymerization is equal to the inverse of the critical concentration. In the case of actin and tubulin assembly, a true equilibrium may not be reached because of the occurrence of polymerization-coupled nucleotide hydrolysis, but the steady-state reaction is characterized by a critical subunit concentration if equation 1 is valid.

As the concentration of subunit is increased above the critical concentration, polymers elongate but the steady-state concentration of the free subunit will be constant and equal to the critical concentration. The critical concentration may thus be defined as either the minimum concentration of subunit at which polymer exists or the steady state concentration of subunit. The critical concentration has most commonly been determined by measuring the polymer content as a function of the total protein concentration and using the extrapolated protein concentration at zero polymer content as the critical concentration.[3-12]

Experiments designed to compare different protein preparations, or to test the effects of specific agents on polymerization, must be able to distinguish changes in the polymerization equilibrium constant (or steady-state free energy change), and hence the critical subunit concentration, and the fraction of protein that is active subunit. In the simplest case, the concentration of inactive protein will be a constant fraction of the protein present. This will not in general be the case, however, if a reversible inhibitor or promoter of assembly is present (see below). For the simple case we can write the

543

following conservation of mass equation:

$$P = (1 - f)T - S_c \tag{2}$$

where T is the total protein concentration, P is the polymer (weight) concentration, S_c is the true critical subunit concentration, and f is the fraction of total protein that is not participating in the polymerization reaction for any reason. This relationship is plotted in FIGURE 1 for different values of the critical concentration and the fraction of active

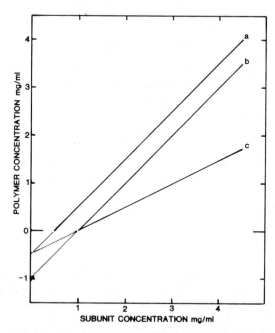

FIGURE 1. Determination of the critical subunit concentration from measurements of polymer concentration as a function of total subunit protein. Three different types of experimental results are indicated. The extrapolated portion of the curves are indicated by dotted lines. Curves a and b indicate samples with different critical concentrations and the same fraction of active protein. Curves a and c indicate samples with the same critical concentration and a different fraction of active protein. Sample b has a higher critical concentration than sample c, and a higher fraction of active protein. The critical concentration for samples a and c is 0.5 mg/ml, and for sample b is 1 mg/ml. Samples a and c contain 100% active subunit, whereas sample c contains 50% active subunit.

protein. It is clear that the true critical concentration is equal to the negative of the intercept on the polymer axis at zero total protein concentration. The critical concentration is equal to the extrapolated protein concentration at zero polymer content only when f = 0. Clearly f = 0 only in the limiting, and experimentally unobtainable, situation of 100% pure, active subunit protein.

　　Whereas the extrapolated subunit concentration at zero polymer content is a useful characterization of a preparation, in that it indicates the minimum concentration of protein needed to make polymer, it is not a measure of the true critical concentration.

The term critical concentration should be reserved for that number that reflects the equilibrium, or steady state, energetics of assembly, and not the minimum concentration for nucleation, (which should be called just that to avoid confusion).

It is also possible to determine the critical concentration from the concentration of unpolymerized protein at steady state.[1] In this analysis the critical concentration is equal to the extrapolated concentration of unpolymerized protein at zero total protein. This procedure is not common, however, because most physical techniques used to assay polymerization, such as viscosity and turbidity, measure the polymer concentration.

In most cases the amount of polymer is not measured directly but is inferred from a physical measurement such as viscosity or turbidity. If the measurements are known to depend only upon the weight concentration of polymer, a constant of proportionality may be used to obtain the actual concentration. If a constant of porportionality is not available, relative values of the critical concentrations in terms of subunit concentration can be obtained by correcting the protein concentration at zero polymer for any difference in the fraction of active protein between the experimental samples. This may be done numerically by correcting the experimental plots for any difference in their slopes.[13,14] This correction can also be quickly performed graphically as indicated in FIGURE 2. For the graphical procedure a line is drawn parallel to a standard line with an intercept on the polymer axis equal to the experimental line (dashed line in FIGURE 2). The intercepts of these parallel lines on the total protein axis are proportional to the true critical concentrations. Note that the standard line must be the one with the greatest observed slope, because this reflects the greatest fraction of active subunit protein.

The critical concentration is the concentration of subunit at which the net assembly rate equals zero, and kinetic measurements may thus be used to determine the critical concentration. To use this procedure, it is necessary to know the functional dependence of assembly rate on subunit concentration. Whereas this is generally assumed to be a linear function, there is reason to believe that this is a simplification for the case of actin and tubulin assembly that involve polymerization-coupled nucleotide triphosphate hydrolysis.[15,16] Second, this procedure does not indicate the fraction of active protein present, and kinetic measurements will not yield valid critical concentrations unless this value is known from independent experiments.

In determining critical subunit concentrations, it is important to also consider the precision of the results. The true critical concentration is obtained by extrapolation to the polymer concentration at zero total protein. This extrapolation is substantially longer than the commonly used extrapolation to the protein concentration at zero polymer. As a result the measured critical concentration is very sensitive to errors in the determination of the slope of the plot. Note that the higher the minimum concentration for assembly, the longer and more inaccurate is the determination of the true critical concentration. Comparisons of the critical concentrations under various conditions must be treated with caution.[17-21]

Subunit will appear to be inactive if it is present below its critical concentration. If two fractions of subunit have different critical concentrations (this could be because of physical differences in the subunits or because of the presence of an inhibitor or promoter of assembly), then a biphasic plot of polymer versus total protein is expected (FIGURE 3). The critical concentration of each component can be obtained by resolving the data into two components as shown in FIGURE 3 and appropriate extrapolation of the data for each component.

If a reversible inhibitor or promoter of polymerization is present, the fraction of active subunit will be a function of the concentration of the agent and protein. For a simple inhibitor of assembly that binds to a single site on the subunit, the fraction of

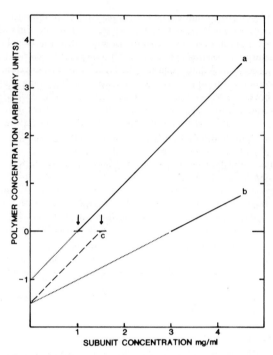

FIGURE 2. Graphical method for comparing critical subunit concentrations. In this case it is assumed that a physical method is used for determining polymer concentration, but the conversion factor for obtaining actual concentration of polymer is not known. Curve a is the standard, which has the greatest slope and therefore reflects the greatest fraction of active protein. The relative critical concentration for sample b is obtained by drawing a line parallel to curve a from its intercept on the polymer axis (curve c, dashed line). The intercepts of curves a and c on the total protein axis (arrows) are proportional to the true critical subunit concentrations of samples a and b.

inactive protein is given by

$$f = K_i I T_a \tag{3}$$

where K_i is the equilibrium constant for binding of the inhibitor, I is the inhibitor concentration, and T_a is the concentration of inhibitor-free, active subunit. If the inhibitor is a constant concentration, i, of the protein concentration, then equation 3 becomes

$$P = T - iK_iT^2/(1+iK_iT) - S_c \tag{4}$$

For a reversible promoter of assembly with an equilibrium binding constant, K_p, at a constant fraction, p, of the protein concentration, equation 4 becomes

$$P = pK_pT2/(1+pK_{pT}) - S_c \tag{5}$$

This relationship is plotted in FIGURE 3 for comparison with the case where two components are present. More complicated results may be expected, of course, if the interaction of inhibitor or promoter and subunit is more complex than assumed. Even for the simplest case, however, it is apparent that it may be difficult to distinguish between an agent that alters the critical subunit concentration for polymerization (and hence the free energy of assembly) or acts stoichiometrically as a reversible activator or inhibitor of assembly (and hence change the fraction of protein able to polymerize). The problems of determining the critical concentration by a long range extrapolation as discussed above are, of course, even more severe if the proper extrapolation is nonlinear as indicated in FIGURE 3. Given the error inherent in these types of measurements, it cannot be assumed that the data will clearly distinguish between the different possibilities indicated in FIGURE 3, and care must be used in interpreting such data.

In addition to the critical subunit concentration, the other important parameters of microtubule assembly are those that describe the kinetics of assembly and disassembly. It has been reported that the rate of assembly is a linear function of subunit concentration,[22,23] as predicted by a simple nucleation-condensation model for helical polymerization.[1] As originally developed, however, this model does not incorporate either polymerization-coupled nucleotide hydrolysis or the role of assembly cofacters

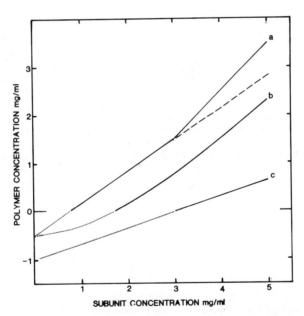

FIGURE 3. Effect of heterogenity of subunit activity on critical concentration analysis. Curve a (solid line) indicates data that could be obtained for a mixture of subunits with two different critical concentrations. In this case it is assumed that two thirds of the subunit has a critical concentration of 0.5 mg/ml, whereas the remaining third has a critical concentration of 1 mg/ml. The second component (curve c) is resolved by subtracting the extrapolated values of the first component (dashed line) from the data at high protein concentrations. Curve b indicates the data expected for the case where a simple promoter of assembly is present. In this example the product of the promoter equilibrium binding constant, K_p, and the fraction of promoter present, p, is assumed to be 0.25, whereas the critical concentration for the polymerization reaction is fixed at 0.5 mg/ml. Note that the proper extrapolation to obtain the critical concentration is nonlinear in this case.

such as MAPs, or intermediates such as oligomers. This model can therefore only be considered as an approximation of the actual reactions occurring.

During the polymerization of tubulin, exchangable, subunit-bound GTP is converted to tightly bound GDP in the microtubule.[24,25] This "polymerization-coupled" nucleotide hydrolysis is important in the assembly of tubulin and actin, and is necessary, for example, for the phenomenon of subunit treadmilling.[26]

In addition to guanine nucleotide, microtubule assembly involves nonmicrotubule proteins as cofactors. These microtubule-associated proteins (MAPs) appear to promote microtubule formation by stabilizing intersubunit bonds.[21] In the absence of microtubule assembly, MAPs promote the formation of MAP-tubulin oligomers, which may be intermediates in the assembly of the microtubule.[16,27,28]

In the basic nucleation-condensation model, the rate of polymerization is described by equation 6:

$$dp/dt = k_+ MS - k_- M \tag{6}$$

where k_+ and k_- are the assembly and disassembly rate constants, M is the molar concentration of assembly and disassembly sites, and S is the free subunit concentration.

This equation is a simplification of the actual reactions occurring because it ignores the hydrolysis of GTP, the association and dissociation of MAPs, and the possible existence of more than one type of assembly or disassembly site. A model for microtubule assembly that specifically takes these factors into account has been proposed.[29] The model may be simplified by several assumptions that reduce microtubule assembly to a single, unidirectional cycle. This cycle consists of the addition of a GTP-containing subunit to the end of the microtubule, the hydrolysis of the bound GTP and the release of inorganic phosphate, the dissociation of GDP-containing subunit from the microtubule, and finally the regeneration of free GTP-containing subunit by exchange or transphosphorylation. As a consequence of this mechanism the ratio of GTP to GDP-containing subunits will depend upon the relative rates of assembly and disassembly and the GTP hydrolysis rate. A more exact model based upon similar principles has recently been presented.[30]

When the proposed mechanism of polymerization-coupled GTP hydrolysis is taken into account, equation 6 becomes

$$dp/dt = k_{+T} MS - k_{-D} M/(1 + k_{+T} S/V) \tag{7}$$

where k_{+T} is the association rate constant for GTP-containing subunits, k_{-D} is the dissociation rate constant for GDP-containing subunits, and V is the rate of hydrolysis of GTP that applies after subunit addition to the microtubule. No similar equations are available to our knowledge for the role of MAPs or oligomers in assembly.

From equation 7 it is possible to calculate the critical subunit concentration by setting the assembly rate to zero. This yields equation 8:

$$S_c = ((1 + 4k_{-D}/V)^{.5} - 1)/2k_{+T}/V \tag{8}$$

There has been interest recently in subunit flow or treadmilling through actin filaments and microtubules. This phenomenon is a consequence of a difference in critical concentrations at opposite ends of the polymer, and depends upon the hydrolysis of nucleotide triphosphate.[26] In the nucleation-condensation model, the critical concentration is equal to the ratio of the disassembly to assembly rate constants, but it is not clear how nucleotide hydrolysis might alter the critical concentration.

According to equation 8 subunit flow through the polymer will occur if either the rate of hydrolysis differs at opposite ends of the polymer, or if the activation energy for assembly-disassembly differs at opposite ends. The activation energy, of course, determines the magnitude of both the assembly and disassembly rate constants. This possibility is consistent with the data of Bergen and Borisy,[23] which can be fit by equation 7 by varying the rate constants at each end of a microtubule, but keeping the ratio of the assembly and disassembly rates the same. It is not necessary to propose that hydrolysis of GTP somehow alters the rate constants for assembly or disassembly, and does so to a different degree at opposite ends of the polymer.

Thermodynamic principles appear to require that if subunit treadmilling is occurring then subunits at opposite ends of the polymer must differ at steady state. The most likely difference is in the proportion of guanosine triphosphate (GTP) to guanosine diphosphate (GDP) that is present (this in turn may be viewed from equation 8 as reflecting a difference in the turnover rate of subunits by assembly-disassembly). For such a difference to exist at steady state the rate of hydrolysis of GTP must be comparable to the subunit association rate. Note that models that propose a very slow rate of hydrolysis, and the existence of a steady-state GTP cap at the ends of the microtubule, do not appear consistent with the existence of subunit treadmilling. Evidence for a GTP cap has been obtained at high subunit concentrations,[31] but this is not inconsistent with the present model. Direct evidence that the nucleotide composition of the ends must be taken into account has been provided by Engelborghs and Van Houtte.[32]

Strictly speaking, at least five rate constants must be used to describe microtubule assembly. These are the assembly constants for GTP tubulin and for GDP tubulin, the disassembly constants for GTP tubulin and GDP tubulin, and a rate constant for the hydrolysis of GTP. Additional rate constants may also be needed to describe the role of MAPs or oligomers. Even in the absence of MAPs or oligomers, structural considerations suggest that more than a single type of assembly-disassembly reaction may occur.[29]

If MAPs dissociate rapidly, we would expect that this would control the disassembly rate of subunits at all concentrations, not just at low concentrations. Kinetic experiments would thus yield the dissociation rate of MAPs, but the kinetics would not necessarily be different than predicted by equation 6. The participation of oligomers in assembly, however, raises the possibility of cooperative interactions occurring between subunits. Such interactions may explain some of the kinetic (*e.g.* "catastrophic" disassembly) and steady state properties (*e.g.* irreversible response to drugs) of microtubules.

Evidence for the participation of oligomers in microtubule growth has grown recently.[33,34] If oligomers are linear aggregates that add to microtubules by end addition then the effect on the kinetics of assembly will probably be minimal. This is a simple consequence of the fact that if n subunits form an oligomer, the molar concentration of reacting species is reduced by n, but this is exactly compensated for because each assembly step adds n subunits. Whereas the kinetics of assembly may be little changed by participation of oligomers, the properties of microtubules so formed may be very different. In particular it is likely that growth of microtubules by oligomer addition will result in assembly involving at least three different reactions, rather than one as usually assumed.[29] The difference may be visualized by comparing the process of helical growth, in which each reaction step is identical to the prior step, and what we term cylindrical growth (which is the process by which a brick chimney is made), in which there are at least three different reaction steps required to form each new layer of the cylinder. We have termed these the primary reaction (the first addition to a new layer), the secondary reaction (the next eleven additions to a thirteen-sided microtubule), and the tertiary reaction (the final addition to complete a layer). As a consequence, different reaction steps may dominate assembly-disassembly and steady-state processes.

In conclusion, I would like to suggest the following rules for the presentation and discussion of microtubule assembly data. First of all, it should be made clear how critical subunit concentrations are determined; such measurements should always be accompanied by estimates of the fraction of active protein present. Treatments that alter the extent of microtubule assembly should be classified by their effects on the critical concentration and on the fraction of active protein. The complication that an agent may act as a reversible promoter or inhibitor of assembly may need to be considered in the above analysis. For any agent, it should be determined if identical levels of assembly are obtained if the agent is added either during assembly or to fully formed microtubules. Agents that do not produce reversible microtubule behavior need to be further characterized by determining their effects on the critical concentration and the fraction of active protein when added to free subunit and fully formed microtubules. Kinetic data should take into account the probability that assembly and disassembly rates are not a linear function of subunit concentration. Because kinetic measurements do not indicate the fraction of active protein present, they must be accompanied by steady-state critical concentration measurements to be quantitatively interpreted. Finally, in model building, consideration should be given to the likely complexity of microtubule assembly, including the possible involvement of oligomers, the possible existence of more than a single reaction step, and the effects of cooperative interactions between subunits mediated by MAPs.

REFERENCES

1. Oosawa, F. and S. Asakura. 1975. Thermodynamics of the polymerization of protein. Academic Press. London.
2. Oosawa, F. & M. Kasai. 1962. J. Mol. Biol. 4: 10–21.
3. Herzog, W. & K. Weber. 1978. Eur. J. Biochem. 91: 249–254.
4. Karr, T. L., A. E. Podrasky & D. L. Purich. 1979. Proc. Natl. Acad. Aci. USA 76: 5475–5479.
5. Karr, T. L., D. Kristofferson & D. L. Purich. 1980. J. Biol. Chem. 255: 1853–1185.
6. Keller, T. C. S. & L. I. Rebhun. 1982. J. Cell Biol. 93: 788–796.
7. Maccioni, R. B. & N. W. Seeds. 1978. Arch. Biochem. Biophys. 185: 262–271.
8. Pantaloni, D., M. F. Carlier, C. Simon & G. Batelier. 1981. Biochemistry 20: 4709–4716.
9. Sandoval, I. V. & P. Cuatrecasas. 1978. Eur. J. Biochem. 91: 151–161.
10. Sloboda, R. D. & J. L. Rosenbaum. 1975. Biochemistry 18: 48–55.
11. Waxman, P. G., A. A. DelCampo, M. C. Lowe & E. Hamel. 1981. Eur. J. Biochem. 120: 129–136.
12. Webb, W. C. & L. Wilson. 1980. Biochemistry 19: 1933–2001.
13. Regula, C. S., J. R. Pfeiffer & R. D. Berlin. 1981. J. Cell Biol. 89: 45–53.
14. Sternlicht, H. & I. Ringel. 1979. J. Biolog. Chem. 254: 10540–10550.
15. Weisenberg, R. C. 1981. Cell Motility 1: 485–497.
16. Zackroff, R. V., R. C. Weisenberg & W. J. Deery. 1980. J. Mol. Biol. 139: 641–660.
17. Engelborghs, E., K. A. H. Heremans, L. C. M. DeMaeyer & J. Hoebeke. 1976. Nature (London) 259: 686–688.
18. Berkowitz, S. A. & J. Wolff. 1981. J. Biol. Chem. 256: 11216–11223.
19. Carlier, M. F. & D. Pantaloni. 1978. Biochemistry 17: 1908–1915.
20. Keates, R. A. B. 1981. Can. J. Biochem. Cell Biol. 59: 353–360.
21. Murphy, D. B., K. A. Johnson & G. G. Borisy. 1977. J. Mol. Biol. 117: 33–52.
22. Johnson, K. A. & G. G. Borisy. 1979. J. Mol. Biol. 117: 1–52.
23. Bergen, L. G. & G. G. Borisy. 1980. J. Cell Biol. 84: 141–150.

24. WEISENBERG, R. C., W. J. DEERY & P. J. DICKINSON. 1976. Biochemistry **15:** 4248–4254.
25. DAVID-PHEUTY, T., H. P. ERICKSON & D. Pantoloni. 1977. Proc. Natl. Acad. Sci. USA **74:** 5372–5376.
26. WEGNER, A. 1976. J. Mol. Biol. **108:** 139–150.
27. MANDELKOW, E., A. HARMSEN & E. MANDELKOW. 1980. Nature (London) **287:** 595–599.
28. PALMER, G. R., D. C. CLARK, P. M. BAYLEY & D. B. SATTELLE. 1982. J. Mol. Biol. **160:** 641–658.
29. WEISENBERG, R. C. 1980. J. Mol. Biol. **139:** 660–677.
30. CARLIER, M. F., T. L. HILL & Y. CHEN. 1984. Proc. Natl. Acad. Sci. USA **80:** 7520–7523.
31. BONNE, D. & D. PANTALONI. 1982. Biochemistry **21:** 1075–1081.
32. ENGELBORGHS, Y. & A. VAN HOUTTE. 1981. Biophys. Chem. **14:** 195–202.
33. KRAVIT, N. G., C. S. REGULA & R. D. BERLIN. 1984. J. Cell Biol. **99:** 188–198.
34. BURNS, R. G. & K. ISLAM. 1984. FEBS Letters **173:** 67–74.

Nucleation of Microtubule Assembly

Experimental Kinetics, Computer Fitting of Models, and Observations on Tubulin Rings[a]

HAROLD P. ERICKSON AND WILLIAM A. VOTER

Department of Anatomy
Duke University Medical Center
Durham, North Carolina 27710

INTRODUCTION: NUCLEATION AND ELONGATION

FIGURE 1 shows a typical assembly reaction as a function of time, plotted both on a linear and a semilogarithmic scale. The reaction comprises two phases, termed nucleation and elongation. During the final half to two-thirds of the reaction, the elongation phase, the rate of the reaction decreases as the plateau value is approached. The log plot gives a straight line, indicating a second order reaction. During this time the number of polymerization sites (microtubule ends) is constant, and the reaction decreases as the pool of subunits is used up. During the nucleation phase the direct plot is concave upward, showing an accelerating reaction. The number of polymerization sites increases as nuclei are formed spontaneously from the pool of subunits. Elongation, of course, is also occurring during the nucleation phase. Because nucleation is strongly concentration dependent (see below), it virtually ceases after about 20–50% of the subunit pool is used up. Note that there appears to be a lag phase at the very beginning of assembly during which no significant change in turbidity is observed. Electron microscope specimens prepared during this time always show a small number of microtubules, suggesting that nucleation and assembly are occurring at too low a level to register a turbidity. The apparent lag phase might be due to a separate mechanism, but we treat it as part of the nucleation phase.

All of the experimental work presented here used microtubule-associated protein (MAP)-free tubulin, purified by several cycles of assembly and phosphocellulose chromatography. Our goal is to understand the simplest assembly system before trying to include the complexities of MAPs and rings. Details of methods and analysis are presented in our recent paper.[1]

A TWO-DIMENSIONAL (2-D) SHEET IS THE BASIC POLYMER IN NUCLEATION AND GROWTH

The microtubule is a helical polymer of subunits. In their classic treatment of biological assembly, Oosawa and Kasai[2] showed that formation and growth of polymers too small to form an intact helix could be very unfavorable. The critical step in nucleation was closure of the first turn of the helix. The model seems to apply reasonably well to actin assembly, where the helix is completed with three subunits. Because microtubules are helical polymers, it might be thought that this model would

[a]This work was supported by NIH Grant GM28553.

552

also apply to the unfavorable nucleation that is observed in *in vitro* assembly. For the microtubule lattice, however, a very large number of subunits would be required to form an intact helix. At least 40 subunits would be needed to close one turn of the three-start helix, and 66 subunits are needed if growth were by way of the five-start helix. These numbers are almost an order of magnitude larger than the observed concentration dependence of tubulin assembly (see below), so the critical event in nucleation must involve polymers much too small to form an intact helix.

In our earliest study of microtubule assembly[3], we used electron microscopy to visualize the polymers formed during the time-course of assembly. These experiments showed that assembly began by formation and growth of 2-D polymers, small pieces of the microtubule wall. At the earliest assembly times these sheets had 3–5 protofilaments; later the sheets had grown wider, to 8–12 protofilaments. Eventually the polymers were almost all converted into intact cylindrical microtubules that continued to grow longer. Sheets with 13 or more protofilaments were very rarely seen, suggesting that they closed to form intact cylinders when they reached their normal

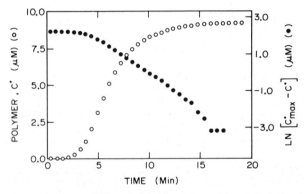

FIGURE 1. The total polymer determined from turbidity is plotted on a linear scale (O) and a logarithmic scale (●). At this tubulin concentration ($11.1\mu M$) nucleation continues until about 6 min, after which polymerization is by elongation alone. The straight line of the log plot after 6 min is consistent with a second order polymerization onto a constant number of microtubule ends. (W. A. Voter & H. P. Erickson.[1] With permission from the *Journal of Biological Chemistry*.)

complement of 13 protofilaments. Micrographs showed that sheets were curved exactly like a piece of the microtubule wall so closure would be expected as soon as the wall had grown to the proper circumference.

Open sheets are often thought to be grossly different from intact microtubules, but at the molecular level, the two forms are virtually identical. The lattice and bonding pattern are only different for the small fraction of subunits at an edge or a discontinuity. In some assembly conditions, intact microtubules are rarely formed, apparently because inversions of protofilaments in the growing sheet make closure impossible.[4] The large sheets produced in these conditions leave about the same critical concentration of free subunits, demonstrating that they are close to microtubules in stability. Often the large sheets are stable for long periods of time in a mixture with intact microtubules. If closure of the helical lattice were a crucial step in nucleation or assembly, the open sheets would be significantly less stable than, and should be replaced by, the intact microtubules. Closure to form the helical lattice is thermodynamically favorable and will occur when possible, but it is not an essential step in assembly. Microtubule assembly is essentially a 2-D polymerization.

The proposed pathway of assembly, in which small 2-D polymers are precursors to large sheets or eventually intact microtubules, has been confirmed by several laboratories for a variety of different tubulins and assembly conditions. FIGURE 2 shows micrographs of small 2-D polymers prepared during the course of our kinetics analysis discussed below. We consider this pathway, with small sheets as intermediates, to be well established. An important corollary of this pathway is that nucleation of microtubules is a question of 2D, not helical assembly.

A THERMODYNAMIC ANALYSIS—DEFINITION OF PATHWAY OF ASSEMBLY AND CRITICAL NUCLEUS

A first step in understanding nucleation is to ask whether nucleation makes sense in terms of simple principles of thermodynamics. Specifically, are the small 2-D polymers sufficiently stable to be formed in the quantity needed to nucleate the number of microtubules eventually formed, and to form and grow spontaneously from the pool of subunits? The answer to both questions is a qualified yes.[5]

It is perhaps worthwhile to restate a fundamental principle: assembly of protein subunits must proceed through a series of bimolecular reactions. The simultaneous collision of three or more subunits to form a trimer or larger polymer is far too rare an event to play any role in assembly. Therefore every oligomer must be assembled through a sequence of smaller intermediates, starting with the dimer. Some oligomers could be formed by pairwise reaction of preformed intermediates, but in the simplest case, the intermediates will grow by stepwise addition of single subunits. The sequence of these intermediates and the bimolecular reactions that link them are called the pathway of assembly.

The least stable polymer is called the critical nucleus. We will use the symbol n to designate the number of subunits in the critical nucleus (see reference 1 for a discussion of alternative definitions of nucleus). The principle of biomolecular reactions says that each polymer must originate from smaller intermediates. More specifically, each large polymer must have passed through the critical nucleus stage. Thus the concentration of the critical nucleus, the lowest concentration of any intermediate, acts as a bottleneck in the pathway of assembly.

A crucial step for the thermodynamic analysis was to separate the free energy of assembly into three terms: the longitudinal and lateral bond energies, and a term called the entropic energy. This third term accounts for the entropy lost when a subunit is polymerized, that is, it is the free energy required to immobilize a subunit in the polymer. Separating this entropic free energy as an explicit term was essential to go from the qualitative treatment of Oosawa and Kasai[2] to a quantitative estimate of concentrations of the various small polymers.[5]

Our analysis was based on one experimentally determined parameter; the equilibrium subunit concentration, $C_c = 10^{-5}$ M; and three adjustable parameters: the longitudinal and lateral bond energies, and the entropic free energy. We initially set the entropic energy term to 15 kcal/mol, the lowest value that could be justified theoretically, and found that small polymers were extremely unfavorable. Using that value we had to conclude that spontaneous nucleation could not occur.

We eventually decided that a much lower value had to be used for this entropic energy parameter, even if it was difficult to justify theoretically. More recently Jencks[6] has shown that a value of 2–5 kcal/mol appears appropriate for a variety of protein-protein and protein-ligand interactions. With a value of 2.1 kcal/mol for the entropic free energy and a $C_c = 10 \ \mu M$, our calculations showed that spontaneous nucleation and growth were thermodynamically favored for a starting subunit concentration greater than a certain value (termed the critical supersaturation, and equal to about $4 \times C_c$ for these parameters).

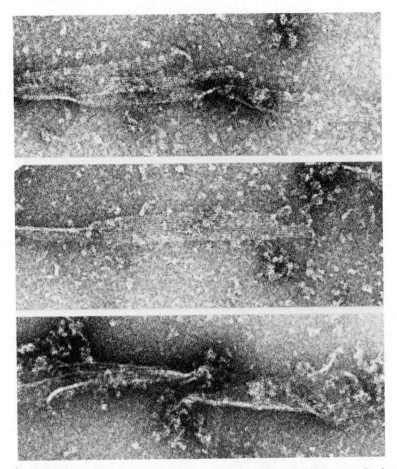

FIGURE 2. Electron micrographs of a negatively stained specimen from an early time in assembly. Note that the polymers are small sheets, sometimes anastomosing or interconnected into small bundles. Magnification 150,000X.

In the pathway of assembly deduced from this study the best candidate for the critical nucleus is a two-filament sheet. For subunit concentrations above the critical supersaturation, all two-filament sheets are sufficiently stable to exist and serve as intermediates in the pathway of assembly. More importantly, the stability of the sheets increases as each pair of subunits is added. This means that all two-filament sheets should spontaneously grow longer.

Lateral growth requires nucleation of a new protofilament along the side of the growing sheet. As discussed below this should be proportional to the length of the sheet, so it should eventually become favorable as the two-filament sheets elongate spontaneously.

A KINETIC ANALYSIS

The rate-limiting step in nucleated polymerization is the addition of a subunit to the critical nucleus, converting it to a stable seed that will spontaneously elongate. Implicit here is the reasonable assumption that rate constants are the same for each subunit addition. Because the critical nucleus is the least stable intermediate, its low concentration limits the overall rate of polymerization. Thus an experimental determination of the kinetics of polymerization can give information on the size of the critical nucleus.

Log-Log Plots of Reaction Time Versus Concentration

The simplest indication of nucleus size is obtained by plotting the log of a characteristic reaction time (we use $t_{10\%}$, the time to reach 10% of plateau assembly) versus the log of starting subunit concentration. Oosawa and Kasai[2] showed that, subject to certain conditions, this plot should be a straight line with a slope of $(n + 1)/2$. The most important restriction is that the back reaction (loss of a subunit from a polymer) must be negligible compared to the forward (elongation) reaction. This will be true when the free subunit concentration is much higher than C_c.

The importance of this restriction is not always realized in experimental applications. It has been safely ignored in studies of actin assembly, where the starting subunit concentration is 10 to 50 times C_c. For tubulin assembly, where the starting concentration is only 2 to 5 times C_c, the Oosawa-Kasai treatment must be modified. We have found by computer simulations that the log-log plot remains linear in this lower concentration range, but the slope changes to $(n + 1.5)/2$. This is a relatively small error for the large nucleus ($n = 6$ to 7) found in our recent analysis, but it would be important in cases where the nucleus is smaller.

The log-log plot for our complete set of assembly kinetics[1] is shown in FIGURE 3. The points fit a straight line over the entire concentration range of 6.7 to 19μM. The slope of the curve is 3.77, from which we deduce a nucleus size of $n = 6$, using the $(n + 1.5)/2$ rule (n would be 6.5 using the $(n + 1)/2$ rule). Probably the most important conclusion is that the nucleus is much larger than the trimer found for actin assembly. This is consistent with the observation that the assembly time has a very strong concentration dependence: the time to plateau varied 50-fold over the 3-fold concentration range investigated.

Fitting the Kinetics to the Wegner-Engel Model

A much better analysis of nucleation can be obtained if the entire time course of assembly at the different starting concentrations is fit to model curves. The complete pathway of assembly involves a sequence of second order equations for each subunit addition in the nucleation and elongation phases. The very large set of coupled differential equations describing these reactions is simple to set up for numerical integration, but it is extremely expensive in computer time. The definitive treatment of this problem is the work of Wegner and Engel,[7] who introduced certain simplifying assumptions that reduced the problem to a pair of coupled differential equations. These are rapidly solved even on a small computer.

There are two assumptions involved in the Wegner-Engel analysis. First the critical nucleus is assumed to be in rapid equilibrium with the pool of subunits. In this case all steps before the nucleus are invisible to the kinetics, and the system can be described as if the nucleus were in equilibrium with free subunits. Thus the concentration of the nucleus will vary as the nth power of the subunit concentration. The second assumption is that each step of elongation after nucleation has the same forward and reverse rate constants. This is very reasonable for subunit addition to a growing helix, and should also apply to some aspects of 2-D polymerization.

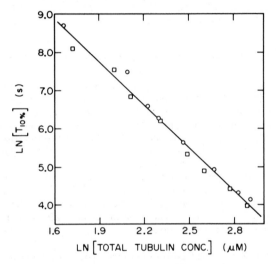

FIGURE 3. Log-log plot of tenth-time versus total starting protein concentration. The points were from two data sets, and the solid line was generated by linear regression. The slope of -3.77 indicates a nucleus size of 6 subunits. (W. A. Voter & H. P. Erickson.[1] With permission from the *Journal of Biological Chemistry*.)

The analysis is based on experimental data, a set of curves for the time course of polymerization at different starting subunit concentrations. Theoretical curves are generated by computer for these same starting concentrations using different values for the nucleus size, n. It is generally found that only one value of n will give a good fit, and this is taken to be the size of the nucleus under the experimental conditions.

The Wegner-Engel analysis has been applied extensively to studies of actin assembly by several labs and for a variety of assembly conditions.[7-11] The conclusion for most assembly conditions is that $n = 3$. In some conditions $n = 2$ or 4 gives a better fit. In all cases, the fit of the theoretical curves to the experimental data is extremely good. Our study[1] was the first to apply this approach to microtubule assembly. Important modifications of the theory, based on the pathway of assembly presented above, were required to obtain a good fit to the experimental kinetics.

Simple Homogeneous Nucleation

For our initial analysis of microtubule kinetics, we used the same model that had worked for actin: rapid equilibrium of a nucleus of n subunits, followed by elongation steps with identical forward and reverse kinetics. Obviously this ignores the complexity of 2-D polymerization discussed above, but it is a good starting point. The best fit, shown in FIGURE 4, was obtained for a nucleus size $n = 6$.

There are two striking differences in the curve fitting between actin and tubulin. First, the microtubule nucleus is much larger than that of actin. The nucleus size is determined largely by fitting the mid-points of the curves for the entire concentration range, and essentially confirms the conclusion from the log-log plot. Second, the fit of each curve to the experimental kinetics was almost perfect for actin, but for microtubules it is not very good. The computer curves rise too early and fall off too slowly, so they never match the "explosive" character of the experimental kinetics. The poor fit suggests that microtubule assembly is more complex than a single step of nucleation followed by identical steps of elongation.

A Nucleation Model Specific for 2-D Assembly

It was not surprising that this simplest model did not work, because it was based on the pathway of assembly of a simple helical polymer. We therefore modified the Wegner-Engel analysis to account for 2-D growth of the microtubule wall. The model is shown diagrammatically in FIGURE 5. There are now two steps of nucleation. The first step is formation of the primary nucleus, assumed to be a two-filament sheet

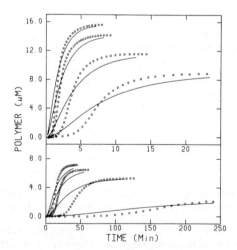

FIGURE 4. Experimental kinetics of tubulin polymerization (open circles) and an attempt to fit the data by a simple (one-step) homogeneous nucleation model. Tubulin concentrations ranged from 6.7 to 19.0 μM. Note the change in time scale between panels. The solid curves were generated by the Wegner-Engel equations for a nucleus size $n = 6$. The curves fit the half-times reasonably well for most concentrations, but the curves do not have the sharp sigmoid shape of the experimental data. (W. A. Voter & H. P. Erickson.[1] With permission from the *Journal of Biological Chemistry*.)

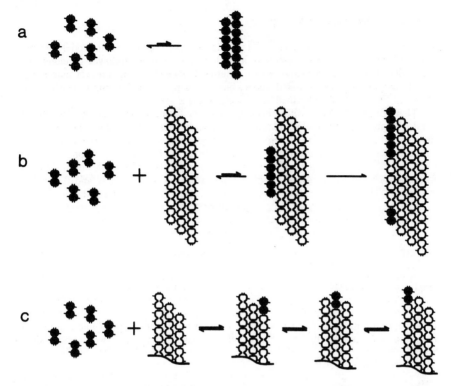

FIGURE 5. Our two-step nucleation model for the 2-D polymerization of the microtubule wall. 6-S tubulin dimers are indicated as dumbbell-shaped subunits with lateral and longitudinal bonds protruding. The most recently added subunits are shown in black. a: The primary nucleation mechanism is assumed to be formation of a two-filament sheet. The nucleus is in rapid (compared to the subsequent elongation steps) equilibrium with the pool of free subunits. The heptamer shown here is the nucleus size determined by our computer fitting. b: Lateral growth is assumed to require the formation of a short protofilament, in equilibrium with free subunits, on the side of an existing sheet. Once the protofilament reaches a critical size, indicated here as three subunits, it is stable and grows rapidly to the length of the sheet. c: Elongation is assumed to involve a second order addition of subunits onto the end of each protofilament. (W. A. Voter & H. P. Erickson.[1] With permission from the *Journal of Biological Chemistry*.)

consisting of n subunits. This is essentially the same as the homogeneous nucleation of actin. As discussed above this two-filament sheet can elongate spontaneously like the helix of actin. In order to make the sheet grow wider, however, we have postulated a second nucleation step, to initiate a new protofilament. This nucleus, formed on the edge of an existing sheet, consists of a single short filament of m subunits. Elongation is assumed to occur at the ends of all protofilaments.

There are now two parameters, n and m in the model, and one might think that this would be sufficient to fit any collection of data. That was not the case. The best fit, shown in FIGURE 6, was markedly improved from the simpler model but still is not perfect. Specifically, the explosive character of the experimental kinetics (starting very slowly, accelerating very rapidly, and then tapering off sharply) is still not matched

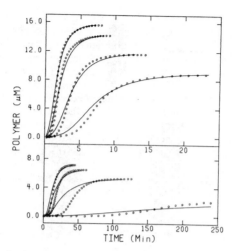

FIGURE 6. The best fit of experimental kinetics to our two-step nucleation model (FIGURE 5). The curves were obtained for primary nucleus size $n = 7$, and secondary nucleus (new protofilament) size $m = 3$. (W. A. Voter & H. P. Erickson.[1] With permission from the *Journal of Biological Chemistry*.)

completely by the computer curves. The best fit was obtained for $n = 7$ and $m = 3$. Changing either of these values by ± 1 made the fit somewhat worse and changes of ± 2 gave a distinctly poorer fit.

Heterogeneous Nucleation

Ferrone et al.[12,13] used a model of heterogeneous nucleation to explain the kinetics of Hb-S assembly. This is also a two-step nucleation but somewhat different from our model in FIGURE 5. Our model was actually based on their derivation, but we restricted the second nucleation to take place only along the edge of an existing sheet, thus initiating a new protofilament. In our model this secondary nucleation is proportional to the total length of lateral edge, which in turn is proportional to the square root of the total polymer (assuming a constant ratio of length to width of the 2-D polymers). In the Ferrone model secondary nucleation is facilitated along all parts of existing polymers. The secondary nucleation is therefore directly proportional to the total mass of existing polymer, rather than to the square root as in our model. In both two-step models, especially in the unconstrained heterogeneous nucleation of Ferrone et al., assembly is autocatalytic and can be very explosive with time. The reason is that once a few polymers are nucleated and elongate, they spawn a large number of secondary nuclei. The increased nucleation sharply accelerates the overall polymerization until the concentration of free subunits drops too low to support further nucleation.

We tried fitting the microtubule kinetics to a heterogeneous nucleation model similar to that of Ferrone et al., and obtained an even better fit than with our 2-D assembly model (FIGURE 7). The best fit was obtained for nucleus size 6 for both the primary and secondary nucleus. The structure of the presumed secondary nucleus is, however, not clear. One attractive candidate is the "hook decorated microtubules," in which a curved microtubule wall grows off an existing microtubule or sheet. Burton and Himes[4] have demonstrated this type of structure for a variety of assembly conditions, and we have seen complex bundles of sheets, probably related to these, in our own preparations (FIGURE 2, and unpublished observations of embedded material). This phenomenon is especially prominent at high protein concentrations.

Overall the unconstrained heterogeneous nucleation model is not as satisfying as our two-step model, in spite of the fact that it gives a somewhat better fit to the kinetics. Our two-step model was based on the heterogeneous nucleation model already developed by Ferrone et al.[12,13] but we modified the secondary nucleation in a very specific way to generate lateral growth of the 2-D polymer. The unconstrained heterogeneous nucleation does not account for lateral growth. It is possible that the formation of hook decorations has a large effect on the kinetics, masking the lateral growth step. This might be explained by a three stage model, incorporating secondary nucleation for lateral growth and tertiary nucleation for hook decorations, but we would then have so many parameters that the model would lose credibility.

Summary of Nucleation Models and Conclusions

The strongest conclusion is that the size of the critical nucleus, under our assembly conditions, is about 6–7 subunits. This number is obtained from the slope of the log-log plot of assembly time versus concentration and from the fit of every model tested to the

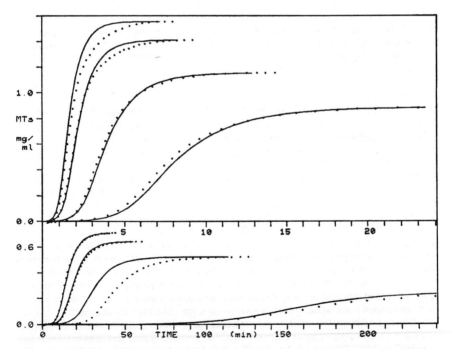

FIGURE 7. The best fit of experimental kinetics to the unconstrained heterogeneous nucleation model of Ferrone et al.[12,13] Both the primary and secondary nuclei were six subunits for this best fit. Note that the computer curves have a sharper sigmoid shape than those in FIGURE 6 and give a somewhat better fit to the explosive kinetics of assembly.

experimental kinetics. We assume that this nucleus comprises two protofilaments of three and four subunits. This is consistent with our thermodynamic analysis, which showed that any two-filament sheet could elongate spontaneously and therefore serve as a nucleus. According to our thermodynamic analysis the smallest two-filament sheet, a trimer or tetramer, should serve as a nucleus. How do we explain our finding that we need a heptamer? Polymers with less than 6–7 subunits are apparently less stable than would be predicted by simple addition of bonds. We note that there is a similar problem for actin nucleation: the simplest model would predict a dimer for the critical nucleus (because the third subunit closes the helix), but experimentally it is found that $n = 3$–4.[7-10] Apparently nucleation is more complex than the simplest ball and stick model in both cases.

A second strong conclusion is that a simple model of homogeneous nucleation, which gave a perfect fit to experimental kinetics of actin assembly, gives a very poor fit for the microtubule data. The fit is greatly improved by a two-step nucleation mechanism, so we conclude that a secondary nucleation is necessary. The second step may be nucleation of a new protofilament along the edge of a sheet or heterogeneous nucleation of new sheets along all polymer surfaces. The former model is attractive to us because it is designed specifically for the 2-D assembly of the microtubule lattice. The latter model, unconstrained heterogeneous nucleation, gives a better fit to the experimental data and can also be rationalized by structures seen in electron microscope studies.

Further study is needed to explore two questions. First, why is the primary nucleus as large as it is? We need to know if the value $n = 7$ is obtained for a wide range of assembly conditions or if it can be much larger or smaller. Second, what is the nature of the secondary nucleation? Is it nucleation of each new protofilament in the growing 2-D polymer, or is it nucleation of new polymers on all exposed polymer surfaces? Again, experiments with different assembly conditions should help decide whether heterogeneous nucleation is dominating the assembly kinetics.

THE ROLE OF RINGS IN MICROTUBULE ASSEMBLY

Previous Experiments and Models with Tubulin + MAPS

Ten years ago we obtained micrographs of microtubule sheets that had ring polymers attached to the ends and sides.[14] We drew two conclusions from these structures. First, rings were identified as protofilaments that curved to form coils or circles when they were not bonded in the microtubule sheet. Second, rings were postulated to be intermediates in assembly, perhaps required for nucleation.

The identification of rings as coiled protofilaments was based on images in which individual protofilaments could be traced from the microtubule wall into the the ring. Several labs have presented images of disassembling microtubules showing a similar continuity of protofilaments and rings.[15-17] A number of ring structures have been described, but we believe it is now well established that they are all composed of tubulin subunits in a modified (curved) protofilament bonding pattern.

Our second conclusion was that rings were an intermediate in assembly. This conclusion was based on images showing rings attached to small sheets, apparently uncoiling and straightening out as they were incorporated into the growing microtubule lattice. In addition, we found that 6-S tubulin separated from rings by Sepharose chromatography would not assemble unless the ring fraction was added back. We now understand that this apparent requirement for rings was actually due to the MAPs. MAPs were bound to the rings and were separated exclusively into the ring fraction by our chromatography. In the buffer conditions used at that time MAPs were essential for assembly, so the apparent requirement for rings was actually for the MAPs that were bound to them. Regardless, the images of rings attached to the growing sheets remain highly suggestive that in this system they are incorporated as preformed intermediates, rather than dissociating first to subunits.

This interpretation has been supported by recent work of Pantaloni et al.[18] who used radiolabeled tubulin to show that "tubulin incorporated in the initial stages of assembly came predominantly from rings (85–90%). This result indicates that oligomers directly issued from rings are the first intermediates in the assembly process and suggests that at the beginning of polymerization incorporation of isomers or fragments of rings proceeds at a faster rate than their dissociation into dimers." The conclusion that rings, or fragments issued directly from them, can be incorporated directly into growing microtubules is thus well supported for the conditions of these two studies.

It is now clear, however, that rings are not always obligatory intermediates in assembly. They are especially important in assembly of MAP-enriched tubulin, as in the two studies mentioned above. In most other assembly systems, rings are rarely seen attached to growing microtubule polymers, and in some conditions, rings may be completely absent both before and during assembly. Most of these assembly systems are MAP free. Frequently, rings are not formed in the absence of MAPs or, as discussed below, rings that do form may have a different structure that makes them incapable of being incorporated directly into microtubules. Our interpretation that rings are intermediates in assembly is thus valid only for certain assembly conditions.

Present Experiments with MAP-free Tubulin

In our experiments with MAP-free tubulin, we found that assembly kinetics varied considerably depending on some details of tubulin preparation. We eventually pinned the differences down to the presence or absence of rings. This preliminary exploration led to several interesting observations.

The formation of rings is affected by several factors in the handling of the protein. Magnesium concentration is very important: 5 mM magnesium could support ring formation, but few or no rings were obtained at 0.5 mM magnesium. Temperature affects both the formation and stability of rings, but we have not yet explored this effect in detail. The most important variable that could be easily controlled was tubulin concentration. In our standard assembly buffer (3.4 M glycerol, 5 mM magnesium sulfate, 0.05 M morpholino-ethane-sulfonic acid, pH 6.6, 1 mM GTP) rings were formed when microtubules were disassembled (at 0°C) at a tubulin concentration greater than 5 mg/ml.

A remarkable characteristic of these rings is that once formed they appeared to be very stable. Thus if microtubules were disassembled at >5 mg/ml, rings would form, and these would remain stable for hours even if they were diluted to less, to 1 mg/ml. This is clearly not an equilibrium situation because rings were never formed spontaneously at this low protein concentration. We have preliminary observations that show rings disassembling very slowly (half-life 2–3 hours in dilute solution at 8°C). The rings appear to be stable in dilute solution because their disassembly is very slow.

Based on these observations, we have two protocols for tubulin preparation. In both protocols tubulin from the freezer is thawed, assembled at 37°C, and pelleted in the centrifuge. This removes inactive tubulin, generally 10–30% of the total. (1) Ring-rich tubulin is prepared by putting the pellet on ice for 20–30 minutes before adding cold buffer to resuspend the protein. The protein concentration in this pellet is very high

(20–40 mg/ml), and rings apparently form as the microtubules disassemble. When buffer is subsequently added to resuspend and dilute the protein these rings remain for several hours at 0°C. (2) Ring-free tubulin is prepared by resuspending the pellet in fresh buffer at 37°C to a protein concentration less than 5 mg/ml. This procedure ensures that the concentration of free tubulin is always <5 mg/ml. This dilute suspension is then placed on ice to disassemble the microtubules. Electron microscopy routinely showed a complete absence of rings in tubulin prepared by this protocol.

TABLE 1 summarizes some experiments showing that the kinetics of assembly were always slower when rings were present. The effect on assembly kinetics was especially pronounced at high protein concentrations. It seemed that ring-rich preparations had an additional lag time of 7–10 minutes. We eventually concluded that the rings in our MAP-free preparation were of a special form that could not be incorporated intact into the microtubule and were relatively stable, requiring perhaps 10 minutes to disassemble. There was no indication of preferential incorporation of ring tubulin, so we believe the rings probably dissociate to free subunits before they can contribute to assembly.

This is in striking contrast to the case of MAP-enriched tubulin where rings or fragments derived directly from them are preferentially incorporated. The difference is

TABLE 1. The Effect of Rings on Assembly Kinetics[a]

Tubulin Concentration (μM)	Rings Present?	Tenth Time (min)
11.4	yes	18.4
10.8	no	7.2
16.9	yes	9
16.3	no	1.6

[a]Two tubulin stock solutions (25 μM) were prepared—one with and one without rings as described in the text. Dilutions were made from each stock solution, and assembly runs were performed. The tenth time is the time required for the turbidity to reach 10% of its maximum value. Note that these assembly times are 7–11 minutes longer for tubulin preparations enriched in rings.

probably due to the role of MAPs in the previous studies. Ring polymers of tubulin plus MAPs are frequently open coils or fragments, which could be incorporated into the microtubule lattice simply by uncoiling and straightening out. The rings in our present study were formed from purified tubulin in a Mg^{2+}-glycerol buffer and are quite different in structure. These are almost all intact double rings,[19] which are probably stabilized by the multiple contacts between the inner and outer rings. These rings cannot uncoil until the intact circle is broken, and this first break may be the rate-limiting step in disassembly. Once the circle is broken, disassembly may be very fast, because there are no MAPs to stabilize them. This model is clearly a speculation, but it is one possible explanation for the retardation of assembly by MAP–free rings.

In the kinetics and model-building studies presented above, the complication of rings was avoided by using 0°C tubulin preparations that had no rings. We believe that the tubulin in these samples is exclusively 6-S tubulin subunits, and that both nucleation and elongation occur directly from the pool of subunits.

In conclusion, we would like to raise a warning that metastable rings are likely to exist in standard tubulin preparations of other labs. The possible involvement of metastable rings should be considered whenever the preparation has involved disassembly of a pellet at high protein concentration. We have demonstrated here that these rings strongly affect assembly kinetics, and there is considerable evidence that rings behave differently from free subunits in drug and nucleotide binding. It is therefore important to know whether rings are present.

Sedimentation analysis is probably not a good assay for the presence of rings. MAP-stabilized rings are known to dissociate in the ultracentrifuge,[14,20] and it is very likely that MAP-free rings also dissociate much more rapidly at high pressure. Thus if the protein concentration is below the critical concentration for ring formation the ultracentrifuge might not show the presence of the metastable polymers. This could explain a long-standing contradiction in the work of Frigon and Timasheff.[21] Their ultracentrifuge observations were consistent with a rapid equilibrium of rings and subunits, with a critical concentration for ring formation of 5–15 mg/ml at different magnesium concentrations. Electron microscopy, however, showed abundant rings in samples diluted to 0.1 to 1.0 mg/ml, an order of magnitude below where they could be detected in the ultracentrifuge. The rings seen by microscopy of these dilute preparations were certainly real, but they apparently dissociated in the centrifuge too rapidly to be detected.

The presence of metastable rings should be considered whenever the tubulin preparation has involved disassembly of a microtubule pellet at high protein concentration. They can apparently be avoided if the pellet is suspended to <5 mg/ml before disassembly. We suggest that the best assay for the presence of rings is carefully controlled electron microscopy.

REFERENCES

1. VOTER, W. A. & H. P. ERICKSON. 1984. J. Biol. Chem. **259:** 10430–10438.
2. OOSAWA, F. & M. KASAI. 1962. J. Mol. Biol. **4:** 10–21.
3. ERICKSON, H. P. 1974. J. Cell Biol. **60:** 153–167.
4. BURTON, P. R. & R. H. HIMES. 1978. J. Cell Biol. **77:** 120–133.
5. ERICKSON, H. & D. PANTALONI. 1981. Biophys. J. **34:** 293–309.
6. JENCKS, W. P. 1981. Proc. Natl. Acad. Sci. USA **78:** 4046–4050.
7. WEGNER, A. & J. ENGEL. 1975. Biophys. Chem. **3:** 215–225.
8. WEGNER, A. & P. SAVKO. 1982. Biochemistry **21:** 1909–1913.
9. TOBACMAN, L. S. & E. D. KORN. 1983. J. Biol. Chem. **258:** 3207–3214.
10. COOPER, J. A., E. L. BUHLE JR., S. B. WALKER, T. Y. TSONG & T. D. POLLARD. 1983. Biochemistry **22:** 2193–2202.
11. FRIEDEN, C. & D. W. GODDETTE. 1983. Biochemistry **22:** 5836–5843.
12. FERRONE, F. A., J. HOFRICHTER, H. R. SUNSHINE & W. A. EATON. 1980. Biophys. J. **32:** 361–380.
13. FERRONE, F. A., J. HOFRICHTER & W. A. EATON. 1985. J. Mol. Biol. **183:** 611–631.
14. ERICKSON, H. P. 1974. J. Supramol. Struct. **2:** 393–411.
15. WARNER, F. D. & P. SATIR. 1973. J. Cell Sci. **12:** 313–326.
16. KIRSCHNER, M. W., R. C. WILLIAMS, M. WEINGARTEN & J. C. GERHART. 1974. Proc. Natl. Acad. Sci. USA **71:** 1159–1163.
17. BORDAS, J., E.-M. MANDELKOW & E. MANDELKOW. 1983. J. Mol. Biol. **164:** 89–135.
18. PANTALONI, D., M. F. CARLIER, C. SIMON & G. BATELIER. 1981. Biochemistry **20:** 4709–4716.
19. VOTER W. A. & H. P. ERICKSON, 1979. J. Supramol. Struct. **10:** 419–431.
20. MARCUM, J. M. & G. G. BORISY. 1978. J. Biol. Chem. **253:** 2852–2857.
21. FRIGON, R. P. & S. N. TIMASHEFF. 1975. Biochemistry **14:** 4559–4566.

Dynamics of Tubulin and Calmodulin in the Mammalian Mitotic Spindle

J. RICHARD McINTOSH,[a] WILLIAM M. SAXTON,[a] DEREK
L. STEMPLE,[a] ROGER J. LESLIE,[a] AND
MICHAEL J. WELSH[b]

[a]Department of Molecular, Cellular, and Developmental Biology
University of Colorado
Boulder, Colorado 80309
and
[b]Department of Anatomy
University of Michigan
School of Medicine
Ann Arbor, Michigan 48109

Microtubules (MTs) have attracted interest and experimental attention in part because they are involved in numerous cellular processes and in part because they are intriguingly dynamic. Many studies with light and electron microscopes have shown that MTs can change their location and extent of polymerization, depending on the physiological state of the cell. Efforts to study this dynamism and the cellular factors that control it have, however, approached the problem either indirectly or with low time resolution. For example, one can look at the changes in the birefringent retardation of the mitotic spindle as a result of anaphase or of some experimental treatment. Such observations are, however, indirect assays of tubulin assembly, because birefringence depends on the orientation and bunching of spindle MTs in addition to the amount of polymer and of other birefringent components present. One can directly study the effects of a given cellular perturbation on MT assembly by fixation and subsequent immunofluorescence, but such investigations are necessarily of low time resolution. MT assembly can be studied quickly and directly *in vitro*, but the systems of buffers and the proteins used for chemical analysis may not be an accurate reflection of the state within the cell. To study the assembly of MTs in their natural context, one wants a way to look directly at the polymerization behavior of tubulin in cells.

Fluorescence microscopy of dye-tagged molecules, or fluorescent analogue cytochemistry, is a powerful way to follow the polymerization dynamics of specific proteins *in vivo*.[1] The requirements of the method are well defined: 1) a fluorophore-conjugated protein that is indistinguishable from the native macromolecule by all available functional tests, 2) a way to introduce the labeled protein into appropriate cells, 3) sufficient fluorescence signal to permit localization of the labeled protein at adequate space resolution, and 4) a device for following the fluorophore that can quantify the observed fluroescence. This approach to the study of tubulin was first used by Keith, Feramisco, and Shelanski[2] who recognized the suitability of dichlorotriazinyl aminofluorescein (DTAF) as a fluorophore to label tubulin. They showed that DTAF tubulin will form cytoplasmic fibers that resembled MTs in their distribution and response to colchicine. Wadsworth and Sloboda[3] confirmed these findings in a study of DTAF tubulin microinjected into sea urchin eggs, noting particularly the speed with which the fluoresecent tubulin analogue would incorporate into spindle-like structures.

Our laboratory has been investigating tubulin dynamics of living cells in collaboration with E. D. Salmon of the University of North Carolina (see his paper, this volume, p. 580). We have obtained additional evidence that DTAF tubulin is a good analogue for native mammalian brain tubulin *in vitro*[4] and have shown that when this protein analogue is injected into living cells, the fluorescent fiber arrays that form are essentially identical to the distribution of total tubulin polymer as seen in fixed cells by immunofluorescence.[5] A quantitative analysis of both the rates of fluorescence incorporation and of fluorescence redistribution after photobleaching have confirmed the previous impression of highly dynamic tubulin behavior in the mitotic spindle,[6] and have demonstrated a rather rapid turnover in the interphase MT arrays. Details of this work have been published elsewhere.[4-7] In this contribution we summarize the most important of our findings about tubulin and discuss plausible interpretations of the results. The behavior of tubulin will be compared with that of a less-prevalent spindle protein, calmodulin. The similarities and differences between the behaviors of these two spindle components are informative with regard to some aspects of spindle structure and to potentially powerful ways to learn more about the organization and function of the mitotic machinery.

BEHAVIOR OF DTAF-TUBLIN *IN VITRO*

Our conditions for conjugation of DTAF to MTs and for the subsequent purification of polymerization-active tubulin give a dye/protein molar ratio of about one.[4] One-dimensional sodium dodecyl sulfate (SDS) gel electrophoresis shows that there is little protein aggregation in the final material, and that 90 to 100% of the dye is covalently bound to protein (it runs with the protein, not at the dye-front). When labeled protein is microinjected into cells, there is no fluorescence in the interphase nucleus, although unconjugated fluorophore injected into cytoplasm rapidly stains chromatin. We infer that all the dye, including the small fraction that runs at the dye-front, is tightly bound to tubulin under physiological conditions. Two-dimensional gel electrophoresis shows that both alpha and beta tubulin are labeled, beta about twice as strongly as alpha. The fluorescent protein has a slightly more acidic isoelectric point than unlabeled material, consistent with the conjugation of DTAF to amine groups on the protein.

DTAF-tubulin assembles and disassembles in the presence of heat-stable brain mircotubule-associated proteins (MAPs) at rates indistinguishable from those of unlabeled tubulin plus MAPs at equivalent concentrations.[4] Pure DTAF-tubulin plus MAPs form true MTs, as seen by electron microscopy of negatively stained material. It adds to the ends of sea urchin flagellar MTs with the bias in polymerization rates characteristic of unlabeled brain tubulin: faster at the distal end of the flagellum. Both labeled and unlabeled tubulin add to MTs initiated by centrosomes at their ends distal to the centrosome. In neither of the above experiments is there an observable tendency for the DTAF-tubulin to stain the walls of existing MTs, suggesting that the fluorescence of MTs formed in the presence of labeled tubulin is a result of DTAF-subunits incorporated into the MT wall, not simply bound to the polymer's surface.

BEHAVIOR OF DTAF-TUBULIN *IN VIVO*

DTAF-tubulin microinjected into living mammalian cells incorporates into fibrous arrays that strongly resemble the cell's MT complexes.[5,6] Immunofluroescent pictures

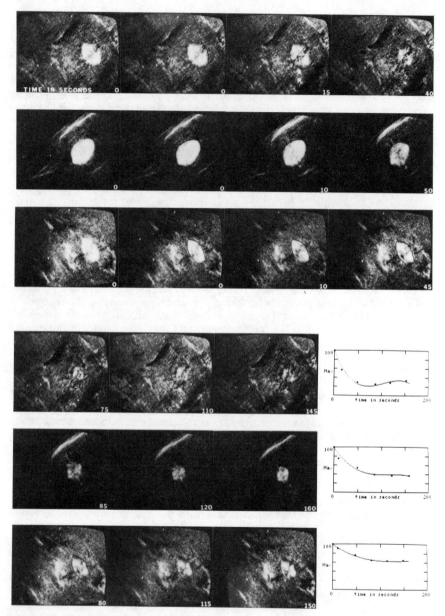

FIGURE 1. Cultured mammalian cells, strain PtK_1, viewed with polarization optics. The images shown are photographs taken on 35 mm film from a video monitor displaying computer-averaged images of 16 successive video frames (about one-half sec of data). The top three rows of pictures are different cells, the first injected with 10 μM TC complex, the second with 10 μM DTAF-TC complex, and the third with 10 μM colchicine and bovine serum albumin at a mass concentration equal to the tubulin in the previous injections. The next three rows of pictures show the further development of birefringence with time. The points for the graphs shown at the end of each

of cells fixed after equilibration with injected tubulin show that the image of rhodamine-labeled antitubulin, which stains all the MTs, and the fluorescein fluorescence of the injected tubulin are very similar. The DTAF-tubulin image changes in expected ways upon treatment of the cells with nocodazole or taxol, and the time-dependent images of fluorescent tubulin over the cell cycle are consistent with the statement that the cell uses DTAF-tubulin as it does its own protein.

CONTROLS FOR DTAF-TUBULIN INCORPORATION INTO MTS *IN VIVO*

These observations do not exclude the possibility that the fluorescent tubulin analogue binds *in vivo* to the walls of existing MTs, staining their surfaces rather than incorporating into the walls of the cellular polymers. Whereas the evidence cited above from experiments *in vitro* makes this suggestion unlikely, it is not impossible. The ideal way to test this possibility would be to find a mutant cell in which all tubulin was missing under nonpermissive conditions, for example, an amber mutant in all relevant tubulin genes combined with a temperature sentitive supresser mutation. Unfortunately no such strain of injectable cells exists. We are therefore in the process of conducting several approximations to this experiment. One is based on the fact that tubulin forms a complex with colchicine that is slow to dissociate.[8] One can form tubulin-colchicine (TC) complexes *in vitro,* separate them from unbound colchicine by gel filtration chromatography, and then inject them into cells to ask whether they will interfere with cellular physiology. Colchicine probably works to block MT assembly by forming a TC complex that adds to MT ends, thereby both distorting the normal lattice and reducing the rate of addition of further subunits.[9,10] The blocking of MT assembly *in vivo* by injected TC complexes should therefore be an indication of the extent to which they compete with endogenous tubulin for end addition. By this logic, a comparison of the efficacy with which TC and DTAF-TC interfere with endogenous tubulin assembly is a measure of the capacity of DTAF-tubulin to incorporate into MTs *in vivo*.

We have made TC complexes and DTAF-TC complexes and injected them into mitotic mammalian cells. Either complex induces a rapid reduction in spindle birefringence (FIGURE 1). Spindle birefringence is slightly affected by injection of buffer containing bovine serum albumin, presumably a result of the perturbation caused by the injection itself. The birefringence is equivalently affected by injection of BSA plus 10 μM colchicine, suggesting that unbound colchicine leaves the cell faster than it can bind to the endogenous tubulin. Further, most of the cells that were injected with colchicine plus BSA entered an apparently normal anaphase, whereas only 10% of the cells injected with TC or DTAF-TC complex continued with mitosis. We conclude that the DTAF-TC is essentially as good at blocking the addition of endogenous tubulin as is the unlabeled TC complex. These data suggest that the cell does not distinguish between injected tubulin and DTAF-tubulin.

---

picture series were obtained by drawing a boundary around the spindle birefringence in the first image of each cell, then integrating the light intensity within and outside the boundary. The outside intensity was used to normalize all the pictures in one series to the same brightness level, then the inside intensity was used to estimate total spindle birefringence. The same boundary was aligned by eye on each image in the series to be sure that the same area was used to define the spindle in each case. The curves are cubics fit by a method of minimum sum squared error. The cell injected with colchicine and BSA entered anaphase, whereas the others did not. × 725.

GEOMETRY OF INCORPORATION OF DTAF-TUBULIN *IN VIVO*

One might hope that changes in the distribution of fluorescence during the incorporation of DTAF-tubulin might be informative about the cellular loci most active in tubulin assembly. So far, this hope has not been realized. The spindle seems to increase in fluorescence over its entire volume, and the fluorescence of the interphase MT array is buried in the fluorescence of unincorporated tubulin until the equilibration process is well advanced. We cannot say from the data in hand where subunits add to an MT *in vivo*.

RATES OF INCORPORATION OF DTAF-TUBULIN MICROINJECTED INTO CELLS

The behavior of DTAF-tubulin injected into interphase mammalian cells is rather what one might expect. The fluorescence is initially concentrated at the site of injection, then it spreads throughout the cytoplasm at a rate appropriate for diffusion. The cytoplasm is uniformly bright for several minutes, but over 3 to 20 min, a fibrous fluorescence appears, consistent with the notion that DTAF-tubulin is slowly incorporating into cellular polymers. It is difficult to say when such a reaction is complete, but our best estimates of its rate suggest a half-time of 20 ± 10 min ($n = 11$). In mitotic cells, the rate of incorporation is dramatically faster. The spindle becomes brighter than background before the newly injected protein has had time to diffuse across the cell. The mean half-time for fluorescence incorporation into 6 metaphase cells was 18 ± 14 sec and for 10 prometaphase asters was 14 ± 7 seconds.

Whereas these differences in rate are sufficiently marked to suggest an interesting difference between mitosis and interphase in the tubulin turn-over mechanisms, the injection experiments may be criticized on the grounds that the transient increase in tubulin concentration that follows injection may cause nonphysiological behavior. This criticism motivates a different approach to the problem in which there will be no concentration transient at the time of the experiment.

RATES OF DTAF-TUBULIN TURNOVER AS SEEN BY FLUORESCENCE REDISTRIBUTION AFTER PHOTOBLEACHING

We have used the method of fluorescence redistribution after photobleaching (FRAP) to circumvent the problems resulting from tubulin concentration transients immediately following injection. In these experiments cells are injected prior to study and allowed the time necessary for the injected protein to equilibrate with the endogenous tubulin pools. One then brings the preparation back to a microscope, equipped with a stage incubator to assure physiological temperatures, and finds previously injected cells by their fluorescence. A microbeam of laser light is imaged for a brief time (0.5–0.05 sec) on a part of the fluorescent cell, for example one spindle pole, and pictures are taken at successive times with a sensitive video camera to record the patterns and rates of fluorescence redistribution (FIGURES 2 and 3). In such experiments there is no change in tubulin concentration at the time of the experiment, only a change in the visibility of some of the tubulin (bleached versus unbleached). The reappearance of fluorescence is due entirely to the motion of unbleached fluorophores, because the bleaching reaction is irreversible. With this approach one eliminates some of the concerns relevant to injection-incorporation experiments, but one introduces a new set of problems that requires a new set of controls, as described below.

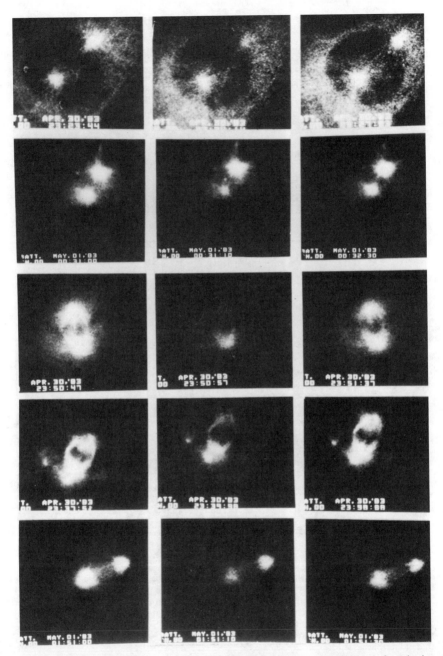

FIGURE 2. Mammalian cells injected with DTAF-tubulin at least one-half hour before viewing to permit equilibration of the fluorescent analogue with the cellular tubulin pool. Each cell is shown before, immediately after, then about a minute after photobleaching by a microbeam of argon laser light at 488 nm. The rows of pictures are prophase, prometaphase, metaphase, early anaphase, and mid-to-late anaphase. Different extents of bleaching and various rates of recovery are shown, but in every case, the fluorescence returns. × 1200.

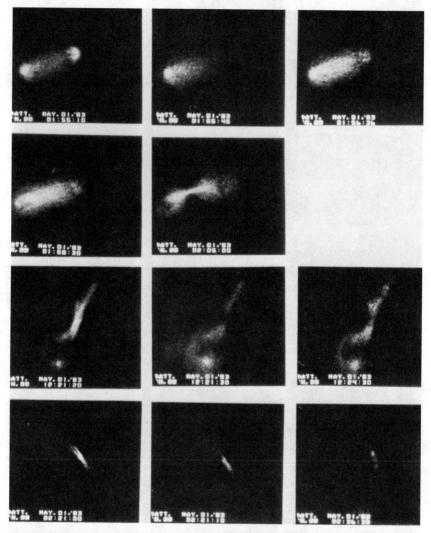

FIGURE 3. A set of images similar to that shown in FIGURE 2. The anaphase of the first five images is the same as that shown at the end of FIGURE 2. These pictures demonstrate that a cell bleached twice will complete anaphase and cytokinesis. The different times of recovery suggest that the rate of fluorescence redistribution becomes slower in later anaphase. The images in the next two rows are of telophase midbodies and show that the trend to slower fluorescence redistribution continues into late mitosis. Midbody fluorescence redistribution is about as slow as that of interphase. × 1200.

We have measured the rates of FRAP of DTAF-tubulin in spindles and in the interphase MT complex of both PtK and BSC cells. The mean half-time for FRAP of metaphase spindles (indistinguishable between the two cell types) is 11 ± 6 sec ($n = 19$). For prometaphase asters, the half-time is 14 ± 10 sec ($n = 10$). For interphase cells, the mean half-time is 200 ± 85 seconds. Thus our studies with FRAP confirm the difference in rates of tubulin turnover between mitotic and interphase cells, suggesting that this is a real feature of tubulin metabolism. The reasons and mechanisms for the difference are yet to be determined.

GEOMETRY OF DTAF-TUBULIN TURNOVER AS SEEN BY FRAP

If there were a substantial movement of MTs during the time that a photobleach was recovering, one would expect to see a translation of the bleached spot during its recovery. We have looked in both mitotic and interphase cells, but find no evidence for such a movement. In the spindle, the fluorescence comes back uniformly into the region that has been bleached.[5] Therefore cellular MTs do not show any evidence of treadmilling, that is, the steady-state addition of subunits to one end and removal from the other.[11] Spindle fluorescence recovery is fast enough that it is not possible to exclude a slow treadmilling of MTs during metaphase, but one can say with confidence that treadmilling is not the mechanism by which the spindle rapidly recovers fluorescence. In interphase on the other hand, the bleached spot sometimes moves a little during recovery. Examination of many cases shows that the spot occasionally moves in toward the centrosome and sometimes moves away. Bidirectional movement is confusing to a model of cellular tubulin turnover based on treadmilling and suggests that things are going on which we do not fully understand. Our observations on spot movement will be published in detail elsewhere.

CONTROLS FOR MT DISRUPTION BY FLUORESCENCE BLEACHING

The mitotic rate of FRAP is so fast that one is concerned about the possibility of some sort of experimental artifact. A simple explanation would be that rapid FRAP is a repair artifact of laser-induced MT destruction. Several lines of evidence suggest, however, that the process of bleaching does not disrupt MTs. Polymer made *in vitro* from pure DTAF-tubulin and heat-stable MAPs may be bleached to 80% of its initial fluorescence without change in its length distribution, as seen by negative staining.[4] MTs bleached *in vitro* to negligible fluorescence will continue to elongate by addition of DTAF-tubulin at their ends, but they do not incorporate fluorescence into their surfaces at a measureable rate. We infer that bleached MT ends retain their polymerization activity, but that no new sites for subunit addition are opened up along an MT wall by the bleaching process. Microtubule structure is not disrupted by bleaching *in vivo* as seen either with polarization optics,[6] by immunofluorescence with antitubulin, or by electron microscopy.[5] Microtubule function is not disrupted by bleaching *in vivo*, as seen both by the continuation of normal anaphase in cells bleached two or three times during chromosome movement (FIGURE 3) and by the persistence of particle saltation along originally fluorescent MTs that have been bleached during interphase.

A CHEMICAL STUDY OF TUBULIN PHOTOBLEACHING

The polymerization activity of DTAF-tubulin *in vitro* seems to be much more sensitive than the structure of MTs to damage by photobleaching *in vitro*. By trial and error we have identified conditions in which bleaching causes no visible damage to tubulin's capacity to polymerize, but there are circumstances in which the fluorescent analogue loses its activity upon irradiation. When pure DTAF-tubulin at 1 mg/ml plus 20 μg/ml unlabeled MAPs in a conventional piperazinediethanesulfonic acid (PIPES) polymerization buffer is bleached in a 1 cm path-length cuvette, using the 488nm line from a 2 W argon ion laser at full power (about 650 mW), approximately 5 min is required to obtain a bleach to 80% fluorescence. This treatment kills the polymerization activity of the tubulin, regardless of whether the protein was in polymer or monomer form at the time of bleaching. Electrophoresis of the irradiated material shows that some of the tubulin has become cross-linked to MAPs and other tubulin molecules, suggesting that MAP inactivation might contribute to the loss of activity. Fresh MAPs added after bleaching do not, however, restore polymerization, and all assembly activity is lost when only 20% of the tubulin is bleached. These results suggest that some sort of mobile intermediate is formed during bleaching that can damage tubulin molecules lying at a considerable distance from the bleaching event. Sheetz and Koppel[12] have shown that the bleaching of a fluorophore attached to the outside of a red blood cell membrane will induce cross-linking of proteins that reside on the inner surface of the membrane, also suggesting that the active species in the bleach-induced cross-linking reaction can diffuse before it acts.

Photobleaching is thought to involve the formation of free radicals,[13] so we have experimented with several additives to try to reduce the loss of polymerization activity experienced during bleaching of DTAF tubulin. Free radical scavengers, such as ascorbic acid, reduce the amount of damage done by a given energy of irradiation, but they also reduce the amount of photobleaching. When a preparation of DTAF tubulin is made 4 mM in ascorbate, about five times more energy is required to achieve equivalent bleaching. When this energy is delivered, however, the resulting 20% bleach again induced a complete loss of polymerization activity. Sulfhydryl-reducing agents, such as mercaptoethanol, on the other hand, decrease the rate of loss of activity without affecting the rate of photobleaching. The cellular reducing agent glutathione is even more affective in protecting polymerization activity. In 5 mM glutathione, there is also a reduction in the extent of protein cross-linking, as seen by gel electrophoresis, suggesting that the formation of inter- and intra-molecular disulfide bonds may contribute to the mechanisms of tubulin inactivation and cross-linking.

There are suggestions in the literature that the amount of damage to proteins during photobleaching depends on the rate at which the bleaching energy is delivered to the fluorophore.[12] We therefore constructed a capillary flow cell whose inner diameter was about the same as the cross-section of the laser beam in our apparatus for microirradiation. DTAF tubulin was passed through the tube just fast enough that the time any one fluorophore would be exposed to the beam was about 0.5 sec, the longest time of irradiation in our cellular bleaching experiments. Sufficient material was then passed though the tube to permit its subsequent analysis for cross-linking and loss of polymerization activity. When DTAF tubulin was bleached by about 20% during a 0.5 sec irradiation, the loss of polymerization activity was substantially less than that seen in a 5 min bleach by 20 percent. Fast bleaching by 20% in glutathione leaves most of the DTAF tubulin active. When pure DTAF tubulin is diluted with unlabeled tubulin by 1:10, about the dilution achieved in an injection experiment, the resulting mixture may be bleached fast in glutathione with no observable loss in polymerization activity. Because these are the conditions that best approximate the experiments on FRAP *in vivo,* we think it likely that photobleaching of some of the injected fluorescent analogue does not seriously perturb tubulin assembly in the cell. Our experiments do not yet tell us, however, whether an individual DTAF-tubulin molecule may be bleached without losing its polymerization activity.

The reduction in photo-induced damage associated with fast bleaching at high power densities is consistent with a model in which protein damage is a two-step process. Our working hypothesis is that when the fluorophore is bleached: a comparatively long-lived intermediate is created. The intermediate can diffuse, and when it encounters a protein, it promotes oxidation of protein sulfhydryls, causing both denaturation and cross-linking. We suggest that the intermediate is itself photo-labile, so if the light intensity is very high, there is a good probability that the intermediate will be destroyed before it has a chance to act on a protein to which it has diffused. Such a two-step model makes several predictions that we are in the process of testing.

POSSIBLE MECHANISMS FOR RAPID, UNIFORM MT TURNOVER IN MITOSIS

The rapid rates of both incorporation and FRAP of DTAF-tubulin in mitotic cells suggest that there is something special about spindle MTs as compared with those of interphase. One possibility is that the concentration of MT ends in the spindle is much larger than that found in interphase. There is no rigorous study of MT length distributions available for mammalian spindles, but the data available for *Dictyostelium,*[14] for the alga *Ochromonas,*[15] and for the fungus *Puccinia*[16] suggest that at metaphase there is an essentially uniform distribution of MT lengths. One would expect, therefore, that there are plus ends of MTs scattered approximately uniformly throughout the metaphase spindle. This allows one to understand the uniformity of fluorescence incorporation into spindles, even if subunit exchange is confined to MT ends. In principle, the end exchange of subunits could also account for the fast rate of MT turnover, but a quantitative analysis of the rates observed here, using known disassembly rate constants and a model permitting subunit exchange at ends only, shows that simple condensation polymerization is unlikely to account for fast mitotic FRAP.[6] To fit the data, one must assume a mean MT length of less than a micrometer, and the structure data on spindle tubule length exclude this possibility. A model based on treadmilling would permit subunit exchange at the observed rates, but makes the strong prediction of bleach spot movement during recovery, a prediction that is not born out by experiment.

Inoue[17] has suggested that MT subunits in solution could exchange with material in the polymer wall at sites other than the polymer ends. One can imagine an "endotubulase" that catalyses this exchange at different rates during different times of the cell cycle and fits all the data available. There is, however, compelling evidence that the major site of subunit addition to MTs *in vitro* is at the polymer ends.[18,19] It would therefore be appealing to find a model to account for fast MT turnover that is based on the assumption of subunit exchange at polymer ends.

Such a model has recently been proposed by Mitchison and Kirschner.[20,21] They have shown that MTs *in vitro* display a dynamic instability that leads some MTs to elongate slowly while others rapidly disappear. They interpret their observations in the light of data on the rates of addition and loss for tubulin subunits bound to either guanosine triphosphate (GTP) or guanosine diphosphate (GDP).[22] In this model, MTs are stable and grow as long as their ends are occupied by tubulin with GTP bound. If GTP hydrolysis catches up with subunit addition, generating a GDP-tubulin end, then the fast off-rate for this molecular species causes the MT to deploymerize rapidly to nothing. For MTs grown from a centrosome, fiber disappearance unveils a polymer initiation site, so a new MT can start to grow. Thus, a labile but constant MT array, like a metaphase spindle, can be thought to exist in rapid turnover as a result of occasional, catastrophic disassembly of a few MTs and continuous replacement of lost polymer by slow MT regrowth.

The Mitchison and Kirschner model would account for our data, so long as one assumes that the disappearance and reappearance of different MTs occurs asynchronously. The recovery of a bleached spot would then result from the rapid, occasional disappearance of individual bleached MTs, followed by their replacement, one by one, with MTs containing unbleached tubulin. The model also allows one to understand the stability of MTs attached to chromosomes, because they would have their plus ends capped by binding to the kinetochores, thereby preventing their fast disassembly. The model predicts that kinetochore MTs will show slow FRAP, a possibility that we are currently testing. The model also allows one to understand how the spindle might concentrate proteins on the kinetochore fibers, because they would be the oldest MTs in the metaphase spindle. If a particular MAP, for example, a calmodulin binding MAP, was slow in binding to MTs, then it would become concentrated on the more stable spindle fibers, because the others would not stay around long enough for appreciable binding to occur.

The difference in MT lability between mitosis and interphase can be understood in this model by the differences in the number of MTs initiated by the centrosomes at each cell cycle stage.[23] This single change would alter the mean number of GTP-tubulin subunits at the end of each MT, and thus the probability that a particular MT would suffer a catastrophic disassembly. The idea is thus very appealing in the face of information currently available. There are in addition several specific predictions of the model, so we should be able soon to test its validity more directly.

THE DYNAMICS OF SPINDLE CALMODULIN

Of the few spindle proteins other than tubulin that have so far been identified, calmodulin (CaM) is one the most interesting, because it is known to regulate so many biological processes.[24] CaM has been identified as a spindle protein by immunofluorescence[25] and by fluorescent analogue cytochemistry.[26] We have therefore undertaken to measure its rate of turnover in the mitotic spindle.

We have derivatized CaM with either tetramethyl rhodamine[26] or DTAF. Both fluorochromes yields a CaM analogue that is active by the criteria of showing a calcium-dependent shift in electrophoretic mobility[24] and of conferring a calcium-dependence on the specific activity of brain phosphodiesterase. Both analogues incorporate into mitotic spindles within seconds of injection into a mitotic cell. Neither shows specific localization in interphase other than being excluded from the nucleus.

All of our quantitative work on the rate of CaM turnover *in vivo* has been done by measuring the FRAP of DTAF-CaM during metaphase and anaphase. With the data in hand (8 metaphase cells and 2 anaphase), we cannot distinguish between the two mitotic stages. At metaphase, DTAF-CaM is distributed in a spindle-like morphology with a marked concentration at the poles and a characteristic staining of the kinetochore spindle fibers[25] (FIGURE 4). Following fluorescence bleaching, there is first a rapid recovery of background, presumably due to diffusion, then a slower rise in the specific fluorescence above background (FIGURE 4). We have used the same sort of methods to quantify the rate of CaM FRAP as those that were applied for tubulin. The mean half-time for CaM fluorescence redistribution is 42 ± 32 sec, about three times slower than the rate for tubulin. Preliminary results suggest that this rate is a result of two separate processes: a comparatively rapid FRAP on the spindle fibers and a slower FRAP in the vicinity of the pole.

The difference between turnover rates of CaM and tubulin is at first surprising, given the low molecular weight of the former and the intuitive sense that CaM will simply be bound to one or more spindle proteins, whereas tubulin will be incorporated into an MT wall. There are numerous possible explanations for the difference, and it seems too soon to speculate on its significance. In current work we are trying to measure the FRAP rate of the kinetochore MTs to see whether these more stable fibers turnover at the same slower rate as the CaM, and we are planning to measure CaM

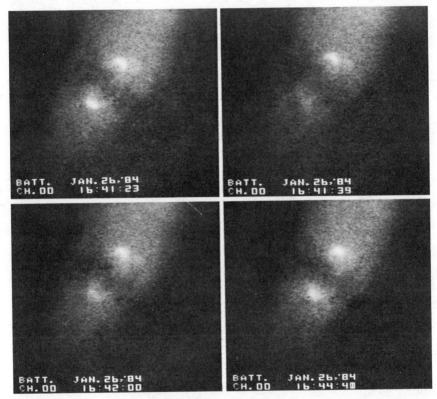

FIGURE 4. A PtK$_1$ cell injected with DTAF-calmodulin about 1 hr before viewing. Images were prepared as in FIGURE 2. The bleaching and redistribution of fluorescence is clear. × 1700.

FRAP with better time and space resolution to see if it confirms the heterogeneity in recovery, indicating the existence of several binding compartments.

SIGNIFICANCE OF SPINDLE COMPONENT TURNOVER

We hope that further work on the turnover of MTs will allow us to identify the significant pathways by which tubulin assembles and disassembles in cells. For example, if we can learn the site of subunit loss from the kinetochore MTs during anaphase, we will determine something quite important for specifying the location and mechanism of the spindle motors. Additional work on FRAP rates of other spindle components may help us to learn about their interactions with one another. It seems likely that interacting proteins will display similar turnover rates, and we may with FRAP analysis be able to group spindle proteins into classes that represent functional units.

The fact that CaM turns over more slowly than tubulin suggests that CaM does not bind directly to all the MTs of the half spindle. If the rate of CaM FRAP turns out to be slower even than that of the kinetochore MTs, then there will be a suggestion that at least some of the spindle CaM is bound not to MTs but to some other structural component of the spindle. Through a continuation of this kind of work, we hope to be able to learn a good deal about the interactions between different spindle components and to make progress toward understanding the mechanisms of mitosis.

REFERENCES

1. TAYLOR, D. L., P. A. AMATO, K. LUBY-PHELPS & P MCNEIL. 1984. Fluorescence analogue cytochemistry. Trends Biochem. Sci. **9:** 88–91.
2. KEITH, C. H., J. R. FERAMISCO & M. SHELANSKI. 1981. Direct visualization of fluorescein-labeled microtubules *in vitro* and in microinjected fibroblasts. J. Cell Biol. **88:** 234–240.
3. WADSWORTH, P. & R. D. SLOBODA. 1983. Microinjection of fluorescent tubulin into dividing sea urchin eggs. J. Cell Biol. **97:** 1249–1254.
4. LESLIE, R. J., W. M. SAXTON, T. J. MITCHISON, B. NEIGHBORS, E. D. SALMON & J. R. MCINTOSH. 1984. Assembly properties of fluorescent labeled tubulin *in vitro* before and after fluorescence bleaching. J. Cell Biol. **99:** 2146–2156.
5. SAXTON, W. M., D. L. STAMPLE, R. J. LESLIE, E. D. SALMON, M. ZAVORTINK & J. R. MCINTOSH. 1984. Tubulin dynamics in cultured mammalian cells. J. Cell Biol. **99:** 2177–2186.
6. SALMON, E. D., R. J. LESLIE, W. M. SAXTON, M. L. KAROW & J. R. MCINTOSH. 1984. Spindle microtubule dynamics in sea urchin embryos. J. Cell Biol. **99:** 2164–2176.
7. SALMON, E. D., W. M. SAXTON, R. J. LESLIE, M. L. KAROW & J. R. MCINTOSH. 1984. Diffusion coefficient of fluorescein-labeled tubulin in the cytoplasm of embryonic cells of the sea urchin. J. Cell Biol. **99:** 2157–2164.
8. WILSON L. & J. BRYAN. 1974. Biochemical and pharmacological properties of microtubules. Adv. Cell Biol. **3:** 21–71.
9. MARGOLIS, R. L. & L. WILSON. 1977. Addition of colchicine-tubulin complex to microtubule ends: the mechanism of substoicheometric poisoning. Proc. Natl. Acad. Sci. USA **74:** 3466–3470.
10. STERNLICHT, H. & I. RINGEL. 1979. Colchicine inhibition of microtubule assembly via copolymer formation. J. Biol. Chem. **254:** 10540–10550.
11. MARGOLIS, R. L. & L. WILSON. 1978. Opposite end assembly and disassembly of microtubules at steady state *in vitro*. Cell **13:** 1–8.
12. SHEETZ, M. P. & D. E. KOPPEL. 1979. Membrane damage caused by irradiation of fluorescent concanavalin A. Proc. Natl. Acad. Sci. USA **76:** 3314–3317.
13. WEBB, W. W., L. S. BARAK, D. W. TANK & E- S WU. 1980. Molecular mobility on the cell surface. Biochem. Soc. Symp. **46:** 191–205.
14. MCINTOSH, J. R., U- P ROOS, B. NEIGHBORS & K. L. MCDONALD. 1985. Architecture of the microtubule component of mitotic spindles from *Dictyostelium discoideum*. J. Cell Sci. **75:** 93–129.
15. TIPPIT, D. H., L. PILLUS & J. D. PICKETT-HEAPS. 1980. Organization of the spindle microtubules in *Ochromonas dancia*. J. Cell Biol. **87:** 531–545.
16. TIPPIT, D. H., C. T. FIELDS, K. L. O'DONNELL, J. D. PICKETT-HEAPS & D. L. MCLAUGHLIN. 1984. The organization of microtubules during anaphase and telophase spindle elongation in the rust fungus *Puccinia*. Eur. J. Cell Biol. **34:** 34–44.

17. INOUE, S. & H. RITTER. 1975. Dynamics of mitotic spindle organization and function. *In* Molecules and Cell Movement. S. Inoue & R. E. Stephens, Eds.: 3–30. Raven Press. New York.

18. BRYAN, J. 1976. A quantitative analysis of microtubule elongation. J. Cell Biol. **71:** 749–767.

19. BERGEN, L. G. & G. G. BORISY. 1980. Head-to-tail polymerization of microtubules *in vitro*. Electron microscopic analysis of seeded assembly. J. Cell Biol. **84:**141–150.

20. MITCHISON, T. & M. KIRSCHNER. 1984. Microtubule assembly nucleated by isolated centrosomes. Nature (London) **312:** 232–237.

21. MITCHISON, T. & M. KIRSCHNER. 1984. Dynamic instability of microtubule growth. Nature (London) **312:** 237–241.

22. CARLIER, M-F., T. L. HILL & Y. CHEN. 1984. Interference of GTP hydrolysis in the mechanism of microtubule assembly: an experimental study. Proc. Natl. Acad. Sci. USA **81:** 771–775.

23. SNYDER, J. A. & J. R. McINTOSH. 1975. Initiation and growth of microtubules from mitotic centers in lysed cells. J. Cell Biol. **67:** 744–760.

24. MEANS, A. R., J. S. TASH & J. G. CHAFOULEAS. 1982. Physiological indications of the presence, distribution, and regulation of calmodulin in eukaryotic cells. Physiol. Rev. **62:** 1–39.

25. WELSH, M. J., J. R. DEDMAN, B. R. BRINKLEY & A. R. MEANS. 1979. Tubulin and calmodulin. Effects of microtubule and microfilament inhibitors on localization in the mitotic apparatus. J. Cell. Biol. **81:** 641–655.

26. ZAVORTINK, M., M. J. WELSH & J. R. McINTOSH. 1983. The distribution of calmodulin in living cells. Exp. Cell. Res. **149:** 375–385.

Microtubule Dynamics in Mitotic Spindles of Living Cells[a]

PATRICIA WADSWORTH[b] and E. D. SALMON

Department of Biology
University of North Carolina
Chapel Hill, North Carolina 27514

Observations on living, dividing cells,[1-3] together with more recent experiments with microtubules (MTs) assembled *in vitro* (for review see reference 4), suggest that MTs are dynamic structures and that an understanding of this dynamic behavior is critical for an understanding of mitosis. Early work on living cells clearly demonstrated the labile equilibrium between the MT-containing spindle fibers and a cellular pool of tubulin subunits. In addition, a role for this assembly and disassembly in chromosome movement was proposed.[2,5] These studies showed that the assembly of spindle MTs and their subsequent disassembly are intimately linked to the mitotic process. The pathways and regulation of tubulin association and dissociation with spindle MTs, however, were not resolved. A variety of morphological studies, performed using immunofluorescence and electron microscopy, have revealed that spindles are composed, in part, of a bipolar arrangement of MTs that radiate from the spindle poles and overlap at the metaphase plate.[6-12] Such techniques, however, yield only static images of fixed stained cells, so the contribution of MTs to chromosome movement has remained unclear.

In an effort to investigate spindle MT dynamics, we have developed techniques to introduce fluorochrome-labeled tubulin subunits into living, dividing cells to monitor tubulin behavior. This work has been carried out in collaboration with the laboratory of J. R. McIntosh (this volume p. 566). Using this approach, termed fluorescence analogue cytochemistry,[13] the distribution and dynamics of the tubulin subunit can be investigated in living cells, avoiding the complexities of *in vitro* conditions. Similar methods were initially used by Keith *et al.*,[14] who observed the dynamic behavior of the interphase MT array in cultured cells. Subsequently, Wadsworth and Sloboda[15] demonstrated the rapid polymerization of fluorescent tubulin into MT arrays in sea urchin eggs during mitosis. These preliminary studies suggested the utility of these techniques to examine the behavior of tubulin in living cells.

For our experiments, bovine brain tubulin has been covalently modified with the fluorochrome 5-(4,6-dichlorotriazin-2-yl) amino fluorescein (DTAF) and the resulting DTAF tubulin shown to retain the native characteristics of the tubulin molecule. (Details of the labeling procedures can be found in reference 16 and J. R. McIntosh, this volume p. 566). Following microinjection into suitable cells, the DTAF-tubulin has been shown to coassemble with all cytoplasmic MT containing structures in living cells.[17,18] Thus the exogenous labeled subunits serve as "tracers" of the behavior of the endogenous tubulin pool.

In initial experiments, mitotic cells were microinjected with DTAF tubulin, and

[a]This work was supported by NIH GM 24364.
[b]Present address: Department of Zoology, University of Massachusetts, Amherst, Massachusetts 01003.

the pattern of fluorescence followed using low-light-level video microscopy. When the cells were observed just after injection, one would expect that only the most favored sites for MT assembly would have incorporated the fluorescent tubulin subunits. Such experiments, however, revealed that spindles, and interphase arrays, appeared fluorescent in a uniform manner.[17] No preference for particular assembly sites was observed at the resolution of the light microscope. These experiments also showed that the incorporation of fluorescent tubulin into spindles was very rapid, appearing complete within 30–60 seconds after injection into mammalian cells[17] and sea urchin embryos.[15,18] However, these initial surprising results provided only a rough estimate of the rate of incorporation and did not necessarily reflect steady state tubulin exchange with MTs. For example, the observed rapid incorporation of fluorescent tubulin into MTs in injected cells could result from a promotion of MT assembly by the increase in the total tubulin concentration following microinjection. Another experimental approach is required to measure the rate and pattern of tubulin incorporation under steady state MT assembly conditions.

The sensitive and quantitative technique of fluorescence recovery after photobleaching (FRAP),[19–21] originally developed to study the mobility of cell surface components, has now been used to measure steady state MT dynamics in living cells. DTAF tubulin is microinjected into cells and allowed to equilibrate with the endogenous tubulin pool. FRAP analysis is then performed on fluorescent MTs at steady state. In a typical FRAP experiment, a brief pulse of laser light is used to bleach irreversibily the DTAF-tubulin fluorophores in a localized region of the cell. Because MTs are not depolymerized by the bleaching intensities and durations we used, recovery of fluorescence measures the turnover of unbleached DTAF-tubulin subunits for bleached subunits.[17,18] Control experiments indicate that FRAP, performed under appropriate conditions, reliably measures the mobility of fluorescent molecules without damage to living cells.[16–18,21]

We have used either video imaging techniques or photometric techniques to record FRAP.[22,23] In the former, fluorescence images of the recovery process are used to determine whether spatial rearrangements of bleached regions occur during recovery and to estimate the rate of fluorescence recovery. In the latter method, quantitative information is obtained from a photomultiplier, interfaced with an Apple computer, which monitors continuously the recovery process. For both techniques, incomplete recovery indicates that a fraction of the bleached fluorophores are immobile during the time of the experiment.

SPINDLE FRAP

Dividing cells of the sea urchin, *Lytechinus variegatus,* have been used to study spindle MT dynamics with the FRAP technique. Cells, microinjected about 30 minutes before mitosis to allow equilibration of the injected and endogenous tubulin, incorporate DTAF tubulin into spindle and astral fibers at mitosis.[18,22] The distribution of fluorescence during mitosis has been shown to correlate well with the pattern of spindle birefringence, seen with polarized light microscopy.[18] When the fluorescent spindle fibers were locally photobleached (FIGURE 1), and FRAP measured using video techniques, recovery was found to be rapid, nearly complete, and apparently uniform throughout the bleached region.[18,22]

We have now examined the recovery of fluorescence in the sea urchin spindle in more detail using photon-counting, photomultiplier recording techniques.[20–22] For these experiments, the initial fluorescence is measured using an attenuated laser

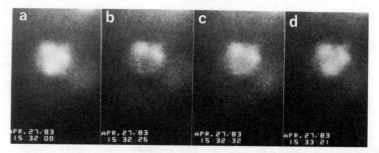

FIGURE 1. Fluorescence recovery after photobleaching (FRAP) of a metaphase tripolar spindle in a first division embryo of *Lytechinus variegatus*. The cell was injected with DTAF-labeled tubulin before spindle formation. The spindle appears uniformly fluorescent before bleaching (**a**). The lower pole was photobleached (**b**) using the 488 nm line from an argon ion laser (Spectrophysics) and a bleaching spot diameter of 4.2 μm. Recovery is shown at 6 sec (**c**) and 55 sec (**d**) after bleaching. Magnification, 165×. (Modified from Salmon.[22] With permission from Cold Spring Harbor Laboratory Publications.)

microbeam to excite the fluorophores along a cylindrical region through the central half-spindle. A photon-counting photomultiplier, coupled to a computer is used to sample and store the data. The fluorescence in the selected region is then photobleached using a brief pulse (100–300 msec) of unattenuated laser light (approximately 10^5 times more intense than the measurement beam). Recovery of fluoresence is monitored as a function of time after bleaching, using the attenuated laser beam.

Analysis of recovery curves indicates that FRAP of spindle MTs consists of at least two major kinetic phases (FIGURE 2). The first phase is due to the diffusion of tubulin subunits through the bleached region.[18] The kinetics of this diffusion, which can be independently measured by bleaching the cytoplasm outside of the MT-containing spindle region, are in good agreement with video measurements of the rate of tubulin diffusion in the sea urchin cytoplasm.[23] The second phase of recovery occurs exponentially and is about 20 times slower. It is due to the incorporation of unbleached DTAF-tubulin subunits into spindle fiber MTs within the bleached region. The first-order rate constant of this incorporation phase was determined by plotting $1n(\overline{F}^\infty - \overline{F}(t))$ where $\overline{F}^\infty$ is the average, recovered fluorescence at approximately 190 sec after bleaching (FIGURE 2). For the sea urchin spindle, recovery of bleached fluorescence is nearly complete (%R = 88 +/− 11) with a first order rate constant, k = 0.045 +/− .013, and the corresponding half-time, $t_{1/2}$ = 16 +/− 4.5 sec (12 cells analyzed). These results are in good agreement with previous measurements from video images.[18]

We have also examined spindle FRAP in mammalian tissue culture cells, strain BSC 1, using these photometric techniques. A typical recovery curve and corresponding log plot are shown in FIGURE 3. The rate of FRAP is rapid, with an average rate constant, k = .017 +/− .004, and average half-time, $t_{1/2}$ = 42 +/− 11 sec at 32°C (17 cells analyzed). The average percentage of recovery, %R = 71 +/− 18 percent. Again, recovery consists of a diffusional phase and a subsequent incorporation phase (FIGURE 3). For these mammalian spindles, however, the data for the incorporation phase are less well fit by a single exponential than the data for the sea urchin spindle. Rather, our preliminary results on BSC 1 spindles suggest that incorporation occurs at two different rates. Previous work, on a variety of spindle types, has shown that kinetochore

MTs (KMTs) are differentially stable to cooling, whereas nonkinetochore MTs are depolymerized.[22–26] The biphasic incorporation phase observed here may be due to different rates of exchange of fluorescent tubulin in stable, KMTs and labile, nonkinetochore spindle MTs. Many of the BSC 1 spindles examined showed two rates of incorporation, although the contribution of each to recovery varied. The faster of the

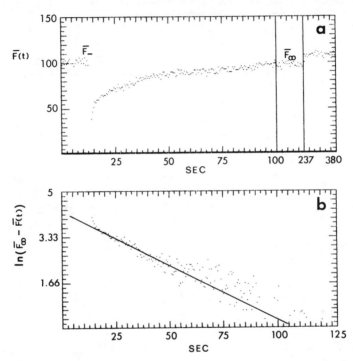

FIGURE 2. a: Computer records of FRAP in a sea urchin metaphase spindle obtained by photometric techniques.[20–22,41] Fluorescence excitation and bleaching were produced by a laser microbeam (488 nm) with a circular Gaussian profile by initially focusing the laser beam onto the field diaphram plane of the epi-illuminator[18] using a spherical focusing lens (17.5 mm focal length). During the measurement phase the laser was attenuated by a factor of approximately 5 × 10^{-4} compared with the bleaching intensity. Spindle fluorescence was bleached by 50–60% within a 100 msec exposure to the laser beam. The number of photon counts was recorded every 0.5 sec using an EMI 9863A photomultiplier, C-10 photon counter (EMI), and an Apple computer to sample and store the data. The first 5 fluorescence values were used to normalize the fluorescence data. A total of 382 samples were taken for this experiment; samples 0–100, 227–237, and 367–382 are plotted. $\overline{F}$ = average, normalized, initial fluorescence; $\overline{F}^{\infty}$ = average, normalized, final fluorescence. **b:** The incorporation phase of fluorescence recovery was analyzed according to the perturbation-relaxation function:

$$\overline{F}^{\infty} - \overline{F}(t) = (\overline{F}^{\infty} - \overline{F}(o))e^{-kt}$$

by plotting $\ln(\overline{F}^{\infty} - \overline{F}(t))$ versus time. A straight line was fit by eye through the data points. This line was used to calculate the following parameters: k, from the slope of the line; and percent recovery, %R = $(\overline{F}^{\infty} - \overline{F}(o))/(\overline{F} - \overline{F}(o))$, where $\overline{F}(o)$ was determined from the value of $(\overline{F}^{\infty} - \overline{F}(t))$ on the straight line at the time of bleaching (t = 0).

two rates, which accounts for the majority of the total recovery, is the value we report here for the BSC 1 spindles (see FIGURE 3). Close examination also suggests that sea urchin spindles may contain two incorporation rates (FIGURE 2). In these spindles, however, the slower phase is minor and not observed in the majority of spindles. These spindle-specific differences may be due to the proportion of differentially stable KMTs[24-26] to total MTs in these different spindles.

To further investigate the two incorporation rates in FRAP, we have examined BSC 1 spindle behavior as a function of temperature. Mitotic cells held at 23°C continue through mitosis, although much more slowly than those at 37°C.[27] When spindles were photobleached at 23°C (FIGURE 4), the half-time for recovery was

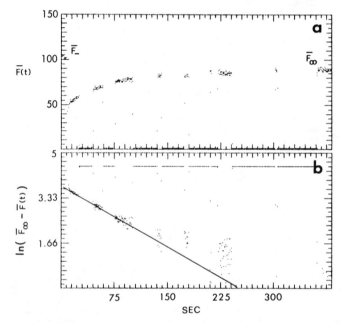

FIGURE 3. **a:** Computer record of FRAP in the mitotic spindle of a mammalian cell, strain BSC 1. Experimental conditions are as in FIGURE 2. Gaussian bleaching beam diameter = 4.5 μm; duration of bleach = 100 msec. The number of photon counts was recorded every 0.5 sec; a total of 377 samples were taken. **b:** Computer generated plot of ln $(\bar{F}^\infty - \bar{F}(t))$ from the data in 3a. Best fit line was used to calculate k, $t_{1/2}$, and %R as described in the legend to FIGURE 2.

prolonged over that seen at 32°C. At lower temperatures, the rate of FRAP was considerably reduced. A linear, Arrhenius relationship between ln k and 1/T was found for temperatures between 23°C and 32°C (correlation coefficient = .990). By linear regression analysis, the rate of FRAP at 37°C was calculated to have a half-time, $t_{1/2}$ = 24 sec (TABLE 1). The percentage of the initial bleached fluorescence that was recovered during the experiment was also reduced at lower temperatures (TABLE 1).

Several explanations for the reduced rate of FRAP are possible. For example, a fraction of the non-KMTs may be preferentially depolymerized at lower tempera-

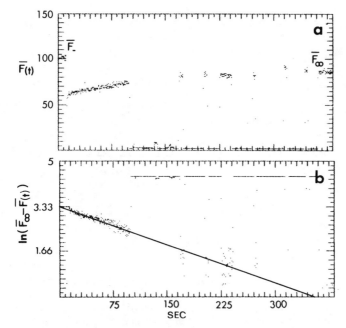

FIGURE 4. **a:** Computer record of FRAP in the mitotic spindle of a BSC 1 cell at 23°C. Experimental conditions as in FIGURE 2. Bleaching beam diameter = 2.5 μm; duration of bleach = 100 msec. The number of photon counts was recorded every 0.5 sec, and a total of 377 samples were taken. **b:** Computer-generated plot of ln $(\bar{F}^{\infty} - \bar{F}(t))$ using the data in **4a**. Best fit line was used to calculate k, $t_{1/2}$, and %R, as described in legend to FIGURE 2.

tures,[24-26] and the more stable KMTs may represent a higher percentage of the MT population in the spindle. Thus, the rate of FRAP would be reduced if the KMTs recover fluorescence more slowly. In addition, cooling may reduce the rate of tubulin turnover by directly affecting protein interactions or by indirect affects on other cellular processes, such as the level of cellular ATP.[28]

TABLE 1. Rate and Percent of FRAP in BSC 1 Spindles Depend on Temperature

T (°C)[a]	$t_{1/2}^{b}$	% R[b]	Number of Cells
20	144 ± 18	47 ± 26	5
23	107 ± 9	39 ± 20	4
26	75 ± 33	53 ± 25	7
32	42 ± 11[c]	71 ± 18	17
37	24[d]	—	—

[a]T(°C) was maintained using a Sage Air Curtain Incubator for 32°, and the room temperature was adjusted for measurements at 20°, 23°, and 26° C.

[b]Calculated from computer generated plots of ln $(\bar{F}^{\infty} - \bar{F}(t))$, as described in FIGURE 2b.

[c]The $t_{1/2}$ for recovery at 32° has now been determined by linear regression analysis of plots of ln $(\bar{F}^{\infty} - \bar{F}(t))$, see reference 41.

[d]Calculated from preceding data by linear regression analysis of a plot of ln k vs $1/T$(°K).

TABLE 2. Comparison of FRAP at Metaphase in Sea Urchin, BSC 1, and Newt Spindles

Cell Type	Physiological Temperature (°C)	k^a (sec^{-1})	$t^a_{1/2}$ (sec)	%R^a
Sea Urchin (*L. variegatus*)	23–25	.045	16	88
BSC 1 mammalian cells	37	.029^b	23.6^b	—
Newt (*Taricha granulosa*)	23	.008	86.6	73

^{a}Calculated from computer plots of ln $(\overline{F}^\infty - \overline{F}(t))$ as described in FIGURE 2b.
^{b}Calculated from data in TABLE 1 as described in the text.

TABLE 2 summarizes our FRAP measurements on three different cell types at physiological temperatures. In all cases, cells are microinjected with DTAF tubulin and allowed at least 5–10 minutes to equilibrate. The spindles, which appear uniformly fluorescent, are photobleached in one-half spindle approximately midway between the chromosomes and the pole. In all three cell types, a large percentage (73–88%) of the bleached fluorescence is recovered during the time of the experiment. This measurement shows that most of the bleached fluorescence is "mobile" and that a small but consistent proportion is not recovered in all spindles examined. Much greater variability is observed in the rate of FRAP, with spindles in newt lung epithelial cells having a half-time for recovery approximately five times slower than the sea urchin at the same physiological temperature. At present, no explanation for this difference in rate of FRAP is available, but is likely due to spindle specific differences. Interestingly, a notable difference in the two spindle types is the pole-to-pole length, suggesting that MT length may influence FRAP. Other differences, however, such as the percentage of KMTs in the spindle may contribute to the different rates of FRAP.

The rates of FRAP obtained photometrically for the sea urchin and BSC 1 cells are comparable to FRAP rates obtained previously using video imaging techniques.[17,18] For the sea urchin, these values are in close agreement: 19 +/− 6.2 sec using video techniques and 16 +/− 4.5 sec for the photometric measurements. For the BSC 1 cells, however, the rate of FRAP measured with video techniques (13 +/− 7 sec) was faster than the rate we estimated photometrically (24 sec). This difference may occur because both the diffusion and incorporation phases of recovery are included in the video measurements, which would lead to an underestimate of $t_{1/2}$ in the more slowly recovering BSC 1 spindles. In addition, subtle differences in culture conditions, temperature, and extent of bleaching may also contribute to the observed difference.

Finally, these photometric techniques have been used to estimate how rapidly and completely the microinjected fluorescent protein equilibrates with spindle MTs. Saxton *et al.*[17] showed that metaphase spindles in BSC 1 and PtK$_1$ mammalian cells are labeled to apparent equilibrium levels, as determined by examination of video micrographs, within approximately one minute of injection of DTAF-labeled tubulin. Is DTAF-tubulin incorporation into spindle MTs complete within this one minute period of fluorescence incorporation? To answer this question, we have used photometric techniques to measure the fluorescence in the central half-spindle of a metaphase BSC 1 cell within 10 minutes of injection, cooled the cell to 4°C for 10 minutes to induce MT depolymerization, then rewarmed the preparation for 10 minutes to induce MT repolymerization to steady state, and remeasured the fluorescence in the same spindle. Spindle fluorescence values, before and after cooling and rewarming, were nearly identical (for example: 4610 average counts/.5 sec and 4936 average counts/.5 sec respectively). This result shows that DTAF tubulin incorporates rapidly and nearly completely throughout the MTs of the spindle fibers in unbleached cells and provides additional evidence that the rapid rates of FRAP measured here are not merely the result of laser-induced damage to the MTs and subsequent repair (references 17, 18, and McIntosh, this volume p. 566).

NON-STEADY STATE TUBULIN DISSOCIATION RATE

If the steady state tubulin incorporation rate into spindle MTs is as rapid as the FRAP results indicate, than an equivalent or faster rate of MT depolymerization should occur when assembly is abruptly blocked at steady state. This prediction has been demonstrated previously using colchicine-like drugs to abruptly block MT assembly and reveal the intrinsic rate of non-steady state tubulin dissociation from spindle MTs.[29] For these experiments, high concentrations of the drug colchicine were microinjected into sea urchin eggs at metaphase, and the rate of MT disassembly was determined from the loss of spindle birefringence retardation (BR). Such experiments revealed an exponential loss of BR with a characteristic rate constant of 0.11 sec and corresponding half-time of 6.5 sec for MT depolymerization.[29] Moreover, the loss of BR occurred uniformly throughout the spindle and asters.

More recently, we have used tubulin colchicine (TC) complex, which contains no free colchicine, as the depolymerizing agent.[30-32] Again, rapid spindle disassembly was observed after microinjection, but at final intracellular concentrations much lower than required for colchicine alone (TABLE 3). Saxton et al. have also found similar results following microinjection of TC into mitotic mammalian cells.[33] These experiments demonstrate that the TC complex is the active intermediate in colchicine-induced MT disassembly. They also confirm that TC acts as a substoichiometric MT poison in vivo,[30-32] as expected from the cooperative character of MT polymerization.

IMPLICATIONS OF MT DYNAMICS FOR THE MECHANISM OF MITOSIS

Our current knowledge of the MT polymer-subunit equilibrium, based largely on in vitro measurements, strongly suggests that MTs elongate by the addition of subunits to the polymer ends.[4,6,34-38] Subunit addition is favored at one end of the polymer,

TABLE 3. Summary of Colchicine and TC Injection Experiments for Embryos of *Lytechinus variegatus*

Injected Substance	Intracellular Concentration (μM)	Loss of Spindle BR
Colchicine	>100	rapid, nearly complete
	10–100	slower, incomplete
	<10	marginal or none
Tubulin	1–5	none
TC[a]	>2	rapid, nearly complete
	.1–2	slower, incomplete

[a]Colchicine to tubulin ratio = .81

referred to as the plus end.[37] This discrepancy in the rate of elongation from the two polymer ends results in MTs that possess an intrinsic growth polarity. Furthermore, it has been demonstrated that MTs, assembled to steady state from tubulin and MT-associated proteins purified from mammalian brain, can assemble at one end while simultaneously disassembling from the other end, at an equal but opposite rate, a process known as treadmilling.[36]

Is the rate of FRAP observed here consistent with a turnover mechanism that occurs solely by simple equilibrium tubulin subunit exchange at the ends of MTs? At steady state, the mean number of dimers, m, replaced at the end of an MT by such a mechanism, in a given time period, can be calculated according to the equation $m = 2(kd\ t/\pi)^{1/2}$.[38] The *in vivo* dissociation constant has been determined from our measurements of MT disassembly following colchicine injection[29] and corresponds to greater than 180 and probably 992 dimers/sec for the sea urchin spindle. These values are calculated from the measured rate constant and the average estimated length of spindle MTs measured from electron micrographs (greater than 1 μm and probably 5.5 μm). Thus, at most, 159 dimers can be replaced from an MT end in 20 sec, a reasonable average $t_{1/2}$ for FRAP in these cells. This value corresponds to about 10% of the dimers in a 1 μm long MT and 2% of the dimers in a 5 μm long MT and is clearly insufficient to account for the observed rate of FRAP in living cells.[18]

An alternative mechanism to account for the observed FRAP rates is that cells use an efficient treadmilling mechanism.[36] Indeed, it has been suggested that treadmilling of MTs in the half-spindle, toward the spindle poles, is responsible for chromosome movement.[39,40] If such a mechanism were used, it predicts that a continual flow of MTs would occur in spindles. Subunits would add at their plus ends, located distal to the poles in the region of the metaphase plate, and disassemble from their minus ends, proximal to the poles.[10-12] Thus a poleward movement of bleached regions would be expected. Initial FRAP experiments strongly suggested that no poleward translocation of the bleached region occurs during the recovery process (references 17, 18, and McIntosh, this volume p. 566).

We have further investigated the question of bleach translocation during recovery using two different FRAP experiments.[41] In the first experiment, a modified FRAP system, capable of producing bar bleach patterns across the half-spindle, normal to the pole-to-pole spindle axis, was used. Because the bar bleach extends across the entire

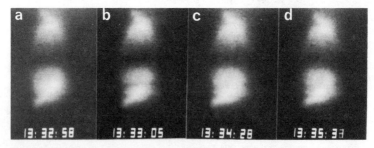

FIGURE 5. Video micrographs[23] of FRAP in a metaphase BSC 1 spindle microinjected with DTAF-tubulin. The spindle was photobleached in the lower half-spindle using an argon ion laser beam initially focused through a cylindrical lens to the field diaphram plane of the epi-illuminator to produce a bar pattern of about 1.6 μm width at the specimen plane. Bleach duration was 100 msec. Cell is shown before (**a**), immediately after bleaching (**b**), and after 82 (**c**) and 146 (**d**) sec of recovery.

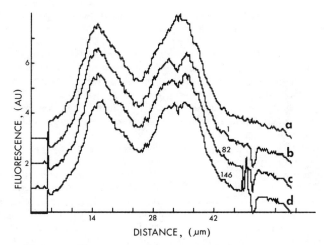

FIGURE 6. Video line scans[29] along the pole-to-pole axis of the spindle shown in FIGURE 5. Video voltages and video micrographs were recorded simultaneously. The line scans have been offset along the Y axis for clarity. Curve a, before bleaching, shows the distribution of fluorescence intensity due to the fluorescent spindle MTs. Curve b, just after bleaching, shows the localized decrease in fluorescence as a result of bleaching. Recovery of fluorescence at 82 (c) and 146 (d) sec after bleaching are shown. AU = arbitrary units.

half-spindle, recovery cannot occur by lateral displacement of unbleached MTs into the bleached spot. An example of such an experiment is shown in FIGURE 5. The bleached region recovers fluorescence without any visible translocation either toward or away from the spindle pole.

In addition to visual examination of video micrographs, quantitative information about the recovery of slit bleaches can be obtained by sampling the video voltages along a line through the spindle pole-to-pole axis.[29] These video line scans and the video micrographs are recorded simultaneously, before the bleach and at intervals during the recovery process, to prevent photobleaching from continual epi-illumination. A set of video line scans is shown in FIGURE 6, which correspond to the video micrographs in FIGURE 5. This experiment demonstrates two important points. First, the bleached region recovers fluorescence without shifting in position along the pole-to-pole axis, as expected from visual inspection of video micrographs of recovery. Secondly, the recovery of fluorescence is only about 70% complete, confirming our results using a photomultiplier recording of spot bleaches (FIGURES 2 and 3). Taken together, this information strongly suggests that a continual flow of subunits toward the spindle pole is not occurring during mitosis for 70% of the MT polymer in the half-spindle. We have also been unable to detect any translocation of that fraction of bleached fluorescence that is not recovered during the time course of these experiments. Although higher resolution analysis is needed to demonstrate unequivically this latter point, all of our analysis to date supports the idea that recovery occurs without a translocation of the bleached region.

In another approach to test poleward flow of tubulin in the half-spindle, the rate of FRAP was measured for circular spot bleaches of two different diameters using our photomultiplier recording method.[41] If treadmilling of tubulin were occurring, one would expect different recovery rates for the two different size bleaches.[20] The rate of

FRAP should be proportional to the diameter of the bleached region if recovery of fluorescence occurs by flow of unbleached regions of MTs through the bleached region. The rate of FRAP, measured using spots of 2.8 or 4.5 μm, was very similiar (t = 43 +/− 13 sec and 41 +/− 10 sec, respectively at 32°C), showing that recovery of fluorescence is independent of the size of the bleached region.

Recently, Mitchison and Kirschner[42,43] have proposed a model for steady state MT dynamics in which MTs exist in rapid growing and rapid shortening phases that interconvert rarely. This end-dependent "dynamic instability" was observed using MAP-free MTs assembled *in vitro*. Their data show that, at apparent equilibrium, some MTs continue to elongate at the expense of other MTs that depolymerize "catastrophically". Rapid, asynchronous MT elongation and depolymerization could be in accord with the rapid rates of tubulin turnover measured in spindle fibers of living mitotic cells using FRAP. This model is attractive in that it combines the rapid, dynamic MT behavior seen *in vivo*,[2,5,17,18,29] the well documented end-dependent behavior measured *in vitro*,[34-38] and the stocastic instability proposed for GTP-tubulin "caps" at MT ends.[44] The dynamic instability model, however, implies that new MTs would regrow from nucleating sites left unoccupied by the catastrophic disassembly of an MT. Because in mitotic cells, nucleation is thought to occur primarily at the spindle poles,[8-11] this model suggests that a directionality to recovery would be seen, from the pole toward the equator, when large areas are bleached.[17,18,41] This result has not been observed in populations of MTs in living cells. If dynamic MT depolymerization and repolymerization were highly asynchronous, however, then directionality to fluorescence recovery might be difficult to detect. Further analysis of the kinetic parameters of the dynamic instability model in living cells is required to resolve these issues.

The data presented here are consistent with the dynamic equilibrium model for mitosis[2,5] in which exchange is not restricted to the ends of the MTs. As is the case for other mitotic models, no definitive proof for the dynamic equilibrium model is currently available. Many of the observations on living mitotic cells, however, are consistent with dynamic microtubule behavior during mitosis. For example, the rapid disassembly and reassembly of spindles in response to various perturbations, the "Northern Lights" behavior of spindle fibers, and the changes in spindle fiber orientation during anaphase, all well documented in living cells, suggest dynamic MT behavior *in vivo*.[1-3] Whereas it is difficult to imagine that polymers can insert subunits along their length, this ability may be related to proteins associated with the surface of an MT. For example, end-dependent and treadmilling behavior have been described in preparations of MTs containing up to 25% MAPs.[36] Such molecules may restrict the addition and loss of tubulin dimers to the MT ends by stabilizing the MT surface lattice.[45,46] Perhaps the same MAP molecules that are associated with the surface of *in vitro* assembled neuronal MTs are not associated with spindle MTs. Unique spindle MAPs,[47] or indeed the absence of MAPs, may permit subunit insertion all along the length of the spindle MTs or regulate the catastrophic behavior of MTs.[42,43] If such were the case, one would predict that the well-characterized neuronal MAPs (such as MAP-2) might reduce the rate of FRAP by altering the mechanism of subunit exchange with MTs, a possibility that we are now investigating.

ACKNOWLEDGMENTS

We appreciate the stimulating discussions and advice of Dick McIntosh, Bill Saxton, Roger Leslie, Ken Jacobson, Lans Taylor, Tom Hays, Mike Caplow, George Langford, and Shinya Inoué.

REFERENCES

1. NICKLAS, R. B. 1975. Chromosome movement: Current models and experiments on living cells. In Molecules and Cell Movement. S. Inoué & R. E. Stephens, Eds.: 97–117. Raven Press. New York.

2. INOUÉ, S. & H. RITTER JR. 1975. Dynamics of mitotic spindle organization and function. In Molecules and Cell Movement. S. Inoué & R. E. Stephens, Eds.: 3–30. Raven Press. New York.

3. BAJER, A. S. & J. MOLE-BAJER. 1975. Lateral movements in the spindle and the mechanism of mitosis. In Molecules and Cell Movement. S. Inoué & R. E. Stephens, Eds.: 77–96. Raven Press. New York.

4. PURICH, D. L. & D. KRISTOFFERSON. 1984. Microtubule assembly: A review of progress, principles and perspectives. Adv. Protein Chem. 36: 133–211.

5. INOUÉ, S. & H. SATO. 1967. Cell motility by labile association of molecules. J. Gen. Physiol. 50: 259–292.

6. MCINTOSH, J. R., W. Z. CANDE & J. A. SNYDER. 1975. Structure of physiology of the mammalian mitotic spindle. In Molecules and Cell Movement. S. Inoué and R. E. Stephens, Eds.: 31–76. Raven Press. New York.

7. FUGE, H. 1977. Ultrastructure of mitotic cells. Int. Rev. Cytol. 6(Suppl). 1–58.

8. INOUÉ, S. 1981. Cell division and the mitotic spindle. J. Cell. Biol. 91(pt2):131s–147s.

9. MCINTOSH, J. R. 1982. Mitosis and the cytoskeleton. In Developmental Order: Its origin and regulation. 77–115. Alan R. Liss, Inc., New York.

10. EUTENEUER, U. & J. R. MCINTOSH. 1981. Structural polarity of kinetochore microtubules in PtK1 Cells. J. Cell Biol. 89: 338–345.

11. TELZER, B. R. & L. T. HAIMO. 1981. Decoration of spindle microtubules with dynein: evidence for uniform polarity. J. Cell Biol. 89: 373–378.

12. EUTENEUER, U., W. T. JACKSON & J. R. MCINTOSH. 1982. Polarity of spindle microtubules in Haemanthus endosperm. J. Cell Biol. 94: 644–653.

13. TAYLOR, D. L., P. A. AMATO, K. LUBY-PHELPS & P. MCNEIL. 1984. Fluorescence analogue cytochemistry. Trends Biochem. Sci. 9: 88–91.

14. KEITH, C. H., J. R. FERAMISCO & M. SHELANSKI. 1981. Direct visualization of fluorescein-labeled microtubules in vitro and in microinjected fibroblasts. J. Cell Biol. 88: 234–240.

15. WADSWORTH, P. & R. D. SLOBODA. 1983. Microinjection of fluorescent tubulin into dividing sea urchin eggs. J. Cell Biol. 97: 1249–1254.

16. LESLIE, R. J., W. M. SAXTON, T. J. MITCHISON, B. NEIGHBORS, E. D. SALMON & J. R. MCINTOSH. 1984. Assembly properties of fluorescent labeled tubulin in vitro before and after fluorescence bleaching. J. Cell Biol. 99: 2146–2156.

17. SAXTON, W. M., D. L. STEMPLE, R. J. LESLIE, E. D. SALMON, M. ZAVORTINK & J. R. MCINTOSH. 1984. Tubulin dynamics in cultured mammalian cells. J. Cell Biol. 99: 2175–2186.

18. SALMON, E. D., R. J. LESLIE, W. M. SAXTON, M. L. KAROW & J. R. MCINTOSH. 1984. Spindle microtubule dynamics in sea urchin embryos: Analysis using a fluorescein-labeled tubulin and measurements of fluorescence redistribution after laser photobleaching. J. Cell Biol. 99: 2165–2174.

19. PETERS, R. 1981. Translational diffusion in the plasma membrane of single cells as studied by fluorescence microphotolysis. Cell Biol. Int. Rep. 5: 733–760.

20. AXELROD, D., D. E. KOPPEL, J. SCHLESINGER, E. ELSON & W. W. WEBB. 1976. Mobility measurement by analysis of fluorescence photobleaching recovery kinetics. Biophys. J. 16: 1055–1069.

21. JACOBSON, K., E. ELSON, D. KOPPEL & W. WEBB. 1983. International workshop on the application of fluorescence photobleaching to problems in cell biology. Fed. Proc. Fed. Am. Soc. Exp. Biol. 42: 72–79.

22. SALMON, E. D. 1984. Tubulin dynamics in microtubules of the mitotic spindle. In Molecular biology of the cytoskeleton. G. G. Borisy, D. Cleveland & D. Murphy, Eds.: 99–109. Cold Spring Harbor Laboratory Publications.

23. SALMON, E. D., W. M. SAXTON, R. J. LESLIE, M. L. KAROW & J. R. MCINTOSH. 1984. Diffusion coefficient of fluorescein-labeled tubulin in the cytoplasm of embryonic cells of the sea urchin. J. Cell Biol. 99: 2157–2164.

24. BRINKLEY, B. R. & J. CARTWRIGHT JR. 1975. Cold-labile and cold-stable microtubules in the mitotic spindle of mammalian cells. Ann. N.Y. Acad. Sci. **253:** 428–439.
25. SALMON, E. D. & D. A. BEGG. 1980. Cold-stable microtubules in kinetochore fibers of insect spermatocytes during anaphase. J. Cell Biol. **85:** 853–865.
26. RIEDER, C. L. 1981. The structure of the cold-stable kinetochore fiber in metaphase PtK_1 cells. Chrom. (Berl.) **84:** 145–158.
27. RIEDER, C. L. 1981. Effects of hypothermia (20–25°C) on mitosis in PtK_1, Cells. Cell Biol. Int. Rep. **5:** 563–573.
28. BERSHADSKY, A. D. & V. I. GELFAND. 1981. ATP-dependent regulation of cytoplasmic microtubule disassembly. Proc. Natl. Acad. Sci. USA **78:** 3610–3613.
29. SALMON, E. D., M. MCKEEL & T. HAYS. 1984. Rapid rate of tubulin dissociation from microtubules in the mitotic spindle *in vivo* measured by blocking polymerization with colchicine. J. Cell Biol. **99:** 1066–1075.
30. MARGOLIS, R. L. & L. WILSON. 1977. Addition of colchicine-tubulin complex to microtubule ends: the mechanism of substoichiometric poisoning. Proc. Natl. Acad. Sci. USA **74:** 3466–3470.
31. STERNLICHT, H. & I. RINGEL. 1979. Colchicine inhibition of microtubule assembly via copolymer formation. J. Biol. Chem. **254:** 10540–10550.
32. DEERY, W. J. & R. C. Weisenberg. 1981. Kinetic and steady-state analysis of microtubules in the presence of colchicine. Biochemistry **20:** 2316–2324.
33. SAXTON, W. M., D. L. STEMPLE & J. R. MCINTOSH. 1984. Microinjection of colchicine bound tubulin into tissue culture cells: Colchicine acts as a substoichiometric microtubule poison *in vivo*. J. Cell Biol. **99:** 38a.
34. BINDER, L. I., W. L. DENTLER & J. L. ROSENBAUM. 1975. Assembly of chick brain tubulin and flagellar microtubules from chlamydomonas and sea urchin sperm. Proc. Natl. Acad. Sci. **72:** 1122–1126.
35. JOHNSON, K. A. & G. G. BORISY. 1977. Kinetic analysis of microtubule self-assembly *in vitro*. J. Mol. Biol. **117:** 1–31.
36. MARGOLIS, R. L. & L. WILSON. 1978. Opposite end assembly and disassembly of microtubules at steady state *in vitro*. Cell **13:** 1–8.
37. BERGEN, L. G. & G. G. BORISY. 1980. Head-to-tail polymerization of microtubules *in vitro*. Electron microscopic analysis of seeded assembly. J. Cell Biol. **84:** 141–150.
38. ZEEBERG, B., R. REID & M. CAPLOW. 1980. Incorporation of radioactive tubulin into microtubules at steady state: experimental and theoretical analysis of diffusional and directional flux. J. Biol. Chem. **255:** 9891–9899.
39. MARGOLIS, R. L., L. WILSON & B. I. KEIFER. 1978. Mitotic mechanism based on intrinsic microtubule behavior. Nature (London) **272:** 450–452.
40. MARGOLIS, R. L. & L. WILSON. 1981. Microtubule treadmills—possible molecular machinery. Nature (London) **293:** 705–711.
41. WADSWORTH, P. & E. D. SALMON. 1985. Analysis of the treadmilling model during metaphase of mitosis using fluorescence redistribution after photobleaching. J. Cell Biol. **102:** In press.
42. MITCHISON, T. & M. KIRSCHNER. 1984. (a) Microtubule assembly nucleated by isolated centromeres. Nature (London) **312:** 232–237.
43. MITCHISON, T. & M. KIRSCHNER. 1984. (b) Dynamic instability of microtubule growth. Nature (London) **312:** 237–242.
44. CARLIER, M. -F., T. L. HILL & Y. CHEN. 1984. Interference of GTP hydrolysis in the mechanism of microtubule assembly: an experimental study. Proc. Natl. Acad. Sci. USA **81:** 771–775.
45. MURPHY, D. B., K. A. JOHNSON & G. G. BORISY. 1977. Role of tubulin-associated proteins in microtubule nucleation and elongation. J. Mol. Biol. **117:** 33–52.
46. SLOBODA, R. D. & J. L. ROSENBAUM. 1979. Decoration and stabilization of intact, smooth-walled microtubules. Biochemistry **18:** 48–55.
47. IZANT, J. G., J. A. WEATHERBEE & J. R. MCINTOSH. 1983. A microtubule associated protein antigen unique to mitotic spindle microtubules in PtK_1 cells. J. Cell Biol. **96:** 424–434.

Control of Microtubule Assembly-Disassembly in Lysed-Cell Models

WILLIAM J. DEERY[a]

Department of Cell Biology
Baylor College of Medicine
Houston, Texas 77030

INTRODUCTION

Most eukaryotic cells contain a network of microtubules (MTs) that extends throughout the cytoplasm, commonly referred to as the cytoplasmic microtubule complex (CMTC) (for review, see reference 1). Assembly of many MTs occurs from the centrosome or MT organizing center (MTOC) and proceeds in an organized fashion to the cell periphery in daughter cells following mitosis or in cells recovering from colcemid block.[1] The factors responsible for the precise spatial distribution of MTs and the intracellular localization of these factors remain to be elucidated. The centrosome is one site that may contain initiating factors required for MT assembly,[2–4] although there are conditions where MT assembly can occur free of the centrosome both in lysed cells[5,6] and in intact cells.[7–10]

Permeabilized mammalian cell systems have been used to demonstrate that centrosomes and kinetochores could nucleate the assembly of purified, exogenous tubulin (for review, see reference 11). For example, when Triton X-100 extracted 3T3 cells are incubated in tubulin dimer that lacks the ability to self-nucleate, MT assembly occurs predominantly from the organizing centers and is similar to that observed *in vivo*.[5,12] Recently, we have used another detergent system, Brij-58, in which the ultrastructure and permeability properties have been characterized by Schliwa and coworkers[13,14] and Cande *et al.*[15] Unlike the Triton-extracted cells, Brij-lysed cells retain endogenous, unpolymerized tubulin, and this has allowed us to examine the characteristics of MT assembly *in situ* from endogenous, unpurified MT protein.[6,16] Random assembly occurs at pH 6.9 and requires nucleotide triphosphates, whereas site-associated MTs are favored at a pH above neutrality in the presence of ATP.

Since the discovery that Ca^{++} inhibits MT assembly *in vitro*,[17] subsequent studies have indicated that several factors such as MAPs and calmodulin[18–24] can influence the degree of Ca^{++} sensitivity. In addition, ATP and Ca^{++}/calmodulin have been shown to decrease the ability of stable tubule-only proteins (STOPs) to stabilize MTs at depolymerizing temperatures.[25] These are only some of many findings that have implicated intracellular Ca^{++} in the regulation of MT assembly *in vivo*.

Incubation of cells in Triton X-100 solubilizes most of the plasma membrane, and unpolymerized tubulin is readily diluted and extracted, creating a permeabilized cell system for studying the MT-subunit dissociation reaction.[14,26,27] The cells are also well extracted for subsequent immunofluorescent studies. We observe a number of cytoplasmic MTs that do not disassemble upon either cold treatment or various

[a]This research was supported by National Institutes of Health Grant CA 23022.

593

dilution conditions, but do disassemble in the presence of micromolar Ca^{++} concentrations. Previously, calmodulin has been localized on kinetochore-to-pole MTs in the mitotic spindle.[28] Indirect immunofluorescence studies using the Triton-lysed system[27] indicate that calmodulin is also associated with cytoplasmic MTs and the centrosomes.

MATERIAL AND METHODS

Cell Culture and Lysis

Swiss mouse fibroblast (3T3) cells were grown on glass coverslips for 48 h to 90–95% confluency in plastic petri dishes containing Dulbecco's modified Eagle's medium supplemented with 10% (for 3T3 cells) or 5% (for SV3T3 cells) fetal calf serum at 10% CO_2 and were used in all experiments.

Immediately before lysis, cells on coverslips were removed from petri dishes and washed for 15 s at room temperature in 0.08 M piperazinediethane sulfonic acid (PIPES) buffer, pH 6.9 containing 10 mM ethylene glycol bis (β-aminoethylether)-N,N,N',N'-tetraacetic acid (EGTA), 1 mM MgCl$_2$, and 0.1 mM guanosine triphosphate (GTP) (extraction buffer) essentially as described by Schliwa et al.[14] Cell lysis in Brij-58 extraction buffer and MT reassembly studies were then carried out as described previously in detail.[6,16] Cell lysis in Triton X-100 extraction buffer and subsequent MT disassembly studies were carried out as described elsewhere in detail.[27] When Ca^{++} was present, the free Ca^{++} concentration was calculated using the Ca^{++}-EGTA buffer system described previously.[20,22]

Immunofluorescent Staining and Microscopy

Either immediately after cells were lysed and washed or after their exposure to various MT assembly/disassembly buffer conditions, cells were fixed for 30 min at 25°C with 3% formaldehyde in PIPES-reassembly buffer (RB) containing 1% dimethyl sulfoxide (DMSO) and processed for indirect immunofluorescence using sheep antibody to tubulin or calmodulin and fluorescein-tagged rabbit-anti-sheep IgG as described previously in detail.[1] Monospecific antibody to calmodulin was prepared in sheep against native rat testis protein and was purified on a calmodulin affinity column.[29] Cells were examined with a Leitz Orthoplan microscope equipped with epi-illumination, and photographs were recorded on Tri-X Pan film (Kodak).

Gel Electrophoresis

Cells near confluence in 100 mm plastic dishes were lysed as described above and collected in 200 μl boiling sample buffer with a rubber policeman. Cell extracts were prepared for electrophoresis and applied to 0.1% sodium dodecyl sulfate (SDS) polyacrylamide slab gels consisting of a 5% pH 6.8 stacking gel and either a 6.3% or 10% polyacrylamide pH 8.8 resolving gel.[30] After staining and destaining, the gels were scanned at 525 nm in a Quick Scan Jr. (Helena Laboratory), and the relative area under each peak was calculated by the gel scan integrator.

RESULTS

Retention of Endogenous Tubulin

After cells are exposed for 3–3.5 min to 0.08 M PIPES-RB containing 0.15% Brij-58 and 1 mM Mg^{++} – 10 mM EGTA, holes are present throughout the plasma membrane with a mean diameter of 95 nm as determined by scanning electron microscopy. This diameter is approximately four times that of a MT. The relative extent to which tubulin in the depolymerized state is maintained within the cytoplasm after lysis was examined by SDS gel electrophoresis. Cells were pretreated with 0.16 μM colcemid for 3.5 h to depolymerize most cytoplasmic MTs and either were not lysed (FIGURE 1A) or were lysed with 0.05% Triton X-100 for 90 s and maintained in a colcemid block for an additional 15 min (FIGURE 1B) or with 0.15% Brij-58 for 3.5 min and maintained in a colcemid block for an additional 40 min (FIGURE 1C), followed by gel electrophoresis. Using the protein band that migrates with a molecular weight characteristic of actin as an internal reference, the ratio of the protein in the tubulin doublet region to actin is essentially the same in unlysed and Brij-lysed samples, whereas in the Triton X-100 lysed cells the ratio is reduced by 30 percent. These data indicate that the unpolymerized cytoplasmic tubulin remains associated with the cytoplasm and is not readily permeable to the Brij-lysed cell membranes.

Characteristics and Nucleotide Requirements of Endogenous Tubulin Reassembly

After colcemid treatment for 3–3.5 h *in vivo,* followed by lysis and incubation in colcemid-RB for 10–40 min, essentially no MTs were present (FIGURE 2A). When these cells, which were exposed to colcemid for 10 min (FIGURE 2B) or 40 min (FIGURE 2C) after lysis, are reversed from colcemid block and incubated in 1 mM GTP-RB for 30 min, MTs reassemble randomly throughout the cytoplasm. Although some assembly is associated with the centrosome, for the most part MTs appear to be free-ended. Under these conditions a nucleotide triphosphate is required for initiation, because submicromolar concentrations of GTP or millimolar guanosine diphosphate (GDP) did not induce MT assembly (FIGURE 2D). MT assembly was induced *in situ* in the presence of millimolar concentrations of adenosine triphosphate (ATP) (FIGURE 4A) or uridine triphosphate (UTP) (FIGURE 4B) plus GDP, apparently by way of transphosphorylation of GDP as reported *in vitro.*[31,32] Like GTP-induced assembly at pH 6.9, MTs are primarily not associated with the centrosomes. Assembly in UTP is similar to that in GTP, whereas MTs are significantly longer in ATP.

The lengths of free MTs were measured with a micrometer ocular as described elsewhere.[5,12] At least 10 MTs were measured per cell, and in each case 10 cells were selected. The mean length of free-ended MTs assembled in GTP or UTP plus GDP was 5 μm, whereas MTs assembled in ATP plus GDP have a mean length of 17 μm (FIGURE 3). There is evidence that the increased lengths in ATP are due to the phosphorylation of a MT protein(s). MTs assembled in the nonhydrolyzable ATP analog, AMPPNP ([B,G-imidoadenosine-5'-triphosphate], which is reported to contain low levels [0.1%] of contaminating ATP), are much shorter and similar in average length to those assembled in GTP or UTP (FIGURE 3C). When the chromatographically purified analog AMPPCP was used (1 mM), however, polymerization did not occur (FIGURE 4C) except when 1–5 μM ATP was added. It should be noted that, in the presence of ATP or UTP, GDP stimulates rather than inhibits assembly; without

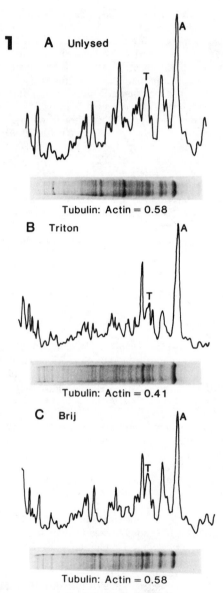

FIGURE 1. SDS polyacrylamide gel electrophoresis (PAGE) of proteins from unlysed and lysed 3T3 cells. Unlysed cells (**A**), cells lysed for 1.5 min in 0.05% Triton X-100 (**B**), or cells lysed for 2.5 min in 0.15% Brij-58 (**C**) were prepared for PAGE as described in MATERIALS AND METHODS. (Deery and Brinkley.[6] With permission from *The Journal of Cell Biology*.)

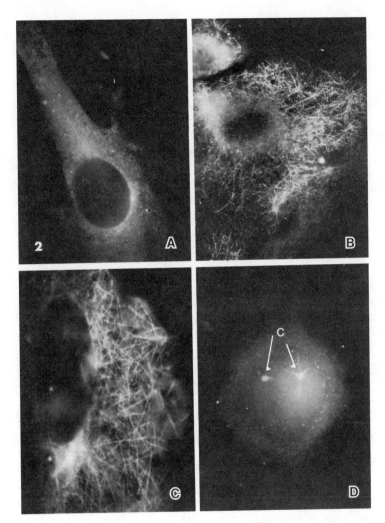

FIGURE 2. Microtubule reassembly from endogenous tubulin in the presence of GTP and inhibition of assembly by GDP. **A:** Cells were pretreated with colcemid *in vivo*, lysed in Brij extraction buffer, and exposed to colcemid-RB for 10–40 min at 25°C. × 1,000. **B:** Lysed cells were incubated for 25 min at 37°C in GTP-RB after a 10-min or **C:** 40-min colcemid block following lysis. × 1,000 and × 1,385, respectively. **D:** Lysed cells were incubated in RB containing 1 m*M* GDP. C = centrosomes. × 1,385. Magnification reduced by 35% in **A–D**. (Deery and Brinkley.[6] With permission from *The Journal of Cell Biology*.)

added GDP, for example, in the presence of these triphosphates alone, no assembly is observed.

MT assembly *in vivo* appears to occur not at random but in association with MTOCs. Because previous *in vitro* studies have indicated that MT nucleation can be retarded at pH values above neutrality,[33,34] the effect of increasing the pH > 6.9 was

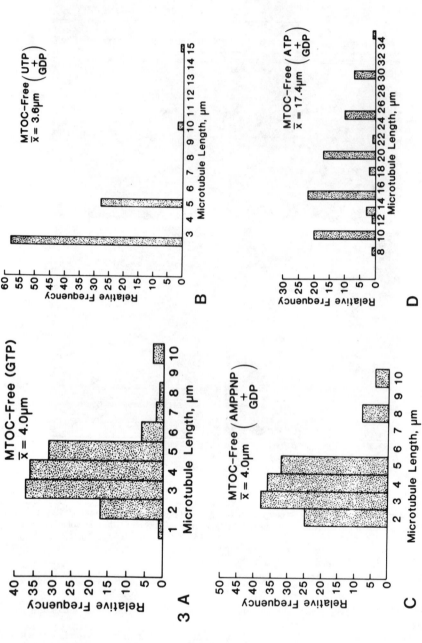

FIGURE 3. Effects of nucleotides on microtubule lengths. The lengths of free MTs were measured with a micrometer ocular as previously described.[5,12] At least 10 MTs were measured per cell, and in each case 10 cells were selected.[6]

examined. When lysed cells containing colcemid-depolymerized tubulin were incubated at 37°C between 7.0–7.6 in the presence of GTP, MT assembly continued to be predominantly free of the centrosome. If the lysed cells, however, were preincubated at 25°C for 10 min in PIPES buffer at pH 7.6 containing 1.5 mM ATP (with or without 0.02 mM 8-Bromo-cAMP) and subsequently incubated at 37°C for 20–30 min at 1

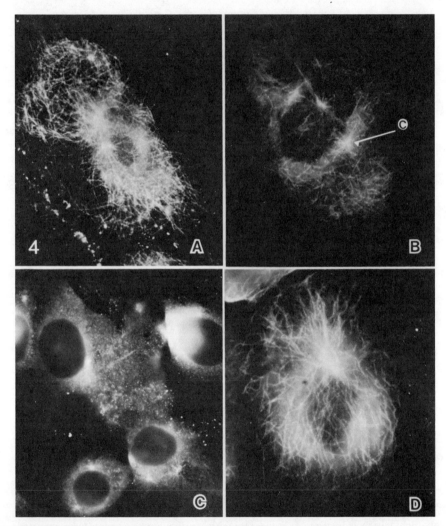

FIGURE 4. Effects of nucleotides and pH on the pattern of microtubule assembly. Lysed cells pretreated with colcemid as described in FIGURE 2 were incubated for 25 min at 37°C in 1 mM ATP-0.3 mM GDP-RB, pH 6.9 (**A**); 1 mM UTP-0.3 mM GDP-RB, pH 6.9. (**B**); 1 mM AMPPCP-0.3 mM GDP-RB, pH 6.9. (**C**); or 1 mM ATP-0.3 mM GDP-0.01 mM 8-Bromo-cAMP-RB, pH 7.6. (**D**). C = centrosome. (**A**) × 760. (**B**) × 990. (**C**) × 760. (**D**) × 990. (Deery and Brinkley.[6] With permission from *The Journal of Cell Biology.*)

mM ATP-0.2 mM GDP, a more organized pattern of assembly was observed. As shown in FIGURE 4D, MTs extended from a central focus (centrosome) outward to the cell periphery. It should be pointed out, however, that some MTs continued to assemble randomly. Thus, in this system, a minimal condition for site-associated MTs requires both high pH and ATP.

Localization of Calmodulin on Microtubules

A complete CMTC was observed (FIGURE 5A) when cells were first fixed in formaldehyde and followed by lysis in cold acetone as described previously.[1] Earlier studies from this laboratory have found calmodulin to be associated predominantly with a cold-stable population of MTs of the mitotic apparatus.[28] The failure to observe anticalmodulin staining of any cytoplasmic MTs in previous studies may have been at least partially related to the protocol of fixation prior to lysis, because this generally yields high fluorescent background and diffuse staining throughout the cytoplasm of interphase cells (FIGURE 5B). When cells were extracted before fixation, however, a more definitive immunofluorescent staining pattern was obtained. When cells were lysed in 0.5% Triton-10 mM EGTA-RB for 90 s and subsequently washed in 4% polyethylene glycol (PEG) to stabilize MTs, the CMTC was stained by calmodulin antibody (FIGURE 5D) and was similar to that observed with antitubulin (FIGURE 5C). MT staining was not obtained with antibody that did not bind to the calmodulin-affinity column, or with anticalmodulin that was preabsorbed for 1 h at 25°C and overnight at 2°C with a 60-fold molar excess of purified calmodulin (not shown).

Behavior of CMTC to Depolymerizing Conditions

Cold-stable MTs associated with the mitotic apparatus have been observed previously by electron microscopy[35–37] and indirect immunofluorescence. Thus, the stability properties of the CMTC were examined at low temperatures. When cells were lysed at room temperature and subsequently incubated for 10–15 min in RB at 0–4°C, some MTs appeared to depolymerize (FIGURE 6A). Numerous MTs, however, were resistant to the cold dilution and seemed to be closely associated with the plasma membrane at their distal ends as viewed by phase-contrast microscopy (data not shown). A similar pattern was observed if living cells were first incubated in medium at 0–4°C for 40 min followed by fixation, or lysis then fixation. It is interesting to note here that upon incubation of lysed cells in ATP-RB, cytoplasmic MTs become more sensitive to cold or warm dilution (data not shown). A similar ATP effect has also been reported by others.[38,39] When lysed cells containing only the cold-resistant MTs were prepared for anticalmodulin immunofluorescence, the staining pattern appeared the same as that observed with antitubulin (FIGURE 6B).

Another way to examine a dilution effect on the CMTC is to subject living cells to hypotonic swelling.[40] When cells were treated with hypotonic medium (3 H$_2$O:1 medium) for 15–20 min followed by fixation and lysis or lysis in Triton-RB then fixation, many cytoplasmic MTs were found to be disrupted, although some persisted (FIGURE 6C). The hypotonic-resistant tubules (and the centrosomes) also stained with anticalmodulin (FIGURE 6D) and were similar in number to those diluted after lysis. Recently, others[41] have also found calmodulin to be localized in the centrosome.

The CMTC can be effectively disrupted by treatment of living cells for several hours with the MT-inhibitor drug, colcemid (for example, see reference 1). Following incubation of living cells with colcemid, essentially all the cytoplasmic MTs depolymerized; frequently, a few tubules could be found associated with the centrosome (FIGURE 6E). When cells treated in this fashion were incubated with anticalmodulin, resistant MTs and the centrosomes were decorated in addition to some diffuse

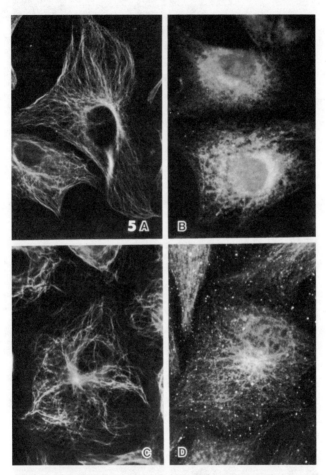

FIGURE 5. Anticalmodulin staining of 3T3 interphase microtubules. Cells were fixed with 3% formalin in PBS and lysed for 6.5 min in $-20°C$ acetone followed by either antitubulin (**A**) or anticalmodulin (**B**) immunofluorescence. **C** and **D** show cells lysed in 0.5% Triton-10 mM EGTA-RB for 90 s, washed in 4% PEG-RB, and prepared for antitubulin and anticalmodulin staining, respectively. (**A**) × 465; (**B**) × 480; (**C**) × 550; (**D**) × 550. (Deery et al.[27] With permission from *The Journal of Cell Biology.*)

cytoplasmic staining (FIGURE 6F). Thus, these disassembly experiments provide evidence that the calmodulin antibody staining was indeed associated with MTs, as indicated by the sensitivity of the staining pattern to three different depolymerizing conditions.

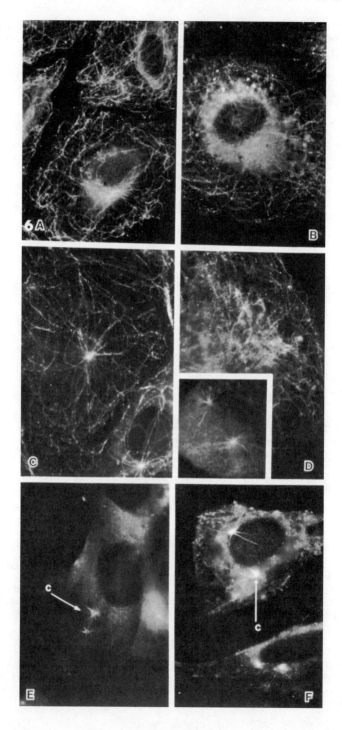

Response of Cytoplasmic Microtubules to Ca^{++}

If intracellular levels of Ca^{++} are involved in the regulation of MT assembly-disassembly, then the CMTC should be sensitive to Ca^{++} concentrations in the micromolar range, particularly if calmodulin is associated with the tubules.[20,25] After lysis in Triton-RB and incubation at 25–37°C in reassembly buffer containing 11 μM free Ca^{++}, the CMTC was disrupted and disassembly in many cases was nearly complete within 7 min (FIGURE 7A). This sensitivity was similar to previous findings of Schliwa et al.[14] Consistent with in vitro studies,[42–44] depolymerization appeared to occur from the polymer ends, starting in many cases at the distal ends at the cell periphery and proceeding toward the centrosome. A number of free-ended MTs segments, however, could be observed in the cytoplasm, apparently a result of disassembly from both ends. Although disassembly along the polymer wall cannot be ruled out in some cases (for example, see reference 45), depolymerization was more extensive at 60 μM Ca^{++}, and disassembly from opposite ends was more pronounced (FIGURE 7C). It is important to note that the extent of disassembly at all Ca^{++} concentrations that were studied varied from cell to cell. At 110 μM Ca^{++}, however, disassembly was essentially complete in most cells (FIGURE 7E). As was observed under the various disassembly conditions shown in FIGURE 6, upon Ca^{++} disassembly, the anticalmodulin staining pattern followed the antitubulin staining of MTs (FIGURE 7 **B, D, F**).

DISCUSSION

Detergent-lysed cell systems have made it possible to subject various intracellular components to defined, controlled buffer conditions such that factors influencing the MT assembly or disassembly process can be examined. Previous studies have shown, for example, that purified, exogenous tubulin will assemble from kinetochores and centrosomes in a fashion similar to that observed in vivo. Other permeabilized cell systems have been used to help elucidate the role of factors (for example, nucleotides and Ca^{++}) in MT-mediated processes such as chromosome movement (for review, see reference 11) and intracellular transport of various cytoplasmic elements.[46–48] In the present study, experiments using two lysed-cell systems, Brij-58 and Triton X-100, have revealed several interesting characteristics pertaining to MT assembly and disassembly, respectively. The endogenous, depolymerized tubulin is retained in Brij-lysed cells; the average length and organization of MTs subsequently reassembled appear to be influenced by nucleotides and pH. Using Triton-lysed cells, a cold/dilution-stable population of MTs is observed; MTs are disrupted, however, by micromolar Ca^{++}, and calmodulin appears to be associated with cytoplasmic MTs.

The observation that cytoplasmic tubulin in the unpolymerized state (in the

---

FIGURE 6. Anticalmodulin staining of 3T3 interphase microtubules resistant to cold/dilution or hypotonic treatment. Cells were lysed in Triton-10 mM EGTA-RB for 90 s, incubated in RB at 2°C for 12 min, fixed, and stained with antitubulin (**A**) or anticalmodulin (**B**). Note the presence of microtubules at the cell membrane. C and D show cells exposed for 20 min to 1 medium: 3 H$_2$O at 37°C, lysed, and stained with either antitubulin or anticalmodulin, respectively. Note the staining of centrosomes (**D** inset). Colcemid (0.3 μM) was added to cells in medium for 2.5 h, and cells were then prepared for either antitubulin (**E**) or anticalmodulin (**F**) staining. Centrosomes (**C**) are clearly stained with anticalmodulin. (**A**) × 600; (**B**) × 705; (**C**) × 735; (**D**) × 740; (**E**) × 735; (**F**) × 705. (Deery et al.[27] With permission from The Journal of Cell Biology.)

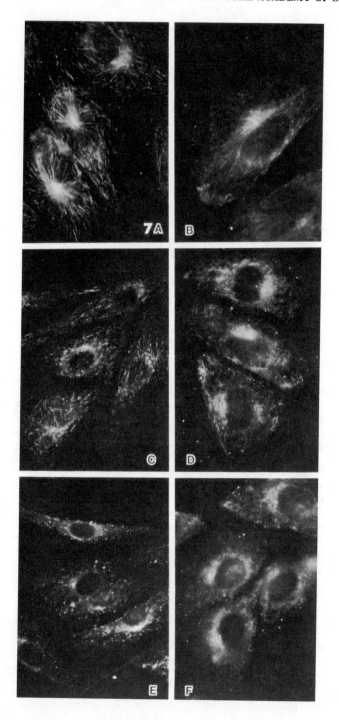

presence of colcemid, for example) is well retained for a relatively long time after lysis in Brij-58 is intriguing. The significant retention raises some doubt as to whether tubulin exists as a "soluble" cytoplasmic protein and strongly suggests that, at least in this lysis system, the tubulin is in an aggregated state and/or may associate with the cytoplasmic matrix or microtrabeculae.[49] It is interesting that several reports have presented evidence for the presence of particulate tubulin or nonmicrotubular aggregates of tubulin in nondividing surf clam eggs,[50] transformed 442 rat kidney cells,[51] and neuroblastoma cells.[52]

Endogenous tubulin polymerization *in situ* has a nucleotide triphosphate requirement; assembly is induced by GTP, ATP, or UTP as observed for MT protein *in vitro*. Although GDP alone does not induce assembly, it is required for ATP or UTP polymerization. Because the nonhydrolyzable ATP analog AMPPCP plus GDP fails to promote assembly, apparently GDP can be transphosphorylated by ATP or UTP *in situ*, as reported *in vitro*.[31,32] At pH values that are optimal for polymerization *in vitro*, reassembly in all cases is attenuated at the centrosomes, in contrast to the organized pattern observed from centrosomes *in vivo*. Because exogenous MAP-depleted tubulin assembles predominantly from centrosomes in Triton-lysed cells,[5,12] it is possible that following lysis, endogenous tubulin might be associated with assembly-promoting factors such as MAPs, which allow for spontaneous, self-nucleating processes to occur as described *in vitro*.[53-55] Such factors can effectively lower the critical concentration of tubulin for nucleation and assembly at sites other than MTOCs. Other studies[7,56] suggest that a similar mechanism may influence the assembly of free MTs in living cell following treatment with the MT assembly-promotor drug, taxol.

We have found conditions where the reassembly of tubulin in lysed cells yields an organized pattern similar to that observed *in vivo*, alkaline pH and ATP. Although these conditions appear to favor site-associated MTs, we are not certain that all cytoplasmic MTs grow from the MTOCs in these models. Indeed, some MTs may assemble free and then become aligned as a result of association with either MTs extending from MTOCs, or other cytoskeletal components or determinants unaffected by colcemid. Ultimately, such interactions would lead to a highly organized CMTC. On the other hand, results by Karsenti *et al.*[10] have recently indicated that some cytoplasmic MTs in cultured cells are not associated with the centrosome.

We observed a significant increase in the average length of MTs assembled in ATP versus those polymerized in GTP or UTP. The increased lengths are reduced, however, when ATP-induced assembly occurs in the presence of nonhydrolyzable ATP analogs. Because ATP and UTP are similar in their ability to induce assembly by way of transphosphorylation of GDP,[31] these observations strongly suggest that phosphorylation of tubulin and/or MAPs is involved in the length differences. Phosphorylation of tubulin and associated proteins *in vitro* by a MAP-kinase has been reported[57,58] and other studies have shown that the kinase is dependent on cyclic adenosine monophosphate (cAMP).[53,59,60] Although the mechanism and significance of phosphorylation on MT assembly are not clear at this time, several studies have demonstrated a stimulatory effect of cAMP on MT growth in several transformed cell types in culture[61,62] and in a lysed-cell system.[12] In addition to the ATP effect, MTs will reassemble randomly in cells subjected to more acidic media.[8] In light of these

---

FIGURE 7. Calcium-induced disassembly of CMTC. Cells were lysed in Triton-10 m*M* EGTA-RB for 90 s, washed for 30 s in 1 m*M* EGTA, then 11 μ*M* Ca^{++}-RB each, and then incubated for 7 min in RB solutions containing either 11 (**A** and **B**), 60 (**C** and **D**), or 110 μ*M* (**E** and **F**) free Ca^{++}. Antitubulin staining is shown in **A**, **C**, and **E**; anticalmodulin is shown in **B**, **D**, and **F**. (**A**) × 480; (**B**) × 445; (**C**) × 355; (**D**) × 555; (**E**) × 355; (**F**) × 555. (Deery *et al.*[27] With permission from *The Journal of Cell Biology*.)

observations, one possible explanation for increased MT lengths and organization in the presence of ATP and alkaline pH is that phosphorylation and alkaline pH may decrease the rate and number of spontaneous nuclei formed by inhibiting or attenuating the nucleation reaction, while subsequently enhancing the kinetics of elongation. Thus, the data presented here indicate that intracellular pH and the phosphorylation state of proteins involved in tubulin polymerization should be considered in the control of organized MT assembly. It is also interesting to speculate that cell cycle-related changes in these parameters might facilitate the formation of discrete MT complexes.

The results from the Triton-lysed cell system indicate that calmodulin is associated with both cytoplasmic MTs and the centrosomes. In addition to 3T3 cells, antibodies to calmodulin also stain cytoplasmic MTs of PtK, bovine kidney (MDBK), African Green monkey kidney (BSC-1), HeLa, and human lung (WI-38) (data not shown). The indirect immunofluorescent staining pattern observed using anticalmodulin is sensitive to various MT depolymerizing conditions. Furthermore, calmodulin is located on MTs reassembling upon recovery from these conditions. The CMTC can be disrupted when exposed to micromolar Ca^{++} concentrations, and our data as well as others[14,18,20,22,25,28] suggest that this sensitivity may be mediated by calmodulin because an antibody to calmodulin stains these MTs.

We have also observed that some cytoplasmic MTs are relatively insensitive to dilution and cold-induced disassembly. Stability of MTs to cold temperatures *in vivo* and *in vitro* has been previously reported. A number of MTs whose ends are associated with the kinetochores and the poles (centrosomes), as well as interzonal tubules in the mitotic apparatus of cultured cells, resist disassembly at 0–4°C.[28,35,36] These MTs also stain with calmodulin antibodies. A similar stability from the interaction of a nontubulin protein(s) (STOPs) with the MT has been observed for a population of MTs isolated from rat brain.[39,63,64] Whether similar factors are responsible for cold-stable spindle and cytoplasmic MTs remains to be elucidated. Because many of the stable MTs appear to be associated with the centrosome and either the kinetochore at mitosis[35] or the plasma membrane during interphase,[40] it is conceivable that they may be specifically involved in chromosome movements as well as cell shape changes and membrane receptor mobility.

It is clear that calmodulin can increase the sensitivity of MTs to Ca^{++} in certain cases *in vitro*,[19,20,21,25] and presently it appears that calmodulin may mediate disassembly through an interaction(s) with proteins that associate with MTs, for example, MAPs[18,19,65] or STOPs.[25,64] Tubulin has not been clearly shown to bind calmodulin. Furthermore, calmodulin does not increase the intrinsic Ca^{++} sensitivity of MTs assembled in the absence of MAPs.[19] As previously mentioned, the association of calmodulin with MTs does not appear to require Ca^{++}, because anticalmodulin stains MTs after lysis and incubation in buffer containing 10 mM EGTA. A low Ca^{++} requirement is practical in that it allows calmodulin to associate with MTs without necessarily inducing disassembly. Precedent exists for the association of calmodulin with other proteins, for example phosphorylase kinase, in a Ca^{++}-independent fashion.[66]

Previous studies using a similar lysed-cell system indicated that various anticalmodulin drugs inhibit Ca^{++}-induced disassembly and provided evidence for a calmodulin-MT association.[14] Our work, however, as well as others[22,25] have shown that anticalmodulin drugs do not necessarily interfere with Ca^{++} sensitivity of tubulin purified *in vitro* or in lysed cells. Together, these data suggest that calmodulin may associate with MTs in a Ca^{++}-independent manner but only affect the state of polymerization in the presence of micromolar Ca^{++}. Interestingly, such a mechanism has also been proposed by Sobue *et al.*[67] for the calmodulin regulation of actin-myosin interaction in the presence of caldesmon and myosin light-chain kinase.

ACKNOWLEDGMENTS

I am grateful to Dr. B. R. Brinkley and Dr. A. R. Means for their valuable discussions, to Ms. Linda Wible for the assistance with cell cultures, and to Ms. Pat Williams for helping prepare this manuscript.

REFERENCES

1. BRINKLEY, B. R., S. H. FISTEL, J. M. MARCUM & R. L. PARDUE. 1980. Int. Rev. Cytol. **63:** 59–95.
2. BERNS, M. W., J. B. RATTNER, S. BRENNER & S. MEREDITH. 1977. J. Cell Biol. **72:** 351–367.
3. GOULD, R. R. & G. G. BORISY. 1977. J. Cell Biol. **73:** 601–615.
4. SNYDER, J. A. 1980. Cell Biol. Int. Rep. **4:** 859–868.
5. BRINKLEY, B. R., S. M. COX, D. A. PEPPER, L. WIBLE, S. L. BRENNER & R. L. PARDUE. 1981. J. Cell Biol. **90:** 554–562.
6. DEERY, W. J. & B. R. BRINKLEY. 1983. J. Cell Biol. **96:** 1631–1641.
7. DEBRABANDER, M., G. GEUENS, R. NUYDENS, R. WILLEBRORDS & J. DEMEY. 1981. Proc. Natl. Acad. Sci. USA **78:** 5608–5612.
8. DEBRABANDER, M., G. GEUENS, R. NUYDENS, R. WILLEBRORDS & J. DEMEY. 1982. Cold Spring Harbor Symp. Quant. Biol. Vol. XLVI: 227–240.
9. WEHLAND, J. & I. V. SANDOVAL. 1983. Proc. Natl. Acad. Sci. USA **80:** 1938–1941.
10. KARSENTI, E., S. KOBAYASHI, T. MITCHISON & M. KIRSCHNER. 1984. J. Cell Biol. **98:** 1763–1776.
11. SNYDER, J. A. 1981. *In* Mitosis/Cytokinesis. A. M. Zimmerman & A. Forer, Eds.: 301–325. Academic Press. New York.
12. TASH, J. S., L. LAGACE, D. R. LYNCH, S. M. COX, B. R. BRINKLEY & A. R. MEANS. 1981. Conf. Cell Proliferation Cold Spring Harbor (Book A) **8:** 1171–1185.
13. SCHLIWA, M. 1980. *In* Procedings of the 38th Meeting of the Electron Microscopy Society of America. G. W. Bailey, Ed.: 814–817. Claitor's. Baton Rouge, La.
14. SCHLIWA, M., U. EUTENEUER, J. C. BULINSKI & J. G. IZANT. 1981. Proc. Natl. Acad. Sci. USA **78:** 1037–1041.
15. CANDE, W. Z., K. MCDONALD & R. L. MEEUSEN. 1981. J. Cell Biol. **88:** 618–629.
16. DEERY, W. J. & B. R. BRINKLEY. 1982. *In* Gene Regulation. B. W. O'Malley, Ed.: 327–341. Academic Press. New York.
17. WEISENBERG, R. C. 1972. Science **177:** 1104–1105.
18. REBHUN, L. I., D. JEMIOLO, T. KELLER, W. BURGESS & R. KRETSINGER. 1980. *In* Microtubules and Microtubule Inhibitors. M. DeBrabander & J. DeMey, Eds.: **3:** 243–252. Elsevier, North Holland Biomedical Press.
19. LEE, Y. C. & J. WOLFF. 1982. J. Biol. Chem. **257:** 6306–6310.
20. MARCUM, J. M., J. R. DEDMAN, B. R. BRINKLEY & A. R. MEANS. 1978. Proc. Natl. Acad. Sci. USA **75:** 3771–3775.
21. NISHIDA, E., H. KUMAGAI, I. OHTSUKI & H. SAKAI. 1979. J. Biochem. **85:** 1257–1266.
22. PERRY, G., B. R. BRINKLEY & J. BRYAN. 1981. *In* Cell and Muscle Motility. R. Dowben & J. Shay, Eds.: **2:** 73–84. Plenum Press. New York.
23. BERKOWITZ, S. A. & J. WOLFF. 1981. J. Biol. Chem. **256:** 11216–11223.
24. KELLER, T. C. S., D. K. JEMIOLO, W. H. BURGESS & L. I. REBHUN. 1982. J. Cell Biol. **93:** 797–803.
25. JOB, D., E. H. FISCHER & R. L. MARGOLIS. 1981. Proc. Natl. Acad. Sci. USA **78:** 4679–4682.
26. BERSHADSKY, A. D. & V. I. GELFAND. 1981. Proc. Natl. Acad. Sci. USA **78:** 3610–3613.
27. DEERY, W. J., A. R. MEANS & B. R. BRINKLEY. 1984. J. Cell Biol. **98:** 904–910.
28. WELSH, M. J., J. R. DEDMAN, B. R. BRINKLEY & A. R. MEANS. 1979. J. Cell Biol. **81:** 624–634.
29. CHAFOULEAS, J. G., J. R. DEDMAN, R. P. MUNJAAL & A. R. MEANS. 1979. J. Biol. Chem. **254:** 10262–10267.

30. AMES, G. F. L. 1974. J. Biol. Chem. **249:** 634–644.
31. PENNINGROTH, S. M. & M. W. KIRSCHNER. 1977. J. Mol. Biol. **115:** 643–673.
32. WEISENBERG, R. C., W. J. DEERY & P. DICKINSON. 1976. Biochemistry **15:** 4248–4254.
33. GASKIN, F., C. R. CANTOR & M. L. SHELANSKI. 1974. J. Mol. Biol. **89:** 737–758.
34. REGULA, C. S., J. R. PFEIFFER & R. D. BERLIN. 1981. J. Cell Biol. **89:** 45–53.
35. BRINKLEY, B. R. & J. CARTWRIGHT. 1975. Ann. N.Y. Acad. Sci. **253:** 428–439.
36. EICHENLAUB-RITTER, U. & A. RUTHMANN. 1982. Chromosoma (Berlin) **85:** 687–706.
37. LAMBERT, A. M. & A. BAJER. 1977. Cytobiologie **15:** 1–23.
38. BERSHADSKY, A. D. & V. I. GELFAND. 1981. Proc. Natl. Acad. Sci. USA **78:** 3610–3613.
39. MARGOLIS, R. L. & C. T. RAUCH. 1981. Biochemistry **20:** 4451–4458.
40. BRINKLEY, B. R., S. M. COX & D. A. PEPPER. 1980. *In* Testicular Development Structure and Function. A. Steinberger & E. Steinberger, Eds.: 305–314. Raven Press. New York.
41. WILLINGHAM, M. C., J. WEHLAND, C. B. KLEE, N. D. RICHERT, A. V. RUTHERLAND & I. H. PASTAN. 1983. J. Histochem. Cytochem. **31:** 445–461.
42. KARR, T. L., D. KRISTOFFERSON & D. L. PURICH. 1980. J. Biol. Chem. **255:** 11853–11856.
43. SUMMERS, K. & M. W. KIRSCHNER. 1979. J. Cell Biol. **83:** 205–217.
44. WEISENBERG, R. C. & W. J. DEERY. 1981. Biochem. Biophysics Res. Commun. **102:** 924–931.
45. SALMON, E. D. & S. M. WOLNIAK. 1984. Cell Motility **4:** 155–167.
46. STEARNS, M. E. & R. L. OCHS. 1982. J. Cell Biol. **94:** 727–739.
47. CLARK, T. G. & J. L. ROSENBAUM. 1982. Proc. Natl. Acad. Sci. USA **79:** 4655–4659.
48. FORMAN, D. S. 1982. Exp. Cell Res. **141:** 139–147.
49. WOLOSEWICK, J. J. & K. R. PORTER. 1979. J. Cell Biol. **82:** 114–149.
50. WEISENBERG, R. C. 1972. J. Cell Biol. **54:** 266–278.
51. RUBIN, R. W. & R. H. WARREN. 1979. J. Cell Biol. **82:** 103–113.
52. OLMSTED, J. B. 1981. J. Cell Biol. **89:** 418–423.
53. SLOBODA, R. D., W. L. DENTLER & J. L. ROSENBAUM. 1976. Biochemistry **15:** 4497–4505.
54. MURPHY, D. B., K. A. JOHNSON & G. G. BORISY. 1977. J. Mol. Biol. **117:** 33–52.
55. BERKOWITZ, S. A., J. KATAGIRI, H. K. BINDER & R. C. WILLIAMS. 1977. Biochemistry **16:** 5610–5617.
56. SIMONE, L. D., S. L. BRENNER, L. J. WIBLE, D. S. TURNER & B. R. BRINKLEY. 1981. J. Cell Biol. **9:** 337a.
57. EIPPER, B. A. 1974. J. Biol. Chem. **249:** 1398–1406.
58. RAPPAPORT, L., J. F. LeTERRIER, A. VIRION & J. NUNEZ. 1976. Eur. J. Biochem. **62:** 539–549.
59. GOODMAN, D. B., P. H. RASMUSSEN, F. DiBELLA & C. E. GUTHROW. 1970. Proc. Natl. Acad. Sci. USA **67:** 652–659.
60. SOIFER, D., A. LASZLO, K. MACK, J. SCOTTO & L. SICONOLFI. 1975. Ann. N.Y. Acad. Sci. **253:** 598–610.
61. DiPASQUALE, A. M., J. McGUIRE, G. MOELLMANN & S. J. WASSERMAN. 1976. J. Cell Biol. **71:** 735–748.
62. HSIE, A. & T. PUCK. 1971. Proc. Natl. Acad. Sci. USA **68:** 358–361.
63. WEBB, B. C. & L. WILSON. 1980. Biochemistry **19:** 1993–1998.
64. JOB, C., C. T. RAUCH, E. H. FISCHER & R. L. MARGOLIS. 1982. Biochemistry **21:** 509–515.
65. LEE, Y. C. & J. WOLFF. 1984. J. Biol. Chem. **259:** 1226–1230.
66. COHEN, P., A. BURCHELL, J. G. FOULKES, P. T. W. COHEN, T. VANAMAN & A. NAIRN. 1978. FEBS Lett. **92:** 287–293.
67. SOBUE, K., K. MORIMATO, M. INU, K. KANDA & S. KAKIUCHI. 1982. Biomed. Res. **3:** 188–196.

Contribution of Microtubules to Cellular Physiology: Microinjection of Well-Characterized Monoclonal Antibodies into Cultured Cells

JÜRGEN WEHLAND, HEINZ C. SCHRÖDER,[a,b] AND
KLAUS WEBER

Max Planck Institute for Biophysical Chemistry
D-3400 Goettingen, Federal Republic of Germany

INTRODUCTION

Microtubules perform multiple cellular functions that have been studied in detail using antimitotic drugs. As the action of these drugs is not always understood and possible side effects cannot be ruled out, we supposed that an independent molecular approach would be helpful. Therefore we started with microinjection of monoclonal tubulin antibodies into living cells. Our results emphasize that microinjection of specific antibodies is a powerful tool with which to study the physiological function of a known and characterized cytoplasmic antigen. The detailed characterization of one monoclonal tubulin antibody (clone YL 1/2) that was used for microinjection studies showed that its epitope is located on the carboxyterminal amino acid residues of α tubulin. As the monoclonal YL 1/2 antibody recognizes specifically the tyrosinated form of α tubulin, we recently became interested in the posttranslational modification of tubulin due to the enzymatic tyrosination of the α subunit. We have developed a rapid purification of the enzyme tubulin-tyrosine ligase by immunoaffinity chromatography. Because the physiological function of the tubulin modification is still unknown, antibodies against the enzyme were microinjected. Preliminary results on 3T3 cells are discussed.

PERTURBATION OF MICROTUBULAR FUNCTIONS INDUCED BY INJECTION OF A RAT MONOCLONAL TUBULIN ANTIBODY INTO LIVING CELLS

Two types of antibodies would be of interest in studying the functions of microtubules in cultured cells by microinjection. The first type resulting in a depolymerization of microtubules could provide effects comparable to those seen in studies using microtubule-depolymerizing drugs. The second type of antibody could recognize epitopes of the tubulin molecule exposed on the surface of intact microtu-

---

[a]H.C. Schröder was a recipient of a Liebig-Stipendium of the Fonds der Chemischen Industrie.
[b]Present address: Institute for Physiological Chemistry, University of Mainz, D-6500 Mainz, Federal Republic of Germany.

bules without introducing depolymerization. Such an antibody could affect microtubular functions without a loss of microtubules, but by a displacement of microtubule associated proteins (MAPs), or by the masking of sites recognized by other cellular structures or proteins on the tubulin molecule. Certainly type-two antibodies seem of more interest in proving the role of microtubules in intracellular organelle movement and the interaction with other cellular structures such as intermediate filaments. The rat monoclonal tubulin antibody YL 1/2, isolated by J.V. Kilmartin using yeast tubulin,[1] is a type-two antibody. When YL 1/2 antibody was injected into cultured fibroblasts, cytoplasmic microtubules were readily decorated by the injected antibody and, dependent on the antibody concentration microtubules were redistributed (FIGURE 1). At low antibody concentrations microtubules were decorated but maintained their normal cytoplasmic distribution. At higher concentrations the antibody induced bundling and perinuclear aggregation of microtubules, but no depolymerization. That YL 1/2 did not interfere with the polymerization-depolymerization process could also be verified by *in vitro* polymerization studies using purified brain microtubule protein.[2] The YL 1/2 epitope is exposed on the microtubule surface, readily accessible to the antibody, and saturable. In electron micrographs, YL 1/2 bound to microtubules is easily recognized. When antibody-injected cells were examined by video time-lapse microscopy, no visible intracellular motions were detected. Such cells also lacked locomotion, indicating that the *in situ* decoration and reorganization induced by the injected tubulin antibodies was sufficient to effect some microtubular functions. Further injection experiments showed that reorganization of microtubules interferred with the organization and location of the Golgi complex and with the distribution of intermediate filaments. Furthermore the antibody when injected into interphase cells affected the formation of normal spindle microtubules and therefore arrested such cells in mitosis (for details, see references 2 and 3). In general these results showed that microinjection of antibodies is a powerful tool with which to study the contribution of specific structural proteins to cellular physiology. With respect to microtubules, Blose *et al.* recently showed that intermediate filaments collapsed in living cells microinjected with polyclonal and monoclonal tubulin antibodies.[4]

CHARACTERIZATION OF THE EPITOPE RECOGNIZED ON TYROSINATED α TUBULIN BY THE RAT MONOCLONAL ANTIBODY YL 1/2

Currently, epitopes for most monoclonal antibodies remain unknown, as a disproportional effort in time is involved in such work. A detailed characterization of an epitope, however, can be very rewarding for an understanding of antibody specificity. The major reason to attempt such a characterization for YL 1/2 was the early observation that the antibody recognized α tubulin only when it carried a carboxyterminal tyrosine. If this residue was removed by carboxypeptidase, α tubulin was not recognized by YL 1/2.[2]

TABLE 1 summarizes the currently established contributions of particular amino acids to the YL 1/2 epitope.[5] The availability of purified tubulin ligase (see below) allowed us to prepare α tubulin carrying in the last position phenylalanine, tyrosine (4-hydroxyphenylalanine), or DOPA (3,4-dihydroxy-phenylalanine). All three derivatives were recognized in Western blots by YL 1/2. A competitive enzyme-linked immunosorbent assay (ELISA) revealed that the phenylalanine derivative had an 8-fold higher affinity than the natural tyrosine derivative, which is assumed to be the original antigen. In line with this observation, a further hydroxyl group on the benzene

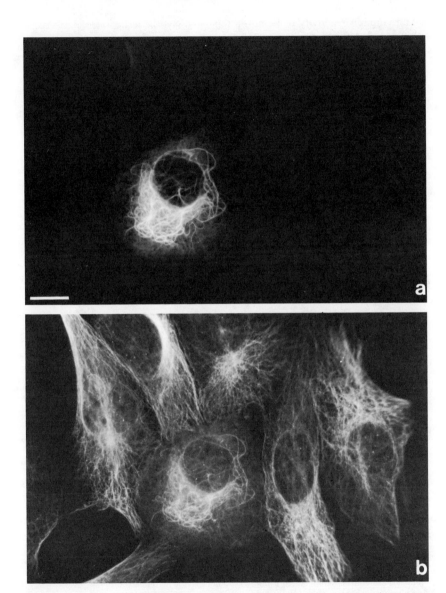

FIGURE 1. Double immunofluorescence microscopy of Swiss 3T3 fibroblasts injected with the rat monoclonal YL 1/2 antibody (12 mg IgG/ml in the injection solution). Three hours after the injection of YL 1/2, cells were fixed and processed for double indirect immunofluorescence using polyclonal rabbit tubulin antibody followed by a mixture of fluorescein-conjugated goat anti-rat IgGs and rhodamine-conjugated goat anti-rabbit IgGs. **a**: This portion shows the distribution of injected YL 1/2 and **b** the distribution of microtubules in both uninjected cells and the injected cells. Note the bundling and perinuclear aggregation of microtubules in the cell previously injected with YL 1/2 antibody. Bar indicates 10 μm.

TABLE 1. Immunoreactivity of Some Proteins and Synthetic Peptides with YL 1/2 Antibody[a]

Proteins and Peptides	Concentration for 50% Competition	Reactivity on Western Blot
Carboxypeptidase A-treated tubulin (-Glu-Glu-Glu-Glu)	none[c]	−
Tyr tubulin derivative (-Glu-Gly-Glu-Glu-Tyr)	0.04 μM	+
Phe tubulin derivative (-Glu-Gly-Glu-Glu-Phe)	0.005 μM	+
DOPA tubulin derivative (-Glu-Gly-Glu-Glu-DOPA)	0.2 μM	+
rec A protein (-Thr-Asn-Glu-Asp-Phe)	0.8 μM	+
catalytic subunit of protein kinase (-Glu-Phe-Ser-Glu-Phe)	none[d]	+
rabbit muscle aldolase (-Ser-Asn-His-Ala-Tyr)	none[c]	−
actin (-His-Arg-Lys-Cys-Phe)	none[c]	−
performic acid-oxidized actin (-His-Arg-Lys-CySO$_3$-Phe)	not tested[e]	+
Gly-Glu-Glu-Glu-Gly-Glu-Glu-Tyr	1.5 μM	
Gly-Glu-Glu-Glu-Gly-Glu-Glu-Glu	none[b]	
Glu-Tyr	500 μM	
Asp-Phe	150 μM	
Glu-Trp	3000 μM	
Tyr-Glu	none[b]	
Val-Tyr	none[b]	
Val-Phe	none[b]	
Asp-Phe-OMe	none[b]	
Asp-Phe-NH$_2$	none[b]	
(Cys-Phe)$_2$	none[b]	
CySO$_3$-Phe	150 μM	
γ-Glu-Tyr	500 μM	
γ-Glu-Phe	150 μM	
Gly-Tyr	none[b]	
Glu-Gly-Tyr	5500 μM	
Glu-Glu-Tyr	7.5 μM	
Ala-Phe-Pro-Leu-Glu-Phe (ACTH-carboxyterminal peptide)	100 μM	

[a]Immunoreactivity was assayed by competitive ELISA with tyrosinated α tubulin. The carboxyterminal sequences of proteins are given in parentheses.
[b]The highest concentration tested (10 mM) did not give 50% competition.
[c]The highest concentration tested (10 μM) did not give 50% competition.
[d]The highest concentration tested (4 μM) gave no competition.
[e]Insoluble in the buffers used. For further details see reference 5.

ring, as in DOPA, reduced the affinity about 20-fold. Further insight into the epitope requirements was obtained with various synthetic peptides using a competitive ELISA. The combined results defined three major requirements.

First, the carboxyterminal residue must be a phenylalanine derivative. Hydroxylation decreases the activity, and a size increase of the ring system to tryptophan results in a drastically lowered affinity (see, for instance, Glu-Tyr versus Glu-Trp).

Second, the carboxyterminal phenylalanine or tyrosine must carry the free carboxylate group, as the corresponding ester or amide are ineffective (see Asp-Phe versus Asp-Phe-OMe or Asp-Phe-NH$_2$).

Third, the penultimate residue should carry a negatively charged side chain, although some tolerance is observed. Thus glutamic acid (the naturally occurring residue), aspartic acid, and cysteic acid all qualify, and γ-Glu-Tyr seems as effective as α-Glu-Tyr. That these dipeptide sequences must occur at the very carboxyl end and carry the free α-carboxylate group is shown by several observations, such as that Glu-Tyr is active in competitive ELISA but not Tyr-Glu. Although α tubulin contains two interior Asp-Tyr sequences and three interior Glu-Phe sequences, it is not recognized by YL 1/2 unless it is tyrosinated at the carboxyl end. Similarly β tubulin with one Glu-Glu-Tyr sequence, in addition to one Glu-Tyr and two Glu-Phe sequences again occurring in interior positions,[6,7] is always inactive with YL 1/2, and so is muscle actin, which has three corresponding dipeptide sequences in interior positions.[8]

Transition from the dipeptide Glu-Tyr to the tripeptide Glu-Glu-Tyr increases the affinity 60-fold. Thus, in line with the sequence of tyrosinated α tubulin, a glutamic acid, or possibly only a further negative charge in the third position from the carboxyl end, strengthens the binding considerably. This contribution seems to explain why the tripeptide Glu-Gly-Tyr, which lacks a negative charge in the penultimate residue, is recognized, although very weakly under our assay conditions. The octapeptide covering the carboxyl end of tyrosinated α tubulin has only a 5-fold higher affinity than the corresponding tripeptide. Thus any residue prior to the last three should only have a relatively small contribution to the strength of the epitope. It is generally believed that epitopes cover about 5 to 7 amino acid residues, which occur either in sequence (linear or continuous epitope) or arise from different parts of the polypeptide chain (conformational or discontinuous epitope) (see reference 9). Within this model, the YL 1/2 epitope seems a continuous epitope where the last three residues of tyrosinated α tubulin are particularly important for affinity. We note, however, that even the octapeptide is still 40-fold less active than tyrosinated α tubulin. This situation is, however, not without precedence, as it seems that due to conformational flexibility, short peptides can be immunologically less active than the corresponding sequence fixed in the parental polypeptide (see, for instance, reference 10). Thus the YL 1/2 epitope of tyrosinated α tubulin belongs to the increasing list of linear epitopes thought to be prime antigenic targets because of their segmental mobility as revealed by X-ray crystallographic data on several proteins well characterized in their epitopes (see references 11 and 12). Such sites are often situated in loops, and these structures occur in several proteins at the two ends of the molecule.

Use of point mutants is a valuable tool to delineate the amino acid requirements of an epitope. We found that once a preliminary characterization is obtained, a computer data bank of protein sequences offers an alternative possibility for delineating linear epitopes. Specifying only a carboxyterminal tyrosine or phenylalanine, preceded by a glutamic or aspartic acid, we did such a search early in 1984 (TABLE 2). Proteins sharing these requirements included several procaryotic polypeptides that were difficult to obtain and the *Escherichia coli* rec A protein that was commercially available. Rec A protein contained the three carboxyterminal residues[13] of the YL 1/2 epitope. The few eukaryotic polypeptides where fit was given with the last two residues included the hormone ACTH, which we had already found to be recognized by YL 1/2 in the ELISA test. An additional candidate was the catalytic subunit of the cyclic adenosine monophosphate (cAMP)-dependent protein kinase from bovine heart.[14] Western blots showed that both rec A and the kinase were easily recognized by YL 1/2, whereas actin and aldolase, which fulfill the epitope requirement solely in the last position, were inactive. Given the various sequences, we conclude that, at least for YL

1/2, a fit of the last two residues is sufficient to obtain a positive immune blot as seen with the catalytic subunit. This conclusion was further explored. The carboxyterminal sequence of actin (-His-Arg-Lys-Cys-Phe)[8] with its three positively charged residues differs distinctly from that of tyrosinated α tubulin (-Glu-Gly-Glu-Glu-Tyr),[6,7] which is very acidic. As, however, our studies with synthetic peptides showed that the dipeptide cysteic acid-phenylalanine is a slightly better competitor than glutamic acid-tyrosine, we used performic acid-oxidized actin and obtained a decoration in Western blots that was clearly absent when normal actin is used. These results are particularly striking, because the cross-reaction between tyrosinated α tubulin and oxidized actin involves not a single identical amino acid residue, but arises from the chemical specifications of the epitope discussed above: the negative charge in the penultimate position and a carboxyterminal phenylalanine or tyrosine residue carrying the carboxylate group.

TABLE 2. Proteins with Carboxyterminal Sequences Fitting the Minimal Requirements of the YL 1/2 Epitope[a]

Protocatechuate 3,4-deoxygenase, α chain, *Pseudomonas aeruginosa*	(-Val-Phe-Phe-Asp-Phe)
Penicillinase precursor, *Staphylococcus aureus* PC-1	(-Val-Met-Lys-Glu-Phe)
Regulatory protein cII, bacteriophage λ	(-Ile-Gln-Met-Glu-Phe)
Helix-destabilizing protein, bacteriophage T7	(-Glu-Asp-Gly-Asp-Phe)
Rec A protein, *Escherichia coli*	(-Thr-Asn-Glu-Asp-Phe)
72K DNA-binding protein, adenovirus 2 and 5	(-Asn-Pro-Phe-Asp-Phe)
ACTH (adrenocorticotropic hormone)	(-Phe-Pro-Leu-Glu-Phe)
Rabbit β_2 microglobulin	(-Trp-Asp-Arg-Asp-Tyr)
Catalytic subunit of cAMP-dependent protein kinase (bovine heart)	(-Glu-Phe-Ser-Glu-Phe)

[a]The carboxyterminal sequences of proteins are given in parentheses. Of this list of proteins only rec A and the kinase were tested in immune blots (see text). For further details see reference 5.

The positive reaction of YL 1/2 with rec A protein and the catalytic subunit in immunoblots was followed up by competitive ELISA. As expected from the fit with the epitope in the three last positions, rec A protein was only 20 times less active than tyrosinated α tubulin (see TABLE 1). Thus this procaryotic protein is an excellent antigen. Unexpectedly, however, the kinase revealed no inhibition in ELISA either because the carboxyterminal residues are not available to YL 1/2 in the native conformation of the enzyme, or because the concentrations used were too low to observe an effect. The catalytic subunit must have, therefore, at least a 1000-fold lower affinity for YL 1/2 than tyrosinated α tubulin. Thus, given the conditions of the previous microinjection studies it is very unlikely that the physiological effects and morphological alterations observed arose from an interaction of YL 1/2 with a related kinase rather than with microtubules.

As the epitope of YL 1/2 becomes understood,[5] slight "defects" in absolute specificity of this monoclonal antibody to tyrosinated α tubulin are recognized. This is most likely not a peculiarity of YL 1/2. Similar "defects" can be expected for many good monoclonal antibodies currently used once they are subjected to a scrutiny similar to that given to YL 1/2. After a certain degree of epitope characterization is reached by synthetic peptides the use of a protein sequence data bank can be helpful. Conformational epitopes, however, cannot be approached this way, and even for linear epitopes there are severe limitations, as only a fraction of eukaryotic proteins are known by sequence. When microinjection studies are done, a new order of specificity problems arises as any potential, although very minor component, additionally recognized by the antibody, can perturb the conclusions. Similar arguments hold in principle for antibodies with which the function of cell surface antigens is studied and for antibodies used in medical applications. Western blots of total cell-free extracts, particularly when done only on one cell type, are not sufficiently sensitive to detect very minor cross-reacting components. Thus we encourage the additional use of those methods that result in an enrichment of such components, that is, immunoaffinity chromatography and immunoprecipitation. Currently there is no general substitute for microinjection of antibodies to explore the contribution of a particular structural protein to cellular physiology. Thus it is particularly important to recognize the limits of antibody specificity and to control for them.

TUBULIN-TYROSINE LIGASE (TTL)

Several mechanisms are thought to regulate the multiple cellular functions of microtubules. Besides factors that might specifically and locally influence the polymerization and/or depolymerization of microtubules, the heterogeneity of tubulin and of microtubule-associated proteins could be responsible for possibly distinct microtubular functions. In addition, the reversible posttranslational modifications of tubulin itself might rapidly modulate microtubular functions. The β subunit can be phosphorylated[15,16] and the carboxyterminal tyrosine of the α subunit is subject to a cycling mechanism (see reference 17).

Tyrosination of α tubulin first described in 1974[18] has since been studied in some detail. The carboxyterminal Tyr of α tubulin is encoded by the mRNA.[6] It can be removed by a presumptive tubulin tyrosine carboxypeptidase,[19,20] which has not been characterized in detail. A Tyr residue can again be covalently attached to the carboxyterminal Glu residue of detyrosinated α tubulin by an adenosine triphosphate (ATP)-dependent cellular enzyme termed tubulin-tyrosine ligase (TTL).[18,21] Even though a variety of suggestions have been made (see reference 17), the physiological function of TTL is essentially not known. One possible way to the function of TTL is to microinject highly specific antibodies to the enzyme into living cells and search for perturbation of cellular functions.

Brain is a suitable source for TTL. Purification is rather cumbersome, as several steps are required and the amount of enzyme obtained is rather small. A previous procedure obtained nearly homogeneous TTL by ion exchange chromatography and two affinity chromatography steps using immobilized ATP as well as immobilized tubulin.[22] Such highly purified ligase is unfortunately rather unstable.[23] When we repeated this purification procedure with pig brain we found the ligase to be contaminated by actin and tubulin. Introduction of some modifications, however, (see TABLE 3) provided a suitable scheme with which nearly homogeneous ligase could be obtained.[24] Even though we were able to overcome the instability of the enzyme activity by inclusion of glycerol in the buffers, a 7000-fold purification was necessary, and the yield was only 10% compared to the total activity present in the starting material. Nevertheless at a protein concentration of about 20 $\mu g/ml$, purified ligase could be

TABLE 3. Purification of Tubulin-Tyrosine Ligase from Pig Brain[a]

Fraction	Purification Step	Total Protein (mg)	Total Enzyme Activity (U)	Specific Activity (U/mg)	Yield (percent)	Purification (-fold)
I.	First warm supernatant	9,420	1,040	0.11	100	1
II.	DEAE-cellulose	1,472	997	0.68	96	6
III.	Sepharose 4B-sebacic acid hydrazide-ATP	21.0	349	16.6	34	150
IV.	Glycerol gradient	9.1	380	42	37	380
V.	Phosphocellulose	0.14	107	762	10.3	6,930
VI.	Tyrosyl-aminohexyl-Sepharose 4B	0.05	16.2	324	1.6	2,950

[a] A preparation procedure, starting from 1,000 g of pig brain, is given. For the detailed purification procedure see reference 24.

stored in the presence of 20% glycerol for more than one year at −70°C without noticeable loss in activity.

The sedimentation coefficient of purified ligase determined by glycerol gradient centrifugation was 3.2 S, and sodium dodecyl sulfate (SDS) gels showed an apparent molecular weight of 40,000. Thus the ligase behaves as a monomeric globular protein. Enzyme mixed with purified tubulin sediments at 7.3 S, indicative of the formation of a 1 to 1 complex.[24] These physical-chemical parameters are in agreement with previous reports.[22,25]

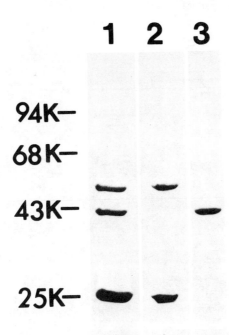

FIGURE 2. Analysis of immunoaffinity purified TTL on a 10% SDS gel stained with Coomassie blue. Mouse monoclonal antibodies specific for TTL were coupled to Sepharose, and the affinity matrix was incubated with crude brain extracts. (For details of immunoaffinity chromatography see reference 24.) After extensive washing finally with 1.5 M NaCl, the matrix was stripped directly with SDS sample buffer (lane 1) or first with 3 M MgCl$_2$ followed by SDS sample buffer (lane 2). Lane 3 represents the fraction containing nearly homogeneous TTL eluted from immobilized TTL antibodies with 3 M MgCl$_2$.

As TTL was initially available in rather small amounts, the production of monoclonal antibodies in mice was indicated. The four independently isolated IgG$_1$ clones reacted with TTL in precipitation assays and Western blots but did not inhibit the enzyme activity. The monoclonal antibodies to brain TTL had a moderately broad cross-species reactivity among mammals and detected the enzyme activity also in nonneuronal tissues such as liver.[24]

Monoclonal antibody (clone LA/C4) isolated from ascites fluid allowed a rapid and convenient purification of TTL by immunoaffinity chromatography. FIGURE 2 gives the details of this large-scale purification that allowed nearly quantitative

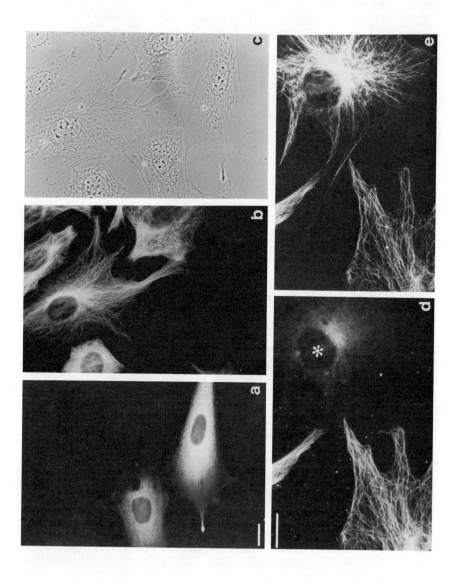

recovery. The 37°C high-speed supernatant obtained after microtubule formation in the crude brain extract was applied to the immunoaffinity resin. The resin was washed repeatedly with ligase stabilization buffer containing 10% glycerol and then centrifuged through 1 M sucrose to remove aggregates that had formed during the incubation. After a final wash with 1.5 M NaCl, TTL could be released with 3 M MgCl$_2$. The presence of 10–20% glycerol during the purification provided highly active enzyme. TTL was more than 95% pure. Starting with 1 kg of brain we obtained about 1 to 1.2 mg of enzyme.[24]

Immunoaffinity-purified TTL was used to raise antibodies in rabbits. Positive sera were affinity-purified on immobilized TTL. These antigen affinity-purified rabbit antibodies inhibited the activity of ligase in enzyme assays. They also revealed a broad cross-reactivity. They recognized ligase on immunoblots derived from SDS gels of extracts of various cells and tissues and inhibited ligase activity present in crude cell extracts. Both assays showed that the affinity-purified rabbit antibodies reacted with ligase present in crude extracts of Swiss 3T3 mouse fibroblasts, the cell line that we used for microinjection experiments (see below). In indirect immunofluorescence microscopy, we did not observe any specific location of ligase in the cytoplasm of cultured cells. Although this may be due to the small amount of ligase present in these cells, alternative explanations are not excluded.

When Swiss 3T3 mouse fibroblasts were injected with polyclonal ligase antibodies and examined 12 hours later by immunofluorescence microscopy using the YL 1/2 rat monoclonal tubulin antibody specific for the tyrosinated form of α tubulin, no microtubules were detected in the injected cells. Polyclonal tubulin antibodies, however, revealed that the cells previously injected with ligase antibodies had a normal microtubular display (FIGURE 3). These results suggest that the posttranslational addition of Tyr to the c terminus of α tubulin can be inhibited *in vivo* in Swiss 3T3 cells by injected ligase antibodies, and therefore a tubulin-tyrosine carboxypeptidase must remove the Tyr residue as it is encoded by the mRNA.[6] A newly isolated mouse monoclonal antibody that inhibited ligase activity also interfered with the posttranslational tyrosination of α tubulin *in vivo* when injected into Swiss 3T3 cells.

Based on two peptide antibodies specifically reacting only with either tyrosinated or detyrosinated α tubulin, separate populations of microtubules have been proposed.[26] Whether such subsets of microtubules within the same cell imply different functions is unknown. Our microinjection experiments with ligase antibodies indicate, however, that at least in the 3T3 fibroblasts that we have used, all cytoplasmic microtubules can lose the carboxyterminal tyrosine. These initial results encourage us to examine whether the absence of the tyrosine in α tubulin interferes with a physiological function of microtubules.

FIGURE 3. Double immunofluorescence microscopy of Swiss 3T3 cells injected with affinity-purified rabbit TTL antibodies (2 mg IgG/ml in the injection solution). Twelve hours after injection, cells were fixed and processed for double immunofluorescence: (a–c) using YL 1/2 followed by a mixture of fluorescein-conjugated goat antirabbit and rhodamine-conjugated goat antirat IgGs; (c), corresponding phase contrast of cells in (a) and (b); (d, e) using a mixture of YL 1/2 antibody and polyclonal sheep antitubulin antibody followed by a mixture of fluorescein-conjugated rabbit antirat IgGs and rhodamine-conjugated guinea pig antisheep IgGs. (a) and (d) were viewed with optics selective for fluorescein and the same fields with optics selective for rhodamine (b and e). Asterisk in d marks the cell previously injected with TTL antibodies. Note that the YL 1/2 antibody specific for the tyrosinated form of α tubulin does not recognize microtubules in the cell previously injected with ligase antibody (a and b), whereas polyclonal tubulin antibodies reveal a normal microtubular display in the cell previously injected with ligase antibody (compare d and e). Bars indicate 10 μm.

ACKNOWLEDGMENTS

The microinjection studies with the YL 1/2 antibody and the initial characterization of this antibody were done in collaboration with Dr. I.V. Sandoval and Dr. M.C. Willingham at the National Institutes of Health, Bethesda, Md. (see references 2 and 3). We thank Dr. M. Osborn for critically reading the manuscript.

[NOTE ADDED IN PROOF. Our predictions on the epitope requirement of the YL 1/2 antibody were confirmed for another cross-reacting protein from clam and sea urchin eggs as well as mouse cells: the small subunit of the ribonucleotide reductase having the carboxyterminal sequence—Leu-Asp-Ala-Asp-Phe. Immobilized YL 1/2 enabled the purification of the reductase by immuno affinity chromatography (STANDART et al. 1985. J. Cell Biol. **100:** 1968–1976; THELANDER et al. 1985. J Biol. Chem. **260:** 2737–2741). The reductase is already released from the affinity matrix by 1M KCl, suggesting weak binding in agreement with the weak antigenic activity of the dipeptide Asp-Phe. In line with our results on the epitope characterization of YL 1/2, cross-reacting antibodies are not so unusual and can be generated once amino acid sequence homology is found for different proteins. The neurofibrillary staining with α-MSH (α-melanocyte-stimulating hormone) antibodies is due to similar amino terminal amino acid sequences present in the middle sized neurofilament protein and α-MSH (SHAW et al. 1985. FEBS Lett. **181:** 343–346). When antibodies were elicited using a synthetic octa- or decapeptide sharing six consecutive amino acids with a viral protein (hepatitis B virus polymerase) and rabbit myelin basic protein, such antibodies reacted with both proteins (FUJINAMI & OLDSTONE. 1985. Science **230:** 1043–1045).]

REFERENCES

1. KILMARTIN, J. V., B. WRIGHT & C. MILSTEIN. 1982. Rat monoclonal anti-tubulin antibodies derived by using a new nonsecreting rat cell line. J. Cell Biol. **93:** 576–582.
2. WEHLAND, J., M. C. WILLINGHAM & I. V. SANDOVAL. 1983. A rat monoclonal antibody reacting specifically with the tyrosylated form of α-tubulin. I. Biochemical characterization, effects on microtubule polymerization *in vitro* and microtubule polymerization and organization *in vivo*. J. Cell Biol. **97:** 1467–1475.
3. WEHLAND, J. & M. C. WILLINGHAM. 1983. A rat monoclonal antibody reacting specifically with the tyrosylated form of α-tubulin. II. Effects on cell movement, organization of microtubules, and intermediate filaments, and arrangement of Golgi elements. J. Cell Biol. **97:** 1476–1490.
4. BLOSE, S. H., D. I. MELTLZER & J. R. FERAMISCO. 1984. 10 nm filaments are induced to collapse in living cells microinjected with monoclonal and polyclonal antibodies against tubulin. J. Cell Biol. **98:** 847–858.
5. WEHLAND, J., H. C. SCHRODER & K. WEBER. 1984. Amino acid sequence requirements in the epitope recognized by the α-tubulin specific rat monoclonal antibody YL 1/2. EMBO J. **3:** 1295–1300.
6. VALENZUELA, P., M. QUIROGA, J. ZALDIVER, W. J. RUTTER, M. W. KIRSCHNER & D. W. CLEVELAND. 1981. Nucleotide and corresponding amino acid sequences encoded by α- and β-tubulin mRNAs. Nature (London) **289:** 650–655.
7. PONSTINGL, H., E. KRAUHS, M. LITTLE & T. KEMPF. 1981. Complete amino acid sequence of α-tubulin from porcine brain. Proc. Natl. Acad. Sci. USA **78:** 2757–2761.
8. VANDEKERCKHOVE, J. & K. WEBER. 1979. The complete amino acid sequence of actins from bovine aorta, bovine heart, bovine fast skeletal muscle, and rabbit slow skeletal muscle. Differentiation **14:** 123–133.
9. VAN REGENMORTEL, M. H. V. 1984. *In* Hybridoma Technology in Agricultural and

Veterinary Research. N. Stern & H. R. Gamble, Eds.: 42–80. Rowman and Allenheld. Totowa, N.J.

10. SACHS, D. H ., A. N. SCHECHTER, A. EASTLAKE & C. B. ANFINSEN. 1972. An immunologic approach to the conformational equilibria of polypeptides. Proc. Natl. Acad. Sci. USA **69:** 3790–3794.

11. WESTHOF, E., D. ALTSCHUH, D. MORAS, A. C. BLOOMER, A. MONDRAGON, A. KLUG & M. H. V. VAN REGENMORTEL. 1984. Correlation between segmental mobility and the location of antigenic determinants in proteins. Nature (London) **311:** 123–126.

12. TAINER, J. A., E. D. GETZOFF, H. ALEXANDER, R. A. HOUGHTON, A. J. OLSON, R. A. LERNER & W. A. HENDRICKSON. 1984. The reactivity of anti-peptide antibodies is a function of the atomic mobility of sites in a protein. Nature (London) **312:** 127–134.

13. SANCAR, A., C. STACHELEK, W. KONIGSBERG & W. D. RUPP. 1980. Sequences of the rec A gene and protein. Proc. Natl. Acad. Sci. USA **77:** 2611–2615.

14. SHOJI, S., D. C. PARMELEE, R. D. WADE, S. KUMAR, L. H. ERICSSON, K. A. WALSH, H. NEURATH, G. L. LONG, J. G. DEMAILLE, E. H. FISCHER & K. TITANI. 1981. Complete amino acid sequence of the catalytic subunit of bovine cardiac muscle cyclic AMP-dependent protein kinase. Proc. Natl. Acad. Sci. USA **78:** 848–851.

15. EIPPER, B. A. 1974. Rat brain tubulin and protein kinase activity. J. Biol. Chem. **249:** 1398–1406.

16. SANDOVAL, I. V. & P. CUATRECASAS. 1976. Protein kinase associated with tubulin: affinity chromatography and properties. Biochemistry **15:** 3424–3432.

17. THOMPSON, W. C. 1982. The cyclic tyrosination/detyrosination of alpha tubulin. Methods Cell Biol. **24:** 235–255.

18. BARRA, H. S., C. A. ARCE, J. A. RODRIGUEZ & R. CAPUTTO. 1974. Some common properties of the protein that incorporates tyrosine as a single unit into microtubule proteins. Biochem. Biophysics Res. Commun. **60:** 1384–1390.

19. ARGARANA, C. E., H. S. BARRA & R. CAPUTTO. 1978. Release of [14C] tyrosine from tubulinyl-[14C] tyrosine by brain extract. Separation of a carboxy peptidase from tubulin tyrosine ligase. Mol. Cell Biochem. **19:** 17–22.

20. KUMAR, N. & M. FLAVIN. 1981. Preferential action of a brain detyrosinating carboxypeptidase on polymerized tubulin. J. Biol. Chem. **256:** 7678–7686.

21. RAYBIN, D. & M. FLAVIN. 1975. An enzyme tyrosylating alpha-tubulin and its role in microtubule assembly. Biochem. Biophysics Res. Commun. **60:** 1384–1390.

22. MUROFUSHI, H. 1980. Purification and characterization of tubulin-tyrosine ligase from porcine brain. J. Biochem. **87:** 979–984.

23. FLAVIN, M., T. KOBAYASHI & T. M. MARTENSEN. 1982. Tubulin-tyrosine ligase from brain. Methods Cell Biol. **24:** 257–263.

24. SCHRÖDER, H. C., J. WEHLAND & K. WEBER. 1985. Purification of brain tubulin-tyrosine ligase by biochemical and immunological methods. J. Cell Biol. **100:** 276–281.

25. RAYBIN, D. & M. FLAVIN. 1977. Enzyme which specifically adds tyrosine to the α-chain of tubulin. Biochemistry **16:** 2189–2194.

26. GUNDERSEN, G. G., M. H. KALNOSKI & J. C. BULINSKI. 1984. Distinct populations of microtubules: tyrosinated and nontyrosinated alpha-tubulin are distributed differently *in vivo*. Cell **38:** 779–789.

Purification of Tubulin by Fast-Performance Liquid Chromatography[a]

FELICIA GASKIN AND SUKLA ROYCHOWDHURY

Oklahoma Medical Research Foundation
Oklahoma City, Oklahoma 73104

Fast performance liquid chromatography (FPLC) was used to purify tubulin from porcine microtubule protein (MTP) prepared as previously described using two cycles of assembly-disassembly.[1] Pharmacia Mono Q (anion exchanger) column chromatography is a fast and reproducible way to purify as little as 1 mg or as much as 100 mg tubulin. The procedure is done at 4–8°C, and the purified tubulin is competent to assemble with 0.02 mM taxol or in 4 M glycerol plus 10% dimethyl sulfoxide (DMSO). The purification scheme is described at the end of this paper. FIGURE 1 demonstrates the chromatography profile of a typical experiment that took 85 min to purify 38 mg tubulin. This tubulin fraction contains trace enzyme activities, that is, <0.1 mU/mg/min nucleoside diphosphate kinase (NDPK), 0.1 nmoles/mg/min Mg-ATPase, and <0.2 mU/mg/min myokinase activities. Using the same assay conditions, MTP contains 60 mU/mg/min NDPK, 7.0 nmoles/mg/min Mg-ATPase, and 87.5 mU/mg/min myokinase. The NDPK and myokinase activities were done as described in references 2 and 3. ATPase activity was determined using ATPγ^{32}P and published procedures in 0.1 M 2-N(morpholino)ethanesulfonic acid (MES), 1 mM MgCl$_2$, and 1 mM ethylene glycol bis (β-aminoethyl ether)-N, N, N', N'-tetracetic acid (EGTA), pH 6.6, 37°C.[4] FIGURE 2 demonstrates that a high molecular weight (HMW) component copurifies with tubulin on Mono Q columns and that this component can be removed by Mono S (cation exchanger) chromatography. Tubulin comes off in the void on this latter column. Alternatively, MTP can be put on the Mono S column, followed by the Mono Q column. Tubulin and some microtubule-associated proteins (MAPs) come off in the void from the Mono S column, and this fraction is treated as in steps 4–10 in the purification scheme for the Mono Q column.

In our purification scheme, guanosine triphosphate (GTP) and the major (MAPs) come off in the void on the Mono Q column. Tubulin-nucleotide is isolated in the tubulin fraction using MTP containing 0.5 mM GTP that was incubated 20 min at 4°C, 30 min at 37°C, and 10 min at 4°C. The nucleotide/tubulin ratio was determined by precipitating the protein with 50% ethanol at pH 5.6 and measuring the nucleotide concentration at A$_{252}$. The protein contained a nucleotide/tubulin ratio of 2.07. This number is similar to the nucleotide/tubulin ratio of 1.87 that we find for pelleted microtubules that were assembled in 0.5 mM GTP. Both the Mono Q purified tubulin and pelleted microtubules contained 0.77 moles labeled nucleotide/tubulin dimer. Thus most of the exchangeable nucleotide appears to stay bound to tubulin during Mono Q chromatography. In one step it is possible to separate active tubulin nucleotide from the main enzyme contaminants, MAPs, and unbound nucleotide.

The purification of tubulin from MTP using Mono Q, Mono S columns is as

[a]This work was supported in part by NIH Grant NS 19224.

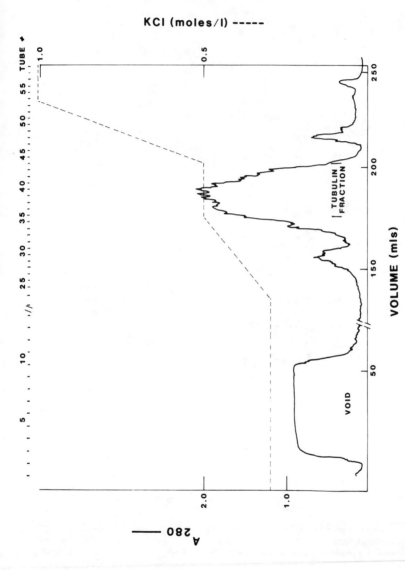

FIGURE 1. Chromatography profile of MTP (70.5mg/49.5mls) on the 10mm × 10cm Pharmacia Mono Q column at 8°C in 0.1 M MES, 0.5 mM MgCl$_2$, 1 mM EGTA, pH 6.6. The tubulin fraction (tubes 36–43) contains 38 mg protein. PAGE on this tubulin fraction shows one high molecular weight component that can be removed by Mono S chromatography (See FIGURE 2). The void contains 13.5 mg of nontubulin proteins (based on PAGE) and unbound GTP and GDP. Tubes 29–30 and 33–35 contain 1.4 and 1.8 mg protein (mainly tubulin), respectively. Tubes 47–48 contains 1.2 mg tubulin. Total recovery of protein from the column was 85 percent.

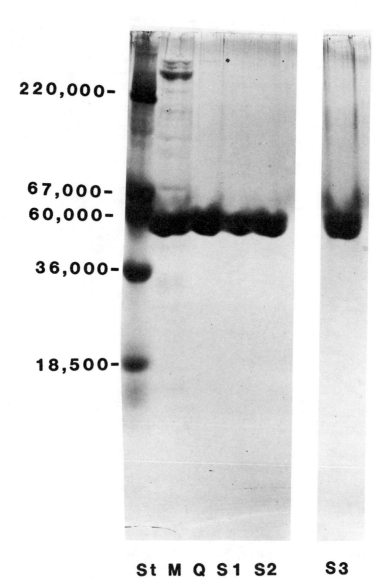

FIGURE 2. Polyacrylamide gel electrophoresis of MTP and FPLC-purified tubulin. M is MTP (11 μg). Q is the major tubulin fraction (7.9 μg) from the Mono Q column. The HMW component that copurifies with tubulin is designated by (·). S1, S2, and S3 are Mono Q, Mono S purified tubulin at 3.6, 7.5, and 15 μg, respectively. St are the following standards, ferritin (220,000 and 18,500), albumin (67,000), catalase (60,000), and lactate dehydrogenase (36,000). A 2.5–27% acrylamide exponential gradient gel was run in tris-borate buffer with 0.1% SDS, 0.1 mM EDTA, 1 mM mercaptoethanol pH 8.5 as described previously.[5]

follows: 1) Dilute MTP with 0.1 M MES, 0.5 mM MgCl$_2$, 1 mM EGTA, pH 6.6 buffer to 1.5–2.5 mg/ml. 2) Add GTP to $2 \times 10^{-5} M$. 3) Centrifuge the protein at 100,000 g, 30 min, 4°C. 4) Add 3 M KCl in buffer so the final concentration is 0.3 M KCl. 5) Filter the protein through a 0.45 micron millipore filter. 6) Inject onto Mono Q at 1.25 mls/min for the 5mm × 5cm column and 3.0 mls/min for the 10mm × 10cm column. We have injected up to 18 mg and 110 mg MTP on the small and larger Mono Q columns, respectively. Based on these amounts, it should be possible to use at least 200 mg MTP on the 16mm × 20cm column. 7) Elute with 0.3 M KCl in buffer for 15 to 60 mls depending on the amount of MTP injected and the size of the column. The void contains MAPs and no tubulin based on polyacrylamide gel electrophoresis (PAGE). 8) Run a linear gradient to 0.5 M KCl in buffer at 0.4 mls/min (30 min) or 1.6 mls/min (25–30 min) for the small and larger columns, respectively. 9) Tubulin begins to come off at 0.45 to 0.48 M KCl. Keep the elutant at 0.5 M KCl until the tubulin fraction is eluted (see FIGURE 1). Recovery of tubulin is 65–80 percent. PAGE shows a HMW component in the tubulin fraction (see FIGURE 2). 10) Use dialysis or Sephadex G-25 chromatography to remove the KCl. 11) The HMW component in the tubulin fraction from the Mono Q column binds to the Mono S column and tubulin comes off in the void. 12) Alternatively, MTP in buffer can be put on the Mono S column. Tubulin and several MAPs come off in the void. The void from the Mono S column is treated as in Step 4 and the same procedure is followed through Step 10.

ACKNOWLEDGMENTS

We wish to thank C. Ray Powers Jr., Duane Compton, and Steve Graham for excellent technical assistance.

REFERENCES

1. SHELANSKI, M. L., F. GASKIN & C. R. CANTOR. 1973. Microtubule assembly in the absence of added nucleotides. Proc. Natl. Acad. Sci. USA **70:** 765–768.
2. Biochemica Information I. 1973. 144–145. Boehringer Mannheim.
3. Biochemica Information II. 1975. 108–109. Boehringer Mannheim.
4. GASKIN, F., S. B. KRAMER, C. R. CANTOR, R. ADELSTEIN & M. L. SHELANSKI. 1974. A dynein-like protein associated with microtubules. FEBS Lett. **40:**281–286.
5. GASKIN, F. 1981. *In vitro* microtubule assembly regulation by divalent cations and nucleotides. Biochemistry **20:**1318–1322.

Hydrophobic Interactions of Tubulin

JOSÉ MANUEL ANDREU

Unidad de Biomembranas
Instituto Inmunología
Consejo Superior de Investigaciones Científicas
28006 Madrid, Spain

Tubulin purified from mammalian brain is a water soluble dimer with sedimentation coefficient 5.8 S^1 and molecular weight close to 100,000.[2] It is able, however, to associate to phosphatidylcholine vesicles where it is apparently inserted into the lipid bilayer.[3] Tubulin has large regions capable of hydrophobic interactions that were probed with the mild detergents octylglucoside and sodium deoxycholate.[4] The interaction of tubulin with these amphiphiles has been examined in detail, and the effects of binding on the hydrodynamic, conformational, and functional properties of this protein have now been explored. Binding was measured by equilibrium gel chromathography[5] employing radiolabeled detergents. A total of 93 ± 10 deoxycholate monomers and $\geqslant 60$ monomers of octylglucoside were bound to tubulin cooperatively and reversibly. Binding could also be monitored by quenching of the intrinsic protein fluorescence that was clearly different from the spectral changes induced by denaturation by sodium dodecylsulfate. Massive binding of the mild detergents was not observed in the presence of 3.4 M glycerol, which is known to cause preferential protein hydration and to favor self-association and stabilization of tubulin.[6] The far ultraviolet circular dichroism of tubulin showed changes induced by both mild detergents that indicated moderate unfolding and were very different from dodecylsulfate denaturation. A marked increase in degradation was induced by the mild detergents when tubulin was subjected to controlled proteolysis by trypsin, thermolysin or *Staphylococcus aúreus* V8 protease. Octylglucoside induced tubulin self-association, as shown by the sedimentation velocity pattern of FIGURE 1A (upper profile). The bimodal protein pattern was apparent from the beginning of the run, and the trough never touched the baseline, suggesting a fast equilibrating protein self-association reaction.[7] Removal of octylglucoside by gel chromathography generated a normal tubulin sedimentation pattern. The tubulin-deoxycholate complex (FIGURE 1B, upper profile) sedimented as a single peak and more slowly than the protein without detergent. Sedimentation equilibrium measurements of tubulin in the presence of deoxycholate close to its critical micelle concentration indicated an apparently monodisperse system and a molecular weight of $144,000 \pm 5,000$. Therefore the decrease in sedimentation rate induced by deoxycholate was not due to dissociation of the tubulin heterodimer, but to a marked increase in its frictional coefficient. This was verified by diffusion measurements, and the results are shown in TABLE 1, which summarizes the properties of tubulin bound to the detergents. Octylglucoside and deoxycholate were found to inhibit the binding to tubulin of 2-methoxy-5 (2',3',4'-trimethoxyphenyl)tropone, a reversibly binding probe for the colchicine site,[8] reducing the number of sites available for interaction. Octylglucoside also inhibited microtubule assembly. Release of octylglucoside from the saturated tubulin-detergent complex rendered the protein able to bind colchicine ligands and to assemble into microtubules again. It is concluded that cytoplasmic tubulin shows a reversible transition between the water soluble state and amphipathic detergent-bound forms.

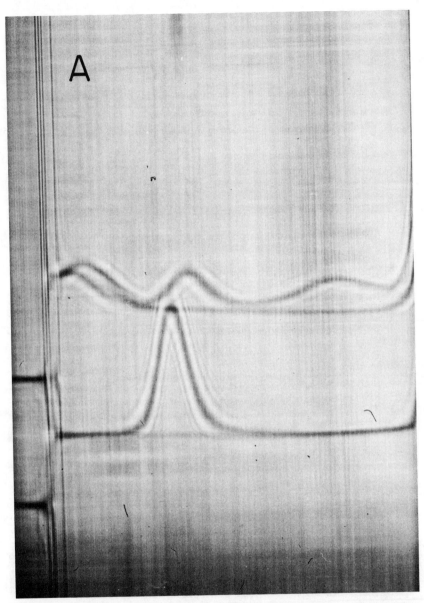

FIGURE 1. Effects of the binding of mild detergents on the sedimentation velocity pattern of tubulin. Tubulin was purified from calf brain and equilibrated in 10 mM sodium phosphate buffer 0.1 mM GTP pH 7.0.[1,8] It was loaded into two double sector cells with regular and wedge sapphire windows, placed in an AnD rotor, and centrifuged at 56,000 rpm in a Beckman Model E analytical ultracentrifuge equipped with electronic speed control, photoelectric scanner, and RTIC unit. Schlieren pictures shown were taken 56 minutes after reaching speed. **A:** Upper profile: 7 mg/ml tubulin in buffer containing 50 mM octylglucoside. Lower profile: an aliquot of the same protein without detergent (25°C, bar angle 65°). First peak to the left in the upper profile is present both in the sample and reference channels and is due to sedimentation of the detergent. The second peak sedimented at a rate close to the tubulin heterodimer (in the lower profile) and the third peak at $s_{20,w} = 11.9$ S.

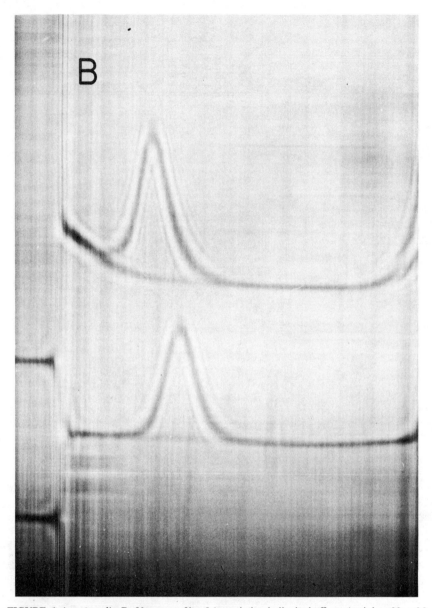

FIGURE 1 (*continued*). **B:** Upper profile: 5.1 mg/ml tubulin in buffer containing 20 mM deoxycholate. Lower profile: without detergent (20°C, bar angle 60°). It has to be noted that the skewed appearance of the tubulin-deoxycholate peak is due to the sedimentation of detergent micelles that is also observed in the reference channel. This profile became symmetrical after baseline correction or in velocity runs employing the photoelectric scanner. Sedimentation rates were corrected for the viscosity and density of the detergent solutions.

TABLE 1. Hydrodynamic Characteristics of the Tubulin-Detergent Complexes

Detergent	Maximal Binding (g detergent/g protein)	Sedimentation Velocity Pattern	Sedimentation Coefficient	Molecular Weight	f/f_{min} [c]	$D_{20,w}^{o} \times 10^7$ (cm^2/s) [d]
Octylglucoside	0.18	bimodal	5 and 12 S			
Deoxycholate	0.39 ± 0.04	single symmetrical peak	$s_{20,w}^{o}$ = 4.8 ± 0.3 S	144,000 ± 5,000[a] (139,000 ± 4,000) for dimer or 68,000 ± 2,000 for monomer)	1.86	3.9 ± 0.2 (3.2 ± 0.4 for dimer or 6.4 ± 0.9 for monomer)
None	—	single symmetrical peak	$s_{20,w}^{o}$ = 5.6 ± 0.2 S	100,000[b]	1.35	5.6 ± 0.5 (5.2 ± 0.2)

[a]Measured by sedimentation equilibrium. (Values in parenthesis were calculated from the binding data for the detergent bond dimer and monomer.)
[b]From the amino acid sequence of porcine tubulin.[2]
[c]Calculated from $s_{20,w}^{o}$ and M_r.
[d]Measured in the ultracentrifuge employing a synthetic boundary cell. (Values in parenthesis were calculated for the detergent bond dimer and monomer from $s_{20,w}^{o}$ and the molecular weights.)

REFERENCES

1. NA, G. C. & S. N. TIMASHEFF. 1982. Methods Enzymol. **85:** 393–408.
2. PONSTINGL, H., E. KRAUHS, M. LITTLE, T. KEMPF, R. HOFER-WARBINEK & W. ADE. 1982. Cold Spring Harbor Symp. Quant. Biol. **46:** 191–197.
3. KUMAR, N., R. D. KLAUSNER, U. N. WEINSTEIN, R. BLUMENTAL & M. FLAVIN. 1981. J. Biol. Chem. **256:** 5886–5889.
4. ANDREU, J. M. 1982. EMBO J. **1:** 1105–1110.
5. HUMMEL, J. P. & W. J. DREYER. 1962. Biochim. Biophys. Acta **63:** 530–532.
6. NA, G. C. & S. N. TIMASHEFF. 1981. J. Mol. Biol. **151:** 165–178.
7. PRAKASH, V. & S. N. TIMASHEFF. 1983. Anal. Biochem. **131:** 232–235.
8. ANDREU, J. M., M. J. GORBUNOFF, J. C. LEE & S. N. TIMASHEFF. 1984. Biochemistry **23:** 1742–1752.

An Analysis of Tubulin Molecular Weight by Equilibrium Chromatography: Further Evidence for Tubulin Oligomers

N.G. KRAVIT AND R.D. BERLIN

Department of Cell Physiology and Biophysics
University of Connecticut Health Center
Farmington, Connecticut 06032

We have previously demonstrated[1] that phosphoscellulose-purified tubulin (PC tubulin) at room temperature does not behave as simple α-β dimers. Results from ultrafiltration, nondenaturing gel electrophoresis, and chemical cross-linking indicate that at room temperature, tubulin dimers participate in an indefinite self-association leading to the formation of an equilibrium mixture of oligomers. In order to further define the characteristics of oligomerization, we studied the behavior of PC tubulin by equilibrium chromatography.[2] This technique measures the degree of protein partitioning or adsorption into porous resins; the effective molecular weight (MW) of a protein can then be determined by comparing its partitioning with those of marker proteins.

Calibration curves for Sepharose 4B, Sepharose 6B-CL, Biogel A 0.5m, and Sephadex G-75 are shown in FIGURE 1. The correspondence between either the partition coefficient (K_{av}) or the column parameter (K_p—see legend to FIGURE 1) and log MW for each resin is comparable to that obtainable by column chromatography. The apparent molecular weight of PC tubulin at 3–5 mg/ml is also marked in FIGURE 1. FIGURE 1B shows that on Sepharose 4B, the molecular weight of PC tubulin is ~150K. This value is significantly greater than 110K and therefore supports our previous conclusion that tubulin oligomerizes. We expected, however, to find an even greater average molecular weight, that is, ~300K, as predicted by ultrafiltration. One likely explanation was suggested by the experiments shown in FIGURE 1A. These experiments derived the molecular weight of PC tubulin under the same conditions as those employed in the Sepharose 4B experiment, but used Sephadex G-75 and Biogel A 0.5m. On these resins, PC tubulin had a molecular weight of ~60 and ~66K respectively, well below that of a dimer. Indeed, Sephadex G-75 should have excluded tubulin altogether.

We therefore hypothesize an interaction between PC tubulin and resin. Because recovery from all columns run was ~100%, this interaction must be reversible. Such an interaction would increase the amount of tubulin retained on the column and thereby increase the K_{av} or K_p, leading to a decrease in the apparent MW.

A reversible interaction should also be strongly affected by changing the resin-to-protein ratio. Accordingly, we changed that ratio by varying the agarose content of the resin matrix. The results are summarized in the first part of Table 1. An MW consistent with oligomer formation was clearly shown using Sepharose 4B, but it is readily apparent that as the agarose content of the resin increased, the apparent MW of PC tubulin decreased.

We also varied the resin-to-protein ratio by changing the protein concentration. These experiments are summarized in the second part of TABLE 1. As the concentration

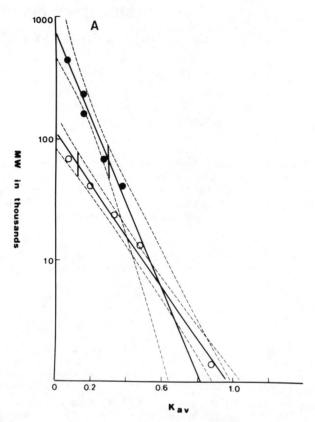

FIGURE 1. Calibration curves for equilbrium chromatography on four different resins. The molecular weight markers were vitamin B12 (MW = 1.4K), RNase I (MW = 12.7K), soybean trypsin inhibitor (MW = 22.7K), ovalbumin (MW = 40K), bovine serum albumin (MW = 67K), aldolase (MW = 158K), catalase (MW = 232K), ferritin (MW = 445K), and thyroglobulin (MW = 669K). All proteins used were freshly prepared from lyophilized powders shortly before use, except for PC tubulin, which was freshly prepared as previously described.[1] Column procedure: All steps were performed at room temperature. Tips for a Gilson P1000 pipetteman were plugged with Pyrex wool and filled with ~1.2 ml of resin. The resin was washed exhaustively with 20 mM NaP$_i$ – 100 mM glutamate, pH 6.75 (PG buffer) prior to each run. Just before adding sample, the top third of the column was stirred and allowed to resettle. Two ml of sample was then percolated through the resin, fully equilibrating the resin with sample, and the run-off fraction was collected into a preweighed tube. The protein in the column was then eluted with 2 ml of PG buffer, and this elulate was also collected into a tared tube. The tubes were reweighed to determine the exact volumes collected, and the protein concentration applied to the column as well as of the two fractions were measured by OD$_{280}$ or by Lowry's method.[3] Protein recoveries were consistently between 97 and 104 percent. The solid lines represent best fits as calculated by linear regression, and the dotted lines represent 95% confidence limits. Molecular

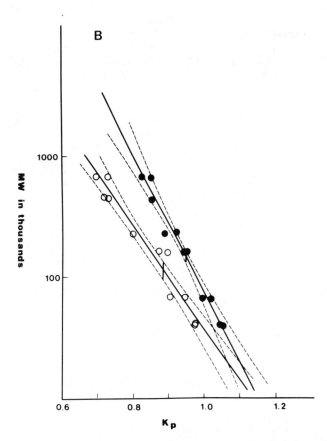

FIGURE 1. (*Continued*) weight markers are shown as open or closed circles, and the values obtained for PC tubulin are marked by line segments. **A:** Calibration curves for Sephadex G-75 (O) and Biogel A 0.5m (●). Void volumes (V_os) were determined as above using Dextran blue 2000, which had been repurified over a Biogel A1.5m column. Only the breakthrough fraction was used as a V_o marker. The internal volume (V_I) was measured using potassium chromate as the marker. K_{av}s were calculated as follows:

$$K_{av} = \frac{\dfrac{V_e P_e}{P_a} - V_O}{V_I}$$

where V_e = volume of eluate, P_e = protein concentration of eluate, and P_a = protein concentration applied to column. **B:** Calibration curves for Sepharose 4B (●) and Sepharose 6B CL (O). Given that log MW is a linear function of K_{av}, it can readily be shown that log MW is also a linear function of $V_e P_e/P_a = K_p$, as long as all determinations of K_p on a given resin were performed using the same internal and void volumes, that is, using the same column, as was done in this study.

of PC tubulin increased, so did its apparent MW. By contrast, aldolase showed no concentration-dependent variation in its elution behavior (data not shown).

Thus, the behavior of PC tubulin during equilibrium chromatography was consistent with interactions of tubulin dimers both with themselves to form oligomers, and with agarose and dextran resins. The latter interactions can lead to artefactually lowered estimates of the apparent MW of PC tubulin from methods employing agarose and dextran resins. In particular, tubulin-resin interactions may become extremely

TABLE 1. The Molecular Weight of PC tubulin as a Function of Concentration and Resin Composition[a]

Resin Type	Composition	[PC Tubulin]	Apparent MW
Sepharose 4B	4% agarose	4.6 mg/ml	150 K
Sepharose 6B	6% agarose	4.3	120 K
Biogel 0.5 m	10% agarose	3.3	66 K
Sephadex G-75	cross-linked dextran	3.0	60 K
Sepharose 6B	6% agarose	0.3 mg/ml	44 K
		0.5	50 K
		1.1	69 K
		2.2	95 K
		4.6	120 K

[a]All values shown were determined using the columns and the standard curves of FIGURE 1.

important during column chromatography, in which resin-to-protein ratios are typically much higher than those used in equilibrium chromatography.

REFERENCES

1. KRAVIT, N. et al., 1984. J. Cell Biol. **99:** 184.
2. ACKERS, G. 1970. Adv. Protein Chem. **24:** 343.
3. LOWRY, O. et al. 1951. J. Biol. Chem. **193:** 265.

Dideoxyguanosine Nucleotides and Microtubule Assembly

ERNEST HAMEL AND CHII M. LIN

Laboratory of Pharmacology and Experimental Therapeutics
Developmental Therapeutics Program
Division of Cancer Treatment
National Cancer Institute
National Institutes of Health
Bethesda, Maryland 20892

We have been studying interactions of ribose-modified analogs of guanosine diphosphate (GDP) and guanosine triphosphate (GTP) with tubulin in reactions with and without microtubule-associated proteins (MAPs). The most interesting analogs were 2′,3′-dideoxyguanosine 5′-diphosphate (ddGDP) and triphosphate (ddGTP), for ddGDP was the only diphosphate able to support microtubule assembly. Only ddGTP was active without MAPs at low tubulin concentrations and low ionic strengths. Here we shall contrast the properties of MAP-dependent microtubules formed with ddGDP or with ddGTP to those formed with GTP in 0.1 M 2-(N-morpholino)ethanesulfonate-0.5 mM MgCl$_2$.

With 1.0 mg/ml of tubulin, the minimum nucleotide concentration required for microtubule assembly was 10 μM GTP, 5 μM ddGTP, but 200 μM ddGDP. In a dilutional study with 0.4 mM nucleotide, the critical concentration for assembly was 0.09 mg/ml with GTP and only 0.04 mg/ml with ddGTP or ddGDP. MgCl$_2$ added to reaction mixtures had little effect on GTP-dependent assembly, was moderately stimulatory with ddGTP, and inhibited ddGDP-dependent polymerization. (By contrast, Mg^{2+} was required for MAP-independent polymerization with ddGDP.) No nucleotide breakdown occurs during assembly with ddGDP, but both GTP and ddGTP are hydrolyzed to diphosphates in reactions initially closely coupled to microtubule assembly. Exchangeable site GDP is displaced by all three nucleotides during polymerization, and microtubules formed with [α-^{32}P]GTP contain 0.95 mol of [α-^{32}P]GDP per mol of tubulin, with [β-^{32}P]ddGTP 1.16 mol of [β-^{32}P]ddGDP, and with [β-^{32}P]ddGDP 0.84 mol of [β-^{32}P]ddGDP. Microtubules formed with the three nucleotides differ in their sensitivities to temperature and Ca^{2+}. With GTP, microtubule assembly requires a reaction temperature of 20–25°, and significant disassembly occurs at 15–20°. Assembly and disassembly with ddGTP occur at much lower temperatures: 10–15° and 0–5°, respectively. With ddGDP, assembly occurs at a higher temperature than with GTP (25–30°), but disassembly occurs at a lower temperature (5–10°). The GTP tubules are most labile to Ca^{2+}, complete disassembly occurring at 2 mM CaCl$_2$. With ddGDP, complete disassembly required 10 mM Ca^{2+}, whereas ddGTP tubules were stable even in 10 mM Ca^{2+}. Microtubule lengths were measured to evaluate the nucleotides' relative abilities to support nucleation (FIGURE 1). Whereas the average tubule formed with ddGTP was less than one-third the length of GTP tubules, the length of tubules formed with ddGDP was similar to that with GTP. In an attempt to gain insight into a possible role for nucleotide hydrolysis in the flux of tubulin subunits in microtubules, treadmilling was examined with GTP, ddGTP, and ddGDP, both with and without Mg^{2+} (FIGURE 2). Nucleotide flux could be readily demonstrated with GTP tubules, with and without Mg^{2+}, but only minimal

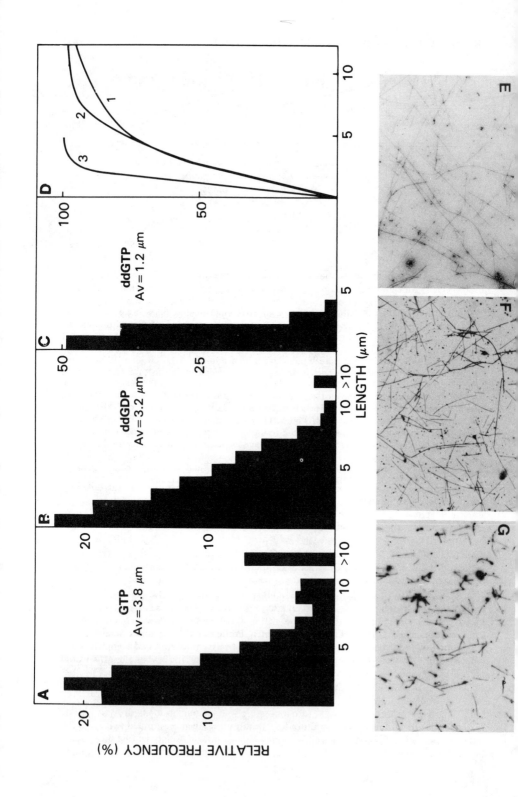

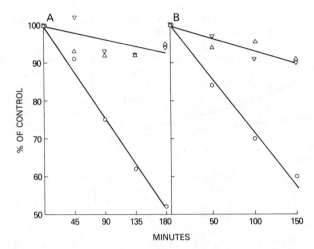

FIGURE 2. Evaluation of flux of tubulin subunits through microtubules formed with GTP, ddGTP, and ddGDP. Each 0.5 ml reaction mixture initially contained 1.6 mg/ml tubulin, 0.4 mg/ml of heat-treated MAPs, radiolabeled nucleotide at 0.2 mM, 0.1 M MES (pH 6.6), and, if indicated, 0.5 mM MgCl$_2$. Polymerization reactions were followed turbidimetrically for one and one-half hr at 37°, well into the plateau phase under all conditions. At this point a 10-fold excess of the appropriate nonradioactive nucleotide was added to each reaction mixture, and nucleotide equilibration was assumed to be complete after another 5 min at 37°. At this point (zero time), and at the times indicated in the FIGURE, 0.1 ml aliquots were centrifuged in a 37° Ti 50 Beckman rotor for 30 min at 35,000 rpm. The pellets were washed three times in 0.1 M MES (pH 6.6) at 37° containing, if appropriate, 0.5 mM MgCl$_2$. They were then dissolved in 0.12 ml of 8 M urea. The radioactivity and protein content of the resulting solutions were determined, and all data were normalized against the zero time values for each nucleotide. Symbols in both panels are as follows: O, the nucleotide was GTP (initially [α-^{32}P]GTP); $\triangle$, the nucleotide was ddGTP (initially [β-^{32}P]ddGTP); $\triangledown$, the nucleotide was ddGDP (initially [β-^{32}P]ddGDP). **A:** Tubulin flux in the presence of 0.5 mM MgCl$_2$. **B:** Tubulin flux in the absence of exogenous Mg^{2+}.

FIGURE 1. Microtubule length distributions with 0.4 mM GTP, ddGDP and ddGTP. Reaction mixtures (0.1 ml) containing 1.0 mg/ml tubulin, 0.33 mg/ml heat-treated MAPs, 0.1 M 2-(N-morpholino)ethanesulfonate (MES) (pH 6.4), 0.5 mM MgCl$_2$, and 0.4 mM nucleotide were incubated for 30 min at 37°. Aliquots were diluted 4-fold with an isothermic solution of 50% sucrose containing 0.1 M MES and 0.5 mM MgCl$_2$. Portions of these mixtures were placed on 200-mesh carbon-coated copper grids, which were negatively stained with 0.25% uranyl acetate and examined in the electron microscope. High power micrographs confirmed the formation of microtubules, and length measurements were made on micrographs with a magnification of 5250 (see panels E–G) by using a Zeiss Videoplan 2 image analyzer. **A:** Length distribution of microtubules formed with GTP. **B:** Length distribution of microtubules formed with ddGDP. **C:** Length distribution of microtubules formed with ddGTP. **D:** Comparison of cumulative length distributions of microtubules formed with the three nucleotides. Curve 1, GTP; curve 2, ddGDP; curve 3, ddGTP. **E:** Micrograph of microtubules formed with GTP. **F:** Micrograph of microtubules formed with ddGDP. **G:** Micrograph of microtubules formed with ddGTP. Magnifications: × 2800.

changes in radiolabeled nucleotide content occurred with the analogs (despite the greatly increased number of microtubule ends in ddGTP tubules).

Even though the analogs support vigorous assembly reactions, we could detect little binding of ddGTP, virtually no binding of ddGDP to unassembled tubulin, and little displacement of exchangeable site GDP without polymerization. The molar ratio of GDP required for complete inhibition of assembly was 7:1 with GTP, 2:1 with ddGTP, and 1:50 with ddGDP, in agreement with the binding studies. GTP hydrolysis was not inhibited, and even stimulated, by ddGTP, whereas GTP in a molar ratio to ddGTP of 1:10 inhibited ddGTP hydrolysis about 50% (despite the fact that ddGTP is hydrolyzed more rapidly than GTP).

Our observations with dideoxyguanosine and other analogs lead to the following conclusions: 1) Any ribose modification reduces nucleotide affinity for the exchangeable site, but the ability of nucleotides to promote nucleation is a property distinct from their affinity for tubulin. Polymerization with ddGDP and with ddGTP MAPs probably derives from the potent activity of ddGTP in nucleation. 2) The simple mechanism tubulin · ddGTP $\rightleftharpoons$ microtubule · ddGDP $\rightleftharpoons$ tubulin · ddGDP is excluded by the different properties of microtubules formed with ddGTP and with ddGDP. 3) Findings with nucleotide analogs should be extended only with caution to the interaction of GTP with tubulin.

The GTP Concentration Modulates the Association Rate Constant for Microtubule Assembly

KHALID ISLAM AND ROY G. BURNS

Biophysics Section
Department of Physics
Imperial College of Science and Technology
London SW7 2BZ, England

Guanosine triphosphate (GTP) binds to a readily exchangeable E-site on the tubulin (Tu) dimer. Microtubule (MT) assembly affects the site and renders it relatively nonexchangeable, and also promotes the hydrolysis of bound GTP to guanosine diphosphate (GDP). Unfractionated cold-dissociated MT protein is, though, a mixture of Tu dimers, free MAP-2, and MAP-2:Tu oligomers, and both Tu dimers and the

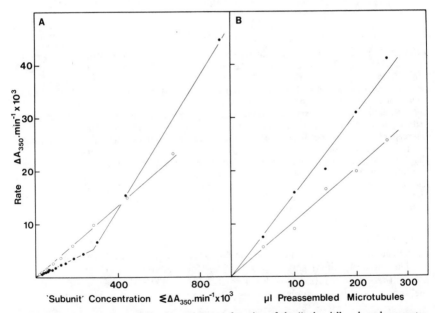

FIGURE 1. Rate of microtubule elongation as a function of the "subunit" and seed concentration. Three times microtubule protein, 0.92 mg · ml^{-1} containing $\approx$0.5 mole GDP · mole^{-1} Tu, and an oligomeric:total Tu ratio of $\approx$0.6, was assembled at 37°C with 20 (O) and 500 (●) μM GTP. The assembly kinetics were measured turbidimetrically at A$_{350_{\text{nm}}}$,[3] and the final extent of assembly was approximately the same at the two GTP concentrations.[1]
A: A pseudo first-order reaction plot of the instantaneous rate (ΔA_{350} · min^{-1}) as a function of the depleting "subunit" concentration ($\Sigma \Delta A_{350}$ · min^{-1}), showing that the assembly at 20 μM GTP conforms to a single pseudo first-order reaction, whereas the assembly at 500 μM GTP is described by the summation of two such reactions.

MAP-2:Tu oligomers can participate directly in MT assembly.[1] The E-site on the oligomers is also partially protected[2] such that GTP differentially binds to dimers and oligomers.

The assembly kinetics of three times recycled chick brain MT protein have been examined as a function of the GTP concentration. The protein was prepared by pelleting two times recycled MTs through a 30% sucrose cushion,[1] cold dissociating,

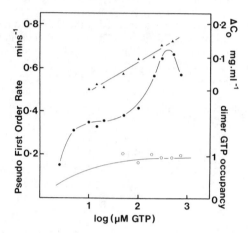

FIGURE 2. Rate of assembly, GTP-dimer occupancy, and the oligomeric critical concentration as a function of the GTP concentration. The pseudo first-order rate constant was determined for three times MT protein at 37°C as a function of the GTP concentration (see FIGURE 1A for details). The kinetics are biphasic at 500 μM GTP (FIGURE 1A and reference 1) yielding two values for the apparent pseudo first-order rate constant: an initial rate (●) representing dimer + oligomer addition, and a second rate (○) representing dimer addition. The difference between the critical concentration for dimer and oligomer addition (ΔC_o:▲) was determined from the intercept of the initial rate. In addition, the GTP occupancy of the Tu dimers (———) has been computed from the total Tu concentration (6.56 μM), the total GDP concentration (3.28 μM), the GTP concentrations, and published values for k_dGTP and k_dGDP (0.1 and 0.2 μM respectively[5]). This calculation underestimates the dimer GTP occupancy as it assumes that all Tu dimers are available and that there is an equilibrium between the Tu dimers and GDP. However, ≈60% of the Tu is oligomeric,[1] and following the gel filtration pretreatment, most of the GDP is bound to the relatively nonexchangeable oligomer E-site.

and then eluting through a Sephadex G-50 column, equilibrated in reassembly buffer.[3] Such protein contains ≈0.5 mole · mole^{-1} GDP, bound primarily to the oligomers.

The assembly kinetics at 20 μM GTP conform, following nucleation, to a single pseudo first-order reaction. The kinetics at 500 μM GTP, however, are described by the summation of two such reactions (FIG. 1A), corresponding to the addition of

---

B: The initial rate of elongation (measured during the lag period of unseeded preparations) as a function of the concentration of preassembled MT seeds. The seed protein (1.05 mg · ml^{-1}) was preassembled to equilibrium at 50 μM GTP, and then increasing concentrations challenged at 25 and 500 μM GTP with 250 μl of unassembled MT protein (1.04 mg · ml^{-1}) prewarmed to 37°C in a final volume of 500 μl. This procedure, which does not yield absolute values for k_{+1} and k_{-1}, was necessary as preassembled sheared MTs are inefficient seeds for oligomer addition.

dimers and oligomers and to dimers alone.[1] Extrapolation of the oligomeric phase yields a value for the difference in the critical concentrations ($C_o^{oligo}-C_d^{dimer}$).

The faster pseudo first-order rate at 500 μM GTP is not due to enhanced nucleation. The phase describing dimer addition is slower than that at 20 μM GTP, indicating less nucleation at the higher GTP concentration (FIGURE 1A). Similarly, the initial rate on seeding with preassembled MTs is faster at 500 μM GTP (FIGURE 1B). The association rate constant (k_{+1}) is therefore higher at 500 than at 20 μM GTP.

This higher value of k_{+1} is not due to increasing the GTP-Tu dimer concentration, as the dimer GTP occupancy exceeds 80% by 20 μM GTP (FIGURE 2). The increase must therefore reflect enhanced oligomer addition resulting from GTP binding to the relatively inaccessible E-site on the oligomers.

Analysis of the pseudo first-order rate constant as a function of the GTP concentration shows that it is approximately constant between 10–50 μM GTP, increases markedly above 50 μM, and attains a maximum at 500–600 μM (FIGURE 2). This increase is accompanied by a rise in $C_o^{oligo}-C_d^{dimer}$, and as k_{-1}^{dimer} is approximately zero for unphosphorylated MT protein,[3,4] both the k_{+1} and k_{-1} for oligomer addition must be affected.

The results demonstrate that the GTP concentration alters the kinetics of MT assembly by modulating the oligomeric rate constants. This modulation is observed at GTP concentrations within the expected physiological range.[4] The multi-subunit nature of the oligomer permits a number of integer steps in the control of the oligomeric rate constants: the GDP:GTP content of the oligomers appears to play an intimate role in regulating MT assembly.

REFERENCES

1. BURNS, R. G. & K. ISLAM. 1984. FEBS Lett. **173:** 67–74.
2. WEISENBERG, R. C., W. J. DEERY & P. J. DICKINSON. 1976. Biochemistry **15:** 4248–4254.
3. BURNS, R. G., K. ISLAM & R. CHAPMAN. 1984. Eur. J. Biochem. **141:** 609–615.
4. BURNS, R. G. & K. ISLAM. 1986. Ann. N.Y. Acad. Sci. **466:** 340–356. This volume.
5. JACOBS, M. & M. CAPLOW. 1976. Biochem. Biophys. Res. Commun. **68:** 127–135.

The Carboxyterminal Region of Tubulin Regulates Its Assembly into Microtubules

LUIS SERRANO,[a] RICARDO B. MACCIONI,[b]
AND JESÚS AVILA[a]

[a]Centro de Biología Molecular (CSIC-UAM)
Universidad Autónoma
Canto Blanco
28049 Madrid, Spain

[b]Department of Biophysics/Genetics
University of Colorado Health Sciences Center
Denver, Colorado 80262

The main characteristic of tubulin is its capacity to self assemble into microtubules. Since the initial report of Weisenberg,[1] the assembly *in vitro* of microtubules has been extensively studied. It is known that brain microtubules assembled *in vitro* contain other proteins than tubulin known as microtubule-associated proteins (MAPs),[2-4] which facilitate microtubule polymerization. Tubulin depleted of MAPs is unable to assemble except under certain conditions involving the addition of cosolvents or polycations that may substitute the requirements of MAPs, or the addition of a high concentration of tubulin (*i.e.* reference 5). As a first step to study how MAPs or polycations facilitate tubulin polymerization, we have located the tubulin site for the interaction of the two most studied MAPs: MAP-2 and tau factor. To study such interaction, the tubulin molecule was cleaved by limited proteolysis with subtilisin, and the interaction of the resulting fragments with MAPs was studied. Subtilisin cleaves tubulin subunits yielding a related pattern, a large fragment (48 kD) containing the aminoterminal and a small fragment (4 kD) containing the carboxyterminal fragment. Binding experiments with MAP-2 and tau factor (the latter experiment done in collaboration with E. Montejo) indicated that the 48 kD fragment did not bind to MAPs, whereas the 4 kD fragment did (FIGURE 1). More important than that, however, we found that after removal of the carboxyterminal regions, the cleaved tubulin can assemble in a way similar to that of tubulin in the presence of MAPs (FIGURE 2). Also, similar results have been found for the assembly of tubulin in the presence of polycations (polylysine) or that of subtilisin-digested tubulin. The morphology of the polymers assembled from undigested or subtilisin-digested tubulin were not identical. Indeed, in the latter case hooked microtubules were found (insert in FIGURE 2). These results suggest that the carboxyterminal is involved in the regulation of the assembly of tubulin and in determining the final shape of the polymerized structure. Reports related to this work have been previously published.[6,7]

REFERENCES

1. WEISENBERG, R. C. 1972. Science **177:** 1104–1105.
2. MURPHY, D. B. & G. G. BORISY. 1975. Proc. Natl. Acad. Sci. USA **72:** 2696–2700.
3. SLOBODA, R. D., W. L. DENTLER & J. L. ROSENBAUM. 1976. Biochemistry **15:** 4497–4505.
4. CLEVELAND, D. W., S. Y. HWO & M. W. KIRSCHNER. 1977. J. Mol. Biol. **116:** 207–225.
5. TIMASHEFF, S. N. & L. M. GRISHAM. 1980. Annu. Rev. Biochem. **49:** 565–591.
6. SERRANO, L., J. AVILA & R. MACCIONI. 1984. Biochemistry **23:** 4675–4681.
7. SERRANO, L., J. DE LA TORRE, R. MACCIONI & J. AVILA. 1984. Proc. Natl. Acad. Sci. USA **81:** 5989–5993.

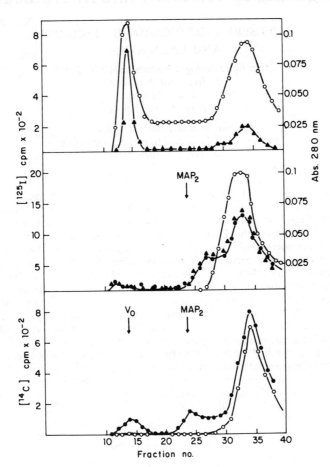

FIGURE 1. Binding of MAPs to undigested and digested tubulin fragments. Upper panel: Unlabeled (O) or labeled [^{125}I] tubulin (▲) was added to a mixture of MAPs (final concentration of tubulin, 2 mg/ml; final concentration of MAPs, 0.4 mg/ml), and the mixture was chromatographed on a Sepharose 4B column (0.3 × 21). Middle: [^{125}I]subtilisin-digested tubulin[6] was mixed with (▲) or without (●) MAPs in the same conditions as above. The position of unlabeled subtilisin-digested tubulin (2 mg/ml) is indicated (O). Lower panel: 0.1 ml of tubulin labeled in the carboxyterminal with [^{14}C]tyrosine (2 mg/ml; final specific activity 40–170 cpm/µg) was digested with subtilisin and the 4 kD carboxyterminal region purified as indicated.[6] This region (50 µl, 5 µg) was incubated for 15 min at 25° with MAP-2 (50 µl, 30 µg) and chromatographed on a Sepharose 4B column as previously indicated. O = 4 kD; ● = 4 kD with MAP-2.

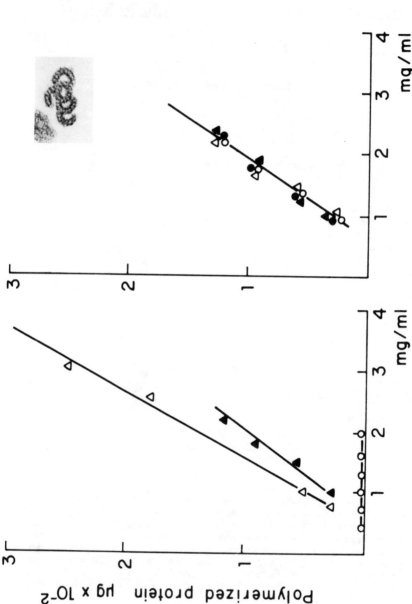

FIGURE 2. Polymerization of undigested and digested tubulin in the presence of MAPs or polylisine. Left panel shows the effect of increasing concentrations of undigested tubulin incubated in the absence ($\bigcirc$) or presence of MAPs ($\bullet$) at a ratio 1/10 w/w or polylysine ($\blacktriangle$) at a ratio 1/50 w/w, in w/w ($\triangle$), tau factor ($\bullet$) at a ratio of 1/8 w/w, polylysine ($\blacktriangle$) at a ratio 1/50 w/w, and in the absence of any of them ($\bigcirc$). The polymerized protein at each concentration was quantified following the Lowry method. The volume of each sample was 0.15 ml. Right panel shows the effect of the concentration of subtilisin-digested tubulin in its assembly in the presence of MAP-2 at a ratio 1/10

Tubulin Domain Structure Studied by Limited Proteolysis and Antibody Labeling

EVA-MARIA MANDELKOW,[a] MARILLE HERRMANN,[b]
AND URSULA RÜHL

Max Planck Institute for Medical Research
Biophysics Department
D-6900 Heidelberg
Federal Republic of Germany

We have studied the substructure of the tubulin monomers by a combination of controlled proteolysis and antibody labeling. When native tubulin is briefly digested with some proteases, one observes fragments accounting for roughly one-third and two-thirds of the molecule, in agreement with earlier studies.[1] The two subunits (α or β) show different stabilities with respect to enzymatic cleavage. For example, trypsin preferentially splits α tubulin into M_r = 36 and 14 kD components, whereas chymotrypsin splits β tubulin into 31 and 20 kD fragments (FIGURE 1). By eluting the fragments from gels and partial sequencing, we determined[2,3] that the main tryptic cleavage site of α tubulin occurs at arg339, and the main chymotryptic cleavage of β tubulin is at tyr281 (FIGURE 2). Thus the small fragments represent the C-terminal parts of the molecules, in contrast to an earlier report.[4]

When comparing tubulin with other known proteins,[5,8] it appears that all features related to nucleotide binding are confined to the N-terminal fragment. This includes the regions of residues 58–80, 142–148, 180–183, and 242–246. In particular, there are striking similarities with the nucleotide-binding domain of glutathione reductase.[5] This suggests that the proteolytic fragments of tubulin closely correspond to functional domains.

Tubulin and its cleavage products were identified by their reaction with high-affinity antibodies specific for α and β tubulin (FIGURE 1). They were raised against electrophoretically purified monomers that were coupled to keyhole limpet hemocyanin and injected into rabbit knee lymph nodes. The antibodies were purified by affinity chromatography and characterized by radioimmunoassay (half maximal binding around 10^{-9}–$10^{-7}M$), blotting, immunofluorescence, and electron microscopy. About 30% of the antibodies were subunit-specific, whereas 70% cross-reacted with both tubulin chains. There was no cross-reactivity with MAPs.

In spite of immunization with intact tubulin chains, the antibodies are rather specific for one of the domains. Thus the antibodies raised against α tubulin recognize predominantly the large domains; they are either specific for α tubulin or cross-react with the large domain of β tubulin. On the other hand, the antibodies obtained with β tubulin recognize mainly the small domains (β-specific and β-cross-reacting fractions). This means that the antibodies can be used to discriminate between the tubulin chains as well as between the domains generated by the proteases. The apparent lack of

[a]Present address: Max Planck Unit for Structural Molecular Biology, Ohnhorststr. 18, D-2000 Hamburg 52, Federal Republic of Germany.
[b]Present address: Universitätsklinikum Charlottenburg, Chirurgische Klinik, Spandauer Damm 130, D-1000 Berlin 19, Federal Republic of Germany.

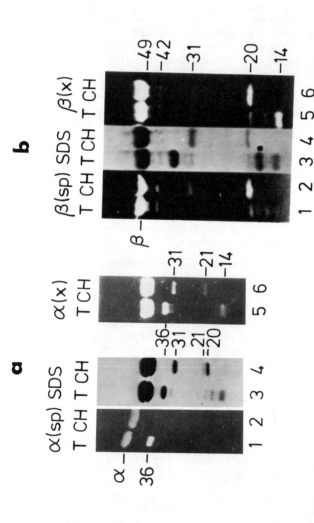

FIGURE 1. Sodium dodecyl sulfate (SDS) gels (12.5%) of proteolytic digests of phosphocellulose-purified tubulin and immunoblots with chain-specific and cross-reacting antibodies. **a:** Lane 3: Tryptic digest of PC tubulin with prominent bands at 36 and 14 kD. Lane 4: Chymotryptic digest with prominent fragments at 31 and 20 kD. Lane 1: Blot of tryptic digest with α-specific antibody, showing label on α tubulin and the 36 kD fragment, but not at 14 kD. Lane 2: Blot of chymotryptic digest with α-specific antibody, showing label on α tubulin only. Lane 5: Blot of tryptic digest with α-cross-reacting antibody, showing label on α and β tubulin and the 36 kD and 14 kD fragments of α tubulin. Lane 6: Blot of chymotryptic digest with α-cross-reacting antibody, showing label on α and β tubulin, on 31 kD fragment. (β), but not on 20 kD fragment of β tubulin. Cross-reactivity of α antibody is mainly between large N-terminal domains of α and β tubulin (lanes 5, 6). **b:** Lanes 3 and 4: SDS gel of tryptic and chymotryptic digests. The band at 18 kD (*lane 3) is soybean trypsin inhibitor. Lane 1: Blot of tryptic digest with β-specific antibody, showing label mainly on β tubulin and some minor fragments. Lane 2: Blot of chymotryptic digest with β-specific antibody, showing label on β tubulin and a main band at 20 kD. Lane 5: Blot of tryptic digest with β-cross-reacting antibody, showing bright staining of α and β tubulin and at 14 kD. Lane 6: Blot of chymotryptic digest with β-cross-reacting antibody, showing staining mainly of α and β tubulin and at 20 kD. Cross-reactivity of β antibody is between the small C-terminal domains of α and β tubulin (lanes 5, 6).

antigenicity of one domain upon immunization with the whole chain is not simply due to lack of antigenic determinants, as shown by the cross-reacting antibodies. A possible explanation is the removal of sites of the "silent" domains of the antigen, for example due to aggregation or degradation. The complementarity of the domains in terms of antigenicity may be related to the structure of the tubulin dimer.[3]

Evidence for different stabilities of domains comes from experiments in which tubulin is incubated with SDS for different times and at different temperatures, then run on gels and blotted with the various antibody fractions. One observes spontaneous degradation of α and β tubulin, with α being the more labile one. The major fragments generated by spontaneous degradation are similar to the domains of enzymatic cleavage or variants thereof. In particular, the small domain of α tubulin and the large domain of β tubulin are more labile than their complementary domains. Thus the domains that show low stability are the same as those that fail to evoke an antigenic response.

Approximate locations of antigenic regions have been determined from the immunoblots of the enzymatic cleavage products, using our polyclonal antibodies as

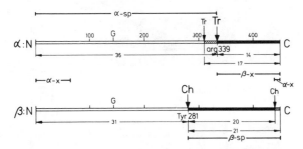

FIGURE 2. Diagram of α- and β-tubulin chains, their main proteolytic fragments, and antibody specificities. Trypsin cleaves mainly α tubulin at arg339; chymotrypsin cleaves mainly β tubulin at tyr281. Secondary cleavage sites are also indicated. The solid bars show the C-terminal domains.

well as several commercial monoclonal ones (Amersham, Seralab). The results were compared with predictions based on hydrophilicity.[6] The main epitopes of our α-cross-reacting antibodies are within about 8 kD of the N-terminus. Epitopes of the β-cross-reacting antibodies are within 12 kD of the C-terminus. In contrast to our antibodies, all commercial ones bind to the C-terminal domain, either at the C-terminus itself (*e.g.* YL ½, Seralab, see reference 9) or within a few kDal of the C-terminus (e.g. Amersham anti-α and anti-β).

Electron microscopy shows that microtubules are disrupted by antibodies or F_{ab} fragments. A variety of breakdown products are observed (FIGURE 3). They include "10nm" fibers (consisting largely of MAPs), structureless sheets (mostly tubulin), thick tapered fibrous bundles (tubulin + MAPs), and wispy fibers (probably MAP oligomers). The "10 nm" fibers are distinct from protofilaments and may represent a remnant of the MAP skeleton of microtubules after destruction of the tubulin core by the antibodies. Disruption of microtubules is also observed after microinjection of antibodies into living cells,[7] in contrast to other antibodies described in the literature. It suggests that some of the sites recognized by our antibodies are involved in intersubunit bonding.

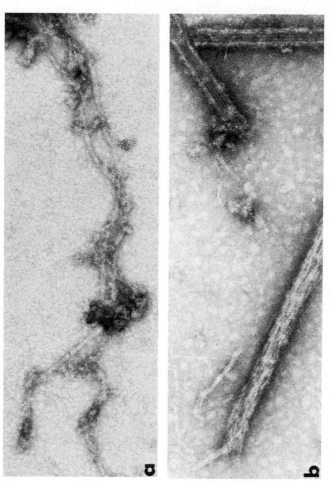

FIGURE 3. a: Microtubule protein polymerized in reassembly buffer, then incubated for 10 min with β-specific F_{ab} fragments. Microtubules disappear within minutes. Early breakdown products include fibers of about 10 nm width composed of 2–4 subfibers about 3.5 nm wide. **b:** Microtubule protein polymerized in stabilizing buffer (with 25% glycerol), then incubated for 15 minues with β-specific F_{ab} fragments. Disassembly is slowed down, compared to glycerol-free buffer. One observes intact microtubules and fibrous bundles with tapered ends and an irregular repeat of 40–50 nm. Note short wispy fibers in the background. Magnification 135,000 X.

REFERENCES

1. BROWN, H. R. & H. P. ERICKSON. 1983. Arch. Biochem. Biophys. **220:** 46–51.
2. MANDELKOW, E.-M., M. HERRMANN, & U. RÜHL. 1985. J. Mol. Biol. **185:** 311–327.
3. KIRCHNER, K. & E.-M. MANDELKOW. 1985. EMBO J. **4:** 2397–2402.
4. MACCIONI, R. B. & N. W. SEEDS. 1983. Biochemistry **22:** 1567–1572.
5. PAI, E. F. & G. E. SCHULZ. 1983. J. Biol. Chem. **258:** 1752–1757.
6. HOPP, T. & K. WOODS. 1981. Proc. Natl. Acad. Sci. USA **78:** 3824–3828.
7. FÜCHTBAUER, A., M. HERRMANN, E.-M. MANDELKOW & B. M. JOCKUSCH. 1985. EMBO J. **4:** 2807–2814.
8. STERNBERG, M. J. E. & W. R. TAYLOR. 1984. FEBS Lett. **175:** 387–392.
9. WEHLAND, J., M. C. WILLINGHAM & I. V. SANDOVAL. 1983. J. Cell Biol. **97:** 1467–1475.

Microtubule Assembly and Disassembly Studied by Synchrotron X-Ray Scattering and Cryoelectron Microscopy

E. MANDELKOW,[a,c] E.-M. MANDELKOW,[a,c]
AND J. BORDAS[b]

[a]Max Planck Institute for Medical Research
Heidelberg, Federal Republic of Germany
[b]MRC-SERC Biology Support Laboratory
Daresbury, Warrington, England

The role of tubulin oligomers in microtubule assembly and disassembly has been investigated by two methods that combine high structural resolution with fast time slicing. One is time-resolved X-ray scattering, which yields the average behavior of an assembling protein solution and the structures of the predominant reaction partners.[1,2] The other is cryoelectron microscopy of hydrated samples frozen in amorphous ice. This avoids staining, drying, or chemical fixation; it preserves the particles in their native state and opens a way to study time-dependent processes, because freezing leads to very rapid physical fixation.[3]

Oligomers (= protofilament fragments, *ca* 40–60nm long) play a twofold role in microtubule assembly. They are the units from which rings are formed at low temperature, but they are also important for microtubule nucleation at elevated temperature[1,4] and possibly elongation.[5] This implies that rings break apart into oligomers and dimers prior to microtubule assembly at 37°C, and that microtubules break apart into oligomers and dimers prior to ring assembly at 4°C (FIGURE 1). The functional distinction between active oligomers (leading to microtubules) and inactive ones appears to be related structurally to the degree of curvature: active oligomers are straight, inactive ones are coiled. Thus, conditions that favor the coiled conformation (as measured by the stability of rings) tend to inhibit microtubule assembly. Examples are low temperature, or mM guanosine diphosphate (GDP), or Ca^{++} at high temperature. A special case is that of stabilizing agents such as glycerol, which preserves not only microtubules but also rings; the slow release of oligomers from rings is in part responsible for the slower assembly of microtubules in glycerol. Conversely, activation of oligomers is achieved by high temperature and guanosine triphosphate (GTP) (*cf.* reference 6).

Rings are convenient indicators of oligomer conformation in the presence of brain microtubules-associated proteins (MAPs), but oligomers exist even when no rings are formed, for example in phosphocellulose-purified tubulin.[1] The equilibrium between dimers and oligomers must be rapid because oligomers are observed immediately after Airfuge centrifugation. X-ray patterns during a cycle of assembly and disassembly are similar in the presence and absence of MAPs if one disregards the MAP-dependent phases of initial ring dissolution and final ring reassembly. In particular, there is a similar oligomer- dependent stage of nucleation. These experiments were performed in

[c]Present address: Max Planck Unit for Structural Molecular Biology, Ohnhorststr. 18, D-2000 Hamburg 52, Federal Republic of Germany.

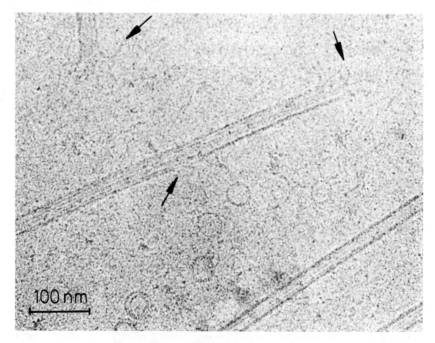

100 nm

FIGURE 1a

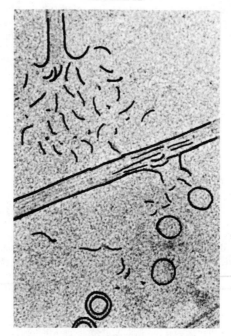

FIGURE 1b

usual assembly buffers (*e.g.* 0.1*M* piperazine-*N*, *N*-bis-(2-ethane sulfonic acid) (PIPES) or 2-(*N*-morpholino)ethane sulfonic acid (MES), 1 m*M* Mg^{++} GTP), that is, without high concentrations of glycerol or Mg^{++}. The driving force for assembly comes from the high concentration of tubulin (10–40mg/ml). It appears that one of the factors limiting the assembly of this protein at lower concentrations is the smaller number of oligomers. Assembly *in vivo* is possibly controlled by the availability of oligomers that may be stored in an active and inactive form.

Microtubule assembly is a nonhelical process, even though the final structure appears cylindrically symmetric at low resolution. This follows from structural

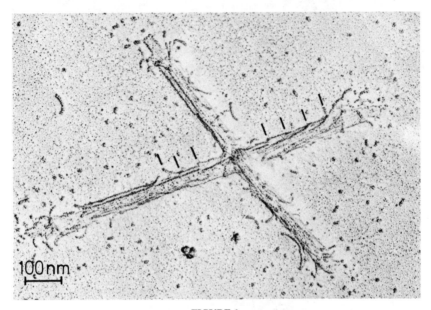

FIGURE 1c

FIGURE 1. a: Microtubules frozen rapidly during disassembly in a thin layer of amorphous ice. **b:** The same as **a,** with main features redrawn. Note coiled protofilaments fraying from ends, tubulin oligomers in background nearby (protofilament fragments = microtubule breakdown products), and newly formed single and double rings further away. Microtubules also seem to disintegrate from within (arrow), suggesting a combination of endwise disassembly and internal breakage. **c:** Similar preparation of disassembling microtubules visualized by glycerol spraying and rotary shadowing with platinum-carbon. Note protofilament fragments in background and disintegration of protofilaments along the length of a microtubule, with indications of a 50–60nm repeat.

considerations (the B lattice of dimers prevents helical symmetry),[7,8] as well as from the participation of the oligomers in assembly. Thus helical models of the type developed for actin[9] do not seem applicable to microtubules, although the basic distinction between nucleation and elongation still holds. FIGURE 2 is a diagram of different pathways of tubulin assembly that accounts for the observed polymorphism.[4,10] The reaction starts with the lateral association ("accretion") of oligomers, which leads to the closure of a microtubule cylinder. This is the nucleus in the strict sense, because it can grow by elongation only. The relative rate and extent of accretion and elongation can be varied, for example, by changing the equilibrium between

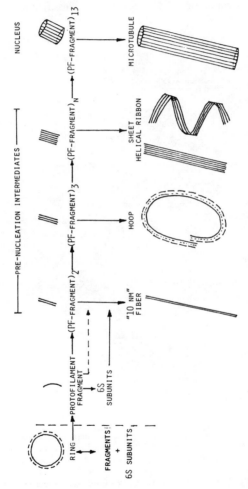

FIGURE 2. Model of tubulin assembly into microtubules and polymorphic forms.

oligomers and dimers. In the usual reassembly buffers, accretion is rapid and results mainly in closed microtubules. If elongation is favored, one finds polymorphic forms, all of which may be regarded as variants of incomplete microtubule walls. Thus the polymorphs are not assembly intermediates, but rather transient overshoot aggregates;

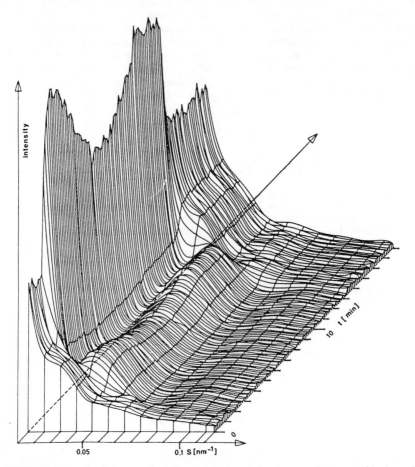

FIGURE 3. Example of time-resolved X-ray experiment showing overshoot assembly (experiment T47/08/83). The pattern of the initial cold solution arises from a mixture of rings, subunits, and oligomers. After the temperature jump to 37°C, one observes rapid assembly of labile microtubules or polymorphic forms, followed by partial disassembly and slow reassembly of stable microtubules. The final part of the traces shows the disassembly after a reverse temperature jump to 0°C.

they disappear during a redistribution phase.[10] This type of overshoot may be termed structural, because its prominent feature is the generation of different assembly forms. The simplest case is a nonequilibrium length distribution of otherwise regular particles, as observed, for example, with actin.[9] Tubulin aggregates tend to be more polymorphic because accretion and elongation take place in different directions (two-dimensional assembly).

There is a second type of overshoot that appears as a transient maximum in the average degree of polymerization (FIGURE 3). This may be termed assembly overshoot, because its prominent feature is the instability of the aggregates formed during the initial phase, irrespective of their structure (these depend on the relative rates of accretion and elongation, as above.) Assembly overshoot requires different pools of protein, one of which is ready for rapid assembly (*e.g.* oligomers activated by GTP and/or enriched in MAPs), but forms unstable aggregates (*e.g.* following GTP hydrolysis), whereas the slow phase results in stable structures. If the particles generated during the rapid phase are stabilized, the overshoot behavior is converted into biphasic assembly kinetics without transient disassembly. The phases may be dissected by controlling the levels of nucleotide, metal ions, pH, and ionic strength.

REFERENCES

1. BORDAS, J., E.-M. MANDELKOW & E. MANDELKOW. 1983. J. Mol. Biol. **164:** 89–135.
2. MANDELKOW, E., A. HARMSEN, E.-M. MANDELKOW & J. BORDAS. 1980. Nature (London) **287:** 595–599.
3. MANDELKOW, E.-M. & E. MANDELKOW. 1985. J. Mol. Biol. **181:** 123–135.
4. MANDELKOW, E., E.-M. MANDELKOW & J. BORDAS. 1983. TIBS **8:** 374–377.
5. WEISENBERG, R. C. 1980. J. Mol. Biol. **139:** 660–677.
6. CARLIER, M. F. & D. PANTALONI. 1978. Biochemistry **17:** 1908–1915.
7. MCEWEN, B. & S. EDELSTEIN. 1980. J. Mol. Biol. **139:** 123–145.
8. MANDELKOW, E.-M., R. SCHULTHEISS, R. RAPP, M. MÜLLER & E. MANDELKOW. 1986. J. Cell Biol. In press
9. OOSAWA, F. & S. ASAKURA. 1975. Thermodynamics of the polymerisation of protein. Academic Press. London.
10. MANDELKOW, E., R. SCHULTHEISS & E.-M. MANDELKOW. 1984. J. Mol. Biol. **177:** 507–529.

ATP-Dependent Gelation Contraction of Microtubules *in Vitro*

CHRISTOPHER CIANCI, DAVID GRAFF,
BAOCHONG GAO, AND RICHARD C. WEISENBERG

Department of Biology
Temple University
Philadelphia, Pennsylvania 19122

Adenosine triphosphate (ATP) will induce gelation contraction of third-cycle calf brain microtubules assembled *in vitro*.[1] After ATP addition to steady state microtubules, gelation contraction is observable within 30 minutes and is essentially complete within 60 minutes. The resulting gel may occupy less than 25% of its original volume. ATP specificity is demonstrated by the inability of contraction to occur in adenosine diphosphate (ADP), inosine triphosphate (ITP), cytidine triphosphate (CTP), or in nonhydrolyzable ATP analogues.

Electron microscopy has revealed that the gel contains microtubules that radiate out from dark-staining focal centers, forming structures similar in appearance to mitotic spindles and asters (FIGURE 1). Some microtubules seem to pass through these centers, and others interconnect one aster-like structure with another. Bundles of microtubules are observed, although these could be a result of sample preparation. Immunofluorescence has confirmed the presence of a network of aster-like structures in the contracted gel. The focal centers stain with antitubulin, are cold stable, and are similar in protein composition to microtubules.

Time-lapse microscopy has shown that aster-like structures move relative to each other. Experiments done in collaboration with Robert Allen and Shinya Inuoe have demonstrated particle movement in the gel. These particles can be seen in electron micrographs. They are roughly spherical and often appear to be associated with microtubules. The particles are probably the same material that comprises the focal centers.

Microtubule gelation contraction needs only those proteins that copurify with tubulin after three cycles of assembly and disassembly. We now have obtained a phosphocellulose-purified preparation that also undergoes gelation contraction upon addition of ATP (FIGURE 2). This preparation is reduced in microtubule-associated proteins and requires 5% dimethyl sulfoxide (DMSO) for assembly. Sodium dodecyl sulfate polyacrylamide gel electrophoresis (SDS-PAGE) of the phosphocellulose-purified proteins reveals the loss of the high molecular weight proteins. There is a minor band of 145K molecular weight (MW) and a more prominent 74k MW band in the gelled protein. Two proteins of 107K and 98k MW are significantly increased in the nongelled protein fraction. Poorly resolved low molecular weight proteins are also present in the preparation. In both the phosphocellulose protein and third-cycle protein preparations, the processes of gelation contraction and the morphology of the resulting gels appear to be similar. The ATPase activity of the phosphocellulose protein preparation was 4.4 nanomoles of ATP hydrolyzed/min per mg of protein. This is approximately one half of that in the third-cycle preparation, which is 9.3 nanomoles of ATP hydrolyzed/min per mg of protein. Several experiments have shown that cyclic AMP stimulates the ATP-induced gelation contraction of relatively inactive prepara-

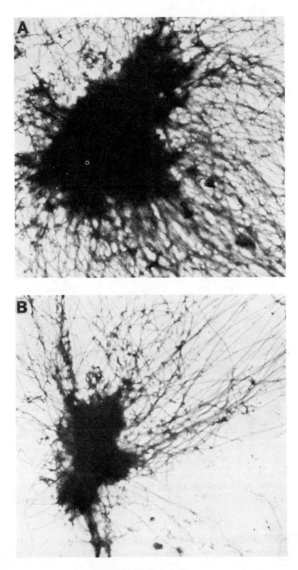

FIGURE 1, A-B

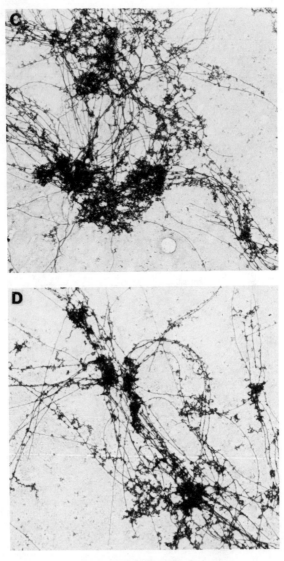

FIGURE 1, C-D

FIGURE 1. Electron micrographs of contracted microtubule gels negatively stained with 0.5% uranyl acetate on carbon coated grids. Before staining, the samples were washed in a microtubule stabilizing medium (0.1 M 2-(N-morpholino)ethane sulfonic acid (MES), 50% glycerol, 0.5 mM MgCl$_2$, and 1.0 mM ethylene glycol bis (β-aminoethyl ether)-N,N,N',N'-tetraacetic acid (EGTA) at pH 6.6). **A** and **B** are contracted third cycle microtubule gels; magnification 8,000 $\times$ and 6,500 $\times$, respectively. **C** and **D** are contracted microtubule gels from a preparation obtained by purification of third cycle protein on a phosphocellulose column; magnification 5,000 $\times$ and 4,500 $\times$, respectively.

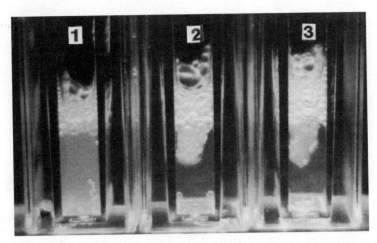

FIGURE 2. Contracted microtubule gels of phosphocellulose-purified protein. The protein was assembled in 5% DMSO with various GTP concentrations. Cell 1, 0.5 mM GTP; cell 2, 1.0 mM GTP; and cell 3, 1.5 mM GTP. Two mM ATP was added to each sample 30 minutes after inducing polymerization, and the samples were photographed after an additional hour.

tions. Whereas it has been established that ATP is a requirement for microtubule gelation contraction, the role of ATP has yet to be determined.

ATP-dependent microtubule-based movement and the formation of aster-like structures may reflect cellular microtubule behaviors. Gelation contraction of purified microtubules could be related to *in vivo* processes such as mitotic spindle assembly, neurite extension, and axonal transport.

REFERENCE

1. WEISENBERG, R. C. & C. CIANCI. 1984. J. Cell Biol. **99:** 1527–1533.

Segregation of an Antigenically Unique Subset of Alpha-Tubulin Subunits into Cold-Stable Microtubules during *in Vitro* Microtubule Assembly[a]

WILLIAM C. THOMPSON

Department of Human Biological Chemistry and Genetics
The University of Texas Medical Branch
Galveston, Texas 77550

We have previously shown that a unique monoclonal antitubulin antibody (Ab 1-6.1) binds to only a subset of cytoplasmic microtubules or microtubule segments within individual mouse or human fibroblastic cells.[1] The heterogeneity of the cytoplasmic microtubule complex is especially evident when detergent-extracted cytoskeletons are double stained by indirect immunofluorescence using both the monoclonal Ab 1-6.1 and a polyclonal rabbit antitubulin reagent (FIGURE 1). As shown in this figure, individual microtubules and microtubule segments within the cytoplasmic complex are stained well by Ab 1-6.1, whereas neighboring microtubules remain unstained.

It will be important to define the nature of the epitope recognized by Ab 1-6.1 in order to begin to understand how such antigenically different microtubule subsets are generated and whether they may have unique functions within the cell. As one approach to localizing the Ab 1-6.1 epitope region on the tubulin molecule, I planned to assay tubulin fragments by immunoblotting procedures. Mouse brain tubulin was the first choice for these studies, because the immunofluorescence study had already shown the antibody to bind to some mouse tubulin, including neuronal tubulin. Mouse brain tubulin was isolated by two cycles of assembly in the presence of glycerol and disassembly in its absence. When the twice-cycled mouse brain tubulin was resolved by sodium dodecyl sulfate polyacrylamide gel electrophoresis (SDS-PAGE), western blotted to nitrocellulose, and immunostained using Ab 1-6.1 and a peroxidase conjugated second antibody, little or no Ab 1-6.1 antibody was bound to the mouse tubulin (FIGURE 2). Ab 1-6.1 did bind to the alpha-tubulin subunits of twice-cycled tubulin from several other species, however.

To determine the fate of Ab 1-6.1 immunoreactive tubulin during the reassembly of mouse brain microtubules *in vitro,* samples of the homogenate and of the various pellets and supernatant fractions during microtubule protein purification were assayed by immunoblotting as in FIGURE 2. The bulk of the immunoreactive tubulin was found to have segregated into cold stable microtubules, which were normally discarded as the C1P fraction during the cyclic purification protocol. In a second type of experiment, a mouse brain homogenate was fractionated in the cold by centrifugation for 2.5 hours at 200,000 × g to pellet the 30S rings, and free 6S tubulin was precipitated from the cold supernatant by vinblastine sulfate. When the two tubulin fractions were analyzed by immunoblotting, tubulin containing the Ab 1-6.1 epitope was found to have segregated with the 30S ring fraction. These results may indicate that Ab 1-6.1 immunoreactive

[a]This work was supported by DHHS Grant GM 33505.

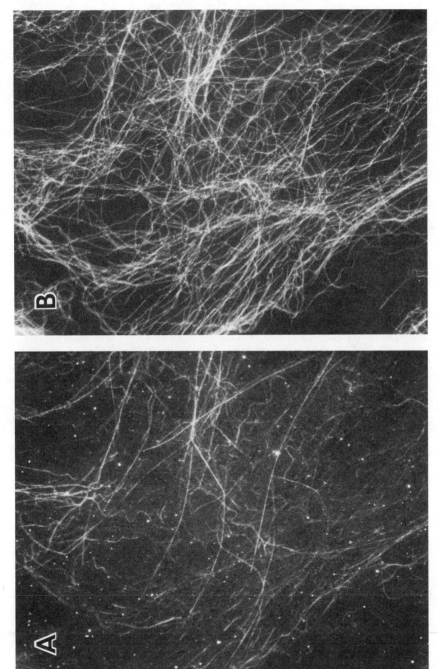

FIGURE 1. Monoclonal antitubulin Ab 1-6.1 stains a subset of cytoplasmic microtubules in a large flat mouse embryo cell. Coverslip cultures of mouse embryo cells were detergent-extracted in a microtubule stabilizing buffer, and the resulting cytoskeletons were then formaldehyde-fixed, extracted with cold methanol, and double stained by indirect immunofluorescence. **A:** Monoclonal Ab 1-6.1 (rhodamine); **B:** polyclonal antitubulin antibodies from rabbit (fluorescein).

tubulin preferentially interacts with MAPs conferring cold stability. The presence of the immunoreactive subspecies of tubulin in the purified C2S fractions may depend upon whether the immunoreactive tubulin or the particular MAPs are in excess in the initial homogenate.

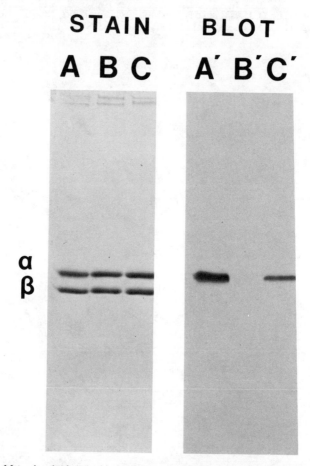

FIGURE 2. Monoclonal Ab 1-6.1 binds to the alpha subunit of twice-cycled brain tubulin from several species, but not from mouse brain. Brain microtubule proteins, purified by two cycles of temperature-dependent microtubule assembly and disassembly, were resolved by SDS-PAGE, transferred to nitrocellulose, and immunostained using Ab 1-6.1 and a peroxidase-conjugated second antibody. A: Chick brain microtubule proteins; B: mouse brain microtubule proteins; C: rabbit brain microtubule proteins.

Immunoblotting of chick brain tubulin fragments generated by formic acid or CNBr indicate that the Ab 1-6.1 epitope is located in the large N-terminal acid fragment and probably within the N-terminal third of the alpha subunit. In a recent report, an antigenic subset of cytoplasmic microtubules has also been detected by antibodies to a second epitopic region, the C-terminus of alpha-tubulin molecules lacking a terminal tyrosine residue.[2]

REFERENCES

1. THOMPSON, W. C., D. J. ASAI & D. H. CARNEY. 1984. J. Cell Biol. **98:** 1017–1025.
2. GUNDERSON, G. G., M. H. KALNOSKI & J. C. BULINSKI. 1984. Cell **38:** 779–789.

Direct Visualization of Steady State Microtubule Dynamics *in Vitro*

DAVID KRISTOFFERSON, TIM MITCHISON,
AND MARC KIRSCHNER

Department of Biochemistry and Biophysics
University of California
San Francisco, California 94143

Microtubule (MT) steady state dynamics have been understood using the treadmilling model of Margolis and Wilson.[1] In this model steady state MTs grow at one end and simultaneously depolymerize at the other end. The assembly and disassembly rates are balanced, and the polymer mass and number concentration remain constant.

Recently, treadmilling has been challenged by the dynamic instability model of Mitchison and Kirschner.[2] In this model the majority of the MTs grow at a moderate rate at both ends. This set of growing MTs is balanced at steady state by the very rapid disassembly of a minority of the MTs. The difference between the growing or shrinking MT states may be due to the presence or absence of tubulin (Tb)-guanosine triphosphate (GTP) subunits at the polymer ends. Tb-GTP ends have slow dissociation rates and tend to grow at steady state, whereas TB-guanosine diphosphate (GDP) ends have high dissociation rates and tend to depolymerize rapidly. Whereas polymer mass is maintained in this model, the number concentration should decrease with time.

All published experimental tests of *in vitro* treadmilling have thus far relied on the measurement of radioisotope incorporation into steady state microtubules. As such, they are indirect tests, and it would be preferable to use an experimental method that allows direct visualization of the two ends. Using biotinylated tubulin, we have monitored steady state microtubule dynamics directly. Microtubules are polymerized first from unmodified tubulin and then incubated with biotinylated tubulin to add biotinylated segments to the polymer ends. These segments are visualized by immuno-fluorescence using Texas red-streptavidin (or anti-biotin antibody), whereas the entire microtubule length is revealed by antitubulin monoclonal and fluorescein-conjugated

TABLE 1. Steady State Length Distribution Data for Microtubules in Buffer A[a]

	Steady State Time Points			
	0 min	10 min	20 min	30 min
Average MT Length (μm)	32.0 (0.8)	38.0 (1.0)	47.7 (1.3)	50.8 (1.5)
Average + End Length[b] (μm)	11.7 (0.2)	24.3 (0.5)	33.7 (0.8)	40.4 (1.1)
Average − End Length (μm)	2.4 (0.1)	4.1 (0.2)	5.9 (0.2)	8.4 (0.5)

[a]Buffer A includes 80 mM piperazine-N,N-bis-(2-ethane sulfonic acid) (PIPES), 1 mM MgCl$_2$, 1 mM EGTA, 1 mM GTP, 10 mM acetyl phosphate, and 2 IU acetate kinase. Phosphocellulose (PC)-Tb concentration is 5.4 mg/ml. Assembly at 37°C initiated with MT seeds polymerized in Buffer B (see TABLE 2). 500 MTs were measured for each time point.

[b]The + end length is defined as the length of the longer of the two biotinylated segments on the microtubule ends. The + segments are over three times longer than − end segments, so the possibility for confusion is minimal. In the event that a microtubule is labeled at only one end or completely saturated with biotinylated-Tb, this is also considered a + end length. Errors reported (in parentheses) are the standard error of the mean (s/$\sqrt{n}$).

TABLE 2. Steady State Length Redistribution Data for Microtubules in Buffer B[a]

	Steady State Time Points			
	0 hr	1 hr	2 hr	3 hr
Average MT length (μm)	14.0 (0.4)	14.4 (0.4)	15.3 (0.4)	17.0 (0.4)
Average + End length (μm)[b]	4.4 (0.1)	5.4 (0.2)	5.7 (0.2)	7.1 (0.2)
Average − End Length (μm)	1.67 (0.07)	1.76 (0.07)	1.73 (0.08)	2.24 (0.11)

[a]Buffer B includes 30% (v:v) glycerol, 80 mM PIPES, 10 mM MgCl$_2$, 1 mM ethylene glycol bis (β-aminiethyl ether)-N,N,N',N'-tetraacetic acid (EGTA), 1 mM GTP, 10 mM acetyl phosphate, and 2 IU acetate kinase. PC-Tb concentration is 1.5 mg/ml, and the assembly is done at 37°C. The dimeric Tb at steady state is a mixture of about 50% biotinylated and 50% unmodified Tb, which in control experiments has been shown to reduce the amount of total polymer formation by only 7% relative to pure unmodified Tb. 500 MTs were measured for each time point.

[b]The + end length and indicated errors (in parentheses) are defined in TABLE 1.

secondary staining. Aliquots for immunofluorescence are taken at steady state, and lengths of the segments at the polymer ends are measured. Data for the MTs polymerized under two different conditions are given in TABLES 1 and 2. In TABLE 1, a drastic increase in the average length and a consequent 40% drop in MT-number concentration occurs over only a 30-minute time course. Growth occurs off both ends of most MTs, whereas some MTs undergo a catastrophic depolymerization. Although the magnitude of the steady state redistribution is drastically reduced in glycerol buffer (TABLE 2), the same qualitative effects are observed. Both ends of the majority of the MTs grow, whereas the MT number concentration drops by about 18% over three hours. We anticipate that a similar qualitative effect will be observed using Tb with MAPs because a decrease in MT number concentration has already been detected by Kristofferson and Purich.[3] The data presented favor the dynamic instability model over treadmilling. Dynamic instability provides a new interpretation for several unclear problems in microtubule dynamics, such as the dependence of the Tb dissociation constant on the free Tb concentration. It provides a means by which MT drugs may be incorporated into, and then block, growing MTs, leading to a loss of the Tb-GTP cap and subsequent disassembly.

REFERENCES

1. MARGOLIS, R. L. & L. WILSON. 1978. Cell **13**: 1–8.
2. MITCHISON, T. & M. KIRSCHNER. 1984. Nature (London) **312**: 237–242.
3. KRISTOFFERSON, D. & D. L. PURICH. 1981. Arch. Biochem. Biophys. **211**: 222–226.

Microtubule-Dependent Intracellular Motility Investigated with Nanometer Particle Video Ultramicroscopy (Nanovid Ultramicroscopy)

M. DE BRABANDER, G. GEUENS, R. NUYDENS,
M. MOEREMANS, AND J. DE MEY

Division of Cellular Biology and Chemotherapy
Department of Life Sciences
Janssen Pharmaceutica
B-2340 Beerse, Belgium

The movement of intracellular organelles in vertebrate tissue-cultured cells, resulting in dynamic subcellular organization and polarity by seemingly guided and purposeful transport and mutual encounters, has been the subject of relatively few thorough studies. The motility of most optically detectable organelles is saltatory. It is fundamentally different from Brownian motion and displays both similarities and essential differences with other types of intracellular organelle transport such as axonal transport, pigment granule movement, and cytoplasmic streaming in plant cells. Saltations are dependent on the presence of, and occur along, microtubules. The molecular motor is unknown and may be located in the organelle membrane, or in the microtubule wall, or in both.

We developed a new approach of general applicability to probe the role of various proteins in subcellular transport. Colloidal gold particles of 20 or 40 nm diameter microinjected into living cells are invisible to the eye in the light microscope. Individual particles, can, however, easily be discerned using transmitted light at high numerical aperture and video contrast enhancement (FIGURE 1). They appear as black dots on a white background. In epipolarization microscopy they appear as brightly staining dots on a background consisting of the interference reflection picture of the cells (FIGURE 1).

Ultrastructural observation of the same cells prepared as whole mounts or in thin sections, confirmed that individual 20 nm particles could be resolved in the light microscope. The injected particles were not within membrane-lined vesicles, but free in the cytoplasmic matrix often in the vicinity of microtubules. Uncoated particles stabilized with polyethylene glycol, or particles coated with albumin were transported along microtubules in a saltatory, ATP-dependent fashion at the same velocity and with the same frequency as endogenous organelles (FIGURE 1). Because subunit flux within microtubules is a possible mechanism, we injected gold particles coupled to a monoclonal antitubulin antibody.[1] Many of these particles assumed fixed positions for several hours, often forming linear arrays. The observations show that gold particles having the same size as the smallest cellular membrane-lined organelles are transported by a microtubule-associated mechanism, as has been described for larger (250–500 nm) polystyrene and other beads microinjected in axons[2] or cultured cells.[3] Microtubule treadmilling does not appear to be involved.

The possibility of following 20–40 nm particles, and probably even smaller ones that can be coupled to many proteins within living cells, provides a tool of wide

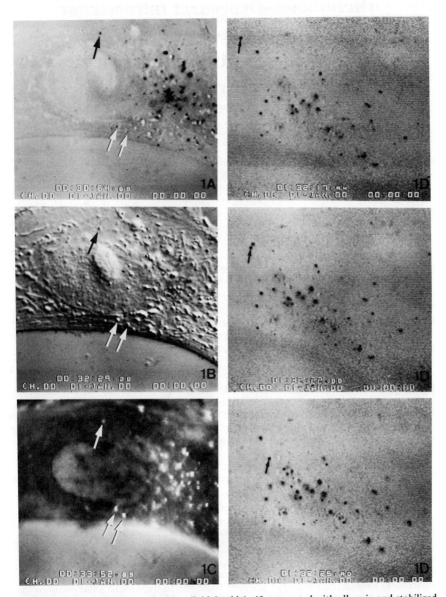

FIGURE 1. A cell was injected with colloidal gold (~40 nm coated with albumin and stabilized with polyethylene glycol) at time −30 min and sequentially viewed with transmitted light (**A**) differential interference contrast (**B**), and epipolarization (**C**). A few particles that are coincident in the three pictures are denoted by arrows (magnification × 1.840). **D**: The rapid linear movement of a particle (arrow) is followed in time. Other particles move also or disappear by the change of focal plane (magnification × 1.760). This figure reduced by 10 percent.

applicability for studying the fate and behavior of these proteins. We wish to call this new method nanometer particle video ultramicroscopy or nanovid ultramicroscopy.

REFERENCES

1. KILMARTIN, J. V., B. WRIGHT & C. MILSTEIN. 1982. Rat monoclonal antitubulin antibodies derived by using a new nonsecreting rat cell line. J. Cell Biol. **93:** 576–582.
2. ADAMS, R. J. & D. BRAY. 1983. Rapid transport of foreign particles microinjected into crab axons. Nature (London) **303:** 718–720.
3. BECKERLE, M. C. 1984. Microinjected fluorescent beads exhibit saltatory motion in tissue culture cells. J. Cell Biol. **98:** 2126–2132.

Centrosomes Are Required for the Assembly of a Bipolar Spindle in Animal Cells[a]

GREENFIELD SLUDER,[b] FREDERICK J. MILLER,[b] AND
CONLY L. RIEDER[c]

[b]Worcester Foundation for Experimental Biology
Shrewsbury, Massachusetts 01545
[c]Wadsworth Center for Research
New York State Department of Health
Albany, New York 12201

There are two conflicting views on the origin of spindle bipolarity. The traditional view holds that specialized structures (centrosomes) form the poles of the spindle and therefore determine the spindle axis.[1,2] During spindle assembly, chromosomes are aligned by the poles and not vice versa. This view has been challenged by observations[3-5] that suggest that chromosomes alone are sufficient to organize a bipolar spindle. These observations have been interpreted by some to indicate that lateral interactions between the condensed chromosomes and/or between bundles of kinetochore microtubules provide the forces that align the chromosomes and organize the spindle structure (reviewed in reference 2).

We critically tested the ability of chromosomes in a mitotic cytoplasm to organize a bipolar spindle in the absence of centrosomes. Sea urchin eggs were treated with 5 × 10^{-6} M colcemid for 7–9 min before fertilization, to block future microtubule assembly. Fertilization events were normal except that a sperm aster was not formed and the pronuclei remained up to 70 μm apart. After nuclear envelope breakdown, individual eggs were irradiated with 366 nm light to photochemically inactivate the colcemid. A functional haploid, bipolar spindle was immediately assembled in association with the male chromosomes. By contrast, the female pronucleus in most of these eggs remained as a small nonbirefringent hyaline area throughout mitosis (FIGURE 1b–d). High voltage electron microscopy of serial semithick sections from individual eggs, previously followed *in vivo,* revealed that the female chromosomes were randomly distributed within the remnants of the nuclear envelope. No microtubules were found in these pronuclear areas even though the chromosomes were well condensed and possessed prominent kinetochores with well-developed coronas (FIGURES 2–4).

In the remaining eggs, a weakly birefringent monaster was assembled in the female pronuclear area (FIGURE 1e). These monasters lacked centrioles, contained an abundance of radially arrayed membranes, and possessed small patches of pericentriolar-like material into which numerous radially oriented microtubules terminated. Chromosomes were monooriented around the monaster: only the kinetochores facing the

[a]This work was supported by NIH GM 30758 (to G. Sluder), NSF PCM-8402441 (to G. Sluder), NCI PO 30-12708 (to WFEB), Biotechnological Resource related Grant RR02157 (to C.L. Rieder), and PHS 01219 (to the New York State Department of Health).

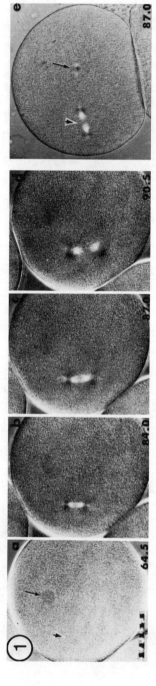

FIGURE 1. Development of a zygote with separate pronuclei: polarization optics. **a:** After nuclear envelope breakdown but before irradiation. Male pronucleus indicated by arrowhead; female pronucleus indicated by arrow. **b–c:** After irradiation with 366 nm light. Spindle is assembled with male chromosomes. Female chromosomes remain in a small hyaline area. **d:** Male spindle in anaphase; female chromosomes still in a small hyaline area. **e:** Different cell with female monaster indicated by arrow and a male spindle shown by arrowhead. Minutes after fertilization shown in lower corner of each frame. 10 μm per scale division.

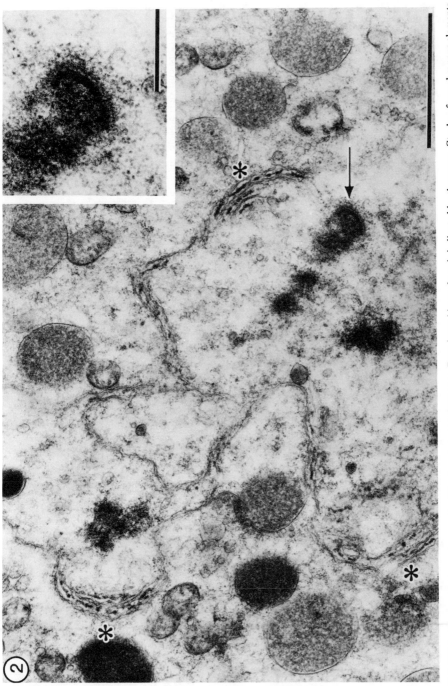

FIGURE 2. Survey electron micrograph of the female pronuclear area. Prominent kinetochore is indicated by the arrow. Stacks of membranous elements are indicated by the asterisks. Note the numerous yolk granules in and around the nuclear area. Inset: higher magnification of the kinetochore indicated by the arrow. Bar = 2.0 μm; inset bar = 0.5 μm.

monaster center were attached to microtubules. The sister kinetochores, facing away from the monaster, lacked microtubules.

These observations demonstrate that chromosomes in a mitotic cytoplasm are not able to organize a bipolar spindle in the absence of a spindle pole or even in the presence of a monaster. In fact, they do not even assemble kinetochore microtubules in

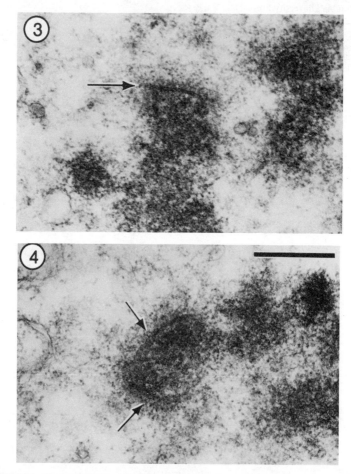

FIGURES 3, 4. Electron micrographs of other sections from the same female pronuclear area. Note that the well-developed kinetochores (arrows) are devoid of microtubules. FIGURE 4, bar = 0.5 μm.

the absence of a spindle pole, and kinetochore microtubules form only on kinetochores facing the pole when a monaster is present. This study also provides the first direct experimental proof for the longstanding paradigm that the sperm provides the centrosomes used in the development of the sea urchin zygote.

REFERENCES

1. MAZIA, D. 1961. Mitosis and the physiology of cell division. *In* The Cell. J. Brachet & A. E. Mirsky, Eds.: 77–412. Academic Press. New York.
2. NICKLAS, R. B. 1971. Mitosis. *In* Advances in Cell Biology. D. M. Prescott, L. Goldstein & E. H. McConkey. Eds.: 275–297. Academic Press. New York.
3. SCHRADER, F. 1953. Mitosis, the movement of chromosomes in cell division. 2nd Edition. Columbia University Press. New York.
4. DIETZ, R. 1966. The dispensability of the centrioles in the spermatocyte divisions of *Pales ferruginea* (Nematocera). Chromosomes Today **1:** 161–166.
5. KARSENTI, E., J. NEWPORT & M. KIRSCHNER. 1984. Respective roles of centrioles and chromatin in the conversion of microtubule arrays from interphase to metaphase. J. Cell Biol. **99:** 47s–54s.

Incorporation and Turnover of Labeled Exogenous Tubulin in the Mitotic Spindles of HeLa Cells and *Chaetopterus* Oocytes

DENNIS GOODE AND VIDYA SARMA

Department of Zoology
University of Maryland
College Park, Maryland 20742

A directional flux or treadmilling of tubulin subunits within microtubules (MTs) has been proposed to be one aspect of the mechanism of mitotic movements.[1,2] However, no direct evidence for treadmilling in mitotic spindle MTs has been obtained.

The incorporation of tubulin into mitotic spindles *in situ* was studied by incubating permeablized mitotic cells in solutions containing tubulin labeled with [³H]GTP or dichlorotriazinylamino fluorescein (DTAF). If treadmilling of spindle MTs occurs, exogenous tubulin and guanosine tryphosphate (GTP) should become incorporated into spindle MTs and replace the endogenous subunits until a constant level is reached. Metaphase HeLa cells or *Chaetopterus* oocytes were lysed in a microtubule-assembly buffer plus 1% Nonidet P-40, 1 mg/ml 100,000 × g supernatant beef-brain tubulin, 5×10^{-5} M cold GTP, and 20 μCi/ml [³H]GTP. After different periods of incubation, mitotic spindles were isolated in 2 M glycerol-containing assembly buffer, rinsed, and separated from unbound counts by centrifugation through a 4 M glycerol cushion. Pelleted spindles were dissolved in 1 M NaOH, and ³H and protein content were determined. ³H counts per mg protein increase linearly for about 15 min and then reach a plateau or steady state in both *Chaetopterus* oocytes (FIGURE 1) and HeLa cells. Addition of 10 mM CaCl$_2$ blocks incorporation. Little or no [³H]GTP is incorporated if exogenous tubulin or lysed cells are omitted from the assembly mixture.

To measure the loss rate of [³H]GTP tubulin from mitotic spindles, cells were incubated in tubulin + [³H]GTP for 20 min, and a 20-fold excess of cold GTP (2 mM) was added. Samples were removed after incubation for different periods, and spindles were isolated and counted as described above. [³H]GTP is lost from spindles at a rate of about 16%/min until a new steady state is reached in about 8 minutes. These results are consistent with an incorporation and turnover of [³H]GTP tubulin in spindle MTs of these lysed cell models.

The location of this newly incorporated tubulin in the spindle was investigated by incorporating fluorescent DTAF tubulin into mitotic spindles of these lysed cell types. In both cases, a short pulse (2–5 min) appears to label MTs near metaphase chromosomes, and longer exposures label the entire spindle.

The rates of incorporation, saturation, and turnover that we see by [³H]GTP and fluorescent tubulin incorporation *in situ* are faster than observed with brain MTs at steady state *in vitro*,[3] but are in the range of the rates of spindle fiber formation in prophase and spindle MT reassembly after cooling.[4]

674

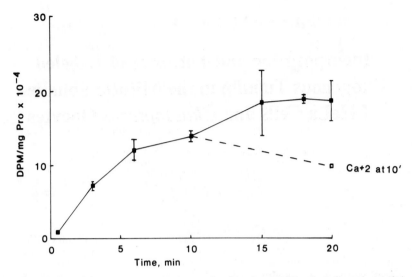

FIGURE 1. Metaphase *Chaetopterus* eggs were incubated at 34°C for the indicated periods in a solution containing calf-brain tubulin and [³H]GTP. Four m*M* CaCl₂ was added to one tube in each experiment at 10 min (------).

REFERENCES

1. MARGOLIS, R. L., L. WILSON & B. I. KIEFFER. 1978. Mitotic mechanism based on intrinsic microtubule behavior. Nature (London) **272:** 450–452.
2. GOODE, D. 1981. Microtubule turnover as a mechanism of mitosis and its possible evolution. BioSystems **14:** 271–287.
3. MARGOLIS, R. L. & L. WILSON. 1978. Opposite end assembly and disassembly of microtubules at steady state *in vitro*. Cell **13:** 1–8.
4. GOODE, D. 1973. Kinetics of microtubule formation after cold disaggregation of the mitotic apparatus. J. Mol. Biol. **80:** 531–539.

Tubulin-Colchicine Interactions and Polymerization of the Complex[a]

JOSE MANUEL ANDREU[b] AND SERGE N. TIMASHEFF[c]

[b]*Unidad de Biomembranas*
Instituto de Inmunologia
Consejo Superior de Investigaciones Cientificas
28006 Madrid, Spain

[c]*Graduate Department of Biochemistry*
Brandeis University
Waltham, Massachusetts 02254

INTRODUCTION

In a study of the binding of [H^3]colchicine to cultured cells, Taylor[1] in 1965 estimated that binding to a critical three to five percent fraction of the set of cellular binding sites inhibited the correct formation of the mitotic spindle. It was suggested that the colchicine (COL) receptors were the elements that constitute microtubules. The COL receptor, tubulin, was purified and shown to bind approximately one molecule of COL with high affinity.[2] Following the discovery of the conditions for assembly of microtubules *in vitro*,[3] Margolis and Wilson[4,5] reported that the tubulin-drug complex adds to microtubule ends, preventing polymer growth and the tubulin flux in steady-state microtubules. It has been shown that the binding of tubulin-COL to microtubule ends is reversible and that the binding constant is of the same order of magnitude as that for the addition of unliganded tubulin.[6] The detailed molecular mechanism of COL binding to tubulin, however, and the events leading to the inhibition of growth at microtubule ends remained largely unknown.

A number of studies have demonstrated that the binding of COL to tubulin is a complex process. It is slow and difficult to reverse and the unoccupied binding sites denature rapidly. These characteristics seriously hampered any attempt at rigorous characterization of the binding of COL to tubulin by equilibrium techniques. (See a discussion on this in reference 8.) In a kinetic study of this interaction, Garland[7] showed that the faster phase of COL binding to tubulin conformed to a two-step mechanism, namely a fast reversible binding, followed by a slow conformational change:

$$TB + COL \xrightleftharpoons{K_1} TB \cdot COL \xrightleftharpoons[k_{-2}]{k_2} (TB \cdot COL)'$$

This resulted in the formation of a fluorescent essentially stable complex ($k_{-2} = (5 - 9) \times 10^6 \text{ s}^{-1}$) with a high apparent binding affinity, $K_{overall} = K_1 k_2 / k_{-2} = 10^7 \text{ M}^{-1}$ at 37°C.

[a]This work was supported by United States Public Health Service Grants CA-16707 and GM-14603 (to S. N. Timasheff), Fogarty International Fellowship TW 02983 (to J. M. Andreu), and the Fondo de Investigaciones Sanitarias y Comision Asesora de Investigaciones (to program No. 42032 of CSIS research, to J. M. Andreu, Brandeis publication number 1576).

In order to resolve the complex tubulin-COL interactions into simpler ones, we designed equilibrium probes for the COL binding site and examined the effects of liganding on the structure of purified tubulin and the self-assembly of the protein.

BINDING OF MONO- AND BIFUNCTIONAL
ANALOGUES OF COLCHICINE

The structures of the ligands employed are shown in FIGURE 1. From a critical examination of the literature, it became apparent to us that the active parts of the

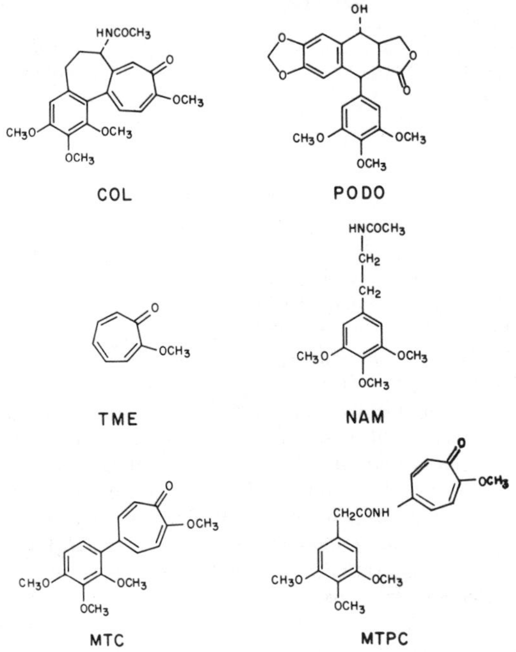

FIGURE 1. Structural formulas of ligands. COL: colchicine; PODO: podophyllotoxin; NAM: N-acetylmescaline; TME: tropolone methyl ether; MTC: 2-methoxy-5-(2,3,4-trimethoxyphenyl)-2,4,6-cycloheptatriene-1-one; MTPC: 2-methoxy-5-[(3-(3,4,5-trimethoxyphenyl)propiony-lamino]-2,4,6-cycloheptatriene-1-one.

(COL) molecule were probably its trimethoxyphenyl and tropolone rings, the direct contribution of the middle ring being minor. The contribution of the trimethoxyphenyl ring was first probed with [14C] mescaline. This was found to bind to one site on tubulin with an apparent equilibrium constant of 10^4 M^{-1}. This binding, however, was not inhibited by podophyllotoxin (PODO), which shares the trimethoxyphenyl ring with colchicine, suggesting that the protonated amino group of mescaline was contributing to the nonspecific binding of this compound. We turned, therefore, to N-acetylmesca-line (NAM). Its binding was not detectable at 25°C by the Hummel and Dreyer[9] gel chromatography technique employed. At 37°C, however, this ligand bound to tubulin

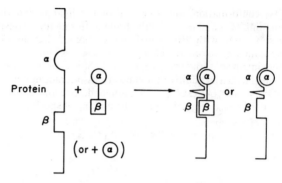

FIGURE 2. Model reaction for the binding of a bifunctional ligand ($\alpha - \beta$) to two subsites on the protein. The binding of the α moiety induces in the protein a conformational change that brings into proper geometric alignment the β binding locus, but it does not affect the conformation of the β locus itself.

weakly, with an apparent equilibrium constant of 430 ± 250 M^{-1}. This interaction was inhibited specifically by colchicine.[8]

The contribution of the tropolone ring to COL binding was probed first with tropolone, following the report of its interaction with tubulin.[10] Although tropolone was found to bind to tubulin, all detailed studies had to be carried out with the closer COL analogue, tropolone methyl ether (TME), because the development of tropolone fluorescence was found to be due to complexing with Mg^{2+} ions, and not to binding to tubulin, as no fluorescence was observed in the absence of Mg^{2+}.[11] Because Mg^{2+} is normally present in tubulin solutions, this circumstance may lead to complications. On the other hand, TME does not interact with magnesium.[11] The binding of [^{3}H]TME to tubulin was found to attain equilibrium within 10 minutes at 37 °C, but required 120 minutes at 0°. At 0°C, it bound reversibly to 0.95 ± 0.20 sites per tubulin dimer with an apparent equilibrium constant, $K_b = (2.2 \pm 0.2) \times 10^3$ M^{-1}. The interaction was favored at lower temperatures, with $\Delta H°_{app} = -8.3 \pm 1.0$ kcal mol^{-1} and $\Delta S°_{app} = -15.2 \pm 3.6$ eu. Thus, $\Delta G°_{app}$ varied from -4.2 to -3.5 kcal mol^{-1} between 0°C and 37°C. This interaction was found to be specific of the COL site in mutual inhibition experiments.[8]

The binding constants of the two parts of COL point to weak interactions between each of them and tubulin, the sum of their standard free energy changes being considerably smaller than that of COL binding (at 37°C, $\Delta G°_{app} = -3.6$ kcal mol^{-1} for TME, -3.7 kcal mol^{-1} for NAM, and -10.2 kcal mol^{-1} for COL). Furthermore, the binding of TME and NAM in the presence of each other was found to be unaffected by the other molecule, indicating an absence of cooperation (either positive or negative) between them, meaning that both binding subsites exit on the tubulin molecule. The slowness of the binding of COL, however, as well as that of TME suggest that this reaction is accompanied by a conformational change in the protein.

A detailed thermodynamic and mechanistic analysis of these results was carried out in terms of the concept of a bifunctional ligand.[8] The model, shown in FIGURE 2, represents such a bifunctional ligand, $\alpha - \beta$, capable of interacting with a bifocal site on the protein. The analysis was based on three assumptions. The bindings of α and β are independent, which is consistent with the absence of cooperation between NAM and TME binding. The bindings are not perturbed by the covalent attachment $\alpha - \beta$, consistent with the knowledge[12] that a COL analogue consisting only of the two rings of interest is as strong a microtubule inhibitor as COL. The changes in the internal and external mobility, conformation, and solvation of the ligands and the protein are similar for the binding of α and $\alpha - \beta$ to the protein; this again is consistent with the experimental results and with theoretical expectations.[13] Let us decompose the apparent binding standard free energy change, $\Delta G^\circ_{obs} = -RT\ln K_{app}$ into two contributions, ΔG°_{int}, the intrinsic standard free energy change of the formation of the protein-ligand bond, which is additive for the various interacting parts of the ligand molecule, and $\Delta G^\circ_{na} = \Delta G^\circ_{obs} - \Delta G^\circ_{int}$, which is nonadditive for the parts. Then, if ΔG^i expresses the standard free energy change of the binding of species i, we have

$$\Delta G^\alpha_{obs} = \Delta G^\alpha_{int} + \Delta G^\alpha_{na} \tag{1}$$

Assumptions 1 and 2 give

$$\Delta G^{\alpha-\beta}_{obs} = \Delta G^\alpha_{int} + \Delta G^\beta_{int} + \Delta G^{\alpha-\beta}_{na} \tag{2}$$

Assumption 3 is

$$\Delta G^{\alpha-\beta}_{na} = \Delta G^\alpha_{na} \tag{3}$$

Combining these relations we obtain

$$\Delta G^\beta_{int} = \Delta G^{\alpha-\beta}_{obs} - \Delta G^\alpha_{obs} \tag{4}$$

and

$$\Delta G^\beta_{na} = \Delta G^\alpha_{obs} + \Delta G^\beta_{obs} - \Delta G^{\alpha-\beta}_{obs} \tag{5}$$

ΔG_{na} contains the contributions from the changes in mobility and conformation, and an immutable contribution, ΔG°_c, the cratic-free energy change that stems from the change of the entropy of mixing when the number of particles is reduced during binding,

$$\Delta G^\circ_c = -T\delta S_{mix} = -RT\ln X_{p\ell}/X_p X_\ell \tag{6}$$

where X_i is the mole fraction of species i, and the subscripts p, pℓ, and ℓ mean unliganded protein, liganded protein and ligand, respectively. At room temperature, ΔG°_c has a value of 2.4 kcal mol^{-1}. Thus, elimination of the covalent bond from $\alpha-\beta$ should result in an unfavorable contribution to ΔG° of 2.4 kcal mol^{-1}. The analysis of the binding data on COL and its two parts, given in TABLE 1, shows that ΔG°_{na} for the COL system can be accounted for totally by the cratic contribution.

This treatment has also made it possible to analyze the temperature dependence of COL binding to tubulin.[8] Colchicine and TME have opposite dependencies. The

TABLE 1. The Contributions of the Tropolone and Trimethoxyphenyl Rings to the Binding of Colchicine

| | | Free Energy Change | |
| | | Value (kcal mol^{-1}) | |
Ligand	Nature of Contribution	25°C	37°C
Colchicine[a]	$\Delta G^{\alpha-\beta}_{obs}$	-9.4 ± 0.5	-10.2 ± 0.2
Tropolone methyl ether	ΔG^α_{obs}	-3.9 ± 0.4	-3.6 ± 0.6
N-acetyl-mescaline	ΔG^β_{int}	-5.5 ± 0.5	-6.6 ± 0.4
N-acetyl-mescaline	ΔG^β_{obs}	$\gtrsim -3.4$	-3.7 ± 0.5
N-acetyl-mescaline	ΔG^β_{na}	$\gtrsim +2.1$	$+2.9 \pm 0.5$

[a]Values derived from the literature (see reference 8).

binding of TME is weakly exothermic, which suggests hydrogen bond formation. The derived curve for NAM indicated a markedly endothermic interaction, consistent with the transfer of the nonpolar trimethoxyphenyl ring from an aqueous medium into a hydrophobic crevice on the tubulin molecule.

This thermodynamic analysis predicts that bifunctional analogues of COL that contain only the trimethoxyphenyl and tropolone rings connected by simple covalent arms should bind to tubulin with high affinity and probably in a reversible manner. This prediction was tested by a direct binding study[14] of the COL analogue, 2-methoxy-5(2,3,4-trimethoxyphenyl)-2,4,6-cycloheptatrien-1-one (MTC).[12] FIGURE 3 shows the binding isotherm of MTC to tubulin at 25°C. Using a variety of techniques (Hummel and Dreyer gel chromatography, generation of ligand fluorescence, protein fluorescence quenching, and ligand difference absorption spectroscopy), binding was found to one site with an apparent equilibrium constant, $K_b = (4.6 \pm 0.4) \times 10^5 \, M^{-1}$ at 25°C. Binding was inhibited by COL and podophyllotoxin, and MTC was found to be

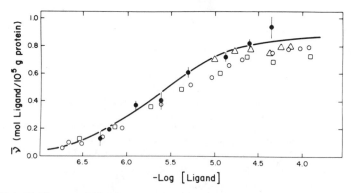

FIGURE 3. Binding isotherm of MTC to tubulin in PG buffer (10 mM sodium phosphate-0.1 mM GTP), pH 7.0, 25°C. Solid circles (●) are column measurements at a protein concentration of $(0.5–1.2) \times 10^{-5} \, M$. Open symbols are binding measurements from ligand fluorescence at (□) 1.8×10^{-6}, (○) 5.3×10^{-6}, and (△) $3.6 \times 10^{-5} \, M$ protein. The solid line is a fit to the column measurements ($K_b = 4.8 \times 10^5 \, M^{-1}$; n = 0.88) obtained from a Scatchard plot of these data.

an inhibitor of colchicine binding. By contrast to colchicine, binding and dissociation were rapid, and binding could be readily detected, even at low temperatures. A van't Hoff analysis of fluorometric binding titrations at different temperatures gave the thermodynamic parameters $\Delta H°_{app} = -1.6 \pm 0.7$ kcal mol^{-1}, $\Delta S°_{app} = 20.5 \pm 2.5$ e.u., and $\Delta C°_p \simeq 0.$[14] Analysis in terms of the bifunctional ligand model developed for COL showed that the third assumption of that model is not valid for this compound, because in MTC, there is free rotation about the bond connecting the two rings. This requires introduction into equation 3 of an additional term, $\Delta G_{i\cdot rot.}$, which describes the unfavorable contribution of the loss of internal rotation, so that equation 5 becomes

$$\Delta G^{\beta}_{na} = \Delta G^{\alpha}_{obs} + \Delta G^{\beta}_{obs} - \Delta G^{\alpha-\beta}_{obs} + \Delta G^{\alpha-\beta}_{i\cdot rot.} \qquad (7)$$

For a compound such as MTC, $\Delta G°_{i\cdot rot.}$ can be estimated to have a value of 1–3 kcal mol^{-1},[15] giving for ΔG_{na} a value between 1.5 and 3.5 kcal mol^{-1}, which again corresponds to the cratic-free energy change contribution. Furthermore, the observed values of ΔH°_{app} and ΔS°_{app}[14] were fully consistent with the above-described analysis of the contributions of the two independent moieties, NAM and TME.[8]

A different analogue of COL, 2-methoxy-5-[3-(3,4,5-trimethyloxyphenyl)propionyl]amino-2,4,6-cycloheptatrien-1-one(MTPC), was synthesized[14] with the purpose of probing the flexibility of the COL binding site, because in this compound the two rings are mutually arranged in a geometry different from MTC. The interaction of MTPC with tubulin was weak and could not be detected except for the inhibition of the rate of binding of [^{3}H]COL and of microtubule assembly at high concentrations of the analogue. In the competition experiments, MTPC behaved similarly to the previously characterized monofunctional COL analogues. It is likely that MTPC is a ligand conformationally inactivated by the presence of the propionamide spacer and binds weakly to the COL site through one of its rings only.

Simple bifunctional COL analogues such as MTC constitute convenient equilibrium high affinity probes for the COL binding site of tubulin and are attractive compounds for the testing of the fast and reversible inhibition of microtubules in living cells.

CONSEQUENCES OF BINDING, PROPERTIES OF THE SUBSITES, AND MECHANISM OF BINDING

The binding of COL and its analogues to tubulin was found to induce changes in several relevant properties both of the ligands and of the protein. COL and MTC became fluorescent on binding to tubulin, whereas in free solution their fluorescence is barely detectable.[10,14,16] No significant enhancement of fluorescence was observed with the single ring analogues, TME and NAM, the weakly active analogue MTPC and PODO.[14,16] Therefore, the promotion of fluorescence is linked to the simultaneous binding to tubulin of both the tropolone and trimethoxyphenyl rings when these are connected covalently. Such a fluorescence enhancement of COL is not observed in organic solvents, except at very low temperatures.[17] This enhancement is not promoted by the interaction of COL with micelles of the detergents octylglucoside, sodium deoxycholate and sodium dodecyl sulfate (J. M. Andreu, unpublished). This suggests restriction of intramolecular motions of COL and MTC, specific to their binding to tubulin, as a major factor in the generation of fluorescence, rather than simply their transfer into an apolar environment. Such a restriction could affect the torsion of the biaryl bond and also the conformations of the nonplanar tropolone ring. In this context, it has been reported[18] that the 340 nm circular dichroic band of COL becomes reduced on binding to tubulin, leading to the proposal that a conformational change occurs in the ligand.[18] Taking as a base the X-ray crystallographic structure of colchicine,[19] it has been suggested[18] that the tropolone ring undergoes an isomerization that gives a smaller dihedral angle with the trimethoxyphenyl ring and that might result in extended conjugation and fluorescence.

The natures of the interactions of the two rings with their corresponding subsites on tubulin were probed by means of difference absorption spectroscopy.[16] While the perturbations of ligand absorption due to binding to tubulin subsites were weak and difficult to analyze in the case of the monofunctional ligands, they were readily observable with COL, MTC, and PODO. The tubulin-PODO difference spectrum, shown in FIGURE 4A, was mimicked by difference spectra of the alkaloid in methanol and chloroform versus an aqueous solution and was found to be most likely related to the trimethoxyphenyl ring. This indicated directly that PODO binds through the trimethoxyphenyl ring and that the binding subsite of this ring, common to COL and PODO, provides an environment more similar to the organic solvents than to water.

This conclusion supports the hydrophobic nature of the binding of trimethoxyphenyl ligands to tubulin, which we have proposed on strictly thermodynamic grounds.[8]

On the other hand, the COL-tubulin and MTC-tubulin difference spectra, shown in FIGURE 4B, could be ascribed mainly to perturbations of the tropolone ring absorption in the 300 to 400 nm region. Related changes could be observed in the formamide or methanol versus water solution spectra of COL, suggesting the possible participation of hydrogen bonding by the carbonyl group of the tropolone ring in the binding of COL to tubulin. It has to be noted that a peak at 387 nm, similar to the one present in these spectra, is generated by the dimerization of COL at high concentrations.[20] This suggests a possible stacking interaction with aromatic amino acids of tubulin, because Lambeir and Engelborghs[26] have shown that the apparent thermody-

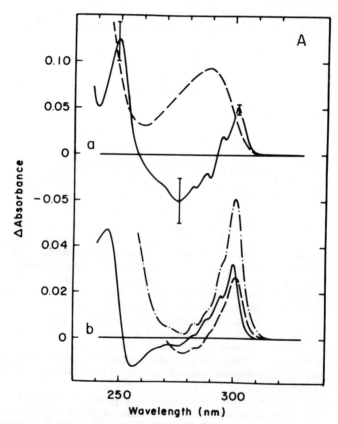

FIGURE 4A. Difference absorption spectra generated by the interaction of ligands with tubulin. Podophyllotoxin. (a) (——) Equilibrium difference spectrum of a 50 μM podophyllotoxin and 16 μM tubulin solution versus 50 μM podophyllotoxin and 16 μM tubulin in separate solutions. (---) Absorption spectrum of 50 μM podophyllotoxin alone reduced 8 times. (b) Difference spectra of 100 μM podophyllotoxin in various solvents: (——) methanol versus 10 mM sodium phosphate buffer, pH 7.0; (·-·) chloroform versus phosphate buffer; (---) formamide versus phosphate buffer.

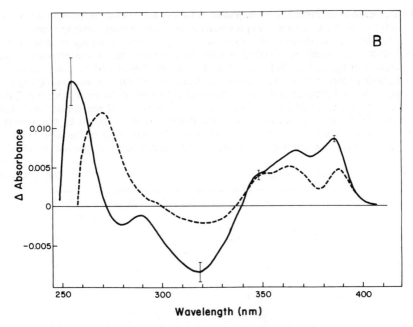

FIGURE 4B. Difference absorption spectra generated by the interaction of ligands with tubulin. COL and MTC. (-----) Equilibrium difference spectrum of a 16 μM COL–10 μM tubulin solution versus 16 μM COL and 10 μM tubulin in separate solutions; (——) equilibrium difference spectrum of 31 μM MTC–4 μM tubulin versus ligand and protein in separate solutions.

namic parameters calculated for the first event of COL binding to tubulin are coincident with those of COL dimerization.[20] The simplest interpretation of the strong quenching of tubulin fluorescence by COL and MTC would also be binding close to a protein fluorophore, most probably a tryptophan.[16] Nevertheless, both hydrogen bonding or stacking interactions at the tropolone subsite can be accommodated within the apparent thermodynamic characteristics of the TME-tubulin interaction. All this type of information is relevant to the elucidation of the properties of the subsites of the COL binding site of tubulin. It is obviously not sufficient to discriminate between positions in the current controversy on the localization of the COL binding site probed by various approaches.[21–24] Major questions that remain unanswered are whether COL binds to α tubulin, to β tubulin, indistinctly to either of them, or at the $\alpha - \beta$ subunit interface in the heterodimer.

Turning now to the effects of liganding on the properties of the protein, none of the ligands induced any aggregation of tubulin under standard conditions. Whereas PODO and NAM binding did not display any evidence of a conformational change in tubulin, binding of COL, MTC, and TME was accompanied by similar small perturbations of the circular dichroism spectrum of tubulin. Although these small dichroism perturbations could not be readily interpreted in terms of changes in secondary structure, they pointed to the binding of the tropolone ring as being the source of the previously postulated conformational change of tubulin,[7] which has now been most likely observed in our studies.[16] COL is known to induce a weak GTPase activity in tubulin.[25] The stoichiometric tubulin-COL complex GTPase activity was characterized for our system and found to have a Michaelis constant, $K_m = 7 \pm 2 \ \mu M$ and a maximal rate of $0.012 \pm 0.001 \ min^{-1}$.[d] The turnover was proportional to the saturation of the COL site.[16] MTC was found also to induce GTPase activity. In this case the enzyme activity was determined as a function of MTC concentration at saturating substrate. The maximal rate of guanosine triphosphate (GTP) hydrolysis was $0.0098 \pm 0.0005 \ min^{-1}$, and the enzyme activation followed, although not exactly, the binding of MTC monitored by fluorescence.[14] Neither of the single ring analogues, nor PODO, induced this activity, although the binding of TME seems to induce the proper conformation. This could be explained by the weakness of the TME-tubulin interaction and the rapidity of dissociation of the complex, relative to the turnover of the GTPase activity of stable tubulin-COL. The COL-induced GTPase was found to be dependent on the chemical nature of buffer components and to be raised 4-fold by the addition of $3.4 \ M$ glycerol to phosphate buffer.[27]

Analysis of the above-described results has led to the conclusions that (1) the binding of TME to tubulin induces a conformational change; (2) this conformational change is closely related to that induced by the binding of COL; (3) the first event in the binding of COL to tubulin is probably the binding of the tropolone end of the drug; (4) this first event induces a conformational change in the protein, which brings the existing trimethoxyphenyl binding site into proper position for this ring to fall into place. The equilibria involved are summarized in the following scheme:

$$
\begin{array}{ccc}
TB + COL \xrightleftharpoons{1} TB \cdot COL & & \\
\alpha \ (slow) \Updownarrow \qquad \qquad \Updownarrow 2' \ (slow) & & \\
TB^* + COL \xrightleftharpoons{\beta} TB^* \cdot COL \xrightleftharpoons{2''} TB^* : COL &
\end{array}
$$

$$
\begin{array}{cc}
TB + TME \xrightleftharpoons{3} TB \cdot TME & \\
\alpha \ (slow) \Updownarrow \qquad \qquad \Updownarrow 4 \ (slow) & \\
TB^* + TME \xrightleftharpoons{\beta'} TB^* \cdot TME &
\end{array}
$$

$$
TB + NAM \xrightleftharpoons{5} TB \cdot NAM
$$

TB $\cdot$ *COL and TB*:COL represent complexes in which only one or both tubulin-ligand contacts are made, respectively.

Our conclusion that the first event in the binding of COL to tubulin is the binding of the tropolone ring[8,16] is strongly supported by the kinetic analysis of the first event by Lambeir and Engelborghs,[26] because the thermodynamic parameters that they have calculated for the first step of COL binding ($\Delta H^{\circ}_{app} = -7.9 \ kcal \ mol^{-1}$,

---

[d]There is an error in Andreu et al.[14] The correct number is reported in Andreu & Timasheff,[16] and the COL- and MTC-induced GTPase activities are similar.

$\Delta S°_{app} = -15.1$ e.u.) are strikingly coincident with our measured values for the interaction of TME with tubulin ($\Delta H°_{app} = -8.3$ kcal mol^{-1}, $\Delta S°_{app} = -15.2$ e.u.).[8] Then, in the scheme, it is possible to equate equilibrium 3 with equilibrium 1; that is, equilibrium 1 is the binding of the tropolone ring of COL to tubulin. Proceeding along the scheme, equilibrium 2' is the conformational change of liganded tubulin, and equilibrium 2'' is the intramolecular (first order) binding of the trimethoxyphenyl ring of COL to the altered tubulin. Analysis of all the equilibrium constants available resulted in the following values of the standard free energy changes of the various steps of the COL binding to tubulin:

$\Delta G° (1) = -3.2$ kcal mol^{-1} (binding of the TME ring of COL)
$\Delta G° (2') = -0.2$ kcal mol^{-1} (conformational change in the protein)
$\Delta G° (2'') = -6.8$ kcal mol^{-1} (binding of the trimethoxyphenyl ring of COL)

These values have led to the conclusion[16] that, although the tropolone methyl ether end of COL binds first to tubulin and imparts some stability to the slowly forming isomer, Tb*, it is the subsequent rapid binding of the trimethoxyphenyl ring, covalently bonded to the tropolone ring, that provides the strong free energy change that thermodynamically drives the reaction to completion and permits to lock kinetically the liganded protein in the new conformation. This product has the properties of being a GTPase and of inducing tubulin to polymerize in a structure different from microtubules, as discussed in the next section.

POLYMERIZATION OF THE TUBULIN-COL COMPLEX AND ITS RELATION TO MICROTUBULE ASSEMBLY

The bindings of TME, NAM, MTC, and MTPC were all accompanied by an inhibition of microtubule formation that was weak in the case of the single ring analogues[8] and substoichiometric in MTC, just as in COL.[14] These results were just as expected, but the basic question of why the binding of a ligand to the COL site of tubulin results in the inhibition of microtubule assembly remained unexplored in our earlier studies.

The general agreement that the mechanism of the inhibition of microtubules by COL involves a direct binding of the tubulin-COL complex to microtubule ends prompted us to examine the self-association properties of this complex. In sedimentation velocity experiments it was shown that addition of COL induces a small enhancement of tubulin polymerization into double ring structures in the presence of 16 mM MgCl$_2$.[27] When the tubulin-COL complex, however, was placed into the microtubule assembly medium, heating to 37°C resulted in a strong increase in turbidity,[28] which was fully reversible by cooling to 10°C or by the addition of Ca^{2+} ions. This reaction also required the presence of Mg^{2+} ions and GTP. These observations indicated the formation of large polymers with properties similar to microtubules. Electron microscopic examination, however, showed that no microtubules were present in the system. Glutaraldehyde-fixed samples showed amorphous structures, as well as fibrous or sheetlike large aggregates,[27,28] or spirals under different buffer conditions.[29]

Prior to carrying out a quantitative examination of this polymerization phenomenon, the adequacy of the light scattering (turbidity) technique was thoroughly checked out. The wavelength dependence of turbidity was found to follow the reciprocal second power of the wavelength ($\Delta A \propto \lambda^{-2.1}$), rather than the reciprocal third power ($\Delta A \propto \lambda^{-3}$) required for the Berne theory[30] to be applicable, as is true in microtubule assembly.[31] The simultaneous determination of the critical concentration, C_r, by turbidity, sedimentation of the protein, and sedimentation of the ^{3}H-labeled bound COL[28] yielded the same value of C_r, permitting the use of turbidity in a totally empirical way for the determination of that parameter.

This polymerization was then characterized in terms of its dependence on the solution environment and compared with microtubule assembly. The results are summarized in TABLE 2. It was found that this polymerization is inhibited by Ca^{2+} ions, 10^{-4} M $CaCl_2$ being effective. Binding of GTP is required for polymerization; guanosine diphosphate (GDP) is incapable of inducing polymerization, although growth may occur when there is an excess of GDP over GTP, but with a higher critical concentration than in the absence of GDP. Polycations and microtubule-associated

TABLE 2. Properties of the Polymerization of the Tubulin-COL Complex Compared with Microtubule Assembly *in vitro*[a]

Property	Tubulin-COL Polymerization	Microtubule Polymerization
Type of polymerization	nucleated condensation	nucleated condensation treadmill mechanism
Morphology of polymer	associated filaments (tentative)	helical tube
Thermodynamics of the growth reaction		
ΔG^0_{app} (kcal mol^{-1}; 37°C)	−7.1	−7.0 (−5.9)[b]
ΔH^0_{app} (kcal mol^{-1}; 37°C)	19.4	2.15
ΔS^0_{app} (eu; 37°C)	85	30
$\Delta C^0_{p,app}$ (cal deg^{-1} mol^{-1})	−1,200	−1,500
$\Delta\nu_{H^+}$	0.94	0.86
$\Delta\nu_{Mg^{2+}}$	0.96	0.78
Qualitative effects of		
Ca^{2+}	inhibition	inhibition
Ionic strength	inhibition	inhibition
Polycations	no enhancement; lag suppression	enhancement
Role of nucleotides		
GTP binding	required	required
GTP hydrolysis rate (minute^{-1}): by soluble protein	0.02 (0.07)[c]	not detected
By protein in polymer	0.02 (0.07)[c] not coupled to polymerization	0.13 induced by polymerization; proceeding for one cycle at an intrinsic rate of 0.25
GDP effects	inhibition	inhibition

[a]Polymerization of tubulin-COL was in PG, 16 mM MgCl$_2$ buffer, pH 7.0, except where indicated. Microtubule assembly was performed in the same buffer, containing 3.4 M glycerol and 1 mM EGTA, except where indicated and in studies from other laboratories.
[b]Without glycerol.
[c]With 3.4 M glycerol. For details, see reference 27.

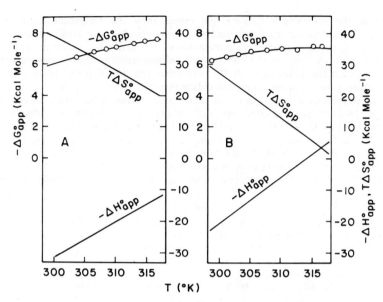

FIGURE 5. (A) Enthalpy and entropy contributions to the apparent standard free energy change of polymerization of tubulin-COL. The points are the experimental results and the solid lines the parameters resulting from the van't Hoff analysis. The solvent was PG, 16 mM MgCl$_2$ buffer, pH 7.0. (B) Enthalpy and entropy contributions to the apparent standard free energy change of polymerization of tubulin during microtubule assembly [data of Lee & Timasheff (1977) obtained in PG, 16 mM MgCl$_2$, 1 mM EGTA, 3.4 M glycerol buffer, pH 7.0].

proteins (MAPs) were shown to have essentially no effect on the strength of the polymerization, although they did affect the rate.

The effects of temperature, Mg^{2+} ions, and H^{+} ions on the growth reaction were determined from their effect on the critical concentration, because the polymer growth constant, K_g, is given by $K_g = C_r^{-1}$. A van't Hoff analysis of the temperature dependence of the polymerization gave results strikingly similar to microtubule formation, as shown in FIGURE 5. This reaction is characterized by a negative heat capacity change and a positive entropy change. The effects of H^{+} and Mg^{2+} were examined in terms of the linked function theory of Wyman.[32] The variation of the equilibrium constant with ligand activity, a_x, is determined directly by the difference between the preferential bindings of ligand to the product and the reactant:

$$(\partial \ln K_g / \partial \ln a_x)_{a_{j \neq x}} = (\partial m_x / \partial m_p)^{\text{polymer}}_{\mu_{x'} m_{i \neq p}} - (\partial m_x / \partial m_p)^{\text{protomer}}_{\mu_{x'} m_{i \neq p}} \qquad (8)$$

where m_x and m_n are the molal concentrations of the ligand and the tubulin heterodimer. When the ligand concentration is low enough to neglect the effects of the displacement of bound water, the slope of a plot of log K_g versus log a_x gives the change in the number of ligand molecules bound to the protein during the reaction. Application of this analysis showed that addition of each tubulin-COL complex kinetic unit to the growing polymer is accompanied by the binding of one magnesium ion and one proton,[27] similar to microtubule assembly.[31]

Two essential differences were found from microtubule assembly.[27] First, the

GTPase activity of the tubulin-COL complex was found to be independent of the state of aggregation. This indicates that the E-GTP site in this polymer is exposed to solvent and that the tubulin-COL complex is present in a conformational state somehow related to that of tubulin assembled into microtubules, whether it is polymerized or not. The second difference was that, contrary to what is known of microtubule assembly, the tubulin-COL complex polymerization is weakened by the presence of 3.4 M glycerol.[27] This, however, was found to be a secondary effect, because glycerol increases fourfold the rate of hydrolysis of GTP by the complex, and the product, GDP, is an inhibitor of the polymerization.[27] The reversibly binding analogue MTC has been found to induce a cooperative tubulin polymerization that is cold reversible, does not occur in the absence of GTP, requires Mg^{2+}, and is inhibited by Ca^{2+},[14] similarly to the polymerization of the tubulin-COL complex.

Based on these observations, we have proposed a mechanism for microtubule inhibition by COL that essentially encompasses and reconciles various seemingly disagreeing reports in the literature. We have proposed that both the polymerization of tubulin-COL in the "wrong geometry" and the inhibition of microtubule assembly are due to a distortion in the geometry of the normal bonds between protomers.[27] The reversible binding of a tubulin-COL complex to the end of a microtubule is statistically favored by a factor of about 5 over a normal tubulin molecule.[26] According to our scheme, it would add on with a geometry slightly different from an unliganded tubulin protomer. For example, the lateral interaction geometry may be altered from the tube-forming one (angle of 152° between vicinal protomers) to a sheet-forming one (bond angle of 180° between protomers). Further addition of a few unliganded protomers in the vicinity of the drug-complexed one can occur until newly added ones fall out of proper register, stopping growth.[27] This would generate a statistical "cap" as proposed by Margolis and Wilson,[4,5] stemming from a low degree of copolymerization, proposed by Sternlicht and Ringel.[33] It is interesting to note that whereas the copolymerization of tubulin-COL with tubulin in the formation of microtubules is very limited, tubulin can be incorporated into the tubulin-COL polymers to a large extent, as has been recently shown by Saltarelli and Pantalon.[34] A COL-mediated alteration of the $\alpha - \beta$ subunit contacts in tubulin can be inferred from the effect of this drug on the dissociation of the dimeric protein.[35] If this alteration leads to a tightening of tubulin molecules, COL binding could abolish the protein flexibility necessary for the autosteric transition needed in microtubule formation and lead to the formation of the polymers with the wrong geometry.[27]

REFERENCES

1. TAYLOR, E. W. 1955. J. Cell Biol. **25:** 145–168.
2. WEISENBERG, R. C., G. G. BORISY & E. W. TAYLOR. 1968. Biochemistry **7:** 4466–4479.
3. WEISENBERG, R. C. 1972. Science **177:** 1196–1197.
4. MARGOLIS, R. L. & L. WILSON. 1977. Proc. Natl. Acad. Sci. USA **74:** 3466–3470.
5. MARGOLIS, R. L. & L. WILSON. 1978. Cell **13:** 1–8.
6. LAMBEIR, A. & Y. ENGELBORGHS. 1980. Eur. J. Biochem. **109:** 619–624.
7. GARLAND, D. L. 1978. Biochemistry **17:** 4266–4272.
8. ANDREU, J. M. & S. N. TIMASHEFF. 1982. Biochemistry **21:** 534–543.
9. HUMMEL, J. M. P. & W. J. DREYER. 1962. Biochim. Biophys. Acta **63:** 530–532.
10. BHATTACHARYYA, B. & J. WOLF. 1974. Proc. Natl. Acad. Sci. USA **71:** 2627–2631.
11. ANDREU, J. M. & S. N. TIMASHEFF. 1982. Biochim. Biophys. Acta **714:** 373–377.
12. FITZGERALD, T. J. 1976. Biochem. Pharmacol. **25:** 1383–1387.
13. STEINBERG, I. Z. & H. A. SCHERAGA. 1963. J. Biol. Chem. **238:** 172–181.

14. ANDREU, J. M., M. J. GORBUNOFF, J. C. LEE & S. N. TIMASHEFF. 1984. Biochemistry 23: 1742–1752.
15. GLASSTONE, S. 1940. Textbook of Physical Chemistry. p. 875. Van Nostrand. Princeton, N.J.
16. ANDREU, J. M. & S. N. TIMASHEFF. 1982. Biochemistry 21: 6465–6476.
17. LETERRIER, F. & F. RIEGER. 1975. CRSSA Travail Scientifique 5: 224–225.
18. DETRICH, M. W., R. C. WILLIAMS, T. L. MACDONALD, L. WILSON & D. PUETT. 1981. Biochemistry 20: 5999–6005.
19. MARGULIS, T. N. 1974. J. Am. Chem. Soc. 96: 899–902.
20. ENGELBORGHS, Y. 1981. J. Biol. Chem. 256: 3276–3278.
21. SCHMITT, M. & D. ATLAS. 1976. J. Mol. Biol. 102: 743–758.
22. MORRIS, N. R., M. M. LAI & C. E. OAKLEY. 1979. Cell 16: 437–442.
23. CABRAL, F., M. E. SOBEL & M. M. GOTTESMAN. 1980. Cell 20: 29–36.
24. SERRANO, L., J. AVILA & R. B. MACCIONI. 1984. J. Biol. Chem. 259: 6607–6611.
25. DAVID-PFEUTY, T., C. SIMON & D. PANTALONI. 1979. J. Biol. Chem. 256: 3279–3282.
26. LAMBEIR, A. & Y. ENGELBORGHS. 1981. J. Biol. Chem. 256: 3279–3282.
27. ANDREU, J. M., T. WAGENKNECHT & S. N. TIMASHEFF. 1982. Biochemistry 22: 1556–1566.
28. ANDREU, J. M. & S. N. TIMASHEFF. 1982. Proc. Natl. Acad. Sci. USA 79: 6753–6756.
29. SALTARELLI, D. & D. PANTALONI. 1982. Biochemistry 21: 2996–3006.
30. BERNE, B. J. 1974. J. Mol. Biol. 89: 756–758.
31. LEE, J. C. & S. N. TIMASHEFF. 1977. Biochemistry 16: 1754–1764.
32. WYMAN, J. 1964. Adv. Protein Chem. 19: 224–285.
33. STERNLICHT, M. & I. RINGEL. 1979. J. Biol. Chem. 148: 627–632.
34. SALTARELLI, D. & D. PANTALONI. 1983. Biochemistry 22: 4607–4614.
35. DETRICH III, H. W., R. C. WILLIAMS & L. WILSON. 1982. Biochemistry 21: 2392–2400.

Kinetics and Steady State Dynamics of Tubulin Addition and Loss at Opposite Microtubule Ends: The Mechanism of Action of Colchicine[a]

LESLIE WILSON AND KEVIN W. FARRELL

Department of Biological Sciences
University of California
Santa Barbara, California 93106

INTRODUCTION

During the past several years our laboratory has been investigating the dynamics of tubulin addition and loss at the opposite ends of steady state microtubules *in vitro*. We have also been interested in understanding the mechanisms of action of chemical agents that alter the dynamics of tubulin addition and loss. Our approach has taken advantage of the fact that microtubules, when assembled to a plateau level *in vitro*, are not in simple equilibrium at either end with the tubulin free in solution. Rather, a continuous net incorporation of tubulin into the polymer mass and a precisely balanced net loss of tubulin from the polymer mass occur at apparent equilibrium (*e.g.,* see reference 1). Because the net uptake and loss of tubulin occurs at microtubule ends and at different locations, we have interpreted this behavior as indicating that there is a net gain of tubulin subunits at one end of each microtubule (called the net assembly, or A end) and a balanced net loss of the tubulin at the opposite end (the net disassembly or D end).

We recently developed a double isotope procedure for labeling microtubule A and D ends differentially that has enabled us to study the net gain and loss of tubulin at both microtubule ends simultaneously under a variety of conditions.[2,3] The double label procedure has been coupled to a filtration-collection assay that involves rapid stabilization of microtubules and collection of the labeled polymers on glass fiber filters.[4] These procedures have been used to analyze the net rates of tubulin addition and loss at opposite microtubule ends under steady state and non-steady state conditions in control microtubule populations and after treatment with a number of drugs that affect microtubule assembly and disassembly dynamics.

The purpose of this report is to describe a number of properties of the bovine brain microtubule protein system used in our laboratory, and to set forth the evidence that the isotope incorporation methods we use with the system reveal the dyamics of net tubulin addition and loss at the two microtubule ends. We also describe the results of our studies on the action of one drug, colchicine, which slows the rates of tubulin addition and loss at the two ends of the microtubules differentially.

[a]This work was supported by USPHS Grant NS13560 from the Institute of Neurological and Communicative Diseases and Stroke (L. Wilson), by PHS Grant CA36389, awarded by the National Cancer Institute, DHHS (L. Wilson), and by USPHS Grant GM26732, from the Institute of General Medical Sciences (K.W. Farrell).

METHODS FOR ANALYSIS OF TUBULIN ADDITION AND LOSS

We have used bovine brain microtubules purified in the absence of glycerol by three cycles of warm assembly and cold disassembly for many of our recent studies (reference 5, modified as in reference 6). The microtubules contain a high proportion of microtubule-associated proteins (25% microtubule-associated proteins, 75% tubulin) and have a critical subunit concentration at steady state of 0.32 to 0.35 mg/ml (total protein) at 30°C. The protein solutions used in all experiments contain a guanosine triphosphate (GTP) regenerating system consisting of 10 mM acetyl phosphate and 0.1 IU/ml acetate kinase.[7]

The method for labeling steady state microtubules differentially with ^{3}H and ^{14}C-labeled nucleotides has been described previously.[2,3] The microtubules are constructed so that they are labeled only with ^{14}C at the D ends, and a mixture of ^{14}C and ^{3}H at the A ends (see diagram in FIGURE 1). Labeled GTP binds exchangeably to soluble tubulin, and the labeled nucleotide becomes nonexchangeable upon incorporation into the microtubules.[8,9] Validity of the method depends upon the ability of radiolabeled guanine-nucleotide-tubulin complexes to become incorporated predominantly at microtubule A ends and be lost predominantly at D ends at steady state by a treadmilling mechanism. (See next section for further discussion and the APPENDIX for an analysis of possible errors due to "wrong end" addition).

When microtubule protein is polymerized to steady state in the presence of [14C]GTP and a GTP regenerating system, the microtubules become uniformly labeled with [14C]guanine nucleotide. Upon pulsing with [3H]GTP, the [3H]guanine nucleotide becomes incorporated at the operationally defined A ends and serves as the marker for the steady state rate of A end net tubulin addition, whereas the [14C]guanine continues to be added at A ends and lost from D ends in an unaltered fashion. The rate of steady state D end tubulin loss is determined by monitoring [14C]guanine nucleotide loss after addition of 25-fold excess unlabeled GTP. The labeled microtubules are collected on glass fiber filters for analysis of label incorporation. Steady state rates of tubulin incorporation at A ends are calculated from linear regression lines of the rates of [3H]GTP uptake from 5 to 45 min after label addition and are corrected for the concentration of microtubule A ends in solution (microtubule number concentration). Mean microtubule lengths are obtained by analysis of negative stain electron micrographs of microtubule suspensions using a MOP-3 image processer to accumulate and process length data. Data are corrected throughout the [3H]nucleotide pulse for the presence of short microtubules, which, if present, would become labeled at both ends with tritium. Similarly, steady state rates of tubulin loss at microtubule D ends are calculated using linear regression analysis of the rates of [14C]GTP loss after chasing with excess unlabeled GTP (corrected for the concentration of microtubule ends in solution). Loss rate data are also corrected for the presence of short microtubules that would lose all [14C]guanine nucleotide during the GTP chase.

The filter procedure for determining labeled GTP incorporation into and loss from microtubules during assembly and steady state treadmilling was modeled after the technique developed by Maccioni and Seeds,[10] and has been described in detail by Wilson et al.[9] Briefly, aliquots of radiolabeled microtubule suspensions are diluted into 4 ml of a microtubule stabilizing buffer at 30° C, consisting of 30% glycerol (v/v), 10% dimethylsulfoxide (v/v), 5.0 mM adenosine triphosphate (ATP) in 100 mM 2-(N-mopholino) ethane sulfonic acid (MES), 1 mM ethylene glycol bis (β-aminoethylether)-N, N, N' N'-tetracetic acid (EGTA), and 1 mM MgSO$_4$, pH, 6.75, which stops tubulin uptake and loss. The stabilized microtubules are filtered through Whatman GF/F glass fiber filters, washed with additional stabilizing buffer, and processed for determination of radioactivity. Filters retain the microtubules quantitatively.

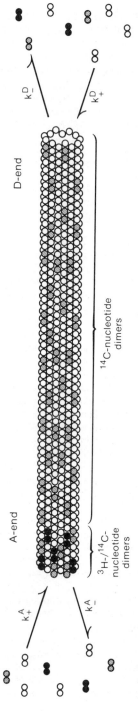

FIGURE 1. A schematic diagram of microtubules differentially labeled with ^{14}C and ^{3}H at opposite ends. Subscripts "+" and "−" refer to association rate constants and dissociation rate constants, respectively. Superscripts refer to operationally defined A and D microtubule ends. Double circles represent tubulin dimers. Open circles represent unlabeled dimers; hatched circles represent dimers labeled with [14C]guanine-nucleotide; closed circles represent dimers labeled [3H]guanine-nucleotide. Shown is a steady state microtubule that had been assembled in the presence of [14C]GTP, becoming labeled with [3H]guanine nucleotide-containing dimers at the A ends after a [3H]GTP pulse. The microtubules are labeled with ^{14}C and ^{3}H at A ends, and ^{14}C only at D ends.

ANALYSIS AND INTERPRETATION OF STEADY STATE FLUX DATA
OBTAINED WITH THE DOUBLE-LABEL PROCEDURE

A typical experiment in which double-labeled steady state microtubules were used to determine tubulin addition and loss at opposite microtubule ends is shown in FIGURE 2 in its entirety, and in FIGURES 3 and 4 in expanded form to illustrate critical portions of the experiment. The experiment was also used to determine the effects of colchicine on steady state tubulin addition and loss (discussed in a later section). Similar to results obtained by turbidimetry (data not shown), analysis of [^{14}C]nucleotide incorporation during initial polymerization of the microtubules revealed that the microtubules

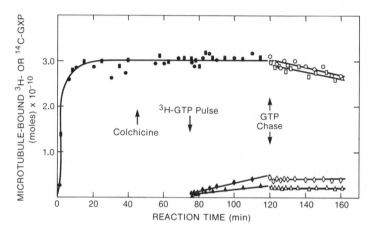

FIGURE 2. Effect of colchicine on the incorporation of tubulin at A ends and loss of tubulin at D ends of steady state microtubules: double label procedure. Two identical portions of a bovine brain microtubule protein solution, 2.96 mg/ml total protein, were polymerized to steady state at 30°C in reassembly buffer containing [^{14}C]GTP and a GTP regenerating system. At 45 min (arrow), 0.2 μM colchicine was added to one portion, and incubation was continued ([^{14}C]guanine nucleotide data: circles = untreated; squares = colchicine-treated). A [^{3}H]GTP pulse was added to both microtubule suspensions at 75 min (arrow) ([^{3}H]guanine nucleotide pulse data: untreated suspension = filled diamonds; colchicine-treated = filled triangles). A 25-fold excess unlabeled GTP chase was initiated at 119 min (arrow), and was continued for an additional 45 min (open symbols denote chase conditions). Data on the ordinate reflect the quantity of bound nucleotide per filter disc (40 μl of microtubule suspension containing 0.049 mg of tubulin in microtubules was applied to each disc).

reached a steady state plateau after 20 min of incubation at 30°C. Addition of a [^{3}H]GTP pulse at 75 min resulted in a 2–3 min initial burst of label incorporation, followed by a linear incorporation of label (see expanded version of the pulse data, FIGURE 3). We believe the initial burst of label incorporation does not reflect the flux incorporation of tubulin at microtubule A ends, but rather reflects equilibrium exchange of tubulin and/or another rapid exchange reaction that reaches equilibrium quickly. The rate of net tubulin addition at microtubule ends in the microtubule suspension was determined using the linear regression line of the data beginning after the initial burst, and was calculated to be 1.1/sec (66 molecules of tubulin/min/microtubule).

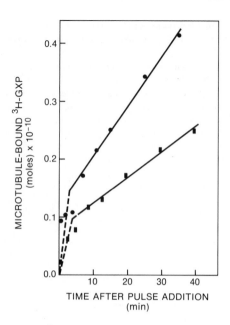

FIGURE 3. Incorporation of [³H]guanine nucleotide into untreated and colchicine-treated microtubules at plateau. Data represent the pulse portion of the experiment shown in FIGURE 2 with expanded scales. Circles = control microtubules; squares = colchicine-treated microtubules.

Addition of 25-fold excess unlabeled GTP as a chase for both isotopes was performed 44 min after initiation of the [³H]GTP pulse. At D ends of the microtubules (¹⁴C data, FIGURE 4), a linear rate of label loss occurred that continued to the end of the experiment. The rate of net tubulin loss was 1.14/sec (68 molecules of tubulin/min/microtubule). This value is indistinguishable from that obtained with the tritium label for the rate of net tubulin addition at A ends of the microtubules, and demonstrates that the microtubules were indeed at steady state, with net A end tubulin gain balanced precisely by the net rate of tubulin loss.

Evidence That the Majority of the Tritium-Guanine Nucleotide Incorporation Occurs at Microtubule A Ends and Carbon-14-Guanine Nucleotide Loss Occurs at D Ends by a Flux Mechanism

One kind of evidence that ³H-labeled tubulin adds predominantly at A ends and that ¹⁴C-labeled tubulin is lost predominantly from D ends by a flux mechanism derives from analysis of pulse-chase data (see FIGURES 2, 4). Zeeberg et al.[11] determined that if equilibrium exchange were solely responsible for tritium incorporation during a [³H]GTP pulse, only 41.4% of the tritium-labeled dimers incorporated during a pulse would be retained in the microtubules during a chase of equal duration. This would occur if there were a single population of microtubules in suspension, and both ends of the microtubules were at equilibrium with the soluble subunit pool. If the extent of tritium retention in an equal-duration chase were significantly greater than 41%, the data would be consistent with two possible interpretations. The data could mean that (1) the tritium-label incorporation occurred predominantly by a flux mechanism, or that (2) length redistributions were occurring in the microtubule population of the kind observed recently by Mitchison and Kirschner, that is, relatively slow elongation of the majority of the microtubules offset by relatively rapid shortening of a minority of the microtubules.[12,13] Evidence will be presented in a later section showing that microtubule length redistribution, if it occurs, is below detectable levels in the microtubule population used in these studies. Thus, retention of most of the tritium from a pulse during a chase of equal duration would indicate that the label is adding to the microtubules by a flux mechanism.

The addition of excess unlabeled GTP, which permits measurement of [14C]guanine nucleotide loss at microtubule D ends, also serves as a chase for the [3H]guanine nucleotide incorporated at A ends. At A ends of untreated steady state microtubules after addition of a GTP chase (FIGURE 4), there was an initial loss of 3H that amounted to 19% of the tritium pulse. This was followed in less than 3 min by a stable plateau lasting at least 40 min during which no detectable label loss occurred. These data are inconsistent with a purely equilibrium exchange mode of label incorporation during the [3H]GTP pulse, and indicate that labeled dimer incorporation into preformed microtubules occurred predominantly by a steady state treadmilling reaction. If one assumes that the initial burst of [3H]nucleotide uptake during a pulse and burst of label loss during a chase predominantly represented equilibrium exchange of tubulin or a kinetically rapid exchange of guanine nucleotide at microtubule ends, and excludes the

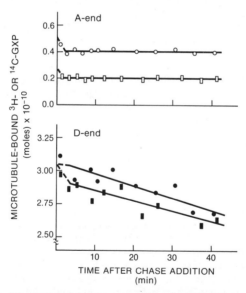

FIGURE 4. Loss of radiolabeled tubulin from A and D ends of untreated and colchicine-treated microtubules. Data from the chase portion of the experiment shown in FIGURE 2 (expanded scales). Top panel: [3H]nucleotide data, that is, A-end label loss (open circles = untreated microtubules; open rectangles = colchicine-treated microtubules). Bottom panel:[14C]nucleotide data; D-end label loss (closed circles = untreated microtubules; closed rectangles = colchicine-treated microtubules). The intercept positions on the ordinate for each data set represent the radiolabel incorporation immediately prior to addition of the chase.

initial 2–3 min of pulse and chase data in the calculation made above, then virtually all of the label incorporated during the linear phase of a pulse was retained in the chase. These data strongly support the conclusion that the tritium-label incorporation represented by the linear phase of the data represents the flux incorporation of tubulin at A ends. Similarly, the carbon-14 label loss represented by the linear phase of the data must represent the flux loss of tubulin at microtubule D ends.

In addition, it is clear that during the period of chase, the rates of carbon-14 loss and tritium loss were very different (FIGURE 4), which would not occur if labeling during the construction of the double-labeled microtubules had occurred predominantly by an equilibrium exchange mechanism. Further, an upward deviation from linearity was clearly apparent when the data shown in FIGURE 3 were plotted as a function of the square root of time (data not shown). This indicates that flux incorporation of label was occurring during the pulse.[11] Finally, in experiments in which double labeled microtubules were diluted from steady state to obtain apparent off-rate constants, the tubulin labeled with [14C]guanine nucleotide was lost between two and three times faster than the tubulin labeled with [3H]nucleotide (data not shown for this experiment, but see reference 2 and 6, and the data described below). This is consistent only with a flux mechanism as being responsible for labeled tubulin incorporation. If an equilibrium exchange mode predominated, the loss rates for the two labels would have been equal.

Can the Steady State Isotope Incorporation Data Be Interpreted in Terms of the Presence of Small Numbers of Rapidly Depolymerizing Microtubules Offset by Large Numbers of Growing Microtubules (No Treadmilling of Individual Microtubules)?

Mitchison and Kirschner have recently obtained evidence that under certain conditions microtubules are able to coexist in growing and shortening populations that interconvert infrequently,[12,13] and have raised the question[13] whether length redistributions in the microtubule population, rather than tubulin flux, might account for the isotope incorporation and loss data. It appears, however, very unlikely that microtubule length redistributions can be responsible for the exchange data obtained in double-label experiments such as the one shown in FIGURES 2–4.

First, an important prediction of the Mitchison and Kirschner model is that the majority of the microtubules should be elongating at the rate revealed by the [3H]guanine nucleotide incorporation data (i.e., 1.1/sec). Over the time period of 45–165 min after assembly initiation (FIGURE 2), this flux rate would translate into a mass redistribution equivalent to approximately 5 μm of microtubules length, well within detectable limits. In the experiments described in FIGURES 2–4, however, we observed no evidence of a microtubule length redistribution. At 45 min after initiation of polymerization (30 min prior to addition of the [3H]GTP pulse), the mean length of the microtubule population was 14.1 μm. At the end of the unlabeled GTP chase, 165 min after initiation of assembly, the mean length was 13.4 μm—effectively unchanged. Equally important, the distribution of microtubule lengths did not change (data not shown).

Therefore, these results exclude a dynamic instability mechanism, as proposed by Mitchison and Kirschner.[13] Only two known mechanisms could account for incorporation of radiolabeled tubulin subunits into microtubules at mass steady state without a length redistribution: equilibrium exchange or treadmilling. The reasons for excluding equilibrium exchange as being a major influence have been discussed above. Thus, the most reasonable conclusion is that the linear uptake of [³H]GTP represents a flux incorporation rate of 1.1/sec by the treadmilling mechanism.

Also, the explanation for our isotope exchange data advanced by Mitchison and Kirschner is incompatible with previously published pulse-chase data[1] obtained using a single-label procedure with sheared bovine brain microbutubles at steady state (FIGURE 5). In this experiment, bovine brain microtubules were assembled to steady state at 30°C, sheared to a mean length of 3.5 μm, pulsed for 30 min with [³H]GTP, then chased with a 30-fold excess of unlabeled GTP. The microtubule A ends were elongating at a rate of 0.7 μm per hour based upon the tritium incorporation rate in a pulse experiment with a separate but identical microtubule preparation. The microtubules retained all incorporated tritium for the first 90 min of the chase (equivalent to approximately 2 μm of microtubule length), then began to lose the tritium. The

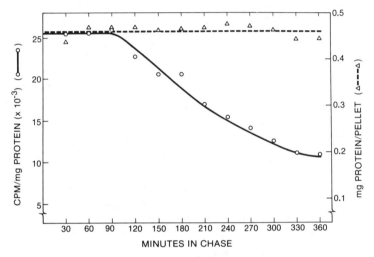

FIGURE 5. Demonstration of subunit flux in microtubules by a pulse-chase experimental procedure. Sheared steady state bovine brain microtubules (mean length, 3.5 μm) were pulsed with [³H]GTP for 30 minutes. Microtubules were then chased with a 30-fold excess of unlabeled GTP for the times indicated. Label and protein retained in the polymer were assayed in microtubule pellets after centrifugation through 50% sucrose cushions.[1]

experiment was terminated after 360 min, at which time approximately 60% of the label had been lost. It seems reasonable to assume that the remaining label resided in the microtubules that were greater than 4.2 μm in length (approximately 40% of the microtubule population). The essentially complete retention of tritium for the initial 90 min of chase followed by the gradual loss of the majority of label during the next 270 min is incompatible with the microtubule behavior observed by Mitchison and Kirschner. The majority of the microtubules would have had to accumulate tritium by growing at both their ends for 30 min, retain the label quantitatively for an additional 90 min, then suddenly undergo a phase transition and begin depolymerizing slowly. This kind of behavior seems highly unlikely, and would clearly be evident at steady state in double label experiments in which net uptake and loss rates are continuously monitored. We conclude that if length rearrangements are occurring in the population of microtubules used in this study, or if a rare switching event results in the rapid disassembly of a subset of microtubules, the contribution of the phenomenon to the isotope data we have obtained is below detectable levels.

EFFECT OF COLCHICINE ON TUBULIN ADDITION AND LOSS AT OPPOSITE MICROTUBULE ENDS AT STEADY STATE

The double-label experiment described previously (FIGURES 2–4) was also used to determine the effects of colchicine on tubulin addition and loss at microtubule A- and D ends. Addition of 0.2 μM colchicine to one portion of the steady state microtubule suspension 45 min after initiation of polymerization produced little immediate effect. This concentration had been predetermined in other experiments (not shown) to be that required for 50% inhibition of A end net tubulin uptake. Colchicine inhibited the initial burst of label incorporation by approximately 33%, as determined by extrapolation of the linear regions of the data to zero time of incubation. Colchicine inhibited the linear incorporation of label by 49% to 0.57/sec (34 molecules of tubulin/min/ microtubule) (FIGURE 3).

Colchicine at 0.2 μm exerted little effect on the rate of label loss at D ends (FIGURE 4, ^{14}C data). There was a small initial burst of label loss followed within a minute or two by a linear rate of label loss that continued for the remainder of the experiment. The rate of net tubulin loss at D ends, determined using the linear region of the data beginning after the initial burst of label loss, was 0.93/sec (56 molecules of tubulin/ min/microtubule). This rate is approximately 82% of the rate of label loss at the D ends of untreated microtubules, and indicates that under the conditions used, colchicine substantially inhibited A end tubulin addition while minimally affecting D end tubulin loss.

At A ends of colchicine-treated microtubules in the chase portion of the experiment shown in FIGURE 4, there was an initial loss of label that amounted to 26% of the tritium pulse followed within 2 min by a stable plateau during which no detectable loss of label occurred. These results are similar to those obtained with untreated microtubules and indicate that label incorporation at A ends of colchicine-treated microtubules, while reduced in rate, occurred predominantly by a treadmilling reaction (see also reference 6).

ANALYSIS AND INTERPRETATION OF KINETIC DATA OBTAINED BY DILUTION-INDUCED DISASSEMBLY OF DOUBLE-LABELED MICROTUBULES: DIFFERENTIAL KINETIC STABILIZATION OF OPPOSITE MICROTUBULE ENDS BY COLCHICINE

Determination of Apparent Rate Constants

The apparent molecular off-rate constants at the A and D ends of microtubules and the effects of drugs on the apparent rate constants can be estimated by dilution-induced disassembly measurements of double-label microtubules. The values obtained in such experiments, however, should be interpreted cautiously because the initial rate of tubulin loss from microtubules upon dilution from steady state is not a linear function of the initial free tubulin concentration.[14,15] When diluted into microtubule protein concentrations below the steady state critical subunit concentration, the initial rate of tubulin loss at both microtubule ends is faster than predicted by the linear relationship of Oosawa and Kasai.[16] For example, between a 10-fold dilution and a 100-fold dilution, the initial rate of dimer loss for bovine brain microtubules increased by 300% rather than the 11% expected on the basis of a linear rate plot (see below).

Results of a typical experiment carried out by a single 100-fold dilution of untreated microtubules appears in FIGURE 6. This experiment was carried out similarly to the experiment described in FIGURE 2, with the microtubules being diluted 60 min after the [³H]GTP pulse rather than being chased undiluted with unlabeled GTP. The depolymerization kinetics at A and D ends for the first 60 sec after dilution

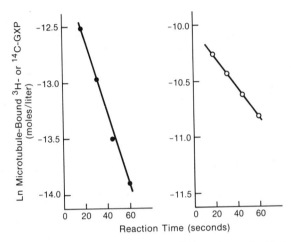

FIGURE 6. Dilution-induced disassembly: left panel, [³H]guanine nucleotide loss (microtubule A ends); right panel, [¹⁴C]guanine nucleotide loss (microtubule D ends). Double-labeled microtubules were constructed as described in the legend to FIGURE 2 except that the microtubules were sheared by passage twice through a 25-guage needle prior to addition of the [³H]GTP (4.98 mg/ml total microtubule protein; mean length, 4.7 μm). The microtubules were labeled at their A ends with tritium for a distance of 0.21 μm. No microtubules were fully labeled with tritium. Microtubules were diluted 100-fold into reassembly buffer (300 μl diluted to 3 ml) 60 min after initiation of the [³H]GTP pulse. Samples were stabilized at 30°C at 15 second intervals, collected on glass fiber filters, and assayed for retained radiolabel. It took approximately 15 s to process experimental samples; thus, it was not possible to obtain data points prior to 15 seconds. Approximately 35% of the microtubule polymer mass had depolymerized in the control sample during the initial 15 seconds after dilution (40% of the tritium label remained at 15 seconds, and 65% of the carbon-14 label remained). Closed circles = tritium loss (A ends); open circles = carbon-14 loss (D ends). Solid lines represent theoretical plots derived from fitting first-order decay curves to the data by least-squares nonlinear analysis. Apparent off rate constants were calculated using the linear regression lines of the data.

closely approximated first-order loss reactions from which the initial rates of tubulin loss were calculated. It has been proposed that the first-order approximation of such data reflect an exponential loss of microtubule ends.[17,18] The apparent off-rate constant at the microtubule D ends was 105/sec, approximately 2.6 times faster than the apparent off rate at the A ends. A typical dilution series showing the nonlinear relationship between the initial rate of dimer loss at each microtubule end and the initial free tubulin concentration is shown in FIGURE 7. In this experiment, the data was approximately linear to 20-fold dilution, but deviated significantly at both microtubule ends at greater dilutions.

The observation that the D ends of the bovine brain microtubules we have been studying exhibit apparent rate constants that are two to three times faster than the A end constants has been reported previously from this laboratory,[2,3,6,14] and indicates that under certain growth conditions, the D ends of these microtubules could grow faster than the A ends. Thus, the end of the microtubule operationally defined as the D end using steady state isotope incorporation data could be the plus end (not the minus end as concluded by Bergen and Borisy[19]) if the end designation based upon growth of tubulin onto flagellar seeds were applied to this steady state system.[20] The relationship

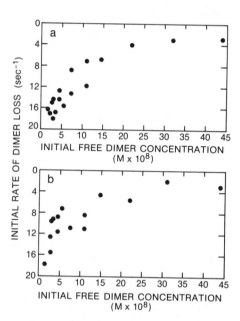

FIGURE 7. Dependence of the initial rate of dilution-induced microtubule depolymerization on the initial free tubulin dimer concentration. Suspensions of double-labeled bovine microtubules (3.1 mg/ml total protein) were serially diluted (5–100-fold) into reassembly buffer containing 0.1 mM GTP at 30°C. The initial rates of tubulin loss at microtubule A ends (top panel) and microtubule D ends (bottom panel) were derived from disassembly plots such as shown in FIGURE 6 (data from reference 14, FIGURE 2).

between the A and D ends operationally defined at steady state, and the plus and minus ends as defined structurally by formation of tubulin hooks[21,22] or by dynein decoration[23] remains to be determined.

Because the initial rate of tubulin loss is not a simple linear function of the initial free tubulin concentration, the rate constants measured by dilution analysis in the presence and absence of drugs such as colchicine (see below) should be considered to be apparent rate constants. Because the degree of deviation from nonlinearity is similar for the initial loss rate from A and D ends, the relative values of the apparent rate constants for the two ends should be valid. Similarly, the effects of drugs at the two ends should also be valid in terms of the effects of the drug on one end relative to the other.

Effects of Colchicine on the Kinetics of Tubulin Loss at Opposite Microtubule Ends

The precise molecular mechanisms by which colchicine inhibits polymerization at microtubule ends has remained highly controversial, certainly due in part to the complex nature of the mechanism, but also to the use of different microtubule protein preparations in different laboratories, and to differences in the conditions employed to study the mechanism of inhibition (discussed in reference 6; see also references 19 and 24–27). For example, when brain microtubule protein is assembled in the presence of colchicine-tubulin complexes, the rate and extent of microtubule assembly is reduced, and the drug-tubulin complexes become extensively incorporated into the polymers at the tubulin-colchicine (TC) complex to free tubulin ratios determined by the ratio of TC complex to free tubulin present initially in solution.[6,24] Studies in which the rates of microtubule elongation onto flagellar axonemal seeds were measured using brain microtubule protein in the presence and absence of TC complexes indicated that the drug reduced the rate constants for tubulin addition equally at the two microtubule ends without affecting the rate constants for tubulin loss.[19]

By contrast, when free colchicine is added to preformed steady state bovine brain microtubules, the drug produces a potent substoichiometric inhibition of tubulin addition at microtubule A ends. Inhibition is brought about by the binding of a very few TC complexes to A ends, without detectable incorporation of colchicine into the microtubule lattice.[26] At the colchicine concentration that inhibits net tubulin incorporation at A ends by 50%, there is no tubulin loss at A ends, and the microtubules continue to depolymerize at the D ends at rates similar to control microtubules (see FIGURE 3). These results differ substantially from those of Ringel and Sternlicht,[24] and Bergen and Borisy,[19] and are consistent with the idea that when colchicine is added to preformed steady state microtubules, inhibition of net tubulin A end addition occurs by a kinetic capping mechanism, in which both the rate of tubulin loss and the rate of tubulin gain at A ends are inhibited by the drug, with little or no effect at the D ends.

Dilution studies carried out previously in which bovine brain microtubules pulse-labeled with ^{3}H[GTP] at their A ends were treated with colchicine and then diluted indicated that the drug kinetically slowed the rate of tubulin loss from A ends as compared with loss from A ends of untreated microtubules (FIGURE 8 and see reference 26). More recently, we have constructed double-labeled microtubules containing incorporated TC complexes at a number of different TC complex to tubulin ratios and have determined the effects of the incorporated TC complexes on the apparent off-rate constants at both microtubule ends.[6] Untreated microtubules and microtubules containing TC complexes were diluted 30-fold into reassembly buffer at 30°C. Apparent dissociation rate constants were derived from untreated and TC-complex-containing microtubule disassembly curves (e.g., see FIGURE 6), and were plotted both for A end and D end loss as a function of the initial TC complex to free tubulin ratio in solution (FIGURE 9). In marked contrast to the results of Bergen and Borisy,[19] we found that the dissociation rate constants at both A and D microtubule ends were reduced by the incorporated TC complexes. Moreover, the A ends of the TC complex-containing microtubules were substantially more sensitive than the D ends at all TC complex to free tubulin ratios. For example, at a TC complex to free tubulin ratio in solution of 0.04 (a molar ratio of 0.02 moles TC complex to free tubulin in the polymers), the A end dissociation rate constant was reduced by 97%, whereas the D end was reduced by only 5 percent. Substantial D end inhibition was observed as the

LOSS OF ASSEMBLY-END LABEL

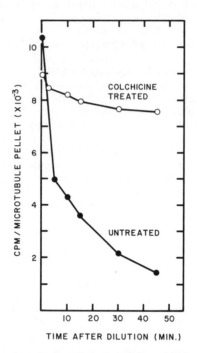

FIGURE 8. Kinetic capping of microtubule A ends by colchicine. Bovine brain microtubule protein (2.45 mg/ml) was assembled to steady state at 30°C in the presence of a GTP-regenerating system, and pulsed for 1 hr with [^{3}H]GTP to label the A ends. The microtubule suspension was split into two portions, and incubation of one portion was continued for an additional 15 min with 100 μM colchicine in the presence of 2.5 mM unlabeled GTP. The drug-treated microtubule suspension was diluted 5-fold into buffer containing 100 μM colchicine and 2.5 mM GTP, and incubated for the times indicated (open circles). Untreated microtubules were similarly diluted, but into colchicine-free buffer. Remaining tritium-labeled nucleotide in the microtubules was determined after collection of the microtubules through 50% sucrose.[9] Control microtubules, initially 8 μm in length and labeled for 0.8 μm at the A ends, had lost 85.9% of the assembly end label 45 min after dilution. The colchicine-treated microtubules lost 15.1% of the label, or 0.12 μm of length. This experiment was initially presented in a confusing manner, in that the specific activity of label remaining in the microtubules rather than the total quantity of label remaining was reported. Whereas the conclusion remains unchanged qualitatively, the original representation led to an overestimate of the degree of label retention (see reference 26, FIGURE 1).

TC complex to free tubulin ratios in the microtubules was increased. For example, an approximately 80% reduction in the off-rate constant at D ends was observed at a ratio of 0.05 moles TC complex to free tubulin in the polymers.

The differential, or polar poisoning of microtubules by incorporated TC complexes is in good agreement with previous results obtained in this laboratory using independent methods. For example, when colchicine[26] or TC complexes[3] were added to steady state microtubules at concentrations that reduced net A end addition almost completely, dimer loss from D ends continued as measured by decreases in polymer mass. A similar result was found in the experiment described in FIGURE 3, with a colchicine concentration that reduced A end tubulin uptake by 50 percent. The differential poisoning of microtubule ends might also explain how large numbers of TC complexes become incorporated into microtubules prepared by assembling microtubule protein in the presence of TC complexes, but not when colchicine or TC complexes are added to preformed microtubules at steady state. Incorporation of TC complexes along with tubulin into assembling microtubules could occur by coassembly of the TC complexes and tubulin at the relatively refractory D ends, whereas A ends remain kinetically stabilized by the drug (no gain nor loss of tubulin). At or near steady state, at least during early times after colchicine addition, the free tubulin concentration would be below that required for growth at D ends, and TC complex incorporation could not occur.

It is unclear why such significant differences in colchicine poisoning are obtained at steady state with bovine brain microtubules and with microtubules composed of brain tubulin and axonemal seeds. One possibility is that the heterologous axonemally seeded

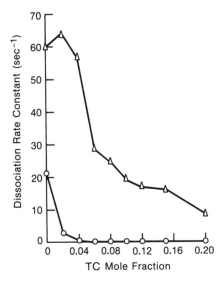

FIGURE 9. Apparent dissociation rate constants for microtubule A ends (circles) and D ends (triangles) as a function of the ratio of the initial TC complex to free tubulin in solution. Microtubule protein solutions containing different ratios of TC complex and free tubulin were assembled to steady state at 30°C in the presence of 0.1 mM GTP and a GTP regenerating system (total protein concentration at each ratio, including the added TC complex, was between 2.0 and 2.5 mg/ml). Untreated and TC-complex-containing microtubules (0.5 ml) were diluted 30-fold into reassembly buffer containing 0.1 mM GTP at 30°C. Apparent dissociation rate constants were determined from disassembly curves (*e.g.,* see FIGURE 6) by nonlinear regression analysis and corrected for the polymer number concentration. The apparent rate constants are shown as a function of the initial TC complex to free tubulin ratio in solution. At a TC complex to free tubulin ratio in solution of 0.02, the polymers actually contained a ratio of TC complex to free tubulin of 0.01. These microtubules were constructed by incubating 0.03 mg/ml (0.03 μM) TC complex with 1.96 mg/ml of drug-free microtubule protein (14.7 μM tubulin); total microtubule protein concentration was 2.0 mg/ml. The microtubules composed of other ratios of TC complex and free tubulin were prepared in a similar fashion, but with increasing TC-complex concentrations in solution.

system differs from the homologous system in a significant way; indeed, the calculated off-rate constants of the heterologous system are much larger than those of the homologous system.

The polar poisoning effect of incorporated TC complexes leads to a dramatic enhancement of the kinetic differences between opposite microtubule ends. For example, the apparent dissociation rate constants of untreated microtubules differed by a factor of approximately 3 (FIGURE 9), whereas at a TC complex to free tubulin ratio in microtubules of 0.02, the difference was nearly 60-fold. Such microtubules would exhibit extremely biased polar growth in cells, and it is interesting to consider the possibility that colchicine-like molecules exist in cells that regulate microtubule growth polarity by a similar differential kinetic stabilization of microtubule ends.

In summary, we have used a double-label isotope procedure to radiolabel microtubule A and D ends differentially, and have used the procedure to characterize partially the dynamics of tubulin addition and loss at opposite microtubule ends. We conclude that the isotope exchange data we have obtained reflect the steady state addition of tubulin at one microtubule end and loss from the opposite end (flux, or treadmilling), and that the data cannot be attributed to length redistributions of the kind recently described by Mitchison and Kirschner.[12,13] We also conclude that labeled guanine-nucleotide uptake and loss occurs predominantly by a flux mechanism and not by simple equilibrium exchange in the microtubule system described in this report. It is reasonable to assume that the reactions giving rise to microtubule flux (treadmilling) *in vitro* are important in microtubule function, and that understanding the flux mechanism and how the mechanism is used by cells are important goals for continued study.

We have found that colchicine produces a differential kinetic stabilization of microtubule ends, with A ends far more sensitive to inhibition of tubulin addition and loss than D ends, and suggest the possibility that naturally occurring colchicine-like molecules may affect microtubule growth in cells by a similar mechanism. Finally, we believe that the dissimilar microtubule behaviors observed in different laboratories are due to the use of different experimental conditions and distinct microtubule protein compositions, and are revealing the wide range of *in vitro* and perhaps *in vivo* capabilities of microtubules. Attention to these differences may lead to new insights regarding the mechanism and regulation of microtubule assembly and function in cells.

APPENDIX

Analysis of Errors in the Double Labeling Protocol Due to Labeling of Microtubule D Ends During a Tritium Pulse

The extent to which [³H]guanine nucleotide becomes incorporated at the D ends of steady state bovine brain microtubules is clearly low, but has not yet been quantified experimentally. The extent to which the D-end incorporation introduces errors into the determination of A- and D-end rate constants is the subject of this analysis.

Definitions

The microtubule (MT) A end is the operationally defined net assembly end, and the MT D end is the net disassembly end at steady state. GXP indicates that the form of the guanine nucleotide (*i.e.*, GTP, GDP) is not specified.

Consider the situation in which MTs are assembled to steady state in the presence of [^{14}C]GTP. The MTs become labeled throughout their lengths with [^{14}C]GXP, that is, at both A and D ends. At steady state, trace amounts of a second label, [^{3}H]GTP, are added. As a result of the treadmilling reaction, MT A ends become labeled both with ^{3}H and ^{14}C. In addition, some of the [^{3}H]guanine nucleotide will also be incorporated at MT D ends by equilibrium exchange.[11] Upon dilution of the double-labeled MTs, the observed initial rate of [^{3}H]label loss, $R_{obs,^{3}H}$, will have two components: loss of the ^{3}H from MT A ends, plus some contribution from loss of ^{3}H at MT D ends, that is,

$$R_{obs,^{3}H} = R_{A,^{3}H} + R_{D,^{3}H} \qquad (1)$$

where $R_{A,^{3}H}$ and $R_{D,^{3}H}$ are the initial rates of loss of [^{3}H]label from MT A and D ends respectively. Writing these components in terms of the individual rate constants,

$$R_{obs,^{3}H} = (k_{+,A} \cdot P - k_{-,A}) + j(k_{+,D} \cdot P - k_{-,D}) \qquad (2)$$

where the subscripts A and D refer to MT A and D ends, respectively, and + and − designate association and dissociation rate constants, respectively. P is the initial concentration of unassembled tubulin protomer, and j is the fraction of MT D ends that contributes to the observed [^{3}H]GXP loss rate, and $0 < = j < = 1$. When $j = 0$, the observed rate of [^{3}H]label loss represents loss from A ends alone. For this situation to occur, no incorporation of [^{3}H]GXP takes place at the MT D ends during the steady state pulse with [^{3}H]GTP. When $j = 1$, pulse labeling of D ends with [^{3}H]GTP is so extensive that the observed loss of [^{3}H]label represents the combined loss from both MT ends. Such a situation would occur in microtubules in which the treadmilling reaction was not occurring (*e.g.*, when the s parameter of Wegner was 0^{28}). The actual value of j will be determined by factors such as the polarity of the MT treadmilling reaction and the relative magnitudes of the A- and D-end rate constants.

Differentiating equation 2 with respect to P gives:

$$dR_{obs,^{3}H}/dP = k_{+,A} + j \cdot k_{+,D} \qquad (3)$$

but $dR_{obs,^{3}H}/dP$ represents the association rate constant derived from experimentally measured rates of loss of [^{3}H]label, $k_{+}{}^{3}H$. Thus,

$$k_{+,^{3}H} = k_{+,A} + j \cdot k_{+,D} \qquad (4)$$

From equation 2, when $P = 0$,

$$R_{obs,^{3}H} = -(k_{-,A} + j \cdot k_{-,D}) \qquad (5)$$

but, when $P = 0$, $R_{obs,^{3}H} = -k_{-,^{3}H}$, the dissociation rate constant derived from the experimentally measured rates of loss of [^{3}H]label. Thus,

$$k_{-,^{3}H} = (k_{-,A} + j \cdot k_{-,D}) \qquad (6)$$

Equations 4 and 6 indicate what is intuitively obvious: that the apparent MT A-end association and dissociation rate constants, derived from the initial rates of [^{3}H]label loss, will be overestimations of the actual A-end on- and off-rate constants by an amount dependent upon the rate of loss of [^{3}H]label from MT D ends.

To determine the initial rates of dimer loss from MT D ends, the initial [^{3}H]nucleotide loss rate is subtracted from the initial [^{14}C]nucleotide loss rate (which represents dimer loss from both MT ends combined). Thus,

$$R_{obs,D} = R_{obs,^{14}C} - R_{obs,^{3}H} \qquad (7)$$

where $R_{obs,D}$ is the initial rate of dimer loss from MT D ends.

Writing equation 7 in terms of the on- and off-rate constants for each MT end,

$$R_{obs,D} = (k_{+,A} \cdot P - k_{-,A} + k_{+,D} \cdot P - k_{-,D})$$

$$-(k_{+,A} \cdot P - k_{-,A} + j \cdot k_{+,D} \cdot P - j \cdot k_{-,D}) \qquad (8)$$

thus,

$$R_{obs,D} = (k_{+,D} \cdot P - k_{-,D}) \cdot (1 - j) \qquad (9)$$

differentiating equation 9 with respect to P,

$$dR_{obs,D}/dP = k_{+,D}(1 - j) \qquad (10)$$

but $dR_{obs,D}/dP = k_{+,^{14}C}$, the MT D-end association rate constant derived from radionucleotide loss. Therefore,

$$k_{+,^{14}C} = k_{+,D}(1 - j) \qquad (11)$$

Also, when $P = 0$, from equation 9,

$$R_{obs,D} = -k_{-,D}(1 - j) \qquad (12)$$

but when $P = 0$, $R_{obs,D} = -k_{-,^{14}C}$, the apparent MT D-end dissociation constant derived from radiolabel loss rates. Therefore,

$$k_{-,^{14}C} = k_{-,D}(1 - j) \qquad (13)$$

Again, equations 11 and 13 indicate the intuitively obvious: that the MT D-end association and dissociation rate constants, derived from the radiolabel data, will underestimate the actual values of the rate constants by an amount equal to $j \cdot k_{+,D}$ and $j \cdot k_{-,D}$, respectively.

Under steady state conditions, the flux rate of incorporation of radiolabeled dimers, F, measured after the initial rapid burst of label uptake, is a function of the actual association and dissociation rate constants,

$$F = k_{+,A} \cdot P_c - k_{-,A} \qquad (14)$$

where P_c is the steady state protomer concentration.

By contrast, the steady state flux rate calculated from the rate constants derived from radiolabel loss, F', is

$$F' = k_{+^{3}H} \cdot P_c - k_{-,^{3}H} \qquad (15)$$

Substituting equations 4 and 6 into equation 15,

$$F' = (k_{+,A} \cdot P_c - k_{-,A}) + j \cdot (k_{+,D} \cdot P_c - k_{-,D})$$

but at steady state,

$$(k_{+,D} \cdot P_c - k_{-,D}) = -(k_{+,A} \cdot P_c - k_{-,A})$$

Thus,

$$F' = (k_{+,A} \cdot P_c - k_{-,A}) \cdot (1 - j)$$

Also,

$$(k_{+,A} \cdot P_c - k_{-,A}) = F,$$

therefore,

$$F' = F \cdot (1 - j) \tag{16}$$

or,

$$j = 1 - (F'/F) \tag{17}$$

Thus, j, the extent to which D-end loss of [³H]nucleotide introduces errors in the determinations of A- and D-end rate constants, can be measured experimentally by comparing the empirically determined steady state flux rate, F, with the steady state flux rate calculated from rate constants determined using the double-label protocol, according to equation 17. Once j has been determined, the values for the association and dissociation rate constants at each MT end can be corrected by substituting for j into equations 4, 6, 11, and 13.

For example, if incorporation of the [³H]label occurred only by the treadmilling reaction (*i.e.* all the label on the MT A ends), then j = 0 and F' = F. As treadmilling incorporation of the [³H]label becomes less significant (*i.e.* as the Wegner s parameter approaches 0), the [³H]GXP will increasingly label both A- and D-MT ends. As a result, j → 1 and both F', F → 0.

ACKNOWLEDGMENTS

We greatfully acknowledge the expert technical assistance of Mr. Herbert Miller, and thank Dr. Mary Ann Jordan, Dr. David J. Asai, Dr. Roger Leslie, Dr. John Kahonu Gathuru and Ms. Susan Overton for helpful and stimulating discussions regarding the analysis of this work.

REFERENCES

1. MARGOLIS, R. L. & L. WILSON. 1981. Nature (London) 293: 705–711.
2. FARRELL, K. W. & M. A. JORDAN. 1982. J. Biol. Chem. 257: 3131–3138.
3. JORDAN, M. A. & K. W. FARRELL. 1983. Anal. Biochem. 130: 41–53.
4. WILSON, L., K. B. SNYDER, W. C. THOMPSON & R. L. MARGOLIS. 1982. Methods Cell Biol. 24: 159–169.
5. ASNES, C. F. & L. WILSON. 1979. Anal. Biochem. 98: 64–73.
6. FARRELL, K. W. & L. WILSON. 1984. Biochemistry 23: 3741–3748.
7. MACNEAL, R. K., B. C. WEBB & D. L. PURICH. 1977. Biochem. Biophysics Res. Commun. 74: 440–447.
8. WEISENBERG, R. C., W. J. DEERY & P. J. DICKINSON. 1976. Biochemistry 15: 4248–4254.
9. MARGOLIS, R. L. & L. WILSON. 1978. Cell 13: 1–8.
10. MACCIONI, R. B. & N. W. SEEDS. 1978. Arch. Biochem. Biophys. 185: 262–271.
11. ZEEBERG, B., R. REID & M. CAPLOW. 1980. J. Biol. Chem. 255: 9891–9899.
12. MITCHISON, T. & M. KIRSCHNER. 1984. Nature (London) 312: 232–237.
13. MITCHISON, T. & M. KIRSCHNER. 1984. Nature (London) 312: 237–242.

14. FARRELL, K. W., R. H. HIMES, M. A. JORDAN & L. WILSON. 1983. J. Biol. Chem. **258:** 14148–14156.
15. CARLIER, M-F., T. L. HILL & Y-D. CHEN. 1984. Proc. Natl. Acad. Sci. USA **81:** 771–775.
16. OOSAWA, F. & S. ASAKURA. 1975. Thermodynamics of the Polymerization of Protein. Academic Press. London.
17. KARR, T. L. & D. L. PURICH. 1979. J. Biol. Chem. **254:** 10885–10889.
18. KRISTOFFERSON, D., T. L. KARR & D. L. PURICH. 1980. J. Biol. Chem. **255:** 8567–8572.
19. BERGEN, L. & G. G. BORISY. 1983. J. Biol. Chem. **258:** 4190–4194.
20. BERGEN, L. & G. G. BORISY. 1980. J. Cell Biol. **84:** 141–150.
21. HEIDEMANN, S. R. & J. R. MCINTOSH. 1980. Nature (London) **286:** 517–519.
22. EUTENEUER, U. & J. R. MCINTOSH. 1981. J. Cell Biol. **89:** 338–345.
23. HAIMO, L. T., B. R. TELZER & J. L. ROSENBAUM. 1979. Proc. Natl. Acad. Sci. USA **76:** 5759–5763.
24. STERNLICHT, H. & I. RINGEL. 1979. J. Biol. Chem. **254:** 10540–10550.
25. LAMBEIR, A. & Y. ENGELBORGHS. 1980. Eur. J. Biochem. **109:** 619–624.
26. MARGOLIS, R. L., C. RAUCH & L. WILSON. 1980. Biochemistry **19:** 5550–5557.
27. DEERY, W. J. & R. C. WEISENBERG. 1981. Biochemistry **20:** 2316–2324.
28. WENGER, A. 1976. J. Mol. Biol. **108:** 139–150.

Kinetic and Thermodynamic Aspects of Tubulin-Ligand Interactions: Binding of the Colchicine Analog 2-Methoxy-5-(2′, 3′, 4′-Trimethoxyphenyl) Tropone[a]

YVES ENGELBORGHS[b] AND THOMAS J. FITZGERALD[c]

[b]University of Leuven
Laboratory of Chemical and Biological Dynamics
Celestijnenlaan 200 D
B-3030 Leuven, Belgium

[c]The Florida Agricultural and Mechanical University
College of Pharmacy
Tallahassee, Florida 32307

INTRODUCTION

The study of the binding of colchicine and a number of analogues have already revealed many details about the peculiar binding process involved. The combined binding of podophyllotoxin and tropolone proved the existence of two partial binding sites: one for the trimethoxyphenyl (A) ring and one for the 2-methoxytropone (C) ring.[1] The A site also binds mescaline[2] independently from the occupation of the B site with tropolone methyl ether.[3] As discussed by Andreu and Timasheff,[3] bifunctional ligands that span the two subsites have a larger free energy decrease upon binding as compared to the binding of the two subligands together, because the entropy decrease, due to the loss of freedom upon binding, counts only once.

The studies of the binding kinetics of colchicine[4,5] allowed the dissection of the binding process in two parallel phases, each phase showing a nonlinear concentration dependence due to the presence of two steps: a fast initial binding and a slow conformational change of the complex. The change of the colchicine conformation in the complex was demonstrated independently by circular dichroism studies,[6] and the change of the protein conformation is evidenced by the induced GTPase activity[7,8] and the increased stability of the tubulin dimer.[9] A comparison of the thermodynamic parameters of the initial binding of colchicine[5] and of the equilibrium binding of tropolone lead Andreu and Timasheff[10] to the conclusion that colchicine binds initially by way of the tropolone ring.

The biphasic nature of the colchicine binding kinetics was interpreted as a proof for the existence of two tubulin species or two conformations in slow equilibrium.[5]

Among the different colchicine analogues, the bicyclic 2-methoxy-5-(2′,3′,4′-trimethoxyphenyl) tropone [AC] is of special interest because it lacks the central ring closure and therefore can reveal information about the role of this central part in binding and kinetics. Its efficiency as a microtubule inhibitor was shown by Fitzger-

[a]Financial support from the Belgian National Fund for Scientific Research (project FKFO 2.0074.83) and from the Research Fund of the University of Leuven (project OT/X/32) is acknowledged.

709

ald.[11] The structure of the analogue is similar to that of colchicine, as shown by Rossi *et al.*[12] Equilibrium binding studies of this analogue have been done by Andreu *et al.*[13] and by Bane *et al.*[14] The latter authors also studied the binding kinetics at low AC concentrations (up to 0.3 mM) where the observed pseudo first-order rate constant was linearly dependent on the AC concentration. Therefore a dissection of the binding in two steps was not possible. In this work we present a kinetic study of the same analogue at higher concentrations (up to 4 mM) where pronounced deviations from linearity are indeed observed. In this way, information about the two steps of the binding mechanism could be obtained. The data for the second phase are of sufficient accuracy, so that, here too, a complete analysis is possible. These results point to a qualitative similarity between the behavior of the acyclic analogue and colchicine itself, but remarkable quantitative differences are observed. The initial fast binding of AC cannot simply be identified with the binding of its tropolone ring, as was the case with colchicine itself.[5,10]

MATERIAL AND METHODS

Microtubule protein was purified from pig brain homogenates according to the method of Shelanski *et al.*[15] and modified as previously described.[16] Glycerol was added up to 1 M in the homogenization buffer and raised to 4 M prior to the first polymerization to increase its yield. The second polymerization cycle was done without glycerol. This preparation contained about 15% microtubule-associated proteins. This protein solution was stored under liquid nitrogen. Pure tubulin was separated from the microtubule associated proteins by chromatography on phosphocellulose (Whatman P11)[17] and concentrated by centrifugation on centriflo^R cones (Amicon). The tubulin $\cdot$ GTP$_2$ and free nucleotide content was determined by two-component analysis using the measured absorption at 278 and 255 nm, and the following extinction coefficients: for tubulin 1.2 (mg/ml)$^{-1}$ cm^{-1} at 278 nm;[9] and 0.65 (mg/ml)$^{-1}$ cm^{-1} at 255 nm (own calibration with Sephadex G25 purified tubulin $\cdot$ GTP$_2$ complex); and for GTP 12.17 and 7.66 mM^{-1} cm^{-1} at 255 nm and 278 nm, respectively. GTP was obtained from Boehringer. For AC, an extinction coefficient of 18.8 mM^{-1} cm^{-1} at 350 nm[14] was used that corresponded with our own determination.

AC was prepared and purified as previously described[11] and dissolved in dimethyl-sulfoxide (DMSO). In all experiments the final concentration of DMSO was adjusted to 10% before mixing.

The kinetics of the binding are measured in a stopped flow instrument, specially designed for fluorescence measurements and built in the laboratory. An Hanovia mercury-xenon 200-Watt arc lamp was used. The excitation monochromator was set at the 365 nm mercury line. The optical pathway was perpendicular to the flow direction and was 2 mm deep and 8 mm wide. Emission was collected over a wide angle, using a Kodak Wratten filter 2B (cutoff at 395 nm). The dead time of the instrument was determined with the reaction of N-bromosuccinimide with N-acetyltryptophanamide[18] and was found to be about 3 milliseconds.

All kinetic experiments were done in a buffer at pH 6.4, I = 0.1 M with the following composition: 50 mM morpholinoethane-sulfonic acid, 70 mM NaCl, 1 mM magnesium acetate, 1 mM ethylene glycol bis (β-aminoethylether)-N,N,N',N'-tetra-acetic acid, 1 mM NaN$_3$, 0.1 mM guanosine triphosphate (GTP), 5% (v/v) DMSO. Only for the preparation of microtubule protein, the previously indicated glycerol concentrations were used in the absence of DMSO.

All fittings were done using a nonlinear least squares fitting program based on the

Marquardt algorithm.[19] No corrections were made for possible dimerization of AC, such as with colchicine[20,5] because these corrections were rather small.

RESULTS

FIGURE 1 shows the time course of the appearance of fluorescence (inverted) upon the binding of AC to tubulin. The upper drawing shows the experimental data and fitted curve at 28°C and 2 mM AC, and the lower curve at 0.2 mM (same temperature). The pronounced decrease of the signal/noise ratio is largely due to the decrease of the signal as a consequence of the strong inner filter effects in the presence of the large excess of free AC molecules. The experimental curves (200 points) are

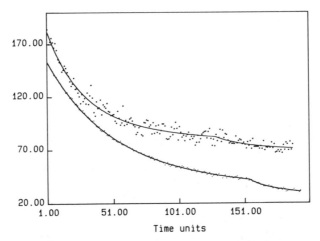

FIGURE 1. Fluorescence increase (inverted) upon the binding of AC to pure tubulin. Both curves are measured at 28°C. The experimental points and the fitted curve are displayed. Upper curve at 2 mM AC, 1 ms/time unit before and 5 ms/time unit after the inflection point. Lower curve at 0.2 mM AC, 5 ms/time unit before and 25 ms after the inflection point. Protein concentration: 1.5 mg/ml. The data are fitted for a sum of two exponential decays.

directly fitted for five parameters:

$$F = F_1 + F_2 \cdot \exp(-k \cdot t) + F_4 \cdot \exp(-k' \cdot t)$$

where F_1 is the fluorescence at equilibrium, F_2, F_4 are the amplitudes of the two phases, and k, k' are the (observed) rate constants of the two phases.

Goodness of fit was judged from the magnitude of the residual least squares sum and from the spreading of the residuals. FIGURES 2 and 3 show the concentration dependence of the observed rate constants for the two phases. Both show a very pronounced deviation from the simple pseudo first-order concentration dependence. Each curve is fitted for the following equation, previously derived for the two-step binding mechanism:

$$T + AC \underset{}{\overset{K_1}{\rightleftharpoons}} TAC \underset{k_{-2}}{\overset{k_2}{\rightleftharpoons}} TAC^*$$

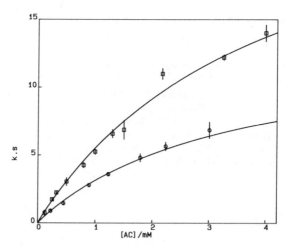

FIGURE 2. Nonlinear dependence of the observed rate constants of the first phase on the total AC concentration at 17.3°C (upper curve) and at 5°C (lower curve). Protein concentration is between 1.5 mg/ml and 2 mg/ml. Error bars are standard deviations for six experiments or more; the continuous line is the weighted least squares fit.

where K_1 is the association constant for the initial binding, and k_2, k_{-2} are the rate constants for the conformational change of the complex. The values of k_{-2} were determined by Bane et al.[14] and can be neglected in these conditions. The rate equation is therefore

$$d[TAC^*]/dt = k_2 \cdot ([T]_{tot} - [TAC^*]) \frac{K_1[AC]}{1 + K_1[AC]}$$

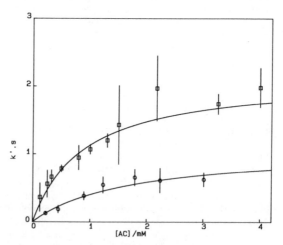

FIGURE 3. Nonlinear dependence of the observed rate constant of the second phase on the total concentration of AC at 17.3°C (upper curve) and at 5°C (lower curve). Protein concentration is between 1.5 and 2 mg/ml.

At low AC concentration this equation reduces to

$$d[TAC^*]/dt = ([T]_{tot} - [TAC^*]) \cdot k_2 \cdot K_1 \cdot [AC]$$

so that the observed rate constant depends linearly on the AC concentration as was indeed found by Bane et al.[14] The calculated bimolecular rate constant is then $k_2 \cdot K_1$, and therefore the temperature dependence gives, in fact, $E_a + \Delta H^\circ_{(1)}$, the sum of the activation energy of the conformational change and the standard enthalpy change of the initial binding. The individual parameters can thus only be obtained from the nonlinear concentration dependence at high AC concentrations, where the full equation has to be applied. The experiments were repeated at five different temperatures, which allowed the determination of activation energy for the conformational change from the Arrhenius plot and the thermodynamic parameters for the initial

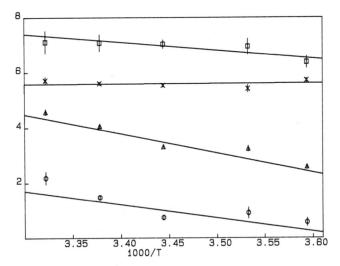

FIGURE 4. Van 't Hoff plot for the initial binding equilibrium (K_1) and Arrhenius plot for the on-rate constant (k_2) of the subsequent conformational change for the two phases. ($\square$) lnK_1 for the second phase; ($\times$) lnK_1 for the first phase; ($\triangle$) lnk_2 for the first phase; ($\bigcirc$) lnk_2 for the second phase.

binding from the van 't Hoff plot (FIGURE 4), and this for the two phases. The obtained parameters are listed in TABLES 1 and 2. For the sake of completeness the data on k_{-2} obtained by Bane et al.[14] are added.

DISCUSSION

The results of these experiments clearly show the existence of two phases and a nonlinear concentration dependence of the observed rate constants in each phase for the binding of AC to tubulin, just as in the case of colchicine binding. Attempts to fit the concentration dependence of the observed rate constants for the two phases by a sequence of two conformational equilibria, as proposed for ATP binding to myosin by Trybus and Taylor,[21] were unsuccessful. The concentration dependence of the two

TABLE 1. Thermodynamic and Kinetic Parameters of the Overall Binding Equilibrium[a]

	Andreu et al.[13]	Bane et al.[14]	This work
$K_b/M^{-1} \times 10^{-5}$	4.8 ± 0.3^b	3.5^c	2.9 ± 0.15^d
$\Delta H°/(kJ \cdot mol^{-1})$	-6.7 ± 3	-28.4	-30 ± 7
$\Delta S°/(J \cdot mol^{-1}K^{-1})$	85 ± 8	15.0	22 ± 26
$E_{a(tot)}/(kJ \cdot mol^{-1})$	—	55.5 ± 1.6	57 ± 3

[a]For the calculation of our values, the data on k_{-2} of Bane et al.[14] were used. The errors are calculated as the sum of the errors of individual contributions.
[b]25°C
[c]37°C
[d]23°C

observed rate constants, which, in this case, are the roots of a quadratic equation, could never be fitted with the same individual rate constants. It is also apparent that the concentration dependence of the second phase is much more pronounced than in the sequential case.

Using our data and the values for k_{-2} from Bane et al.[14] the overall binding constant can be calculated, as well as the thermodynamic parameters for the overall equilibrium (at least for the first phase). The equilibrium data, obtained in this way, were compared with those of other authors in TABLE 1. Similar values are found for the three data sets, although for the thermodynamic data a slightly better agreement was found between the present data and those of Bane et al.

The initial binding of the analogue is remarkably different from the behavior of colchicine (TABLE 2): whereas the initial binding of colchicine is exothermic ($\Delta H°/ kJ \cdot mol^{-1} = -33.5 \pm 8$), it is almost athermic for the first phase and endothermic for the second phase. The thermodynamic parameters for the initial binding of AC are also clearly different from those of tropolone.[3] This could mean that AC binds initially by way of its trimethoxyphenyl ring, but it cannot be excluded that already at the initial binding, both rings contribute, especially if the higher flexibility of the AC molecule is taken into account.

TABLE 2. Thermodynamic and Kinetic Parameters of the Individual Steps[a]

	AC		Colchicine
	1st phase	2nd phase	1st phase[d]
[a]K_1/M^{-1}	273 ± 13	1188 ± 200	220
$\Delta H°/(kJ \cdot mol^{-1})$	-1 ± 1.6	24 ± 7	-33 ± 12
$\Delta S°/(J \cdot mol^{-1} \cdot K^{-1})$	43 ± 7	142 ± 14	-63 ± 40
k_2/s^{-1}	58.5 ± 0.4	4.4 ± 0.4	0.3
$E_a/(kJ \cdot mol^{-1})$	58 ± 2	39 ± 4	100 ± 5
$k_{-2}/s^{-1 b,c}$	0.074		
$E_a/(kJ \cdot mol^{-1})^c$	87 ± 12		

[a]The data for k_{-2} are obtained from Bane et al.,[14] and are attributed to the first phase, as they are determined from protein fluorescence enhancement upon release of AC. Protein fluorescence quenching upon binding of AC occurs only in the first phase.
[b]23°C
[c]Data from Bane et al.[14]
[d]Data from Lambeir and Engelborghs[5]

The rate constant for the conformational change of the initial complex to the final state is much larger than for colchicine. The activation energy is also much lower. By determining the overall observed biomolecular rate constant and its activation energy, Bane *et al.* calculated the activation energy of the global process $T + AC \rightleftharpoons TAC^*$ with $E_{a,tot} = E_a + \Delta H°_{(1)}$ at least for the first phase. Due to the athermicity of the initial binding, the overall activation energy is very close to the activation energy of the conformational change (E_a) itself. This is not true for colchicine, where the initial binding is strongly exothermic. The higher activation energy for the conformational change of the initial to the final tubulin-colchicine complex can be attributed to two effects: a contribution for the deformation of colchicine in both directions, or a steric effect from the protein for the entrance to and the exit from the final state, which could be smaller for AC. In the absence of data on the conformational changes of the small molecules themselves, these two possibilities cannot be distinguished. The different activation energies for the two phases point to a different contribution from the protein. The second phase remarkably shows a higher affinity in the initial binding but a much slower rate constant for the conformational change of the complex, despite a smaller activation energy.

The presence of the two phases is still a puzzling fact, and the question arises whether the two phases are linked to the presence of the subsites. Here we describe several models where the possibility is considered that colchicine and/or AC initially bind in two different ways: one way with the tropolone first, the second way with the trimethoxyphenyl ring first, to their respective binding sites. Bane *et al.* made the interesting observation that in protein fluorescence quenching, only the first phase is observed. An explanation might be that in the second phase, the fast initial binding of the tropolone ring already results in the correct orientation for energy transfer, so that the quenching would be too fast for measuring. Assuming that both initial complexes would be able to convert into the final complex with different rate constants (k_2 and k_3), the following equations would describe such a model:

$$k_{obs} = \frac{k_2 \cdot K_1 \cdot [AC] + k_3 \cdot K_2 \cdot [AC]}{1 + K_1 \cdot [AC] + K_2 \cdot [AC]}$$

or

$$k_{obs} = \frac{k_2 \cdot K_2 + k_3 \cdot K_2}{(K_1 + K_2)} \times \frac{(K_1 + K_2) \cdot [AC]}{1 + (K_1 + K_2) \cdot [AC]}$$

It is clear that the last equation is analogous to the one used before, and therefore is able to describe the nonlinear concentration dependence, but not the presence of two exponential phases. If one of both initial complexes would not be able to convert into the final complex, this would simply mean that in the above equations, $k_3 = 0$ and would not change the conclusions. Another model could be designed where the two partial sites could be saturated independently (more likely with AC than colchicine), but where, for example only the singly saturated sites could convert. For this model the following equation can be derived:

$$k_{obs} = \frac{(k_2 \cdot K_1 + k_3 \cdot K_2) \cdot [AC]}{1 + (K_1 + K_2) \cdot [AC] + K_1 \cdot K_2^* \cdot [AC]^2}$$

where K_2^* is the binding of AC in site 2 with site 1 already occupied, which might influence the binding at the other site. Again, this model does not describe a biphasic reaction, and it predicts a decrease of k_{obs} at very high concentrations, where the term

$K_1 \cdot K_2^* \cdot [AC]^2$ becomes very large. Although this has not been observed, it would probably only occur at even higher concentrations than the ones studied here. From these considerations it is clear that all models with a fast interconversion of species can account for the nonlinear concentration dependence of the observed rate constant, but cannot explain the presence of two phases. Two possibilities remain to explain these phases: the existence of two major tubulin species, or of two conformations of tubulin that interconvert very slowly. Here again a difference is apparent between colchicine and AC. With colchicine the amplitude ratio is strongly temperature dependent; this is not the case for AC binding, where 60% (+10%) of the total fluorescence increase occurs in the first phase, almost independently from temperature. Other indications for two states in the case of the tubulin-colchicine complex come from the kinetics of the salt-induced dissociation of the tubulin-colchicine complex.[22]

CONCLUSIONS

From the combined studies, we can now make the following conclusions. 1) Both colchicine and AC bind to tubulin in two, and only two, steps. 2) The kinetics are biphasic, pointing to the presence of two tubulin species or two slowly interconverting conformers. The initial binding of AC in the fast phase shows a lower affinity than in the slow phase. The rate of the subsequent conformational change is, however, higher. 3) The thermodynamic parameters for the initial binding of AC are different from those of colchicine. We are not sure whether AC binds first through its tropolone ring. 4) The activation energy for the conformational change of the initial complex is much lower for tubulin-AC than for tubulin-colchicine, due to the higher flexibility of AC. 5) For the calculation of the binding rate constant at high colchicine or AC concentrations, as can be necessary for *in vivo* perturbation studies, the full nonlinear expression for the observed rate constant has to be used.

ACKNOWLEDGMENTS

The authors wish to thank Dr. J. Andreu (Madrid) for encouragement and for a sample of 5 mg AC, which allowed a feasibility study. Peter Verlinden and Michel Daniels provided technical assistence. Y.E. is a senior research associate of the Belgian National Fund.

REFERENCES

1. CORTESE, F., B. BHATTACHARYYA & J. WOLFF. 1977. J. Biol. Chem. **252:** 1134–1140.
2. HARRISSON, C. M. H., B. M. PAGE & H. M. KEIR. 1976. Nature (London) **260:** 138–139.
3. ANDREU, J. M. & S. TIMASHEFF. 1982. Biochemistry **21:** 534–543.
4. GARLAND, D. L. 1978. Biochemistry **17:** 4266–4272.
5. LAMBEIR, A. & Y. ENGELBORGHS. 1981. J. Biol. Chem. **256:** 3279–3282.
6. DETRICH III, H. W., R. C. WILLIAMS JR., T. L. MACDONALD, L. WILSON & D. PUETT. 1981. Biochemistry **20:** 5999–6005.
7. DAVID-PFEUTY, T., C. SIMON & D. PANTALONI. 1979. J. Biol. Chem. **254:** 11696–11702.
8. ANDREU, J. M. & S. N. TIMASHEFF. 1981. Arch. Biochem. Biophys. **211:** 151–157.
9. DETRICH III, H. W. & R. C. WILLIAMS. 1978. Biochemistry **17:** 3900–3907.
10. ANDREU, J. M. & S. N. TIMASHEFF. 1982. Biochemistry **21:** 6465–6476.
11. FITZGERALD, T. J., 1976. Biochem. Pharmacol. **25:** 1383–1387.

12. ROSSI, M., J. LINK & J. C. LEE. 1984. Arch. Biochem. Biophys. **231:** 470–476.
13. ANDREU, J. M., M. J. GORBUNOFF, J. C. LEE & S. N. TIMASHEFF. 1984. Biochemistry **23:** 1742–1752.
14. BANE, S., D. PUETT, T. L. MACDONALD & R. C. WILLIAMS JR. 1984. J. Biol. Chem. **259:** 7391–7398.
15. SHELANSKI, M. L., F. GASKIN & C. R. CANTOR. 1973. Proc. Natl. Acad. Sci. USA **70:** 765–768.
16. ENGELBORGHS, Y., L. C. M. DE MAEYER & N. OVERBERGH. 1977. FEBS Lett. **80:** 81–85.
17. WEINGARTEN, M. D., A. H. LOCKWOOD, S. -Y. HWO & M. W. KIRSCHNER. 1975. Proc. Natl. Acad. Sci. USA **72:** 1858–1862.
18. PETERMAN, B. F., 1979. Anal. Biochem. **93:** 442–444.
19. BEVINGTON, P. R. 1969. Data Reduction and Error Analyses for the Physical Sciences. McGraw-Hill. New York.
20. ENGELBORGHS, Y. 1981. J. Biol. Chem. **256:** 3276–3278.
21. TRYBUS, K. M. & E. W. TAYLOR. 1982. Biochemistry **21:** 1284–1294.
22. IDE, G. & Y. ENGELBORGHS. 1981. J. Biol. Chem. **256:** 11684–11687.

Interactions of Vinblastine and Maytansine with Tubulin[a]

RICHARD F. LUDUENA,[b] WENDY H. ANDERSON,[b]
VEENA PRASAD,[b] MARY ANN JORDAN,[c]
KATHLEEN C. FERRIGNI,[b] MARY CARMEN ROACH,[b]
PAUL M. HOROWITZ,[b] DOUGLAS B. MURPHY,[d]
AND ARLETTE FELLOUS[e]

[b]Department of Biochemistry
University of Texas Health Science Center
San Antonio, Texas 78284

[c]Department of Biological Sciences
University of California
Santa Barbara, California 93106

[d]Department of Cell Biology and Anatomy
John Hopkins University School of Medicine
Baltimore, Maryland 21205

[e]Unité de Recherche sur la Grande Thyroide
et la Regulation Hormonale
Institute National de la Santé et de la Recherche Medicale
F94279 Bicentre, France

INTRODUCTION

The *Vinca* alkaloids, vinblastine and vincristine (FIGURE 1), are 9-ringed compounds purified from the Madagascar periwinkle *Vinca rosea*.[1] They bind to tubulin with high affinity and prevent microtubule assembly.[2] Clinically, vinblastine is the drug of choice to treat Hodgkin's disease and vincristine to induce remission of acute lymphocytic leukemia.[3,4] Maytansine (FIGURE 1) is a macrocyclic ansa macrolide isolated from African plants of the genera *Maytenus* and *Putterlickia*.[5,6] It also binds tightly to tubulin and blocks microtubule assembly.[2] Although it has been found to be active against a variety of cancers, maytansine's toxicity is too high for it to be a useful therapeutic tool. The reason we are considering maytansine and the *Vinca* alkaloids together in the same article is that, despite their structural dissimilarity, they appear to bind to the same site or sites on the tubulin molecule.

Interestingly, other than the fact that they both inhibit microtubule assembly, maytansine's effects on the tubulin molecule are profoundly different from those of vinblastine.

[a]This work was supported by Grants CA26376 and GM23476 from the National Institutes of Health (to R.F.Luduena).

THE BINDING OF VINBLASTINE AND MAYTANSINE TO TUBULIN

Vinblastine binding to a variety of tubulins has been measured in different ways (TABLE 1). The reported dissociation constants range from 0.16 μM to 45 μM, a difference of more than two orders of magnitude. Perhaps these differences may be explained in part by the observation of Wilson *et al.*[7] that freezing tubulin reduced its

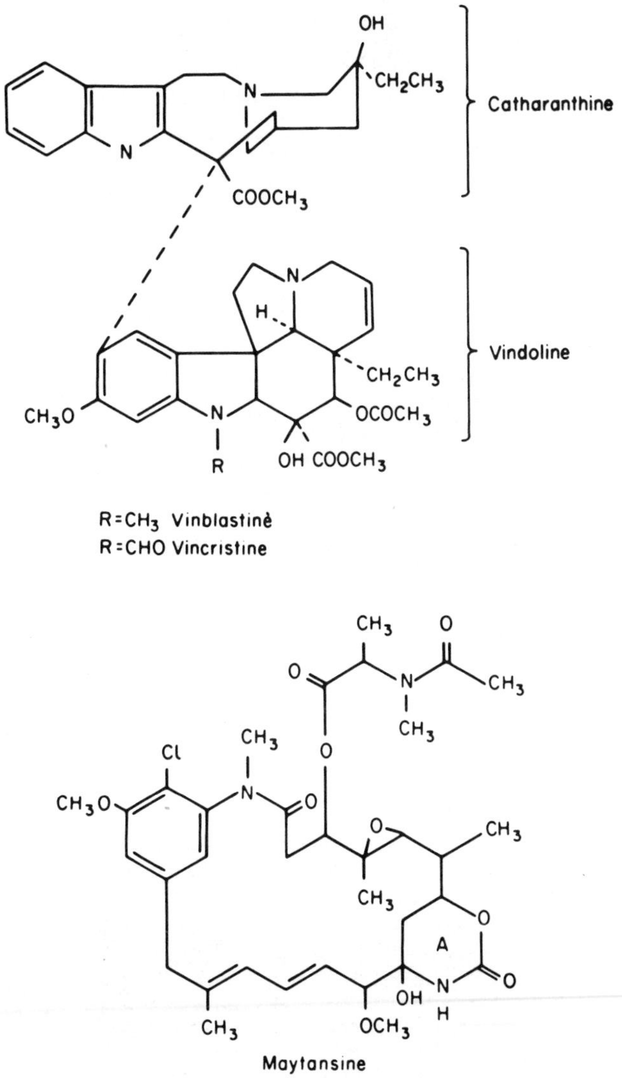

FIGURE 1. Structures of the *Vinca* alkaloids (left) and maytansine (right).

TABLE 1. Vinblastine Binding to Tubulin as Reported in the Literature

Source of Tubulin	Methodology	Number of Binding Sites	K_d (μM)	References
Chick Brain	Sephadex gel, E[a], S[b]	1.8	2.2	Wilson et al.[7]
	Gel filtration, E	1.9	ND[c]	Wilson et al.[7]
Pig Brain	Filter disc, NE[d], S	0.42	0.19	Owellen et al.[8]
Rat Brain	Filter disc, NE, S	0.74	0.16	Bhattacharyya and Wolff[9]
Calf Brain	Fluorescence, E, CV[e]	2	42	Lee et al.[10]
	Gel filtration, E, S	1.8	45	Lee et al.[10]
Sea urchin sperm	Filter disc, NE	ND	10	Wilson et al.[7]
	Gel filtration, E	2.2–2.3	ND	Wilson et al.[7]
Sea urchin egg	Sedimentation, E, S	2.3	4.2	Wilson et al.[11]

[a]E = equilibrium technique.
[b]S = Scatchard; number of sites and binding affinity determined by Scatchard analysis.[13]
[c]ND = Not determined.
[d]NE = Nonequilibrium technique.
[e]CV = Continuous variation; number of sites determined by continuous variation method of Job.[15]

affinity for vinblastine by a factor of about twenty. Perhaps the vinblastine binding sites on tubulin decay in an incremental manner, gradually losing their affinity for vinblastine, in contrast to the colchicine binding site that decays in an all-or-none fashion.[2]

As shown in TABLE 1, estimates of the number of high-affinity vinblastine binding sites on tubulin have varied from 0.42 to 2.3. In evaluating these data, however, it is important to remember that, with a ligand such as vinblastine, which binds reversibly to tubulin, the stoichiometry of binding can only be measured accurately by methods in which the binding is in equilibrium. The filter disc method is not one of these, but the others mentioned in TABLE 1 are. When only results obtained by equilibrium methods are considered, the stoichiometry of binding ranges from 1.8 to 2.3, suggesting the presence of two vinblastine binding sites on the tubulin molecule.

We should stress, however, that accurate measurement of the number of vinblastine binding sites is technically difficult. Even certain equilibrium techniques may give potentially misleading results. For example, Wilson et al.[7] obtained stoichiometries of 1.9 binding sites in chick brain tubulin and 2.2–2.3 in sea urchin sperm tubulin using equilibrium gel filtration.[12] Only single concentrations of [3H]vinblastine, however, were used in each experiment. Consequently the binding affinities could not be simultaneously determined. It is theoretically possible, therefore, that the known low-affinity binding sites could have contributed to the measured stoichiometry. One can be much more confident about binding stoichiometries if they are determined by testing a series of vinblastine concentrations and the resulting data subjected to Scatchard analysis.[13] This type of analysis was employed on data obtained by equilibrium gel filtration by Lee et al.[10] when they found 1.8 binding sites for vinblastine in calf brain tubulin and by Wilson et al.[11] to find 2.3 binding sites on sea urchin egg tubulin. The elegant data in these two reports fit the Scatchard plot extremely closely and probably constitute the strongest evidence for two binding sites. Using a different technique, the equilibrium Sephadex gel method,[14] Wilson et al.[7] found 1.8 sites in chick brain tubulin by Scatchard analysis. The scatter of the data, however, was such that some of the points could have fit a bimodal curve, suggesting a more complex model consisting of one binding site of moderately high affinity and one of very high affinity.

A very different experimental approach is based on the fact that vinblastine inhibits the intrinsic fluorescence of tubulin.[10] This permits the measurement of the binding stoichiometry by the continuous variation method of Job[15] in which the fluorescence of a series of tubulin concentrations is measured. Enough vinblastine is then added to each sample to give a total tubulin plus vinblastine concentration of 5 μM, and the fluorescence is measured again. After making appropriate corrections for internal filter effects, the change in fluorescence is plotted as a function of the mole fraction of ligand. The result is a rising and falling curve whose peak should be at a value for mole fraction of ligand corresponding to the maximal binding. Thus, for Lee et al.[10] the curves peaked at a value of 0.67, suggesting two binding sites. In our hands, using bovine brain tubulin, the intersection was at 0.4 (FIGURE 2), implying one binding site. In our experiment, the tubulin was prepared by a cycle of assembly followed by phosphocellulose chromatography,[16] a procedure not used in preparing any of the tubulins whose binding to vinblastine was measured in the reports summarized in TABLE 1. Conceivably, this isolation procedure could have permitted or induced the decay of one high-affinity binding site and not the other. It is interesting that when Prakash and Timasheff used the continuous variation method to measure the binding of vinblastine's analogue, vincristine, to tubulin, their data could not distinguish between a one-site and a two-site model.[17] Vinblastine and vincristine are so similar (FIGURE 1), that it is highly unlikely that a binding site for one would not also bind the

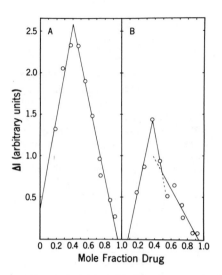

FIGURE 2. Continuous variation analysis of the binding of maytansine (A) and vinblastine (B) to tubulin. Microtubule protein was prepared from bovine cerebra by the method of Fellous *et al.*[16] Tubulin was purified by chromatography on phosphocellulose using the buffer of Fellous *et al.*[16] Protein was determined by amino acid analysis on a Durrum D-500 Amino Acid Analyzer using norleucine as an internal standard. To each of a series of 2.0 ml tubulin solutions were added microliter amounts of either maytansine (A) or vinblastine (B) such that the total of the tubulin and drug concentrations was about 5 μM. Samples were placed in fluorescence cells in a Perkin-Elmer Model 44 Spectrophotofluorimeter and excited at 299 nm. The emission at 334 nm was determined before and after drug addition. The quenching due to addition of the drug was corrected for dilution and internal absorbtion.[10] The quenching is plotted as a function of the mole fraction of drug in the drug-tubulin mixture.

other. Thus, if there are two binding sites for vinblastine on the tubulin molecule, there should also be two for vincristine. The results of Prakash and Timasheff, therefore, underline the difficulties of accurately determining the stoichiometry of vinblastine binding to tubulin.

In summary, the preponderance of the evidence is that there are two high-affinity binding sites for vinblastine on the tubulin molecule. Because tubulin consists of two similar subunits, it is not surprising that there should be two sites. The fact that the affinities of the two sites should apparently be identical suggests that the portions of the tubulin molecule that constitute the binding sites are regions of high homology between the α and β subunits. In addition to the high-affinity sites, vinblastine binds to a number of low-affinity sites on the tubulin molecule.[11,18] The binding has a strong ionic component[9] and could arise from a nonspecific interaction between vinblastine, which is a base, and tubulin, which is an acidic protein.

Maytansine binding to tubulin has been measured directly using [³H]maytansine and also by its suppression of tubulin's intrinsic fluorescence (FIGURE 3). The dissociation constants estimated by these two approaches are, respectively, 0.67[19] and 0.9–6 μM. Maytansine will competitively inhibit the binding to tubulin of either [³H]vinblastine[20] or [³H]vincristine[19] with K_i's of 0.5 and 0.4 μM, respectively. Also, the binding of maytansine is competitively inhibited by vincristine.[19] These results indicate that maytansine must occupy at least one of vinblastine's binding sites on the tubulin molecule. The number of maytansine binding sites, however, is not clear. In the two studies where maytansine was found to competitively inhibit the binding of *Vinca* alkaloids to tubulin,[19,20] the molar ratios of the alkaloids to tubulin was very low and it may have been difficult to detect the presence of a second vinblastine site. Interestingly, when the continuous variation method is applied to maytansine's binding to tubulin, a stoichiometry of one binding site is obtained (FIGURE 2). It is not impossible that maytansine may have only one high-affinity binding site on the tubulin molecule, whereas vinblastine may have two identical sites. As we shall argue shortly, maytansine is likely to have a binding site that overlaps that of vinblastine. Thus, even if each subunit of tubulin has identical vinblastine binding sites, the portion of the maytansine binding site where vinblastine does not bind could be present only in one subunit and be replaced in the other subunit by a region to which maytansine cannot bind. On the other hand, we should remember that if maytansine binds at a second site without affecting the intrinsic fluorescence of tubulin, this binding would be undetectable in an experiment such as that shown in FIGURE 2. Similarly, if one vinblastine site has vanished during the isolation procedure, as suggested above, it is quite possible that the missing vinblastine site could also have bound to maytansine.

THE EFFECTS OF VINBLASTINE AND MAYTANSINE ON THE TUBULIN MOLECULE

The effects of vinblastine on the tubulin molecule have been studied using a variety of approaches. Vinblastine causes changes in the ultraviolet spectrum of tubulin suggesting the movement of aromatic chromophores to a less polar environment.[10]

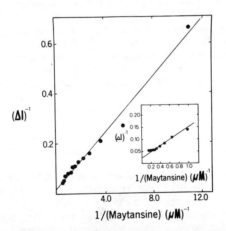

FIGURE 3. Double reciprocal plot of the effect of maytansine on the intrinsic fluorescence of tubulin. To 2.2 ml of tubulin (0.35 mg/ml) in a fluorescence cell at 25°C were added microliter amounts of 0.2 m*M* maytansine. Excitation was at 299 nm. After each addition, the emission at 334 nm was measured. Inset: To 2.0 ml of tubulin (0.39 mg/ml) in a fluorescence cell at 15°C were added microliter amounts of 1.03 m*M* maytansine. Excitation and emission were at 299 nm and 334 nm, respectively. The graphs show the reciprocal of the change in intensity of emission as a function of the maytansine concentration.

Other than this, vinblastine's effects seem to stabilize tubulin's conformations against decay rather than to alter it.

For example, vinblastine will not change tubulin's circular dichroism spectrum but will prevent spontaneous changes in it.[10,21] Also, vinblastine stabilizes the binding of guanosine triphosphate (GTP) and colchicine to tubulin[22-24] and stabilizes the $\alpha\beta$ heterodimer for intersubunit cross-linking by dimethyl-3,3'-(tetramethylenedioxy) dipropionimidate.[25] Vinblastine strongly inhibits the alkylation of both α and β subunits by iodo[14C]acetamide and N-ethyl[14C]maleimide, suggesting that certain sulfhydryl groups in tubulin become less accessible in the presence of vinblastine.[26] Vinblastine also inhibits nonspecific intermolecular cross-link formation by N,N'-ethylene-bis(iodoacetamide) (EBI).[26]

EBI is a potentially useful probe for specific effects of tubulin ligands because it can form two intrachain cross-links in the β subunit of tubulin.[27,28] The formation of one of these, designated the β^* cross-link, is suppressed by colchicine, podophyllotoxin, and nocodazole, but is enhanced by vinblastine.[26] This enhancement need not be a conformational effect because it could be due to vinblastine's inhibition of competing nonspecific cross-linking reactions. EBI also forms another cross-link, designated β^s, in a reaction that is slowed down, by about 70%, by vinblastine.[28] It is tempting to speculate that at least one of the sulfhydryls involved in the β^s cross-link is located close to the vinblastine binding site.

The effects of maytansine on the tubulin molecule have not been studied as extensively as those of vinblastine, but the effects that have been observed are almost all different from vinblastine's. Maytansine does not stabilize colchicine binding[29] and, in the presence of either GTP or podophyllotoxin, has no effect on the alkylation of tubulin by iodo[14C]acetamide.[30] In the absence of any other ligand, maytansine inhibits alkylation by about 10 percent.[28] This is very different from vinblastine, which always inhibits alkylation of tubulin, even in the presence of GTP, by as much as 66 percent.[26] In fact, when the two drugs are tested together, maytansine abolishes vinblastine's inhibitory effect on allylation.[30] Again, in contrast with vinblastine, maytansine, in the presence of either GTP or podophyllotoxin, has no effect on the formation of nonspecific intermolecular cross-links by EBI,[30] although in the absence of other ligands, maytansine inhibits nonspecific cross-link formation almost as well as does vinblastine.[28] A simple model to account for these different effects is that vinblastine affects a large area on the tubulin molecule, which includes the colchicine binding site and numerous sulfhydryls; maytansine, on the other hand, would affect only a small area, including very few sulfhydryls. These areas would be small enough to be contained within the areas affected by GTP and podophyllotoxin; this would explain why the small inhibitory effects of maytansine on alkylation by iodoacetamide and on nonspecific cross-linking by EBI disappear in the presence of GTP and podophyllotoxin.

When the effects of maytansine and vinblastine on the intrachain cross-linking of tubulin by EBI are compared, one similarity and one difference are observed. Maytansine enhances β^* formation to exactly the same extent as does vinblastine.[30] Because maytansine, unlike vinblastine, does not inhibit any competing reaction, its enhancement of β^* formation is likely to be due to a maytansine-induced conformational change. In contrast to the formation of β^*, that of β^s is completely inhibited by maytansine, not just slowed down, as is the case in the presence of vinblastine.[28] Several models are possible, but one very simple one is that maytansine and vinblastine have overlapping binding sites and that one of the β^s sulfhydryls is located in that portion of the maytansine site to which vinblastine does not bind, but close enough to the vinblastine binding site that its rate of alkylation can be diminished by vinblastine.

EFFECTS OF VINBLASTINE AND MAYTANSINE ON TUBULIN AGGREGATION

The most striking difference between vinblastine and maytansine is their effect on tubulin polymerization. Although both drugs are potent inhibitors of microtubule assembly, vinblastine induces an alternative form of tubulin polymerization, whereas maytansine does not.[2] Vinblastine can cause pure tubulin to aggregate into crystals;[31] similar crystals are seen in cells treated with vinblastine or vincristine.[32-35] The crystals are made of a double-helical "macrotubule," which consists of two spirals about 180–200 A apart.[35-37] Each spiral consists of a filament of about the same diameter as the microtubule protofilament and is apparently made of tubulin molecules joined longitudinally.[36,37] In certain preparations of tubulin, vinblastine will induce formation of loose aggregates of spirals rather than crystals.[35,36,38-40] These spirals are morphologically similar to those seen in crystals.[41] In addition, vinblastine will cause tubulin to form rings and semicircles whose dimensions suggest that they are abortive spirals.[41] Vinblastine-induced polymerization of tubulin is strongly stimulated by the presence of microtubule-associated proteins (MAPs) that appear to lower the critical tubulin concentration for assembly, exactly as they do in normal microtubule assembly.[40-42] Interestingly, individual MAPs have strikingly different effects on vinblastine-induced polymerization; in the presence of tau, vinblastine causes tubulin to polymerize into clusters of long spirals with high turbidity, whereas microtubule-associated protein 2 (MAP-2) causes formation of isolated short spirals of low turbidity.[40,41] Vinblastine-induced tubulin polymerization appears to be a universal phenomenon occurring with tubulins prepared from sources as disparate as chordate brains, sea urchin eggs, and avian erythrocytes.[32-43] The effect may be most spectacular in erythrocyte tubulin, where addition of 20 μM vinblastine to a solution of 10 μM tubulin can precipitate 96% of the tubulin.[43]

The molecular explanation for vinblastine-induced tubulin polymerization is potentially complicated because vinblastine has at least one, and probably two, high-affinity and several low-affinity binding sites on the tubulin molecule. It is therefore not immediately obvious which site or sites mediate this aggregation. Based on the fact that the concentration of vinblastine required for aggregation is much higher than that required to inhibit normal microtubule assembly, Bhattacharyya and Wolff have proposed that assembly inhibition is mediated by the high-affinity and aggregation by the low-affinity sites.[9] We must remember, however, that inhibition of microtubule assembly by colchicine, podophyllotoxin, and vinblastine requires only that a few drug molecules bind to the end of a microtubule.[44-46] By contrast, vinblastine-induced aggregation might require that each molecule of tubulin in the aggregate be bound to the drug. Such a model would explain why induction of aggregation needs a higher concentration of vinblastine than does inhibition of microtubule assembly. Perhaps the most telling evidence is that when sea urchin egg tubulin is induced to form crystals by addition of [³H]vinblastine, the molar ratio of vinblastine to tubulin is exactly 1:1.[47] We have observed the same ratio in spiral polymers of tau, tubulin, and [³H]vinblastine.[48] This strongly suggests that vinblastine-induced aggregation involves only one high-affinity site. It is certainly conceivable, however, that binding of vinblastine to the low-affinity sites could stimulate further aggregation, if only by neutralizing the mutually repulsive negative changes on the tubulin molecule.

A simple hypothesis to summarize the foregoing argument is as follows. When vinblastine binds to the high-affinity sites, it induces or stabilizes in tubulin a conformation that is only partly compatible with normal microtubule assembly. Thus, a tubulin-vinblastine complex is sufficiently similar to an unliganded tubulin molecule that it can bind to a microtubule end, but it is sufficiently dissimilar that an unliganded tubulin molecule cannot bind to it. In other words, if we imagine that an unliganded tubulin molecule has two lateral and two longitudinal sites where it can interact with other tubulin molecules, perhaps vinblastine alters a lateral and a longitudinal site permitting addition of the complex to the microtubule but inhibiting further addition of unliganded tubulin. If we now assume that the vinblastine-induced alteration in these tubulin-tubulin interaction sites forbids only normal tubulin-tubulin interactions but permits abnormal ones, with other tubulin-vinblastine complexes, we have the following situation. At low vinblastine concentrations, microtubule assembly is inhibited, but the ratio of liganded to unliganded tubulin is too low to permit aggregation of tubulin-vinblastine complexes. At high vinblastine concentrations, the ratio of liganded to unliganded tubulin is high enough to permit aggregation of these complexes.

The contrast of vinblastine with maytansine is a substantial one. Even at millimolar concentrations, and in the presence of MAPs, maytansine does not induce tubulin aggregation.[29] In fact it is a potent inhibitor of this aggregation.[49] Thus, 0.5 μM maytansine inhibits by 97% polymerization of a mixture of 2.5 μM tau, 10 μM tubulin, and 39 μM vinblastine. These numbers raise the possibility that maytansine is not acting by displacing bound vinblastine, but by "poisoning" substoichiometrically the formation of vinblastine-induced spirals. Conceivably, a maytansine-tubulin complex could add on to the end of a spiral and inhibit further growth.

DRUGS THAT INHIBIT MICROTUBULE ASSEMBLY—A HYPOTHESIS

A variety of drugs are now known that inhibit microtubule assembly *in vitro* and *in vivo*. Recent discoveries suggest, however, that some of these drugs do not inhibit tubulin polymerization per se, but rather channel it in a different direction, the end-product being something other than a microtubule. It is now clear that, under the appropriate conditions, these drugs can permit or even induce formation of aberrant tubulin polymers. Vinblastine's ability to generate tubulin spirals under a wide range of conditions have already been discussed. We now have evidence that colchicine, podophyllotoxin, and nocodazole also allow formation of aberrant tubulin polymers under certain conditions.

Colchicine is a potent inhibitor of microtubule assembly, but it has recently been reported that the purified tubulin-colchicine complex can polymerize to give ribbon- or sheet-like structures under the same conditions used for normal microtubule assembly.[50-52] The thermodynamic properties and the effects of calcium and magnesium are similar to those found in normal microtubule assembly.[50-52] We have recently found that, in the presence of MAP-2 that has been cleaved with chymotrypsin, colchicine will induce formation of a large amount of an amorphous polymer (FIGURE 4). The polymerization proceeds at 37°C and also at 0°C and is dependent on the colchicine concentration.[53] Besides inducing formation of tubulin-colchicine polymers, colchicine will permit formation of the tubulin-vinblastine spirals and, in fact, will bind tightly to them.[47,54]

In contrast to colchicine, podophyllotoxin does not appear to induce formation of any tubulin polymer. In fact, it will inhibit both types of colchicine-induced polymerization described above.[53,55] Like colchicine, however, podophyllotoxin permits formation of the tubulin-vinblastine complex and binds tightly to it as well, suggesting that the binding of a tubulin molecule to podophyllotoxin is not incompatible with that tubulin molecule forming part of a polymer.

Nocodazole is similar to podophyllotoxin in that it inhibits the colchicine-induced tubulin polymerization mediated by chymotrypsin-treated MAP-2.[53] Also, like podophyllotoxin, nocodazole does not inhibit vinblastine-induced tubulin polymerization. The most surprising finding about the interaction of nocodazole with tubulin is that it can induce massive polymerization of erythrocyte tubulin (TABLE 2). In one experiment, for example, treatment of 10 μM erythrocyte tubulin with 20 μM nocodazole precipitated 93% of the tubulin. The nocodazole-induced aggregate appears to be amorphous when examined by electron microscopy.

Maytansine presents a strong contrast to the other drugs. Not only does it not

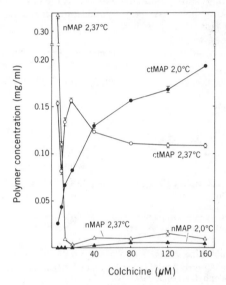

FIGURE 4. Effect of colchicine concentration on the formation of tubulin aggregates in the presence of normal and chymotrypsin-treated MAP-2. Aliquots (250 ml) of tubulin (1.0 mg/ml) were incubated for 30 min either at 0°C (0, Δ) or at 37°C (0, Δ) in the presence of either normal MAP-2 (Δ, Δ) or chymotrypsin-treated MAP-2 (0, 0), and in the presence of the indicated concentrations of colchicine, MAP-2 concentration was 0.30 mg/ml. Samples were centrifuged at 39,000 × g for 40 min at either 4°C (0, Δ) or 30°C (0, Δ). Protein was determined in the pellets by the method of Lowry *et al.*[57] The chymotrypsin-treated MAP-2 was prepared by incubating the MAP-2 for 8 min at 37°C with 0.7 mg/ml chymotrypsin. Proteolysis was stopped by addition of 2 mM phenylmethane-sulfonyl fluoride. Electrophoretic analysis and reconstitution with microtubules indicated that the promoter domain was still functional.

induce any type of aggregate formation, but unlike podophyllotoxin, where maytansine has been tested, it potently prevents formation of all kinds of vinblastine-induced tubulin polymers (TABLE 3). Also noteworthy is its interaction with erythrocyte tubulin. This is an unusual isotype of tubulin having what appears to be an inherently greater ability to self-assemble than does the better-studied brain tubulin. As we have seen, erythrocyte tubulin undergoes massive aggregation in the presence of vinblastine and even nocodazole. Also, erythrocyte tubulin polymerizes at 0°C to give disc-like structures.[56] As shown in TABLE 2, maytansine inhibits the 0°C polymerization of erythrocyte tubulin, but podophyllotoxin has a much smaller effect.

TABLE 2. Effect of Drugs on the Polymerization of Brain and Erythrocyte Tubulins[a]

Experiment 1: Erythrocyte Microtubule Protein, 1.5 mg/ml

Addition	Temperature	Polymer Concentrations (mg/ml)	Polymerization (Percent of Control)
None[b]	0°C	0.32	100
Maytansine, 20 μM[b]	0°C	0.06	17
Podophyllotoxin, 20 μM[b]	0°C	0.24	75
None[c]	37°C	0.97	100
Maytansine, 20 μM[c]	37°C	0.27	28
Podophyllotoxin, 20 μM[c]	37°C	0.95	97

Experiment 2: Erythrocyte Tubulin, 1.0 mg/ml, 37°C[d]

Addition	Polymer Concentration (mg/ml)	Polymerization (Percent of Control)
None	0.11	100
Nocodazole, 20 μM	0.93	824
Podophyllotoxin, 20 μM	0.07	65
Maytansine, 20 μM	0.05	40

Experiment 3: Microtubule Protein, 1.0 mg/ml, 37°C[e]

Source of Protein	Addition	Polymer Concentrations (mg/ml)	Polymerization (Percent of Control)
Erythrocyte	None	0.45	100
	Vinblastine, 20 μM	0.90	201
	Nocodazole, 20 μM	0.82	182
	Podophyllotoxin, 20 μM	0.29	65
Brain	None	0.27	100
	Vinblastine, 20 μM	0.17	62
	Nocodazole, 20 μM	0.15	56
	Podophyllotoxin, 20 μM	0.13	49

[a]All experiments were done on tubulin or microtubule protein fractions from chickens. The buffer used in every case was 0.1 M piperazinediethanesulfonic acid (PIPES), pH 6.94, 1 mM GTP, and 0.1 mM MgCl$_2$. Samples of microtubule protein were clarified by centrifugation prior to the experiment. Pure tubulin was prepared by running microtubule protein on phosphocellulose. Polymers were collected by centrifugation.

[b]Samples incubated at 0°C for 2 hours and polymers collected by centrifugation at 4°C for 40 min at 27,000 × g.

[c]Samples incubated 30 min at 37°C and then 30 min at 30°C. Polymers were collected by centrifugation at 30°C for 40 min at 27,000 × g.

[d]Samples incubated 30 min at 37°C. Polymers collected by centrifugation at 30°C for 40 min at 39,000 × g.

[e]Samples incubated 45 min at 37°C. Polymers collected by centrifugation at 30°C for 40 min at 39,000 × g.

We can perhaps summarize all of these data in the following hypothesis. It is likely that the tubulin molecule, when free in solution, has a different conformation than when it forms part of a microtubule. In the latter case, its conformation is one where its four tubulin-tubulin interaction sites are oriented so that the tubulin can interact with other tubulin molecules to create the microtubule. When drugs such as colchicine, podophyllotoxin, nocodazole, or vinblastine bind to tubulin they induce or stabilize in the tubulin molecule a conformation that is similar and perhaps, in some cases,

TABLE 3. Effect of Maytansine on the Polymerization of Tubulin in the Presence of Vinblastine and Colchicine[a]

Temperature	MAP	Drug	Maytansine (μM)	Polymer Concentration (mg/ml)	Percent of Control
37°C[b]	Tau[c]	Vinblastine, 40 μM	0	0.83	100
			20	0.13	15
	nMAP-2[d]	Vinblastine, 40 μM	0	0.60	100
			20	0.02	3
0°C[b]	Tau[c]	Vinblastine, 40 μM	0	0.80	100
			20	0.02	3
	nMAP-2[d]	Vinblastine, 40 μM	0	0.13	100
			20	0.05	40
37°C	ctMAP-2[e]	Colchicine, 100 μM	0	0.11	100
			50	0	0

[a] Aliquots (250 ml) of phosphocellulose-purified bovine brain tubulin (1.0 mg/ml) were incubated at the indicated temperatures for 30 min in the presence of the indicated drugs and MAPs. Samples were centrifuged and protein determined as described in the legend to FIGURE 4.
[b] All incubations were done in duplicate.
[c] Concentration was 0.15 mg/ml.
[d] Concentration was 0.30 mg/ml. nMAP-2 = undigested MAP-2.
[e] ctMAP-2 = MAP-2 digested with chymotrypsin.

identical to its conformation in the intact microtubule. Therefore, at least some of the tubulin-tubulin interaction sites are unaffected by the ligand, and the tubulin-ligand complex can easily add on to a growing microtubule. The presence of the bound ligand, however, inhibits further addition of unliganded tubulin, either because the ligand itself sterically impedes access to a tubulin-tubulin interaction site or because the ligand-induced conformational change has altered the nature or location of some of these interaction sites, so that a tubulin molecule cannot add on to a tubulin-ligand complex and still create a microtubule. Thus, these drugs can poison microtubule assembly.

Let us assume, however, that the tubulin-tubulin interaction sites, though shifted in position or partly covered by the ligand, are still available on the surface of the tubulin molecule. This molecule, therefore, under the appropriate conditions, could interact at these sites with other liganded or unliganded tubulin molecules to form polymers. These polymers cannot be microtubules, however, either because the tubulin molecules that form them are binding to each other at angles incompatible with being in a microtubule or because one or two of the normal tubulin-tubulin interactions cannot occur in the presence of ligand. It is difficult to predict the morphology of such polymers. It is reasonable to speculate that if the altered tubulin-tubulin interactions are still tight and highly specific, the resulting polymer would be a regular one as is the vinblastine-induced spiral. If the interactions are weak and less specific, the resulting polymer will be amorphous, as are the colchicine- and nocodazole-induced polymers. It is also difficult to predict the precise conditions that could lead to polymerization of tubulin-ligand complexes. As we have seen, polymerization of colchicine-tubulin complexes is facilitated by an artificially altered MAP-2, whereas unaltered MAP-2 has no effect. Likewise, nocodazole-tubulin complex polymerization is dependent upon the presence of an unusual tubulin, namely, that of the erythrocyte. In summary, it appears that for some drugs, substoichiometric inhibition of microtubule assembly and induction of nonmicrotubule polymers are two aspects of the same phenomenon. Binding of a ligand will prevent assembly by microtubules but permit formation of alternative structures.

Maytansine is apparently an exception to this rule, because it does not induce the formation of any kind of tubulin polymer and, unlike podophyllotoxin, strongly inhibits formation of the tubulin-vinblastine polymer and the cold-stable erythrocyte tubulin discs. Its effects on the tubulin molecule, as far as we can tell, appear to be quite different from those of vinblastine. It may be that maytansine induces a conformational change in the tubulin molecule that is different from those that occur in the presence of the other drugs; perhaps this conformational change is one that abolishes most of the tubulin-tubulin interactions. Conversely, perhaps maytansine prevents whatever conformational change occurs in the tubulin molecule as part of the assembly pathway. In either case, a tubulin molecule bound to maytansine could simply be unable to polymerize, no matter what the conditions are. It is, however, conceivable that a tubulin-maytansine complex could itself form polymers under conditions that have not yet been tested. It will be of interest to examine the effects of maytansine on the conformation of tubulin and also its effects on treadmilling. The data presently available, however, suggest that maytansine is closer than any of the other antitubulin drugs to being a pure inhibitor of tubulin polymerization.

ACKNOWLEDGMENT

We gratefully acknowledge the skillful technical assistance of Phyllis T. Smith.

REFERENCES

1. GOODMAN, L. S. & A. GILMAN. 1975. The Pharmacological Basis of Therapeutics. Macmillan Publishing Co. New York.
2. LUDUENA, R. F. 1979. *In* Microtubules. K. Roberts & J. S. Hyams, Eds.: 65–116. Academic Press. London.
3. AMIN AUR, R. J. 1974. *In* Advances in Acute Leukemia. F. J. Cleton, D. Crowther & J. S. Malpas, Eds.: 95–114. American Elsevier Publishing Co., New York.
4. ULTMANN, J. E. & D. D. NIXON. 1969. *In* Leukemia and Lymphoma. J. F. Holland, P. A. Miescher & E. R. Jaffe, Eds.: 152–179. Grune and Stratton. New York.
5. KUPCHAN, S. M., Y. KOMODA, W. A. COURT, G. J. THOMAS, R. M. SMITH, A. KARIM, C. J. GILMORE, R. C. HALTIWANGER & R. F. BRYAN. 1972. J. Am. Chem. Soc. **94:** 1354–1356.
6. KUPCHAN, S. M., Y. KOMODA, A. R. BRANFMAN, A. T. SNEDEN, W. A. COURT, G. J. THOMAS, H. P. H. HINTZ, R. M. SMITH, A. KARIM, G. A. HOWIE, A. K. VERMA, Y. NAGAO, R. G. DAILEY, V. A. ZIMMERLY & W. C. SUMMER. 1977. J. Org. Chem. **42:** 2349–2357.
7. WILSON, L., K. M. CRESWELL & D. CHIN. 1975. Biochemistry **14:** 5586–5592.
8. OWELLEN, R. J., D. W. DONIGIAN, C. A. HARTKE, R. M. DICKERSON & M. J. KUHAR. 1974. Cancer Res. **34:** 3180–3186.
9. BHATTACHARYYA, B. & J. WOLFF. 1976. Proc. Natl. Acad. Sci. USA **73:** 2375–2378.
10. LEE, J. C., D. HARRISON & S. N. TIMASHEFF. 1975. J. Biol. Chem. **250:** 9276–9282.
11. WILSON, L., A. N. C. MORSE & J. BRYAN. 1978. J. Mol. Biol. **121:** 225–268.
12. HUMMEL, J. P. & W. J. DREYER. 1962. Biochim. Biophys. Acta **63:** 530–537.
13. SCATCHARD, G. 1949. Ann. N.Y. Acad. Sci. **51:** 660–672.
14. HIROSE, M. & Y. KANO. 1971. Biochim. Biophys. Acta **251:** 376–379.
15. JOB, P. 1928. Ann. Chem. (Paris) **9:** 113–203.
16. FELLOUS, A., J. FRANCON, A. M. LENNON & J. NUNEZ. 1977. Eur. J. Biochem. **78:** 167–174.
17. PRAKASH, V. & S. N. TIMASHEFF. 1983. J. Biol. Chem. **258:** 1689–1697.
18. WILSON, L. 1975. Ann. N.Y. Acad. Sci. **253:** 213–231.
19. MANDELBAUM-SHAVIT, F., M. K. WOLPERT-DE FILIPPES & D. G. JOHNS. 1976. Biochem. Biophysics Res. Commun. **72:** 47–54.
20. BHATTACHARYYA, B. & J. WOLFF. 1977. FEBS Lett. **75:** 159–162.
21. VENTILLA, M., C. R. CANTOR & M. SHELANSKI. 1975. Biochemistry **11:** 1554–1561.
22. TAN, L. P. & J. R. LAGNADO. 1975. Biochem. Soc. Trans. **3:** 121–124.
23. GARLAND, D. & D. C. TELLER. 1975. Ann. N.Y. Acad. Sci. **253:** 232–238.
24. WILSON, L. 1970. Biochemistry **9:** 4999–5007.
25. LUDUENA, R. F., E. M. SHOOTER & L. WILSON. 1977. J. Biol. Chem. **252:** 7006–7014.
26. LUDUENA, R. F. & M. C. ROACH. 1981. Biochemistry **20:** 4444–4450.
27. LUDUENA, R. F. & M. C. ROACH. 1981. Biochemistry **20:** 4437–4444.
28. ROACH, M. C. & R. F. LUDUENA. 1984. J. Biol. Chem. **259:** 12063–12071.
29. BHATTACHARYYA, B. & J. WOLFF. 1977. FEBS Lett. **75:** 159–162.
30. LUDUENA, R. F. & M. C. ROACH. 1981. Arch. Biochem. Biophys. **210:** 498–504.
31. NA, G. C. & S. N. TIMASHEFF. 1982. J. Biol. Chem. **257:** 10387–10391.
32. BENSCH, K. G. & S. E. MALAWISTA. 1969. J. Cell Biol. **40:** 95–107.
33. BENSCH, K. G., R. MARANTZ, H. WISNIEWSKI & M. SHELANSKI. 1969. Science **165:** 495–496.
34. BRYAN, J. 1971. Exp. Cell. Res. **66:** 129–136.
35. MARANTZ, R. & M. SHELANSKI. 1970. J. Cell Biol. **44:** 234–238.
36. ERICKSON, H. P. 1975. Ann. N.Y. Acad. Sci. **253:** 51–52.
37. FUJIWARA, K. & L. G. TILNEY. 1975. Ann. N.Y. Acad. Sci. **253:** 27–50.
38. WARFIELD, R. K. N. & G. B. BOUCK. 1974. Science **186:** 1219–1221.
39. MONACO, G., P. CALISSANO & D. MERCANTI. 1977. Brain Res. **129:** 265–274.
40. LUDUENA, R. F., A. FELLOUS, J. FRANCON, J. NUNEZ & L. MCMANUS. 1981. J. Cell Biol. **89:** 680–683.

41. LUDUENA, R. F., A. FELLOUS, L. MCMANUS, M. A. JORDAN & J. NUNEZ. 1984. J. Biol. Chem. **259:** 12890–12898.
42. DONOSO, J. A., K. M. HASKINS & R. H. HIMES. 1979. Cancer Res. **39:** 1604–1610.
43. LUDUENA, R. F., M. C. ROACH, M. A. JORDAN & D. B. MURPHY. 1985. J. Biol. Chem. **260:** 1257–1264.
44. MARGOLIS, R. L. & L. WILSON. 1977. Proc. Natl. Acad. Sci. USA **74:** 3466–3470.
45. MARGOLIS, R. L. & L. WILSON. 1978. Cell **13:** 1–8.
46. WILSON, L., M. A. JORDAN, A. MORSE & R. L. MARGOLIS. 1982. J. Mol. Biol. **159:** 125–149.
47. BRYAN, J. 1972. Biochemistry **11:** 2611–2616.
48. LUDUENA, R. F. & M. C. ROACH. Unpublished results.
49. FELLOUS, A., R. F. LUDUENA, V. PRASAD, M. A. JORDAN, W. ANDERSON, R. OHAYON & P. T. SMITH. 1985. Cancer Res. **45:** 5004–5010.
50. SALTARELLI, D. & D. PANTALONI. 1982. Biochemistry **21:** 2996–3006.
51. ANDREU, J. M. & S. N. TIMASHEFF. 1982. Proc. Natl. Acad. Sci. USA **79:** 6753–6756.
52. ANDREU, J. M., T. WAGENKNECHT & S. N. TIMASHEFF. 1983. Biochemistry **22:** 1556–1566.
53. FELLOUS, A., V. PRASAD, R. F. LUDUENA & M. A. JORDAN. 1983. J. Cell Biol. **97:** 200a.
54. PALANIVELU, P. & R. F. LUDUENA. 1982. J. Biol. Chem. **257:** 6311–6315.
55. SALTARELLI, D. & D. PANTALONI. 1983. Biochemistry **22:** 4607–4614.
56. MURPHY, D. B. & K. T. WALLIS. 1983. J. Biol. Chem. **258:** 8357–8364.
57. LOWRY, D. H., N. H. ROSEBROUGH, A. L. FARR & R. J. RANDALL. 1951. J. Biol. Chem. **193:** 26.

Taxol: Mechanisms of Action and Resistance[a]

S. B. HORWITZ, L. LOTHSTEIN, J. J. MANFREDI,
W. MELLADO, J. PARNESS, S. N. ROY, P. B. SCHIFF,
L. SORBARA, AND R. ZEHEB

Department of Molecular Pharmacology
Albert Einstein College of Medicine
Bronx, New York 10461

INTRODUCTION

Taxol was isolated in 1971 from the stem bark of the plant, *Taxus brevifolia,* a member of the yew family.[1] It and related compounds have also been found in the leaves, stems, and roots of a variety of other *Taxus* species.[2] In experimental murine systems, taxol has demonstrated antileukemic and tumor inhibitory properties and is presently being tested in Phase I clinical trials in humans. This drug is a complex ester consisting of a taxane derivative with an oxetan ring, an unusual chemical structure whose biological activity had not been previously studied.

Taxol has become an important tool for studying the structure and function of microtubules. It complements other plant alkaloids such as colchicine, podophyllotoxin, and the vinca alkaloids, as a probe for elucidating the complex mechanisms involved in the polymerization and depolymerization of microtubules. The following discussion will review the research that has been done in our laboratory to study the mechanism of action and of resistance to taxol.

EFFECTS OF TAXOL IN CELLS

Taxol is a potent inhibitor of cell replication that blocks cells in the late G_2 or M phases of the cell cycle.[3] The drug does not have a primary effect on DNA synthesis. The addition of taxol to synchronized cells at the beginning of S phase in no way alters the progression of the cells through this phase of the cell cycle.

As seen by tubulin immunofluorescence and electron microscopy, taxol-treated cells exhibit an unusual interphase microtubule cytoskeleton. Such cells not only maintain their cytoplasmic microtubules, but in addition develop prominent bundles of microtubules. Incubation of cells with taxol results in the stabilization of cellular microtubules to depolymerization by cold and by antimitotic agents such as steganacin and colchicine.[3,5] The presence of bundles of microtubules is characteristic of the cytoskeleton of taxol-treated cells and has been reported in a number of cell lines.[3,6,7–10,11] These bundles tend not to be associated with the microtubule organizing center; in taxol-treated PtK2 cells, essentially none of the microtubule bundles are

[a]Research that originated in the authors' laboratory was supported by USPHS Grants CA 15714 and Gm 29042 and by American Cancer Society Grant CH-86.

associated with the centrosome.[6] An association of microtubule bundles with the endoplasmic reticulum, however, has been observed in various cell lines. This was first seen in HeLa cells,[3] and is being studied extensively in various cell types of primary mouse dorsal root ganglion-spinal cord cultures in which taxol induces abnormal microtubule arrays and dramatically increases the number of cytoplasmic microtubules (FIGURE 1). Such arrays include microtubules aligned along the endoplasmic reticulum and hexagonal groupings of microtubules in the cytoplasm.[12,13]

The reorganization of cellular microtubules, in the presence of taxol, to form microtubule bundles is both a concentration-dependent and time-dependent phenomenon. Extraction of taxol-treated cells with the nonionic detergent Triton X-100 clearly indicates that the taxol-induced bundles remaining in the extracted preparations consist of parallel arrays of cellular microtubules. Within the parallel arrays, there appear to be some cross-bridges between adjacent microtubules.[14]

Although the process by which taxol induces the formation of microtubule bundles is not clear, an intact cell with normal energy levels is necessary for the cytoskeletal reorganization that is required for the formation of microtubule bundles. For example, microtubule bundles are never observed in detergent-extracted cytoskeletons that are incubated with taxol, although taxol binds to these cytoskeletons. In addition, the depletion of adenosine triphosphate (ATP) in cells after treatment with sodium azide results in a lack of microtubule bundle formation, although, again, taxol binds to such cells.[11] Therefore, it is likely that there are specific cellular conditions, perhaps including a "bundling factor," that are necessary for the microtubule cytoskeleton reorganization observed in the presence of taxol.

Experimental evidence indicates that the cellular receptor for taxol is the microtubule, the polymerized form of tubulin. The specific binding of taxol to a cell saturates in a concentration-dependent manner.[11] Pretreatment of cells with colchicine or vinblastine results in the inhibition of taxol binding. When such cells were examined by immunofluorescence with antibodies against tubulin, it was evident that the pretreatment had completely depolymerized the microtubule cytoskeleton. The cellular receptor for taxol was lost and saturable binding was no longer observed. Saturable binding also was not seen in sheep erythrocytes,[14] which like most mammalian red blood cells do not contain tubulin.[15] Further evidence that the cellular receptor for taxol is the polymer form of tubulin comes from experiments in which the binding of taxol to detergent extracted cytoskeletons was observed. The extracted cytoskeleton contains no unassembled tubulin dimer, only microtubules. When the detergent extraction was done in the presence of calcium, however, specific binding of taxol was not observed. Calcium induces microtubule depolymerization.[14]

EFFECTS OF TAXOL ON PURIFIED TUBULIN *IN VITRO*

Taxol has unusual effects on the assembly of calf brain microtubules *in vitro* that correlate with our knowledge of the action of the drug in cells. In contrast to other antimitotic agents that inhibit microtubule polymerization, taxol enhances both the rate and yield of microtubule assembly by shifting the equilibrium between the tubulin dimer and polymer in favor of the polymer.[4] The drug, however, induces the formation of microtubules that are shorter than those observed in the absence of the drug, indicating that taxol is responsible for increasing the number of nucleation events at the start of microtubule assembly.[16] Additionally, taxol alters the kinetics of tubulin assembly. Whereas there is normally a three to four minute lag period prior to observing the advent of assembly when following this process by changes in turbidity, assembly appears to begin instantly on the addition of taxol.

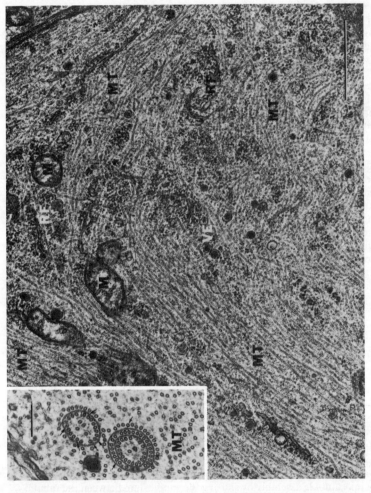

FIGURE 1. Electron micrograph illustrating an unusual abundance of microtubules (MT) coursing through the cytoplasm near an exiting process of a neuron in a 13-day fetal mouse dorsal-root ganglion explant exposed to taxol (1 μM) for 6 days (+ nerve growth factor), after an initial period of development for more than 2 weeks in control culture medium. The MTs appear in various orientations interspersed with foci of vesicles (VE), mitochondria (M), and ribosomal formations (RF). Original magnification × 40,000; reduced by 2 percent. Inset: Transverse section through concentric ordered arrays of MTs alternating with layers of macromolecular material in a portion of neuritic extension near the soma. Connections between some MTs and these nonmembranous lamellae appear at various points in these complexes (*e.g.*, arrows). Some nearby MTs appear to be deployed in various linear and other groupings. Original magnification × 80,000; reduced by 2 percent. Scale bar: 0.2 μm. (Masurovsky *et al.*[13] With permission from IBRO and Pergamon Press.)

The drug has a unique capacity to assemble tubulin into its polymer form. The critical concentration of microtubule protein required for assembly is reduced in the presence of taxol, and the microtubules formed are stable to depolymerization by calcium or cold. Taxol also assembles tubulin under conditions in which polymerization would not normally occur. For example, assembly occurs in the absence of microtubule-associated proteins, exogenously added guanosine triphosphate (GTP), organic buffers, or even warm temperatures.[17–20] In addition to studying calf brain tubulin, we have reported that taxol, in the absence of GTP, induces the assembly of calcium stable microtubules from flagellar tubulin solubilized from sea urchin sperm and dramatically reduces the critical concentration of protein required for polymerization.[21]

All the evidence suggests that there is a taxol binding site on the microtubule. For example, when taxol is added to steady state microtubules assembled in the absence of the drug, the microtubules become resistant to depolymerization by calcium. In addition, treadmilling of *in vitro* microtubules is reduced in the presence of taxol.[17,18,22] The taxol binding site is distinct from the exchangeable GTP binding site and the binding sites for colchicine or podophyllotoxin and vinblastine.[17,18] Although antimitotic drugs such as colchicine do not bind at the taxol site, they can inhibit taxol binding if added prior to assembly. This probably is due to an inhibition of assembly by colchicine that results in an absence of microtubules, the target for taxol binding.[23]

Maximal effects of taxol are observed when the concentration of drug is stoichiometric with the tubulin dimer concentration. The preparation and use of [³H]taxol has made it clear that the drug binds specifically and reversibly to assembled microtubules *in vitro* and the stoichiometry of binding approaches one mole of taxol bound per mole of tubulin dimer in the polymer.[24]

The enhancement of both the rate and yield of microtubule assembly by taxol makes this drug distinct from all other antimitotic agents whose mechanism of action has been analyzed. The drug exhibits specificity for the tubulin-microtubule system; taxol does not influence actin polymerization or bind to intermediate filaments or DNA.[24]

MATERIAL AND METHODS

Taxol was obtained from the National Cancer Institute. All drugs were dissolved at a concentration of 10 mM in dimethylsulfoxide and stored at $-20°C$. The macrophage-like cell line J774.2 was maintained and used to follow the effects of drugs on cell growth as previously described.[25] The preparation of microtubule protein (MTP) from calf brain and the *in vitro* microtubule assembly assay have been published.[26,27]

Growth Curves

Cells were plated (1×10^5 cells/ml) in complete medium containing a range of drug concentrations in 60 mm tissue culture dishes in a final volume of 4 milliliters. After 72 hours, the medium was removed. Cells were suspended with the aid of a rubber policeman in 2 ml Dulbecco's modified Eagle's medium (DME) containing 0.02% ethylenediamine tetraacetic acid (EDTA) and counted in a Coulter counter.

Binding of [³H]Taxol to Cells

Total and nonspecific binding of [³H]taxol (0.3 μM final concentration; 0.032 Ci/mmol) were measured according to the method described by Manfredi et al.[11]

Isolation of Cell Membranes and their Analysis by Electrophoresis

Cells growing in 100 mm tissue culture dishes were metabolically labeled with [^{3}H]glucosamine (34.6 Ci/mmol; 10 μCi/ml) in 5 ml of complete medium for 16 hr at 37°C in a humid CO_2 incubator. Cells were washed three times with 10 ml ice-cold phosphate buffered saline and centrifuged at 500 × g for 30 sec; plasma membranes were prepared by the method of Atkinson and Summers[28] with some modification. Briefly, the cells were resuspended in two volumes of homogenizing buffer (10 m*M* Tris-HCl, pH 8.0, 75 m*M* sucrose, 25 m*M* MgCl$_2$, 1.5 m*M* EDTA, 5 m*M* dithiothreitol, 0.15 *M* NaCl, 0.15 *M* KCl and the protease inhibitors phenylmethylsulfonyl fluoride, leupeptin, benzamidine, aprotinin, and pepstatin, each at 10 μg/ml). Cells were ruptured in a Dounce homogenizer using 20 strokes; the extent of breakage was followed by light microscopy. The unbroken cells and nuclei were removed by centrifugation at 500 × g for 60 sec at 4°C. The supernatant was mixed with 70% sucrose (w/v) solution (10 m*M* Tris-HCl, pH 7.5, containing 5 m*M* MgCl$_2$, 0.5 m*M* EDTA, 1 m*M* dithiothreitol, 0.15 M NaCl, 0.15 *M* KCl, and the protease inhibitors phenylmethylsulfonyl fluoride, leupeptin, benzamidine, aprotinin, and pepstatin, each at 10 μg/ml) (Buffer A), and the final density of sucrose was adjusted to 40.5 percent. This sucrose solution was layered at the bottom of a previously prepared discontinuous sucrose gradient consisting of equal volumes of 20%, 34%, and 40% sucrose (w/v) in Buffer A (total volume 17 ml) and centrifuged in a SW 28.1 rotor (Beckman) for 16 hr at 26,000 rpm at 4°C. The opaque band, at the "34/40" interface was removed and centrifuged at 15,000 rpm for 30 min at 4°C. The pellet was resuspended in a minimum volume of 10 m*M* Tris-HCl, pH 7.5 and stored in liquid N$_2$ for future use.

Protein concentration in the membrane fraction was measured by the method of Lowry *et al.*[29] with bovine serum albumin as the standard. Sodium dodecylsulfate-polyacrylamide gel electrophoresis (SDS-PAGE) of membrane proteins was performed according to the procedure described by Laemmli.[30]

RESULTS AND DISCUSSION

Structure-Activity Relationships

There has been considerable interest in the potential use of taxol as an antitumor agent primarily because of its unusual mechanism of action and interesting chemistry. The formulation of taxol for clinical trials and its use in humans has been severely hampered because of the extreme insolubility of the drug in aqueous solvents. In an attempt to improve the solubility characteristics of the drug, information about the chemical structure of the drug as it relates to biological activity has been sought. A number of natural and semisynthetic congeners of taxol (FIGURE 2) have been studied.[31] Two parameters of each compound have been examined: cytotoxicity toward the macrophage-like cell line, J774.2, and the ability to promote microtubule assembly *in vitro* in the absence of exogenous GTP. Experiments have indicated that both an intact taxane ring and an ester side chain at position C-13 are required for cytotoxicity. Small alterations in the structure of the drug, such as loss of the acetyl group at C-10 (10-deacetyltaxol) or a change of the *N*-acyl substituent (cephalomannine) do not have major effects on activity (TABLE 1).

The semisynthetic derivative of taxol, 2'-7-diacetyltaxol, did not promote microtubule assembly *in vitro,* although it maintained some of its cytotoxicity. To dissect the roles of the acetyl groups at C-7 and C-2', 2'-acetyltaxol and 7-acetyltaxol have been synthesized.[32] There are three hydroxyl groups in taxol but each is of differing activity.

The C-1 hydroxyl group is tertiary, and is therefore unreactive to normal conditions of acetylation, whereas the C-7 hydroxyl group is hindered. The C-2' hydroxyl group is unhindered, and acetylation of taxol with acetic anhydride in pyridine resulted in a very high yield of 2'-acetyltaxol. To prepare 2'-7-diacetyltaxol, acetylation of taxol was carried out in the presence of acetic anhydride using carbodiimide/4-dimethylamino-pyridine to regenerate acetic anhydride.[33] Selective deacetylation of the diacetyltaxol derivative in mild base resulted in the formation of 7-acetyltaxol. Taxol and its acetates were isolated by high performance liquid chromatography (HPLC) (FIGURE 3).

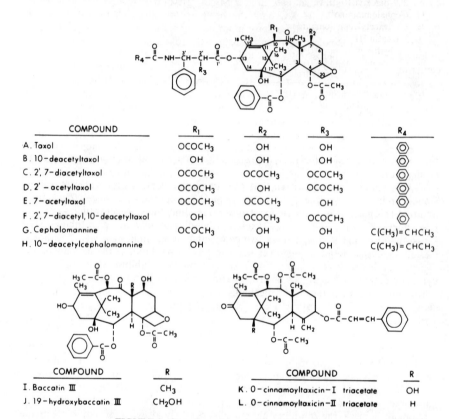

COMPOUND	R_1	R_2	R_3	R_4
A. Taxol	OCOCH$_3$	OH	OH	⬡
B. 10-deacetyltaxol	OH	OH	OH	⬡
C. 2', 7-diacetyltaxol	OCOCH$_3$	OCOCH$_3$	OCOCH$_3$	⬡
D. 2'-acetyltaxol	OCOCH$_3$	OH	OCOCH$_3$	⬡
E. 7-acetyltaxol	OCOCH$_3$	OCOCH$_3$	OH	⬡
F. 2', 7-diacetyl, 10-deacetyltaxol	OH	OCOCH$_3$	OCOCH$_3$	⬡
G. Cephalomannine	OCOCH$_3$	OH	OH	C(CH$_3$)=CHCH$_3$
H. 10-deacetylcephalomannine	OH	OH	OH	C(CH$_3$)=CHCH$_3$

COMPOUND	R
I. Baccatin III	CH$_3$
J. 19-hydroxybaccatin III	CH$_2$OH

COMPOUND	R
K. O-cinnamoyltaxicin-I triacetate	OH
L. O-cinnamoyltaxicin-II triacetate	H

FIGURE 2. Structure of taxol and related taxanes.

Although 2'-acetyltaxol and 2',7-diacetyltaxol inhibited the growth of cells in culture, they did not induce microtubule assembly *in vitro*. The cytotoxic activity of these two taxol derivatives suggests that these compounds may be converted intracellularly to either taxol or unknown taxol metabolites. No information is available on the metabolic fate of taxol or any related taxanes. The 7-acetyltaxol, however, is very similar to taxol in both its effects on the replication of cells and on microtubule polymerization *in vitro*. The lack of *in vitro* activity observed with both 2'-acetyltaxol and 2',7-diacetyltaxol suggests that the C-2' hydroxyl group on the ester side chain is either involved in a specific interaction with microtubules, alters the conformation of

TABLE 1. Activity of Taxol and Taxol Congeners

Compound	Induction of Microtubule Assembly	Relative Cytotoxicity
A. Taxol[a]	Yes	+ + + + +
B. 10-deacetyltaxol[a]	Yes	+ + +
C. 2′,7-diacetyltaxol[a]	No	+
D. 2′acetyltaxol[b]	No	+ + +
E. 7-acetyltaxol[b]	Yes	+ + + +
F. 2′,7-diacetyl,10-deacetyltaxol[a]	No	+ + .
G. Cephalomannine[a]	Yes	+ + + +
H. 10-deacetylcephalomannine[a]	Yes	+ +
I. Baccatin III[a]	No	—[c]
J. 19-hydroxybaccatin III[a]	No	—[c]
K. 0-cinnamoyltaxicin-I-triacetate[a]	No	—[c]
L. 0-cinnamoyltaxicin-II-triacetate[a]	No	—[c]

[a]Data from Parness *et al.*[31]
[b]Data from Mellado *et al.*[32]
[c]No activity at 10 μM.

the drug to an inactive form, or inhibits its interaction with microtubules by steric hinderance.

These studies, demonstrating that the properties of taxol and 7-acetyltaxol are similar, indicate that a free hydroxyl group at C-7 is not required for *in vitro* activity and that this position is available for structural modifications. Recent studies in which the effects of taxol and 7-acetyltaxol on the disassembly of microtubules were measured also indicated that these two compounds have similar activity.[34] Although 7-acetyltaxol has no advantage over taxol in terms of solubility, the introduction of charged or hydrophilic groups at C-7 could improve solubility while maintaining cytotoxic activity. These types of investigations could lead to the development of an antitumor drug with the same mechanism of action as taxol but with improved solubility.

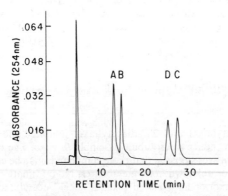

FIGURE 3. Isolation of taxol and taxol acetates by reversed phase HPLC. 2.5 nmoles of each compound in dimethylsulfoxide were injected on an analytical reverse phase HPLC column (Ultrasphere-ODS 4.6 × 250 mm) and eluted with $CH_3OH:H_2O$ (75:25 v/v) at 0.7 ml/min. A = taxol; B = 2′-acetyltaxol; C = 2′,7-diacetyltaxol; D = 7-acetyltaxol.

Taxol Resistant Cells

Drug-resistant cell lines were developed over a one year period by growing the murine macrophage-like cell line J774.2 in the presence of stepwise increases in the concentration of taxol, colchicine, or vinblastine.[47] The concentration of taxol was raised to 50 μM (J7/TAX-50), vinblastine to 1 μM (J7/VBL-1), and colchicine to 20 μM (J7/CLC-20). The cells were maintained continuously in these drug concentrations. J7/TAX-50 cells were analyzed after growth in the absence of taxol for 240 days (J7/TAX-50/OD-240). The J7/TAX-50 cells became partially dependent on taxol for normal growth, and removing taxol from the growth medium resulted in some cell death and unusual morphology until the cell population stabilized after being maintained in the absence of the drug for 7–10 days.

The growth inhibitory effects of taxol, colchicine, vinblastine, and bleomycin on the parental and drug-resistant cell lines were measured after a 72 hr period (TABLE 2). The J7/TAX-50 cells were approximately 800-fold resistant to taxol and also cross-resistant to colchicine (58-fold) and vinblastine (43-fold). In addition, an approximate 4-fold increase in sensitivity to bleomycin was detected. The J7/CLC-20 and J7/VBL-1 cell lines were approximately 550-fold and 250-fold resistant, respectively, to colchicine and vinblastine. Both of these resistant cell lines demonstrated some cross-resistance to each other and to taxol, but like J7/TAX-50, indicated a small but definite increased sensitivity to bleomycin. J7/TAX-50 cells grown in the absence of the drug for eight months (J7/TAX-50/OD-240) exhibited a sensitivity to taxol, colchicine, and vinblastine that was very similar to that of the parental drug-sensitive cell line.

Although the increased sensitivity to bleomycin is not currently understood, this drug has a mechanism of action that does not involve the tubulin-microtubule system; it binds to and degrades DNA.[35,36] In contrast to taxol, colchicine, and vinblastine, which are hydrophobic molecules, bleomycin is a hydrophilic compound that is extremely soluble in aqueous solvents. This latter difference may be important for drug resistance because bleomycin may cross the plasma membrane by a mechanism quite distinct from that used by taxol.

To determine if the taxol-resistant cell line has an altered ability to bind the drug, the steady state accumulation of taxol was measured using radiolabeled drug. As had been previously reported,[11] a steady state accumulation of taxol was obtained in J774.2 cells within 45 minutes. The total accumulation of taxol in J7/TAX-50 was reduced by approximately 90% compared to the parental line (FIGURE 4). In J7/TAX-50/OD-240, the steady state accumulation of the drug remained 30% below that of

TABLE 2. Drug Sensitivity of J774.2 and Resistant Cell Lines

Cell Line	ED_{50} (μM)[a]			
	Taxol	Vinblastine	Colchicine	Bleomycin
J7	0.06	0.03	0.04	1.3
J7/TAX-50[b]	50.0[c] (833)[d]	1.3(43)	2.3(58)	0.37(.28)
J7/TAX-50/OD-240	0.1(1.7)	0.03(1)	0.05(1.5)	0.20(.15)
J7/VBL-1	2.5(42)	7.4(247)	1.9(48)	0.74(.57)
J7/CLC-20	3.9(65)	0.5(17)	22.0(550)	0.32(.15)

[a]Drug concentration that inhibits cell division by 50% after 72 hrs.
[b]Cells maintained in 50 μM taxol during cross-resistance experiments.
[c]Maximum solubility of drug in medium is 50 μM.
[d]Numbers in parenthesis equal ratio of ED_{50} for resistant cell line to that for J774.2.

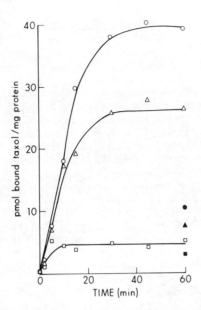

FIGURE 4. Binding of [³H]taxol to J7, J7/TAX-50, and J7/TAX-50/OD-240 cells. Confluent 35 mm plates of cells were incubated with 2 ml of Dulbecco's modified Eagle's medium containing 20% horse serum containing 0.3 μM [³H]taxol at 37° for the indicated times. Cells were washed three times with ice-cold phosphate buffered saline, and lysed with 1 ml 1N NaOH for 16 hrs. An aliquot of cell lysate was neutralized with an equal volume of glacial acetic acid and [³H] radioactivity determined. O = J7; △ = J7/TAX-50; and □ = J7/TAX-50/OD-240 cells. The closed symbols denote the binding of 0.3 μM [³H]taxol in the presence of a 100-fold excess of unlabeled drug. Each point represents the average of 4 determinations.

J774.2. A 100-fold excess of unlabeled drug diluted in each case the accumulation of labeled drug, indicating that cells have a specific binding site for taxol.

Membranes were prepared from J7, J7/TAX-50, and J7/TAX-50/OD-240 cells, and the membrane proteins at the "34/40" interface on a sucrose gradient were analyzed by SDS-PAGE. Silver staining of such 7.5% acrylamide gels clearly indicated a protein with an approximate molecular weight of 135,000 that was present in the membranes prepared from the taxol-resistant cells, but was not detectable in the parental cell line or cells grown in the absence of the drug for 8 months. When these cell lines were metabolically labeled with [³H]glucosamine and the membrane proteins at the "34/40" interface examined by autoradiography after SDS-PAGE, a 135,000 molecular weight glycoprotein could be visualized only in the J7/TAX-50 cells (FIGURE 5). This 135,000 molecular weight glycoprotein could also be metabolically radiolabeled with [³H]leucine, [³⁵S]methionine or [³²P]orthophosphate. Phosphoamino acid analysis of the membrane phosphoglycoprotein indicated that phosphorylation was at serine and threonine residues.

The presence of the phosphoglycoprotein in the plasma membrane of taxol resistant cells correlated well with resistance to the drug. Resistance to the growth inhibitory properties of taxol and the presence of the membrane phosphoglycoprotein is dependent on the presence of the drug in the medium indicating the unstable nature of taxol resistance. Similar but not identical membrane phosphoglycoproteins were found

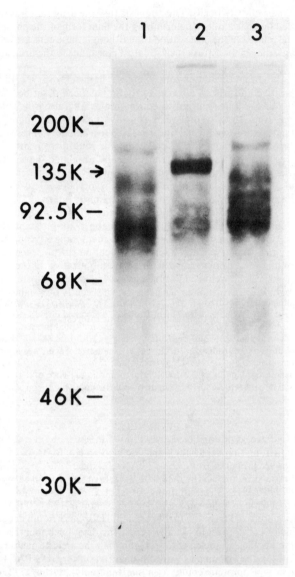

FIGURE 5. SDS-polyacrylamide gel profiles of membranes from cells metabolically labeled with [³H]glucosamine. Cells were labeled, and membranes were isolated as described under METH-ODS. Samples from the "34/40" interface, each containing 2.7 × 10⁴ cpm, were used for analysis. Lane 1 = J7 (15 μg protein); Lane 2 = J7/TAX-50 (10 μg protein); and Lane 3 = J7/TAX-50/OD-240 (17 μg protein). The gel (7.5%) was fixed in 10% acetic acid, 5% methanol (v/v) incubated with "Enhance" (New England Nuclear) for 60 min, washed with water for 60 min, dried, and exposed for one month to Kodak SB-5 film. Molecular weight standards (Amersham) were (from top to bottom): ¹⁴C-labeled methylated myosin, phosphorylase B, bovine serum albumin, ovalbumin and carbonic anhydrase.

in J7/CLC-20 and J7/VBL-1 cells. This suggests that the drug that was used to select each resistant cell line has a role in determining the final form of the phosphoglycoprotein. This may be related to the observation that although these cells are cross-resistant to different drugs, they demonstrate their greatest resistance to the drug against which they have been selected.

A two- to three-fold taxol-resistant Chinese hamster ovary cell line that is temperature sensitive and has an altered α tubulin has been described.[37] Analysis of tubulin from J7/TAX-50 has not indicated any major alterations as compared with tubulin prepared from the parental taxol-sensitive cells. Much more likely is the possibility that our taxol-resistant cell line is related to a category of resistant cells that have been described as demonstrating pleiotropic or multidrug resistance.[38-41] Such cells are cross-resistant to drugs that are structurally unrelated, have altered permeability properties, contain high molecular weight glycoproteins in their plasma membrane, and often possess alterations in their karyotypes such as the inclusion of double-minute chromosomes or homogeneously staining regions.[42-45] Although the taxol-sensitive J774.2 cell line does not have any double-minute chromosomes, double-minute chromosomes are clearly visible in the taxol-resistant cell line. The existence of the double-minute chromosomes, the membrane glycoprotein, and the expression of drug resistance are all closely dependent on the presence of taxol in the growth medium and may be associated with gene amplification in the drug-resistant cell lines. The presence of amplified DNA sequences in Chinese hamster cells and their correlation with multidrug resistance has been shown recently by Roninson *et al.*[46] The function of the 135,000 M_r phosphoglycoprotein in the membranes of taxol-resistant J774.2 cells and its role in drug resistance is of major interest.

ACKNOWLEDGMENT

The authors are grateful to Dr. E. Masurovsky for valuable discussions.

REFERENCES

1. WANI, M. C., H. L. TAYLOR, M. E., WALL, P. COGGON & A. T. MCPHAIL. 1971. J. Am. Chem. Soc. **93:** 2325–2327.
2. MILLER, R. W., R. G. POWELL, C. R. SMITH JR., E. ARNOLD & J. CLARDY. 1981. J. Org. Chem. **46:** 1469–1474.
3. SCHIFF, P. B. & S. B. HORWITZ. 1980. Proc. Natl. Acad. Sci. USA **77:** 1561–1565.
4. SCHIFF, P. B., J. FANT & S. B. HORWITZ. 1979. Nature (London) **277:** 665–667.
5. CROSSIN, K. L. & D. H. CARNEY. 1981. Cell **27:** 341–350.
6. DEBRABANDER, M., G. GEUENS, R. NUYDENS, R. WILLERBRORDS & J. DEMEY. 1981. Proc. Natl. Acad. Sci. USA **78:** 5608–5612.
7. DEBRABANDER, M. 1982. Cell Biol. Int. Rep. **6:** 901–915.
8. ALBERTINI, D. F. & J. I. CLARK. 1981. Cell Biol. Int. Rep. **5:** 387–397.
9. BRENNER, S. L. & B. R. BRINKLEY. 1982. Cold Spring Harbor Symp. Quant. Biol. **46:** 241–254.
10. TOKUNAKA, S., T. M. FRIEDMAN, Y. TOYAMA, M. PACIFICI & H. HOLTZER. 1983. Differentiation **24:** 39–47.
11. MANFREDI, J. J., J. PARNESS & S. B. HORWITZ. 1982. J. Cell Biol. **94:** 688–696.
12. MASUROVSKY, E. B., E. R. PETERSON, S. M. CRAIN & S. B. HORWITZ. 1981. Brain Res. **217:** 392–398.
13. MASUROVSKY, E. B., E. R. PETERSON, S. M. CRAIN & S. B. HORWITZ. 1983. Neurosci. **10:** 491–509.
14. MANFREDI, J. J. & S. B. HORWITZ. 1984. Pharm. Ther. **25:** 83–125.

15. DUSTIN, P. 1984. *Microtubules*. 2nd Ed. Springer Verlag. Berlin.
16. SCHIFF, P. B. & S. B. HORWITZ. 1981. *In* Molecular Actions and Targets for Cancer Chemotherapeutic Agents. A. C. Sartorelli, J. S. Lazo & J. R. Bertino, Eds. Vol. 2: 483–507. Bristol-Myers Cancer Symposium. Academic Press. New York.
17. SCHIFF, P. B. & S. B. HORWITZ. 1981. Biochemistry **20:** 3247–3252.
18. KUMAR, N., 1981. J. Biol. Chem. **256:** 10435–10441.
19. THOMPSON, W. C., L. WILSON & D. L. PURICH. 1981. Cell Motility **1:** 445–454.
20. HAMEL, E., A. A. DEL CAMPO, M. C. LOWE & C. M. LIN. 1981. J. Biol. Chem. **256:** 11887–11894.
21. PARNES, J., C. F. ASNES & S. B. HORWITZ. 1983. Cell Motility **3:** 123–130.
22. CAPLOW, M. & B. ZEEBERG. 1982. Eur. J. Biochem. **127:** 319–324.
23. HORWITZ, S. B., J. PARNESS, P. B. SCHIFF & J. J. MANFREDI. 1982. Cold Spring Harbor Symp. Quant. Biol. **46:** 219–226.
24. PARNESS, J. & S. B. HORWITZ. 1981. J. Cell Biol. **91:** 479–487.
25. HORWITZ, S. B., G. H. CHIA, C. HARRACKSINGH, S. ORLOW, S. PIFKO-HIRST, J. SCHNECK, L. SORBARA, M. W. SPEAKER, E. W. WILK & O. M. ROSEN. 1981. J. Cell Biol. **91:** 798–802.
26. SHELANSKI, M. L., F. GASKIN & C. R. CANTOR. 1973. Proc. Natl. Acad. Sci. USA **70:** 765–768.
27. GASKIN, F., C. R. CANTOR & M. L. SHELANSKI. 1974. J. Mol. Biol. **89:** 737–758.
28. ATKINSON, P. H. & D. F. SUMMERS. 1971. J. Biol. Chem. **246:** 5162–5175.
29. LOWRY, O. H., N. J. ROSEBROUGH, A. L. FARR & R. J. RANDALL. 1951. J. Biol. Chem. **193:** 265–275.
30. LAEMMLI, U. K. 1970. Nature (London) **277:** 680–685.
31. PARNESS, J., D. G. I. KINGSTON, R. G. POWELL, C. HARRACKSINGH & S. B. HORWITZ. 1982. Biochem. Biophysics Res. Commun. **105:** 1082–1089.
32. MELLADO, W., N. F. MAGRI, D. G. I. KINGSTON, R. GARCIA-ARENAS, G. A. ORR & S. B. HORWITZ. 1984. Biochem. Biophysics Res. Commun. **124:** 329–336.
33. HASSNER, A. & V. ALEXANIAN. 1978. Tetrahedron Lett. **46:** 4475–4478.
34. LATASTE, H., V. SENILH, M. WRIGHT, D. GUENARD & P. POTIER. 1984. Proc. Natl. Acad. Sci. USA **81:** 4090–4094.
35. BURGER, R. M., J. PEISACH & S. B. HORWITZ. 1981. Life Sci. **28:** 715–727.
36. ROY, S. N. & S. B. HORWITZ. 1984. Cancer Res. **44:** 1541–1546.
37. CABRAL, F., I. ABRAHAM & M. M. GOTTESMAN. 1981. Proc. Natl. Acad. Sci. USA **78:** 4388–4391.
38. CARLSEN, S. A., J. E. TILL & V. LING. 1977. Biochim. Biophys. Acta **467:** 238–250.
39. RIORDAN, J. R. & V. LING. 1979. J. Biol. Chem. **254:** 12701–12705.
40. BIEDLER, J. L. & R. H. F. PETERSON. 1981. *In* Molecular Actions and Targets for Cancer Chemotherapeutic Agents. A. C. Sartorelli, J. S. Lazo & J. R. Bertino, Eds.: Vol. 2: 453–482. Bristol-Myers Cancer Symposium. Academic Press. New York.
41. BECK, W. T., M. C. CIRTAIN & J. L. LEFKO. 1984. Mol. Pharmacol. **24:** 485–492.
42. BIEDLER, J. L. & B. A. SPENGLER. 1976. Science **191:** 185–187.
43. GRUND, S. H., S. R. PATIL, H. O. SHAH, P. G. PAUW & J. K. STADLER. 1983. Mol. Cell. Biol. **3:** 1634–1647.
44. KUO, T., S. PATHAK, L. RAMAGLI, L. RODRIGUEZ & T. C. HSU. *In* Gene Amplification. R. T. Schimke, Ed.: 53–57. Cold Spring Harbor Laboratory. New York.
45. ROBERTSON, S. M., V. LING & C. P. STANNERS. 1984. Mol. Cell. Biol. **4:** 500–506.
46. RONINSON, I. B., H. T. ABELSON, D. E. HOUSMAN, N. HOWELL & A. VARSHAVSKY. 1984. Nature (London) **309:** 626–628.
47. ROY, S. N. & S. B. HORWITZ. 1985. Cancer Res. **45:** 3856–3863.

A Mechanism of Cellular Resistance to Drugs that Interfere with Microtubule Assembly[a]

FERNANDO R. CABRAL, RICHARD C. BRADY, AND
MATTHEW J. SCHIBLER

Department of Medicine
Division of Endocrinology
University of Texas Medical School
Houston, Texas 77225

INTRODUCTION

Several years ago we began an investigation of microtubule function in mammalian cells. Our approach was to isolate mutants of Chinese hamster ovary (CHO) cells with defective microtubules in which a clear biochemical alteration in tubulin could be demonstrated. These strains could then be used to determine what cellular functions are affected by the defective microtubules. In order to use this approach, a method was needed to isolate microtubule mutants at high frequency. The route upon which we eventually decided was to select CHO cells resistant to the cytotoxic effects of drugs known to interact with microtubules. This methodology has allowed us to isolate a number of CHO mutants with well-defined alterations in tubulin that share a number of properties that will be described in this article. In addition, many of these mutants have been shown to be conditional for growth, allowing us to select revertants and to study the consequences to the cell having defective microtubules. Although it was not one of our primary goals, the isolation of these mutants has forced us to consider why these cells are drug-resistant and how the drugs must affect microtubule assembly. It is these later considerations that will be emphasized in this communication.

MUTANT SELECTION

The ability of a number of naturally occurring plant alkaloids to poison microtubule assembly has been recognized for some time.[1] These agents can be grouped into two classes based on their effects on microtubules. The majority of the drugs inhibit microtubule assembly *in vitro* and lead to the disappearance of cytoplasmic and spindle microtubules when cells are exposed to these agents *in vivo* or in culture.[1] Thus far, only taxol has been shown to actually promote microtubule assembly and stabilize those structures both *in vitro* and *in vivo*.[2] In general, the microtubule destabilizing drugs have been shown to bind to free tubulin dimers in solution,[1] whereas taxol appears to bind stoichiometrically with tubulin in the polymerized but not the nonpolymerized state.[2]

[a]These studies were funded in part by Grants GM29955 from the NIH and CD-154 from the American Cancer Society. F. R. Cabral is the recipient of a Jr. Faculty Research Award from the American Cancer Society.

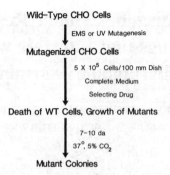

FIGURE 1. Selection scheme for the isolation of drug-resistant CHO mutants. Wild-type cells are mutagenized with ethyl methanesulfonate (EMS) or ultraviolet (UV) light to a 20% survival rate, and are then allowed to recover for three days. The mutagenized cells are then plated onto 100 mm tissue culture dishes at a density of 5×10^5 cells/dish in 20 ml of α-minimal essential medium (MEM) containing 10% fetal bovine serum and an appropriate concentration of the selecting drug. Under these conditions, most of the cells die, but any drug-resistant cells in the population continue to grow and form visible colonies after 7–10 days at 37°.

We have used the cytotoxic properties of these drugs to isolate drug-resistant CHO cells. Because the drugs are known to bind to tubulin or microtubules with high affinity, the expectation is that some of the resistant cells will have altered microtubules. An example of our selection scheme is shown in FIGURE 1. All of our selections are limited to a single step in order to avoid problems in assigning a biochemical defect to the phenotype of the mutant cells. As expected, based on earlier work by Ling and his associates in CHO cells[3] and by other groups in other cells types,[4,5] the most frequently derived mutants obtained by this selection are those that have an altered ability to accumulate the drugs intracellularly. These permeability mutants have the interesting property of being resistant not only to the selecting drug, but to a wide variety of other hydrophobic drugs as well.[4-6] Thus, permeability mutants may be easily recognized by their cross-resistance to drugs, such as puromycin, which have mechanisms of action unrelated to microtubule function. Positive identification of a tubulin mutant may be made by looking for electrophoretic changes in the mobility of α and β tubulin using two-dimensional gel electrophoresis.[7] In this way, we have found a number of microtubule mutants in CHO cells (TABLE 1). These mutants represent only a small percentage of all the drug-resistant cells we have isolated, the bulk of the remainder being permeability mutants. The microtubule mutants share a number of properties that will now be discussed.

TABLE 1. Mutants with Altered Tubulin

Selecting Drug	Alteration	Number Isolated
colchicine	β tubulin	1
colcemid	α or β tubulin	1/1
griseofulvin	β tubulin	2
maytansine	α tubulin	2
taxol	α or β tubulin	9/9

MUTATIONS AFFECTING α OR β TUBULIN CAN LEAD TO RESISTANCE TO ANY MICROTUBULE ACTIVE DRUG

A priori one might expect that a drug-resistant microtubule mutant should have an alteration in the drug-binding site on tubulin and, thus, demonstration of an alteration in either α or β tubulin should define the drug-binding subunit. Whereas several mutants of this type have been reported,[8] the vast majority of mutants isolated in our own laboratory do not appear to have an alteration in the drug-binding site. This conclusion is based on an inability to measure altered drug binding in the mutants we have tested, on the cross-resistance patterns of the mutants (see below), and on the fact that resistance to any particular drug may be conferred by an alteration in either α or β tubulin. This last point is illustrated in FIGURE 2 by three taxol-resistant mutants, Tax 2-4, Tax 8-2, and Tax 11-6 (unpublished). Compared to the wild-type pattern, these mutants display extra spots corresponding to altered tubulin subunits. For example, strains 8-2 and 11-6 have extra spots corresponding to an altered α tubulin (arrowheads, FIGURE 2C, D), whereas strain 2-4 has an extra spot corresponding to an altered β tubulin (arrowhead, FIGURE 2B). These extra spots have been shown to represent electrophoretic variants of tubulin by peptide mapping experiments. Similar results have been obtained for cells resistant to colcemid (unpublished). These results point out the danger in assuming that an alteration in a particular tubulin subunit in a drug-resistant mutant defines the subunit that possesses the drug-binding site.

MICROTUBULE MUTANTS ARE FREQUENTLY CONDITIONAL FOR GROWTH

Although our mutants were selected solely on the basis of their resistance to the cytotoxic effects of microtubule active drugs, a significant number of them are temperature-sensitive[11,12] or drug-dependent[9] for growth. This property has allowed us to isolate revertants of the mutants (see below) and to study the effects of altered microtubule assembly on the physiology of the cells. In all the mutants we have examined thus far, the tubulin defects result in aberrant spindle microtubule assembly but leave the cytoplasmic microtubules morphologically and functionally intact. The mitotic defects that result from the disruption of spindle microtubule assembly are most easily studied in cells requiring taxol for normal growth (references 9, 10, and unpublished work).

Some of these mutants display a partial requirement for taxol for growth (*e.g.* Tax-11, FIGURE 3), whereas others appear to be fully dependent on the drug for cell division (*e.g.* Tax-18, FIGURE 3). This growth behavior is also reflected in the morphology of these cells in the presence or absence of the drug. FIGURE 4A shows the morphology of Tax-18 cells in the presence of taxol, whereas FIGURE 4B shows similar cells deprived of taxol for two days. Note that the latter cells appear much larger and flatter and instead of a single nucleus, have a number of oddly shaped micronuclei. A similar result is obtained for Tax-11 shown in FIGURES 4C and D except that the taxol-deprived cells appear to have a mixture of normal and abnormal morphologies. This mixture is not the result of subclone heterogeneity because recloning the cells results in the same appearance. Rather, this mutant probably has a less severe defect than Tax-18, and statistically, some of the cells exhibit the defect at any given time, whereas others do not. Examination of Tax-18 and other mutants of this type using immunofluorescence microscopy with tubulin antibodies reveals normal cytoplasmic

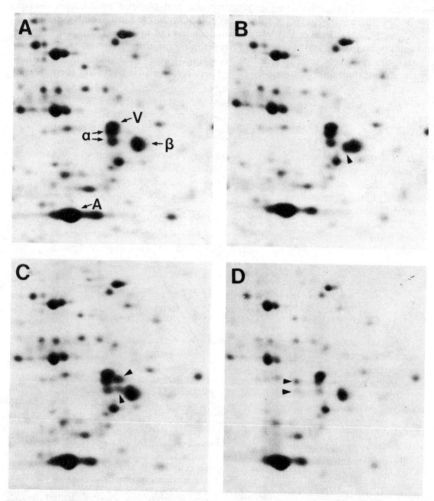

FIGURE 2. Two-dimensional gel autoradiograms of wild-type (**A**) and taxol-resistant cells (**B–D**). Cells were grown to approximately 70% confluency in 24-well dishes, labeled for 30 min with [³⁵S]methionine, lysed in hot sodium dodecyl sulfate (SDS), and run on two-dimensional gels. Only a portion of the autoradiogram of these gels is shown. **A** = wild-type cells; **B** = Tax 2-4; **C** = Tax 8-2; and **D** = Tax 11-6. The arrows in panel **A** point out the positions of actin (**A**), vimentin (**V**), α tubulin (α), and β tubulin (β). For reasons we do not understand, α tubulin migrates as two species of unequal apparent molecular weight on our gels. The two species do not represent distinct gene products, however, because single gene mutations alter the mobility of both spots. The arrowheads in panels **B–D** indicate the presence of tubulins with altered electrophoretic mobility that are not seen in the wild-type cells.

microtubules but an accumulation of cells with prometaphase-like spindles.[10] Metaphase and anaphase spindles are not seen, nor are midbodies, consistent with a lack of successful cytokinesis. Both immunofluorescence and electron microscopic studies suggest the absence of interpolar spindle microtubules in the mutant cells deprived of taxol.[10] As a consequence of this lesion, chromosomes fail to segregate during mitosis, and cell division is blocked. The cells do not remain in mitosis, however, and continue through the cell cycle several times before dying. These observations indicate that spindle microtubules are not necessary for short-term cell viability, but their role in

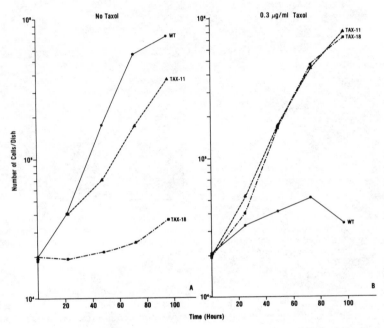

FIGURE 3. Growth of taxol-resistant cells in the presence (**B**) and absence (**A**) of taxol. A constant number of cells (2×10^4) were plated into each of 5 wells of a 24-well tissue culture dish in 1 ml of complete medium containing 0 or 0.3 μg/ml of taxol. At each time point, the medium from an appropriate well was transferred to a Coulter counter vial, and the attached cells on the dish were removed with 1 ml of 0.25% trypsin—1mM ethylenediaminetetraacetic acid (EDTA) in phosphate-buffered saline. These trypsinized cells were then combined with cells removed with the original media, 8 ml of Isoton was added, and the cells were counted in a Coulter model ZBI cell counter. Note that the taxol-resistant cells exhibit slower (Tax-11) or virtually no growth (Tax-18) in the absence of taxol (**A**), but both strains grow equally well when the medium is supplemented with 0.3 mg/ml of taxol (**B**).

chromosome segregation and cytokinesis make them crucial for the survival of nonquiescent cells.

In many respects, the mutation in these cells mimics the effects that are seen when normal cells are treated with low concentrations of microtubule disrupting drugs. The drugs preferentially inhibit the assembly of interpolar spindle microtubules, arrest cells in prometaphase, disrupt chromosome organization, prevent cytokinesis, and lead to the formation of large, multinucleated cells.[1] This suggests that the mutation in the

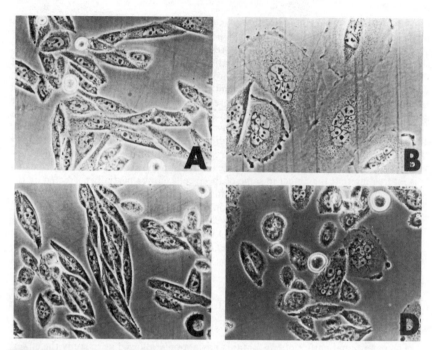

FIGURE 4. Morphology of Tax-18 **(A,B)** and Tax-11 **(C,D)** in the presence **(A,C)** and absence **(B,D)** of 0.2 μl/ml taxol. Cells were seeded onto 60 mm dishes in the presence or absence of the drug and photographed 2 days later using a Nikon Diaphot inverted phase microscope equipped with a Nikon DL 40X objective. Approximate magnification: 200X.

taxol requiring mutants is one that hinders the assembly of the microtubules or makes them more labile. Further evidence for this idea comes from cross-resistance studies described in the next section.

COLCEMID-RESISTANT MUTANTS ARE OFTEN MORE SENSITIVE TO TAXOL, AND TAXOL RESISTANT MUTANTS ARE MORE SENSITIVE TO COLCEMID

In contrast to the permeability mutants described earlier, microtubule mutants do not exhibit cross-resistance to a wide variety of diverse drugs. They do, however, exhibit cross-resistance to other microtubule active drugs. For example, a colcemid-resistant β-tubulin mutant, Cmd-4, isolated several years ago was shown to be cross-resistant to colchicine and vinblastine, but only slightly more resistant to griseofulvin.[13] Subsequently, we have found that this cell line is hypersensitive to taxol relative to the wild-type parental cells; this property can be used in the isolation of revertants (unpublished work). Similarly, a maytansine-resistant α tubulin mutant we recently isolated is cross-resistant to vinblastine, colcemid, and griseofulvin, but is hypersensitive to taxol.[14] By contrast, many of our taxol resistant mutants (reference 9 and unpublished work) are hypersensitive to microtubule-disrupting drugs. An example of this is shown in FIGURE 5. Similar results to these have been described in other CHO mutants[15,16] and in tubulin mutants of *Aspergillus nidulans*.[17]

In general, our drug-resistant CHO mutants are only two to threefold resistant to the selecting drug, and cells selected for resistance to a microtubule-disrupting drug are cross-resistant to other microtubule-disrupting drugs, but are more sensitive to a microtubule-stabilizing drug, taxol. Conversely, cells selected for resistance to taxol are more sensitive to microtubule-disrupting drugs. Although these statements are generally true, there are a number of clear and often perplexing exceptions that are as yet unexplained. In any case, the cross-resistance patterns make it unlikely that an alteration in the drug-binding site on tubulin is responsible for the drug resistance. In the case of the taxol dependent (and resistant) mutants, this conclusion seems especially valid because the cells are unable to divide in the absence of any added drug. It is interesting that the defect in some of these cells can be partially corrected by the relatively nonspecific microtubule-stabilizing agent, dimethylsulfoxide (DMSO). FIG-URE 6 shows growth curves for Tax-18 in the presence and absence of taxol and DMSO. As may be seen, DMSO improves the growth of the mutant, but taxol is better. It is unlikely that taxol and DMSO bind to the same site, yet both correct the defect in these cells (although to varying degrees), again arguing against a defect in the taxol-binding site.

MUTANT SUBUNITS MUST ASSEMBLE IN ORDER TO CONFER DRUG RESISTANCE

The conditional nature of our mutants has not only allowed us to study the effects of the mutations on microtubule function, but also to select revertants of the drug-resistance phenotype.[12] This was initially done to show that the altered tubulin was responsible for the drug-resistance seen in the mutant cells, but during the course of these studies it became clear that isolation of revertants provides a way of obtaining assembly-defective tubulins. For example, Cmd-4 is a colcemid-resistant mutant of CHO cells that has an altered β tubulin (β^*) and is temperature-sensitive (ts) for growth. Assuming that drug resistance and the ts phenotype were both conferred by the β^* tubulin, we selected Cmd-4 revertants able to grow at the nonpermissive temperature. As expected, many of these revertants were no longer colcemid-resistant and no longer expressed the β^* tubulin.[12] In one revertant, however, there was another alteration in β^* tubulin (to give β^{**} tubulin), which caused the β^{**} tubulin to migrate to yet another position on two-dimensional gels. The β^{**} tubulin was found to be unstable and incapable of assembling into microtubules.[12] The revertant has wild-type properties because its microtubules are composed of only wild-type subunits, implying that the β^* tubulin must assemble into microtubules in order to confer drug resistance. Recently, we have isolated several more revertants with this general phenotype (work in progress) showing that prevention of the assembly of mutant subunits is a fairly common mechanism of reversion. These results argue against the possibility that the β^* tubulin confers resistance by sequestering the drug or altering it in some way.

DRUG RESISTANCE BEHAVES AS A CODOMINANT TRAIT

Examination of whether a genetic trait behaves in a recessive or dominant manner can often provide an indication of the nature of the mutation. For example, recessive mutants are often lacking a functional gene product, whereas dominant mutants often

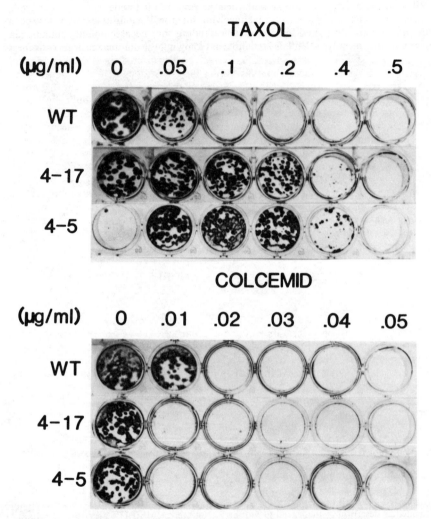

FIGURE 5. Drug resistance of two taxol-resistant mutants. Approximately 100 cells of each strain were plated into individual wells of a 24-well tissue culture dish containing varying concentrations of taxol or colcemid. The cells were allowed to grow for seven days, the medium was removed, and the cells were then stained with a solution of one percent methylene blue in water. Note that the two taxol resistant strains are only a few fold more resistant to taxol than are the wild-type (WT) cells, and that the strains are more sensitive to colcemid than are wild-type cells. Strain 4-5 does not grow without taxol and thus falls into the class of taxol-dependent mutants described in the text. In order to test its sensitivity to colcemid, 0.05μl/ml of taxol was included in the medium. Wild-type cells containing this level of taxol did not show an altered sensitivity to colcemid.

have a gene product with an altered or greater activity than the wild-type gene product. Whereas these generalities work well when the gene product under study is an enzyme, they work much less well in systems involving macromolecular assembly where gene dosage effects and a multiplicity of genes coding for the assembling subunits can complicate the analysis. Still, we thought that knowing the dominance or recessiveness

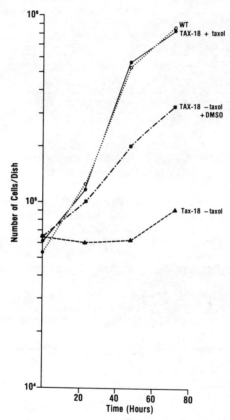

FIGURE 6. Growth of Tax-18 in the presence or absence of DMSO. Cells were plated and counted as described in the legend to FIGURE 3. Note that DMSO (1%) partially corrects the defect in Tax-18 but is not as effective as taxol. The taxol is added to the culture media as a solution in DMSO, but DMSO at the levels introduced with taxol has no visible effect on the growth or morphology of the cells. Only concentrations of DMSO above 0.5% are effective.

of our mutants would provide another constraint in devising a mechanism to explain the drug-resistance of the cells.

Testing dominance in CHO cells is easily accomplished by fusing the mutant cells with a wild-type strain containing 8-azaguanine and ouabain resistance markers.[18,19] The hybrid cells may then be selected in HAT medium (medium containing hypoxanthine, aminopterin, and thymidine) containing ouabain, and tested for drug resistance. For all mutants we have tested thus far, the drug resistance of the hybrids is

TABLE 2. Characteristics of Microtubule Mutants

1. Two- to threefold resistance to the selecting drug.
2. Mutants resistant to one microtubule destabilizing drug are cross-resistant to others but are hypersensitive to taxol. Taxol-resistant mutants are often hypersensitive to microtubule-destabilizing drugs.
3. Existence of drug-dependent mutants.
4. Mutations mimic drug effects.
5. Mutations in α or β tubulin (or MAPs?) can confer resistance to any given microtubule active drug.
6. Mutant subunits must assemble to confer drug resistance.
7. Drug resistance behaves codominantly.

intermediate between the wild-type and the mutant cells. Thus, drug resistance in these mutants is said to be codominant or incompletely dominant. An explanation of this behavior is given in the next section.

THE MECHANISM OF DRUG-RESISTANCE IN MICROTUBULE MUTANTS

The properties that we have just described for CHO microtubule mutants isolated in our laboratory are summarized in TABLE 2. Whereas exceptions can be found, most of our mutants conform to these generalities very well. Once we recognized the existence of these common traits, we began to try to formulate a mechanism that could explain them. Our current working model is diagrammed in FIGURE 7. As with all models, it is not perfect and cannot explain all our observations, especially the

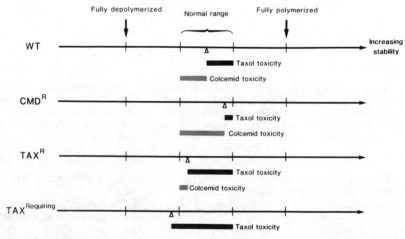

FIGURE 7. The mechanism of drug resistance in CHO mutants with altered tubulin. See text for details.

sometimes confusing cross-resistance patterns we see in some mutants. Rather, it is meant to simplify the existing data and suggest experiments to help understand how cells become resistant to drugs that affect microtubule assembly.

It is well known that polymerized microtubules in living cells are in a dynamic equilibrium or steady state with $\alpha\beta$ heterodimers.[1] Assume that this equilibrium is tightly regulated by the cell, and so the stability of the microtubules (as influenced by tubulin concentration, calcium levels, and availability or activity of MAPs) must be maintained within fairly narrow limits in order for microtubules to be able to carry out their myriad of functions. We might think of stability by considering a stability line as shown in FIGURE 7, where stability increases from left to right. The vertical hash marks indicate the permissible range of stability for continued microtubule function. Wild-type cells, on the average, reside near the midpoint of this range. If a depolymerizing drug (*e.g.* colcemid) is given to wild-type cells, those cells are shifted to the left along the stability line. An amount of drug sufficient to shift them outside the normal range (hashed bar, top line) is toxic because the microtubules will be too unstable (or unable to assemble sufficiently) to carry out their functions. Similarly, addition of a stabilizing drug (*e.g.* taxol; solid bar, top line) will be toxic if it shifts the cells too far to the right such that the microtubules become hyperstable.

When thought of in these terms, a mutation conferring colcemid resistance is one that produces more stable microtubules (cells are shifted to the right along the stability line). More colcemid is thus needed to shift the cells outside the normal range (hashed bar, line 2). This scheme predicts that such mutants should be cross-resistant to any agent that destabilizes microtubules but should be more sensitive to taxol that stabilizes the microtubules. As discussed above, this is, in fact, observed. It also explains why only two to threefold resistance is obtained. Mutations leading to greater resistance will shift the cells outside the normal range and thus will themselves be lethal. As will be discussed, this is what we believe has happened in the case of taxol-dependent mutants. The scheme also explains the codominant nature of the mutations and the fact that alterations in α or β tubulin (or MAPs) can lead to resistance to any given drug.

Mutations conferring taxol resistance are the opposite of what has just been discussed. These shift the cells to the left, that is, their microtubules are less stable, and therefore they can tolerate higher levels of taxol, but lower levels of depolymerizing drugs. One extreme example of this class of mutations is the taxol-requiring mutants. The lesion in these cells has destabilized the microtubules to such an extent that they are now outside the normal range and unable to survive under normal conditions. The addition of a small amount of an exogenously supplied stabilizing agent, taxol, however, shifts the cells back into the normal range and they can now survive as taxol-dependent cells. One would predict, however, that addition of too much taxol would still be toxic, because it would make the microtubules hyperstable. This was observed.[9]

The usefulness of any hypothesis lies not only in its ability to explain data, but also in its ability to make predictions and suggest new experiments. One such prediction is that colcemid-dependent mutants should exist. Until now it has been difficult to test this prediction because of the low frequency of tubulin mutants among colcemid-resistant cells. Recently, however, it has been reported in Gottesman's laboratory[20] that a colcemid-dependent mutant was created by transferring multiple copies of a mutant β-tubulin gene from Cmd-4 into wild-type cells. Presumably, the introduction of high levels of mutant β tubulin into these cells has hyperstabilized the microtubules to such an extent that an exogenously supplied destabilizer, colcemid, is required to keep microtubule stability within acceptable limits. Consistent with this interpretation, the colcemid-dependent mutant is very taxol sensitive.

It is our hope that this model is accepted in the spirit in which it is presented: not as dogma, but rather as a framework for understanding mutant properties that have been observed. We expect that the model will evolve into more sophisticated forms as more data is generated that requires explanation, and as further predictions of the model are tested.

REFERENCES

1. DUSTIN, P. 1978. Microtubules. Springer-Verlag. Berlin.
2. HORWITZ, S. B., J. PARNESS, P. B. SCHIFF & J. J. MANFREDI. 1982. Cold Spring Harbor Symp. Quant. Biol. **XLVI:** 219–226.
3. LING, V. & L. H. THOMSON. 1974. J. Cell. Physiol. **83:** 103–116.
4. DANO, K. 1973. Biochim. Biophys. Acta **323:** 466–483.
5. INABA, M., H. KOBAYASHI, Y. SUKURAI & R. K. JOHNSON. 1979. Cancer Res. **39:** 2200–2203.
6. BECH-HANSEN, N. T., J. E. TILL & V. LING. 1976. J. Cell. Physiol. **88:** 23–32.
7. CABRAL, F. & G. SCHATZ. 1979. Methods Enzymol. **56G:** 602–613.
8. LING, V., J. E. AUBIN, A. CHASE & F. SARANGI. 1979. Cell **18:** 423–430.
9. CABRAL, F. 1983. J. Cell Biol. **97:** 22–29.
10. CABRAL, F., L. WIBLE, S. BRENNER & B. R. BRINKLEY. 1983. J. Cell Biol. **97:** 30–39.
11. CABRAL, F., I. ABRAHAM & M. M. GOTTESMAN. 1981. Proc. Natl. Acad. Sci. USA **78:** 4388–4391.
12. CABRAL, F., I. ABRAHAM & M. M. GOTTESMAN. 1982. Mol. Cell. Biol. **2:** 720–729.
13. CABRAL, F., M. SOBEL & M. M. GOTTESMAN. 1980. Cell **20:** 29–36.
14. SCHIBLER, M. J. & F. CABRAL. 1985. Can. J. Biochem. Cell Biol. **63:** 503–510.
15. WARR, J. R., D. J. FLANAGAN & M. ANDERSON. 1982. Cell Biol. Int. Rep. **6:** 455–460.
16. GUPTA, R. S., T. K. W. HO, M. R. K. MOFFAT & R. GUPTA. 1982. J. Biol. Chem. **257:** 1071–1078.
17. OAKLEY, B. R. & N. R. MORRIS. 1981. Cell **24:** 837–845.
18. JHA, K. K. & H. L. OZER. 1976. Somatic Cell Genet. **2:** 215–223.
19. PONTECORVO, G. 1975. Somatic Cell Genet. **1:** 397–400.
20. CABRAL, F., M. SCHIBLER, I. ABRAHAM, C. WHITFIELD, R. KURIYAMA, C. MCCLURKIN, S. MACKENSEN & M. M. GOTTESMAN. 1984. *In* Molecular Biology of the Cytoskeleton. G. G. Borisy, D. Cleveland & D. Murphy, Eds.: 305–317. Cold Spring Harbor Press. New York.

Tubulozole: A New Stereoselective Microtubule Inhibitor[a]

M. DE BRABANDER,[b] G. GEUENS,[b]
R. NUYDENS,[b] R. WILLEBRORDS,[b] M. MOEREMANS,[b]
R. VAN GINCKEL,[b] W. DISTELMANS,[b] C. DRAGONETTI,[c]
AND M. MAREEL[c]

[b]Laboratory of Cellular Biology and Chemotherapy
Department of Life Sciences
Janssen Pharmaceutica Research Laboratories
B-2340 Beerse, Belgium

[c]Laboratory of Experimental Cancerology
Department of Radiotherapy and Nuclear Medicine
University Hospital
B-9000 Gent, Belgium

INTRODUCTION

Microtubule inhibitors have been of paramount importance in elucidating mechanisms of assembly and functional aspects of microtubules both in the test tube and in living cells and organisms. Colchicine, whose effects on mitosis were known long before microtubules were known as such, has played an irreplaceable role in the rapid progression of the field.[1] It was the first substance that was shown to bind with high affinity to tubulin subunits, to inhibit the association of subunits into microtubules, and to inhibit the assembly or destroy microtubule systems in many cell types. These are now the standard criteria for any compound to be ranked as a bona fide microtubule inhibitor. A number of other plant alkaloids fulfill these criteria to various degrees of specificity: vinca-alkaloids, podophylotoxin, rotenone, steganacin, and maytansine. They all have in common a number of methoxy groups that are crucial for their activity. Sulfhydryl reagents and physical agents such as high pressure and low temperature are also known to affect microtubule assembly.[1] They also affect many other systems, however, and thus cannot be regarded as specific inhibitors.

A class of new synthetic inhibitors, the benzimidazole carbamates, of which nocodazole is the most widely used, was introduced around 1975.[2,3] They had been used as anthelmintic, antifungal, or antineoplastic agents. Although these drugs apparently bind to the colchicine site on tubulin they often show a remarkable difference in species specificity. One of the major advantages of nocodazole appeared to be its very fast penetration in cells and binding to tubulin and its equally fast reversibility. This allows one, in particular, to study more accurately the dynamics of microtubule assembly and disassembly in the living cell.

In particular, because of the activity against invasion and metastasis formation by malignant cells, we conduct a continuing systematic search for new potential microtubule inhibitors. Recently, we found a new stereoselective microtubule inhibitor:

[a]This work was supported by a Grant from the I.W.O.N.L., Brussels, Belgium.

tubulozole (proposed non-proprietary name). The compound exists as a *cis*-isomer called hereafter tubulozole-C and as a *trans*-isomer called tubulozole-T (FIGURE 1). The *cis*-isomer appears to be a potent and specific microtubule inhibitor, the *trans*-isomer being inactive at 100 times higher concentrations. In the following we will give a condensed description of our findings thus far. More detailed information is available in reference 4 and in several papers that are in press.

EFFECTS ON TUBULIN POLYMERIZATION *IN VITRO*

Tubulozole-C inhibits the polymerization of dog brain tubulin in a dose-dependent way (FIGURE 2). Both the rate of assembly and the total amount of polymer are decreased. In parallel experiments, tubulozole-C was more active than nocodazole and

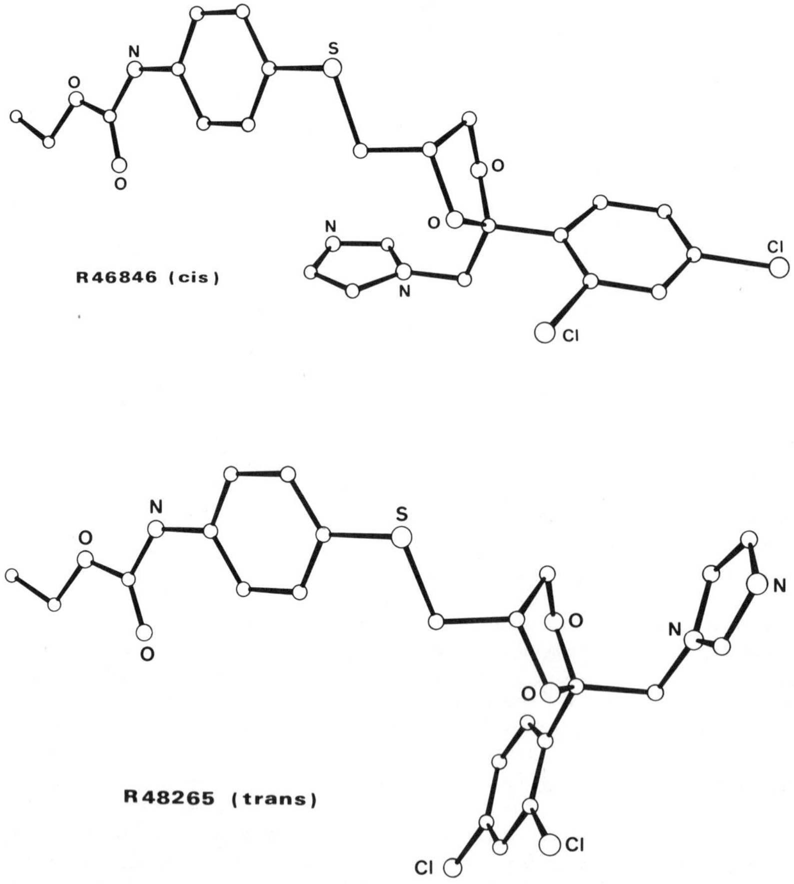

FIGURE 1. Three-dimensional representation of tubulozole-C (R 46 846) and tubulozole-T (R 48 265) based on X-ray data of related compounds and quantum chemical calculations. Obtained from H. Moereels, Janssen Pharmaceutica.

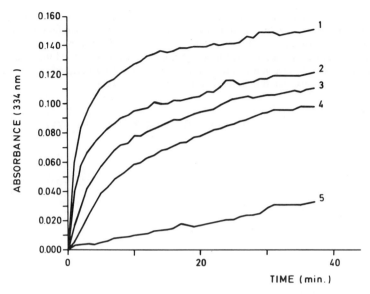

FIGURE 2. The effect of tubulozole-C on dog brain tubulin polymerization. Concentrations of tubulozole-C: $1 = 0$; $2 = 10^{-7} M$; $3 = 5 \times 10^{-7} M$; $4 = 10^{-6} M$; $5 = 5 \times 10^{-6} M$. Dog brain tubulin was prepared by one polymerization-depolymerization cycle in buffer A (0.1 M piperazinedi-ethane sulfonic acid (PIPES); 1 mM MgSO$_4$; 1 mM ethylene glycol bis (β-aminoethyl ether)-N,N,N',N'-tetraacetic acid (EGTA) pH 6.94). The final protein concentration in the assay was 2.4 mg/ml. Polymerization was initiated at 37°C with 1 mM guanosine triphosphate (GTP), and followed turbidimetrically at 334 nm. The rate of polymerization was determined by measuring the slope of the linear part of the curve.

colchicine (FIGURE 3). Tubulozole-T produced a very slight inhibition only at high concentrations. This could be accounted for by the contamination of the batch used with less than 1% of tubulozole-C.

PRIMARY AND SECONDARY EFFECTS ON CULTURED CELLS

To investigate the effects on cellular microtubules, cells were treated with the compound and processed for immunocytochemistry using affinity-purified antitubulin antibodies and the peroxidase-antiperoxidase procedure[5] or embedded for electron microscopy. Transformed (MO$_4$) and nontransformed (MO) C$_3$H mouse embryonal cell lines and PTK$_2$ cells were used. In cells treated with tubulozole-C ($1.8 \times 10^{-6} M$ or higher) virtually all cytoplasmic and mitotic microtubules are depolymerized within 30–60 min (see *e.g.* FIGURE 4). At concentrations between $1.8 \times 10^{-6} M$ and $1.8 \times 10^{-7} M$ partial disassembly was seen.

Tubulozole-T did not affect the microtubule system at concentrations up to $1.8 \times 10^{-5} M$. The secondary effects of tubulozole-C were identical to those produced by nocodazole or colchicine. Interphase cells lost their polarized shape by the appearance of undulating membranes all around the cell periphery. Saltatory organelle motility was arrested. Golgi elements, lysosomes, and mitochondria were dispersed randomly within the cytoplasm. The centrosome too became dislodged from its normal perinuclear location. Prolonged treatment resulted in the accumulation of intermediate filament coils and annulate lamellae. Mitotic cells failed to assemble a spindle and were arrested in a rounded state. The chromosomes were dispersed in the cytoplasm, and chromatid separation did not occur. After about 5–6 h, the chromatids were enveloped separately in a new nuclear envelope. Multiple irregular cleavage furrows appeared, and finally the cells flattened out on the substratum without cell division having occurred. These postmitotic cells usually contained multiple irregular nuclei. Tubulozole-T failed to induce any of these alterations. This confirms that the secondary effects described, which are also induced by other microtubule inhibitors,[2] are thus indeed specific consequences of microtubule disassembly and not due to nonspecific side effects.

Both tubulozole-C and -T, at the same concentration ($1.8 \times 10^{-5}\ M$), induced mitochondrial alterations: swelling, loss of cristae, and clearing of the matrix. The mitochondrial damage is thus not due to microtubule dissassembly and apparently not involved in the microtubule inhibitory effect of the *cis*-isomer. It provides a clear-cut

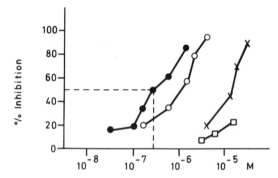

FIGURE 3. Dose-response curves for tubulozole-C (●), tubulozole-T (□), nocodazole (○), and colchicine (x), on the inhibition of dog brain tubulin polymerization. Each point is the mean value of three experiments.

example of how an isomer not showing the specific effect can nevertheless produce the same side effects.

The effects of tubulozole on microtubule integrity were rapidly reversible. After washing the cells with regular medium, normal cytoplasmic microtubule complexes and spindles are reconstructed within 30–60 minutes.

EFFECTS ON MALIGNANT NEOPLASMS *IN VIVO*

One of the major therapeutic applications of microtubule inhibitors is the treatment of leukemias and solid malignant tumors. Tubulozole-C clearly prolongs the survival of mice bearing transplantable tumors (TABLE 1).[4] Its activity compares favorably with that of reference drugs in clinical use such as cyclophosphamide and vincristine, the latter having the same mechanism of action as tubulozole-C. Particularly, solid tumors appeared more susceptible to tubulozole. Moreover, tubulozole is also active on a vinca-resistant subline of leukemia P-388 (Ashirawa and Horimoto, Kyowa Hakko, Japan; personal communication).

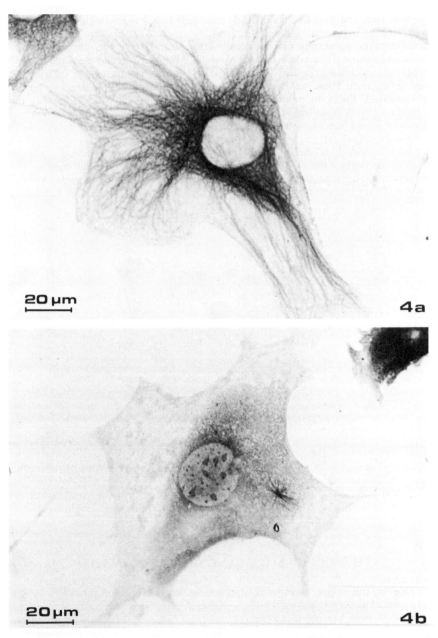

FIGURE 4. Microtubule network of an untreated MO cell (**a**) and a cell treated with 1.8×10^{-6} *M* tubulozole-C for 4 h (**b**). In the treated cell, only a few microtubules are seen radiating from a dislocated centrosome. Peroxidase-antiperoxidase staining (Original magnification × 675; reduced by 4 percent.)

TABLE 1. Effects of Optimal Dosages of Tubulozole, Cyclophosphamide (CPA), and Vincristine Sulfate (VCR) on MO_4 Sarcoma, L_{1210} Leukemia, and TA_3-HA Mammacarcinoma

Drug	Dose (mg/kg)	Scheduled Day of Dosing	Tumor	Inoculum Site	Median Survival Time Percent	No. of Survivors >60 Days
VCR	1	1, 5, 9, 13	L_{1210}	i.p.	186	0
	0.5				200	0
	1	1, 5, 9, 13	MO_4	i.p.	111	0
	0.5				111	0
	1	1, 5, 9, 13	TA_3-HA	i.p.	228	0
	0.5				162	0
CPA	160	1, 5, 9, 13	L_{1210}	i.p.	220	0
	80				227	0
	160	1, 5, 9, 13	MO_4	i.p.	148	0
	80				136	0
	160	1, 5, 9, 13	TA_3-HA	i.p.	164	0
	80				127	0
	40				105	0
Tubulozole	160	1, 5, 9, 13	L_{1210}	i.p.	88	0
	80				188	0
	40				150	0
	20				125	0
	80	1, 5, 9, 13	MO_4	i.p.	189	0
	40				148	0
	20				119	0
	10				100	0
	160	1, 5, 9, 13	TA_3-HA	i.p.	>500	5
	80				>500	6
	40				>500	5
	20				146	1

Six mice per group are injected intraperitoneally (i.p.) each either with 1×10^6 MO_4 cells, 1×10^6 L_{1210} cells, or 1×10^4 TA_3-HA cells. Treatment with tubulozole (suspension) was performed i.p. on days 1, 5, 9, and 13 after tumor inoculation. The median survival time of the treated animals was expressed as a percentage of the median survival time of the control animals (MST %). When MST was 125% or higher, the compound was considered active. More than 4 long-term survivors (for over 60 days) per group were expressed as MST >500 percent.

In order to see whether the antineoplastic effect *in vivo* was due to inhibition of microtubule assembly in the cancer cells, mice were inoculated with L 1210 leukemia cells and treated with a single dosage of tubulozole-C. Malignant cells and normal leukocytes were recovered from the peritoneal cavity and processed for electron microscopy. Microtubules were disassembled in malignant cells for up to 48 h after treatment with 80–160 mg/kg of tubulozole-C. Dispersal of Golgi elements in interphase cells and of chromosomes in the arrested mitotic cells were seen as clear-cut functional indicators of microtubule disturbance. Normal leukocytes residing in the same cavity retained a well-organized microtubule system and normal organelle topography even in animals treated with 640 mg/kg. Accumulation, however, of C-mitosis in the bone marrow was seen after treatment.

Treatment of mice with tubulozole-T did not have any effect on tumor growth.

EFFECTS ON DIRECTIONAL MIGRATION AND MALIGNANT INVASION

The malignant character of a tumor is essentially dependent on the propensity of cancer cells to invade the surrounding tissue, to penetrate into blood and lymphatic vessels, to penetrate the vessel walls and to invade distant organs where they grow as secondary metastases. Mareel *et al.*[6] have shown during recent years that malignant invasion of chick heart fragments in organotypical culture by malignant cancer cells

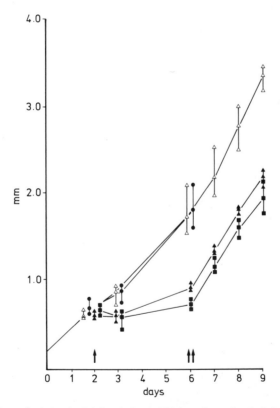

FIGURE 5. Effects of tubulozole-C (■), tubulozole-T (●), and nocodazole (▲) (all 1 μg/ml) on directional migration of MO_4 cells. Aggregates of MO_4 cells having a diameter of 0.2–0.3 mm were explanted on a coverslip in a Leighton tube. The mean diameter of the circular area covered by MO_4 cells that had migrated radially from the aggregate was used as an index of directional migration. Drugs or solvents were added (↑) two days after explantation and removed (↑↑) six days after explanation. Nocodazole and tubulozole-C arrest directional migration in a reversible way. Tubulozole-T has no effect.

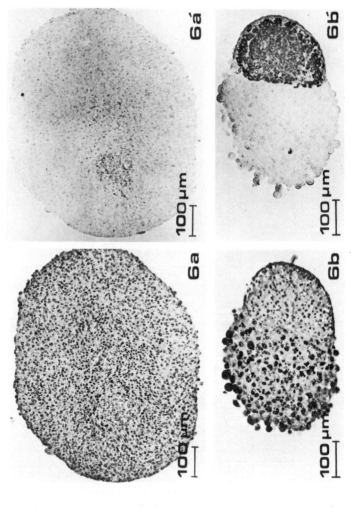

FIGURE 6. Effects of tubulozole-C and -T (1.8×10^{-6} M) on malignant invasion. MO_4-cell aggregates (0.3 mm diameter) were confronted with precultured fragments of nine-day-old embryonic chick cardiac muscle (0.4 mm diameter) in individual cultures on a gyratory shaker.[7] Drugs were added at the onset of the confrontation. Confronting pairs were fixed after four days. Consecutive 8 μm thick sections were stained with hamatoxylin-eosin (**a, b**) and with an antiserum against embryonic chick heart (**a', b'**).[8] In cultures incubated with tubulozole-T (**a, a'**), MO_4 cells have entirely replaced the heart tissue. In cultures incubated with tubulozole-C, MO_4 cells and the heart fragment are separated by a sharp demarcation line ($\times$ 100).

can be arrested by microtubule inhibitors and not by other cytotoxic or cytostatic antineoplastic drugs. They have suggested that this is probably due to the inhibition of directional cell migration by microtubule inhibitors. Therefore, the effects of both isomers of tubulozole on directional cell migration and on malignant invasion *in vitro* and *in vivo* were investigated. FIGURE 5 shows that both nocodazole and tubulozole-C arrest directional migration of cancer cells in culture, whereas tubulozole-T is completely inactive. FIGURE 6 shows that malignant invasion *in vitro* is also arrested by tubulozole-C but not by the *trans*-isomer.

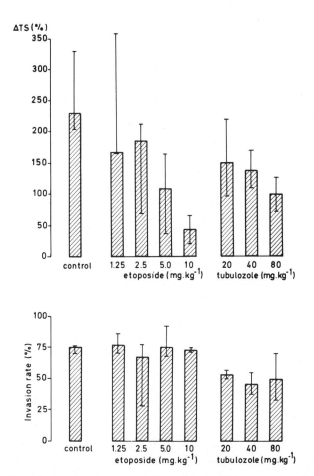

FIGURE 7. Effect of tubulozole-C and etoposide on growth and invasion of malignant cells *in vivo*. Fragments of MO_4 tumors ($\pm$ 1 mm^3) are transplanted under the renal capsule in Swiss mice. The growth in diameter of the tumors, after seven days, was expressed as a percentage of the initial value (ΔTS %). The invasion rate is determined on sections as the depth of invasion into the kidney expressed as a percentage of the total tumor thickness. Animals were treated with saline or with the indicated drug dosages every two days, starting from day one after implantation. Etoposide, a cytostatic drug not acting on microtubules inhibits the growth of the tumors but does not affect invasion. Tubulozole-C significantly inhibits invasion even at dosages that have a smaller effect on tumor growth. The histogram gives median and extreme values.

In order to investigate the potential of tubulozole-C to inhibit malignant invasion *in vivo*, an assay recently developed by Distelmans *et al.*[9] was used. Small tumor fragments were transplanted under the renal capsule of mice. After 6–7 days, measurements were made of the size of the tumor and of the depth of penetration into the kidney tissue. The percentage of the total tumor thickness that penetrated into the kidney was taken as a measure of the invasion rate. FIGURE 7 shows that potent inhibition of tumor growth by etoposide, which does not affect microtubules, has no effect on the invasion rate, whereas this characteristic is clearly inhibited by tubulozole-C.

CONCLUSIONS

Tubulozole-C appears to be a novel type of specific microtubule inhibitor. Structurally it is related to the benzimidazole carbamates by its carbamate moiety, which is essential for the activity of both types of compounds (our unpublished structure activity studies). Tubulozole-C inhibits tubulin polymerization *in vitro* and in cells. It also produces the same secondary effects as classical microtubule inhibitors. The possible binding of the compound to tubulin is currently being investigated.

The most interesting aspect of the compound is that it exists in two isomeric forms: one inactive and the other active on microtubules. This property should be useful whenever questions arise about the specificity of existing microtubule inhibitors in affecting any physiological response. Lumicolchicine has been used often for that purpose.[1] Lumicolchicine, however, is not a true isomer of colchicine; the seven-membered tropolone moiety is changed into a conjugated five-membered and four-membered ring system. Lumicolchicine may thus not only have lost the tubulin binding activity, but also unknown side effects of colchicine. This is less likely to be the case when using stereoisomers.

REFERENCES

1. DUSTIN, P. 1978. Microtubules. Springer-Verlag. Heidelberg.
2. DE BRABANDER, M. J., R. VAN DE VEIRE, F. AERTS, M. BORGERS & P. A. J. JANSSEN. 1976. The effects of methyl[5-(2-thienylcarbonyl)-1H-benzimidazol-2-yl]carbamate, (R 17 934; NSC 238159), a new synthetic antitumoral drug interfering with microtubules, on mammalian cells cultured *in vitro*. Cancer Res. **36:** 905–916.
3. HOEBEKE, J., G. VAN NIJEN & M. DE BRABANDER. 1976. Interaction of nocodazole (R 17 934), a new antitumoral drug, with rat brain tubulin. Biochem. Biophysics Res. Commun. **69:** 319–324.
4. VAN GINCKEL, R., M. DE BRABANDER, W. VANHERCK & J. HEERES. 1984. The effects of tubulozole, a new synthetic microtubule inhibitor, on experimental neoplasms. Eur. J. Cancer Clin. Oncol. **20**(1): 99–105.
5. DE BRABANDER, M., J. DE MEY, M. JONIAU & G. GEUENS. 1977. Immunocytochemical visualization of microtubules and tubulin at the light- and electron-microscopic level. J. Cell Sci. **28:** 283–301.
6. MAREEL, M. M. K. & M. J. DE BRABANDER. 1978. Effect of microtubule inhibitors on malignant invasion *in vitro*. J. Natl. Cancer Inst. **61:** 787–792.
7. MAREEL, M., J. KINT & C. MEYVISCH. 1979. Methods of study of the invasion of malignant C_3H-mouse fibroblasts into embryonic chick heart *in vitro*. Virchows Arch. B **30:** 95–111.
8. MAREEL, M. M., G. K. DE BRUYNE, F. VANDESANDE & C. DRAGONETTI. 1981. Immunohistochemical study of embryonic chick heart invaded by malignant cells in three-dimensional culture. Invasion Metastasis **1:** 195–204.
9. DISTELMANS, W., R. VAN GINCKEL, W. VANHERCK & M. DE BRABANDER. 1985. The kidney invasion test: An assay allowing macroscopic quantification of malignant invasion *in vivo*. Invasion Metastasis **5:** 170–184.

Drugs with Colchicine-like Effects that Specifically Disassemble Plant but Not Animal Microtubules[a]

A. S. BAJER AND J. MOLÈ-BAJER

Department of Biology
University of Oregon
Eugene, Oregon 97403

INTRODUCTION

Studies on the effects of different microtubule (MT) disassembling drugs have demonstrated which processes are dependent upon the presence of MTs. An important example of such dependence is chromosome movement in mitosis. The final effect, total disassembly of the MTs of the mitotic spindle (c-mitosis), can be achieved through different pathways and does not require a highly specific drug.

Temporally controlled direct disassembly of MTs is necessary, however, to understand the mechanism involved in such MT-dependent cellular events. Such disassembly of MTs is especially difficult in living cells of higher plants due to a lack of highly specific drugs. Colchicine, which binds specifically to animal tubulin, has a very low affinity for plant tubulin[1-3] and unphysiologically high doses are required to obtain the desired effect (100–1000 × higher than for animal cells). Recent studies of Morejohn and Fosket[1,4] indicate that the herbicides amiprophos-methyl [0-methyl-0-(4-methyl-6-nitrophenyl)-N-isopropyl-phosphorothioamidate, abbreviated APM] and oryzalin [4(Dipropylamino)-3,5-dinitrobenzenesufonamide, synonyms: El 119, Dirimal, Ryzelan, Surflan] bind specifically to higher plant tubulin *in vitro* in the same way as colchicine does to animal tubulin.

We report here the effects of both drugs on chromosome movements *in vivo* and correlate them with sequential changes in MT behavior in the mitotic spindle. Our material, naked natural protoplasts of *Haemanthus* endosperm, has the largest mitotic spindle we know, with unmatched clarity. The size of the *Haemanthus* spindle allows us to follow spatial arrangements of MTs within the spindle at the light microscopic level. The long duration of all mitotic stages as compared to all other objects permits the temporal dissection of any stage of mitosis. The absence of centrioles eliminates some of the control of MT nucleation and arrangement superimposed by these organelles on the mitotic cycle.

The data presented here concern the selective disassembly of MTs at different stages of anaphase, which is the most suitable stage for these types of studies. They also point to the continuous assembly of MTs as *sine qua non* condition for chromosome movements in anaphase.

[a]This work was supported by NSF research Grants (PCM-8310016 and DCB-8310016).

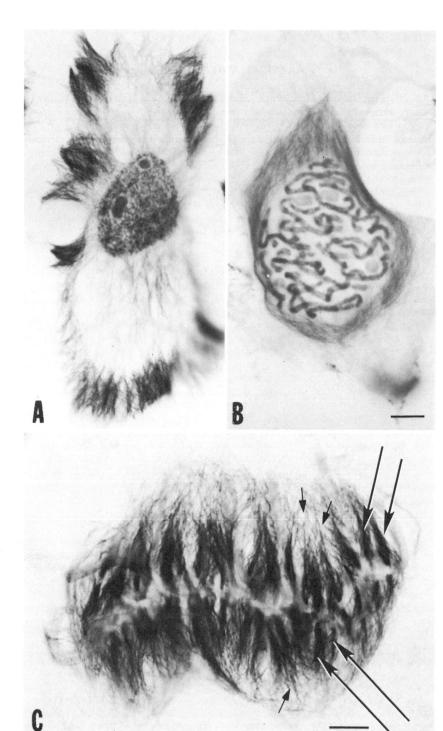

MATERIAL AND METHODS

Endosperm of *Haemanthus katherinae* Bak. (globe lily, a south African plant) and primary cultures of heart endothelium from *Xenopus leavis* were used as material. Observations on chromosome movements *in vivo* were correlated with the arrangements of MTs after immunogold stain (IGS) for tubulin as previously described.[5] Monospecific affinity column purified rabbit antitubulin antibody against dog brain tubulin and antibody against Rose β tubulin were used as primary antibodies. Media and the technique for *Xenopus* heart culture were the same as for the newt.[6]

The technique for studies on endosperm *in vivo*[7] was modified. Cells were sandwiched between layers of 0.5% Phytagar (Gibco) with 3.5% glucose, and 1% Gelrite[8] with 3.5% glucose. Oryzalin was dissolved in buffered glucose or mannitol medium for *Haemanthus* ($0.05 M$ citric acid, sodium citrate buffer, pH 5.2) and in culture medium for *Xenopus;* in both cases the media contained 1% ethanol. For APM, 1% dimethylsulfoxide (DMSO) was substituted for 1% ethanol. Controls showed no influence of ethanol or DMSO at these concentrations. The first detectable effects of ethanol on mitosis in endosperm occurred in 5% ethanol. The concentrations of oryzalin and APM used ranged from 0.1 mM to 1 nM. Preparations in all concentrations were fixed 2, 5, 10, 15, 20, 30, 45 min, and 1 and 2 hr after incubation in the drug solutions.

Mitosis in living cells was followed and recorded either with a Dage 67 Newvicon video camera and a Sony TVO 9000 video cassette recorder, or cells were observed visually under low magnification and their progress monitored. The latter technique does not permit measurement of chromosome movements, but allows a rapid assessment of the drug's action. At chosen stages of mitosis (mostly anaphase), cells were perfused with different concentrations of oryzalin or APM and either followed *in vivo* or fixed and processed for IGS. Perfusion often results in rounding of cells; consequently more cells are lost during IGS processing than in controls. Over 500 cells were followed with the latter technique, and over 50 recorded under high magnification. The data reported here concern mostly short term (up to 1 hr) effects of the drugs and are based on over 8000 mitotic cells studied with IGS method. Studies on longer effects and on their reversal by taxol will be published elsewhere.

RESULTS

There are two extreme cell populations in *Haemanthus* endosperm, a feature that, according to our knowledge, has not been reported: large cells in which the spindle can be easily flattened and small compact cells that do not flatten easily (in our hands).

FIGURE 1. Normal (control) mitosis in endosperm of *Haemanthus katharinae*. IGS method. Bar = 10 μm on **B** for **A,** and **B** and on **C** for **C.**
A: Interphase. MTs form an irregular mesh throughout the cytoplasm. A denser accumulation of MTs is seen at the cell periphery. Such arrangement is characteristic for cells fixed 30–60 min after preparation.
B: Prophase. First stages of formation of the mitotic spindle: MTs form a "clear zone" around the nucleus. Note the bushy outlines of the clear zone. The cytoplasm is nearly devoid of MTs.
C: Metaphase. Kinetochores are aligned in one plane at the equator. The spindle is an aggregate of individual kinetochore fiber complexes. Each kinetochore fiber complex is a "fir tree" composed of the kinetochore fiber proper (long arrows) and non-kinetochore MTs, which form the branches (short arrows).

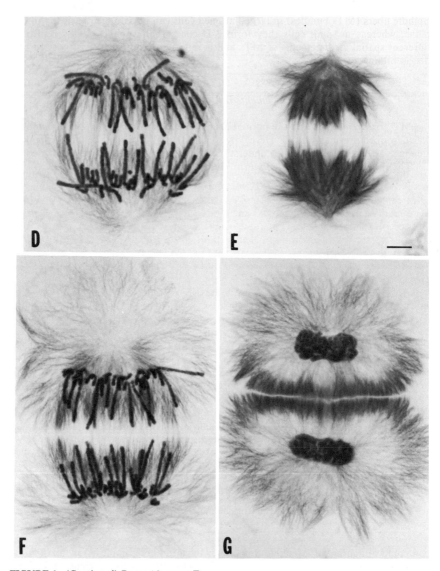

FIGURE 1. (Continued) Bar = 10 μm on **E.**

D and **E:** Midanaphase. Cells in **D** and **E** are in the same stage; **D** is a flattened cell and **E** an unflattened one. The "MT side branches of the fir trees" elongate considerably, extend into the interzone between the trailing chromosome arms, and form the bushy outlines of each half-spindle. Asters begin to form (arrows). Only remnants of the interzonal MTs are present.

F: End of midanaphase. Well developed asters. Comparison with **D** and **E** illustrates the most variable features between different cells, that is, the development of asters and organization of the interzone.

G: Telophase. Two daughter nuclei and phragmoplast with a cell plate are formed. MTs begin to reorganize. In time they will be arranged as in **A**. The uniform distribution characteristic for early telophase is already disturbed, and a slightly denser ring of MTs at the cell periphery is present.

Spindle fibers (MTs bundles) in large flattened cells can be arranged basically in one plane, whereas in small cells they form 3-D cones (half spindles). This results in different spatial arrangements of MTs and is reflected in different sensitivity to MT disassembling agents.

Control: MTs Arrangements in Untreated Endosperm Spindles

MT arrangement during the normal course of endosperm mitosis in *Haemanthus* was described previously.[5,9,10-12] Here only some essential features of spindle organization in control cells, pertinent for understanding the drug action, will be mentioned.

The mitotic spindle of *Haemanthus* endosperm has a pinnate structure.[11,12] It is

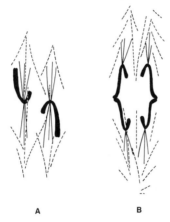

A   B

FIGURE 2. Schematic drawing of a segment of a dissociated mitotic spindle. Continuous lines are kinetochore MTs that form the "trunk" of the microtubular "fir tree." Dotted lines are MTs that form the side branches. Both types of MTs intermingle, and kinetochore fibers proper may contain some MTs from the side branches of neighboring "fir trees."
A: Metaphase. Side branches do not reach the equatorial plate.
B: Midanaphase. Side branches elongate in the direction of the equator. They reach beyond the kinetochores and stretch between the trailing chromosome arms. In later stages they stretch into the interzone and associate with the remnants of interzonal MTs.

composed of aggregates of single units—kinetochore fiber complexes shaped like fir trees. Within a kinetochore fiber complex, the MTs attached to the kinetochore form the trunk of the fir tree and non-kinetochore MTs are the branches (FIGURE 1C). The "fir tree" organization of kinetochore fibers and whole half-spindles become increasingly pronounced until mid-, or occasionally, late anaphase. The side branches, which in earlier stages do not reach even kinetochores, elongate toward the equator, reach beyond kinetochores and penetrate the interzone (FIGURE 1D, E, and F). In late anaphase they have been called polar MTs,[5] although they arise earlier.[11,12] Their ends laterally associate with the remnants of interzonal MTs. A schematic drawing illustrates this process (FIGURE 2). The "fir tree" structure of kinetochore fibers is best seen in flat cells (FIGURE 1C) and as a bushy outline of half-spindles in compact spindles, where individual fibers are not resolved (FIGURE **D–F**). After midanaphase "the trunk" (kinetochore fiber proper) shortens, and the "fir tree" is composed mostly of branches pointing toward the equator.

Effects of APM and Oryzalin

Interphase

Interphase MTs are sensitive to the action of both oryzalin and APM. Within a few minutes (at 0.10 μM and higher) the irregular mesh of MTs disappears and groups of MTs radiating from a common point (center), chromosome-free spindles and extra phragmoplasts, appear in excessive numbers (FIGURE 7D, E). They develop also at 10 nM after a longer time (15 min).

The effect of APM and oryzalin at 100 nM (and higher) on chromosome movements and the arrangements of MTs (and many times cell shape) can often be

TABLE 1. Effects on MT Spindle Organization APM (Amiprophos-methyl)

Concentration	Time	Cells Affected	Cells with Undetected Changes	Total
0.1 μM	2′	89 (38%)	145 (62%)	234
	10′	659 (93%)	49 (7%)	708
	20′	1227 (98%)	26 (2%)	1257
	30′	128 (98%)	3 (2%)	131
	45′	68 (94%)	4 (6%)	72
1 μM	10′	487 (94%)	29 (6%)	526
	20′	215 (100%)	0	215
	2 hr	256 (98%)	6 (2%)	262
				3392
		Oryzalin		
0.1 μM	5′	1784 (96%)	86 (4%)	1870
	10′	233 (96%)	11 (11%)	244
	20′	436 (99%)	5 (1%)	441
	1 hr	1297 (100%)	0	1297
				3684

detected after 30 sec and in most cells about 1 min after the drug perfusion (TABLE 1). We could not detect any effects on the chromosomal cycle. Changes occurring during anaphase are most characteristic. The effects on other stages of mitosis are easier to follow by comparison to anaphase.

Observations in Vivo

The actions of oryzalin and APM on the living mitotic spindle are very similar. The effects of the latter seem to be more gradual and easier to control. Therefore further description will mostly concern APM.

At nanomolar concentration, the effect of APM is not very pronounced. Anaphase is not prolonged. Chromosome movement, especially in later stages of anaphase, however, tends to be asynchronous, and such abnormalities as nondisjunction appear.

At 10 nM concentration, chromosome movement slows down immediately after

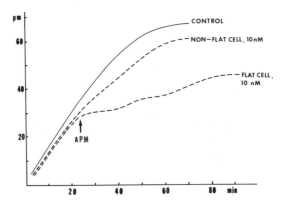

FIGURE 3. Effect of APM *in vivo*. Anaphase chromosome movement in 10 n*M* APM as compared to the control cell. An arrow marks the start of APM perfusion. The movement of chromosomes in a flattened cell is affected more than in an unflattened cell.

perfusion (FIGURE 3). The rate of chromosome movement is 1.5–3 × slower (depending on the cell) than in the majority of normal cells. The only other clear effect is the asynchrony of chromosome movements in anaphase, an effect that is rarely observed under the influence of any other drug. Metaphases enter into anaphase at this concentration.

Starting from a concentration of 0.1 μ*M* and up, both APM and oryzalin stop chromosome movement almost instantaneously, that is, as soon as we can measure it (within 30 sec). These are the only drugs we know that arrest chromosome migration before any effects on the structure of the spindle can be detected. The arrest of chromosome movement may be followed in some cells by the movement of chromosomes back toward the equator. FIGURE 4 shows two examples of chromosome

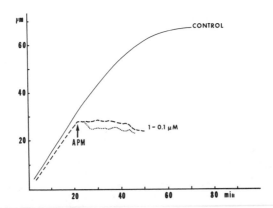

FIGURE 4. Effect of APM *in vivo*. Effect of 1–0.1 μ*M* concentration of APM on anaphase chromosome movement. Each curve is a compound curve from 10 cells. Continuous line: control; interrupted line: arrest of chromosome movements without subsequent movement toward the equator; dotted line: arrest followed by a "backward" movement of chromosomes. The arrow marks the start of APM perfusion.

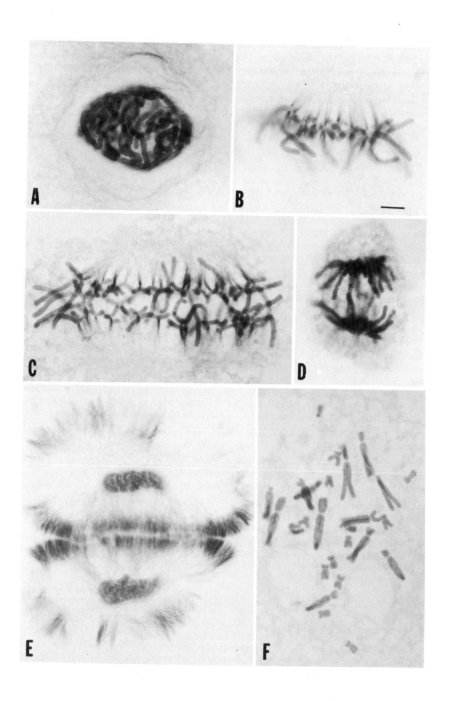

behavior—one where the chromosomes oscillate irregularly after stopping, and another where they move backwards.

The arrest of chromosome movements takes place independently of the stage of anaphase.[13] Because the chromosomal cycle is not inhibited, the arrested anaphase chromosome groups form nuclei that very often fuse. Prometaphases and metaphases do not enter anaphase. In prophase nuclei, the condensation of chromosomes proceeds, and breaking of the nuclear envelope takes place at a time characteristic for normal mitosis. Most of these observations, however, were not carried on long enough to state whether, after some specific MT rearrangement, mitosis was not completed. During preliminary, long-range observations, depending on the stage, either restitution nuclei or typical c-mitoses were formed (FIGURE 5F and 6G).

Changes in MTs Arrangement under the Influence of Drugs

After 2 min treatment with APM (100 nM and up), chromosome movement at any stage of anaphase is arrested. A detailed analysis of MT arrangement in the mitotic spindle at this time did not reveal any differences from untreated cells at the same stage. All classes of MTs are retained and there is no morphological evidence of any changes.

Between 5 and 10 min after the perfusion, a slight change in MT arrangement takes place. In early stages of anaphase the mitotic spindle is well developed— pronounced half spindles as well as kinetochore fibers (many of them with diverging MTs, a characteristic feature of active fibers) are present (FIGURE 5C). The side branches of MT "fir trees" (kinetochore fiber complexes) are, however, missing. In some cells one can also see a tendency of MTs for the kinetochore fibers proper (trunks of the tree) to form tight bundles (e.g., FIGURE 5C). In compact spindles, where one cannot resolve individual kinetochore fibers, the outlines of the half spindles are missing the normal bushy appearance (FIGURES 6C, D, 7B).

In later stages of anaphase, in addition to the morphological changes noted above, polar MTs (which normally elongate along the trailing chromosome arms and into the interzone at this stage) become shorter compared to controls (FIGURES 6C, D, 7B). In many cells at this stage, interzonal MTs are missing.

FIGURE 5. Effect of oryzalin. Bar = 10 μm on **B.**

A: Prophase. 100 nM for 20′. Comparable stage to FIGURE 1B. Only few MTs are left in the clear zone.

B: Metaphase. 100 nM 10′. Only MTs of kinetochore fibers proper persist. They form tight bundles that in some cells are still divergent (in respect to the poles). The side branches of "fir trees" are disassembled. Compare with FIGURE 1C.

C: Early anaphase. 100 nM 15′. Similar to 2B; only kinetochore fibers proper remain. They are still divergent. Compare with FIGURE 1D.

D: Late anaphase. 100 nM for 20′. All MTs in this cell are disassembled. This does not take place in all cells simultaneously, and remnants of MTs, especially at the polar region, may still occasionally be seen after 20′.

E: Telophase. 100 nm for 10′. Compare with FIGURE 1G. MTs in the proximity of the nucleus are disassembled, and only those that form a dense accumulation at the cell periphery and adjacent to the cell plate persist. This arrangement of MTs reminds us of that in control cells during the telophase-interphase transition; it develops usually about one hour after sister nuclei formation. Oryzalin accelerates this process at least three times.

F: Oryzalin. 100 nM for 2 hours. Cells which at the onset of oryzalin action were in prophase and prometaphase form a classical c-mitosis. In this flattened cell all 27 chromosomes are visible.

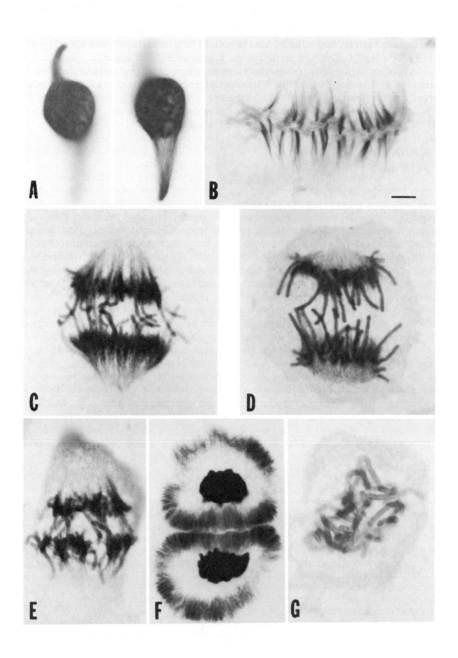

In all stages of anaphase after 10–20 min treatment with the drug, the MTs of kinetochore fibers proper are zipped (laterally associate starting adjacent to the kinetochores) and form tight bundles that persist for a long time. Polar MTs are much shorter (FIGURES 6E, 7C). With a prolonged exposure (20 up to 120 min) polar MTs disassemble further, and bundled kinetochore MTs (kinetochore fibers proper) gradually shorten until only stubs are left at the kinetochores (FIGURE 6E).

In conclusion, the changes described above are very consistent and characteristic of large flat and half flat cells. The following sequence of events occurs: instantaneous arrest of chromosome movement without any apparent change in the arrangement of MTs, followed by the disassembly of side branches of the "fir trees." This is followed by the disassembly of polar MTs proceeding from their distal ends, accompanied by the increased lateral interaction (zipping) of MTs in kinetochore fibers proper (trunk of the tree). The bundles of kinetochore MTs persist for a long time until they finally disassemble (FIGURE 5D).

In case of small, compact, and unflattened cells, the course of events is basically the same: chromosome movement is instantly arrested and a c-mitosis is formed. There is, however, an additional step before the final disassembly, which may require much more time than in flattened cells. The half-spindles in untreated small cells form tightly packed cones. This packing, judging from IGS penetration, becomes even tighter under the influence of the drugs. The first effect observed in compact cells is disappearance of bushy outlines of the half-spindles. Then the cones (the half-spindles) form a pointed structure that elongates and may curve along the cell periphery before its final disassembly (FIGURE 7B, C). Similar elongation of a few MTs before disassembly may occur in large half flat cells, if the individual kinetochore fibers are close enough to each other.

In prometaphase and metaphase (which are also arrested under the influence of these drugs), changes of MT arrangement are similar to those in anaphase: non-kinetochore MTs disassemble, whereas kinetochore MTs (kinetochore MTs proper) zip together, forming a difficult to disassemble unit (FIGURES 5B, and 6B).

A tendency of MTs to associate laterally and elongate is also seen in prophase (FIGURE 6A). Very often the clear zone has two exceptionally long and pointed "polar regions" or is multipolar; in the latter case, polar regions are also long and closely packed.

Telophases, after the formation of the cell plate, are least affected by the drugs. The phragmoplast at this stage is a rather rigid structure that does not disassemble

---

FIGURE 6. Effect of APM. Bar = 10 μm on **B**.
A: Prophase. 100 nM for 20'. Unflattened cell at two optical levels. Elongated and pointed clear zone is composed of tightly bundled MTs that are considerably resistant to further disassembly.
B: Metaphase. 100 nM for 20'. All MTs are disassembled, except tightly bundled kinetochore MTs that represent "the trunks of the fir trees."
C: Midanaphase. 100 nM for 10'. Anaphase is arrested. Polar MTs, which in control cells reach the ends of the trailing chromosome arms (FIGURE D and E), are disassembled at their distal ends, whereas their proximal ends are more tightly associated.
D: Late anaphase. 100 nM for 10'. Similar changes as in C, but disassembly is more advanced as demonstrated by shorter polar MTs. The bushy outlines of the half-spindles are also absent. Compare control FIGURES E and F.
E: Midanaphase. 100 nM 30'. Only short stubs of kinetochore fibers proper remain after longer action of APM. They bundle with remnants of polar MTs.
F: Telophase. 100 nM for 10'. All MTs are disassembled except the resistant bundles of MTs at the cell periphery and adjacent to the cell plate (compare control FIGURES 1G and 2E).
G: C-mitosis. 100 nM for 30'. No MTs are detectable.

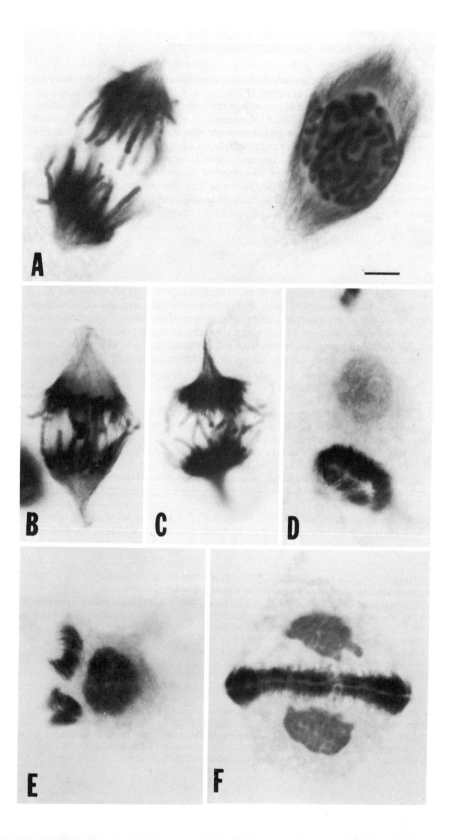

easily. After a prolonged exposure, however, it disassembles and only a narrow ring of tightly bound MTs remains (FIGURE 7F).

In conclusion, there are three important features characterizing the action of oryzalin and APM on mitosis in *Haemanthus:* (1) a stepwise and (2) selective disassembly of spindle MTs and (3) in some stages, acceleration of MT turnover without altering its pattern. The latter feature we can see in interphase and telophase. FIGURE 7D and E shows MT configurations in the cytoplasm after a 10 min APM treatment. Such arrays are found in untreated cells only about one hour after the preparation. The arrangement of MTs in telophase 10 min after perfusion with APM (FIGURE 6E) corresponds to the telophase arrangements in much later stages (about 1 hour).

Effect on Animal Mitosis

Exposure of *Xenopus* cells to oryzalin ranged from a few minutes to several hours.[14] There was no noticeable effect on mitosis in physiological concentrations of the drug. At 50 μM, which we consider an unphysiological concentration, the only disturbance occasionally observed was some desynchronization of anaphase. Anaphase chromosome movements in such cells were within normal range with respect to time, velocity, and the distance covered by chromosomes. The structure of the spindle and arrangement of MTs were indistinguishable from those in untreated anaphases.

Reversibility

The effects of even brief exposure to oryzalin are irreversible. Even after 5 min in 10^{-7} M oryzalin, followed by perfusion with a normal medium, spindles from prometaphase-anaphase do not recover, and a c-mitosis follows. Only in a very small number of cells (which at the time of perfusion were in late prophase just before breaking of the nuclear envelope) kinetochore MTs start to form after 1–2 hours. After 1 hr in normal medium following exposure to the drugs, only 31 (2.5%) of 1266 mitoses screened showed short MTs at the kinetochores. Because the chromosomal cycle is not affected and restitution nuclei are forming within 1–2 hr (depending on the stage), it is difficult to state whether the MTs would recover after a longer time.

FIGURE 7. Effect of oryzalin and APM. Bar = 10 μm on **A.**

A: Different sensitivity of MTs in different stages of mitosis in unflattened cells. Oryzalin, 100 nM for 10'. Anaphase movements are arrested and polar MTs are disassembled. The MTs of the clear zone form tighter bundles and still persist, in contrast to those in flattened cells that disassemble (FIGURE 2A).

B and C: Late anaphase in small unflattened cells (B, 100 nM for 10' and F, 100 nM for 20'). In these very compact half-spindles, anaphase is instantly arrested, but the half-spindles are resistant to the disassembly. The bushy outlines of the half-spindles are absent, and polar MTs are shortened considerably. Some MTs of each half-spindle elongate and curve along the cell periphery. After longer action of APM, these MTs will be disassembled. This demonstrates that the final disassembly may be proceeded by a temporary elongation.

D and E: Interphase. APM 100 nM for 10'. The MTs that formed an irregular meshwork are disassembled. Tight bundles of laterally associated MTs in cytoplasm arise.

F: Telophase. APM 100 nM for 20'. After a longer action, all MTs are disassembled except a narrow ring of tightly bound MTs around the cell equator, which persist for a long time.

DISCUSSION

Higher Plant Tubulin

Several factors make studies on the biochemistry of plant tubulin difficult. The main one is the lack of a plant tissue comparable to neuronal cells that are an abundant source of tubulin. Therefore, the isolation of plant tubulin in quantities sufficient for polymerization is not as simple as for neurotubulin.

Plant tubulin copolymerizes with neurotubulin.[15] Only with the use of a potent MT polymerizing agent, taxol, however, were Morejohn and Fosket[16] able to polymerize plant tubulin from Paul's Scarlet rose in sufficient quantity to make polypeptide maps. It appears that the β chains of plant and animal tubulin are similar, whereas the α chains differ considerably (see also reference 15). That may explain the somewhat different properties of plant and animal tubulin in response to drugs.

Specific Inhibitors of Higher Plant Microtubules

Colchicine

This plant alkaloid has been the most extensively studied mitotic poison[17] and has been widely used to produce plant polyploids. The molecular action of colchicine on animal tubulin in vitro is better understood than that of any other drug. Colchicine-tubulin binding is a two-step reaction: fast reversible followed by slow, resulting in the formation of the tight bond.[14,18] This may explain the difficulty in the reversal of colchicine action.

The mechanism of interaction of colchicine with plant tubulin and MTs is not clear.[19] Colchicine binds at much lower ratio to higher plant tubulin than to neurotubulin.[2,3] It inhibits higher plant mitosis and results in classical c-mitosis only at concentrations that are 10–1000 times higher than in animal cells. Concentrations as high as 0.2% have been used to produce polyploids[20] in plants.

APM and Oryzalin

Oryzalin is widely used in the USA as a herbicide, and APM is still in the experimental stage. The most extensively studied herbicide and MT inhibitor chemically related to oryzalin is trifluralin. It appears, however, that trifluralin adsorbs to glass,[21] and all data on this herbicide, including those on Haemanthus,[22] should be reevaluated.

Both APM and oryzalin are chemically unrelated to colchicine. It was initially thought that these drugs disassemble plant MTs through the release of Ca^{2+} or interfere with the Ca^{2+} transport.[23,24] Because the inhibition of polymerization of plant tubulin in vitro, however, occurs at concentrations 10 times lower and in living cells 100 times lower than those required to interfere with Ca^{2+} transport, this is most likely not the case. Similarly, there is no evidence that both drugs affect only plant but not animal membranes (because MTs in animal cells are not affected by their action, Morejohn and Pepper, unpublished results; Mole-Bajer, unpublished results). On the contrary, Hertel et al.[23] demonstrated that APM at high concentrations (10 μM and up) affects Ca^{2+} uptake by mitochondria of both plant and animal origin. APM binds specifically to plant tubulin and inhibits, in a concentration-dependent manner, the nucleation step of polymerization[4] of plant, but not animal, tubulin in vitro. It is not yet known where or how it binds to the tubulin molecule; its binding site is, however, different from that for colchicine. Oryzalin also binds specifically to plant tubulin *in vitro,* inhibits its polymerization, and partly depolymerizes taxol stabilized MTs.[25] It is possible, therefore, that these two herbicides, oryzalin and APM, are "the colchicines" of the plant kingdom.

Lateral Association and MT Stability in Vivo

The present observations demonstrate the existence of a very subtle balance between MT assembly-disassembly and lateral association in living cells. Disassembly of non-kinetochore MTs always precedes lateral association of the kinetochore MTs, independent of the nature of the factor causing disassembly (*e.g.* preparatory stress, cold, and drugs). The MT bundle is a "molecular superstructure" very resistant to further disassembly. We argue that the thermodynamics of such a superstructure are different from that of a single MT *in vitro*. This prevents (in our experiments) the rapid disassembly of the entire microtubular system by APM or oryzalin. The mechanism of resistance to depolymerization is unknown and may involve several molecular factors such as stable tabule-only polypeptide, (STOP (see Margolis *et al.,* this volume, p. 306)), bundling, or linking proteins and other endogenous mechanisms (*e.g.* see references 26–31). We have never been able, except by UV-microbeam (unpublished results), to disassemble *Haemanthus* MTs rapidly.

Differential Sensitivity of MTs to Disassembly

There is a different sensitivity of *Haemanthus* MTs to the drug's action (a) at different stages of mitosis, (b) between different classes of MTs at the same stage, and (c) along the same MT. In general, newly formed MTs or those that have both ends free are more easily disassembled. Thus, the variation of response of MTs to any experimental treatment depends on their intrinsic stability, spatial arrangement within the cell at the time of the experiment, and a variety of still unknown factors.

(a) MTs are more resistant to the action of oryzalin and APM in prophase and telophase than in any other stage. In the latter case they form a tightly packed structure that is absent in prophase, although the interaction between MTs increases.

(b) The most sensitive MTs from prometaphase until late anaphase are the side branches of kinetochore fiber complexes (polar MTs in later stages) and other nonchromosomal MTs (interzonal MTs before, but not after the cell plate formation). MTs of kinetochore fibers proper zip together and are resistant to further disassembly.

(c) Polar MTs in late anaphase disassemble first from their distal (distal to the pole) ends, whereas phragmoplast MTs (derived from polar MTs) in telophase disassemble from their proximal (proximal to the pole) end (FIGURE 6F). The change in sensitivity of MT ends occurs within a few minutes during anaphase-telophase transition.

Effects in Vivo

The lack of biochemical data on the interaction of drugs with plant tubulin prevents the interpretation of their molecular action on *Haementhus* mitosis *in vivo*. Their mode of action draws attention, however, to several possible factors involved in the regulation of assembly-disassembly and lateral association. We believe that the concentrations of colchicine that are required to see effects on living cells (within mM^{32} and not μM range[13]) are too high for the drug to be exerting one specific effect. The minimum effective range reported for *Haemanthus* was 0.5 mM.[33] There is no effect of colchicine on mitosis in *Haemanthus* endosperm at μM concentrations for several hours, and anaphase cannot be arrested instantly at any concentration. We could not reverse the effects of colchicine, and it takes several weeks to reverse the effect on root tips of *Haemanthus*. In contrast to colchicine, the arrest of anaphase by APM and oryzalin is instantaneous. The experiments on recovery, that is, reversibility of reaction *in vivo*, have not been successful. The fact that the elongation of MTs (bundled only) still proceeds in the presence of drugs and after the arrest of mitosis may indicate that, as *in vitro*, the nucleation step is inhibited. Both kinetochore and phragmoplast MTs are presumably anchored at their $(+)$ ends,[34,35] which may indicate that their elongation occurs at the $(-)$ end. The molecular action of APM and oryzalin, similar to colchicine, may be a two-step reaction.[14,18] The first step may be a rapid binding to tubulin, which is followed by a slow conformational change of tubulin molecule. This may either prevent nucleation of new MTs or make MTs very unstable and prevent the incorporation of the changed subunits. Only further studies can determine whether these interpretations are correct.

The stepwise and selective disassembly of spindle MTs by oryzalin and APM makes it possible to draw conclusions concerning the involvement of different classes of MTs in anaphase chromosome movement. There is no chromosome movement when the side branches of kinetochore fiber complexes (MT "fir tree") or, in later stages, polar MTs, shorten and disassemble. The movement stops, however, even before such changes occur. Our conclusion is that movement is arrested at the moment the assembly of side branches stops. Our methods do not allow us to follow MT assembly-disassembly directly. Our data are, however, consistent with the assumption that a continuous assembly of MTs (most likely of side branches of kinetochore fiber complexes, *i.e.*, MT "fir trees") is necessary for chromosome migration to occur. Thus the shortening of kinetochore MTs is not sufficient for the progress of anaphase, and the pinnate organization of the spindle[11,12] is necessary for its normal function. The assembly is always related to disassembly, and these two processes determine the degree of lateral association in an unknown manner.

Interzonal MTs are also very sensitive to disassembly. Because this is the most variable class of MTs in endosperm cells and their abundance or nearly total absence does not influence the rate of anaphase chromosome movement or the distance chromosomes cover in control cells, we conclude that they are not essential for the movement to occur.

SUMMARY

Mitosis is arrested in *Haemanthus* endosperm by amiprophos-methyl and oryzalin at a concentration of 100 nM, and anaphase chromosome movements are modified at 10 nM. Prolonged exposure to these drugs results in a classical c-mitosis. Anaphase chromosome movement is arrested within less than 30 seconds without any detectable change of MT arrangement, as shown by the immunogold staining method. In the next phase of amiprophos-methyl and oryzalin action, however, all non-kinetochore MTs, and especially those most sensitive to these drugs, polar microtubules, which form abundantly during anaphase, disassemble. Kinetochore microtubules form tight bundles that are very resistant to further drug action and often elongate before they finally disassemble. Because these drugs inhibit MT assembly *in vitro* in a concentration-dependent manner, we conclude that MT assembly is required for chromosome movements in anaphase and that the elongation of polar MTs is necessary for the progress of anaphase. No effect of the drugs on mitosis was detected, even at the saturated level, in the tissue culture of the frog *Xenopus*.

ACKNOWLEDGMENTS

It is a pleasure to thank Dr. C. Cypher (Hutchinson Cancer Center, Seattle, Washington) for critical comments and stimulating discussions in different stages of this work. Monospecific affinity column-purified rabbit antitubulin antibody against dog brain tubulin was a generous gift from Dr. J. De Mey, Janssen Pharmaceutica, Beerse, Belgium, and antibody against Rose β tubulin was a generous gift from Dr. L. C. Morejohn, University of Minnesota, St. Paul, Minnesota. We are also thankful for APM from Dr. C. A. Anderson, Chemagro, Mobay Chemical Corporation, Kansas City, Missouri, and oryzalin from Dr. G. Evans, Agricultural Research Division, E. I. Lilly Company, Greenfield Laboratories, Greenfield, Indiana.

REFERENCES

1. MOREJOHN, L. C. & D. E. FOSKET. 1984. Taxol-induced rose polymerization *in vitro* and its inhibition by colchicine. J. Cell Biol. **99:** 141–147.
2. MOREJOHN, L. C., T. E. BUREAU, L. P. TOCCHI & D. E. FOSKET. 1984. Tubulins from different higher plant species are immunologically nonidentical and bind colchicine differently. Proc. Natl. Acad. Sci. USA **81:** 1440–1444.
3. MOREJOHN, L. C., T. E. BUREAU, L. P. TOCCHI & D. E. FOSKET. 1986. Resistance of plant microtubules to colchicine results from a low affinity interaction of colchicine and tubulin. Manuscript submitted.
4. MOREJOHN, L. C. & D. E. FOSKET. 1984. Inhibition of plant microtubule polymerization *in vitro* by the phosphoric amide herbicide amiprophos-methyl. Science **224:** 874–876.
5. DE MEY, J., A. M. LAMBERT, A. S. BAJER, M. MOEREMANS & M. DE BRABANDER. 1982. Visualization of microtubules in interphase and mitotic plant cells of *Haemanthus* endosperm with the immuno-gold staining (IGS) method. Proc. Natl. Acad. Sci. USA **79:** 1898–1902.
6. BAJER, A. S. 1982. Functional autonomy of monopolar spindle and evidence for oscillatory movement in mitosis. J. Cell Biol. **93:** 33–48.
7. MOLÈ-BAJER, J. & A. BAJER. 1968. Studies of selected endosperm cells with the light and electron microscope. La Cellule **67:** 257–265.
8. SHUNGU, D., M. VALIANT, V. TUTLANE, E. WEINBERG, B. WEISSBERGER, L. KOUPAL, H. GADEBUSH & E. STAPLEY. 1983. Gelrite as an agar substitute in bacteriological media. Appl. Environ. Microbiol. **46:** 840–845.
9. BAJER, A. S. & J. MOLÈ-BAJER. 1982. Asters, poles and transport properties within spindle-like microtubule arrays. Cold Spring Harbor Symp. Quart. Biol. **46:** 263–283.
10. MOLÈ-BAJER, J. & A. BAJER. 1983. The action of taxol on mitosis. Modification of microtubule arrangements of the mitotic spindle. J. Cell Biol. **96:** 527–540.
11. MOLÈ-BAJER, J. & A. S. BAJER. 1985. Pinnate organization of the mitotic spindle in endosperm of a higher plant *Haemanthus* demonstrated by experimental disassembly of microtubules. Yamada Conference on Motility. Nogoya, Sept. 1984. Tokyo University Press. In press.
12. BAJER, A. S. & J. MOLÈ-BAJER. 1986. Reorganization of microtubules in endosperm cells and cell fragments of the higher plant *Haemanthus in vivo*. J. Cell Biol. **102:** 263–281.
13. MOLÈ-BAJER, J. 1983. Anaphase arrest and selective disassembly of spindle microtubules in *Haemanthus* endosperm by oryzalin. J. Cell Biol. **97:** 187a Abst. # 708.

14. FITZGERALD, T. J. & D. G. MAYFIELD. 1976. Effect of colchicine on polymerization of tubulin from rats, mice, hamsters and guinea-pigs. Experientia **32:** 83–84.
15. YADOV, N. S. & P. FILNER. 1983. Tubulin from cultured tobacco cells: Isolation and identification based on similarities to brain tubulin. Planta **157:** 46–52.
16. MOREJOHN, L. C. & D. E. FOSKET. 1982. Higher plant tubulin identified by self-assembly into microtubules *in vitro*. Nature (London) **297:** 426-428.
17. DUSTIN, P. 1978. Microtubules. Springer-Verlag. Berlin. 1–452.
18. FITZGERALD, T. J. 1976. Molecular features of colchicine associated with antimitotic activity and inhibition of tubulin polymerization. Biochem. Pharmacol. **25:** 1385–1387.
19. HART, J. W. & D. D. SABNIS. 1976. Colchicine binding activity in extracts of higher plants. J. Exp. Bot. **27:** 1353–1360.
20. EIGSTI, O. J. & P. DUSTIN. 1957. Colchicine. *In* Agriculture, Medicine, Biology and Chemistry. The Iowa State College Press. Ames, Iowa. 470.
21. STACHAN, S. D. & F. D. HESS. 1982. Dinitroaniline herbicides adsorb to glass. J. Agric. Food Chem. **30:** 389–391.
22. JACKSON, W. T. & D. A. STETLER. 1973. Regulation of mitosis. IV. An *in vitro* and ultrastructural study of effects of trifluralin. Can. J. Bot. **51:** 1513–1518.
23. HERTEL, C., H. QUADER, D. G. ROBINSON & D. MARME. 1980. Antimicrotubular herbicides and fungicides affect Ca^2+ transport in plant mitochondria. Planta **149:** 336–340.
24. UPADYAYA, M. K. & L. D. NOODEN. 1977. Mode of dinitroaniline herbicide action. Plant Physiol. **66:** 1048–1052.
25. MOREJOHN, L. C., T. E. BUREAU & D. E. FOSKET. 1983. Oryzalin binds to plant tubulin and inhibits taxol-induced microtubule assembly *in vitro*. J. Cell Biol. **97:** 211a #803.
26. ALBERTINI, D. F., B. HERMAN & P. SHERLINE. 1984. *In vivo* and *in vitro* studies on the role of HMW-MAPs in taxol-induced microtubule bundling. Eur. J. Cell Biol. **33:** 134–143.
27. HEIDEMANN, S. R. & P. T. GALLAS. 1980. The effect of taxol on living eggs of *Xenopus leavis*. Dev. Biol. **80:** 489–494.
28. LOCKWOOD, A. A. 1979. Molecules in mammalian brain that interact with the colchicine site on tubulin. Proc. Natl. Acad. Sci. USA **76:** 1184–1188.
29. KUMAGAI, H. & H. SAKAI. 1983. A porcine brain protein (35K protein) which bundles microtubules and its identification as glyceraldehyde 3-phosphate dehydrogenase. J. Biochem. (Tokyo) **93:** 1259–1269.
30. SABNINS, D. D. & J. W. HART. 1982. Microtubule proteins and P-proteins. *In* Encyclopedia of Plant Physiology. D. Boulter & B. Parthier Eds.: **14:** 401–437. Springer Verlag. Berlin.
31. SHERLINE, P., K. SCHIAVONE & S. BROCATO. 1979. Endogenous inhibitor of colchicine binding in rat brain. Science **205:** 593–595.
32. MOLÈ-BAJER, J. 1958. Cine-micrographic analysis of c-mitosis in endosperm. Chromosoma **9:** 332–358.
33. SCHMIT-BENNER, A. C. 1981. Contribution a l'etude des mechanismes mitotiques: dynamique de la polymerisation des microtubules analysee *in vivo*, en immunofluorescence et en ultrastructure dans les cellules de plantes (albumen de Clivia et Haemanthus). These, L'Universite Louis Pasteur de Strasbourg.
34. EUTENEUER, U. & J. R. McINTOSH. 1980. Polarity of midbody and phragmoplast microtubules. J. Cell Biol. **87:** 509–515.
35. EUTENEUER, U. & J. R. McINTOSH. 1981. Structural polarity of kinetochore microtubules in PtK cells. J. Cell Biol. **89:** 338–345.

New Antineoplastic Agents with Antitubulin Activity

JANENDRA K. BATRA,[a] CHII M. LIN,[a] ERNEST HAMEL,[a]
LEONARD JURD,[b] AND LARRY J. POWERS[c]

[a]Laboratory of Pharmacology and Experimental Therapeutics
Developmental Therapeutics Program
Division of Cancer Treatment
National Cancer Institute
National Institutes of Health
Bethesda, Maryland 20892

[b]Western Regional Research Center
Agricultural Research Service
United States Department of Agriculture
Berkeley, California 94710

[c]SDS Biotech Corporation
Painesville, Ohio 44077

The GTPase activity of tubulin can be profoundly affected by many agents, including all antitubulin drugs we have examined. We have exploited this property of tubulin to develop a relatively simple screening assay to detect new antitubulin agents among antineoplastic drugs whose mechanism of action is uncertain. As demonstrated in FIGURE 1, antimitotic agents can be distinguished from drugs with no known effect on tubulin by mere inspection of an autoradiogram prepared after thin-layer separation of guanosine diphosphate (GDP) and guanosine triphosphate (GTP). In 1.0 M monosodium glutamate there is significant hydrolysis of GTP that is associated with the formation of sheets of protofilaments. The antimitotic drugs either inhibit or stimulate GTP hydrolysis, even though all antimitotic drugs except taxol inhibit glutamate-induced polymerization. In 0.1 M glutamate, GTP hydrolysis by tubulin is minimal, but a few drugs still induce a GTPase reaction.

FIGURE 2 presents the chemical structures of a number of new compounds that we have found to have substantial antitubulin activity *in vitro*. These drugs have also demonstrated varying degrees of antileukemic activity (murine P388) *in vivo* in screening tests at the National Cancer Institute.

Combretastatin, isolated by G. R. Pettit from the South African tree *Combretum caffrum*, is structurally reminiscent of the A and C rings of colchicine as well as the bicyclic compound 2-methoxy-5-(2', 3', 4'-trimethoxyphenyl)tropone (MTPT) synthesized by T. J. Fitzgerald. Combretastatin, like colchicine and MTPT, stimulates tubulin-dependent GTP hydrolysis. The drug inhibits mitosis in L1210 cells in culture and microtubule assembly (plateau level reduced over 50% at 4–5 μM combretastatin). It competes with colchicine (apparent K_i, 1.1 μM) but not vinblastine in binding to tubulin.

A large number of derivatives of 6-benzyl-1,3-benzodioxole have been synthesized. Many possess significant antitubulin activity *in vitro*, and the most active compounds inhibit mitosis in L1210 cells. Structurally, they are similar to podophyllotoxin, colchicine, and steganacin. The benzylbenzodioxole derivatives inhibit tubulin-dependent GTP hydrolysis, as does podophyllotoxin, whereas steganacin and colchicine stimulate the reaction. We thus consider these compounds most analogous to podophyllotoxin. The most potent compound inhibits colchicine binding to tubulin (apparent K_i, 0.6 μM) and has no effect on the binding of vinblastine. Inhibition of the plateau of microtubule assembly is over 50% with 5–10 μM drug. Active derivatives possess a 1–3 carbon substituent as Rl and a methoxy group as R2. Compounds with additional methoxy groups on the benzyl moiety have little activity.

Derivatives of 4-cyano-5,6-biphenylpyridazin-3-one have no obvious structural analogy to known antitubulin agents. They stimulate tubulin-dependent GTP hydrolysis and have no effect on interactions of colchicine or vinblastine with the protein. The most active derivatives inhibit assembly (turbidity plateau) over 50% at 5–10 μM. *In vitro* antitubulin effects do not correlate well with cytotoxicity or antimitotic activity in L1210 cells. Active derivatives all contain the nitrile and phenyl moieties, but no

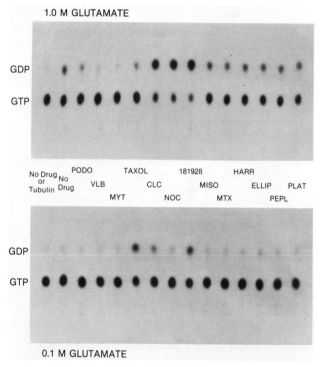

FIGURE 1. Tubulin-dependent GTP hydrolysis as a screening test for antimitotic drugs. Each 50 μl reaction mixture contained 1.0 mg/ml of tubulin, unless indicated, glutamate at the indicated concentration, 10% (v/v) dimethylsulfoxide, 50 μM [α-^{32}P]GTP, and the indicated drug at 0.1 mM. Antimitotic drugs: podophyllotoxin (PODO), vinblastine (VLB), maytansine (MYT), taxol, colchicine (CLC), nocodazole (NOC), and NSC 181928. Drugs with other modes of action: misonidazole (MISO), methotrexate (MTX), harringtonine (HARR), ellipticine (ELLIP), pepleomycin (PEPL), and *cis*-diaminedichloroplatinum (PLAT). Incubation was for 20 min at 37°. An aliquot of each reaction mixture was mixed with 25% acetic acid and spotted on a polyethyleneimine-cellulose thin-layer sheet. Chromatography was in 1.0 M KH$_2$PO$_4$ (pH unadjusted).

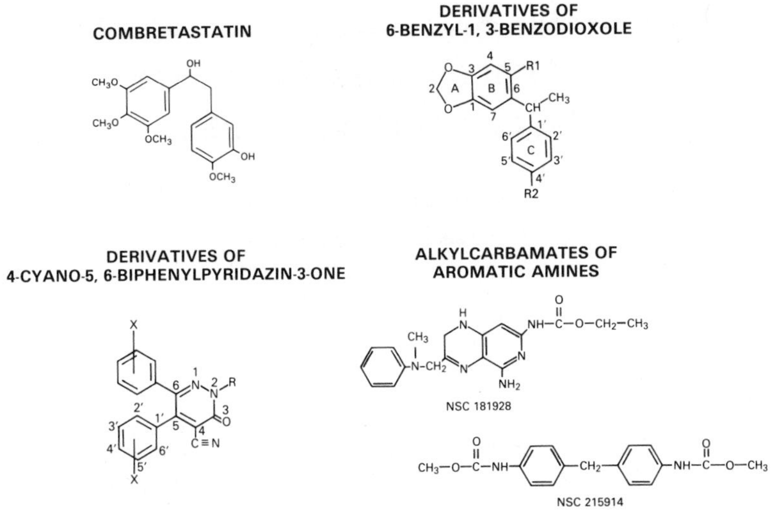

FIGURE 2. New antineoplastic compounds with antitubulin activity *in vitro*.

structure-function pattern is obvious in substituents on the phenyl rings or at position two.

Two alkylcarbamates of aromatic amines, structurally reminiscent of nocodazole, were found to have significant antitubulin activity *in vitro*. These drugs, NSC 181928 (synthesized by C. Temple) and the bis-carbamate NSC 215914, like nocodazole, stimulate tubulin-dependent GTP hydrolysis, inhibit the binding of colchicine but not vinblastine to tubulin, and inhibit microtubule assembly. NSC 181928 almost completely inhibited polymerization at 10 μM, but NSC 215914 was only partially inhibitory at this concentration. Recent reports of carbamates of other amines with antitubulin activity indicate that it should be worthwhile to survey the many carbamates, particularly those with antineoplastic activity, in the NCI collection for antitubulin properties.

Molecular Modeling of Microtubule Inhibitors and the Colchicine Binding Site on Tubulin

MICHAEL N. LIEBMAN

Department of Pharmacology
Mount Sinai School of Medicine of the
City University of New York
New York, New York 10029

We have been developing and applying a variety of computational approaches to examine the relationship between molecular structure and biological function in systems that range in size from biologically active small molecules to enzymes, protein assemblies, and receptors. The ideal data for this analysis include the three-dimensional structure of each of the components: the enzyme/protein/receptor; the ligand, for example, substrates or inhibitors, that bind to the macromolecule; and the complex formed between these components. Rarely are more than one of these components available. Such is the case for mitotic spindle poisons that interact with tubulin. To analyze such a system, we apply techniques of molecular modeling that include structural superposition, representation of molecular electrostatic potentials, van der Waals surfaces, and heuristic modeling of the receptor/binding pocket. We have applied these techniques to study the interaction of tubulin with drugs that bind to the colchicine binding site and act to inhibit the formation of microtubules and to promote the disassembly of the microtubules. We address two specific questions.

What Is the Pharmacophore That Permits Structurally Nonhomologous Molecules to Be Functionally Homologous, and How Specific Is It?

Modern analysis of the structure-function determinants of recognition and reactivity involves the comparison of properties computed for a series of drug molecules. The trimethoxyphenyl group is common to both classes of moledules in this study: structural analogues of colchicine (*e.g.* colcemid and isocolchicine); and functional analogues of colchicine (*e.g.* podophyllotoxin). This structural fragment was used in topological equivalencing of these molecules, with superposition achieved through incorporation of the analysis of functional determinants. This analysis (FIGURE 1) has suggested the source of specificity and reactivity observed in a large series of molecules. As a result, activity in the tubulin binding has been predicted for a series of structurally variable molecules previously untested in this system, and these predictions are currently being tested experimentally through collaborative studies.[1]

What Can We Learn about the Nature of the Tubulin Binding Pocket Based only on the Knowledge of the Structures of the Drug Molecules?

We examined the environments, found in the crystalline state, of a series of functionally related molecules to define the determinants complementary to those

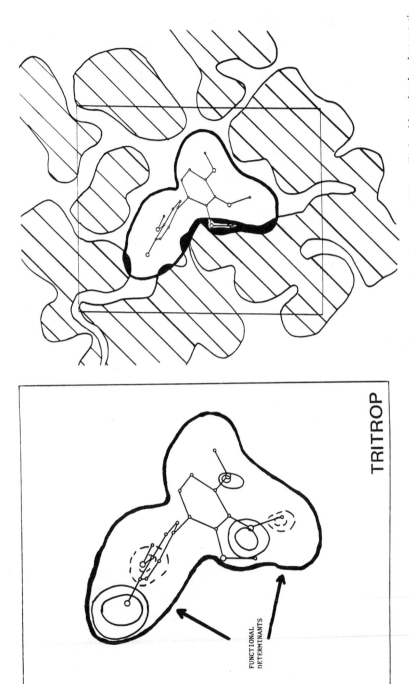

TRITROP

FUNCTIONAL
DETERMINANTS

FIGURE 1. Left: Projection of the van der Waals surface of 2-methoxy-5-(2'3'4'trimethoxyphenol) tropone (thick outline), with molecular electrostatic potential projected for both negative (thin, solid) and positive (thin, dashed) regions. Right: Projection of van der Waals surface of the molecular environment of the modeled binding site as computed from X-ray structure.[1,2]

revealed in the molecules by the analysis above. This approach uses the stability of intermolecular interactions within the crystal lattice and assumes that the fundamental physical forces and interactions (*e.g.* complementation of electrostatic potential surfaces and van der Waals surfaces) are representative of both *in vitro* and *in vivo* complementarity.[2]

This analysis (FIGURE 1) has examined the interaction of the open-ring analogue of colchicine with 29 symmetry-related molecules that form an environment within 6A of any given molecule and reveals the degree of complementation of the environment to those features assigned as specificity determinants above. We expect to use the amino acid sequence and experimental data concerning the composition of the tubulin binding site to conformationally arrange this sequence to reproduce the computed properties and thus yield a model for the binding site in tubulin. This model of the colchicine binding site will be further refined through collaboration with experimentalists.

REFERENCES

1. LIEBMAN, M., 1985. Prog. Clin. Biol. Res. **172B:** 285.
2. LIEBMAN, M., 1986. J. Mol. Graphics. In press.

The Role of the B-Ring of Colchicine on Taxol-Induced Tubulin Polymerization

B. BHATTACHARYYA, G. GHOSHCHAUDHURI, S.
MAITY, AND B. B. BISWAS

Department of Biochemistry
Bose Institute
Calcutta, India

Colchicine and its analogues have been used extensively as a tool for the *in vivo* study of microtubule-dependent processes because they bind tubulin, the subunit of microtubules. This binding of colchicine to tubulin is extremely specific and has high affinity. In order to have this colchicine-like activity, the analogue must have the trimethoxybenzene (A-ring) and methyoxytropone (C-ring) systems combined into a single molecular entity by the appropriate carbon-carbon single bond. Thus, the tetramethoxybicyclic compound 2-methoxy-5 (2′,3′,4′trimethoxyphenyl)tropone possess the minimal molecular features required for colchicine-like biological activity. This interesting analogue is a potent mitotic inhibitor, binds tubulin at the same site where colchicine binds, and inhibits normal microtubule assembly.

We observed that goat brain tubulin polymerization induced by taxol is inhibited by colchicine in a concentration-dependent manner, and half maximal inhibition occurs when only 13% of the total tubulin in the mixture is complexed with colchicine (FIGURE 1). It has been further observed that the polymerization reaction occurred with simultaneous loss of colchicine binding (data not shown). To our surprise, however, 2-methoxy 5(2′,3′,4′trimethoxyphenyl)tropone and desacetamido colchicine, which are known to bind tubulin at the same site where colchicine binds, have been found to be unable to inhibit taxol-induced assembly (FIGURE 2). Although the rate of polymerization is slightly decreased in the presence of each analogue, the final values of polymerization are almost unaffected. Identical results are obtained if tubulin is first incubated with the drug for sixty minutes before the polymerization was initiated with the addition of guanosine triphosphate (GTP) and taxol. This preincubation is not necessary in order to see the effect of these two colchicine analogues, because unlike colchicine, these two analogues bind tubulin even at 4°C, and the binding is almost instantaneous at 37°C. Moreover, these two colchicine analogues were incorporated into polymers when incubated in the presence of taxol (data not shown). Furthermore, preformed taxol-induced polymers of tubulin have been found to bind these two colchicine analogues (TABLE 1). To determine the stoichiometry, preformed taxol-induced polymers were incubated with 10 μM drug at 37°C for 30 minutes, and the complex was pelleted by warm centrifugation (25°C) at 120,000 × g for 30 minutes. The pellets were suspended in cold (0°C) buffer, and the fluorescence emission intensity at 430 nm was determined to calculate the amount of drug bound per mole of tubulin (TABLE 1).

One possible explanation for the inability of the colchicine tubulin complex to polymerize in the presence of taxol might be that colchicine induced a conformational change on tubulin molecules, which is unfavorable for the taxol-induced polymerization. An alternative mechanism might be that the colchicine binding site and the tubulin-tubulin interaction site on taxol-induced polymers shared a "common region"

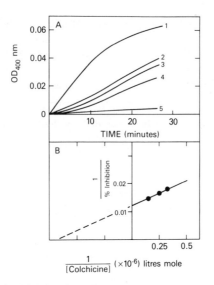

FIGURE 1. Inhibition of taxol-induced tubulin assembly by colchicine.
Panel A: 14 μM tubulin was incubated with different concentrations of colchicine at 37°C for 1.5 hour in assembly buffer. Polymerization reaction was then initiated by the addition of taxol (10 μM), and the assembly was studied spectrophotometrically at 400 nm. Colchicine concentrations: curve 1, 0; curve 2, 3 μM; curve 3, 4 μM; curve 4, 6 μM; and curve 5, 9 μM.
Panel B: Inhibition of the initial rate of turbidity development compared to control (no colchicine) was plotted against each colchicine concentration in the form of a double reciprocal plot.

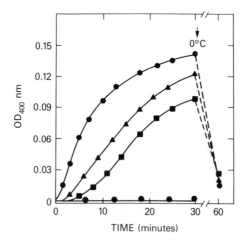

FIGURE 2. Polymerization of tubulin by taxol in the presence of colchicine analogues. 14.5 μM tubulin was incubated with 10 μM analogue at 37°C for 1.5 hour in assembly buffer. To each of the analogue-treated samples, taxol was added to a final concentration of 15 μM, and the polymerization was studied spectrophotometrically at 400 nm: Samples are: control (●——●); desacetamido colchicine (▲——▲); 2-methoxy-5(2′,3′,4′trimethoxy phenyl) tropone (■——■); and colchicine (●——●).

TABLE 1. Analysis of Taxol-Induced Polymers-Analogue Complexes[a]

Analogues	Mole of Analogue Bound per Mole of Tubulin
Desacetamido colchicine	0.40–0.43
2-methoxy-5(2',3',4'trimethoxy phenyl)tropone	0.35–0.39

[a]14.4 μM PC-tubulin was polymerized to steady state at 37°C for 60 minutes with taxol in assembly buffer. These polymers were further incubated with analogue (100 μM) at 37°C for 30 minutes and then pelleted by warm centrifugation at 25°C for 30 minutes. Pellet was suspended in cold (0°C) assembly buffer, and stoichiometry was determined from fluorescence emission spectra at 430 nm.

on the tubulin molecule. Because colchicine and its structural analogue desacetamido colchicine differ on the substituent present in the B-ring, we propose that the B-ring substituent-binding domain of the tubulin molecule is probably the "common region," which is also needed for the tubulin-tubulin interaction in the case of taxol-induced assembly.

Regulation of Neuronal Adenylate Cyclase by Microtubule Proteins[a]

MARK M. RASENICK

Department of Physiology and Biophysics
University of Illinois College of Medicine
Chicago, Illinois 60680

Microtubule-disrupting drugs have been shown to increase adenylate cyclase activity in synaptic membranes prepared from rat cerebral cortex. Activation of the enzyme by hydrolysis-resistant guanosine triphosphate (GTP) analogues and NaF is augmented by colchicine or vinblastine ($EC_{50} = 5 \times 10^{-7}M$), whereas basal and Mn^{++}-stimulated (reflecting catalytic-moiety activation) activities are unchanged.[1] These findings suggest that the microtubule-disrupting drugs increase the "coupling" between the GTP-binding protein that stimulates adenylate cyclase (Ns) and the catalytic moiety of that enzyme. When these membranes are treated with colchicine or vinblastine and subsequently washed, activity of the Ns protein is released from the membranes into the supernatants. This release of Ns activity is indicative of facilitated Ns-catalytic moiety coupling and is similar to that observed in the homologous cGMP phosphodiesterase cascade from retinal rod outer segments.[2]

Incubation of tubulin ($EC_{50} = 0.5~\mu g/ml$; 10 nM) with synaptic membranes inhibits guanyl nucleotide and NaF activated adenylate cyclase by 50–60% without altering basal and Mn^{++} activation of the enzyme (FIGURE 1). The effects of this tubulin incubation are sustained even after membranes are washed. Furthermore, treatments that diminish the amounts of membrane-associated tubulin increase Ns-mediated adenylate cyclase activation.

The photoaffinity GTP analogue (P^3azidoanilido)-P^1-5' GTP (AAGTP) labels both the 42 kD Ns and the 40 kD inhibitory GTP-binding protein (Ni) on the synaptic membranes.[3] Under conditions where microtubule disrupting drug treatment and subsequent washing releases Ns activity into the supernatant, AAGTP-labeled 42 kD protein is also released from those membranes (FIGURE 2). This labeled 42 kD protein is retained by an agarose-tubulin affinity column, and when eluted from the column, the protein displays Ns activity (reconstitution of Ns-deficient adenylate cyclase). GTP-dependent (Ni mediated) inhibition of adenylate cyclase is unaltered by colchicine or vinblastine treatment, and the 40 kD AAGTP labeled protein is not released from the synaptic membrane by these agents.

Although microtubule disrupting drugs increase cyclic adenosine monophosphate (cAMP) accumulation in intact cells,[4] colchicine or vinblastine enhancement of adenylate cyclase in broken cell preparations and the release of AAGTP-labeled Ns has been observed only in membranes from tissues of neural origin. These data lead to the proposal of a hypothetical Ns-tubulin interaction similar to that suggested for actin and fibronectin.[5] The physiological significance of a putative Ns-tubulin interaction (association/dissociation) might be to regulate intracellularly the amount of cAMP produced in response to a given neurotransmitter.

[a]This work was supported by Air Force Office of Scientific Research Grant 83-0249.

794

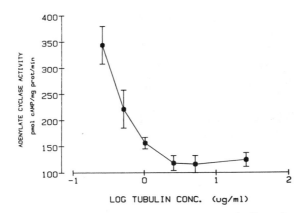

FIGURE 1. Inhibition of Ns-mediated adenylate cyclase by tubulin. Synaptic membranes were washed twice in 20 mM Tris HCl (pH7,5) containing 1 mM dithiothreitol (DTT), 0.3 mM phenylmethyl sulfonyl fluoride (PMSF) and 1 mM ethylene diamine tetraacetic acid (EDTA). Following this, membranes were resuspended in the above buffer containing 5 mM MgCl [no EDTA], linoleic acid (10 μg/ml), phosphatidylcholine (1 mg/ml), and the indicated tubulin (phosphocellulose purified from rat brain) concentration and incubated for 10 minutes at 37°C followed by 15 minutes on ice. Following this, guanylyl inidodiphosphate (Gpp[NH]p) (10 μM) was added, and the tubes were incubated at 30°C for 10 minutes followed by the addition of adenosine triphosphate (ATP) and a 10-minute assay incubation. Adenylate cyclase activity in the absence of added tubulin plus 10 μM GppNHp was 326 pmol cAMP/mg protein/minute. Basal adenylate cyclase activity (in the absence of added GppNHp) was 36 and 47 pmol cAMP/mg protein/min respectively in the absence or presence of 5 μg/ml added tubulin.

Tubulin incubation did not inhibit basal or Mn^{++}-stimulated adenylate cyclase; thus, only Ns-mediated adenylate cyclase appears to be involved. The "IC$_{50}$" for this effect is about 10 nM, and the effect is maximal at 40 nM tubulin. The ratio of microtubule protein:total synaptic membrane protein required to achieve this effect is about 1:500. Although the nature of tubulin interaction with these membranes is unclear, the effects of added tubulin are maintained after the membranes are washed.

One interpretation of these results might be that addition of tubulin to these membranes results in increased tubulin-Ns association and subsequently, decreased Ns-mediated adenylate cyclase coupling. The possibility of a trace impurity in the tubulin preparation having these effects, or of tubulin acting indirectly on Ns, cannot be ruled out.

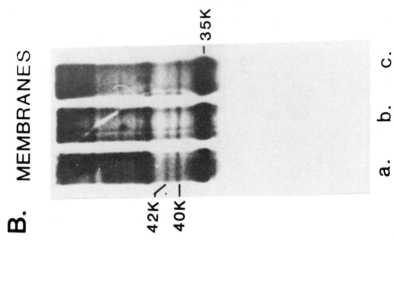

REFERENCES

1. RASENICK, M. M., P. J. STEIN & M. W. BITENSKY. 1981. Nature (London) **294:** 560.
2. STEIN, P. J., M. M. RASENICK & M. W. BITENSKY. 1982. Prog. Retinal Res. **1:** 222.
3. RASENICK, M. M., G. L. WHEELER, M. W. BITENSKY, C. M. KOSACK, R. L. MALINA & P. J. STEIN. 1984. J. Neurochem. **43:** 1447.
4. ZOR, U. 1983. Endocrin. Rev. **4:** 1.
5. ALI, I. & R. HYNES. 1977. Biochim. Biophys. Acta **471:**16.

FIGURE 2. Photoaffinity labeling and colchicine or vinblastine mediated increase in Ns mobility.

A: Synaptic membranes were incubated with 1 μM [^{3}H]AAGTP at 30°C for 20 minutes followed by 20 minutes of UV photolysis on ice. The reaction was quenched with 4 mM DTT, and the membranes were then incubated at 30°C for 10 minutes with 5 μM of either colchicine (lanes b and d) or lumicolchicine (lanes a and c). Following this, the membranes were washed three times with a low ionic strength buffer (2 mM HEPES, pH 7.4, 1 mM MgCl, 2 mM DTT), and the proteins released from the membranes were electrophoresed on 10% polyacrylamide gels and radioautographed. Lanes a and b represent Coomassie blue staining patterns, and lanes c and d are radiofluorographs.

B: Synaptic membranes were treated as above except that [^{32}P]AAGTP was used. Following photolysis, the membranes were treated with H$_2$O (lane a), 0.1 μM vinblastine (lane b) or 10 μM vinblastine (lane c), and washed as above. These radioautographs represent labeling on the membranes rather than on the supernatants.

Similar experiments using unlabeled AAGTP with subsequent adenylate cyclase assays rather than autoradiography show a vinblastine-mediated loss of Ns activity subsequent to washing without a concomitant loss of Ni activity. Under conditions where colchicine or vinblastine enhance "coupling" of the Ns protein with the adenylate cyclase catalytic moiety, washing membranes with buffer releases about 50% of the Ns activity. This is borne out by comparing [^{3}H]DPM in the 42 kD bands from **A,** which are 7,445 in lane d and 4,343 in lane c.

Although the coupling of Ns is enhanced by microtubule disrupting drugs, that of Ni is not. Similarly, the 40 kD GTP-binding protein is not released from the membrane under conditions where Ns appears to be released.

Interactions of Microtubules with Neuronal Intermediate Filaments *in Vitro*[a]

ROBLEY C. WILLIAMS JR.

Department of Molecular Biology
Vanderbilt University
Nashville, Tennessee 37235

INTRODUCTION

The formation of an integrated cytoskeleton from its known individual filamentous components (microtubules, intermediate filaments, and actin filaments) requires the presence of connecting elements, or bridges, between those components. Bridges between filaments have been described, both morphologically and biochemically, in a number of cell types.[1-7] In the special case of neurons, morphological studies[8-18] involving a variety of preparative techniques have established that microtubules are structurally bridged to each other, to intermediate filaments (here called neurofilaments for brevity), and to membrane-bounded organelles by filamentous "arms." Lasek and coworkers[19,20] have established by an extensive series of transport studies that microtubules and neurofilaments are also functionally bridged *in vivo*. Evidence published recently has clearly implicated microtubule associated proteins (MAPs) as being among the proteins that compose the interfilament bridging entities.[21-25]

Detailed understanding of the role of interfilament linkages in the organization of the cytoskeleton requires knowledge of their structure and functional properties at the molecular level. The necessary investigations of interfilament linkage *in vitro* have been largely confined to studies of the linking of actin filaments by MAPs (reviewed in reference 26) and to the studies of the linking of microtubules to neurofilaments that are the main subject of this article. The numerous other interactions that must be of importance in organizing the cytomatrix are thus far untouched.

INVESTIGATIONS *IN VITRO*

Work through December 1984 has been described in a previous volume of these Annals.[27] The reader will find much detail there. Discussion here will be restricted to a brief review of those findings and a description of some recent results.

Berkowitz *et al.*[28] had noticed that microtubules prepared from bovine brain by a modification of the assembly/disassembly method of Shelanski *et al.*[29] were substantially contaminated by filaments of about 10 nm diameter and by proteins that comigrated with the three polypeptides of mammalian neurofilaments. The contami-

[a]This work was supported by Grants GM 25638 and GM 29834 of the National Institutes of Health.

nation seemed best explained by the hypothesis that neurofilaments could adhere to microtubules under the preparative conditions employed. Runge et al.[30] investigated this possibility in detail. Making use of neurofilaments prepared from bovine brain by a combined centrifugal and gel-filtration technique,[31] they found, by falling ball viscometry, that a mixture of microtubules and neurofilaments consistently displayed an apparent viscosity one to two orders of magnitude larger than the viscosity of microtubules or neurofilaments alone. The presence of the high viscosity corresponded to the presence in solution of a rapidly sedimenting complex of the two filament types. These results were interpreted to mean that an associated complex could form in vitro between microtubules and neurofilaments. The complex was found to be easily disrupted by shear and by low temperatures, indicating that it is held together by noncovalent interactions. Most aspects of this study were confirmed by Minami et al.,[32] who found similar concentration-, temperature-, and shear-dependencies of viscosity and reported that microtubules and neurofilaments cosediment and form gels at sufficiently high concentrations. They did not observe an apparent requirement for adenosine triphosphate (ATP) that had been reported by Runge et al.[30] This apparent requirement has since been shown to be an artifact of the presence of contaminating enzymes in the preparation of neurofilaments from bovine brain.[33]

In addition to moderately large amounts of many different enzymic activities, neurofilaments prepared from brain also have firmly attached to them a number of proteins that comigrate with the high molecular weight MAPs. Neurofilaments prepared from spinal cord by the technique of Delacourte et al.[34] are relatively free of these large proteins. Aamodt and Williams[35] found that the combination of such filaments with microtubules made without MAPs failed to result in formation of a complex. MAPs were prepared from microtubule protein by chromatography on phosphocellulose.[36] MAPs became bound to neurofilaments when the two structures were incubated together, a result in agreement with findings of Leterrier et al.[37] When the MAP-coated neurofilaments were incubated with microtubules formed from pure tubulin, a viscous complex formed. Evidently, some component of the MAP preparation can serve as a bridge between the two types of filaments. (Controls showed that the presence of bound MAPs on the neurofilaments did not induce formation of a complex among those filaments alone.)

Chromatography of the MAP fraction by gel-filtration on Bio-Gel A-15m yielded a broad elution profile.[27] Analysis of individual fractions by sodium dodecyl sulfate (SDS)-polyacrylamide gel electrophoresis showed that individual polypeptides were not well separated from each other. The fact that a gel-filtration medium with a nominal exclusion limit of 15 million daltons had to be employed to obtain separation implies that the MAPs are present in solution not as single polypeptides, but rather as complexes of several molecules held together by noncovalent interactions. This notion that MAPs exist, in vitro at least, as components of particles is reinforced by the difficulty of separating them from each other in a native state.[38–40] When fractions from the gel-filtration chromatogram were added to a solution of microtubules and neurofilaments and the resulting mixture was assayed by falling-ball viscometry, it was found[27,41] that the fractions with the greatest viscosity-inducing capacity (and, hence, with the most efficient bridging or linking entities) were those with the largest apparent Stokes' radius. Evidently the largest MAP-containing particles are the most effective linkers.

The fraction containing these particles was found to contain chiefly the high molecular weight MAPs, MAP-1 and MAP-2,[42] and tubulin. Consequently, one or more of these proteins is likely to compose the bridging entity. In an attempt to ascertain whether MAP-2 alone might be the active molecule, a preparation made according to the heat-treatment method of Kim *et al.*[43] was tested and found to be devoid of viscosity-inducing activity. Thus, either MAP-2 alone cannot bridge microtubules and neurofilaments or the high temperatures involved in the preparation inactivate it.

Preliminary experiments[44] in which the covalent cross-linker BS³ [45] was applied to

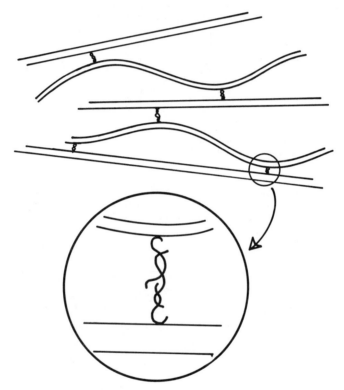

FIGURE 1. Schematic cartoon of the microtubule-neurofilament bridging particle. The straight lines represent microtubules, and the wavy filaments represent neurofilaments. Bridges are pictured as being present at intervals along their lengths. An expanded view of a single bridge (inside the circle) shows that it is composed of one or more copies of each of several proteins (among the likeliest are MAP-1, MAP-2, and tubulin). The protein that makes contact with the wall of the microtubule is pictured as making contact at its other end with proteins of the bridging particle, and not with the wall of the neurofilament. This last proposed geometry is consistent with the available data but is not yet demonstrated.

MAPs show that all of the high molecular weight MAPs and tubulin are cross-linked together. The linking proceeds efficiently over a large range of concentrations of protein and of BS³, indicating that the MAPs are indeed nearest neighbors in solution. Neither the gel-filtration studies nor the cross-linking studies so far establish whether one, or more than one, type of associated MAP-particle exists.

DISCUSSION

The current notion of how the interaction occurs is summarized schematically in FIGURE 1. Microtubules are held to neurofilaments by a number of bridges. Each bridge is composed of several polypeptide chains. It is quite possible that no one protein touches both the microtubule wall and the neurofilament. Rather, the protein that touches the microtubule at one end may be inserted at its other end into the particle, where it is held by noncovalent association to other proteins, one of which makes contact with the neurofilament. It is clear that this kind of scheme is more flexible than arrangements in which bridging is imagined to be carried out by a single molecule that has only a microtubule-binding end and a neurofilament-binding end.

The chief drawback of these studies of bridging *in vitro* is that they do not yield the kind of geometric specificity provided by intracellular localization studies. When a bridging entity is found, we do not know from which part of the cell it came, or whether it is properly oriented in the complexes that are formed. The chief advantage of these studies is that they provide unequivocal evidence that the bridging entities found do, in fact, hold filaments together. Furthermore, they give an idea of the *strength* of the interactions and allow one, at least potentially, to experiment easily with biochemical changes that might serve to regulate the interactions. A secondary advantage is that the successful formation of an interfilament complex assures that all the components necessary for interfilament bridging are indeed present in the active fraction. Hence, we can say with fair certainty that microtubules can be bridged to neurofilaments by entities composed of nothing more than the high molecular weight MAPs and tubulin. If the MAPs are indeed involved in connecting microtubules to intermediate filaments, their name could be thought to be too restrictive to evoke their function. The fact that they are seen as microtubule-associated must reflect their preferred association *in vitro*. A name like cytoskeletal linking proteins (CLIPs) would be more appropriate.

In the course of this work, no entity has been isolated that links microtubules to each other or neurofilaments to each other. The reason for this lack is not clear, although it seems possible that the appropriate particle becomes solubilized under the preparative conditions employed. If a multiprotein particle is involved in these bridges, it may have some polypeptides in common with those of the microtubule-neurofilament bridge.

REFERENCES

1. WOLOSEWICK, J. J. & K. R. PORTER. 1976. Am. J. Anat. **147:** 303–324.
2. PORTER, K. R., R. BYERS & M. H. ELLISMAN. 1979. *In* The Neurosciences: Fourth Study Program. F. O. Schmitt & F. G. Worden, Eds.: 703–722. MIT Press. Cambridge, Massachusetts.
3. SCHLIWA, M., J. VAN BLERKOM & K. B. PRYZWANSKY. 1982. Cold Spring Harbor Symp. Quant. Biol. **46:** 51–67.
4. WANG, E., R. K. CROSS & P. W. CHOPPIN. 1979. Proc. Natl. Acad. Sci. USA **76:** 5719–5723.
5. LAZARIDES, E. & W. J. NELSON. 1982. Cell **31:** 505–508.
6. BAINES, A. J. 1983. Nature (London) **301:** 377–378.
7. BRANTON, D., C. COHEN & J. TYLER, 1981. Cell **24:** 24–32.
8. WUERKER, R. B. & S. L. PALAY. 1969. Tissue Cell **1:** 387–402.
9. BERTOLINI, B., G. MONACO & A. ROSSI. 1970. J. Ultrastruc. Res. **33:** 173–186.
10. WUERKER, R. B. 1970. Tissue Cell **2:** 1–9.
11. SMITH, D. S. 1971. Philos. Trans. Roy. Soc. London Ser. B. **261:** 395–405.
12. SMITH, D. S., U. JALFORS & M. L. CAYER. 1977. J. Cell Sci. **27:** 235–272.

13. ELLISMAN, M. H. & K. R. PORTER. 1980. J. Cell Biol. **87:** 464–479.
14. HODGE, A. J. & W. J. ADELMAN JR. 1980. J. Ultrastruc. Res. **70:** 220–241.
15. RICE, R. V., P. F. ROSLANSKY, N. PASCOE & S. M. HOUGHTON. 1980. J. Ultrastruc. Res. **71:** 303–310.
16. METUZALS, J., V. MONTPETIT & D. F. CLAPIN. 1981. Cell Tissue Res. **214:** 455–482.
17. HIROKAWA, N. 1982. J. Cell Biol. **94:** 129–142.
18. SCHNAPP, B. J. & T. S. REESE. 1982. J. Cell Biol. **94:** 667–679.
19. HOFFMAN, P. N. & R. J. LASEK. 1975. J. Cell Biol. **66:** 351–366.
20. LASEK, R. J., J. A. GARNER & S. T. BRADY. 1984. J. Cell Biol. **99:** 212s–221s.
21. PAPASOZOMENOS, S. CH., L. AUTILIO-GAMBETTI & P. GAMBETTI. 1981. J. Cell Biol. **91:** 866–871.
22. HIROKAWA, N., M. A. GLICKSMAN & M. B. WILLARD. 1984. J. Cell Biol. **98:** 1523–1536.
23. BLOOM, G. S. & R. B. VALLEE. 1983. J. Cell Biol. **96:** 1523–1531.
24. VALLEE, R. B. & G. S. BLOOM. 1984. Mod. Cell Biol. **3:** 21–75.
25. VALLEE, R. B., G. S. BLOOM & W. E. THEURKAUF. 1984. J. Cell Biol. **99:** 38s–44s.
26. POLLARD, T. D., S. C. SELDEN & P. MAUPIN. 1984. J. Cell Biol. **99:** 33s–37s.
27. WILLIAMS JR, R. C. & E. A. AAMODT. 1985. Ann. N. Y. Acad. Sci. **455:** 509–524.
28. BERKOWITZ, S. A., J. KATAGIRI, H. K. BINDER & R. C. WILLIAMS JR. 1977. Biochemistry **16:** 5610–5617.
29. SHELANSKI, M. L., F. GASKIN & C. R. CANTOR. 1973. Proc. Natl. Acad. Sci. USA **70:** 765–768.
30. RUNGE, M. S., T. M. LAUE, D. A. YPHANTIS, M. R. LIFSICS, A. SAITO, M. ALTIN, K. REINKE & R. C. WILLIAMS JR. 1981. Proc. Natl. Acad. Sci. USA **78:** 1431–1435.
31. RUNGE, M. S., W. W. SCHLAEPFER & R. C. WILLIAMS JR. 1981. Biochemistry **20:** 170–175.
32. MINAMI, Y., H. MUROFUSHI & H. SAKAI. 1982. J. Biochem. (Tokyo) **92:** 889–898.
33. AAMODT, E. A. & R. C. WILLIAMS JR. 1984. Biochemistry **23:** 6031–6035.
34. DELACOURTE, A., G. FILLIATRIAU, F. BOUTTEAU, G. BISERTE & J. SCHREVEL. 1980. Biochem. J. **191:** 543–546.
35. AAMODT, E. A. & R. C. WILLIAMS JR. 1984. Biochemistry **23:** 6023–6031.
36. WEINGARTEN, M. D., A. H. LOCKWOOD, S.-Y. HWO & M. W. KIRSCHNER. 1975. Proc. Natl. Acad. Sci. USA **72:** 1858–1862.
37. LETERRIER, J. F., R. K. LIEM & M. L. SHELANSKI. 1982. J. Cell Biol. **95:** 982–986.
38. VALLEE, R. B., M. J. DiBARTOLOMEIS & W. E. THEURKAUF. 1981. J. Cell Biol. **90:** 568–576.
39. KUZNETSOV, S. A., V. I. RODINOV, V. I. GELFAND & V. A. ROSENBLAT. 1981. FEBS Lett. **135:** 237–240.
40. VALLEE, R. B. & S. E. DAVIS. 1983. Proc. Natl. Acad. Sci. USA **80:** 1342–1346.
41. AAMODT, E. A. 1984. Ph.D. Dissertation. Vanderbilt University, Nashville, Tennessee.
42. SLOBODA, R. D., S. A. RUDOLPH, J. L. ROSENBAUM & P. GREENGARD. 1975. Proc. Natl. Acad. Sci. USA **72:** 177–181.
43. KIM, H., L. I. BINDER & J. L. ROSENBAUM. 1979. J. Cell Biol. **80:** 260–276.
44. HITT, A. 1985. Personal communication.
45. STAROS, J. V. 1982. Biochemistry **21:** 3950–3955.

Interaction of Actin Filaments with Microtubules Is Mediated by Microtubule-Associated Proteins and Regulated by Phosphorylation[a]

S. CHARLES SELDEN[b] AND THOMAS D. POLLARD[c]

Department of Cell Biology and Anatomy
The Johns Hopkins University
School of Medicine
Baltimore, Maryland 21205

INTRODUCTION

At the present time, the best evidence for interactions between actin filaments and microtubules comes from biophysical experiments with purified actin and microtubule proteins.[1-6] In cells, actin filaments are occasionally found in close proximity to microtubules,[7-9] and agents that depolymerize actin filaments can inhibit the microtubule-dependent movement of vesicles in neurites,[10-12] but there is really not yet direct evidence for physical association of the two polymers in living cells. In this paper we will first review briefly the status of the biochemical studies on the system and then present new reconstitution experiments demonstrating that MAP-2 can cross-link actin filaments to microtubules. Further, we show that phosphorylation of MAP-2 inhibits its cross-linking activity.

When actin filaments and unfractionated microtubules are mixed together *in vitro,* the viscosity of the solution is higher than the sum of the viscosities of the separate components.[1] This is true for viscosity measured at high shear rates in an Ostwald viscometer and for the apparent viscosity measured at low shear rates in a miniature falling-ball viscometer that amplifies differences between samples.[13]

Pure tubulin microtubules do not form high viscosity complexes with pure actin, but isolated microtubule-associated proteins (MAPs) can interact in some way with actin filaments to produce a high viscosity complex.[1] Both MAP-2[2,4,5] and tau[2,3,5] form high viscosity complexes with actin filaments and aggregate actin filaments into bundles. Sattilaro *et al.*[4] also demonstrated binding of MAP-2 to actin filaments with a pelleting assay. Cross-linking of actin filaments by purified MAPs has recently been confirmed by Arakawa and Frieden[6] using a completely different technique—fluorescence photobleaching recovery. By labeling both the actin and the MAPs in different experiments, they were able to demonstrate that actin filaments are immobilized in the presence of MAPs and that part of the MAPs is bound to the actin filaments.

Reversible phosphorylation of MAPs may be responsible, in part, for the regulation

[a]This work was supported by NIH Research Grant GM-26132 and an NIH Postdoctoral Fellowship to S.C. Selden.
[b]Current address: Revlon Biotechnology Research Center, Rockville, Maryland 20850.
[c]To whom correspondence should be addressed.

of actin-microtubule interaction. First, it was found that mixtures of actin and unfractionated microtubule protein have a lower viscosity in the presence of adenosine triphosphate (ATP) and other nucleoside phosphates than in their absence.[1] Nishida et al.[3] confirmed this observation and suggested that the ATP itself was not the inhibitor, but rather that ATP-dependent phosphorylation of the MAPs inhibits their binding to actin filaments. They preincubated unfractionated microtubule protein (containing protein kinases) with ATP and cyclic adenosine monophosphate (cAMP) before heating to isolate heat stable MAPs. The MAPs incubated with [^{32}P]ATP incorporated ^{32}P and had lower actin cross-linking activity compared with untreated MAPs. We extended this work using a chemical assay for phosphate bound to MAPs and found that there is an inverse relationship between the level of MAP phosphorylation and actin filament cross-linking activity in the absence of microtubules.[5] Using successive treatments with kinases and protein phosphatases, we showed that the effect of MAP phosphorylation on actin gelation is reversible. This strengthens the possibility that phosphorylation of MAPs might regulate their interaction with actin filaments.

All of this work is consistent with a simple model where actin filaments bind to MAPs that project from the surface of microtubules. Both MAP-2 and tau seem to be able to cross-link actin filaments, but it is not clear whether one or both or neither of these MAPs might bind the actin to microtubules. The major problem with the previous work is that all of the experiments with isolated MAPs were done with actin filaments alone; the actin-microtubule system had never been reconstituted from purified components. Here we demonstrate how to reconstitute the system and show that MAP-2, but not tau, can cross-link microtubules to actin filaments.

METHODS

All reagents and protocols for isolating the proteins are described by Selden and Pollard,[5] with the following clarifications and additions. In these experiments we used "conventional" rabbit skeletal muscle actin.[14] Microtubule protein was isolated from hog brain, using two cycles of thermal polymerization/depolymerization. The microtubule protein was loaded onto a 4 × 15 cm column of diethylaminoethyl (DEAE) Sephadex A-50 in 100 mM piperazinediethane sulfonic acid (PIPES) (pH 6.95), 2 mM ethylene glycol-bis(β-aminoethyl ether) N,N,N',N'-tetraacetic acid (EGTA), 2 mM MgCl, 0.1 mM guanosine triphosphate (GTP), and 0.5 mM dithiothreitol (DTT). The column was washed with two column volumes of the same buffer to elute tau peptides, then two column volumes of 0.3 M KCl in column buffer to elute other MAPs, and with 0.8 M KCl in column buffer to elute pure tubulin. The tubulin was desalted on a 2.5 × 8 cm column of Sephadex G-25 equilibrated with DEAE column buffer, then frozen drop-wise in liquid nitrogen. Thawed tubulin was cleared by centrifugation for 15 min in an Eppendorf desk-top centrifuge at 4°C, made 1 mM in GTP and 7% in DMSO, then heated at 37°C for 30 minutes to polymerize the tubules. The tubules were pelleted by centrifugation in a Ti50 rotor at 35,000 rpm, 30°C, for 20 minutes. The pellets were resuspended in two-pellet volumes of 0.1 M PIPES pH 6.95, 2 mM EGTA, 2 mM MgCl$_2$, 1 mM GTP and 7% DMSO using cooling on ice and careful sonication. The suspension was cleared by centrifugation in an Eppendorf centrifuge at 4°C for 15 minutes. This procedure produces readily polymerizable tubulin at 60 to 90 mg per ml.

To produce unfractionated MAPs, twice-cycled microtubule protein was made 0.8 M in NaCl, 80 mM in 2-mercaptoethanol and heated for 5 minutes in a boiling water bath. After centrifugation for 30 minutes at 35,000 rpm in a Ti50 rotor at 5°C, the

MAPs in the clear supernatant were precipitated with 2 M ammonium sulfate. The MAPs were resuspended in a small volume of distilled water by sonication and dialyzed against 100 mM KCl and 4 mM HEPES, pH 6.5.

MAP-2 and tau fractions were obtained by chromatography of the unfractionated MAPs on a 2.5 × 15 cm column of DEAE-Sephadex A-50 in 100 mM PIPES, 2 mM EGTA, 0.5 mM DTT. Tau was in the flow-through volume and was concentrated by precipitation with 2 M ammonium sulfate, centrifugation, and dialysis as above. The rest of the MAPs were eluted with a step to 0.4 M KCl. After concentration by precipitation with 2 M ammonium sulfate, followed by dialysis against 100 mM KCl, 50 mM N-2-hydroxyethylpiperazine-N'-ethanesulfonic acid (HEPES) pH 6.5, 2 mM MgCl$_2$, 2 mM EGTA, 0.5 mM DTT, this MAP fraction was further purified by gel filtration on a 1.5 × 60 cm column of Biogel A 1.5 M (200–400 mesh). The leading portion of the major peak was predominantly MAP-2, with a few minor bands of other high molecular weight peptides.

MAPs of high or low phosphate content were obtained by the method of Nishida et $al.$[2] by incubating twice-cycled microtubule protein for 1 hour in 33 mM PIPES, 12 mM 2-(N-morpholino)ethane sulfonic acid (MES), 5 mM MgCl$_2$, 75 mM KCl (pH 6.6) plus 2 mM EGTA to produce low phosphate MAPs, or plus 2 mM EGTA with 1 mM ATP and 5 μM cAMP to produce high phosphate MAPs. This material was fractionated as above to yield high and low phosphate tau or MAP-2.

Falling-ball viscometry[13] was performed at 37°C in 100 μl capillary pipettes held at an 80° angle using 0.64 mm diameter stainless steel balls. The reconstitution buffer consisted of 25 mM KCl, 80 mM PIPES, 2 mM MgCl$_2$, 1 mM GTP, 2 mM EGTA, 0.2 mM CaCl$_2$, 0.5 mM DTT, 0.01 mM ATP, and 0.9 M DMSO, ph 6.6. Incubation time was 20 minutes.

The presence of actin filaments and microtubules was checked by electron microscopy of samples from the viscometers and negatively stained with uranyl acetate. For measurement of microtubule lengths, we removed 50 μl samples from capillary viscometers within 30 minutes of making viscosity measurements and fixed the material in an equal volume of 40% glycerol in 60% microtubule polymerization buffer with 8% glutaraldehyde, and 2% tannic acid with gentle hand vortexing. Samples were later diluted into 9 volumes of distilled water; a drop was placed on a glow-discharged, carbon-coated grid for 30 seconds, and rinsed for 5 seconds in polymerization buffer, twice for 5 seconds in distilled water, then inverted for 30 seconds on a drop of 2% aqueous uranyl acetate, wicked off and air dried. Electron micrographs were made on a Zeiss EM10A at 1.6, 5, and 10 × 10^3 diameters magnification. Negatives were printed on 8 × 10 inch paper for a further 2.6 × magnification. All microtubules on an individual print were measured, 3 prints per sample; between 50 and 100 tubules were measured per sample. Measurements were made using a Hewlett-Packard digitizing table and program. Tubule length was doubled if one end was off the print, tripled if both ends were not printed.

RESULTS

Polypeptide Composition of Protein Fractions

Both tubulin and actin were free of contaminating polypeptides (FIGURE 1) even on greatly overloaded gels (not shown). Tau preparations consisted of at least five polypeptides with apparent molecular weights of 50 to 68. The mobility of these bands was the same in preparations with 1.9 and 0.5 moles of phosphate per 65,000 g. The

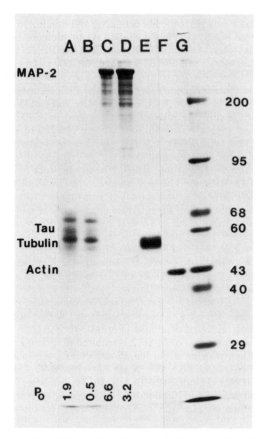

FIGURE 1. Polyacrylamide gel electrophoresis in sodium dodecyl sulfate of the protein fractions used in the experiments. The phosphate contents in moles per mole are given at the bottom. (A, B) Tau. (C, D) MAP-2. (E) Tubulin. (F) Actin. (G) Molecular mass standards with their values in thousands of daltons given on the right.

MAP-2 preparations consisted of a major band of about 280 kD and a variety of minor polypeptides with higher mobility. The band pattern was the same with 6.6 and 3.2 moles of phosphate per 300,000 g.

Reconstitution of a High Viscosity Network from Purified Components

The buffer composition proved to be the critical factor in reconstituting a high viscosity complex of actin filaments and microtubules using pure actin, pure tubulin and unfractionated MAPs. We were constrained by several factors in varying the components in the solution. First, we wanted to approximate physiological conditions, but more importantly we needed to use conditions where both actin and tubulin polymerized fully in the absence of MAPs, so that the addition of MAPs did not substantially change the polymer concentrations (and hence the viscosity). We used 2 mM MgCl$_2$ to promote polymerization of both actin and tubulin. In addition, we used a zwitterionic-sulfonate buffer (PIPES) and DMSO to promote tubulin polymerization.[15] We found that when the concentrations of KCl and PIPES gave ionic strengths between 200 and 250 mM, a mixture of 6.4 μM actin, 600 μg/ml MAPs, and 60 μM tubulin formed solutions with apparent viscosities above 40 poise. In FIGURE 2, this effect is shown in an experiment where we varied the PIPES concentration from 70 to 115 mM in the presence of 20 mM KCl, 2 mM MgCl$_2$, and 1 mM EGTA as the other major ionic species. For subsequent experiments we used 20 to 30 mM KCl with 75 to 85 μM PIPES for an approximate total ionic strength of 200 to 210 mM. Note that mixtures of actin alone with MAPs can also have a high apparent viscosity, but this occurs at lower ionic strengths (FIGURE 2).

MAPs are required to reconstitute a high viscosity network of actin filaments and microtubules (FIGURE 3). Several lines of evidence are consistent with the conclusion that MAP-2 can physically cross-link actin filaments to microtubules.

First, the apparent viscosity of the mixtures depends on the concentrations of actin (FIGURE 4), tubulin (FIGURE 3), and MAPs (FIGURE 4) as expected for an interactive, three-component system.

Second, under the conditions of the reconstitution assay, mixtures of actin with pure tubulin microtubules have a much lower apparent viscosity than the same concentration of tubulin plus MAPs (FIGURE 3). For example, with 40 μM tubulin and 6 μM actin the apparent viscosity was 10 times higher with MAPs than without MAPs. Mixtures of actin filaments and pure tubulin microtubules can have apparent viscosities greater than the sum of the viscosities of the two components, but this requires very high concentration of tubulin (FIGURE 3).

Third, a lower concentration of MAP-2 is required to form a gel with a mixture of actin and tubulin than with an equal length concentration of actin filaments alone (FIGURE 4). For 58 μM actin (9.5 × 10^{10} m per liter), less MAP-2 is required to yield a given viscosity than with 32 μM actin (5.2 × 10^{10} m per liter). Even less MAP-2, however, is required with 32 μM actin plus 127 μM tubulin (4.8 × 10^{10} m per liter), a mixture with approximately the same total polymer length concentration (10 × 10^{10} m

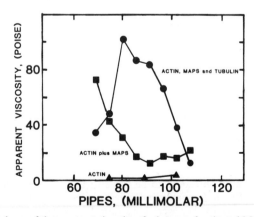

FIGURE 2. Dependence of the apparent viscosity of mixtures of actin and MAPs ± tubulin upon the concentration of PIPES buffer. Conditions: 2 mM MgCl$_2$, 20 mM KCl, 1 mM EGTA, 1 mM GTP, 900 mM DMSO, 0.5 mM dithiothreitol, pH 6.6, PIPES 70 to 115 mM; 6.4 μM actin; ± 60 μM purified tubulin; 37°C. Samples were incubated for 20 min before measuring viscosity. Actin alone (▲); Actin plus MAPs (■); Actin, tubulin, and MAPs (●).

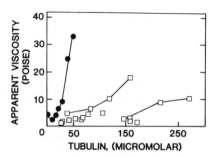

FIGURE 3. Dependence of the apparent viscosity of actin and actin plus MAPs on the concentration of tubulin. Conditions: Reconstitution buffer; 37°C; 6 μM actin plus 600 $\mu g/ml$ MAPs (●); or 6 μM actin (□). Connected squares represent data from a single series on a given day. Unconnected points are related data from other experiments. Samples with tubulin alone or tubulin with 600 $\mu g/ml$ MAPs had apparent viscosities less than one poise.

per liter) as 58 μM actin (9.5 × 10^{10} m per liter). Thus, although MAP-2 can cross-link actin filaments to each other, it is even more effective cross-linking actin filaments to microtubules.

Fourth, the effect of MAP-2 on the viscosity is not attributable to an effect on the length of the microtubules. In fact, the mean length of microtubules is inversely proportional to the MAP-2 concentration (FIGURE 5). This makes it even more remarkable that MAP-2 increases the viscosity of mixtures of actin and microtubules, because shorter tubules should not form a continuous network as readily as long

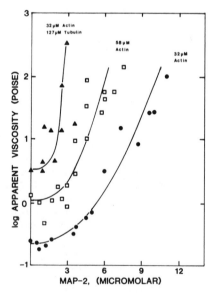

FIGURE 4. Dependence of the apparent viscosity of actin, and actin plus tubulin on the concentration of purified MAP-2. Conditions: Reconstitution buffer; 37°C; 32 μM actin (●); 58 μM actin (□); 32 μM actin with 128 μM tubulin (▲).

tubules. Similarly, the mean length of microtubules is inversely proportional to the tubulin concentration (FIGURE 5). The presence of actin had no effect on the length distribution of 35 μM tubulin ± MAP-2 (data not shown).

Effect of Phosphorylation of MAP-2 on Cross-Linking of Actin Filaments and Microtubules

It was established previously that phosphorylation of MAP-2 inhibits its ability to cross-link actin filaments to each other, and the development of the reconstitution assay has made it possible to show that phosphorylation of MAP-2 also inhibits the cross-linking of actin filaments and microtubules (FIGURE 6). In these experiments we tested only two levels of phosphorylation, 3.2 and 6.6 moles phosphate per 300,000 g, but even with this small difference, the MAP-2 with more phosphate was clearly less effective in raising the viscosity of actin alone and actin with tubulin.

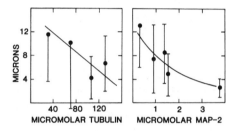

FIGURE 5. Dependence of microtubule length on the concentrations of tubulin and MAP-2. Conditions: 26 mM KCl; 76 mM PIPES (pH 6.6); 2.6 mM EGTA, 2 mM MgCl$_2$; 1 mM GTP; 900 mM DMSO. Samples were polymerized at 37° in a capillary viscometer for 40 min before preparing for EM as described in METHODS. (Left) Pure tubulin; (Right) 35 μM tubulin + MAPs. Points are mean length ± 1 SD.

Tau Does Not Cross-Link Actin Filaments and Microtubules

Although tau can cross-link actin filaments,[2,3,5] it is not an effective cross-linker of actin filaments and microtubules (FIGURE 7). In this experiment, we compared the effect of tau on the viscosity of actin and actin plus tubulin. Less tau is required to produce a given viscosity with 58 μM actin than 32 μM actin, and at both actin concentrations tau can produce a gel. By contrast, a mixture of 32 μM actin with 127 μM tubulin (the same polymer length concentration as 58 μM actin) does not gel even at the highest concentrations of tau that we tested (FIGURE 7). Note that at high concentrations of tau, the viscosity of actin plus tubulin is considerably lower than actin alone. Thus tubulin can inhibit the cross-linking of actin filaments by tau.

DISCUSSION

The reconstitution experiments described here show that MAP-2 is necessary and sufficient for actin filaments and pure tubulin microtubules to form a complex with a high apparent viscosity. Phosphorylation of MAP-2 inhibits its effectiveness in

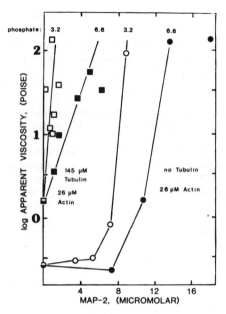

FIGURE 6. Phosphorylation of MAP-2 inhibits cross-linking of actin filaments and microtubules. Dependence of the apparent viscosity on the concentration of MAP-2 for 26 μM actin (●, ○), and 26 μM actin with 145 μM tubulin (■, □). The MAP-2 had 3.2 (open symbols) or 6.6 (closed symbols) moles of phosphate per 300,000 g. Conditions: Reconstitution buffer, 37°C.

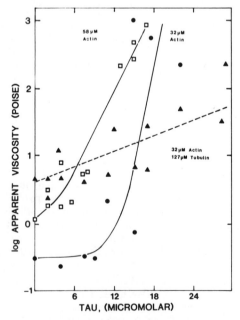

FIGURE 7. Dependence of the apparent viscosity of actin filaments and actin filaments with microtubules upon the concentration of tau. Conditions: Reconstitution buffer; 37°, 32 μM actin (●); 58 μM actin (□); and 32 μM actin with 127 μM tubulin (▲).

reconstituting the high viscosity complex. These reconstitution experiments are required to show which of the many molecules in crude MAP fractions can promote the interaction of actin and tubulin. For example, tau can bind to microtubules and can also cross-link actin filaments, but it is ineffective in reconstituting the actin filament-microtubule complex.

Our interpretation of the reconstitution experiments is that MAP-2 cross-links actin filaments to the microtubules by virtue of its high-affinity binding site for microtubules and second lower-affinity binding site for actin filaments. This conclusion is based on the apparent viscosities of various mixtures of the three components. It is remarkable that MAP-2 results in a higher apparent viscosity of the mixtures of actin filaments and microtubules, because MAP-2 also makes the microtubules shorter, a result that, by itself, would make the viscosity lower.

The conclusion that MAP-2 cross-links actin filaments and microtubules must be qualified, because the falling ball method is not a quantitative assay for non-Newtonian materials such as actin filaments and microtubules.[13] It is useful for detecting the presence of some sort of filamentous network, but the apparent viscosities are not a quantitative measure of the extent of cross-linking. Consequently, more work using rheometers that can measure viscosity and elasticity quantitatively will be required to characterize the physical properties of these samples. Then it should be possible to establish whether the two types of polymers are actually connected together in a continuous network as we have speculated here. Because MAP-2 is a large molecule that extends radially from the surface of microtubules, it is a plausible connector between microtubules and actin filaments. The actin binding site is probably on the radial projection.[16] Because MAP-2 can also cross-link actin filaments to each other, we expect that there is a second low-affinity actin binding site, perhaps in the tubulin-binding domain. On the other hand, tau may fail to cross-link actin filaments and microtubules because its actin binding sites are buried when it binds to microtubules.

Phosphorylation of MAP-2 inhibits both its ability to cross-link actin filaments to each other and to microtubules. The most likely mechanism is that dephosphorylation reduces the affinity of MAP-2 for actin, but this has not been investigated. Undoubtedly, the situation is complex due to the multiple phosphorylation sites on MAP-2 and the well-documented effects of phosphorylation on the interaction of MAP-2 with tubulin.[17]

The experiments with purified proteins establish the potential for interactions between actin filaments and microtubules in living cells and regulation of the association by phosphorylation of MAP-2. The linking together of these two major polymer systems should, at the very least, contribute to the structural integrity of the cytoplastic matrix. Further, actin filaments bound to microtubules might in some circumstances provide the substrate for myosin-powered, microtubule dependent movements.

SUMMARY

We have reconstituted high viscosity networks of actin filaments and microtubules from purified actin, tubulin, and MAPs. MAP-2 can effectively cross-link actin filaments and microtubules, presumably because a low affinity actin binding site is available even when it is bound tightly to microtubules. Phosphorylation of MAP-2 inhibits cross-linking of actin filaments and microtubules. Tau is not an effective cross-linker of actin and microtubules even though it can interact with each polymer individually.

REFERENCES

1. GRIFFITH, L. M. & T. D. POLLARD. 1978. J. Cell Biol. **78:** 958–965.
2. GRIFFITH, L. M. & T. D. POLLARD. 1982. J. Biol. Chem. **257:** 9143–9151.
3. NISHIDA, E., T. KUWAKI & H. SAKAI. 1981. J. Biochem. (Tokyo) **90:** 575–578.
4. SATTILARO, R. F., W. L. DENTLER & E. L. LECLUYSE. 1981. J. Cell Biol. **90:** 467–473.
5. SELDEN, C. S. & T. D. POLLARD. 1983. J. Biol. Chem. **258:** 7064–7071.
6. ARAKAWA, T. & C. FRIEDEN. 1984. J. Biol. Chem. **259:** 11730–11734.
7. GAWADI, N. 1974. Cytobios **10:** 17–35.
8. FORER, A. & O. BEHNKE. 1972. J. Cell Sci. **11:** 491–519.
9. POLLARD, T. D., S. C. SELDEN & P. MAUPIN. 1984. J. Cell Biol. **99:** 33s–37s.
10. ISENBERG, G. & J. V. SMALL. 1978. Cytobiologie **16:** 326–344.
11. SCHWARTZ, J. H. 1980. Sci. Am. **242:** 152–171.
12. BRADY, S. T., R. J. LASEK, R. D. ALLEN, H. L. YIN & T. P. STOSSEL. 1984. Nature (London) **310:** 58–59.
13. POLLARD, T. D. & J. A. COOPER. 1982. *In* Methods in Enzymology. D.W. Frederiksen & L.W. Cunningham, Eds. **85:** 211–233. Academic Press. New York.
14. SPUDICH, J. A. & S. WATT. 1971. J. Biol. Chem. **246:** 4866–4871.
15. OLMSTED, J. B. & G. G. BORISY. 1975. Biochemistry **14:** 2996–3002.
16. SATTILARO, R. F. & W. L. DENTLER. 1982. *In* Biological Functions of Microtubules and Related Structures. Proceedings of the Oji International Seminar. H. Saki, H. Mori & G.G. Borisy, Eds. pp: 297–309. Academic Press. Japan.
17. JAMESON, L., T. FREY, B. ZEEBERG, F. DALLDORF & M. CAPLOW. 1980. Biochemistry **19:** 2472–2479.

Isolation of Microtubule-Secretory Granule Complexes from the Anglerfish Endocrine Pancreas[a]

W.L. DENTLER AND K.A. SUPRENANT

Department of Physiology and Cell Biology
University of Kansas
Lawrence, KS 66045

INTRODUCTION

Ultrastructural analysis of a variety of tissues has established that microtubules are linked to organelle membranes by filamentous microtubule-membrane bridges,[1-5] and recent light microscopic studies have revealed that organelles are transported in close association with microtubules in intact cells and in cell extracts.[6-11] An attractive possibility is that the bridges bind specific organelles to microtubules and mediate microtubule-associated organelle transport. The identification and characterization of the microtubule-membrane bridges should help us understand the role of microtubules in organelle positioning and transport.

To study the bridges, methods must be developed to stabilize them and to isolate microtubule-organelle complexes for analysis *in vitro*. In the present study, we identify bridges that link secretory granules to microtubules in intact islet cells, develop methods to stabilize the bridges so that complexes of microtubules and secretory granules can be isolated for analysis *in vitro*, and identify possible islet microtubule-associated proteins (MAPs) that might be responsible for the microtubule-secretory granule linkages. Pancreatic cells were chosen because earlier pharmacological and morphological studies indicated that microtubules were associated with secretory granules and were necessary for normal secretory activities.[12,13] The anglerfish was chosen because milligram quantities of pure islet tissue could be obtained without manipulation or digestion of the endocrine tissue.[14]

MATERIAL AND METHODS

Isolation of Microtubule-Secretory Granule Complexes

Pancreatic islets were dissected from freshly killed anglerfish and maintained in stabilization buffer (SB: 100 mM 2-(N-morpholino)ethane sulfonic acid (MES), 1 mM MgSO$_4$, 2 mM ethylene glycol bis (β-aminoethyl ether)-N, N, N', N'-tetraacetic acid (EGTA), 250 mM sucrose, 1 mM guanosine triphosphate (GTP), 1 mM dithiothreitol (DTT), 0.1 mM leupeptin, 0.3 mM phenyl methylsulfonyl fluoride (PMSF), pH 6.4 with NaOH) at room temperature for up to 30 min prior to decapsulation.[14] Decapsulated islets were homogenized at room temperature in a

[a]This research was supported by Grants from the NIH (AM 21672 and GM 32556), Biomedical Sciences and General Research Grants funded by the University of Kansas (W.L.Dentler) and the American Cancer Society (PF-2415) (K.A.Suprenant).

siliconized glass-teflon homogenizer in 1 ml SB for every 100 mg islet tissue, and the homogenate was centrifuged at 600 × g for 5 minutes. The pellet was resuspended in fresh SB and recentrifuged at 600 × g. The supernatants from the 600 × g centrifugations were combined and centrifuged at 17,000 × g for 5 min to pellet microtubule-secretory granule complexes.

The 17,000 × g pellet was resuspended in SB and made 5 μM in taxol[15] [stored as a 0.5 mM stock in 100% dimethyl sulfoxide (DMSO)] or 1 M in hexylene glycol to stabilize the microtubules. Although electron microscopy revealed that complexes stabilized by each agent were nearly identical, most experiments were carried out with taxol. The suspension was then centrifuged for 2 min in a Fisher microfuge at 13,000 × g, and the pellet was resuspended in SM + 5 μM taxol. Pellets and supernatants from the microfuge centrifugation were examined by dark field microscopy, and the procedure was repeated until the pellets were composed primarily of microtubule-secretory granule complexes and the supernatants were composed of free secretory granules (see RESULTS and FIGURE 3). Generally, this procedure was repeated four to six times.

The complexes were stable and could be stored in SM at 4°C overnight without any noticeable change in the ultrastructure of the bridges or release of the secretory granules from the microtubules. In some experiments, the preparations were incubated overnight in SM with gold-labeled monoclonal tubulin antibodies (see RESULTS and FIGURE 6).

Isolation of Microtubule Protein

In vitro Assembly and Disassembly

Anglerfish islets were collected and incubated in SB at 4°C for no longer than 30 minutes. Islets were then decapsulated and homogenized at 4°C with a glass-teflon homogenizer (presoaked in 100 mM EGTA) in polymerization buffer (PM: 100 mM piperazinediaminetetraacetic acid (PIPES), 1 mM MgSO₄, 1 mM EGTA, 250 mM sucrose, 1 mM GTP, 1 mM adenosine triphosphate (ATP), 1 mM DTT, 0.1 mM leupeptin, 0.3 mM PMSF, pH 6.6 with NaOH) at a ratio of 0.75 ml PM per gram tissue. (Generally 400–600 mg of islet tissue were homogenized in ~0.45–0.60 ml PM). The homogenate was centrifuged at 40,000 × g for 10 min (4°C), and the supernatant was removed and centrifuged at 40,000 × g for an additional 60 min (4°C). The supernatant was removed, made 1 mM in GTP and ATP, and incubated at 35°C for 45 min to assemble microtubules. The microtubules were pelleted at 40,000 × g for 45 min (30°C). The supernatant (called 1X super) was saved for sodium dodecyl sulfate polyacrylamide gel electrophoresis (SDS-PAGE) or for taxol-microtubule assembly (see below). The microtubule-containing pellets were resuspended in warm PM (¼ of the initial homogenate volume), the microtubules were depolymerized by cooling on ice for 30 min, and the suspensions were centrifuged at 27,000 × g for 20 min (4°C) to pellet cold-stable material. The supernatant was removed, fresh GTP and ATP were added, and the microtubules were assembled and pelleted a second time. The 2X microtubule pellet was then prepared for SDS-PAGE or fixed for electron microscopy.

Taxol Microtubules

Islet tissue was decapsulated, homogenized, and the homogenate centrifuged as described above (in vitro assembly/disassembly). The 40,000 × g supernatant from the homogenate or the 1X super (see above) was then made 5 μM in taxol and incubated at 35°C for 30 min.[15,16] The microtubules were pelleted through 0.5 M sucrose in PM (0.5 ml) at 40,000 × g for 60 min (4°C) and were prepared for electron microscopy or for SDS-PAGE. SDS-PAGE revealed that the composition of microtubules obtained from the 40,000 × g supernatant were essentially identical to those assembled from the 17,000 × g supernatant after sedimentation of microtubule-secretory granule complexes.

Dark Field Light Microscopy

Dark field light microscopy was carried out essentially as previously described.[17] For routine examination, a Zeiss Standard microscope was equipped with a 40X N.A. 0.65 objective, darkfield ultracondenser (N.A. 1.2–1.4), and a Zeiss 75 watt xenon lamp. Photographs were taken with a Nikon F camera, 10X ocular, and Tri X film developed in Diafine.

Electron Microscopy

Intact decapsulated islets were fixed for 1 hr at room temperature in 1% glutaraldehyde in PM, and postfixed for 1 hr at 4°C in 1% OsO_4 in PM. Tissue was then stained for 1–2 hr in 1% aqueous uranyl acetate, dehydrated in ethanol and propylene oxide, and embedded in Epon-Araldite or Spurr resin.

For some experiments, tissue was fixed for 1–5 min in glutaraldehyde-PM, extracted 5–30 min in SM + 0.1% Nonidet P-40 to extract soluble cytoplasm, and returned to glutaraldehyde-PM for 1–2 hr prior to postfixation with OsO_4.[18]

The microtubule-secretory granule complexes were fixed in suspension by dilution of the sample with an equal volume of 2% glutaraldehyde in PM for 1–12 hours, and pelleted at 13,000 × g. Alternatively, samples were pelleted at 13,000 × g, and the pellets were overlayed with 1% glutaraldehyde in PM. All pellets were postfixed in 0.5% OsO_4, stained 1–12 hr in 1% uranyl acetate, dehydrated, and embedded in Epon-Araldite or Spurr resin.

In vitro assembled microtubules and microtubule-secretory granule complexes were negatively stained with 1% aqueous uranyl acetate on freshly glow discharged carbon-coated formvar films.

Electrophoresis

Proteins were separated in slab gels according to the procedure of Laemmli.[19] Gels were run for 10–15 hr at constant voltage and were fixed and stained using the procedure of Fairbanks *et al.*[20]

RESULTS

Association between Microtubules and Secretory Granules In Situ

Thin sections of intact anglerfish islet cells revealed numerous microtubules, and most of the microtubules were linked to secretory granule membranes by thin filaments similar in appearance to the high molecular weight MAPs associated with *in*

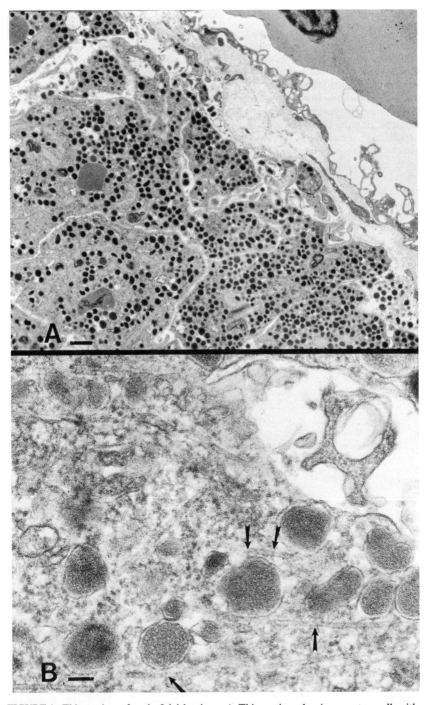

FIGURE 1. Thin sections of anglerfish islet tissue. **A.** Thin sections showing secretory cells with a darkly staining nucleated erythrocyte (NE) in the bloodstream. **B.** Higher magnification of an area similar to that shown in **A.** Microtubules course through the cytoplasm and often contact secretory granule membranes by thin bridges (arrows). **A:** bar = 1 μM. **B:** bar = 0.1 μM.

vitro assembled brain microtubules[21,22] (FIGURE 1). These microtubule-secretory granule complexes were more easily seen in partially extracted islets that had been briefly washed with detergent during fixation (FIGURE 2). In these cells, MAP-like filaments extended from the surfaces of both microtubules and from the secretory granule membranes. In both intact and extracted cells, virtually every microtubule was linked by thin bridges to one or more secretory granules although there were numerous secretory granules that did not appear to be linked to microtubules.

Although these results clearly showed that secretory granules were linked to islet microtubules *in situ,* it was not possible to identify the specific types of granules to which microtubules were attached because the anglerfish islets were composed of several different types of endocrine cells whose secretory granules contain different hormones. The granules associated with microtubules did, however, have a structure similar to those that contain insulin and somatostatin (B. Noe, Emory University, personal communication).

Isolation of Microtubule-Secretory Granule Complexes

To study the associations between microtubules and secretory granules it was necessary to develop methods to stabilize both the microtubules and the microtubule-membrane bridges. In initial experiments, we discovered that the microtubule-secretory granule complexes were stable and could be identified in cell homogenates. Moreover, the complexes could be pelleted from the soluble cell extract and remained intact upon resuspension.

Microtubules were stabilized by adding $1 M$ hexylene glycol[23,24] or $5\ \mu M$ taxol[15] to the buffers in which the pelleted complexes were resuspended. (The addition of stabilizing agents to the cell homogenate would have induced microtubule assembly from soluble tubulin.) When resuspended complexes were examined by electron microscopy, the results with each stabilizing agent was essentially the same: each microtubule observed by negative stain (FIGURES 3 and 4) or in thin section (FIGURES 5 & 6) contained one or more secretory granules bound along its length. Thin bridges linked the microtubules and organelle membranes in many specimens (FIGURES 4B, 5, 6). Although microtubule-secretory granule complexes were stabilized by each stabilizing agent, taxol was used in most experiments because electron microscopy revealed that hexylene glycol partially disrupted some secretory granule membranes.

The major contaminants of the microtubule-secretory granule complex preparations were free secretory granules. The most successful method for purifying complexes from free granules involved successive centrifugations with a microfuge and analysis of the pellets and supernatants by darkfield light microscopy (FIGURE 3). Each centrifugation resulted in the separation of individual secretory granules that remained in the supernatant from the large microtubule-secretory granule complexes that pelleted. The ratio of secretory granule-microtubule complexes to free secretory granules increased with the number of washes. The greatest enrichment of complexes was obtained after four to six centrifugations.

The microtubule-secretory granule linkages were stable to repeated centrifugation and resuspension. In some experiments, the purified complexes were incubated overnight in the presence of a gold-labeled monoclonal tubulin antibody (a gift from J. Kilmartin). As shown in FIGURE 6, the granules were not displaced by the binding of the tubulin antibody, and the complexes remained intact throughout the incubation period. These results show that the bridges linking microtubules to secretory granule membranes are very stable, making purification to homogeneity a likely possibility.

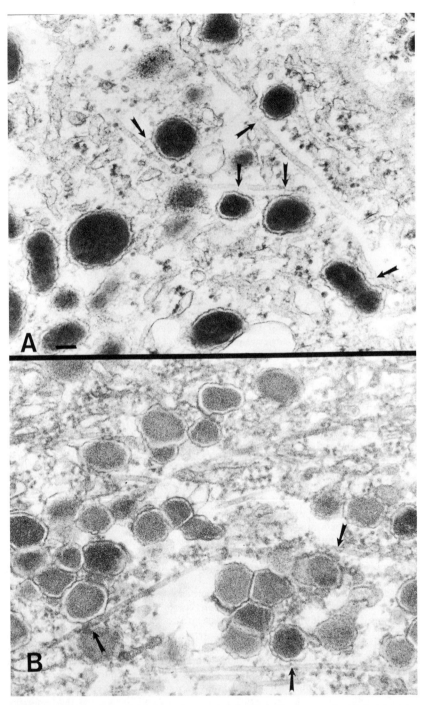

FIGURE 2. Thin sections of islet tissue that was briefly fixed, then extracted with detergent (see MATERIAL AND METHODS). This procedure briefly extracted some of the soluble cytoplasm without disrupting the (morphological) integrity of membranes. The microtubules, thin microtubule-membrane bridges (arrows) as well as MAPs distributed along the microtubules are more easily observed in these preparations. Filamentous material (MeAPS?) can also be seen to extend from secretory granule membranes (arrowheads). **A, B:** bar = 0.1 μM.

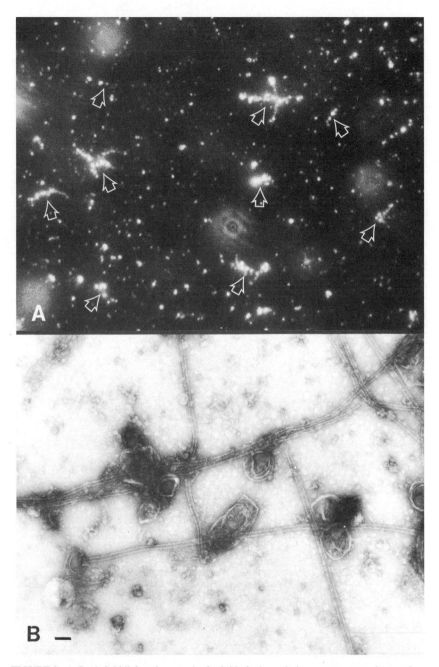

FIGURE 3. A. Dark field light micrograph of a field of microtubule-secretory granule complexes (large arrows) and free secretory granules (arrowheads). **B.** Negatively stained complex obtained from a preparation similar to that shown in the dark field micrograph. Each of the microtubules is bound to one or more secretory granule membranes. Thin sections of this preparation are shown in FIGURE 4. **B:** bar = 0.1 μM.

Morphology of Microtubule-Secretory Granule Complexes

Each negatively stained complex contained one or more microtubules coated with intact secretory granules or membrane vesicles. Although some microtubules contained only one granule (see FIGURE 4B), most microtubules were linked to two or more

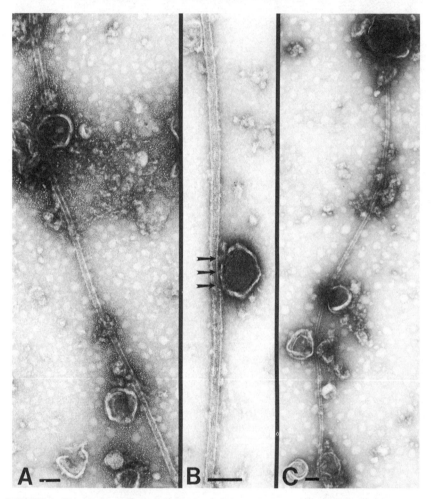

FIGURE 4. Negatively stained purified microtubule-secretory granule complexes. Membrane-bound secretory granules are bound to the microtubules by short microtubule-membrane bridges, shown by arrows in **B**. Bars = 0.1 μM.

granules (FIGURES 4A and C). Short bridges bound each secretory granule membrane to a microtubule in the isolated complexes (FIGURE 3B and 4–6). The membranes were linked to the microtubules by two or three bridges, but linkages mediated by only one bridge were occasionally observed.

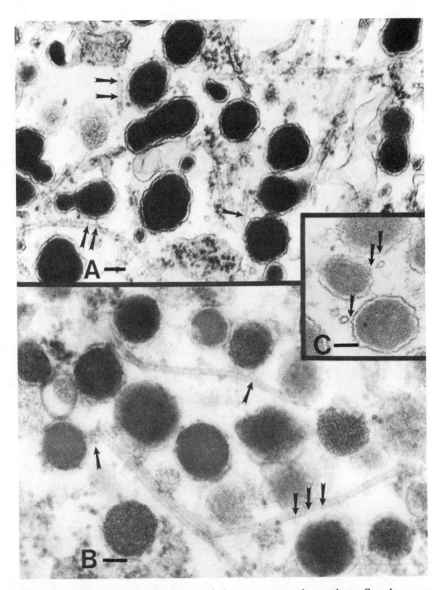

FIGURE 5. Thin sections of isolated microtubule-secretory granule complexes. Complexes were pelleted prior to fixation. Bridges linking microtubules to secretory granule membranes are shown by arrows. Numerous filamentous structures (MeAPs?) are also associated with the secretory granule membranes. Bars = 0.1 μM.

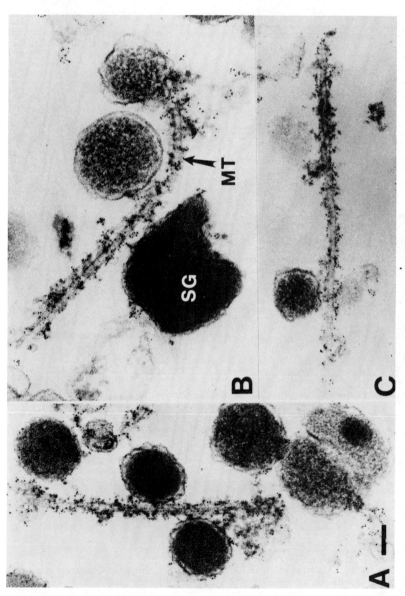

FIGURE 6. Thin sections of isolated microtubule-secretory granule complexes fixed in solution prior to pelleting. Microtubules and secretory granules are spread further apart than in the pelleted preparations (FIGURE 5), and associations are somewhat more easily observed. These complexes were incubated overnight in SB with colloidal gold-labeled yeast antitubulin antibodies (a gift of J. Kilmartin), and the electron-dense dots on the microtubules are colloidal gold particles. The complexes were subsequently fixed, pelleted, and processed for electron microscopy. The complexes appeared identical to those fixed immediately after isolation, indicating that the isolated complexes are stable in the buffers with taxol. Bar = 0.1 μM.

Thin sections of the complexes permitted positive identification of the types of organelles that were attached to microtubules and revealed that the microtubules were only linked to secretory granules and, occasionally, a clathrin coat (see FIGURE 5B). These results showed that (1) the microtubule-membrane bridges tightly linked secretory granules to microtubules *in vivo* (if the interactions were weak or apparent artifacts of electron microscopy the organelles would not have copurified with one another) and (2) the associations were specific (because secretory granules were the only organelles commonly observed to bind to the microtubules *in situ* and *in vitro*).

Reactivation of Granule Movement

To study the nature of the microtubule-membrane associations, isolated microtubule-secretory granule complexes were mixed with various solutions and observed with dark field light microscopy. The secretory granules remained tightly bound to microtubules in the presence of 0.01–5 mM ATP, in buffers adjusted to pH 6.4–7.5, and in the presence of SB plus 50 mM KCl. Some, but not all, secretory granules were dissociated from the microtubules in the presence of up to 0.5 M NaCl, which indicated that the microtubule-membrane bridges were firmly bound to the microtubules. The exposure of *in vitro* assembled pancreatic microtubules to similar salt concentrations released many of the MAPs (see below).

The bridges linking secretory granules to microtubules were, however, very flexible. In the presence of ATP, the granules were observed to move along or pull away from the microtubules and then snap back, implying the presence of an elastic component. They were rarely seen to detach from any single microtubule.

In several experiments, the granules were observed to move along the microtubules. Movements occurred equally well in mM ATP and GTP, and the effects of other nucleotides have not been tested. One curious feature was the presence of small hoops of filaments that were slightly smaller in diameter than microtubular marginal bands found in the anglerfish nucleated erythrocytes. Particle movements occurred bidirectionally on the hoops, and two particles attached to a single hoop could move in opposite directions at the same time. Presumably, these hoops were composed of microtubules because microtubules were the only filamentous structures seen in negatively stained preparations. We presume that the hoops are transient structures (and not marginal bands) because we never observed them in electron micrographs of thin sectioned or negatively stained complexes.

Proteins Associated with Islet Microtubules
and Microtubule-Secretory Granule Complexes

Our previous studies[17] showed that MAPs were necessary to link microtubules to secretory granules *in vitro* but, because we used a heterologous system, the identity of MAPs that linked microtubules to secretory granules *in vivo* was not possible. To identify the pancreatic MAPS, we compared the polypeptides in isolated complexes with the MAPs that copurified with *in vitro* assembled pancreatic microtubules, with the rationale that proteins found in all fractions would be potential components of the microtubule-membrane bridges.

As shown in FIGURE 7, numerous proteins copurified with tubulin through two cycles of *in vitro* assembly and disassembly. Prominent among these proteins were those of M_r 45,000, 90,000, 105,000 150,000, 220,000, and 300,000. Further cycles of

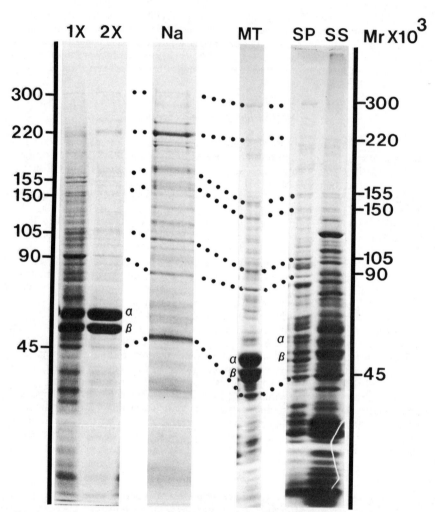

FIGURE 7. SDS-PAGE of microtubule-secretory granule complexes (SP, SS) and *in vitro* assembled islet microtubules (1X, 2X, MT). The dotted lines indicate polypeptides of similar relative molecular weight (determined by running standard protein mixtures in other lanes on each gel) and are drawn between the major MAP proteins found in microtubules assembled *in vitro* by one and two cycles of assembly and disassembly (1X, 2X) and one taxol assembly (MT). Polypeptides released by NaCl from taxol-stabilized 2X microtubules are shown in Na. Microtubule-secretory granule complexes were fractionated into detergent-soluble (SS) and detergent insoluble (SP) fractions. Microtubules were found in the SP fraction (see RESULTS).

assembly/disassembly were not possible because a maximum of 800 mg of islet tissue was all that could be obtained for any single experiment.

The twice-cycled *in vitro* assembled microtubules were decorated with short MAP-like filaments (FIGURE 8). Based on the similarity in appearance of the projections with those previously shown in thin-sectioned brain microtubules assembled *in vitro*,[21,22] we tentatively identified them as being composed of one or more of the MAPs that copurified with the islet microtubules.

The MAPs were also identified by stabilizing the twice-polymerized islet microtubules with taxol and then removing the MAPs with 0.6M NaCl.[16] The microtubules were pelleted after this extraction, and most of the nontubulin proteins that copurified with microtubules remained in the supernatant (FIGURE 7).

A second method to isolate microtubules used taxol to induce microtubule assembly from islet homogenates or "17,000 × g supernatants." Electron microscopy of thin sections revealed that these preparations contained a greater amount of membrane and amorphous material than microtubule preparations purified by cycles of assembly/disassembly. The "taxol microtubules" were coated with MAP filaments and contained the principal polypeptides of M_r 45,000, 70,000, 90,000, 105,000, 150,000, 220,000, and 300,000, that copurified with cycled microtubules (FIGURE 6). Because these polypeptides copurified with microtubules using each of these two assembly methods, we tentatively identify them as pancreatic islet MAPs.

Purified complexes were analyzed by SDS-PAGE to identify MAPs in the microtubule-secretory granule complexes. Because the complexes contained many secretory granule proteins as well as MAPs, the complexes were separated into detergent-soluble and detergent-insoluble fractions by treatment with 1% Triton X-100. Detergent-treated samples were then centrifuged at 13,000 × g for 5 min to pellet the detergent-insoluble material. Electron microscopy revealed that the pellets contained microtubules with MAP-like filaments as well as a large amount of amorphous material believed to be insoluble secretory granule contents.

As shown in FIGURE 7, both detergent soluble and insoluble fractions contained numerous polypeptides from M_r 300,000 to 20,000. The small peptide hormones (M_r <10,000) were run off these gels. The detergent-insoluble material contained polypeptides that comigrated with alpha and beta tubulins as well as those that comigrated with the M_r 300,000, 220,000, 155,000, 150,000, 105,000, 90,000, and 45,000 MAPs that copurified with *in vitro* assembled islet microtubules.

DISCUSSION

Although we have begun to understand the mechanisms that regulate microtubule assembly *in vitro* and *in vivo*, we have comparatively little understanding of microtubule function. Early pharmacological studies revealed that agents that depolymerize or superstabilize microtubules also inhibited secretion,[12,25] axoplasmic transport,[26,27] and pigment granule movements.[28] Recent morphological studies of living cells and cell fractions showed that microtubules were intimately associated with organelle movements,[6-10,29] and examination of fixed and sectioned cells revealed many examples of microtubules that were directly linked to organelle membranes by thin bridges.[1-5] Genetic analysis of mitochondrial and nuclear movements in *Aspergillis* revealed that microtubule-mediated organelle movements were selective, that is, nuclear motility was more dependent on the presence of microtubules than were mitochondrial movements.[30,31] Taken together, these studies show that microtubules are essential for

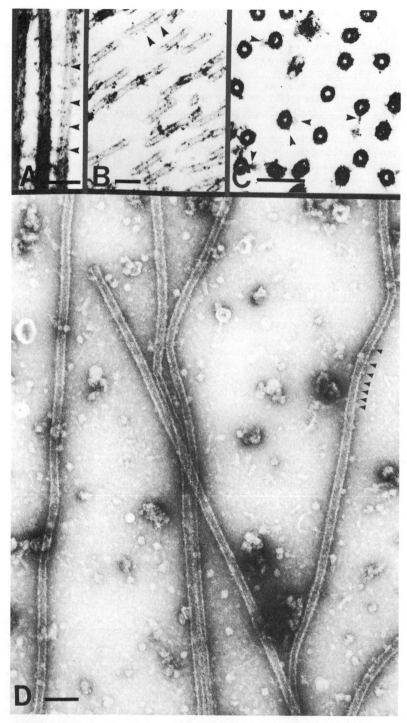

FIGURE 8. Electron microscopy of *in vitro* assembled microtubules. **A:** Negatively stained microtubules showing periodic MAP-like projections. **B:** Thin sections revealing the MAP-projections. Bar = 0.1 μM.

the movements of some organelles and that microtubule-dependent movements may be mediated by microtubule-membrane bridges.

Since the microtubule-membrane bridges were similar in appearance to the high molecular weight MAPs that copurified with brain microtubules,[21,22] initial studies were designed to determine if the MAPs could link organelles to microtubules *in vitro*. Babiss and colleagues[11,32,33] were the first to use negative staining and electron microscopy to assay associations between microtubules and viral particles and suggested that MAPs were necessary for viral particles to bind to microtubules *in vitro*. Sherline *et al.*[34] developed a centrifugation assay and reported that MAPs could associate with pituitary secretory granules *in vitro*.

Suprenant and Dentler[17,23] used dark field light microscopy to directly observe solutions containing microtubules and endocrine pancreatic secretory granules and presented direct evidence that secretory granules bound only to MAP-containing microtubules and not to MAP-free microtubules *in vitro*. The dark field observations were confirmed by electron microscopic observations of fixed and sectioned material that showed the MAP filaments to link microtubules to secretory granule membranes. Their studies also revealed that the associations were modulated: cyclic adenosine monophosphate (AMP) increased the number of associations, possibly by stimulating MAP phosphorylation, and ATP released the granules from microtubules. Although these studies clearly revealed that MAPs could link microtubules to secretory granules, they were limited by their use of a heterologous system. There was no evidence to prove that the proteins that linked microtubules and secretory granules in living islet cells were similar to the brain MAPs used in these studies.

The results presented here show that the bridges that link cytoplasmic microtubules with secretory granules *in vivo* can be stabilized and that organelle-microtubule complexes can be isolated for *in vitro* analysis. In addition to showing that the linkages are relatively stable, we have also shown that they are selective, because secretory granules are the principal organelle associated with the microtubules. Studies of the microtubule-membrane associations can now begin to address the specificity of association as well as the role of the bridges in organelle positioning and movement.

Proteins That Compose the Bridges

We have identified polypeptides in the preparations of microtubule-secretory granule complexes that comigrate, in one-dimensional SDS-PAGE, with MAPs that copurify with islet microtubules. Chief among these polypeptides are those of M_r 300,000, 220,000, 155,000, 150,000, 105,000, 90,000, and 45,000. We have also shown that one or more of these polypeptides compose the filaments associated with *in vitro* assembled islet microtubules. Positive identification of specific proteins that make up the microtubule-membrane bridges will depend on further purification of the bridges, production of antibodies that localize specific MAPs in the microtubule-membrane complexes, and cross-linking studies similar to those carried out with ciliary microtubules and membranes.[35]

How Do Secretory Granules Move along the Isolated Microtubules?

Two types of movements were observed in the isolated microtubule-secretory granule complexes. In one type of movement, secretory granules stretched away from a microtubule and then moved back to their starting place on the microtubule, as if the membranes and microtubules were linked by an elastic bridge. In the other type of movement, secretory granules moved along the microtubules for some distance and did not return to their starting place. In both cases the granules rarely were observed to detach or attach to a microtubule. It is not known if we have simply failed to identify solution conditions necessary to detach or attach microtubules and secretory granules or if other cytoplasmic factors or cytoskeletal components might be necessary for this process *in vivo*. We expect that future studies of the isolated microtubule-secretory granule complexes will reveal the mechanisms responsible for motility and for the attachment and release of the microtubule-membrane bridges.

Although we have no detailed understanding of the mechanisms that regulate the binding of microtubules to membranes, we can consider several general models for this association. The diagram in FIGURE 9 summarizes the types of associations that might link microtubules and membranes. One possibility, shown in FIGURE 9A, is that MAPs attached to microtubules bind to phospholipids in the nearby membrane. The associations could be dependent on local concentrations of charged groups on membranes, and the specificity for the associations may require specific microtubule-associated proteins. A second possibility, in 9C, is that membrane-associated proteins (MeAPs) attach to the microtubules. The specificity for this interaction would depend on the presence of specific MeAPS for specific microtubules or other cytoskeletal structures. The third possibility, shown in FIGURE 8B, is that specific MAPs are designed to recognize specific MeAPs. The major purpose for considering these models is to determine whether the focus for studies of microtubule-membrane interactions should be on the MAPs, the MeAPs, or the associations between MAPs and MeAPs.

Models A and B are the most consistent with data obtained from our *in vitro* studies showing that secretory granules bound to MAP-containing microtubules.[17] The model shown in FIGURE 9C is inconsistent with our data because secretory granules did not attach to MAP-free microtubules. At present, we have no direct data to determine whether model A or model B is the best description of the associations between microtubules and endocrine pancreatic secretory granules. We have, however, found filamentous structures attached to secretory granule membranes[17] (FIGURE 2) which may bind to MAPs and form the microtubule-membrane bridges; these membrane-associated filaments may be the MeAPs that bind to MAPs. It is also possible that the MAP-MeAP interactions may be mediated by a third component. One possibility for this third component is actin, which has been shown to bind to membranes and to MAPs[36-39] *in vitro*.

In addition to our studies of the endocrine pancreas, several studies have recently been focused on the *in vitro* associations of microtubules with vesicles isolated from squid axoplasm. Gilbert and Sloboda[10] recently reported that membrane vesicles isolated from squid axoplasm could attach to and move along microtubules in extruded squid axoplasm. They also reported that the vesicles attached to and moved along dialyzed sperm tail outer doublet microtubules from which most of the MAPs and dynein had been removed (Gilbert & Sloboda, personal communication). In another study, Sheetz *et al.*[8] reported that extracts from squid axoplasm can mediate the attachment and movement of plastic spheres and glass coverslips along *in vitro* assembled MAP-free microtubules. One interpretation of these studies is that the MeAPs are directly responsible for binding vesicles or other particles to the microtubules, but because MAPs are present in soluble cell extracts (as evidenced by their ability to coassemble with microtubules), another equally plausible explanation is that MAPs as well as MeAPs are present in the soluble extracts. Identification of the MAPs and MeAPs in the squid axoplasm as well as those in the pancreatic microtubule-membrane complexes should yield valuable information on the nature of microtubule-membrane interactions.

Based on the results reported here, the purification of microtubule-membrane

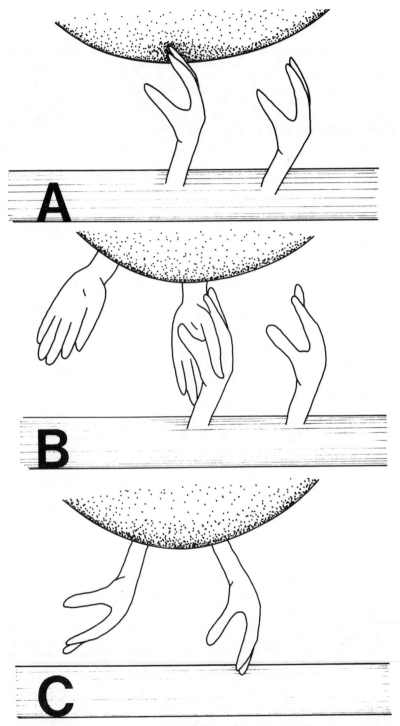

FIGURE 9. Diagram showing the types of specific interactions one might expect to find between microtubules and membranes. See DISCUSSION.

complexes from pancreatic as well as other cell types can now be initiated. With the development of these methods, we eventually hope to find the cytoplasmic equivalent of a flagellum—a motile machine that we can isolate and dissect. Investigation of these homologous microtubule-membrane systems will lead to the identification of the bridge proteins, the discovery of specific microtubule- and membrane-associated proteins that interact with one another, and finally, to the discovery of mechanisms that transport organelles along microtubules *in vivo.*

ACKNOWLEDGMENT

The taxol used in these studies was generously supplied by the Natural Products Branch, Division of Cancer Treatment, National Cancer Institute.

REFERENCES

1. DENTLER, W. L. 1981. Microtubule-membrane interactions in cilia and flagella. Int. Rev. Cytol. **72:** 1–47.
2. RAINE, C. S., B. GHETTI & M. L. SHELANSKI. 1971. On the association between microtubules and mitochondria within axons. Brain Res. **84:** 389–393.
3. SCHNAPP, B. J. & T. S. REESE. 1982. Cytoplasmic structure in rapid-frozen axons. J. Cell Biol. **94:** 667–679.
4. SMITH, D. S., U. JARLFORS & B. F. CAMERON. 1975. Morphological evidence for the participation of microtubules in axonal transport. Ann. N.Y. Acad. Sci. **253:** 472–506.
5. SMITH, D. S., U. JARLFORS & M. L. CAJER. 1977. Structural cross-bridges between microtubules and mitochondria in central axons of an insect (*Periplaneta americana*). J. Cell Sci. **27:** 235–272.
6. ALLEN, R. D., J. METUZALS, I. TASAKI, S. T. BRADY & S. GILBERT. 1982. Fast axonal transport in squid giant axon. Science **218:** 1127–1129.
7. HAYDEN, J. H. & R. D. ALLEN. 1984. Detection of single microtubules in living cells: Particle transport can occur both directions along the same microtubule. J. Cell Biol. **99:** 1785–1793.
8. SHEETZ, M. P., B. J. SCHNAPP, R. D. VALE & T. S. REESE. 1984. Organelle transport in squid axoplasm: Inhibition by drugs that collapse proton gradients. J. Cell Biol. **99:** 118a.
9. WEISENBERG, R. C. & R. D. ALLEN. 1984. ATP-dependent motility *in vitro* by isolated calf brain microtubule proteins. J. Cell Biol. **99:** 239a.
10. GILBERT, S. P. & R. D. SLOBODA. 1984. Bidirectional transport of fluorescently labeled vesicles introduced into extruded squid axoplasm. J. Cell Biol. **99:** 112a.
11. WEATHERBEE, J. A. 1981. Membranes and cell movement: interactions of membranes with the proteins of the cytoskeleton. *In* International Review of Cytology, Supplement 12 A. L. Muggleton-Harris, Ed.: 113. Academic Press. New York.
12. MALAISSE, W. J., F. MALAISSE-LAGAE, E. VANOBBERGHEN, G. SOMERS, G. DEVIS, M. RAVAZZOLA & L. ORCI. 1975. Role of microtubules in the phasic pattern of insulin release. Ann. N.Y. Acad. Sci. **253:** 630–652.
13. ORCI, L., A. A. LIKE, M. AMHERDT, B. BLONDEL, Y. KANAZAWA, E. B. MARLISS, A. E. LAMBERT, C. B. WOLLHEIM & A. E. RENOLD. 1973. Monolayer cell culture of neonatal rat pancreas: an ultrastructural and biochemical study of functioning endocrine cells. J. Ultrastruc. Res. **43:** 270–297.
14. NOE, B. D., C. A. BASTE & G. E. BAUER. 1977. Studies of proinsulin and proglucagon biosynthesis and conversion at the cellular level. I. Fractionation procedure and characterization of the subcellular fractions. J. Cell Biol. **74:** 578–588.
15. SHIFF, P. B. & S. B. HOROWITZ. 1980. Taxol stabilizes microtubules in mouse fibroblast cells. Proc. Natl. Acad. Sci. USA **77:** 1561–1566.

16. VALLEE, R. B. 1982. A taxol-dependent procedure for the isolation of microtubules and microtubule-associated proteins. J. Cell Biol. **92:** 435–442.
17. SUPRENANT, K. A. & W. L. DENTLER. 1982. Association between endocrine pancreatic secretory granules and *in vitro* assembled microtubules is dependent upon microtubule-associated proteins. J. Cell Biol. **93:** 164–174.
18. DENTLER, W. L. 1977. Fine structural localization of phosphatases in cilia and basal bodies of *Tetrahymena pyriformis*. Tissue Cell **9:** 209–222.
19. LAEMMLI, U. K. 1970. Cleavage of structural proteins during the assembly of bacteriophage T4. Nature (London) **227:** 680–685.
20. FAIRBANKS, G., T. L. STECK & D. F. H. WALLACH. 1971. Electrophoretic analysis of the major polypeptides of the human erythrocyte membrane. Biochemistry **10:** 2606–2617.
21. DENTLER, W. L., S. GRANETT & J. L. ROSENBAUM. 1975. Ultrastructural localization of the high molecular weight proteins (MAPs) associated with *in vitro* assembled brain microtubules. J. Cell Biol. **65:** 237–241.
22. SLOBODA, R. D., W. L. DENTLER & J. L. ROSENBAUM. 1976. Microtubule-associated proteins and the stimulation of tubulin assembly *in vitro*. Biochemistry **15:** 4497–4505.
23. SUPRENANT, K. A. & W. L. DENTLER. 1979. Pancreatic secretory granules are associated with microtubules. Biol. Bull. **157:** 398.
24. SUPRENANT, K. A. & W. L. DENTLER. 1980. Microtubule-associated proteins are necessary for the associations of microtubules with organelles. Fed. Proc. Fed. Am. Soc. Exp. Biol. **39:** 1878.
25. BOYD III, A. E., W. E. BOLTON & B. R. BRINKLEY. 1982. Microtubules and beta cell function: effect of colchicine on microtubules and insulin secretion *in vitro* by mouse beta cells. J. Cell Biol. **92:** 425–434.
26. LASEK, R. J. 1980. The dynamic ordering of neuronal cytoskeletons. Neurosci. Res. Program Bull. **19:** 7–30.
27. SMITH, R. S. 1980. The short term accumulation of axonally transported organelles in the region of localized lesions of single myelinated axons. J. Neurocytol. **9:** 39–65.
28. MURPHY, D. B. & L. G. TILNEY. 1974. The role of microtubules in the movement of pigment granules in teleost melanophores. J. Cell Biol. **61:** 757–779.
29. TRAVIS, J. L., J. F. X. KENEALY & R. D. ALLEN. 1983. Studies on the motility of the foraminifera. II. The dynamic microtubular cytoskeleton of the reticulopodial network of *Allogromia laticollaris*. J. Cell Biol. **97:** 1668–1676.
30. OAKLEY, B. R. & N. R. MORRIS. 1980. Nuclear movement is b-tubulin dependent in *Aspergillus nidulans*. Cell **19:** 255.
31. OAKLEY, B. R. & J. E. REINHART. 1984. Nucleii and mitochondria move by different mechanisms in *Aspergillus nidulans*. J. Cell Biol. **99:** 49a.
32. BABISS, L. E., R. B. LUFTIG, J. A. WEATHERBEE, R. R. WEIHING, U. R. RAY & B. N. FIELDS. 1979. Reovirus serotypes 1 and 3 differ in their *in vitro* association with microtubules. J. Virol. **30:** 863–874.
33. LUFTIG, R. B. 1982. Animal virus-cytoskeleton interactions. J. Theor. Biol. **99:** 173–191.
34. SHERLINE, P., Y. C. LEE & L. S. JACOBS. 1977. Binding of microtubules to pituitary secretory granules and secretory granule membranes. J. Cell Biol. **72:** 380–389.
35. DENTLER, W. L., M. M. PRATT & R. E. STEPHENS. 1980. Microtubule-membrane interactions in cilia. II. Photochemical cross-linking of bridge structures and the identification of a membrane-associated dynein-like ATPase. J. Cell Biol. **84:** 381–403.
36. GRIFFITH, L. M. & T. D. POLLARD. 1978. Evidence for actin filament-microtubule interaction mediated by microtubule-associated proteins. J. Cell Biol. **78:** 958–965.
37. NISHIDA, E., T. KUWAKI & H. SAKAI. 1981. Phosphorylation of microtubule-associated proteins (MAPs) and pH of the medium control interaction between MAPs and actin filaments. J. Biochem. (Tokyo) **90:** 575–578.
38. SATTILARO, R. F., W. L. DENTLER & E. L. LeCLUYSE. 1981. Microtubule-associated proteins and the organization of actin filaments *in vitro*. J. Cell Biol. **90:** 467–473.
39. SATTILARO, R. F. & W. L. DENTLER. 1982. The association of MAP-2 with microtubules, actin filaments, and coated vesicles. *In* Biological Functions of Microtubules and Related Structures. Proceedings of the Oji International Seminar. H. Sakai, H. Mohri & G. G. Borisy, Eds.: 297–309. Academic Press. New York.

Membrane Tubulin[a]

C.S. REGULA, P.R. SAGER, AND R.D. BERLIN

Department of Physiology
University of Connecticut Health Center
Farmington, Connecticut 06032

The physiological role of cytoplasmic microtubules can be portrayed largely in terms of membrane phenomena. In addition to capping, the sorting of membrane proteins during endocytosis, and the maintenance of surface polarity, secretory and vesicle transport events can be interpreted as resulting from interactions of microtubules and membranes. This view of microtubule function has been established mainly by the effects of antimicrotubule agents on function, but quantitative studies have indicated a strong correlation between the state of microtubule assembly (or disassembly) and the functional perturbation. In some instances, structural bridges between microtubules and membrane-bounded organelles, such as mitochondria, are demonstrable.[2] At least in mammalian cells, however, direct contacts of microtubules with plasma membrane or with secretory vesicles are unusual. Because microtubules are major determinants of cell shape, membrane curvature, and the organization of other elements of the cytoskeleton, various indirect mechanisms by which microtubules affect membrane function have also been hypothesized. Although such hypotheses have proved to have predictive value (*e.g.,* in establishing the direction and timing of ligand-receptor movement under various conditions,[3]) it has also become clear that tubulin itself binds to phospholipid bilayers.[4,5] Indeed, there is considerable evidence suggesting that tubulin may be associated with biological membranes (reviewed in reference 1). Much of this has been obtained in studies of brain membranes and is complicated by the abundance of soluble brain tubulin (see reference 6 for example). In addition, the amount of membrane-associated tubulin is increased with the postmortem interval prior to membrane isolation.[7] Such evidence has cast a pall of skepticism on the significance or even existence of membrane tubulin. A curious dilemma has been forced on the investigator of membrane tubulin: the closer the identity of a membrane constituent is to soluble tubulin, the more likely to be considered an artifact of adsorption; the greater the difference between the membrane constituent and soluble tubulin, the more likely to be considered an unrelated protein.

The latter may be illustrated by studies of ciliary membrane. This membrane contains a PAS-positive doublet of approximately 55 K daltons. Stephens has shown clearly that it is distinguishable from axonemal tubulins by amino acid composition, tryptic peptide maps, and charge-shift electrophoresis. After Triton solubilization, the α and β constituents of this protein may be purified in nonstoichiometric amounts. This protein is selectively incorporated into reconstituted ciliary membranes.[8] No comparable evidence for a membrane tubulin in mammalian cells has been available so far. It is noteworthy and consistent with a unique membrane species of tubulin that a small fraction of tubulin is synthesized *in vitro* on membrane-bound polysomes, but neither Soifer and Czosnek[9] nor Cleveland *et al.*[10] demonstrated a difference in one-dimensional V8 protease maps of the tubulin formed.

We report here the demonstration of a hydrophobic tubulin isolated from a crude

[a]This was supported by NIH Grant CA 15544.

membrane fraction of bovine brain. The strategy employed was a selective sedimentation of a macromolecular complex from Triton X-100 membrane extracts, followed by partitioning of the complex with a two-phase Triton X-114 system that allows separation of hydrophobic, integral membrane proteins from complex mixtures. The identity of the hydrophobic tubulin was established by electrophoretic immunoblots using mouse monoclonal α- and β-tubulin antibodies and selective chemical cleavage procedures. We believe, thus, to have answered the dilemma posed above, establishing the existence of an integral membrane protein with yet clearly demonstrable compositional similarities to soluble tubulin.

METHODS

Membranes

The procedure of Jones and Matus[11] was modified slightly and applied to bovine brain homogenate. A phosphate-glutamate buffer (PGEP) was employed and modified from Regula *et al.* (1 mM ethylene glycol bis (β-ammoethyl ether)-N,N,N',N'-tetraacetic acid (EGTA), saturated phenylmethyl sulfonyl fluoride, phosphate-glutamate, pH 6.75[12]). An intermediate-speed pellet (9000 g for 20 min) was obtained and subjected to hypotonic lysis. Following gradient centrifugation, a crude membrane fraction containing plasma, synaptic, and some mitochondrial membranes was obtained.

Isolation of Triton-Soluble Proteins

The membrane fraction was diluted in PGEP and 1% Triton X-100 and stirred 60 min at 4°. The mixture was then centrifuged at 100,000 g for 60 min at 4°. This step was designed to sediment postsynaptic densities (PSD) and other detergent insoluble structures.[13]

The supernatant was made 1 mM guanosine triphosphate (GTP), 0.5 mM MgCl$_2$, and 5 μM Taxol and incubated at 30° for 30 min, and then centrifuged at 100,000 g for 30 min at 30°. The resulting translucent pellet was taken up in PGEP.

Phase Separation of Integral Membrane Proteins

Bordier's technique for separating membrane proteins in solutions of Triton X-114 was employed.[14] With this member of the Triton series, a single phase is present at 4° and phase separation occurs above 20°. Hydrophilic proteins are found entirely in the aqueous phase, whereas integral membrane proteins are present in the Triton-condensed phase. The method has been validated with bacterial, erythrocyte,[14] and viral[15] membrane proteins. We modified Bordier's procedure slightly to ensure minimal contamination of the Triton-condensed phase by proteins in the aqueous phase.

To a protein solution of 0.5–1.0 mg/ml, Triton X-114 in a 10% precondensed stock solution in 10 mM TrisCl pH 7.4, 150 mM NaCl was added to a final concentration of 1.4% Triton X-114. The solution was held 30–60 min at 4° and then centrifuged for 3 min at 13,000 g in the cold (no phase separation) to remove particulates. The amount of protein recovered in this pellet was <10% of that found subsequently in the Triton-condensed phase. The latter was obtained by warming to 30° for 3 min and centrifuging for 3 min at 300 g to separate phases. Bordier calls for reextraction of the aqueous phase. We, however, routinely backwashed the Triton phase twice with aqueous buffer to remove contaminating hydrophilic proteins.

Gel Electrophoresis

For routine analysis, a 7.5% polyacrylamide slab gel was employed in an SDS system.[16] Peptides formed after chemical cleavage were electrophoresed on 10–18% linear gradient gels in urea sodium dodecyl sulfate (SDS) with high cross-linking to reduce the loss of small peptides from the gel.[17]

Immunoblotting

Transfer from 7.5% acrylamide slabs to nitrocellulose was performed as described,[18] modified by addition of 0.1% SDS to the blotting buffer. Antibody incubation proceeded for 12–16 h at room temperature; binding was detected with a goat antimouse IgG and IgM antibody-peroxidase conjugate subsequently reacted with 4-chloro-1-naphthol. All antibodies were used at 1:1000 dilutions.

Chemical Cleavage

The procedures used were specifically developed for treatment of proteins in gel pieces after their prior separation on one-dimensional gels. Cyanogen bromide (methionine)[19] and N-chlorosuccinimide (tryptophan)[20] cleavages were performed as described. Gels were silver-stained according to Wray et al.[21]

RESULTS

FIGURE 1 is a representative gel illustrating the protein composition of the pellet formed from the Triton X-100 supernatant after warm incubation in the presence of Taxol-GTP, and its subsequent partition into the aqueous and Triton X-114 condensed phases.

Lane a, for comparison, is a heavily loaded sample of bovine microtubule protein taken through three cycles of assembly and disassembly as described previously.[12] None of these proteins was partitioned into a condensed detergent phase (data not shown). Lane b was the Taxol-GTP sedimented protein. A 55 K doublet is prominent. The heavy band at roughly 46 K is a constant finding but was usually less dense than the 55 K doublet. Lane c is of the aqueous phase derived from an amount of protein equal to that shown in lane b, and lanes d and e are successive washes of the Triton X-114 condensed phase. Lane f is the washed Triton phase. As expected, the patterns of aqueous and Triton phases (lanes c and f) are complementary with several bands unique to both. Most importantly, the Triton phase contains a 55 K doublet. The doublet bands are virtually identical in intensity. We emphasize that none of the tubulin obtained by assembly/disassembly procedures was extractable into the Triton phase. The relative amounts of doublet material obtained from membrane present in aqueous and Triton phases were similar but not stoichiometric.

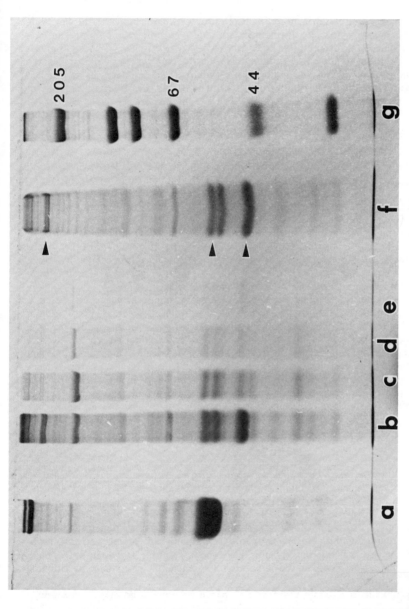

FIGURE 1. SDS-polyacrylamide gel electrophoresis (PAGE) showing protein composition of material pelleted from Triton X-100 membrane extract by incubation in Taxol-GTP and its subsequent phase partition according to the method of Bordier.[14] Lane a: bovine microtubule protein derived from three cycles of assembly/disassembly[12]; lane b: whole Taxol-GTP pellet; lane c: aqueous phase proteins, derived from an aliquot of Taxol-GTP pellet equal to lane b; lanes d, e: successive washes of Triton X-100 condensed phase; lane f: condensed phase after washes; lane g: protein standards. The 7.5% acrylamide gel was run in the Laemmli system[16] and was stained with Coomassie blue.

Immunoblotting

Electrophoretograms of Triton (T) and aqueous (A) phase material, the clarifying pellet (P) obtained prior to phase separation, and control microtubule protein (M) were transferred to nitrocellulose paper and probed with a mouse monoclonal anti-α (FIGURE 2) or anti-β (FIGURE 3) tubulin. Specific staining is seen in all samples; the faint staining seen below the major anti-β-tubulin bands (FIGURE 3) is directed presumably to a tubulin breakdown product. These data clearly establish tubulin-specific epitopes in the 55 K doublet of the Triton-phase material.

Chemical Cleavage

To further compare the phase-separated immunoreactive doublets (which we will call membrane and aqueous tubulins henceforth), the tubulin bands were subjected to chemical cleavage by cyanogen bromide or N-chlorosuccinimide. FIGURE 4 illustrates the results for the α tubulins. As before, M, T, and A refer to microtubule protein and the tubulin of Triton X-114 and aqueous phases respectively. The patterns of M and T are not distinguishable. Surprisingly, the A pattern showed clear differences. The basic pattern of M and T is present. In addition, however, there are several other bands marked by arrows that are also present. Additional bands could represent a second, comigrating protein. That this is probably not the case is suggested by the absence of the uppermost bands shown in the M and T patterns.

FIGURE 5 illustrates the results of N-chlorosuccinimide cleavages of β and α tubulins. Again, the α patterns are similar. The β patterns, however, are not. T, A, and M are all different, but T and A are more similar, showing both additional bands and a difference in relative band intensities. These data suggest, but do not prove, that microtubule and Triton-phase enriched tubulins have identical α chains and that differences in their hydrophobic character are derived from differences in the β chain. The data also suggest that there is a less hydrophobic tubulin (which partitions into the aqueous phase) that is nonetheless derived from membrane and is different in both α and β chains. These possibilities clearly warrant additional purification and analysis.

DISCUSSION

A readily separable hydrophobic tubulin is revealed by these studies. The isolation procedure was designed to minimize contaminating unbound tubulin but did not require its complete exclusion: the principal discriminatory step, partitioning into the Triton X-114 condensed phase, completely excluded soluble tubulin (prepared by cyclic assembly/disassembly). Such partitioning provides a strong operational criterion for integral membrane proteins. All intergral membrane proteins, to our knowledge, can be partitioned into a Triton X-114 condensed phase. It is to be anticipated, however, that should a component of a protein be sufficiently hydrophilic, partitioning may be frustrated.

It is difficult to estimate the extent to which tubulin, as isolated by assembly/disassembly procedures, need be modified to behave as an integral membrane protein. The binding of this tubulin to phospholipid bilayers has been noted[4,5] and at the lipid phase-transition temperature, can be integrated into the phospholipid bilayer. Tubulin also binds various nonionic detergents.[22] The quantity bound is less than that reported for integral membrane proteins but more than for typical hydrophilic proteins. This intermediate character is consistent with the role of hydrophobic bonding in microtu-

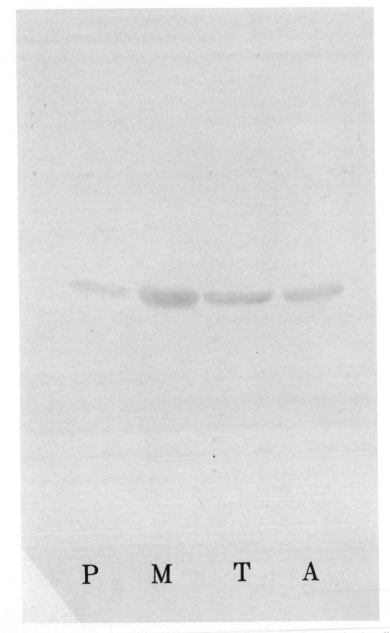

FIGURE 2. Immunoblot of SDS-PAGE probed with mouse α-tubulin monoclonal antibody. Detection was as described in METHODS. A: aqueous phase; T: Triton phase; M: microtubule protein; P: pellet formed by clarifying spin prior to phase separation. The Coomassie patterns were similar to those shown for the pertinent lanes in FIGURE 1. The anti-α-tubulin was obtained from Amersham, Arlington Heights, Il. and was raised to chick brain tubulin.

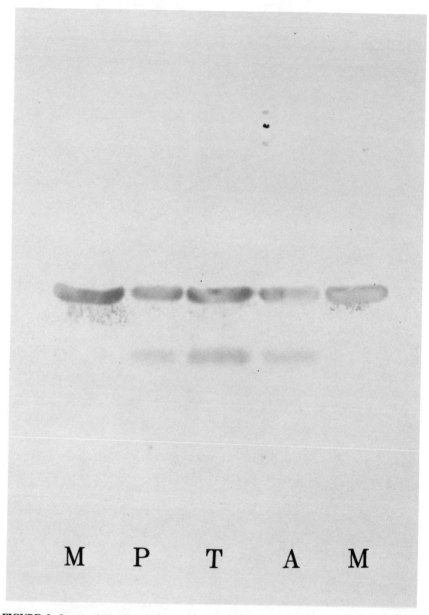

FIGURE 3. Immunoblot of SDS-PAGE probed with mouse β-tubulin monoclonal antibody. Lanes are labeled as in FIGURE 2. The mouse monoclonal was obtained from Amersham and was also raised against chick brain tubulin.

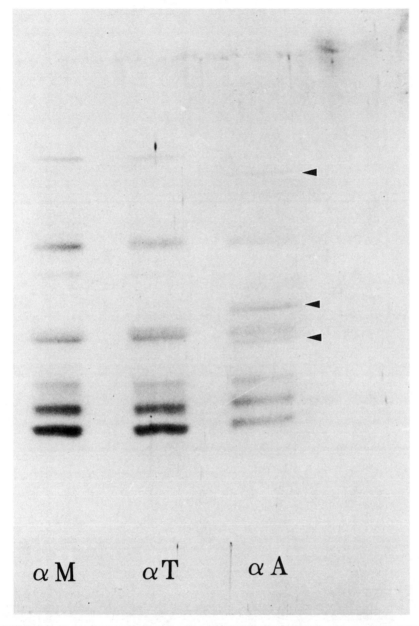

FIGURE 4. Cyanogen bromide cleavage of α tubulins. One-dimensional peptide maps of proteins cleaved *in situ*.[19] αM, αT, and αA refer to the slower moving band of the putative tubulin doublet. M, T, and A refer to tubulin obtained by assembly/disassembly or to membrane-extracted tubulin partioned into Triton X-114 or aqueous phases. The 10–18% linear acrylamide slab gel was run in the urea/SDS system of Hashimoto *et al.*[17] and silver-stained according to Wray *et al.*[21]

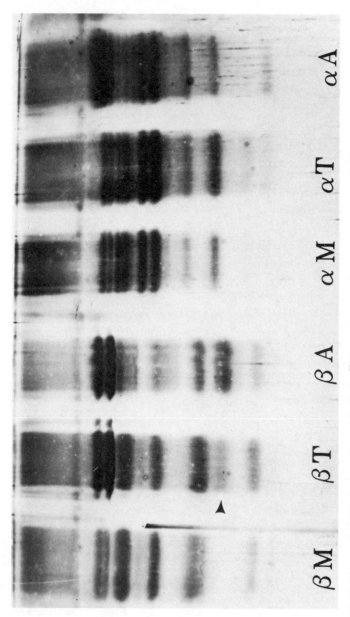

FIGURE 5. *N*-chlorosuccinimide cleavage of α and β tubulins.[20] Nomenclature and gel system are as in FIGURE 4 except that a straight 10% acrylamide slab was used.

bule assembly. It should be emphasized, however, that such bonding in polymer structures, reflected in cold and pressure lability for example, does not generally lend hydrophobic character, as operationally defined here, to the constitutent subunits. It is possible, however, that relatively minor structural modifications can render tubulin hydrophobic. The results of chemical cleavage of the hydrophobic tubulin suggest that such modifications occur in the primary sequence of the β chain.

Although we expected that the Triton X-100 membrane extract would contain contaminating and/or adsorbed soluble tubulin, we also expected it to be identical to the tubulin isolated by assembly/disassembly. Instead, peptide mapping showed the aqueous-phase tubulin to be different from both hydrophobic and microtubule tubulins. The cleavage experiments are not conclusive, however. Contributions from contaminating proteins are possible. As indicated above, however, some peptide bands were actually missing, and it is significant that microtubules were not formed from the Taxol-GTP incubation mixture. Under our conditions, Triton X-100 did not inhibit microtubule assembly (data not shown). Assembly-competent tubulin, therefore, could not have been present in the Triton X-100 extract. Thus, a second tubulin species, in addition to the hydrophobic tubulin, may be extractable by Triton. It is tempting to speculate that such a tubulin could serve as a linkage element between an anchored integral membrane tubulin and microtubules or other associated proteins.

Whether the entire material sedimented from the membrane extract (after prior, cold centrifugation) is a unique macromolecular complex is unknown. It is probable that at least some nontubulins are specifically associated with tubulin. These potential TAPs (tubulin-associated proteins) obviously require further investigation. One should consider, however, whether a hydrophilic tubulin could be pulled into the detergent phase by association with other hydrophobic proteins. This question can be definitively answered when purified hydrophobic tubulin becomes available. It is already clear, however, that such a "piggy-back" mechanism is unlikely. First, in several preparations the hydrophobic tubulin constituted the bulk of the hydrophobic material, that is, there was insufficient "other" protein to provide carrier. Second, there was always aqueous tubulin in amounts that were independent of nontubulins present in the Triton phase. Finally, in other membrane systems, integral and peripheral membrane proteins are partitioned as expected. This, of course, is in addition to evidence from chemical cleavage studies that indicated a unique β chain of the hydrophobic tubulin.

Further structural studies of the hydrophobic tubulin are in progress. The development of specific monoclonal antibodies to this tubulin should open the way to studies of its distribution and functional significance.

REFERENCES

1. BERLIN, R. D., J. M. CARON & J. M. OLIVER. 1979. *In* Microtubules. K. Roberts & J. S. Hyams, Eds. Chapter **10**: 443–485. Academic Press. New York.
2. SMITH, D. S., U. JARLFORS & M. L. CAYER. 1977. J. Cell Sci. **27**: 255–272.
3. OLIVER, J. M. & R. D. BERLIN. 1982. Philos. Trans. Roy. Soc. London Ser. B **299**: 215–235.
4. CARON, J. M. & R. D. BERLIN. 1979. J. Cell Biol. **81**: 665–671.
5. KLAUSNER, R. D., N. KUMAR, J. N. WEINSTEIN, R. BLUMENTHAL & M. FLAVIN. 1981. J. Biol. Chem. **256**: 5879–5885.
6. STROCCHI, P., B. A. BROWN, J. D. YOUNG, J. A. BONAVENTRE & J. M. GILBERT. 1981. J. Neurochem. **37**: 1295–1307.
7. CARLIN, R. K., D. J. GRAB & P. SIEKEVITZ. 1982. J. Neurochem. **38**: 94–100.
8. STEPHENS, R. E. 1983. J. Cell Biol. **96**: 68–75.
9. SOIFER, D. & H. H. CZOSNEK. 1980. J. Neurochem. **35**: 1128–1136.

10. CLEVELAND, D. W., M. W. KIRSCHNER & N. J. COWAN. 1978. Cell **15:** 1021–1031.
11. JONES, D. H. & A. I. MATUS. 1974. Biochim. Biophys. Acta **356:** 276–287.
12. REGULA, C. S., J. R. PFEIFFER & R. D. BERLIN. 1981. J. Cell Biol. **89:** 45–53.
13. COHEN, R. S., F. BLOMBERG, K. BERZINS & P. SIEKEVITZ. 1977. J. Cell Biol. **74:** 1:81–203.
14. BORDIER, C. 1981. J. Biol. Chem. **256:** 1604–1607.
15. WARREN, G., C. FEATHERSTONE, G. GRIFFITHS & B. BURKE. 1983. J. Cell Biol. **97:** 1623–1628.
16. LAEMMLI, U. K. 1970. Nature (London) **227:** 680–685.
17. HASHIMOTO, F., T. HORIGOME, M. KANABAYASHI, K. YOSHIDA & H. SUGANO. 1983. Anal. Biochem. **129:** 192–199.
18. TOWBIN, H., T. STAEHELIN & J. GORDON. 1979. Proc. Natl. Acad. Sci. USA **76:** 4350–4354.
19. LONSDALE-ECCLES, J. D., A. M. LYNLEY & B. A. DALE. 1981. Biochem. J. **197:** 591–597.
20. LISCHWE, M. A. & D. OCHS. 1982. Anal. Biochem. **127:** 453–457.
21. WRAY, W., T. BOULIKAS, V. P. WRAY & R. HANCOCK. 1981. Anal. Biochem. **118:** 197–203.
22. ANDREU, J. M. 1982. EMBO J. **1:** 1105–1110.

Association of Microtubules with Membrane Skeletal Proteins

BARBARA L. FACH, SUSAN F. GRAHAM,
AND ROBERT A. B. KEATES

Department of Chemistry and Biochemistry
University of Guelph
Guelph, Ontario N1G 2W1, Canada

Microtubule associated proteins (MAPs) have previously been identified as proteins that coassembled with tubulin *in vitro* with or without the help of taxol.[1,2] There is a risk that MAPs that link to other structural elements may remain associated with parts of the cell that do not disassemble along with the microtubules on cooling or extraction, and be lost in the preliminary centrifugation before *in vitro* reassembly is attempted. We have therefore examined methods for isolation of the original cellular microtubules by stabilization instead of by reassembly. Glycerol (6.7 M) was used as a stabilizing medium, and appears to limit the rate of subunit exchange.[3] The microtubules thus isolated can be purified by density gradient methods. Stabilized microtubules analyzed on gels showed additional bands not present in reassembled preparations. One of these bands was identified as actin and a pair of bands at 235 and 240 kD as the membrane associated protein, fodrin.[4]

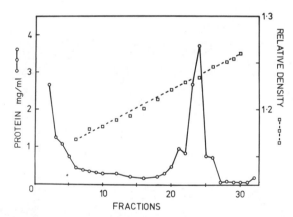

FIGURE 1. Bovine cerebral cortex (20 g) was homogenized (all steps at 22°) in 100 ml buffer containing 0.1 M 2-(N-morpholine) ethane sulfuric acid (MES) K^+ pH 6.6, 1 mM ethylene glycol bis (β-aminoethyl ether)-N, N, N', N'-tetraacetic acid (EGTA), 2 mM Mg^{2+} acetate, 6.7 M glycerol, 0.1% Nonidet P-40, and 0.01 mg/ml each of antipain, aprotinin, and leupeptin. The homogenate was centrifuged at 10,000 × g for 20 minutes. After removal of floating lipid, the supernatant was centrifuged again at 120,000 × g for 60 minutes. The pellet was resuspended in 20 ml buffer, and the 120,000 × g centrifugation repeated. The pellet was resuspended in 12 ml buffer yielding 15 ml suspension. Centrifugation at 2000 × g for 5 min removed undispersed particles, and 3 ml was layered on a linear gradient 6.7 M glycerol to 2.2 M sucrose. This was centrifuged at 25,000 rpm for 3 h (Beckman SW 28 rotor). Fractions of 1 ml were collected and analyzed for protein.

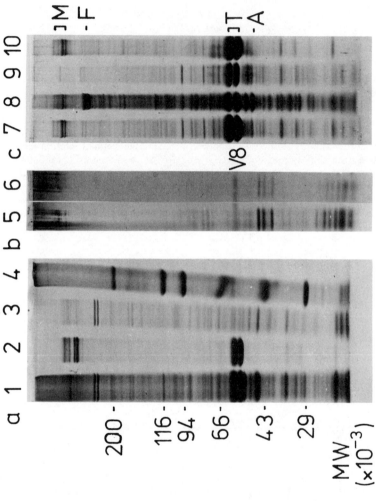

FIGURE 2. Analysis of microtubule protein by polyacrylamide gel electrophoresis. **a.** 4–12% gradient gel. Lane 1, gradient purified microtubules; lane 2, *in vitro* reassembled cerebral cortex microtubules; lane 3, bovine brain membrane fodrin isolated according to Glenney;[4] lane 4, markers. **b.** 6–14% gradient gel. Lane 5, membrane α-fodrin digested with *Staphylococcus aureus* V8 protease; lane 6, microtubule associated 240 kD polypeptide digested with *S. aureus* V8 protease. **c.** 4–12% gradient gel. Lane 7, supernatant after cold disassembly of stabilized microtubules; lane 8, pellet after cold disassembly of stabilized microtubules; lane 9, supernatant after incubation of protein from lane 7 at 37° in the presence of 1 mM GTP; lane 10, pellet after reassembly of protein from lane 7 (compare composition with lane 2). M = 280–350 kD MAPs; F = fodrin; T = tubulin; A = actin; V8 = S. aureus V8 protease.

Glycerol-stabilized microtubules were recovered from density gradients (FIGURE 1). Analysis by gel electrophoresis showed a prominent pair of bands at 235 and 240 kD, similar to fodrin, isolated from brain membrane (FIGURE 2a). The identity of fodrin was confirmed by limited protease digestion (FIGURE 2b). Electron microscopy showed many microtubules in tight clusters and bundles. Densitometry of electrophoresis gels at various stages of isolation showed little change in the tubulin:fodrin ratio, (20:1 in homogenate, 16:1 in second microtubule pellet), but a decreasing content of actin. Microtubules containing fodrin were disassembled in cold glycerol free buffer. After centrifugation at 100,000 × g for 1 h at 2°, most of the fodrin sedimented, but around 80% of the tubulin and 280–350 kD MAPs appeared in the supernatant. The supernatant was incubated for 30 min at 37° in the presence of 1 mM guanosine triphosphate (GTP). Reassembled microtubules were collected by centrifugation and were similar in composition to microtubules prepared by standard *in vitro* reassembly procedures (FIGURE 2c).

Our method of isolation is similar to that of Kirkpatrick[5] but with glycerol instead of hexylene glycol as a stabilizing agent. One important difference is that the microtubules we isolate remain cold labile, and can be disassembled and reassembled *in vitro* when the glycerol is removed. A distinctive feature of the stabilized microtubules is the presence of the membrane-associated protein, fodrin, in proportions greater than the 280–350 kD MAPs. The relatively low content of MAP in our preparation is a result of the very high recovery of microtubules (6.4 mg per g brain) compared to the *in vitro* recycling procedure (0.7–1.3 mg per g brain). This suggests that we have recovered two subsets of microtubules, one of which is lost in the standard procedure. Fodrin is largely lost in the standard reassembly isolation procedure for microtubule protein during the preliminary centrifugation of the homogenate. Reassociation of fodrin with *in vitro* reassembled microtubules has already been demonstrated by Ishikawa *et al.*[6] Fodrin is now known to be associated with the cell in general, and is not unique to brain tissue.[7] Thus we may have found a link between two cytoskeletal systems that may be of general significance.

REFERENCES

1. DENTLER, W. L., S. GRANETT & J. L. ROSENBAUM. 1975. J. Cell Biol. **65:** 237–241.
2. VALLEE, R. 1982. J. Cell Biol. **92:** 435–442.
3. KEATES, R. A. B. 1980. Biochem. Biophysics Res. Commun. **97:** 1163–1169.
4. GLENNEY, J. R., P. GLENNEY, M. OSBORN & K. WEBER. 1982. Cell **28:** 843–854.
5. KIRKPATRICK, J., L. HYAMS, V. L. THOMAS & P. M. HOWLEY. 1970. J. Cell Biol. **47:** 384–394.
6. ISHIKAWA, M., H. MUROFUSHI & H. SAKAI. 1983. J. Biochem. (Tokyo) **94:** 1209–1217.
7. GLENNEY, J. R. & P. GLENNEY. 1983. Cell **34:** 503–512.

Membrane Tubulin Receptors in Human Leukemic Cells

ROBERT W. RUBIN

Department of Anatomy and Cell Biology
University of Miami Medical School
Miami, Florida 33101

We have reported that certain leukemic cell types (such as the CCRF-CEM line) possess tubulin as a major constituent on their cell surfaces. These cells are marked by an extreme sensitivity to cytolysis induced by several tubulin binding drugs.[1] In addition, unrelated drugs such as cytochalasin D and stelazine inhibit Con A-induced capping in these cells and prevent the cytolytic response to tubulin binding drugs. The addition of metabolically labeled, high speed supernatants prepared from cell homogenates, to live CEM cells followed by extensive washing results in the specific binding of tubulin to the surface. Two-dimensional gel video quantitation of radio fluorographs indicates that over 97% of the total bound protein is tubulin. The binding of excess CEM tubulin in this manner also completely inhibits Con A capping.

The 2-D gels reveal that the alpha-tubulin monomer is a more basic isoelectric variant than the bulk of the total cellular tubulin or of purified brain tubulin.[2] This binding is saturable and inhibited by the addition of excess unlabeled high speed supernatant, but not by the addition of 100-fold excess of three-cycled brain tubulin. This binding is also inhibited by the preaddition of wheat germ agglutinin (WGA), but not by the preaddition of peanut agglutinin, Con A, or treatment with neuraminidase. Neuraminidase pretreatment, however, prevents the WGA inhibition. Pretreatment of CEM cells with the lowest concentration of trypsin shown to remove surface tubulin also inhibits the binding of this unusual surface tubulin. Thus, the data taken together suggest that there exists on these cells a sialic acid containing a glycoprotein receptor that binds an isoform of tubulin with great selectivity. Labeled high speed supernatants, from nontumorigenic lines that do not possess iodinatable surface tubulin, when applied to CEM cells, do not result in the binding of any tubulin. Such normal cells likewise do not bind CEM membrane tubulin. We conclude therefore that cells must have both a surface exposed tubulin receptor and a variant form of tubulin in order to express surface tubulin. The presence of this receptor and its ligand on the surface of cells may confer drug sensitivity and affect lateral plasma membrane protein mobility.

REFERENCES

1. RUBIN, R. W., M. QUILLEN, J. J. CORCORRAN, R. GANAPATHI & A. KRISHAN. 1982. Cancer Res. **42:** 1384.
2. QUILLEN, M., C. CASTELLO, A. KRISHAN & R. W. RUBIN. 1985. Cell Biol. **101:** 1–10.

Local Control of the Axonal Cytoskeleton

STEVEN R. HEIDEMANN,[a] HARISH C. JOSHI,[a]
AND R.E. BUXBAUM[b]

[a]Department of Physiology
[b]Department of Chemical Engineering
Michigan State University
East Lansing, Michigan 48824

The ability of isolated growth cones to respond to their environment and grow independently of the cell body[1,2] suggests that, in addition to cell body dependent processes,[3,4] local mechanisms operating within the neurite may organize cytoskeletal assembly. Our investigations suggest one such mechanism: a complementary force interaction within growing neurites, tension on actin filaments supported by compression on microtubules (MTs), integrating MT assembly with the actin-based motility of the growth cone.

PC-12 cells treated with nerve growth factor (NGF) and cyclic adenosine monophosphate (cAMP) extend more neurites faster, and with a shorter lag period than after treatment with NGF alone.[5] We found that neurites of PC-12 cells regenerated in medium with both NGF and cAMP (NC cells) are more stable to retraction induced by treatments that depolymerize MTs than neurites regenerated in NGF only (N cells).[6] These results indicate that MTs within NC neurites are more

TABLE 1. Retraction of PC-12 Neurites upon Laser Transection

Treatment	Retracts within 2 min	No Retraction within 2 min	Lifts off dish
None (n = 27)	24	1	2
25 µg/ml taxol (n = 13)	0	11[a]	2
0.4 µg/ml cytochalasin D (n = 31)	0	28[a]	3
0.1 mM dinitrophenyl (DNP) +			
0.1 mM Arsenate (n = 6)	5	0	1

[a]Although 2 minutes was chosen as the experimental time point, due to the speed of retraction of untreated neurites, these neurites were stable for as long as observed, up to 30 minutes.

stable than MTs in N cells. As suggested by Bray,[7] external force could provide a link between MT stability and the greater rate and extent of PC-12 neurite growth in NC conditions. The tension on the neurite[7] may normally be supported by the compression of MTs. The disassembly of MTs disturbs this equilibrium, and the neurite retracts in response to the tension. The tubulin subunit/polymer equilibrium is shifted away from assembly by compression.[8] An increase in MT stability, relative to a zero force situation, would be required to enable MTs to bear compression without depolymerization.

Preliminary experiments on two retraction phenomena support the hypothesis that MTs are under compression and also suggest that the actin network of PC-12 neurites are under tension. TABLE 1 shows that 24 of 27 neurites transected by a laser beam retract both proximal and distal portions within two minutes. Taxol rescues neurites from retraction, consistent with the notion that stabilizing MTs stiffens them as a compressive element, supporting the neurite even while severed. The retraction of severed neurites is unaffected by combined glycolytic and electron transport poisons, suggesting that retraction is not an active process. The role of actin in neurite adhesion and motility[9] and the lack of intermediate filaments within PC-12 neurites[10] suggested that actin would be the tensile component within neurites.

In support of this speculation, we found that neurites treated with cytochalasin D prior to laser transection are rescued from retraction. Cytochalasin disrupts actin networks,[11] therefore diminishing its ability to bear tension.[12] Consequently, we suggest that cytochalasin rescues neurites from retraction by dissipating the tension within the neurite. If so, cytochalasin-treated neurites should no longer require MTs a support element. TABLE 2 shows that both N and NC neurites are rescued from MT depolymerization-induced retraction by cytochalasin, confirming similar observations on neuroblastoma.[13] Release of tension by cytochalasin should relieve the MTs of some compression load, decreasing the critical concentration of tubulin for assembly.[8] If these forces were of an appropriate magnitude to affect MT assembly *in vivo*, the decrease in the critical concentration should be observable as a stabilization of MTs to threshold depolymerization stimuli. Indeed, N and NC neurites rescued from the

TABLE 2. Effect of Cytochalasin D on Microtubule Depolymerization-Induced Retraction of PC-12 Neurites Regenerated for 24 Hours

Treatment	NFG Only[a]	NGF and cAMP[a]
0.1 μg/ml Noc[b] 15 min	42/681	—
0.2 μg/ml Noc 15 min	—	18/224
2 μg/ml cytochalasin D only, 20 min	214/256	234/220
cytochalasin D, 20 min, Noc 15 min	166/179	173/207
cytochalasin D, 16 min, Noc 15 min	175/193	208/217

[a]Numerators refer to the number of neurites in the surveyed region after treatment, and denominators refer to the number of neurites in the same region prior to treatment.
[b]Noc-nocodazole

threshold dose (but not high doses) of nocodazole that caused MT depolymerization without cytochalasin now contain an apparently normal array of MTs.

REFERENCES

1. SHAW, G. & D. BRAY. 1977. Exp. Cell. Res. **104:** 55–62.
2. SEELEY, P. J. & L. A. GREENE. 1983. Proc. Natl. Acad. Sci. USA **80:** 2789–2793.
3. LASEK, R. J. 1982. Proc. R. Soc. Lond. Ser. B **299:** 313–327.
4. SOLOMON, F. 1981. J. Cell Biol. **90:** 547–553.
5. GUNNING, P. W., G. E. LANDRETH, M. BOTHWELL & E. SHOOTER. 1981. J. Cell Biol. **89:** 240–245.
6. HEIDEMANN, S. R., H. C. JOSHI, A. SHECHTER, J. R. FLETCHER & M. BOTHWELL. 1985. J. Cell Biol. **100:** 916–927.
7. BRAY, D. 1982. Filopodial Contraction and Growth Cone Guidance. *In* Cell Behavior. R. Bellair, A. Curtis & G. Dunn, Eds. Cambridge University Press.
8. HILL, T. L. & M. W. KIRCHNER. 1982. Int. Rev. Cytol. **78:** 1–125.
9. LANDIS, S. C. 1983. Ann. Rev. Physiol. **45:** 567–580.
10. LUCKINBILL-EDDS, L., C. VAN HORN & L. A. GREENE. 1979. J. Neurocytol. **8:** 493–511.
11. SCHLIWA, M. 1982. J. Cell Biol. **92:** 79–91.
12. ROSEN, S. L. 1971. Fundamental Principles of Polymeric Materials for Practicing Engineers. 161–162. Barnes and Noble. New York.
13. SOLOMON, F. & M. MAGENDANTZ. 1981. J. Cell Biol. **89:** 157–161.

Modification of Microtubule-Microfilament Interactions during Neurite Outgrowth

DAVID A. SPERO, EDWARD T. BROWNING,
AND FRED J. ROISEN

Departments of Anatomy and Pharmacology
University of Medicine and Dentistry of New Jersey
Rutgers Medical School
Piscataway, New Jersey 08854

We have shown previously[1-3] that a mixture of bovine brain gangliosides (BBG) applied to Neuro-2a neuroblastoma (Neuro-2a) increased neurite outgrowth in a dose-dependent manner, stimulated the formation of a complex cytoskeleton consisting of a network of microfilament bundles, and induced a highly ordered burst in surface activity characterized by the formation of microvilli and ruffled membranes. Furthermore, we demonstrated that BBG produced long, thin aberrant neurites in Neuro-2a cells grown in the presence of cytochalasin D (cyto-D) at a concentration known to disrupt microfilament function.[2] These neurites contained microtubules and intermediate filaments and lacked microfilaments. By contrast, neurite initiation, but not growth, was observed after 20 h ganglioside exposure in the presence of the microtubule-disruptive agent colcemid (0.25 μg/ml). These results suggested that microtubules were sufficient to support neurite growth under microfilament-limited conditions, but that microfilaments could not support neuritogenesis in the absence of microtubules. To examine further the role that microtubules play in neurite outgrowth, the microtubule promoting and stabilizing effects of taxol were examined under microfilament-limited conditions. Neuro-2a cells were grown in minimum essential medium (MEM) supplemented with 10% heat-inactivated fetal bovine serum (FBS), 75 mg/dl additional $NaHCO_3$, 10 mg/dl gentamicin and 0.1 mM nonessential amino acids. Cultures were exposed to cyto-D (2 μg/ml) and either BBG (19% GM_1, 44% GD_{1a}, 16% GD_{1b}, 20% GT_{1b}, 250 μg/ml) or taxol (1 μM) for 24 h, fixed with 2.5% glutaraldehyde in a cytoskeletal stabilizing buffer,[4] and processed for scanning electron microscopy (SEM), transmission electron microscopy (TEM), and whole-cell transmission electron microscopy (WCTEM). [2] Neurons grown in serum-deprived medium (to stimulate neurite production) or solely with BBG formed thick, highly arborized processes (FIGURE 1A–D). By contrast, cells treated simultaneously with BBG and cyto-D produced long, thin, unbranched neurites that lacked typical growth cones (FIGURE 1E). These aberrant neurites grew in a circular pattern, contained densely packed, longitudinally oriented microtubules and were deficient in microfilaments (FIGURE 2C,G). The simultaneous application of taxol and cyto-D resulted in similar long, thin, unbranched neurites that were packed with microtubules (FIGURE 2D,H). Control cells treated either with taxol or cyto-D did not form processes, but exhibited numerous microvilli. Our recent studies[5] demonstrate that primary neurons from eight-day-old embryonic chick cerebral cortex treated with BBG or taxol under similar microfilament-limiting conditions produced long, thin aberrant neurites that contained an abundance of microtubules. As an initial step in studying the molecular basis for the differences underlying the aberrant neurite outgrowth, we have employed

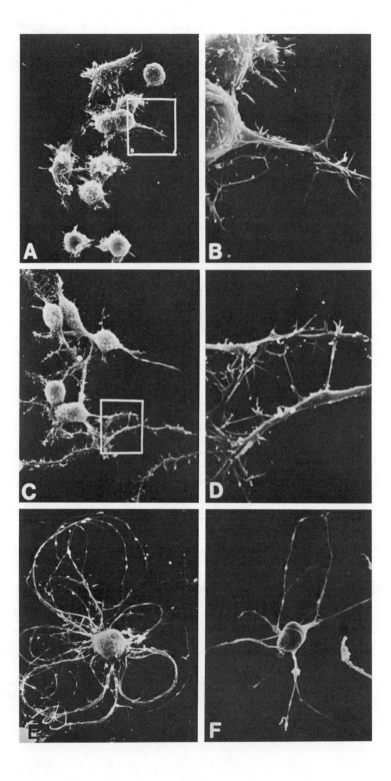

two-dimensional electrophoresis to investigate protein synthesis. Cells were incubated in low-methionine MEM containing 2% FBS and 0.1 mCi [^{35}S]methionine for 4 h under the appropriate treatment conditions, and electrophoresis was carried out as described previously.[6] Visual inspection of autoradiograms from cells treated simultaneously with BBG and cyto-D revealed a decrease in the incorporation of radiolabel into most proteins except the tubulins. These studies demonstrate that cytoskeletal disruptive agents alter the pattern of neurite outgrowth in both transformed neuronal cell lines and cultures of primary neurons by modifying the assembly and synthesis of major cytoskeletal proteins. These results support the hypothesis that the precise coordination of the activities of cytoskeletal elements is essential for normal neurite outgrowth. Furthermore, they suggest that in the absence of microfilaments, microtubules provide the motive force for neurite extension, whereas microfilaments appear essential for neurite initiation and arborization.

REFERENCES

1. ROISEN, F. J., H. BARTFELD, R. NAGELE & G. YORKE. 1981. Ganglioside stimulation of axonal sprouting *in vitro*. Science **214**: 577–578.
2. SPERO, D. A. & F. J. ROISEN. 1984. Ganglioside-mediated enhancement of the cytoskeletal organization and activity in Neuro-2a neuroblastoma cells. Dev. Brain Res. **13**: 37–48.
3. ROISEN, F. J., D. A. SPERO, S. J. HELD, G. YORKE & H. BARTFELD. 1984. Ganglioside induced surface activity and neurite formation of Neuro-2a neuroblastoma cells. *In* Ganglioside Structure, Function, and Biomedical Potential. R. Ledeen, R. Yu, M. Rapport & K. Suzuki, Eds.: 499–511. Plenum Press. New York.
4. LUFTIG, R., P. MCMILLIAN, J. WEATHERBEE & R. WEIHING. 1977. Increased visualization of microtubules by an improved fixation procedure. J. Histochem. Cytochem. **25**: 175–187.
5. SPERO, D. A. & F. J. ROISEN. 1984. Taxol and ganglioside induced aberrant neurite outgrowth from neuroblastoma and CNS neurons cultured in microfilament-limited conditions. Neurosci. Abs. **10**(1):37A.
6. GROPPI, V. & E. BROWNING. 1980. Norepinephrine-dependent protein phosphorylation in intact C-6 glioma cells. Mol. Pharmacol. **18**: 427–437.

FIGURE 1. A–F: Scanning electron micrographs of Neuro-2a cells, 24 h *in vitro*. **A** and **B**: Control cells maintained in MEM containing 10% FBS. **C** and **D**: BBG treatment (250 μg/ml) produced long, highly branched neurites. **E**: Simultaneous treatment with BBG (250 μg/ml) and cyto-D (2 μg/ml). Neurite outgrowth under these conditions resulted in numerous aberrant processes with reduced branching and circular growth patterns. **F**: Simultaneous treatment with taxol (1 μM) and cyto-D (2 μg/ml). These cells also had long, thin neurites that lacked branches and exhibited circular outgrowth. Original magnification: **A**: × 800; **B**: × 2100; **C**: × 800; **D**: × 2400; **E**: × 900; **F**: × 900. A–F reduced by 12 percent.

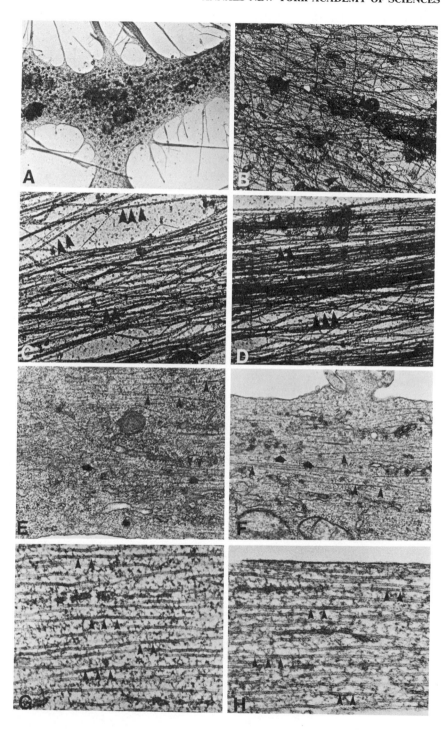

FIGURE 2. A–D: Whole-cell transmission electron micrographs (WCTEMs) of Neuro-2a neurites. A and B: Neurites from control cells grown in serum-deprived medium for 48 hours. The cytoskeleton consisted of a rich network of intermediate filaments (arrows) and microtubules (arrowheads). C: Neurites treated with BBG (250 μg/ml) and cyto-D (2 μg/ml) for 48 h exhibited a cytoskeleton composed solely of numerous microtubules (arrowheads) oriented longitudinally. D: Neurites treated with taxol (1 μM) and cyto-D (2 μg/ml) for 24 h also had an abundance of longitudinally oriented microtubules (arrowheads). E–H: TEMs of Neuro-2a neurites. E: Neurites produced by 48 h serum-depleted medium served as controls. Microtubules (arrowheads) and intermediate filaments (arrows) were prominent. F: BBG-treated neurites also possessed microtubules (arrowheads), intermediate filaments (arrows), and an abundance of other organelles. G: Simultaneous treatment with BBG (250 μg/ml) and cyto-D (2 μg/ml) for 24 h produced an abundance of microtubules (arrowheads) and a paucity of other organelles. H: Simultaneous treatment with taxol (1 μM) and cyto-D (2 μg/ml) resulted in neurites that contained numerous microtubules (arrowheads). Original magnification: A: × 7600; B: × 44,100; C and D: × 64,000; E and F: × 44,000; G and H: × 69,000. A–H reduced by 25 percent.

Mitochondrial Localization of a Microtubule-Related Protein[a]

RADHEY S. GUPTA AND THOMAS J. VENNER

Department of Biochemistry
McMaster University
Hamilton, Canada L8N 3Z5

In earlier studies from this laboratory, stable mutants of Chinese hamster ovary (CHO) cells that exhibited increased resistance to the microtubule (MT) inhibitor podophyllotoxin (Pod^R mutants) have been isolated.[1] Two-dimensional (2-D) gel electrophoretic analyses of total cellular proteins showed that a large number (>20) of independently isolated Pod^R mutants, which exhibited specific cross-resistance or collateral sensitivity towards other MT inhibitors,[2,3] were specifically altered in a protein designated P_1, of approximately 63,000 dalton (see FIGURE 1A and **B**) (references 2 and 4 and unpublished results). The protein P_1 has been shown to be MT-related by a number of different criteria,[2–4] and studies on its quantitation in CHO cells show that it is a major cellular protein present in roughly equimolar amount with $(\alpha + \beta)$ tubulins.[2–4]

To understand the possible role of P_1 in MT structure and functions, specific antibodies to P_1 have been raised in rabbits. In 2-D immunoblots of total cellular proteins from CHO cells, the P_1 antibody reacts specifically with only the P_1 protein and its mutant form M_1 (FIGURE 1C–**F**). The P_1 antibody specifically cross-reacted with a protein of identical molecular weight in cells from various species examined, including human (HeLa and diploid fibroblasts), monkey (vero), mouse (3T3 and LTK$^-$), chicken (chicken embryo fibroblasts, CEF), Syrian hamster (BHK), and Chinese hamster (CHO), indicating its presence in all mammalian and vertebrate species.[5] Immunofluorescence studies using P_1 antibody and a fluorescent isothiocyanate (FITC)-conjugated goat anti-rabbit antibody showed that in interphase cells from various species (*viz.* CHO, 3T3, HDF, CEF), the P_1 antibody specifically reacted with bead and/or string-shaped structures present in discontinuous arrays (see FIGURE 2A for results of CEF cells). The concentration of these bead/string-shaped structures was highest in the perinuclear region. To understand the possible relationship of these structures to MTs, labeling of the same cells (previously labeled with anti-P_1 and FITC-conjugated antibodies) with tubulin antibodies and a rhodamine-conjugated second antibody was carried out. From the representative results of these studies (FIGURE 2**B**), it is evident that the overall distribution of the bead/string-shaped structures in cells was strikingly similar to the distribution of MTs. As a result, their discontinuous arrays appear to be joined by MTs, and in many places MTs appear to be emanating from these bead/string-shaped structures (see arrow in FIGURE 2**B**). Based upon their shape and codistribution with cellular MTs, the structures stained by P_1 antibody appeared very similar to mitochondria.[6–8] To ascertain this, live human diploid fibroblasts (HDFs) were stained with the fluorescent dye rhodamine 123, which binds specifically to mitochondria.[6,8] Subsequently, the same cells were fixed and stained with P_1 antibody. The striking similarity of the two fluorescent patterns

---

[a]This work was supported by the Medical Research Council of Canada.

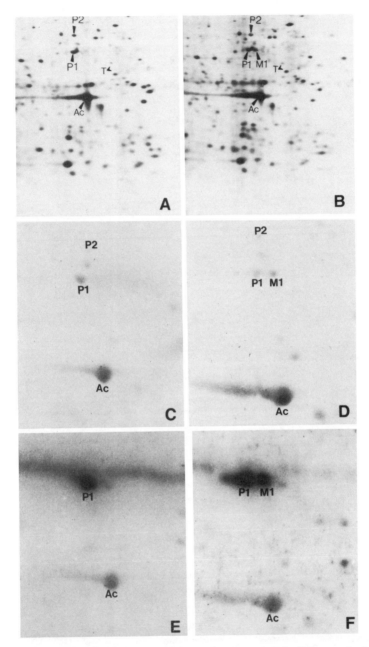

FIGURE 1. A, B: Two-dimensional gel electrophoretic pattern of total cellular proteins from the parental Pod^S and a Pod^{RII} mutant cell line. The positions of the spots corresponding to P$_1$, P$_2$, M$_1$, actin(Ac), and tubulins (T) are indicated in the gels. **C, D:** The Pod^S and Pod^{RII} cells were labeled with low concentrations of [^{35}S]methionine, and after electrophoresis in 2-D gels, the proteins were electroblotted on nitrocellulose and then exposed for 4 h on Kodak X-Omat R film. Only small portions of the fluorographs corresponding to **E** and **F** below are shown. **E, F:** The above blots were treated sequentially[12] with a 1:100 dilution of the P$_1$ antiserum followed by ^{125}I-labeled (F(ab')$_2$ fragment of donkey anti-rabbit IgG (Amersham Corp., Ill.) and then exposed as above.

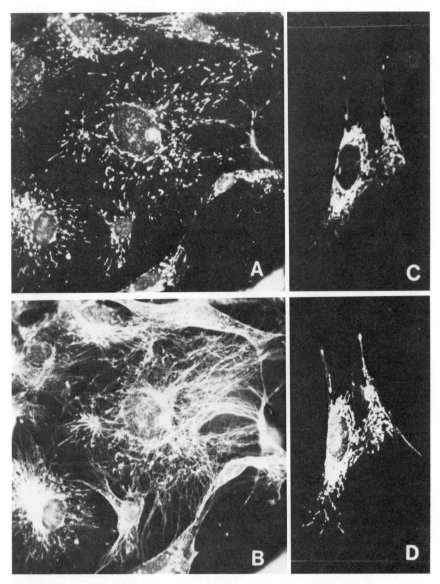

FIGURE 2. A, B: Immunofluorescent labeling of CEF cells with anti-P_1 (**A**) and both anti-P_1 and antitubulin (**B**) antibodies. **C, D:** Fluorescent staining of mitochondria in human diploid fibroblasts with rhodamine 123 and subsequent staining of the same cells with anti-P_1 antibody and FITC-conjugated goat anti-rabbit IgG.

(FIGURE 2C and D) provides strong evidence that the bead/string-shaped structures to which the P_1 antibodies bind are indeed mitochondria.

In view of its MT-related nature, the exclusive location of P_1 on cellular mitochondria is unexpected and surprising. These observations could be understood only if a chemical and functional linkage exists between these two important cellular structures with involvement of P_1 in the linkage. The existence of a linkage between MTs and mitochondria has previously also been suggested by others.[7,9] Earlier studies with podophyllotoxin show that the drug inhibits MT assembly by first binding to the free tubulin dimers that (i.e. drug-tubulin complex) then cap the growing ends of MTs.[10] In this context, our observation that the Pod^R mutants are specifically affected in P_1 suggests that the growing ends of MTs, where the drug-tubulin complex binds, should be located either on or in the vicinity of mitochondria, so that an alteration in P_1 could prevent or affect binding of the drug-tubulin complex to growing ends of MTs. Because the dynamic structure of MTs is dependent upon continued supply of chemical energy, and mitochondria are responsible for cellular energy production, the observed association and inferred linkage between these two structures may not be coincidental. One possibility is that mitochondria, by providing an attachment site for the growing ends of MTs, as inferred above, and a source of needed chemical energy, serve as cellular sites where MT assembly occurs. This suggestion is consistent with our observation that MTs in many places appear to be emanating from mitochondria-like structures and that the highest density of mitochondria in interphase cells is observed in the perinuclear region, from where MTs also appear to originate.[11] The MTs-mitochondria association/linkage also provides a possible mechanism for the efficient transport and distribution of chemical energy to different parts of cells. Further studies on the relationship between these two structures should prove of great importance in understanding their cellular roles.

REFERENCES

1. GUPTA, R. S. 1981. Somatic Cell Genet. **7:** 59–71.
2. GUPTA, R. S., T. K. W. HO, M. R. K. MOFFAT & R. GUPTA. 1982. J. Biol. Chem. **257:** 1071–1078.
3. GUPTA, R. S. 1983. Cancer Res. **43:** 505–512.
4. GUPTA, R. S. & R. GUPTA. 1984. J. Biol. Chem. **259:** 1882–1890.
5. GUPTA, R. S., T. VENNER & A. CHOPRA. 1985. Can. J. Biochem. Cell Biol. **63:** 489–502.
6. JOHNSON, L. V., M. L. WALSH & L. B. CHEN. 1980. Proc. Natl. Acad. Sci. USA **77:** 990–994.
7. BALL, E. H. & S. J. SINGER. 1982. Proc. Natl. Acad. Sci. USA **79:** 123–126.
8. SUMMERHAYES, I. C., D. WONG & L. B. CHEN. 1983. J. Cell Sci. **61:** 87–105.
9. SMITH, D. S., U. JÄRLFORS & M. L. CAYER. 1977. J. Cell Sci. **27:** 255–272.
10. WILSON, L., K. ANDERSON & D. CHIN. 1976. Cell Motility. R. Goldman, R. Pollard & J. Rosenbaum, Eds.: 1051–1064. Cold Spring Harbor. New York.
11. OSBORN, M. & K. WEBER. 1976. Proc. Natl. Acad. Sci. USA **73:** 867–871.
12. TOWBIN, H., T. STAEHELIN & J. GORDON. 1979. Proc. Natl. Acad. Sci. USA **76:** 4350–4354.

Microtubule Function in Human Blood Polymorphonuclear Leukocytes: Analysis Through Heat-Induced Lesions[a]

STEPHEN E. MALAWISTA

Department of Internal Medicine
Yale University School of Medicine
New Haven, Connecticut 06510

The carefully controlled and timed application of heat to human blood polymorphonuclear leukocytes (PMN) induces the formation of membrane-bounded, anucleate, granule-poor cytoplasmic fragments, which retain motile functions of the parent cell, including chemotaxis and phagocytosis. Fragment formation appears to represent an uncoupling of motile machinery concentrated in the protopod (leading front, lamellipodium), followed by its exodus from the bulk of the cell.[1,2] After heat treatment, microtubules are lacking both in the motile fragments, which we have called cytokineplasts (CKP), and in the nonmotile residual cell bodies.[2] Their absence from these preparations may be instructive with regard to the proper role of microtubules both in directed locomotion and in the cellular integrity of PMN. Moreover, because the heat-induced lesion seems directed at a specific centrosomal structure and not confined to this one cell type, appropriate brief heat may prove generally useful as a probe of centrosomal function.[3]

Effects of Heat on PMN: Anchoring and Uncoupling

When adherent PMN are heated to 45°C for 9 minutes, the delicate veil that ornaments the surface of normal PMN (FIGURE 1a) slips off it like a sleeve (uncoupling, FIGURE 1b), and moves away from the motionless residual cell body (FIGURE 1c). The latter is unusually adherent to substrate (anchoring), which allows mutual distancing to occur. With time, the connecting filament breaks and seals itself, freeing the fragment (FIGURE 1d).

Free Fragments (Cytokineplasts): Chemotaxis without Microtubules

These fragments retain multiple motile properties of the parent cell, including adherence, spreading, random locomotion, chemotaxis (directed locomotion), and phagocytosis.[1,2] FIGURES 2 and 3 show chemotaxis of individual PMN and fragment, respectively, toward erythrocytes freshly destroyed by laser microirradiation. Like the parent PMN, fragments also respond chemotactically to zymosan-activated serum

[a]This work was supported in part by Grants from the National Institutes of Health (AM-10493, AM-19742, AM-07107) and by the Arthritis Foundation and its Connecticut Chapter.

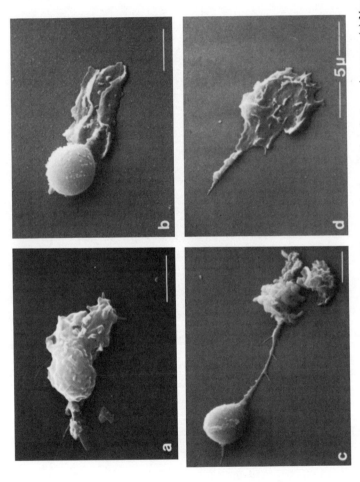

FIGURE 1. Various stages of fragment development in human blood PMN are seen in the scanning electron microscope. (**a**) Normal PMN. The leading front (protopod) is at right, the tail (uropod) at left. Ruffled membrane is integrated over the leading front (right) and cell body. (**b**) After heating. (**b–d**) Ruffled membrane is uncoupled from the rounded-up cell body, whose surface is smooth or tufted. (**c**) With mutual distancing, the connection between cell body and fragment is attenuated. (**d**) A free fragment (cytokineplast) is seen. Bars, ~5 μm. (S. E. Malawista, & A. De Boisfleury Chevance.[2] With permission from *The Journal of Cell Biology*.)

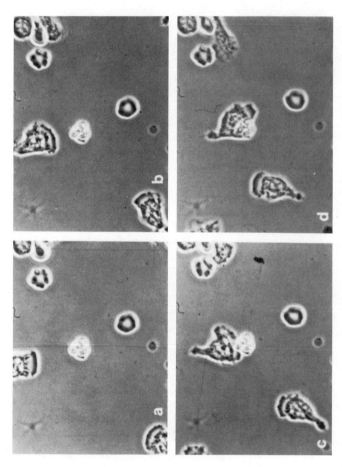

FIGURE 2. Chemotactic behavior of an intact human blood PMN is seen in time lapse cinematography. (**a**) After laser destruction of the central erythrocyte (zero time), local PMN approach from above and two corners. (**b**–**d**) One strikes and wraps itself around the target. Phase contrast, approximately × 750. (S. E. Malawista & A. De Boisfleury Chevance.[2] With permission from *The Journal of Cell Biology*.)

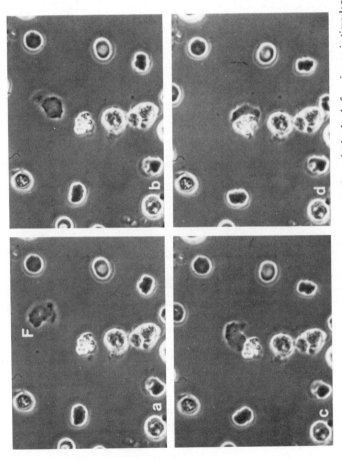

FIGURE 3. Chemotactic behavior of an anucleate cytoplasmic fragment (cytokineplast), made the day before, is seen in time-lapse cinematography. (**a–d**) Response of the fragment (labeled F in FIGURE 3a) is like that of intact PMN (FIGURE 2). Targets lose their chemoattraction after several minutes; this cytokineplast responded to three more fresh laser targets over the ensuing 46 minutes. Phase contrast, approximately × 750. (S. E. Malawista & A. De Boisfleury Chevance.[2] With permission from *The Journal of Cell Biology*.)

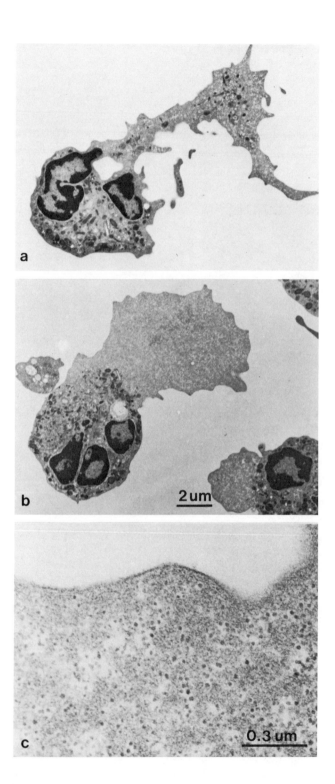

a

b   2um

c   0.3 um

(C5a) and to the formyl peptide, N-formylmethionylleucylphenylalanine (fMet-Leu-Phe).[4] Their functional longevity exceeds that of the parent PMN.[2,5]

In the transmission electron microscope, fragments contain what would be expected in a structure derived primarily from the organelle-excluding hyaline peripheral cytoplasm (FIGURE 4);[2,6] notable by their absence are both centrioles and microtubules. Microfilaments are present and, by immunofluorescence, both actin and myosin are present as well.[6] On stimulation with fMet-Leu-Phe, fragments mobilize calcium (measured by Quin-2 fluorescence) and polymerize actin (measured on gels, as cytoskeleton-associated actin) like the parent PMN.[6a] In addition to actin and myosin, the fact of function dictates that the rather circumscribed portion of the cell represented by the fragment will also bear all the other contractile proteins—known and as yet unknown—as well as the sensing and transducing apparatus necessary for directed locomotion.

Role of Microtubules in Chemotaxis by PMN

Through these fragments we can address the role of microtubules in chemotaxis in a way that avoids the ambiguities attendant upon the use of exogenous drugs such as colchicine. In assays that depend upon migration of PMN across a micropore filter, colchicine seems to interfere with chemotaxis.[7,8] The general interpretation of such data has been that arrays of microtubules, organized from the centrosome, provide direction to cell movement.[9-11] However, colchicine-treated PMN viewed microscopically still point in the right direction[12] and, although they show wider angles of turn than untreated cells, they continue to move toward the chemotactic source.[13] Allan and Wilkinson[13] concluded that just as microtubules are not essential for phagocytosis,[14-16] they are also not important for the spatial detection of gradients, nor for locomotion towards their sources; they are important for accurate turning by leukocytes and for maintaining the shape and polarity of the moving and phagocytosing cell. The work on fragments supports that view: CKP are capable of directed locomotion and phagocytosis even though they are derived from microtubule-free hyaloplasm that is concentrated at the leading front and have no centrosome. The demonstration that this material is by itself sufficient for directed locomotion allows us to interpret the role of microtubules in chemotaxis more precisely[2]: we suggest that they serve to stabilize and orient the "baggage" (nucleus, granuloplasm) carried by the potentially autonomous motile apparatus; in short, they prevent fishtailing.

Residual Cell Body: Lack of Microtubules Associated with Alterations in Pericentriolar Osmiophilic Material

The initial clue to the likely mechanism of uncoupling came from immunofluorescent studies of the monocytes in our leukocyte preparations (PMN being, at that time,

---

FIGURE 4. Control and heated human blood PMN are seen in the transmission electron microscope. (**a**) In this thin section of a spread control cell, cut parallel to the substrate, granuloplasm is generally pervasive; its exclusion from the relatively narrow leading front is best seen in phase contrast (see FIGURE 2). (**b**) After heat there is a rather discreet boundary between a relatively large area that lacks organelles, and the rounded, nucleated, granule-rich cell body. (**c**) Fragments at greater magnification contain filaments best seen subjacent to the cell membrane, but not microtubules. The numerous round dense bodies are glycogen. Approximately (**a**) × 4,500, (**b**) × 4,500, and (**c**) × 54,000. (S. E. Malawista & A. De Boisfleury Chevance.[2] With permission from *The Journal of Cell Biology*.)

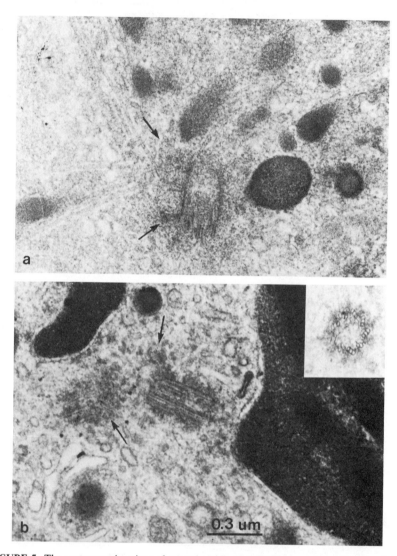

FIGURE 5. The centrosomal regions of control and heated PMN are seen in the transmission electron microscope. (**a**) Control PMN. Microtubules radiate especially from pericentriolar osmiophilic material (arrows). (**b**) Heated PMN. Centriolar structure remains intact (insert) and its paranuclear location is maintained, but pericentriolar osmiophilic material (arrows) appears condensed and either fragmented or clotted, and microtubules are sparse. Approximately × 50,000. (S. E. Malawista & A. De Boisfleury Chevance.[2] With permission from *The Journal of Cell Biology.*)

poor specimens for immunofluorescence), with antibodies directed against pericentrio-lar osmiophilic material, and against microtubules.[3] After heat, the demonstration of both structures by immunofluorescence was severely compromised, even though putative paired centrioles were still visible in phase contrast. These findings in monocytes suggested that uncoupling in PMN might be related to heat-induced centrosomal damage mediated through resultant interference with the apparent constraints upon the membrane imposed by the assembly-disassembly of microtu-bules.[9,17]

When the ultrastructure became available, this interpretation was supported by the finding of similar heat-induced lesions both in the fast-moving PMN and eosinophil, both of which fragment in these preparations, and in the more sluggish monocyte, which (under these conditions) does not fragment.[2] Heated cells have two consistent abnormalities: (a) microtubules are sparse; and (b) the pericentriolar osmiophilic material from which microtubules ordinarily radiate appears condensed and either fragmented or clotted (FIGURE 5). Structure of the centriole itself appears preserved (insert), and its paranuclear location is maintained.

Similar aggregation of pericentriolar osmiophilic material has been described in Chinese hamster ovary (CHO) cells fixed after 15 min at 45.5°C;[18] in CHO cells, that temperature produces a time-dependent (and cell-cycle dependent) mitotic delay. These and other effects of heat on dividing CHO cells have also been ascribed to impaired nucleation of microtubules by centrosomes.[19] The functional implication for the postmitotic PMN is that constraints emanating from the centrosome may now be extended to include maintenance of the motile machinery as an integral part of the cell.

Heat as a Probe of Centrosomal Function

Although we have not ruled out additional effects of heat (*e.g.,* direct inactivation of microtubule protein), the heat-induced lesion of pericentriolar osmiophilic material appears reasonably specific, as evidenced by the persistence of motile function in fragments,[1,2] by the reversibility of antimitotic effects in CHO cells,[18] and by the general preservation of ultrastructural integrity.[2,6,18] Thus, appropriate brief heat treatment may prove generally useful as a probe of centrosomal function, just as colchicine and other agents that interfere with the assembly of microtubules by binding to tubulin have been used successfully to produce a more distal lesion.

REFERENCES

1. KELLER, H. U. & M. BESSIS. 1975. Chemotaxis and phagocytosis in anucleated cytoplasmic fragments of human peripheral blood leukocytes. Nouv. Rev. Fr. Hematol. **15:** 439–446.
2. MALAWISTA, S. E. & A. DE BOISFLEURY CHEVANCE. 1982. The cytokineplast: purified, stable, and functional motile machinery from human blood polymorphonuclear leuko-cytes. J. Cell Biol. **95:** 960–973.
3. MALAWISTA, S. E., A. CHEVANCE DE BOISFLEURY, R. MAUNOURY & M. BESSIS. 1983. Heat as a probe of centrosomal function: a phase-contrast and immunofluorescent study of human blood monocytes. Blood Cells **9:** 443–448.
4. DYETT, D. E., S. E. MALAWISTA, G. VAN BLARICOM, D. A. MELNICK & H. L. MALECH. 1985. Functional integrity of cytokineplasts: specific chemotactic and capping responses. J. Immunol. **135:** 2090–2094.
5. MALAWISTA, S. E., G. VAN BLARICOM & S. B. CRETELLA. 1985. Cytokineplasts from

human blood polymorphonuclear leukocytes: lack of oxidase activity and extended functional longevity. Inflammation **9:** 99–106.

6. MALAWISTA, S. E., A. CHEVANCE DE BOISFLEURY, M. LESOURD & G. C. GODMAN. 1982. Ultrastructure of the cytokineplast (stable functional motile machinery) from human polymorphonuclear leukocytes. Clin. Res. **30:** 560a.

6a. DYETT, D. E., S. E. MALAWISTA, P. H. NACCACHE & R. I. SHA'AFI. 1986. Stimulated cytokineplasts from human polymorphonuclear leukocytes mobilize calcium and polymerize actin; cytoplasts made in cytochalasin B retain a defect in actin polymerization. J. Clin. Invest. **77:** 34–37.

7. CANER, J. E. Z. 1965. Colchicine inhibition of chemotaxis. Arthritis Rheum. **8:** 757–763.

8. MALECH, H. L., R. K. ROOT & J. I. GALLIN. 1979. Structural analysis of human neutrophil migration: centriole, microtubule, and microfilament orientation and function during chemotaxis. J. Cell Biol. **75:** 666–693.

9. OLIVER, J. M. 1978. Cell biology of leukocyte abnormalities. Am. J. Pathol. **93:** 221–260.

10. GALLIN, J. I., E. K. GALLIN, H. L. MALECH & E. B. CRAMER. 1978. Structural and ionic events during leukocyte chemotaxis. *In* Leukocyte Chemotaxis: Methods, Physiology, and Clinical Implications. J. I. Gallin & P. G. Quie, Eds.: 123–141. Raven Press. New York.

11. BANDMANN, U., L. RYDGREN & B. NORBERG. 1974. The difference between random movement and chemotaxis. Effects of antitubulins on neutrophil granulocyte locomotion. Exp. Cell Res. **88:** 63–73.

12. ZIGMOND, S. H. 1977. The ability of polymorphonuclear leukocytes to orient in gradients of chemotactic factors. J. Cell Biol. **75:** 606–616.

13. ALLAN, R. B. & P. C. WILKINSON. 1978. A visual analysis of chemotactic and chemokinetic locomotion of human neutrophil leukocytes. Exp. Cell Res. **111:** 191–203.

14. MALAWISTA, S. E. & P. T. BODEL. 1967. The dissociation by colchicine of phagocytosis from increased oxygen consumption in human leukocytes. J. Clin. Invest. **46:** 786–796.

15. PENNY, R., D. A. GALTON, J. T. SCOTT & V. EISEN. 1966. Studies on neutrophil function. I. Physiological and pharmacological aspects. Br. J. Haematol. **12:** 623–632.

16. MALAWISTA, S. E. 1975. Microtubules and the mobilization of lysosomes in phagocytizing human leukocytes. Ann. N.Y. Acad. Sci. **253:** 738–749.

17. RUDOLPH, S. A. & S. E. MALAWISTA. 1980. Inhibitors of microtubule assembly potentiate hormone-induced cyclic AMP generation in human leukocytes. *In* Microtubules and Microtubule Inhibitors 1980. M. De Brabender & J. De Mey, Eds.: 481–495. North Holland, Amsterdam.

18. BARRAU, M. D., G. R. BLACKBURN & W. C. DEWEY. 1978. Effects of heat on the centrosomes of Chinese hamster ovary cells. Cancer Res. **38:** 2290–2294.

19. COSS, R. A., W. C. DEWEY & J. R. BAMBURG. 1982. Effects of hyperthermia on dividing Chinese hamster ovary cells and on microtubules *in vitro*. Cancer Res. **42:** 1059–1071.

The Function of Microtubules in Directional Cell Movement[a]

URSULA EUTENEUER AND MANFRED SCHLIWA

Department of Zoology
University of California
Berkeley, California 94720

INTRODUCTION

Many cell types are capable of persistent, directional locomotion in the absence of known external stimuli such as chemical signals, substate properties, or electric fields. Essentially all locomoting cells possess an external polar organization characterized by the presence of an anterior lamellipod of variable shape and complexity, and a posterior uropod or tail from which retraction fibers may extent. This external asymmetry seems to be complemented in many instances by an internal axis defined by the position of two prominent cell organelles, namely, the nucleus and the centrosome/ microtubule complex. The finding of an apparent relationship between cell polarity and the position of the centrosome, postulated a century ago,[1] received support from recent experimental studies that correlate shifts in centrosome position with changes in the cell axis and/or changes in the direction of cell locomotion.[2-6] Even though such a correlation between cell polarity and centrosome position could not be established in all instances,[7,8] these and other observations have led to the widely held belief that cell polarity and the direction of cell locomotion are determined by the centrosome, which transmits the necessary information to the rest of the cytoskeleton by way of the microtubules associated with it. It needs to be emphasized, however, that, whereas these studies present evidence for a positive correlation between centrosome position and the establishment of a cell axis or the direction of cell locomotion, it is difficult to determine cause and effect. This question is central to the issue of cell polarity during directional locomotion, but it can not be resolved on the basis of the available evidence.

Here we make use of two model systems, primary cultures of fish epidermal keratocytes and human polymorphonuclear leukocytes (PMNs), to study the involvement of the centrosome/microtubule system in the development and maintenance of polar cell organization and directional movement. Fish keratocytes are highly motile in culture, migrating at a rate of up to 1 μm/second. Interestingly, small fragments derived from the anterior lamella are also capable of directional movement in a manner indistinguishable from that of intact cells. Many of these fragments contain neither microtubules nor centrioles, nor any other major cell organelle, and yet they are capable of persistent, directional locomotion in the absence of the centrosome/ microtubule complex. The second model system, human PMNs, allowed us to obtain some information on the causal relationship between the position and dynamics of the centrosome/microtubule complex and the contractile machinery of the cell. When these cells are treated with the potent tumor promoter, 12-0-tetradecanoyl-phorbol-

[a]This work was supported in part by a CRCC Grant from the University of California and by NIGMS Grant 31041.

acetate (TPA), they flatten and spread on the coverslip and remain nonmotile.[9] In the majority of these cells, the centrosome undergoes splitting[10] into two or even three separate microtubule asters. The asters of TPA-treated cells are in constant, rapid motion through the cytoplasm. We studied the motility of centrosome/microtubule complexes using differential drug treatments and obtained evidence for an interaction between the actin-based cytoplasmic network and the microtubules in the asters. This interaction may give rise to the movement of asters observed under these conditions and may help determine the positioning of centrosomes.

LOCOMOTION OF FISH KERATOCYTES

Single epidermal keratocytes, when emigrating from explants of epidermal sheets, adopt a distinctive morphology. The cells extend a broad, flat lamellipodium from the cell anterior that may span more than 180° of the cell perimeter (FIGURE 1). All major cell organelles, including the nucleus, endoplasmic reticulum, mitochondria, Golgi apparatus, and the bundles of cytokeratin filaments are contained in the globular cell body located at the cell posterior. All locomoting cells have this characteristic canoe shape[11] also observed in certain amphibian cells.[12] Keratocytes move foreward perpendicular to their long axis and change their direction by making smooth, gradual turns, rather than abruptly retracting the existing lamella and extending a new one elsewhere. Their average speed of movement is 0.5 μm/sec, and occasionally reaches over 1 μm/second. They are thus among the fastest moving eukaryotic cells known. Migratory cells usually persist in their general direction of movement for many cell diameters, deviating less than 20% from the shortest possible route (a straight line between beginning and end point of movement). As demonstrated by fluorescence microscopy, actin filaments form a delicate criss-cross pattern in the lamella (FIGURE 1). Neither the bundles of cytokeratin filaments nor the microtubules extend into the lamellar region, but instead are confined to the perinuclear space. Microtubules, in particular, appear wrapped around the nucleus, as shown by the micrographs taken in two different planes of focus (FIGURE 1). Centrioles are also found in the bulbous cell body where they occupy a position close to the nucleus (FIGURE 2). Apparently, they are not surrounded by an extensive aster of microtubules, a finding consistent with the rather random orientation of microtubules as seen by immunofluorescence microscopy.

The presence of an intact system of microtubules is not required for the movement of keratocytes. Depolymerization induced by nocodazole or colcemid treatment in the cold (0°C for 30–45 min), followed by rewarming in the presence of the drug, changes neither the velocity nor the persistency of keratocyte locomotion. The cells completely round up during cold treatment and then develop their characteristic migratory morphology in the absence of microtubules. The movement of several cells treated in this way was recorded, and the cells were then processed for immunofluorescence microscopy with antibodies against tubulin. The course of 4 cells over an 18 min time interval and corresponding phase contrast and fluorescence micrographs for one of the cells are shown in FIGURE 3. Inhibitor-treated cells have precisely the same shape and move with the same degree of persistency as their untreated counterparts. This also applies to their behavioral repertoires, in particular, the avoidance reaction known as contact inhibition after collision with other cells (see diagram in FIGURE 3). Neither fluorescence microscopy (FIGURE 3) nor electron microscopy (not shown) demonstrate the presence of intact microtubules. Occasionally, one or two fluorescent dots, probably corresponding to the centrioles, may be detected. Thus, the movement of intact keratocytes appears to be independent of the presence of microtubules.

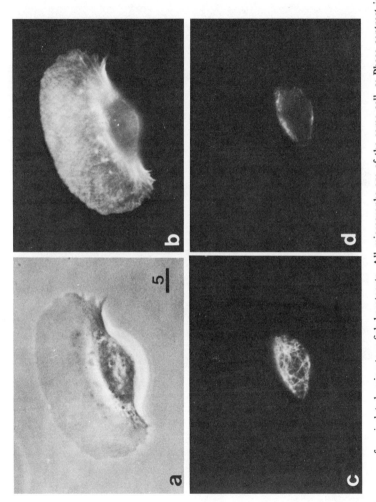

FIGURE 1. Light microscopy of an isolated, migratory fish keratocyte. All micrographs are of the same cell. **a:** Phase contrast image, showing the characteristic canoe shape, the broad, flat lamellipodium, and the bulbous cell body. **b:** Staining for actin with rhodamine-phalloidin. A delicate criss-cross pattern is revealed within the lamella. **c:** Staining for microtubules with an antibody against tubulin. Microtubules are confined to the perinuclear space. **d:** Microtubule staining at a focal plane that bisects the nucleus, showing how microtubules are wrapped around the nucleus. The dimensions of the scale bars in this and all following micrographs are given in micrometers.

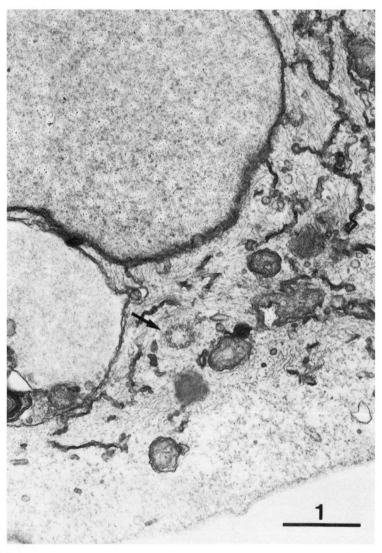

FIGURE 2. Electron micrograph of a thin section that passes through the bulbous cell body. A centriole (arrow) is located close to the nucleus. Whorls of cytokeratin filaments are interspersed between a variety of membraneous organelles.

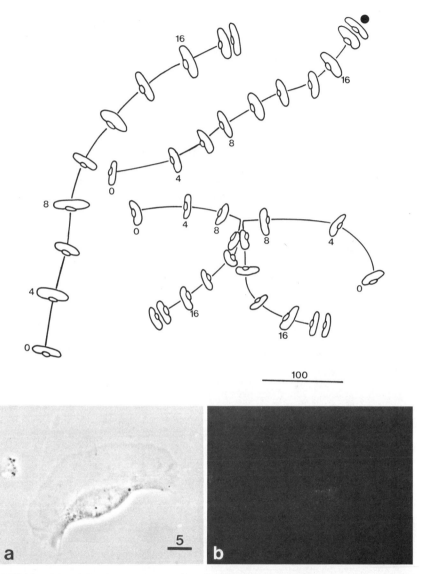

FIGURE 3. Movement of intact keratocytes in the absence of microtubules. Top: Tracing of the paths of 4 living keratocytes over a time period of 20 min after treatment with 6 μg/ml nocodazole for 45 min at 0° C, followed by rewarming in the presence of nocodazole. The two lower cells collide after 9 min and show the typical behavior of contact inhibition. The cell marked by a filled dot is shown in **a** in phase contrast and in **b** after staining with antitubulin. No microtubules are present. Numbers represent time in minutes.

LAMELLAR CELL FRAGMENTS

Additional support for the idea that keratocyte locomotion does not require intact microtubules comes from the observation that small cell fragments derived from the anterior lamella can locomote with the same degree of persistency as intact cells. In their attempt to break loose from a fresh explant, keratocytes at the periphery of such an explant extend lamellae that may elongate and finally detach (pinch off), whereas the cell body is still held in the cell cluster by intercellular junctions. Alternatively, lamellar fragments may be generated by cutting off lamellae with fine glass needles. These fragments, which represent less than 10% of the total cell volume, may move in a pattern indistinguishable from that of intact cells (FIGURE 4), and they may do so in the presence or absence of colcemid. FIGURE 5 shows the course of a small fragment that collides with an intact cell and then changes direction. Immunofluorescence microscopy shows that this fragment, like many others studied in the same way, does not contain a single piece of microtubule, whereas neighboring intact cells display the characteristic circumnuclear microtubule whorls.

Morphological differences between intact cells and lamellar fragments are illustrated in FIGURES 6 and 7. High voltage electron microscopy clearly shows an almost complete absence of membraneous organelles from the lamella in both intact cells and fragments. One noteworthy feature in both is the presence of fibrils, or small bundles of filaments oriented perpendicular to the direction of movement at the cell posterior (FIGURES 6a, b, 7c).

CENTROSOME MOTILITY IN TUMOR PROMOTER-TREATED PMNs

An involvement of the centrosome/microtubule complex in establishing and maintaining directionality of locomotion has repeatedly been proposed for leukocytes. Perhaps the most graphic example of a correlation between cell movement and centrosome position is found in newt eosinophils, where the position of the centrosome is clearly visible by light microscopy as a granule-excluding cytoplasmic domain that almost invariably is located between the leading cell edge and the nucleus.[4] Such high visibility of the centrosome/microtubule complex is not observed in other leukocytes during locomotion, and their behavior is therefore more difficult to study. We found, however, that treatment of human PMNs with TPA induces, and renders visible, an unprecedented form of astral motility whose study provided some clues as to the relationship between centrosome/aster complexes and actin-based cytoplasmic networks.

When randomly locomoting PMNs in culture are treated with 10–100 ng/ml TPA for 20–30 min, they assume a "fried egg" morphology and remain stationary (FIGURE 8), but show constant ruffling activity around the cell perimeter. TPA also induces centrosome splitting[9] in 60–70% of the cells, whereby the two centrioles of the centrosome may separate and form two independent aster systems. This condition may persist as long as the TPA is present, thus distinguishing this phenomenon from transient (3–5 min) splitting induced in the initial phase of chemokinesis.[10] Time-lapse video microscopy shows that the centrosome/aster complexes, whether splitting has occurred or not, are in constant, rapid motion through the cytoplasm. Though the centrosome/asters *per se* are not visible by light microscopy, their movement can easily be followed in time-lapse recordings as the displacement of cytoplasmic domains relative to the rest of the cytoplasm. These moving "entities" may carry with them some of the cytoplasmic granules while pushing aside others, and they may deform and

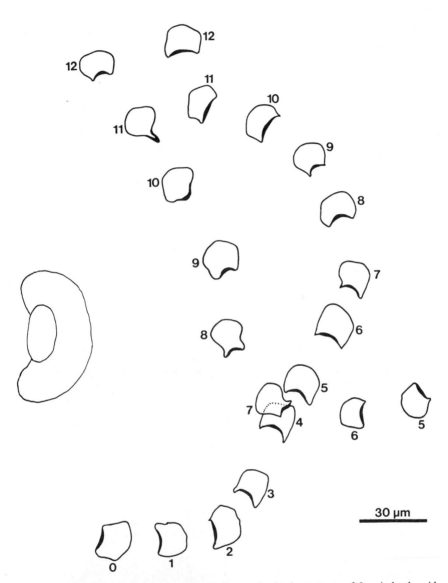

FIGURE 4. Tracings of two anucleate fragments moving in the presence of 5 µg/ml colcemid over a time period of 12 minutes. The patterns of movement of these fragments is identical to that of intact cells. An average-sized intact cell is shown near the left-hand margin for comparison. Numbers represent time in minutes.

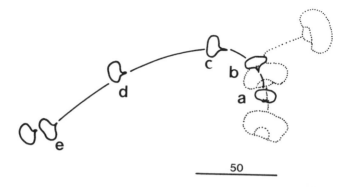

50

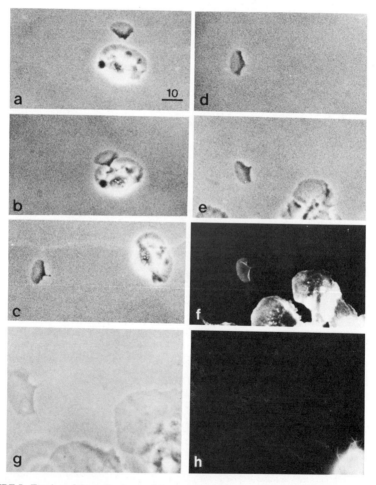

10

FIGURE 5. Tracing of the path of a small fragment (top) and corresponding light micrographs of the live fragment (**a–e**). This preparation was processed for fluorescence microscopy. **f:** Rhodamine-phalloidin staining, shown at the same magnification as **a–e**. **g, h:** Higher magnification of the fragment after processing for immunofluorescence shown in phase contrast (**g**) and antitubulin fluorescence (**h**). No microtubules are present in the fragment, whereas intact cells nearby show the characteristic circumnuclear whorls.

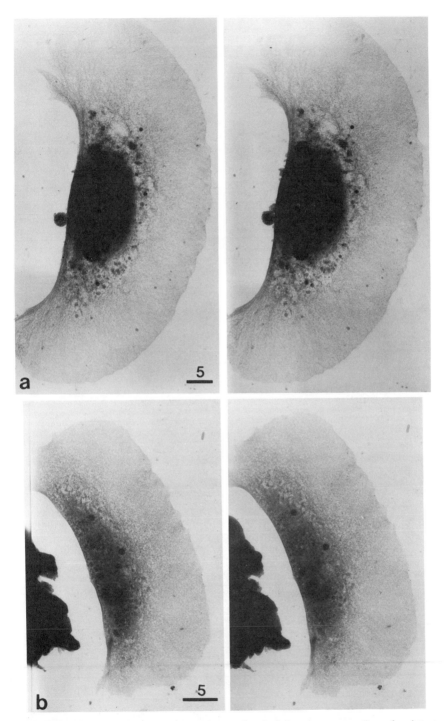

FIGURE 6. Stereo high voltage electron micrographs of whole mount preparations of an intact keratocyte (**a**) and a lamellar fragment (**b**). Note the absence of membranous or vesicular organelles in the lamella of both. Interestingly, just as in intact cells, cell organelles of fragments, if present, are confined to the cell posterior.

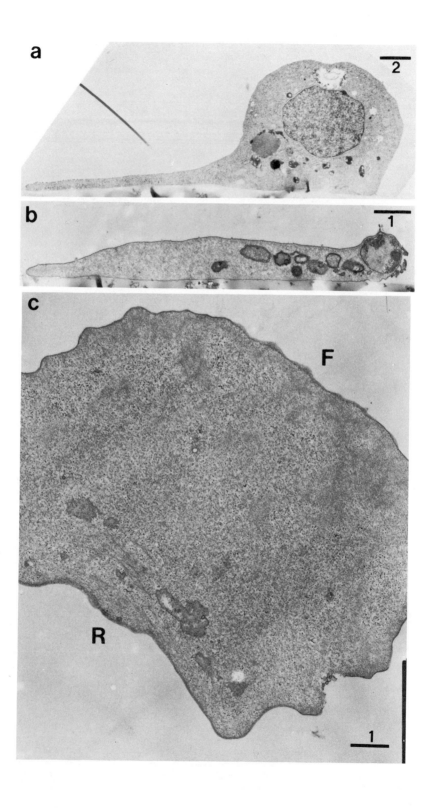

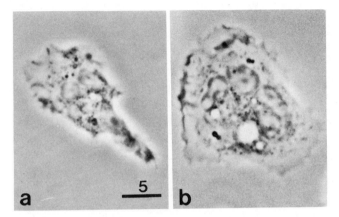

FIGURE 8. Phase contrast micrographs of living human PMNs. Cell in random locomotion (a) and after treatment with 50 ng/ml TPA for 20 min (b).

displace the nucleus. Their identification as centrosme/aster complexes was confirmed by correlative light and high voltage electron microscopy. Cells attached to finder grids were first videotaped to follow the movement of presumptive centrosome/aster complexes, lysed in microtubule-stabilizing medium, and processed for whole-mount high-voltage electron microscopy. In all cases examined, the precise position of the microtubule asters could be predicted with remarkable accuracy from the recording of the live cell before fixation. Although, as a rule, these movements are very difficult to document in a series of still photographs, in one fortunate case, shown in FIGURE 9, a clear vesicle was trapped near, or somehow stayed associated with, one of the split centrosomes so that it served as a convenient marker for its rather elaborate movement through the cytoplasm. Thus, once splitting has occurred, the asters are by no means stationary, but rather are in constant, rapid motion through the cytoplasm, reaching speeds of up to 0.6 μm/second.

MECHANISM OF CENTROSOME SPLITTING AND ASTER MOTILITY

Centrosome motility in TPA-treated cells was used as a model system to elucidate how asters might interact with the rest of the cytoskeleton. Instead of using a visual assay (time-lapse light microscopy) or tedious high-voltage electron microscopy to assess the outcome of experimental treatments (both these approaches would severely limit sample size), we have used immunofluorescence microscopy to monitor the position of centrioles. In this way, the parameter tested was reduced to the assessment

FIGURE 7. a: Transverse section through the central portion of an intact keratocyte, showing the thin lamellipod and the bulbous organelle-containing cell body. **b:** Comparable transverse section of a lamellar fragment. The intracellular organelles enclosed in this fragment are located at the posterior. **c:** Third section from the substrate in a series of horizontal sections of a small lamellar fragment that was known to have moved in a directional fashion before fixation. Very few membrane-bounded intracellular organelles are seen. F and R designate front and rear, respectively.

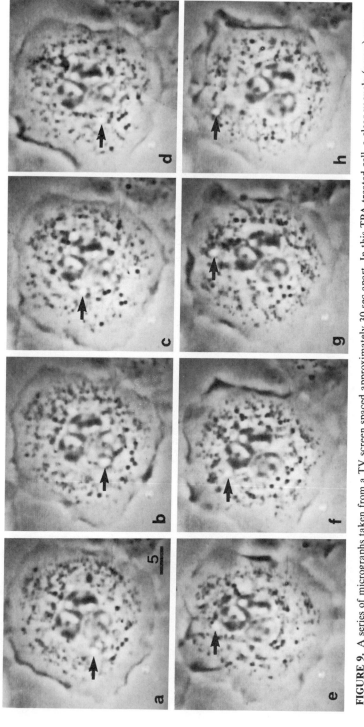

FIGURE 9. A series of micrographs taken from a TV screen spaced approximately 30 sec apart. In this TPA-treated cell, a clear vacuole (arrow) was closely associated with one of the motile aster centers and thus could serve as an excellent marker for its vigorous movement through the cytoplasm.

of splitting versus no splitting, but a large number of cells could be surveyed. Wherever possible, this assay was supplemented by video microscopy or high-voltage electron microscopy. Although PMNs are a poor substrate for immunofluorescence microscopy with tubulin antibodies (tests over many years with many different antibodies and a large variety of staining protocols did not yield consistent results), an antibody kindly provided by Marc Kirschner[13] was found to consistently stain the centrioles in PMNs (as well as microtubules in other cell types in the same preparations, FIGURE 10a and b). Although some staining protocols gave reasonably good results in some of the PMNs (one of the best examples is shown in FIGURE 10c), the procedures worked too

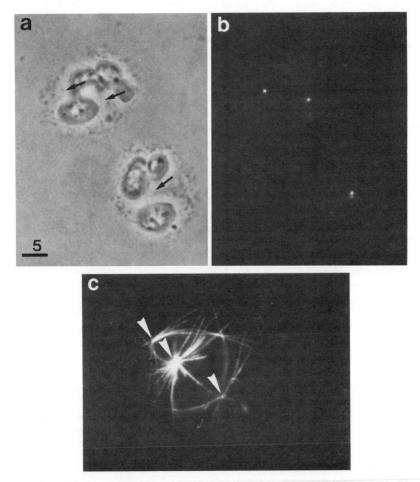

FIGURE 10. Phase contrast (**a**) and corresponding immunofluorescence micrograph (**b**) of TPA-treated PMNs prepared with a tubulin antibody that consistently stains centrioles. In one cell, the two centrioles are still together in the cell center; in the other cell, the centrosome has undergone splitting (arrows). **c**: One of the very best examples of a PMN stained with antitubulin in which the microtubules are actually labeled. In this cell, three asters are present (arrowheads).

inconsistently to allow their use as an assay. Staining of just the centrioles clearly gave the best results.

The effects of microtubule and actin-active compounds (nocodazole and cytochalasin D, respectively), as well as the recovery from these treatments, on the frequency of centrosome splitting, are summarized in TABLE 1. Inhibitor treatments were done as complementary pairs where the compound was applied to cells either before or after TPA treatment. The effectiveness of nocodazole in depolymerizing all PMN microtubules was confirmed by thin-section and whole-mount electron microscopy. In agreement with reports by others,[7] centrioles in randomly locomoting cells are in close proximity to each other and are located close to the nucleus in the cell center (TABLE 1, 92% not split). TPA consistently induces centrosome splitting in 60–70% of the cells. The outcome of the inhibitor treatments depends on the time at which they are administered. Nocodazole treatment for 20 min will prevent centrosome splitting when added before TPA. If the sequence is reversed, that is, if nocodazole is used after TPA has induced centrosome splitting, the splitting frequency is similar to that of cells

TABLE 1. Effects of Nocodazole (NOC) and Cytochalasin D (CD) on the Frequency of Centrosome Splitting in TPA-Treated Human PMNS.[a]

	Percent split	Percent not split	?[c]	Number
Random	4.1	92.0	3.9	263
TPA	66.0	28.4	5.6	697
NOC—TPA	6.0	85.0	9.0	154
TPA—NOC	76.5	23.0	1.5	171
CD—TPA	0.5	97.0	2.5	147
TPA—CD	24.0	73.0	3.0	191
NOC—TPA—REC[b]	53.7	38.7	7.6	257
CD—TPA—REC	52.6	39.0	8.3	399
TPA—CD—REC	55.0	37.1	7.9	355

[a]TPA treatment was for 25 min at 50 ng/ml. Nocodazole treatment was for 20 min at 1 μg/ml. Cytochalasin D treatment was for 25 min at 2 μg/ml.
[b]REC refers to recovery from nocodazole or cytochalasin D, respectively, for 45 min in the presence of TPA.
[c]The column marked by "?" is the percentge of cells that could not be scored.

treated with TPA alone. In both cases microtubules are completely absent, and no aster motility is observed by time-lapse video analysis of living cells. From this observation, it can be concluded that microtubules are required for splitting to occur upon TPA treatment, and that microtubule disassembly does not affect centrosome position once splitting has occurred.

The effect of cytochalasin D is rather remarkable. Not only does it affect the integrity of actin-based networks and, as a consequence, actin-dependent activities such as ruffling and cell spreading, it also has profound effects on microtubule organization, the frequency of centrosome splitting, and the behavior of centrosome/microtubule complexes. Cytochalasin D not only prevents centrosome splitting when applied before TPA treatment, it also induces reversal of splitting (i.e., convergence of separated microtubule asters) in TPA-treated cells. The two centrioles are found in the geometric center of the cell under both conditions, the major difference being the absence of TPA-induced cell spreading if cytochalasin D is applied before TPA. In live cells treated with cytochalasin D after TPA, the aster center undergoes only minimal excursions (less than 3 μm) from its central position; cytoplasmic granules are radially aligned around it and saltate along radial pathways. Thus centrosome splitting depends on an intact actin network, and it is reversed when actin is disrupted in cells with split centrosomes. Both the effects of nocodazole and cytochalasin D are reversible when the compounds are removed; within 45 min, the percentage of cells with split centrosomes returns to a level approaching that of PMNs treated with TPA alone (TABLE 1).

High-voltage electron microscopy of detergent-extracted whole mount preparations graphically illustrates the dramatic changes in cytoskeletal organization induced by cytochalasin D. FIGURE 11a shows a cell treated with TPA alone for 25 minutes. The separated asters are located on opposing sides of the nucleus, and the cell body is pervaded by an intricate network composed largely of actin filaments. Microtubules frequently are gently curved or bent, closely approaching the cell margin or meandering through the cell cortex. The cytochalasin D-treated cell shown in FIGURE 11b has undergone reversal of centrosome splitting. The two centrioles, surrounded by an aster of straight microtubules that point at the cell perimeter at more or less right angles, are located in the cell center. The actin network is literally shredded and absent throughout many portions of the cell body.

The effects of the various experimental treatments on centrosome splitting, microtubule organization, and actin network integrity are diagrammatically summarized in FIGURE 12.

DISCUSSION

We have examined the dynamics of the microtubule system in two different model systems in order to learn more about their possible involvement in directional cell locomotion. The experiments on fish keratocytes provide evidence for directional locomotion of cells and small fragments derived from the anterior lamella in the complete absence of microtubules. The findings on keratocytes differ from those on fibroblasts in culture, where microtubules are believed to play an important role in both the development of an asymmetric locomotory morphology and the subsequent expression of directional motility.[14-17] Their disassembly by colchicine or colcemid leads to a loss of asymmetry and locomotion. By contrast, colcemid or nocodazole-induced microtubule disassembly neither alters the shape of intact keratocytes, nor does it interfere with their ability to locomote in a highly persistent fashion. Thus microtubules are not required for keratocytes to choose and maintain the direction of movement, nor are they essential to organize the cytoplasm for motility. Not only is the force for movement generated and regulated in the lamellar region, but also the cellular control mechanism for directionality is localized in the same cytoplasmic domain. This conclusion is supported by the observation of autonomous, directional migration of small lamellar fragments that consist essentially of a fraction of the "motor" domain. Cell fragments of similar size have previously been generated from fibroblasts,[18] but these "microplasts" do not translocate, but rather display the same stereotypic behavior, such as ruffling or blebbing, over and over again. Large, motile fragments of microtubule and centriole-free lamellipods of leukocytes, on the other hand, are capable of directed locomotion and chemotaxis.[19,20]

Whereas a number of experiments suggest an involvement of microtubules in directional cell movements,[2-6,16,17] it is intriguing to speculate, on the basis of the observations on fish keratocytes, that the role of microtubules lies not in the determination of directional migration, but merely in some supportive function. As mentioned in the INTRODUCTION, previous experiments did not allow us to establish with certainty whether the centrosome/microtubule complex determines cell polarity, or whether it is lined up passively by the activity of some other cellular

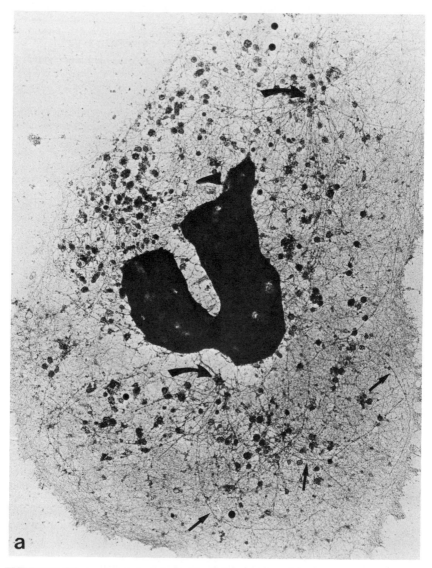

FIGURE 11. Whole mount preparations of detergent-extracted PMNs after treatment with TPA alone (**a**), and after TPA treatment followed by 2 μg/ml cytochalasin D for 30 min. In **a**, the separated centrioles are located on opposing sides of the nucleus (large arrows). Gently curved microtubules closely approach the cell perimeter (arrows). A tight actin network extends across the cell. In **b**, the centriole pair is located in the cell center (double arrow), and microtubules radiate from the pericentriolar area towards the cell boundary (arrows). The actin network is completely disrupted.

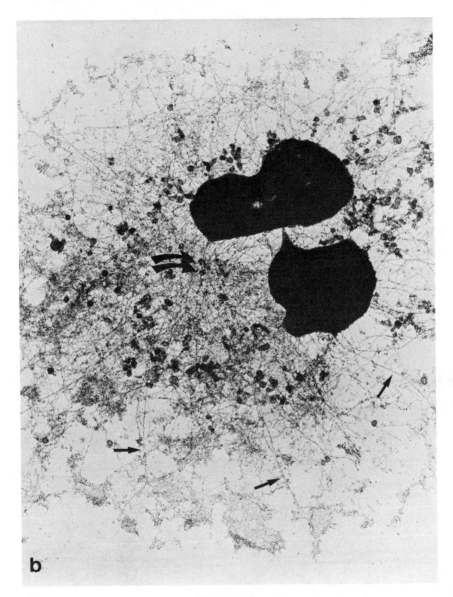

FIGURE 11b.

component. Whereas a correlation between directionality of movement and centrosome position is widely accepted, the cell models used so far did not allow for separation of cause and effect. We feel that the model system used here, TPA-treated human PMNs, may shed some light on this relationship. Even though the findings were obtained on essentially nonpolar, stationary cells, they are of relevance for consider-

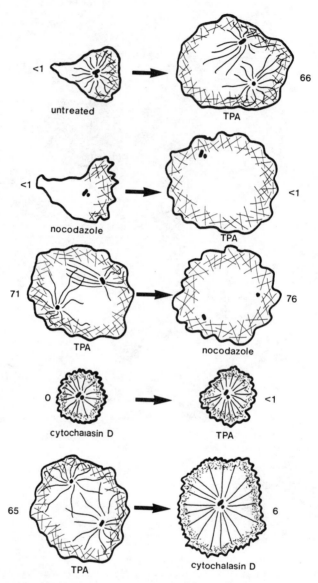

FIGURE 12. Schematic diagram summarizing the effects of the various regimes of inhibitor treatments on the distribution of centrioles (short, fat bars), microtubules (medium lines), and cortical actin (cross-hatched and stippled areas). Numbers are the percentages of cells with split centrosomes found under the various experimental conditions.[22]

ations of the role of microtubules in migratory cells. The results with the different regimes of inhibitor treatments strongly suggest an involvement for an actin-containing cellular component in the positioning and activity of the centrosome/microtubule complex, presumably by way of interaction with the microtubules radiating from the centrosomes. Microtubules are required for splitting to occur, and an actin-containing, possibly cortical network both initiates centrosome splitting and supports centrosome motility later on. If impaired by cytochalasin D, splitting is reversed, and the resulting single aster exhibits little motile activity.

On the basis of these observations, the following scheme to explain centrosome splitting and aster mobility is proposed. Activation of the cell cortex by TPA or chemoattractants[10] induces cell spreading that is likely to exert a force on the microtubules extending into the peripheral filament network, thus pulling the microtubule aster apart and leading to two, occasionally three, independent asters. As long as the cortex remains active, this condition is maintained. It is reversed upon return to a normal migratory morphology[10] or experimental disruption of the actin network (this paper). Thus, we propose that the cell cortex influences the position of the centrosome/microtubule complex. We hypothesize that this interaction, whose molecular basis remains to be determined, also takes place in polar, migratory, cells, where the cell cortex, notably the extending anterior lamellipod, plays a dominant role. The proposed dominance of the cortex, however, does not mean that microtubules are dispensable and do not take part in directional locomotion of many cell types, keratocytes being an exception, perhaps. Rather, the emphasis is shifted from a cell component that sets up polarity to an element that helps reinforce polar organization, set up by the cell cortex. Trinkaus[21] discusses a scenario where microtubules become passively aligned by the cortex, help stabilize cortically determined asymmetry, and thus, in a process of mutual reinforcement, establish cell polarity. In this way the centrosome-aster complex tends to exert a constraining influence and serves to stabilize the bulk of the cell.

ACKNOWLEDGMENTS

We thank M. Kirschner and T. Wieland for gifts of antitubulin and rhodamine-phalloidin, respectively.

REFERENCES

1. VAN BENEDEN, E. 1883. Arch. Biol. **4:** 265–638.
2. ALBRECHT-BUEHLER, G. & A. BUSHNELL. 1979. Exp. Cell. Res. **120:** 111–118.
3. MALECH, H. L., R. K. ROOT & J. I. GALLIN. 1977. J. Cell Biol. **75:** 666–693.
4. KOONCE, M. P., R. A. CLONEY & M. W. BERNS. 1984. J. Cell Biol. **98:** 1999–2010.
5. GOTLIEB, A. I., L. MCBURNIE MAY, L. SUBRAHMANYAN & V. I. KALNINS. 1981. J. Cell Biol. **91:** 589–594.
6. KUPFER, A., D. LOUVARD & S. J. SINGER. 1982. Proc. Natl. Acad. Sci. USA **79:** 2603–2607.
7. ANDERSON, D. C., L. J. WIBLE, B. J. HUGHES, C. W. SMITH & B. R. BRINKLEY. 1982. Cell **31:** 719–729.
8. OLIVER, J. M. & R. D. BERLIN. 1982. Int. Rev. Cytol. **74:** 55–94.
9. SCHLIWA, M., K. B. PRYZWANSKY & G. G. BORISY. 1983. Eur. J. Cell Biol. **32:** 75–85.
10. SCHLIWA, M., K. B. PRYZWANSKY & U. EUTENEUER. 1982. Cell **31:** 705–717.
11. GOODRICH, H. B. 1924. Biol. Bull. **46:** 252–262.

12. BEREITER-HAHN, J., R. STROHMEIER, I. KUNZENBACHER, K. BECK & M. VOTH. 1981. J. Cell Sci. **52:** 289–311.
13. SPIEGELMAN, B. M., M. A. LOPATA & M. W. KIRSCHNER. 1979. Cell **16:** 253–263.
14. GOLDMAN, R. D. 1971. J. Cell Biol. **51:** 752–767.
15. ALBRECHT-BUEHLER, G. 1977. Cell **12:** 333–342.
16. VASILIEV, J. M., & I. M. GELFAND. 1976. Cell Motility **1:** 279–304.
17. GAIL, M. H. & C. BOONE. 1970. Biophys. J. **10:** 980–993.
18. ALBRECHT-BUEHLER, G. 1980. Proc. Natl. Acad. Sci. USA **77:** 6639–6643.
19. KELLER, H. U. & M. BESSIS. 1975. Nature (London) **258:** 723–724.
20. MALAWISTA, S. E. & A. CHEVANCE DE BOISFLEURY. 1982. J. Cell Biol. **95:** 960–973.
21. TRINKAUS, J. P. 1984. Cells Into Organs. The Forces That Shape the Embryo. 2nd ed. Prentice-Hall. Englewood Cliffs.
22. EUTENEUER, U. & M. H. SCHLIWA. 1985. J. Cell Biol. **101:** 96–103.

Role of Microtubules in the Formation of Carotenoid Droplet Aggregate in Goldfish Xanthophores[a]

T. T. TCHEN,[b] ROBERT D. ALLEN,[d] SZE-CHENG J. LO,[c]
THOMAS J. LYNCH,[c] ROBERT E. PALAZZO,[c]
JOHN HAYDEN,[d] GARY R. WALKER,[c]
AND JOHN D. TAYLOR[c]

[b]Department of Chemistry
[c]Department of Biological Sciences
Wayne State University
Detroit, Michigan 48202
and
[d]Department of Biological Sciences
Dartmouth University
Hanover, New Hampshire 03756

INTRODUCTION

Microtubules exist both in cellular appendages such as flagella and cilia and in the main body of cells. In this paper, we shall restrict our discussion to the latter, where microtubules serve a dual role of maintaining, and/or determining, cell shape and participating in motility, particularly intracellular motility. Although the mitotic spindle, containing massive amounts of microtubules and involved in chromosome translocation, has been known for decades, it is relatively recent that microtubules were recognized for their involvement in organelle translocation in a variety of cells (see review by Schliwa[1]). In the case of lower vertebrate pigment cells, both structural and motile roles have been documented. When melanocytes from goldfish were induced in organ cultures by adrenocorticotropin (ACTH), melanocyte stimulating hormone (MSH), or cyclic adenosine monophosphate (cAMP), they primarily assumed a bipolar morphology. If colchicine were present in the medium, however, they assumed an irregular flattened morphology with no dendrites.[2] Further, in many melanophores, there exists in the dendrites a ring of microtubules running parallel to the dendrite and just beneath the plasma membrane (see references 1 and 3). It is therefore clear that microtubules are essential in the formation and/or maintenance of the dendritic shape of these cells. The importance of microtubules in pigment organelle translocation has also been amply demonstrated, particularly in the erythrophores of the squirrel fish and the melanophores of the angelfish, by Porter and Schliwa and their coworkers (see reference 1).

We began our studies on pigment organelle translocation in lower vertebrate chromatophores with the rationale that, because this is a reversible and regulated (hormonal and/or neural) translocation in contrast to organelle translocation in almost

[a]This work was supported by Grants from the NIH (AM13724, T.T. Tchen) and the Michigan Comprehensive Cancer Center (T.J.Lynch).

887

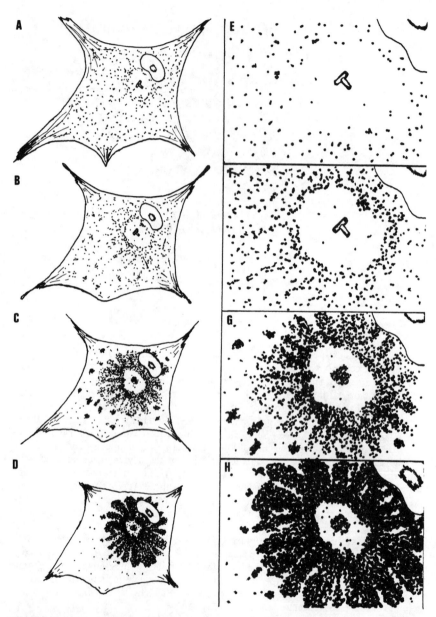

FIGURE 1. Schematic representation of the aggregation of carotenoid droplets. **A–D:**Fully dispersed to fully aggregated states, induced by withdrawal of ACTH. The individual carotenoid droplets are actually much smaller than the size of the dots. **E–H:** Representation of centriolar area from **A–D.** The centrioles (the T-shaped structures) and the carotenoid droplets are both drawn oversize. In the fully dispersed state (**E**), the region around the centriole actually is less densely populated with carotenoid droplets compared to elsewhere in the cytoplasm. As aggregation proceeds, (**F**), the carotenoid droplets migrate to the centriole region but leave a zone of exclusion around the centriole. Further aggregation (**G**) forms a hollow shell of carotenoid droplets surrounding the centriole, which has a few attached carotenoid droplets. All of the carotenoid droplets in the hollow still exhibit oscillatory and/or saltatory motion at this stage.

all other systems (see Bagnara and Hadley[3]), there are more experimental handles for studying the biochemical basis of organelle translocation in the cell. For example, because cAMP causes pigment organelle dispersion in most lower vertebrate pigment cells, (see reference 4) including the goldfish xanthophores,[5] one can ask whether there is a cAMP-dependent protein kinase (cAMP-dPK) acting on the organelles and/or the cytoskeleton. The choice for the xanthophores of the common goldfish was based on the practical consideration that, in order to carry out biochemical studies, one must be able to isolate pigment cells in sufficient purity and quantity. In this respect, the goldfish xanthophores offer three advantages: they are hardy to conditions of cell dissociation from the skin, they are abundant in the skin (approximately 1%), and they are from a cheap and readily available fish. In practice, we can isolate in one day, using 100 fish, 1–2 ml packed hormone-responsive xanthophores of approximately 75% purity. Although this is a pitifully small amount to most biochemists, it has allowed judiciously chosen biochemical studies with correlated microscopic (light, fluorescent, and electron) studies. In this paper, we shall summarize the results leading to the conclusion that the microtubules play an essential role in the formation of the aggregate of the pigment organelles, carotenoid droplets. Almost all of the data are as yet unpublished except in the form of Ph.D. dissertations.[6–8]

RESULTS

In xanthophores with dispersed carotenoid droplets, these droplets are seen by AVEC-DIC microscopy to be in constant oscillatory motion, whereas in xanthophores with aggregated pigment, they are located around the centriole region and exhibit no motion. When the pigment in a cell with dispersed pigment is induced to aggregate by the removal of ACTH, the carotenoid droplets begin to form a hollow shell around the centriole as well as clumps in the more peripheral areas of the cell. Within the hollow of the shell, there is essentially no carotenoid droplets. The carotenoid droplets in the shell, which actually form distinct lobes, are, however, not static. Instead, they undergo continuous oscillatory and/or saltatory motions, and individual droplets move into the hollow area from time to time. Some of these adhere to a structure, presumably the centriole, and eventually build up a small aggregate of a few dozen droplets in the middle of the hollow. Elsewhere in the cell, as more and more carotenoid droplets have moved to the large aggregate, smaller clumps of carotenoid droplets become more evident. In some cells, clumps are seen abutting the nucleus on the opposite side of the centriole surrounded by the large aggregate. Finally, the carotenoid droplets in the aggregate become static. Indeed, the whole cell appears to be frozen with essentially no noticeable motion of any visible organelles. The above summarizes the video studies that unfortunately cannot be reproduced here. Instead, various phases of aggregation are illustrated in FIGURE 1.

As the aggregation process involves self-association of carotenoid droplets and the formation of the main aggregate surrounding the centriole in the form of a hollow shell with a small aggregate in the center, it is pertinent to ask what holds or attracts these

---

FIGURE 1. *(Continued)* Clumps of carotenoid droplets are now clearly seen in the more peripheral portion of the cytoplasm. Final stage of aggregation (**H**): The hollow shell (often with lobed structure) now consists of tightly packed immobile carotenoid droplets. The central carotenoid droplet aggregate on the centriole is larger than in earlier stages. Occasional clumps seem to be blocked by the nucleus from joining the main aggregate (represented by clumps above nucleus in **D**). These figures depict focus planes across the centriole. By moving the focal plane up and down, it was established that the rings of carotenoid droplets in **C, D, G,** and **H** are actually cross-sections of a hollow sphere. Notice also the rounding up of the cell as a whole.

carotenoid droplets. This brings us to the question, what is the structure of the cytoskeleton and how does it relate to the carotenoid droplet aggregate? FIGURE 2 shows the ultrastructure of the cytoskeleton and the patterns of distribution of the major cytoskeletal elements in xanthophores with aggregated carotenoid droplets. It is clear that the area occupied by the large carotenoid droplet aggregate contains only tubulin but is essentially devoid of actin and intermediate filaments. It thus seems reasonable to assume that there is an association of carotenoid droplets with the microtubules, to form the shell of carotenoid droplets, and with the tubulin of the centrioles, to form the small central aggregate. The empty space between them may correspond to the region between the centrioles and the amorphous materials of the microtubule organizing center from which the microtubules radiate.

If the large aggregate depends on carotenoid droplet-microtubule interaction, it might be expected that agents capable of disrupting microtubules would affect its structure. Indeed, preliminary studies indicate that treatment with colchicine for even less than one hour at room temperature causes the compact large aggregate to assume a less compact and irregular shape in many xanthophores.

The nature of the carotenoid droplet-microtubule interaction is not clearly understood. Preliminary results suggest, however, that a carotenoid droplet protein, p57, (protein of 57 kD molecular mass) may be the agent serving both the association of carotenoid droplets to each other and to microtubules. Previously, we have shown that this protein is the main target of cAMP-dependent phosphorylation and that it can be phosphorylated at multiple sites, up to 6 or 7 phosphates per molecule of pp57 (phosphorylated form of p57).[6,9] Recent studies showed that the phosphorylation of p57 and dephosphorylation of pp57 precedes, respectively, the dispersion and aggregation of carotenoid droplets.[14] Also, the minimal concentration of cAMP required for carotenoid droplet dispersion is the same as that for p57 phosphorylation.[14] A VEC-DIC microscopy showed that within two minutes of addition of cAMP ($10^{-3}M$) to xanthophores with aggregated pigment, the carotenoid droplets have changed from the "frozen" state and begin to disperse.[8] These results suggest that self-association of carotenoid droplets requires that this protein be in the unphosphorylated state. Indications that carotenoid droplet-microtubule association also requires that this protein be in the unphosphorylated state came from biochemical studies on the Triton-insoluble cytoskeletons of xanthophores with dispersed or aggregated pigment. When these cytoskeletons from xanthophores were incubated with γ-[^{32}P]ATP and subjected to 2-D fluorographic analysis,[10,11] essentially the same pattern of phosphorylated proteins was obtained regardless of the initial state of pigment distribution. When cytoskeletons of xanthophores with aggregated pigment were incubated with γ-[^{32}P]ATP and cAMP, however, several additional proteins were phosphorylated, one of which appears by 2-D fluorography as a streak. The location of this streak on the 2-D fluorogram closely resembles that of pp57 of carotenoid droplets (FIGURE 3).

We therefore isolated labeled pp57 from both such cytoskeleton preparations and from carotenoid droplets by preparative 1D-electrophoresis[12] and carried out peptide mapping according to Cleveland.[13] The results are as follows. First, both pp57 samples are relatively resistant to chymotrypsin, and only the highest concentration gave significant preteolysis. The patterns of labeled peptides obtained from the two samples are identical. Second, both pp57 samples are quite sensitive to papain digestion, giving rise to numerous labeled peptides of intermediate molecular weight. Again, the patterns are identical with both samples. Finally, with V8 protease digestion, the two samples gave identical patterns of labeled high molecular weight fragments.

Differences, however, were observed in the patterns of labeled low molecular weight fragments.[8] We conclude that these two pp57s are indeed the same protein with perhaps differences in one or two sites of phosphorylation. In preliminary experiments,

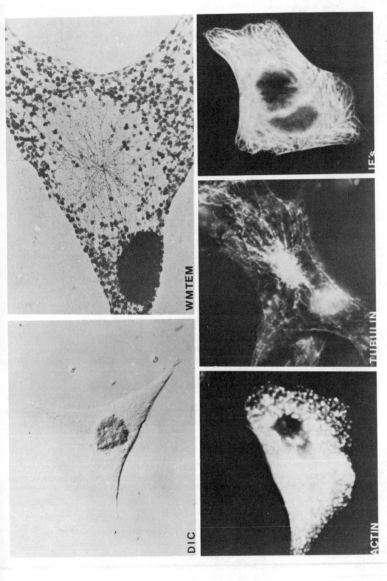

FIGURE 2. Cytoskeleton of xanthophore with aggregated pigment. DIC: light micrograph; WMTEM: whole mount transmission electron micrograph of Triton-insoluble cytoskeleton; ACTIN, TUBULIN, and IF's: immunofluorescence using antiactin, antitubulin, and antiintermediate filament protein respectively. The micrographs demonstrate clearly that the space occupied by the carotenoid droplet aggregate is rich in microtubules, but essentially devoid of actin and intermediate filaments (the two round dark areas). The punctate fluorescence seen with antiactin is characteristic of the xanthophores as normal stress-fiber patterns are seen in nonpigment cells on the same cover slip. These fluorescent structures have been shown to correspond exactly to normal stress-fiber patterns are seen in nonpigment cells on the same cover slip. These fluorescent structures have been shown to correspond exactly to phase-dense particles that in turn have been shown to correspond exactly to the electron-opaque particles seen in WMTEM. It should be pointed out that the small number of these particles seen in WMTEM near the centriole may not be within, but are above and/or below the carotenoid droplet aggregate.

it was found that if the xanthophores were treated with ACTH for only one minute before Triton extraction, the resulting cytoskeleton, upon incubation with cAMP and γ-[^{32}P]ATP, yielded only 10% of pp57 as obtained with cytoskeleton from xanthophores with aggregated pigment.[8] Our interpretation is as follows. In the aggregate, carotenoid droplets associate with each other and with microtubules by way of p57 (unphosphorylated). When the carotenoid droplets and most of their proteins are removed by Triton extraction, some of the p57 that served to bind the carotenoid

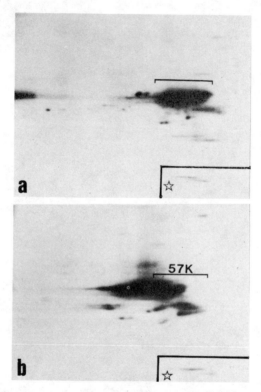

FIGURE 3. Two-dimensional fluorograms of ^{32}P-labeled pp57 from ^{32}P-labeled whole cell and from Triton-insoluble cytoskeletons treated with cAMP and γ-[^{32}P]ATP. Corresponding areas (including the cluster of major cytoskeletal proteins) from 2-D fluorograms are shown in **3a** (^{32}P-labeled whole cell) and **3b** (from cytoskeleton). The dark streaks are due to pp57.[6,8] Underexposed autoradiograms show only pp57 (star insets) in the form of two streaks with molecular mass approximately 57 D and isoelectric points 4.5–4.9.

droplets to the microtubules are left behind with the cytoskeletons. Subsequent treatment of such cytoskeleton with cAMP activates bound cAMP-dPK, which phosphorylates this residual p57. Because p57 has multiple phosphorylation sites, one or two of these sites may be masked when the p57 is bound to the microtubules. Consequently, when ^{32}P-labeled pp57 were digested to small polypeptides by an appropriate enzyme, differences between pp57 from cytoskeletons and from carotenoid droplets were revealed. When such xanthophores are treated with ACTH, the hormone's second messenger activates cAMP-dPK, which rapidly phosphorylates p57 and causes dissociation of carotenoid droplets from each other and from the cytoskeleton.

DISCUSSION

We have presented strong circumstantial evidence that the formation of the carotenoid droplet aggregate requires their self-association and association with microtubules and is suggestive of evidence that this latter association occurs between microtubules and a carotenoid droplet protein (p57) in the unphosphorylated state. Unpublished preliminary results, however, suggest that the microtubules are not involved, at least not directly, in the translocation of the carotenoid droplets. Thus, the role of microtubules in the goldfish xanthophores is apparently quite different from that in other pigment cells, as reviewed by Schliwa.[1]

In a broader perspective, one must bear in mind that there is enormous diversity among organisms and among different cell types in any higher organism. This diversity includes cell shape, composition of the cytoskeleton, and the type of cellular and intracellular motility. With respect to the function(s) of the microtubules in organelle translocation, there are insufficient data at present to draw any general conclusions. The complexity of organelle translocation has been thoughtfully reviewed recently by Schliwa,[1] and it is clear that different translocation processes depend on different cytoskeletal components. In the case of the role of microtubules in determining cell shape, it is clear that microtubules play an important role in dendritic cells, but its role in other cells, such as liver cells, is essentially unknown. The function(s) of the microtubules thus remains an important, fertile, and underexplored area of research.

REFERENCES

1. SCHLIWA, M. 1984. Mechanisms of intracellular organelle transport. *In* Cell and Muscle Motility. J. W. Shay, Ed.: Vol. 5: 1–82. Plenum Publishing Corp. New York.
2. FOOTE, C. D. & T. T. TCHEN. 1967. Studies on the mechanism of hormonal induction of the melanoblast-melanocyte transformation in organ culture. Exp. Cell. Res. **47:** 596–605.
3. BAGNARA, J. T. & M. E. HADLEY. 1973. Chromatophores and Color Change. Prentice Hall, Inc. Englewood Cliffs, N.J.
4. NOVALES, R. R. 1983. Cellular aspects of hormonally controlled pigment translocations within chromatophores of poikilothermic vertebrates. Am. Zool. **23:** 559–568.
5. WINCHESTER, J. D., F. NGO, T. T. TCHEN & J. D. TAYLOR. 1976. Hormone-induced dispersion or aggregation of carotenoid-containing smooth endoplasmic reticulum in cultured xanthophores from the goldfish, *Carassius auratus* L. Endocr. Res. Comm. **3:** 335–342.
6. LYNCH, T. J. 1982. Protein phosphorylation during pigment dispersion in goldfish xanthophores. Dissertation. Wayne State University. Detroit, Mich.
7. WALKER, G. R. 1984. An immunological study of the goldfish *Carassius auratus* L. chromatophore cytoskeleton: The identification of two intermediate filament peptides. Dissertation. Wayne State University. Detroit, Mich.
8. PALAZZO, R. E. 1984. Organelle translocation in xanthophores of *Carassius auratus* L.: Corrective ultrastructure and biochemistry. Dissertation. Wayne State University. Detroit, Mich.
9. LYNCH, T. J., S. J. LO, J. D. TAYLOR & T. T. TCHEN. 1981. Characterization of and hormonal effects on subcellular fractions from xanthophores of the goldfish *Carassius auratus* L. Biochem. Biophysics Res. Commun. **102:** 127–134.

10. O'FARREL, P. H. 1975. High resolution two-dimensional electrophoresis of proteins. J. Biol. Chem. **254:** 40007–40021.
11. LASKEY, R. A. 1980. The use of intensifying screens or organic scintillators for visualizing radioactive molecules resolved by gel electrophoresis. *In* Methods in Enzymology. L. Grossman & K. Moldave, Eds.: Vol. 65:. 363–371.
12. LAEMMLI, U. K. 1970. Cleavage of structural proteins during the assembly of the head of bacteriophage T4. Nature (London) **227:** 680–685.
13. CLEVELAND, D. W., S. G. FISCHER, M. W. KIRSCHNER & U. K. LAEMMLI. 1977. Peptide mapping by limited proteolysis in sodium dodecyl sulfate and analysis by gel electrophoresis. J. Biol. Chem. **254:** 12670–12678.
14. LYNCH, T. J., J. D. TAYLOR & T. T. TCHEN. 1986. Regulation of pigment organelle translocation. I. Phosphorylation of the organelle-associated protein, p 57. J. Biol. Chem. In press.

The Cytomatrix Regulates "Resolute" Transport in Erythrophores

MARK E. STEARNS,[a,c] LESTER I. BINDER,[b,d]
AND MIN WANG[a,c]

[a]Department of Anatomy
Georgetown University
Washington, D.C. 20007
and
[b]Department of Biology
University of Virginia
Charlottesville, Virginia 22903

INTRODUCTION

Extensive HVEM studies of whole mount cultured cells have revealed that microtubules, intermediate filaments, and microfilaments constitute the principle cytoskeletal elements. Using both gluteraldehyde and fast freezing techniques, Porter and colleagues[1-4] have demonstrated that the above types of filaments are cross-linked by, and suspended in, a network of fine filaments termed the microtrabecular lattice. The three-dimensional lattice appeared to form a major component of the cytomatrix and was proposed to play a major role in determining cell shape and mediating intracellular transport.[1,4,5] With regard to transport, the cytomatrix was thought to be dynamic and exhibit dramatic changes in its organization in response to specific cellular signals.[2,5] For example, minute increases in intracellular calcium levels increased the thickness of the cytomatrix filaments that cross-linked organelles with microtubules in both chromatophores[2,3] and in cultured neuroblastoma cells.[4] These physiological increases in calcium ($>10 - {}^6M$) produced pigment aggregation in chromatophores[4] and arrested organelle motion in neurons.[6] Predictably, reduced calcium levels ($<10 - {}^6M$) had the opposing effect and restored transport.

These morphological studies presented intriguing problems as to possible mechanisms by which the lattice might mediate transport. For example: What are the factors and proteins regulating motion? Do different components of the cytomatrix have specific functions? And how is the polarity of transport controlled? In attempting to answer such questions we have focused our attention on microtubules and the microtubule-associated lattice.

Microtubules compose one of the principal cytoskeletal elements in most nonmuscle cell systems, and the majority of visible transport appears to be microtubule directed. Historically, thin section studies of microtubules have shown they are coated with "wispy" filaments[7] that are occasionally observed to cross-link microtubules with the surfaces of organelles.[8] Presumably, the "wispy" filaments compose a major part of

[c]Present address: Department of Pharmacology, Fox Chase Cancer Center, 7701 Burholme Avenue, Philadelphia, Pennsylvania 19111.
[d]Present address: Department of Cell Biology and Anatomy, University of Birmingham, Birmingham, Alabama.

895

the three-dimensional latticework. The HVEM studies of nervous tissue[9] and neuro-blastoma[6] have clearly shown that microtubules were coated with a lattice of cytomatrix filaments. The cytomatrix formed a cross-linking interface between the tubules and the motile organelles.

HVEM studies of squirrel fish erythrophores have unveiled several interesting properties of the microtubule-associated cytomatrix. In erythrophores, microtubules and their associated lattice were the main components that mediated the transport of a single organelle, the pigment granule.[1,2,5,10,11] Pigment granules appeared to be suspended in the cross-linking cytomatrix and aligned in rows along radially arrayed microtubules that emanated from a centrally located cell center. In live cells it was observed that in response to catecholamines and increased intracellular calcium levels, the pigment rapidly aggregated at rates of 10 to 20 μM/second. McNiven, Wang, and Porter[12] have reported that following pigment migration to the cell center, individual granules return to the approximate vicinity from which they originated prior to the onset of aggregation. How the cytomatrix accomplished this remarkable process was not known! One possible clue to the mechanism of action of the cytomatrix has been demonstrated by HVEM studies of freeze-substituted erythrophores. When pigment material was removed during osmium fixation in methoxyethanol, the entire cytomatrix network was revealed for comparison in dispersed and aggregated cells. The authors discovered that part of the cytomatrix, termed the α cytomatrix, comigrated with the aggregating pigment, and part, the β cytomatrix, remained associated with the microtubules.[3] The most reasonable interpretation of these results was that the α cytomatrix aggregated and expanded by interacting with the microtubules and/or the β cytomatrix. They proposed that the cytomatrix functioned as a unit structure that exerted positional and directional control over the motion of granules. It remains to be discovered if one or both of these cytomatrices contain structural and molecular information for regulating the spatial distribution of granules.

In this paper, we report results from morphological, physiological, and immunocytochemical studies of pigment transport in erythrophores. We have recently published a review on chromatophores[4] that should provide the reader with a thorough background of the literature pertaining to this article.

RESULTS

In Situ *Studies of Nocodazole Treated Erythrophores*

Light microscopy studies showed that erythrophores treated with 5 μM nocodazole for 1 hr at 4°C initially were somewhat damaged, as large clear areas appeared in the cytoplasm (FIGURE 1). All intracellular motility was discontinued, and the pigment granules were held in the dispersed state. Saltatory motion was also discontinued. Upon warming to 32 to 37°C in fresh media the pigment still remained dispersed, and spontaneous pigment pulsation was not initiated. Erythrophores recovered a normal ability to transport their pigment by about 30 min incubation. By comparison, erythrophores exposed to low temperatures (4°C for 1 hr) and warmed to 37°C immediately exhibited both pigment pulsation and saltatory motion after several min at 32°C.

If nocodazole-treated cells (5 μM for 1 hr at 4°C) were incubated in TPS solutions containing 0.4 μM nocodazole at 32°C, erythrophores recovered an ability to spontaneously pulse their pigment, but saltatory motion was not reinitiated (FIGURE 1). Pigment aggregation, however, was stimulated (FIGURE 2) at rates about one-third normal speed by $10^{-7} M$ epinephrine, 10^{-5} MCa^{+2}, 0.08 M K^+, and 660 mOsm sucrose

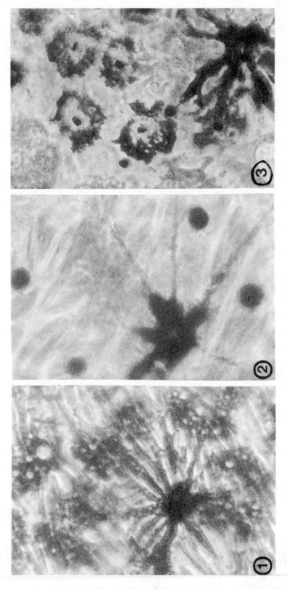

FIGURES 1–3. Pigment granule aggregation and dispersion in nocodazole-treated erythrophores *in situ* (5 μM for 1 hr at 4°C). Erythrophores were the principle chromatophore visible, but a melanophore and xanthophores were also visible. FIGURE 1 shows erythrophores after 5 min at 32°C in media containing 0.4 μM nocodazole. FIGURE 2 shows erythrophores where aggregation was induced by 10^{-7} M epinephrine. FIGURE 3 shows redispersion stimulated with caffeine. Magnification × 1800.

solutions. Here, aggregation occurred in a progressive, resolute fashion that involved an initial patching of the pigment into peripherally located aggregates (peripheral pigment patching) followed by the aggregation of pigment to the cell center. During aggregation, some pulsation of the granules occurred in the central region over about one-third of the cell's diameter.

The aggregation event was reversed with caffeine, reduced calcium levels ($10^{-6}M$), and 0.2 M Na$^+$ solutions (FIGURE 3). The dispersion involved a uniform redistribution of the pigment so that pigment granules were uniformly redispersed throughout the cell. The motion was resolute in nature, and in the presence of nocodazole (0.4 μM) saltatory motion was not reinitiated. Usually it was possible to induce several cycles of pigment aggregation and dispersion without the aid of microtubules, although eventually about 20 percent of the granules became immotile and remained dispersed.

Electron microscopy studies with serial thin section and serial thick section stereo HVEM techniques have revealed that the microtubules were completely removed from the nocodazole-treated cells (compare FIGURES 4, 5). The microtubules were not reassembled following prolonged incubations in the presence of 0.4 μM nocodazole or following pigment aggregation and dispersion. Similar results were observed in studies of whole-mount cultured erythrophores treated with nocodazole (1 μM for 1 hr at 4°C) where the microtubules were completely removed (compare FIGURES 6, 7). A three-dimensional cross-linking cytomatrix consisting of 3 to 9 nm cytomatrix filaments remained (arrowheads), which interconnected the pigment granules (FIGURE 7). In contrast to erythrophores *in situ*, the cultured cells did not aggregate their pigment in response to epinephrine, 660 mOsm sucrose solutions or $10^{-5}M$ Ca^{+2}, but instead they exhibited a partial aggregation or peripheral pigment patching (FIGURE 7). HVEM studies of cultured cells exposed to $10^{-5}M$ Ca^{+2} revealed that the patched pigment granules were extensively cross-linked with 5 to 15 nm filaments (arrowheads, FIGURE 7).

Digitonin Permeabilized Cells In Situ

Erythrophores permeabilized on scales with 0.01% digitonin still continued saltatory motion and retained an ability to aggregate and disperse their pigment. We are currently in the process of investigating the physiological requirements for pigment saltation, and pigment aggregation and dispersion, in nocodazole-treated cells *in situ*. These studies should demonstrate which physiological factors serve to regulate the cytomatrix-dependent activities. In the meantime we have continued investigating the process of resolute motion in intact cultured erythrophores.

Perfusion Studies of Digitonin Permeabilized Cells In Vitro

Our initial attempts to characterize the role of the cytomatrix proteins in pigment transport have involved partially lysing cultured erythrophores with low levels of digitonin (0.001%) in a microtubule stabilizing buffer containing 0.1M piperazine-diethanesulfonic acid (PIPES), 2 mM ethylene glycol bis(β-aminoethyl ether)-N,N,N',N'-tetraacetic acid (EGTA), 2 mM Mg^{2+}, 1.4% PEG 6000, pH 7.2.[13] Small holes of about 50 nm diameter were opened in the plasma membrane, yet the cells were still motile (FIGURE 8). These cell models have permitted the rapid perfusion of ions, small molecules, proteins, and drugs. The main benefit of the approach was that it

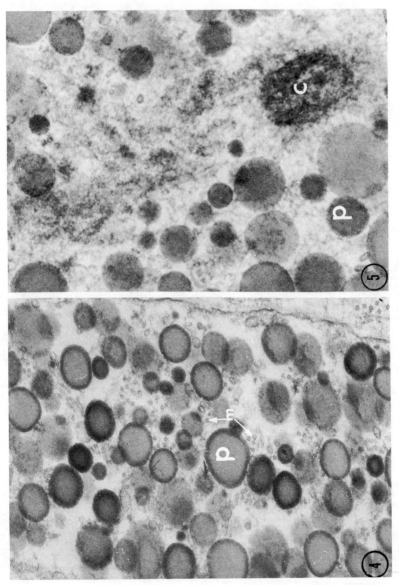

FIGURES 4, 5. The figures compare HVEM cross section images of intact and nocodazole-(5 μ*M* for 1 hr at 4°C) treated erythrophores *in situ.* Microtubules were visible in FIGURE 4, but only granular material remained in FIGURE 5. P = pigment; C = centriole; m = microtubule. Magnification × 53,000.

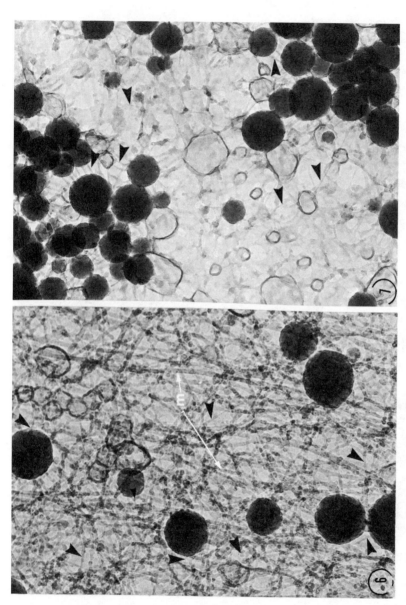

FIGURES 6, 7. The figures compare whole-mounted images of parts of intact and nocodazole-treated (1 μM for 30 min) erythrophores cultured on formvar, carbon-coated gold grid. FIGURE 6 demonstrates that pigment granules (P) and microtubules (m) were cross-linked by cytomatrix filaments (arrowheads). FIGURE 7 shows that once the microtubules were completely removed, the pigment granules formed "PPP," and their surfaces were cross-linked by numerous 5 to 15 nM filaments (arrowheads). Magnification × 45,000.

enabled direct and rapid experimental manipulation of the pigment granules and associated cytomatrix structures.

Studies showed that exposure of the cultured erythrophores to nocodazole (1 μM for 1 hr at 4 °C) either prior to or following permeabilization resulted in a complete inhibition of all observable transport. The disassembly of the microtubules was the only recordable change (see FIGURE 7). In comparison, antimicrofilament agents, 0.1% DNase I and 10 μM cytochalasin B and D had no effect on transport or microtubule organization.

The data reported here summarizes our investigations of the effects of vanadate (VO_3^{+5}) and erythro-hydroxynonyl-adenosine, EHNA, on pigment pulsation in the

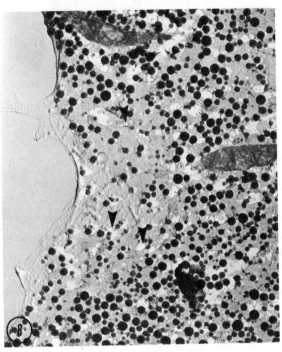

FIGURE 8. Part of a whole-mounted digitonin permeabilized erythrophore (0.001%) lightly coated with Pt/Pd to reveal the small holes (50 nM arrowheads) in the plasma membrane.

digitonin permeabilized cell models. In cultured erythrophores, pigment aggregation occurred in solutions containing epinephrine ($10^{-7}M$), or $10^{-5}M$ Ca^{+2}. The process was inhibited by 50 and 100 μM vanadate and 100 μM EHNA but 20 μM vanadate and 20 to 50 μM EHNA did not prevent aggregation. Cells recovered from the vanadate and EHNA effects with washes of fresh solutions. Alternatively, if cells in the aggregated state were permeabilized with digitonin and asked to disperse their pigment, dispersion was induced with caffeine or cyclic adenosine monophosphate (cAMP). The response was inhibited with 50 μM and 100 μM VO_3^{+5} and 20 μM EHNA. Again, the effects of the drugs were reversed with washes of fresh media, and cells dispersed their pigment. Dispersed cells reaggregated their pigment in the presence of epinephrine or $10^{-5}M$ Ca^{+2}. Following this aggregation step, the cells usually became inactive, and redispersion did not occur. We have yet to carry out the above types of studies on erythrophores *in situ* or more importantly, on nocodazole-treated erythrophores *in situ* (see below).

In several studies, erythrophores were co-cultured with ciliated epithelial cells isolated from rat epididymis. Here we found that dynein ATPase inhibitors used at concentrations of 50 μM vanadate and 100 μM EHNA inhibited both ciliary beating and pigment granule transport (aggregation and dispersion) in erythrophores. Saltatory motion and resolute transport were completely inhibited. Washing out the reagents with fresh media resulted in the recovery of transport processes and ciliary beating.

We have examined the importance of nucleotides, calcium, and calmodulin in transport. Our results showed exogenously supplied adenosine triphosphate (ATP) promoted pigment dispersion but was not required for aggregation to occur. Other nucleotides such as adenosine diphosphate (ADP), adenylyl-imidodiphosphate (AMP-PNP), guanosine triphosphate (GTP), and uridine triphosphate (UTP) did not stimulate aggregation or dispersion at 0.5 to 1.0 mM levels. If cells were pretreated for several hours with metabolic inhibitors such as $2 \times 10^{-5}M$ dinitrophenol (DNP) and 0.01% NaN$_3$ and then permeabilized with 0.001% digitonin, they refused to disperse their pigment in the absence of exogenously supplied ATP. Again, other nucleotides did not support dispersion in place of ATP. Aggregation, on the other hand, readily occurred without the addition of ATP or other nucleotides. The data suggested that dispersion was the energy-consuming event. Both stages required temperatures above 25°C and occurred at rates comparable to intact cells. Studies with calmodulin inhibitors, trifluoperazine, showed that ATP-driven dispersion was not stopped with 25 μM TFP and aggregation was inhibited with TFP.

Monoclonal Antibody Studies in Digitonin-Permeabilized Cells

Dispersed Cells

Immunofluorescent studies showed that MAP-2 was diffusely distributed in the cytoplasm of dispersed erythrophores fixed with 2.9% paraformaldehyde, 0.25% gluteraldehyde, and 0.5% triton X-100, pH 7.2 (FIGURE 9). Control studies showed that the diffuse staining was not produced by aldehyde fixation but indeed represents MAP-2 that is normally not bound to the microtubule surface. If dispersed cells were lysed with 0.01% digitonin and/or fixed with −20°C MeOH for immunostaining, MAP-2 was found coating the nucleus, the centrally located centrosomal complex and the surfaces of microtubules. The diffuse staining MAP-2 component observed in aldehyde fixed cells was removed. The same labeling patterns were obtained in cells lysed with digitonin solutions containing 1 mM ATP and/or 10 μM Ca^{+2} (FIGURES 9 and 10). Similarly, in cells exposed to 1 μM taxol for 1 hr, the antibody labeling patterns showed that the microtubules and centrosomal complex were heavily coated with MAP-2. The results suggested that MAP-2 forms part of a diffusely organized cytomatrix and that MAP-2 also can bind microtubules *in vivo*.

Aggregated Cells

Using antibodies raised against three different MAP-2 epitopes, we have found that MAP-2 antibody preferentially labeled a cytomatrix component associated with aggregating pigment granules. During pigment aggregation, the MAP-2 comigrated with the pigment to the cell center (FIGURE 11). FIGURE 12 shows the corresponding

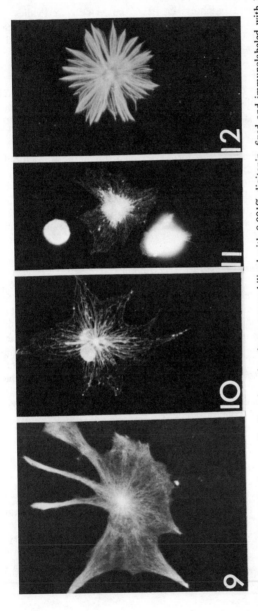

FIGURES 9–12. Immunofluorescent images of cultured erythrophores permeabilized with 0.001% digitonin, fixed and immunolabeled with MAP-2 (FIGURES 9–11) and tubulin (FIGURE 12) monoclonal antibodies that were raised against Chinese Hamster brain microtubule proteins. The erythrophores were fixed in the dispersed state (FIGURES 9, 10) or the aggregated state (FIGURES 11, 12). FIGURE 9 shows that aldehyde fixation in the presence of 0.5% triton X-100 preserved more of the diffusely distributed antigenic material, then was retained with direct −20 °C MeOH fixation (FIGURE 10). In FIGURE 10 only the MAP-2 bound to the microtubules: the centrosomal complex and the nucleus was preserved. FIGURE 11 and 12 show that MAP-2 localized at the cell center, but tubulin remained uniformily arranged following aggregation. Compare FIGURES 9 and 11. Magnification: FIGURE 9 × 2200; FIGURE 10, × 1800; FIGURE 11, × 1500; FIGURE 12, × 1700.

microtubule pattern in an aggregated cell. With pigment granule dispersion, MAP-2 redispersed uniformly from the cell center. If the microtubules were removed with nocodazole (1 μM for 1 hr) and cells exposed to solutions (e.g. 10^{-5} M Ca+2) that stimulate peripheral pigment patching, or PPP, the MAP-2 antibodies labeled a component that associated with the PPPs. Recently, immunoelectron microscopy studies have indicated that MAP-2 binds to the pigment surface at the initial phase of pigment coagulation (data not shown). Thus MAP-2 may form an important cross-link between the granule surfaces and other cytomatrix components. (See the recent review of Stearns[14] and a recent paper by Stearns and Binder.[15])

In contrast to MAP-2, we have found that MAP-1[10] and tau monoclonal antibodies (reference 16 and unpublished data) consistently labeled the microtubules and centrosomal complex of erythrophores regardless of the state of pigment distribution, the composition of solutions, the composition of the lysis media ($+/-$ATP, Ca^{2+}), or the fixation methods used in preparing cells for immunofluorescent studies. These data have suggested that MAP-1 and tau form part of a nonmotile microtubule-bound cytomatrix component. Presumably, MAP-1 and tau have a distinct function from MAP-2. We anticipate that microinjection of MAP antibodies should further help identify the involvement of distinct MAP molecules in cytomatrix-regulated transport events. We exercise caution in our interpretation of the tau and MAP-1 studies because the antibodies stained intermediate filaments and microtubules in epithelial cells (unpublished data).

Monoclonal Antibodies against Erythrophore Cytomatrix Proteins

We have recently developed methods for purifying erythrophores using discontinuous sucrose gradient centrifugation techniques. Taxol-stabilized microtubules and their associated microtubule-associated proteins (MAPs) were subsequently isolated from the purified erythrophores. Silver-stained gels showed that the microtubule pellets contained tubulin plus a wide range of proteins including several high molecular weight proteins of about 270–360 kD. The MAP component was separated from the taxol microtubules with 0.5 M KCl and 0.1% triton X-100, and the tubules were pelleted with centrifugation at 39,000 g for 30 minutes. The supernatant extract was denatured with sodium dodecyl sulfate (SDS) and injected subcutaneously into BALB/c mice. So far we have raised seven monoclonal antibodies that secrete antibodies against the cytomatrix material of erythrophores. The antibodies have been classified into two types based on whether the corresponding antigen comigrates with pigment granules during pulsation or remains permanently associated with a nonmotile microtubule-linked, cytomatrix component. The former were termed pigment-associated transporters (PATs) and the latter erythrophore-microtubule-associated proteins (EMAPs). We have identified 5 PAT antibodies and 2 EMAP antibodies. Initial characterizations of these proteins indicated we were working with antibodies that specifically stained the cytomatrix of erythrophores and did not stain the cytomatrix in melanophores or xanthophores (in preparation). Preliminary results showed that these antibodies were absorbed with a MAPs-extract that coisolated with, and was salt extracted from, the taxol microtubules of erythrophores. Immunoelectron microscopic analysis should locate the position and structural properties of these proteins during pigment transport.

DISCUSSION

We believe that microtubule-dependent transport involves interactions of organelle(s) with a three-dimensional cytomatrix largely found coating the microtubule surfaces.[4,6] In erythrophores, the two main types of motion supported by the cytomatrix were saltatory transport and resolute transport. In dispersed erythrophores, the saltatory motion of organelles along the microtubule surface apparently was powered by cross-linking filaments that extended from the tubule surface. With resolute motion, we believe that the microtubules guide transport produced by a loosely associated cytomatrix component that specifically cross-links the pigment granule surfaces.

Although other cell systems exhibit both types of motion,[3] erythrophores provided an elegant system where the two types of transport were distinguishable on the basis of there dependence on the presence or absence of microtubules. The earliest studies of chromatophores showed that colchicine treatment stopped, or at least slowed, the rate of pigment transport in chromatophores.[11] More recent studies of cultured melanophores[5] and erythrophores[10] exposed to antimitotic drugs (nocodazole and colchicine for 60 min at 4°C) have clearly shown that with removal of the microtubules, saltatory motion was discontinued. Resolute motion was minimized so that dispersed cells were merely competent to form "peripheral pigment patches" when stimulated with epinephrine. These results suggested that a microtubule-independent contractile component was responsible at least for the resolute transport of pigment granules.[5,10] The data further indicated that microtubules were required for the cytomatrix to produce pigment granule aggregation to the cell center. We have observed a similar phenomena in cultured erythrophores that were exposed to nocodazole. Whole-mount electron microscopic images indicated that pigment granules were definitely suspended in a cross-linking network of filaments.[3]

In contrast to the above results we have recently found that erythrophores exposed to nocodazole ($5 \mu M$ for 1 hr at 4°C) *in situ* responded to catecholamines and methylxanthines. The cells transported their pigment to and from the cell center at rates about one-third normal speed. Physiological studies showed that shifts in Ca^{+2} levels could regulate pigment translocations. In erythrophores, the regulatory role of calcium and calmodulin was of potential significance, because shifts of calcium levels were known to produce changes in pigment distribution.[2] Our work here with TFP indicated calmodulin might regulate the Ca^{+2}-triggered motility events. Increases of calcium (10^{-5}) inhibited saltatory transport, whereas reduced calcium levels ($10^{-7}M$) supported motion. Calcium-dependent calmodulin functions might therefore influence MAP-2–dependent transport processes. For example, epinephrine induced increases in calcium levels could stimulate calmodulin, and calmodulin in turn might cause MAP-2 dissociation from tubules. The end result would be increased MAP-2 cross-linking of pigment granule surfaces with motile cytomatrix components. Electron microscopy confirmed that the cytomatrix was the main cytoskeletal component remaining in nocodazole-treated cells. Taken together, the morphological, immunocytochemical, and physiological data strongly implicated the cytomatrix as the main propagator of resolute transport. We believe that the cytomatrix might possess molecular and structural information that responds to specific cellular signals in order to determine the spatial distribution of organelles, the polarity of transport, and rate and extent of organelle motion. Microtubules somehow serve to facilitate the process, perhaps by increasing the rate and extent of transport.

Perfusion Studies of Permeabilized Live Erythrophores

Our studies with digitonin-opened erythrophores basically demonstrated that saltatory and resolute transport in erythrophores were ATP-consuming processes.[13] Micromolar levels of vanadate and EHNA inhibited transport, indicating one or more ATPases were directly or indirectly involved in transport. The levels of vanadate used also reversibly inhibited ciliary beating, a result that was contradictory to results of Buckley et al.,[17] which showed that vanadate microinjected into ciliated epithelial cells did not inhibit saltatory motion at micromolar levels that stopped ciliary beating. An immediate explanation for this discrepancy of results may reside with an inability to adequately control final vanadate levels using the microinjection approach. By comparison, in perfused cells, it was possible to accurately control vanadate or EHNA levels and reverse the effects of vanadate or EHNA. Unfortunately, until the specificity of vanadate or EHNA can be shown or until the specific ATPase can be isolated, the results are merely suggestive of a dynein-like ATPase involvement.

Does Saltatory Motion or Organelle Streaming Depend on the Cytomatrix?

Saltatory motion definitely required microtubules, and it was reversibly inhibited when the microtubules were removed. Recently, Vale et al.[18,19] and Schnapp et al.[20] have reported that axoplasmic exudates contain filamentous elements (e.g. microtubules) that can support both saltatory motion and organelle streaming, or a nonsaltatory type of organelle transport. The saltatory motion required the presence of an unspecified amount of microtubule-bound cytoplasmic material and generally occurred in the more intact axoplasmic exudates. Saltations were especially exhibited by large organelles such as mitochondria. Depending on the size of the vesicle monitored, a nonsaltatory continuous form of motion, termed streaming, occurred on isolated filaments at rates of 5 to 60 μm/second. This motion was ATP-dependent but Ca^{+2} insensitive.

In attempts to identify the specific proteins that powered transport, Vale et al.[18] isolated a crude supernatant fraction that contained an ATP-dependent, vanadate-sensitive ATPase that produced either directed crawling of microtubules on a glass substrate, or organelle streaming on microtubules. The ATPase was either associated with the organelle membrane or was found in a soluble fraction. The soluble ATPase appeared to bind 6S microtubules or latex beads to activate transport.[18] Their results (see Schnapp et al., this volume) indicated that some form of uncontrolled, continuous organelle motion occurred as a function of both an unidentified ATPase and microtubules. Interestingly, the streaming was insensitive to Ca^{+2}, and fairly high vanadate levels (50 to 100 μM) were needed to inhibit motion. Perhaps a unique cytoplasmic ATPase distinct from axonemal dynein actually powers organelle streaming.

Note that in contrast to the above studies of isolated squid axon microtubules, many studies with permeabilized cell models have indicated that a calcium-sensitive, vanadate-inhibited, ATP-dependent, microtubule-bound cytomatrix component was involved in microtubule-directed saltatory transport.[13,21–23] Other studies on intact erythrophores have also indicated calcium, ATP-regulated transport and a dynein-like ATPase-mediated transport.[1,2,24]

Monoclonal Antibody Studies

The MAP-2 antibody-labeling results here indicated that MAP-2 composes part of the motile cytomatrix material. Unfortunately it was somewhat difficult to interpret the immunofluorescent images in dispersed cells, as the labeling was rather diffuse in aldehyde-fixed cells and not bound to the surfaces of any specific structures such as microtubules (except in lysed or MeOH-fixed cells). It was only with the onset of pigment aggregation that MAP-2 was found to bind the surfaces of the granules.

Immunogold electron microsocpy images have elegantly supported this interpretation.[15]

Our studies have represented an initial effort to characterize the proteins regulating microtubule-directed transport. Ultimately, we feel that the production of monoclonal antibodies against cytomatrix proteins (e.g. the motile cytomatrix) will allow characterization of the major proteins involved in transport in erythrophores. The basic strategy was to screen for antibodies that inhibited transport in perfused cells or in microinjected erythrophores. The approach should eventually yield information on proteins (e.g. an ATPase) specifically involved in functional aspects of organelle transport.

Because the pigment granules are the sole organelle being transported in erythrophores and because microtubules are the main cytoskeletal element directing the transport, we should be able to determine potential roles of these specific proteins in transport. In future studies we anticipate that the identification of α-cytomatrix components will at the very least permit more enlightened speculation on potential cellular transport mechanisms.

SUMMARY

Light microscopic studies have indicated that most microtubule-directed transport is either saltatory or resolute in nature. The latter form of transport is an intriguing phenomenon, because it commonly involves the unidirectional bulk motion of an organelle(s) such as chromosomes in dividing cells or pigment granules in chromatophores. We have investigated the ultrastructural and biochemical basis for the resolute transport of pigment in chromatophores. Light and EM studies of erythrophores in situ have clearly shown that when the microtubules were completely removed with nocodazole, resolute transport continued and was stimulated by aggregating and dispersing agents. Light and electron microscopic studies of cultured erythrophores permeabilized with digitonin indicated that resolute motion was produced by a cytomatrix of 3 to 7 nm filaments. Immunofluorescent analysis with several monoclonal antibodies raised against MAP-2 further demonstrated that MAP-2 was an important component of the contractile cytomatrix that powers pigment aggregation and dispersion. We conclude that a microtubule-associated cytomatrix normally produces resolute pigment transport in chromatophores.

REFERENCES

1. LUBY, K. J. & K. R. PORTER. 1980. The control of pigment migration in isolated erythrophores of Holocentrus ascensionus (Osbeck). I. Energy Requirements. Cell 21: 130–138.

2. LUBY-PHELPS, K. J. & K. R. PORTER. 1982. The control of pigment migration in isolated erythrophores of Holocentrus ascensionus (Osbeck). II. The role of calcium. Cell 29: 441–450.

3. PORTER, K. R., M. BECKERLE & M. McNIVEN. 1983. The cytoplasmic matrix. In Modern Cell Biology. Vol. 2. J. R. McIntosh, Ed.: 259–302. Alan R. Liss. New York.

4. STEARNS, M. E. 1984. Cytomatrix in chromatophores. J. Cell Biol. 99: 144s–151s.

5. SCHLIWA, M. & U. EUTENEUER. 1978. A microtubule independent component may be involved in granule transport in pigment cells. Nature (London) 273: 556–558.

6. STEARNS, M. E. 1982. High voltage electron microscopy studies of axoplasmic transport in neurons: A possible regulatory role of divalent cations. J. Cell Biol. 92: 765–776.

7. OCHS, R. L. & P. R. BURTON. 1980. Distribution and selective extraction of filamentous

components associated with axonal units of crayfish nerve cord. J. Ultrastruc. Res. **73:** 169–182.

8. SMITH, D. S., U. JARLFORS & B. F. CAMERON. 1975. Morphological evidence for the involvement of microtubules in axoplasmic transport. Ann. N.Y. Acad. Sci. **253:** 472–506.

9. ELLISMAN, M. H. & K. R. PORTER. 1980. Microtrabecular structure of the axoplasmic matrix: visualization of cross-linking structures and their distribution. J. Cell Biol. **87:** 464–479.

10. BECKERLE, M. C. & K. R. PORTER. 1983. Analysis of the role of microtubules and actin in erythrophore intracellular motility. J. Cell Biol. **96:** 354–362.

11. GREEN, L. 1968. Mechanism of movements of granules in melanocytes of *Fundulus heteroclitus*. Proc. Natl. Acad. Sci. USA **59:** 1179–1186.

12. MCNIVEN, M. A., M. WANG & K. R. PORTER. 1984. Microtubule polarity and the direction of pigment transport reverse simultaneously in surgically severed melanophore arms. Cell **37:** 753–765.

13. STEARNS, M. E. & R. L. OCHS. 1982. A functional *in vitro* model for studies of intracellular motility in digitonin permeabilized erythrophores. Cell Biol. Int. Rep. **94:** 727–739.

14. STEARNS, M. E. & M. WANG. 1986. Resolute transport is a function of the cytomatrix in erythrophores. Submitted to J. Cell Biol.

15. STEARNS, M. E. & L. I. BINDER. 1986. MAP-2 is associated with pigment transport in erythrophores. Submitted to Cell Motility.

16. STEARNS, M. E., M. PAYNE & L. I. BINDER. 1983. MAP-2 and MAP-1 involvement in pigment transport in erythrophores. J. Cell Biol. **97:** (2, Pt. 2) 201a.

17. BUCKLEY, I. & M. STEWART. 1983. Ciliary but not saltatory movements are inhibited by vanadate microinjected into living cultured cells. Cell Motility **3:** 167–184.

18. VALE, R. D., B. J. SCHNAPP, T. S. REESE & M. P. SHEETZ. 1985. Organelle, bead, and microtubule translocations promoted by soluble factors from the squid giant axon. Cell **40:** 559–569.

19. VALE, R. D., B. J. SCHNAPP, T. S. REESE & M. P. SHEETZ. 1985. Movement of organelles along filaments dissociated from the axoplasm of the squid giant axon. Cell **40:** 449–454.

20. SCHNAPP, B. J., R. D. VALE, M. P. SHEETZ & T. S. REESE. 1985. Single microtubules from squid axoplasm support bidirectional movement of organelles. Cell **40:** 455–462.

21. CLARK, T. G. & J. L. ROSENBAUM. 1984. Energy requirements for pigment aggregation in *Fundulus* melanophores. Cell Motility **4:** 431–441.

22. CLARK, T. G. & J. L. ROSENBAUM. 1982. Pigment particle translocation in detergent-permeabilized melanophores of *Fundulus heteroclitus*. Proc. Natl. Acad. Sci. USA **79:** 4655–4659.

23. FORMAN, D. S. 1982. Vanadate inhibits saltatory organelle movement in a permeabilized cell model. Exp. Cell Res. **141:** 139–147.

24. BECKERLE, M. C. & K. R. PORTER. 1982. Inhibitors of dynein activity block intracellular transport in erythrophores. Nature (London) **295:** 701–703.

Microtubules and the Mechanism of
Directed Organelle Movement

BRUCE J. SCHNAPP,[a] R.D. VALE,[a] M.P. SHEETZ,[b]
AND T.S. REESE[a]

[a]Laboratory of Neurobiology
Marine Biological Laboratory
National Institutes of Health
Woods Hole, Massachusetts 02543

[b]Department of Cell Biology and Physiology
Washington University Medical School
St. Louis, Missouri 63110

Our understanding of the mechanisms that cells use to translocate vesicular cytoplasmic organelles has been dramatically extended during the past few years as a result of work from several laboratories. It had long been suspected that microtubules have a role in this process (for a review see reference 1), but only in the past two years has it been demonstrated that single microtubules directly interact with vesicular organelles to generate movement.[2-4] Indeed the molecular components of this interaction are beginning to emerge.[5]

In this report we will focus on the results of our work on axoplasm extruded from a giant axon from the nervous system of the squid. Our aims are to review the evidence that bidirectional transport of organelles results from a direct interaction between organelles and single microtubules and to report on our first attempts at defining the molecular components that mediate this interaction.

MICROTUBULES AS THE SUBSTRATE FOR ORGANELLE MOVEMENT

The advances made in understanding organelle transport at a molecular level are a consequence of two recent innovations that came together at the Marine Biological Laboratory in the summer of 1981. On the one hand, the application of video processing to light microscopy[6,7] enabled structures ten or more times smaller than the resolution limit (about 200 nm) of light optics to be detected. At the same time, it was appreciated that the largest giant axon in the squid, which was originally recognized for its utility in studying the electrical properties of excitable membranes, is also useful in the investigation of cytoplasmic structure.[8] Examination in the video microscope of axoplasm extruded from the giant axon revealed anterograde as well as retrograde organelle movement lasting for many hours.[9] Organelle movement continued in the presence of an adenosine triphosphate (ATP)-containing buffer, and under these conditions single filaments with associated moving organelles would sometimes separate from the edges of the extruded cytoplasm.[10]

We experimented with buffer conditions for promoting the separation of filaments that supported organelle transport. In a 1:1 dilution of buffer X [buffer X was devised[11] to match closely the intracellular compartment of squid axons], large numbers of apparently single filaments consistently separated from the bulk axoplasm and settled on the surface of a coverglass producing extensive fields of isolated filaments (FIGURE

909

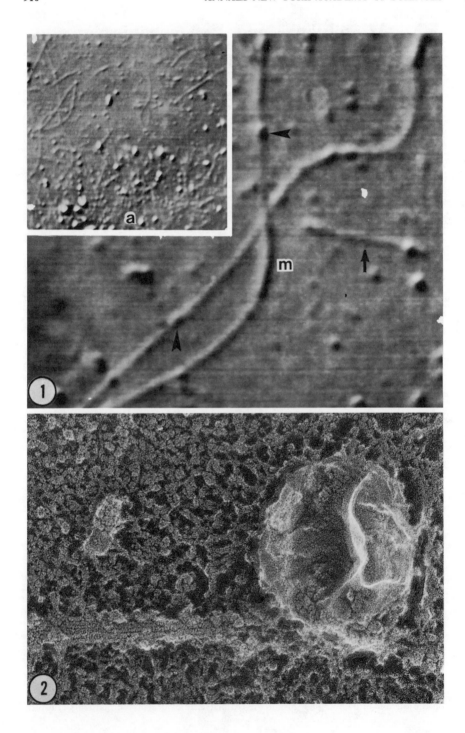

1, inset). Vesicular organelles, derived from the bulk axoplasm, entered the fluid phase and were observed to collide with, attach to, move along, and eventually dissociate from the filaments.[12] Because of the similarity between organelle movement along these isolated filaments and organelle transport in the intact axoplasm, we concluded that these components isolated from the bulk axoplasm represent the minimal machinery for rapid axonal transport. Directed organelle movement along these filaments ruled out mechanisms for organelle transport, such as streaming, which invoke an organized cytomatrix.[12]

To determine the structure of these transport filaments, a technique was devised to identify single filaments, first by video microscopy while they were transporting organelles, and subsequently the same filament by electron microscopy after rapid freezing, freeze-drying, and rotary shadowing.[3] All transport filaments, including those that supported the bidirectional movement of organelles, had a diameter and substructure indicative of a single microtubule (*e.g.* FIGURES 1 and 2). The identity of transport filaments as single microtubules was confirmed by exposing dissociated preparations of axoplasm to an antibody to alpha tubulin, which was then stained with an immunofluorescent secondary antibody; there was a one-to-one correspondence between transport filaments and labeled, tubulin-containing filaments.[3] When rhodamine-phalloidin was used to localize actin in preparations of dissociated axoplasm, single actin filaments were visible, but they were not associated with transport filaments.[13]

CHARACTERISTICS OF ORGANELLE MOVEMENT ALONG ISOLATED AXONAL MICROTUBULES

Although the process of directed organelle movement in intact cells has been recognized for some time, inferences regarding the underlying molecular mechanisms have been hampered by the possibility that organelle movement is influenced by the surrounding cytoplasmic components. For example, effects of treatments or drugs on the cytoplasmic organization cannot be distinguished from direct effects on the transport machinery. By contrast, organelle movement along the isolated axoplasmic microtubules should reflect directly underlying molecular events.

Organelles move continuously, in a consistent direction, at an average velocity of 2.2 μm/sec along isolated axoplasmic microtubules. This result brings to mind a similar experimental situation in which the velocity of myosin-coated beads moving along actin filaments was characteristic for different myosin molecules.[14] Thus, the fact that all axoplasmic organelles move at the same velocity under the dissociated

---

FIGURE 1. Video micrograph of three transport filaments (each is a single microtubule) that have separated from the bulk axoplasm. All three filaments were transporting organelles. An electron micrograph of the filament indicated by the arrow is shown in FIGURE 2. Mitochondrion (m) and vesicular organelles (arrowheads) are associated with the filaments. × 5,500. Inset: Low magnification view of a field of isolated transport filaments (arrowheads) that have separated from the bulk axoplasm (a).

FIGURE 2. Electron micrograph of one of the transport filaments with an attached vesicular organelle from FIGURE 1 (filament indicated by arrow in FIGURE 1). Specimen was prepared by rapid freezing, freeze drying, and rotary shadowing with Pt-Ta-Ir. A map of AuPd squares on the surface of the coverglass enabled the identification of individual filaments by light and electron microscopy.[3] Transport filaments are 24 nm in diameter and have a substructure indicative of a single microtubule. × 15,000.

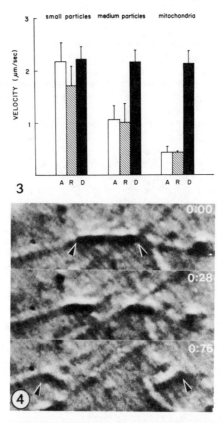

FIGURE 3. The velocity of organelle movement along axoplasmic microtubules in intact, extruded axoplasm (*A*, anterograde direction; *R*, retrograde direction) compared to velocities in dissociated axoplasm (*D*) along isolated microtubules like those illustrated in FIGURE 1. Small particles measure 0.2 μm in diameter (the actual size of some of these is likely to be less than this, perhaps as small as 50 nm); medium particles measure between 0.2 and 0.5 μm; mitochondria are recognized as tubular structures up to several microns in length (see FIGURE 1, for example). In the intact axoplasm there is a negative relationship between size and velocity, and all but the smallest particles move in a discontinuous or saltatory manner. Along isolated microtubules (filled bars), all organelles move continuously at the same velocity of 2.2 μm/second. Note that this is the same velocity characterizing the movement of the smallest organelles in the intact axoplasm.

FIGURE 4. Sequence of three video micrographs taken at 0, 0.28, and 0.76 seconds. A single mitochondrion becomes attached to two microtubules (arrowheads). At 0.28 seconds the organelle splits as a consequence of the two ends moving in opposite directions. Both halves continue to move normally (arrowheads at 0.76 seconds). × 8,000.

conditions suggests that this movement is mediated by a single type of molecular motor.

The continuous movement of organelles along isolated microtubules contrasts with the discontinuous movement of the larger organelles in intact axoplasm. In fact, organelle movement had been considered saltatory[15] because it was visualized by light microscopic techniques that detected only the larger organelles making discontinuous movements through intact cytoplasm. The smaller organelles (less than 100 nm in diameter, perhaps as small as 50 nm), which are only detected after video processing the light microscopic image, clearly move continuously in the intact axoplasm.[16,12] That the larger organelles move discontinuously and more slowly in the intact axoplasm (FIGURE 3) is presumably a consequence of impeding interactions with the surrounding cytoplasm. The decline of organelle velocity with size suggests that the impeding interactions are greater for larger organelles; in fact, small organelles appear to move freely through the cytoplasmic matrix.

Organelles have more than one motor attached to their surfaces. This can be seen directly in mitochondria that frequently attach to two microtubules at the same time (FIGURE 4). Multiple attachments also occur between vesicular organelles and microtubules.[12] Nevertheless, organelles move in a consistent direction along microtubules and rarely, if ever, reverse. Among thousands of observations of moving organelles, only two or three examples of possible reversals have been noted, and two of these involved mitochondria. On the other hand, the ends of mitochondria caught between two points on a single looped microtubule can move simultaneously in opposite directions (FIGURE 5). Although these examples do not constitute reversals of movement, they do indicate that a single mitochondrion has the capacity to move in both directions on the same microtubule.

Single microtubules support the bidirectional movement of organelles,[3] and two organelles moving in opposite or the same direction along the same microtubule often pass each other without an obvious collision (FIGURE 6). These results suggest that microtubules have more than one track along their length that can serve as a substrate for organelle movement.

ORGANELLE MOVEMENT RECONSTITUTED FROM PURIFIED COMPONENTS

A reconstituted system was developed that provided further evidence that microtubules are not simply serving as a support for some other cytoplasmic filament but are, in fact, interacting directly with organelles. The reconstituted system combined a pure fraction of taxol-polymerized, MAP-free microtubules from squid optic lobes with an organelle fraction purified from axoplasm extruded from the giant axon. Organelle movement was observed along these purified microtubules. Organelles purified from squid axoplasm also translocate along axonemal microtubules[17] and along microtubules polymerized from bovine tubulin in the absence of any microtubule-associated protein.[18]

A soluble supernatant fraction from squid axoplasm significantly increased the frequency of organelle movements along purified, MAP-free microtubules.[20] The rate of these movements was the same as those in the native preparation, but movements were primarily unidirectional on individual microtubules. Furthermore, not only did the organelles move along microtubules in the presence of this supernatant fraction, but the microtubules moved along the glass coverslip, even in the absence of organelles (FIGURE 7). The translocator appeared to act on microtubules through the formation of a fixed attachment to the glass, because a glass coverslip previously treated with soluble supernatant that was then washed off still supported microtubule movement. Carboxylated beads, which bear a net negative charge in aqueous solution, also move along purified microtubules at the same rate that the microtubules move along the glass coverslip (0.4 μm/sec), provided that the beads have been previously treated with the supernatant fraction.

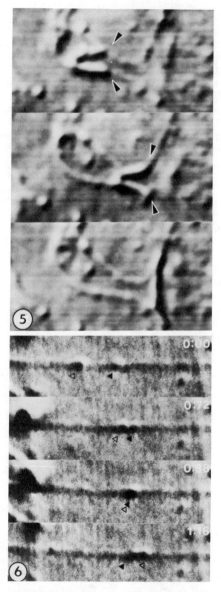

FIGURE 5. Sequence of video micrographs showing a single mitochondrion with its two ends (arrowheads) caught on a single-looped microtubule. Both ends move as the mitochondrion translates from left to right. The two ends move in opposite direction with respect to the orientation of the microtubule. × 8,000.

FIGURE 6. Sequence of video micrographs taken lasting 1.16 seconds. A smaller organelle (filled arrowhead) momentarily stops and is overtaken by a larger particle (open arrowhead). Neither organelle dissociates from the microtubule. Organelles also pass each other moving in opposite directions.[3] These observations indicate that a single microtubule has multiple tracks for organelle movement. × 8,000.

FIGURE 7. Sequence of video micrographs showing taxol-polymerized microtubules from squid optic lobe in the presence of a soluble cytoplasmic fraction from squid giant axon. All microtubules in this field are moving along the glass coverslip at an average velocity of 0.4 μm/second.

Because the microtubule-translocating activity was inhibited by exposing the supernatant fraction to trypsin or temperatures above 60°C, it seemed that the translocating activity is a protein. Because the microtubule movement required ATP and was inhibited by adenyl imidodiphosphate (AMP-PNP), the translocator protein appeared to be an ATPase. Finally, microtubule movement could be inhibited by precoating the glass coverslip with polylysine. Thus, we concluded that a soluble, cytoplasmic translocator protein attaches to the negatively charged surface of the coverslip. Only those translocators that have an appropriate orientation to the tubulin

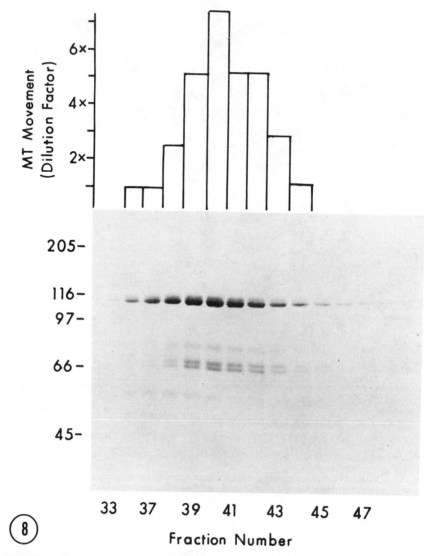

FIGURE 8. The microtubule-translocating activity associated with fractions eluted from a hydroxyapatite column. In this experiment, kinesin was first affinity purified by binding to microtubules in the presence of AMP-PNP.[5] Microtubule translocating activity of each fraction was assayed by determining the maximum dilution that still supports microtubule movement. The peak of activity coelutes with a protein complex consisting of a 110 kD polypeptide, a 60–65 kD doublet, and a faint 80 kD peptide.

lattice of the microtubule would then generate directed forces because only these molecules would bind to the microtubule surface. A similar argument has been invoked to explain how beads coated with myosin could make directed movements along actin filaments.[14]

The microtubule movement on a glass coverslip provided a functional assay that enabled the purification of a protein associated with the microtubule translocating activity (FIGURE 8[5]). Similar proteins have been purified from squid optic lobe, squid axoplasm, and bovine brain.[5] The protein, which has been named kinesin, consists of major polypeptides of 110 K daltons, 60–65 K daltons, and 80 K daltons. This protein elutes from a gel filtration sizing column with an apparent molecular mass of 600–700 K daltons. FIGURE 9 is a schematic diagram illustrating our view of the role of kinesin in promoting movement. The overall molecular weight of kinesin, as well as the molecular weight of its polypeptides, is different from dynein, the ATPase responsible for the interaction between outer doublet microtubules in cilia.

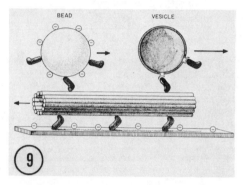

FIGURE 9. Schematic representation of binding of kinesin oligopeptide to a bead, vesicle, and the glass substrate. In each case kinesin interacts with a microtubule, exerting a power stroke directed toward the plus end of the microtubule. Specific receptors, indicated on the surface of the vesicle, might bind kinesin in a configuration that produces the higher rate of movement indicated by the longer arrow.[20]

Kinesin induces the movement of carboxylated beads in only one direction along microtubules.[5,20] Because microtubules are oriented unidirectionally in axons with their plus ends toward the nerve terminals and their minus ends toward the cell body,[19] it was important to determine the direction of the force stroke generated by kinesin. This has been accomplished by assaying bead movement along arrays of microtubules polymerized from centrosomes. In the presence of purified kinesin, beads move from the minus (near the basal bodies) to the plus ends of the microtubules, corresponding to the anterograde direction in intact axons.[18] It remains unclear how movement is induced in the direction opposite to that induced by kinesin, whether by modification of kinesin or by a second translocator factor that could have an affinity for certain classes of organelle.

REFERENCES

1. SCHLIWA, M. 1984. Mechanisms of intracellular organelle transport. *In* Cell and Muscle Motility. J. W. Shaw, Ed.: **5**: 1–82. Plenum Publishing Co. New York.

2. HAYDEN, J. H. & R. D. ALLEN. 1984. Detection of single microtubules in living cells: particle transport can occur in both directions along the same microtubule. J. Cell Biol. **99:** 1783–1793.

3. SCHNAPP, B. J., R. D. VALE, M. P. SHEETZ & T. S. REESE. 1985. Single microtubules from squid axoplasm support bidirectional movement of organelles. Cell **40:** 455–462.

4. ALLEN, R. D., D. G. WEISS, J. H. HAYDEN, D. T. BROWN, H. FUJIWAKE & M. SIMPSON. 1985. Gliding movement of and bidirectional organelle transport along single native microtubules from squid axoplasm: evidence for an active role of microtubules in cytoplasmic transport. J. Cell Biol. **100:** 1736–1752.

5. VALE, R. D., T. S. REESE & M. P. SHEETZ. 1985. Identification of a novel translocator, kinesin, involved in microtubule-based motility. Cell **42:** 39–50.

6. INOUE, S. 1981. Video image processing greatly enhances contrast, quality and speed in polarization-based microscopy. J. Cell Biol. **89:** 346–356.

7. ALLEN, R. D., N. S. ALLEN & J. L. TRAVIS. 1981. Video-enhanced differential interference contrast (AVEC-DIC) microscopy: a new method capable of analyzing microtubule-related movement in the reticulopodial network of *Allogromia laticollaris*. Cell Motility **1:** 291–302.

8. LASEK, R. J. 1974. Biochemistry of the squid giant axon. *In* A Guide to the Laboratory Use of the Squid. J. M. Arnold, W. C. Summers & P. L. Gilbert, Eds.: 69–74. Marine Biological Laboratory. Woods Hole, Mass.

9. BRADY, S. T., R. J. LASEK & R. D. ALLEN. 1982. Fast axonal transport in extruded axoplasm from squid giant axon. Science **218:** 1129–1131.

10. ALLEN, R. D., D. T. BROWN, S. P. GILBERT & H. FUJIWAKE. 1983. Transport of vesicles along filaments dissociated from squid axoplasm. Biol. Bull. **165:** 523.

11. BRADY, S. T., R. J. LASEK, R. D. ALLEN, H. L. YIN & T. P. STOSSEL. 1984. Gelsolin inhibition of fast axonal transport indicates a requirement for actin microfilaments. Nature (London): **310:** 56–58.

12. VALE, R. D., B. J. SCHNAPP, T. S. REESE & M. P. SHEETZ. 1985. Movement of organelles along filaments dissociated from the axoplasm of the squid giant axon. Cell **40:** 449–454.

13. SCHNAPP, B. J., M. P. SHEETZ, R. D. VALE & T. S. REESE. 1984. Filamentous actin is not a component of transport filaments isolated from squid axoplasm. J. Cell Biol. **99**(2): 351a.

14. SHEETZ, M. P. & J. SPUDICH. 1983. Movement of myosin-coated fluorescent beads on actin cables *in vitro*. Nature (London) **303:** 31–35.

15. REBHUN, L. I. 1972. Polarized intracellular particle transport: saltatory movements and cytoplasmic streaming. Int. Rev. Cytol. **32:** 93–137.

16. ALLEN, R. D., J. METUZALS, I. TASAKI, S. T. BRADY & S. P. GILBERT. 1982. Fast axonal transport in squid giant axon. Science **218:** 1127–1128.

17. GILBERT, S. P., R. D. ALLEN & R. D. SLOBODA. 1985. Translocation of vesicles from squid axoplasm on flagellar microtubules. Nature **315:** 245–248.

18. SCHNAPP, B. J., R. D. VALE, T. MITCHISON, T. S. REESE & M. P. SHEETZ. 1985. Translocator protein (kinesin) for rapid axonal transport mediates anterograde movement of beads along purified microtubules. Soc. Neurosci. (abstract).

19. HEIDEMANN, S. R., J. M. LANDERS & M. A. HAMBORG. 1981. Polarity orientation of axonal microtubules. J. Cell Biol. **91:** 661–665.

20. VALE, R. D., B. J. SCHNAPP, T. S. REESE & M. P. SHEETZ. 1985. Organelle, bead, and microtubule translocations promoted by soluble factors from the squid giant axon. Cell **40:** 559–569.

Changes in the Extent of Microtubule Assembly Can Regulate Initiation of DNA Synthesis[a]

DARRELL H. CARNEY,[b] KATHRYN L. CROSSIN,[b,d]
REBECCA BALL,[c] GERALD M. FULLER,[b]
THOMAS ALBRECHT[c] AND WILLIAM C. THOMPSON[b]

[b]Departments of Human Biological Chemistry and Genetics
and [c]Microbiology
The University of Texas Medical Branch
Galveston, Texas 77550

INTRODUCTION

Several lines of evidence indicate that the cytoplasmic microtubule (MT) complex may play a regulatory role in cell proliferation. MT depolymerization can modulate cell surface events and either enhance or inhibit mitogenic signals in several different cell types. For example, MT disruption by colchicine prevents initiation of DNA synthesis in lymphocytes by concanavalin A and alters the distribution of concanavalin A on the surface of these cells.[1] Colchicine also inhibits events leading to initiation of DNA synthesis in neuroblastoma cells,[2] hepatocytes,[3,4] and some cultured fibroblasts.[3] By contrast, MT depolymerization initiates DNA synthesis[5] or enhances the effects of serum or purified growth factors in many types of fibroblasts.[6-11]

We have recently reported that MT depolymerization is sufficient to initiate DNA synthesis in serum-free primary cultures of mouse, chick, and human fibroblasts.[5] These studies demonstrated that MT depolymerization was in itself able to generate an initiation signal. Furthermore, MT stabilization by taxol prevented both the drug-induced depolymerization and the initiation of DNA synthesis, indicating that MT depolymerization was necessary for the MT-disrupting drugs to initiate DNA synthesis. In similar experiments, we have shown that taxol inhibits initiation of DNA synthesis by epidermal growth factor (EGF) and thrombin,[12] suggesting that there might be a universal mechanism for initiating DNA synthesis that involves MT disruption or rearrangements. In the present studies we have expanded this idea and found that oncogenic DNA viruses also initiate host-cell DNA synthesis by a mechanism that is inhibitable by taxol.

The nature of this microtubule-related event remains to be determined. Recent studies have shown that in addition to stabilizing preexisting MTs,[13] taxol promotes assembly of tubulin into tubules[14-16] as well as inhibits the disassembly process.[17] Thus, the normal treadmilling or equilibrium[18] between free tubulin and tubules is blocked,

---

[a]This research is supported by NIH Grants AM-25807 to D.H. Carney, AI-42557 to T. Albrecht, and GM-33505 to W.C. Thompson. D.H. Carney is the recipient of a Research Career Development Award from the NCI (CA-00805), and R. Ball is a James C. McLaughlin fellow.
[d]Current address: Department of Development of Molecular Biology, Rockefeller University, New York, N.Y. 10021.

and the pool of free tubulin is depleted. These findings have raised the possibility that the inhibitory effects of taxol might be quite complex, involving shifts in the tubulin-MT equilibrium in additon to stabilization of structural MT complexes. Adding more confusion to this story, recent studies have demonstrated a number of microtubule subsets within a single cell. Using specific monoclonal antibodies, we have shown that at least two different classes of microtubules[19] exist within single cells. It has also been possible to define other classes of microtubules based on drug, temperature, or pressure sensitivity.[20-22] Thus, it may be possible that only one or more of these subsets is involved in initiation.

In the present studies we have correlated changes in the immunofluorescent patterns of the cytoplasmic MT complexes with initiation of DNA synthesis by various concentrations of colchicine and taxol. We found that colchicine and taxol affect DNA synthesis in a dose-dependent manner corresponding to the amount of MT disruption that is visually detected in the cytoplasm or quantitated by direct antibody binding to fixed cytoskeletons. Interestingly, only a small fraction of the total microtubules appear to be involved in initiation. Thus, these results demonstrate that slight changes in the extent of microtubule assembly or changes in specific functional groups of microtubules may initiate intracellular signals leading to cell proliferation.

METHODS

Cell Preparation and Thymidine Incorporation

Primary cultures of mouse embryo (ME) cells were prepared as described previously.[23] Briefly, cells were isolated from the body walls of 9- to 13-day-old mouse embryos. Primary ME cells were grown in a 1:1 mixture of Dulbecco modified Eagle's medium and Ham's F12 (DV/F12) medium (Gibco) supplemented with 10% calf serum (Irvine Scientific). These primary cultures were subcultured into 35 mm dishes at a density of approximately 6.8×10^4 cells per cm^2. After 18 hr, the media was changed to serum-free media, and the cells were further incubated for 2 days. At this time, the cells were 90–95% quiescent as measured by flow microfluorimetric analysis. At this time, various concentrations of colchicine, growth factors, or virus were added with or without taxol (provided by Dr. John Douros and Dr. Matthew Suffness, National Cancer Institute). Colchicine and taxol were dissolved and diluted in ethanol and added to cultures in 20 or 10 μl volumes, respectively. DNA synthesis was measured in these experiments by the amount of thymidine [methyl-^{3}H]-(New England Nuclear, 20 Ci/mmole) incorporated into acid precipitable material during a 6 hr incubation from 20 to 26 hr after drug addition. At 26 hr the cells were rinsed in phosphate-buffered saline (PBS), then 5 times in 10% trichloroacetic acid (TCA). The TCA-precipitable material was dissolved in KOH, neutralized and counted in 10 ml Beckman Ready Solv MP scintillation fluid.

Immunofluorescence

Cells were grown as described above on glass coverslips at a density of 1×10^4 cells/cm^2. Following the indicated treatment, the cells were prepared for fluorescence as described by Fuller et al.[24] The cells were fixed in 3% formaldehyde, treated with ice-cold acetone then incubated with monospecific rabbit antibody to tubulin. The coverslips were rinsed and incubated with rhodamine-conjugated antirabbit immuno-globulin (Meloy Laboratory), rinsed extensively and mounted on slides in 9:1

PBS:glycerol and examined using a Leitz Orthoplan microscope equipped with ploemopack epifluorescent illumination (850X). Alternative procedures were used for double-label immunofluorescence of cytoskeleton shown in FIGURE 3. These methods are described in detail.[19]

RESULTS

MT Stabilization by Taxol Inhibits Initiation of DNA Synthesis by Colchicine, Growth Factors, and Oncogenic Viruses

As shown in FIGURE 1, long-term thymidine incorporation is stimulated approximately 70–80% at colchicine concentrations above 0.1 μM in nonproliferating serumfree cultures of ME cells. Previous studies have shown that this amount of long-term

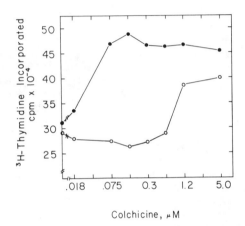

Colchicine, μM

FIGURE 1. Effect of taxol pretreatment on colchicine-initiated DNA synthesis. Quiescent cultures of ME cells were treated with various concentrations of colchicine with or without a one-hour pretreatment with 10 μg/ml taxol. DNA synthesis was measured using a six-hour [³H]thymidine pulse as described in METHODS. ● = colchicine; O = colchicine plus taxol.

stimulation correlates with a four- to five-fold increase in the number of labeled nuclei.[5] Thus, this stimulation appears to represent a true stimulation of DNA synthesis. The presence of 10 μg/ml of taxol (11.7 μM) prevents the initiation of DNA synthesis by colchicine concentrations up to about 0.6 μM colchicine and inhibits approximately 20% of the stimulation at higher colchicine concentrations (FIGURE 1). Thus, colchicine can initiate DNA synthesis in the presence of taxol, however, the half-maximal concentration for the colchicine effect is increased about 20-fold. It should be noted that in these cells, taxol treatment at 10 μg/ml does not appear to affect transport of thymidine, glucose or amino acids, nor incorporation of amino acids into protein.[5] We have also observed that taxol does not simply delay the onset of DNA synthesis. Thus, it appears that the inhibitory effects of taxol represent a true inhibition of the initiation signal.

As shown in TABLE 1, taxol at 10 μg/ml also inhibits the initiation of DNA synthesis by thrombin and EGF. In previous studies, we have shown that 10 μg/ml

TABLE 1. Effect of Taxol on Stimulation of Thymidine Incorporation by Thrombin and Epidermal Growth Factor

	Stimulation of Thymidine Incorporation over Controls (CPM × 10⁻³)		Percent Inhibition
	−Taxol	+Taxol (10 μg/ml)	
Thrombin (0.5 μg/ml)	133	48	64
Thrombin (1.0 μg/ml)	118	38	68
EGF (2.5 ng/ml)	151	77	49
EGF (5.0 ng/ml)	165	83	50

Nonproliferating serum-free cultures of ME cells were pretreated for one hr with or without taxol (10 μg/ml) prior to addition of thrombin or EGF. Thymidine incorporation was determined as described in METHODS and compared to parallel control cultures not treated with growth factors.

taxol does not affect binding of thrombin or EGF to the surface of these cells or the internalization and degradation of these factors.[12] This concentration of taxol blocked about 50 to 70% of the growth-factor–initiated DNA synthesis (TABLE 1). Because this is approximately the same concentration of taxol required to block 50% of the initiation of DNA synthesis by colchicine,[5] these results suggest that growth factors initiate DNA synthesis by a mechanism involving MT depolymerization or rearrangements.

Cytomegalovirus (CMV) as well as other oncogenic DNA viruses initiate one round of host cell DNA synthesis prior to establishing a virulent or permissive infection.[25,26] As shown in Figure 2, CMV stimulates [³H]thymidine incorporation (6 hr incubations) in ME cells approximately 100 percent over control cultures. This stimulation is approximately equal to that observed following addition of growth factors. As shown, taxol addition (10 μg/ml) at the time of infection blocks 60 to 70%

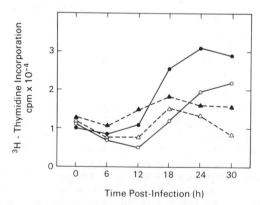

FIGURE 2. Effect of taxol on CMV-stimulated [³H]thymidine incorporation. Nonproliferating serum-free cultures of ME cells were incubated for one hour with CMV in the presence (O) or absence of taxol (●) at 10 μg/ml or with control-cell lysate in the presence (△) or absence (▲) of taxol. After virus adsorption, the cells were rinsed and returned to serum-free medium with or without taxol. [³H]thymidine (1 μCi/ml) was incubated with cells for 6 hr beginning at the indicated times, and incorporation into trichloroacetic acid-precipitable cellular DNA was determined as described.[25]

of the CMV-stimulated DNA synthesis. This suggests that oncogenic DNA viruses, like growth factors, initiate DNA synthesis through a common mechanism involving MT disruption or rearrangements.

In an attempt to determine if thrombin, EGF and cytomegalovirus were causing a depolymerization similar to that seen with colchicine, we treated quiescent populations of mouse cells with thrombin or cytomegalovirus for various periods, fixed the cells and examined the immunofluorescent pattern of the microtubules (data not presented). In all cases, we noted very little change in the total microtubules. At 2–4 hours, there was a slight loss of microtubules, and by 8 hours the microtubules appeared to be brighter with perhaps thicker bundles. Although preliminary, these results indicate that growth factors and viruses may cause slight alterations in the cytoplasmic microtubule complex, but they do not appear to cause a total disruption such as that seen with colchicine or other microtubule-disrupting drugs.

That partial depolymerization may be sufficient to initiate DNA synthesis raises two possibilities; one that slight shifts in the equilibrium between free tubulin and polymerized tubules is enough to trigger initiation, or second, that alterations of only a particular class or subset of microtubules is involved. We have previously shown that within single cells there are at least two antigenically distinct classes of microtubules (see FIGURE 3). Other subsets of microtubules that exist within single cells may also be involved.[19–22,27] To determine how much microtubule depolymerization was needed to initiate DNA synthesis or whether a specific group of microtubules needed to be depolymerized, we have examined the effects of various concentrations of colchicine and taxol on initiation and compared these effects with the amount of MT depolymerization.

Effect of Taxol on Initiation of DNA Synthesis and MT Disruption by Colchicine

Indirect immunofluorescence of the cytoplasmic MTs following treatment with taxol and colchicine in experiments parallel to FIGURE 1 showed that high colchicine concentrations were able to overcome the stabilization of MTs by taxol (FIGURE 4). FIGURE 4a shows control cells treated with 10 μg/ml taxol. These cells exhibit increased numbers of MTs and shorter MTs having multiple nucleation sites. Cells treated with 0.6 μM colchicine for 2 hr after a 1 hr pretreatment with 10 μg/ml taxol are indistinguishable from taxol-treated control cells (FIGURE 4b). FIGURES 4c and 4d illustrate the pattern seen at concentrations of colchicine that began to stimulate DNA synthesis even in the presence of taxol (compare FIGURE 1). Cells in FIGURE 4c were treated with 1.2 μM colchicine following taxol treatment. These cells retain discernible MT structures that extend from the nucleus to about halfway to the cell periphery. Beyond this point, the immunofluorescent pattern becomes diffuse and very faint (note cell periphery marked by arrows). Cells treated with 5 μM colchicine (FIGURE 4d) subsequent to taxol treatment have diffuse staining with intact MT structures only near the nucleus. These results suggest that taxol inhibits initiation of DNA synthesis maximally only when it stabilizes MTs throughout the cytoplasm to the cell periphery.

To further determine whether the observed inhibitory effects of taxol were related to MT stabilization and how much stabilization was required, we measured the amount of DNA synthesis initiated by 1 μM colchicine in the presence of various taxol concentrations. As shown in FIGURE 5, inhibition begins at about 2.5 μg/ml taxol (2.9 μM) and is maximal (80% inhibition) at 20 μg/ml taxol. In this case, the extent of inhibition is dose-dependent for taxol with half-maximal inhibition observed at about 8–10 μg/ml taxol.

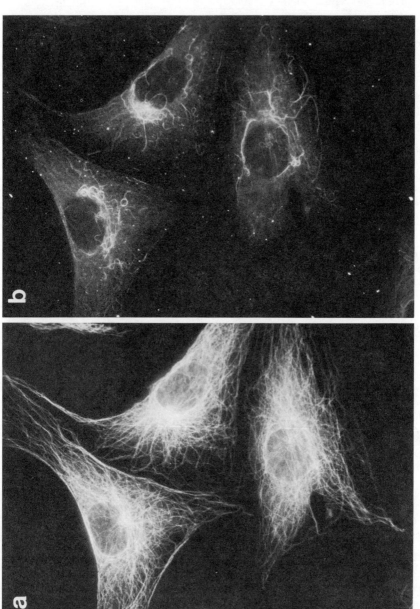

FIGURE 3. Demonstration of an antigenic subset of MTs and MT segments within single ME cells by monoclonal antitubulin 1-6.1. Cells were extracted in MT-stabilizing buffer, the cytoskeletons fixed and double stained using affinity-purified rabbit antitubulin antibodies and mouse monoclonal Ab 1-6.1 as described in detail.[19] Panel **a** shows the entire cytoplasmic MT array visualized by rabbit polyclonal antitubulin, panel **b** the Ab 1-6.1 subset. Ab 1-6.1 is an IgG$_1$, which binds the alpha subunit of tubulin (see Thompson p. 660, this volume).

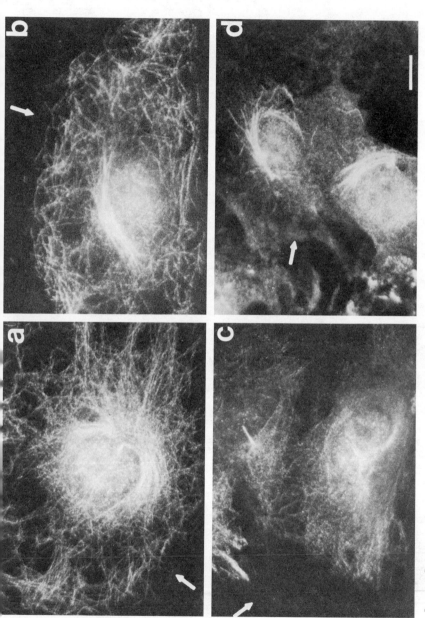

FIGURE 4. Immunofluorescent visualization of the ability of colchicine to disrupt microtubules in the presence of 10 μg/ml of taxol. Coverslip cultures of ME cells were treated for one hour with 10 μg/ml taxol followed by a two-hour treatment with indicated concentrations of colchicine and were then prepared for immunofluorescence as described in METHODS. **a:** Control culture (taxol alone); **b:** 0.6 μM colchicine; **c:** 1.2 μM colchicine; and **d:** 5 μM colchicine. Bar in panel represents 10 μm. Arrows are placed to indicate the cell periphery.

FIGURE 6 illustrates a parallel immunofluorescence experiment with various concentrations of taxol in the presence of 1 μM colchicine. FIGURE 6a and 6b represent cells treated with 0.15 $\mu g/ml$ and 0.6 $\mu g/ml$ taxol, respectively, followed by a two-hour exposure to 1 μM colchicine. As shown in FIGURE 5, both these taxol concentrations are below the critical concentrations required for inhibition of DNA synthesis. As in the cells treated with taxol and high colchicine concentrations, these cells show intact MT structures around the nuclear area, but diffuse cytoplasmic staining to the cell periphery. Cells treated with 2.5 $\mu g/ml$ taxol (FIGURE 6c), the lowest concentration at which inhibition of colchicine-initiated DNA synthesis is observed, exhibit extensive MTs that appear to fill the cytoplasm. FIGURE 6d illustrates cells exposed to 10 $\mu g/ml$ taxol, a concentration that caused 50% of maximal inhibition of DNA synthesis (FIGURE 5). These cells display the characteristic taxol pattern similar to those in control cells never exposed to colchicine. It is noteworthy that half-maximal inhibition occurs at a taxol to colchicine ratio of approximately 13 to 1 (see FIGURES 1 and 5).

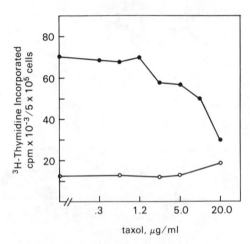

FIGURE 5. Effect of taxol concentration on initiation of DNA synthesis by colchicine. Quiescent cultures of ME cells were pretreated with various concentrations of taxol prior to addition of 1 μM colchicine. DNA synthesis was measured as described in FIGURE 1. ● = taxol + 1 μM colchicine; O = taxol.

Thus, there is a dose-dependent correlation between the extent of MT assembly and the magnitude of the initiation of DNA synthesis, suggesting that changes in the state of the MT-tubulin equilibrium may be responsible for the mitogenic properties of MT-disrupting drugs. Alternatively, because initiation of DNA synthesis was inhibited only when MTs were stabilized throughout the cytoplasm, it could be that a specific subset of MTs near the periphery of the cell is directly depolymerized or rearranged to allow entry of these cells into a proliferative cycle.

To better quantitate the effects of colchicine and taxol on confluent cultures of mitogenically active cells, we metabolically labeled monoclonal antibody 1-1.1 (originally cloned by Dr. David Asai at the University of California, Santa Barbara) with [35S]methionine and bound this antibody to MT stabilized cytoskeletons (Ball et al., manuscript in preparation). As shown in FIGURE 7, with taxol pretreatment there was a significant increase in the amount of polymerized tubulin as determined by Ab 1-1.1

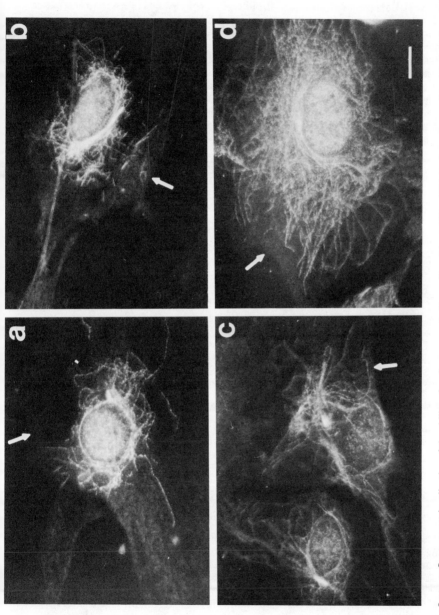

FIGURE 6. Immunofluorescent visualization of the ability of various concentrations of taxol to stabilize MTs in the presence of 1 μM colchicine. Cultures were prepared as described in FIGURE 4 and pretreated with taxol for one hour, followed by a two-hr treatment with 1 μM colchicine. **a:** 0.15 μg/ml taxol; **b:** 0.6 μg/ml taxol; **c:** 2.5 μg/ml taxol; **d:** 10.0 μg/ml taxol. Bar in panel represents 10 μm. Arrows are placed to indicate the cell periphery.

binding. In these experiments, nonspecific Ab 1-1.1 binding as determined in the presence of a 100-fold excess of unlabeled antibody was approximately 1000 counts per minute (CPM) per well. Therefore, the differences observed between control cells and cells treated with 1.2 μM colchicine would represent a 60 to 70% depolymerization of the microtubules. It is important to note that the difference between 0.1 and 10 $\mu g/ml$ taxol at 1.2 μM colchicine would represent less than a 15% shift in the amount of polymerized tubulin. At these concentrations, however, there was no inhibition of colchicine initiation at 0.1 $\mu g/ml$ taxol and half-maximal inhibition at 10 $\mu g/ml$ (FIGURE 5). Similarly, if one compares the effects of colchicine concentration on 10 $\mu g/ml$ taxol, inhibition is overcome between 0.6 and 1.2 $\mu g/ml$ colchicine where there is perhaps a 20 to 25% decrease in polymerized tubulin. Thus, these results confirm that slight changes in the total tubulin-MT equilibrium are sufficient to initiate DNA synthesis.

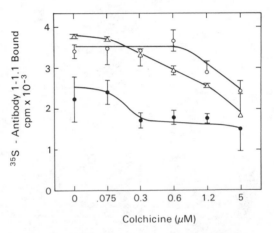

FIGURE 7. Effect of taxol and colchicine on the extent of MT polymerization. Nonproliferating confluent cultures of ME cells were pretreated for one hr without (●) or with taxol at 0.1 $\mu g/ml$ (△) or taxol at 10 $\mu g/ml$ (O) and then for an additional two hr with colchicine at the indicated concentration. Following treatment, the cells were extracted with MT-stabilizing buffer and fixed as described in FIGURE 3, then incubated with labeled [^{35}S]methionine Ab 1-1.1, a β-tubulin specific IgM that has been shown to bind to all cytoplasmic MTs in these ME cells.[19]

DISCUSSION

We recently demonstrated that MT depolymerization by colchicine and other drugs is sufficient to initiate DNA synthesis in serum-free quiescent cultures of primary embryonic fibroblasts.[5] In those studies we also showed that stabilizing the MTs with taxol prior to colchicine addition blocked initiation of DNA synthesis, indicating that MT depolymerization itself was necessary for these drugs to initiate proliferative events. We have also now shown that taxol can block initiation of DNA synthesis initiated by highly purified growth factors[12] and by oncogenic DNA viruses such as CMV. Thus, it appears there is a common mechanism for initiating DNA synthesis that involves MT depolymerization or rearrangements. In the current study we have compared the extent of MT assembly in drug-treated cells as visualized by indirect immunofluorescence microscopy with initiation of DNA synthesis to determine whether initiation can be correlated with either depolymerization of specific groups of MTs or with a general shift in the equilibrium between free tubulin and polymerized microtubules.

Pretreatment of cultures for one hour with 10 μg/ml taxol prevented initiation of DNA synthesis by concentrations of colchicine up to about 0.6 M, but higher colchicine concentrations up to 5 μM overcame the taxol inhibition and initiated DNA synthesis to levels about 80% as high as cultures never exposed to taxol. Similarly, if cultures were pretreated for one hour with various concentrations of taxol followed by addition of colchicine to 1 μM, only concentrations of taxol at or above 2.5 μg/ml showed any inhibition. This suggests that the dose-dependent stimulation or inhibition by these drugs with opposing effects might correlate with either a particular shift in the tubulin-MT equilibrium or with the depolymerization of specific MT structures.

Immunofluorescent studies of cells treated with various concentrations of colchicine and taxol showed that at drug concentrations where DNA synthesis was inhibited, MTs appear to be stabilized throughout the cytoplasm extending nearly to the cell periphery. At lower taxol to colchicine ratios where DNA synthesis is initiated, MTs in the perinuclear region remain intact; those MTs near the periphery of the cell, however, were completely depolymerized as judged by the diffuse immunofluorescent staining. Thus, as the taxol to colchicine ratio decreases, MT depolymerization appears to occur first near the cell periphery. Only at very low taxol to colchicine ratios is MT depolymerization complete. These results, therefore, indicate that initiation of DNA synthesis does not require complete MT depolymerization. Indeed, MTs near the nucleus are stabilized at taxol to colchicine ratios that initiate DNA synthesis, and maximal inhibition of initiation seems to occur only when MT structures extend throughout the cytoplasm to the cell periphery.

These results suggest two possible mechanisms for the involvement of MTs in initiating proliferative events. First, it might be possible that a subset of MTs near the cell surface mediate normal mitogenic signals generated by growth-factor interaction with specific cell surface receptors. In this case, drug-induced MT depolymerization near the cell periphery might mimic the normal mitogenic signal. Indeed other studies have suggested the involvement of MTs in modulating surface assemblies,[1] and several recent reports have indicated that MT depolymerization enhances the mitogenic effects of serum and various growth factors.[6-11] We have also shown that MT stabilization by taxol can inhibit initiation of DNA synthesis by thrombin, EGF,[12] and now by oncogenic DNA viruses. Because these initiations involve cell-surface interaction, it is possible that all of these effects are mediated by a subset of MTs near the cell surface.

A second possibility that could explain our present results as well as those mentioned above is that perturbation of the equilibrium between free tubulin and polymerized MTs is a part of the mitogenic signal. In this case, the critical taxol to colchicine ratio for inhibition of initiation could simply reflect the total effect of these drugs on the pool of free tubulin or a related pool of nucleotides or ions. Colchicine binds to free tubulin[28,29] and to tubulin monomers at the growing end of MTs,[30] preventing further polymerization. Taxol apparently binds to tubulin at a site distinct from the colchicine site,[15] and both promotes MT assembly[14-16] and inhibits disassembly, resulting in a depletion of the free tubulin pools. Thus, at ratios of taxol to colchicine at which taxol no longer inhibits initiation of DNA synthesis, it is possible that taxol is not as effective in preventing colchicine-induced disassembly and the resulting shift in equilibrium toward depolymerization. In our immunofluorescent studies, taxol stabilization of microtubules near the nucleus and perhaps halfway to the periphery did not inhibit initiation. Furthermore, quantitating the effects of these opposing drugs on the amount of polymerized tubulin left in extracted cytoskeletons showed that shifts of as little as 15% of the total tubulin could determine whether or not DNA synthesis was initiated. Thus, if the initiation of proliferative events by MT depolymerization is the result of a shift between tubulin and intact MTs, then the extent of this shift might be quite small, involving only a fraction of the total tubulin or might only require this shift in equilibrium in MTs near the periphery.

A shift in the tubulin-MT equilibrium could lead to initiation in several ways. First, as described above, if a particular subset of MTs near the periphery of the cell were involved in modulating surface receptors, then depolymerization of these MTs might lead to receptor signal activation. We have recently discovered that approximately 50% of the thrombin receptors are anchored to the cytoskeletons (Carney, Bradley, and Thompson, manuscript in preparation). Second, recent studies have shown that as MTs are depolymerized by colchicine or vinblastine, there is an accompanying decrease in the pools of adenosine monophosphate (AMP) and guanosine monophosphate (GMP) and an increase in the uridine triphosphate (UTP) pool.[31] Glucosamine depletes cells of UTP and appears to block initiation of DNA synthesis by colchicine. Thus, changes in these nucleotide pools that accompany MT disruption may be involved in initiation. A third possibility is that free tubulin dimers themselves are involved in initiating DNA synthesis. For example, recent studies have shown that lowering the free tubulin concentration will increase tubulin synthesis and transcription of tubulin mRNA.[32,33] Moreover, if phosphocellulose-purified tubulin is microinjected into single cells, tubulin synthesis is depressed.[34] In similar experiments, we have shown that following microinjection of phosphocellulose-purified tubulin into ME cells, DNA synthesis is initiated (Carney, Kay, and Thompson, manuscript in preparation). Because this tubulin does not appear competent to polymerize into MTs,[34] these results suggest that shifting the MT-tubulin equilibrium or increasing the free tubulin concentration may be sufficient to initiate proliferative events in these cells.

SUMMARY

We have shown that MT depolymerization by colchicine and other drugs is sufficient to initiate DNA synthesis in serum-free cultures of embryonic fibroblasts and that stabilization of MTs with taxol inhibits this initiation. Growth factors and oncogenic DNA viruses also initiate DNA synthesis by a taxol-sensitive mechanism that appears to require MT depolymerization or rearrangements. Because we have shown that microtubule heterogeneity exists within single fibroblastic cells, we have carried out a series of experiments to determine the extent of microtubule disruption necessary to initiate DNA synthesis. We have compared the effects of various concentrations of colchicine and taxol on initiation of DNA synthesis with their effects on cytoplasmic MT complexes as visualized by indirect immunofluorescence microscopy and quantitated by direct binding of radiolabeled monoclonal antibody to cytoskeletons. The opposing effects of these drugs on MTs shows that there is a correlation between the extent of MT depolymerization and initiation of DNA synthesis. Initiation of DNA synthesis by colchicine in the presence of taxol is half-maximal when taxol and colchicine are added to cultures at a ratio of about 13 to 1. At this drug ratio, taxol stabilizes MTs near the nucleus, but MTs near the cell periphery are depolymerized. Maximal inhibition of DNA synthesis by taxol occurs only at taxol to colchicine ratios where MTs extend throughout the cytoplasm to the cell periphery. Thus, depolymerization of a small fraction of total MTs, particularly those near the periphery, may be sufficient to initiate proliferative events.

ACKNOWLEDGMENTS

The authors wish to acknowledge John Douros and Matthew Suffness at the NCI for providing us with taxol, John W. Fenton II for gifts of highly purified human α thrombin, and Helen Amato for excellent secretarial assistance.

REFERENCES

1. EDELMAN, G. 1976. Surface modulation in cell recognition and cell growth. Science **192:** 218–226.
2. BAKER, M. E. 1976. Colchicine inhibits mitogenesis in C1300 neuroblastoma cells that have been arrested in Go. Nature (London) **262:** 785–786.
3. WALKER, P. R., A. L. BOYNTON & J. F. WHITFIELD. 1977. The inhibition by colchicine of the initiation of DNA synthesis by hepatocytes in regenerating rat liver and by cultivated WI-38 and C3H10T 1/2 cells. J. Cell. Physiol. **93:** 89–98.
4. WALKER, P. R. & J. F. WHITFIELD. 1978. Inhibition by colchicine of changes in amino acid transport and initiation of DNA synthesis in regenerating rat liver. Proc. Natl. Acad. Sci. USA **75:** 1394–1398.
5. CROSSIN, K. L. & D. H. CARNEY. 1981. Evidence that microtubule depolymerization early in the cell cycle is sufficient to initiate DNA synthesis. Cell **23:** 61–71.
6. FRIEDKIN, M., A. LEGG & E. ROZENGURT. 1980. Enhancement of DNA synthesis by colchicine in 3T3 mouse fibroblasts stimulated with growth factors. Exp. Cell Res. **129:** 23–30.
7. FRIEDKIN, M., A. LEGG & E. ROZENGURT. 1979. Antitubulin agents enhance the stimulation of DNA synthesis by polypeptide growth factors in 3T3 mouse fibroblasts. Proc. Natl. Acad. Sci. USA **76:** 3902–3912.
8. MCCLAIN, D. M. & G. M. EDELMAN. 1980. Density-dependent stimulation and inhibition of cell growth by agents that disrupt microtubules. Proc. Natl. Acad. Sci. USA **77:** 2748–2753.
9. OTTO, A. M., M. O. ULRICH, A. ZUMBE & L. JIMENEZ DE ASUA. 1981. Microtubule disrupting agents affect two different events regulating the initiation of DNA synthesis in Swiss 3T3 cells. Proc. Natl. Acad. Sci. USA **78:** 3063–3067.
10. OTTO, A., A. ZUMBE, L. GIBSON, A. M. KUBLER & L. JIMENEZ DE ASUA. 1979. Cytoskeleton-disrupting drugs enhance effect of growth factors and hormones on initiation of DNA synthesis. Proc. Natl. Acad. Sci. USA **76:** 6435–6438.
11. TENG, M., J. C. BARTHOLOMEW & M. J. BISSELL. 1977. Synergism between anti-microtubule agents and growth stimulants in enhancement of cell cycle traverse. Nature (London) **268:** 739–741.
12. CROSSIN, K. L. & D. H. CARNEY. 1981. Microtubule stabilization by taxol inhibits initiation of DNA synthesis by thrombin and epidermal growth factor. Cell **27:** 341–350.
13. SCHIFF, P. B. & S. B. HORWITZ. 1980. Taxol stabilizes microtubules in mouse fibroblast cells. Proc. Natl. Acad. Sci. USA **77:** 1561–1565.
14. SCHIFF, P. B., J. FANT & S. G. HORWITZ. 1979. Promotion of microtubule assembly *in vitro* by taxol. Nature (London) **277:** 665–667.
15. SCHIFF, P. B. & S. B. HORWITZ. 1981. Taxol assembles tubulin in the absence of exogenous guanosine 5'-triphosphate or microtubule-associated proteins. Biochemistry **20:** 3247–3252.
16. THOMPSON, W. C., L. WILSON & D. L. PURICH. 1981. Taxol induces microtubule assembly at low temperature. Cell Motility **1:** 445–454.
17. WILSON, L., H. P. MILLER, K. W. FARRELL, K. B. SNYDER, W. C. THOMPSON & D. L. PURICH. 1984. Taxol stabilization of microtubules *in vitro:* Dynamics of tubulin addition and loss at opposite microtubule ends. Biochemistry **24:** 5254–5262.
18. MARGOLIS, R. L. & L. WILSON. 1978. Opposite end assembly and disassembly of microtubules at steady state *in vitro.* Cell **13:** 1–8.

19. THOMPSON, W. T., D. J. ASAI & D. H. CARNEY. 1984. Heterogeneity among microtubules of the cytoplasmic microtubule complex detected by a monoclonal antibody to alpha tubulin. J. Cell Biol. **98:** 1017–1025.
20. BRINKLEY, B. R., S. M. COX & D. A. PEPPER. 1980. Structures of the mitotic apparatus and chromosomes after hypotonic treatment of mammalian cells *in vitro.* Cytogenet. Cell Genet. **26:** 165–174.
21. DEERY, W. J. & B. R. BRINKLEY. 1982. Induction of microtubule assembly *in situ* from unpolymerized tubulin pools in SV-40 transformed 3T3 cells. *In* Gene Regulation Cetus-UCLA Symposium on Molecular and Cellular Biology. B. W. O'Mally & C. F. Fox, Eds.: **26:** 327–341. Academic Press. N.Y.
22. SALMON, E. D. & D. A. BEGG. 1980. Functional implications of cold-stable microtubules in kinetochore fibers of insect spermatocytes during anaphase. J. Cell Biol. **85:** 853–865.
23. CARNEY, D. H., K. C. GLENN & D. D. CUNNINGHAM. 1978. Conditions which affect initiation of animal cell division by trypsin and thrombin. J. Cell. Physiol. **95:** 13–22.
24. FULLER, G. M., B. R. BRINKLEY & J. M. BOUGHTER. 1975. Immunofluorescence of mitotic spindles by using monospecific antibody against bovine brain tubulin. Science **187:** 948–950.
25. ST. JEOR, S. C., T. ALBRECHT, F. D. FUNK & F. RAPP. 1974. Stimulation of cellular DNA synthesis by human cytomegalovirus. J. Virol. **13:** 353–362.
26. GERSHON, D., L. SACHS & E. WINOCOUR. 1966. The induction of cellular DNA synthesis by simian virus 40 in contact-inhibited and x-irradiated cells. Proc. Natl. Acad. Sci. USA **56:** 918–925.
27. GUNDERSEN, G. G., M. H. KALNOSKI & J. C. BULINSKI. 1984. Distinct populations of microtubules: tyrosinated and nontyrosinated alpha tubulin are distributed differently *in vivo.* Cell **38:** 779–789.
28. WILSON, L. 1975. Microtubules as drug receptors: pharmacological properties of microtubule protein. Ann. N.Y. Acad. Sci. **253:** 213–231.
29. WILSON, L. 1970. Properties of colchicine binding protein from chick embryo brain. Interactions with vinca alkaloids and podophyllotoxin. Biochemistry **9:** 4999–5007.
30. WILSON, L., J. R. BAMBURG, S. B. MIZEL, L. M. GRISHAM & K. M. CRESWELL. 1974. Interaction of drugs with microtubule proteins. Fed. Proc. Fed. Am. Soc. Exp. Biol. **33:** 158–166.
31. CHOU, I-N, J. ZEIGER & E. RAPAPORT. 1984. Imbalance of total cellular nucleotide pools and mechanism of the colchicine-induced cell activation. Proc. Natl. Acad. Sci. USA **81:** 2401–2405.
32. BEN-ZEEV, A., S. R. FARMER & S. PENMAN. 1979. Mechanisms of regulatory tubulin synthesis in cultured mammalian cells. Cell **17:** 319–325.
33. CLEVELAND, D. W., M. A. LOPATA, P. SHERLINE & M. W. KIRSHNER. 1981. Unpolymerized tubulin modulates the level of tubulin mRNAs. Cell **25:** 537–546.
34. CLEVELAND, D. W., M. F. PITTENGER & J. R. FERAMISCO. 1983. Elevation of tubulin levels by microinjection suppresses new tubulin synthesis. Nature (London) **305:** 738–740.

Microtubule-Dependent Reticulopodial Surface Motility: Reversible Inhibition on Plasma Membrane Blebs

SAMUEL S. BOWSER AND CONLY L. RIEDER

Wadsworth Center for Laboratories and Research
New York State Department of Health
Albany, New York 12201

Reticulopodia of the foraminiferan protozoan *Allogromia* display rapid (up to 11 μm/sec) and bidirectional saltatory transport of plasma membrane surface markers (polystyrene microspheres; 0.45 to 14 μm diameter).[1,2] We have recently used correlative videomicroscopic and electron microscopic methods to investigate the structural basis of this phenomenon within highly flattened reticulopodia.[3,4] Our results indicate (1) that reticulopodial surface transport is energy-dependent; (2) that it occurs only along that area of plasma membrane that directly overlies cytoplasmic fibrils previously shown[5] to be involved in the bidirectional streaming of intracellular organelles; (3) that these fibrils are composed of bundled microtubules and an associated fibrillar matrix that provides morphological links between the microtubules and membranous cytoplasmic components (*i.e.*, membrane-bound organelles and the plasma membrane); (4) that surface marker particles and intracellular organelles remain bound to microtubules, by way of the matrix material, after the reticulopodial membrane has been removed by treatment with nonionic detergent; and (5) that reticulopodial surface transport is inhibited by agents (*e.g.* cold, colchicine) that induce microtubule disassembly, but not by cytochalasins B and D. Together these findings support our hypothesis that surface motility in *Allogromia* is mediated by labile cytoplasmic microtubules.[4]

In this communication we present new experimental evidence that further strengthens the above hypothesis. Regions of the reticulopodial surface form blebs (FIGURE 1A) when *Allogromia* are treated with a calcium-depleted seawater substitute made hypotonic with D_2O (3:1, v/v). Electron microscopy (FIGURE 1B) of such blebs reveals that the plasma membrane becomes physically dissociated from the underlying microtubule cytoskeleton (which appears *in vivo* as a phase-dense axial core, *c.f.* FIGURES 1A and B). The transport of intracellular organelles and extracellular microspheres is unaffected by this treatment along those areas of reticulopods lacking blebs. Cytoplasmic organelles even continue to saltate within the blebs along the axial microtubule core. Organelles that become detached from the axial core, however, undergo rapid Brownian motion within the fluid environs of the bleb (not illustrated). By contrast, the saltatory transport of membrane surface markers ceases at the membrane blebs (FIGURE 2). Upon reaching a bleb, small (0.45 μm diameter) surface markers may undergo random (Brownian) motion on the surface of the blebs; larger (6 μm diameter) surface markers do not show this random motion. Finally, small microspheres displaying Brownian motion on blebs can reassociate with unblebbed regions and once again undergo rapid saltatory transport (FIGURE 2).

We interpret these observations to indicate that, on blebs, components of the plasma membrane that are associated with the surface markers become reversibly uncoupled from the underlying microtubular cytoskeleton. That small microspheres

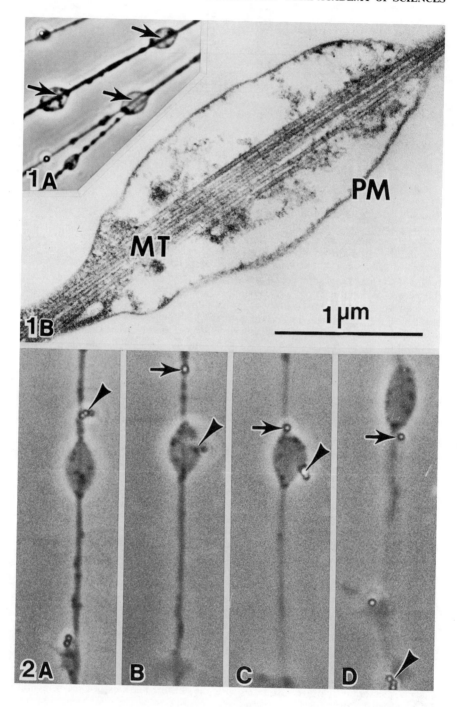

can undergo Browninan motion on the bleb surface, whereas larger microspheres are essentially motionless, directly attests to the fluidity of the blebbed membrane. These observations therefore provide additional evidence that both intracellular and extracellular (surface) motility in reticulopods require an intimate association between the transported membrane elements (*i.e.*, organelles or plasma membrane patches labeled with microspheres) and the microtubule cytoskeleton.

REFERENCES

1. BOWSER, S. S., *et al.* 1984. Cell Biol. Int. Rep. **8:** 1051.
2. BOWSER, S. S. & R. A. BLOODGOOD. 1984. Cell Motility **4:** 305.
3. BOWSER, S. S. 1984. Ph.D. Thesis. State University of New York at Albany.
4. BOWSER, S. S. & C. RIEDER. 1984. Can. J. Biochem. Cell Biol. **63:** 608.
5. TRAVIS, J. L., *et al.* 1983. J. Cell Biol. **97:** 1668.

FIGURE 1. Plasma membrane blebs induced by treating *Allogromia* reticulopodia with a calcium-depleted seawater substitute made hypotonic with D_2O. Phase-contrast light microscopy reveals that bidirectional streaming of cytoplasmic granules continues along a phase-dense core within membrane blebs (arrows, FIGURE 1A). Thin-section electron microscopy (FIGURE 1B) demonstrates that the plasma membrane (PM) has lifted away from the microtubule cytoskeleton (MT) within these blebbed regions. Note the subplasmalemmal densities (especially at PM) and fibrillar matrix associated with the microtubules. Magnification: 1A, × 200; 1B, × 72,000.

FIGURE 2. Sequential phase-contrast light micrographs showing polystyrene microspheres (0.45 µm diameter) in motion on an *Allogromia* reticulopod. Arrowhead points to a cluster of microspheres that encounters a plasma membrane bleb, whereupon it undergoes random movements (FIGURE 2B and C). After 30 seconds it reassociates with the reticulopod on the opposite side of the bleb and resumes rapid, saltatory transport (FIGURE 2D). Another microsphere (arrow) similarly encounters the bleb, undergoes Brownian motion, and reassociates with the opposite reticulopodial surface (FIGURES 2B–D). (FIGURE 2D has been shifted slightly with respect to the other figures.) Magnification: × 333. Time (in seconds): (A) 0; (B) 15; (C) 30; (D) 60.

Intracellular Transport in *Reticulomyxa*

URSULA EUTENEUER,[a] KENT L. McDONALD,[b]
MICHAEL P. KOONCE,[a] AND MANFRED SCHLIWA[a]

[a]*Department of Zoology*
[b]*Department of Botany*
University of California
Berkeley, California 94720

We have recently isolated a giant multinucleate freshwater amoeba closely resembling *Reticulomyxa*[1] and have begun to analyze the structural and biochemical basis of the spectacular form of organelle movements it displays. *Reticulomyxa* possesses a branched cell body consisting of several to many sausage-shaped strands, each approximately 50 μm in diameter. Within these thick strands, vigorous and frequently countercurrent cytoplasmic streaming is prominent. An extensive and highly branched reticulate network of fine filamentous strands extends from this cell body for distances of up to several centimeters. This fine network participates in food capture transport to the cell body. Particles and inclusions such as mitochondria, food vacuoles, and cytoplasmic droplets travel bidirectionally along these fine network strands at rates of up to 25 μm/second. The characteristics of movement, as well as the overall morphology of the reticulate network, are similar to certain *Foraminifera* (*e.g.*, reference 2). Immunofluorescence, transmission, and high voltage electron microscopy demonstrate the presence of a pervasive microtubule system throughout the network and within the cell body. There is an apparent correlation between the diameter of a strand and the number of microtubules. The finest strands, measuring about 100 nm, contain only a single microtubule. A correlative video light-microscopic and high-voltage electron microscopic analysis of the same strands demonstrates that particles can move bidirectionally in these fine strands containing single microtubules. A single microtubule is therefore sufficient to support movement in both directions,[3] a condition also found in other studies.[4,5] Thin sections of reticulopodal strands fixed according to a modified fixation protocol[6] demonstrate a striking, parallel colocalization of microtubules and fine, 6 nm filaments (FIGURE 1a) along the long axes of the strands. Transverse sections show the filaments to be interspersed between the microtubules (FIGURE 1b). Whereas the identity of these filaments as F-actin has yet to be demonstrated by heavy meromyosin decoration, double labeling experiments for fluorescence microscopy with antibodies against tubulin (kindly provided by M. Kirschner) and rhodamine-phalloidin (kindly provided by T. Wieland) strongly suggest that they are. These experiments show a striking correspondence of the labeling patterns (FIGURE 2), indicating that the phalloidin-stainable component is located in very close proximity to the microtubules.

Initial biochemical analyses of cytoplasmic extracts of *Reticulomyxa* suggest the presence of a factor or factors capable of cross-linking microtubules into large, needle-shaped bundles that are easily visible in the light microscope. When bovine brain microtubules are carried through two cycles of assembly/disassembly in the presence of amoeba extract, large microtubule bundles are formed. These bundles contain several minor components besides tubulin, including a high molecular weight component derived from the extract. Further experiments are directed towards characterizing the bundling factor and other proteins involved in supporting saltatory organelle movements.

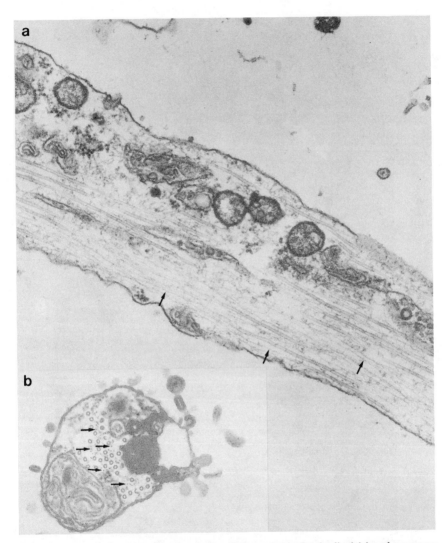

FIGURE 1. Electron micrographs of reticulopodial strands in a longitudinal (a) and transverse section (b) prepared with the osmium-ferricyanide procedure.[6] Numerous 6 nm filaments (arrows) are oriented parallel to, and interspersed between, the microtubules. (Original magnification × 60,000; reduced by 32 percent.)

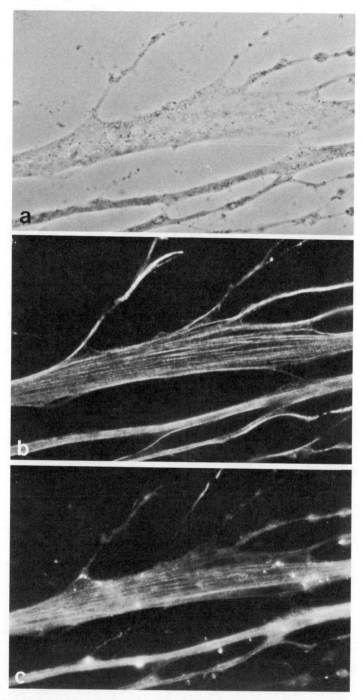

FIGURE 2. Fluorescence microscopy of double-labeled reticulopodial strands. Phase contrast (**a**), tubulin immunofluorescence (**b**), and rhodamine-phalloidin staining (**c**) of several strands fixed with glutaraldehyde, followed by permeabilization with 0.1% Triton X-100. Note the remarkable congruence of tubulin and phalloidin staining within the flattened strand. (× 550).

REFERENCES

1. NAUSS, R. N. 1949. Bull. Torrey Bot. Club **76:** 161–174.
2. JAHN, T. L. & R. A. RINALDI. 1959. Biol. Bull. **117:** 100–118.
3. KOONCE, M. P. & M. SCHLIWA. 1985. J. Cell Biol. **100:** 322–326.
4. HAYDEN, J. H. & R. D. ALLEN. 1984. J. Cell Biol. **99:** 49a.
5. SCHNAPP, B. J., M. P. SCHEETZ, R. D. VALE & T. S. REESE. 1984. J. Cell Biol. **99:** 351a.
6. M.DONALD, K. 1984. J. Ultrastruc. Res. **86:** 107–118.

Intracellular pH Shift Initiates Microtubule-Mediated Motility during Sea Urchin Fertilization[a]

GERALD SCHATTEN, TIMOTHY BESTOR,
RON BALCZON, JOHN HENSON, AND
HEIDE SCHATTEN

Department of Biological Science
Florida State University
Tallahassee, Florida 32306

INTRODUCTION

The sea urchin egg at fertilization is a unique model for investigating the ionic regulation of microtubule assembly and microtubule-mediated motility (reviewed in reference 1). Though the unfertilized egg does not have any assembled microtubules, within five minutes of insemination, a radial monaster assembles, nucleated by the incorporated sperm centriole.[2] This sperm aster is responsible for the movements leading to the fusion of the sperm and egg nuclei during fertilization.[3,4]

In addition to the well-displayed microtubule-mediated movements of the sperm and egg nuclei, the ionic sequence of egg activation is well understood.[5] An intracellular calcium transient initiates a sodium:proton exchange that directly results in a shift in intracellular pH.

RESULTS

When the intracellular pH is reduced to unfertilized values after the formation of the sperm aster, and immediately prior to the migration of the female pronucleus with 10 mM sodium bicarbonate at pH 6.5 (FIGURE 1), the expected microtubule-mediated motions are not observed. In FIGURE 1, in which external pH recorded with a 1.2 mm pH microelectrode within 1 mm of the studied egg is electronically superimposed on the differential interference contrast video image, sperm incorporation (FIGURE 1A) and the formation of the sperm aster are observed (FIGURE 1C). At the time expected for the egg nuclear migration, when the pH is reduced, no motions are observed (FIGURE 1D). Movements are not detected during the subsequent 90 minutes while the intracellular pH is maintained at unfertilized values (FIGURE 1E).

When these cells are permitted to undergo the normal shift in intracellular pH by removal of the sodium bicarbonate (FIGURE 1E), the migration of the egg nucleus is seen to occur within five minutes (FIGURE 1F), and the nuclei move to the egg center by fifteen minutes later (FIGURE 1G). Cleavages into 2 (FIGURE 1H) and 4 (FIGURE 1I) cells occur at an hour and one-and-a-half hours after recovery, respectively. Antitubu-

[a]Support of this research by the National Institutes of Health and the National Science Foundation is gratefully acknowledged.

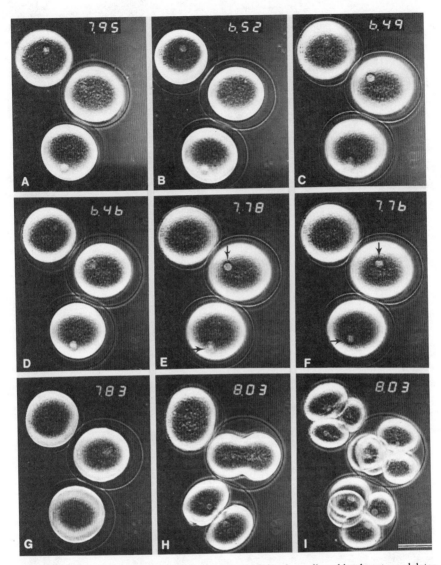

FIGURE 1. Suppression of the cytoplasmic alkalinization by sodium bicarbonate and later recovery. Sperm incorporation, formation of the sperm aster, and the central movement of the sperm nucleus have occurred when 10 mM sodium bicarbonate, pH 6.5, is added at 6.5 minutes postinsemination. **A:** The expected migrations of the egg nuclei are not observed, and further development is arrested. **B–E:** At 90 minutes the bicarbonate is removed, (**E**), and the external pH elevated with sea water to about pH 8.0. Immediately, the sperm aster forms and the egg nuclei (**E, F** arrows) migrate by five minutes (**F**). Cleavages to two (**H**) and four (**I**) cells are noted at one and one-and-a-half hours after the removal of the bicarbonate. Insert: extracellular pH, top right.

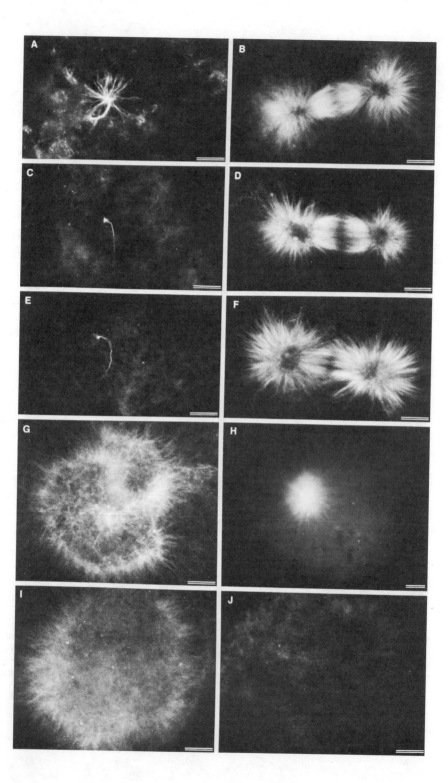

lin immunofluorescence microscopy of fertilization, artificial activation, and manipulations with intracellular pH support the conclusion that an intracellular pH shift from about 6.8 to about 7.2 will include microtubule assembly and with it microtubule-mediated motility. In FIGURES 2A and 2B, the radial sperm aster that forms within five minutes of insemination and the mitotic apparatus are depicted in untreated eggs that have undergone the normal pH shift. In FIGURE 2C the cytoplasmic pH has been secondarily suppressed with sodium bicarbonate, and in FIGURE 2E the pH shift accompanying fertilization has been blocked by insemination in the absence of sodium ions. The cytoplasmic microtubules are unable to assemble, and only the sperm axoneme is detected. Interestingly, if these inhibited cells are later permitted to undergo the cytoplasmic alkalinization, the sperm aster, streak and mitotic apparatus form during subsequent culturing. FIGURES 2D and 2F are the mitotic apparatus of cells incubated in bicarbonate or sodium-free medium for 90 minutes and then allowed to recover for an additional hour.

It is of importance that unfertilized eggs in which the intracellular pH change is artificially induced with the divalent ionophore A23187 (FIGURES G and H), which induces both an intracellular calcium release as well as the pH shift, or with 10 mM HCl (FIGURE 2I), which induces solely the shift in intracellular pH, results in the appearance of monasters that move the female pronucleus to the egg center. Ionophore activation in sodium-free medium, which triggers a calcium release but does not lead to a change in intracellular pH, does not result in any microtubule assembly (FIGURE 2J).

DISCUSSION

In this study, the shift in intracellular pH is found to be essential for the formation and functioning of the egg microtubules during fertilization.[6] When the pH shift at fertilization is blocked or suppressed within five minutes of insemination, neither the microtubules comprising the sperm aster nor their mediated motions are observed. If the pH shift is reduced after the formation of the sperm aster, the microtubules disassemble and the nuclei are not moved. If fertilized eggs in which the pH shift is prevented, or later induced, undergo the pH shift, microtubules assemble and nuclear movements and later development occur normally, but on a schedule delayed by the length of time during which the pH was arrested (FIGURE 1). Artificial elevation of the intracellular pH of unfertilized eggs induces both the assembly of microtubules

---

FIGURE 2. Antitubulin immunofluorescence microscopy of fertilization and artificial activation. Though the unfertilized egg is devoid of assembled microtubules, the radial sperm aster forms within five minutes of insemination (**A**); note the incorporated sperm axoneme. The streak and the mitotic apparatus (**B**) are observed during the first cell cycle. If the cytoplasmic alkalinization is prevented or secondarily suppressed with sodium bicarbonate (**C**), sodium acetate or sodium free medium (**E**), the expected microtubule-containing structures are not observed. Only the incorporated sperm axoneme is apparent at fifteen minutes postinsemination. If these inhibited cells are later permitted to undergo the cytoplasmic alkalinization (recovery from sodium-free medium, sodium acetate, or sodium bicarbonate), the sperm aster, streak and mitotic apparatus form during subsequent culturing. Figures D and F are the mitotic apparatus cells incubated with bicarbonate (**D**) or sodium free medium (**F**) for 90 minutes and then allowed to recover for an additional hour. Artificial activation with the divalent ionophore A23187 (**G, H**) or NH_4Cl (**I**) results in the appearance of a monaster. Ionophore activation in sodium-free medium does not result in any microtubule assembly (**J**).

nucleated by typical maternal centers and the movement of the egg nucleus to the cell center. These results indicate that the shift in intracellular pH during egg activation regulates the assembly and with it the functioning of the microtubules during fertilization.

REFERENCES

1. SCHATTEN, G. 1984. The supramolecular organization of the cytoskeleton during fertilization. *In* Subcellular Biochemistry. D. B. Roodyn, Ed.:**10:** 359–453. Plenum Press. New York.
2. BALCZON, R. & G. SCHATTEN. 1983. Microtubule-containing detergent-extracted cytoskeletons in sea urchin eggs from fertilization through cell division: Antitubulin immunofluorescence microscopy. Cell Motility **3:** 213–226.
3. SCHATTEN, G. & H. SCHATTEN. 1981. Effects of motility inhibitors during sea urchin fertilization. Exp. Cell Res. **135:** 311–330.
4. SCHATTEN, G., H. SCHATTEN, T. H. BESTOR & R. BALCZON. 1982. Taxol inhibits the nuclear movements during fertilization and induces asters in unfertilized sea urchin eggs. J. Cell Biol. **94:** 455–465.
5. STEINHARDT, R., R. ZUCKER & G. SCHATTEN. 1977. Intracellular calcium release at fertilization in the sea urchin egg. Dev. Biol. **58:** 185–196.
6. SCHATTEN, G., T. BESTOR, R. BALCZON, J. HENSON & H. SCHATTEN. 1985. Intracellular pH shift leads to microtubule assembly and microtubule-mediated motility during sea urchin fertilization: correlations between elevated intracellular pH and microtubule activity and depressed intracellular pH and microtubule disassembly. Eur. J. Cell Biol. **36:** 116–126.

Microtubules in Mouse Oocytes, Zygotes, and Embryos during Fertilization and Early Development: Unusual Configurations and Arrest of Mammalian Fertilization with Microtubule Inhibitors[a]

GERALD SCHATTEN, CALVIN SIMERLY, AND
HEIDE SCHATTEN

Department of Biological Science
Florida State University
Tallahassee, Florida 32306

INTRODUCTION

The goal of fertilization is the union of the parental genomes in most animals when the microtubule-containing cytoskeleton forming within the activated egg participates in the motility necessary for the cytoplasmic migrations of the sperm and egg nuclei (reviewed in reference 1). To explore the activity of egg cytoplasmic microtubules during mammalian fertilization and early development, we have performed anti-tubulin immunofluorescence[2] and transmission electron microscopy on mouse oocytes and zygotes throughout fertilization and have explored the effects of colcemid, griseofulvin, and nocodazole.

RESULTS

In the unfertilized oocyte, microtubules were predominantly found in the arrested meiotic spindle[3,4] attached to the cell surface (FIGURE 1A). A dozen cytoplasmic microtubule-containing asters assemble at the time for sperm incorporation. These asters are initially nucleated by maternal microtubule organizing centers (MTOCs) and are frequently found in direct association with the pronuclear surfaces (FIGURE 1B). The incorporated sperm axoneme and the midbody from the meiotic spindle persists (FIGURES 1B and C) as the male and female pronuclei move to the egg center. These asters enlarge, filling the cytoplasm with a dense microtubule-containing matrix; the pronuclei are embedded within the matrix center (FIGURES 1D–F). At the completion of first interphase, the matrix disassembles from the interior towards the periphery, and a sheath of microtubules surrounds the adjacent but still separate pronuclei.

Pronuclear fusion is not observed; rather, following nuclear envelope breakdown at mitosis, the adjacent paternal and maternal chromosome sets first meet at the

[a]Support of this research by the National Institutes of Health and the National Science Foundation is gratefully acknowledged.

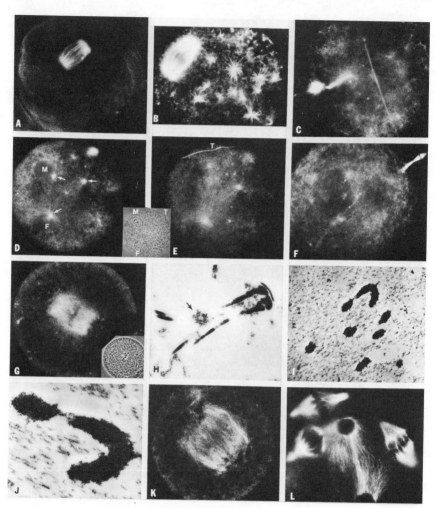

FIGURE 1. Microtubules in mouse oocytes and zygotes during fertilization and early development. **A:** Unfertilized oocyte. Microtubules are found predominantly in the anastral barrel-shaped meiotic spindle. **B:** Cytoplasmic asters. At the time for sperm incorporation, over a dozen cytoplasmic asters appear in the egg cytoplasm before sperm incorporation. **C:** Sperm incorporation. The sperm axoneme is apparent at the sperm incorporation. **D–E:** Pronucleate egg with cytoplasmic asters (arrows). The cytoplasmic asters enlarge as the pronuclei form and asters are frequently found in association with the male and female pronuclei. Inset: phase contrast micrograph; M: Male pronucleus, F: female pronucleus, T: incorporated sperm axoneme. **F:** Late pronucleate egg with cytoplasmic microtubule-containing matrix. **G:** Metaphase. Note barrel-shaped anastral spindle. Inset: phase contrast micrograph. **H–J:** Transmission electron microscopy. **H:** The incorporated sperm axoneme with an embedded centriole (arrow) is apparent and is found in the cytoplasmic region devoid of microtubules, though numerous parallel microtubules are found in the spindle. **I–J:** Numerous parallel microtubules are found in the spindle region. **K:** Metaphase. **L:** Third Division. The mitotic spindles now are typical with well-focused spindle poles.

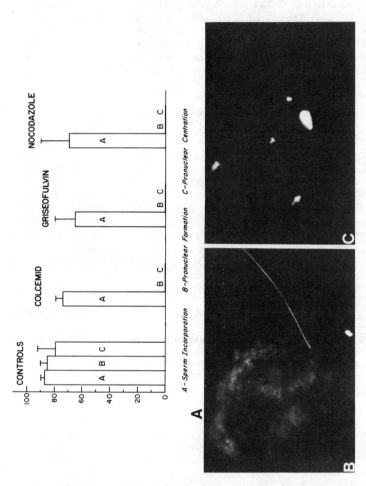

FIGURE 2. Microtubule inhibitors prevent pronuclear formation and apposition, but not sperm incorporation during mammalian fertilization. **A:** Effects of colcemid, griseofulvin and nocodazole. **B:** Antitubulin immunofluorescence of colcemid-treated egg ten hours postinsemination. **B:** Antitubulin immunofluorescence localizes the incorporated sperm axoneme; cytoplasmic microtubules have been disassembled. C:Hoechst DNA fluorescence reveals the incorporated sperm nucleus and the maternal chromosomes that are scattered along the egg cortex. Pronuclear development is arrested in treated eggs.

metaphase plate. The metaphase mitotic apparatus emerges from this perinuclear microtubule sheath and is barrel-shaped and anastral, with broad mitotic poles reminiscent of plant spindles (FIGURE 1G). The incorporated sperm centriole does not appear to participate in the nucleation of mitotic microtubules (FIGURES 1H–K). At anaphase, sparse microtubules appear; at telophase, interzonal microtubules form; and after cleavage, the mitotic spindle is replaced by monasters extending from each blastomere nucleus. Third and later divisions are characterized by mitotic spindles typical for higher cells with narrow mitotic poles and fusiform shapes (FIGURE 1L).

Colcemid (50 μM), griseofulvin (100 μM), and nocodazole (10 μM) inhibit the formation of these microtubules and prevent the movements leading to pronuclear reunion (FIGURE 2A). The meiotic spindle is disassembled in the presence of these inhibitors, and maternal chromosomes are scattered throughout the oocyte cortex (FIGURES 2B and C).

DISCUSSION

The migrations leading to the union of the sperm and egg nuclei at the mouse egg center require the formation of cytoplasmic microtubules as in other animal systems. Inhibition of these microtubules prevents this movement and surprisingly prevents the normal decondensation of the sperm nucleus and meiotic chromosomes. Microtubules in the mouse egg during fertilization appear to be required for the purpose of separation and alignment of the maternal meiotic chromosomes and for the movement of the male and female pronuclei from the cortex into close apposition at the egg center. In contrast to those in most other animals, however, the microtubules are nucleated by numerous maternal sources rather than a single monaster organized by the incorporated sperm centriole, and centrioles may not be paternally inherited during mammalian fertilization.[5,6]

REFERENCES

1. SCHATTEN, G. 1984. The supramolecular organization of the cytoskeleton during fertilization. *In* Subcellular Biochemistry. D. B. Roodyn, Ed.:**10:** 357–451.
2. BRINKLEY, B. R., F. S. FISTEL, J. M. MARCUM & R. L. PARDUE. 1980. Microtubules in cultured cells: Indirect immunofluorescent staining with tubulin antibody. Int. Rev. Cytol. **63:** 59–95.
3. SZOLLOSI, D., P. CALARCO & R. P. DONAHUE. 1972. Absence of centrioles in the first and second meiotic spindles of mouse oocytes. J. Cell Sci. **11:** 521–541.
4. CALARCO-GILLAM, P. D., M. C. SIEBERT, R. HUBBLE, T. MITCHISON & M. KIRSCHNER. 1983. Centrosome development in early mouse embryos as defined by an auto-antibody against pericentriolar material. Cell **35:** 621–629.
5. SCHATTEN, G., C. SIMERLY & H. SCHATTEN. 1985. Microtubule configurations during fertilization, mitosis and early development in the mouse and the requirement for egg microtubule-mediated motility during mammalian fertilization. Proc. Natl. Acad. Sci. USA **82:** 4152–4156.
6. SCHATTEN, H., G. SCHATTEN, D. MAZIA, R. BALCZON & C. SIMERLY. 1986. Behavior of centrosomes during fertilization and cell division in mice and sea urchins. Proc. Natl. Acad. Sci. USA **83:** 105–109.

Automaton Model of Dynamic Organization in Microtubules

STUART R. HAMEROFF,[a] STEVEN A. SMITH,[b] AND
RICHARD C. WATT[a]

[a]*Department of Anesthesiology*
University of Arizona Health Sciences Center
Tucson, Arizona 85724

[b]*Computing Division*
Los Alamos National Laboratories
Los Alamos, New Mexico 87545

Microtubules (MT) participate in a wide range of organized dynamical activities that apparently involve information processing. Cellular automata are dynamical systems that can generate and process patterns and information by way of rules based on lattice neighbor interactions.[1-3] Lattice structure and apparent information processing capabilities have suggested an MT-automaton behavior based on coherent dipole oscillations within MT subunits.[4-6]

Various models of dynamic activities in biomolecules predict functional organization and regulation. Among these, Frohlich has theorized that nonlinear dipole excitations among subunits of biomolecular arrays result in coherent oscillations (10^{-10} to 10^{-11} sec fluctuations), lower frequency metastable states, long-range cooperativity and order, and coupling of protein conformational states to dipole oscillations within protein hydrophobic regions.[7,8] Individual subunit conformation within a lattice array would thus depend on various factors (GTP, Ca^{++}, primary structure, lattice neighbor electrostatic interactions) influencing oscillation phase. Based on MT geometry, we have calculated MT-lattice neighbor electrostatic influences on dipole oscillations as rules for an MT-automaton computer simulation (FIGURE 1). Dimer states at each generation can be determined by neighbor states at the previous generation.

$$\text{state} = \alpha, \text{ if } \sum_{i=1}^{n} f(y) > 0, \text{ state} = \beta, \text{ if } \sum_{i=1}^{n} f(y) < 0,$$

where $n = 7$ as the number of neighbors, and $f(y)$ the force from the "ith" neighbor in the y direction.

$$f(y) \propto \frac{\sin \theta}{r^2}, \sin \theta = \frac{y}{r}; f(y) \propto \frac{y}{r^3};$$

thus states may be determined by $\sum_{i=1}^{n} y/r^3$ and relative values calculated from known MT structure.

FIGURE 2 shows one sequence from an "opened" MT automaton simulation in which 13 MT protofilaments are arranged horizontally. Each generation represents one Frohlich oscillation period (~ one nanosecond). "a" represents an alpha-tubulin dipole oscillation phase, whereas dots connected by lines represent beta-tubulin oscillation phases. Generations 22, 28, 34, and 40 from a particular sequence demonstrate stable patterns (during even-numbered generations) and a traveling kink pattern (~80 meters/sec), which leaves an altered "wake" (? memory). These traveling patterns are known as gliders in automaton nomenclature and have been shown by von Neumann to be capable of universal computing.

Stable and moving patterns of an MT-subunit oscillation phase could function as specific sites of ion or protein binding and transport, orchestration of biomolecular activities, as well as stored information or memory. Interaction and interference of these patterns could process information as in computers and provide "real time" biological information processing. Although the MT-automaton rules and patterns described here are somewhat arbitrary, they demonstrate that MT-automaton activities are feasible and are of potential importance in dynamic organization.

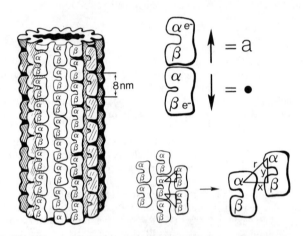

FIGURE 1. Left: MT structure from X-ray diffraction crystallography. Right top: MT-tubulin dimer subunits comprised of α and β monomers. Electron occupancy of either monomer may correlate with coherent dipole oscillations and are represented as "a" for α-monomer electron occupancy and dots connected by lines for β-monomer electron occupancy. Center bottom and right: hexagonal relations lead to neighbor rules based on lattice distances and electrostatic interactions. For neighbors shown (lower right), x = 5 nm, y = 4 nm, r = 6.4 nm.

REFERENCES

1. VON NEUMANN, J. 1966. Collected Works. A. H. Taub, Ed. **5:** 288.
2. Cellular Automata. 1984. Proceedings of an Interdisciplinary Workshop, Los Alamos. D. Farmer, T. Toffoli & S. Wolfram, Eds. North Holland, Amsterdam.
3. WOLFRAM, S. 1984. Cellular automata as models of complexity. Nature (London) **311:** 419–424.
4. SMITH, S. A., R. C. WATT & S. R. HAMEROFF. 1984. Cellular automata in cytoskeletal lattices. Physica **10D:** 168–174.
5. HAMEROFF, S. R., S. A. SMITH, R. C. WATT. 1984. Nonlinear electrodynamics in cytoskeletal protein lattices. *In* Nonlinear Electrodynamics in Biological Systems. W. R. Adey & A. F. Lawrence, Eds.: 567—583. Plenum Press. New York.

FIGURE 2. Generations 22, 28, 34, and 40 from an MT-automaton simulation. Stable patterns have evolved during even-numbered generations. A traveling "glider" has left an altered "wake," which may be likened to a computer operation.

6. HAMEROFF, S. R. & R. C. WATT. 1982. Information processing in microtubules. J. Theor. Biol. **98:** 549–561.
7. FROHLICH, H. 1970. Long range coherence and the action of enzymes. Nature (London) **228:** 1093.
8. FROHLICH, H. 1975. The extraordinary dielectric properties of biological materials and the action of enzymes. Proc. Natl. Acad. Sci. USA **72:** 4211.

Microtubular Screw Symmetry: Packing of Spheres as a Latent Bioinformation Code

DJURO L. KORUGA

Molecular Machines Research Unit
University of Belgrade
11000 Belgrade, Yugoslavia
and
University of Arizona Health Sciences Center
Tucson, Arizona 85724

INTRODUCTION

Microtubules (MT) participate in bioinformation processes such as memory and learning.[1-2] The precise role of MT in these functions is not well understood. If MT are primary information processors, then their subunits should be arranged by sphere-packing symmetry derived from information coding laws.

THE PACKING OF SPHERES

Oh $(\bar{6}/4)$ symmetry group describes face-centered-cubic sphere packing and derives information coding laws.[3] Hexagonal packing of protein monomers independent of the Oh $(\bar{6}/4)$ symmetry group has been used to explain the form patterns of viruses, flagella, and MT.[4] Because hexagonal packing and face-centered-cubic packing of spheres have equal density, I use both to explain MT organization.

Hexagonal packing may be used by fixing conditions. This is possible if the centers of spheres lie on the surface of a cylinder (with radius equal to the Oh $(\bar{6}/4)$ unit sphere) and if (and only if) the sphere values in the axial direction (lattice) of the cylinder by order of sphere packing is the same as in the dimension in which the face-centered-cubic packing is done.

Because $\bar{6}$-fold symmetry axis of Oh group is inverse, there must be two kinds of spheres (white and black) on the cylinder surface, but linked such that they have the dimension value in which the face-centered-cubic packing is done. This packing on the cylinder surface leads to "screw symmetry" (FIGURE 1).

From coding theory, the symmetry laws of α- and β-tubulin subunits lead to the conclusion that 13 protofilaments passes one of the best known binary error-correcting codes [K_1 (13, 2^6, 5)] with 64 code words.

Symmetry theory further suggests that on the surface of a circular cylinder in axial direction there must be a code of length of 24 monomer subunits. If the coding efficiency is used as a criterion of transmission, then 6-binary dimers of K_1 code should be coded to give a 4-dimer ternary sequence of K_2 [24, 3^4, 13] code. This code may result from interaction between 24 tubulin monomers and high molecular weight proteins. Under the influence of Ca^{2+}-calmodulin, binary dimers of K_1 code give dimer ternary sequence of K_2 code. In this way, K_1 and K_2 codes, which result from the property of the Oh $(\bar{6}/4)$ symmetry group, lead a K (B^6T^4) transmission bioinforma-

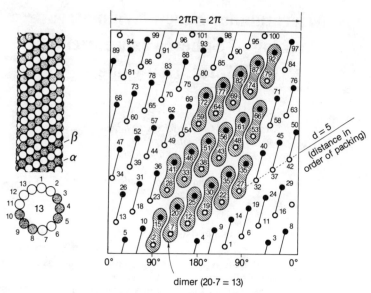

FIGURE 1. Because the optimal number of spheres for information processing is 11, 12, or 13, we show here the solution for the packing of 13 spheres. The distance between spheres in order of packing is 5. Sphere indices in axial direction of dimers (white-black) must be 13 because each dimer represents the dimension of face-centered-cubic packing. From information theory, this means that the arrangement of white and black spheres corresponds to a binary code of length n = 13 and distance d = 5 with 64 code words.

tion code, which may combine with guanosime triphosphate (GTP) and ions for intracellular dynamic and organizational activities.

CONCLUSION

Symmetry theory and MT structures lead to the conclusion that packing of tubulin subunits is equal to information coding. This means that microtubules possess a code

TABLE 1. K-function Code

	Code	
Digital Sum	$K_1[13, 2^6, 5]$	$K_2[24, 3^4, 13]$
0	18	19
1	32	16^+ 16^-
2	10	10^+ 10^-
3	4	4^+ 4^-
4	—	1^+ 1^-
All	$2^6 = 64$	$3^4 = 81$

Coding relations between K_1 and K_2 codes show that 18 binary words of K_1 code yield 19 ternary words of K_2 code; there exists one "nonsense" word. Further, we see that 14 binary words are ambiguous with ternary words. Finally two more "nonsense" (stop) and ambiguities are seen in coding of amino acids in the genetic code. Thus, MT may code for dynamic information processing in some elements like DNA codes for genetic structure.

system that can provide both memory and intracellular dynamic activities. The latent bioinformation K code has been identified and in that way showed that microtubules along with DNA and RNA are unique cell structures that possess a code system.

ACKNOWLEDGMENTS

I would like to thank Dr. Radmila Mileusnic, University of Belgrade and Dr. Stuart Hameroff, University of Arizona for helpful and stimulating discussion. I thank Pat Kime for excellent secretarial assistance.

REFERENCES

1. MILEUSNIC, R., S. P. R. ROSE & P. TILLSON. 1980. Learning and chick brain tubulin. J. Neurochem **34:** 1007–1014.
2. CRONLY-DILLON, J., D. CARDEN & C. BIRKS. 1974. The possible involvement of brain microtubules in memory fixation. J. Exp. Biol. **61:** 443–454.
3. SLOANE, H. J. A. 1984. The packing of spheres. Sci. Am. **250:** 116–125.
4. ERICKSON, R. D. 1973. Tubular packing of spheres in biological fine structure. Science **181:** 705–716.

Conference Summary

B. R. BRINKLEY

Department of Cell Biology and Anatomy
Schools of Medicine and Dentistry
University of Alabama in Birmingham
Birmingham, Alabama 35294

Ten years ago, the New York Academy of Sciences sponsored a conference entitled The Biology of Cytoplasmic Microtubules, which attracted scientists from many laboratories throughout the world. Since that time, knowledge and interest in microtubules and related cytoskeletal components has increased exponentially, resulting in the need for a second conference in 1984. In comparing my notes with the excellent summary of the 1974 conference by E.W. Taylor, it is apparent that investigators are still probing many of the same fundamental questions of a decade ago, but with an accent on molecular biology and the accompanying technology of gene cloning, sequencing, and monoclonal antibodies. Whereas the emphasis of the first conference was focused largely on tubulin and microtubules *in vitro,* the emphasis today is on microtubules *in vivo* using lysed cell models and even living cells in culture. Indeed, the capacity to apply cytochemistry to living cells as described by several participants of this conference, opens many new vistas in microtubule research. A major focus of the present conference was on the molecular biology of microtubules. Indeed, a missing link of a decade ago, namely knowledge of tubulin genes and genetics is now in place. The field is moving rapidly, and the areas of tubulin genetics, structure, evolution, nucleotide interaction, assembly-disassembly, regulation, and microtubule-associated proteins (MAPs) are being aggressively persued by many laboratories. More certain than ever is the notion that microtubules play a key role in motility, cytoplasmic transport, secretion, cell morphogenesis, sensory transduction, and membrane and cell surface properties. The evidence that they may be intermediates in signal transduction and initiation of DNA synthesis (Carney *et al.*) embraces even more noble functions including growth control and hormone action. In fact, the functional implications seem almost endless, and yet, the mechanisms of how microtubules actually carry out even a single function in the cell are as elusive as ever.

The current state of knowledge of microtubules from gene to assembled array was discussed in this conference and is outlined in FIGURE 1. The heterogeneity of α and β subunits suggested by the existence of multiple protein bands on SDS gels and by other earlier techniques is now explained by the existence of a multigene family for tubulin as shown by analysis of genomic DNA by Southern hybridization with cDNA probes.[1] The gene-cloning data has now been elegantly supported by genetic analysis in *Aspergillus* by Morris and coworkers, *Chlamydomonas* (Silflow *et al.*), and in mammalian cells (Cabral *et al.*). Although the number of tubulin mutants is still relatively small, the approach is powerful and some exciting new information on tubulin gene regulation is at hand. Interestingly, most of the mutations have involved the β subunit, and these have interfered with important cellular functions such as sporulation in *Aspergillus* (Morris *et al.*). Little and coworkers reported that animal tubulin genes are genetically very stable, with a greater heterogeneity between the amino acid sequence of yeast and *Physarum* than between birds and mammals. The latter underwent evolutionary divergence about 280 million years ago and yet show 99% sequence homology between the α subunits and a 95% homology between the β

subunit. The low mutation rates in animal tubulin, especially the α subunit, suggests that rigid contraints have been placed on the tubulins in these organisms due to their involvement in critical functions such as mitosis, meiosis, and embryogenesis.

The expression of the tubulin genes appear to vary considerably to different organisms. Silflow and coworkers reported that *Chlamydomonas* contained two α and two β genes, both of which were active and coordinately expressed throughout the life cycle of the organisms. The two β genes encode for identical β tubulins, and the same is true for the α, but the reason for the duality of expression was not yet apparent. In most animal cells, tubulin genes are developmentally expressed. *Drosophila* contain 4 α and 4 β genes with β-2 and β-3 being selectively expressed during development. The β-3 gene is expressed during embryogenesis and β-2, a testis specific gene, is activated during meiosis.

Using a rat α-tubulin cDNA probe, Ginzberg and coworkers identified 15–20 copies of tubulin genes in the rat genome but were uncertain as to how many were actually functional. Indeed, Cowan and coworkers found an unusually large number of

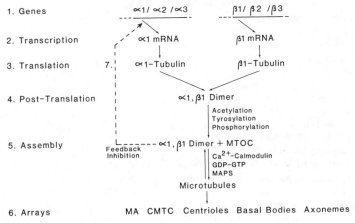

FIGURE 1. Tubulin synthesis and microtubule assembly involves at least seven discrete steps. All are discussed in this volume.

tubulin pseudogenes in the mammalian genome. Two identical tubulin pseudogenes are carried by gorillas and chimpanzees, two primates separated by four million years of evolution. How and why large numbers of tubulin pseudogenes have been maintained in the genome of animal cells remains unclear.

Gene expression and tubulin synthesis appears to be autoregulated in most eukaryotic organisms as reported by Cleveland and coworkers. It has long been known that tubulin protein levels and/or synthetic rate increases with progression of cells through the cell cycle.[2–5] Work by Ben-Ze'ev et al.[6] and Cleveland et al.[7] suggested that tubulin mRNA levels may be regulated by the amount of tubulin monomer in the cytoplasm. Indeed, microinjection of tubulin monomer into cultured cells to increase the level of cytoplasmic tubulin yielded equivalent results.[8] Cleveland and coworkers (see Lau et al.) have further dissected the molecular events underlying tubulin autoregulation by injecting cloned chicken β-tubulin genes into cultured mouse L cells. When an intracellular pool of tubulin was increased by drug-induced microtubule depolymerization, the chicken gene was down-regulated coordinately with the endogenous mouse α- and β-tubulin gene.

These interesting results indicate that cells can monitor the amount of unpolymerized tubulin in their cytoplasm and adjust the rate of synthesis accordingly. Such autoregulation may be unique to the tubulins and does not appear to be operational for other cytoskeletal proteins, including actin. It remains to be seen how and why such an elegant feedback regulation is confined to tubulin and microtubules. Farmer and coworkers reported that autoregulation of tubulin synthesis may be important in development. Obviously, the notion of a feedback loop in tubulin synthesis has other interesting implications. For example, To what extent can other synthetic events such as DNA synthesis (see Carney et al.) be influenced by the ratio of dimer/polymer in the cytoplasm? Or how does autoregulation relate to the critical concentration of tubulin needed for microtubule assembly at microtubule organizing centers (MTOCs)?

Another area of active research involves tubulin regulation by post-translational modification of the α and β subunits. Rosenbaum and coworkers presented an excellent example of how cytoplasmic α tubulin can be modified by this mechanism to become axoneme tubulin in regenerating flagella of Chlaymdomonas. Shortly after experimental removal of flagella, cytoplasmic α-1 tubulin is transported up the axoneme where it is acetylated to become α-3 before being incorporated into the axonemal microtubules.

In view of the highly conserved nature of microtubules and their subunit proteins, it is surprising that microtubules are involved in such a plethora of physiological functions. As noted by Soloman in these proceedings, one possible explanation is that functional specialization is facilitated by the noncovalent association of a family of rather minor components known as microtubule-associated proteins. MAPs assemble with microtubules in vitro and promote the assembly of purified tubulin (see Vallee for review). Some of the best known MAPs include MAP-1, MAP-2 with molecular masses of approximately 300,000 kD, tau protein 55,000–68,000 kD, and a family of widely distributed MAPs of 120,000 kD and 210,000 kD. The availability of libraries of monoclonal antibodies made against MAPs and tau has resulted in a better appreciation of the distribution of these proteins, and several new MAPs have recently been discovered. Reports from several laboratories made it abundantly clear that MAPs are indeed organ, tissue, and cell specific as shown beautifully by indirect immunofluorescence (Binder et al., Drubin et al., Matus and Riederer, Olmsted et al., Bloom et al., and Wiche et al.). Through immunocytochemical techniques, it is now apparent that MAPs are indeed bound to microtubules in living cells (Drubin et al.).

In addition to the more familiar MAPs, several new species have been discovered. Matus and coworkers have identified MAP-3 consisting of two native peptides with a molecular mass of 180,000 kD and MAP-5 consisting of a broad band migrating between MAPs 1 and 2 and a major breakdown product that migrates below MAP-3. Olmsted's group has identified yet another, MAP-4, a 210,000 kD protein found in a variety of mammalian cell lines and a large number of organs and tissues.

Obviously, the list of MAPs is growing and it is painfully clear that a better system of nomenclature is needed. The current system of numbering MAPs 1,2,3 . . . has limited application and leads to confusion and redundancy. Terms such as tau, tap, and STOPs are somewhat more descriptive but further confuse the literature. Clearly, this conference accentuated the need for a revised, universally acceptable system of nomenclature. A system that encompasses either functional or distributional parameters would be handy. Obviously, information on the MAP genes and their sequences would also be helpful in this regard. MAP and tau genes are being cloned, but little information on the molecular biology of MAPs was presented at this conference.

In addition to the expanding list of traditional MAPs, a number of other proteins maintain close association with microtubules and appear to play a significant role in regulating microtubule function. These include calmodulin, a cyclic adenosine monophosphate (cAMP)-dependent kinase, and a calcium/calmodulin-dependent kinase as reported by DeLorenzo and coworkers (see Vallano et al.). These proteins appear to be intermediates in the overall scheme of calcium regulation of microtubule assembly-disassembly. The calcium/calmodulin kinase appears to play a functional role in the phosphorylation of MAP-2 and possibly tubulin. Although the role of calmodulin in microtubule biology is not yet defined, its spatial proximity to microtubules of the mitotic spindle and cytoplasmic microtubule complex (Deery) suggests an important regulatory function. It is hoped that the biochemical sequala from calcium release to microtubule response should soon be known.

How microtubules are assembled and disassembled in cells is still an active and controversial area of research, and the experimental strategies encompass both *in vitro* and *in vivo* systems, as represented in this volume by the laboratories of Weisenberg, Caplow, Pantaloni and Carlier, Wilson, Burns, Sternlicht, Purich, Engelborghs, and Wolff. The bulk of the work centers around tubulin/nucleotide interactions and the nature of critical concentrations (Weisenberg). Of considerable interest is the kinetics of tubulin addition to, and loss from, the growing or shrinking microtubule, following a flurry of interest in treadmilling brought on by the experiments of Margolis and Wilson,[9] and Bergen and Borisy.[10] Thoughts on how tubulin dimers add to and leave the polymer has become more complex and mystifying. The fact that guanosine triphosphate (GTP) hydrolysis is not mechanistically coupled to tubulin polymerization led Pantaloni and Carlier to conclude that the ends of growing microtubules are capped by GTP tubulin. At the critical concentration, a small GTP cap exists that maintains the microtubule at steady state. As the GTP is hydrolyzed to guanosine diphosphate (GDP), the cap gets smaller. When tubules lose their GTP cap, they become unstable and are depolymerized. According to this model, microtubules exist in two states: capped (stabilized, but slowly growing) and uncapped (unstable and rapidly depolymerizing).

Over the years, several laboratories have succeeded in initiating microtubule assembly from discrete MTOCs in permeabilized cells (see Deery for review). Recently, however, Kirschner's group[11] have succeeded in isolating a functional MTOC, the centrosome, from cultured mammalian cells. Examination of the isolated centrosomes by electron microscopy showed that it consisted of a pair of centrioles surrounded by pericentriolar material. When the isolated centrosomes were incubated with pure brain tubulin, microtubules grew from the center outward in a radial pattern much like that seen in living cells or lysed cell models. The shape of the assembly curve, determined by counts of microtubules/centrosome as a function of tubulin concentration, indicated that a finite, saturable number of initiation sites existed on each centrosome. Moreover, microtubules that were nucleated by the centrosome contained 13 protofilaments, whereas those that assembled spontaneously contained mostly 14 protofilaments.

The most interesting results came when Mitchinson and Kirschner diluted the saturated centrosome preparation with buffer.[11] Following the dilution of tubulin from a steady state concentration of 14 μM to a concentration of 7.5 μM, the number of microtubules decreased by 40 percent. The average length of the microtubules, however, increased by 40 percent. Thus, at tubulin levels well below the steady state concentration, some microtubules were shrinking, whereas others were growing. This unexpected phenomena termed a microtubule catastrophe[12] has many interesting implications if indeed it occurs in living cells. Thus, microtubule asters would be composed of a dynamic array of simultaneously growing and shrinking microtubules.

Mitchinson and Kirschner[13] also demonstrated this phenomena using *in vitro* assembly conditions and concluded that their results agreed with the GTP capping model of Pataloni and Carlier. When the tubulin-GTP cap of an individual microtubule is depleted by hydrolysis to tubulin-GDP, the microtubule would completely disassemble. The free nucleation site would then begin to initiate the assembly of a new microtubule. If the "catastrophe" is real, it certainly adds a new wrinkle to the dynamics of microtubules.

Tubulin is a receptor for numerous drugs. Where would the field be today without colchicine or taxol? Several new studies are highlighted in this volume, including studies of vinblastine and maytansine binding (Luduena *et al.*), tubulin-colchicine interactions (Wilson and coworkers, and Andreu) and a new stereo-selective microtubule inhibitor, tubulazole (De Brabander and coworkers). Such drugs continue to serve as important probes in the study of microtubules, but after 30 years of investigation, it is surprising that we still lack a firm knowledge of how colchicine binds to tubulin.

An obvious highlight of the conference were the reports that microtubules could be envisioned and analyzed kinetically in individual living cells. Emphasis on microtubules in living cells began several years ago when monospecific, and later monoclonal, antibodies to tubulin and MAPs became available for use as immunofluorescent probes. The technique of indirect immunofluorescence not only confirmed the presence of microtubules in the mitotic apparatus, but revealed an elaborate cytoplasmic microtubule complex (CMTC) in the cytoplasm of interphase cells. As cells entered mitosis, the CMTC disappeared and was replaced by the mitotic apparatus. When mitosis was completed, the CMTC reappeared. Recent advances in optics, microinjection techniques, morphometrics, and computerized image analysis has enabled some investigators to examine microtubule arrays in living cells. McIntosh and coworkers injected fluorescently labeled tubulin (DTAF-tubulin) into living cells and found that it was rapidly incorporated into microtubules, resulting in images identical to those seen in fixed, permeabilized cells. Likewise, when DTAF-labeled calmodulin was injected, the pattern of incorporation was essentially identical to that seen by indirect immunofluorescence. Through the use of these fluorescent analogues, combined with a technique for measuring fluorescence distribution after photobleaching (FRAP), McIntosh's group was successful in measuring tubulin dynamics in the spindle and CMTC of sea urchin eggs and mammalian cells. The half-time for spindle FRAP was 19 seconds. Moreover, the half-time for redistribution of fluorescent tubulin after photobleaching was 18 times shorter in mitotic cells than it was in interphase cells. No evidence for treadmilling of subunits was seen, and the rate of exchange in the spindle was too fast to be explained by simple end-dependent tubulin exchange. These exciting new findings, tempered by the possibility of artifact or misinterpretation, suggest that current models for tubulin assembly/disassembly may not apply to living cells.

The idea of observing microtubule dynamics in real time in a living cytoplasm is overwhelmingly appealing. Two laboratories[14,15] recently published remarkable video images of organelle movement along individual microtubules dissociated from the axoplasm of the squid giant axon. Organelles moved in the same or opposite directions along single microtubules like cars on a monorail sometimes passing other granules en route or switching from one microtubule to another. Vale and coworkers[15] presented evidence that a soluble factor from supernatants of axon preparations is required, along with adenosine triphosphate (ATP), for organelle movement.

The field of microtubule research was launched by a few electron micrographs of mitotic spindles and flagella taken over a quarter of a century ago. What followed was an avalanche of investigations into the biochemistry, pharmacology, and cell biology of microtubules, leading to an era of molecular biology and fluorescent analogue cytochemistry of living cells. We have indeed come full circle. The new techniques enable us to once again look into cells and visualize single microtubules, but this time, in the unique environment of living cytoplasm surrounded by a myriad of regulatory molecules and interacting cytoskeletal proteins. Perhaps now a small part of the puzzle can be solved.

We are indeed grateful to The New York Academy of Sciences for recording a significant era of the microtubule history in the proceedings of these two excellent conferences. Our appreciation is also extended to David Soifer for his untiring organizational efforts.

REFERENCES

1. CLEVELAND, D. W., S. H. HUGHES, E. STUBBLEFIELD, M. W. KIRSCHNER & H. E. VARMUS. 1981. J. Biol. Chem. **256:** 3130–3134.
2. FOREST, G. L. & R. R. KLEVECZ. 1972. J. Biol. Chem. **247:** 3147–3152.
3. NOLAND, B. J., B. A. WALTERS, R. A. TOBEY, J. M. HARDIN & G. R. SHEPHERD. 1974. Exp. Cell Res. **85:** 234–238.
4. KLEVECZ, R. R. & G. L. FOREST. 1975. Ann. N. Y. Acad. Sci. **253:** 292–303.
5. LAWRENCE, J. M. & H. N. WHEATLEY. 1975. Cytobios **13:** 167–179.
6. BEN-ZE'EV, A., S. R. FARMER & S. PENMAN. 1979. Cell **17:** 319–325.
7. CLEVELAND, D. W., M. A. LOPATA, P. SHERLINE & M. W. KIRSCHNER. 1981b. Cell **25:** 537–546.
8. CLEVELAND, D. W., M. F. PITTENGER & J. R. FERAMISCO. 1983. Nature (London) **305:** 738–740.
9. MARGOLIS, R. L. & L. WILSON. 1978. Cell **13:** 1–8.
10. BERGEN, L. G. & G. G. BORISY. 1980. J. Cell Biol. **84:** 141–150.
11. MITCHISON, T. & M. KIRSCHNER. 1984a. Nature (London) **312:** 232–237.
12. MCINTOSH, J. R. 1984. Nature (London) **312:** 196–197.
13. MITCHISON, T. & M. KIRSCHNER. 1984b. Nature (London) **312:** 237–241.
14. ALLEN, R. D., D. G. WEISS, J. H. HAYDEN, D. T. BROWN, H. FUJIWAKE & M. SIMPSON. 1985. J. Cell. Biol. **100:** 1736–1752.
15. VALE, R. D., T. S. REESE & M. SHEETZ. 1985c. Cell **41:** 39–50.

Index of Contributors

Subject Index